Classical and Quantum Information

Classical and Quantum Information

Dan C. Marinescu

Gabriela M. Marinescu

ELSEVIER

AMSTERDAM • BOSTON • HEIDELBERG • LONDON
NEW YORK • OXFORD • PARIS • SAN DIEGO
SAN FRANCISCO • SINGAPORE • SYDNEY • TOKYO

Academic Press is an imprint of Elsevier

Academic Press is an imprint of Elsevier
30 Corporate Drive, Suite 400, Burlington, MA 01803, USA
The Boulevard, Langford Lane, Kidlington, Oxford, OX5 1GB, UK

Library of Congress Cataloging-in-Publication Data
Marinescu, Dan C.
Classical and quantum information / Dan C. Marinescu, Magdalena Marinescu.
p. cm.
Includes bibliographical references and index.
ISBN 978-0-12-383874-2 (hardback)
1. Quantum computers. 2. Data processing–Technological innovations. I. Marinescu, Gabriela M. II. Title.
QA76.889.M363 2010
004.1–dc22
2010038738

British Library Cataloguing-in-Publication Data
A catalogue record for this book is available from the British Library.

For information on all Academic Press publications
visit our Web site at *www.elsevierdirect.com*

Typeset by: diacriTech, India

Printed and bound by CPI Group (UK) Ltd, Croydon, CR0 4YY

Transferred to digital print 2012

To Vera Rae

Contents

Preface

I want to know God's thoughts; the rest are details.
Albert Einstein

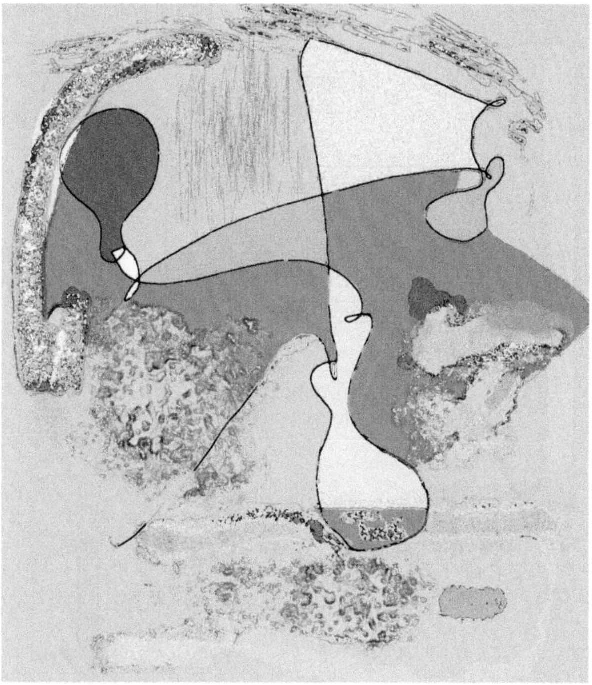

A new discipline, quantum information science, has emerged in the last two decades of the twentieth century at the intersection of physics, mathematics, and computer science. Quantum information processing is an application of quantum information science that covers the transformation, storage, and transmission of quantum information; it represents a revolutionary approach to information processing.

We have witnessed the development of microprocessors, high-speed optical communication, and high-density storage technologies, followed by the widespread use of sensors, and more recently, multi- and many-core processors and spintronics technology. We are now able to collect humongous amounts of information, process the information at high speeds, transmit the information through high-bandwidth and low-latency channels, store it on digital media, and share it using numerous applications built around the World Wide Web. Thus, the full cycle at the heart of the information revolution was closed (Figure P.1 [284]), and this revolution became a reality that profoundly affects our daily life.

Now, at the beginning of the twenty-first century, information processing is facing new challenges: Heat dissipation, leakage, and other physical phenomena limit our ability to build increasingly faster and, implicitly, increasingly smaller solid-state devices; it is very difficult to ensure the security of our communication; we are overwhelmed by the volume of information we are bombarded

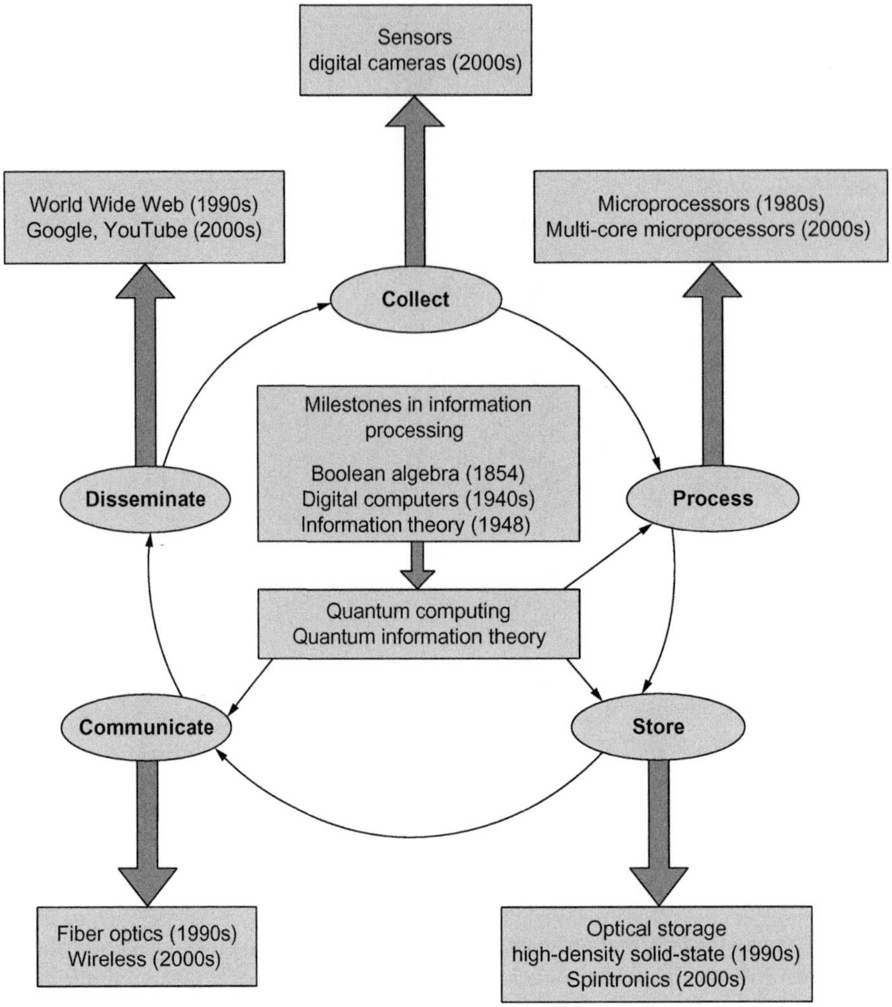

FIGURE P.1

Our ability to collect, process, store, communicate, and disseminate information has increased considerably during the last two decades of the twentieth century. The 1980s was the decade of microprocessors; advances in solid-state technologies allowed the increase of the number of transistors on a chip by three orders of magnitude and a substantial reduction of the cost of a microprocessor. In the 1990s, we have seen major breakthroughs in optical storage, high-density solid-state storage technologies, fiber optics communication, and the widespread acceptance of the Word Wide Web. The first decade of the twenty-first century is the decade of sensors, rapid information dissemination, and multi-core microprocessors.

with, and it is increasingly difficult to extract useful information from the vast ocean of information surrounding us.

Information, either classical or quantum, is physical; this is the mantra repeated throughout the book. Therefore, we must understand the physical processes that affect the state of the systems used to carry information. The physical processes for the storage, transformation, and transport of classical information are governed by the laws of classical physics, which limit our ability to process information increasingly faster using present-day solid-state technologies. The speed of charge carriers in semiconductors is finite; to increase the speed of the device, we have to pack the logic gates as tightly as possible.

The heat dissipated by a device increases with the clock rate to the power of 2 or 3, depending on the solid-state technology. Heat removal is a hard problem for densely packed devices; the heat produced by a solid-state device is proportional to the number of gates and, thus, to the volume of the device. If we pack the gates into a sphere, the heat dissipated is proportional to the volume of the sphere and can be removed through the surface of the sphere; while the amount of the heat increases as the cube of the radius, our ability to remove it only increases as the square of the radius of the sphere. We are thus limited in our ability to increase the speed and density of classical circuits.

These facts provide serious motivation to search for an alternative physical realization of computing and communication systems. Scientists are now exploring revolutionary means to overcome the limitations of computing and communication systems based on the laws of classical physics. Quantum and biological information processing provide a glimpse of hope in overcoming some of the limitations we mentioned and could revolutionize computing and communication in the third millennium. DNA computing together with quantum computing and quantum communication are the most promising avenues explored nowadays. While significant progress has been made in understanding the properties of quantum information, fundamental questions regarding biological information are still waiting for answers. For example, how do we explain the semantic aspect of biological information? Or, how is information from a damaged region of the brain recovered?

Quantum information is information stored as a property of a quantum system—e.g., the polarization of a photon or the spin of an electron. Quantum information can be transmitted, stored, and processed following the laws of quantum mechanics. Several physical embodiments of quantum information are possible; for example, quantum communication involves a source that supplies quantum systems in a given state, a noisy channel that "transports" the quantum system, and the recipient that receives and decodes the quantum information. The source could be a laser producing monochromatic photons, the channel could be an optical fiber, and the recipient a photocell; the source could also be an ion trap controlled by laser pulses, the channel a series of trapped ions, and the receiver a photo detector reading out the state of the ions via laser-induced fluorescence [275]. The diversity of the processes and technologies to process quantum information gives us hope that practical applications of quantum information will emerge sooner rather than later.

The physical processes for photons, ion traps, quantum dots, nuclear magnetic resonance (NMR), and other quantum systems are very different and could distract us from the goal of discovering the common properties of quantum information independent of its physical support. To study the properties of quantum information, we use an abstract model that captures the critical aspects of quantum behavior; this model, quantum mechanics, describes the properties of physical systems as entities in a finite-dimensional Hilbert space. Therefore, quantum information theory requires a basic understanding of quantum mechanics and familiarity with the mathematical apparatus used by quantum mechanics and information theory.

Quantum information has special properties: The state of a quantum system cannot be measured or copied without disturbing it; the quantum state of two systems can be *entangled*; the two-system ensemble has a definite state, though neither individual system has a well-defined state of its own; and we cannot reliably distinguish non-orthogonal states of a quantum system. Charles Bennett noted that

"Speaking metaphorically, quantum information is like the information in a dream: attempting to describe your dream to someone else changes your memory of it, so you begin to forget the dream and remember only what you said about it" [38].

The properties of quantum information are remarkable and could be exploited for information processing. In quantum computing systems an exponential increase in parallelism requires only a linear increase in the amount of space needed. Thus, in principle, a quantum computer will be able to solve problems that cannot be solved with today's computers; reversible quantum computers avoid logically irreversible operations and can, in principle, dissipate arbitrarily little energy for each logic operation. Quantum information theory allows us to design algorithms for quantum key distribution and for quantum teleportation. Eavesdropping on a quantum communication channel can be detected with very high probability.

Decoherence, the randomization of the internal state of a quantum computer due to interactions with the environment, is a major problem in quantum information processing; quantum computers rely on undisturbed evolution of quantum coherence. Quantum error correction allows reliable communication over noisy quantum channels, provided that the channels are not too noisy. We should caution the reader that the complexity of the circuits involved in quantum error correction is far beyond today's technological possibilities; a fault-tolerant implementation of Shor's quantum-factoring algorithm would most likely require thousands of physical qubits, at least two orders of magnitude more qubits than the systems reported in the literature have been able to harness. It may be possible, though, to resort to techniques that exploit the specific properties of individual physical realizations of quantum devices to manage the complexity of the quantum circuits for fault-tolerant systems. Fault-tolerant quantum computing still requires many more years of research.

Quantum information processing involves several areas including quantum algorithms, quantum complexity theory, quantum information theory, quantum error-correcting codes, quantum cryptography, and quantum reliability. This book covers basic concepts in quantum computing, quantum information theory, and quantum error-correcting codes.

Classical information theory is a mathematical model for the transfer, storage, and processing of information based on the laws of classical physics. In the late 1940s, Claude Shannon proved that it is possible to reliably transmit information over noisy classical communication channels; this discovery triggered the search for classical error-correcting codes, and the first codes were discovered by Richard Hamming in the early 1950s. Error correction is a critical component of modern technologies for reliable transfer, storage, and processing of classical information. Quantum information theory combines classical information theory with quantum mechanics to model information-related processes in quantum systems. The foundations of quantum information theory were established in the late 1980s by Charles Bennett and others, and the interest in quantum information increased dramatically in the mid-1990s after Peter Shor and Andrew Steane showed that quantum error correction is feasible and, together with Robert Calderbank, demonstrated that good quantum error-correcting codes exist.

New discoveries add to the excitement of quantum information science. Topological quantum computing, proposed by Kitaev in 1997 and further developed by Friedman, Kitaev, Larsen, and Wang, has the potential to revolutionize fault-tolerance. In 2005, Grover discovered the fixed-point quantum search. In 2008, Smith and Yard showed that communication is possible over zero-capacity quantum channels, and in 2009, Hastings provided an answer to one of the most important open questions

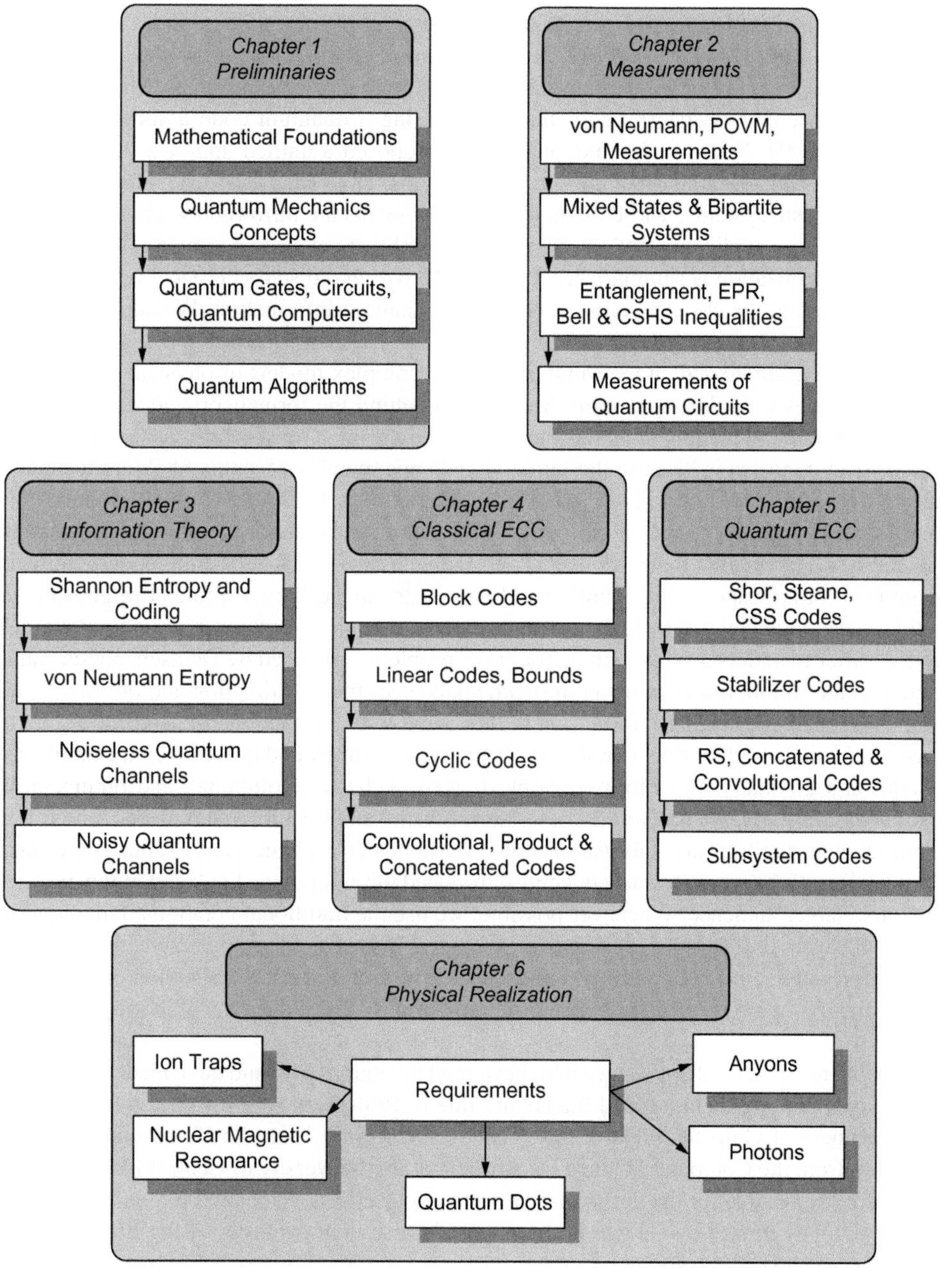

FIGURE P.2

Organization of the book at a glance.

in quantum information theory, showing that the minimum entropy output of a quantum communication channel is not additive. In 1999, Knill, Laflamme, and Viola reformulated quantum error correction and proposed to view quantum error-correcting codes as subsystems where the information resides in noiseless subspaces rather than considering a quantum code a subspace of a larger Hilbert space; in 2004, Kribs, Laflamme, and Poulin proposed a unified approach to quantum error correction and extended the concept of noiseless subsystems. Their work led to the introduction of operator quantum error subsystems. These theoretical developments are mirrored by advances in quantum communication—e.g., applications of quantum cryptography are close to commercialization.

The book's organization is summarized in Figure P.2. We first discuss classical concepts and then, gradually, we move to the corresponding concepts for quantum information. We adopted this philosophy for several reasons. First, the classical concepts are easier to grasp. Natural sciences develop increasingly more accurate and, at the same time, more complex models of physical reality; the level of abstraction makes it harder to develop the intuition behind the formalism, and it is more difficult to master the mathematical apparatus the models are based on. The second reason why we discuss first the classical concepts is because the targeted audience members for this book are not physicists familiar with quantum mechanics, but the larger population of scientists, engineers, students, or ordinary people puzzled by the "strange" properties of quantum information. Some of them are familiar with the classical information theory concepts and with classical error-correcting codes; for them, the significant leap is to transpose their intuition and knowledge to a different frame of reference.

We follow the same philosophy in the presentation of quantum algorithms; we analyze first quantum oracles, the easier to understand algorithms for "toy" problems proposed by Deutsch, Jozsa, Bernstein, and Vazirani, and Simon, followed by an in-depth analysis of phase estimation and of Grover's search algorithm. The chapter covering information theory starts with thermodynamic and Shannon entropy and classical channels, and we then introduce von Neumann entropy and quantum channels. We discuss linear codes and gradually move to more sophisticated cyclic, convolutional, and other families of classical codes; similarly, we analyze first the Shor, Steane, and CSS quantum error-correcting codes before introducing stabilizer and subsystem codes. We hope that the numerous examples will facilitate the understanding of the more abstract concepts introduced throughout the book and will make the book accessible to a larger audience. Whenever possible, we use the traditional notations in the literature or in the original papers that introduced the basic concepts. This required a careful selection of characters and fonts; for example, a 2^n-dimensional Hilbert space is denoted as $\mathcal{H}_{2^n}$, Shannon entropy is H, the parity-check matrix of a code is H, the Hadamard transform is $\mathbf{H}$, and the transfer matrix of a Hadamard gate is H.

We are indebted to several colleagues who have read the manuscript and have made many constructive suggestions. Among them, special thanks are due to Professors Dan Burghelea from the Mathematics Department at Ohio State University, Eduardo Mucciolo from the Physics Department, and Pawel Wocjan from the Computer Science Department at the University of Central Florida. Of course, the authors are responsible for the errors that, in spite of our efforts, may still be found in the text.

The artwork was created by George Dima, a gifted artist, concertmaster of the Bucharest Philharmonic, and accomplished creator of computer-generated graphics (see http://picasaweb.google.com/degefe2008). We express our thanks to Patricia Osborn, Gavin Becker, Lisa Lamenzo, and the editorial staff from Elsevier for their suggestions.

CHAPTER

Preliminaries

1

No amount of experiments can ever prove me right; a single experiment can prove me wrong.
Albert Einstein

What is Information?

Carl Friederich von Weizsäcker's answer, "Information is what is understood," implies that information has a sender and a receiver who have a common understanding of the representation and the means to convey information using some properties of the physical systems [446]. He adds, "*Information has no absolute meaning; it exists relatively between two semantic levels*" [447].

Once asked the question "What is time?" Richard Feynman answered: "*Time is what happens when nothing else happens.*" Unfortunately, history did not record Feynman's answer to the question "What is information?" and thus we do not have a crisp, witty, and insightful answer to a question central to the twenty-first century science. Indeed, the questions *What is information?* and *What is its relationship with the physical world?* become more important as we try to better understand physical phenomena at the quantum scale and the behavior of biological systems.

It is easy to understand why there is no simple answer to the question we posed at the beginning of this section; like matter and energy, information is a primitive concept. Thus, it is rather difficult to rigorously define it. Informally, we can state that information abstracts properties of and allows us to distinguish among objects/entities/phenomena/thoughts; information is a common denominator for the very diverse contents of our material and spiritual world. There is a common expression of information as strings of bits, regardless of the objects/entities/processes/thoughts it describes. Moreover, these bits are independent of their physical embodiment. Information can be expressed using pebbles on the beach, mechanical relays, electronic circuits, and even atomic and subatomic particles.

Classical information is information encoded to some property of a physical system obeying the laws of classical physics. Classical information is transformed using logic operations. Classical gates implement logic operations and allow for processing of classical information with classical computing devices.

Quantum information is information encoded to some property of quantum particles and obeys the laws of quantum mechanics. Quantum information is transformed using quantum gates, the building blocks for quantum circuits, which, in turn, can be assembled to build quantum computing and communication devices. The societal impact of information increases if the physical embodiments of bits and gates become smaller and we need less energy to process, store, and transmit information. This justifies our interest in quantum information.

This Book

This book covers topics in quantum computing, quantum information theory, and quantum error correction, three important areas of quantum information processing. Quantum information theory and quantum error correction build on the scope, concepts, methodology, and techniques developed in the context of their close relatives, classical information theory and classical error-correcting codes. It seems natural to follow the historical evolution of the concepts, and in this book, we first introduce the classical version of the concepts and techniques that are often simpler and easier to grasp. We then discuss in detail the significant leaps forward necessary to apply the concepts and techniques to quantum information.

Information theory is a mathematical model for transmission and manipulation of classical information. Quantum information theory studies fundamental problems related to the transmission of quantum information over classical and quantum communication channels such as the entropy of quantum systems, the capacity of classical and quantum channels, and the effect of noise, fidelity, and optimal information encoding. Quantum information theory promises to lead to a deeper understanding of fundamental properties of nature and, at the same time, support new and exciting applications.

Error-correcting codes allow us to detect and then correct errors during the transmission of classical information over classical channels and to build fault-tolerant computing and communication systems that obey the laws of classical physics. Quantum error-correcting codes exploit the fundamental properties of quantum information investigated by quantum information theory and play an important role in the fault-tolerance of quantum computing and communication systems. Quantum error-correcting codes are critical for the practical use of quantum computing and communication systems.

The first chapter of the book provides basic concepts from mathematics, quantum mechanics, and computer science necessary for understanding the properties of quantum information. Then we discuss the building blocks of a quantum computer, quantum circuits, and quantum gates and survey some of the properties of quantum algorithms. Figure 1.1 provides a structured view of the topics covered in this chapter: (1) the mathematical apparatus used by quantum mechanics; (2) the fundamental ideas of quantum mechanics; (3) and the circuits and algorithms for quantum computing devices.

1.1 ELEMENTS OF LINEAR ALGEBRA

Familiarity with complex numbers, algebraic structures such as groups, Abelian groups, and fields [57] and linear algebra [169] is required to understand the mathematical formalism of quantum mechanics. A review of algebraic structures used in coding theory is given in Section 4.4; in this section we review concepts such as vector spaces, inner product, norm, distance, orthogonality, basis, orthonormal basis, dimension of a vector space, linear transformation and matrices, eigenvectors and eigenvalues, and trace.

A *vector space* is an algebraic structure consisting of:

1. An Abelian group $(V, \ )$ whose elements, v_i , are called "vectors" and whose binary operation, " ", is called *addition*.
2. A field, $\mathbb{F}$, of numbers whose elements are called "scalars"; we restrict $\mathbb{F}$ to be either $\mathbb{R}$ (the field of real numbers) or $\mathbb{C}$ (the field of complex numbers).
3. An operation called "multiplication with scalars" and denoted by " ", which associates to any scalar, $c \ \mathbb{F}$, and vector, $v_i \ V$, a new vector, $v_j \ c \ v_i \ V$. $\mathbb{F}$ acts linearly on V: if $a, b \ \mathbb{F}$ and $u, v \ V$, then $a \ (u \ v) \ a \ u \ a \ v$ and $(a \ b) \ u \ a \ u \ b \ u$.

Assume that $\mathbb{F} \ \mathbb{C}$; it is easy to show that $\mathbb{C}^{m \ n}$, the set of all matrices $A \ [a_{ij}]$, with the entries $a_{ij} \ \mathbb{C}, 1 \ i \ n, 1 \ j \ m$, is a vector space where addition of two matrices, $A \ [a_{ij}]$ and $B \ [b_{ij}]$, is defined as $A \ B \ [a_{ij} \ b_{ij}]$; the inverse with respect to addition of $A \ [a_{ij}]$ is $A \ [\ a_{ij}]$; and the identity element is $E \ [0]$.

A set, $\mathcal{B}$, of vectors is called a *basis* in V if (1) every vector, $v \ V$, can be expressed as a linear combination of vectors from $\mathcal{B}$, and (2) the vectors in $\mathcal{B}$ are linearly independent. The *dimension of a vector space* is the cardinality of $\mathcal{B}$. We consider only *finite-dimensional* vector spaces, and in this case, the cardinality is the number of elements of $\mathcal{B}$. An n-dimensional vector space will be denoted as V_n.

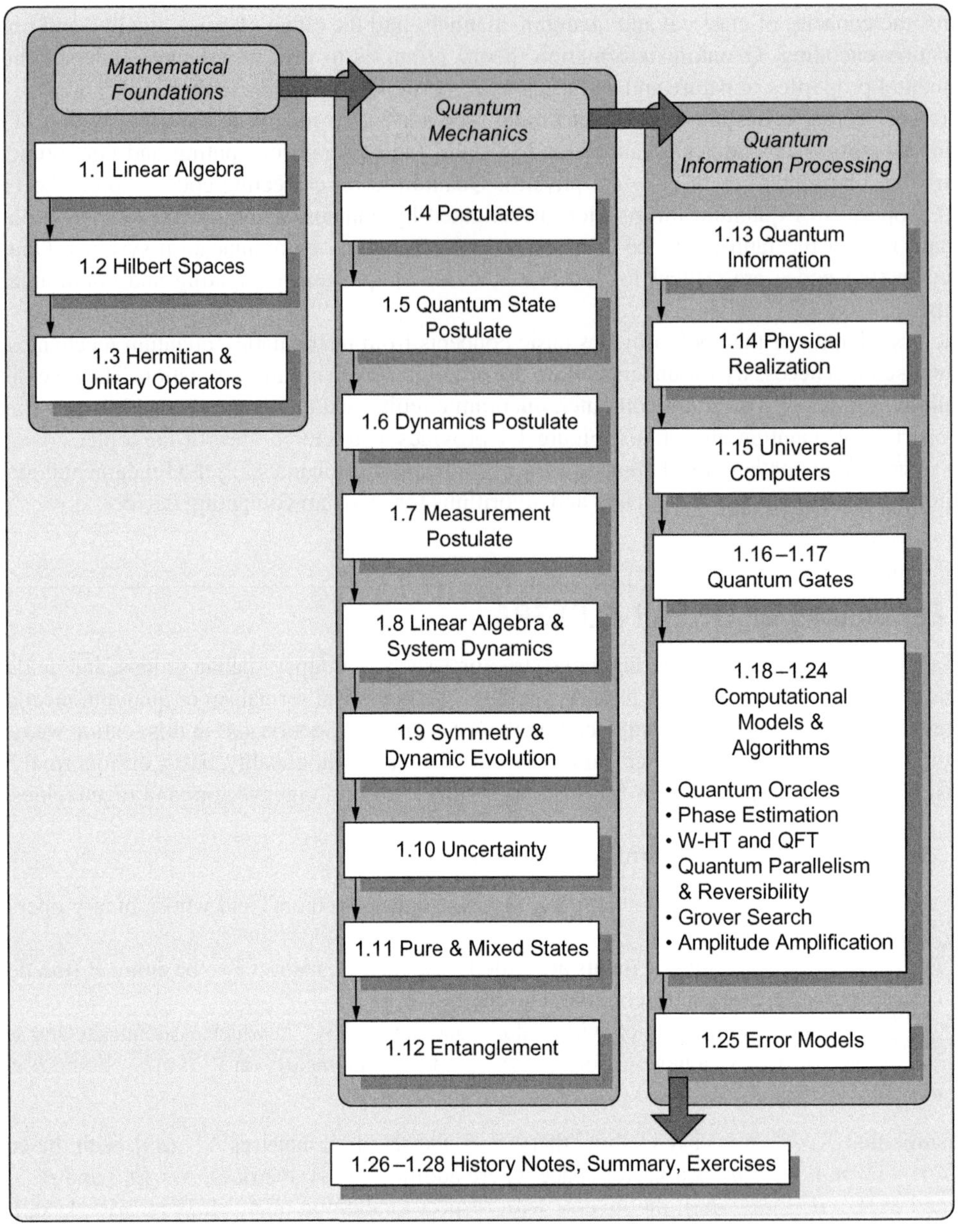

FIGURE 1.1

The organization of Chapter 1 at a glance.

An *inner product* in the vector space, V_n, over the field, $\mathbb{F}$, is a mapping, $g : V_n \quad V_n \quad \mathbb{F}$, with several properties; $v_i, v_j, v_k \quad V_n$, and $c \quad \mathbb{F}$:

1. Obeys the addition rule in V_n:

$$g(v_i \quad v_j, v_k) \quad g(v_i, v_k) \quad g(v_j, v_k) \text{ and } g(v_i, v_j \quad v_k) \quad g(v_i, v_j) \quad g(v_i, v_k).$$

2. Obeys the multiplication, with a scalar rule in V_n:

$$g(c \quad v_i, v_j) \quad c \quad g(v_i, v_j) \text{ and } g(v_i, c \quad v_j) \quad c \qquad g(v_i, v_j),$$

   with c , the complex conjugate of c when $\mathbb{F} \quad \mathbb{C}$.
3. Satisfies the following relations:

$$g(v_i, v_j) \quad g(v_j, v_i) \text{ if } \mathbb{F} \quad \mathbb{R} \quad \text{and} \quad g(v_i, v_j) \quad g \ (v_j, v_i) \text{ if } \mathbb{F} \quad \mathbb{C}.$$

4. The inner product is nondegenerate, $g(v_i, v_i) \quad 0$ and $g(v_i, v_i) \quad 0$, if and only if $v_i \quad 0$.

To simplify the notation, the inner product $g(v_i, v_j)$ will be written as v_i, v_j , and we shall use this notation from now on.

If an inner product in V is provided, then the *norm*, v , of the vector, $v \quad V_n$, is the square root of the inner product of the vector with itself:

$$v \qquad \overline{v, v} \ .$$

The *distance*, $d(v_i, v_j)$, of the two vectors, $v_i, v_j \quad V_n$, is

$$d(v_i, v_j) \qquad v_i \quad v_j \qquad \overline{(v_i \quad v_j), (v_i \quad v_j)} \ .$$

Two vectors, $v_i, v_j \quad V_n$, are *orthogonal* if

$$v_i, v_j \qquad 0.$$

The vectors of $b_1, b_2, \ldots b_n \quad V_n$ form an *orthonormal basis* if the inner product of any two of them is zero, $b_i b_j \quad 0, \quad (i,j) \quad 1, n \ , i \quad j$, and the norm of a vector is equal to unity, $b_i \qquad b_i, b_i \quad 1, \ i \quad 1, n$.

A *linear operator*, $\mathbf{A}$, between two vector spaces, V and W, over the field, $\mathbb{F}$, is any mapping from V to W, $\mathbf{A} : V \quad W$, linear in its inputs

$$\mathbf{A} \qquad \underset{i}{} c_i v_i \qquad \underset{i}{} c_i \mathbf{A}(v_i).$$

The identity operator, $\mathbf{I}$, maps $v \quad V_n$ to itself, $\mathbf{I}(v) \quad v$. A *linear transformation* is a linear operator, with $V \quad W$.

The *dual* of a vector space V, denoted as V , is the set of all scalar-valued linear maps, $\varphi : V \quad \mathbb{F}$. If we define the addition and scalar multiplications in V as

$$(\varphi \quad \phi)(v) \quad \varphi(v) \quad \phi(v) \quad \text{and} \quad (c\varphi)(v) \quad c\varphi(v),$$

with $v \quad V, c \quad \mathbb{F}$ and $\varphi, \phi \quad V$, then the dual is also a vector space over the field $\mathbb{F}$. If V is an n-dimensional vector space, so is V , and if the vectors $b_1, b_2, \ldots, b_j, \ldots, b_n$ form a basis for V_n, then V is also an n-dimensional vector space, and the vectors $b^1, b^2, \ldots, b^i, \ldots, b^n$ defined by the property

$$(b^i, B_j) \quad \delta_{ij} \quad \begin{cases} 1 & \text{if} \quad i \quad j, \\ 0 & \text{if} \quad i \quad j \end{cases}$$

form a basis for V . An inner product in V provides an isomorphism—i.e., an invertible linear operator, $\mathbf{A} : V \quad V$, $\mathbf{A}(v)(w) \quad v, w$.

If we choose a basis, $\mathcal{B} \quad b_1, b_2, \ldots, b_n$, then the linear transformation A is represented by an $n \quad n$ *matrix*, $A \quad [a_{ij}]$, $1 \quad i, j \quad n$. Let the vectors, v and w, be

$$v \qquad v_i b_i \quad \text{and} \quad w \qquad w_i b_i,$$
$$\qquad i \qquad\qquad\qquad i$$

and $w \quad Av$. Then $w_i \quad {}^{n}_{j\ 1} a_{ij} v_j$, which can be written as

$$\begin{matrix} w_1 \\ w_2 \\ \vdots \\ w_n \end{matrix} \qquad \begin{matrix} a_{11} & a_{12} & \ldots & a_{1n} \\ a_{21} & a_{22} & \ldots & a_{2n} \\ \vdots & \vdots & \vdots & \vdots \\ a_{n1} & a_{n2} & \ldots & a_{nn} \end{matrix} \qquad \begin{matrix} v_1 \\ v_2 \\ \vdots \\ v_n \end{matrix} ,$$

with the right side of the equality representing the product of the $n \quad n$ matrix whose elements are a_{ij}'s, with the $n \quad 1$ matrix whose elements are the v_i's.

Addition and multiplication of matrices satisfy standard algebraic laws:

$$A \quad B \quad B \quad A \qquad A \quad (B \quad C) \quad (A \quad B) \quad C \quad A(B \quad C) \quad AB \quad AC$$
$$A(BC) \quad (AB)C \quad (A \quad B)C \quad AC \quad BC \qquad AI \quad IA \quad A,$$

with I, the identity matrix, an $n \quad n$ matrix with main diagonal elements equal to one and off-diagonal elements equal to zero. Nonzero matrices do not always have inverses, and the product of two matrices is in general noncommutative, $AB \quad BA$.

The *determinant* of the $n \quad n$ matrix A, $\det(A)$, is a number calculated from the elements of matrix A; it vanishes if and only if the matrix represents a linear transformation that is not one to one. The determinant can be written as a polynomial,

$$\begin{matrix} a_1 & a_2 & a_3 \\ b_1 & b_2 & b_3 \\ c_1 & c_2 & c_3 \\ \vdots & \vdots & \vdots & \ddots \end{matrix} \qquad \sum_{ijk} \epsilon_{ijk} \quad a_i b_j c_k \qquad ,$$

where $(ijk\ldots)$ is a permutation of indices $1, 2, 3, \ldots$ and

$$\epsilon_{ijk} \quad \begin{cases} 1 & \text{for even permutations,} \\ 1 & \text{for odd permutations.} \end{cases}$$

The determinants of two $n \times n$ matrices, A and B, have the following property:

$$\det(AB) = \det(A)\det(B).$$

An *eigenvector*, v, of a linear transformation $\mathbf{A}$ is a nonzero vector such that

$$\mathbf{A}v = \lambda v.$$

The scalar, λ, is called an *eigenvalue* corresponding to the eigenvector v of $\mathbf{A}$. The previous expression can also be written as

$$\mathbf{A}v = \lambda \mathbf{I} v.$$

Thus:

$$(\mathbf{A} - \lambda \mathbf{I})v = 0.$$

This equation can be transformed to a matrix equation by choosing a basis, $\{b_i\}$, for V_n. With respect to this basis, v can be expressed as

$$v = \sum_i c_i b_i.$$

Then,

$$(\mathbf{A} - \lambda \mathbf{I}) \sum_i c_i b_i = 0,$$

where the coefficients must satisfy the equation

$$\sum_i (\mathbf{A} - \lambda \mathbf{I})_{i,j} c_i = 0$$

for any fixed j. A nontrivial solution exists only if the determinant

$$\det(\mathbf{A} - \lambda \mathbf{I}) = 0.$$

The scalar λ for which that happens is an eigenvalue of $\mathbf{A}$. This condition becomes:

$$\begin{vmatrix} (a_{11} - \lambda) & a_{12} & a_{13} & \cdots & a_{1n} \\ a_{21} & (a_{22} - \lambda) & a_{23} & \cdots & a_{2n} \\ a_{31} & a_{32} & (a_{33} - \lambda) & \cdots & a_{3n} \\ \vdots & \vdots & \vdots & \ddots & \\ a_{n1} & a_{n2} & a_{n3} & \cdots & (a_{nn} - \lambda) \end{vmatrix} = 0.$$

This is called the *characteristic equation*. The characteristic equation above is a polynomial of degree n in λ, where n is the dimension of the vector space. If $\mathbb{F}$ is either $\mathbb{R}$ or $\mathbb{C}$, then the polynomial has n

possibly complex numbers as roots, and by the "fundamental theorem of algebra," can be expressed as a product of linear factors:

$$(\lambda_1 \quad \lambda)(\lambda_2 \quad \lambda)(\lambda_3 \quad \lambda) \quad (\lambda_n \quad \lambda) \quad 0.$$

If $\mathbb{F} \quad \mathbb{C}$, then the n roots, $\lambda_1, \lambda_2, \lambda_3, \ldots, \lambda_n$, are the *eigenvalues* of the operator and are independent of the basis chosen to represent the operator as a matrix. If $\mathbb{F} \quad \mathbb{R}$, then the real roots are eigenvalues.

If the characteristic equation has less than n distinct roots, there are *multiple roots*; the root λ of multiplicity larger than one is said to be *multiple*. The *multiplicity m of a root λ_i* is the number of times the factor $(\lambda_i \quad \lambda)$ appears in the product above. For a multiple eigenvalue it is possible to have more than one eigenvector, all linearly independent, and the corresponding linear space is of dimension $\quad m$.

If there are multiple eigenvalues of **A** and if for each multiple eigenvalue the multiplicity, m, is equal to the dimension, d, then it is possible to find a basis, $b_1, b_2, \ldots, b_j, \ldots, b_n$, for **A** such that each basis vector is an eigenvector of **A**:

$$\mathbf{A}b_1 \quad \lambda_1 b_1, \quad \mathbf{A}b_2 \quad \lambda_2 b_2, \quad \mathbf{A}b_3 \quad \lambda_3 b_3, \ldots, \mathbf{A}b_n \quad \lambda_n b_n.$$

Then **A**, with respect to the basis $b_1, b_2, \ldots, b_j, \ldots, b_n$, can be expressed by the diagonal matrix

$$\begin{matrix} \lambda_1 & 0 & 0 & & 0 \\ 0 & \lambda_2 & 0 & & 0 \\ 0 & 0 & \lambda_3 & & 0 \\ \vdots & \vdots & \vdots & \ddots & \\ 0 & 0 & 0 & & \lambda_n \end{matrix} .$$

Example. Let $\mathbb{F} \quad \mathbb{C}$, and consider the matrix

$$A \quad \begin{matrix} 1 & 2i \\ 2i & 2 \end{matrix}$$

representing a linear transformation, $A : \mathbb{C} \quad \mathbb{C}^2$. λ is an eigenvalue of A if the determinant of the matrix

$$A \quad \lambda I \quad \begin{matrix} 1 \quad \lambda & 2i \\ 2i & 2 \quad \lambda \end{matrix}$$

is zero. The resulting characteristic polynomial

$$\lambda^2 \quad \lambda \quad 6 \quad 0$$

has the roots $\lambda_1 \quad 2$ and $\lambda_2 \quad 3$. Hence, the eigenvectors of A satisfy one or the other of the following systems of equations:

$$\begin{cases} x_1 \quad 2iy_1 & 2x_1 \\ 2ix_1 \quad 2y_1 & 2y_1 \end{cases} \quad \text{or} \quad \begin{cases} x_2 \quad 2iy_2 & 3x_2 \\ 2ix_2 \quad 2y_2 & 3y_2 \end{cases},$$

where x_1, x_2 and y_1, y_2 represent the components corresponding to the basis vectors used to represent matrix A.

The equations can be rewritten as

$$\begin{matrix} x_1 & 2iy_1 \\ ix_1 & 2y_1 \end{matrix} \quad \begin{matrix} 0 \\ 0 \end{matrix} \quad \text{or} \quad \begin{matrix} 2x_2 & iy_2 \\ 2ix_2 & y_2 \end{matrix} \quad \begin{matrix} 0 \\ 0 \end{matrix},$$

Solving these systems of equations, we obtain the eigenvectors $(1, \ \frac{1}{2}i)$ and $(1, 2i)$. These eigenvectors are not unique; $(\lambda, 2i\lambda)$ are also eigenvectors. These eigenvectors can be used as a new basis, and the transformed matrix A has a diagonal form

$$\begin{matrix} 2 & 0 \\ 0 & 3 \end{matrix} \ \lambda,$$

with respect to this basis. Now we review several important properties of square matrices with elements from $\mathbb{C}$ useful for the proof of the proposition presented at the end of this section.

The *trace* of a square matrix, $A \quad a_{ij}$, $1 \quad i,j \quad n$, $a_{ij} \quad \mathbb{C}$, is the sum of the elements on the main diagonal of A, $\text{tr}(A) \quad a_{11} \quad a_{22} \qquad a_{nn}$. From this definition, it is easy to prove several properties of the trace:

1. The trace is a linear map. Given a scalar $c \quad \mathbb{R} \quad \mathbb{C}$, then $\text{tr}(A \quad B) \quad \text{tr}(A) \quad \text{tr}(B)$ and $\text{tr}(cA) \quad c\text{tr}(A)$. As a consequence, if $A(t)$ is a matrix-valued function and $\frac{d}{dt}[A(t)]$ denotes the matrix whose entries are the derivatives of the entries of $A(t)$, then

$$\frac{d}{dt} \, tr[A(t)] \quad \text{tr} \ \frac{d}{dt}[A(t)] \ .$$

2. The trace is invariant to transposition, $\text{tr}(A) \quad \text{tr}(A^T)$. Indeed, the diagonal elements, a_{ii}, of a square matrix, $A \quad [a_{ij}]$, are invariant to transposition.
3. The trace is not affected by the order of the matrices in a product of two matrices:

$$\text{tr}(AB) \quad \text{tr}(BA).$$

If $A \quad a_{ij}$ and $B \quad b_{ij}$, $1 \quad i,j \quad n$, the diagonal elements of the products, AB and BA, are $\sum_{i\ 1}^{n} a_{ki}b_{ik}$ and $\sum_{i\ 1}^{n} b_{ki}a_{ik}$, respectively thus, $\text{tr}(AB) \quad \text{tr}(BA)$. Consequently, if U is a square $n \quad n$ matrix and it is invertible, then

$$\text{tr}(U^{\ 1}AU) \quad \text{tr}(A).$$

Given three $n \quad n$ matrices, A, B, and C, we have $\text{tr}(ABC) \quad \text{tr}(CAB) \quad \text{tr}(BCA)$. Another consequence, if $\lambda_1, \lambda_2, \ldots, \lambda_n$ are the eigenvalues of an operator $\mathbf{A}$ with associated square matrix A, then

$$\text{tr}(A) \qquad \sum_{i\ 1}^{n} \lambda_i.$$

4. If A is a square matrix and A is its complex conjugate (i.e., the entry of index ij is the complex conjugate of the entry ji), then

$$\mathrm{tr}(A\ \ A)\quad 0.$$

Indeed,

$$\mathrm{tr}(A\ \ A) \qquad \sum_{1\ \ i,j\ \ n} a_{ij}a_{ji} \qquad \sum_{1\ \ i,j\ \ n} a_{ij}\ ^{2}.$$

It follows immediately that

$$\mathrm{tr}(A\ \ A)\quad 0 \qquad A\quad 0.$$

From properties 1–4, it follows that *the set of n n square matrices, with elements from $\mathbb{C}$, form a vector space with the inner product defined as*

$$A, B \qquad \mathrm{tr}(A\ \ B).$$

5. The determinant and the trace are related:

$$\det(I\quad \epsilon A)\quad 1\quad \epsilon\quad \mathrm{tr}(A)\quad o(\epsilon^2).$$

1.2 HILBERT SPACES AND DIRAC NOTATIONS

In this section we recast the linear algebra concepts using the formalism introduced by Paul Dirac for quantum mechanics [125]. Hilbert spaces could be of infinite dimension, but we shall only discuss finite-dimensional Hilbert spaces.

An *n-dimensional Hilbert space*, $\mathcal{H}_n$, is an n-dimensional vector space over the field of complex numbers, $\mathbb{C}$, equipped with an inner product. In this case, the isomorphism, A, considered previously, will be called *Hermitian conjugation* and denoted as $()^{\dagger}$. The state of a physical system will be represented either as a vector in $\mathcal{H}_n$ and referred to as a *ket vector* or, equivalently, by its Hermitian conjugate and referred to as a *bra vector*. In this vector space, distances and angles between vectors can be measured.

An example of a Hilbert space of dimension n is $\mathbb{C}_n$ when equipped with the inner product:

$$a, b \qquad \sum_{i\ \ 1}^{n} a_i b_i\,, \text{ with } a \quad (a_1, a_2, \ldots, a_n) \text{ and } b \quad (b_1, b_2, \ldots, b_n).$$

This Hilbert space has a canonical orthonormal base:

$$(1,0,\ldots,0,0), (0,1,0,\ldots,0,0), \ldots, (0,0,\ldots,1,\ldots,0), \ldots, (0,0,\ldots,0,1).$$

All n-dimensional Hilbert spaces are isomorphic with $\mathbb{C}_n$ by isomorphisms that identify the inner products (isometries). More precisely, if $v_1, v_2, \ldots, v_i, \ldots, v_n$ is an orthonormal base, then the unique linear operator, $\mathbf{A} : \mathcal{H}_n \quad \mathbb{C}_n$, defined by $\mathbf{A}(v_i) \quad (0,0,\ldots,1,\ldots,0)$ and extended linearly to all $v_i \quad \mathcal{H}_n$, is such an isometry. Conversely, for any isometry $\mathbf{A} : \mathcal{H}_n \quad \mathbb{C}_n$, $v_i \quad \mathbf{A}^{\ 1}(0,0,0,\ldots,1,\ldots,0)$ provides an orthonormal basis in $\mathcal{H}_n$. In view of these remarks, we can restrict ourselves to $\mathbb{C}_n$ as the "standard" representation of an n-dimensional Hilbert space, $\mathcal{H}_n$.

Given two Hilbert spaces, $\mathcal{H}_m$ and $\mathcal{H}_n$, over the same field, $\mathbb{F}$, their *tensor product,* $\mathcal{H}_m \quad \mathcal{H}_n$, is an $(m \quad n)$-dimensional Hilbert space, $\mathcal{H}_{mn}$. If $\ e_1\ ,\ e_2\ ,\ldots,\ e_m\ $ is an orthonormal basis in $\mathcal{H}_m$ and $\ f_1\ ,\ f_2\ ,\ldots,\ f_n\ $ is an orthonormal basis in $\mathcal{H}_n$, then $\ e_i \quad f_j\ , 1 \quad i \quad m,\ 1 \quad j \quad n$, is an orthonormal basis in $\mathcal{H}_{mn}$.

Quantum mechanics uses a special notation for vectors; instead of the vector v, we write $\ v\ $ and call it the ket vector representation of v; instead of the adjoint $v^\dagger$, we write $\ v\ $ and call it the bra vector representation of v. With this convention, the inner product $(v,w) \quad (w^\dagger v)$ can also be written as $\ w \quad v\ $.

The canonical orthonormal basis of $\mathbb{C}^n$ can be expressed as ket vectors as

$$0\ ,\ \ 1\ ,\ldots,\ \ i\ ,\ldots,\ \ n \quad 1\ ,$$

or, as bra vectors as

$$0\ ,\ \ 1\ ,\ldots,\ \ i\ ,\ldots,\ \ n \quad 1\ .$$

We will also use the matrix representation of each ket vector $\ i\ $ in this canonical representation as a column vector, with a 1 in the $i^{th} \quad 1$ row and 0 in all the others. For example,

$$0 \quad \begin{matrix} 1 \\ 0 \\ \vdots \\ 0 \\ \vdots \\ 0 \end{matrix}\ ,\ \ 1 \quad \begin{matrix} 0 \\ 1 \\ \vdots \\ 0 \\ \vdots \\ 0 \end{matrix}\ ,\ldots,\ i \quad \begin{matrix} 0 \\ 0 \\ \vdots \\ 1 \\ \vdots \\ 0 \end{matrix}\ ,\ldots,\ n \quad 1 \quad \begin{matrix} 0 \\ 0 \\ \vdots \\ 0 \\ \vdots \\ 1 \end{matrix}\ ,$$

and the matrix representation of the bra vector $\ i\ $ as a row vector, with 1 in the $i^{th} \quad 1$ position and 0 in all the others:

$$0 \quad 1\ 0\ \ldots\ 0\ \ldots 0\ ,$$

$$1 \quad 0\ 1\ \ldots\ 0\ \ldots 0\ ,$$

$$\vdots$$

$$i \quad 0\ 0\ \ldots\ 1\ \ldots 0\ ,$$

$$\vdots$$

$$n \quad 1 \quad 0\ 0\ \ldots\ 0\ \ldots 1\ .$$

We have:

$$i \quad j \quad \delta_{i,j} \quad \begin{cases} 0 & \text{if } i \quad j, \\ 1 & \text{if } i \quad j, \end{cases} \quad 0 \quad (i,j) \quad n \quad 1,$$

where $\delta_{i,j}$ is the *Kronecker delta* symbol.

An n-dimensional ket, ψ , can be expressed in this basis as a linear combination of the orthonormal basis ket vectors

$$\psi \quad \alpha_0 \; 0 \quad \alpha_1 \; 1 \qquad \alpha_i \; i \qquad \alpha_{n \;\; 1} \; n \quad 1 \,,$$

where $\alpha_0, \alpha_1, \ldots, \alpha_i, \ldots, \alpha_{n \;\; 1}$ are complex numbers.

The vector, ψ , can be expressed in different bases. For example, if $n \quad 2$ when the vector is expressed in the basis 0 , 1 as

$$\psi \quad \alpha_0 \; 0 \quad \alpha_1 \; 1 \,,$$

then in a new basis, x , y , defined by

$$x \quad \frac{1}{\overline{2}}(\; 0 \qquad 1 \;), \quad y \quad \frac{1}{\overline{2}}(\; 0 \qquad 1 \;) \,,$$

the vector ψ is expressed as

$$\psi \quad \frac{1}{\overline{2}}(\alpha_0 \quad \alpha_1) \; x \quad \frac{1}{\overline{2}}(\alpha_0 \quad \alpha_1) \; y \,.$$

For each ket vector ψ , there is a *dual*, the bra vector denoted by ψ . As mentioned earlier, the bra and ket vectors are related by Hermitian conjugation:

$$\psi \quad (\; \psi \;)^{\dagger}, \quad \psi \quad (\; \psi \;)^{\dagger}.$$

The bra vector ψ is expressed as a linear combination of the orthonormal bra vectors:

$$\psi \quad \alpha_0 \; 0 \quad \alpha_1 \; 1 \qquad \alpha_i \; i \qquad \alpha_{n \;\; 1} \; n \quad 1 \,,$$

where $\alpha_0, \alpha_1, \ldots, \alpha_i, \ldots, \alpha_{n \;\; 1}$ are the complex conjugates of $\alpha_0, \alpha_1, \ldots, \alpha_i, \ldots, \alpha_{n \;\; 1}$.

In matrix representation, a ket vector is expressed as the column matrix

$$\psi \qquad \begin{matrix} \alpha_0 \\ \alpha_1 \\ \vdots \\ \alpha_i \\ \vdots \\ \alpha_{n \;\; 1} \end{matrix} \,,$$

and a dual bra vector is expressed as the row matrix

$$\psi \quad \alpha_0 \;\; \alpha_1 \;\; \ldots \;\; \alpha_i \;\; \ldots \;\; \alpha_{n \;\, 1} \; .$$

Recall that the inner product of two vectors, (ψ , φ), is a complex number. The inner product denoted as $\psi \;\; \varphi$ has the following properties:

1. The inner product of a vector with itself is a non-negative real number. Indeed,

$$\psi \;\; \psi \qquad \alpha_0 \;\; \alpha_1 \;\; \alpha_2 \qquad \alpha_{n \;\, 1} \begin{matrix} \alpha_0 \\ \alpha_1 \\ \alpha_2 \\ \vdots \\ \alpha_{n \;\, 1} \end{matrix}$$

$$\alpha_0 \;\; ^2 \qquad \alpha_1 \;\; ^2 \qquad \alpha_2 \;\; ^2 \qquad\qquad \alpha_{n \;\, 1} \;\; ^2,$$

with $\alpha_i \;\; ^2$, the square of the modulus of the complex number α_i,

$$\alpha_i \;\; ^2 \qquad e(\alpha_i) \;\; ^2 \qquad =m(\alpha_i) \;\; ^2.$$

Thus, $\psi \;\; \psi \qquad \mathbb{R}$, and

$$\psi \;\; \psi \; \begin{cases} 0 & \text{if } \psi \qquad (00\ldots0), \\ > 0 & \text{otherwise.} \end{cases}$$

2. Linearity. If ψ , φ , $\xi \qquad \mathcal{H}_n$ and $(a,b,c) \quad \mathbb{C}$, then

$$\psi \;\; (c \;\; \varphi \;) \quad c \; \psi \;\; \varphi \; ;$$

$$(a \; \psi \qquad b \; \varphi \;) \;\; \xi \qquad a \; \psi \;\; \xi \qquad b \; \varphi \;\; \xi \; ;$$

$$\xi \;\; (a \;\; \psi \qquad b \;\; \varphi \;) \quad a \; \xi \;\; \psi \qquad b \; \xi \;\; \varphi \; .$$

3. Hermitian symmetry:

$$\psi \;\; \varphi \qquad \varphi \;\; \psi \; .$$

Let **A** and **B** be two linear, operators represented as $m \quad n$ and $p \quad q$ matrices:

$$A \quad \begin{matrix} a_{11} & a_{12} & \ldots & a_{1n} \\ a_{21} & a_{22} & \ldots & a_{2n} \\ a_{31} & a_{32} & \ldots & a_{3n} \\ \vdots & \vdots & & \vdots \\ a_{m1} & a_{m2} & \ldots & a_{mn} \end{matrix} \quad \text{and} \quad B \quad \begin{matrix} b_{11} & b_{12} & \ldots & b_{1q} \\ b_{21} & b_{22} & \ldots & b_{2q} \\ b_{31} & b_{32} & \ldots & b_{3q} \\ \vdots & \vdots & & \vdots \\ b_{p1} & b_{p2} & \ldots & b_{pq} \end{matrix} \; .$$

Then, their *tensor product,* $\mathbf{A} \otimes \mathbf{B}$, is an $mp \times nq$ matrix defined as

$$A \otimes B = \begin{pmatrix} a_{11}B & a_{12}B & \dots & a_{1n}B \\ a_{21}B & a_{22}B & \dots & a_{2n}B \\ a_{31}B & a_{32}B & \dots & a_{3n}B \\ \vdots & \vdots & & \vdots \\ a_{m1}B & a_{m2}B & \dots & a_{mn}B \end{pmatrix}.$$

Here, $a_{ij}B$, $1 \le i \le m$, $1 \le j \le n$, is a submatrix whose entries are the products of a_{ij} and all the elements of matrix B. Consistent with this definition, the tensor product of two-dimensional vectors (2×1 matrices), (a,b) and (c,d), is the four-dimensional vector (a 4×1 matrix)

$$\begin{pmatrix} a \\ b \end{pmatrix} \otimes \begin{pmatrix} c \\ d \end{pmatrix} = \begin{pmatrix} ac \\ ad \\ bc \\ bd \end{pmatrix}.$$

Example. The tensor product of vectors $(|0\rangle, |1\rangle) \in \mathcal{H}_2$:

$$|0\rangle \otimes |1\rangle = \begin{pmatrix} 1 \\ 0 \end{pmatrix} \otimes \begin{pmatrix} 0 \\ 1 \end{pmatrix} = \begin{pmatrix} 0 \\ 1 \\ 0 \\ 0 \end{pmatrix}.$$

An $m \times n$ matrix obtained as a product of an $m \times 1$ matrix (a ket vector $|\psi\rangle$) and a $1 \times n$ matrix (a bra vector $\langle\varphi|$) is sometimes called the *outer product*. For example, let $|\psi\rangle, |\varphi\rangle \in \mathcal{H}_4$ be

$$|\psi\rangle = \alpha_0|0\rangle + \alpha_1|1\rangle + \alpha_2|2\rangle + \alpha_3|3\rangle,$$

$$\langle\varphi| = \beta_0\langle 0| + \beta_1\langle 1| + \beta_2\langle 2| + \beta_3\langle 3|.$$

Then the outer product, $|\psi\rangle\langle\varphi|$, is the matrix

$$|\psi\rangle\langle\varphi| = \begin{pmatrix} \alpha_0 \\ \alpha_1 \\ \alpha_2 \\ \alpha_3 \end{pmatrix} \begin{pmatrix} \beta_0 & \beta_1 & \beta_2 & \beta_3 \end{pmatrix} = \begin{pmatrix} \alpha_0\beta_0 & \alpha_0\beta_1 & \alpha_0\beta_2 & \alpha_0\beta_3 \\ \alpha_1\beta_0 & \alpha_1\beta_1 & \alpha_1\beta_2 & \alpha_1\beta_3 \\ \alpha_2\beta_0 & \alpha_2\beta_1 & \alpha_2\beta_2 & \alpha_2\beta_3 \\ \alpha_3\beta_0 & \alpha_3\beta_1 & \alpha_3\beta_2 & \alpha_3\beta_3 \end{pmatrix}.$$

We conclude with the observation that in the case of Hilbert spaces any root of the characteristic polynomial of a linear transformation A is an eigenvalue. If an eigenvalue has multiplicity 1, the eigenvector is unique up to multiplication with a scalar, but if we insist that the eigenvector is unitary (of length 1), then the eigenvector be unique up to a phase factor. As we shall see in Sections 1.5, 1.6, and 1.7, vectors in Hilbert space represent the states of a quantum system; the transformation of states and the measurements are represented by linear operators.

1.3 HERMITIAN AND UNITARY OPERATORS: PROJECTORS

A *linear operator* **A** maps vectors in a space $\mathcal{H}_n$ to vectors in the same space $\mathcal{H}_n$. Given the vectors ψ , φ $\quad \mathcal{H}_n$, the transformation performed by the linear operator **A** can be described as

$$\varphi \quad \mathbf{A} \; \psi \; .$$

With respect to an orthonormal basis chosen once and for all, the linear operator **A** on $\mathcal{H}_n$ can be represented by the matrix

$$A \quad [a_{ij}], \quad \text{with} \quad a_{ij} \quad \mathbb{C}, \; 1 \quad i,j \quad n.$$

The *adjoint* of a linear operator **A** is denoted as $\mathbf{A}^\dagger$. The matrix $A^\dagger$ describing the adjoint operator $\mathbf{A}^\dagger$ is the transpose conjugate matrix of A,

$$A^\dagger \quad [a_{ji}], \quad 1 \quad i,j \quad n.$$

Throughout this book, the term *operator* means *linear operator*, and we shall use the terms operator **A** and matrix A interchangeably. An operator acts on the left of a ket vector and on the right of a bra vector. Moreover,

$$[\mathbf{A} \;\; \varphi \;]^\dagger \qquad \varphi \; {}^\dagger \mathbf{A}^\dagger \qquad \varphi \;\; \mathbf{A}^\dagger.$$

An operator, **A**, is *normal* if $\mathbf{A}\mathbf{A}^\dagger \quad \mathbf{A}^\dagger\mathbf{A}$. An operator **A** is *Hermitian (self-adjoint)* if $\mathbf{A} \quad \mathbf{A}^\dagger$. Clearly, a Hermitian operator is normal. The sum of two Hermitian operators, $\mathbf{A} \quad \mathbf{A}^\dagger$ and $\mathbf{B} \quad \mathbf{B}^\dagger$, is Hermitian

$$(\mathbf{A} \quad \mathbf{B})^\dagger \quad \mathbf{A}^\dagger \quad \mathbf{B}^\dagger \quad \mathbf{A} \quad \mathbf{B},$$

while their product is not necessarily Hermitian. The necessary and sufficient condition for the product of two Hermitian operators to be Hermitian is $\mathbf{AB} \quad \mathbf{BA}$ or $\mathbf{AB} \quad \mathbf{BA} \quad 0$. The expression

$$[\mathbf{A},\mathbf{B}] \quad \mathbf{AB} \quad \mathbf{BA}$$

is called the *commutator* of **A** and **B**. Thus, *the product of two Hermitian operators is a Hermitian operator if and only if their commutator is equal to 0.*

We now introduce a class of Hermitian operators of special interest in quantum mechanics; a *projector*, **P**, is a Hermitian operator, with the property, $\mathbf{P}^2 \quad \mathbf{P}$. The outer product of a state vector φ

with itself, is a projector: $\mathbf{P}_\varphi \quad \varphi \ \varphi$. Two projectors, $\mathbf{P}_i$ and $\mathbf{P}_j$, are *orthogonal* if, for *every* state $\psi \quad \mathcal{H}_n$, the following, equality holds:

$$\mathbf{P}_i\mathbf{P}_j \ \psi \quad 0.$$

This condition is often written as

$$\mathbf{P}_i\mathbf{P}_j \quad 0.$$

A set of orthogonal projectors, $\mathbf{P}_0, \mathbf{P}_1, \mathbf{P}_2, \ldots$, is *complete/exhaustive* if

$$\underset{i}{\mathbf{P}_i} \quad I.$$

It is easy to see that

$$(\mathbf{P}_\varphi)^\dagger \quad (\ \varphi \ \varphi\ )^\dagger \quad \varphi \ \varphi \quad \mathbf{P}_\varphi.$$

If φ is normalized, $\varphi \ \varphi \quad 1$, then $\mathbf{P}^2_\varphi \quad \mathbf{P}_\varphi$:

$$(\mathbf{P}_\varphi)^2 \quad (\mathbf{P}_\varphi)^\dagger(\mathbf{P}_\varphi) \quad (\ \varphi \ \varphi\ )(\ \varphi \ \varphi\ ) \quad \varphi \ \varphi \quad \mathbf{P}_\varphi.$$

Properties of Hermitian Operators

Hermitian operators enjoy a set of remarkable properties; some of them, *P1–P5*, are discussed next.

P1. The eigenvalues of a Hermitian operator in $\mathcal{H}_n$ are real.

Proof. We calculate the inner product, $\psi \ \mathbf{A} \ \psi$, where λ_ψ is the eigenvalue corresponding to the eigenvector ψ of $\mathbf{A}$:

$$\psi \ \mathbf{A} \ \psi \quad \psi \ [\mathbf{A} \ \psi\ ] \quad \psi \ \lambda_\psi \ \psi \quad \lambda_\psi \ \psi \ \psi\ .$$

If we take the adjoint of $\mathbf{A}$, $\psi \quad \lambda_\psi \ \psi$, and use the fact that $\mathbf{A}$ is Hermitian, $\mathbf{A} \quad \mathbf{A}^\dagger$, it follows that $\psi \ A \quad \lambda_\psi \ \psi$. We now rewrite the inner product:

$$\psi \ \mathbf{A} \ \psi \quad [\ \psi \ \mathbf{A}] \ \psi \quad \lambda_\psi \ \psi \quad \psi \quad \lambda_\psi \ \psi \ \psi\ .$$

It follows that

$$\lambda_\psi \ \psi \ \psi \quad \lambda_\psi \ \psi \ \psi \qquad (\lambda_\psi \quad \lambda_\psi) \ \psi \ \psi \quad 0, \qquad \psi \quad \mathcal{H}_n.$$

Since $\psi \ \psi \quad 0$, it follows that $\lambda_\psi \quad \lambda_\psi$; thus, the eigenvalue λ_ψ is a real number. ■

P2. Two eigenvectors, ψ and φ , of the Hermitian operator $\mathbf{A}$ *with the distinct eigenvalues, $\lambda_\psi \quad \lambda_\varphi$, are orthogonal, $\psi \ \varphi \quad 0$.*

Proof. We calculate the inner product, $\psi \ \mathbf{A} \ \varphi$, and use the fact that λ_φ is an eigenvalue of $\mathbf{A}$ corresponding to the eigenvector φ , thus, $\mathbf{A} \ \varphi \quad \lambda_\varphi \ \varphi$:

$$\psi \ \mathbf{A} \ \varphi \quad \psi \ [\mathbf{A} \ \varphi\ ] \quad \psi \ \lambda_\varphi \ \varphi \quad \lambda_\varphi \ \psi \ \varphi\ .$$

The eigenvalue λ_ψ corresponds to the eigenvector ψ ; thus, $\mathbf{A}\ \psi\quad \lambda_\psi\ \psi$. If we take the adjoint of this expression and use the fact that $\mathbf{A}$ is Hermitian, $\mathbf{A}\quad \mathbf{A}^\dagger$, it follows that $\psi\ A\quad \lambda_\psi\ \psi$. We can now rewrite the inner product:

$$\psi\ \mathbf{A}\ \varphi \qquad [\ \psi\ \mathbf{A}]\ \varphi \qquad \lambda_\psi\ \psi \qquad \varphi \qquad \lambda_\psi\ \psi\ \varphi\ .$$

It follows that

$$\lambda_\varphi\ \psi\ \varphi \qquad \lambda_\psi\ \psi\ \varphi \qquad \lambda_\psi \quad \lambda_\varphi.$$

This is possible if and only if $\psi\ \varphi\quad 0$. ■

P3. For every Hermitian operator **A** *there exists a basis of orthonormal eigenvectors such that the matrix A is diagonal in that basis and its diagonal elements are all the eigenvalues of* **A**.

Proof. Let ψ_i , $1\quad i\quad n$, be an eigenvector of **A** and λ_i the corresponding eigenvalue. If the eigenvalues are distinct, $\lambda_i\quad \lambda_j$ for $i\quad j$, $1\quad i,j\quad n$, we construct the orthonormal basis as follows: we select an eigenvector ψ_j from a subspace that is orthogonal to the subspace spanned by $(\ \psi_1\ ,\ \psi_2\ ,\ldots,\ \psi_i\ )$.
Then:

$$A \qquad \begin{matrix} \lambda_1 & 0 & 0 & \\ 0 & \lambda_2 & 0 & \\ 0 & 0 & \lambda_3 & \\ \vdots & \vdots & \vdots & \ddots \\ & & & & \lambda_n \end{matrix} \qquad .$$

If the eigenvalues, λ_j, are degenerate (not distinct), then there are many bases of eigenvectors that diagonalize matrix A. If λ is an eigenvalue of multiplicity $K > 1$, the set of corresponding eigenvectors generates a subspace of dimension K, the eigenspace corresponding to that λ; the eigenspaces corresponding to different eigenvalues are orthogonal. Assume that λ is a degenerate eigenvalue, and the corresponding eigenvectors are ψ_{λ_1} and ψ_{λ_2} :

$$\mathbf{A}\ \psi_{\lambda_1} \qquad \lambda\ \psi_{\lambda_1} \qquad \text{and} \qquad \mathbf{A}\ \psi_{\lambda_2} \qquad \lambda\ \psi_{\lambda_2}\ .$$

Then, $a_1, a_2\quad \mathbb{C}$:

$$\mathbf{A}(a_1\ \psi_{\lambda_1} \qquad a_2\ \psi_{\lambda_2}\ ) \quad \lambda(a_1\ \psi_{\lambda_1} \qquad a_2\ \psi_{\lambda_2}\ ).$$

We can say that there is a whole subspace spanned by the vectors, ψ_{λ_1} and ψ_{λ_2} , the elements of which are eigenvectors of **A**, with eigenvalues λ. ■

P4. The necessary and sufficient condition for two Hermitian operators, $\mathbf{A}_1$ *and* $\mathbf{A}_2$*, to have the same set of eigenvectors is that they commute with each other:*

$$[\mathbf{A}_1, \mathbf{A}_2] \quad \mathbf{A}_1\mathbf{A}_2 \quad \mathbf{A}_2\mathbf{A}_1 \quad 0.$$

P5. The eigenvectors of a projector operator, $\mathbf{P}_\psi\quad \psi\ \psi$ *, where* $\psi\ \psi\quad 1$*, are either perpendicular or collinear to the vector* ψ *and their eigenvalues are* 0 *and* 1*, respectively.*

Proof. Assume φ is an eigenvector of $\mathbf{P}_\psi$ corresponding to the eigenvalue λ:

$$\mathbf{P}_\psi \quad \varphi \qquad \lambda \quad \varphi$$

or

$$\psi \quad \psi \quad \varphi \qquad \lambda \quad \varphi .$$

The inner product, $\psi \quad \varphi$, is a number γ:

$$\gamma \quad \psi \qquad \lambda \quad \varphi .$$

This implies that

$$\lambda \quad 0 \text{ if } \gamma \quad 0, \text{ when } \psi \text{ and } \varphi \text{ are perpendicular}$$

or

$$\lambda \quad 1 \text{ if } \gamma \quad 1, \quad \text{when} \quad \psi \text{ and } \varphi \text{ are parallel and normalized.}$$ ■

An operator **A** is *unitary* if $\mathbf{A}_n\mathbf{A}_n^\dagger \quad \mathbf{A}_n^\dagger\mathbf{A}_n \quad \mathbf{I}_n$. Clearly, a unitary operator is normal. The product of unitary operators is unitary, but the sum is not; "product" in this case means composition, not to be confused with tensor product.

A unitary operator **A** *preserves the inner product*; thus, it preserves the distance in a Hilbert space. If ψ_1 , ψ_2 , φ_1 , $\varphi_2 \quad \mathcal{H}_n$, $\varphi_1 \quad \mathbf{A} \quad \psi_1$, and $\varphi_2 \quad \mathbf{A} \quad \psi_2$, then:

$$\varphi_1 \quad \varphi_2 \qquad \psi_1 \quad \psi_2 .$$

Indeed, the inner product, $\varphi_1 \quad \varphi_2$, can be written as

$$\varphi_1 \quad \varphi_2 \qquad [\mathbf{A} \quad \psi_1 \;]^\dagger \mathbf{A} \quad \psi_2 \qquad \psi_1 \quad \mathbf{A}^\dagger\mathbf{A} \quad \psi_2 \qquad \psi_1 \quad \psi_2 .$$

It follows immediately that a unitary operator **A** preserves the norm of a vector:

$$[\; \varphi \quad A^\dagger] \, [A \quad \varphi \;] \qquad \varphi \quad [A^\dagger A] \quad \varphi \qquad \varphi \quad \varphi .$$

If **A** is a unitary operator, then we can undo its action (i.e., a unitary operator is invertible and $\mathbf{A}^{\;1} \quad \mathbf{A}^\dagger$).

Every normal operator has a complete set of orthonormal eigenvectors. If e_i is an eigenvector (eigenstate) of the operator **A** and λ_i is the associated eigenvalue, then we can write

$$\mathbf{A} \quad e_i \qquad \lambda_i \quad e_i .$$

The *spectral decomposition* of a normal operator **A** is

$$\mathbf{A} \qquad \underset{i}{} \; \lambda_i \, \mathbf{P}_i,$$

with $\mathbf{P}_i \quad e_i \quad e_i$, the projector corresponding to the eigenvector e_i .

Proof. Note that $(\mathbf{A} \quad \mathbf{A}^{\dagger})/2$ and $(\mathbf{A} \quad \mathbf{A}^{\dagger})/2i$ are Hermitian and commute. By *P3* and *P4* there exists an orthonormal base, $e_1 \ , \ e_2 \ , \ldots, \ e_n$, of eigenvectors simultaneously for $(\mathbf{A} \quad \mathbf{A}^{\dagger})/2$ and $(\mathbf{A} \quad \mathbf{A}^{\dagger})/2i$. Hence, they are also eigenvectors for

$$\mathbf{A} \qquad \frac{\mathbf{A} \quad \mathbf{A}^{\dagger}}{2} \quad i\frac{\mathbf{A} \quad \mathbf{A}^{\dagger}}{2i}$$

Let $\lambda_1, \lambda_2, \ldots, \lambda_n$ be the eigenvalues corresponding to the eigenvectors $e_1 \ , \ e_2 \ , \ldots, \ e_n$:

$$\mathbf{A} \qquad \lambda_i \mathbf{P}_i, \text{ with } \mathbf{P}_i \qquad e_i \ \ e_i \ .$$
$$i$$

Indeed, both $\mathbf{A}$ and $\ _i \lambda_i \mathbf{P}_i$ have the same matrix representation with respect to the base $e_1 \ , \ e_2 \ , \ldots, \ e_n$. ■

Example. Choose an orthonormal basis in $\mathcal{H}_2$. Let $\mathbf{A}$ be a 2 2 matrix in this basis:

$$A \quad \begin{matrix} a_{11} & a_{12} \\ a_{21} & a_{22} \end{matrix} \ .$$

Calculate its eigenvalues, λ_1 and λ_2, possibly equal. Find linearly independent solutions of the two linear equations:

$$A \begin{matrix} x_1 \\ y_1 \end{matrix} \qquad \lambda_1 \begin{matrix} x_1 \\ y_1 \end{matrix} \quad \text{and } A \begin{matrix} x_2 \\ y_2 \end{matrix} \qquad \lambda_2 \begin{matrix} x_2 \\ y_2 \end{matrix} \ .$$

Define

$$v_1 \qquad \begin{matrix} \dfrac{x_1}{x_1{}^2 \quad y_1{}^2} \\ \dfrac{y_1}{x_1{}^2 \quad y_1{}^2} \end{matrix} \quad \text{and} \quad v_2 \qquad \begin{matrix} \dfrac{x_2}{x_2{}^2 \quad y_2{}^2} \\ \dfrac{y_2}{x_2{}^2 \quad y_2{}^2} \end{matrix} \ .$$

The vectors, v_1 and v_2 , form another basis for $\mathbf{A}$, such that we can express $\mathbf{A}$ as

$$\mathbf{A} \quad \lambda_1 \ v_1 \ \ v_1 \qquad \lambda_2 \ v_2 \ \ v_2 \ .$$

Now we introduce the concept of positive operators and define the square root and the modulus of a positive operator; then we describe the canonical decomposition of an invertible operator [354].

Positive Operators

An operator $\mathbf{V} : \mathcal{H}_n \qquad \mathcal{H}_n$ is a *positive operator*, $\mathbf{V} > 0$, if:

$$\mathbf{V}\varphi \ \ \varphi > 0, \qquad \varphi \quad \mathcal{H}_n, \ \ \varphi \qquad 0.$$

The eigenvalues of a positive-definite operator are real and positive.

An operator is *positive semi-definite* if $\varphi \quad 0 \;\; \mathbf{V}\varphi \;\; \varphi \qquad \varphi \;\; \mathbf{V} \;\; \varphi \qquad 0$ and the eigenvalues of $\mathbf{V}$ are real and non-negative; thus, $\mathrm{tr}V \quad 0$. A positive semi-definite operator is self-adjoint. If two positive operators commute, then their product is a positive operator.

If $\mathbf{V}$ is a positive semi-definite operator, there exists a unique positive semi-definite operator, $\mathbf{Q}$, such that $\mathbf{Q}^2 \quad \mathbf{V}$, denoted also by $\mathbf{Q} \qquad \overline{\mathbf{V}}$. Indeed, there exists an orthonormal base in $\mathcal{H}_n$ consisting of the eigenvectors of $\mathbf{V}$; with respect to this basis, $\mathbf{V}$ can be represented as a diagonal matrix, $\mathrm{diag}\,(p_1, p_2, \ldots, p_n)$. Since $\mathbf{V}$ is positive, $p_i \quad 0$; hence, they have unambiguously defined positive square roots. Define

$$\mathbf{Q} \qquad \overline{\mathbf{V}} \text{ to be represented by diag } \quad \overline{p_1}, \;\; \overline{p_2}, \ldots, \;\; \overline{p_n} \;.$$

The *modulus* of any operator, $\mathbf{V}$, is defined as

$$\mathbf{V} \qquad \overline{\mathbf{V}^\dagger \mathbf{V}} \quad 0,$$

and it is always positive. One can also consider $\overline{VV^\dagger} \quad 0$, but, unless $\mathbf{V}$ is normal, this is not the same as $\overline{V^\dagger V} \quad 0$.

The operator, $\mathbf{Q}$, is *invertible* if

$$\mathbf{Q} \;\; \mathbf{Q} \quad ^{1} \quad \mathbf{I}.$$

If an operator $\mathbf{Q}$ *is invertible, then it can be written as a product of a positive-definite operator,* $\mathbf{V}$*, and a unitary operator,* $\mathbf{U}$*:*

$$\mathbf{Q} \quad \mathbf{VU}, \quad \text{with} \quad \mathbf{V} \quad \mathbf{Q} \qquad \overline{\mathbf{QQ}^\dagger} \quad \text{and} \quad \mathbf{U} \quad \mathbf{Q} \;\; ^{1}\mathbf{Q}.$$

From the definition, it follows immediately that $\mathbf{V}$ is positive as the modulus of $\mathbf{Q}$. To check that U is unitary, observe that $\mathbf{Q}$ and $\mathbf{Q} \quad ^{1}$ are self-adjoint; hence, $\mathbf{Q} \quad ^{\dagger} \qquad \mathbf{Q}$.
Then we have

$$\mathbf{UU}^\dagger \quad \mathbf{Q} \;\; ^{1}\mathbf{QQ}^\dagger(\;\mathbf{Q} \;\; ^{1})^\dagger \quad \mathbf{Q} \;\; ^{1}\mathbf{QQ}^\dagger \;\; \mathbf{Q} \;\; ^{1} \quad \mathbf{Q} \;\; ^{1} \;\mathbf{Q} \;\; \mathbf{Q} \;\; \mathbf{Q} \;\; ^{1} \quad \mathbf{I}.$$

To prove that this decomposition is unique, we assume that

$$\mathbf{Q} \quad \mathbf{V}_1\mathbf{U}_1 \quad \text{and} \quad \mathbf{Q} \quad \mathbf{V}_2\mathbf{U}_2.$$

Then, we use the fact that $\mathbf{U}_1$ and $\mathbf{U}_2$ are unitary, $\mathbf{U}_1\mathbf{U}_1^\dagger \quad \mathbf{U}_2\mathbf{U}_2^\dagger \quad \mathbf{I}$, to get

$$\mathbf{QQ}^\dagger \quad [\mathbf{V}_1\mathbf{U}_1][\mathbf{V}_1\mathbf{U}_1]^\dagger \quad \mathbf{V}_1\mathbf{U}_1\mathbf{U}_1^\dagger\mathbf{V}_1^\dagger \quad \mathbf{V}_1\mathbf{V}_1^\dagger \quad \mathbf{V}_1^2$$

and

$$\mathbf{QQ}^\dagger \quad [\mathbf{V}_2\mathbf{U}_2][\mathbf{V}_2\mathbf{U}_2]^\dagger \quad \mathbf{V}_2\mathbf{U}_2\mathbf{U}_2^\dagger\mathbf{V}_2^\dagger \quad \mathbf{V}_2\mathbf{V}_2^\dagger \quad \mathbf{V}_2^2.$$

It follows that $\mathbf{V}_1 \quad \mathbf{V}_2$ and immediately that $\mathbf{U}_1 \quad \mathbf{U}_2$.

Another useful property of a positive operator $\mathbf{V}$ regards the trace of the product of $\mathbf{V}$ and a unitary operator $\mathbf{U}$. The maximum with respect to $\mathbf{U}$ of the trace of the product $\mathbf{VU}$ is

$$\mu \quad \max_{\mathbf{U}}[\mathrm{tr}\,(\mathbf{VU})] \quad \mathrm{tr}\,(\mathbf{V}) \quad \text{if}\ (\mathbf{VU}) \quad 0.$$

We show this only for the case when $\mathbf{V}$ is invertible and $\mathbf{V} \quad \mathbf{V} \quad \mathbf{U}$. Let $p_1, p_2, \ldots, p_i, \ldots, p_n \quad 0$ be the eigenvalues of $\mathbf{V}$:

$$\mathbf{V} \quad \mathrm{diag}\,[p_1, p_2, \ldots, p_n].$$

The elements of $U \quad [u_{ij}]$, the matrix associated with the unitary operator $\mathbf{U}$, cannot be larger than unity, $u_{ij} \quad 1$. It follows that

$$\mathrm{tr}\,(\mathbf{VU}) \qquad \underset{i}{\quad} p_i u_{ii} .$$

Thus, the maximum is obtained when $u_{ii} \quad 1$ and $\mathbf{U} \quad \mathbf{I}$; then $\mu \quad \mathrm{tr}\,(\mathbf{V})$, as stated.

A useful relation is the *Schwarz inequality* for the inner product of two operators (see Problem 1.1 at the end of this chapter):

$$\mathrm{tr}\ \mathbf{VV}^{\dagger}\ \mathrm{tr}\ \mathbf{QQ}^{\dagger} \qquad \mathrm{tr}\ \mathbf{VQ}^{\dagger} \ ^{2}.$$

Last, we consider a Hilbert space, $\mathcal{H}_n$, with an orthonormal basis $e_1 , e_2 , \ldots, e_n$. Consider now a vector, φ , in the extended Hilbert space $\mathcal{H}_n \quad \mathcal{H}_n$:

$$\varphi \qquad \overset{n}{\underset{i\ 1}{\quad}} e_i \quad e_i .$$

Let $\mathbf{V}$ be any operator in $\mathcal{H}_n$ and let $\mathbf{V}^T$ be its transpose in $\mathcal{H}_n$. If we want to apply the operator $\mathbf{V}$ defined on $\mathcal{H}_n$ to the state vector φ defined on $\mathcal{H}_n \quad \mathcal{H}_n$, it is easy to see that we have to write

$$(\mathbf{V} \quad \mathbf{I}) \ \varphi \qquad \mathbf{I} \quad \mathbf{V}^T \quad \varphi$$

or

$$\overset{n}{\underset{i\ 1}{\quad}} [\mathbf{V}\ \ e_i\ ] \quad e_i \qquad \overset{n}{\underset{i\ 1}{\quad}} e_i \quad \mathbf{V}^T \ e_i .$$

We have now concluded the survey of the mathematical apparatus required to describe information embodied by a quantum system. As we shall see in Chapter 6, the physical processes for the storage, transformation, and transport of information for photons, ion traps, quantum dots, nuclear magnetic resonance (NMR), and other quantum systems are very different and could distract us from the goal of discovering the common properties of quantum information independent of its physical support. To study the properties of quantum information, we have to resort to an abstract model that captures the essence of quantum behavior; this model, quantum mechanics, describes the properties of physical systems as objects in a finite-dimensional Hilbert space.

1.4 POSTULATES OF QUANTUM MECHANICS

A model of a physical system is an abstraction based on *correspondence rules* that relate the entities manipulated by the model to the physical objects or systems in the real world. Once such rules are established, we can operate only with the abstractions according to a set of *transformation rules*. To ensure the usefulness of the model and its ability to describe physical reality, we have to *validate* the model and compare its prediction with the physical reality. To ensure expressiveness, the ability of the model to describe the physical system, the correspondence and the transformation rules must be kept as simple as possible, but, at the same time, complete—in other words, capable of capturing the relevant properties of the physical system and of its dynamics, its evolution in time.

Distinguishability and system dynamics require the model to abstract the concepts of *observable* and of *state* of the physical object. An observable is a property of the system state that can be revealed as a result of some physical transformation. The state at time t is a synthetic characterization of the object that could be revealed by the measurement of relevant observables at time t.

The model must also abstract the concept of *measurement*; it should describe the relation between the state of the object before and after the measurement and specify how to interpret the results of a measurement, how to map the range of possible results to abstractions. In the physical world, we often have to deal with a collection of physical objects. If $A, B, C, \ldots$ are the abstractions of the objects $a, b, c, \ldots$, respectively, we need another transformation rule to specify how to construct $A, B, C, \ldots$, the abstraction corresponding to the collection $a, b, c, \ldots$. Last, but not least, we need transformation rules to describe the system dynamics, the evolution of the system in time.

Quantum mechanics is a model of the physical world at all scales; it describes more accurately than classical physics systems at the atomic and subatomic scale. This model allows us to abstract our knowledge of a quantum system, to describe the state of single and composite quantum systems, the effect of a measurement on the system's state, and the dynamics of quantum systems. A quantum state summarizes our knowledge about a quantum system at a given moment in time, it allows us to describe what we know, as well as, what we do not know, about the system. An impressive number of experiments have produced results consistent with the prediction of quantum mechanics and so far there is no experimental evidence to disprove it; thus, we shall use this model to study the properties of quantum information.

The correspondence and transformation rules are captured by the *postulates of quantum mechanics* (Figure 1.2). We find it useful to expand the traditional three postulates of quantum mechanics, the state postulate, the dynamics postulate, and the measurement postulate, to emphasize some aspects important for quantum information processing:

1. A quantum system, $\mathcal{Q}$, is described in an n-dimensional Hilbert space, $\mathcal{H}_n$, where n is finite. The Hilbert space $\mathcal{H}_n$ is a linear vector space over the field of complex numbers with an inner product. The dimension, n, of the Hilbert space is equal to the maximum number of reliably distinguishable states the system $\mathcal{Q}$ can be in.
2. A state ψ of the quantum system $\mathcal{Q}$ corresponds to a direction (or ray) in $\mathcal{H}_n$. In Section 1.11, we shall see that the most general representation of a quantum state is any density operator over an n-dimensional Hilbert space with n finite. The density operator is Hermitian, has non-negative eigenvalues, and has a trace equal to unity.

1. A PHYSICAL SYSTEM IS REPRESENTED BY A HILBERT SPACE.

2. A STATE OF THE SYSTEM IS A RAY IN THIS SPACE.

3. THE SPONTANEOUS EVOLUTION OF THE SYSTEM IN ISOLATION IS DESCRIBED BY A CERTAIN UNITARY TRANSFORMATION IN THIS HILBERT SPACE.

4. TWO OR MORE SYSTEMS ARE REPRESENTED BY THE TENSOR PRODUCT OF THE HILBERT SPACES REPRESENTING EACH COMPONENT SYSTEM.

5. A MEASUREMENT OF A QUANTUM SYSTEM CORRESPONDS TO A PROJECTION OF ITS STATE INTO ORTHOGONAL SUBSPACES. THE SUM OF THESE PROJECTIONS IS ONE.

FIGURE 1.2

The postulates of quantum mechanics.

3. When the internal conditions and the environment of a quantum system are completely specified and no measurements are performed on the system, the system's evolution is described by a unitary transformation in $\mathcal{H}_n$ defined by the Hamiltonian operator. A unitary transformation $\mathbf{U}$ is linear and preserves the inner product. The spontaneous evolution of an unobserved quantum system with the density matrix, ρ, is

$$\rho \quad \mathbf{U}\rho\mathbf{U}^{\dagger},$$

with $\mathbf{U}^{\dagger}$, the adjoint of $\mathbf{U}$.

4. Given two independently prepared quantum systems, $\mathcal{Q}$ described in $\mathcal{H}_n$ and $\mathcal{S}$ described in $\mathcal{H}_m$, the bipartite system consisting of both $\mathcal{Q}$ and $\mathcal{S}$ is described, in a Hilbert space, $\mathcal{H}_n \quad \mathcal{H}_m$, the tensor product of the two Hilbert spaces.

5. A measurement of the quantum system $\mathcal{Q}$ in the state $| \psi \rangle$ described in $\mathcal{H}_n$ corresponds to a resolution of $\mathcal{H}_n$ to orthogonal subspaces, $\{\mathcal{H}^j\}$, and a projection of the system's state to these subspaces, $\{\mathbf{P}_j\}$, such that the sum of the projections is $\sum \mathbf{P}_j = 1$. The measurement produces the result, j, with the probability

$$\text{Prob}(j) = \| \mathbf{P}_j | \psi \rangle \|^2.$$

The state after the measurement is

$$| \varphi \rangle = \frac{\mathbf{P}_j | \psi \rangle}{\| \mathbf{P}_j | \psi \rangle \|} = \frac{\mathbf{P}_j | \psi \rangle}{\sqrt{\text{Prob}(j)}}.$$

Manipulation of coherent quantum states is at the heart of quantum computing and quantum communication. A quantum computation involves a single entity and consists of unitary transformations of the quantum state. Quantum communication involves multiple entities and involves the transmission of quantum states over noisy communication channels.

1.5 QUANTUM STATE POSTULATE

> A state is a complete description of a physical system. In quantum mechanics, a state $| \psi \rangle$ of a system is a vector—in fact, a direction (ray) in the Hilbert space $\mathcal{H}_n$.

Consider a canonical base, $\{|0\rangle, |1\rangle, \ldots, |k\rangle, \ldots, |(n-1)\rangle\} \in \mathcal{H}_n$; the state, $|\psi\rangle \in \mathcal{H}_n$, can be written as a linear combination of basis states:

$$| \psi \rangle = \sum_{k=0}^{n-1} \alpha_k | k \rangle.$$

The coefficients

$$\alpha_k = \langle k | \psi \rangle$$

are complex numbers and represent the *probability amplitudes*; the probability of observing the state $|k\rangle$ is $p_k = |\alpha_k|^2$.

By convention, state vectors are assumed to be normal(ized) (i.e., $\langle \psi | \psi \rangle = 1$). Therefore:

$$\sum_{k=0}^{n-1} |\alpha_k|^2 = 1.$$

The *length* of a bra vector, $\langle \psi |$, or of the corresponding ket vector, $| \psi \rangle$, is defined as the square root of the positive number, $\langle \psi | \psi \rangle$. For a given state, the bra or ket vector representing it is defined only as a direction, and its length is undetermined up to a factor; the factor is chosen so that *the vector length is usually set equal to unity.* Even then the vector could be undetermined because it can be multiplied by a quantity of modulus 1. Such a quantity, the complex number, $e^{i\gamma}$, where γ is real and is called a *phase factor*.

The *inner product* of two state vectors, ψ and φ, represents the generalized "angle" between these states and gives an estimate of their *overlap*.

The inner product, $\psi \quad \varphi \quad 0$, defines orthogonal states; the implication of $\psi \quad \varphi \quad 1$ is that ψ and φ are parallel—in fact, one and the same state.

The inner product of two state vectors is a complex number, but the square of the inner product $\psi \quad \varphi \quad ^2$, a real number, can be interpreted as a quantitative measure of the "relative orthogonality" between these states.

Superpositions of Quantum States

We assume that the state of a dynamical quantum system at a particular time corresponds to a vector ψ ; if this state is the result of a superposition of other states, φ and ξ, it can be represented by the linear expression

$$\psi \quad \alpha \ \varphi \quad \beta \ \xi \ .$$

The *superposition principle*: every vector (*ray*) in the Hilbert space corresponds to a possible state; given two states, φ and ξ, we can form another state as a superposition of these two states, $\alpha \ \varphi \quad \beta \ \xi$. This is a characteristic of the Hilbert space, which contains all possible superpositions of its vectors, and is suited for the description of *interference effects*.

The state superposition has several properties derived from the properties of linear transformations:

1. Symmetry: the order of the state vectors in the superposition is not important:

$$\psi \quad \alpha \ \varphi \quad \beta \ \xi \quad \beta \ \xi \quad \alpha \ \varphi \ .$$

2. Each state in the superposition can be expressed as a superposition of the other states:

$$\varphi \quad \frac{1}{\alpha}(\ \psi \quad \beta \ \xi \ ).$$

3. The superposition of a state with itself results in the original state:

$$\alpha_1 \ \varphi \quad \alpha_2 \ \varphi \quad (\alpha_1 \quad \alpha_2) \ \varphi \ .$$

If $\alpha_1 \quad \alpha_2 \quad 0$, there is no superposition and the two components cancel each other by an interference effect. If $a_1 \quad a_2 \quad a_3$, we assume that the result of the superposition is the original state itself, and we can conclude that if the ket (bra) vector corresponding to a state is multiplied by any nonzero complex number, the resulting ket (bra) vector will correspond to the same state.

There is a *fundamental difference* between a quantum and a classical superposition. For example, a superposition of a membrane vibration state with itself results in a different state with a different magnitude of the oscillation; the magnitude of such a classical oscillation has no correspondence in any physical characteristic of a quantum state. A classical state with amplitude of oscillation zero everywhere is a membrane at rest. No corresponding state exists for a quantum system since a zero ket vector corresponds to no state at all.

Quantum states have several other properties:

- A state is specified by the direction of a ket vector; the length of the vector is irrelevant.
- The states of a dynamical system are in one-to-one correspondence with all the possible orientations of a ket vector.
- The directions of the ket vectors ψ and (ψ) are not distinct.
- When a state ψ is the result of a superposition of two other states, the ratio of the complex coefficients α and β effectively determines the state ψ .

A quantum state is a *ray in a Hilbert space*; a ray is an *equivalence class* of vectors that differ by multiplication by a nonzero complex scalar. We can choose an element of this class (for any nonvanishing vector) to have unit norm $\psi \;\; \psi \quad 1$. For such a normalized vector, we can say that ψ or $e^{i\gamma} \;\; \psi$, where $e^{i\gamma} \quad 1$ describe the same physical state. The phase factor $e^{i\gamma}$ becomes physically significant when it appears in a superposition state $\alpha \;\; \varphi \quad e^{i\gamma}\beta \;\; \xi$ as a *relative phase*.

Transformation of Quantum States

If we apply a transformation to a quantum system in the state ψ (e.g., rotate it, or let it evolve for some time Δt), the system evolves to a different state, φ . The physical interaction of a quantum system (e.g., the interaction of an atom with a magnetic field) is represented in our formalism by an *operator*.

Consider a canonical base, $0 ,\ldots, \; j ,\ldots, \; k ,\ldots, \; (n \quad 1) \quad \mathcal{H}_n$. A Hermitian operator, $\mathbf{A} \quad \mathbf{A}^{\dagger}$, applied to the state $\psi \quad {}_k\alpha_k \;\; k$, with $\alpha_k \quad k \;\; \psi$, produces a new state φ :

$$\varphi \quad \mathbf{A} \;\; \psi \quad \text{or} \quad \varphi \quad \psi \;\; \mathbf{A}^{\dagger} \quad \psi \;\; \mathbf{A}.$$

It follows that

$$j \;\; \varphi \quad j \;\; \mathbf{A} \;\; \psi \quad j \;\; \mathbf{A} \quad \underset{k}{\quad} \alpha_k \;\; k \quad \underset{k}{\quad} j \;\; \mathbf{A} \;\; k \; \alpha_k \quad \underset{k}{\quad} j \;\; \mathbf{A} \;\; k \;\; k \;\; \psi \; .$$

The expressions, $j \;\; \varphi$, give the amount that each basis state, j , contributes to the state φ ; this amount is given in terms of a linear superposition of $k \;\; \psi$, the probability amplitudes in each basis state of the original state, ψ . The numbers, $j \;\; \mathbf{A} \;\; k$, tell how much of each amplitude, $k \;\; \psi$, goes to the sum for each j ; these coefficients are the components, A_{jk}, of the matrix, $A \quad [A_{jk}]$, associated with the linear operator **A**:

$$A_{jk} \quad j \;\; \mathbf{A} \;\; k \; .$$

Once we have determined the matrix A associated with the operator **A** for one basis, we can calculate the corresponding matrix for another basis; the matrix can be transformed from one *representation* to another.

The Qubit

Consider a quantum system in a two-dimensional Hilbert space, a *qubit*. The state of a qubit, $\psi \quad \mathcal{H}_2$, can be expressed using the canonical base, 0 and 1 , as

$$\psi \quad \alpha_0 \;\; 0 \quad \alpha_1 \;\; 1 , \quad \alpha_0, \alpha_1 \quad \mathbb{C}, \quad \alpha_0 \;\; ^2 \quad \alpha_1 \;\; ^2 \quad 1.$$

Recall that using polar coordinates we can express an arbitrary complex number, $z = x + iy$, as $z = r\cos\theta + ir\sin\theta$ (where $x = r\cos\theta$, $y = r\sin\theta$, and $i = \sqrt{-1}$). Then using Euler's identity $e^{i\theta} = \cos\theta + i\sin\theta$, we can express z as $z = re^{i\theta}$.

It follows immediately that the state of a qubit can be defined using four real parameters, $r_0, \delta_0, r_1, \delta_1$, and substituting $\alpha_0 = r_0 e^{i\delta_0}$ and $\alpha_1 = r_1 e^{i\delta_1}$:

$$| \psi \rangle = r_0 e^{i\delta_0} | 0 \rangle + r_1 e^{i\delta_1} | 1 \rangle .$$

The only measurable quantities for the state $| \psi \rangle$ are the probability amplitudes $| \alpha_0 |^2$ and $| \alpha_1 |^2$; thus, multiplying the state by an arbitrary global phase, $e^{i\gamma}$, has no observable consequences; indeed,

$$| e^{i\gamma} \alpha_0 |^2 = (e^{i\gamma} \alpha_0)^* (e^{i\gamma} \alpha_0) = (e^{-i\gamma} \alpha_0^*)(e^{i\gamma} \alpha_0) = \alpha_0^* \alpha_0 = | \alpha_0 |^2 .$$

A similar derivation can be carried out for $| \alpha_1 |^2$; if we multiply the expression of the state by $e^{-i\delta_0}$, we get a new expression for the state of the qubit:

$$| \psi \rangle = r_0 e^{i\delta_0} e^{-i\delta_0} | 0 \rangle + r_1 e^{i(\delta_1 - \delta_0)} | 1 \rangle = r_0 | 0 \rangle + r_1 e^{i\varphi} | 1 \rangle .$$

We notice that now the state depends only on three real parameters, r_0, r_1, and $\varphi = \delta_1 - \delta_0$. We express the state $| \psi \rangle$ in Cartesian coordinates as

$$| \psi \rangle = r_0 | 0 \rangle + (x + iy) | 1 \rangle$$

and observe that the normalization constraint $\langle \psi | \psi \rangle = 1$ requires that

$$| r_0 |^2 + | x + iy |^2 = 1.$$

This condition can be rewritten as

$$r_0^2 + x^2 + y^2 = 1.$$

This is the equation of unit sphere in a real three-dimensional space with the Cartesian coordinates x, y, r_0. Thus, the state $| \psi \rangle$ is represented by a vector connecting the center to a point on the unit sphere. If we use polar coordinates (Figure 1.3a), then the three-dimensional Cartesian coordinates can be expressed as

$$x = r\sin\theta\, \cos\varphi,\ y = r\sin\theta\, \sin\varphi,\ z = r\cos\theta .$$

But $r = 1$; thus,

$$| \psi \rangle = z | 0 \rangle + (x + iy) | 1 \rangle = \cos\theta | 0 \rangle + \sin\theta\, (\cos\varphi + i\sin\varphi) | 1 \rangle = \cos\theta | 0 \rangle + e^{i\varphi} \sin\theta | 1 \rangle .$$

The state requires only two parameters, θ and φ. We also notice that when $\theta = 0$, then $| \psi \rangle = | 0 \rangle$, and when $\theta = \pi/2$, then $| \psi \rangle = e^{i\varphi} \sin\theta | 1 \rangle$; this suggests that $0 \leq \theta \leq \pi/2$ and $0 \leq \varphi \leq 2\pi$ may generate all the points on the unit sphere.

To show that we can only consider the upper hemisphere of the so-called *Bloch sphere*, we consider two opposite points: $| \psi \rangle$, with polar coordinates $(1, \theta, \varphi)$ and $| \psi' \rangle$, with polar

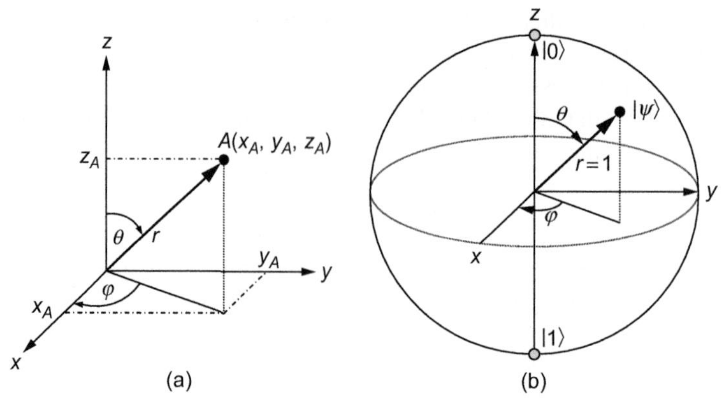

FIGURE 1.3

(a) The Cartesian coordinates of the point $A(x_A, y_A, z_A)$, with polar coordinates (r, θ, φ), are $x_A \quad r\sin\theta\cos\varphi$, $y_A \quad r\sin\theta\sin\varphi$, and $z_A \quad r\cos\theta$. (b) The Bloch sphere representation of a qubit in the state ψ .

coordinates $(1, \theta\ ) \quad (\pi - \theta, \varphi \quad \varphi \quad \pi)$. It is easy to see that the two states differ only by a phase factor of $-1 \quad e^{-i\pi}$, and we recall that multiplication by $e^{i\gamma}$, γ, has no observable consequence:

$$\psi \quad \cos(\pi - \theta)\ 0 \quad e^{i(\varphi \quad \pi)}\sin(\pi - \theta)\ 1 \quad -\cos\theta\ 0 - e^{i\varphi}\sin\theta\ 1 \quad - \psi\ .$$

Thus, when we use $\theta \quad \theta/2$ we can represent the state of a qubit as a point on the Bloch sphere (Figure 1.3b) as

$$\psi \quad \cos\frac{\theta}{2}\ 0 \quad e^{i\varphi}\sin\frac{\theta}{2}, \quad 0 \quad \theta \quad \pi,\ 0 \quad \varphi \quad 2\pi.$$

Examples of Quantum Mechanical Operators

We present now a few basic, quantum operators:

1. The *Pauli spin* (rotation) operators for spin-½ particles, $\sigma_I, \sigma_x, \sigma_y, \sigma_z$, also denoted as I, X, Y, Z, are:

$$\sigma_I \quad \begin{matrix} 1 & 0 \\ 0 & 1 \end{matrix}, \ \sigma_x \quad \begin{matrix} 0 & 1 \\ 1 & 0 \end{matrix}, \ \sigma_y \quad \begin{matrix} 0 & -i \\ i & 0 \end{matrix}, \text{ and } \sigma_z \quad \begin{matrix} 1 & 0 \\ 0 & -1 \end{matrix}.$$

2. The *Hadamard* operator, H, rotates the basis with an angle of 45 degrees:

$$H \ \psi \quad \frac{1}{\overline{2}} \quad \begin{matrix} 1 & 1 \\ 1 & -1 \end{matrix}.$$

3. The *rotations around the x,y,z axes*, with an angle θ, are carried out by the following unitary operators:

$$\mathcal{R}_x(\theta) \quad \cos\frac{\theta}{2}\sigma_I \quad i\sin\frac{\theta}{2}\sigma_x \qquad \begin{matrix} \cos\frac{\theta}{2} & i\sin\frac{\theta}{2} \\ i\sin\frac{\theta}{2} & \cos\frac{\theta}{2} \end{matrix} \quad ,$$

$$\mathcal{R}_y(\theta) \quad \cos\frac{\theta}{2}\sigma_I \quad i\sin\frac{\theta}{2}\sigma_y \qquad \begin{matrix} \cos\frac{\theta}{2} & \sin\frac{\theta}{2} \\ \sin\frac{\theta}{2} & \cos\frac{\theta}{2} \end{matrix} \quad ,$$

and

$$\mathcal{R}_z(\theta) \quad \cos\frac{\theta}{2}\sigma_I \quad i\sin\frac{\theta}{2}\sigma_z \qquad \begin{matrix} e^{\ \ i\frac{\theta}{2}} & 0 \\ 0 & e^{i\frac{\theta}{2}} \end{matrix} \quad .$$

4. The *rotation around an axis connecting the origin with the point* $A(x_A, y_A, z_A)$ on the Bloch sphere, with an angle θ, is

$$\mathcal{R}_A(\theta) \quad \cos\frac{\theta}{2}\sigma_I \quad i\sin\frac{\theta}{2}(x_A\sigma_x \quad y_A\sigma_y \quad z_A\sigma_z).$$

5. The *inversion (parity)* operator Π creates a new state by reversing the sign of all coordinates.
6. The *ip* operator in a two-dimensional Hilbert space, $\mathcal{H}_2$, written as a linear combination of orthonormal basis vectors is

$$\mathbf{A} \quad 0 \ 1 \qquad 1 \ 0 \ .$$

When applied to a ket state vector, $\psi \quad \alpha_0 \ 0 \quad \alpha_1 \ 1$, or the bra ψ , we obtain, respectively,

$$\mathbf{A} \ \psi \qquad (\ 0 \ 1 \qquad 1 \ 0 \ )(\alpha_0 \ 0 \quad \alpha_1 \ 1 \ ) \quad \alpha_0 \ 1 \quad \alpha_1 \ 0 \ ,$$

$$\psi \ \mathbf{A} \quad (\alpha_0 \ 0 \quad \alpha_1 \ 1 \ )(\ 0 \ 1 \qquad 1 \ 0 \ ) \quad \alpha_0 \ 1 \quad \alpha_1 \ 0 \ .$$

7. The z-component of the angular momentum, $\mathbf{J}_z$. This operator is defined in terms of the rotation operator for an infinitesimally small angle $\Delta\theta$ around the z-axis,

$$\mathbf{R}_z(\Delta\theta) \quad 1 \quad \frac{i}{\hbar}\mathbf{J}_z\Delta\theta,$$

with $\hbar \quad h/2\pi$, the reduced Planck constant. Among the new states resulting from application of this rotation operator, some are the same as the initial state, except for a phase factor. The phase is proportional to the angle $\Delta\theta$, thus:

$$\mathbf{R}_z(\Delta\theta) \ \ \psi_0 \qquad 1 \quad \frac{i}{\hbar}\ \mathbf{J}_z \ \Delta\theta \qquad \psi_0 \qquad e^{im\Delta\theta} \quad \psi_0 \qquad (1 \quad im\Delta\theta) \ \ \psi_0 \ .$$

When we compare this with the definition of $\mathbf{J}_z$ above, we see that

$$\mathbf{J}_z \ \ \psi_0 \qquad m\hbar \ \ \psi_0 \ ,$$

where $m\hbar$ is the amount of the z-component of the angular momentum. This expression can be interpreted in the following way: If we operate with $\mathbf{J}_z$ on a state with a definite angular momentum about the z-axis, we get the same state multiplied by a factor $m\hbar$. The state ψ_0 is an eigenvector of the operator, J_z, with the eigenvalue $m\hbar$.

8. The *displacement* operator, $\mathbf{D}_x(L)$, by distance, L, along the x-axis. A small displacement, Δx along x, transforms a state ψ to ψ , where

$$\psi \quad \mathbf{D}_x(\Delta x) \quad \psi \quad 1 \quad \frac{i}{\hbar}\, \mathbf{p}_x\, \Delta x \quad \psi\ .$$

When Δx 0, the, ψ should become the initial state ψ —i.e., $\mathbf{D}_x(0)$ 1. For infinitesimally small Δx, the change of $\mathbf{D}_x$ from its value 1 should be proportional to Δx; the proportionality quantity, $\mathbf{p}_x$, is the x-component of the *momentum operator*.

1.6 DYNAMICS POSTULATE

> The dynamics of a quantum system is specified by a Hermitian operator, **H**, called the Hamiltonian; the time evolution of the system is described by the Schrödinger equation:
>
> $$i\hbar \tfrac{d}{dt} \quad \psi(t) \qquad \mathbf{H} \quad \psi(t)\ ,$$
>
> where $\hbar$ is the reduced Planck's constant.

More precisely, $\psi(t)$ $e^{\;it/\hbar}\mathbf{H}$ $\psi(0)$, with $\psi(0)$, the initial state of the system at t 0. The Schrödinger equation of a finite dimensional quantum system is in fact a *coupled system*[1] *of linear differential equations*.

The Schrödinger equation represents the relation between the states of a quantum system observed at different instants of time. When we make an observation (measurement) on a dynamical quantum system, the state of the quantum system is changed; it is projected with a certain probability onto one of the basis vectors, i.e., it "collapses." However, between observations, the evolution of the quantum system is expected to be governed by equations of motion and that makes the state at one time determine the state at a later time (causality is assumed to apply in a similar way as in classical physics). The equations of motion will apply as long as the quantum system is left undisturbed by any observation.

The Schrödinger equation, as written above, gives the general law for the variation with time of the vector corresponding to the state at any time. The operator, $\mathbf{H}(t)$, the Hamiltonian, is a real, linear operator, characteristic of the dynamical system. The Hamiltonian is a special operator: it describes the complete dynamics of a quantum system under time evolution, and it also determines the energy eigenstates (equilibrium states) of the system.

The Hamiltonian matrix, H $[H_{ij}]$, contains all the physics behind the actions that, when applied to the quantum system, cause it to change; thus, the matrix may depend on time. If we know this matrix, we have a complete description of the behavior of the system with respect to time. The Hamiltonian **H**,

[1] A set of partial differential equations is a *coupled system* if the same functions $f(x)$ and/or their derivatives appear in several of the equations, so that solving one equation depends on the solutions of another.

as the total energy, takes different forms for different physical systems. For example, the energy of a body of mass m in a *simple harmonic motion* is given by

$$\frac{p^2}{2m} \quad \frac{m\omega^2 x^2}{2},$$

where x, p, and ν are, respectively, the position, momentum, and the oscillation frequency of the body at a given moment in time, and $\omega \quad 2\pi\nu$. A body executes a *simple harmonic motion* when the force that tries to restore the body to its equilibrium (rest) position is proportional to the displacement of the body.[2]

To describe the state of a quantum system, we need to select a set of basis states and to express through the matrix, H_{ij}, the physical laws that apply to that particular system. The rule is to find the Hamiltonian corresponding to a particular physical situation, such as the interaction with a magnetic field, with an electric field, and so on. For nonrelativistic phenomena and for some other special cases, there are very good approximations. For example, the form of the Hamiltonian containing the kinetic energy and the Coulomb interaction between nuclei and electrons in atoms is a very good starting point to describe chemical phenomena.

The Hamiltonian of a quantum system is a Hermitian operator:

$$\mathbf{H} \quad \mathbf{H}^\dagger \qquad H_{ij} \quad H_{ji}.$$

This property is required by the condition that the total probability that an isolated quantum system is in *some* state (any state) does not change in time and by the fact that H, the energy observable, must have real eigenvalues. If our system is an isolated particle, as time goes on we will find it in the very same state, $\psi \quad {}_i\alpha_i \;\, i$, with $\; i \;$ an orthonormal basis in $\mathcal{H}_n$; the probability to find it *somewhere* at time, t, is

$$\underset{i}{} \quad \alpha_i(t) \;\; ^2 \quad 1.$$

This probability must not vary with time, though the individual probability amplitudes vary with time.

Next, we consider some simple quantum systems and describe them using the Hamiltonian.

A System with One Basis State

A hydrogen atom at rest can be described to a good approximation with only one basis state, which is an energy eigenvector of the hydrogen atom. Since the atom is at rest, we assume that the external conditions do not change in time; thus, the Hamiltonian, $\mathbf{H}$, is independent of time and

$$i\hbar \frac{d\alpha_1}{dt} \quad H_{11}\,\alpha_1.$$

---

[2]A typical example is the motion of an object on a spring when it is subject to an elastic restoring force, $F \quad kx$, with k the spring constant and x the displacement of the body; the motion is sinusoidal in time.

The system is described with one differential equation for the probability amplitude α_1, with H_{11}, as constant (in time), and its solution is

$$\alpha_1 \quad (\text{const})\, e^{\;\frac{i}{\hbar} H_{11} t}.$$

The solution expresses the time dependence of a state characterized by a definite total energy, $E \quad H_{11}$, of the system.

A System with Two Basis States

Let us assume the basis states are 0 and 1, and the system is in the state ψ. If the probability amplitude of being in the state 0 is $\alpha_0 \quad 0 \;\psi$ and the probability amplitude of being in the state 1 is $\alpha_1 \quad 1 \;\psi$, then the state vector ψ can be written as

$$\psi \qquad 0 \;\; 0 \;\; \psi \qquad 1 \;\; 1 \;\; \psi$$

or

$$\psi \quad \alpha_0 \;\; 0 \quad \alpha_1 \;\; 1 \;.$$

If we assume that the system changes its state at any given moment in time, the coefficients, α_0 and α_1, will change in time according to the equations

$$i\hbar \frac{d\alpha_0}{dt} \quad H_{11}\alpha_0 \quad H_{12}\alpha_1$$

and

$$i\hbar \frac{d\alpha_1}{dt} \quad H_{21}\alpha_0 \quad H_{22}\alpha_1.$$

Solution: We have to make some assumptions about the H matrix in order to solve these equations. Consider the following two hypotheses.

Stationary states hypothesis: We assume that once the system is in one of the states, 0 or 1, there is no chance that it could transition to the other state. In this case, the matrix elements expressing transitions from one state to another are $H_{12} \quad 0$ and $H_{21} \quad 0$, and the equations become

$$i\hbar \frac{d\alpha_0}{dt} \quad H_{11}\alpha_0,$$

$$i\hbar \frac{d\alpha_1}{dt} \quad H_{22}\alpha_1.$$

The solutions of these differential equations for the probability amplitudes are

$$\alpha_0 \quad (\text{const})\, e^{\;\frac{i}{\hbar} H_{11} t},$$

$$\alpha_1 \quad (\text{const})\, e^{\;\frac{i}{\hbar} H_{22} t}.$$

These are the amplitudes for *stationary* states, with energies $E_0 \quad H_{11}$ and $E_1 \quad H_{22}$, respectively. These two states are separated by an energy barrier. If the two states, 0 , and 1 , are symmetrical, the two energies are equal:

$$E_0 \quad E_1 \quad E.$$

Nonstationary states hypothesis: Let us assume that there is a small probability (amplitude) that the system could transition from one state to the other and thereby "tunnel" through the energy barrier separating the two (symmetrical) states. That means

$$H_{12} \quad H_{21}^{} \quad H_{21} \qquad \mathcal{E}$$

up to a phase factor.

The initial two equations for the probability amplitudes become

$$i\hbar \frac{d\alpha_0}{dt} \quad E\alpha_0 \quad \mathcal{E}\alpha_1,$$

$$i\hbar \frac{d\alpha_1}{dt} \quad E\alpha_1 \quad \mathcal{E}\alpha_0.$$

We solve this system of two differential equations. First, we take the sum of the two equations

$$i\hbar \frac{d}{dt}(\alpha_0 \quad \alpha_1) \quad (E \quad \mathcal{E})(\alpha_0 \quad \alpha_1),$$

then we take the difference of the two equations

$$i\hbar \frac{d}{dt}(\alpha_0 \quad \alpha_1) \quad (E \quad \mathcal{E})(\alpha_0 \quad \alpha_1).$$

The solutions of these new differential equations are, respectively,

$$\alpha_0 \quad \alpha_1 \quad a\, e^{\ \frac{i}{\hbar}(E \quad \mathcal{E})t}$$

and

$$\alpha_0 \quad \alpha_1 \quad b\, e^{\ \frac{i}{\hbar}(E \quad \mathcal{E})t}.$$

The integration constants, a and b, are chosen to give the appropriate initial conditions for a particular system. By adding and subtracting the last two equations, we get the probability amplitudes

$$\alpha_0(t) \quad \frac{a}{2} e^{\ \frac{i}{\hbar}(E \quad \mathcal{E})t} \quad \frac{b}{2} e^{\ \frac{i}{\hbar}(E \quad \mathcal{E})t},$$

$$\alpha_1(t) \quad \frac{a}{2} e^{\ \frac{i}{\hbar}(E \quad \mathcal{E})t} \quad \frac{b}{2} e^{\ \frac{i}{\hbar}(E \quad \mathcal{E})t}.$$

These solutions have a physical interpretation. If $b \quad 0$,

$$\alpha_0(t) \quad \alpha_1(t) \quad \frac{a}{2} e^{\ \frac{i}{\hbar}(E \quad \mathcal{E})t}.$$

The two probability amplitudes are equal and have the same frequency, $\nu \quad (E \quad \mathcal{E})/\hbar$, in the exponent. We can say that the system is in a state of definite energy, $(E \quad \mathcal{E})$, at this frequency, in a *stationary* state, when the amplitudes, α_0 and α_1, for the system in state 0 and, respectively, 1 are equal.

If $a \quad 0$,

$$\alpha_0(t) \quad \frac{b}{2} e^{\quad \frac{i}{\hbar}(E \quad \mathcal{E})t},$$

$$\alpha_1(t) \quad \quad \frac{b}{2} e^{\quad \frac{i}{\hbar}(E \quad \mathcal{E})t},$$

and

$$\alpha_0(t) \quad \quad \alpha_1(t).$$

This is another possible stationary state. This time the two amplitudes have the frequency $(E \quad \mathcal{E})/\hbar$. We say that the system is in a state of definite energy $(E \quad \mathcal{E})$ if the two amplitudes are equal, but of opposite sign (i.e., $\alpha_0 \quad \quad \alpha_1$).

Now, let us assume that at $t \quad 0$ the system is in the state 0 . Then:

$$\alpha_0(0) \quad \frac{a \quad b}{2} \quad 1,$$

$$\alpha_1(0) \quad \frac{a \quad b}{2} \quad 0,$$

with the result

$$a \quad b \quad 1.$$

The amplitudes become

$$\alpha_0(t) \quad e^{\quad \frac{i}{\hbar} Et} \quad \frac{e^{\frac{i}{\hbar}\mathcal{E}t} \quad e^{\quad \frac{i}{\hbar}\mathcal{E}t}}{2},$$

$$\alpha_1(t) \quad e^{\quad \frac{i}{\hbar} Et} \quad \frac{e^{\frac{i}{\hbar}\mathcal{E}t} \quad e^{\quad \frac{i}{\hbar}\mathcal{E}t}}{2},$$

and we can rewrite them as

$$\alpha_0(t) \quad e^{\quad \frac{i}{\hbar}Et} \cos \frac{\mathcal{E}t}{\hbar},$$

$$\alpha_1(t) \quad e^{\quad \frac{i}{\hbar}Et} \sin \frac{\mathcal{E}t}{\hbar}.$$

The probability that the system is found in the state 0 at time t is

$$\alpha_0(t)^{\;2} \quad \cos^2 \frac{\mathcal{E}t}{\hbar}.$$

The probability that the system is in the state 1 at time t is

$$\alpha_1(t) \quad ^2 \quad \sin^2 \frac{\mathcal{E}t}{\hbar}.$$

At the initial moment t 0, the probability that the system is in the state 1 is zero increases to one and continues to oscillate between zero and one in time. The probability that the system is in the state 0 is one at the initial moment, t 0, decreases to zero, and then oscillates between one and zero in time. We say that the magnitude of the two probability amplitudes varies *harmonically*[3] with time. The probability of finding the system is in one of the two states varies back and forth between the magnitudes of the two individual probabilities.

1.7 MEASUREMENT POSTULATE

> The numerical outcome of a measurement of observable A of a quantum system in the state $\varphi \quad \mathcal{H}_n$ is an eigenvalue λ_i of the operator **A** used to measure observable A (Figure 1.4); after the measurement, the quantum state of the system is an eigenvector a_i of **A** corresponding to the eigenvalue λ_i.

This postulate is sometimes called the *postulate of collapse* because as we shall see later a measurement projects, or collapses, the state of the quantum system being measured.

Another formulation of this postulate is that *mutually exclusive measurement outcomes correspond to orthogonal projection operators (projectors)* $\mathbf{P}_0, \mathbf{P}_1, \ldots$. From the definition of a complete set of orthogonal projectors it follows that the number of projectors in a complete orthogonal set must be *less than*, or *equal to*, the dimension of the Hilbert space. The measurement postulate can be reformulated in terms of completeness of a set of projectors: *A complete set of orthogonal projectors specifies an exhaustive measurement.*

The resulting state after applying the transformation given by $\mathbf{P}_i$ to a quantum system in the state φ is

$$\psi \quad \mathbf{P}_i \quad \varphi \; .$$

The probability of obtaining a measurement outcome, λ_i, is the norm of the resulting state

$$p(\lambda_i) \qquad \psi \quad ^2 \quad \mathbf{P}_i \quad \varphi \quad ^2 \quad (\mathbf{P}_i \quad \varphi \;)^\dagger \; \mathbf{P}_i \quad \varphi \qquad \varphi \;\; \mathbf{P}_i^\dagger \mathbf{P}_i \quad \varphi \qquad \varphi \;\; (\mathbf{P}_i)^2 \quad \varphi \; .$$

But $(\mathbf{P}_i)^2 \quad \mathbf{P}_i$; thus,

$$p(\lambda_i) \qquad \varphi \;\; \mathbf{P}_i \quad \varphi \; .$$

---

[3]A harmonic variation is expressed in complex exponential form as $Ae^{i(\omega t \quad \epsilon)}$, where $\omega \quad 2\pi\nu$ is the angular velocity.

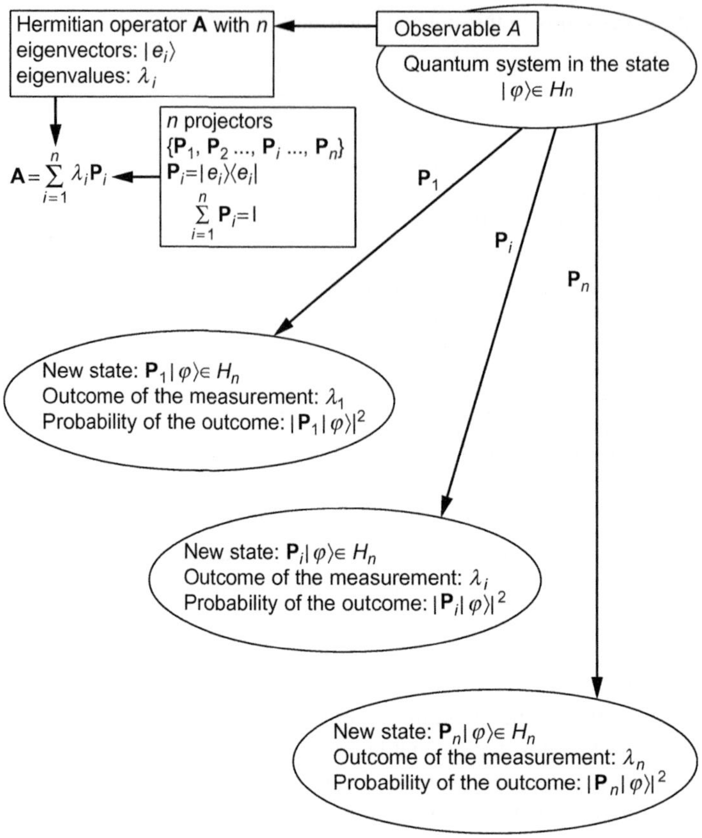

FIGURE 1.4

The measurement of observable A of a quantum system in the state φ $\mathcal{H}_n$. The operator $\mathbf{A}$ corresponding to observable A is Hermitian and has n eigenvalues, λ_i $\mathbb{R}$, 1 i n. The corresponding eigenvectors e_i , 1 i n, form an eigenbasis. The set of n projectors $\mathbf{P}_i$ is complete if $\,^n_i\,_1 \mathbf{P}_i$ $\mathbf{I}$; the state of the system after the transformation due to $\mathbf{P}_i$ is ψ_i $\mathbf{P}_i$ φ , the outcome of the measurement is λ_i, and the Prob(λ_i) $\mathbf{P}_i$ φ ^{2}.

The completeness of the set of projectors implies that the total probability for all possible measurement outcomes is $\;_i p(\lambda_i)$ 1. It follows that

$$\varphi \;\mathbf{P}_i\; \varphi \quad 1.$$
$$i$$

After the measurement, the state ψ becomes a normalized pure quantum state defined as

$$\psi \qquad \frac{\psi}{\varphi \;\mathbf{P}_i\; \varphi} \qquad \frac{\mathbf{P}_i \;\varphi}{\varphi \;\mathbf{P}_i\; \varphi}.$$

Example. Consider a two-dimensional Hilbert space, $\mathcal{H}_2$, where we have chosen the orthogonal basis vectors x and y. We define two possible state vectors of a system in this space:

$$\varphi_A \quad \alpha_x \; x \quad \alpha_y \; y \quad \text{and} \quad \varphi_B \quad \beta_x \; x \quad \beta_y \; y .$$

The initial state of the system could be ψ_A, with probability p, or ψ_B, with probability $1 \quad p$; we say that the system is a *mixed* ensemble of quantum states.

We perform a large number, N, of measurements corresponding to the projectors

$$\mathbf{P}_x \quad x \; x \quad \text{and} \quad \mathbf{P}_y \quad y \; y .$$

We wish to predict, n_x, the number of times out of the total number N of measurements, when we expect to obtain the measurement outcome corresponding to basis vector x. We use the conditional probabilities, Prob($x \quad \varphi_A$) and Prob($x \quad \varphi_B$), to express n_x as

$$\begin{aligned} n_x \quad & N[\text{Prob}(\ \varphi_A\ )\ \text{Prob}(x \quad \varphi_A\ ) \quad \text{Prob}(\ \varphi_B\ )\ \text{Prob}(x \quad \varphi_B\ )] \\ & N[p \ \ \varphi_A \ \ \mathbf{P}_x \ \ \varphi_A \quad (1 \quad p)\ \varphi_B \ \ \mathbf{P}_x \ \ \varphi_B\ ] \\ & N[p \ \ \alpha_x \ ^2 \quad (1 \quad p) \ \ \beta_x \ ^2]. \end{aligned}$$

The ratio, n_x/N, is bounded from below by the smaller of $\alpha_x \ ^2$ and $\beta_x \ ^2$ as $0 \quad p \quad 1$.

Let us assume that the initial state of the system is a *coherent superposition* of the states, φ_A and φ_B, corresponding to the *pure* state $\varphi(\gamma_A, \gamma_B)$, with

$$\begin{aligned} \varphi(\gamma_A, \gamma_B) \quad & \gamma_A \ \ \varphi_A \quad \gamma_B \ \ \varphi_B \\ & (\gamma_A \alpha_x \quad \gamma_B \beta_x) \ \ x \quad (\gamma_A \alpha_y \quad \gamma_B \beta_y) \ \ y , \end{aligned}$$

and where γ_A and γ_B are chosen such that the state $\varphi(\gamma_A, \gamma_B)$ is normalized, $\gamma_A \ ^2 \quad \gamma_B \ ^2 \quad 1$.

In this case, the probability of a measurement outcome corresponding to basis vector x is n_x/N

$$\begin{aligned} n_x/N \quad & \varphi(\gamma_A, \gamma_B) \ \ \mathbf{P}_x \ \ \varphi(\gamma_A, \gamma_B) \\ & \gamma_A \alpha_x \quad \gamma_B \beta_x \ ^2 . \end{aligned}$$

We notice that in certain cases it is possible to choose γ_A and γ_B such that $\gamma_A \alpha_x \quad \gamma_B \beta_x \quad 0$ and then, $n_x \quad 0$, though $\alpha_x \ ^2, \ \beta_x \ ^2 > 0$. This phenomenon is known as *destructive interference*.

There is an important distinction between *coherent superpositions* (of the type that produce a single pure state) and *incoherent admixtures* (of the type that produce a mixed ensemble of quantum states).

We will now summarize several important properties of the quantum observables and quantum operators related to the concepts discussed in this section:

1. An *observable* in quantum mechanics is a Hermitian operator with a complete set of eigenvectors. This set of eigenvectors can be chosen to be mutually orthogonal; they form a basis.
2. The eigenvalues of a Hermitian operator are real numbers.
3. Eigenvectors of a Hermitian operator corresponding to different eigenvalues are mutually orthogonal.
4. If two Hermitian operators commute they have a common basis of orthonormal eigenvectors (an eigenbasis). If they do not commute, then no common eigenbasis exists.
5. A complete set of commuting observables is the minimal set of Hermitian operators with a unique common eigenbasis.
6. The eigenvalues of a unitary operator are complex numbers of modulus 1.
7. Eigenvectors of a unitary operator corresponding to different eigenvalues are mutually orthogonal.
8. In general, an operator must be normal for the property of orthogonality of eigenvectors to hold.
9. If **A** is a Hermitian operator and A is an observable of a system, then the measurement of observable A of the system in the state φ leaves the system in a state that is an eigenvector a of **A**, and the probability of this outcome is $a \;\; \varphi \;\; ^2$.

1.8 LINEAR ALGEBRA AND SYSTEMS DYNAMICS

Linear algebra allows us to describe the evolution in time of classical as well as quantum systems. Consider first a discrete-time, finite-dimensional, nondeterministic classical system; the state transitions occur at distinct time instances $t_1, t_2, \ldots, t_k, \ldots$; the state space is of dimension n. The state of the system at time t_k is a stochastic vector, $\sigma \quad p_1^k, p_2^k, \ldots, p_n^k$, with $\;_i p_i^k \quad 1$ and p_i^k, the probability of the system being in the state i, $1 \quad i \quad n$, at time t_k. The dynamics of the system is captured by a directed graph G with vertices corresponding to the states; the arcs correspond to transitions and are labeled with the transition probabilities. This graph is characterized by the *adjacency matrix* $A \quad [a_{ij}]$, with a_{ij}, the probability of the transition from state i to state j. The adjacency matrix A is a *row stochastic matrix*, and the sum of all elements in a row is equal to 1.

For example, in the graph in Figure 1.5, the state of the system is given by the vector $\sigma^T(t_k) \quad (1/2\ 1/4\ 1/4)$, and the adjacency matrix is

$$A \quad \begin{matrix} 1/4 & 1/2 & 1/4 \\ 1/6 & 1/3 & 1/2 \\ 7/12 & 1/6 & 1/4 \end{matrix} \quad V.$$

If the system is in the state σ_{t_k} at time t_k, then the state at time $t_k \;\; 1$ will be $\sigma_{t_k \;\; 1} \quad A\sigma_{t_k}$:

$$\sigma_{t_k \;\; 1} \quad \begin{matrix} 1/4 & 1/2 & 1/4 \\ 1/6 & 1/3 & 1/2 \\ 7/12 & 1/6 & 1/4 \end{matrix} \quad \begin{matrix} 1/2 \\ 1/4 \\ 1/4 \end{matrix} \quad \begin{matrix} 5/16 \\ 7/24 \\ 19/48 \end{matrix} \; .$$

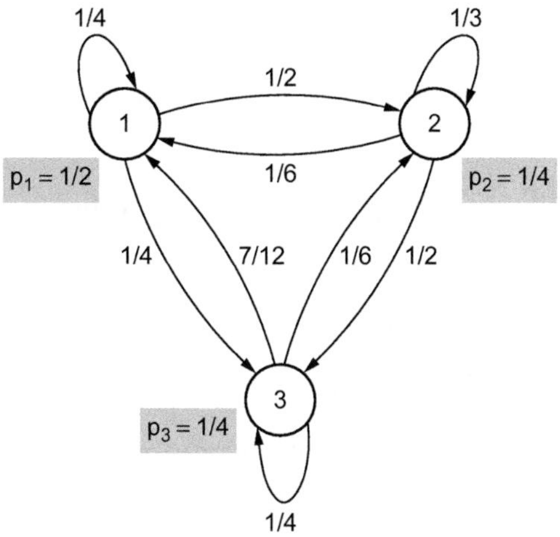

FIGURE 1.5

Systems dynamics. The directed graph, G, with adjacency matrix A.

The next state of a classical Markovian discrete-time stochastic system depends only on the current state. The system could reach a state σ_i from two distinct initial states, σ_a and σ_b; the memoryless property does not allow us to distinguish which path was taken to reach σ_i and makes it impossible to return to the initial state; the system is *nonreversible*. Only if A is a permutation matrix, a matrix with one nonzero element in each row and each column, is the discrete-time system reversible. The description of a classical stochastic system with a continuous-time and a possibly infinite state space is considerably more complex and will not be discussed.

We have been concerned with the dynamics of classical systems where probabilities are real numbers, 0 $\; p_i, p_j \;$ 1, and $\;_i p_i \;$ 1; as such probabilities can only increase when added, $p_i \;\; p_j \;\; p_i$ and $p_i \;\; p_j \;\; p_j$. What if the probabilities are functions of complex numbers (e.g., they are the square of the modulus of complex numbers, $q_i \;^2 \;\; \alpha_i \;\; i\beta_i \;^2$), and we replace the condition $\;_i p_i \;\; 1$ with $_i \; q_i \;^2 \quad _i (\alpha_i^2 \;\; \beta_i^2 \quad 1$?

With these rules in place, we notice that $q_1 \;\; q_2 \;^2$ can be smaller than $q_1 \;^2$ or $q_2 \;^2$. For example, if $q_1 \;\; 3 \;\; 5i$ and $q_2 \;\; -4-4i$, then $q_1 \;^2 \;\; 9 \;\; 25 \;\; 34$, $q_2 \;^2 \;\; 16 \;\; 16 \;\; 32$, and $q_1 \;\; q_2 \;^2 \;\; (3-4) \;\; i(5-4) \;^2 \;\; 1 \;\; 1 \;\; 2$.

The arcs in the directed graphs are now labelled with complex numbers, and we now require that the adjacency matrix be unitary, rather than row stochastic. If q_{ij} is the label of the arc from state i to state j, then the adjacency matrix is

$$Q \quad Q[q_{ij}] \quad \text{and} \quad QQ^{\dagger} \quad I.$$

Then:

$$\sigma_{t \;\; 1} \quad Q \;\; \sigma_t \qquad \sigma_t \quad \sigma_{t \;\; 1} \;\; Q.$$

Indeed, $\sigma_{t \;\; 1} \quad Q \;\; \sigma_t$ implies that $\sigma_{t \;\; 1} \quad \sigma_t \;\; Q^{\dagger}$ and $\sigma_{t \;\; 1} \;\; Q \quad \sigma_t \;\; Q^{\dagger}Q \quad \sigma_t$.

A logical question we address next is why we choose the transition probabilities for a discrete-time quantum system to be functions of complex numbers. The answer is that *the new probability rules allow us to capture the wave-like behavior of quantum systems while the real-valued probabilities describe classical, particle-like behavior.*

The Wave-Particle Duality and the New Probability Rule

One of the greatest discoveries leading to the formulation of the wave equation of Schrödinger in 1926 was the *wave-particle duality principle* formulated by Louis de Broglie in 1923; this principle expresses states that atomic and subatomic particles (matter) and photons (energy) exhibit both wave-like and particle-like properties. The wavelength λ and the momentum p of any form of matter are related: $\lambda \quad h/p$, with h, Planck's constant.

To grasp the physical significance of the probability rules, we consider the double-slit experiments discussed [148]. When the experiment is performed with bullets and when both slits are open, the number of bullets $n(x)$ collected at a spot x between the two slits is the sum, $n(x) \quad n_{up}(x) \quad n_{low}(x)$, with $n_{up}(x)$, the number of bullets when only the upper slit is open and $n_{low}(x)$ is the number of bullets when only the lower slit is open. If N bullets are shot and $p(x) \quad n(x)/N$ is the probability of a bullet hitting spot x with both slits open, and $p_{up}(x) \quad n_{up}(x)/N$ and $p_{low}(x) \quad n_{low}(x)/N$ correspond to the two probabilities with only one slit open, then the usual probability rule applies: $p(x) \quad p_{up}(x)$ $p_{low}(x)$.

If the double slit experiment is performed with waves, then in some spots the amplitude $a(x)$ of the waves is reinforced, $a(x) > \max(a_{up}, a_{low})$, while in other spots it is diminished, $a(y) < \min(a_{up}(y), a_{low}(y))$. This is due to the phenomenon of *interference* characteristic to wave-like behavior.

In conclusion, *state transformation in a Hilbert space corresponds to application of a unitary operator to the current state, and the probability of a state is the modulus of a complex number.* For example, in $\mathcal{H}_2$ a system in the state

$$\varphi \quad \alpha_0 \;\; 0 \quad \alpha_1 \;\; 1$$

has the probability $p_0 \quad \alpha_0 \;^2$ to be in the state 0 and the probability $p_1 \quad \alpha_1 \;^2$ to be in the state 1, with $\alpha_0 \;^2 \quad \alpha_1 \;^2 \quad 1$. This system will be the embodiment of a unit of quantum information, called a qubit.

1.9 SYMMETRY AND DYNAMIC EVOLUTION

Symmetry plays an important role in classical as well as quantum physics. The mathematical concept that describes the symmetry of a physical system or of a physical object is the concept of a *group*. An abstract group is a set G equipped with a map, $g : G \quad G \quad G$ and $(g)^{\;1} : G \quad G$, and an element, $e \quad G$, so that the familiar axioms discussed in Section 4.4 hold. A *subgroup* G of the group G is a subset, such that $g_1, g_2 \quad G$ implies $g_1 \; g_2^{\;1} \quad G$.

The set of symmetries of a physical system form a group; one can compose two symmetries, consider the inverse of each symmetry, and there always exists the "identity" as an obvious symmetry. Most of these groups belong to a remarkable class of groups, the *Lie groups*. An abstract Lie group satisfies the axioms of a group and, in addition, is a "smooth manifold"[4] such that both the composition and the inverse are given by differentiable functions. The most familiar group is $GL(n,F)$, the group of symmetries of an n-dimensional vector space over the field F of real or complex numbers.

Almost all Lie groups can be described as subgroups of $GL(n,\mathbb{R})$ or $GL(n,\mathbb{C})$. For example, $SL(n,\mathbb{R})$, the group of symmetries of $\mathbb{R}^n$ that preserves the n-dimensional volume, is the subgroup of $GL(n,\mathbb{R})$ consisting of matrices with a determinant equal to 1. As another example, $O(n,\mathbb{R})$, the group of symmetries of the vector space $\mathbb{R}^n$ that preserves the angles and the distances, is a subgroup of matrices, $\mathbf{A}$, with real elements, and $\mathbf{A}\mathbf{A}^T \quad \mathbf{I}$. Also, $U(n)$, the group of symmetries of the Hilbert space $\mathbb{C}^n$, is the subgroup of matrices $\mathbf{A}$ in $GL(n,\mathbb{C})$, with $\mathbf{A}\mathbf{A}^\dagger \quad \mathbf{I}$.

Lie algebra is a mathematical object that is mathematically, rather than conceptually, simpler than a Lie group. To each Lie group one can associate a Lie algebra; if the Lie group is a group of symmetries, then the Lie algebra is usually called *infinitesimal symmetries.* Each finite-dimensional Lie algebra determines a unique (connected and simply connected) Lie group.

A Lie algebra is a vector space, V, over a field, F, together with a binary operation $[,]$ called a Lie bracket with the following properties:

1. Bilinearity:

$$[\alpha u \quad \beta v, w] \quad \alpha[u,w] \quad \beta[v,w],\ [w,\alpha u \quad \beta v] \quad \alpha[w,u] \quad \beta[w,v], \qquad \alpha,\beta \quad F,\ u,v,w \quad V.$$

2. The Lie bracket $[u,u]$:

$$[u,u] \quad 0, \qquad u \quad V.$$

3. Jacobi identity:

$$[u,[v,w]] \quad [v,[w,u]] \quad [w,[u,v]] \quad 0, \qquad u,v,w \quad V.$$

The operation defined by the Lie bracket is not in general associative; $[[x,y],z]$ need not equal $[x,[y,z]]$.

The Lie algebra of $GL(n,F)$ is denoted by $gl(n,F)$ and is the vector space of all $n \quad n$ matrices, with elements from the field F and bracket $[A,B] \quad A \ B \quad B \ A$. The Lie algebra of $SL(n,\mathbb{R})$ and $O(n,\mathbb{R})$ are subspaces of $n \quad n$ matrices A, with real elements and $\mathrm{tr}(A) \quad 0$, respectively, and $A \quad A^T \quad 0$ (skew symmetric matrices). The Lie algebra of $U(n)$, denoted by $u(n)$, consists of $n \quad n$ matrices with complex elements that satisfy $A \quad A^\dagger \quad 0$, equivalently with matrices A such that (iA) is a Hermitian matrix.

An element in the Lie algebra defines a smooth one-parameter family of elements in the Lie group parameterized by $t \quad \mathbb{R}$, which for $t \quad 0$ is exactly e. Any smooth one-parameter family of elements in the group, which for $t \quad 0$ is e, is an element in the Lie algebra of the group. If an $n \quad n$-matrix A is an element of $gl(n,F)$, then the one-parameter family is

$$e^{tA} \quad I \quad \frac{tA}{1!} \quad \frac{(tA)^2}{2!} \qquad .$$

[4]A smooth manifold is a geometrical object that is locally the same as the Euclidian space $\mathbb{R}^n$ [268].

This series is convergent for any t to an invertible matrix. If $\mathbf{A}(t)$ is a one-parameter family of elements in $gl(n,F)$, then

$$\left.\frac{d\mathbf{A}(t)}{dt}\right|_{t=0}$$

is a matrix, not necessarily invertible, and hence, an element in $gl(n,F)$.

The noncommutativity of the Lie group is reflected in the bracket operation; more precisely if

$$A = \left.\frac{d\mathbf{A}(t)}{dt}\right|_{t=0} \quad \text{and} \quad B = \left.\frac{d\mathbf{B}(t)}{dt}\right|_{t=0},$$

then

$$\lim_{t \to 0} \frac{\mathbf{A}(t)\mathbf{B}(t)\mathbf{A}(t)^{-1}\mathbf{B}(t)^{-1} - \mathbf{I}}{t^2} = [A,B].$$

In particular,

$$[A,B] = \lim_{t \to 0} \frac{e^{tA}e^{tB}e^{-tA}e^{-tB} - I}{t^2}.$$

The equality can be verified by applying the l'Hôspital rule from calculus.

Usually the elements of a Lie algebra of a Lie group of symmetries (i.e., infinitesimal symmetries) define conservation laws (first observed by Nöther [14]).

If an observable $\mathbf{A}$ (a self-adjoint operator) is invariant by the dynamics of a quantum system defined by the Hamiltonian $\mathbf{H}$ (i.e., all eigenstates and eigenvalues are preserved by the dynamics), which means $\mathbf{A}e^{-\frac{i}{\hbar}\mathbf{H}t} = e^{-\frac{i}{\hbar}\mathbf{H}t}\mathbf{A}$ (Section 1.6); one obtains (by derivation at $t = 0$) $[\mathbf{H},\mathbf{A}] = 0$. This holds even in the more general case when the dynamics is defined by the time-dependent Hamiltonian, $\mathbf{H}(t)$; precisely, $[\mathbf{H}(t),\mathbf{A}] = 0$.

1.10 UNCERTAINTY PRINCIPLE AND MINIMUM UNCERTAINTY STATES

In classical physics *nondeterminism* is due to *uncontrolled causes that are recognized to exist and that, if better known, would make the predictions better.* The quantum state postulate reveals the critical role of nondeterminism in quantum theory; the uncertainty principle shows that it has broader implications on our ability to observe the properties of a quantum system and makes us wonder if its nature is fundamentally different from the role it plays in classical physics.

In quantum physics nondeterminism means that a *precise knowledge at the quantum level is impossible.* The uncertainty principle introduced by Werner Heisenberg in 1927 is at the heart of the special nature of the nondeterminism of quantum mechanics. Uncertainty is an intrinsic property of quantum systems. The accuracy of measured values of physical properties, such as position and momentum along the axis used to measure the position, is limited; the precise knowledge of both the position and the momentum is forbidden in a quantum system. This limitation cannot be avoided, as shown by many experiments performed over the years. Einstein, who doubted that "God is playing dice," questioned the truth of such an indeterminacy.

The *uncertainty principle*: consider canonically conjugated observables x, the position of a particle, and p_x, the momentum of the same particle at position x; x and p_x are three-dimensional vectors, and we consider their projections along the same direction, d. Then Δx, the uncertainty in determining the projection of the position, and Δp_x, the uncertainty in determining the projection of the momentum at position x along the same direction, are constrained by the inequality:

$$\Delta x \, \Delta p_x \geq \frac{\hbar}{2},$$

where $\hbar = h/2\pi$ is the reduced Planck's constant.

Let us assume that we are interested in two observables, A and B, of a quantum particle that are associated with the Hermitian operators, $\mathbf{A}$ and $\mathbf{B}$, respectively. We prepare two disjoint sets, $\mathcal{S}_1$ and $\mathcal{S}_2$, each set consisting of a large number of quantum systems in the identical state $|\varphi\rangle$. For the systems in $\mathcal{S}_1$, we measure first the observable A and then the observable B on all particles. We obtain the "same" value for the observable A, while we notice a large standard deviation of the observable B. For the systems in $\mathcal{S}_2$, we measure first the observable B and then the observable A. Now we obtain the same value for the observable B, while we notice a large standard deviation of the observable A.

Call $\Delta\mathbf{A}$ and $\Delta\mathbf{B}$ the standard deviations of the measurements of observables A and B, respectively. The uncertainty principle can be expressed as

$$\Delta\mathbf{A} \, \Delta\mathbf{B} \geq \frac{1}{2} \left| \langle \varphi | [\mathbf{A},\mathbf{B}] | \varphi \rangle \right|,$$

with $[\mathbf{A},\mathbf{B}]$, the commutator of the two operators, $\mathbf{A}$ and $\mathbf{B}$. If A and B are noncommutative quantum observables, then $[\mathbf{A},\mathbf{B}] > 0$. It follows that $\Delta\mathbf{A}\Delta\mathbf{B} > 0$; thus, there is a minimum level of uncertainty that cannot possibly be removed.

On the other hand, *when two operators commute, they can be diagonalized simultaneously, and we can measure the eigenvalues of one of them without disturbing the eigenvectors of the other.* This property is important for a class of quantum error-correcting codes, the stabilizer codes, discussed in Section 5.12.

The quantum mechanical uncertainty relation $\Delta x \, \Delta p_x \geq \frac{\hbar}{2}$ has a classical counterpart based on the wave phenomenon. An acoustic signal with intensity $s(t)$ cannot have precise timing and precise pitch; the two must satisfy the inequality

$$\Delta t \, \Delta\omega \geq \frac{1}{2},$$

where

$$(\Delta t)^2 = \int t^2 s(t)dt - \left(\int ts(t)dt \right)^2 \text{ and } (\Delta\omega)^2 = \int \omega^2 S(w)d\omega - \left(\int \omega S(\omega)d\omega \right)^2.$$

In this expression, $S(\omega) = \mathcal{F}(s(t))$ is the Fourier transform of the function $s(t)$ and $\omega = (2\pi)/f$, with $f = 1/T$, the frequency and T the period of the signal.

Peres observes [324] that in the case of acoustic signals we can have *approximate* values for the time and the frequency; musical scores indicate both time and frequency and allow musicians to produce sounds that capture the information generated by the composer. It seems quite reasonable to consider also approximate values for the observables of a quantum system. For example, we can consider a

Gaussian distribution of the position and wavelength, $\lambda \quad h/p$, of a wave packet with mean x and variance σ, and, respectively, p. Then the wave function

$$\phi(x) \quad (\pi\sigma^2) \;^{1/4} \exp \frac{x \quad x}{2\sigma^2} \quad \frac{1}{\hbar} ipx$$

is a minimum uncertainty wave packet with

$$\Delta x \quad \frac{1}{\overline{2}}\sigma, \quad \Delta p \quad \frac{\hbar}{\overline{2}}\sigma, \quad \text{and} \quad \Delta x \, \Delta p \quad \frac{\hbar}{2}\sigma^2.$$

Minimum uncertainty wave functions can be extended to noncommuting operators other than position and momentum [324]. Coherent superposition states, introduced in Section 1.7 and discussed in more depth in Section 3.16, are minimum uncertainty states.

1.11 PURE AND MIXED QUANTUM STATES

So far, our discussion has focused on a class of states of a quantum system, described by Dirac's ket and bra vectors, or, by a *wave function* in a finite-dimensional Hilbert space. These states are called *pure quantum states*; a pure state can be expressed as a superposition of the basis vectors of an orthonormal basis, $\mathcal{B} \quad b_1 \,, \; b_2 \,, \ldots, \; b_n \quad \mathcal{H}_n$, as

$$\psi \quad \sum_{i \; 1}^{n} \alpha_i \; b_i \,, \quad \text{with} \quad \sum_{i \; 1}^{n} \; \alpha_i \;^2 \quad 1.$$

A pure state that is a linear combination of two or more component states is sometimes called a *coherent superposition*. According to the postulates of quantum mechanics, discussed in Section 1.4, the evolution of a closed quantum system can be completely described as a unitary transformation of pure states in a Hilbert space.

For a qubit in a coherent superposition state, there is always a basis in which any measurement of the qubit will produce the same result. For example, the state

$$\psi \quad \frac{1}{\overline{2}}(\, 0 \quad 1 \,)$$

represents a qubit that has a 50% probability to produce either 0 or 1 as a result of a measurement. If we rotate the basis by 45 degrees by applying a Hadamard transformation, H, the state of the qubit becomes

$$H \;\; \psi \quad \frac{1}{\overline{2}} \begin{matrix} 1 & 1 \\ 1 & 1 \end{matrix} \quad \frac{1}{\overline{2}} \begin{matrix} 1 \\ 1 \end{matrix} \quad \begin{matrix} 1 \\ 0 \end{matrix} ,$$

and all measurements of the resulting state 0 produce the same result, 0.

We introduce a new characterization of the state of a quantum system by means of the density operator. We can associate every vector, $\psi \quad \mathcal{H}_n$, with a matrix $A \quad \mathbb{C}^{n \quad n}$; indeed, $\mathbb{C}^{n \quad n}$, the set of

$n \quad n$ matrices with complex elements, is an inner-product vector space; thus, there is an isomorphism from a Hilbert space to $\mathbb{C}^{n \quad n}$. The *density matrix of a pure state*,

$$\psi \qquad \overset{n}{\underset{i \quad 1}{}} \alpha_i \;\; b_i \;, \qquad \text{with} \qquad \overset{n}{\underset{i \quad 1}{}} \;\; \alpha_i \;\; ^2 \quad 1,$$

is given by the outer product

$$\rho \qquad \psi \quad \psi \qquad \begin{matrix} \alpha_1\alpha_1 & \alpha_1\alpha_2 & \dots & \alpha_1\alpha_n \\ \alpha_2\alpha_1 & \alpha_2\alpha_2 & \dots & \alpha_2\alpha_n \\ \vdots & \vdots & & \vdots \\ \alpha_n\alpha_1 & \alpha_n\alpha_2 & \dots & \alpha_n\alpha_n \end{matrix} \quad .$$

The density matrix of a pure state is a Hermitian operator; indeed

$$\rho^\dagger \qquad (\;\psi \quad \psi\;)^\dagger \qquad \psi \quad \psi \qquad \rho.$$

The trace of the density matrix of a pure state is equal to 1:

$$\mathrm{tr}(\rho) \qquad \overset{n}{\underset{i \quad 1}{}} \alpha_i\alpha_i \qquad \overset{n}{\underset{i \quad q}{}} \;\; \alpha_i \;\; ^2 \quad 1.$$

A *mixed state* is a statistical ensemble, $(\;\psi_1\;, p_1), (\;\psi_2\;, p_2), \dots, (\;\psi_n\;, p_n)$, with p_i, a probability in the classical sense, $\overset{n}{\underset{i}{}}\,_1 p_i \quad 1$, rather than a probability amplitude. The *density matrix of a mixed state*, $\overset{n}{\underset{i}{}}\,_1 p_i \;\; \psi_i \quad \mathcal{H}_n$, is defined as

$$\rho \qquad \overset{n}{\underset{i \quad 1}{}} p_i \;\; \psi_i \;\; \psi_i \;,$$

where more than one factor p_i is greater than zero. The density matrix of a mixed state is a Hermitian operator, and its trace is one. Different mixtures of pure states could have the same density matrix; for example, the mix of the pair of states

$$\frac{1}{\sqrt{2}}(\;0 \qquad 1\;) \quad \text{and} \qquad \frac{1}{\sqrt{2}}(\;0 \qquad 1\;),$$

with probability $p \quad 1/2$, and the mix of pair of states $\;0\;$ and $\;1\;$, with probability $p \quad 1/2$, have the same density matrix; See also Section 2.3 where we show that the density matrix allows us to distinguish pure states from mixed (impure) states: $\mathrm{tr}(\rho^2) \quad 1$ for a pure state and $\mathrm{tr}(\rho^2) < 1$ for a mixed state. Pure states are represented as points on the surface of the *Bloch sphere*, while mixed states are points inside the *Bloch ball*. An *incoherent mixture* remains a mixture no matter what basis we choose to describe it.

The density matrix plays an important role in quantum information theory. It provides an answer to the question "How much information can we acquire about a quantum state?" *Pure states* are characterized by maximal knowledge or minimal ignorance; in principle, there is nothing more to be learned about a quantum system in a pure state [44]. Whenever we can only attribute probabilities to possible states, or when we are allowed to observe only a subsystem of a composed system, we cannot acquire maximum information about the entire quantum system and we say that the system is in a *mixed state.* The density operator, ρ, allows us to distinguish pure from mixed quantum states; ρ is a Hermitian operator and $\mathrm{tr}(\rho) \quad 1$.

1.12 ENTANGLEMENT AND BELL STATES

We have discussed a number of intriguing properties of quantum states, and we add to this list the entanglement, a phenomenon without a classical counterpart. Quantum mechanical systems have a unique property: A bipartite system, a system composed of two subsystems, can be in a pure state for which it is not possible to assign a definite state to each of its component subsystems. This strong correlation of quantum states is called *entanglement*; entangled states are pure states of bipartite quantum systems.

Erwin Schrödinger discovered the phenomenon of entanglement,[5] and in 1935 he made a crucial observation:

> *Total knowledge of a composite system does not necessarily include maximal knowledge of all its parts, not even when these are fully separated from each other and at the moment are not in uencing each other at all.*

Charles Bennett and Peter Shor comment on the effect of entanglement for quantum information processing [44]:

> *Classical information can be copied freely, but can only be transmitted forward in time to a receiver in the sender s forward light cone. Entanglement by contrast cannot be copied, but can connect any two points in space-time. Conventional data-processing operations destroy entanglement, but quantum operations can create it, preserve it and use it for various purposes, notably speeding up certain computations and assisting in the transmission of classical data or intact quantum states (teleportation) from a sender to a receiver.*

According to the postulates of quantum mechanics, the joint state of two or more quantum systems is a vector in a Hilbert defined as the tensor product of the Hilbert spaces $\mathcal{H}_{n_1}, \mathcal{H}_{n_2}, \ldots, \mathcal{H}_{n_k}$ used to represent the individual states of the component systems

$$\mathcal{H}_{n_1 \quad n_2 \quad\quad n_k} \quad \mathcal{H}_{n_1} \quad \mathcal{H}_{n_2} \quad\quad \mathcal{H}_{n_k}.$$

For example, the state of a quantum system consisting of two qubits is a vector in $\mathcal{H}_{2^2} \quad \mathcal{H}_2 \quad \mathcal{H}_2$, with the orthonormal basis $00 \;,\; 01 \;,\; 10 \;,\; 11$, expressed as

$$\psi \quad \alpha_{00} \; 00 \quad \alpha_{01} \; 01 \quad \alpha_{10} \; 10 \quad \alpha_{11} \; 11 \; ,$$

with $\alpha_{00} \;^2 \quad \alpha_{01} \;^2 \quad \alpha_{10} \;^2 \quad \alpha_{11} \;^2 \quad 1$.

[5]Entanglement is the English translation of the German noun Verschränkung, the name used by Schrödinger to describe this phenomenon.

Sometimes the state of a two-qubit system can be factored as the tensor product of the individual states of two qubits. For example, when α_{00} α_{10} $1/2$ and α_{01} α_{11} $i/2$, the state ψ is the tensor product of the states of the two qubits, ψ_1 and ψ_2 :

$$
\begin{aligned}
\psi \quad & \frac{1}{2}[\ 00 \quad i\ 01 \quad 10 \quad i\ 11\ ] \\
& \frac{1}{2}[\ 0 \quad (\ 0 \quad i\ 1\ ) \quad 1 \quad (\ 0 \quad i\ 1\ )] \\
& \frac{1}{2}(\ 0 \quad 1\ ) \quad (\ 0 \quad i\ 1\ ) \\
& \psi_1 \quad \psi_2\ .
\end{aligned}
$$

The individual states of the two qubits are well defined:

$$\psi_1 \quad \frac{1}{\overline{2}}(\ 0 \quad 1\ ) \quad \text{and} \quad \psi_2 \quad \frac{1}{\overline{2}}(\ 0 \quad i\ 1\ ).$$

This factorization is not always feasible. For example, consider a special state of a two-qubit system when

$$\alpha_{00} \quad \alpha_{11} \quad 1/\ \overline{2} \quad \text{and} \quad \alpha_{01} \quad \alpha_{10} \quad 0.$$

The resulting state, β_{00} , is called a *Bell state*, and the pair of qubits is called an *EPR pair*:

$$\beta_{00} \quad \frac{00 \quad 11}{\overline{2}}.$$

There are three other *Bell states*:

$$\beta_{01} \quad \frac{01 \quad 10}{\overline{2}}, \quad \beta_{10} \quad \frac{00 \quad 11}{\overline{2}}, \quad \text{and} \quad \beta_{11} \quad \frac{01 \quad 10}{\overline{2}}.$$

These states form an orthonormal basis and can be distinguished from one another. Bell states are *entangled* states; the four states are called *maximally entangled* states, and β_{11} is an *anticorrelated* state.

Entangled states are never in an ideal form; the source producing the entangled state is affected by noise, and the communication channel used to transfer entangled states or the quantum circuits used to manipulate the states can add noise. Thus, we have to *purify*, or *distill*, partial entanglement. Assume we have N pairs of qubits, each pair in the same entangled but mixed state with density matrix ρ; the ensemble of pairs is in the state given by the N-fold tensor product, $\rho^{\ N}$. Several protocols exist to transform the set of the N pairs of partially entangled pairs to $M < N$ pairs of maximally entangled particles using only local operations and classical communication primitives [39, 40, 118].

Entangled states are affected by decoherence—*the environment conspires to disentangle the quantum systems we have prepared in an entangled state*—and we have to protect the information embodied by entangled quantum states. The fragility of quantum information requires that we encode quantum states and then manipulate them in an encoded form.

After this introduction of basic concepts of quantum physics, we review several aspects of quantum information processing. We start with a discussion of quantum information and then introduce quantum computational devices and quantum algorithms.

1.13 QUANTUM INFORMATION

The unit of classical information is a *bit*, with two possible values, $B \quad 0, 1$; n bits form a *register*, $R \quad 0, 1 \;^{n}$. The state of a system can be characterized by the $m \quad 2^n$ possible configurations of the bits in register R. The vector $(p_1 p_2 \quad p_m)$, with $\;_i p_i \quad 1$, and the $(m \quad m)$ adjacency matrix P, describe the evolution of the system. A reversible transformation of the system state is described by a permutation matrix P, a matrix with only one nonzero element in each row and column; the new state is $P(p_1 p_2 \quad p_m)^T$.

When the physical support of information is a quantum system, we talk about *quantum information*. The unit of quantum information is a *qubit*, abstracted as $\mathcal{C}^2 \quad\; 0 \;,\; 1 \;$; a register of n qubits is $(\mathcal{C}^2) \;^{n}$, and the reversible transformation of the state of a quantum system is described by a *unitary* transformation. The density matrix ρ allows the description of pure as well as mixed states, with the probability of the individual components of the mix described by the vectors $q_1 q_2 \quad q_n$ and $\;_i q_i \quad 1$; ρ is a diagonal matrix, with q_i as its elements; and a reversible transformation, P transforms the density matrix ρ in ρ , with $\rho \quad P \rho P^{\dagger}$.

As expected, properties of quantum systems such as entanglement and superposition, the inability to clone the state of a quantum system, the fact that a measurement is an irreversible process, and the instability of quantum states due to interactions with the environment have a profound effect on the ways we process quantum information. The question we pose now is if this new embodiment of information leads to special attributes that can be exploited in the process of manipulating information.

Superposition

The *wave-particle duality* characterizes the behavior of quantum systems; *interference* occurs as a manifestation of the wave behavior and is a consequence of the superposition of quantum states. Due to superposition, we can compute the 2^n values of a function $f(x)$, with x a binary n-tuple, in one time step using a single copy of a quantum circuit implementing the transformation $f(x)$ of its input x. For example, consider a lengthy computation of a function $f(x)$, where the argument x is one bit and the result is also one bit, $f(0)$ or $f(1)$. We are not interested in the actual value of the function, but only if the function is constant, $f(0) \quad f(1)$, or balanced, $f(0) \quad f(1)$. A sequential classical computer evaluates first $f(0)$, then $f(1)$, and finally compares the two to provide the answer; a parallel classical computer evaluates $f(0)$ and $f(1)$ concurrently, but needs two processors for this task. A quantum computer with the input a superposition of $\; 0 \;$ and $\; 1 \;$ allows us to extract *global information* about the function f and thus to solve Deutsch's problem in a single time step (the time needed to evaluate the function for one value of the argument), and with a single copy of the hardware (Section 1.19).

We are now in a better position to understand Feynman's argument that the exact simulation of a quantum system is only possible with a quantum computer. Assume that a state of the quantum system used for the exact quantum simulation can be expressed using n bits, with n a large number (e.g., $n \quad 10^3$). Such a state is described by $2^n \quad 2^{1000} \quad 10^{333}$ complex numbers. No classical computer is able to store and process this colossal amount of data; on the other hand, the exact simulation of the system can be carried out with a quantum computer, provided that we expect the answer either to be that the resulting state is identical with a reference state or not, or to provide a measure of their distance.

We are capable of simulating the behavior of quantum circuits using classical computers, a problem critical for the study of quantum fault-tolerance [1, 107, 438]. The simulation of quantum circuits is possible when the behavior of the quantum system is confined to a small region of the vast Hilbert space.

Entanglement

The *entanglement* of quantum systems has a profound impact on quantum information processing. Consider a composite quantum system consisting of two subsystems, A and B, with n and m qubits, respectively. The question we pose is if it is possible to reconstruct the state of the joint systems from measurements performed separately on one of the subsystems alone. The answer to this question was given by John Bell [25, 26] who established that the information about the quantum state of a composite system is contained in *non-local correlations* between the two subsystems; these non-local correlations cannot be revealed by any measurement performed on one of the subsystems alone.

We should point out that non-local correlations that cannot be revealed by measurements performed on only one of the component systems are not restricted to quantum phenomena. Think, for example, of two correlated random variables, X and Y, related to the behavior of two systems, A and B, respectively; there are no local measurements that allow either A or B to determine the joint probability density function, $p_{XY}(x,y)$. In addition to non-local correlations, a critical property of the quantum entanglement is that a measurement of one of the two subsystems forces the other one to change its state, as we shall see when we discuss the EPR experiment in Section 2.16.

Entanglement has important implications for quantum error correction, a critical aspect of quantum information processing. Entanglement allows us to detect quantum errors without altering the state of the qubits in a quantum register, and then to correct the errors. First, we entangle the qubits in the register with ancilla qubits prepared in a well-defined state; as a result of the entanglement, the ancilla qubits contain information about the qubits in error. Then we perform a non-demolition measurement of the ancilla qubits to determine the *error syndrome*. The error syndrome tells us if an error has occurred and, if so, which qubits were affected and what type of error was present. An undesirable consequence of entanglement is the occurrence of *spatial* and *time-correlated quantum errors*. Once a qubit is affected by an error, the error can propagate to other qubits correlated to the one in error; a time-correlated error re-occurs at a later time, on the same qubit, after we have corrected its initial occurrence.

Measurements, Preparation, and Extraction of Classical Information from Quantum Information

Classical information is independent of the medium used to transport it. Yet, classical information is often carried by the same types of particles as quantum information (e.g., electrons, or photons). Why should we expect quantum information to be different from classical information?

To answer this question, we should establish if it is possible to freely convert one type of information to another and then recover the original information [230]. We can convert classical to quantum information and then convert back the quantum information to the original classical information. Formally, this process consists of two stages: *preparation*, when the quantum information is generated from the classical one, and *measurements*, when classical information is obtained from the quantum information (Figure 1.6a).

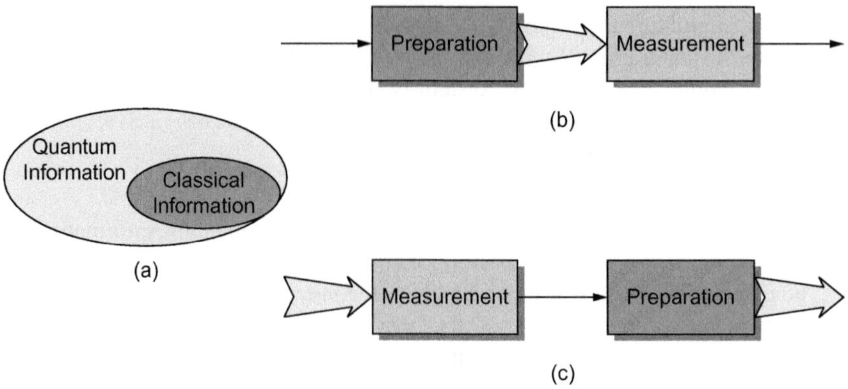

FIGURE 1.6

Classical and quantum information and conversion from one to another. Classical information is represented by thin arrows and quantum information as thick arrows. (a) Classical information can be regarded as a particular form of quantum information. (b) Classical information can be recovered from the quantum information when the preparation phase is followed by a measurement. The conversion path is: **classical** **quantum** **classical**. (c) Quantum information cannot be recovered when the preparation follows the measurement; the measurement is an irreversible process and alters the state of quantum systems. The conversion path in this case is **quantum** **classical** **quantum**.

The remaining question is if we can convert quantum to classical information and then convert the classical information to quantum information indistinguishable from the original one; this means that we should first perform a measurement to extract classical information and then use it to prepare quantum information (Figure 1.6b). The only possibility to compare quantum mechanical systems is in terms of statistical experiments, and this is not possible; because a measurement is an irreversible process, it alters the state of a quantum system. Chapter 2, devoted to quantum measurements, covers the arguments supporting this statement.

We conclude that, indeed, quantum information is qualitatively different from classical information. Even though classical and quantum information can be carried out by the same types of particles, the physical processes are different; in the former case, the physical processes are subject to classical physics, and later, to quantum physics. As we shall see in Chapter 2, some states of a quantum system, the *pure states*, have a classical counterpart, and a measurement allows us to distinguish orthogonal pure states. Therefore, classical information can be regarded as a particular form of quantum information (Figure 1.6c).

Manipulation of Quantum States

Now we turn our attention to the question of whether the states of a quantum system are stable for sufficiently long periods of time to allow the transformations prescribed by a quantum algorithm to progress without hinderance. Unfortunately, only rarely can the state of a quantum system be considered stable; more often, it is unstable. For example, the famous state of Schrödinger's cat is a "superposition" of being at the same time "dead" and "alive" formally described as $\text{cat} \quad \frac{1}{2}(\text{dead} \quad \text{alive})$.

This state is possible in quantum mechanics, but never observed in practice; all the cats we have ever seen, or will ever see, were and will be either dead or alive. The instability of quantum states is due to the interactions of the system with the environment; the correlation between a quantum system and the environment is very strong and leads to the phenomenon of *decoherence*, which hinders quantum information processing.

The quantum information we wish to process is encoded as correlations among the components of the quantum system; the interactions of the environment with the quantum system erode in time these useful correlations, which are transformed to correlations between the quantum system and the environment. The decoherence is only one of the problems we have to address when we think about building quantum computing and communication devices.

Another problem is the *accumulation of errors*, a problem we are familiar with from the study of classical analog circuits; if each analog circuit performing the transformation T_i introduces an error ϵ_i, then, after N steps, instead of the desired transformation, $\;_{i\ 1}^{N} T_i(I)$, the transformation of the input I will be $\;_{i\ 1}^{N} T_i(1 \quad \epsilon_i)(I)$. Similarly, if each quantum gate introduces a small error ϵ, then, after $1/\epsilon$ gates, the cumulated error will be significant enough to affect the result.

In addition to the bit-flip errors we are familiar with from the study of fault-tolerance of classical systems, we have to deal with phase-flip errors and with combinations of bit- and phase-flip errors of a quantum system. This brief discussion motivates the attention paid to quantum error correction, one of the topics of this book.

Physical Embodiment of Quantum Information

The quantum analog of a classical bit is called a *qubit*; following our mantra that"information is physical," we link the abstraction called qubit with the physical reality and provide a glimpse at some of the properties of quantum particles we have to manipulate in order to process quantum information.

Quantum particles have some properties with no classical counterpart. For example, the *spin* is an intrinsic angular momentum[6] of a quantum particle, related to its intrinsic rotation about an arbitrary direction. The spin of a quantum particle can be observed as the result of the interaction of the intrinsic angular momentum of the particle with an external magnetic field **B**. Classical physics operates with the concept of angular momentum arising from a rotation around a well-defined axis of a body.

There are two classes of quantum particles, those with a spin multiple of one-half, called *fermions*, and those with a spin multiple of one, called *bosons*. The spin quantum number of fermions can be $s \quad 1/2$, $s \quad 1/2$, or an odd multiple of $s \quad 1/2$. Electrons, protons, and neutrons are fermions. The spin quantum number of bosons can be $s \quad 1$, $s \quad 1$, $s \quad 0$, or a multiple of $\quad 1$.

A quantum particle such as the electron is not a "body" in the classical sense and does not have a defined axis of rotation. The electron is characterized by a charge with a nonstationary spatial distribution. The variation in time of this charge distribution can be associated with an intrinsic rotation of the electron about directions randomly oriented in space.

One possible embodiment of a qubit is the spin of the electron, the quantum number characterizing the intrinsic angular momentum of the electron; the electron spin has either the value $\quad 1/2$ or $\quad 1/2$ along the measurement axis, regardless of what that axis is. The introduction of this two-valued

---

[6]The intrinsic angular momentum of a quantum particle should be distinguished from its orbital angular momentum.

quantum number for the electron led Pauli to postulate his *exclusion principle*. According to Pauli's exclusion principle, no more than two electrons can occupy the same "orbit," and those two electrons must have antiparallel spins.

Indistinguishability is a principle of quantum mechanics and says that all quantum particles of the same type are alike; for example, we cannot distinguish an electron from another. Therefore, the operation of swapping the position of two electrons in a system with many electrons leaves the system's state unchanged; or, in other words, the operation is symmetric, and it is represented by a unitary transformation acting on the wave function as discussed in Section 1.7. In three dimensions, an exchange of two bosons is represented by an identity operator; the wave function is invariant and we say that the particles obey Bose statistics. The exchange of two fermions in three dimensions changes the sign of the wave function; the particles are said to obey Fermi statistics.

A *photon*,[7] a particle of light, is another important two-state quantum system used to embody a qubit; the quantum information can be encoded as the polarization of the photon. Photons differ from the spin-½ electrons in two ways: (1) they are massless, and (2) they have spin of one. A photon is characterized by its vector momentum (the vector momentum determines the frequency) and its polarization.

Light has a dual nature, wave-like and corpuscular-like; as an *electromagnetic radiation*, light consists of an electric and a magnetic field perpendicular to each other and, at the same time, perpendicular to the direction the energy is transported by the electromagnetic wave; the electric field oscillates in a plane perpendicular to the direction of light, and the way the electric field vector travels in this plane defines the polarization of light.

When the end of the electric field vector oscillates along a straight line, we say that the light is *linearly polarized*. When the end of the electric field vector moves along an ellipse, the light is *elliptically polarized*, and when it moves around a circle, the light is *circularly polarized*. If the light comes toward us and the end of the electric field vector moves around in a counterclockwise direction, we say that the light has *right-hand polarization*; if it moves in a clockwise direction, we say that the light has *left-hand polarization*.

The last embodiment of quantum information we discuss are the *anyons*, quantum particles of interest for topological quantum computing. Anyons are indistinguishable particles defined in two dimensions; they are different from either fermions or bosons. Consider a gas of electrons squeezed between two slabs of semiconductor materials such that the movement of electrons is restricted to two dimensions only. At a very low temperature and in a strong magnetic field the two-dimensional electron gas has a strongly entangled ground (lowest energy) state separated from all other states by an energy gap. This lowest-energy state carries an electric charge that is not an integer multiple of the electron charge and does not have the quantum numbers associated with electrons [344]. The properties of the anyons manifest themselves as the fractional quantum Hall effect (FQHE) discovered by Daniel Tsui, Horst Störmer, and Arthur Gossard [424]; the FQHE is discussed in Section 6.15.

We have only mentioned three possible embodiments of quantum information. Others are discussed in detail in Chapter 6, which covers physical realizations of quantum computing and communication devices. The diversity of potential implementations of quantum information processing devices is a source of optimism and, at the same time, anxiety. On one hand, it gives us great hope that at least

---

[7] The name photon comes from the Greek word "photo," meaning light.

some of the theoretical ideas will ultimately lead to feasible quantum information processing devices; on the other hand, it gives us some feeling about the vastness of the field and the depth of knowledge required to master a discipline involving multiple areas of mathematics, physics, and computer science.

Quantum Information Processing Systems

The last subject of our survey of quantum information covers quantum information processing systems. A quantum computer is a physical device capable of transforming quantum information as required by a quantum algorithm; a quantum communication system transmits either classical or quantum information from one place to another through a quantum channel.

The operation of a quantum computer takes advantage of fundamental principles of quantum physics—superposition, quantum interference, quantum entanglement, and the high dimensionality of the state space of a quantum system—to solve some computationally "hard" problems efficiently. A quantum algorithm aims to increase the probability of obtaining the correct answer by arranging so that all computational paths leading to the right answer interfere constructively, while the computational paths to wrong answers interfere destructively. This strategy has been applied successfully to solving a subset of "hard" computational problems such as factoring large numbers and searching large unstructured databases; the quantum algorithms of Shor [379] and Grover [182] are milestones in quantum computing.

To process information with a quantum system, we first prepare the system in an initial quantum state, then transform this state through a set of operations prescribed by a quantum algorithm, and, finally, measure the resulting state of the system. This sequence of steps resembles the ones followed by a classical device up to the point when we examine the result. The qualitative difference between a classical system outputting the result of a transformation and the measurement of the final state of the quantum system is due to the *randomness of the quantum measurement process.* If the result of the quantum computation is one qubit in state ψ α_0 0 α_1 1 , then a measurement of the qubit will produce the result 0, with probability p_0 α_0 ^{2}, or 1, with the probability p_1 α_2 ^{2}.

A quantum algorithm generates a probability distribution of the results, while a randomizing algorithm for a classical computer uses randomness as part of its logic. There are qualitative differences between randomization algorithms and quantum algorithms. The former are designed with the hope of achieving good performance in the "average case" taken over all possible random choices. The latter exploit the entanglement and superposition of quantum information; the randomness is not part of their logic, but of the physical transformations leading to the result.

Landauer [264] observes that classical computers function based on the intuitive and completely abstract computability theory due to Church, Turing, Post, and Gödel, while "the properties of quantum computers are not postulated in abstracto, but deduced from the laws of physics" [117]. A quantum computer coupled with a quantum communication system performs functions related to quantum cryptography that are not feasible with a classical system; however, a quantum computer can only perform computations that can be carried out by a classical computer, albeit much more efficiently.

A quantum communication channel transmits information using quantum effects when, for example, the information is encoded as the spin of an electron or in the polarization of a photon. Eavesdropping on a quantum communication channel can be detected with very high probability because the effect of intrusion is an irreversible transformation, a measurement of the quantum state.

We conclude that the special properties of quantum information have important consequences for computation and for communication, and we should investigate the physical realization of quantum information processing devices.

1.14 PHYSICAL REALIZATION OF QUANTUM INFORMATION PROCESSING SYSTEMS

A quantum computer is a physical device designed to transform quantum information embodied by the state of a quantum system. The physical processes required to transform the quantum state in a controlled manner are different for ion traps, solid-state, optical, NMR, or other possible physical implementations. Thus, we need first to define the basic requirements to build a quantum computer regardless of the physical phenomena involved. DiVincenzo [130] provides a crisp and clear formulation of five plus two (additional) requirements (discussed in depth in Section 6.1) for a quantum information processing system:

1. A scalable physical system with well-characterized qubits
2. The ability to initialize the state of the qubits
3. Long relevant decoherence times, much longer than the gate operation times
4. A "universal" set of quantum gates
5. A qubit-specific measurement capability

The two additional requirements are related to the ability to communicate:

1. The ability to interconvert "stationary" (memory) and "flying" (communication) qubits
2. The ability to faithfully transmit "flying" qubits between specified locations

These requirements come naturally to mind. They have an immediate correspondent for classical computers. Indeed, we cannot conceive of a state-of-the-art computer built with circuits whose state cannot be controlled or initialized to a desired state. It would be impractical to build a computer unless we have a finite set of building blocks, and it seems obvious that we should have access to the results of a computation.

A *well-characterized* qubit means that the relevant parameters of the physical process, including the Hamiltonian, $\mathbf{H}(t)$, of the system and the coupling between a qubit and the environment and between this qubit and other qubits, must be known. For an implementation of a quantum computer, we should only consider those quantum systems that satisfy this requirement. For example, the superselection rules (SSRs)[8] prohibit entangled states involving different particle numbers; thus, a two-qubit system consisting of two quantum dots and an electron in a superposition state as being on one or the other quantum dot would not satisfy the first requirement [130].

[8] An SSR is a restriction on the allowed local operations on a system, not on its allowed states, and it is associated with a group of physical transformations [24]. Such restrictions could be imposed by the properties of the underlying theory, or arise due to physical restrictions. The operations it applies include unitary transformations, $\mathbf{O}\rho \quad U\rho U^{\dagger}$, and measurements, $\mathbf{O}_r\rho \quad \mathcal{M}_r\rho\mathcal{M}_r^{\dagger}$, with $\quad \mathcal{M}_r\mathcal{M}_r^{\dagger} \quad 1$.

The *scalability* requirement is not only related to the ability to carry out complex transformations, but also to the necessity, discussed later, of having reliable circuits; as we shall see shortly, every "useful qubit" requires a large number of additional qubits, 10 or more, to ensure fault-tolerance.

It is self-evident that we should prepare a qubit in a *well-defined initial state* in order to carry out a quantum computation with predictable results. An n-qubit quantum computer operates on a 2^n-dimensional Hilbert space, $\mathcal{H}_{2^n}$, with the computational basis states $x_1x_2 \quad x_n$, with $x_i \quad 0$ or $x_i \quad 1$. This requirement translates to the fact that *any computational basis state,* $x_1x_2 \quad x_n$ *, should be prepared in at most n steps.* The practical questions are how to initialize a qubit in a state with maximal entropy and how fast the preparation of qubits can be done. A possible solution is to have an independent system to prepare qubits in an initial state either by forcing them to the ground state of the Hamiltonian, a process called "cooling," or by performing a measurement that collapses the state of a qubit to a basis state (e.g., 0). Then, a "qubit conveyor belt" should provide access to the qubits in the initial state whenever they are needed.

Decoherence in its simplest form means that a "pure state," $\varphi \quad \alpha_0 \; 0 \quad \alpha_1 \; 1$ is transformed by interaction with the environment to a "mixed state," with the density matrix $\rho \quad \alpha_0 \; ^2 \; 0 \; 0 \quad \alpha_1 \; ^2 \; 1 \; 1$. The pure state of a qubit is represented by a vector connecting the origin with a point on the surface of the Bloch sphere, while a mixed state of a qubit is represented by a vector connecting the origin to a point inside the Bloch (solid) sphere; the von Neumann entropy of a mixed state is measured by the distance of such a point to the surface of the Bloch sphere, as discussed in Section 2.3.

Since the early days of computing, reliability has been a major concern [79]; therefore, it seems reasonable to ask ourselves if a reliable quantum computer could be built at all, knowing that quantum states are subject to decoherence. The initial thought was that a quantum computation could only be carried out successfully if its duration were shorter than the decoherence time of the quantum computer. As we shall see in Section 5.25, the decoherence time ranges from about 10^4 seconds for the nuclear spin to $10 \; ^9$ seconds for the quantum dot charge. Thus, it seems very problematic that a quantum computer could be built unless we have a mechanism to periodically deal with errors.

Probabilistic classical computer and communication systems can be error corrected and operate effectively in the presence of noise without requiring an exponential precision, while the operation of their analog counterparts is conditioned by a lack of noise and an exponential precision. The analog nature of quantum information and of quantum transformation raised serious questions for the experimental realization of quantum information processing systems. These questions were answered by showing that quantum computer and communication systems are more similar to probabilistic classical systems than to analog ones [40, 245, 246, 247, 380, 402].

Now we know that quantum error-correcting codes could be used to ensure fault-tolerant quantum computing; quantum error correction allows us to deal algorithmically with decoherence. A quantum error-correcting code maps a "logical qubit" onto several "physical qubits" and then encodes these qubits using two classical error-correcting codes to deal with bit-flip and phase-flip errors, as discussed in detail in Chapter 5. There is a significant price to pay to achieve fault-tolerance through error correction—the number of qubits required to correct errors is an order of magnitude larger than the number of "useful" qubits.

In 1996, Shor [382] showed how to perform reliable quantum computations when the probability of a qubit or quantum gate error decays polylogarithmically with the size of the computation, a rather unrealistic assumption. The *quantum threshold theorem* states that arbitrarily long computations can be carried out with high reliability provided that the error rate is below an *accuracy threshold* according to Knill, Laflamme, and Zurek [246]. In 1999, Aharonow and Ben-Or [5] proved that reliable computing is possible when the error rate is smaller than a constant threshold, but the cost is polylogarithmic in time and space. In practice, error correction is successful for a quantum system whose decoherence time is four to five orders of magnitude larger than the *gate time*, the time required for a quantum gate operation.

In the next section, we discuss in more detail the practical requirements regarding universal quantum gates. Here, we only note that the fact that any logic function could be implemented using a small set of universal classical gates had a significant impact on current solid-state technology; it allowed us to reduce the size of classical circuits as predicted by *Moore's law* which states that the number of transistors per chip that yields the minimum cost per transistor doubles every 18 months or so.

An important aspect of a discussion regarding universal quantum gates is that many-body quantum interactions are difficult to control and to analyze. Thus, it is necessary to have a universal set of quantum gates that require at most two-body interactions, and this translates to the ability to implement any unitary transformation using only one-qubit and two-qubit gates. *Any unitary transformation can be approximated arbitrarily well by a quantum circuit consisting of two-qubit* CNOT *gates, and one-qubit* H, T, *and* S *gates.* First, we show that any unitary transformation $\mathbf{U} \in \mathcal{H}_{2^n}$ can be carried out by a set of unitary transformations, $U_1, U_2, \ldots, U_k, \ldots, U_m$, with $m \leq 2^n - 1$, which act only on two or fewer computational basis states. Then, we show that each unitary transformation U_k, $1 \leq k \leq m$, can be expressed exactly as a product of transformations carried out by CNOT and one-qubit gates, and finally, we show that the transformation carried out by a one-qubit gate can be approximated arbitrarily well by the transformations carried out by the H, T, and S gates. As an example, we show how to implement a Toffoli gate with the set of universal quantum gates described above.

Consider a quantum circuit that consists of m gates, $G_k, 1 \leq k \leq m$, which carry out the unitary transformations, $\mathcal{G}_k = e^{i\mathbf{H}_k(t)/\hbar}, 1 \leq k \leq m$, where $\mathbf{H}_k(t)$ is the Hamiltonian describing the evolution of the quantum system when implementing the k gate, and $\hbar$ is the Planck's constant. The simplest modus operandi of the circuit is to consider discrete times $t_1, t_2, \ldots, t_k, \ldots, t_m, t_{m+1}$ when each Hamiltonian is turned on and off. For example, $\mathbf{H}_k$ should be turned on at t_k and turned off before t_{k+1} when $\mathbf{H}_{k+1}$ is turned on.

Recall that the control unit of a classical processor decides which functional units are active and what specific function each unit is expected to carry out at time t. An important question we have to answer is how to control each quantum gate; who plays the role of the "control unit" of a classical computer in a quantum computing setup? The answer is that the individual quantum circuits are controlled by a physical apparatus that regulates the electric and magnetic fields when the quantum gates are implemented on the ion traps or NMR, or that regulates the temperature of the quantum bath for quantum dots, and so on.

We are able to examine the final results as well as partial results of a computation carried out by a classical computer. Classical error correction techniques require the ability to examine some information derived from a sequence of bits stored or transmitted through a communication channel (e.g., the syndrome of linear codes). Therefore, it is natural to require the ability to measure the state of qubits. If the density matrix of a qubit is

$$\rho = p|0\rangle\langle 0| + (1-p)|1\rangle\langle 1| + \alpha|0\rangle\langle 1| + \alpha^*|1\rangle\langle 0|,$$

with $\alpha \in \mathbb{C}$, then an "ideal measurement" should provide an outcome of 0, with probability p, and an outcome of 1, with probability $(1-p)$, regardless of the value of α, regardless of the state of the neighboring qubits, and without changing the state of the quantum computer. In this case, a "nondemolition measurement" leaves the qubit in the state, with ρ 0 0 , after reporting the outcome 0 and leaves it in the state, with ρ 1 1 , after reporting the outcome 1.

We know that a measurement of a quantum state collapses the state to one of the basis states; a measurement is an irreversible process. This poses serious challenges to the design of quantum algorithms, which should avoid irreversible transformations. Quantum error correction requires nondemolition measurements of the error syndrome to preserve the state of the physical qubits.

Lastly, the two additional requirements put forth by DiVincenzo recognize the inherent symbiosis between computation and communication. This is also true for quantum information; indeed, quantum information may be processed or stored at a different location than the one it is collected from, and quantum computers may need to exchange information among themselves. There is a general agreement that photons and optical communication channels are ideal for transporting "flying qubits."

We conclude that a quantum computer is an ensemble consisting of a classical and a quantum component, and the quantum component must satisfy the requirements outlined by DiVincenzo [130]. A quantum computer consists of quantum circuits, and quantum circuits are built by interconnecting quantum gates.

1.15 UNIVERSAL COMPUTERS: THE CIRCUIT MODEL OF COMPUTATION

Informally, a *universal computer* is a single machine able to perform any physically possible computation. The classical theory of computability, as well as the quantum computability theory, admit the existence of universal computers. Moreover, the concept of universal computer developed in the context of the computability theory has immediate practical consequences; it implies that the components used to build such an instrument must also be universal. Indeed, if we could obtain the solution to a problem using a physical computing instrument without having a systematic method to produce that instrument using a set of universal components, then the solution would not necessarily be "computable" in a useful sense [117].

The universality of quantum computers implies *the ability to carry out computations of arbitrary complexity—in other words, computations involving an arbitrary number of operations and an arbitrary amount of storage.* We should be able to maintain the computer in operation for arbitrary long periods of time and provide an arbitrary large number of qubits in a standard state. The last requirement is equivalent to the unlimited "blank tape" of Turing machines.

A classical computing engine is a *deterministic system* evolving from an initial/input state to a final/output state. The input state and all states traversed by the system during its dynamic evolution have some canonical labelling. Moreover, the label of each state can be measured by an external observer. The label of the final state is a deterministic function f of the label of the input state; we say that the engine computes a "function" f. Two classical computing engines are *equivalent* if they compute the same function f given the same labelling of their input and output states [114].

While early computers could only execute a fixed program, the computers we discuss are universal; they implement an *instruction set* and can carry out any computation described by a *program* expressed as a set of instructions from this set. The most popular architecture of a *universal computer* was proposed by John von Neumann [79]; this is a *stored-program computer* with four interconnected components: (1) a control unit, (2) an arithmetic and logic unit (ALU), (3) memory, and (4) an input/output unit.

Modern computers process classical information and obey the laws of classical physics, which, ultimately, limit our ability to increase the speed of solid-state circuits and to make them increasingly smaller. The power dissipation of classical circuits increases as κ^{α}, where κ is the clock rate and $2 \quad \alpha \quad 3$; when we double the speed, the power dissipation may increase by a factor of $2^3 \quad 8$. Heat dissipation for modern computers operating at clock rates of few GHz is an extremely challenging problem and has forced manufacturers of microprocessors to switch their attention from increasing the clock rate to building multicore systems.

A classical universal computer is built from *classical gates* (Figure 1.7); a gate implements a Boolean function $f:\ 0,1\ ^{n} \qquad 0,1\ ^{m}$ (Figure 1.7a). The classical gates used in existing computers perform *irreversible transformations,* and the transformations carried out by these gates produce a substantial amount of "informational junk." This information must be erased, a dissipative process responsible for a substantial amount of heat, and heat removal is a major problem for classical processors.

Classical Gates

Practical considerations require a computer to be built using gates from a small set; in other words, the existence of one or more universal sets of classical gates is required. Formally, we say that a set of classical gates is *universal* if every Boolean function can be implemented using only gates from that family.

We can *use a small number of gates to compute any Boolean function.* For simplicity in our presentation we consider a restricted family of Boolean functions, with n bits as input and a single output bit $f:\ 0,1\ ^{n} \qquad 0,1$, and provide a proof by induction that one can construct a circuit for this class of Boolean functions using a small number of different logical gates. First, we consider the case $n \quad 1$ and realize that we need four circuits to map the two possible inputs to the outputs (Figure 1.8a): (1) one that performs the identity transformation $y \quad x$ (this can be done with a single wire); (2) one that flips the input x, $x \quad 0 \quad y \quad 1$, $x \quad 1 \quad y \quad 0$ (this can be done with a NOT gate); (3) one that produces $y \quad 0$ regardless of the input (this can be done with an AND gate with the two inputs 0 and x); and (4) one that produces $y \quad 1$ regardless of the input (this can be done with an OR gate with the two inputs 1 and x). Then we assume that we can construct the circuit such that it is able to compute a function $f_n(b_1, b_2, \ldots, b_n)$ on n bits, and we wish to prove that by using NOT, AND, and XOR gates we can construct a circuit able to compute $f_{n \quad 1}(b_0, b_1, b_2, \ldots, b_n)$ on $(n \quad 1)$ bits. Call $f^{(0)}_{n \quad 1} \quad f_{n \quad 1}(0, b_1, b_2, \ldots, b_n)$ and $f^{(1)}_{n \quad 1} \quad f_{n \quad 1}(1, b_1, b_2, \ldots, b_n)$; then the circuit in Figure 1.8b will compute $f_{n \quad 1}(b_0, b_1, b_2, \ldots, b_n)$.

To build a classical circuit implementing a Boolean function we need (1) wires; (2) gates of several types, NOT, AND, XOR; (3) the ability to replicate the input, an operation called *fanout*; (4) the ability to interchange two bits, an operation called *swapping*; and last, but not least, (5) a supply of ancilla, or auxiliary bits, e.g., the bits initialized to 0 or to 1 for the circuits in Figure 1.8a.

Reversibility and Computation

Logical reversibility: A Boolean function is reversible if there is a one-to-one relationship between input and output. In the early 1960s, Landauer established that a necessary condition for a computational process to be physically reversible is that the logical function it implements be logically

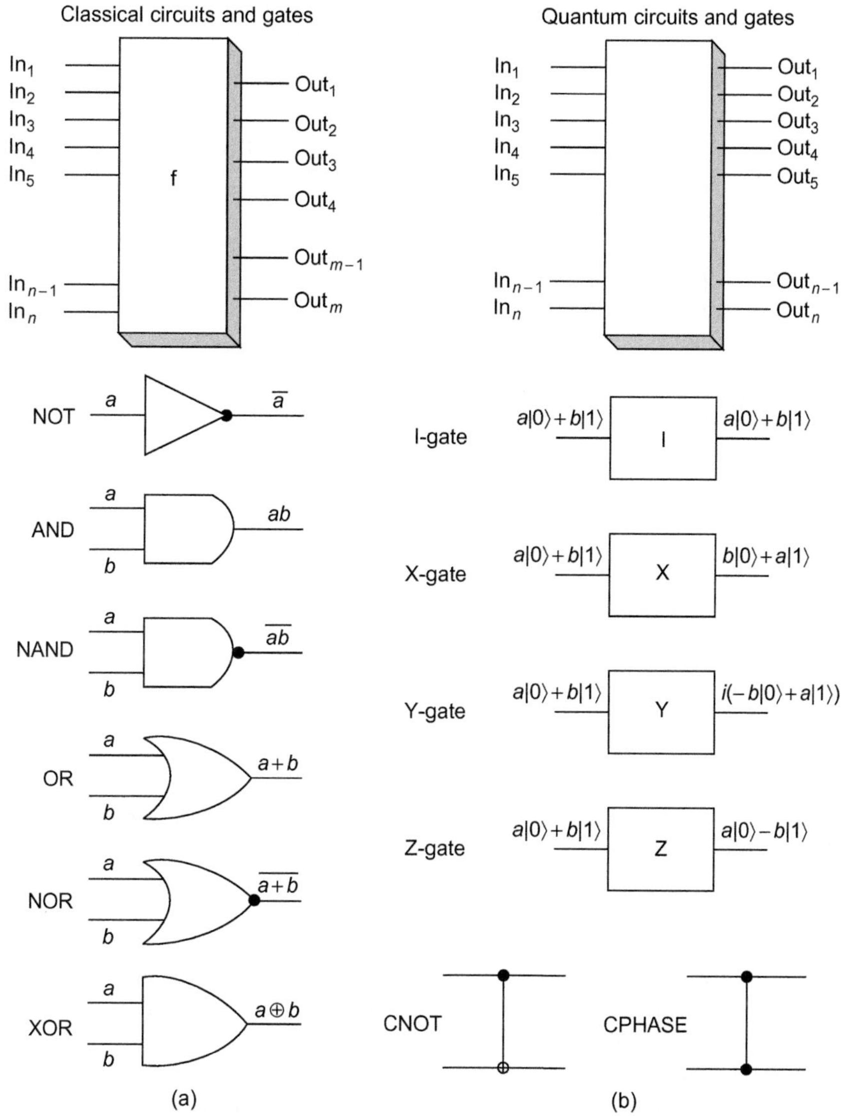

FIGURE 1.7

Classical and quantum circuits and gates. (a) A classical circuit implements a multi-output Boolean function, $f:\ 0,1^{\ n} \quad 0,1^{\ m}$, and is constructed using a finite number of classical gates. The number of inputs, n, and the number of outputs, m, of a classical circuit may be, and often are, different ($n \quad m$). The NOT, AND, NAND, OR, NOR, and XOR classical gates are shown. a and b are Boolean variable; a is the negation of a; ($a \quad b$) is the logical sum aORb, also written as $a \quad b$; ab is the logical product aANDb, also written as $a \quad b$; and $a \quad b$ is the exclusive OR (XOR) of a and b. (b) A quantum circuit performs a unitary operation in the Hilbert space $\mathcal{H}_n$ and consists of a finite collection of quantum gates; each quantum gate implements a unitary transformation on k qubits for a small k and must have the same number of inputs and outputs. The one-qubit gates for the Pauli transformations, $\sigma_I, \sigma_x, \sigma_y$ and σ_z, and the two-qubit gates, CNOT and CPHASE, are shown.

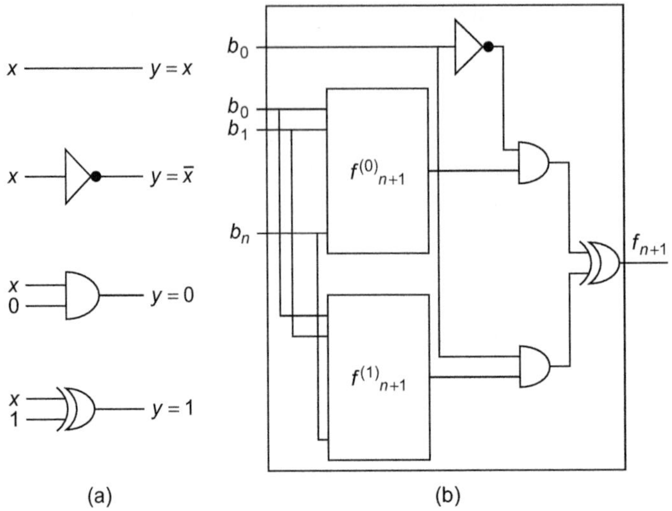

FIGURE 1.8

(a) The four circuits mapping an input x to an output y. The circuits are able to compute a Boolean function $f: \ 0,1 \ ^{n} \quad 0,1$ for $n \quad 1$: (1) a wire for the identity $y \quad x$; (2) a bit-flip implemented with a NOT gate; (3) a circuit that produces $y \quad 0$ regardless of the input, using an AND gate with the two inputs 0 and x; and (4) a circuit that produces $y \quad 1$ regardless of the input, using an OR gate with the two inputs 1 and x. (b) The circuit to compute a function $f_{n \quad 1}(b_0, b_1, b_2, \ldots, b_n)$ on $(n \quad 1)$ bits given that we can construct the circuit able to compute a function $f_n(b_1, b_2, \ldots, b_n)$ on n bits; the circuit uses CNOT, AND, and XOR gates.

reversible [264]; he showed that any irreversible computation may be transformed into a reversible one by embedding it into a larger computation where no information is lost. The next step, Bennett's discovery that computations can be done reversibly [30], led to the investigation of classical reversible gates. Then, in 1980 Toffoli showed that classical reversible gates can be used to construct classical circuits [421].

For example, instead of the classical XOR, one could use the CNOT (controlled-NOT) gate. The CNOT is reversible and has two inputs, a and b; the source, or "control," bit, a, is not changed and affects the setting of the "target" bit b; when $a \quad 1$, b is flipped. Toffoli also showed that we can replace an irreversible NAND gate with a reversible Toffoli gate. The classical Toffoli gate has three inputs (a, b, c) and three outputs $(a, b, (c \quad ab))$, where stands for binary addition (Figure 1.9a). The NAND gate transforms its two inputs, a and b, to their logical product and negates it; its output is $(\overline{ab})$, an expression also written as $(a \quad b)$, or as (a NAND b).

The classical Toffoli gate is universal for reversible Boolean logic: Indeed, we know that the NAND gate is universal; any Boolean function can be implemented using only NANDs. We see immediately that the output of a Toffoli gate, with $c \quad 1$, is $a \quad b$ or aNANDb (Figure 1.9b); thus, the Toffoli gate in this configuration simulates a NAND gate. This simulation is not very effective, as the number of additional bits introduced in a circuit constructed solely with NAND gates increases linearly with the number of gates.

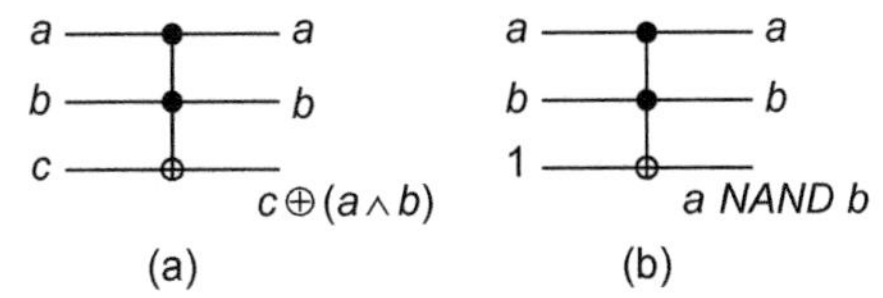

FIGURE 1.9

The classical Toffoli gate is universal for reversible Boolean logic. (a) The gate has three inputs (a, b, c) and three outputs $(a, b, (c \quad (ab)))$, where stands for binary addition, a and b are "control" bits, and c is a "target" bit. The two control bits are unchanged, and the target is flipped when at least one of the control bits is 1. (b) The output of a Toffoli gate, with $c \quad 1$, is $a \quad b$, or aNANDb. The NAND gate transforms its two inputs, a and b, to their logical product and negates it; its output is $\overline{ab}$, also written as $(a \quad b)$, or as $(a$ NAND $b)$.

The Circuit Model of Computation

The circuit model of computation relates the practical implementation of an algorithm with the circuits to compute Boolean functions; in terms of computational power, this model is equivalent to the Turing machine model. This statement requires some qualifications as the circuit model deals with practical implementation (i.e., with finite systems), while the Turing machine model assumes unbounded resources (e.g., an infinite tape). The circuit model should also be able to distinguish between computable and noncomputable functions.

So far we have been concerned with the ability of the circuit model of computation to express an algorithm. To construct the circuit for a particular algorithm, we have to design a protocol telling us what types of classical gates are needed, how to interconnect them, and how to locate the results of the computations. The need for the circuit design protocol leads to a deeper connection between the circuit computational model and the Turing machine computational model. We should find the means to prohibit the ability to "hide" the complexity of an algorithm in the protocol for building the circuit, or even to consider embedding a noncomputable function (e.g., a function for the "halting problem") in the protocol.

To address the problem of hiding the complexity of an algorithm in the computational system, the circuit model of computation uses the concept of *uniform circuit family*. A circuit in the family, denoted as $\mathcal{C}_n$, has n input bits and any number of auxiliary and output bits, when the input is a string x of n bits; the output of the circuit is denoted as $\mathcal{C}_n(x)$. A circuit is *consistent* if when the input is a string of length $m < n$, then $\mathcal{C}_m(x) \quad \mathcal{C}_n(x)$. A circuit family is uniform and is denoted as $\mathcal{C}_n$ if an algorithm for a Turing machine to generate a description of the protocol for the design of the circuit, given n, the size of the input, exists. If such an algorithm for a Turing machine does not exist, then the circuit family is called *nonuniform*. The equivalence of the circuit and the Turing computational models is restricted to uniform circuit families.

We assume that a computing machine, $\mathcal{M}$, computes one function, f, of its input to produce the desired result. To hide the complexity, we could first alter the input and then present it to another machine, $\mathcal{M}$, able to produce the same result in polynomial time. To address this problem, we consider the input as consisting of two parts: (a) a *program/protocol* determining which function the computing machine $\mathcal{M}$ will compute and (b) the actual input for the desired function. We require the existence of an algorithm for a Turing machine to generate a description of the protocol.

The full specification of the state for the Turing machine computational model as well as for the circuit computational model requires the specification of a set of numbers, all expected to be measurable at any instant of time. Quantum physics excludes the existence of physical systems with this property; thus, these computational models are effectively classical, and we have to address the question of quantum computational models from a new perspective.

Computational problems are said to be "hard" if the time to solve the problem is extremely long, regardless of the hardware used. If the time $t(n)$ to carry out a computation with n data elements as input satisfies the condition $t(n) \quad \text{Poly}(n)$, where $\text{Poly}(n)$ is a polynomial in n, then we have a *polynomial time computation*; otherwise, it is an *exponential computation*. Complexity theory, the branch of computer science addressing the question of which problems are hard and which ones are easy, defines "easy" problems as *polynomial time* and "hard" problems as *exponential time.*

In Section 1.13 we outlined the arguments that classical computers are not powerful enough to efficiently solve problems such as factorization of large integers, or "efficient" simulation of physical systems at an atomic and subatomic level. We also discussed the Church-Turing principle, namely, that every function that could be regarded as computable is computable by a Turing machine. A *quantitative version of the Church-Turing principle* relates the behavior of the abstract model of computation provided by the Turing machine concept with the physical computing devices used to carry out a computation. This thesis states that: "Any physical computing device can be simulated by a Turing machine in a number of steps polynomial in the resources used by the computing device" [319]. While no one has been able to find counter-examples for this thesis, the search has been limited to systems that are based on the laws of classical physics. Yet, our universe is essentially quantum mechanical; therefore, there is a possibility that the computing power of quantum computing devices might be greater than the computing power of classical computing devices [381].

1.16 QUANTUM GATES, CIRCUITS, AND QUANTUM COMPUTERS

Quantum Gates

The building blocks of a quantum computer are quantum gates. Each quantum gate implements a unitary transformation on k qubits for a small k (Figure 1.7b). The *fanout*, the ability of a logic gate output to drive a number of inputs of other logic gates to form more complex circuits, is nontrivial for quantum gates; quantum gates implement unitary and, thus, reversible transformations, and are required to have the same number of input and output qubits, a strict rule that we do not impose on classical gates, which may have different numbers of inputs and outputs.

In 1980, Paul Benioff realized that the Hamiltonian time evolution of an isolated quantum system is reversible and could mimic a reversible Boolean computation [27]. A few years later, in 1985, David Deutsch observed that the linearity of the Schrödinger equation implies that *mapping of the basis states uniquely specifies the dynamics of an arbitrary initial state* [114].

A one-qubit gate carries out a unitary transformation, $\mathbf{A}$, of an input state

$$\varphi_{out} \quad \mathbf{A} \ \varphi_{in} \quad \text{or} \quad \begin{matrix} \gamma_0 \\ \gamma_1 \end{matrix} \quad \begin{matrix} a_{11} & a_{12} \\ a_{21} & a_{22} \end{matrix} \quad \begin{matrix} \alpha_0 \\ \alpha_1 \end{matrix}, \ \text{with} \ A \quad \begin{matrix} a_{11} & a_{12} \\ a_{21} & a_{22} \end{matrix}.$$

Example. *One-qubit gates for the Pauli transformations.* The Pauli transformations given by the set, $\mathcal{G}\quad \sigma_I,\sigma_x,\sigma_y,\sigma_z$, are carried out by the quantum gates usually denoted as I, X, Y, and Z, respectively:

$$I \quad \begin{matrix} 1 & 0 \\ 0 & 1 \end{matrix}, \quad X \quad \begin{matrix} 0 & 1 \\ 1 & 0 \end{matrix}, \quad Y \quad \begin{matrix} 0 & i \\ i & 0 \end{matrix}, \quad \text{and} \quad Z \quad \begin{matrix} 1 & 0 \\ 0 & 1 \end{matrix}.$$

Example. *The Hadamard gate, H.* It performs the transformation

$$H \quad \frac{1}{2} \begin{matrix} 1 & 1 \\ 1 & 1 \end{matrix}$$

and can be used to transform one basis to another:

$$(\ 0\ ,\ 1\ ) \qquad \frac{0 \quad 1}{2}, \frac{0 \quad 1}{2} \quad \text{or} \quad \frac{0 \quad 1}{2}, \frac{0 \quad 1}{2} \qquad (\ 0\ ,\ 1\ ).$$

The quantum CNOT gate transforms an arbitrary initial state of two qubits expressed in the basis, (00 , 01 , 10 , 11), as follows:

$$\alpha_0 \ 00 \quad \alpha_1 \ 01 \quad \alpha_2 \ 10 \quad \alpha_3 \ 11 \qquad \alpha_0 \ 00 \quad \alpha_1 \ 01 \quad \alpha_3 \ 10 \quad \alpha_2 \ 11 .$$

In this expression, $\alpha_0, \alpha_1, \alpha_2$, and α_3 are arbitrary complex numbers satisfying the condition $\alpha_0 \ ^2 \quad \alpha_1 \ ^2 \quad \alpha_2 \ ^2 \quad \alpha_3 \ ^2 \quad 1$; the matrix describing this unitary transformation is

$$U_{CNOT} \quad \begin{matrix} 1 & 0 & 0 & 0 \\ 0 & 1 & 0 & 0 \\ 0 & 0 & 0 & 1 \\ 0 & 0 & 1 & 0 \end{matrix}.$$

If, for simplicity, we omit the time-ordered product, then the Hamiltonian of the transformation can be written as

$$U_{CNOT} \quad e^{\frac{i}{\hbar} \quad H(t)dt}.$$

There is not a unique solution to this equation; many types of Hamiltonians can be used to implement the CNOT gate [129].

The CNOT gate has special properties [114], some very useful for quantum error correction [129]. First, it can produce perfectly entangled states from non-entangled states (Figure 1.10a). Second, it can

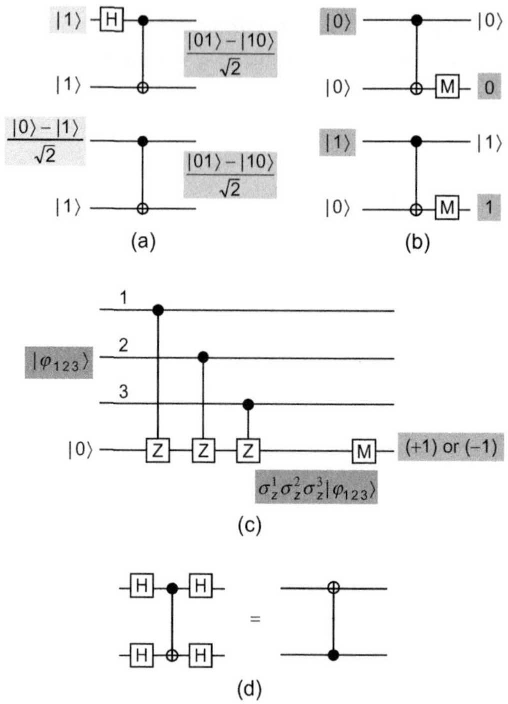

FIGURE 1.10

Properties of the quantum CNOT gate. (a) It produces perfectly entangled states from non-entangled states. When the control qubit is in the state $\frac{0-1}{2}$ (this state can be produced by a Hadamard gate, with the input 1), and if the target qubit is in the state 1 , then the output is in an EPR state, $\frac{01-10}{2}$. (b) It can be used for a "nondemolition measurement" of the control qubit. (c) A more sophisticated nondemolition measurement, M, of the operator $Z^1Z^2Z^3$ applied to the state φ_{123} of a system consisting of three particles, 1,2, and 3. The measurement of the target qubits of the three CNOT gates produces the result in an ancilla qubit initially in the state 0 . (d) When both the control and the target qubits are in a rotated basis, $\frac{0\ \ 1}{2}, \frac{0-1}{2}$, then the role of the control and target qubit are reversed.

be used for a "nondemolition measurement" of the control qubit if this qubit is either in state 0 or 1 , but not in a superposition state; in Figure 1.10b we see that the target qubit ends up in the same state as the control qubit. Recall that the result of a measurement is an eigenvalue of the operator; thus, the result of the measurement of the target qubit is 0 when the control qubit is in state 0 and is 1 when the control qubit is in state 1 . Third, the nondemolition measurement, M, of the operator $Z^1Z^2Z^3$ applied to the state φ_{123} of a quantum system consisting of three particles, 1,2, and 3 (control qubits of three CNOT gates), produces the result in an ancilla qubit initially in the state 0 , as shown in Figure 1.10c. Lastly, when both the control and the target qubits are in a rotated basis, $\frac{0\ \ 1}{2}, \frac{0-1}{2}$, the roles of the source and target qubits are reversed (Figure 1.10d).

Nondemolition Measurements

Such measurements are critical for quantum error correction because they leave the original state of a quantum system unchanged [91]. Circuits similar to the one in Figure 1.10c will be discussed in Section 5.14, but their remarkable properties are outlined now. A critical question for the error correction of a quantum code is the parity of a codeword. In our example we wish to determine the parity of a codeword consisting of the three qubits 1, 2, and 3. The answer to this question is that the parity of qubit 1 is given by the eigenvalue of the transformation Z_1 applied to qubit 1; an eigenvalue of 1 corresponds to even parity and 1 to odd parity of the qubit. Consequently, the eigenvalue of the product, $Z_1 Z_2 Z_3$, gives the parity of the three-bit codeword; 1 implies even parity and 1 odd parity. As noted by DiVincenzo [129], prior to the discovery of the properties of the circuit in Figure 1.10c, it was thought that a measurement of a multi-particle Hermitian operator would require that each particle be measured separatively and could only be done in a demolishing manner [289], and that would have posed tremendous challenges to quantum error correction. These examples convince us that the CNOT gate plays a very important role in quantum information processing.

Multi-Qubit Gates

The Toffoli gate is an example of a three-qubit gate; the unitary transformation performed by the Toffoli gate can be described using the basis states, 000 , 001 , 010 , 011 , 100 , 101 , 110 , 111 as,

$$U_{Toffoli} \quad \begin{matrix} 1 & 0 & 0 & 0 & 0 & 0 & 0 & 0 \\ 0 & 1 & 0 & 0 & 0 & 0 & 0 & 0 \\ 0 & 0 & 1 & 0 & 0 & 0 & 0 & 0 \\ 0 & 0 & 0 & 1 & 0 & 0 & 0 & 0 \\ 0 & 0 & 0 & 0 & 1 & 0 & 0 & 0 \\ 0 & 0 & 0 & 0 & 0 & 1 & 0 & 0 \\ 0 & 0 & 0 & 0 & 0 & 0 & 0 & 1 \\ 0 & 0 & 0 & 0 & 0 & 0 & 1 & 0 \end{matrix} .$$

Multi-qubit gates require simultaneous access to the state of several qubits; intuitively, we expect that *the larger the number of qubits a quantum gate operates on, the more difficult it is to implement that gate;* one-qubit gates are the easiest to implement, two-qubit gates are more complex, and so on. While it may be more expedient to express a transformation using multi-qubit gates such as the Toffoli gate, it seems appropriate to ask whether it is possible to implement multi-qubit gates with simpler, one- and two-qubit gates, a question we shall examine in more detail in the next section when we discuss universality of quantum gates.

As an example, consider the simulation of the three-qubit controlled-U gate of Deutsch (Figure 1.11), where U is a generic unitary transformation and the transformation carried out by this

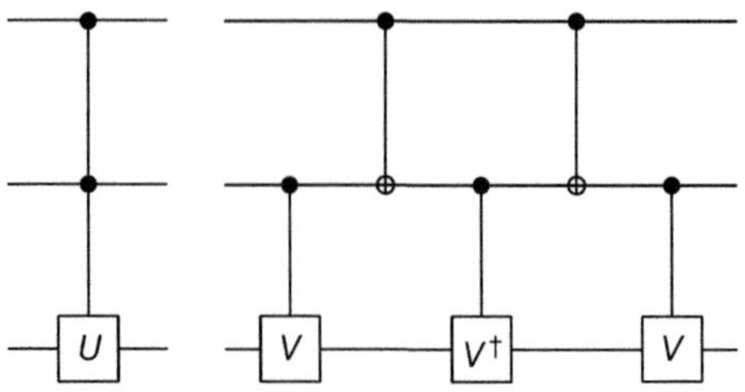

(a) (b)

FIGURE 1.11

(a) The three-qubit controlled-U gate of Deutsch; the unitary transformation, U, is carried out when both control qubits are set to 1 . (b) The simulation of the three-qubit controlled-U using two CNOT gates and three two-qubit controlled-V gates, with V^2 U [395].

gate is U_D:

$$U \quad \begin{matrix} u_{11} & u_{12} \\ u_{21} & u_{22} \end{matrix} \quad \text{and} \quad U_D \quad \begin{matrix} 1 & 0 & 0 & 0 & 0 & 0 & 0 & 0 \\ 0 & 1 & 0 & 0 & 0 & 0 & 0 & 0 \\ 0 & 0 & 1 & 0 & 0 & 0 & 0 & 0 \\ 0 & 0 & 0 & 1 & 0 & 0 & 0 & 0 \\ 0 & 0 & 0 & 0 & 1 & 0 & 0 & 0 \\ 0 & 0 & 0 & 0 & 0 & 1 & 0 & 0 \\ 0 & 0 & 0 & 0 & 0 & 0 & u_{11} & u_{12} \\ 0 & 0 & 0 & 0 & 0 & 0 & u_{21} & u_{22} \end{matrix} .$$

This gate was studied extensively [117, 127, 274, 395], and it was proved that the quantum version of it can be decomposed into simpler parts, while the classical one used for Boolean reversible computations cannot. The decomposition of the Deutsch gate as presented in Figure 1.11 uses two CNOT gates and three two-qubit controlled-V gates [395].

Quantum Circuits

Quantum circuits are collections of *quantum gates* interconnected by *quantum wires*. The actual structure of a quantum circuit, the number and the types of gates, as well as the interconnection scheme are dictated by the unitary transformation, **U**, carried out by the circuit. Though in our description of quantum circuits we use the concepts *input* and *output registers* of qubits, we should be aware that physically, the input and the output of a quantum circuit are not separated as their classical counterparts are; this convention allows us to describe the effect of unitary transformation carried out by the circuit in a more coherent fashion.

In all descriptions of quantum circuits in addition to gates, we see *quantum wires* that move qubits and allow us to *compose* more complex circuits from simpler ones that, in turn, are composed of quantum gates. We compose components by connecting the output of one to the input of another; we also compose operations when the results of an operation are used as input to another. The composition does not affect the quantum states. The quantum wires do not perform any transformations in a computational sense; sometimes we can view them as transformations carried out by the Pauli identity operator σ_I.

Quantum Computers

Quantum computers perform unitary transformations; the input qubits in the state x should be returned to the same state at the end of the computation. A quantum circuit requires a number of *ancilla qubits* to store partial results; the ancilla qubits are initially in a well-defined state, usually 0, and must be returned to the same state at the end of the computation. Figure 1.12 shows a schematic representation of the transformations required by reversibility. The original circuit has two input registers, one in the state $x^{(n)}$ and the other in the state $y^{(m)}$, as well as two output registers. As a result of the transformation, one output register will be in the state $f(x)^{(n)}$ and the other in the state $g(x)^{(m)}$. First, we add two new registers, one for the result and the other for the ancilla qubits, both in the state 0; then, we add CNOT gates for bitwise AND of y and $f(x)$, and finally, we add CNOT gates to reverse the computation and to return the result and ancilla registers to their initial state.

The quantum circuit in Figure 1.13 shows a circuit with an input register of six qubits in the state x. The unitary transformation, **U**, is applied to the register of qubits in the state $y^{(m)}$. The unitary transformation **U** is carried out on the register y if $x_1 \quad x_2 \quad x_3 \quad x_4 \quad x_5 \quad x_6 \quad 1$. The five ancilla qubits used to store partial results are returned to their original state, state 0, at the end of the computation.

In the next section, we discuss one of the requirements formulated by DiVincenzo, the universality of quantum gates. The Solovay-Kitaev theorem states that it is possible to approximate, with a desired level of accuracy, the quantum circuit implementing a quantum algorithm, though no finite set of quantum gates can generate all unitary operations.

1.17 UNIVERSALITY OF QUANTUM GATES: THE SOLOVAY-KITAEV THEOREM

Practical considerations related to fault-tolerance restrict the diversity of quantum gates used to express a quantum algorithm and lead us to the question of universality of quantum gates. In this section we discuss the theoretical aspects of the universality of quantum gates.

We start with the observation that one-qubit quantum gates cannot be universal; indeed, they cannot place two initially unentangled qubits in an entangled state. It is also clear that classical gates cannot be universal for quantum computing because they cannot create a superposition of quantum states.

We know from Section 1.15 that a gate operating on k qubits is represented by a unitary transformation on a 2^k-dimensional Hilbert space. We denote by $SU(d)$ the special unitary group of degree d the multiplicative group of $(d \quad d)$ unitary matrices with determinant equal to 1. The d-dimensional vector space of $(d \quad d)$ unitary matrices is a manifold and can be parameterized by a *continuum* of real parameters. For example, when $k \quad 1$ we have the set of one-qubit gates characterized by unitary transformations, **U**, in $SU(2)$ parameterized by the set of continuous parameters, α, β, and θ:

$$\mathbf{U}_2 \qquad \begin{matrix} e^{i\alpha}\cos\theta & e^{i\beta}\sin\theta \\ e^{\quad i\beta}\sin\theta & e^{\quad i\alpha}\cos\theta \end{matrix} \quad .$$

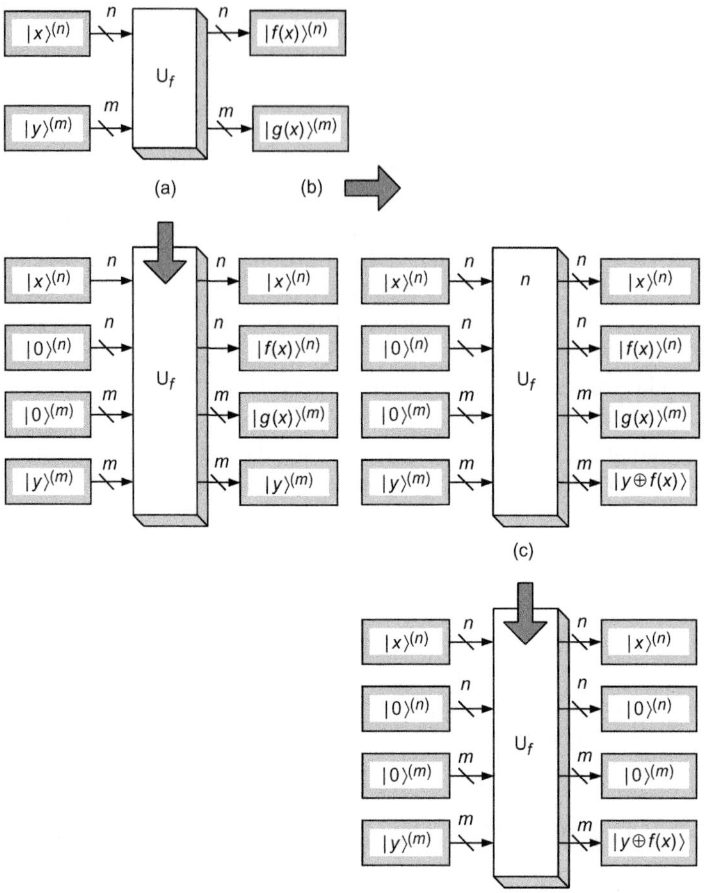

FIGURE 1.12

Schematic representation of a reversible quantum gate array. The circuit carries out the unitary transformation, $\mathbf{U}_f$, and has two input registers, one with n qubits in the state $x^{(n)}$ and the other with m qubits in the state $y^{(m)}$. There are two output registers; after the transformation, one will be in the state $f(x)^{(n)}$ and the other in the state $g(x)^{(m)}$. (a) We add two new registers, one for the result and the other for the ancilla qubits, both in the state 0. (b) We add CNOT gates for bitwise AND of y and $f(x)$. (c) We add CNOT gates to reverse the computation and to set the second and third output registers to zero.

No finite set of quantum gates can generate all unitary operators; thus, to define universality we should consider the *approximate* simulation of a circuit by another one. Cast in terms of quantum computability, rather than seeking computational universality, we wish to *approximate* any quantum algorithm without using too many more gates than in its original description.

Barenco *et al* showed that CNOT and one-qubit gates are sufficient to carry out universal quantum computations [16]; Lloyd established that almost any two-qubit gate belongs to a set of universal gates [274]. The general solution to the problem was given by Solovay in 1995 for $SU(2)$, but not published;

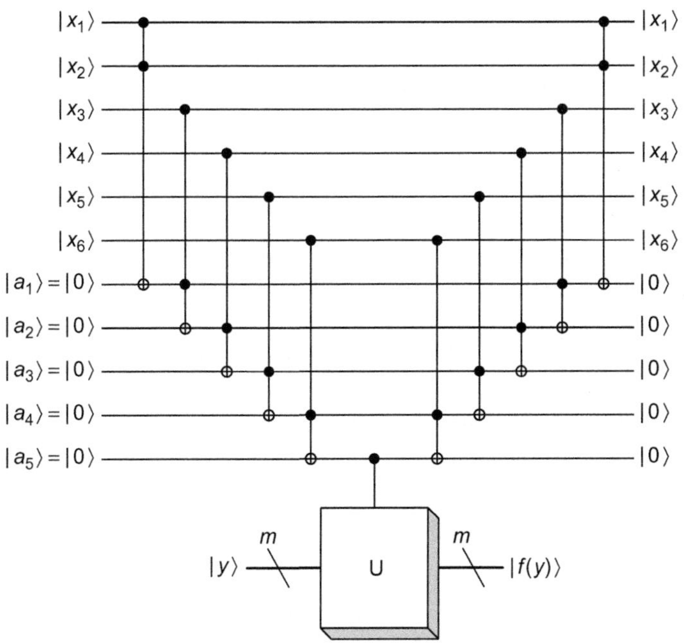

FIGURE 1.13

A reversible quantum circuit. The input register x is in the state x $x_1x_2x_3x_4x_5x_6$; the register y has m qubits. The unitary transformation, $\mathbf{U}$, is applied to y only if the condition x_1 x_2 x_3 x_4 x_5 x_6 1 is satisfied. There are five ancilla qubits used to store partial results; initially, they are in the state 0 and after the transformation $\mathbf{U}$, they are returned to the state 0 . The circuit uses 10 `Toffoli` gates.

in 1997, Kitaev published a review paper [239] and outlined a proof for the general case of $SU(d)$ and then Solovay announced that he had generalized the result for $SU(d)$ as well [111]. According to the Solovay-Kitaev theorem, if a set $\mathcal{G}$ of one-qubit quantum gates generates a dense subset of $SU(2)$, then it is possible to construct good approximations of any gate using short sequences of gates from the set $\mathcal{G}$ of one-qubit gates; in other words, $\mathcal{G}$ is guaranteed to fill out $SU(2)$ quickly.

The problem of finding efficient approximations of quantum gates can also be cast as *quantum compiling* [202]. In classical computing, a compiler transforms a source code to an object code consisting of instructions the target computer was designed to carry out; the translation should take a short time and the object code should run efficiently. By analogy, a quantum compiler expresses a quantum computation as a set of quantum gates, hopefully from a relatively small set. The problem is to define the set of gates and to design efficient translation algorithms. In this context, "efficient" means that the translation takes a short time and, even more importantly, that the resulting string of gates is of minimum length.

A set of gates $\mathcal{B}$ $SU(d)$, called *base gates*, is said to be *computationally universal* if given any gate performing the unitary transformation, $\mathbf{U}$ $SU(d)$, we can find a string consisting of gates from the set $\mathcal{B}$ and their inverses, such that the product of these gates approximates $\mathbf{U}$ with arbitrary precision ϵ; then $\mathcal{B}$ is said to be a dense subset of $SU(d)$.

An important question is the length of the string of basis gates used to approximate **U**, with precision ϵ. Lloyd's construction requires $\mathcal{O}(e^{1/\epsilon})$ gates [274]. The Solovay-Kitaev algorithm runs in a time polynomial in $\log(1/\epsilon)$ and produces strings of length $\mathcal{O}(\log^c(1/\epsilon))$, with c a constant, $3 \quad c \quad 4$. The constant, c, cannot be smaller than 1; indeed, a ball of radius ϵ in $SU(d)$ has a volume proportional to $\epsilon^{d^2 \; 1}$. To approximate every element of $SU(d)$ to a precision ϵ, we need $\mathcal{O}((1/e)^{d^2 \; 1})$ different strings of gates. It has been shown [202] that at least for some universal sets of base gates, $c \quad 1$; in other words, we need only $\mathcal{O}(\log(1/\epsilon))$ gates to approximate any gate to a precision ϵ.

Given an arbitrary projector, **P**, and two states, φ and φ, such that $(\varphi \quad \varphi) \quad \epsilon$, then the probabilities of the two states satisfy the inequality

$$\mathbf{P} \; \varphi \quad ^2 \quad \mathbf{P} \; \varphi \quad ^2 \quad (\; \mathbf{P} \; \varphi \qquad \mathbf{P} \; \varphi \quad)(\; \mathbf{P} \; \varphi \qquad \mathbf{P} \; \varphi \quad) \quad 2\epsilon.$$

A similar inequality holds for a mixed state, with the density matrices ρ and ρ.

The Solovay-Kitaev theorem guarantees that different sets of universal gates can simulate each other exceedingly fast and to a very good approximation.

Theorem. *$\mathcal{G} \quad SU(d)$ is a universal family of gates, where $SU(d)$ is a group of operators in a d-dimensional Hilbert space if:*

1. *$\mathcal{G}$ is closed under the inverse operation, $g \quad \mathcal{G} \quad g^{\;\;1} \quad \mathcal{G}$ and*
2. *$\mathcal{G}$ generates a dense subset of $SU(d)$,*

then there exists a finite sequence of gates from $\mathcal{G}$ such that

$$\mathbf{U} \quad SU(d),\ \epsilon > 0,\ \ g_1, g_2, \ldots, g_q \quad \mathcal{G}: \ \mathbf{U} \quad \mathbf{U}_{g_1, g_2, \ldots, g_q} \qquad \epsilon \ \text{ and } \ q \quad \mathcal{O}(\log^2 1/\epsilon),$$

where $\quad \mathbf{U} \quad$ is the norm of the linear operator, **U**, and $\mathbf{U}_{g_1, g_2, \ldots, g_q}$ is an implementation of **U** using only, q, gates from the set, $\mathcal{G}$.

A proof for the general case, $SU(d)$, is given by Kitaev [239]. An algorithm for compiling an arbitrary one-qubit gate to a sequence of gates from a fixed and finite set is described in [111]. The algorithm is based on the Solovay-Kitaev theorem and runs in $\mathcal{O}(\log^{2.71}(1/\epsilon))$ time and produces an output sequence of $\mathcal{O}(\log^{3.97}(1/\epsilon))$ quantum gates. The algorithm can be used to compile Shor's algorithm, which uses rotations of $\pi/2^k$ using a target set consisting of Hadamard H gates and $\pi/8$ one-qubit gates and CNOT gates.

An important class of quantum circuits are the *stabilizer circuits* used for quantum error-correcting codes. Such circuits consist only of Hadamard, phase, and measurement one-qubit gates and CNOT gates. Unitary stabilizer circuits are also known as *Clifford group circuits*. It turns out that such circuits can be simulated efficiently using classical computers, a very important consideration for the investigation of the fault-tolerance of quantum circuits [1].

Recall that a quantum circuit maps n input qubits to precisely n output qubits, a necessary, but not a sufficient, condition for reversibility. The transfer matrix, U, of a quantum circuit is a *unitary operator* $\mathbf{U}\mathbf{U}^\dagger \quad \mathbf{U}^\dagger\mathbf{U} \quad \mathbf{I}$; when applied to an input register, with n qubits in the state $\psi^{(n)}$, the circuit produces n qubits in the state $\varphi^{(n)} \quad U \; \psi^{(n)}$. It is easy to show that any n-qubit quantum circuit, A, can be simulated by another circuit, B, constructed with only CNOT and H, S, and T one-qubit gates that carry out the following transformations:

$$U_{CNOT} \quad \begin{matrix} 1 & 0 & 0 & 0 \\ 0 & 1 & 0 & 0 \\ 0 & 0 & 0 & 1 \\ 0 & 0 & 1 & 0 \end{matrix}\, , \quad H \quad \frac{1}{2} \begin{matrix} 1 & 1 \\ 1 & 1 \end{matrix}\, , \quad S \quad \begin{matrix} 1 & 0 \\ 0 & i \end{matrix}\, , \quad T \quad \begin{matrix} 1 & 0 \\ 0 & \exp(i\pi/4) \end{matrix}\, ,$$

with $S \quad T^2$ and $S^2 \quad Z$. The process of constructing the circuit B consisting of only CNOT and H, S, and T one-qubit gates that simulates the unitary transformation A of n qubits consists of three stages:

1. Decompose the unitary transformation, A, as a product of unitary transformations, $U_k, 1 \quad k \quad 2^n$, which affect at most two computational basis states.
2. Rearrange the basis states so that each unitary transformation, U_k, affects only one qubit and can thus be carried out by either a two-qubit CNOT gate or by one-qubit gate.
3. Approximate the transformation carried out by any one-qubit gate by the transformation carried out by the three gates in the set H, T, S .

1.18 QUANTUM COMPUTATIONAL MODELS AND QUANTUM ALGORITHMS

A *quantum computational model* specifies the resources needed for a quantum computer, as well as the means to specify and control a quantum computation. To process quantum information, each computational model requires several steps from the set: free-time quantum evolution, controlled-time evolution, preparation, and measurement.

Several quantum computational models have been proposed: the quantum Turing machine, the quantum circuit, the topological quantum computer, the adiabatic, one-way quantum computer, and measurement models (Figure 1.14).

The Quantum Turing Machine Computational Model

This model is a generalization of the classical Turing machine model; it was the first quantum computational model introduced by Benioff in 1980 [27] and further developed by Bernstein and Vazirani in 1997 [50]. Though based on quantum kinematics and dynamics, Benioff's model was classical in the sense that it required specification of the state as a set of numbers measurable at any instant of time. Then, in 1982, Richard Feynman introduced a "universal quantum simulator" consisting of a lattice of spin systems with near-neighbor interactions; Feynman's model could simulate any physical system with a finite-dimensional state space but did not include a mechanism to select arbitrary dynamic laws.

The Quantum Circuit Model

The standard model for a quantum computer is related to the classical circuit model. It was proposed by David Deutsch [115] and further developed by Andrew Yao [464]; this model applies only to uniform families of quantum circuits.

To answer the question of how close is the quantum circuit model to the classical one, we should remember the procedure discussed in Section 1.15 for constructing the Boolean circuit implementing

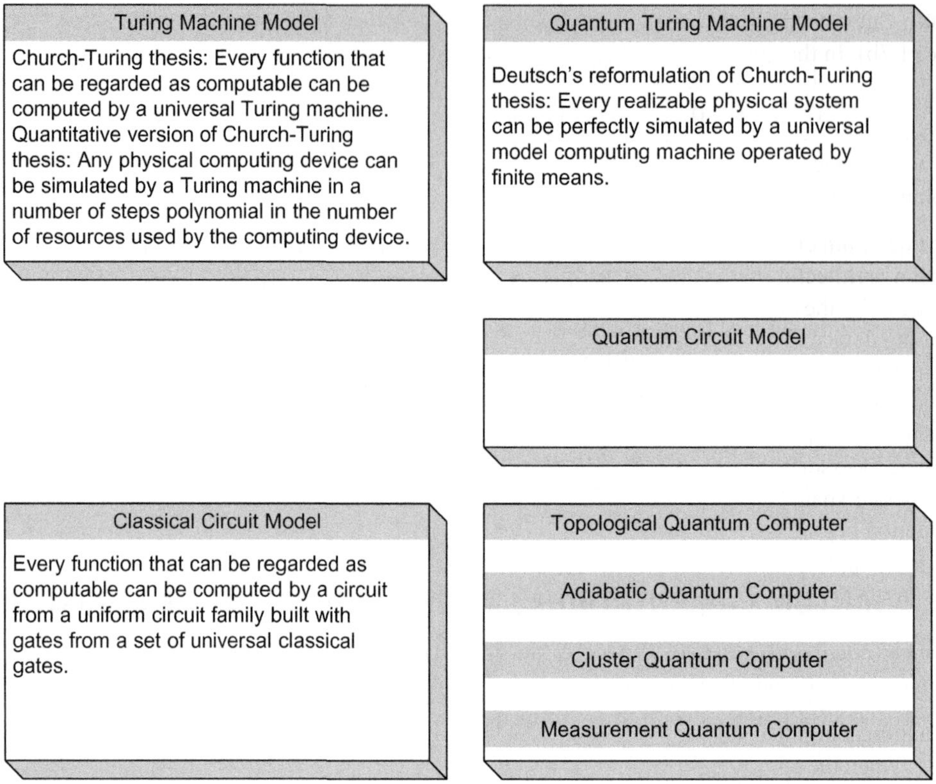
Turing Machine Model
Church-Turing thesis: Every function that can be regarded as computable can be computed by a universal Turing machine.
Quantitative version of Church-Turing thesis: Any physical computing device can be simulated by a Turing machine in a number of steps polynomial in the number of resources used by the computing device.
Quantum Turing Machine Model
Deutsch's reformulation of Church-Turing thesis: Every realizable physical system can be perfectly simulated by a universal model computing machine operated by finite means.
Quantum Circuit Model
Classical Circuit Model
Every function that can be regarded as computable can be computed by a circuit from a uniform circuit family built with gates from a set of universal classical gates.
Topological Quantum Computer
Adiabatic Quantum Computer
Cluster Quantum Computer
Measurement Quantum Computer

FIGURE 1.14

Classical and quantum computational models.

a function, $f: \{0,1\}^n \mapsto \{0,1\}$, and the five elements required for the construction of the circuit. Some of the challenges posed by the quantum circuit model are that the no-cloning theorem prohibits fanout; the decoherence of quantum states makes even the implementation of quantum wires nontrivial; and we can only approximate the function of one-qubit gates with gates from a small set of universal quantum gates.

To capture the finiteness of resources required by a physical realization of a computing device, David Deutsch restated the Church-Turing hypothesis [115]: *"Every realizable physical system can be perfectly simulated by a universal model computing machine operating by finite means."* The much stronger formulation of the Church-Turing hypothesis is not satisfied by a Turing machine $\mathcal{T}$ operating in the realm of classical physics; indeed, the set of states of a classical physical system form a continuum due to the continuity of classical dynamics. The classical Turing machine, $\mathcal{T}$, cannot simulate every classical dynamics system because there are only countable ways of preparing the input for $\mathcal{T}$.

In the quantum circuit model, *a quantum computation is carried out by quantum circuits that transform information under the control of external stimuli.* A quantum circuit operating on n qubits performs a unitary operation in the Hilbert space, $\mathcal{H}_{2^n}$, and consists of a finite collection of quantum gates; each quantum gate implements a unitary transformation on a small number k of qubits (Figure 1.7b). In the quantum circuit model, a computation involving n qubits consists of three stages:

1. Initialization stage, when the quantum computer is prepared in a basis state in $\mathcal{H}_{2^n}$
2. Processing stage, when a sequence of one- and two-qubit gates are applied to the qubits
3. Readout stage, when a measurement of a subset of qubits in the computational basis allows us to obtain the classical information revealing the result of the computation

The quantum circuit model involves an *open* quantum system; the role of the control unit of the von Neumann architecture is played by a classical system that controls the temperature, the pressure, the magnetic field, the electric field, or other parameters specific to the physical realization of the quantum computing device.

A quantum computing engine, $\mathcal{Q}$, consists of two components: a finite processor consisting of n two-state observables (qubits), $p_i \in \mathcal{H}_{2^n}$, and an infinite memory consisting of an infinite sequence of two-state observables (qubits), m_i. The computation consists of steps of finite duration, t_s; during each step, the processor and a finite segment of the memory interact. The state of $\mathcal{Q}$ is a unit vector in a Hilbert space spanned by simultaneous eigenvectors, $|b_i\rangle$, forming *computational basis states.* The computation begins at $t = 0$, and we need to specify the state, $|\psi\rangle$, only at times that are multiples of t_s. The system dynamics are specified by a unitary operator $\mathbf{U}$:

$$|\psi(kt_s)\rangle = \mathbf{U}^k |\psi(0)\rangle .$$

While a classical Turing machine signals the end of a computation when two consecutive states are identical, two states of a reversible computing engine $\mathcal{Q}$ can never be identical. Thus, the processor of the quantum computing engine $\mathcal{Q}$ must have an internal qubit, p_e, initially in the state $|0\rangle$ and set to the state $|1\rangle$ when the computation has finished. This qubit must be periodically observed from the outside without affecting the state of $\mathcal{Q}$; this is possible because this qubit contains only classical information.

We say that a quantum "program" is *valid* if the state of p_e goes to $|1\rangle$ in a finite time; this statement is analogous to the one for a classical Turing machine, $\mathcal{T}$, in which case a "program" is valid if the machine halts after a finite number of steps. We now realize that a classical Turing machine is similar to a quantum computing engine $\mathcal{Q}$ whose evolution in time ensures that at the end of each computational step it remains in a computational basis state provided that it started in one.

In summary, according to the circuit model, a quantum "program" is a sequence of quantum gates applied to a group of qubits at a time. The function of a multi-qubit gate can be simulated with one-qubit and two-qubit gates; multi-qubit gates are more difficult to realize physically.

The Topological Quantum Computer Model

In 1997, Alexei Kitaev [240] proposed anyonic quantum computing as an inherently fault-tolerant quantum computing model. Kitaev showed that a system of non-abelian anyons can efficiently simulate a quantum circuit; in Kitaev's scheme measurements are necessary to simulate some quantum gates. Preskill elaborated on the idea of topological quantum computing, showing that it provides an elegant means to ensure fault-tolerance [341]. Freedman, Larsen, and Wang showed that the measurements can be postponed until the final readout of the results [157]; then they proved that a system of anyons can be simulated effectively by a quantum circuit, thus demonstrating that the topological quantum computing model has similar computing power as the quantum circuit model [158]. Mochon gave a constructive proof that anyonic magnetic charges with fluxes in a nonsolvable finite group can perform universal quantum computations; the gates are based on the elementary operations braiding, fusion, and vacuum pair creation [295].

Recall from Section 1.13 that a fractional quantum Hall effect, FQHE, occurs when a two-dimensional electron gas placed in a strong magnetic field, at very low temperature, behaves as a system of anyons, particles with a fractional charge (e.g., $e/3$, where e is the electric charge of an electron). In a nutshell, *a topological quantum computer braids world-lines by swapping the positions of anyons.* This terse statement needs a fair amount of explaining. A *braid*, or a strand, connects two objects; informally, world-lines are representations of particles as they move through time and space. Particle world-lines form braids in a space with (2 1) dimensions (two space dimensions and one time dimension) (Figure 1.15). The symmetry group generated by the exchanges of n particles is in a one-to-one correspondence with the braid group.

B_n, the braid group on n strands, is an infinite group with an intuitive geometric representation; it shows how n strands can be laid out to connect two groups of n objects (Figure 1.16). The braid group is a multiplicative group; the product of two elements is constructed geometrically by laying the two elements next to one another in the order dictated by the product and connecting the two objects in the middle. To construct the inverse of an element, the left-hand set becomes the right-hand set and vice versa. The identity element connects objects with the same index from the left and right sets, and the strands are straight lines. Matrices form a non-Abelian representation of the braid group.

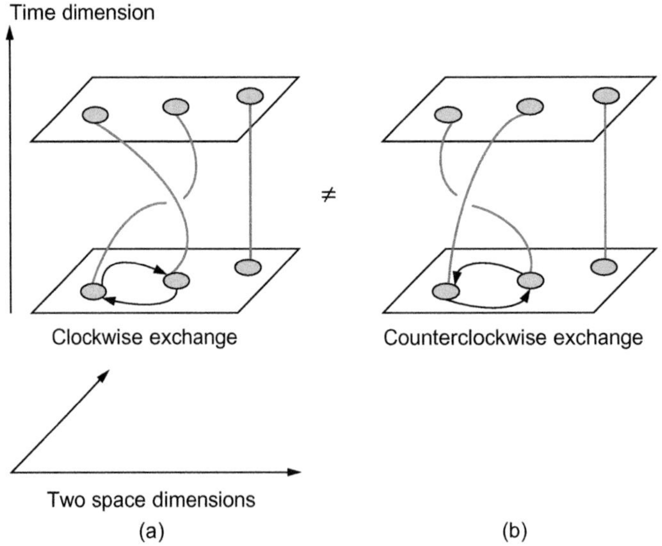

FIGURE 1.15

Particle world-lines form braids in a space, with (2 1) dimensions. (a) The braids corresponding to the clockwise exchange of two particles. φ_{init} α_0 0 α_1 1 is the initial state, and φ_{final} γ_0 0 γ_1 1 is the final state of the system. The symmetry group generated by the exchanges of two particles is in one-to-one correspondence with the braid group, B_2, of 2 2 unitary matrices. (b) The braids corresponding to the counterclockwise exchange of the two particles.

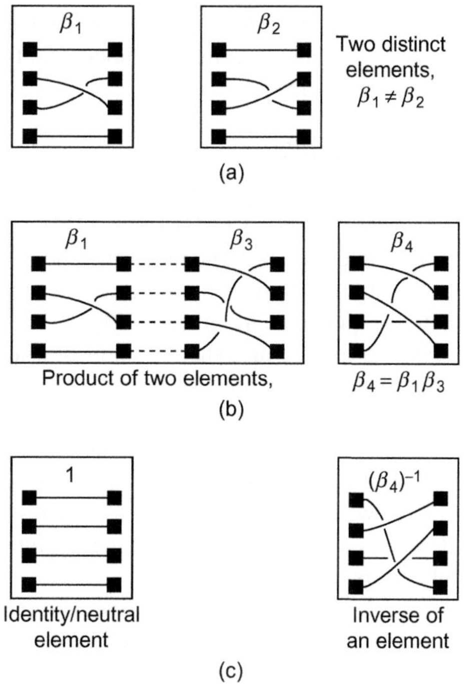

FIGURE 1.16

The braid group, B_4. $\beta_1, \beta_2, \beta_3, \beta_4$, are elements of B_4; each element consists of two sets of four objects (in our case squares), one on the left and one on the right, connected by four strands. Each set of four squares is arranged vertically; a strand connects one object from the left set with an object from the right set. A strand moves from left to right, and knots are not allowed; e.g., the four strands of β_1 connect the objects as follows: $(1, \text{left}) \quad (1, \text{right})$, $(2, \text{left}) \quad (3, \text{right})$, $(3, \text{left}) \quad (2, \text{right})$, and $(4, \text{left}) \quad (4, \text{right})$. (a) $\beta_1 \quad \beta_2$ because the positions of the strands connecting objects 2 and 3 are different; $(2, \text{left}) \quad (3, \text{right})$ is above in β_1 and $(2, \text{left}) \quad (3, \text{right})$ is below in β_2. (b) The braid group is a multiplicative group; the product of two elements is constructed by laying the two elements next to one another in the order dictated by the product and connecting the two objects in the middle. (c) The construction of the inverse of an element: the left set becomes the right set and vice versa. The identity element connects objects with the same index from the left and right sets, and the strands are straight lines; it is easy to see that $(\beta_4)(\beta_4)^{-1} \quad 1$.

The symmetry group generated by the exchanges of any two particles are in one-to-one correspondence with the braid group, B_2, of 2 2 unitary matrices. If $\varphi_{init} \quad \alpha_0 \; 0 \quad \alpha_1 \; 1$ denotes the initial state and $\varphi_{final} \quad \gamma_0 \; 0 \quad \gamma_1 \; 1$ denotes the final state of the system in Figure 1.15a, then the two states are related by the relation:

$$\begin{matrix}\gamma_0 \\ \gamma_1\end{matrix} \quad \begin{matrix} a_{11} & a_{12} \\ a_{21} & a_{22}\end{matrix} \quad \begin{matrix}\alpha_0 \\ \alpha_1\end{matrix} \quad \text{or} \quad \varphi_{final} \quad A \;\; \varphi_{init}\,, \quad \text{with} \;\; A \quad \begin{matrix} a_{11} & a_{12} \\ a_{21} & a_{22}\end{matrix} \,.$$

This relation resembles the one describing the transformation carried out by a one-qubit quantum gate and hints at the fact that the topological quantum computer model and the quantum circuit model are closely related; in fact, they are equivalent [158].

The Adiabatic Quantum Computing Model

The adiabatic quantum computing model was proposed in 2000 by Farhi *et al* [145] who suggested an algorithm to solve optimization problems, such as SATISFIABILITY (SAT); there is now evidence that this algorithm takes an exponential time for some (nondeterministic polynomial time) NP-complete problems. The interest in the adiabatic quantum computing was renewed in 2005 when Aharonov *et al* proved that it is equivalent to the quantum circuit model [3].

An *adiabatic* process is a *quasi-static* thermodynamic process when no heat is transferred; the opposite of an adiabatic process is an *isothermal* process when heat is transferred to maintain the temperature constant. An *adiabatic evolution of a quantum system means that the Hamiltonian is slowly varying;* recall from Section 1.6 that the Hamiltonian operator corresponds to the total energy of the quantum system. The Hamiltonian is a Hermitian operator, and its eigenvector corresponding to the smallest eigenvalue (i.e., to the lowest total energy of the system) is called the *ground state* of the system. A *local* Hamiltonian describes a quantum system where the interactions occur only among a constant, and rather small, number of particles.

The adiabatic approximation is a standard method to derive approximate solutions to the Schrödinger equation when the Hamiltonian is slowly varying. This method is based on a simple idea: If the quantum system is prepared in a ground state and the Hamiltonian varies slowly enough, then, as time goes on, the system will stay in a state close to the ground state of the instantaneous Hamiltonian. This idea is captured by the adiabatic theorem of Born and Fock [65]:

> *A physical system remains in its instantaneous eigenstate if a given perturbation is acting on it slowly enough and if there is a gap between the eigenvalues (corresponding to this eigenstate) and the rest of the Hamiltonian s spectrum.*

Consider a quantum system in the state $\psi(t) \quad \mathcal{H}_n$, with a Hamiltonian, $\mathbf{H}(t)$. The evolution of the system is described by the Schrödinger equation,

$$i\frac{d}{dt} \quad \psi(t) \qquad \mathbf{H}(t) \quad \psi(t) \ .$$

Assume that the Hamiltonian $\mathbf{H}(t)$ is slowly varying:

$$\mathbf{H}(t) \qquad \mathbf{H}(t/T),$$

where T controls the rate of variation of $\mathbf{H}(t)$, and $\mathbf{H}(t/T)$ belongs to a smooth one-parameter family of Hamiltonians, $\mathbf{H}(s), 0 \quad s \quad 1$. The instantaneous eigenstates, $i;s$, and eigenvalues, E_i, of $\mathbf{H}(s)$ are defined as

$$\mathbf{H}(s) \quad i;s \qquad E_i \quad i;s \ .$$

The eigenvalues of $\mathbf{H}(s)$ are ordered

$$E_0(s) \quad E_1(s) \qquad\qquad E_n(s).$$

Call $\psi_0 \quad i \quad 0; s \quad 0$ the ground state of $\mathbf{H}(0)$. The adiabatic theorem says that if the gap between the lowest energy levels $E_1(s) \quad E_0(s) > 0, \quad 0 \quad s \quad 1$, then the state, $\psi(t)$, of the system after an evolution described by the Schrödinger equation will be very close to the ground state of the Hamiltonian $\mathbf{H}(t)$ for $0 \quad t \quad T$ when T is large enough.

The adiabatic quantum computer evolves between an initial state with the Hamiltonian, $\mathbf{H}_{init}$, and a final state with the Hamiltonian, $\mathbf{H}_{final}$. The input data and the algorithm are encoded as the ground state of $\mathbf{H}_{init}$, and the result of the computation is the ground state of $\mathbf{H}_{final}$. The running time of the adiabatic computation is determined by the minimal spectral gap of all Hamiltonians of the form

$$\mathbf{H}(s) \quad (1 \quad s)\mathbf{H}_{init} \quad s\mathbf{H}_{final}, \quad 0 \quad s \quad 1.$$

These Hamiltonians lie on the straight line connecting $\mathbf{H}_{init}$ and $\mathbf{H}_{final}$ [3]. The ground state of the Hamiltonian $\mathbf{H}_{final}$ for the optimization algorithm in [145] was a classical state in the computational basis, and $\mathbf{H}_{final}$ was a diagonal matrix as the solution of a combinatorial optimization problem. This restriction was removed by Aharonov *et al* [3] who require only that the Hamiltonians be *local*. This condition resembles the one imposed on the quantum circuit model—namely, that the quantum gates operate on a constant number of qubits.

The One-Way Quantum Computer Model

In 2001, Raussendorf and Briegel proposed a model based on a special type of entangled states, called *cluster states* [349]. A cluster state refers to a family of quantum states of n-qubit two- or three-dimensional lattice, or even a more general graph, in which each vertex corresponds to a qubit. A specific preparation procedure is applied (e.g., vertices are connected using controlled-Z two-qubit gates).

To process quantum information in this network, it is sufficient to measure the qubits in a certain order, in a certain basis. In a two-dimensional lattice, the quantum information propagates horizontally by measuring the qubits on the wire, while qubits on the vertical connection implement two-qubit gates. The bases used for measurement depend on the results of previous measurements. Any quantum circuit can be implemented on a cluster state [349].

The Quantum Measurement Model

In this model, introduced by Nielsen [308], no coherent dynamical operations are allowed; the model allows only three operations: (1) preparation of qubits in the initial state, 0 ; (2) storage of qubits; and (3) projective measurements (Section 2.7) of up to four qubits at a time in arbitrary bases.

Quantum Algorithms

Once we are convinced that given a function f is able to assemble a quantum circuit capable of evaluating this function, we switch our attention to quantum algorithms. Recall that a computational problem is considered tractable if an algorithm to solve it in a number of steps and requiring storage space polynomial in the size of the input exists. There are classically intractable problems, such as the Travelling Salesman Problem, which are proven to belong to the complexity class *nondeterministic polynomial time* ($\mathcal{NP}$). The expectation that quantum algorithms could lead to efficient solutions of "hard" computational problems proved to be justified.

A number of "toy problems," such as the Deutsch problem, provided a first glimpse of hope. Then, in 1994, Peter Shor found a polynomial time algorithm for the factorization of n-bit numbers on quantum computers [379]. His discovery generated a wave of enthusiasm for quantum computing for two major reasons: the intrinsic intellectual beauty of the algorithm and the fact that efficient integer factorization has important applications.

The security of widely used cryptographic protocols is based on the conjectured difficulty of the factorization of large integers. Like most factorization algorithms, Shor's algorithm reduces the factorization problem to the problem of finding the period of a function, but uses quantum parallelism to find a superposition of all values of the function in one step. Then the algorithm calculates the quantum Fourier transform of the function, which sets the amplitudes to multiples of the fundamental frequency, the reciprocal of the period. To factor an integer, Shor's algorithm measures the period of the function.[9]

In 1996 Grover described a quantum algorithm for searching an unsorted database $\mathcal{D}$ containing N items in a time of order $\sqrt{N}$ [182]; on a classical computer, the search requires a time of order N.

It seems natural to start our analysis of quantum algorithms with the very first algorithms developed to reveal the power of quantum parallelism. These algorithms for "toy" problems proposed by Deutsch, Jozsa, Bernstein and Vazirani, and Simon will be discussed before an in-depth analysis of phase estimation and of the Grover search algorithm.

The algorithms discussed in the next section are constructed around *black boxes* and measurements performed on the output of these black boxes. The algorithms of Shor and Grover use *white boxes* implementing the quantum Fourier transform (QFT) or the quantum Hadamard transform and determine the period of a function, or the angle of the rotation of a state vector.

1.19 DEUTSCH, DEUTSCH-JOZSA, BERNSTEIN-VAZIRANI, AND SIMON ORACLES

In this section, we show that one can construct quantum circuits capable of providing simple answers, in an *efficient* manner, to questions that require very elaborate computations, very much like oracles do.[10] Classical solutions to such problems require either an exponential number of time steps or an exponential number of copies of a circuit able to compute the value of a function. The term "efficient" means that the quantum solution requires a single copy of the quantum circuit and either a single or a linear number of time steps.

In the realm of computing, an *oracle* is an abstraction for a black box that can follow a very elaborate procedure to answer a very complex question with a "yes" or "no" answer. Quantum computing has a natural affinity for oracles because the result of a quantum computation is probabilistic and it is a superposition of all possible results.

Several increasingly complex oracles used in quantum computing are:

- *Deutsch oracle.* Given a function $f: \{0,1\} \mapsto \{0,1\}$, the oracle decides if the function is constant, $f(0) = f(1)$, or balanced, $f(0) \neq f(1)$.

[9]A powerful version of the technique used by Shor is the *phase-estimation algorithm* of Kitaev [238].

[10]The name "oracle" comes form the Latin verb "orare," meaning "to speak." An oracle is a source of wise pronouncements or prophetic opinions with some connection to an infallible spiritual authority.

- *Deutsch-Jozsa oracle.* Given a function $f:\ 0,1\ ^n \quad 0,1$, the oracle decides if the function is constant or balanced. The function is balanced if for half of the arguments it is equal to 0 and for the other half it is equal to 1. If the function is neither constant, nor balanced; the oracle's answer is meaningless.
- *Bernstein-Vazirani oracle.* Given a function $f:\ 0,1\ ^n \quad 0,1$ of the form $f(x) \quad a\ \ x$, where a is a constant vector of 0s and 1s and $a\ \ x$ is the scalar product of the vectors a and x, the oracle determines the value of a in one time step.
- *Simon oracle.* Given a function $f:\ 0,1\ ^n \quad 0,1\ ^{n-1}$, which is 2 1 and periodic with period a, the oracle returns the period in $\mathcal{O}(n)$ measurements.

To illustrate the feasibility of quantum oracles, we present two quantum circuits, one implementing the oracle for the Deutsch problem and the other for the Bernstein-Vazirani problem. Let us first discuss briefly the *Deutsch oracle*. Recall that we could use two copies of a classical circuit to compute in a single time step, $f(0)$ and $f(1)$, and decide if the function, f, is balanced or constant, assuming that we can ignore the time to compare $f(0)$ and $f(1)$. Alternatively, we can use a single copy of the circuit built to compute a binary function f of a binary argument x, $f(x)$, but in this case we need two time steps, one to compute $f(0)$ and the other to compute $f(1)$.

The quantum circuit in Figure 1.17 shows the quantum circuit for the Deutsch problem [283]. It is easy to see that the quantum circuit produces $f(0) \quad f(1)$. We calculate the state of the system after each stage, ξ_i , $1 \quad i \quad 3$, and we conclude that

$$\xi_3 \qquad f(0) \quad f(1) \quad \frac{0 \ - \ 1}{\overline{2}} \ .$$

This expression tells us that by measuring the first output qubit of the circuit in Figure 1.17, we are able to determine $f(0) \quad f(1)$ after performing a single evaluation of the function. If the outcome of a measurement of the first qubit is equal to 0, we conclude that the function is constant, and if it is 1, we conclude that the function is balanced.

It is relatively easy to construct a circuit for the more general oracle capable of solving the Deutsch-Jozsa problem. We use as a model the circuit in Figure 1.17, and the input is a register $x^{(n)}$ of n qubits.

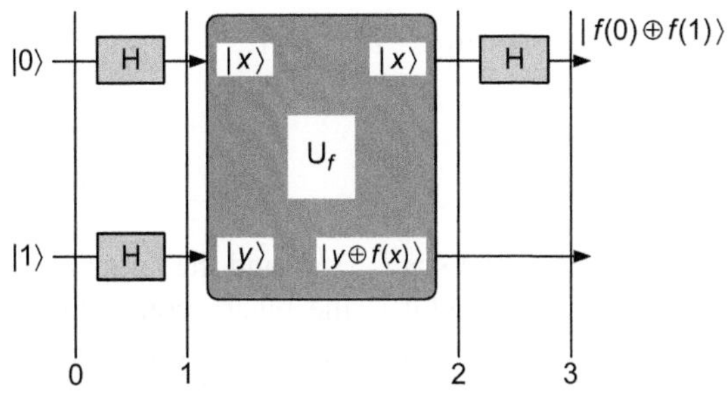

FIGURE 1.17

A quantum circuit for solving the Deutsch problem.

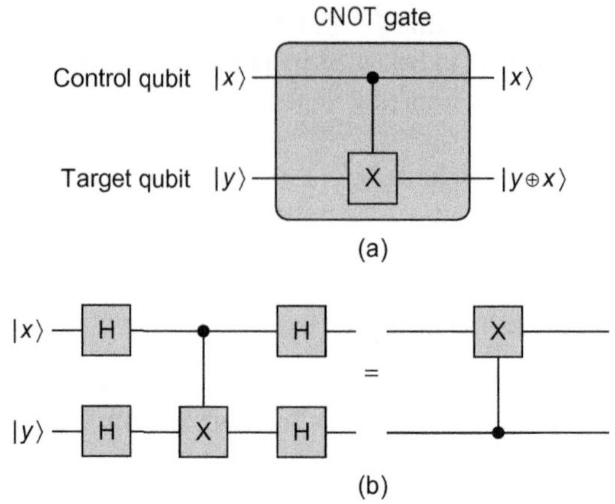

FIGURE 1.18

(a) A `CNOT` gate. (b) `Hadamard` gates on the input and output lines of a `CNOT` gate allow us to swap control and target qubits.

Before discussing the Bernstein-Vazirani oracle, we review the effect of adding two Hadamard gates to the control and target qubits of a CNOT gate (Figure 1.18); when we add two Hadamard gates on each input and output line of a CNOT gate, we swap the control and the target qubits. A detailed analysis of this circuit can be found in [283].

Now we turn our attention to the Bernstein-Vazirani problem. The quantum circuit in Figure 1.19 computes the binary value of the scalar product,

$$z \quad (a \quad x) \qquad {}^{\;n}_{i\;\;1} \; a_i x_i,$$

of two n-dimensional binary vectors,

$$x \quad (x_n x_{n-1} \ldots x_1), \quad x_i \quad 0,1 \;, \quad 1 \quad i \quad n$$

and

$$a \quad (a_n a_{n-1} \ldots a_1), \quad a_i \quad 0,1 \;, \quad 1 \quad i \quad n.$$

Thus,

$$a \quad a_n \quad 2^{n-1} \quad a_{n-1} \quad 2^{n-2} \qquad a_2 \quad 2^1 \quad a_1 \quad 2^0.$$

Initially, the control qubits, $x\ ^{(n)}$, are in the state $0\ ^{(n)}$ and the target qubit, y , is in the state 1 . As a result of the computation, the target qubit is the binary complement, z, of the inner product, $(a \quad x) \quad {}^{n}_{i\;1}\, a_i x_i$. If the inner product has an even number of terms equal to 1, then $z \quad 0$; if the inner product has an odd number of terms equal to 1, then $z \quad 1$. The reason for choosing the initial state of the target qubit to be 1 and obtaining the complement of the inner product $(a \quad x)$ will become

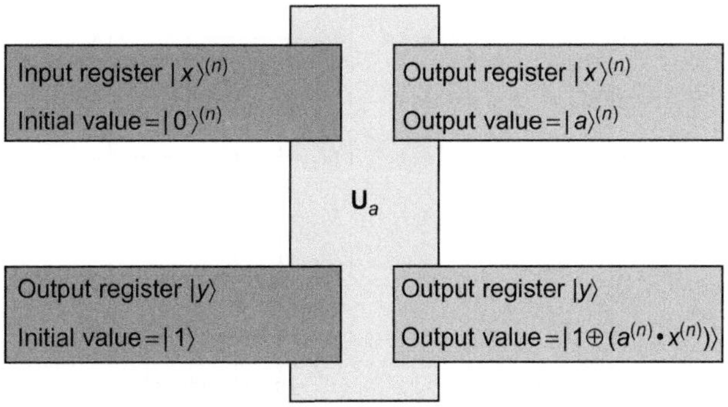

FIGURE 1.19

The black box of the Bernstein-Vazirani oracle performs a unitary transformation, U_a. The expression, $f(x) = a \cdot x$, contains an n-bit vector hardwired into the circuit. The oracle has as the input a register of n qubits, $|x\rangle^{(n)}$, initially in the state $|0\rangle^{(n)}$, and a register, $|y\rangle$, of one qubit, initially set to the state $|1\rangle$. The oracle returns in the output register $|x\rangle^{(n)}$ the n-qubit vector $|a\rangle^{(n)}$ and in the output register $|y\rangle$ the expression $|1 \oplus (a^{(n)} \cdot x^{(n)})\rangle = |1 \oplus \sum_{i=1}^{n} a_i x_i\rangle$.

apparent later when we examine the actual transformation carried out by the circuit. A classical solution to the problem of determining the constant vector, $a = (a_1, a_2, \ldots, a_n)$, requires n computations; a quantum circuit implementing the oracle requires only one copy of the circuit and one time step.

Example. A Bernstein-Vazirani oracle for $n = 6$ and $a = (a_6 a_5 a_4 a_3 a_2 a_1) = 011011$:

$$a = 0 \times 2^5 + 1 \times 2^4 + 1 \times 2^3 + 0 \times 2^2 + 1 \times 2^1 + 1 \times 2^0 = 27.$$

The oracle and the inner working of circuits for this example are shown in Figures 1.20 and 1.21, respectively. The circuit on top in Figure 1.21 has two Hadamard gates on each control qubit and two Hadamard gates on the target line; this circuit performs the same transformation as the one in Figure 1.20 because the Hadamard transformation is unitary, $\mathbf{HH}^\dagger = \mathbf{HH} = \mathbf{I}$. The circuit in the middle of Figure 1.21 has two Hadamard gates on each control and target line of every CNOT; Figure 1.18 shows that in this case we swap control and target qubits. The circuit at the bottom of Figure 1.21 reflects the reversal of the role of control and target qubits; when the input register is $|000000\rangle$, the output register will be $|011011\rangle$, the binary expression of constant $a = 27$; the control qubit will be equal to $|1\rangle$. We want the input target qubit to be equal to 1 to allow the flipping of the qubits corresponding to 1 in the binary expression of the constant vector a.

Given a black box that carries out the transformation, $f_a : \{0,1\}^n \mapsto \{0,1\}^{n-1}$, the problem addressed by the *Simon oracle* is to determine the period, a, with a minimum number of evaluations of function, f_a. The function f_a is said to be *periodic for bitwise modulo* 2 *addition,* an operation denoted as $\oplus$, if there exists an integer, $a \neq 0$, such that

$$f_a(x) = f_a(x \oplus a), \quad 0 \leq x \leq 2^n - 1.$$

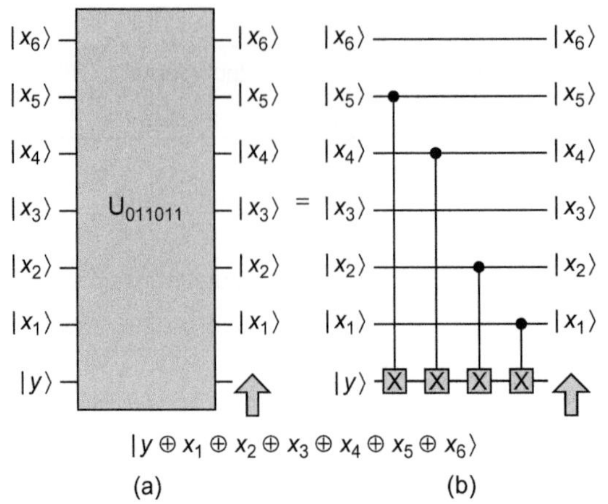

FIGURE 1.20

The Bernstein-Vazirani oracle for n 6 and a 27. (a) The black box. (b) The actual circuit with four CNOT gates. The output qubit, y , is initially in the state 1 and becomes 1 a_6x_6 a_5x_5 a_4x_4 a_3x_3 a_2x_2 a_1x_1 . Here, is addition modulo 2. In binary, a a_6 2^5 a_5 2^4 a_4 2^3 a_3 2^2 a_2 2^1 a_1 2^0, or a 011011.

Then a is called the period of f_a. If the function is periodic with period a, then given two values of the argument, x_i, x_j, $f_a(x_i)$ $f_a(x_j)$ x_i x_j a. Addition modulo 2 is associative and x_i x_i 0. Thus, x_i x_j a a x_i x_j; this suggests a method to determine a: compute x_i x_j, x_j, 0 x_j $2^n - 1, x_j$ x_i for every argument x_i. As we can see in Figure 1.22a, after m evaluations of $f_a(x_i)$, we have eliminated $\binom{m}{2}$ $m(m-1)/2$ possible values of a. The classical solution requires $\mathcal{O}(2^n)$ operations, while the Simon oracle finds the answer in $\mathcal{O}(n)$ operations.

At the heart of the solution provided by the Simon oracle is a measurement of the quantum state of a register of multiple qubits. The Born rule states that a measurement of one qubit in the state ψ α_0 0 α_1 1 produces the outcome 0, with the probability p_0 α_0 ^{2}, and 1, with the probability p_1 α_1 ^{2}. Now we outline an extension of the Born rule for a register of multiple qubits discussed in more detail in Section 2.5. We express the state of a register of (n 1) qubits, ψ $^{(n\ 1)}$, as a superposition of states of two groups of qubits; the first consists of a single qubit, which could be described in the orthonormal base [0 , 1], and the second group consists of n qubits:

$$\psi^{(n\ 1)} \quad \alpha_0 \ 0 \quad \psi_0^{(n)} \quad \alpha_1 \ 1 \quad \psi_1^{(n)}.$$

If we *only* measure the single qubit, the outcome is either 0 or 1. The extended Born rule states that when the outcome of the measurement is 0, the state of the n qubits in the second group is ψ_0 $^{(n)}$, and when the outcome of the measurement is 1, the state of the n qubits is ψ_1 $^{(n)}$.

After this brief digression, we return to the problem solved by the Simon oracle and consider the quantum circuit in Figure 1.22b. This circuit uses the Walsh-Hadamard transform discussed in detail in Section 1.21; for now we only mention that it is the result of applying a Hadamard transformation to each one of the n qubits of an input register.

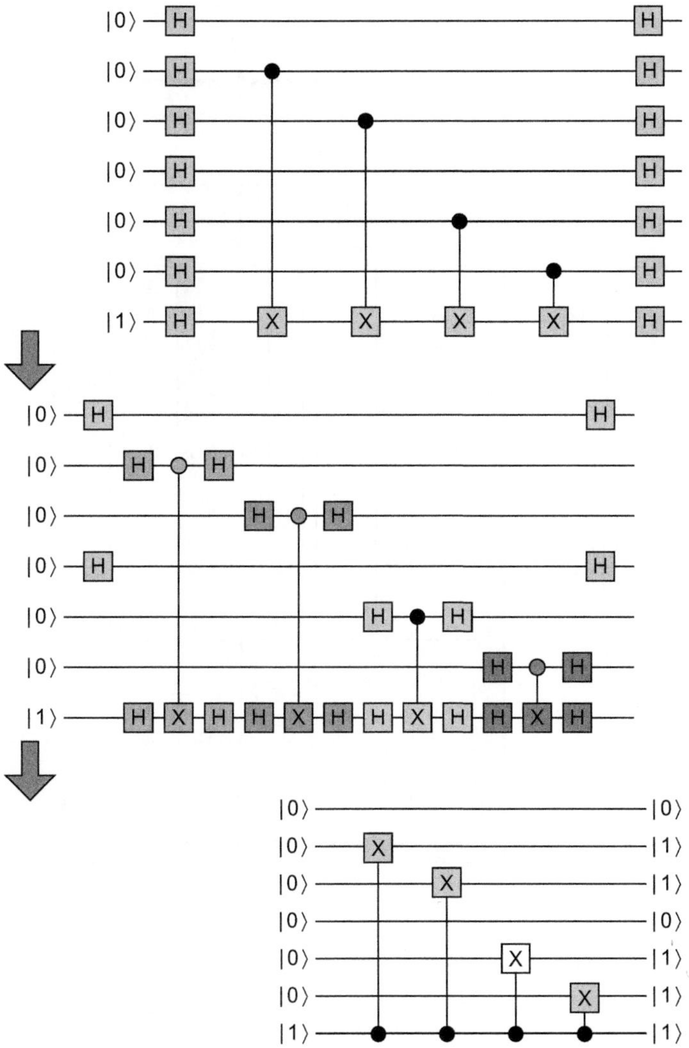

FIGURE 1.21

The inner workings of the Bernstein-Vazirani oracle in Figure 1.20. Top: we add two `Hadamard` gates on each line without altering the function. Middle: we add two `Hadamard` gates on each control and target line of every `CNOT`, and we thereby swap the control and target qubit of every `CNOT` gate. Bottom: the target and control qubits of each `CNOT` gate are swapped. Now all `CNOT` gates share a common control qubit (the original target qubit) initially in the state 1 . Thus, the output six-bit register will contain the binary expression of a 27, the vector 011011 .

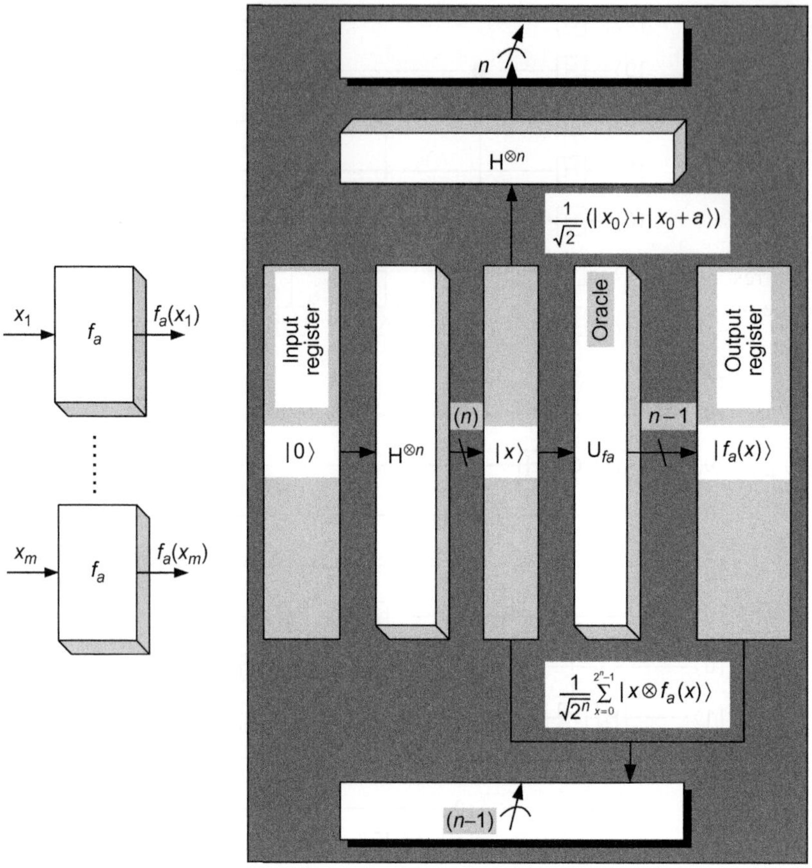

(a) (b)

FIGURE 1.22

Simon's problem: find the period, a 0, of the function, f_a : 0, 1 ^{n} 0, 1 $^{n-1}$. (a) A classical solution requires an exponential number of trials. There are 2^n possible values of a, and after m evaluations of $f_a(x_i)$, we have eliminated $m(m-1)/2$ possible values of a. (b) The quantum solution: apply a Walsh-Hadamard transform to the input register on n qubits in the state 0 $^{(n)}$ to create an equal superposition state; then the oracle performs the unitary transformation, $\mathbf{U}_f$. Now the joint state of the input and the output registers of the oracle is φ ; we measure the output register only. As a result of this measurement, the input register will be left in the superposition state 1/ 2(x_0 x_0 a), with x_0, the value of the argument, x, corresponding to the outcome of the measurement we have observed, $f_a(x_0)$. Finally, apply a Walsh-Hadamard transform to the input register and then perform a measurement. Repeat this procedure $\mathcal{O}(n)$ times to determine the n bits of the period a $(a_1a_2 \ldots a_n)$.

First, we apply a Walsh-Hadamard transform to the input register on n qubits in the state $|0\rangle^{(n)}$ and create an equal superposition state:

$$\frac{1}{2^{n/2}}\sum_{x=0}^{2^n-1}|x\rangle.$$

This state is then transformed by the oracle to the state $|f_a(x)\rangle$. The state of the qubits in the output register of the oracle is an equal superposition of all 2^{n-1} values of the function $f_a(x)$; recall that each value, $f_a(x_i)$, corresponds to a pair of values of the argument, namely, x_i and $x_i \oplus a$. The joint state of the input and the output registers of the oracle is

$$|\varphi\rangle = \frac{1}{2^{n/2}}\sum_{x=0}^{2^n-1}|x\rangle|f_a(x)\rangle = \frac{1}{2^{n/2}}\sum_{x=0}^{2^n-1}|x, f_a(x)\rangle.$$

Now we apply a measurement to the qubits in the output register *only*; all 2^{n-1} values of the function f_a are equally likely to be the outcome of this measurement. Let x_0 be the value of the argument x corresponding to the outcome of the measurement we have observed, $f_a(x_0)$. Then, according to the generalized Born rule, the input register will be left in a superposition corresponding to the two arguments $|x_0\rangle$ and $|x_0 \oplus a\rangle$, namely, in the state

$$\frac{1}{\sqrt{2}}(|x_0\rangle + |x_0 \oplus a\rangle).$$

Unfortunately, from this superposition we cannot identify the two integers, x_0 and $x_0 \oplus a$. If we could, then computing the period would be trivial: $x_0 \oplus (x_0 \oplus a) = a$. Nevertheless, we can use our old trick: apply the Walsh-Hadamard transformation to the input register. Recall that the Hadamard gate transforms a single qubit as follows:

$$\mathbf{H}|x\rangle = \frac{|0\rangle + (-1)^x|1\rangle}{\sqrt{2}} = \frac{1}{\sqrt{2}}\sum_{y\in\{0,1\}}(-1)^{x\cdot y}|y\rangle.$$

Similarly, the Walsh-Hadamard transform of n qubits is

$$\mathbf{H}^{\otimes n}|x\rangle^{(n)} = \frac{1}{\sqrt{2^n}}\sum_{y=0}^{2^n-1}(-1)^{x\cdot y}|y\rangle^{(n)}.$$

When we apply the Walsh-Hadamard transformation to the qubits in the input register, we obtain

$$\mathbf{H}^{\otimes n}\frac{1}{\sqrt{2}}(|x_0\rangle + |x_0 \oplus a\rangle) = \frac{1}{\sqrt{2^{n+1}}}\sum_{y=0}^{2n-1}\left[(-1)^{x_0\cdot y} + (-1)^{(x_0\oplus a)\cdot y}\right]|y\rangle^{(n)}.$$

Call $\mathcal{Y}_0$ the subset of y, such that $a \cdot y = 0$, and $\mathcal{Y}_1$ the subset of y, such that $a \cdot y = 1$. We observe that

$$\sum_{y\in\mathcal{Y}_1}\left[(-1)^{x_0\cdot y} + (-1)^{(x_0\oplus a)\cdot y}\right]|y\rangle^{(n)} = 0.$$

Indeed, when $a \cdot y = 1$, then $(-1)^{(x_0 \oplus a)\cdot y} = (-1)^{x_0 \cdot y}(-1)^{a \cdot y} = -(-1)^{x_0 \cdot y}$. It follows that the input register is in the state

$$\frac{1}{\sqrt{2^{n-1}}} \sum_{y \perp y_0} \left[(-1)^{x_0 \cdot y} - (-1)^{x_0 \cdot y}\right] |y\rangle^{(n)} = \frac{1}{\sqrt{2^{n-1}}} \sum_{y \perp y_0} (-1)^{x_0 \cdot y} |y\rangle^{(n)}.$$

Now we carry a measurement of the n qubits in the input register and obtain with equal probability one of the values of $y \perp y_0$—in other words, a value of y, such that $y \cdot a = 0$. But

$$y \cdot a = 0 \Leftrightarrow \sum_{i=0}^{2^n-1} y_i a_i = 0 \mod 2.$$

Every value of y, with the exception of $y = 0$, allows us to establish a relation among the n bits of a. Thus, the oracle allows us to determine a after $\mathcal{O}(n)$ operations.

We have concluded the analysis of the first group of quantum algorithms and continue our presentation with the discussion of more sophisticated quantum algorithms. Useful information about the transformation carried out by a quantum algorithm is often encoded in the phases of the basis states. We discuss next a very useful procedure to extract this information through quantum phase estimation.

1.20 QUANTUM PHASE ESTIMATION

Quantum phase estimation is the process of determining the phase of the basis states after a transformation in a Hilbert space. Several quantum algorithms and the quantum Fourier transform use quantum phase estimation. We start our discussion with an example that illustrates the basic idea of quantum phase estimation; the problem we pose is to determine *the order of an element* in $GF(q)$, a finite field with q elements. Factoring an integer, N, can be reduced to the following problem: Given an integer a, find the smallest positive integer, r, with the property that $a^r \equiv 1 \mod N$. Indeed, $a^r \equiv 1 \mod N$ implies that $(a^r - 1) = mN$. But $a^r - 1 = (a - 1)(a^{r-1} + a^{r-2} + \ldots + a + 1)$; thus, either $(a - 1)$ or $(a^{r-1} + a^{r-2} + \ldots + a + 1)$ is divided by N. Also, order finding is important for coding theory, as we shall see in Chapter 4. Section 4.4 provides an in-depth discussion of finite fields; here, we simply state that integers modulo a prime number form an algebraic structure with a finite number of elements and two operations (e.g., addition and multiplication) with standard properties. For example, the integers modulo 5 form a finite field with five elements, $\mathbb{Z}_5 = \{0,1,2,3,4\}$; $\mathbb{Z}_5^* = \{1,2,3,4\}$ is the set of nonzero elements in $\mathbb{Z}_5$. The order of an element, $a_i \in \mathbb{Z}_5^*$, is the smallest integer r such that $a_i^r = 1$. In our example, the orders of the four elements $\{1,2,3,4\}$ in $\mathbb{Z}_5^*$ are, respectively, $\{1,4,4,2\}$; indeed,

$$1^1 = 1 \mod 5, \quad 2^4 = 1 \mod 5, \quad 3^4 = 1 \mod 5, \quad 4^2 = 1 \mod 5.$$

The elements of $\mathbb{Z}_5^*$ can be expressed in a canonical base, with $|j\rangle$, the binary expression of the element j (e.g., $|3\rangle = |011\rangle$). We are looking for a linear transformation, $\mathbf{U}_a$, of the basis vectors with the property

$$\mathbf{U}_a |j\rangle = |aj\rangle, \ \mathbf{U}_a^2 |j\rangle = \mathbf{U}_{a^2} |j\rangle = |a^2 j\rangle, \ \ldots, \mathbf{U}_a^{2^n} |j\rangle = \mathbf{U}_{a^{2^n}} |j\rangle = |a^n j\rangle.$$

If r is the order of a, then $\mathbf{U}_a^r = \mathbf{U}_{a^r} = \mathbf{I}$; this implies that the eigenvalues of $\mathbf{U}_a$ are of the form $e^{2\pi i (k/r)}$, with k, an integer, and the eigenvectors of $\mathbf{U}_a$ are

$$\varphi_k \qquad \sum_{j\ 0}^{r-1} e^{-2\pi i \frac{jk}{r}} \ a^j \ .$$

We see that the information about r is encoded in the phase $\omega_k \quad -2\pi i(jk/r)$ of the eigenvector φ_k . This justifies our interest in quantum phase estimation.

Eigenvalue Kickback

We wish to construct a quantum circuit to determine the phase of a unitary transformation. The circuit in Figure 1.23 illustrates the mapping carried out by a controlled gate; $f(x)$ is the transformation applied to the target qubit, and we see that the function, $f(x)$, determines the phase of the control qubit:

$$\begin{aligned} x \ (\ 0 \ - \ 1 \ ) \qquad & x \ (\ f(x) \quad 0 \ - \ f(x) \quad 1 \ ) \\ & x \ (\ f(x) \quad - \ f(x) \quad 1 \ ) \\ & (-1)^{f(x)} \ x \ (\ 0 \ - \ 1 \ ). \end{aligned}$$

Indeed, if $f(x) \quad 0$, then $x \ (\ 0 - \ 1 \ ) \quad x \ (-1)^0(\ 0 \ - \ 1 \ )$ as $0 \quad 0 \quad 0$; similarly, if $f(x) \quad 1$, then $x \ (\ 0 \ - \ 1 \ ) \quad x \ (-1)^1(\ 0 - \ 1 \ )$ as $1 \quad 0 \quad 1$.

Next, we examine controlled-U circuits, which apply unitary transformations, $\mathbf{U}$ and $\mathbf{U}^k$, to their eigenvector, φ , with the eigenphase, ω:

$$\mathbf{U} \ \varphi \quad e^{2i\pi\omega} \ \varphi \quad \text{and} \quad \mathbf{U}^k \ \varphi \quad e^{2i\pi\omega k} \ \varphi \ .$$

The intuition is that $\mathbf{U}^k$, *k successive applications of the transformation* $\mathbf{U}$ *that rotate the eigenvector with an angle ω, result in a rotation with an angle $\omega_k \quad \omega k$.* In Figure 1.24 we see that

$$\begin{aligned} &\mathbf{U}(\ 1 \quad \varphi \ ) && e^{2\pi i\omega} \ 1 \quad \varphi \\ &\mathbf{U}(\ 0 \quad \varphi \ ) && 0 \quad \varphi \\ &\mathbf{U}((\alpha_0 \ 0 \quad \alpha_1 \ 1 \ ) \quad \varphi \ ) && (\alpha_0 \ 0 \quad e^{2\pi i\omega}\alpha_1 \ 1 \ ) \quad \varphi \\ &\mathbf{U}^k(\ 1 \quad \varphi \ ) && e^{2\pi i\omega k} \ 1 \quad \varphi \\ &\mathbf{U}^k(\ 0 \quad \varphi \ ) && 0 \quad \varphi \\ &\mathbf{U}^k((\alpha_0 \ 0 \quad \alpha_1 \ 1 \ ) \quad \varphi \ ) && (\alpha_0 \ 0 \quad e^{2\pi i\omega k}\alpha_1 \ 1 \ ) \quad \varphi \end{aligned}$$

The controlled-U can be any unitary transformation including a controlled phase shift described by

$$\mathbf{R}_m \quad \begin{matrix} 1 & 0 \\ 0 & e^{\frac{2\pi i}{2^m}} \end{matrix} \ .$$

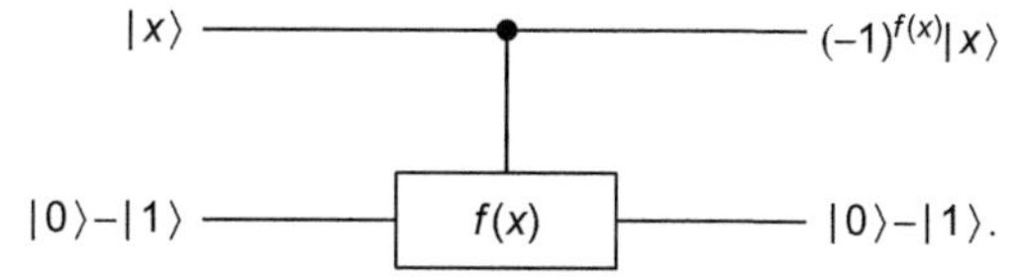

FIGURE 1.23

Eigenvalue kickback. The function, $f(x)$, determines the phase of the control qubit.

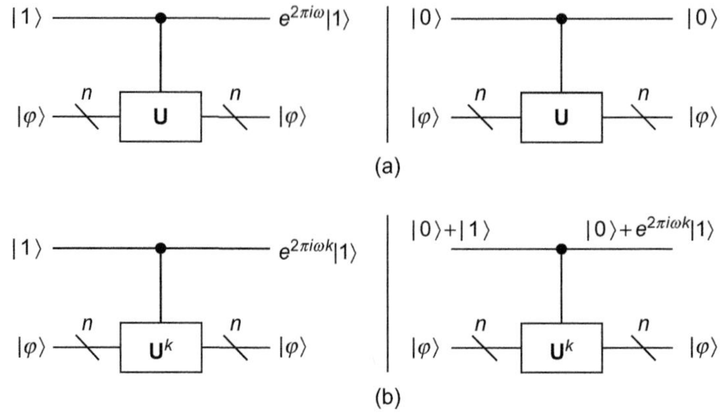

FIGURE 1.24

The eigenvalue of linear transformations for `controlled-U` operations with an eigenvector as the input. (a) The unitary transformation, $\mathbf{U}$, applied to an eigenvector, $\varphi \quad \mathcal{H}_{2^n}$, with an eigenphase, ω. (b) The unitary transformation, $\mathbf{U}^k$.

If the phase is $\omega \quad 2^{-m}$, then the controlled phase shift, $\mathbf{R}_m^{-1}$, rotates the eigenvector φ , with an angle $(2\pi i)/2^m$; when the control qubit is 1 , the eigenvalue is equal to 1.

After a phase shift, we apply to the control qubit a Hadamard gate that carries out a transformation described by

$$\mathbf{H}\frac{0 \quad (-1)^j \; 1}{2} \qquad j \begin{cases} \mathbf{H}\frac{0 \quad 1}{2} & 0 \quad \text{when} \quad j \quad 0, \\ \mathbf{H}\frac{0 \; - \; 1}{2} & 1 \quad \text{when} \quad j \quad 1. \end{cases}$$

Estimation of the Eigenphase

The problem we address now is how to determine the real number, $\omega \quad (0,1)$, the eigenphase of an eigenvector φ expressed as

$$\varphi \qquad \sum_{k \; 0}^{2^n-1} e^{2\pi i\omega k} \; k \; .$$

We observe that

$$\sum_{k \; 0}^{2^n-1} e^{2\pi i\omega k} \; k \qquad 0 \quad e^{2\pi i(2^{n-1}\omega)} \; 1 \qquad 0 \quad e^{2\pi i(2^{n-2}\omega)} \; 1 \qquad\qquad 0 \quad e^{2\pi i\omega} \; 1 \; .$$

We can express ω in binary as

$$\omega \quad \frac{\omega_1}{2} \quad \frac{\omega_2}{2^2} \quad \frac{\omega_3}{2^3} \qquad \frac{\omega_j}{2^j} \qquad ,$$

or in a compact form as

$$\omega \quad 0.\omega_1\omega_2\omega_3 \quad \omega_j \quad .$$

It is easy to see that

$$\begin{array}{ll} 2\omega & \omega_1.\omega_2\omega_3\ldots\omega_j\ldots \\ 2^2\omega & \omega_1\omega_2.\omega_3\ldots\omega_j\ldots \\ \vdots & \\ 2^{n\ \ 1}\omega & \omega_1\omega_2\ldots\omega_{n\ \ 1}.\omega_n\ldots . \end{array}$$

Then we can express

$$\sum_{k\ \ 0}^{2^n\ \ 1} e^{2\pi i\omega k}\ k \quad 0 \quad e^{2\pi i(0.\omega_n\omega_{n\ \ 1}\ldots)}\ 1 \qquad 0 \quad e^{2\pi i(0.\omega_{n\ \ 1}\omega_n\ldots)}\ 1$$
$$0 \quad e^{2\pi i(0.\omega_2\omega_3\ldots)}\ 1 \qquad 0 \quad e^{2\pi i(0.\omega_1\omega_2\ldots)}\ 1\ .$$

Example. We wish to estimate ω $0.\omega_1\omega_2\omega_3$. A Hadamard gate transforms an input state, $(\ 0 \quad e^{2\pi i(0.\omega_1)}\ 1\ )/\ \overline{2}$, to ω_1 (top of Figure 1.25a); indeed, $e^{2\pi i(0.\omega_1)}$ $e^{2\pi i(\omega_1/2)}$ $e^{\pi i\omega_1}$ $(e^{\pi i})^{\omega_1}$ $(\ 1)^{\omega_1}$. Thus ϕ $(\ 0 \quad e^{2\pi i 0.\omega_1}\ 1\ )/\ \overline{2}$ $(\ 0\ \ (\ 1)^{\omega_1}\ 1\ )/\ \overline{2}$ and $H\ \phi$ ω_1 as shown above. When we measure the output of the Hadamard gate, we obtain the real value ω_1 (top of Figure 1.25b).

The circuit at the bottom of Figure 1.25a carries out the transformation

$$\frac{0 \quad e^{2\pi i(0.\omega_3)}\ 1}{\overline{2}} \quad \frac{0 \quad e^{2\pi i(0.\omega_2\omega_3)}\ 1}{\overline{2}} \quad \frac{0 \quad e^{2\pi i(0.\omega_1\omega_2\omega_3)}\ 1}{\overline{2}} \qquad \omega_1 \quad \omega_2 \quad \omega_3\ .$$

A measurement after a controlled-phase-shift is equivalent to a measurement followed by a controlled phase-shift if the outcome (classical information) is equal to 1. The circuit at the bottom of Figure 1.25b adds a measurement as the final stage of the circuit at the bottom of Figure 1.25a and carries out the transformation that produces the real values ω_1, ω_2, and ω_3:

$$\frac{0 \quad e^{2\pi i(0.\omega_3)}\ 1}{\overline{2}} \quad \frac{0 \quad e^{2\pi i(0.\omega_2\omega_3)}\ 1}{\overline{2}} \quad \frac{0 \quad e^{2\pi i(0.\omega_1\omega_2\omega_3)}\ 1}{\overline{2}} \qquad \omega_1\omega_2\omega_3.$$

We return now to the problem of finding the order of the elements of $\mathbb{Z}_5$, expressed as r $2k_2$ k_1. We consider the unitary transformation denoted as $\mathbf{U}_2$ and its successive applications, $\mathbf{U}_2^2$, $\mathbf{U}_2^3$, and $\mathbf{U}_2^4$. The basis vectors are transformed as follows:

	$\mathbf{U}_2$	$\mathbf{U}_2^2$	$\mathbf{U}_2^3$	$\mathbf{U}_2^4$
001	010	100	011	001
010	100	011	001	010
011	001	010	100	011
100	011	001	010	100

The eigenvectors of $\mathbf{U}_2$ are

$$\varphi_j \quad \sum_{k\ \ 0}^{3} e^{\ \ 2\pi i\frac{jk}{4}}\ 2^k \bmod 5$$

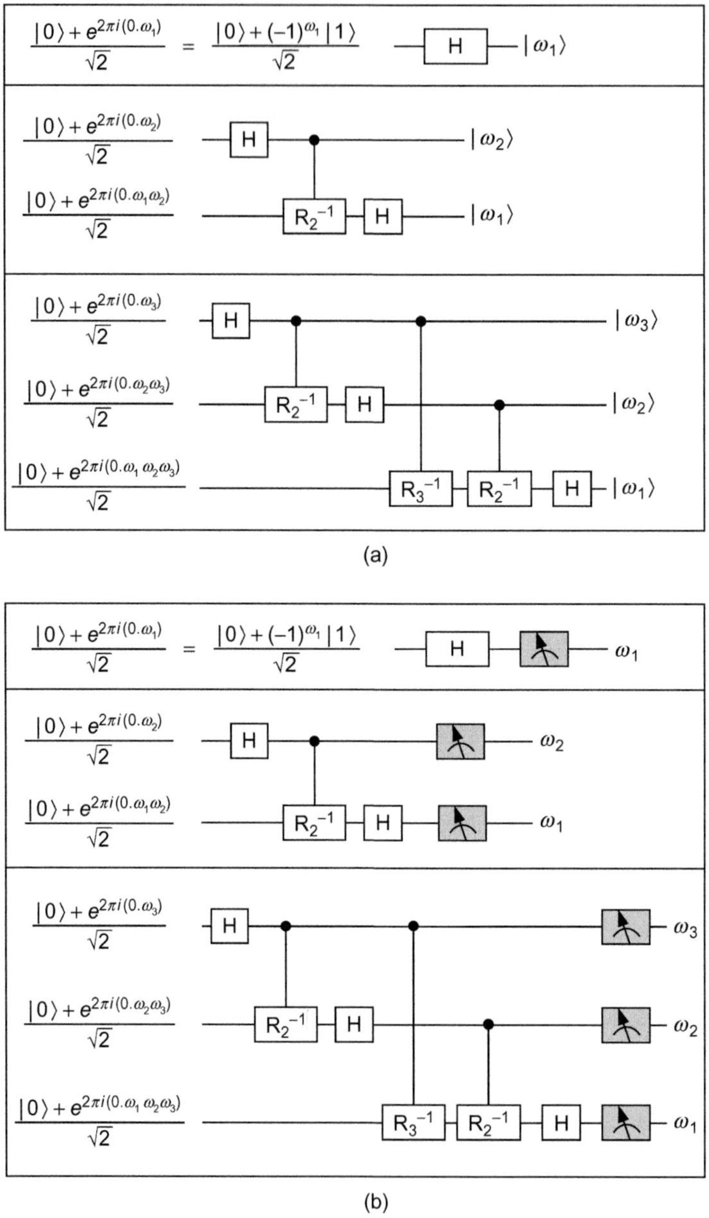

FIGURE 1.25

Quantum circuits to determine ω $0.\omega_1\omega_2\omega_3$. (a) Controlled-phase shifts, $\mathbf{R}_2$ and $\mathbf{R}_3$, followed by a Hadamard transform, $\mathbf{H}$, produce: (top) ω_1 ; (middle) ω_2 and ω_1 ; (bottom) ω_3 , ω_2 , and ω_1 . (b) A measurement after a controlled-phase shift is equivalent to a measurement followed by a controlled-phase shift if the outcome (classical information) is equal to 1. The outcome of the measurements are: (top) ω_1; (middle) ω_2 and ω_1; (bottom) ω_3, ω_2, and ω_1.

or

$$
\begin{array}{lllll}
\varphi_0 & 001 & 010 & 100 & 011 \;, \\
\varphi_1 & 001 & e^{-2\pi i \frac{1}{4}}\; 010 & e^{-2\pi i \frac{2}{4}}\; 100 & e^{-2\pi i \frac{3}{4}}\; 011 \;, \\
\varphi_2 & 001 & e^{-2\pi i \frac{2}{4}}\; 010 & e^{-2\pi i \frac{2}{4}2}\; 100 & e^{-2\pi i \frac{3}{4}2}\; 011 \;, \\
\varphi_3 & 001 & e^{-2\pi i \frac{3}{4}}\; 010 & e^{-2\pi i \frac{2}{4}3}\; 100 & e^{-2\pi i \frac{3}{4}3}\; 011 \;.
\end{array}
$$

Then,

$$\frac{1}{2}(\varphi_0 \quad \varphi_1 \quad \varphi_2 \quad \varphi_3) \quad 001 \;.$$

The information about the order r is encoded in the phase $\omega_j \quad -2\pi i(jk/r)$ of the eigenvector φ_j :

$$
\begin{array}{lll|lll}
\mathbf{U}_2\; \varphi_0 & \varphi_0 & & \mathbf{U}_2(0 \quad 1)\; \varphi_0 & (0 \quad 1)\; \varphi_0 \;, & \\
\mathbf{U}_2\; \varphi_1 & e^{2\pi i \frac{1}{4}} & \varphi_1 & \mathbf{U}_2(0 \quad 1)\; \varphi_1 & 0 \quad e^{2\pi i \frac{1}{4}}\; 1 & \varphi_1 \;, \\
\mathbf{U}_2\; \varphi_2 & e^{2\pi i \frac{2}{4}} & \varphi_2 & \mathbf{U}_2(0 \quad 1)\; \varphi_2 & 0 \quad e^{2\pi i \frac{2}{4}}\; 1 & \varphi_2 \;, \\
\mathbf{U}_2\; \varphi_3 & e^{2\pi i \frac{3}{4}} & \varphi_3 & \mathbf{U}_2(0 \quad 1)\; \varphi_3 & 0 \quad e^{2\pi i \frac{3}{4}}\; 1 & \varphi_3 \;.
\end{array}
$$

We also see that

$$
\begin{array}{lll}
\mathbf{U}_2^2(0 \quad 1)\; \varphi_0 & (0 \quad 1)\; \varphi_0 \;, & \\
\mathbf{U}_2^2(0 \quad 1)\; \varphi_1 & 0 \quad e^{2\pi i \frac{1}{4}2}\; 1 & \varphi_1 \;, \\
\mathbf{U}_2^2(0 \quad 1)\; \varphi_2 & 0 \quad e^{2\pi i \frac{2}{4}2}\; 1 & \varphi_2 \;, \\
\mathbf{U}_2^2(0 \quad 1)\; \varphi_3 & 0 \quad e^{2\pi i \frac{3}{4}2}\; 1 & \varphi_3 \;.
\end{array}
$$

The circuit to find the order of the elements of $\mathbb{Z}_5$ is depicted in Figure 1.26.

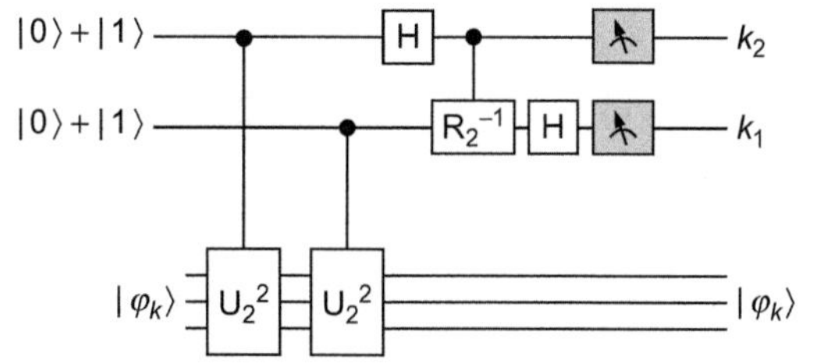

FIGURE 1.26

Quantum circuit to determine the order of an element in $\mathbb{Z}_5$; $r \quad 2k_2 \quad k_1$.

We mentioned that the circuits for quantum phase estimation are also used for the quantum Fourier transform (QFT). Recall that the classical discrete Fourier transform performs the following mapping:

$$j \quad (0,0,\ldots,0,1,0,\ldots,0) \qquad 1, e^{2\pi i \frac{j}{n}}, e^{2\pi i \frac{2j}{n}}, \ldots, e^{2\pi i \frac{(n\ \ 1)j}{n}} \ .$$

Similarly, the QFT discussed in the next section transforms the basis vectors in $\mathcal{H}_{2^n}$ as follows:

$$j \qquad \sum_{k\ \ 0}^{2^n\ \ 1} e^{2\pi i \frac{j}{2^n} k} \ k \ .$$

We see immediately that the circuits in Figure 1.25 implement, in fact, an inverse QFT. There is yet another connection between phase estimation and QFT; we have assumed that the phase is of the form $\omega \quad j/2^k$, with j, an integer, so the next question is if we can approximate ω by ω when this condition is not satisfied. The answer is that the QFT will transform the state as follows:

$$\sum_{k\ \ 0}^{2^n\ \ 1} e^{2\pi i \omega k} \ k \qquad \omega \qquad \sum_{j} \alpha_j \ j \ .$$

The error is bounded:

$$\text{Prob} \quad \frac{j}{2^n} \quad \omega \quad \frac{1}{2^n} \quad \frac{8}{\pi^2} \quad \text{and} \quad \alpha_j \quad \mathcal{O} \quad \frac{1}{j/2^n \quad \omega} \ .$$

We introduce the next two transforms of interest for quantum computing: the Walsh-Hadamard transform used by several quantum algorithms, including the Grover search algorithm, and the QFT used by several algorithms, including the Shor quantum factoring algorithm.

1.21 WALSH-HADAMARD AND QUANTUM FOURIER TRANSFORMS

In this section we first introduce Hadamard matrices and then discuss the Walsh-Hadamard transform and the QFT.

Hadamard Matrix

A Hadamard matrix of order n is an $n \quad n$ matrix, with elements h_{ij}, either $\ 1$ or $\ 1$; a Hadamard matrix of order $2n$ is a $2n \quad 2n$ matrix:

$$H(n) \quad [h_{ij}],\ 1 \quad i \quad n,\ 1 \quad j \quad n \quad \text{and} \quad H(2n) \quad \begin{matrix} H(n) & H(n) \\ H(n) & H(n) \end{matrix} \ .$$

$H(n)$ has the following properties:

1. The product of a Hadamard matrix and its transpose satisfies the following relations:

$$H(n)H(n)^T \quad nI_n \quad \text{and} \quad H(n)^T H(n) \quad nI_n.$$

2. The exchange of rows or columns transforms one Hadamard matrix to another one.

3. The row vectors, $h_1, h_2, \ldots, h_n$, of the matrix, $H(n)$ are pairwise orthogonal:

$$h_k \quad h_l \quad 0, \quad (k,l) \qquad 1,n \;, \quad k \qquad .$$

4. The multiplication of rows or columns by 1 transforms one Hadamard matrix to another one.
5. If $n \quad 2^{q \quad 1}$, then $H(2n)$ can be expressed as the tensor product of q matrices of size 2 2:

$$H(2n) \quad H(2) \quad H(2) \qquad\qquad H(2) \quad H(2).$$

Recall that the tensor product of the $p \quad q$ matrix $A \quad [a_{ij}]$, with the $r \quad s$ matrix $B \quad [b_{kl}]$, is the $(p \quad r) \quad (q \quad s)$ matrix $C \quad [c_{mn}]$, with $c_{mn} \quad a_{ij} \quad b_{kl}$, where i,j,k,l,m,n are the only integers that satisfy the relations $m \quad (i \quad 1)r \quad k$ and $n \quad (j \quad 1)s \quad l$. In other words, C is the matrix obtained by replacing the element, a_{ij}, of matrix A by the $r \quad s$ matrix, $a_{ij} \quad B$.

Let us consider the first property: $H(n)$ and $H(n)^T$ can be written as

$$H(n) \qquad \begin{array}{c} h_1 \\ h_2 \\ \vdots \\ h_n \end{array} \quad , H(n)^T \qquad h_1^T \; h_2^T \; \ldots \; h_n^T \; .$$

The elements of h_k and h_k^T are identical: $h_{ki} \quad h_{ik}$.
Then,

$$H(n)H(n)^T \qquad \begin{array}{cccc} h_1 \;\; h_1 & h_1 \;\; h_2 & \ldots & h_1 \;\; h_n \\ h_2 \;\; h_1 & h_2 \;\; h_2 & \ldots & h_2 \;\; h_n \\ \ldots & \ldots & \ldots & \ldots \\ h_n \;\; h_1 & h_n \;\; h_2 & \ldots & h_n \;\; h_n \end{array} \qquad \begin{array}{cccc} n & 0 & \ldots & 0 \\ 0 & n & \ldots & 0 \\ \ldots & \ldots & \ldots & \ldots \\ 0 & 0 & \ldots & n \end{array} \qquad nI_n.$$

Now multiply the previous equation with $H(n) \quad ^1$:

$$H(n) \quad ^1 H(n)H(n)^T \quad nH(n) \quad ^1 I_n.$$

It follows that:

$$H(n)^T \quad nH(n) \quad ^1 \qquad H(n)^T H(n) \quad nH(n) \quad ^1 H(n) \qquad H(n)^T H(n) \quad nI_n.$$

Properties 2–5 follow immediately. The Hadamard matrices used in quantum computing are normalized:

$$H(n) \quad 2 \quad ^{n/2}[h_{ij}].$$

Example. Hadamard matrices of order 2,4, and 8 are:

$$H(2) \qquad \frac{1}{\overline{2}} \begin{array}{cc} 1 & 1 \\ 1 & 1 \end{array} \; , \; H(4) \qquad \frac{1}{2} \begin{array}{cccc} 1 & 1 & 1 & 1 \\ 1 & 1 & 1 & 1 \\ 1 & 1 & 1 & 1 \\ 1 & 1 & 1 & 1 \end{array} \; ,$$

$$H(8) \quad \frac{1}{2\ \bar{2}} \begin{matrix} 1 & 1 & 1 & 1 & 1 & 1 & 1 & 1 \\ 1 & 1 & 1 & 1 & 1 & 1 & 1 & 1 \\ 1 & 1 & 1 & 1 & 1 & 1 & 1 & 1 \\ 1 & 1 & 1 & 1 & 1 & 1 & 1 & 1 \\ 1 & 1 & 1 & 1 & 1 & 1 & 1 & 1 \\ 1 & 1 & 1 & 1 & 1 & 1 & 1 & 1 \\ 1 & 1 & 1 & 1 & 1 & 1 & 1 & 1 \\ 1 & 1 & 1 & 1 & 1 & 1 & 1 & 1 \end{matrix} .$$

The Walsh-Hadamard Transform

The Walsh-Hadamard transform performs a randomization operation, but it is perfectly reversible. Given n qubits, the transform allows us to construct a quantum mechanical system, with $N \quad 2^n$ states. The Walsh-Hadamard transform, $\mathbf{H} \quad ^n$, rotates each of the n qubits independently:

$$\mathbf{H} \quad ^n \quad \mathbf{H}(N), \quad \text{with} \quad N \quad 2^n.$$

The Walsh-Hadamard transform is used to create an equal superposition of qubits when the input is in the state $00\ldots0$. For example,

$$\mathbf{H}(4) \ 00 \quad \frac{00 \qquad 01 \qquad 10 \qquad 11}{2}$$

and

$$\mathbf{H}(8) \ 000 \quad \frac{000 \qquad 001 \qquad 010 \qquad 011 \qquad 100 \qquad 101 \qquad 110 \qquad 111}{2\ \bar{2}}.$$

If we start in an arbitrary state, $\psi \quad {}_{k\ 0}^{N\ 1} \alpha_k \ k$, and apply the Walsh-Hadamard transform, then the system reaches the state

$$\varphi \quad \mathbf{H} \quad ^n \ \psi \quad \sum_{k\ 0}^{N\ 1} \beta_k \ k \ , \quad \text{with} \quad \beta_k \quad \sum_{j\ 0}^{N\ 1} (\ 1)^p \alpha_j,$$

with $p \quad 0, 1$. For example, in $\mathcal{H}_4$, when $\psi \quad \alpha_0 \ 00 \quad \alpha_1 \ 01 \quad \alpha_2 \ 10 \quad \alpha_3 \ 11$, we have $\beta_0 \quad \alpha_0 \quad \alpha_1 \quad \alpha_2 \quad \alpha_3$, $\beta_1 \quad \alpha_0 \quad \alpha_1 \quad \alpha_2 \quad \alpha_3$, $\beta_2 \quad \alpha_0 \quad \alpha_1 \quad \alpha_2 \quad \alpha_3$, $\beta_3 \quad \alpha_0 \quad \alpha_1 \quad \alpha_2 \quad \alpha_3$.

Recall that the probability of projecting a superposition state,

$$\psi \quad \sum_{k\ 0}^{N\ 1} \alpha_k \ k \ ,$$

onto the basis state, k , is equal to the square of the absolute value of the probability amplitude of that basis state, $p_k \quad \alpha_k \ ^2$. The probability amplitude, α_k, is a complex number characterized by its magnitude and phase. The state, ψ , is characterized by two N-dimensional vectors: the probability amplitude, a vector of complex numbers, and the probability, a vector of real numbers,

$$(\alpha_0,\alpha_1,\ldots,\alpha_{N\ 1}) \quad \text{and} \quad (\alpha_0\ ^2,\ \alpha_1\ ^2,\ldots,\ \alpha_{N\ 1}\ ^2), \quad \text{respectively.}$$

Two states may have the same probability vectors but different probability amplitude vectors. For example, the states

$$\psi_1 \quad \frac{0 \quad 1}{\overline{2}} \quad \text{and} \quad \psi_2 \quad \frac{0 \quad 1}{\overline{2}}$$

have different probability amplitude vectors, $(1/\ \overline{2}, 1/\ \overline{2})$ and $(1/\ \overline{2},\ \ 1/\ \overline{2})$, respectively, but have the same probability vector $(1/2, 1/2)$.

It follows that the Walsh-Hadamard transform of a superposition state of n qubits preserves the magnitude of the probability amplitude of individual components.

Quantum Fourier Transform

The QFT is a unitary operator that transforms the vectors of an orthonormal basis,

$$0\ ,\ 1\ ,\ldots,\ j\ ,\ldots,\ k\ ,\ldots,\ N\quad 1 \qquad \mathcal{H}_N, \text{ with } N \quad 2^n,$$

as follows:

$$j \qquad \frac{1}{\overline{N}} \sum_{k\ 0}^{N\ 1} e^{i2\pi jk/N}\ \ k\ ,\ i \qquad \overline{\ \ 1}.$$

The QFT transforms a state, ψ , of a quantum system to another state, φ :

$$\psi \qquad \sum_{j\ 0}^{N\ 1} \alpha_j\ \ j \qquad \varphi \qquad \sum_{k\ 0}^{N\ 1} \beta_k\ \ k\ .$$

The probability amplitude, β_k, is the *discrete Fourier transforms* of the probability amplitude α_j, $0 \quad j \quad N \quad 1$:

$$\beta_k \qquad \frac{1}{\overline{N}} \sum_{j\ 0}^{N\ 1} \alpha_j e^{i2\pi jk/N}.$$

The binary representation of integers j and k is

$$j \quad j_0 2^{n\ \ 1} \quad j_1 2^{n\ \ 2} \qquad\qquad j_{n\ \ 2} 2^1 \quad j_{n\ \ 1} 2^0 \text{ and } k \quad k_0 2^{n\ \ 1} \quad k_1 2^{n\ \ 2} \qquad\qquad k_{n\ \ 2} 2^1 \quad k_{n\ \ 1} 2^0.$$

Then, the definition of the QFT can be rewritten as

$$j_0 j_1 \ldots j_{n\ 1} \quad \frac{1}{2^{n/2}} \quad \underset{k_0\ (0,1)\, k_1\ (0,1)\quad k_{n\ 1}\ (0,1)}{\ldots} \quad e^{i2\pi j \ {}^{n\ 1}_{m\ 0} k_m 2 \ {}^{m}} \quad k_0 k_1 \ldots k_m \ldots k_{n\ 1} \,,$$

$$j_0 j_1 \ldots j_{n\ 1} \quad \frac{1}{2^{n/2}} \quad \underset{k_0\ (0,1)\, k_1\ (0,1)\quad k_{n\ 1}\ (0,1)\, m\ 0}{\ldots \quad {}^{n\ 1}} \quad e^{i2\pi j k_m 2 \ {}^{m}} \quad k_m \,,$$

$$j_0 j_1 \ldots j_{n\ 1} \quad \frac{1}{2^{n/2}} \underset{m\ 0\quad k_m\ (0,1)}{{}^{n\ 1}} \quad e^{i2\pi j k_m 2 \ {}^{m}} \quad k_m \,.$$

The bit k_m may only take two values, 0 and 1. We note that $e^{i2\pi j k_m}$ 1 when k_m 0; thus:

$$j_0 j_1 \ldots j_{n\ 1} \quad \frac{1}{2^{n/2}} \underset{m\ 0}{{}^{n\ 1}} \quad 0 \quad e^{i2\pi j 2 \ {}^{m}} \quad 1 \,.$$

But

$$0 \quad e^{i2\pi j 2 \ {}^{m}} \quad 1 \quad \begin{matrix} 1 \\ 0 \end{matrix} \quad \begin{matrix} 0 \\ e^{i2\pi j 2 \ {}^{m}} \end{matrix} \quad \begin{matrix} 1 \\ e^{i2\pi (j/2^m)} \end{matrix} \,.$$

The transformation of the input j can be rewritten as

$$j \quad \frac{1}{2^{n/2}} \quad \begin{matrix} 1 \\ e^{i2\pi (j/2^0)} \end{matrix} \quad \begin{matrix} 1 \\ e^{i2\pi (j/2^1)} \end{matrix} \quad \begin{matrix} 1 \\ e^{i2\pi (j/2^2)} \end{matrix} \quad \begin{matrix} 1 \\ e^{i2\pi (j/2^{n\ 1})} \end{matrix} \,.$$

If we denote by $S_{p,q}$, 0 p, q n 1, p q, and the joint transformation of qubits p and q as

$$S_{p,q} \quad \begin{matrix} 1 & 0 & 0 & 0 \\ 0 & 1 & 0 & 0 \\ 0 & 0 & 1 & 0 \\ 0 & 0 & 0 & e^{i\pi/2^{p\ q}} \end{matrix} \,,$$

then we obtain an equivalent expression for the QFT that leads immediately to the quantum circuit for the QFT in Figure 1.27. The QFTs of the n qubits are given by the following transformations:

$$\begin{aligned}
&k_0 && S_{0,n\ 1} S_{0,n\ 2} \ldots S_{0,4} S_{0,3} S_{0,2} S_{0,1} (H \ \ j_0 \ ) \\
&k_1 && S_{1,n\ 1} S_{1,n\ 2} \ldots S_{1,4} S_{1,3} S_{1,2} (H \ \ j_1 \ ) \\
&k_2 && S_{2,n\ 1} S_{2,n\ 2} \ldots S_{2,4} S_{2,3} (H \ \ j_2 \ ) \\
&\vdots && \\
&k_{n\ 3} && S_{n\ 3,n\ 1} S_{n\ 3,n\ 2} (H \ \ j_{n\ 2} \ ) \\
&k_{n\ 2} && S_{n\ 2,n\ 1} (H \ \ j_{n\ 2} \ ) \\
&k_{n\ 1} && (H \ \ j_{n\ 1} \ ).
\end{aligned}$$

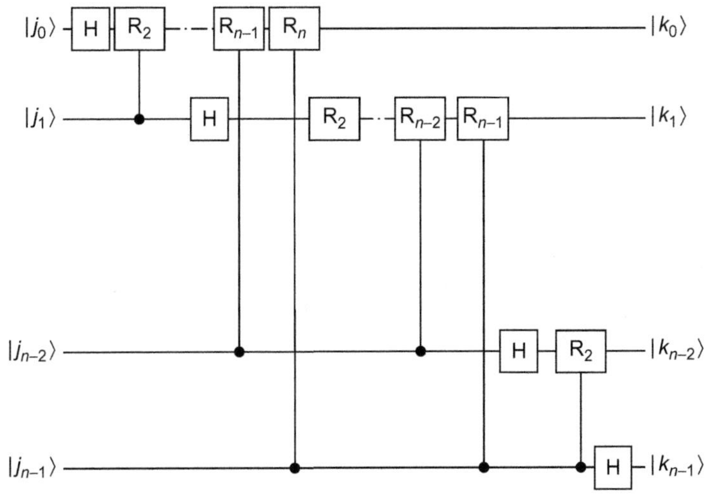

FIGURE 1.27

A circuit for the quantum Fourier transform.

All $S_{p,q}$ transform qubit p as a target and use q as a control qubit. $S_{p,q}$ does not change the value of a qubit; it only changes the phase. If both qubits p and q are equal to 1, then $S_{p,q}$ adds the value π to the phases of qubit p. The one-qubit gates, R_m, in Figure 1.27 represent the unitary transformation

$$R_m \quad \begin{matrix} 1 & 0 \\ 0 & e^{2\pi i/2^m} \end{matrix} \; .$$

This transformation is followed by a bit reversal. A qubit b is transformed to a qubit b , with b, the bit reversal of b. If $b \quad b_0 2^{n-1} \quad b_1 2^{n-2} \qquad b_{n-3}2^2 \quad b_{n-2}2 \quad b_{n-1}$, then $b \quad b_{n-1}2^{n-1}$ $b_{n-2}2^{n-2} \quad b_{n-3}2^{n-3} \qquad b_2 2^2 \quad b_1 2 \quad b_0$.

The circuit shown in Figure 1.27 does not carry out the bit reversal.

1.22 QUANTUM PARALLELISM AND REVERSIBLE COMPUTING

We will now take a closer look at the transformations of quantum states required by a quantum computation. A classical computation evaluates a function, f, for a particular value of the argument x; the outcome of this evaluation is $y \quad f(x)$. The picture is slightly different for a quantum computation, which evaluates a function, $f(\ x \ )$, with $x \quad \mathcal{H}_{2^n}$. The unitary transformation, $\mathbf{U}_f$, required by f maps computational basis states to computational basis states. $\mathbf{U}_f$ is *reversible*; this means that at the end of the computation the n-qubit *input register* is in its initial state x .

The quantum circuit for $\mathbf{U}_f$ should have access to an m-qubit register of *ancilla qubits* in the state a to store partial results of the computation. The circuit in Figure 1.13 illustrates the use of ancilla qubits to store partial results for a reversible controlled gate with multiple control qubits. To simplify

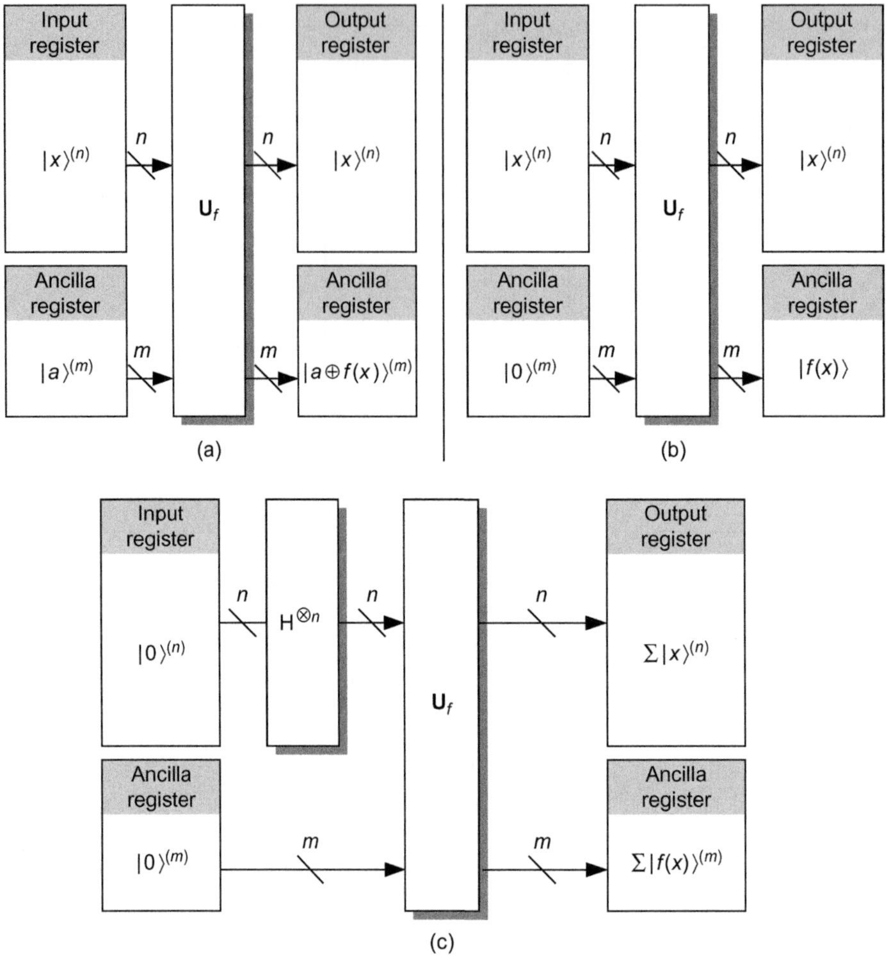

FIGURE 1.28

Quantum parallelism. (a) The unitary transformation, $\mathbf{U}_f$, is applied to an input register of n qubits in the state $x^{(n)}$; the transformation also requires a register of m ancilla qubits in the state $a^{(m)}$. After the transformation, the state of the ancilla qubits is $a \quad f(x)^{(m)}$. (b) When $a^{(m)} \quad 0^{(m)}$, the state of the ancilla qubits is $f(x)^{(m)}$. (c) When $x^{(n)} \quad 0^{(n)}$ and we apply a Walsh-Hadamard transform to the qubits in the input register, the output is a superposition of the values of $f(x)$ for all values of the argument $0 \quad x \quad 2^n - 1$.

the presentation in Figure 1.28(a) we grouped together the ancilla qubits and the second input register shown in Figure 1.12 from Section 1.16. $\mathbf{U}_f$ carries out the following transformation:

$$\mathbf{U}_f \quad x^{(n)} \quad a^{(m)} \qquad x^{(n)} \quad a \quad f(x)^{(m)}.$$

The ancilla qubits must be returned to their initial state at the end of the computation, after recording the result of the computation. When the ancilla qubits are initially in the state $|0\rangle^{(m)}$ (Figure 1.28b), then $|a \oplus f(x)\rangle^{(m)} \mapsto |f(x)\rangle^{(m)}$, and the unitary transformation $\mathbf{U}_f$ returns $|f(x)\rangle^{(m)}$:

$$\mathbf{U}_f\left(|x\rangle^{(n)} \otimes |0\rangle^{(m)}\right) = |x\rangle^{(n)} \otimes |f(x)\rangle^{(m)}.$$

In Section 1.21 we learned that when we apply a Walsh-Hadamard transform to a register of n qubits in the state $|0\rangle^{(n)}$, we obtain an *equal superposition state*, a state in which the probability amplitudes of different states are equal to $\frac{1}{\sqrt{2^n}}$:

$$\mathbf{H}^{\otimes n}|0\rangle^{(n)} = \frac{1}{\sqrt{2^n}}\sum_{i=0}^{2^n-1}|i\rangle.$$

We apply the Walsh-Hadamard transform, $\mathbf{H}^{\otimes n}$, to the input register and the identity transformation, $\mathbf{I}^{\otimes m}$, to the register of ancilla qubits:

$$\mathbf{U}_f\left(\mathbf{H}^{\otimes n} \otimes \mathbf{I}^{\otimes m}\right)\left(|0\rangle^{(n)} \otimes |0\rangle^{(m)}\right) = \frac{1}{\sqrt{2^n}}\sum_{i=0}^{2^n-1}\mathbf{U}_f\left(|i\rangle^{(n)} \otimes |0\rangle^{(m)}\right).$$

We showed earlier that $\mathbf{U}_f\left(|x\rangle^{(n)} \otimes |0\rangle^{(m)}\right) = |x\rangle^{(n)} \otimes |f(x)\rangle^{(m)}$; thus:

$$\mathbf{U}_f\left(\mathbf{H}^{\otimes n} \otimes \mathbf{I}^{\otimes m}\right)\left(|i\rangle^{(n)} \otimes |0\rangle^{(m)}\right) = \frac{1}{\sqrt{2^n}}\sum_{i=0}^{2^n-1}|i\rangle^{(n)} \otimes |f(i)\rangle^{(m)}.$$

The result of the transformation $\mathbf{U}_f$ is a *superposition of the* 2^n values of the function $f(x)$ for each of the 2^n possible values of the argument x (Figure 1.28c). If f is a one-to-one function, then m should be equal to n; sometimes, we require only a "yes" or "no" answer and $m = 1$ (e.g., in the case of several of the algorithms discussed in Section 1.19).

The outcome of a measurement of the output ancilla qubits reveals only one of the values of the function—e.g., $f(x_i)$—but we have no way either to discover the argument x_i or to force the system to produce $f(x_0)$ for a particular argument x_0. Therefore, we have to defer the measurement of the output ancilla qubits; first, we should amplify the amplitude of the projection on the particular basis state corresponding to the argument of interest, or carry out additional computations using as input this superposition state, and only then carry out the measurement to access the result of the computation.

Algorithmically, this means that we have to work harder to get the desired result, but there is a silver lining: the transformation $\mathbf{U}_f$ does the work of a classical system consisting of 2^n copies of the circuit evaluating in parallel the function, $f: \{0,1\}^n \mapsto \{0,1\}^m$, for all possible values of its argument x. This property of quantum systems, referred to as *quantum parallelism*, is one of the reasons for the excitement triggered by the possibility of building a quantum computer. Think about a computation with $n = 100$; classically, to evaluate the value of a function $f(x)$ for all possible arguments $x \in \{0,1\}^{100}$, we would need $2^{100} \approx 10^{33}$ copies of the circuit or, equivalently, 10^{33} time steps, with a single copy of the circuit; a quantum computer can realize this monumental task with a single copy of the circuit and in one time step. A critical facet of any quantum algorithm is to amplify the amplitude of the desired solution, and we shall see this idea very clearly in our discussion of the Grover search algorithm in Section 1.23.

Quantum, as well as classical, algorithms start from an initial state and then cause a set of state transformations of the quantum or of the classical device, which eventually lead to the desired result. Indeed, the first step for any quantum computation is to initialize the system to a state that we can easily prepare; then we carry out a sequence of unitary transformations that cause the system to evolve toward a state that provides the answer to the computational problem.

A quantum operation is a rotation of the state ψ in N-dimensional Hilbert space. Thus, the ultimate challenge is to build up powerful N-dimensional rotations as sequences of one- and two-dimensional rotations. For any quantum algorithm, there are multiple paths leading from the initial to the final state, and there is a degree of interference among these paths. The amplitude of the final state—thus, the probability of reaching the desired final state—depends on the interference among these paths. This justifies the common belief that quantum algorithms are very sensitive to perturbations, and one has to be extremely careful when choosing the transformations the quantum mechanical system is subjected to. Recently, Grover showed that the search algorithm is extremely robust and instead of using the Walsh-Hadamard transform, one can in principle use almost any unitary transformation [185].

If we know how to efficiently compute a function $f(x)$ and its inverse $f^{1}(x)$, then we can efficiently map any input state x to $f(x)$, and the computation is reversible. If U_f is a quantum circuit with the input as well as the output consisting of two registers of n and, respectively, m qubits, then $\mathbf{U}_f[\ a \quad b\ ] \quad (\ a \quad f(b)\ ) \quad b$; when a 0 and b x, the circuit transforms its input 0 x to $f(x)$ x as shown in Figure 1.29a.

If $U_{f\ 1}$ is a quantum circuit with the input as well as the output consisting of two registers n and, respectively, m qubits, then $\mathbf{U}_{f\ 1}[\ c \quad d\ ] \quad c \quad (\ d \quad f^{1}(c)\ )$; when c $f(x)$ and d x, the circuit transforms its input $f(x)$ x to $f(x)$ 0, as shown in Figure 1.29b; indeed, $x \quad f^{1}(f(x)) \quad x \quad x$ 0. When the two circuits work in tandem, they implement the linear transformation, $\mathbf{U}_f\mathbf{U}_{f\ 1}$, mapping the tensor product, 0 x to $f(x)$ x, as shown in Figure 1.29c.

This brief discussion reinforces the realization that the development of quantum algorithms requires a different kind of thinking than required for the development of classical algorithms. We cannot replicate the ability of a classical computation to immediately reveal the result of the evaluation of a function $f(x)$; the superposition state resulting from the application of the unitary transformation $\mathbf{U}_f$ reveals only the form of the solution dictated by the shape of the function f. This partially explains why so few quantum algorithms have been developed so far, a topic addressed by Shor [386].

1.23 GROVER SEARCH ALGORITHM

We now discuss the quantum search algorithm of Grover, a discovery impressive not only due to its simplicity and elegance, but also due to the range of potential applications. Preskill called Grover's algorithm "perhaps the most important new development" in quantum computing.

> *If quantum computers are being used 100 years from now, I would guess they will be used to run Grover s algorithm or something like it,*

Preskill says. The quantum search illustrates the quintessence of a quantum algorithm: *take advantage of quantum parallelism to create a superposition of all possible values of a function and then*

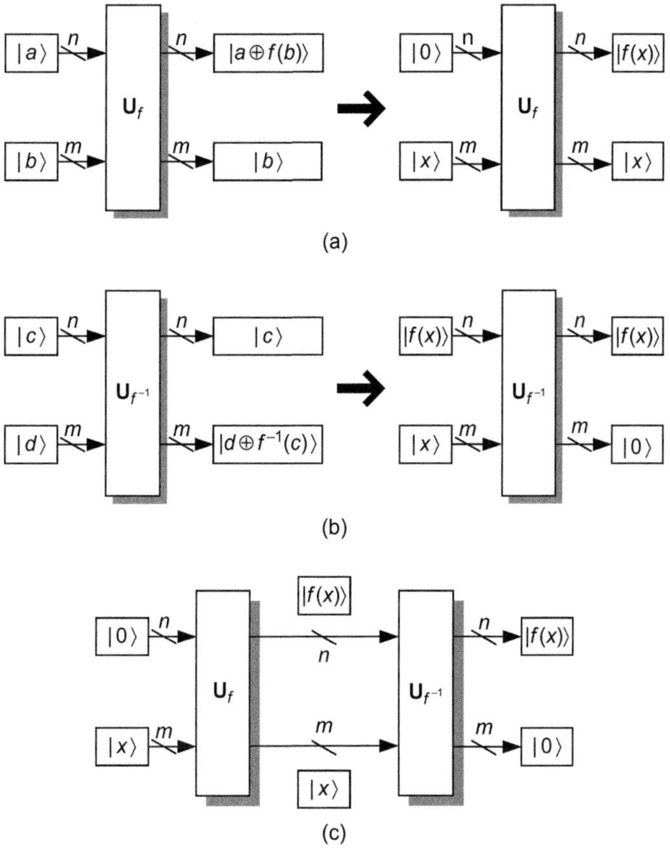

FIGURE 1.29

Reversible computation. (a) The quantum circuit, U_f, maps its input, a b to (a $f(b)$) b ; when a 0 and b x, then $\mathbf{U}_f$(0 x) $f(x)$ x . (b) $U_{f^{-1}}$ maps its input, c d to c (d $f^{-1}(c)$); when c $f(x)$ and d x, then $\mathbf{U}_{f^{-1}}$($f(x)$ x) $f(x)$ 0 . (c) If the two circuits work in tandem implementing the linear transformation $\mathbf{U}_f\mathbf{U}_{f^{-1}}$, then they map the tensor product 0 x to $f(x)$ x .

amplify—i.e., increase the probability amplitude of the solution. It also illustrates the contrast between classical and quantum algorithm strategies: *a classical search algorithm continually reduces the amplitude of nontarget states, while a quantum search algorithm amplifies the amplitude of the target states.* In this context, to amplify means to increase the probability.

Introduction

Grover describes the series of steps that led to his algorithm in [187]; he starts by discretizing the Schrödinger equation and considers a unidimensional case. Grover argues that when we start with a uniform linear superposition and let it evolve in the presence of a potential function it will gravitate toward points at which the potential is lower; therefore, if we want to design an algorithm to reach specific marked states, these marked states should correspond to a lower potential. The algorithm should implement several iterations of the transformations obtained from the evolution of the Schrödinger equation.

The main idea of the quantum search algorithm [183–186] is to rotate the state vector in a two-dimensional Hilbert space defined by an initial and a final (target) state vector. The algorithm is iterative and each iteration causes the same amount of rotation. A more recent version of the algorithm, the *fixed-point quantum search* discovered by Grover [189], could be used for quantum error correction [355], as discussed in Section 5.24.

The speedup of the Grover algorithm is achieved by exploiting both quantum parallelism and the fact that, according to quantum theory, a probability is the square of an amplitude. Bennett *et al.* [42] and Zalka [465] showed that the Grover algorithm is optimal. No classical or quantum algorithm can solve this problem faster than time of order $\overline{N}$.

The Grover search algorithm can be applied directly to a wide range of problems (see, for example, [164]). Even problems not generally regarded as searching problems can be reformulated to take advantage of quantum parallelism and entanglement and lead to algorithms that show a square root speedup over their classical counterparts [277].

The Intuition

We consider a search space, $Q \quad q_i$, consisting of $N \quad 2^n$ items; each item $q_i, 1 \quad i \quad N$, is uniquely identified by a binary n-tuple i, called *the index* of the item. We assume that $M \quad N$ items satisfy the requirements of a query, and we wish to identify one of them.

The classic approach is to repeatedly select an item q_i, decide if the item is a solution to the query, and if so, terminate the search. If there is a single solution ($M \quad 1$), then a classical search algorithm requires $\mathcal{O}(N)$ iterations; in the worst case, we need to examine all N elements; if we repeat the experiment many times, then, on average, we will end up examining $N/2$ elements before finding the desired one.

In this section we provide an intuitive explanation for two transformations at the heart of the Grover algorithm, the *inversion about the mean* in the classical context and the *phase inversion.*

Given a set of integers $Q \quad q_i$, with the mean $q \quad (1/N) \quad {}_{i \; 1}^{N} q_i$, the inversion about the mean, denoted as P, transforms q_i to q_i:

$$P : q_i \quad q_i, \quad \text{with} \quad q_i \quad q \quad (q \quad q_i) \quad 2q \quad q_i.$$

In the example depicted in Figures 1.30a and 1.30b, the mean is $q \quad 49$, and the individual integers are transformed as follows:

34 98 34 64, 66 98 66 32, 47 98 47 51, 63 98 63 35,
54 98 54 44, 28 98 28 70, 42 98 42 56, 58 98 58 40.

The inversion about the mean can be reformulated if we represent the transformation P as a matrix of state transitions, the elements of the set Q as a column vector Q, and the set of new values as the column vector Q :

$$Q \quad 2PQ \quad Q \quad (2P \quad I)Q,$$

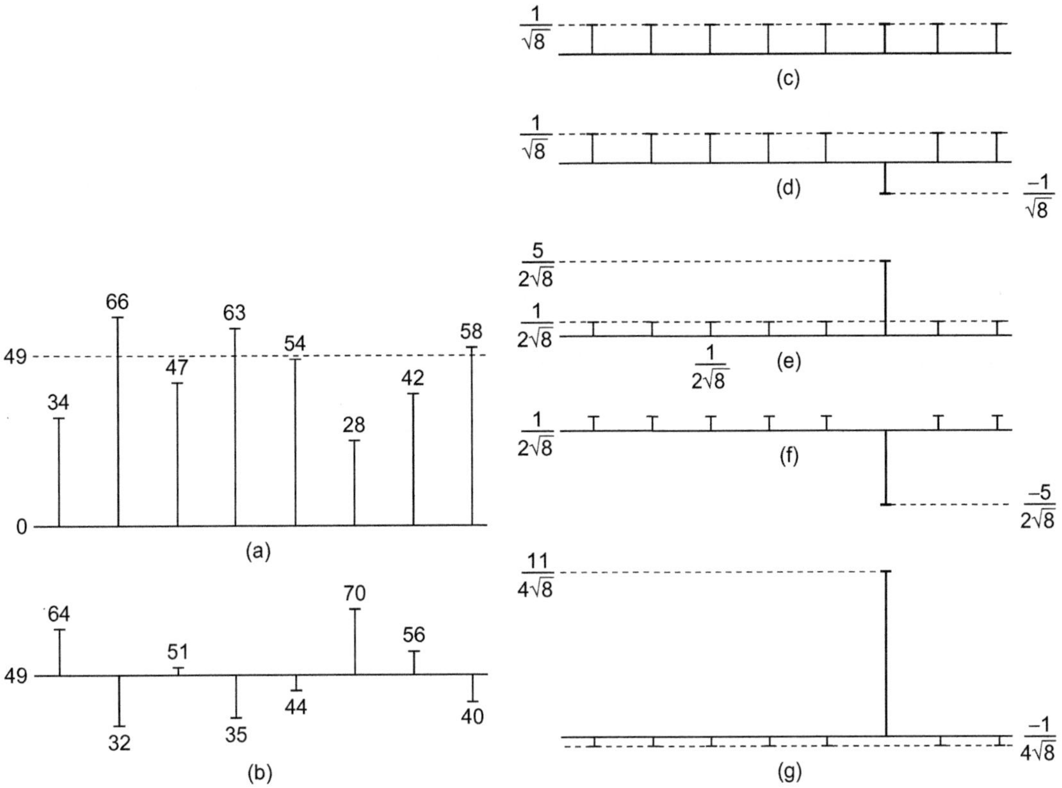

FIGURE 1.30

Graphic illustration of the inversion about the mean and the phase inversion. The amplitude of vertical bars is proportional to the modulus of the numbers. (a) A set of eight integers, $\mathcal{Q} \quad 34, 66, 47, 63, 54, 28, 42, 48$, with the average $q \quad 1/8 \; {}_{i}^{8}\, {}_{1} q_i \quad 49$. (b) Inversion about the mean of the set, $\mathcal{A}$; integer q_i becomes $q_i \quad q \quad (q - q_i) \quad 2q - q_i$. (c) A set of eight complex numbers of equal amplitude $q_i \quad 1/ \; \overline{8}$. (d) Phase inversion of the sixth item, the one we searched for. (e) Inversion about the mean from step (d): $q \quad 3/(4 \; \overline{8})$; it follows that $q_1 \quad q_2 \quad q_3 \quad q_4 \quad q_5 \quad q_7 \quad q_8 \quad 1/(2 \; \overline{8})$ and $q_6 \quad 5/(2 \; \overline{8})$. (f) A second phase inversion of the item we searched for. (g) A second inversion about the new mean from step (e): $q \quad 1/(8 \; \overline{8})$; now $q_1 \quad q_2 \quad q_3 \quad q_4 \quad q_5 \quad q_7 \quad q_8 \quad -1/(4 \; \overline{8})$ and $q_6 \quad 11/(4 \; \overline{8})$.

with

$$P \quad \frac{1}{N} \begin{matrix} 1 & 1 & \ldots & 1 \\ 1 & 1 & \ldots & 1 \\ \vdots & \vdots & & \vdots \\ 1 & 1 & \ldots & 1 \end{matrix} \quad \text{and} \quad Q \quad \begin{matrix} q_1 \\ q_2 \\ \vdots \\ q_N \end{matrix}, \quad Q \quad \begin{matrix} q_1 \\ q_2 \\ \vdots \\ q_N \end{matrix}.$$

Next, we consider the case when we label the N items with complex numbers of equal amplitude, $q_i \quad 1/\ \overline{N}$, and require that after each transformation the sum of the squares of the moduli of the complex numbers be 1; we also assume that there is an oracle capable of inverting the phase of the complex number identifying the item we search for. The inversion about the mean amplifies the amplitude of the item we search for and reduces the amplitude of all other items. When $q_1 \quad q_2 \quad q_3 \quad q_4 \quad q_5 \quad q_7 \quad q_8 \quad 1/(2\ \overline{8})$, the average amplitude after the first phase inversion (Figure 1.30c and 1.30d) is

$$q \quad \frac{7\ \frac{1}{\overline{8}} \qquad \frac{1}{\overline{8}}}{8} \quad \frac{3}{4\ \overline{8}}.$$

Then, the inversion about q (Figure 1.30e) leads to

$$q_1 \quad q_2 \quad q_3 \quad q_4 \quad q_5 \quad q_7 \quad q_8 \quad 2\ \frac{3}{4\ \overline{8}} \qquad \frac{1}{\overline{8}} \quad \frac{1}{2\ \overline{8}}$$

and

$$q_6 \quad 2\ \frac{3}{4\ \overline{8}} \qquad \frac{1}{\overline{8}} \quad \frac{5}{2\ \overline{8}}.$$

A second phase inversion of the item we search for (Figure 1.30f) followed by a second inversion about the new mean,

$$q \quad \frac{7\ \frac{1}{2\ \overline{8}} \qquad \frac{5}{2\ \overline{8}}}{8} \quad \frac{1}{8\ \overline{8}},$$

leads to the situation depicted in Figure 1.30g. We notice the reduction of the amplitudes of the complex numbers labelling the items that do not match our search,

$$q_1 \quad q_2 \quad q_3 \quad q_4 \quad q_5 \quad q_7 \quad q_8 \quad 2\ \frac{1}{8\ \overline{8}} \qquad \frac{1}{2\ \overline{8}} \qquad \frac{1}{4\ \overline{8}},$$

and the further amplification of the amplitude of the item we searched for,

$$q_6 \quad 2\ \frac{1}{8\ \overline{8}} \qquad \frac{5}{2\ \overline{8}} \quad \frac{11}{4\ \overline{8}}.$$

The example in Figures 1.30c–g allows us to refocus the discussion from classical to quantum search; we now view the complex numbers $q_i, 1 \quad i \quad 8$, as the probability amplitudes of the basis states of the canonical base in $\mathcal{H}_8$. The original state of the system is

$$\varphi \quad q_1\ 000 \quad q_2\ 001 \quad q_3\ 010 \quad q_4\ 011 \quad q_5\ 100 \quad q_6\ 101 \quad q_7\ 110 \quad q_8\ 111\ .$$

The final state after two iterations of phase inversion followed by inversion about the mean is

$$\varphi \quad q_1\ 000 \quad q_2\ 001 \quad q_3\ 010 \quad q_4\ 011 \quad q_5\ 100 \quad q_6\ 101 \quad q_7\ 110$$
$$q_8\ 111\ .$$

The item we searched for is labelled 101, and we have amplified the probability amplitude of the state 101 from 1/8 0.125 to $(11/4\ \overline{8})^2$ 121/128 0.97227 in two iterations; thus, we have increased the probability that a measurement of the state, φ , produces the result 101.

Phase Inversion and the Oracle

For simplicity, we assume that the database, $\mathcal{D}$, contains $N\ \ 2^n$ items and only one item satisfies the query. The basic idea of the quantum search is to associate the index of an item, $0\ \ x\ \ 2^n\ \ 1$, with the corresponding basis state, x , of a Hilbert space, $\mathcal{H}_{2^n}$. The canonical base in $\mathcal{H}_{2^n}$ is $0\ ,\ 1\ ,\ 2\ ,\ldots,\ 2^n\ \ 1\ .$

The quantum search algorithm requires an *oracle* to identify the index, x_0, of the item we are searching for. To mark the desired item, the oracle performs a phase inversion. To understand the inner working of the oracle, we consider a function, $f(x)$, such that

$$f(x)\ \ \begin{cases} 0 & \text{if } x\ \ x_0 \quad x \text{ is not a solution to the query,} \\ 1 & \text{if } x\ \ x_0 \quad x \text{ is a solution to the query.} \end{cases}$$

A black box performing a unitary transformation, $\mathbf{U}_f$, for a function, $f\colon\ 0,1\ ^n\ \ \ 0,1\ ,$ is shown in Figure 1.31a. The black box used as an oracle accepts as the input n qubits representing the basis state, x , corresponding to the index, x, and an *oracle qubit*, y ; the oracle qubit is flipped if $f(x)\ \ 1$ (Figure 1.31b). The oracle recognizes the solution, inverts its phase, and produces at the output $(\ \ 1)^{f(x)}[\ x\ \ \ \ y\ ]$. If the oracle qubit is set at $y\ \ \ \ 1$, after the Hadamard gate, H, is applied to it the qubit state becomes $\frac{0\ \ \ 1}{\overline{2}}$. When the oracle recognizes the solution, the oracle qubit is flipped to the state $\frac{1\ \ \ 0}{\overline{2}}$ or $\frac{0\ \ \ 1}{\overline{2}}$. The output of the oracle becomes $[(\ \ 1)^{f(x)}\ \ x\ ]\ \ \ (\ 0\ \ \ \ 1\ )/\ \overline{2}$. The basis state corresponding to the index, x_0, of the item we are searching for is branded as the solution by the oracle, and its phase is inverted; the oracle qubit remains unchanged. To take advantage of the quantum parallelism, we apply as the input to the oracle an equal superposition of all basis states,

$$\psi\ \ \ \frac{1}{\overline{2^n}} \ \ _{x\ \ 0}^{2^n\ \ 1}\ \ x\ .$$

The oracle carries a phase inversion denoted as $\mathbf{O}_{x_0}$; in other words, it inverts the phase of the basis state corresponding to x_0 and leaves all other basis states unchanged. The $(n\ \ \ 1)$ qubit output of the oracle is

$$[\mathbf{O}_{x_0}\ \ \psi\ ]\ \ \ \frac{0\ \ \ 1}{\overline{2}}.$$

Inversion about the Mean

For $\psi\ \ \ \frac{1}{\overline{2^n}}\ \ _{x\ \ 0}^{2^n\ \ 1}\ x$, an equal superposition state, the operator describing the conditional phase shift, or inversion about the mean, is

$$\mathbf{D}_\psi\ \ \ 2\mathbf{P}_\psi\ \ \ \mathbf{I}\ \ 2\ \ \psi\ \ \psi\ \ \ \mathbf{I},$$

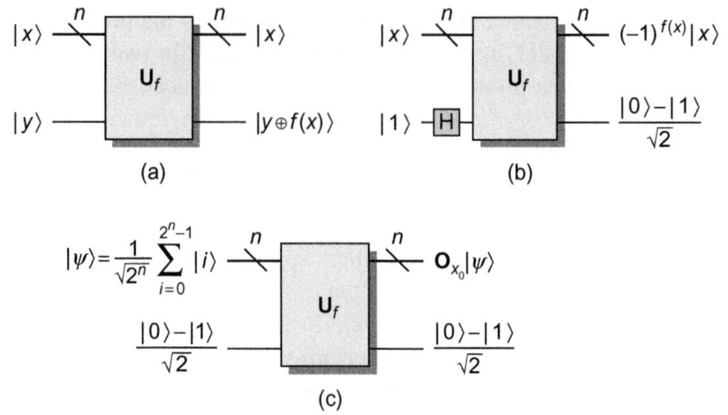

FIGURE 1.31

The oracle for the Grover quantum search algorithm. (a) A black box performing a unitary transformation, $\mathbf{U}_f$, of its n-qubit input, x, with $f: \; 0,1 \;^n \quad 0,1$. (b) The black box used as an oracle carries out a phase shift of the input x when $x \quad x_0$ and $\mathbf{O}_{x_0} \; x_0 \quad (-1)^{f(x_0)} \; x_0$; here, x_0 is the index of the item we are searching for. (c) When we apply ψ, a superposition of all basis states, as the input, the oracle carries a phase inversion denoted as $\mathbf{O}_{x_0}$ and produces $\mathbf{O}_{x_0} \; \psi$ at the output; the oracle qubit remains unchanged.

or

$$\mathbf{D}_\psi \quad 2 \begin{pmatrix} \frac{1}{\sqrt{N}} \\ \frac{1}{\sqrt{N}} \\ \vdots \\ \frac{1}{\sqrt{N}} \\ \vdots \\ \frac{1}{\sqrt{N}} \end{pmatrix} \begin{pmatrix} \frac{1}{\sqrt{N}} & \frac{1}{\sqrt{N}} & \cdots & \frac{1}{\sqrt{N}} & \cdots & \frac{1}{\sqrt{N}} \end{pmatrix} - \begin{pmatrix} 1 & 0 & \ldots & 0 & \ldots & 0 \\ 0 & 1 & \ldots & 0 & \ldots & 0 \\ \vdots & & & & & \\ 0 & 0 & \ldots & 1 & \ldots & 0 \\ \vdots & & & & & \\ 0 & 0 & \ldots & 0 & \ldots & 1 \end{pmatrix}.$$

The matrix representation of the operator, $\mathbf{D}_\psi$, is

$$\mathbf{D}_\psi \quad \begin{pmatrix} \frac{2}{N}-1 & \frac{2}{N} & \ldots & \frac{2}{N} & \ldots & \frac{2}{N} \\ \frac{2}{N} & \frac{2}{N}-1 & \ldots & \frac{2}{N} & \ldots & \frac{2}{N} \\ \vdots & & & & & \\ \frac{2}{N} & \frac{2}{N} & \ldots & \frac{2}{N}-1 & \ldots & \frac{2}{N} \\ \vdots & & & & & \\ \frac{2}{N} & \frac{2}{N} & \ldots & \frac{2}{N} & \ldots & \frac{2}{N}-1 \end{pmatrix}.$$

It is easy to see that $\mathbf{D}_\psi^\dagger \quad [(\; \psi \quad \psi \; - \mathbf{I}]^\dagger \quad \mathbf{D}_\psi$; thus, the inversion about the mean is unitary:

$$\mathbf{D}_\psi \mathbf{D}_\psi^\dagger \quad \mathbf{D}_\psi^\dagger \mathbf{D}_\psi \quad \mathbf{I}.$$

To convince ourselves that individual basis states suffer an inversion about the mean, we consider a system in the state $\xi \quad \mathcal{H}_N$:

$$\xi \qquad \overset{N\ 1}{\underset{i\ 0}{}} \; a_i \;\; i \, .$$

Then,

$$\zeta \quad \mathbf{D}_\psi \;\; \xi \qquad \overset{N\ 1}{\underset{k\ 0}{}} \; b_k \;\; k \, .$$

The probability amplitude, b_k, is

$$b_k \quad 2a \quad a_k, \quad 0 \quad k \quad N \quad 1, \quad \text{with} \quad a \quad \frac{1}{N} \overset{N\ 1}{\underset{i\ 0}{}} \; a_i.$$

Jozsa [224] observes that any state, ξ , can be expressed as a sum of components parallel and orthogonal to $\mathbf{D}_\psi \;\; \xi$, namely, α and β . He shows that given any state ξ , $\mathbf{D}_\psi \;\; \xi$ preserves the two-dimensional space spanned by ξ and ψ .

As a result of the inversion about the mean, the probability amplitudes are redistributed. According to Grover [182]:

> *After the inversion, the amplitude in each state increases/decreases; the amplitude is as much below/above a as it was above/below a before the inversion.*

If the probability amplitudes of all but one basis states are positive, after we apply the inversion about the mean, the probability amplitude of the negative basis state is amplified and becomes positive (Figure 1.32). Intuitively, the inversion about the mean increases the amplitude of the target state by $\frac{2}{\overline{N}}$ at each iteration.

Grover Search Algorithm

Figure 1.33 illustrates the transformations for the Grover quantum search algorithm: We apply an n-dimensional Walsh-Hadamard transform to create an equal superposition of the indices of all items in the database; then we perform $\overline{N}$ Grover iterations, $\mathbf{G}$. A Grover iteration, $\mathbf{G}$, consists of the transformation performed by the oracle, $\mathbf{O}$, followed by an inversion about the mean, $\mathbf{D}_\psi$:

$$\mathbf{G} \quad \mathbf{D}_\psi \mathbf{O} \quad (2 \;\; \psi \;\; \psi \quad \mathbf{I})\mathbf{O}.$$

At each iteration, the phase of the "solution(s)" is rotated with an angle of θ radians; the mean value of the probability amplitudes of the superposition state is the same, but the amplitude of the " solution" is amplified.

One may question why we need to carry out several iterations once the oracle is able to identify the solution. Recall that originally the input to the oracle is an equal superposition of 2^n basis vectors, each with a probability amplitude of $1/\;\overline{2^n}$. For example, if $n \quad 30$, then the probability amplitude of each of the $2^{30} \quad 10^9$ basis vectors in the Hilbert space, $\mathcal{H}_{2^n}$, is very low, of the order $10 \;\; ^4$. It is hard

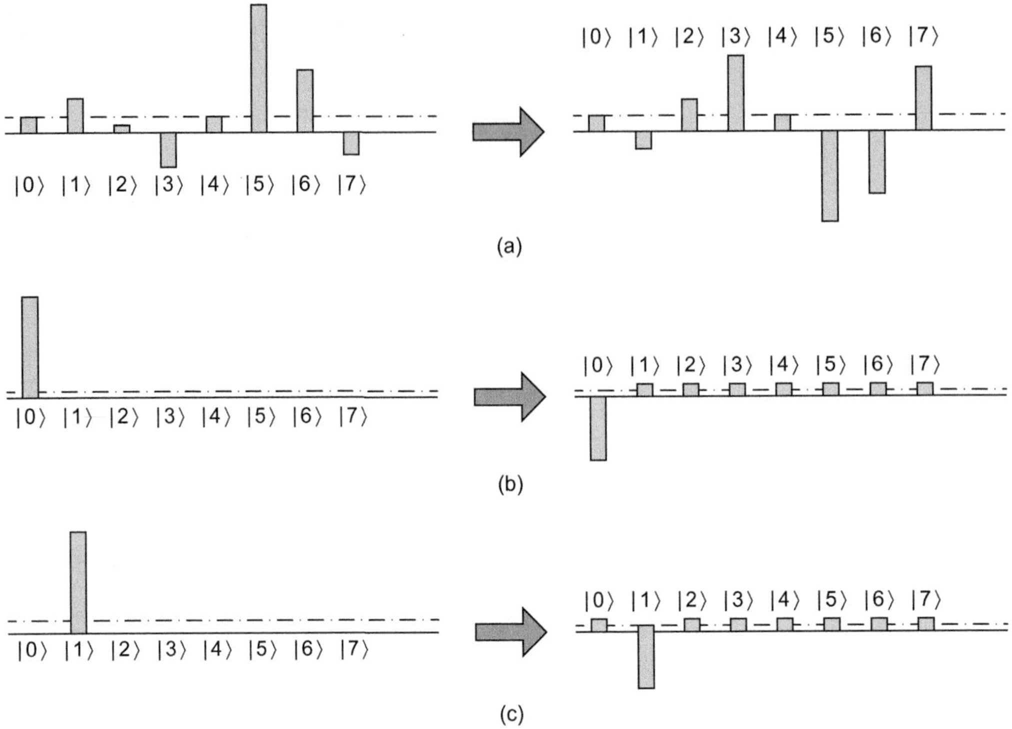

FIGURE 1.32

(a) An example of inversion about the mean in $\mathcal{H}_{2^3}$. (b) and (c) The inversion about the mean of basis vectors 0 and 1 .

to distinguish the solution, which has a very small negative probability amplitude; thus, we need to perform several iterations and amplify the amplitude of the solution.

Geometric Interpretation of One Grover Iteration

First, we review a property of the groups of transformations related to the symmetry operators: The product of two reflections in a two-dimensional Euclidian plane is a rotation (see Section 1.9). Consider two lines, $L1$ and $L2$, at an angle γ (Figure 1.34a); the reflection, $L2$, of line $L2$ about line $L1$ is shown in Figure 1.34b. Then in Figure 1.34c we see a reflection of $L2$ about $L2$; the product of the two reflections is a rotation of $L2$ with an angle 2γ.

We consider a slightly modified search problem and assume that there are, M N, items in the database, $\mathcal{D}$, that satisfy the search criteria. The N items in the database, $\mathcal{D}$, are partitioned in two disjoint subsets: $\mathcal{S}$, the set of items that satisfy the search criteria, and $\mathcal{S}$, the set of items that do not satisfy the search criteria (Figure 1.35a):

$$\mathcal{D} \quad \mathcal{S} \quad \mathcal{S}, \qquad \mathcal{D} \quad N, \qquad \mathcal{S} \quad M, \quad \text{and} \quad \mathcal{S} \quad N-M.$$

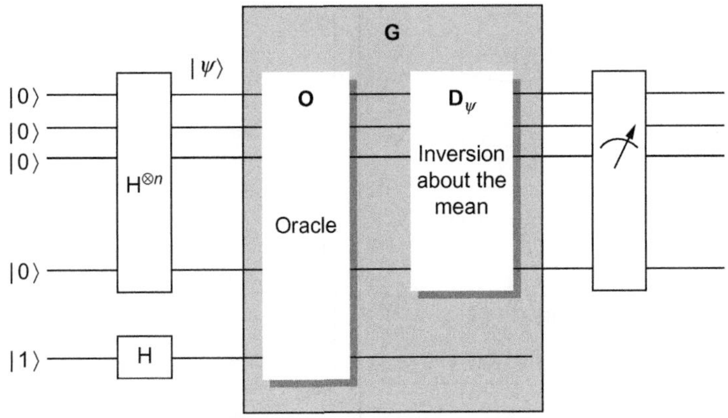

FIGURE 1.33

The schematic representation of the transformations for the Grover quantum search algorithm. An n-dimensional Walsh-Hadamard transform creates ψ , an equal superposition of the indices of all items in the database. Then we perform $\overline{N}$ Grover iterations, $\mathbf{G}$. A Grover iteration, $\mathbf{G}$, consists of the transformation performed by the oracle, $\mathbf{O}$, followed by an inversion about the mean, $\mathbf{D}_{\psi}$. Finally, we measure the result.

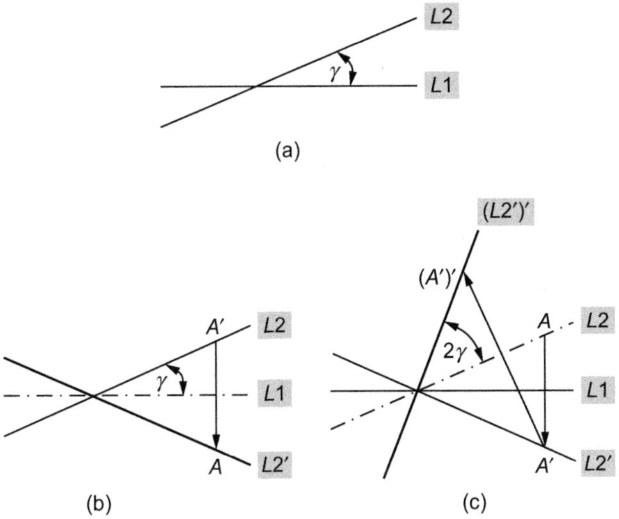

FIGURE 1.34

(a) Two lines, $L1$ and $L2$, at an angle γ in a two-dimensional Euclidean plane. (b) $L2$ is the reflection of line $L2$ about $L1$. (c) ($L2$) is the reflection of line $L2$ about $L2$. The two successive reflections correspond to a rotation with an angle 2γ.

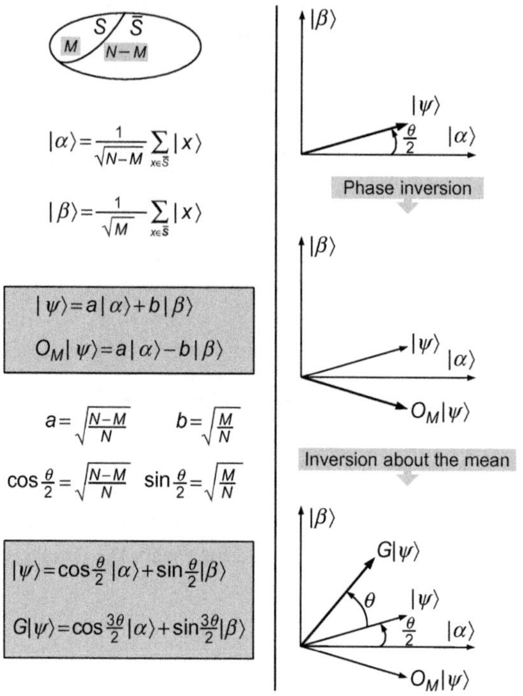

(a) (b)

FIGURE 1.35

Geometric interpretation of one Grover iteration. (a) The N items in the database, $\mathcal{D}$, are partitioned in two disjoint subsets: $\mathcal{S}$ includes items that do not satisfy the search criteria, and $\mathcal{S}$ includes the ones that do. We define two normalized states: α , an equal superposition of the $N - M$ basis states, x , corresponding to indices x $\mathcal{S}$, and β , an equal superposition of the M basis states, x $\mathcal{S}$. (b) Any superposition state can be represented as ψ a α b β . A phase inversion, $\mathbf{O}_M$, produces the state, a α $- b$ β ; thus, it is a reflection of ψ about α . The inversion about the mean, $\mathbf{D}_\psi$, is a reflection of the state, $\mathbf{O}_M$ ψ , about ψ . The product of two reflections is a rotation; an iteration of the Grover algorithm, G ψ , rotates the state, ψ $\cos(\theta/2)$ α $\sin(\theta/2)$ β , by an angle θ toward β and therefore toward the superposition of the states that satisfy the search criteria; thus, $\mathbf{G}$ ψ $\cos\frac{3\theta}{2}$ α $\sin\frac{3\theta}{2}$ β .

We define two states: α is an equal superposition of the $N - M$ basis states, x , corresponding to indices x of items that *do not satisfy the query*, x $\mathcal{S}$; and β is an equal superposition of the M basis states, x , corresponding to indices, x, of items that *satisfy the query*, x $\mathcal{S}$:

$$\alpha \quad \frac{1}{\overline{N - M}} \quad x \quad \text{and} \quad \beta \quad \frac{1}{\overline{M}} \quad x .$$
$$\scriptstyle x \ \mathcal{S} \qquad\qquad\qquad\qquad x \ \mathcal{S}$$

The two states are orthogonal, α β 0; we can then express any state, ψ , as the superposition of the two states (Figure 1.35b) as

$$\psi \quad a \ \alpha \quad b \ \beta , \quad \text{with} \quad a^2 \quad b^2 \quad 1.$$

The probability that a state x corresponding to the index x of an item that does not satisfy the query is $(N \quad M)/N$; similarly, the probability that a state x corresponding to the index x of an item that satisfies the query is $(M)/N$. Thus,

$$a \qquad \overline{\frac{N \quad M}{N}} \quad \text{and} \quad b \qquad \overline{\frac{M}{N}}.$$

In this representation, $\mathbf{O}_M$, the phase inversion of the M solutions to the query carried out by the oracle, transforms the original state, $\psi \quad a \; \alpha \quad b \; \beta$, to $\mathbf{O}_M \; \psi \quad a \; \alpha \quad b \; \beta$. The phase inversion is represented geometrically as a reflection of the state vector ψ about α; the projection on β changes its sign, as seen in Figure 1.35b. But ψ is an equal superposition state; therefore, the inversion about the mean, $\mathbf{D}_\psi$, is a reflection of the state, $\mathbf{O}_M \; \psi$, about ψ. The product of the two reflections is a rotation. To determine the rotation angle, we denote the angle between ψ and α as $\theta/2$; then,

$$\cos\frac{\theta}{2} \qquad \overline{\frac{N \quad M}{N}} \quad \text{and} \quad \sin\frac{\theta}{2} \qquad \overline{\frac{M}{N}}.$$

We can express ψ as

$$\psi \qquad \cos\frac{\theta}{2} \; \alpha \qquad \sin\frac{\theta}{2} \; \beta .$$

From Figure 1.35b, we see that

$$\mathbf{G} \; \psi \qquad \mathbf{D}_\psi \mathbf{O}_M \; \psi \qquad \cos\frac{3\theta}{2} \; \alpha \qquad \sin\frac{3\theta}{2} \; \beta .$$

We conclude that *one Grover iteration rotates the phase of the "solution" with θ radians*. After k Grover iterations, the state is

$$\mathbf{G}^k \; \psi \qquad \cos((2k \quad 1)\theta/2) \; \alpha \qquad \sin((2k \quad 1)\theta/2) \; \beta .$$

Now we answer the question of how many iterations are needed when $M << N$. One iteration rotates the state vector with an angle θ; thus, the number of iterations is

$$n_R \qquad \frac{\pi/2}{\theta} \quad .$$

When $M << N$, θ is a small angle, and we approximate

$$\theta/2 \quad \sin(\theta/2) \qquad \overline{\frac{M}{N}} \qquad \theta \quad 2 \; \overline{\frac{M}{N}}.$$

It follows that

$$n_R \qquad \frac{\pi/2}{2 \; \overline{\frac{M}{N}}} \qquad \pi/4 \; \overline{\frac{N}{M}} \quad .$$

We conclude that the number of iterations is

$$n_R = \mathcal{O}\left(\sqrt{\frac{N}{M}}\right).$$

Example. Consider the case $n = 2$: There are $N = 2^2 = 4$ items in the database, with indices 00, 01, 10, and 11; when $M = 1$, there is a unique solution. The quantum circuit in Figure 1.36 allows us to identify the solution in one iteration. Indeed, in this case the state after the first Walsh-Hadamard transform is

$$|\psi\rangle = \frac{|00\rangle + |01\rangle + |10\rangle + |11\rangle}{2}.$$

The angle, $\theta/2$, between this state and state $|\alpha\rangle$ is

$$\cos\theta/2 = \sqrt{\frac{N-M}{N}} = \frac{\sqrt{3}}{2} \quad \text{and} \quad \theta/2 = \frac{\pi}{6}.$$

One Grover iteration rotates the state, $|\psi\rangle$, with an angle $\theta = \pi/3$. Thus, after one iteration, the angle of the state vector is $\theta/2 + \theta = \pi/2$, and the state vector is along the solution $|\beta\rangle$. Figure 1.37 shows the four circuits of the oracle corresponding to solutions 00, 01, 10, and 11, respectively.

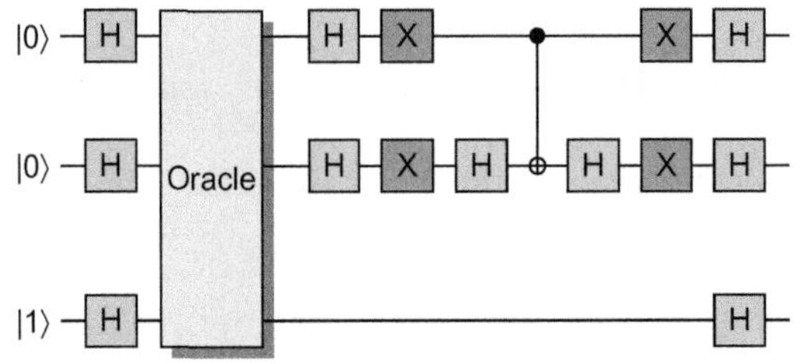

FIGURE 1.36

Two-qubit Grover circuit.

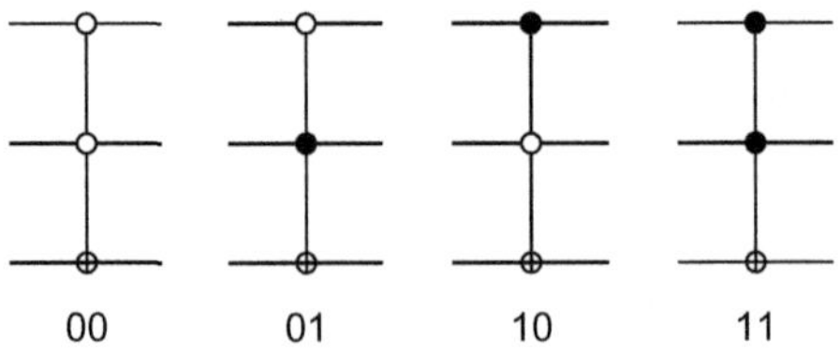

FIGURE 1.37

The circuits of the oracle for the Grover algorithm in Figure 1.36. The four circuits correspond to the solutions 00, 01, 10, and 11, respectively.

Stopping Criteria

The algorithm needs a precise stopping criterion (i.e., the precise number of target states). Without this stopping criterion, the rotation of the current state could overshoot and drift away from the desired solution.

The sophisticated culinary analogy of this process attributed to Kristen Fuchs is described by Gilles Brassard [72]:

> *Quantum search is like cooking a souf e. You put the state obtained by quantum parallelism in a quantum oven and let the answer rise slowly. Success is almost guaranteed if you open the oven at the right time. But the souf e is likely to fall—the amplitude of the correct answer will go to zero—if you open the oven too early. Furthermore, the souf e could burn if you overcook it; strangely, the amplitude of the desired state shrinks after reaching its maximum.*

The Grover algorithm was generalized to the case in which the initial amplitudes are either real or complex and follow any arbitrary distribution [58, 59]. Recently, Grover showed that by replacing selective inversions with selective phase shifts of $\pi/3$ the algorithm preferentially converges to the target state irrespective of the step size or the number of iterations [189]. This strategy and the amplitude amplification are discussed in the next section.

1.24 AMPLITUDE AMPLIFICATION AND FIXED-POINT QUANTUM SEARCH

We first emphasize the amplitude amplifications aspect of the quantum search, then discuss the general formulation of the amplitude amplification problem, and, finally, we present the core ideas of the fixed-point quantum search.

Quantum Search and Amplitude Ampli cation

The Grover quantum search algorithm can be summarized as follows: Consider a quantum mechanical system, with N distinguishable states; all N basis states have initially the same amplitude. We have N objects, each identified by a unique label, x. We map each label x to one of the basis states of the system and iteratively amplify the amplitude of the basis state, τ , corresponding to the solution, $x \quad \tau$. The amplitude amplification is achieved by repeated state transformations as prescribed by the Walsh-Hadamard transform. Finally, we perform a measurement that collapses the system's state to a basis state. With high probability, we end up observing the value τ, the label of the object we are searching for, because the probability amplitude of the corresponding basis state, τ , has been amplified.

The centerpiece of the quantum search algorithm is a binary function, $f(x)$, with the property that $f(x) \quad 1$ if and only if x is the label of the object we are searching for, $f(\tau) \quad 1$. Recall that we can construct a quantum circuit that will evaluate any function that can be evaluated classically; thus, we can construct a circuit to evaluate $f(x)$. If b is an ancilla qubit, we can always construct a circuit to carry out the following state transformation:

$$x, b \qquad x, f(x)\ XOR\ b\ .$$

If the ancilla, b, is initially in the state $b \quad (1/\ \overline{2})(\ 0 \qquad 1\ )$, then the circuit will invert the amplitude of all states when $f(x) \quad 1$.

Soon after Grover discovered the search algorithm, he realized that the Walsh-Hadamard transform can be replaced by almost any other transformation [185]. Let us follow Grover's arguments and notations. Assume that initially the system is in the state γ , and we wish to force it to the state τ through a sequence of unitary transformations **U**. Call $U_{\tau\gamma}$ the amplitude of reaching the state τ starting from the state γ after one application of the unitary transformation **U**:

$$U_{\tau\gamma} \quad \tau \ \mathbf{U} \ \gamma \ .$$

The corresponding probability is $p_{\tau\gamma} \quad U_{\tau\gamma}{}^{2}$. On average we would need $\mathcal{O}(1/ \ U_{\tau\gamma}{}^{2})$ trials (one trial is one application of U) for a success. Grover shows that in fact one can reach the state τ only in $\mathcal{O}(1/ \ U_{\tau\gamma} \ )$ trials. When $U_{\tau\gamma} << 1$, this results in a significant speedup.

Call $\mathbf{I}_x$ the transformation that inverts the amplitude of the basis state x . The corresponding matrix, $I_x \quad [u_{ij}], 1 \quad i,j \quad N$, is a diagonal one, with all diagonal terms equal to 1, except the element corresponding to x:

$$u_{ij} \quad \begin{matrix} 0 & \text{if} & i & j, & \\ 1 & \text{if} & i & j & x, \\ 1 & \text{if} & i & j & x. \end{matrix}$$

When x corresponds to the basis vector, x , the projector, $\mathbf{P}_x \quad x \ x$, is an $N \quad N$ matrix whose only nonzero element is equal to 1, and it is the diagonal element on row x and column x. We see that

$$I_x \quad I \ 2 \ x \ x \ ,$$

with I, the $N \quad N$ identity matrix. For example, for $N \quad 4$ and $x \quad 3$:

$$I_3 \quad \begin{matrix} 1 & 0 & 0 & 0 \\ 0 & 1 & 0 & 0 \\ 0 & 0 & 1 & 0 \\ 0 & 0 & 0 & 1 \end{matrix} \quad \text{and} \quad 3 \ 3 \quad \begin{matrix} 0 & 0 & 0 & 0 \\ 0 & 0 & 0 & 0 \\ 0 & 0 & 1 & 0 \\ 0 & 0 & 0 & 0 \end{matrix} .$$

Thus,

$$\begin{matrix} 1 & 0 & 0 & 0 \\ 0 & 1 & 0 & 0 \\ 0 & 0 & 1 & 0 \\ 0 & 0 & 0 & 1 \end{matrix} \quad \begin{matrix} 1 & 0 & 0 & 0 \\ 0 & 1 & 0 & 0 \\ 0 & 0 & 1 & 0 \\ 0 & 0 & 0 & 1 \end{matrix} \quad 2 \quad \begin{matrix} 0 & 0 & 0 & 0 \\ 0 & 0 & 0 & 0 \\ 0 & 0 & 1 & 0 \\ 0 & 0 & 0 & 0 \end{matrix} \quad \text{or} \quad I_3 \quad I \ 2 \ 3 \ 3 \ .$$

We define transformation **Q** as

$$\mathbf{Q} \quad \mathbf{I}_\gamma \mathbf{U}^{\ 1} \mathbf{I}_\tau \mathbf{U}.$$

Consider the state obtained by applying the inverse of **U** to the target state:

$$\delta \quad \mathbf{U}^{\ 1} \ \tau \ .$$

Now we apply the new transformation, **Q**, to the initial state, γ , and to the state, δ . It is easy to show that:

$$\mathbf{Q}\ \gamma \quad (1 \quad 4\ U_{\tau\gamma}\ ^{2})\ \gamma \quad 2U_{\tau\gamma}\ \delta\ .$$

Indeed,

$$\begin{aligned}
\mathbf{Q}\ \gamma \quad & (\ \mathbf{I}_{\gamma}\mathbf{U}\ ^{1}\mathbf{I}_{\tau}\mathbf{U})\ \gamma \\
& [\ (\mathbf{I}\ \ 2\ \gamma\ \ \gamma\ )\mathbf{U}\ ^{1}(\mathbf{I}\quad 2\tau\ \ \tau\ )\mathbf{U}]\ \gamma \\
& \mathbf{IUU}\ ^{1}\ \gamma \quad \mathbf{IU}\ ^{1}2\ \ \tau\ \ \tau\ \ \mathbf{U}\ \gamma \quad 2\ \gamma\ \ \gamma\ \ \mathbf{U}\ ^{1}\mathbf{IU}\ \gamma \\
& 4\ \gamma\ \ \gamma\ \ \mathbf{U}\ ^{1}\ \ \tau\ \ \tau\ \ \mathbf{U}\ \gamma \\
& \ \gamma \quad 2\mathbf{IU}\ ^{1}\ \ \tau\ U_{\tau\gamma} \quad 2\ \gamma \quad 4\ \gamma\ U_{\tau\gamma}U_{\tau\gamma} \\
& \ \gamma \quad 4\ U_{\tau\gamma}\ ^{2}\ \gamma \quad 2U_{\tau\gamma}(\mathbf{U}\ ^{1}\ \ \tau\ ) \\
& (1 \quad 4\ U_{\tau\gamma}\ ^{2})\ \gamma \quad 2U_{\tau\gamma}\ \delta\ .
\end{aligned}$$

It is left as an exercise for the reader to prove that

$$\mathbf{Q}(\ \delta\ ) \quad 2U_{\tau\gamma}\ \gamma \quad \delta\ .$$

The two equations can be combined as

$$\mathbf{Q}\ \begin{matrix} \gamma \\ \delta \end{matrix} \quad \begin{matrix} (1 \quad 4\ U_{\tau\gamma}\ ^{2}) & 2U_{\tau\gamma} \\ 2U_{\tau\gamma} & 1 \end{matrix} \quad \begin{matrix} \gamma \\ \delta \end{matrix}\ .$$

Thus, *the transformation* **Q** *preserves the two-dimensional vector space spanned by* γ *and* δ , and each application of it rotates a vector in this space by approximately 2 $U_{\tau\gamma}$ radians. The case of interest is when γ and δ are almost orthogonal. Each application of **Q** rotates the system state by an angle 2 $U_{\tau\gamma}$; thus, the initial state γ is transformed into the final state τ after a number of applications of **Q** equal to

$$n_Q \quad \frac{\frac{\pi}{2}}{2\ U_{\tau\gamma}} \quad \frac{\pi}{4\ U_{\tau\gamma}}\ .$$

Grover shows that the result extends to the case when the amplitudes in the states γ and τ are rotated by arbitrary phases instead of being inverted—in other words, when $\mathbf{I}_{\gamma}$ and $\mathbf{I}_{\tau}$ are replaced by arbitrary rotations.

This result shows that quantum search is a robust algorithm and that the Walsh-Hadamard transform on n qubits, $\mathbf{W} \quad \mathbf{H}\ ^{n}$, can in fact be replaced by almost any other unitary transformation.

If we wish to perform an exhaustive search starting from the zero state 0 and use the Walsh-Hadamard transform on n qubits as the unitary transformation, $\mathbf{U} \quad \mathbf{W} \quad \mathbf{W}\ ^{1}$, then **Q** becomes

$$\mathbf{Q} \quad \mathbf{U}_{0}\mathbf{W}\mathbf{I}_{\tau}\mathbf{W}.$$

Note that $\mathbf{W}\mathbf{I}_{0}\mathbf{W}$ is in fact the inversion-about-zero operation described in Section 1.23. Since $\mathbf{I}_{0}$ $\mathbf{I} \quad 2\ \ 0\ \ 0$, it follows that

$$\mathbf{W}\mathbf{I}_{0}\mathbf{W}\ x \quad \mathbf{W}(\mathbf{I} \quad 2\ \ 0\ \ 0\ )\mathbf{W}\ x \quad x \quad 2\mathbf{W}\ \ 0\ \ 0\ \ \mathbf{W}\ x\ .$$

If $X \quad 1/N \quad {}^{N}_{i \;\; 1} x_i$, then we see that the i-th component of $\mathbf{W I_0 W}$ can be expressed as

$$x_i \quad 2X \quad X \quad (X \quad x_i).$$

Amplitude Ampli cation

In 2000 Gilles Brassard, Peter Hoyer, Michele Mosca, and Alain Trapp generalized the amplitude amplification idea [214]. The new process allows us to find a "Good" element $x \quad X$ after an expected number of applications of the unitary transformation $\mathbf{U}$ and of its inverse, $\mathbf{U}^{\;1}$; the number of iterations is proportional to $1/ \quad \bar{a}$, with a the probability of getting a "Good" element x if $\mathbf{U} \quad 0$ is measured, Figure 1.38.

Let $\mathbf{U}$ be a unitary operator in a Hilbert space, $\mathcal{H}_N$, with an orthonormal basis

$$X \qquad 0 \,, \; 1 \,, \ldots, \; N \quad 1 \; .$$

The only condition imposed on $\mathbf{U}$ is to be invertible; thus, $\mathbf{U}$ must not involve any measurements.

If $\chi : X \qquad 0, 1$ is a Boolean function, we say that the basis state x is a "Good" state if $\chi(x) \quad 1$ and x is a "Bad" state" if $\chi(x) \quad 0$. The Boolean function, χ, partitions the Hilbert space $\mathcal{H}_N$ into two subspaces,

$$\mathcal{H}_N \quad \mathcal{H}_{Good} \quad \mathcal{H}_{Bad}, \quad \mathcal{H}_{Good} \quad \mathcal{H}_{Bad} \qquad ,$$

such that

$$\begin{array}{ll} \mathcal{H}_{Good} & \text{subspace spanned by the set of “Good” basis states } x \text{ , such that } \chi(x) \quad 1, \\ \mathcal{H}_{Bad} & \text{subspace spanned by the set of “Bad” basis states } x \text{ , such that } \chi(x) \quad 0. \end{array}$$

Every pure state, $\gamma \qquad \mathcal{H}_N$, has a unique decomposition as a projection onto $\mathcal{H}_{Good}$ and $\mathcal{H}_{Bad}$:

$$\gamma \qquad \gamma_{Good} \qquad \gamma_{Bad} \; .$$

The amplitude amplification is based on an operator $\mathbf{Q}$ defined as

$$\mathbf{Q} \quad \mathbf{Q}(U, \chi, \phi, \varphi) \qquad \mathbf{U S}_0(\phi)\mathbf{U}^{\;1}\mathbf{S}_\chi(\varphi),$$

with ϕ and φ, two angles, such that $0 \quad \phi$ and $\varphi \quad \pi$, and $\mathbf{S}_\chi$, an operator that conditionally changes the amplitudes of "Good" states:

$$\mathbf{S}_\chi(\varphi) \;\; x \qquad \begin{cases} e^{i\varphi} \;\; x & \text{if} \;\; \chi(x) \quad 1, \\ \quad x & \text{if} \;\; \chi(x) \quad 0. \end{cases}$$

Similarly, $\mathbf{S}_0$ amplifies the amplitude by a factor $e^{i\phi}$ if the state is 0 :

$$\mathbf{S}_0(\phi) \;\; 0 \qquad \begin{cases} e^{i\phi} \;\; 0 & \text{if} \;\; \chi(0) \quad 1, \\ \quad 0 & \text{if} \;\; \chi(0) \quad 0. \end{cases}$$

Let us apply the transformation, $\mathbf{U}$, to a system in the state 0 :

$$\Phi \quad \mathbf{U} \; 0 \qquad \Phi_{Good} \qquad \Phi_{Bad}$$

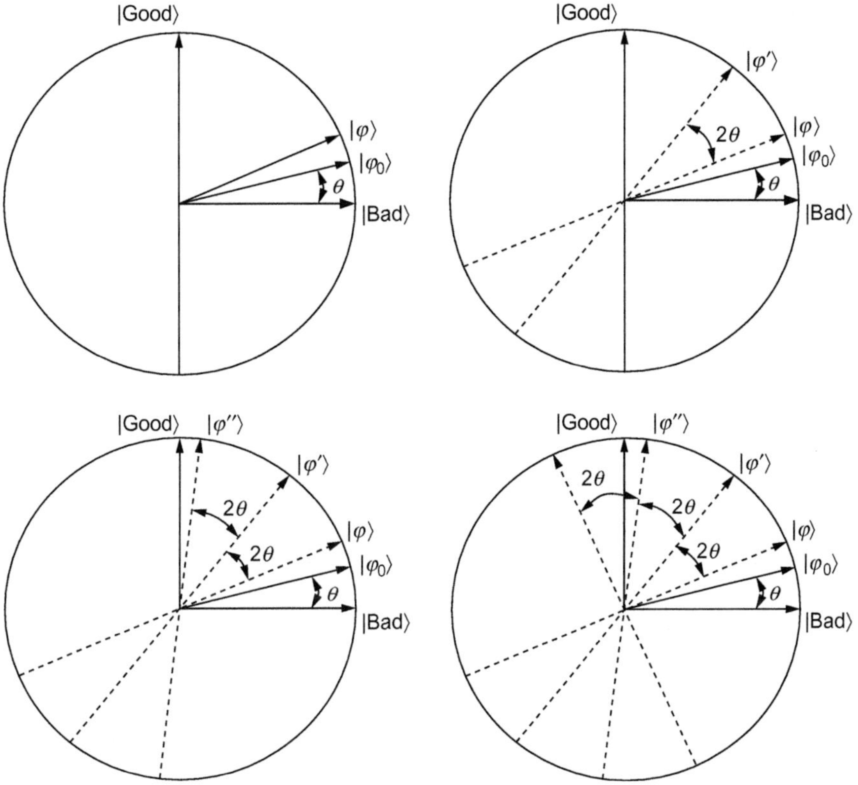

FIGURE 1.38

The effect of amplitude amplification; the original state, φ , is rotated toward the Good state and may overshoot it.

and consider the subspace H_Φ spanned by Φ_{Good} and Φ_{Bad} . As before, a denotes the probability of success:

$$a \quad \Phi_{Good} \quad \Phi_{Good} \ .$$

We also define the angle $\theta \quad \frac{\pi}{2}$, such that

$$\sin^2(\theta) \quad a.$$

Proposition. *The Hilbert subspace, $\mathcal{H}_\Phi$, is stable under the transformation* **Q**. *If* $0 \quad a \quad 1$, *then* $\mathcal{H}_\Phi$ *has dimension* 2*; otherwise, it has dimension 1.*

This proposition is similar to the one we discussed earlier, and it implies the following equations:

$$\mathbf{Q} \quad \Phi_{Good} \quad (1-2a) \quad \Phi_{Good} \quad - \quad 2a \quad \Phi_{Bad} \ ,$$
$$\mathbf{Q} \quad \Phi_{Bad} \quad 2(1-a) \quad \Phi_{Good} \quad (1-2a) \quad \Phi_{Bad} \ .$$

$\mathbf{Q}$ is a unitary operator; thus, if 0 a 1, then $\mathcal{H}_{\Phi}$ has an orthonormal basis consisting of the two eigenvectors of $\mathbf{Q}$, Φ ,

$$\Phi \quad \frac{1}{2} \quad \frac{1}{a} \quad \Phi_{Good} \qquad \frac{i}{1 \quad a} \quad \Phi_{Bad} \quad ,$$

with the corresponding eigenvalues

$$\lambda \qquad \exp(\quad i2\theta).$$

We can express Φ_{Good} and Φ_{Bad} in terms of the eigenvectors

$$\Phi_{Good} \qquad \frac{a}{2}(\; \Phi \qquad \Phi \;)$$

and

$$\Phi_{Bad} \qquad i \; \frac{1 \quad a}{2} (\; \Phi \qquad \Phi \;).$$

Thus,

$$\mathbf{U} \; 0 \qquad \Phi \qquad \Phi_{Good} \qquad \Phi_{Bad} \qquad \frac{i}{2}[\exp(i\theta) \; \Phi \qquad \exp(\quad i\theta) \; \Phi \quad].$$

It follows that after the j-th application of the transformation $\mathbf{Q}$, the state of the system is

$$\mathbf{Q}^{(j)} \; \Phi \qquad \frac{i}{2}(\exp[i\theta(2j \quad 1)] \; \Phi \qquad \exp[\quad i\theta(2j \quad 1)] \; \Phi \;)$$

or

$$\mathbf{Q}^{(j)} \; \Phi \qquad \frac{1}{a}\sin[(2j \quad 1)\theta] \; \Phi_{Good} \qquad \frac{1}{1 \quad a}\cos[(2j \quad 1)\theta] \; \Phi_{Bad} \; .$$

This shows that after we apply m times, with m 0, the transformation $\mathbf{Q}$, a measurement will produce a "Good" state, with the probability p_{Good} $\sin[(2m \quad 1)\theta]^2$. The following proposition follows immediately.

Proposition. *Let* $\mathbf{U}$ *be any unitary transformation that does not perform any measurements and has an initial probability of success equal to* 0 a 1 *and let* $\sin^2(\theta)$ a, 0 θ $\frac{\pi}{2}$. *Let* $\chi : \mathbb{Z}$ 0, 1 *be any Boolean function. Set m* $\frac{\pi}{4\theta}$. *After m applications of* $\mathbf{Q}$, *the resulting state,* $\mathbf{Q}^{(m)}\mathbf{U}$ 0 , *is a "Good" state, with probability at least* $\max(1 \quad a, a)$.

We note that the Grover algorithm is a particular instance of amplitude amplification when:

- the oracle implements the Boolean function, f χ,
- the transformation, $\mathbf{U}$, is the Walsh-Hadamard transform, $\mathbf{W}$ $\mathbf{H}$ ^{n}, on n qubits, and
- the Grover iteration, $\mathbf{W}\mathbf{S}_0\mathbf{W}\mathbf{S}_f$, corresponds to the operator, $\mathbf{Q}$ $\mathbf{U}\mathbf{S}_0\mathbf{U}$ $^1\mathbf{S}_{\chi}$.

This iteration carried out by transformation **Q** can be regarded as a *rotation* in the two-dimensional space spanned by the starting vector, $|\psi_0\rangle = \mathbf{U}|0\rangle = \sum_{x \in X} \alpha_x |x\rangle$, and the state consisting of a uniform superposition of solutions to the search problem. The initial state may be expressed as

$$|\psi_0\rangle = \frac{1}{\sqrt{2}}\left(\sqrt{a}\,|Good\rangle + \sqrt{1-a}\,|Bad\rangle\right).$$

Fixed-Point Quantum Search

The "classical" quantum search algorithm does not guarantee that the state vector does not over-rotate beyond the solution to the search problem, or, in terms of the culinary analogy discussed in Section 1.23, that the souffle is not overcooked. In his 2005 paper [189], Lov Grover shows that by replacing the selective amplitude inversion with selective phase shifts of $\theta = \frac{\pi}{3}$, the algorithm converges to the desired target state irrespective of the amount of rotation of the state vector at each iteration and the number of iterations. The new algorithm is thus extremely robust. Figure 1.39, inspired by Grover's paper [189], summarizes the differences between fixed-point quantum search and amplitude amplification.

The system is initially in the state $|\gamma\rangle$, and we wish it to end in the state $|\tau\rangle$; when we apply the unitary operator, **U**, the state vector always moves closer to the target state.

We define the transformation, **Q**, as

$$\mathbf{Q}_{\tau\gamma} = \mathbf{U}\mathbf{R}_\gamma\mathbf{U}^\dagger\mathbf{R}_\tau U.$$

In this expression, $\mathbf{R}_\gamma$ and $\mathbf{R}_\tau$ are selective phase shifts by $\theta = \frac{\pi}{3}$ of the states, $|\gamma\rangle$ and $|\tau\rangle$, respectively:

$$\mathbf{R}_\gamma = \mathbf{I} - [1 - \exp(i\theta)]\,|\gamma\rangle\langle\gamma|,$$
$$\mathbf{R}_\tau = \mathbf{I} - [1 - \exp(i\theta)]\,|\tau\rangle\langle\tau|.$$

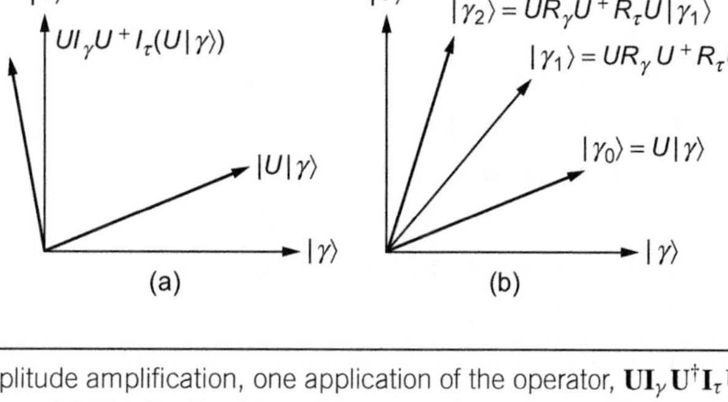

FIGURE 1.39

(a) In the case of amplitude amplification, one application of the operator, $\mathbf{U}\mathbf{I}_\gamma\mathbf{U}^\dagger\mathbf{I}_\tau\mathbf{U}$, could overshoot the desired target state, $|\tau\rangle$. (b) The fixed-point quantum search guarantees that $\mathbf{U}\mathbf{R}_\gamma\mathbf{U}^\dagger\mathbf{R}_\tau\mathbf{U}\,|\gamma\rangle$ always moves toward the target state $|\tau\rangle$.

Proposition. *If when we apply the unitary transformation* $\mathbf{U}$ *the probability of reaching the state* τ *from* γ *is* $p_{\tau\gamma}^{(U)} \quad \mathbf{U}_{\tau\gamma} \quad ^2 \quad (1 \quad \epsilon)$, *then when we apply the transformation* $\mathbf{Q}_{\tau\gamma}$ *to the system in the state* γ , $\mathbf{Q}_{\tau\gamma} \ \gamma \quad \mathbf{U}\mathbf{R}_\gamma\mathbf{U}^\dagger\mathbf{R}_\tau\mathbf{U} \ \gamma$, *the probability of reaching the state* τ *becomes* $(1 \quad \epsilon^3)$; *thus,*

$$\mathbf{U}_{\tau\gamma} \quad ^2 \quad (1 \quad \epsilon) \qquad \mathbf{Q}_{\tau\gamma} \quad ^2 \qquad \mathbf{U}\mathbf{R}_\gamma\mathbf{U}^\dagger\mathbf{R}_\tau\mathbf{U} \quad ^2 \quad (1 \quad \epsilon^3).$$

The proof of this proposition follows from a one-by-one application of the successive transformations implied by $\mathbf{Q}_{\tau\gamma}$. First, we compute $\mathbf{U} \ \gamma$, then we apply to the resulting state the transformation $\mathbf{R}_\tau \quad I \quad (1 \quad \exp(i\theta)) \ \tau \ \tau$, and so on. We use the fact that $\mathbf{U}_{\tau\gamma} \quad \tau \ \mathbf{U} \ \gamma$, $\mathbf{U}_{\tau\gamma} \quad \gamma \ U^{\ 1} \ \tau$ and $\mathbf{U}_{\tau\gamma} \quad ^2 \quad \mathbf{U}_{\tau\gamma}\mathbf{U}_{\tau\gamma}$. It follows that

$$\begin{aligned}\mathbf{Q}_{\tau\gamma} \ \gamma \quad & \mathbf{U}\mathbf{R}_\gamma\mathbf{U}^\dagger\mathbf{R}_\tau\mathbf{U} \ \gamma \\ & [\exp(i\theta) \quad \mathbf{U}_{\tau\gamma} \quad ^2 (\exp(i\theta) \quad 1)^2]\mathbf{U} \ \gamma \quad \mathbf{U}_{\tau\gamma}(\exp(i\theta) \quad 1) \ \tau \ .\end{aligned}$$

Since $\mathbf{U} \ \gamma$ deviates from the desired target state, with the probability $\epsilon \quad 1 \quad U_{\tau\gamma} \quad ^2$, it follows that the probability that this superposition deviates from the desired target state τ is

$$\begin{aligned}q_{\tau\gamma}^{(Q)} \quad & (1 \quad \mathbf{U}_{\tau\gamma} \quad ^2) \quad [\exp(i\theta) \quad \mathbf{U}_{\tau\gamma} \quad ^2 [\exp(i\theta) \quad 1]^2] \quad ^2 \\ & \epsilon \quad \exp(i\theta) \quad (1 \quad \epsilon)[\exp(i\theta) \quad 1]^2] \quad ^2 \\ & \epsilon^3.\end{aligned}$$

Thus, the probability of reaching the state τ after one application of $\mathbf{Q}_{\tau\gamma}$ is

$$p_{\tau\gamma}^{(Q)} \quad 1 \quad \epsilon^3.$$

Figure 1.39b shows that the successive applications of the fixed-point quantum search operator move the state closer to the target state. The fixed-point quantum search transformation can be defined recursively:

$$\begin{aligned}
\mathbf{Q}_{\tau\gamma}^{(0)} \quad & \mathbf{U}, \\
\mathbf{Q}_{\tau\gamma}^{(1)} \quad & \mathbf{Q}_{\tau\gamma}^{(0)}\mathbf{R}_\gamma(\mathbf{Q}_{\tau\gamma}^{(0)})^\dagger\mathbf{R}_\tau\mathbf{Q}_{\tau\gamma}^{(0)} \quad \mathbf{U}\mathbf{R}_\gamma\mathbf{U}^\dagger\mathbf{R}_\tau\mathbf{U}, \\
\mathbf{Q}_{\tau\gamma}^{(2)} \quad & \mathbf{Q}_{\tau\gamma}^{(1)}\mathbf{R}_\gamma(\mathbf{Q}_{\tau\gamma}^{(1)})^\dagger\mathbf{R}_\tau\mathbf{Q}_{\tau\gamma}^{(1)} \\
& \mathbf{U}(\mathbf{R}_\gamma\mathbf{U}^\dagger\mathbf{R}_\tau\mathbf{U})(\mathbf{R}_\gamma\mathbf{U}^\dagger\mathbf{R}_\tau^\dagger\mathbf{U})(\mathbf{R}_\gamma^\dagger\mathbf{U}^\dagger\mathbf{R}_\tau\mathbf{U})(\mathbf{R}_\gamma\mathbf{U}^\dagger\mathbf{R}_\tau\mathbf{U}), \\
\mathbf{Q}_{\tau\gamma}^{(3)} \quad & \mathbf{Q}_{\tau\gamma}^{(2)}\mathbf{R}_\gamma(\mathbf{Q}_{\tau\gamma}^{(2)})^\dagger\mathbf{R}_\tau\mathbf{Q}_{\tau\gamma}^{(2)} \\
& \mathbf{U}(\mathbf{R}_\gamma\mathbf{U}^\dagger\mathbf{R}_\tau\mathbf{U})(\mathbf{R}_\gamma\mathbf{U}^\dagger\mathbf{R}_\tau^\dagger\mathbf{U})(\mathbf{R}_\gamma^\dagger\mathbf{U}^\dagger\mathbf{R}_\tau\mathbf{U}) \\
& (\mathbf{R}_\gamma\mathbf{U}^\dagger\mathbf{R}_\tau\mathbf{U})(\mathbf{R}_\gamma\mathbf{U}^\dagger\mathbf{R}_\tau^\dagger\mathbf{U})(\mathbf{R}_\gamma^\dagger\mathbf{U}^\dagger\mathbf{R}_\tau^\dagger\mathbf{U}) \\
& (\mathbf{R}_\gamma\mathbf{U}^\dagger\mathbf{R}_\tau\mathbf{U})(\mathbf{R}_\gamma^\dagger\mathbf{U}^\dagger\mathbf{R}_\tau^\dagger\mathbf{U})(\mathbf{R}_\gamma^\dagger\mathbf{U}^\dagger\mathbf{R}_\tau\mathbf{U}) \\
& (\mathbf{R}_\gamma\mathbf{U}^\dagger\mathbf{R}_\tau\mathbf{U})(\mathbf{R}_\gamma\mathbf{U}^\dagger\mathbf{R}_\tau^\dagger\mathbf{U})(\mathbf{R}_\gamma^\dagger\mathbf{U}^\dagger\mathbf{R}_\tau\mathbf{U}) \\
& (\mathbf{R}_\gamma\mathbf{U}^\dagger\mathbf{R}_\tau\mathbf{U}), \\
\mathbf{Q}_{\tau\gamma}^{(4)} \quad & \ldots \\
\vdots
\end{aligned}$$

As we can see, the recursion relation is not a simple one, and we cannot derive a closed form expression for the m-th iteration.

We end this introductory chapter with a brief discussion of error models and quantum algorithms.

1.25 ERROR MODELS AND QUANTUM ALGORITHMS

Quantum errors affect the processing, storage, and transmission of quantum information. In this section we discuss models for errors affecting quantum computations described by quantum algorithms. The primary goal of a quantum algorithm is to produce a desired quantum state starting from a classical input state such as 0 . To obtain the result of a computation, we perform a measurement of some or of all the qubits affecting the output state.

A quantum algorithm can be described using a quantum network. A *quantum network* is a space-time diagram of the operations applied by the algorithm to each qubit. The computational state of an algorithm is maintained with arbitrary precision if and only if the error of each quantum gate of the quantum network implementing the algorithm is below a certain threshold. Only if this condition is satisfied could one rely on the results of a quantum computation.

Without loss of generality, we assume that errors occur at *given quantum network locations*, and we consider two classes of errors:

1. *Operational errors.* They may occur during each gate operation and may affect all the qubits the gate is operating on.
2. *Memory errors.* A quantum network is partitioned into qubit time units determined by the maximum execution time of a gate. A memory error occurs if there is a loss of information during a qubit time unit. There is an interesting relationship between parallelism and memory errors. If t is the longest interval of time a qubit can be in storage without any significant loss of information and q is the number of qubits actively involved in a computation (thus, the number of gates executing in parallel), then the minimum number of operations per unit of time must be considerably larger than q/t.

To construct computational quantum error models, we consider three aspects: (1) the type of error operators that may occur at a given location, (2) the nature of the mixture of error operators at a given location, and (3) whether the errors at different locations are independent or not. We make several simplifying assumptions:

- The operators used to transform a qubit are affected by a small number of errors; the error operators lead to either no error, bit-flip, phase-flip, or bit and phase-flip.
- Each error occurs at a given location. The actual behavior of a quantum network can be represented as a sum of networks, with linear error operators at each network location. This *error expansion* is not unique; it gives the correct input–output behavior of the network, yet it is not capable of accurately representing the intermediate states of a qubit. The final state of the computation is obtained by adding the final state of each of the networks.
- *Leakage errors*, manifestations of loss of amplitude in the two-dimensional Hilbert space, are nonexistent. This last assumption is unrealistic for some quantum systems. For example, in an ion trap the amplitude of a level may be lost to other levels that in turn are storing information for different operations [243, 246].

The error analysis is simpler if one stochastically and independently places a standard error at an error location. The stochastic assumption means the state of the environment associated with each element in the error expansion must be orthogonal. We present several quantum computational error models:

- *Independent stochastic errors.* At each error location one makes an independent random choice of either the identity operator or a stochastic combination of linear operators satisfying the unitarity assumption. The probability of an error at that location is equal to the probability of the stochastic combination.
- *Independent errors.* The error expansion is obtained by assigning a quantum operation to each error location. The quantum operation is expressed as a set of linear operators labelled by the states of the environment,

$$e_0 \; I \qquad \underset{i}{} \; e_i \; A_i,$$

with $\mathrm{tr}(A_i) \quad 0, \;\; i \quad 1$. The strength of the error associated with this quantum operation is

$$\underset{i}{} e_i \; A_i \qquad \sup \qquad \underset{i}{} e_i \; A_i \;\; \psi \quad .$$

We cannot talk about the probability of error, only about the error strength of the quantum operation.

- *Quasi-independent stochastic errors.* In this model an error occurs, with the probability p, and the error expansion is obtained as a stochastic sum such that the probability of all terms of the sum that have a nonidentity quantum operation at a given set of k error locations is at most p^k.
- *Quasi-independent errors.* Each summand of an error expansion is associated with a set of failed error locations, and all other error locations are associated with an identity operator.

The following proposition [243] relates quantum algorithms using perfect operations to algorithms with imperfect operations.

Proposition. *There exists a constant δ such that for every $\epsilon > 0$, a quantum algorithm using perfect operations can be converted to an equivalent quantum algorithm with imperfect operations, each with an error at most δ, such that the final error is at most ϵ. The overhead of the converted algorithm is polylogarithmic in $\epsilon^{\ 1}$ and the number of computational steps of the algorithm.*

1.26 HISTORY NOTES

Our knowledge about the physical world continually evolves, stimulated by new discoveries. In the seventeenth century Johannes Kepler formulated the laws of planetary motion, which were empirically adequate but did not explain why the planets move the way they do. Sir Isaac Newton provided an adequate justification for Kepler's laws based on the three laws of physics he had formulated by 1666 and his law of gravitational forces; yet, Newton's mechanics could not answer why the actions at a distance obey an inverse square law. This question could be answered only after Albert Einstein and David Hilbert developed the general relativity theory during the second decade of the twentieth century.[11]

The history of science is ripe with similar examples: The Dutch mathematician Hendrik Antoon Lorentz showed that Maxwell's equations for the electromagnetic field are invariant to the *Lorentz transformations* he had proposed in 1904, but are not invariant under Galilean transformation of Newtonian mechanics; Lorentz's transformations are rather counterintuitive, predicting length contraction and time dilation, while Galilean transformations are more intuitive and easier to comprehend. Only after Einstein postulated that the laws of physics are the same in all inertial frames of reference and that the speed of light is constant, independent of the source, did it become clear that Lorentz transformations are anchored in physical reality.

This brings up fundamental questions about our knowledge: How do we develop theories and models that are better than others? How do we distinguish true knowledge from inadequate knowledge? Is quantum mechanics an accurate description of the physical world? Is it the ultimate description of the physical world?

The discipline of philosophy that studies knowledge is called *epistemology*. If we examine the history of epistemology, we notice a shift from the theories that stressed the absolute, the permanent character of knowledge, to theories that emphasize the relativity, the situation-dependence, the evolution of knowledge and its active interference with the physical world.

The Greek philosopher Plato believed that knowledge is absolute and reflects a set of universal "ideas" or "forms." Aristotle, his disciple, emphasized the role of logic, gathering knowledge through rational reflection, and empiricism, which seeks knowledge through sensory perception. Aristotle's ideas were greatly admired during and immediately after the Renaissance, when empiricism and rationalism were fashionable. During the second half of the nineteenth century, Immanuel Kant introduced the human mind as an active originator of experience, rather than just a passive recipient of perception. Kant believed that knowledge results from organization of perceptual data on the basis of inborn "categories" such as space, time, and causality. These a priori categories are static, even though the basic concepts are subjective. Kant's philosophy attempted to provide philosophical grounds for Newton's mechanics by showing that classical physics is entirely compatible with transcendental conditions for objective knowledge. Niels Bohr was strongly influenced by Kant's philosophy.

The twentieth century was dominated by different flavors of *pragmatism*—e.g., logical positivism, and the "Copenhagen interpretation" of quantum mechanics. The common theme of pragmatism is that knowledge consists of "models" that allow us to approach "problem solving" in the simplest possible fashion. It is implicitly assumed that models abstract properties of the entities they describe and cannot, and should not, capture all properties of these entities. Thus, multiple parallel models, possibly contradictory, may exist at any instant of time, and we should always choose the model that helps us solve the problem at hand. The ultimate truth, the reality behind these models, the so-called "Ding an sich"[12], is not only unattainable, but also meaningless.

---

[11] On 25 November 1915, Einstein submitted his paper *The Field Equations of Gravitation*, which gives the correct field equations for general relativity. Five days earlier, Hilbert had submitted a paper *The Foundations of Physics*, which contained the correct field equations for gravitation. Hilbert's paper contains some important contributions to relativity not found in Einstein's work.

[12] "Ding an sich" is a German expression; its ad litteram translation is "the thing in itself."

While matter and energy preoccupied the minds of philosophers starting with Leucippus and Democritus several hundred years before our era and later preoccupied the minds of many generations of natural scientists, information, per se, became a subject of serious investigation only after the significant technological developments in communication in the late 1940s.

There is little wonder that information is not a central concept in quantum mechanics, or that information theory, as developed by Claude Shannon, is not concerned with the behavior of quantum systems capable of carrying information. The milestones that mark the inception of the information age happened in the second half of the twentieth century: the transistor was invented by William Shockley, John Bardeen, and Walter Brattain, just before Christmas in 1947; the first commercial computer, UNIVAC I, became operational in 1951 [285]; the DNA double-helix structure was discovered by Sir Francis Harry Compton Crick and James Dewey Watson in 1953; the first microprocessor, 4004, was produced by Intel in 1971.

It is fascinating to follow the evolution of the ideas leading to today's quest for processing quantum information using quantum mechanical devices and to see the tremendous progress in quantum information processing we have witnessed since the early 1980s. The last two decades mirror the exciting decades from the beginning of the twentieth century when the basic concepts of quantum mechanics were developed.

The stage was set by the introduction of the quantum theory in 1900 by Max Planck, which triggered a chain of scientific discoveries (Figure 1.40): In 1905, Albert Einstein developed the theory of the photoelectric effect; in 1911, Sir Ernest Rutherford presented the planetary model of the atom; in 1913, Niels Bohr introduced the quantum model of the hydrogen atom; in 1924, Prince Louis de Broglie related the momentum p of a particle with the wavelength λ of the wave associated with the particle, $p \quad \lambda/h$, with Planck's constant $h \quad 6.626 \quad 10^{34}$ Joule-second.

Quantum mechanics has its origins in the 1925 "matrix quantum mechanics" of Heisenberg [206]. Later that year, Max Born and Pascual Jordan used infinite matrices to represent basic physical quantities and developed a complete formalism for quantum mechanics. Inspired by de Broglie's wave-particle duality, Erwin Schrödinger introduced in 1926 the equation for the dynamics of the wave function [366]; the same year Schrödinger and Paul Dirac showed the equivalence of Heisenberg's matrix formulation with Schrödinger's wave function.

The study of information and of the relation of information with physics followed a sinuous path: in 1929, Leó Szilárd pioneered the study of the physics of information while analyzing the Maxwell demon [414]; in 1961, Rolf Landauer showed that the erasure of information is a dissipative process and provided a quantitative characterization of the process, known as Landauer's principle [264].

The study of computability was a response to David Hilbert's Entscheidungs problem formulated in the 1930s; Hilbert asked if there was a mechanical procedure for separating mathematical truths from mathematical falsehoods. While studying this problem Alonzo Church and Stephen Kleene introduced the λ-definable functions in 1936. Church proved that "every effectively calculable function (effectively decidable predicate) is general recursive" [95]. Alan Turing introduced the concept of the Turing machine and proved that every function that could be regarded as computable is computable by a Turing machine [428]. The Church-Turing principle is generally formulated as "every function which can be regarded as computable can be computed by a universal computing machine." In layman's terms, a universal computer is a single machine that can perform any physically possible computation.

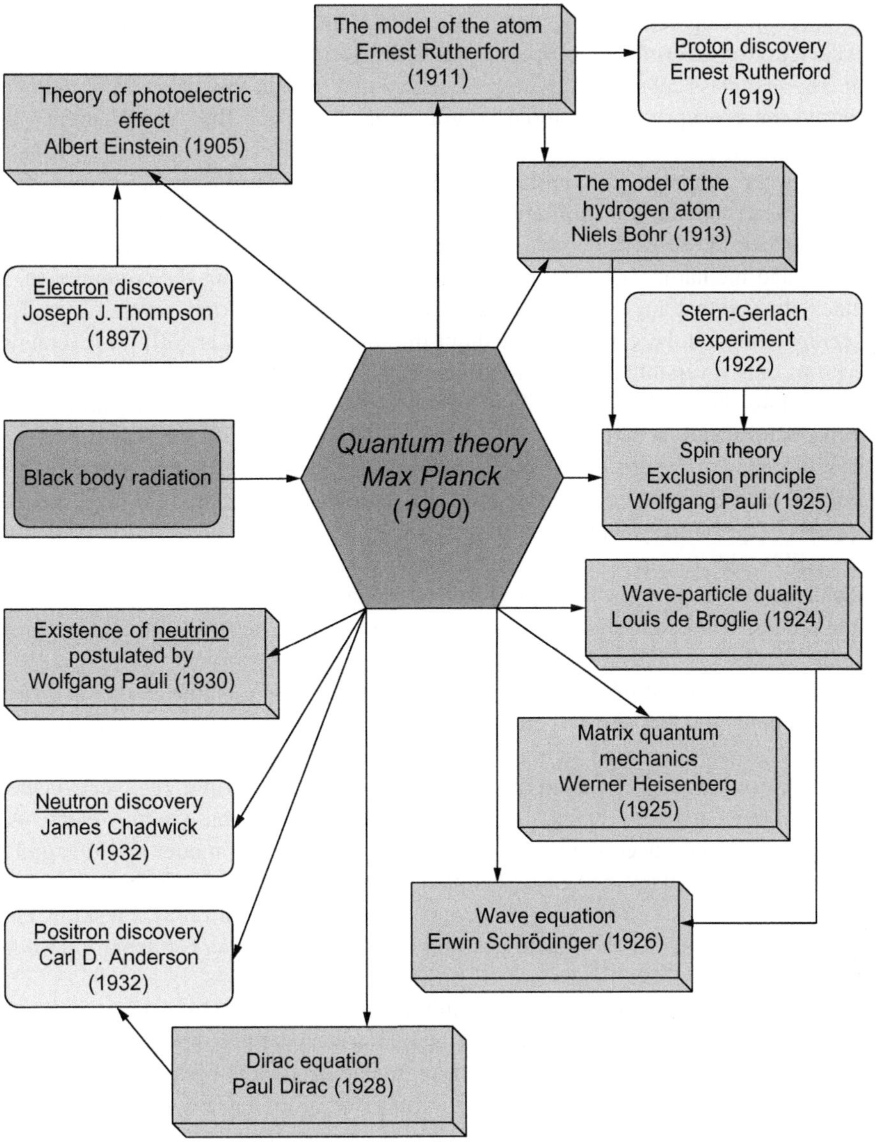

FIGURE 1.40

The chain of great discoveries in quantum physics during the *golden years*, 1900–1932. Theoretical models were developed to explain the results of the experiments. In turn, the theoretical models predicted new properties and phenomena and suggested new experiments to establish the validity of their predictions.

The theoretical developments in the mid-1930s were followed by the practical construction of mechanical devices for information processing and the formulation of basic principles for their construction. The first classical computers were designed and built in the early 1940s; John von Neumann introduced the concept of the "stored-program computer" and the "von Neumann computer architecture" in 1946 [79].

As faster and faster computers were built, more attention was paid to energy dissipation. Theoretical physicists argued that if there is a minimum amount of energy required to perform an elementary logical operation, then the faster a computer becomes, the more power is needed for the operation of the computer, and the harder it is to remove the heat generated during the computation. In 1973, Charles Bennett showed that any computation can be carried out using only reversible steps; thus, in principle a computation can be carried out without dissipating any power [30]. A reversible Turing machine performs only reversible computations as follows: it runs forward all the steps prescribed by the algorithm, outputs the result, and, finally, reverses all the steps to return to its initial state; at the end of this process no energy is dissipated.

The idea of using quantum effects for processing information followed a more direct path; in 1982, Richard Feynman envisioned the quantum computer, a physical device that takes advantage of the distinct properties of quantum systems to process information; Feynman conjectured that only a quantum computer will be able to carry out an "exact simulation" of a physical system [149]. In 1982, Paul Benioff recognized that the time evolution of an isolated quantum system described by the Hamiltonian is a reversible process and can mimic a reversible computation; he developed the idea of a quantum Turing machine [28, 29].

In 1985, David Deutsch introduced the concept of "quantum parallelism" and gave a concrete example of a computation that showed the distinction between classical and quantum computers [114]. In 1994, Peter Shor developed an algorithm for factoring large numbers [379] and generated a wave of excitement for the newly founded discipline of quantum computing. Two years later, in 1996, Lov Grover discovered a quantum search algorithm for an unsorted database with N elements.

Alexei Kitaev [240] introduced the topological quantum computing model in 1997, and in 2000, Michael Freedman, Alexei Kitaev, and Zhenghan Wang showed that the topological quantum computing has a computing power similar to that of the quantum circuit model [158]. Preskill's class notes cover a rigorous presentation of topological quantum computing [344]; an elementary introduction to the subject can be found in [102].

In 1995, David DiVincenzo formulated the requirements for practical implementations of a quantum information processing system [126]. Deutsch, Barenco, and Ekert observed in 1995 that "all computer programs may be regarded as symbolic representations of some of the laws of Physics, specialized to apply to specific processes, therefore the limits of computability coincide with the limits of science itself. If the laws of Physics did not support computational universality, they would be decreeing their own un-knowability" [117].

In the early 1990s, some 40 years after Shannon's development of information theory, Charles Bennett, Gilles Brassard, Richard Jozsa, Asher Peres, Benjamin Schumacher, Peter Shor, William, Wootters, and others introduced the basic concepts of quantum information theory [33–35, 37, 39, 40, 42, 44, 46–48]. Quantum information theory had to reconcile two different views regarding nondeterminism, the one that is the basic tenet of quantum mechanics and the one at the foundation of information theory. The nondeterminism of quantum mechanics reflects our *inability to know precisely the state of atomic or subatomic particles,* a very provoking thought contested bitterly by some, including Albert Einstein. Information theory and classical physics take a different view: *probabilities re ect lack of knowledge.* George Boole categorically declares that "probability is expectation founded on partial knowledge. A perfect acquaintance with all the circumstances affecting the occurrence of an event would change expectation into certainty, and leave neither room nor demand for a theory of probabilities" [64].

1.27 SUMMARY

A theme reverberating throughout the entire book is that *information is physical* and that information and physics are deeply intertwined. Classical and quantum information are anchored in physical reality; they are "engraved" onto some property of the physical system used to transmit, store, and transform information. Classical information we are familiar with is subject to the laws of classical physics, while quantum information obeys the laws of quantum mechanics.

The special properties of quantum information such as superposition and entanglement, the inability to clone quantum states, and the fact that a measurement projects the quantum state pose considerable challenges to quantum information processing and, at the same time, open intriguing possibilities for communication, storage, and computing with quantum information. For example, we can detect with high probability the presence of an intruder on a quantum communication channel, but at the same time, we have to measure the state of a quantum register to detect errors without altering its state.

Quantum information is transformed using quantum gates, the building blocks for quantum circuits which, in turn, can be assembled to build quantum computing and communication devices. A quantum bit, a *qubit*, is a quantum system with two distinguishable states; the state of a qubit is represented by a vector in a two-dimensional Hilbert space, $\mathcal{H}_2$. A quantum computer is a physical device designed to transform quantum information embodied by the state of a quantum system.

The physical processes required to transform the quantum state in a controlled manner are different for ion traps, solid-state, optical, NMR, or other possible physical implementations. The requirements for the physical implementation of quantum devices come naturally to mind. They have an immediate correspondent for classical computers: we cannot conceive a state-of-the-art computer built with circuits whose state cannot be controlled or be initialed to a desired state; it would be impractical to build a computer unless we have a finite set of building blocks; and it seems obvious that we should have access to the results of a computation.

An abstract computation could be viewed as the creation of symbols, the "output," which encode information according to some preexisting conventions in a systematic manner and have abstract properties specified by other symbols, the "input" of the process. Communication is a process whose output symbols are affected not only by the properties specified by the input but also by factors that are not entirely under our control. The "symbols" are physical objects and are therefore subject to the laws of physics.

An oracle is an abstraction for a black box that can follow a very elaborate procedure to answer a very complex question with a "yes" or "no" answer. Quantum computing has a natural affinity for oracles because the result of a quantum computation is probabilistic and it is a superposition of all possible results; increasingly complex oracles are the ones proposed by Deutsch, Deutsch-Jozsa, Bernstein-Vazirani, and Simon.

Quantum algorithms start from an initial state and then cause a set of state transformations of the quantum computer, which eventually lead to the desired result. Indeed, the first step for any quantum mechanical computation is to initialize the system to a state that we can easily prepare; then we carry out a sequence of unitary transformations that cause the system to evolve toward a state that provides the answer to the computational problem. A quantum operation is a rotation of the state ψ in N-dimensional Hilbert space. Thus, the ultimate challenge is to build up powerful N-dimensional rotations as sequences of one- and two-dimensional rotations. For any quantum algorithm, there are multiple paths leading from the initial to the final state, and there is a degree of interference among these paths. The amplitude of the final state (thus, the probability of reaching the desired final state) depends on the interference among these paths. This justifies the common belief that quantum algorithms are very sensitive to perturbations, and one has to be extremely careful when choosing the transformations the quantum mechanical system is subjected to.

Grover introduced a quantum algorithm for searching an unsorted database containing N items in a time of order $\overline{N}$, while on a classical computer the search requires a time of order N. The speedup of the Grover algorithm is achieved by exploiting both quantum parallelism and the fact that, according to quantum theory, a probability is the square of an amplitude. No classical or quantum algorithm can solve this problem faster than time of order $\overline{N}$. The Grover search algorithm can be applied directly to a wide range of problems.

1.28 EXERCISES AND PROBLEMS

Problem 1.1. Let A and B be two linear operators. Prove that

$$\text{tr}\ AA^{\dagger}\ \ \text{tr}\ BB^{\dagger} \qquad \text{tr}\ AB^{\dagger}\ {}^{2}.$$

This is the Schwarz inequality for the operator inner product.
Hint: Assume $A \quad [a_{ij}]$ and $B \quad [b_{ij}]$, $1 \quad i \quad n$, $1 \quad j \quad m$. Then:

$$\text{tr}\ AA^{\dagger} \qquad (a_{11}\ {}^{2} \qquad a_{1m}\ {}^{2}) \ \ (a_{21}\ {}^{2} \qquad a_{2m}\ {}^{2}) \qquad (a_{n1}\ {}^{2} \qquad a_{nm}\ {}^{2})$$

and

$$\text{tr}\ AB^{\dagger} \qquad (a_{11}b_{11} \qquad a_{1m}b_{1m}) \ \ (a_{21}b_{21} \qquad a_{2m}b_{2m}) \qquad (a_{n1}b_{n1} \qquad a_{nm}b_{nm}).$$

Use the Cauchy-Schwarz inequality: if $(\alpha_1, \alpha_2, \quad, \alpha_n, \beta_1, \beta_2, \quad, \beta_n) \quad \mathbb{C}$, then

$$\alpha_1\ {}^{2} \quad \alpha_2\ {}^{2} \qquad \alpha_n\ {}^{2} \quad \beta_1\ {}^{2} \quad \beta_2\ {}^{2} \qquad \beta_n\ {}^{2} \quad \alpha_1\beta_1 \quad \alpha_2\beta_2 \qquad \alpha_n\beta_n\ {}^{2}.$$

Problem 1.2. Show that $\mathbb{C}^{m \quad n}$, the set of all matrices, $A \quad [a_{ij}]$, with the elements, $a_{ij} \quad \mathbb{C}$, is a vector space. The addition of two matrices, $A \quad [a_{ij}]$ and $B \quad [b_{ij}]$, is defined as $A \quad B \quad [a_{ij} \quad b_{ij}]$, the inverse of $A \quad [a_{ij}]$ is $\ A \quad [\ \ a_{ij}]$, and the identity element is $E \quad [0]$.

Problem 1.3. Show that the set of $n \quad n$ square matrices, with elements from $\mathbb{C}$, form an inner product vector space. The inner product of two matrices is defined as

$$A, B \qquad \mathrm{tr}(A \; B),$$

with $A^{\cdot}$ the complex conjugate of the matrix A.

Problem 1.4. Show that opposite points on the Bloch sphere correspond to orthogonal qubit states.

Problem 1.5. Consider a point A on the Bloch sphere, with the coordinates x_A, y_A, z_A, and the vector connecting the origin of the sphere with A. Show that a rotation with an angle θ around this vector is described by:

$$\mathcal{R}_A(\theta) \quad \cos\frac{\theta}{2}\ \sigma_I \quad i\sin\frac{\theta}{2}\ (x_A\sigma_x \quad y_A\sigma_y \quad z_A\sigma_z),$$

Problem 1.6. Show that $0 , 1 , 2 , 3$ form an orthonormal basis in $\mathcal{H}_4$. Construct the operator, Π_4, that permutes circularly the basis in $\mathcal{H}_4$ as follows: $0 \quad 1 , 1 \quad 2 , 2 \quad 3$, and $3 \quad 0$. Construct its matrix representation. Calculate $\Pi_4 \ \psi$ and $\psi \ \Pi_4$, with $\psi \quad \alpha_0\ 0 \quad \alpha_1\ 1 \quad \alpha_2\ 2 \quad \alpha_3\ 3$. *Hint:* Express the canonical basis states as $00 , 01 , 10 , 11$.

Problem 1.7. The controlled-phase CPHASE gate in Figure 1.7b transforms the two input qubits in the states φ_1 and φ_2 as follows:

$$\varphi_1\varphi_2 \qquad (-1)^{\varphi_1\varphi_2}\ \varphi_1\varphi_2 \ .$$

Construct the matrix describing this transformation. Show that this gate is in fact a controlled-Z gate that flips the phase of the target qubit when the control qubit is set; show that the role of the target and control qubits could be reversed.

Problem 1.8. The Fredkin and Toffoli gates in Figure 1.41 operate on three qubits.

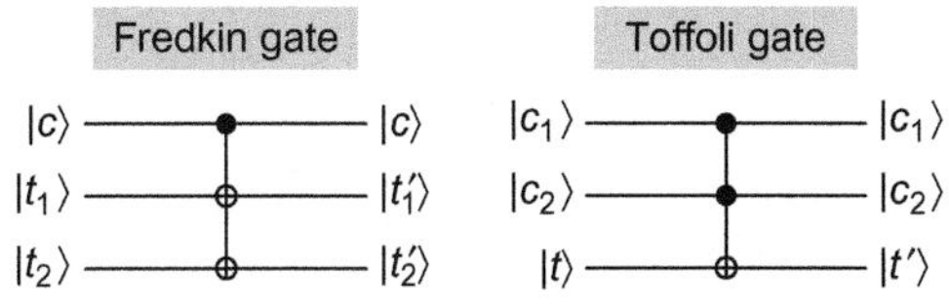

FIGURE 1.41

The Fredkin gate operates as follows: the control qubit, c , is transferred to the output unchanged; when $c \quad 0$, the target qubits, t_1 and t_2 , are transferred to the output unchanged; when $c \quad 1$, the two target qubits, t_1 and t_2 , are swapped. The Toffoli gate operates as follows: the control qubits, c_1 and c_2 , are transferred to the output unchanged; when $c_1 \quad c_2 \quad 0$, the target qubit is transferred to the output unchanged; when $c_1 \quad c_2 \quad 1$, the target qubit is flipped.

Show that the unitary transformations carried out by the Fredkin and Toffoli gates are, respectively:

$$U_{Fredkin} \quad \begin{matrix} 1 & 0 & 0 & 0 & 0 & 0 & 0 & 0 \\ 0 & 1 & 0 & 0 & 0 & 0 & 0 & 0 \\ 0 & 0 & 1 & 0 & 0 & 0 & 0 & 0 \\ 0 & 0 & 0 & 1 & 0 & 0 & 0 & 0 \\ 0 & 0 & 0 & 0 & 1 & 0 & 0 & 0 \\ 0 & 0 & 0 & 0 & 0 & 0 & 1 & 0 \\ 0 & 0 & 0 & 0 & 0 & 1 & 0 & 0 \\ 0 & 0 & 0 & 0 & 0 & 0 & 0 & 1 \end{matrix} \quad \text{and} \quad U_{Toffoli} \quad \begin{matrix} 1 & 0 & 0 & 0 & 0 & 0 & 0 & 0 \\ 0 & 1 & 0 & 0 & 0 & 0 & 0 & 0 \\ 0 & 0 & 1 & 0 & 0 & 0 & 0 & 0 \\ 0 & 0 & 0 & 1 & 0 & 0 & 0 & 0 \\ 0 & 0 & 0 & 0 & 1 & 0 & 0 & 0 \\ 0 & 0 & 0 & 0 & 0 & 1 & 0 & 0 \\ 0 & 0 & 0 & 0 & 0 & 0 & 0 & 1 \\ 0 & 0 & 0 & 0 & 0 & 0 & 1 & 0 \end{matrix} .$$

Hint: The Fredkin gate transforms the basis vectors in $\mathcal{H}_8$ as follows:

000 000 , 001 001 , 010 010 , 011 011 ,
100 100 , 101 110 , 110 101 , 111 111 .

The Toffoli gate transforms the basis vectors in $\mathcal{H}_8$ as follows:

000 000 , 001 001 , 010 010 , 011 011 ,
100 100 , 101 101 , 110 111 , 111 110 .

Problem 1.9. Show that classical Fredkin and Toffoli gates allow *fanout*; in other words, they can replicate one of their inputs at two different outputs, while the quantum Fredkin and Toffoli gates do not violate the *no cloning theorem*. *Hint:* Consider the case when the input of the two quantum gates are in the basis states, 0 and 1 , as well as the case when the input is in a superposition state, e.g., 1/ 2(0 1).

Problem 1.10. Provide an intuitive justification of the reason why a quantum circuit must have the same number of input and output qubits and relate your answer to the reversibility of quantum transformations. *Hint:* The transformation, **U**, carried out by the circuit is unitary, $\mathbf{UU}^{\dagger}$ $\mathbf{U}^{\dagger}\mathbf{U}$ **I**.

Problem 1.11. Construct the transfer matrix of the quantum circuit in Figure 1.42.

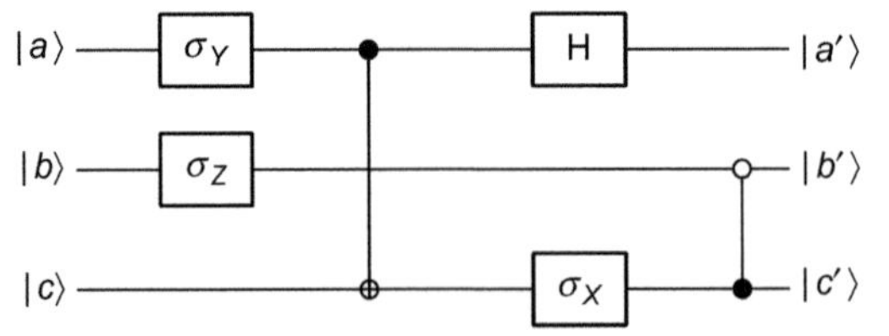

FIGURE 1.42

$\sigma_x, \sigma_y, \sigma_z$ are the transformations given by the Pauli matrices, and H is the Hadamard transform.

Problem 1.12. If $\sigma_I, \sigma_x, \sigma_y$, and σ_z are Pauli matrices and ρ is the density matrix of a qubit, show that

$$\frac{1}{2}\sigma_I \quad \frac{1}{4} \quad \rho \quad \sigma_x \rho \sigma_x \quad \sigma_y \rho \sigma_y \quad \sigma_z \rho \sigma_z \; .$$

Problem 1.13. Construct a quantum circuit able to solve the Deutsch-Jozsa problem.

Problem 1.14. By replacing selective inversions by selective phase shifts of $\pi/3$, the Grover algorithm preferentially converges to the target state irrespective of the step size, or the number of iterations [189]. Discuss the new algorithm and explain its novel features.

Problem 1.15. Show that if $H(n)$ is a Hadamard matrix, then so is

$$H(2n) \quad \begin{matrix} H(n) & H(n) \\ H(n) & H(n) \end{matrix} \; .$$

Hint: Show that

$$H(2n)H(2n)^{\dagger} \quad 2nI,$$

with I, the $n \quad n$ identity matrix.

Problem 1.16. Consider the transformation, **Q**, defined in Section 1.24 as

$$\mathbf{Q} \quad \mathbf{I}_{\gamma}\mathbf{U} \; ^{1}\mathbf{I}_{\tau}\mathbf{U}.$$

Initially the system is in the state γ , and we wish to force it to the state τ through a sequence of unitary transformations **U**. We apply the new transformation **Q** to the initial state γ and to the state $\delta \quad \mathbf{U} \; ^{1} \; \tau$, obtained by applying the inverse of **U** to the target state.

Show that

$$\mathbf{Q}(\; \gamma \;) \quad \gamma \; (1 \quad 4 \; U_{\tau\gamma} \; ^{2}) \quad 2U_{\tau\gamma}(\mathbf{U} \; ^{1} \; \tau \;)$$

and

$$\mathbf{Q}(\mathbf{U} \; ^{1} \; \tau \;) \quad \mathbf{U} \; ^{1} \; \tau \quad 2U_{\tau\gamma} \; \gamma$$

Here, $U_{\tau\gamma}$ is the amplitude of reaching the state τ starting from the state γ after one application of the unitary transformation U:

$$U_{\tau\gamma} \quad \tau \; \mathbf{U} \; \gamma \; .$$

Hint: Use the fact that **U** is unitary $\mathbf{UU} \; ^{1} \quad \mathbf{I}$. Recall also that

$$\gamma \; \gamma \quad 1 \quad \text{and} \quad \tau \; \tau \quad 1$$

and that

$$\mathbf{U}_{\tau\gamma} \quad (\; \tau \; \mathbf{U} \; \gamma \;)^{\dagger} \quad \gamma \; \mathbf{U} \; ^{1} \; \tau \; .$$

We have also shown that $\mathbf{I}_{\gamma} \quad \mathbf{I} \quad \gamma \; \gamma$ and $\mathbf{I}_{\tau} \quad I \quad \tau \; \tau$.

CHAPTER

Measurements and Quantum Information

2

Not everything that can be counted counts; not everything that counts can be counted.
Albert Einstein

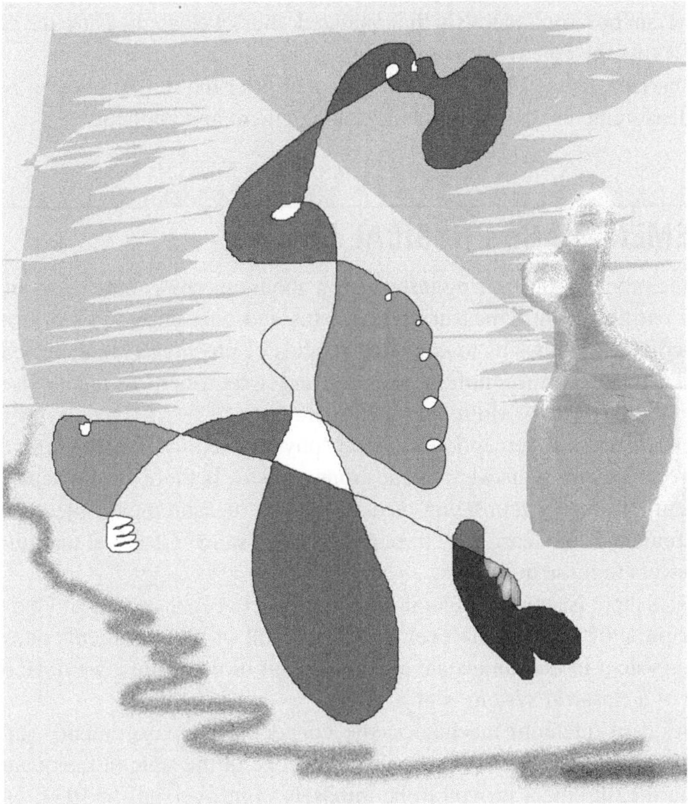

Quantum information processing requires a deeper understanding of the measurement process and its effects on a quantum system; measurements allow us to observe the outcome of transformations carried out by quantum circuits and quantum communication channels. This justifies the important role measurements play in quantum information theory and the extended coverage of measurements of quantum systems in this chapter.

Recall that a quantum measurement is an irreversible process that transforms quantum information to classical information and that there is a strong coupling between the system being measured and the measuring system. The measurements postulate tells us that the outcome of a measurement of an observable, A, of a quantum system is an eigenvalue of the self-adjoint operator, $\mathbf{A}$, associated with the measurement of the observable. Operator $\mathbf{A}$ is a *projector* for observable A of a quantum system. The quantum state immediately after the measurement is an eigenvector of the operator $\mathbf{A}$. We say that a quantum measurement *collapses the state* of a quantum system.

We have seen in Chapter 1 that *we cannot simultaneously measure observables associated with noncommuting operators;* if two operators, $\mathbf{A}$ and $\mathbf{B}$, do not commute, then measuring observable A using projector $\mathbf{A}$ will interfere with the eigenvectors of projector $\mathbf{B}$ and vice versa. Indeed, only two commuting matrices can be simultaneously diagonalized; thus, we can measure the eigenvalues of one of them without affecting the eigenvectors of the other.

Figure 2.1 summarizes the organization of the chapter; after the introduction of basic concepts, we analyze quantum measurements, then discuss the entanglement and applications.

2.1 MEASUREMENTS AND PHYSICAL REALITY

A measurement is the process of acquiring information about our environment, about physical objects, phenomena, or any entities using a measuring apparatus and our senses; this information eventually evolves into knowledge and allows us to construct models of physical systems. Measurements are an important step in the process of scientific discovery; any experiment to assess the correctness of a theoretical model of the physical world requires measurements.

The relationship between our perceptions and the "physical reality" surrounding us is central to the theory of knowledge and requires answers to questions such as, Is there a clear separation between the measuring instrument and the system being measured? Is a measurement objective? Are we able to make "ideal measurements" and learn the "true state" of a system? Classical and quantum physics do not give similar answers to these questions.

In classical physics there is an unambiguous distinction between a system and the measuring instrument. Also, a description of the system developed as a result of measurements does not depend on a measurement of the system in the same state at a later point in time; thus, we have reasons to believe that the description of a classical system is objective.

From the inception of quantum mechanics, the concept of "measurement" and the relationship between the outcome of the measurement and our knowledge of the state of the quantum system prior, during, and after the measurement proved to be intensely controversial. In 1934, Niels Bohr made a stunning observation regarding the measurement of quantum systems:

> *A subsequent measurement to a certain degree deprives the information given by a previous measurement of its significance for predicting the future course of phenomena. Obviously, these facts not only set a limit to the extent of the information obtainable by measurement, but they also set a limit to the meaning which we may attribute to such information. We meet here in a new light the old truth that in our description of nature the purpose is not to disclose the real essence of the phenomena, but only to track down so far as possible, relations between the manifold aspects of our experience.*

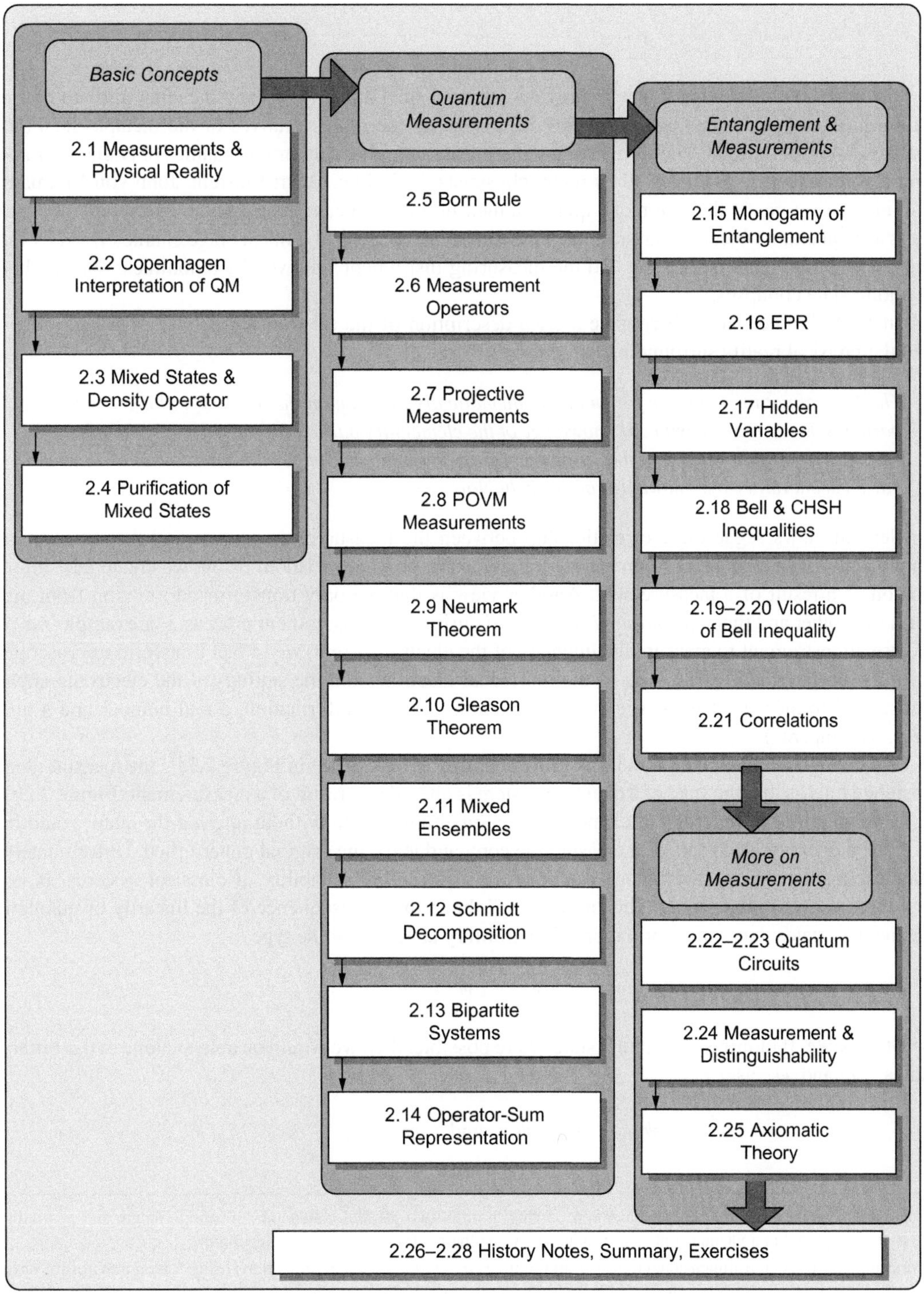

FIGURE 2.1

The organization of Chapter 2 at a glance.

The question of whether a numerical value of a physical quantity has any meaning until an observation has been performed was answered differently by the representatives of the metaphysical and realist schools of thought. The Copenhagen interpretation of the quantum mechanics of Niels Bohr and Werner Heisenberg was contested by many physicists, including Albert Einstein. John von Neumann provided an axiomatic definition of a quantum measurement process; he treats the measurement apparatus as a quantum object and argues that a measurement consists of a linear stage characterized by the joint state of the quantum system and the measuring instrument followed by a nonlinear stage, when this joint state collapses.

In 1958, Werner Heisenberg gave a crisp description of the relationship between our knowledge and the physical reality surrounding us:

> *The laws of nature which we formulate mathematically in quantum theory deal no longer with the particles themselves but with our knowledge of the elementary particles. ... The conception of objective reality... evaporated into the ... mathematics represents no longer the behavior of elementary particles but rather our knowledge of this behavior.*

Several views regarding the relationship between the measurement process and the information obtained as a result of a measurement are possible. The common wisdom is that we create new information as a result of a measurement. Another view is that we only transform information from one form to another and that *information is conserved* during the measurement process. For example, when we use an instrument to measure the intensity of the electric current, we in fact transform microscopic information provided by the state of the individual electrons and the settings of the electronic apparatus, used to measure the electric current, to another type of information, a real number and a unit (e.g., 3.5 ampere[1]).

Projecting an object leads to a loss of information, as we can see in Figure 2.2a[2]; the measurement postulate tells us that the state of a quantum system is altered as a result of a measurement (Figure 2.2b). A naive solution to the problem of measuring a quantum system without altering the state would be to clone a quantum state and then measure the copy and leave the original unperturbed. Unfortunately, state replication, one of the favorite methods to increase the reliability of classical systems, is not possible for quantum systems. The "no-cloning" theorem, a consequence of the linearity of quantum mechanics, states that a transformation of an arbitrary state ψ of the type

$$\psi \quad \psi \quad \psi$$

is not possible. If cloning of quantum states were possible, then we would be able to clone two arbitrary states, ψ and φ, as

$$\psi \quad \psi \quad \psi \quad \text{and} \quad \varphi \quad \varphi \quad \varphi .$$

---

[1]One ampere is the constant current that when owing through two parallel conductors of infinite length and negligible cross-section placed in a vacuum 1 m apart produce between them a force of 2 10 7 Newton/m.

[2]We can reconstruct an n-dimensional object from multiple (n 1)-dimensional projections using Radon transforms, or its complex analog, the Penrose transform; CAT scan is an application of this process.

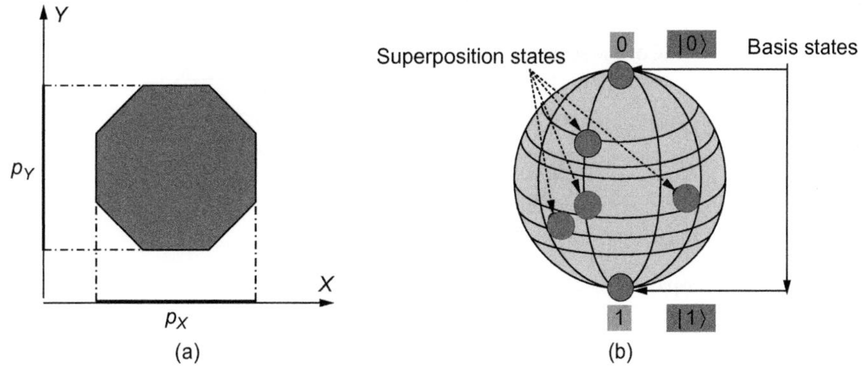

FIGURE 2.2

(a) Information is lost when we project an object; we cannot recognize the shape of a two-dimensional object from a one-dimensional projection. (b) A quantum measurement projects quantum information on a subspace of the original Hilbert space and replaces quantum information with classical information; the measurement of a qubit in a superposition state $|\varphi\rangle = \alpha_0|0\rangle + \alpha_1|1\rangle$ projects it on one of the basis states, $|0\rangle$ or $|1\rangle$, and reveals classical information, 0 or 1.

By linearity, it follows that

$$\alpha|\psi\rangle + \beta|\varphi\rangle \rightarrow \alpha|\psi\rangle \otimes \alpha|\psi\rangle + \beta|\varphi\rangle \otimes \beta|\varphi\rangle.$$

On the other hand, a quantum copy machine would then be able to clone the superposition state, $\alpha|\psi\rangle + \beta|\varphi\rangle$, as

$$\alpha|\psi\rangle + \beta|\varphi\rangle \rightarrow (\alpha|\psi\rangle + \beta|\varphi\rangle) \otimes (\alpha|\psi\rangle + \beta|\varphi\rangle),$$

as shown in Figure 2.3. But we see that

$$\alpha|\psi\rangle \otimes \alpha|\psi\rangle + \beta|\varphi\rangle \otimes \beta|\varphi\rangle \neq (\alpha|\psi\rangle + \beta|\varphi\rangle) \otimes (\alpha|\psi\rangle + \beta|\varphi\rangle).$$

We conclude that a *quantum copy machine able to clone two arbitrary quantum states, $|\psi\rangle$ and $|\phi\rangle$, cannot clone a superposition of the two states, $|\xi\rangle = \alpha|\psi\rangle + \beta|\phi\rangle$.*

Surprisingly, the fact that a measurement alters the state of a quantum system could be very beneficial; one cannot eavesdrop on a quantum communication channel without being detected, and this has important applications in computer security. Quantum key distribution protocols exploit this property of quantum information.

The rejection of nondeterminism, a basic tenet of quantum mechanics, has motivated the development of several alterative models such as the hidden variable theory, discussed at length in Section 2.17. The Bell theorem provides the theoretical foundation for the study of correlations of distant events and enables us to show the nonexistence of "local hidden variables." The Bell inequality is derived using a very simple model of a physical system that makes only two common-sense assumptions: physical properties are independent of observations, the *realism principle*, and the measurement of different physical properties of different objects carried out by different observers at distinct locations cannot

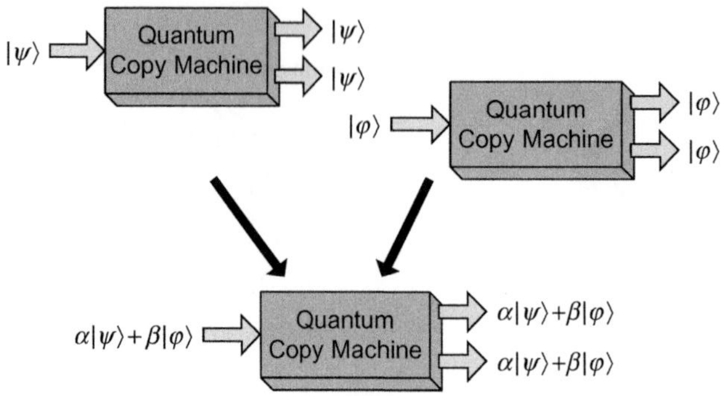

FIGURE 2.3

A quantum copy machine able to clone two arbitrary states, ψ and φ . It could also clone the superposition state, (α ψ β φ). Then the output of the quantum copy machine would be the tensor product of the original superposition state and its clone, (α ψ β φ) (α ψ β φ), different than the value one would expect due to the linearity of quantum state transformations, (α ψ α ψ β φ β φ).

in uence each other, *the locality principle*. John Bell proved that independently of any specific physical theory *there is an upper limit to the correlation of distant events* if the principle of locality is valid; any hidden variable account of the quantum theory applied to quantum systems consisting of distant subsystems must have the separability property; thus, no such hidden variable theory can be correct. Bell showed that a deterministic model based on hidden variables generates results whose averages are identical to those predicted by quantum mechanics, but the hidden variables do not satisfy the separability requirement discussed in Section 2.20.

While no experiment can prove a theory to be right, a single experiment could prove a theory to be wrong; no experiment carried out from the mid-1920s until this day has produced results contradicting the prediction of quantum mechanics. Yet, quantum systems violate Bell inequality, and the only logical conclusion is that one of the two assumptions of the Bell model, or possibly both, must be false. It is hard to accept that physical properties depend on our observations and discount the realism; alternatively, can we discount the locality?

2.2 COPENHAGEN INTERPRETATION OF QUANTUM MECHANICS

Quantum mechanics, perhaps one of the most successful scientific theories in the history of mankind, is in some respects counterintuitive, challenges our imagination, seems to violate fundamental principles of classical physics, and raises many questions that have preoccupied the minds of physicists and philosophers since the early 1930s. Among them: Is there an objective reality? What is the role of the experiment? How accurate can our knowledge about physical objects be? What is the relationship between quantum mechanics and classical physics?

Classical physics is deterministic; we know how a system will evolve in time if we determine the initial state of the physical system by observing its properties at the time we choose to be the initial moment, as well as all external forces acting on the system. Moreover, the act of observing the system does not affect its later behavior. In classical physics there is a clear and unambiguous distinction between the system and the measuring instrument. The description of a physical system is objective; it does not depend on measurements carried out at a later time.

The *Copenhagen interpretation of quantum mechanics* is the name given to a set of principles formulated by Niels Bohr, Werner Heisenberg, Max Born, and several other physicists who spent time at the Institute for Theoretical Physics of the University of Copenhagen[3] in their attempt to answer the questions raised by quantum mechanics. Niels Bohr had a very practical approach to these questions; he believed that quantum mechanics is a natural generalization of classical physics, that classical concepts did not change their meaning, but that their application was restricted. Early on, Bohr observed that his quantum theory of the hydrogen atom leads to results that coincide approximately with the ones provided by classical electrodynamics. He then postulated the so-called *correspondence rule*, a heuristic principle stating that the behavior of systems described by quantum mechanics reproduces classical physics in the limit of large quantum numbers. The correspondence principle stipulates that a new scientific theory should be able to explain the phenomena under the condition when the earlier theory is valid. For example, Einstein's special relativity satisfies the correspondence principle because it reduces to classical mechanics in the limit of velocities small compared to the speed of light; general relativity reduces to Newtonian gravity in the limit of weak gravitational fields; the Laplace theory of celestial mechanics reduces to Kepler's when interplanetary interactions are ignored; Kepler's reproduces Ptolemy's equation in a coordinate system where the Earth is stationary; and statistical mechanics reproduces thermodynamics when the number of particles is large. Paul Feyerabend, another philosopher of science, believes that *succeeding theories like classical and quantum mechanics are incommensurable, and that there is partial lack of rationality in the choice between incommensurable theories.*

Complementarity is a principle formulated by Niels Bohr in 1928. Complementarity re ects the de Broglie duality principle; photons, electrons, and other atomic or subatomic objects sometimes exhibit wave-like behavior, and at other times, particle-like behavior. Sometimes it is possible to observe simultaneously both types of behavior (e.g., in the case of double-slit experiments). Bohr expressed the view that the complementarity of kinematic and dynamic properties of quantum particles can only be observed in mutually exclusive experiments.

The Copenhagen interpretation refers to concepts such as Bohr complementarity and the correspondence principle, Born statistical interpretation of the wave function, and nondeterminism. Almost synonymous with nondeterminism, the Copenhagen interpretation aims to account for the violation of the basic principles that classical physics rests on [475]:

- *Causality:* Every event has a cause.
- *Determinism:* Every later state of a system is uniquely determined by any earlier state.
- *Continuity:* When a process evolves from an initial to a final state it has to go through all intervening states.

[3]The Institute for Theoretical Physics of the University of Copenhagen was named "Niels Bohr Institute" on 7 October 1965 on Niels Bohr's 80th birthday.

- *Conservation of energy:* The energy of a closed system can be transformed into various forms but is never gained, lost, or destroyed.
- *The principle of space and time:* Physical objects exist separately in space and time in such a way that they are localizable and countable. Processes that describe the evolution of physical objects take place in space and time.

Some of the tenets of the philosophy of Bohr are:

1. Theory must rely on experimental practice.
2. Experimental practice assumes a prescientific practice of (1) description that establishes the requirements for a measurement apparatus and defines what scientific experience means, and (2) understanding the role of the environment, the identification of physical objects. This pre-scientific experience defines primitive categories such as duration and change of duration, position of a physical object and change of position, and cause-and-effect relations. These categories are necessary for objective knowledge and must be part of any description of nature.
3. Classical physics is based on exact concepts derived from these categories. The concepts of classical physics allow us to understand and to communicate results of our experiments.
4. The quantization of the energy of a harmonic oscillator (implicitly, of the energy radiated by a blackbody) discovered by Planck requires a revision of the foundation for the use of classical concepts. The use of classical concepts is well defined only when effects of energy quantization are negligible, i.e., when the Planck constant can be neglected. Classical concepts can only be applied to phenomena that are macroscopic manifestations of a measurement performed on the object. They re ect the uncontrollable interaction between a measurement apparatus and the physical object.
5. The quantum mechanical description of an object requires that a separation line be drawn between the description of the object and the measurement apparatus. But, this line of separation is not between the macroscopic instrument and the microscopic object. Parts of the measurement apparatus may sometimes be included in the quantum mechanical description of the physical object.
6. The Schrödinger wave equation has a symbolic character. It cannot provide a visual description of physical reality, but it can be used to predict the outcome of a measurement.

We conclude this section with the words of Richard Feynman in *Lecture Notes on Physics* [148]:

The principle of science, the definition, almost, is the following: The test of all knowledge is the experiment. Experiment is the sole judge of the scientific truth. But what is the source of knowledge? Where do the laws to be tested come from? Experiment itself helps to produce these laws, in the sense that it gives us hints. But also needed is the imagination to create from these hints the great generalizations—to guess at the wonderful, simple, but very strange patterns beneath them all, and then to experiment to check again whether we have made the right guess.

The next topics we address cover the measurements of quantum systems and require a closer examination of the density operator.

2.3 MIXED STATES AND THE DENSITY OPERATOR

The density operator allows us to distinguish pure states from mixed states of a quantum system. *Pure states* are characterized by maximum knowledge; they can be expressed as superpositions of basis vectors of an orthonormal basis. *Mixed states* require a statistical characterization and describe:

1. Ensembles, or statistical mixtures, of pure states; the system can be in any of the pure states, ψ_1 , ψ_2 , ψ_3 ,..., with the probabilities $p_1, p_2, p_3, \ldots$.
2. Composite systems; for example, two systems, $\mathcal{A}$ and $\mathcal{B}$, components of a larger system, $\mathcal{AB}$, in an entangled pure state.

The density operator of a system in a pure state ψ is ρ $\quad \psi \quad \psi$, while the density operator of the ensemble of pure states, ψ_1 , ψ_2 , ψ_3 ,..., with the probabilities $p_1, p_2, p_3, \ldots$, is

$$\rho \quad \sum_{(i)} p_i \; \psi_i \quad \psi_i \, .$$

Given the bipartite system, $\mathcal{AB}$, in the entangled pure state, ψ_{AB} , the density operator of $\mathcal{A}$ captures only the information available to an observer who has infinitely many opportunities to the examine subsystem $\mathcal{A}$; the *reduced density operator of subsystem* $\mathcal{A}$ is defined as

$$\rho_A \quad \mathrm{tr}_B(\; \psi_{AB} \quad \psi_{AB} \;).$$

Example. *A quantum system in a mixed state.* To relate the concept of a mixed state with physical systems, we discuss brie y the Stern-Gerlach experiment. Recall that the spin of an atomic or subatomic particle is the intrinsic angular momentum of the particle. The spin is quantified; the spin quantum number, s, takes discrete values. There are two types of particles: *fermions*, or spin-½ particles, with $s \quad 1/2$; and *bosons*, or integer-spin particles, with $s \quad 0, \quad 1$; for example, the electrons are fermions, while the photons are bosons.

The spin interacts with a magnetic field, and the Stern-Gerlach experiment shows that a beam of hydrogen atoms consists of a mix of atoms with two different spin values. The experimental setup in Figure 2.4 uses a nonuniform magnetic field whose z-axis component is normal to the planar cap; the components of the magnetic field along the x- and y-axis are negligible. The atomic beam entering the apparatus consists of a statistical mixture of hydrogen atoms in the spin states, and , de ected upwards and, respectively, downwards. If the atomic beam consists of N particles and N is the number of atoms de ected upwards, then for very large N, the probability of an atom to have its spin up approaches the ratio, $N \;/N$; similarly, if N is the number of atoms de ected downwards, the probability of an atom to have its spin down approaches the ratio, $N \;/N$.

In the Stern-Gerlach experiment, a *pure spin state* corresponds to a *completely polarized beam*, while a statistical mixture corresponds to either a *partially polarized beam* when the probabilities of possible states are unequal, or to an *unpolarized beam* if the probabilities of the states are equal. The value of an observable cannot be predicted with certainty. Yet, we have to make a distinction between the probabilities associated with a pure state and the probabilities associated with impure states or statistical mixtures. For example, in the case of the spin-½ particles we just discussed, there are

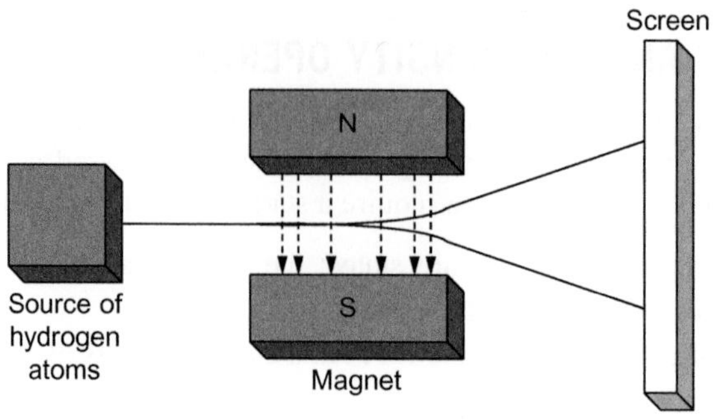

FIGURE 2.4

The hydrogen atomic beam entering the apparatus used in the Stern-Gerlach experiment consists of a statistical mixture of atoms in the spin states, and , de ected upwards and, respectively, downwards.

two possible pure spin states, and , each with density matrix, ρ and ρ , respectively. If $N \quad N \quad N$, then $p \quad N\ /N$, $p \quad N\ /N$, and $p \quad p \quad 1$. The density matrix of the mixed state is $\rho \quad p\ \rho \quad p\ \rho$.

After this qualitative discussion of pure and mixed states, we switch our focus to properties of the density operator; recall from Section 1.11 that the density operator is self-adjoint. It is easy to show that any self-adjoint 2 2 matrix, ρ, can be represented as

$$\rho \quad \frac{1}{2}(\sigma_I \quad r_x\sigma_x \quad r_y\sigma_y \quad r_z\sigma_z) \quad \frac{1}{2}\begin{pmatrix} 1 \quad r_z & r_x \quad ir_y \\ r_x \quad ir_y & 1 \quad r_z \end{pmatrix},$$

with $r_x, r_y, r_z \quad \mathbb{C}$ and $\sigma_I, \sigma_x, \sigma_y, \sigma_z$ and the Pauli matrices

$$\sigma_I \quad \begin{pmatrix} 1 & 0 \\ 0 & 1 \end{pmatrix}, \quad \sigma_x \quad \begin{pmatrix} 0 & 1 \\ 1 & 0 \end{pmatrix}, \quad \sigma_y \quad \begin{pmatrix} 0 & i \\ i & 0 \end{pmatrix}, \quad \sigma_z \quad \begin{pmatrix} 1 & 0 \\ 0 & 1 \end{pmatrix}.$$

Then r_x, r_y, and r_z are equal to

$$r_x \quad \rho_{12} \quad \rho_{21}, \quad ir_y \quad \rho_{21} \quad \rho_{12}, \quad r_z \quad \rho_{11} \quad \rho_{22} \quad \text{if } \rho \quad \begin{pmatrix} \rho_{11} & \rho_{12} \\ \rho_{21} & \rho_{22} \end{pmatrix}.$$

The *trace of the density operator* has the following properties:

$$\text{tr}(\rho) \quad 1, \quad \text{tr}\left(\rho^2\right) \quad \frac{1}{2}\left(1 \quad r^2\right), \quad \text{where} \quad r^2 \quad r_x^2 \quad r_y^2 \quad r_z^2 \quad 1.$$

Proof. We use the previous expression of the density operator and construct the trace of the density operator as the sum of the diagonal elements:

$$\text{tr}(\rho) \quad \frac{1}{2}[(1 \quad r_z) \quad (1 \quad r_z)] \quad 1.$$

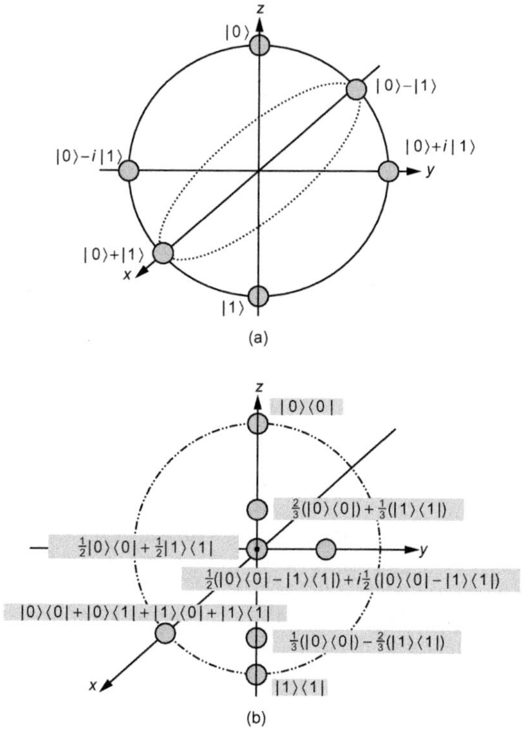

FIGURE 2.5

Bloch sphere and Bloch ball. (a) Pure states are points on the surface of a Bloch sphere. (b) Pure and mixed states are points on the surface of or inside a Bloch ball, respectively.

It is easy to see that

$$\rho^2 \quad \frac{1}{2}\begin{pmatrix} 1 \quad r_z & r_x \quad ir_y \\ r_x \quad ir_y & 1 \quad r_z \end{pmatrix} \frac{1}{2}\begin{pmatrix} 1 \quad r_z & r_x \quad ir_y \\ r_x \quad ir_y & 1 \quad r_z \end{pmatrix} \quad \frac{1}{4}\begin{bmatrix} 1 \quad r^2 \quad 2r_z & 2(r_x \quad ir_y) \\ 2(r_x \quad ir_y) & 1 \quad r^2 \quad 2r_z \end{bmatrix}.$$

Thus,

$$\mathrm{tr}\left(\rho^2\right) \quad \frac{1}{2}\left(1 \quad r^2\right).$$

The condition, $\det(\rho) \quad 0$, becomes

$$\det(\rho) \quad \frac{1}{4}\left[\left(1 \quad r_z^2\right) \quad \left(r_x^2 \quad r_y^2\right)\right] \quad \frac{1}{4}\left(1 \quad r^2\right) \quad 0.$$

Thus, $r^2 \quad 1$. This expression implies that there is a one-to-one correspondence between the possible density matrices of a single qubit and points either on the surface of the Bloch sphere, $r \quad 1$, or within the Bloch ball, $0 \quad r < 1$ (Figure 2.5). ■

Properties of the Density Operator

The density operator has several properties:

1. If a qubit is in a pure basis state, then it can be represented as a point on the Bloch sphere, det(ρ) 0, and its density matrix could only be

$$\rho \quad \begin{pmatrix} 1 & 0 \\ 0 & 0 \end{pmatrix} \quad \text{or} \quad \rho \quad \begin{pmatrix} 0 & 0 \\ 0 & 1 \end{pmatrix}.$$

2. If the density operator of a single qubit has only one nonzero eigenvalue, then the state is a pure state. If the density operator has more than one nonzero eigenvalue, then the state is a mixed state.
3. Infinitely many ensembles of pure states may be characterized by the same density operator.
4. The density operator of an equal mixture of two orthogonal single qubit pure states is

$$\rho_{eq} \quad \frac{1}{2}\begin{pmatrix} 1 & 0 \\ 0 & 1 \end{pmatrix} \quad \rho_{ne},$$

with ρ_{ne}, the density operator of a mixture of states with nonequal probabilities.

Proof. For pure states, r^2 1; thus, det(ρ) 1 r^2 0. The determinant of the density operator of states represented as points on the surface of the Bloch (solid) sphere is equal to zero. Any 2 2 density operator ρ can be diagonalized as:

$$\rho \quad \begin{pmatrix} \alpha_{11} & 0 \\ 0 & \alpha_{22} \end{pmatrix}.$$

Then det(ρ) $\alpha_{11}\alpha_{22}$ 0; the trace of the density operator must be one, tr(ρ) α_{11} α_{22} 1. The two equations have two sets of solutions:

$$\alpha_{11} \quad 1, \alpha_{22} \quad 0 \quad \text{or} \quad \alpha_{11} \quad 0, \alpha_{22} \quad 1.$$

We conclude that the density operator of a pure basis state could only be

$$\rho \quad \begin{pmatrix} 1 & 0 \\ 0 & 0 \end{pmatrix} \quad \text{or} \quad \rho \quad \begin{pmatrix} 0 & 0 \\ 0 & 1 \end{pmatrix}.$$

We notice that the first expression of the density operator corresponds to ρ 0 0 and the second to ρ 1 1 . The density operator of a pure state can be represented as the sum of outer products of basis states. ■

Example. The density matrix of the state φ (1/ $\bar{2}$)(0 1) is

$$\rho_\varphi \quad \varphi \quad \varphi \quad \frac{0\ 0 \quad 0\ 1 \quad 1\ 0 \quad 1\ 1}{2} \quad \frac{1}{2}\begin{pmatrix} 1 & 1 \\ 1 & 1 \end{pmatrix}.$$

The diagonalization of the density operator may have caused a rotation of the basis vectors. If we call the new basis vectors, 0 and 1 , resulting from the rotation of 0 and 1 , respectively, then the density operator of a pure state can only be ρ 0 0 or ρ 1 1 .

Example. An equal mixture, (p_0 p_1 1/2), of the pure states of a single qubit, 0 and 1 , and an equal mixture of the pure states, (1/ $\overline{2}$)(0 1) and (1/ $\overline{2}$)(0 1), have the same density operator.

The density operator of the first mixture is

$$\rho \quad \sum_i p_i \rho_i \quad \frac{1}{2}\rho_0 \quad \frac{1}{2}\rho_1 \quad \frac{1}{2} 0 \; 0 \quad \frac{1}{2} 1 \; 1 \quad \frac{1}{2}\begin{pmatrix} 1 & 0 \\ 0 & 0 \end{pmatrix} \quad \frac{1}{2}\begin{pmatrix} 0 & 0 \\ 0 & 1 \end{pmatrix} \quad \frac{1}{2}\begin{pmatrix} 1 & 0 \\ 0 & 1 \end{pmatrix}$$

and

$$\mathrm{tr}\,(\rho_0) \quad (1 \quad 0) \quad 1 \quad \text{and} \quad \mathrm{tr}\,(\rho_1) \quad (0 \quad 1) \quad 1.$$

The density operator of the equal mixture of the states 1/ $\overline{2}$(0 1) and 1/ $\overline{2}$(0 1) is

$$\rho \qquad \frac{1}{\overline{2}}\begin{pmatrix} 1 \\ 1 \end{pmatrix} \frac{1}{\overline{2}}\begin{pmatrix} 1 & 1 \end{pmatrix} \quad \frac{1}{2}\begin{pmatrix} 1 & 1 \\ 1 & 1 \end{pmatrix}, \quad \mathrm{tr}\,(\rho \;) \quad \frac{1}{2}(1 \quad 1) \quad 1,$$

$$\rho \qquad \frac{1}{\overline{2}}\begin{pmatrix} 1 \\ 1 \end{pmatrix} \frac{1}{\overline{2}}\begin{pmatrix} 1 & \quad 1 \end{pmatrix} \quad \frac{1}{2}\begin{pmatrix} 1 & 1 \\ 1 & 1 \end{pmatrix}, \quad \mathrm{tr}\,(\rho \;) \quad \frac{1}{2}(1 \quad 1) \quad 1,$$

and

$$\rho \quad \sum_i p_i \rho_i \quad \frac{1}{2}\rho \quad \frac{1}{2}\rho \quad \frac{1}{2}\;\frac{1}{2}\begin{pmatrix} 1 & 1 \\ 1 & 1 \end{pmatrix} \quad \frac{1}{2}\;\frac{1}{2}\begin{pmatrix} 1 & 1 \\ 1 & 1 \end{pmatrix} \quad \frac{1}{2}\begin{pmatrix} 1 & 0 \\ 0 & 1 \end{pmatrix},$$

$$\mathrm{tr}(\rho^2) \quad \mathrm{tr}\left[\begin{pmatrix} \frac{1}{2} & 0 \\ 0 & \frac{1}{2} \end{pmatrix}\begin{pmatrix} \frac{1}{2} & 0 \\ 0 & \frac{1}{2} \end{pmatrix}\right] \quad \mathrm{tr}\left[\begin{pmatrix} \frac{1}{4} & 0 \\ 0 & \frac{1}{4} \end{pmatrix}\right] \quad \frac{1}{4} \quad \frac{1}{4} \quad \frac{1}{2} < 1.$$

The trace of the density operator satisfies the condition, $\mathrm{tr}(\rho^2)$ 1, for pure states and, $\mathrm{tr}(\rho^2) < 1$, for mixed states.

Example. Consider two mixtures of states with the nonequal probabilities p_0 1/4 and p_1 3/4.

First, we consider a mixture of the states 0 and 1 ; the density operator is

$$\rho_{ne} \quad \sum_i p_i \rho_i \quad \frac{1}{4}\rho_0 \quad \frac{3}{4}\rho_1 \quad \frac{1}{4}\begin{pmatrix} 1 & 0 \\ 0 & 0 \end{pmatrix} \quad \frac{3}{4}\begin{pmatrix} 0 & 0 \\ 0 & 1 \end{pmatrix} \quad \frac{1}{4}\begin{pmatrix} 1 & 0 \\ 0 & 3 \end{pmatrix}.$$

Now we consider a mixture of the states 1/ $\overline{2}$(0 1) and 1/ $\overline{2}$(0 1); the density operator is

$$\rho_{ne} \quad \sum_i p_i \rho_i \quad \frac{1}{4}\rho \quad \frac{3}{4}\rho \quad \frac{1}{4}\;\frac{1}{2}\begin{pmatrix} 1 & 1 \\ 1 & 1 \end{pmatrix} \quad \frac{3}{4}\;\frac{1}{2}\begin{pmatrix} 1 & 1 \\ 1 & 1 \end{pmatrix} \quad \frac{1}{8}\begin{pmatrix} 4 & 2 \\ 2 & 4 \end{pmatrix}$$

or

$$\rho_{ne} \quad \frac{1}{4}\begin{pmatrix} 2 & 1 \\ 1 & 2 \end{pmatrix}.$$

We see that $\rho_{ne} \quad \rho_{eq}$ and $\rho_{ne} \quad \rho_{eq}$.

We now discuss the defining properties of a density operator; an operator ρ in the Hilbert $\mathcal{H}_n$ can be considered a density operator if and only if it has three properties:

1. It is self-adjoint, $\rho \quad \rho^{\dagger}$.
2. It is non-negative, $\rho \quad 0$.
3. Its trace is equal to 1, $\text{tr}[\rho] \quad 1$.

The following proposition shows that operators with these properties form a convex set.[4]

Proposition. *The set of density operators in a Hilbert space, $\mathcal{H}_n$, is a convex linear subset of $n \quad n$ self-adjoint matrices.*

Proof. Assume that ρ_a and ρ_b are density operators and thereby satisfy the three properties. Given a real λ such that $0 \quad \lambda \quad 1$, we construct the following operator:

$$\rho(\lambda) \quad \lambda\rho_a \quad (1 \quad \lambda)\rho_b.$$

We see immediately that $\rho(\lambda)$ is self-adjoint because ρ_a and ρ_b are self-adjoint:

$$\rho^{\dagger}(\lambda) \quad [\lambda\rho_a \quad (1 \quad \lambda)\rho_b]^{\dagger} \quad \lambda\rho_a^{\dagger} \quad (1 \quad \lambda)\rho_b^{\dagger} \quad \lambda\rho_a \quad (1 \quad \lambda)\rho_b \quad \rho(\lambda).$$

Given any state $\varphi \quad \mathcal{H}_n$, we see that

$$\varphi\ \rho(\lambda)\ \varphi \quad \lambda\ \varphi\ \rho_a\ \varphi \quad (1 \quad \lambda)\ \varphi\ \rho_b\ \varphi \quad 0.$$

Finally,

$$\text{tr}(\rho(\lambda)) \quad \lambda\text{tr}(\rho_a) \quad (1 \quad \lambda)\text{tr}(\rho_b) \quad \lambda \quad (1 \quad \lambda) \quad 1.$$

■

The property of density matrices to form a convex set has important consequences, as we shall see in the next chapters. There are infinitely many ways to express $\rho(\lambda)$ as a convex combination of other states and, thus, infinitely many ways to prepare a mixed state; if we have an observable, M, associated with the self-adjoint operator, $\mathbf{M}$, and wish to determine the expected value for different preparations of the state, ρ_a and ρ_b, these values are indistinguishable. The expected value of the observable $\mathbf{M}$ for the system with the density matrix, $\rho(\lambda)$, is defined in Section 2.10:

$$\mathbf{M} \quad \text{tr}(\mathbf{M}\rho(\lambda)).$$

---

[4]A convex subspace of a vector space is a subset that contains all the points of a straight line connecting any two points in the subset.

As before, $\rho(\lambda) \quad \lambda\rho_a \quad (1 \quad \lambda)\rho_b$. Call $\mathbf{M}_a \quad \mathrm{tr}\,(\mathbf{M}\rho_a)$ and $\mathbf{M}_b \quad \mathrm{tr}\,(\mathbf{M}\rho_b)$ the average values of the measurements for the systems with density matrices ρ_a and ρ_b. Then

$$\mathbf{M} \quad \lambda\,\mathbf{M}_a \quad (1 \quad \lambda)\,\mathbf{M}_b\,.$$

Thus, regardless of how we have prepared a mixed state with the density operator ρ, the average value of the outcome of a measurement of any observable of the system will be the same. On the other hand, a pure state can be prepared in one and only one way; a pure state is an eigenvector of some observable and the outcome of the measurement of that observable will always be equal to 1.

The density operator, ρ, of an ensemble captures only the information available to an observer who has the opportunity to examine infinitely many states of the ensemble. The entropy of a mixed state of non-orthogonal pure states, which are not mutually distinguishable, is given by the von Neumann entropy:

$$S(\rho) \quad \mathrm{tr}\,(\rho \log \rho)\,.$$

When the pure states of the ensemble are orthogonal, they can be treated as classical states and then the entropy is given by the known Shannon expression:

$$H \quad \sum_{(i)} p_i \quad \log p_i.$$

In summary, the density operator is a positive semi-definite, self-adjoint operator with trace equal to unity. The eigenvalues of a positive semi-definite operator are real numbers greater than or equal to zero; thus, the operator must have a positive trace and a non-negative determinant. The trace of the density operator satisfies the condition $\mathrm{tr}(\rho^2) \quad 1$ for pure states and $\mathrm{tr}(\rho^2) < 1$ for mixed states.

2.4 PURIFICATION OF MIXED STATES

Pure states have a particular appeal for quantum information theory; thus, a legitimate question is if given a quantum system, $\mathcal{A}$, in a mixed state we can identify another system, $\mathcal{B}$, such that the bipartite system, $\mathcal{C} \quad \mathcal{AB}$, is in a pure state. System $\mathcal{B}$, called a *reference system*, is only a mathematical construct without a physical support.

The process of finding a reference system given a quantum system, $\mathcal{A}$, in a mixed state, $\rho_{\mathcal{A}}$, is called *purification*. If $\psi_{\mathcal{C}}$ is the pure state of the bipartite system, $\mathcal{C} \quad \mathcal{AB}$, we wish that

$$\rho_{\mathcal{A}} \quad \mathrm{tr}_{\mathcal{B}}\,(\ \psi_{\mathcal{C}} \quad \psi_{\mathcal{C}}\ ).$$

Let $\mathcal{H}_n$ be the Hilbert space for system $\mathcal{A}$, with an orthonormal basis $\ e_i\ $; then the decomposition of the density operator of $\mathcal{A}$ is

$$\rho_{\mathcal{A}} \quad \sum_i p_i\, e_i^A \quad e_i^A\,.$$

The procedure of purification consists of the following steps:

1. Introduce $\mathcal{B} \quad \mathcal{H}_n$, which shares the same space with $\mathcal{A}$. Let $e_j^B \quad \mathcal{H}_n$ be an orthonormal basis for $\mathcal{H}_n$.

2. Define a pure state of the bipartite system

$$\psi_C \quad \sum_i \overline{p_i}\, e_i^A \quad e_i^B .$$

We now compute the reduced density operator for $\mathcal{A}$ for this state as

$$\mathrm{tr}_B(\ \psi_C \quad \psi_C\ ) \quad \sum_i\sum_j \overline{p_i p_j}\, e_i^A \quad e_i^A\ \mathrm{tr}\left(\ e_i^B \quad e_j^B\ \right)$$

$$\sum_i\sum_j \overline{p_i p_j}\, e_i^A \quad e_i^A\ \delta_{ij} \quad \sum_i p_i\, e_i^A \quad e_i^A \qquad \rho_{\mathcal{A}}.$$

A formal definition of purification requires a more general formalism for positive semi-definite operators [444]. First, we review several linear algebra concepts: the *kernel of a linear transformation* $\mathcal{L}: \mathcal{H}_a \quad \mathcal{H}_b$, denoted as Ker($\mathcal{L}$), is the set of all vectors, $\zeta \quad \mathcal{H}_a$, such that $\mathcal{L}(\ \zeta\ )\quad 0$. The *rank of a linear transformation* is the dimension of its range; in our example, rank$(\mathcal{H}_a)\quad \dim(\mathcal{H}_b)$. If $\mathcal{L}(\mathcal{H}_a)$ denotes a linear transformation of the vector space $\mathcal{H}_a$, then supp$[\rho]$, the *support of a normal operator* $\rho \quad \mathcal{L}(\mathcal{H}_a)$, is defined as the subspace of $\mathcal{H}_a$ spanned by the eigenvectors of ρ having a nonzero eigenvalue,

$$\mathrm{supp}[\rho] \quad \mathrm{span}\ \ \xi_i\ : 1 \quad i \quad n, \lambda_i \quad 0\ ,$$

where $\rho \quad \sum_i^n\ _1 \lambda_i\ \xi_i \quad \xi_i$ is a spectral decomposition of ρ. Obviously,

$$\dim(\mathrm{supp}[\rho]) \quad \mathrm{rank}(\rho).$$

Formal De nition of Puri cation

Let $\mathcal{H}_a$ and $\mathcal{H}_b$ be Hilbert spaces and $\rho \quad \mathrm{Pos}(\mathcal{H}_a)$ be any positive semi-definite operator. Then a purification of ρ in $\mathcal{H}_a \quad \mathcal{H}_b$ is any vector $\ \varphi\ $, such that

$$\rho \quad \mathrm{tr}_{\mathcal{H}_b}(\ \varphi \quad \varphi\ ).$$

Let $\Pi_{\mathcal{H}_c}$ be the projection of $\mathcal{H}_a$ onto a subspace $\mathcal{H}_c$. Then,

$$\Pi_{\mathrm{supp}(\rho)} \quad \sum_{i \quad S} \ \xi_i \quad \xi_i\ ,$$

with $\ \xi_i\ : i \quad R\ $, an orthonormal basis in $\mathcal{H}_a$. If $\dim[\mathcal{H}_b] \quad \mathrm{rank}[\rho]$, then there exists a mapping $A \quad \mathcal{L}(\mathcal{H}_a, \mathcal{H}_b)$, such that

$$A^\dagger A \quad \Pi_{\mathrm{supp}(\rho)}.$$

Indeed, $\ \xi_i\ : i \quad R\ $, an orthonormal basis in $\mathcal{H}_a$, exists because $\dim(\mathcal{H}_b) \quad \mathrm{rank}(\rho) \quad R$. Thus,

$$A \quad \sum_{i \quad S} \gamma_i \ \xi_i$$

satisfies the condition $A^\dagger A \quad \Pi_{\mathrm{supp}(\rho)}$.

The Existence of Puri cation

Let $\mathcal{H}_a$ and $\mathcal{H}_b$ be Hilbert spaces and ρ Pos$(\mathcal{H}_a)$ be any positive semi-definite operator in $\mathcal{H}_a$. Then a purification of ρ exists in $\mathcal{H}_a$ $\mathcal{H}_b$ if and only if dim$[\mathcal{H}_b]$ rank$[\rho]$. Moreover, when dim$[\mathcal{H}_b]$ dim$[\mathcal{H}_a]$, a purification of ρ always exists.

Proof. The proof of the two clauses is rather short; when dim$[\mathcal{H}_b]$ rank(ρ), then there exists the mapping A, such that $A^\dagger A$ $\Pi_{\text{supp}(\rho)}$. Given a matrix A, we denote by vec(A) the column vector constructed as follows: We take each row vector of A, transpose it, and stack the vectors in order, on top of one another. Define

$$\psi \quad \text{vec}(\ \overline{\rho}A^\dagger) \quad \mathcal{H}_a \quad \mathcal{H}_b.$$

Then ψ is a purification of ρ because

$$\text{tr}_{\mathcal{H}_b}(\ \psi \quad \psi \ ) \qquad \overline{\rho}A^\dagger A \ \overline{\rho} \quad \rho.$$

Conversely, when ψ $\mathcal{H}_a$ $\mathcal{H}_b$ satisfies the condition $\text{tr}_{\mathcal{H}_b}(\ \psi \quad \psi \ )$ ρ, consider Ψ vec $^{1}(\ \psi \ )$. Then,

$$\rho \quad \text{tr}_{\mathcal{H}_b}(\ \psi \quad \psi \ ) \quad \Psi\Psi^\dagger.$$

Therefore, rank(ψ) rank(ρ). Because the rank of ψ cannot exceed the dimension of $\mathcal{H}_b$, it follows that

$$\text{dim}\,[\mathcal{H}_b] \quad \text{rank}(\rho).$$

■

This property will be used in Chapter 3 when we discuss information transmission through a noisy quantum channel.

Next, we discuss quantum measurements and start our presentation with a review of a postulate of quantum mechanics, the Born rule.

2.5 BORN RULE

Quantum measurements are subject to the Born rule; this fundamental postulate[5] of quantum mechanics is typically formulated as follows: Consider a system in the state ψ $\mathcal{H}_n$ and an observable A. Let $\mathbf{P}_i, 1$ i n be a set of projectors on a_1 , a_1 ,..., a_n , in an orthonormal basis of $\mathcal{H}_n$, where we measure the observable A; the projector, $\mathbf{P}_i$, has a_i as an eigenvector and λ_i as the corresponding eigenvalue. Then the probability for a measurement of the observable A of yielding a particular value λ_i is

$$p(\lambda_i) \qquad a_i \ \psi \quad ^2.$$

[5]Even though no violation of the Born rule has been observed experimentally, it would be desirable to replace this postulate by a derivation based on a set of fundamental assumptions. A. M. Gleason has presented a mathematical motivation for the form of the Born rule [171], and more recently, W. H. Zurek used a mechanism termed environment-assisted invariance, or *envariance*, for a novel derivation of the Born rule [472].

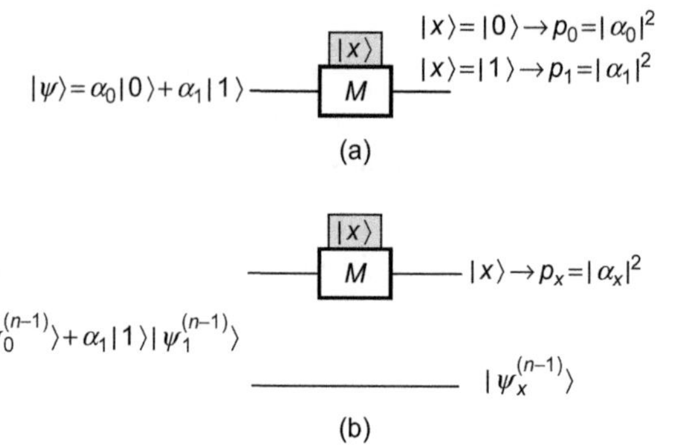

FIGURE 2.6

(a) The measurement of a single qubit in the state, $|\psi\rangle = \alpha_0|0\rangle + \alpha_1|1\rangle$, using a one-qubit measurement gate. (b) We use a one-qubit measurement gate to measure only the first qubit of a register of n qubits in the state $|\psi^{(n)}\rangle = \alpha_0|0\rangle|\psi_0^{(n-1)}\rangle + \alpha_1|1\rangle|\psi_1^{(n-1)}\rangle$.

In Section 1.7 we showed that the probability of an outcome, λ_i, is

$$p(\lambda_i) = \langle\psi|\mathbf{P}_i|\psi\rangle,$$

with $\mathbf{P}_i = |a_i\rangle\langle a_i|$; this expression is equivalent to the Born rule:

$$\langle\psi|\mathbf{P}_i|\psi\rangle = \langle\psi|a_i\rangle\langle a_i|\psi\rangle = (\langle a_i|\psi\rangle)^\dagger(\langle a_i|\psi\rangle) = |\langle a_i|\psi\rangle|^2.$$

The measurement of a single qubit in the state, $|\psi\rangle = \alpha_0|0\rangle + \alpha_1|1\rangle$, using a one-qubit measurement gate is illustrated in Figure 2.6a. The measurement is performed in the orthonormal basis $(|0\rangle, |1\rangle)$. The instrument displays the result, 0, corresponding to the state, $|0\rangle$, with the probability $p_0 = |\alpha_0|^2$, and the result, 1, corresponding to the state, $|1\rangle$, with the probability $p_1 = |\alpha_1|^2$.

The generalized Born rule for a register of n qubits in the state $|\psi^n\rangle \in \mathcal{H}_{2^n}$ when we carry out a measurement of only the first qubit using a one-qubit measurement gate is presented in Figure 2.6b. In this case, we consider two subspaces, $\mathcal{H}_2$ and $\mathcal{H}_{2^{n-1}}$ of $\mathcal{H}_{2^n}$, and express the state of the n-qubit register as

$$|\psi^n\rangle = \alpha_0(|0\rangle \otimes |\psi_0^{(n-1)}\rangle) + \alpha_1(|1\rangle \otimes |\psi_1^{(n-1)}\rangle).$$

Here, $|\psi_0^{(n-1)}\rangle$ and $|\psi_1^{(n-1)}\rangle$ are the eigenvectors of the two projectors in $\mathcal{H}_{2^{n-1}}$. If the result displayed by the instrument is 0 (if the single qubit measured is projected on $|0\rangle$), then the remaining $n-1$ qubits are in the state $|\psi_0^{(n-1)}\rangle$; if the result displayed by the instrument is 1 (when the single qubit measured is projected on $|1\rangle$), then the remaining $n-1$ qubits are in the state $|\psi_1^{(n-1)}\rangle$.

We can also apply the generalized Born rule to measuring the first k qubits of an n-qubit register originally in the state

$$|\psi^n\rangle = \sum_{i=0}^{2^k-1} \alpha_i |i^{(k)}\rangle \otimes |\psi_i^{(n-k)}\rangle.$$

When the result of the measurement performed on the first k qubits is $x = r$, we know that the remaining $(n-k)$ qubits are in the state $|\psi_r^{(n-k)}\rangle$.

2.6 MEASUREMENT OPERATORS

The measurement of an observable of a quantum system in the state, $\psi \quad \mathcal{H}_n$, is characterized by a set of operators, $\mathbf{M}_i$. Call λ_i the set of possible outcomes of the measurement of the quantum system and recall that these outcomes are eigenvalues of the measurement operators and must be real numbers to allow a physical interpretation of the measurement process. Only self-adjoint operators have real eigenvalues; thus, *the measurement operators,* $\mathbf{M}_i$*, must be self-adjoint*:

$$\mathbf{M}_i^\dagger \mathbf{M}_i \quad \mathbf{M}_i^2 \quad \mathbf{M}_i.$$

According to the measurement postulate of quantum mechanics, the probability that the outcome with the index i, λ_i, occurs as a result of the measurement $\mathbf{M}_i$ is

$$p(i) \quad \psi \ \mathbf{M}_i \ \psi \ .$$

The sum of the probabilities of all possible outcomes of the measurement must be 1:

$$\sum_i p(i) \quad \sum_i \ \psi \ \mathbf{M}_i \ \psi \quad 1.$$

This implies that the measurement operators must satisfy the *completeness equation*,

$$\sum_i \mathbf{M}_i \quad \mathbf{I}.$$

The measurement causes the system to change its state; if the state of the system immediately prior to the measurement is ψ , then after the measurement $\mathbf{M}_i$, the state of the system is $\varphi_i \quad \mathbf{M}_i \ \psi$. The *normalized state* of the system is

$$\varphi_i \quad \frac{\mathbf{M}_i \ \psi}{\sqrt{\ \psi \ \mathbf{M}_i \ \psi\ }}.$$

Example. *Consider a single qubit in the state* $\psi \quad \alpha_0 \ 0 \quad \alpha_1 \ 1$. There are only two possible outcomes of a measurement in the canonical basis $0 , 1$; we can only observe the basis states 0 or 1 . The corresponding projectors are

$$\mathbf{P}_0 \quad 0 \ 0 \quad \begin{pmatrix} 1 & 0 \\ 0 & 0 \end{pmatrix} \quad \text{and} \quad \mathbf{P}_1 \quad 1 \ 1 \quad \begin{pmatrix} 0 & 0 \\ 0 & 1 \end{pmatrix}.$$

These two operators are self-adjoint, and $\mathbf{P}_0^\dagger \mathbf{P}_0 \quad \mathbf{P}_0$ and $\mathbf{P}_1^\dagger \mathbf{P}_1 \quad \mathbf{P}_1$. The probability of the outcome 0 corresponding to the basis state 0 is

$$p(0) \quad \psi \ \mathbf{P}_0 \ \psi \quad \alpha_0 \ ^2.$$

Indeed,

$$\mathbf{P}_0 \ \psi \quad \begin{pmatrix} 1 & 0 \\ 0 & 0 \end{pmatrix} \begin{pmatrix} \alpha_0 \\ \alpha_1 \end{pmatrix} \quad \begin{pmatrix} \alpha_0 \\ 0 \end{pmatrix}$$

and

$$p(0) \quad \psi\ (\mathbf{P}_0\ \psi\ ) \quad (\alpha_0 \alpha_1) \begin{pmatrix} \alpha_0 \\ 0 \end{pmatrix} \quad \alpha_0\ ^2.$$

The normalized state of the qubit after applying the projector, $\mathbf{P}_0$, up to a phase, is given by

$$\varphi_0 \quad \frac{\mathbf{P}_0\ \psi}{\overline{\psi\ \mathbf{P}_0\ \psi}} \quad \frac{\alpha_0\ 0}{\sqrt{\alpha_0\ ^2}} \quad 0\ .$$

Similarly, the probability of the outcome, 1, corresponding to the basis state, 1 , is

$$p(1) \quad \psi\ \mathbf{P}_1\ \psi \quad \alpha_1\ ^2.$$

The normalized state of the qubit after applying the projector, $\mathbf{P}_1$, up to a phase, is given by

$$\varphi_1 \quad \frac{\mathbf{P}\ \psi}{\overline{\psi\ \mathbf{P}_1\ \psi}} \quad \frac{\alpha_1\ 1}{\sqrt{\alpha_1\ ^2}} \quad 1\ .$$

The completeness condition is satisfied: $p(0) \quad p(1) \quad \alpha_0\ ^2 \quad \alpha_1\ ^2 \quad 1.$

Next, we discuss projective measurements and then positive operator-valued measures (POVMs).

2.7 PROJECTIVE MEASUREMENTS

In this section we discuss projection measurements whose outcomes are deterministic and introduce the von Neumann axiomatic definition of the measurement process.

When the projectors, $\mathbf{M}_k \quad \mathcal{H}_n$, form an *orthogonal set*, $\mathbf{M}_j\mathbf{M}_k \quad 0$ when $j \quad k$, and if the set is complete, $\sum_k \mathbf{M}_k \quad \mathbf{I}$, then any observable $\mathbf{M}$ of the quantum system has a spectral decomposition,

$$\mathbf{M} \quad \sum_k k\mathbf{M}_k,$$

with $\mathbf{M}_k$, the projector onto the eigenspace of $\mathbf{M}$, with the eigenvalue k. The measurement operator corresponding to a basis vector, $k \quad \mathcal{H}_n$, is $\mathbf{M}_k \quad k\ k$; the operator $\mathbf{M}_k$ is self-adjoint and idempotent:

$$\mathbf{M}_k \quad \mathbf{M}_k^\dagger \quad \text{and} \quad \mathbf{M}_k^2 \quad k\ k\ k\ k \quad k\ k \quad \mathbf{M}_k.$$

To answer the question of when is the outcome of a measurement deterministic, we use the concept *projective measurement*.

Projective measurements. Let $\mathbf{M}_k$ be a set of self-adjoint measurement operators. The set is complete and consists of orthogonal operators and

$$\mathbf{M}_i\mathbf{M}_j \quad \delta_{ij}\mathbf{M}_i,$$

with δ_{ij}, the Kronecker delta function; the operators allow us to conduct projective measurements.

When measuring a quantum system in the state ψ , the probability of observing the outcome with the index k is

$$p(k) \quad \psi\ \mathbf{M}_k\ \psi\ .$$

Projective measurements are repeatable; the outcome observed as a result of projective measurement is deterministic. Indeed, consider a quantum system in the state ψ immediately prior to a measurement carried out using the projective measurement operator, $\mathbf{M}_k$; the state immediately after the measurement, ψ , is

$$\psi \quad \frac{\mathbf{M}_k\ \psi}{p(k)}.$$

If we apply the operator $\mathbf{M}_k$ again, this time to the state ψ , there is no state change. Indeed, the state after applying $\mathbf{M}_k$ for the second time is

$$\psi \quad \frac{\mathbf{M}_k\ \psi}{\sqrt{p\ (k)}} \quad \frac{\mathbf{M}_k \quad \mathbf{M}_k\ \psi}{\sqrt{p\ (k)} \quad p(k)},$$

with

$$p(k) \quad \psi\ \mathbf{M}_k\ \psi \quad \text{and} \quad p\ (k) \quad \psi\ \mathbf{M}_k\ \psi\ .$$

We can now derive the expression for $p\ (k)$:

$$p\ (k) \quad \psi\ \mathbf{M}_k\ \psi \quad \left[\psi\ \frac{\mathbf{M}_k^{\dagger}}{p(k)}\right]\mathbf{M}_k\left[\frac{\mathbf{M}_k}{p(k)}\ \psi\right] \quad \frac{\psi\ \mathbf{M}_k\ \psi}{p(k)}.$$

Then,

$$\sqrt{p\ (k)p(k)} \quad \sqrt{\psi\ \mathbf{M}_k\ \psi} \quad \sqrt{p(k)}.$$

Finally, we see that the state has not been affected by the second measurement:

$$\psi \quad \frac{\mathbf{M}_k \quad \mathbf{M}_k\ \psi}{\sqrt{p\ (k)} \quad p(k)} \quad \frac{\mathbf{M}_k\ \psi}{p(k)} \quad \psi\ .$$

The fact that projective measurements are repeatable has straightforward mathematical and physical interpretations: projecting a quantum state on the basis vectors of the Hilbert space transforms quantum information into classical information; thus, applying again the same operator does not lead to a state change. The outcome observed as a result of such a measurement is deterministic: $p\ (k) \quad \psi\ \mathbf{M}_k\ \psi \quad 1$.

The expected value of the projective measurement $\mathbf{M}$, $\mathbf{E}(\mathbf{M})$, denoted as $\mathbf{M}$, is defined as

$$\mathbf{M} \quad \mathbf{E}(\mathbf{M}) \quad \sum_k kp(k) \quad \sum_k k\ \psi\ \mathbf{M}_k\ \psi \quad \psi\left[\sum_k k\mathbf{M}_k\right]\psi \quad \psi\ \mathbf{M}\ \psi\ .$$

The standard deviation of a projective measurement $\mathbf{M}$ is defined as

$$\mathbf{E}[\mathbf{M} \quad \mathbf{M}\ ]^2 \quad \mathbf{E}\left[\mathbf{M}^2 \quad 2(\mathbf{M})(\ \mathbf{M}\ ) \quad (\ \mathbf{M}\ )^2\right] \quad \mathbf{M}^2 \quad (\ \mathbf{M}\ )^2.$$

Example. Assume that an observable of a qubit in the state, $\psi \quad (0 \quad 1)/\ \overline{2}$, has the two eigenvectors, 0 and 1 , with the eigenvalues 1 and 1, respectively. Then the probabilities of the two outcomes are $p(\ 1) \quad p(\ 1) \quad 1/2$.

The projector operators corresponding to the two eigenvectors, $\mathbf{M}_0$ and $\mathbf{M}_1$, are:

$$\mathbf{M}_0 \quad 0\ 0 \quad \begin{pmatrix} 1 & 0 \\ 0 & 0 \end{pmatrix} \quad \text{and} \quad \mathbf{M}_1 \quad 1\ 1 \quad \begin{pmatrix} 0 & 0 \\ 0 & 1 \end{pmatrix}.$$

The projective measurement of the state ψ produces the eigenvalue (1) corresponding to the eigenvector 0 , with the probability

$$p(\ 1) \quad \psi\ \mathbf{M}_0\ \psi \quad \psi\ 0\ 0\ \psi\ ,$$

and the eigenvalue (1) corresponding to the eigenvector 1 , with the probability

$$p(\ 1) \quad \psi\ \mathbf{M}_1\ \psi \quad \psi\ 1\ 1\ \psi\ .$$

It is easy to see that

$$\mathbf{M}_0\ \psi \quad \begin{pmatrix} 1 & 0 \\ 0 & 0 \end{pmatrix} \frac{1}{\overline{2}} \begin{pmatrix} 1 \\ 1 \end{pmatrix} \quad \frac{1}{\overline{2}} \begin{pmatrix} 1 \\ 0 \end{pmatrix},$$

$$p(\ 1) \quad \psi\ (\mathbf{M}_0\ \psi\ ) \quad \frac{1}{\overline{2}} (1 \quad 1) \frac{1}{\overline{2}} \begin{pmatrix} 1 \\ 0 \end{pmatrix} \quad \frac{1}{2} (1 \quad 1) \begin{pmatrix} 1 \\ 0 \end{pmatrix} \quad \frac{1}{2}.$$

Similarly, $p(\ 1) \quad \psi\ (\mathbf{M}_1\ \psi\ ) \quad 1/2$.

John von Neumann provided an axiomatic definition of a quantum measurement process [441]; his measurement scheme is an ancestor of the quantum decoherence theory and treats the measurement apparatus as a quantum object. According to this axiomatic definition, the measurement consists of two stages (Figure 2.7):

1. The quantum system, S, interacts with a macroscopic measuring apparatus, M, to measure the quantity, Q. This interaction is linear; if at time, t, the quantum system is in a superposition state $\psi \quad \sum_i \alpha_i\ a_i$ and the apparatus is in the state m , the ensemble $S \quad M$ is in the state $\sum_i \alpha_i\ a_i$ m . In this expression, a_i are the eigenvectors of Q corresponding to the possible outcomes of the measurement, q_i (the eigenvalues of Q). At the end of this stage, the system is in the state $\sum_i \gamma_i\ a_i \quad m_i$, with m_i , the state in which M registers the outcome, q_i; the states of the apparatus, m_i , are orthonormal.
2. As a result of a nondeterministic process, the ensemble $S \quad M$ "collapses" from the state $\sum_i \gamma_i\ a_i$ m_i to the state $a_i \quad m_i$ for some i.

In summary, in a projective measurement the measurement operators are self-adjoint and idempotent. The number of such operators is equal to the dimension of the Hilbert space. Since orthogonal measurement operators commute, they correspond to simultaneous observables. A von Neumann measurement is a conditional expectation onto a maximal Abelian subalgebra of the algebra of all bounded operators acting on the Hilbert space [329].

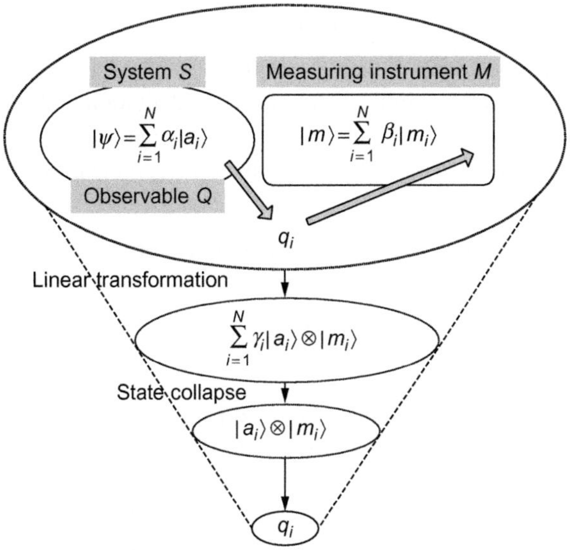

FIGURE 2.7

Two-stage von Neumann-type measurement. We measure the observable, Q, of the quantum system, S, using a macroscopic apparatus, M. At time t, the quantum system is in a superposition state $\psi \quad \sum_i \alpha_i \, a_i$ and the apparatus is in the state m . q_i is the eigenvalue corresponding to the eigenvector a_i of observable Q. At the end of the first phase, the composite system, $S \quad M$, is in the state $\sum_i \gamma_i \, a_i \quad m_i$. Then, as a result of the measurement, the system collapses in the state $a_i \quad m_i$ for some i.

2.8 POSITIVE OPERATOR-VALUED MEASURES (POVMs)

Projective measurements are restrictive and not always possible. Sometimes we destroy a quantum particle in the process of measurement; thus, the repeatability of a projective measurement is violated. For example, a photon may be absorbed by a polarization filter and no longer available for measurements. In such cases where the system is measured only once, the post-measurement state of the system is no longer of interest and the probabilities of the measurement outcomes are the ones that count.

A generalized type of measurement, the *positive operator-valued measure*, or POVM, is concerned only with the statistics of the measurement. The POVM operators are not necessarily orthogonal or commutative and *allow the possibility of measurement outcomes associated with non-orthogonal states.* Recall that measurement operators corresponding to non-orthogonal states do not commute and are therefore not simultaneously observable. The number of POVM operators may differ from the dimension of the Hilbert space, while the number of projective operators is precisely equal to the dimension of the Hilbert space.

Given the set of measurement operators, $\mathbf{M}_i$, describing a measurement performed on a quantum system in the state, ψ , a *POVM* has the elements, $\mathbf{A}_i$, defined by

$$\mathbf{A}_i \quad \mathbf{M}_i^{\dagger}\mathbf{M}_i \quad \text{and} \quad \sum_i \mathbf{A}_i \quad \mathbf{I}.$$

The operators $\mathbf{A}_i$ have several properties, including:

- They are positively defined; recall that a positively defined operator has real and positive eigenvalues. This property follows immediately from the definition of $\mathbf{A}_i$.
- $p(i) \quad \psi\ \mathbf{A}_i\ \psi$. This follows immediately from the fact that $p(i) \quad \psi\ \mathbf{M}_i^\dagger \mathbf{M}_i\ \psi$. The set of operators, $\mathbf{A}_i$, can be used, instead of $\mathbf{M}_i$, to estimate the probabilities of various measurement outcomes.

Not all the elements of a POVM are measurement operators. Only in the case of projective measurements, all the POVM elements $\mathbf{A}_i$, where

$$\mathbf{A}_i \quad \mathbf{P}_i^\dagger \mathbf{P}_i \quad \mathbf{P}_i$$

are the same as the measurement operators.

POVM measurements prevent us from making misidentification errors, but sometimes they may not provide any information at all. POVM allows us to distinguish the cases when the outcome of a measurement identifies with certainty the original state from the ones when the identification is not possible. POVM measurements can be *optimal*, a topic discussed in more depth in Section 3.12; optimality implies that the average information gained as a result of the measurement is maximized.

Example. Consider a qubit in one of two possible states:

$$\psi_1 \quad 0 \quad \psi_1 \quad \begin{pmatrix} 1 \\ 0 \end{pmatrix} \quad \text{or} \quad \psi_2 \quad \frac{(0 \quad 1)}{\overline{2}} \quad \psi_2 \quad \frac{1}{\overline{2}} \begin{pmatrix} 1 \\ 1 \end{pmatrix}.$$

We define a POVM with three operators (elements). The first two are

$$\mathbf{A}_1 \quad \frac{\overline{2}}{1 \quad \overline{2}}\ 1\ 1 \quad \mathbf{A}_1 \quad \frac{\overline{2}}{1 \quad \overline{2}} \begin{pmatrix} 0 \\ 1 \end{pmatrix} \begin{pmatrix} 0 & 1 \end{pmatrix} \quad \frac{\overline{2}}{1 \quad \overline{2}} \begin{pmatrix} 0 & 0 \\ 0 & 1 \end{pmatrix}$$

and

$$\mathbf{A}_2 \quad \frac{\overline{2}}{1 \quad \overline{2}} \frac{(0 \quad 1)(0 \quad 1)}{2} \quad \mathbf{A}_2 \quad \frac{\overline{2}}{2(1 \quad \overline{2})} \begin{pmatrix} 1 \\ 1 \end{pmatrix} \begin{pmatrix} 1 & 1 \end{pmatrix}$$

or

$$\mathbf{A}_2 \quad \frac{\overline{2}}{2(1 \quad \overline{2})} \begin{pmatrix} 1 & 1 \\ 1 & 1 \end{pmatrix}.$$

To ensure completeness, we define a third operator:

$$\mathbf{A}_3 \quad \mathbf{I} \quad \mathbf{A}_1 \quad \mathbf{A}_2 \quad \begin{pmatrix} 1 & 0 \\ 0 & 1 \end{pmatrix} \quad \frac{\overline{2}}{1 \quad \overline{2}} \begin{pmatrix} 0 & 0 \\ 0 & 1 \end{pmatrix} \quad \frac{\overline{2}}{2(1 \quad \overline{2})} \begin{pmatrix} 1 & 1 \\ 1 & 1 \end{pmatrix}.$$

Thus,

$$\mathbf{A}_3 \quad \frac{1}{2(1 \quad \overline{2})} \begin{pmatrix} 2 & \overline{2} & \overline{2} \\ \overline{2} & 2 & \overline{2} \end{pmatrix}.$$

Now we show that the three possible outcomes of the measurement associated with $\mathbf{A}_1, \mathbf{A}_2, \mathbf{A}_3$ lead to the following conclusions regarding the original state of the system:

$\mathbf{A}_1$ the original state was ψ_2
$\mathbf{A}_2$ the original state was ψ_1
$\mathbf{A}_3$ we do not know if the original state was ψ_1 or ψ_2

First, we show that when there is an outcome associated with $\mathbf{A}_1$, the probability that the original state was ψ_1 is $p_{\mathbf{A}_1}(\psi_1)$ 0 and $p_{\mathbf{A}_1}(\psi_2) > 0$; thus, the system could only have been in the state ψ_2. Indeed,

$$\mathbf{A}_1 \psi_1 \quad \frac{\overline{2}}{1 \quad \overline{2}} \begin{pmatrix} 0 & 0 \\ 0 & 1 \end{pmatrix} \begin{pmatrix} 1 \\ 0 \end{pmatrix} \quad \begin{pmatrix} 0 \\ 0 \end{pmatrix}$$

and

$$p_{\mathbf{A}_1}(\psi_1) \quad \psi_1 \, [\mathbf{A}_1 \psi_1] \quad \begin{pmatrix} 1 & 0 \end{pmatrix} \begin{pmatrix} 0 \\ 0 \end{pmatrix} \quad 0.$$

Also,

$$\mathbf{A}_1 \psi_2 \quad \frac{\overline{2}}{1 \quad \overline{2}} \begin{pmatrix} 0 & 0 \\ 0 & 1 \end{pmatrix} \frac{1}{\overline{2}} \begin{pmatrix} 1 \\ 1 \end{pmatrix} \quad \frac{1}{1 \quad \overline{2}} \begin{pmatrix} 0 \\ 1 \end{pmatrix}$$

and

$$p_{\mathbf{A}_1}(\psi_2) \quad \psi_2 \, [\mathbf{A}_1 \psi_2] \quad \frac{1}{\overline{2}(1 \quad \overline{2})} \begin{pmatrix} 1 & 1 \end{pmatrix} \begin{pmatrix} 0 \\ 1 \end{pmatrix} \quad \frac{1}{\overline{2}(1 \quad \overline{2})}.$$

When there is an outcome associated with $\mathbf{A}_2$, the probability that the original state was ψ_2 is equal to zero, $p_{\mathbf{A}_2}(\psi_2)$ 0 and $p_{\mathbf{A}_2}(\psi_1) > 0$, therefore, the system could only have been in the state ψ_1. Indeed,

$$\mathbf{A}_2 \psi_2 \quad \frac{\overline{2}}{2(1 \quad \overline{2})} \begin{pmatrix} 1 & 1 \\ 1 & 1 \end{pmatrix} \frac{1}{\overline{2}} \begin{pmatrix} 1 \\ 1 \end{pmatrix} \quad \frac{1}{2(1 \quad \overline{2})} \begin{pmatrix} 0 \\ 0 \end{pmatrix}$$

and

$$p_{\mathbf{A}_2}(\psi_2) \quad \psi_2 \, [\mathbf{A}_2 \psi_2] \quad \frac{1}{\overline{2}} \begin{pmatrix} 1 & 1 \end{pmatrix} \frac{1}{2(1 \quad \overline{2})} \begin{pmatrix} 0 \\ 0 \end{pmatrix} \quad 0.$$

Also,

$$\mathbf{A}_2 \psi_1 \quad \frac{\overline{2}}{2(1 \quad \overline{2})} \begin{pmatrix} 1 & 1 \\ 1 & 1 \end{pmatrix} \begin{pmatrix} 1 \\ 0 \end{pmatrix} \quad \frac{\overline{2}}{2(1 \quad \overline{2})} \begin{pmatrix} 1 \\ 1 \end{pmatrix}$$

and

$$p_{\mathbf{A}_2}(\ \psi_1\ )\quad \psi_1\ [\mathbf{A}_2\ \psi_1\ ]\quad \begin{pmatrix}1 & 0\end{pmatrix}\frac{\bar{2}}{2(1\quad \bar{2})}\begin{pmatrix}1\\1\end{pmatrix}\quad \frac{1}{\bar{2}(1\quad \bar{2})}.$$

Finally, when there is an outcome associated with $\mathbf{A}_3$, there is a nonzero probability that the original state was either ψ_1 or ψ_2 ; thus, $p_{\mathbf{A}_3}(\ \psi_1\ ) > 0$ and $p_{\mathbf{A}_3}(\ \psi_2\ ) > 0$. Indeed,

$$\mathbf{A}_3\ \psi_1\quad \frac{1}{2(1\quad \bar{2})}\begin{pmatrix}2 & \bar{2} & \bar{2}\\ & \bar{2}\quad 2 & \bar{2}\end{pmatrix}\begin{pmatrix}1\\0\end{pmatrix}\quad \frac{1}{\bar{2}(1\quad \bar{2})}\begin{pmatrix}1 & \bar{2}\\ & 1\end{pmatrix},$$

$$p_{\mathbf{A}_3}(\ \psi_1\ )\quad \psi_1\ [\mathbf{A}_3\ \psi_1\ ]\quad \frac{1}{\bar{2}(1\quad \bar{2})}\begin{pmatrix}1 & 0\end{pmatrix}\begin{pmatrix}1 & \bar{2}\\ & 1\end{pmatrix}\quad \frac{1}{\bar{2}}.$$

Similarly,

$$\mathbf{A}_3\ \psi_2\quad \frac{1}{2(1\quad \bar{2})}\begin{pmatrix}2 & \bar{2} & \bar{2}\\ & \bar{2}\quad 2 & \bar{2}\end{pmatrix}\frac{1}{\bar{2}}\begin{pmatrix}1\\1\end{pmatrix}\quad \frac{1}{\bar{2}(1\quad \bar{2})}\begin{pmatrix}1 & \bar{2}\\ & 1\end{pmatrix},$$

$$p_{\mathbf{A}_3}(\ \psi_2\ )\quad \psi_2\ [\mathbf{A}_3\ \psi_2\ ]\quad \frac{1}{\bar{2}}\begin{pmatrix}1 & 1\end{pmatrix}\frac{1}{\bar{2}(1\quad \bar{2})}\begin{pmatrix}1 & \bar{2}\\ & 1\end{pmatrix}\quad \frac{1}{\bar{2}}.$$

Clearly, 0 $\left(p_{\mathbf{A}_3}(\ \psi_1\ ), p_{\mathbf{A}_3}(\ \psi_2\ )\right)$ 1 and, in this case,

$$p_{\mathbf{A}_3}(\ \psi_1\ )\quad p_{\mathbf{A}_3}(\ \psi_2\ )\quad \frac{1}{\bar{2}}.$$

There is never a mistake in identifying the original state when we apply the POVM measurement operators, $\mathbf{A}_1$ and $\mathbf{A}_2$. This example shows that when we use the POVM element, $\mathbf{A}_3$, there is not enough information for a reliable identification.

A clear exposition of POVM can be found in Peres [325]; POVM has been applied to the analysis of quantum key distribution and the analysis of the eavesdropping strategy by Ekert *et al* [136]. A review of POVM, followed by a discussion of an implementation of POVM for photonic qubits, is offered by Brandt [70, 71].

We conclude the discussion of projective and POVM measurements with the observation that the main difference between them is that the number of available preparations of the system and that of available outcomes may be different from each other and different from the dimension of the Hilbert space. The probability of an outcome, i, is given by $p_i\quad \mathrm{tr}(\mathbf{A}_i\rho)$, in the case of POVM, while for a von Neumann measurement, it is $p_i\quad \mathrm{tr}(\mathbf{P}_i\rho)$.

2.9 NEUMARK THEOREM

The Neumark theorem addresses the problem of extending POVM operators from a smaller to a larger Hilbert space; a consequence of this theorem is that a realizable experimental procedure for a POVM measurement with operators, $\mathbf{A}_i$, always exists.

To show this, we introduce the concept *maximal quantum test*: Consider a quantum system and assume that the maximum number of different outcomes obtainable in a test of the system is N; any test with exactly N different outcomes is a maximal quantum test. For example, when atoms with spin s are used in a Stern-Gerlach experiment, $2s \quad 1$ different outcomes are observed, regardless of the orientation of the magnetic field. The "strong superposition principle" [324] states that any complete set of mutually orthogonal vectors has a physical realization as a maximal quantum test. The Newmark theorem can be formulated as follows [6].

Theorem. *The Hilbert space, $\mathcal{H}_n$, in which the set $\mathbf{A}_i, 1 \quad i \quad n$, of POVM operators is defined can be extended to a larger Hilbert space, $\mathcal{H}_N$, with $N > n$, such that in $\mathcal{H}_N$ there exists a set of orthogonal projectors, $\mathbf{P}_i$, with the two properties:*

- $\sum_i \mathbf{P}_i \quad \mathbf{I}$.
- *$\mathbf{A}_i$ is the result of projecting $\mathbf{P}_i \quad \mathcal{H}_N$ into $\mathcal{H}_n$.*

Proof. We add $N \quad n$ extra dimensions to $\mathcal{H}_n$ by introducing the unit vectors, v_k , orthogonal to each other, $v_j \; v_k \quad \delta_{jk}, (n \quad 1) \quad (j,k) \quad N$, and also orthogonal to all a_i [324]; then the vectors

$$w_i \qquad a_i \qquad \sum_{l \; n \; 1}^{N} \gamma_{il} \, v_l \qquad 1 \quad i \quad N$$

form a complete orthonormal basis in $\mathcal{H}_N$ if and only if

$$w_i \; w_j \qquad \delta_{ij}.$$

This means that

$$a_i \; a_j \qquad \sum_{l \; n \; 1}^{N} \gamma_{il} \, \gamma_{jl} \qquad \delta_{ij}.$$

The unknown coefficients, γ_{il}, can be determined from this system of equations. We notice that we have more equations than unknowns. We also notice that a_i are not arbitrary; they satisfy the closure property $\sum \; a_i \quad a_i \qquad \sum \mathbf{A}_i \quad \mathbf{I}$. We write this system as

$$\sum_{k \; 1}^{n} \alpha_{ik} \alpha_{jk} \qquad \sum_{l \; n \; 1}^{N} \gamma_{il} \, \gamma_{jl} \qquad \delta_{ij} \quad (i,j \quad 1,\ldots,N).$$

We now consider the square matrix of order N:

$$M \qquad \begin{pmatrix} \alpha_{p1} & \ldots & \alpha_{pn} & \gamma_{p(n \; 1)} & \ldots & \gamma_{pN} \\ \alpha_{q1} & \ldots & \alpha_{qn} & \gamma_{q(n \; 1)} & \ldots & \gamma_{qN} \\ \vdots & & \vdots & \vdots & & \vdots \\ \alpha_{N1} & \ldots & \alpha_{Nn} & \gamma_{N(n \; 1)} & \ldots & \gamma_{NN} \end{pmatrix}.$$

The first n columns of the matrix, M, consist of the, α_{ik}, given to us, and the last $N \quad n$ columns are the unknown, γ_{il}. This matrix is unitary; the first n columns can be considered n orthonormal vectors in an N-dimensional space, and the remaining $(N \quad n)$ column vectors can be constructed in infinitely many ways. The N column vectors allow us to construct the orthonormal vectors, w_i , and then the projections of w_i to $\mathcal{H}_n$ give us the a_i . ■

In the general case, the extension from $\mathcal{H}_n$ to $\mathcal{H}_N$ requires the introduction of ancilla qubits. The Neumark theorem proves that an arbitrary POVM, with a finite number of elements can, in principle, be converted to a maximal quantum test by using the ancilla, an auxiliary independently prepared quantum system.

2.10 GLEASON THEOREM

A measurement provides additional information about a quantum system in a pure or in a mixed state, and it affects the density operator of the system. The post-measurement state is characterized by new probabilities, and it is different from the state of the quantum system prior to the measurement.

We now consider N independent versions of a quantum system, each in a different state ψ_i , $1 \quad i \quad N$. The Gleason theorem [171] allows us to compute the ensemble average of the operator, $\mathbf{A}$, after we measure the same observable, A, on all N versions of the system [2].

Theorem. *The ensemble average of the observable, A, is equal to the trace of the product of the measurement operator,* $\mathbf{A}$, *and the density operator,* ρ:

$$\mathbf{A} \quad \mathrm{tr}(\mathbf{A}\rho).$$

Proof. Let a_j denote the set of vectors of an orthonormal basis and assume that these orthogonal vectors are the eigenstates of a self-adjoint operator. The ensemble average, $\mathbf{A}$, of the measurements performed on the observable A is defined as

$$\mathbf{A} \quad \frac{1}{N}\sum_{i \; 1}^{N} \psi_i \; \mathbf{A} \; \psi_i \; .$$

If the system is initially in the state ψ_i , as a result of applying the measurement operator $\mathbf{A}$, it ends up in the state $\varphi_i \quad \mathbf{A} \; \psi_i$. Thus, the ensemble average can be expressed as the average of inner products of the states before and after the measurement

$$\mathbf{A} \quad \frac{1}{N}\sum_{i \; 1}^{N} \psi_i \; \varphi_i \; .$$

The vectors of the orthonormal basis, a_j , and the corresponding projectors, satisfy the completeness condition, $\sum_j a_j \quad a_j \quad \mathbf{I}$. Then we rewrite

$$\psi_i \; \mathbf{A} \qquad \psi_i \; \mathbf{IA} \; \psi_i \qquad \psi_i \left[\sum_{j \; 1}^{N} a_j \quad a_j \; \mathbf{A} \right] \psi_i \qquad \sum_{j \; 1}^{N} \psi_i \; a_j \quad a_j \; \mathbf{A} \; \psi_i \; ,$$

or

$$\psi_i \; \mathbf{A} \; \psi_i \qquad \sum_{j \; 1}^{N} a_j \; \mathbf{A} \; \psi_i \quad \psi_i \; a_j \; .$$

Note that in the previous expression, $\psi_i \; a_j$ is the projection of ψ_i on the basis vector a_j . It follows that

$$\mathbf{A} \quad \frac{1}{N}\sum_{i \;\; 1}^{N}\sum_{j \;\; 1}^{N} a_j \; \mathbf{A} \; \psi_i \quad \psi_i \; a_j \qquad \sum_{j \;\; 1}^{N} a_j \; \mathbf{A} \frac{1}{N}\sum_{i \;\; 1}^{N} \psi_i \quad \psi_i \; a_j \; .$$

When $p_i \quad 1/N$, the density operator ρ is

$$\rho \quad \frac{1}{N}\sum_{i \;\; 1}^{N} \psi_i \quad \psi_i \; .$$

Then the ensemble average $\mathbf{A}$ can be written as

$$\mathbf{A} \quad \sum_{j \;\; 1}^{N} a_j \; \mathbf{A}\rho \; a_j \qquad \mathrm{tr}(\mathbf{A}\rho).$$

To complete the proof, we recall that $\mathrm{tr}(\mathbf{A}\rho) \quad \mathrm{tr}(\rho\mathbf{A})$. ■

Example. Consider a mixed ensemble of quantum states in a two-dimensional Hilbert space, with the orthonormal basis vectors 0 and 1 .

The states and their respective occurrence probabilities in the mixed ensemble are:

State			Probability
ψ_A	$\alpha_0 \; 0$	$\alpha_1 \; 1$	p
ψ_B	$\beta_0 \; 0$	$\beta_1 \; 1$	$1 \quad p$

The density operator of this mixed ensemble of states is

$$\rho \quad p \; \psi_A \quad \psi_A \qquad (1 \quad p) \; \psi_B \quad \psi_B \; .$$

A measurement of this system is specified by the projectors

$$\mathbf{P}_0 \quad 0 \;\; 0 \quad \text{and} \quad \mathbf{P}_1 \quad 1 \;\; 1 \; .$$

The two possible outcomes of a measurement of the system in the mixed state are 0 and 1. In a series of such measurements, we expect to obtain the outcome 0, with the probability:

$$p(0) \quad \mathrm{Prob}(\psi_A)\mathrm{Prob}(0, \psi_A) \quad \mathrm{Prob}(\psi_B)\mathrm{Prob}(0, \psi_B).$$

In this expression, $\mathrm{Prob}(\psi_A)$ is the probability of the state ψ_A in the mix, $\mathrm{Prob}(0, \psi_A)$ is the conditional probability of observing the outcome 0 in the state ψ_A , and it is equal to $\psi_A \; \mathbf{P}_0 \; \psi_A$. Similarly, $\mathrm{Prob}(0, \psi_B)$ is the conditional probability of observing the outcome 0 in the state ψ_B , and it is equal to $\psi_B \; \mathbf{P}_0 \; \psi_B$. However, $\mathrm{Prob}(\psi_A) \quad p$ and $\mathrm{Prob}(\psi_B) \quad 1 \quad p$; thus,

$$p(0) \quad p \; \psi_A \; \mathbf{P}_0 \; \psi_A \qquad (1 \quad p) \; \psi_B \; \mathbf{P}_0 \; \psi_B \; ,$$

with $\text{Prob}(0, |\psi_A\rangle) = |\alpha_0|^2$ and $\text{Prob}(0, |\psi_B\rangle) = |\beta_0|^2$. It follows that

$$p(0) = p|\alpha_0|^2 + (1-p)|\beta_0|^2.$$

According to the Gleason theorem, ensemble-averaged quantities such as $p(0)$ above can be represented in terms of the density operator ρ as

$$\begin{aligned} p(0) &= \text{tr}(\rho \mathbf{P}_0) \\ &= \text{tr}(\mathbf{P}_0 \rho). \end{aligned}$$

Indeed,

$$\begin{aligned} \text{tr}(\rho \mathbf{P}_0) &= \langle 0|\rho \mathbf{P}_0|0\rangle + \langle 1|\rho \mathbf{P}_0|1\rangle \\ &= \langle 0|\rho|0\rangle \\ &= \langle 0|[p|\psi_A\rangle\langle\psi_A| + (1-p)|\psi_B\rangle\langle\psi_B|]|0\rangle \\ &= p\langle 0|\psi_A\rangle\langle\psi_A|0\rangle + (1-p)\langle 0|\psi_B\rangle\langle\psi_B|0\rangle \\ &= p|\alpha_0|^2 + (1-p)|\beta_0|^2 \\ &= p(0). \end{aligned}$$

The density operator plays an important role in quantum information theory, notably in the characterization of the noise inevitable in quantum communication. Consider quantum states affected by noise and characterized by vectors that are not necessarily normalized in a Hilbert space (e.g., $|\tilde{\psi}_i\rangle$ and $|\tilde{\varphi}_i\rangle$). The relationship among quantum states with the same density operator is captured by the following proposition.

Proposition. *Consider two sets of mixed states, $|\tilde{\psi}_i\rangle$ and $|\tilde{\varphi}_i\rangle$, that may not be normalized to unit length; the two sets are extended to have the same number of component states by padding the one with fewer states with additional zero vectors. Let $U = [u_{ij}]$ be a unitary matrix of complex numbers. $|\tilde{\psi}_i\rangle$ and $|\tilde{\varphi}_i\rangle$ generate the same density operator if and only if*

$$|\tilde{\psi}_i\rangle = \sum_j u_{ij}|\tilde{\varphi}_j\rangle.$$

Proof. We only prove one side of the proposition, namely, that if $|\tilde{\psi}_i\rangle = \sum_j u_{ij}|\tilde{\varphi}_j\rangle$, then the two vectors characterize states with the same density operator:

$$\sum_i |\tilde{\psi}_i\rangle\langle\tilde{\psi}_i| = \sum_{i,j,k} u_{ij}u_{ik}^* |\tilde{\varphi}_j\rangle\langle\tilde{\varphi}_k| = \sum_{j,k}\left(\sum_i u_{ki}^\dagger u_{ij}\right)|\tilde{\varphi}_j\rangle\langle\tilde{\varphi}_k|.$$

Thus,

$$\sum_i |\tilde{\psi}_i\rangle\langle\tilde{\psi}_i| = \sum_{j,k}\delta_{k,j}|\tilde{\varphi}_j\rangle\langle\tilde{\varphi}_k| = \sum_j |\tilde{\varphi}_j\rangle\langle\tilde{\varphi}_j|.$$

■

Corollary. *Given a Hilbert space, $\mathcal{H}_n$, of dimension, $n \quad 3$, the only possible measure of the probability of the state associated with a particular linear subspace, $\mathcal{H}_m$, where $m < n$, of the Hilbert space $\mathcal{H}_n$, is the trace of the operator product of the projection operator,* **E**, *that projects $\mathcal{H}_n$ onto $\mathcal{H}_m$, and the density matrix, ρ, of the system, $tr(\rho\mathbf{E})$.*

The corollary makes the assumption that in quantum mechanics we should be able to measure the events and assign probabilities to all possible orthogonal projections in a Hilbert space $\mathcal{H}$. The projectors are unitary, $\mathbf{E}^2 \quad \mathbf{E}$ and self-adjoint $\mathbf{E} \quad \mathbf{E}^\dagger$. The state of a quantum system is a mapping of each projection to a real number between zero and one, the probability of a particular outcome of the measurement:

$$\mathbf{E} \quad p(\mathbf{E}); \quad 0 \quad p(\mathbf{E}) \quad 1.$$

This mapping must have the three properties:

1. $p(\mathbf{0}) \quad 0.$
2. $p(\mathbf{1}) \quad 1.$
3. $\mathbf{E}_1\mathbf{E}_2 \quad 0 \qquad p(\mathbf{E}_1 \quad \mathbf{E}_2) \quad p(\mathbf{E}_1) \quad p(\mathbf{E}_2).$

The first and second properties give the probabilities for the projectors on the null subspace and the entire space, **0** and **1**, respectively; the last property re ects the fact that probabilities assigned to mutually orthogonal spaces must be additive; this makes sense because mutually orthogonal subspaces can be considered exclusive alternatives.

The Gleason theorem is an important result in quantum logic; quantum logic treats quantum events as logical propositions and studies the relationships and structures formed by these events. Formally, a quantum logic is a set of events that is closed under a countable disjunction of countably many mutually exclusive events. The representation theorem in quantum logic shows that these logics form a lattice that is isomorphic to the lattice of subspaces of a vector space with a scalar product.[6] The representation theorem allows us to treat quantum events as a lattice $\mathcal{L} \quad \mathcal{L}(\mathcal{H})$ of subspaces of a real or complex Hilbert space $\mathcal{H}$. The Gleason theorem allows us to assign probabilities to these events. The Gleason theorem [171] also shows that the Born rule for the probability of obtaining a specific outcome of a given measurement follows naturally from the structure formed by the lattice of events in a real or complex Hilbert space.

2.11 MIXED ENSEMBLES AND THEIR TIME EVOLUTION

We now discuss the effect of a measurement of a mixed ensemble on its density operator and show that the time evolution of mixed ensembles can be expressed through the time evolution of its density operator. The density operator, ρ, or the quantum state vector, ψ , are *mathematical objects* that represent our knowledge about the preparation of a physical system. They are not "physical" objects in the way hydrodynamic waves or electromagnetic wavepackets are, though we could be tempted to

[6]A necessary result for the Gleason theorem to be applicable is that the field, K, over which the vector space is defined is either the real numbers, complex numbers, or the quaternions; the definition of the inner product of a nonzero vector with itself will satisfy the requirements to make the vector space a Hilbert space for real numbers, complex numbers, or the quaternions.

interpret them as such. In this mathematical framework, we can think of pure states, ψ , as representing "minimal uncertainty" states, while a density operator, ρ, represents the additional uncertainties, the probabilities, associated with a mixed ensemble.

We wish to express the probability of an outcome of the measurement knowing the density operator of the initial state of a system, and we also wish to determine the density operator of the resulting state when one of the projectors is used for the measurement.

Proposition. *Consider a mixed ensemble of the states,* ψ_i *, with the corresponding probabilities,* p_i*, and a density operator,* $\rho \quad \sum_i p_i \; \psi_i \;\; \psi_i$ *. Call* $p(k)$ *the probability of obtaining the outcome, with the index* k *and* ρ_k *the post-measurement density operator of the mixed ensemble in a measurement of an observable corresponding to the projector* $\mathbf{P}_k$*; then*

$$p(k) \quad \mathrm{tr}(\rho \mathbf{P}_k) \quad \text{and} \quad \rho_k \quad \frac{\mathbf{P}_k \rho \mathbf{P}_k}{\mathrm{tr}(\rho \mathbf{P}_k)}.$$

Proof. $p_i \quad$ Prob(ψ_i) denotes the probability of the state ψ_i ; call Prob(k, ψ_i) the probability of observing the outcome k of a measurement of the state ψ_i and recall that this probability is $\psi_i \; \mathbf{P}_k \; \psi_i$; thus,

$$p(k) \quad \sum_i \mathrm{Prob}(\ \psi_i\ )\ \mathrm{Prob}(k,\ \psi_i\ ) \quad \sum_i p_i \; \psi_i \; \mathbf{P}_k \; \psi_i \quad \sum_i p_i \; \psi_i \; \mathbf{P}_k \mathbf{I} \; \psi_i \ .$$

We use the fact that $\mathbf{I} \quad \sum_j \; j \;\; j$ and obtain

$$p(k) \quad \sum_i p_i \; \psi_i \; \mathbf{P}_k \sum_j (\ j \;\; j\ ) \; \psi_i \quad \sum_j \; j \; \rho \mathbf{P}_k \; j \quad \mathrm{tr}(\rho \mathbf{P}_k).$$

Now we concentrate on the second relationship and denote by $\varphi_{(i,k)}$ the normalized state resulting when the projector, or measurement operator, $\mathbf{P}_k$, is applied to the state, ψ_i of the mixed ensemble. Call $p(i \; k)$ the conditional probability of applying $\mathbf{P}_k$ to the state ψ_i of the mixed ensemble; then the density operator, ρ_k, of the mixed ensemble after the application of $\mathbf{P}_k$ is

$$\rho_k \quad \sum_i p(i \; k) \; \varphi_{(i,k)} \;\; \varphi_{(i,k)} \ .$$

The normalized state resulting when $\mathbf{P}_k$ is applied to the state ψ_i of the mixed ensemble is

$$\varphi_{(i,k)} \qquad \frac{\mathbf{P}_k \; \psi_i}{\sqrt{\psi_i \; \mathbf{P}_k \; \psi_i}}.$$

The relation between conditional probabilities allows us to express

$$p(i \; k) \quad \frac{p(k \; i) p_i}{p(k)},$$

with the probability of outcome with the index k, $p(k) \quad \mathrm{tr}(\rho \mathbf{P}_k)$, and the conditional probability of applying $\mathbf{P}_k$ to the state ψ_i of the mixed ensemble, $p(i \; k) \quad \psi_i \; \mathbf{P}_k \; \psi_i$. Here, $p(k \; i)$ is the conditional probability of applying $\mathbf{P}_i$ to the state ψ_k of the mixed ensemble.

The density operator of the mixed ensemble after the application of $\mathbf{P}_k$ is

$$\rho_k \quad \sum_i p_i \frac{\psi_i \ \mathbf{P}_k \ \psi_i}{\mathrm{tr}(\rho \mathbf{P}_k)} \quad \frac{\mathbf{P}_k \ \psi_i \quad \psi_i \ \mathbf{P}_k}{\psi_i \ \mathbf{P}_k \ \psi_i} \quad \frac{\mathbf{P}_k[\sum_i p_i \ \psi_i \quad \psi_i \ ]\mathbf{P}_k}{\mathrm{tr}(\rho \mathbf{P}_k)}.$$

Thus,

$$\rho_k \quad \frac{\mathbf{P}_k \rho \mathbf{P}_k}{\mathrm{tr}(\rho \mathbf{P}_k)}.$$

■

Mixed ensembles, such as $p_i^A, \ \psi_i^A$, with the density operator ρ_A and $p_i^B, \ \psi_i^B$, with the density operator ρ_B, can be combined. The states, ψ_A and ψ_B the in the combined ensemble occur with the probabilities pp_i^A and $(1 \quad p)p_i^B$, respectively. The density operator ρ of the combined ensemble is

$$\rho \quad p\rho_A \quad (1 \quad p)\rho_B.$$

Here, p and $(1 \quad p)$, with $0 \quad p \quad 1$, are the probability of occurrence in the combined ensemble of the individual mixed ensembles A and B, respectively.

Recall that the Schrödinger equation describes the evolution in time of the mathematical objects describing the system preparation, given that we know the Hamiltonian of the system. When we perform a measurement on a quantum system, we gain new information about that system, and our initial description of it needs to change instantaneously in order to make possible the prediction of the outcomes of further measurements. According to the Schrödinger equation, a unitary time evolution operator, T, can be used to express the evolution in time of a pure quantum state:

$$\psi(t) \quad \mathbf{T}(t,0) \ \psi(0) \quad \text{and} \quad \psi(t) \quad \psi(0) \ \mathbf{T}(0,t).$$

According to its definition, the density operator at time, t, can be written as

$$\begin{aligned} \rho(t) \quad & \sum_i p_i \ \psi_i(t) \quad \psi_i(t) \\ & \sum_i p_i \mathbf{T}(t,0) \ \psi_i(0) \quad \psi_i(0) \ \mathbf{T}(0,t) \\ & \mathbf{T}(t,0) \sum_i p_i \ \psi_i(0) \quad \psi_i(0) \ \mathbf{T}(0,t) \\ & \mathbf{T}(t,0)\rho(0)\mathbf{T}(0,t). \end{aligned}$$

Now, the trace of the density operator can be written as

$$\mathrm{tr}(\rho(t)) \quad \mathrm{tr}(\mathbf{T}(t,0)\rho(0)\mathbf{T}(0,t)) \quad \mathrm{tr}(\mathbf{T}(0,t)\mathbf{T}(t,0)\rho(0)) \quad \mathrm{tr}(\rho(0)),$$

where $\mathbf{T}(0,t)\mathbf{T}(t,0) \quad I$. We see that the trace of the density operator is preserved under Hamiltonian evolution; the Hamiltonian evolution preserves the eigenvalues of the density operator.

2.12 BIPARTITE SYSTEMS: SCHMIDT DECOMPOSITION

Bipartite systems are of interest in quantum information theory because quantum systems interact with one another and with the environment; such interactions affect the information encoded in the quantum state. For example, a quantum system, $\mathcal{A}$, in a pure state may interact with the environment and as a result of this interaction the state of the system may become a mixed state; the system $\mathcal{A}$ may be looked at as part of a bipartite system, $\mathcal{A}$ environment. The formalism to describe bipartite systems is based on *Schmidt decomposition*.

Proposition. *Let* $\psi_{\mathcal{C}}$ *be the state of a bipartite system,* $\mathcal{C}$ $\mathcal{AB}$, *with* $\mathcal{A}$ $\mathcal{H}_A$ *and* $\mathcal{B}$ $\mathcal{H}_B$*; there are two orthonormal bases, called Schmidt bases* e_j^A $\mathcal{H}_A$ *and* e_j^B $\mathcal{H}_B$, *such that*

$$\psi_{\mathcal{C}} \quad \sum_j \lambda_j \, e_j^A \quad e_j^B \, .$$

In this expression, λ_j are called *Schmidt coefficients*, λ_j are real numbers, and $\sum_j \lambda_j^2 \quad 1$.

Proof. When $\dim[\mathcal{H}_A] \quad \dim[\mathcal{H}_B] \quad n$, we can express the state of $\mathcal{C} \quad \mathcal{H}_A \quad \mathcal{H}_B$ as a superposition,

$$\psi_{\mathcal{C}} \quad \sum_{i \; 1}^{n} \sum_{k \; 1}^{n} \gamma_{ik} \, u_i \quad v_k \, ,$$

using the orthonormal bases, $u_j \quad \mathcal{H}_A$ and $v_j \quad \mathcal{H}_B, 1 \quad j \quad n$.

The complex coefficients, $\gamma_{ik}, 1 \quad i,k \quad n$, can be viewed as elements of a matrix $\Gamma \quad [\gamma_{ik}]$. The singular value decomposition theorem states that an $n \quad n$ matrix, Γ, over a field, F, of real or complex numbers can be decomposed as a product of three $n \quad n$ matrices over F: $\Gamma \quad A \Delta B^{\dagger}$, with $A \quad [\alpha_{ij}]$ and $B \quad [\beta_{ij}]$, unitary matrices, and $\Delta \quad [\delta_{jj}]$, a diagonal matrix. This allows us to write the previous expression as

$$\psi_{\mathcal{C}} \quad \sum_{i \; 1}^{n} \sum_{j \; 1}^{n} \sum_{k \; 1}^{n} \alpha_{ij} \delta_{jj} \beta_{jk} \, u_j \quad v_k \, .$$

We denote

$$e_i^A \quad \sum_{j \; 1}^{n} \alpha_{ij} \, u_j \, , \quad e_j^B \quad \sum_{k \; 1}^{n} \beta_{jk} \, v_k \, , \quad \text{and} \quad \lambda_j \quad \delta_{jj}.$$

Recall that u_j form an orthonormal basis in $\mathcal{H}_A$ and that A is unitary. It is easy to see that the vectors, e_i^A, form an orthonormal basis in $\mathcal{H}_A$,

$$e_i^A \, e_k^A \quad \sum_{j \; 1}^{n} \sum_{k \; 1}^{n} \alpha_{ji}^{*} \alpha_{ik} \, u_j \, u_k \quad \delta_{jk},$$

because $\sum_i \; \alpha_{ii} \; ^2 \quad 1$. Similarly, the vectors $\; e_j^B \;$ form an orthonormal basis in $\mathcal{H}_{\mathcal{B}}$. Thus,

$$\psi_{\mathcal{C}} \quad \sum_j \lambda_j \, e_j^A \quad e_j^B \; .$$

The proof for the case when the two subsystems have Hilbert spaces with different dimensions is more elaborate. ■

Consider a bipartite pure state, $\psi_{\mathcal{C}}$, of a composite system, $\mathcal{C} \quad \mathcal{AB}$. The number of nonzero eigenvalues, λ_j, of $\rho^{\mathcal{C}}$, thus, the number of terms in the Schmidt decomposition of $\psi_{\mathcal{C}}$ is called the *Schmidt number of state* $\psi_{\mathcal{C}}$. The Schmidt number is a measure of the degree of entanglement of the two subsystems.

A *separable bipartite pure state,* $\varphi_{\mathcal{AB}}$, over the Hilbert space, $\mathcal{H}_{\mathcal{A}} \quad \mathcal{H}_{\mathcal{B}}$, is a tensor product of two pure states $\zeta_{\mathcal{A}} \quad \mathcal{H}_{\mathcal{A}}$ and $\xi_{\mathcal{A}} \quad \mathcal{H}_{\mathcal{B}}$:

$$\varphi_{\mathcal{AB}} \qquad \zeta_{\mathcal{A}} \qquad \xi_{\mathcal{B}} \; .$$

It follows that the reduced density matrices correspond to the pure states

$$\rho^{\mathcal{A}} \qquad \zeta_{\mathcal{A}} \;\; \zeta_{\mathcal{A}} \quad \text{and} \quad \rho^{\mathcal{B}} \qquad \xi_{\mathcal{B}} \;\; \xi_{\mathcal{B}} \; .$$

The concepts "separability," "correlation," and "entanglement" should be correctly understood in the following sense [342]:

- Two subsystems, $\mathcal{A}$ and $\mathcal{B}$, can be correlated even if they are separable; for example, two spins in the state

$$\varphi_{\mathcal{AB}} \qquad \mathcal{A} \qquad \mathcal{B}$$

  are separable, but correlated since both spins are up. This state can be created by acting individually on the two subsystems; the two subsystems need not be at any instant of time together. We can just send a message to the two individuals preparing the two systems that each one of them should have the spin up.
- Entanglement is a special type of correlation. The two correlated systems are not separable. The entanglement cannot be created locally at the two sites where the subsystems $\mathcal{A}$ and $\mathcal{B}$ are located; the two subsystems $\mathcal{A}$ and $\mathcal{B}$ must interact directly with each other to become entangled. The state

$$\varphi_{\mathcal{AB}} \qquad \frac{1}{2}(\quad \mathcal{A} \qquad \mathcal{B} \qquad \mathcal{A} \qquad \mathcal{B} \;)$$

  requires a unitary transformation to be applied to the two qubits that have to be brought to some place and allowed to interact.

We conclude this discussion with the observation that the Schmidt number provides a clear distinction between bipartite pure states, $\psi_{\mathcal{AB}}$, that are separable and those that are entangled: *A bipartite pure state* $\psi_{\mathcal{AB}}$ *is entangled if its Schmidt number is greater than one; otherwise, it is separable.*

2.13 MEASUREMENTS OF BIPARTITE SYSTEMS

We continue our discussion of bipartite quantum systems and explore the question of how to gather information about one of the subsystems, $\mathcal{A}$, from measurements performed on the bipartite system, $\mathcal{C}$ $\mathcal{AB}$ [44, 307]. The bipartite system may be in a product state when $\psi_{\mathcal{C}}$ $\psi_{\mathcal{A}}$ $\psi_{\mathcal{B}}$, with $\psi_{\mathcal{A}}$, the state of subsystem $\mathcal{A}$ and $\psi_{\mathcal{B}}$, the state of subsystem $\mathcal{B}$, or in a state with some degree of entanglement between the two subsystems when $\psi_{\mathcal{C}}$ $\psi_{\mathcal{A}}$ $\psi_{\mathcal{B}}$ (Figure 2.8). The following proposition allows us to find the average value of an observable of one subsystem only.

Proposition. *Let $\psi_{\mathcal{C}}$ be the current state of a bipartite system, $\mathcal{C}$, consisting of two subsystems, $\mathcal{A}$ and $\mathcal{B}$; we measure an observable, M, of the subsystem $\mathcal{A}$. Then the average value of the corresponding measurement operator,* $\mathbf{M}^{\mathcal{A}}$, *is*

$$\mathbf{M}^{\mathcal{A}} \quad \psi_{\mathcal{C}}\ \mathbf{M}^{\mathcal{A}} \quad \mathbf{I}^{\mathcal{B}}\ \psi_{\mathcal{C}} .$$

This follows immediately if we think of the measurement as applying the operator, $\mathbf{M}^{\mathcal{A}}$ $\mathbf{I}^{\mathcal{B}}$, to the bipartite system; $\mathbf{I}^{\mathcal{B}}$ is the identity operator applied to subsystem $\mathcal{B}$.

Example. *The measurement of a bipartite system of spin-½ particles.* If $\mathcal{C}$ is a system of two entangled particles and each subsystem can be described using the basis vectors and , then $\psi_{\mathcal{C}}$ α_0 α_1 . The expected value of the observable O for one of the two particles, without specifying which one, is

$$O \quad \psi_{\mathcal{C}}\ \mathbf{O} \quad \mathbf{I}\ \psi_{\mathcal{C}} \quad \alpha_0\ ^2(\quad \mathbf{O} \quad) \quad \alpha_1\ ^2(\quad \mathbf{O} \quad) \quad \mathrm{tr}(\rho\mathbf{O}),$$

where the density operator is given by ρ $\alpha_0\ ^2$ $\alpha_1\ ^2$. The operators and are projectors on the basis states and . The result is similar to the case where any of the two particles is detected with the probability $\alpha_0\ ^2$ or $\alpha_1\ ^2$ to be in the state or , respectively.

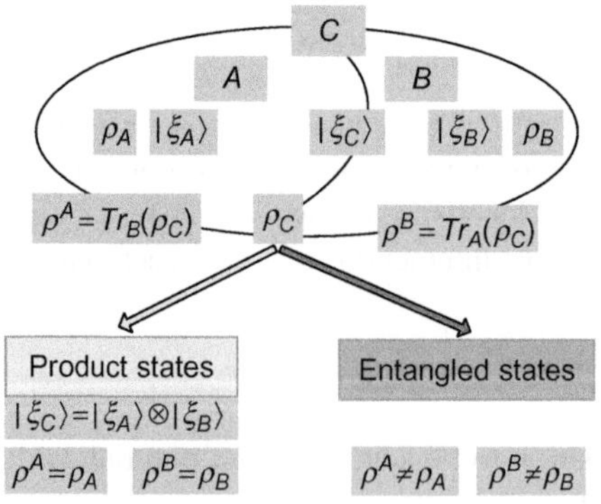

FIGURE 2.8

$\mathcal{C}$ is a bipartite system consisting of systems $\mathcal{A}$ and $\mathcal{B}$. The state and the density operators of the three systems are, respectively, $\xi_{\mathcal{C}}$ and $\rho_{\mathcal{C}}$, $\xi_{\mathcal{A}}$ and $\rho_{\mathcal{A}}$, $\xi_{\mathcal{B}}$ and $\rho_{\mathcal{B}}$. The bipartite system can be either in a product state or in an entangled state. $\mathcal{C}$ is in a product state if $\xi_{\mathcal{C}}$ $\xi_{\mathcal{A}}$ $\xi_{\mathcal{B}}$; in this case, the reduced density operators satisfy the equalities $\rho^{\mathcal{A}}$ $\rho_{\mathcal{A}}$ and $\rho^{\mathcal{B}}$ $\rho_{\mathcal{B}}$.

If we repeat the experiment many times, the result is also similar to that obtained for an ensemble of particles; some in state   and, some in state   .

Let $\mathcal{C} \quad \mathcal{AB}$ be a bipartite system consisting of subsystems, $\mathcal{A}$ and $\mathcal{B}$, and described by the density operator, $\rho_{\mathcal{C}}$. The *partial trace of* $\rho_{\mathcal{C}}$ *over system* $\mathcal{B}$ is

$$\text{tr}_{\mathcal{B}}(\rho_{\mathcal{C}}) \quad \text{tr}_{\mathcal{B}}(a_1 \quad a_2 \qquad b_1 \quad b_2) \qquad a_1 \quad a_2 \, \text{tr}_{\mathcal{B}}(b_1 \quad b_2) \qquad a_1 \quad a_2 \quad b_1 \, b_2 \, ,$$

with a_1 , a_2 any two vectors in the state space of $\mathcal{A}$ and b_1 , b_2 any two vectors in the state space of $\mathcal{B}$. The *reduced density operator of subsystem* $\mathcal{A}$ is defined as $\rho^{\mathcal{A}} \quad \text{tr}_{\mathcal{B}}(\rho_{\mathcal{C}})$.

Proposition. *Let* $\mathcal{C} \quad \mathcal{AB}$ *be a bipartite system in a product state, with* $\rho_{\mathcal{A}}$*, the density operator of subsystem* $\mathcal{A}$*,* $\rho_{\mathcal{B}}$ *the density operator of subsystem* $\mathcal{B}$*, and* $\rho_{\mathcal{C}} \quad \rho_{\mathcal{A}} \quad \rho_{\mathcal{B}}$*. Then the reduced density operator of each subsystem is equal to the density operator of the subsystem*

$$\rho^{\mathcal{A}} \quad \rho_{\mathcal{A}} \quad \text{and} \quad \rho^{\mathcal{B}} \quad \rho_{\mathcal{B}}.$$

Proof. Recall that the trace of the density operator of a system is equal to unity:

$$\text{tr}_{\mathcal{A}}(\rho_{\mathcal{A}}) \quad 1 \quad \text{and} \quad \text{tr}_{\mathcal{B}}(\rho_{\mathcal{B}}) \quad 1.$$

According to the definition of the reduced density operator:

$$\rho^{\mathcal{A}} \quad \text{tr}_{\mathcal{B}}(\rho_{\mathcal{C}}) \quad \text{tr}_{\mathcal{B}}(\rho_{\mathcal{A}} \quad \rho_{\mathcal{B}}) \quad \rho_{\mathcal{A}}\text{tr}_{\mathcal{B}}(\rho_{\mathcal{B}}) \quad \rho_{\mathcal{A}},$$
$$\rho^{\mathcal{B}} \quad \text{tr}_{\mathcal{A}}(\rho_{\mathcal{C}}) \quad \text{tr}_{\mathcal{A}}(\rho_{\mathcal{A}} \quad \rho_{\mathcal{B}}) \quad \rho_{\mathcal{B}}\text{tr}_{\mathcal{A}}(\rho_{\mathcal{A}}) \quad \rho_{\mathcal{B}}.$$

■

This result re ects also our intuition; if indeed both the density operators, $\rho_{\mathcal{A}}$, and the reduced density operator, $\rho^{\mathcal{A}}$, characterize the same state of the system, $\mathcal{A}$, then the average of an observable should be the same regardless of whether it is computed using $\rho_{\mathcal{A}}$ or $\rho^{\mathcal{A}}$. This is true only for product states; if the system is in an entangled state, we expect that the interaction of the two subsystems will affect the outcome of a measurement in a more subtle manner (Figure 2.8).

Proposition. *Let* $M^{\mathcal{A}}$ *be an observable of* $\mathcal{A}$ *and* $M^{\mathcal{C}}$ *be the corresponding observable of* $\mathcal{A}$ *but performed on* $\mathcal{C} \quad \mathcal{AB}$*. Let* $\mathbf{I}^{\mathcal{B}}$ *be the identity operator in the state space of* $\mathcal{B}$ *and* $\rho_{\mathcal{A}}$ *and* $\rho_{\mathcal{C}}$ *be the density operators for the two systems, respectively. Then, the measurement operator for the observable* $M^{\mathcal{C}}$ *is*

$$\mathbf{M}^{\mathcal{C}} \quad \mathbf{M}^{\mathcal{A}} \quad \mathbf{I}^{\mathcal{B}}.$$

Proof. Let φ be any state of $\mathcal{B}$; let m be an eigenstate of $\mathbf{M}^{\mathcal{A}}$, with the eigenvalue m and $\mathcal{P}_m$ the projector with outcome m. Then the spectral decomposition of the measurement operator $\mathbf{M}^{\mathcal{A}}$ for the observable $M^{\mathcal{A}}$ is

$$\mathbf{M}^{\mathcal{A}} \quad \sum_m m\mathcal{P}_m.$$

The projector of the measurement operator $\mathbf{M}^{\mathcal{C}}$ is

$$\mathbf{M}^{\mathcal{C}} \quad \mathbf{M}^{\mathcal{A}} \quad \mathbf{I}^{\mathcal{B}} \quad \sum_m m\mathcal{P}_m \quad I^{\mathcal{B}}.$$

A measurement of the state, $m \quad \varphi$, of $\mathcal{C}$, must produce the result, m, with certainty. The average of the measurements performed on $\mathcal{A}$ and on $\mathcal{C}$ must be the same:

$$\mathrm{tr}(\mathbf{M}^{\mathcal{A}}\rho_{\mathcal{A}}) \quad \mathrm{tr}((\mathbf{M}^{\mathcal{A}} \quad \mathbf{I}^{\mathcal{B}})\rho_{\mathcal{C}}).$$

The partial trace, $\rho^{\mathcal{A}} \quad \mathrm{tr}_{\mathcal{B}}(\rho_{\mathcal{C}})$, is the only function satisfying this equation (see Nielsen and Chuang [307]). ∎

Proposition. *If* $\psi_{\mathcal{C}}$ *is a pure state of the bipartite system* $\mathcal{C} \quad \mathcal{AB}$*, then the eigenvalues of the reduced density operators of the two component systems are identical. Thus, many properties of the two subsystems that are determined by the eigenvalues of the reduced density operators are the same for the two subsystems.*

Proof. The reduced density operator for the subsystem $\mathcal{A}$ can be transformed when we apply the Schmidt decomposition as

$$\rho^{\mathcal{A}} \quad \mathrm{tr}_{\mathcal{B}}(\rho_{\mathcal{C}}) \quad \mathrm{tr}_{\mathcal{B}}(\ \psi_{\mathcal{C}} \quad \psi_{\mathcal{C}}\ ) \quad \sum_{j\ 1}^{n}\lambda_j\, e_j^{\mathcal{A}} \sum_{j\ 1}^{n}\lambda_j\ \ e_j^{\mathcal{A}} \quad \sum_{j\ 1}^{n}\ \lambda_j\ ^2\, e_j^{\mathcal{A}}\ \ e_j^{\mathcal{A}}\ .$$

Recall that the eigenvalues of the density operator are the coefficients of the projectors in the sum; thus, the eigenvalues of $\rho^{\mathcal{A}}$ are $\ \lambda_i\ ^2$. Similarly,

$$\rho^{\mathcal{B}} \quad \mathrm{tr}_{\mathcal{A}}(\rho_{\mathcal{C}}) \quad \mathrm{tr}_{\mathcal{A}}(\ \psi_{\mathcal{C}} \quad \psi_{\mathcal{C}}\ ) \quad \sum_{j\ 1}^{n}\lambda_j\, e_j^{\mathcal{B}} \sum_{j\ 1}^{n}\lambda_j\ \ e_j^{\mathcal{B}} \quad \sum_{j\ 1}^{n}\ \lambda_j\ ^2\, e_j^{\mathcal{B}}\ \ e_j^{\mathcal{B}}\ .$$

∎

Communication over quantum channels involves the transport of quantum particles and could certainly benefit from the formalism described above. This formalism allows us to determine the state of $\mathcal{A}$, the original quantum particle(s) prepared in the state $\varphi_{\mathcal{A}}$ based on observations performed on the bipartite system, $\mathcal{AB}$, in the state $\varphi_{\mathcal{AB}}$. Next, we discuss quantum teleportation, an application of a measurement of one subsystem of a bipartite system.

Quantum Teleportation

Quantum teleportation is the transfer of a quantum state from one quantum system to another. We assume that Alice and Bob are given a pair of entangled particles, called P_1 and P_2, in a maximally entangled state:

$$\psi_{12} \quad \beta_{00} \quad \frac{00 \quad 11}{\ 2}.$$

Eve asks Alice to deliver to Bob a message encoded as the state $\psi_3 \quad \alpha_0\ 0 \quad \alpha_1\ 1$ of particle P_3 (Figure 2.9). Alice follows the following procedure: She first entangles her half of the pair with the particle from Eve; she applies a CNOT to this pair, using the state of P_3 as the control qubit and the state of P_1 as the target qubit. Then Alice performs a joint measurement of her two qubits, whereby she measures the state of P_3 and leaves P_1 unchanged; as a result of this measurement, she obtains one of four values, 00 , 01 , 10 , and 11 , which correspond to the classical information 00, 01, 10, and 11, respectively, with the equal probability, $p \quad 1/4$. Alice sends the results of the measurements,

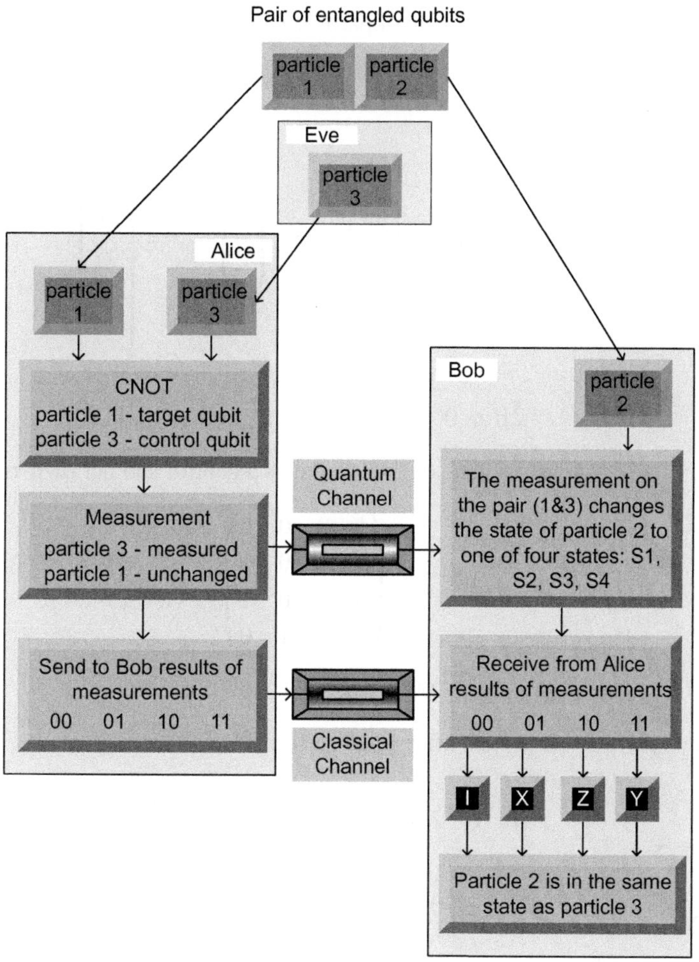

FIGURE 2.9

Quantum teleportation. Initially, Eve s particle is in the state $\psi_C \quad \alpha_0 \, 0 \quad \alpha_1 \, 1$; the pair of particles shared by Alice and Bob is in a maximally entangled state, β_{00} . At the end of the experiment, Bob s half of the entangled pair is in the state ψ_C , while the state of Eve s particle was affected by the quantum measurement and is no longer ψ_C . The change of state of Bob s particle does not occur instantaneously. Alice must use a classical communication channel to transmit the results of her measurements to Bob.

00, 01, 10, or 11, over a classical communication channel; upon receiving the results, Bob applies one of four transformations to the state of P_2: I if the string is 00, X for 01, Z for 10, or Y for 11. After the transformation, the state of P_2 replicates the state of P_3.

Quantum teleportation *does not involve an instantaneous exchange of information;* the state of Bob's qubit after Alice's measurement is not dependent on the state of P_3. No measurement performed by Bob after Alice's measurement and without knowing her result contains definite information about particle P_3. Alice *must use a classical communication channel* to transmit the result of her measurement to Bob; therefore, there is no instantaneous transfer of information. An analysis of this process follows:

Before the measurement ξ , the joint state of particle P_3 and the entangled pair of particles P_1 and P_2, is

$$\xi \quad \psi_3 \quad \psi_{12} \quad \begin{pmatrix} \alpha_0 \\ \alpha_1 \end{pmatrix} \quad \frac{1}{\overline{2}} \begin{pmatrix} 1 \\ 0 \\ 0 \\ 1 \end{pmatrix} \quad \frac{1}{\overline{2}} \begin{pmatrix} \alpha_0 \\ 0 \\ 0 \\ \alpha_0 \\ \alpha_1 \\ 0 \\ 0 \\ \alpha_1 \end{pmatrix},$$

$$\xi \quad (1/\ \overline{2})(\alpha_0\ 000 \quad \alpha_0\ 011 \quad \alpha_1\ 100 \quad \alpha_1\ 111\ ).$$

Alice applies a CNOT to the pair; she uses Eve's qubit as a control and her own as a target:

$$\kappa \quad (G_{CNOT} \quad I)(\ \xi\ ) \quad \begin{pmatrix} 1 & 0 & 0 & 0 \\ 0 & 1 & 0 & 0 \\ 0 & 0 & 0 & 1 \\ 0 & 0 & 1 & 0 \end{pmatrix} \quad \begin{pmatrix} 1 & 0 \\ 0 & 1 \end{pmatrix} (\ \xi\ ),$$

$$\kappa \quad (1/\ \overline{2})(\alpha_0\ 000 \quad \alpha_0\ 011 \quad \alpha_1\ 101 \quad \alpha_1\ 110\ ).$$

ζ , the state of the three-particle system after Alice performs her measurement, is

$$\zeta \quad (H \quad I \quad I)\ \kappa \quad \frac{1}{\overline{2}} \begin{pmatrix} 1 & 1 \\ 1 & 1 \end{pmatrix} \quad \begin{pmatrix} 1 & 0 \\ 0 & 1 \end{pmatrix} \quad \begin{pmatrix} 1 & 0 \\ 0 & 1 \end{pmatrix} \quad \kappa\ ,$$

$$\zeta \quad \frac{1}{2}[\alpha_0(\ 000 \quad 011 \quad 100 \quad 111\ ) \quad \alpha_1(\ 001 \quad 010 \quad 101 \quad 110\ )].$$

We rewrite the expression for ζ as

$$\zeta \quad 1/2[\ 00\ (\alpha_0\ 0 \quad \alpha_1\ 1\ ) \quad 01\ (\alpha_0\ 1 \quad \alpha_1\ 0\ )$$
$$10\ (\alpha_0\ 0 \quad \alpha_1\ 1\ ) \quad 11\ (\alpha_0\ 1 \quad \alpha_1\ 0\ )].$$

From this expression, it follows that when Alice performs a joint measurement of her two qubits, she gets one of the four equally probable results 00 or 01 , or 10 or 11 . As we already know, the measurement forces the pair of qubits 1 and 3 to one of the four basis states, and transforms quantum information to classical information. Then, she sends Bob the results of her measurements, 00, 01, 10, or 11, over a classical communication channel. At the same time, the measurement performed by Alice forces the qubit, P_2, in Bob's possession to change to one of the four states:

1. $\eta_{00} \quad \alpha_0\ 0 \quad \alpha_1\ 1$ when the result is 00.
2. $\eta_{01} \quad \alpha_0\ 1 \quad \alpha_1\ 0$ when the result is 01.
3. $\eta_{10} \quad \alpha_0\ 0 \quad \alpha_1\ 1$ when the result is 10.
4. $\eta_{11} \quad \alpha_0\ 1 \quad \alpha_1\ 0$ when the result is 11.

The density operator for the state ζ is

$$\rho_{\mathcal{C}}(\zeta) \quad \sum_{i\ 1}^{4} p_i \rho_i \quad 1/4(\rho_0 \quad \rho_1 \quad \rho_2 \quad \rho_3)$$

$$1/4\{ 00 \quad 00 \left[(\alpha_0\, 0 \quad \alpha_1\, 1\,)(\alpha_0\, 0 \quad \alpha_1\, 1\,)\right]$$
$$01 \quad 01 \left[(\alpha_0\, 1 \quad \alpha_1\, 0\,)(\alpha_0\, 1 \quad \alpha_1\, 0\,)\right]$$
$$10 \quad 10 \left[(\alpha_0\, 0 \quad \alpha_1\, 1\,)(\alpha_0\, 0 \quad \alpha_1\, 1\,)\right]$$
$$11 \quad 11 \left[(\alpha_0\, 1 \quad \alpha_1\, 0\,)(\alpha_0\, 1 \quad \alpha_1\, 0\,)\right]\}.$$

Bob's qubit is the second of the pair, and the reduced density operator of Bob's qubit is

$$\rho^{\mathcal{B}} \quad \mathrm{tr}_{\mathcal{A}}(\rho_{\mathcal{C}}(\zeta)) \quad 1/4[(\alpha_0\, 0 \quad \alpha_1\, 1\,)(\alpha_0\, 0 \quad \alpha_1\, 1\,)$$
$$(\alpha_0\, 1 \quad \alpha_1\, 0\,)(\alpha_0\, 1 \quad \alpha_1\, 0\,)$$
$$(\alpha_0\, 0 \quad \alpha_1\, 1\,)(\alpha_0\, 0 \quad \alpha_1\, 1\,)$$
$$(\alpha_0\, 1 \quad \alpha_1\, 0\,)(\alpha_0\, 0 \quad \alpha_1\, 1\,)].$$

Then,

$$\rho^{\mathcal{B}} \quad \frac{1}{4}[2\,0 \quad 0\,(\,\alpha_0\,^2 \quad \alpha_1\,^2) \quad 2\,1 \quad 1\,(\,\alpha_0\,^2 \quad \alpha_1\,^2)].$$

But we know that $\alpha_0\,^2 \quad \alpha_1\,^2 \quad 1$; thus,

$$\rho^{\mathcal{B}} \quad \frac{0\ 0 \quad 1\ 1}{2} \quad \frac{1}{2}\left[\begin{pmatrix}1 & 0\\ 0 & 0\end{pmatrix} \quad \begin{pmatrix}0 & 0\\ 0 & 1\end{pmatrix}\right] \quad \frac{1}{2}\begin{pmatrix}1 & 0\\ 0 & 1\end{pmatrix} \quad \frac{1}{2}I.$$

This confirms that the state of Bob's qubit after Alice's measurements, but before Bob has learned the measurement results, is $I/2$, and it is independent on the state of the particle P_3, as stated earlier. Note that Bob's qubit is in a mixed state; indeed,

$$\mathrm{tr}\left(\left(\frac{I}{2}\right)^2\right) \quad \frac{1}{2} < 1.$$

The particles, P_1 and P_3, are in pure states, while P_2 is in a mixed state. Needless to say that the teleportation Gedankenexperiment described in this section does not violate the "no-cloning theorem." The state of P_3 has not been cloned; it has been altered in the measurement process. P_2 acquires the original state of P_3 as a result of teleportation.

2.14 OPERATOR-SUM (KRAUS) REPRESENTATION

The state ψ of a quantum system, with the density matrix, ρ, can be transformed by applying the unitary operator, **U**, or a measurement operator, **M**. Call ρ the density matrix of the quantum system after the transformation and denote by, $\mathcal{E}$, a generic quantum operation, be it a unitary transformation or a measurement, such that

$$\rho \quad \mathcal{E}(\rho).$$

We distinguish open from closed quantum systems; an *open quantum system* consists of a *principal system* in contact and interacting with an *environment*. The ensemble (principal system environment) forms a *closed quantum system*. We consider three cases: (1) a unitary transformation, **U**, is applied to a system in isolation; (2) a measurement of the system; and (3) a unitary transformation, **U**, applied to an *open* system.

When the quantum operation is a unitary transformation **U** applied to a system in isolation (Figure 2.10),

$$\rho \quad \mathcal{E}(\rho) \quad \mathbf{U}\rho\mathbf{U}^{\dagger}.$$

If the quantum operation is a measurement of the quantum system, with the density matrix ρ, and the measurement is described by the collection of measurement operators, $\mathbf{M}_m$, then the probability of the outcome labelled m is $p(m) \quad \psi \; \mathbf{M}_m^{\dagger}\mathbf{M}_m \; \psi$, and the normalized state after the measurement is carried out by the operator $\mathbf{M}_m$ is

$$\varphi \quad \frac{\mathbf{M}_m \; \psi}{\sqrt{\psi \; \mathbf{M}_m^{\dagger}\mathbf{M}_m \; \psi}}.$$

It is easy to show that when the measurement is carried out using the operator $\mathbf{M}_m$ immediately after the measurement:

1. The density matrix of the quantum system is

$$\rho_m \quad \mathcal{E}_m(\rho) \quad \mathbf{M}_m\rho\mathbf{M}_m^{\dagger}.$$

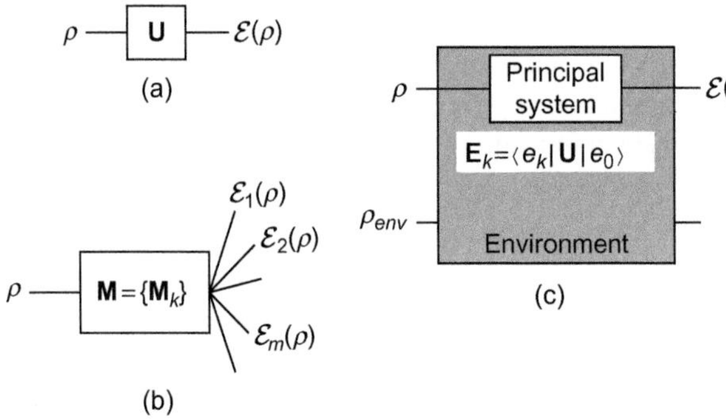

FIGURE 2.10

Quantum operations. (a) An isolated quantum system subject to a unitary transformation, **U**; the density matrix after the operation is $\rho \quad \mathcal{E}(\rho) \quad \mathbf{U}\rho\mathbf{U}^{\dagger}$. (b) A measurement described by the collection of measurement operators, $\mathbf{M}_m$; if the measurement operator is $\mathbf{M}_m$, then the density matrix after the measurement is $\rho_m \quad \mathcal{E}_m(\rho) \quad \mathbf{M}_m\rho\mathbf{M}_m^{\dagger}$. (c) An open quantum system, with the density matrix ρ, that interacts with an environment, with the density matrix ρ_{env}, subject to a unitary transformation, **U**. If $\mathbf{E}_k \quad e_k \; \mathbf{U} \; e_0$ is an operator on the state space of the principal system, with e_0 the initial state and e_k an orthonormal basis in the Hilbert space of the environment, then the density matrix after the measurement is $\mathcal{E}(\rho) \quad \sum_k \mathbf{E}_k\rho\mathbf{E}_k^{\dagger}$.

2. The probability of the outcome m is

$$p(m) \quad \mathrm{tr}(\mathcal{E}_m(\rho)).$$

Lastly, we consider an open system consisting of a *principal system* and the environment Figure 2.10c. The density matrices of the two systems are ρ and ρ_{env}, respectively; we assume that the principal system and the environment are in a product state $\rho \quad \rho_{env}$. Then the effect of the unitary transformation $\mathbf{U}$ on the open system can be represented as

$$\mathcal{E} \quad \mathrm{tr}_{env}\left(\mathbf{U}(\rho \quad \rho_{env})\mathbf{U}^{\dagger}\right).$$

If e_0 is the initial state and e_k is an orthonormal basis in the Hilbert space of the environment, then the density matrix of the environment in the pure state e_0 is $\rho_{env} \quad e_0 \quad e_0$. Call $\mathbf{E}_k \quad e_k \ \mathbf{U} \ e_0$ an operator on the state space of the principal system. Then the effect of the unitary transformation $\mathbf{U}$ on the principal system is

$$\mathcal{E}(\rho) \quad \sum_k e_k \ \mathbf{U}[\rho \quad (e_0 \quad e_0)]\mathbf{U}^{\dagger} \ e_k \ .$$

This can be expressed in a form known as the *operator-sum representation*:

$$\mathcal{E}(\rho) \quad \sum_k \mathbf{E}_k \rho \mathbf{E}_k^{\dagger}.$$

The operator-sum representation can be seen as a transformation in which the original state, ρ, is subject to random transformations,

$$\rho_k \quad \frac{\mathbf{E}_k \rho \mathbf{E}_k^{\dagger}}{\mathrm{tr}(\mathbf{E}_k \rho \mathbf{E}_{\mathbf{k}}^{\dagger})},$$

each occurring with the probability $p_k \quad \mathrm{tr}(\mathbf{E}_k \rho \mathbf{E}_k^{\dagger})$, or

$$\mathcal{E}(\rho) \quad \sum_k p_k \rho_k.$$

The *trace-preserving transformations*, $\mathcal{E}$, have the property that

$$\mathrm{tr}\,(\mathcal{E}(\rho)) \quad 1.$$

This implies that for all density operators ρ, we have

$$\mathrm{tr}(\mathcal{E}(\rho)) \quad \mathrm{tr}\left(\sum_k \mathbf{E}_k \rho \mathbf{E}_k^{\dagger}\right) \quad \mathrm{tr}\left(\sum_k \mathbf{E}_k^{\dagger} \mathbf{E}_k \rho\right).$$

It follows immediately that to be trace-preserving, the operators $\mathcal{E}$ must satisfy the condition

$$\sum_k \mathbf{E}_k^{\dagger} \mathbf{E}_k \quad \mathbf{I}.$$

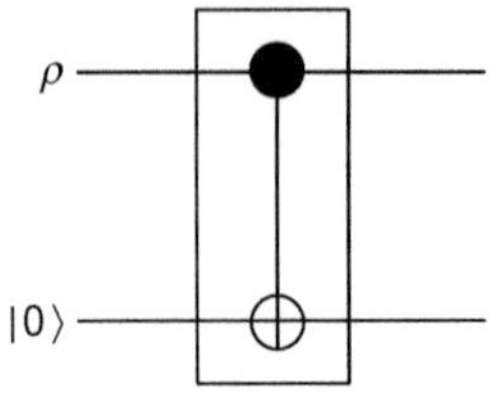

FIGURE 2.11

An open quantum system, with the density matrix ρ, that interacts with an environment, with the density matrix ρ_{env}, subject to a unitary transformation CNOT. The principal system is the control qubit, and the environment is the target qubit, initially in the state 0 .

A *non-trace-preserving* quantum operation is characterized by the inequality

$$\sum_i \mathbf{E}_i^\dagger \mathbf{E}_i \quad \mathbf{I}.$$

It corresponds to a transformation that involves measurement of the quantum system.

Example. *Consider the circuit in Figure 2.11.* The principal system is the control qubit, and the environment is initially in the state 0 ; the basis states of the environment, e_k, are 0_E and 1_E . The unitary transformation $\mathbf{U}$ can be written as

$$\mathbf{U} \quad 0_P 0_E \; 0_P 0_E \quad 0_P 1_E \; 0_P 1_E \quad 1_P 1_E \; 1_P 0_E \quad 1_P 0_E \; 1_P 1_E \, ,$$

where the subscripts E and P refer to the environment and principal system, respectively.

Here, the operation elements are

$$\mathbf{E}_0 \quad 0_E \, \mathbf{U} \, 0_E \quad 0_P \; 0_P \quad \text{and} \quad \mathbf{E}_1 \quad 1_E \, \mathbf{U} \, 1_E \quad 1_P \; 1_P \, .$$

Thus,

$$\mathcal{E}(\rho) \quad \mathbf{E}_0 \rho \mathbf{E}_0^\dagger \;\; \mathbf{E}_1 \rho \mathbf{E}_1^\dagger \quad \mathbf{P}_0 \rho \mathbf{P}_0 \;\; \mathbf{P}_1 \rho \mathbf{P}_1,$$

where $\mathbf{P}_0$ 0 0 and $\mathbf{P}_1$ 1 1 are the projectors.

The operator sum representation *is not unique.* We show that given a system, with the density matrix ρ, two operations, $\mathcal{E}$ and $\mathcal{F}$, with the distinct operation elements $\mathbf{E}_1, \mathbf{E}_2$ and $\mathbf{F}_1, \mathbf{F}_2$, lead to the same state of the system, $\mathcal{E}(\rho) \quad \mathcal{F}(\rho)$. Consider the following operations and elements:

Operation	Operation Element 1	Operation Element 2
$\mathcal{E}$	$\mathbf{E}_1 \quad \frac{1}{\sqrt{2}}\mathbf{I} \quad \frac{1}{\sqrt{2}}\begin{pmatrix} 1 & 0 \\ 0 & 1 \end{pmatrix}$	$\mathbf{E}_2 \quad \frac{1}{\sqrt{2}}\mathbf{X} \quad \frac{1}{\sqrt{2}}\begin{pmatrix} 0 & 1 \\ 1 & 0 \end{pmatrix}$
$\mathcal{F}$	$\mathbf{F}_1 \quad \frac{1}{\sqrt{2}}(\mathbf{E}_1 \quad \mathbf{E}_2) \quad \frac{1}{2}\begin{pmatrix} 1 & 1 \\ 1 & 1 \end{pmatrix}$	$\mathbf{F}_2 \quad \frac{1}{\sqrt{2}}(\mathbf{E}_1 \quad \mathbf{E}_2) \quad \frac{1}{2}\begin{pmatrix} 1 & 1 \\ 1 & 1 \end{pmatrix}$

We start from the definition

$$\mathcal{E}(\rho) \quad \mathbf{E}_1\rho\mathbf{E}_1^\dagger \quad \mathbf{E}_2\rho\mathbf{E}_2^\dagger.$$

Then,

$$\mathcal{F}(\rho) \quad \mathbf{F}_1\rho\mathbf{F}_1^\dagger \quad \mathbf{F}_2\rho\mathbf{F}_2^\dagger \quad \frac{1}{2}[(\mathbf{E}_1 \quad \mathbf{E}_2)\rho(\mathbf{E}_1^\dagger \quad \mathbf{E}_2^\dagger)] \quad [(\mathbf{E}_1 \quad \mathbf{E}_2)\rho(\mathbf{E}_1^\dagger \quad \mathbf{E}_2^\dagger)]$$

or, after expansion, we get

$$\mathcal{F}(\rho) \quad \mathbf{E}_1\rho\mathbf{E}_1^\dagger \quad \mathbf{E}_2\rho\mathbf{E}_2^\dagger \quad \mathcal{E}(\rho).$$

The next question is how can the unitary transformation **U** be applied to an open system when the environment has a very large number of degrees of freedom. The answer, discussed in Section 3.18, is that *when the principal system is specified in a Hilbert space of dimension d, it is sufficient to model the environment in a Hilbert space of dimension d^2.*

We now switch topics and discuss entanglement and its applications for quantum information processing.

2.15 ENTANGLEMENT: MONOGAMY OF ENTANGLEMENT

Entanglement plays a critical role in quantum computing and quantum information theory, as we shall see when we discuss the quantum communication channels and quantum error correction; for example, when two parties share an entangled quantum state, they are able to communicate securely. In this section we present some basic facts about entanglement and introduce the *monogamy of entanglement*, a property of quantum systems with important consequences for many applications, including quantum cryptography where it quantifies how much information an eavesdropper could potentially obtain.

The monogamy of entanglement is the deeper root of our inability to clone quantum states. Figure 2.12 illustrates why the monogamy of entanglement prevents quantum states to be cloned. Consider two maximally entangled quantum states, $\mathcal{A}$ and $\mathcal{B}$. Assume that we have a quantum copy machine able to clone quantum states. If the input to this quantum copy machine is system $\mathcal{A}$, then

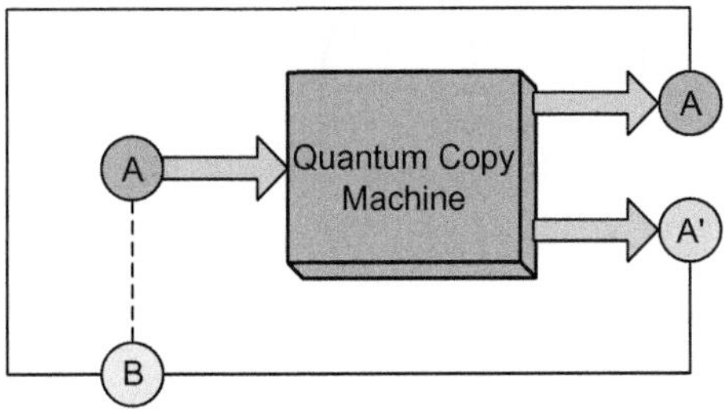

FIGURE 2.12

In violation of the monogamy of entanglement, a quantum copy machine would allow a quantum state $\mathcal{B}$, maximally entangled with another quantum state $\mathcal{A}$, to end up entangled with the two copies of state $\mathcal{A}$.

the output will be the original, system $\mathcal{A}$, and a perfect replica of it, $\mathcal{A}$. Thus, the quantum system, $\mathcal{B}$, would end up being entangled with both systems $\mathcal{A}$ and $\mathcal{A}$, in violation of the monogamy of entanglement. The dotted line represents the original entanglement of $\mathcal{B}$ with $\mathcal{A}$, and the solid lines the entanglement of $\mathcal{B}$ with $\mathcal{A}$ and its clone $\mathcal{A}$.

The joint state of a maximally entangled pair system is a pure state, while individual particles are in mixed state; thus, the state of individual particles cannot be known with certainty and individual particles cannot be cloned. Recall from Section 1.12 that the *Bell states* are *entangled* states of a *Bell pair*:

$$\beta_{00} \quad \frac{00 \quad 11}{\overline{2}}, \qquad \beta_{01} \quad \frac{01 \quad 10}{\overline{2}}, \qquad \beta_{10} \quad \frac{00 \quad 11}{\overline{2}}, \qquad \beta_{11} \quad \frac{01 \quad 10}{\overline{2}}.$$

The Bell states form an orthonormal basis and can be distinguished from one another. All four states are called *maximally entangled* states. The last one is called an *anti-correlated* state.

Proposition. *Bell states are pure states; the joint state of an EPR pair is known exactly (it is a pure state), while the state of either qubit of the pair is not (it is a mixed state).*

Proof. Let us pick one of the Bell states, say β_{10} , and compute its density operator, $\rho_{(\beta_{10})}$. Then we compute the density operator of one of the qubits of the EPR pair, say the second one, $\rho_{(\beta_{10},second)}$. We expect that the traces of the two density operators satisfy the known relations for pure and mixed, respectively, states:

$$\mathrm{tr}\left(\rho^2_{(\beta_{10})}\right) \quad 1 \quad \text{and} \quad \mathrm{tr}\left(\rho^2_{(\beta_{10},second)}\right) < 1.$$

First, we compute the density operator for the pair:

$$\rho_{(\beta_{10})} \quad \beta_{10} \quad \beta_{10} \quad \frac{(\;00 \quad 11\;)}{\overline{2}}\frac{(\;00 \quad 11\;)}{\overline{2}}$$

$$\frac{1}{2}(\;00 \;\; 00 \quad\; 00 \;\; 11 \quad\; 11 \;\; 00 \quad\; 11 \;\; 11\;)$$

$$\frac{1}{2}\left[\begin{pmatrix}1&0&0&0\\0&0&0&0\\0&0&0&0\\0&0&0&0\end{pmatrix} \quad \begin{pmatrix}0&0&0&1\\0&0&0&0\\0&0&0&0\\0&0&0&0\end{pmatrix} \quad \begin{pmatrix}0&0&0&0\\0&0&0&0\\0&0&0&0\\1&0&0&0\end{pmatrix} \quad \begin{pmatrix}0&0&0&0\\0&0&0&0\\0&0&0&0\\0&0&0&1\end{pmatrix}\right]$$

$$\frac{1}{2}\begin{pmatrix}1&0&0&1\\0&0&0&0\\0&0&0&0\\1&0&0&1\end{pmatrix}.$$

Then,

$$\rho^2_{(\beta_{10})} \quad \frac{1}{2}\begin{pmatrix} 1 & 0 & 0 & 1 \\ 0 & 0 & 0 & 0 \\ 0 & 0 & 0 & 0 \\ 1 & 0 & 0 & 1 \end{pmatrix} \frac{1}{2}\begin{pmatrix} 1 & 0 & 0 & 1 \\ 0 & 0 & 0 & 0 \\ 0 & 0 & 0 & 0 \\ 1 & 0 & 0 & 1 \end{pmatrix} \quad \frac{1}{4}\begin{pmatrix} 2 & 0 & 0 & 2 \\ 0 & 0 & 0 & 0 \\ 0 & 0 & 0 & 0 \\ 2 & 0 & 0 & 2 \end{pmatrix}$$

$$\frac{1}{2}\begin{pmatrix} 1 & 0 & 0 & 1 \\ 0 & 0 & 0 & 0 \\ 0 & 0 & 0 & 0 \\ 1 & 0 & 0 & 1 \end{pmatrix} \qquad \mathrm{tr}\left(\rho^2_{(\beta_{10})}\right) \quad \frac{1}{2}(1 \quad 0 \quad 0 \quad 1) \quad 1.$$

Now we compute the reduced density operator and the partial trace for the second qubit of the EPR pair by tracing the first qubit:

$$\rho_{(\beta_{10},second)} \quad \mathrm{tr}_{first}\left(\rho_{\beta_{10}}\right) \quad \frac{\mathrm{tr}_{first}\,(\ 00\ \ 00 \qquad 00\ \ 11 \qquad 11\ \ 00 \qquad 11\ \ 11\ )}{2}$$

$$\frac{0\ \ 0\ \ 0\,0 \qquad 0\ \ 1\ \ 0\,1 \qquad 1\ \ 0\ \ 1\,0 \qquad 1\ \ 1\ \ 1\,1}{2}.$$

But,

$$0\,0 \quad 1, \quad 0\,1 \quad 0, \quad 1\,0 \quad 0, \quad 1\,1 \quad 1$$

and

$$0\ \ 0 \quad \begin{pmatrix} 1 & 0 \\ 0 & 0 \end{pmatrix} \quad \text{and} \quad 1\ \ 1 \quad \begin{pmatrix} 0 & 0 \\ 0 & 1 \end{pmatrix}.$$

It follows that

$$\rho_{(\beta_{10},second)} \quad \frac{0\ \ 0 \qquad 1\ \ 1}{2} \quad \frac{1}{2}\begin{pmatrix} 1 & 0 \\ 0 & 1 \end{pmatrix} \quad \frac{I}{2}$$

$$\rho^2_{(\beta_{10},second)} \quad \frac{1}{4}\begin{pmatrix} 1 & 0 \\ 0 & 1 \end{pmatrix}\begin{pmatrix} 1 & 0 \\ 0 & 1 \end{pmatrix} \quad \frac{1}{4}\begin{pmatrix} 1 & 0 \\ 0 & 1 \end{pmatrix} \quad \mathrm{tr}\left(\rho^2_{(\beta_{10},second)}\right) \quad \frac{1}{4}(1 \quad 1) \quad \frac{1}{2}.$$

To complete the proof of this proposition, we have to repeat the calculation for the first qubit of the β_{10} pair following the pattern presented above. Then we have to redo the calculations for the other three Bell states: β_{00}, β_{01}, and β_{11}. ■

One of the most intriguing properties of quantum information is the shareability of quantum correlations. While classical correlations can be shared among many parties, quantum correlations cannot be shared freely; quantum correlations of pure states are monogamous. As we know from statistics, two random variables, X and Y, correlated with each other, may also be individually correlated with other random variables. In contrast, if two quantum systems, $\mathcal{A}$ and $\mathcal{B}$, are in a maximally entangled pure state, then neither of them can be correlated with any other system in the universe. Otherwise, there is tradeoff between the amount of entanglement of two qubits and the quantum correlation each of the two qubits could share with a third one. It is believed that if the two qubits are as much entangled with each other as it is possible, they cannot be entangled or even classically correlated with another qubit.

Consider three qubits, a, b, c, with the first two in a maximally entangled pure state,

$$\psi_{ab} \quad \frac{1}{\sqrt{2}} 0_a \quad 0_b \quad \frac{1}{\sqrt{2}} 1_a \quad 1_b ,$$

and the third in the state ψ_c [418]. There is no quantum state shared by the three qubits, such that when we remove the third qubit, c, we get the joint state of the qubits a and b, ψ_{ab} , and, at the same time, the three-qubit state does not change when we interchange the qubits a and b. Thus, the joint state of the three qubits is

$$\psi_{abc} \quad \psi_c \quad \left[\frac{1}{\sqrt{2}} 0_a \quad 0_b \quad \frac{1}{\sqrt{2}} 1_a \quad 1_b \right].$$

There is no symmetry between the qubits b and c, b is maximally entangled with a, and c is not entangled with a.

The next question is whether all entangled states are monogamous; for example, we may inquire if entangled mixed states are monogamous. Schumacher introduced the term *shareable quantum states*, and Bennett *et al* [40] gave an example of a mixed entangled state that is shareable rather than monogamous. Consider a noisy quantum channel shared by Alice, Bob, and Eve; with the probability $1/2$, a qubit sent by Alice is transmitted unchanged to Bob, and with the probability $1/2$, the qubit is intercepted and Bob gets a random qubit. When Alice sends to Bob half of a maximally entangled pair, the state shared by Alice and Bob is still entangled, while the state shared by the three is always symmetric with respect to Eve and Bob; thus, Eve is entangled with Alice as well.

2.16 EINSTEIN-PODOLSKI-ROSEN (EPR) THOUGHT EXPERIMENT

"Liebe Gott würfelt nicht,"[7] Einstein's famous pronouncement, re ects Einstein's skepticism shared by some physicists and philosophers regarding the nondeterminism of quantum theory. In 1935, in the now famous paper "Can Quantum-Mechanical Description of Physical Reality be Considered Complete?" [133] Einstein, Podolsky, and Rosen proposed a thought experiment challenging the very foundations of quantum theory. They concluded that we have a choice, either to accept the principle of locality or to accept the postulates of quantum mechanics. In his autobiography, Einstein wrote [134]:

> *The paradox forces us to relinquish one of the following two assertions: (1) the description by means of the wave function ψ is complete, (2) the real states of spatially separated objects are independent of each other.*

Einstein concluded that the quantum mechanical description by means of the wave function is not complete, and he had no doubts that the second assertion is true. Nonlocality would violate one of the basic postulates of the Special Theory of Relativity.[8]

---

[7] "Dear God does not play dice."

[8] The special theory of relativity of Einstein describes the motion of particles moving at close to the speed of light. The two basic postulates of special relativity are: (1) The speed of light is the same for all observers, no matter what their relative speeds; (2) The laws of physics are the same in any inertial (i.e., non-accelerated) frame of reference. This means that the laws of physics observed by a hypothetical observer traveling with a relativistic particle must be the same as those observed by an observer who is stationary in the laboratory.

Before presenting the EPR concepts, we describe a relatively simple experiment known as the ARC, which seems to illustrate that the randomness of the quantum theory is illusory.

The Atomic Radiation Cascade (ARC) Experiment

The ARC source in Figure 2.13 produces two polarization-entangled photons that move in opposite directions. An atom (e.g., Ca^{40}), initially laser pumped to an excited S state, the upper level of the cascade state, emits two photons that appear to have the same polarization. Let x_1 , x_2 , y_1 , and y_2 denote the states of the two photons, with linear polarization along the x- and y-axis, respectively, orthogonal to the z-axis. It can be shown [325] that the state of the entangled system of the two photons can only be

$$\psi \quad (1/\ \bar{2})(\,x_1\ \ x_2\quad y_1\ \ y_2\ ).$$

One of the photons is detected by a polarizing cube[9] located in the immediate vicinity of the source, and the other is re ected by a distant mirror and then detected by a second calcite crystal. Each detector

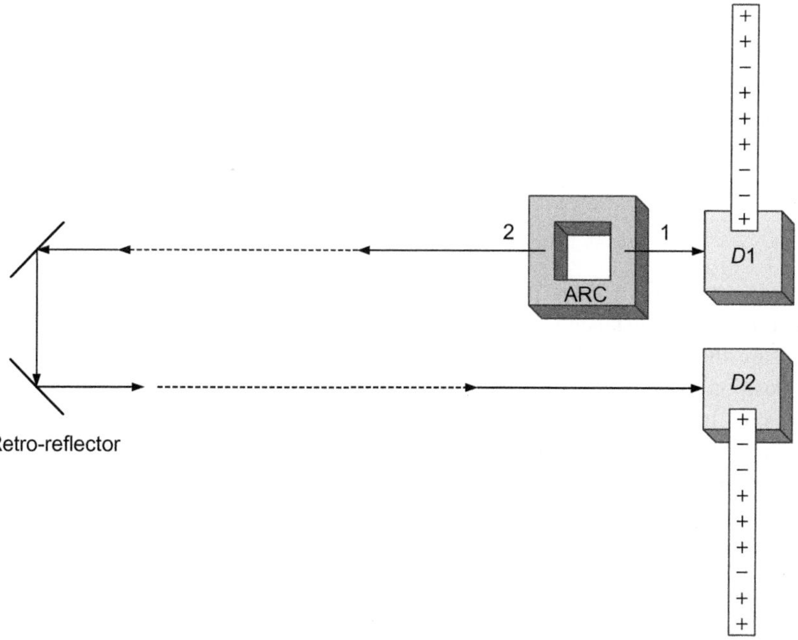

FIGURE 2.13

An ARC source produces two polarization-entangled photons. Detectors $D1$, located in the immediate vicinity of the source, and $D2$, detecting a photon re ected by a distant mirror, are connected to output devices that print one of the symbols or depending on the polarization of the photons. The two paper tapes contain random sequences of s and s, but they are perfectly correlated. The second tape, which looks perfectly random when examined in isolation, is in fact deterministic. We could have predicted with absolute certainty the sequence of symbols on that tape.

[9]A polarizing cube is made of two prisms, with a dielectric film on the sides that are stuck together. The polarizing cube transmits photons with horizontal polarization and re ects photons with vertical polarization.

is connected to a device that prints on a paper tape the symbols and depending on the polarization of the photon it has detected (Figure 2.13). There is a delay between the measurement performed by the upper detector located in the immediate vicinity of the source and the measurement performed by the second detector on the photon whose path is re ected by the distant mirror.

As expected, each paper tape contains a completely random sequence of and symbols. If we compare the two paper tapes, we see that they are perfectly correlated; an observer able to see the first tape can predict with certainty the next symbol to be printed on the second tape. Amazingly, the second tape, which looks perfectly random when examined in isolation, is in fact deterministic. We could have predicted with absolute certainty the sequence of symbols on that tape.

The EPR Experiment

EPR is a thought (Gedanken) experiment; the EPR argument is based on the concept "element of physical reality" defined as follows: *If without in any way disturbing a system we can predict with certainty the value of a physical quantity, then there exists an element of physical reality corresponding to this physical quantity.*

We consider a bipartite system consisting of two particles, P_1 and P_2, prepared in such a state that their total momentum is close to zero and their relative distance is close to L, where L is much larger than the distance that allows the two particles to interact with each other, as in Figure 2.14a. If we denote by γ a normalizable function with a very high and very narrow peak and by x_1, x_2, p_1, p_2 the positions and the momenta of the two particles, then the state of the system is described by an entangled wave function:

$$\psi \quad \gamma(x_1 \quad x_2 \quad L)\gamma(p_1 \quad p_2).$$

We do not know anything about the position of the particles or about their individual momenta. We only know that they are at distance L from one another and that the total momentum is equal to zero.

If we measure the position of the first particle, x_1, the wave function allows us to predict with certainty the position of the second particle, x_2. It is easy to argue that both x_1 and x_2 are elements of the physical reality. Indeed, the measurements of the two systems do not affect each other because the distance L was chosen to be large enough to prevent such interactions. Thus, no change may take place in the second system as a result of the measurement performed in the first system.

We could have measured the momentum of the first particle, p_1, and then we would have been able to predict the momentum of the second one, p_2. Similar arguments indicate that both p_1 and p_2 are elements of the physical reality. Yet, the Heisenberg inequality precludes the simultaneous assignment of precise values to both the position, x_2, and the momentum, p_2, of the second particle because the two operators corresponding to the two measurements of observables do not commute.

Bohm Version of EPR

David Bohm gives a simpler example of the same paradox [63] (Figure 2.14b). Recall that the "spin" is the quantum number characterizing the intrinsic angular momentum of a boson or fermion and that quantum mechanics asserts that at most one component of the spin of a particle may be definite.

Consider a spinless system such as the neutral pion, π^0, which can decay in two spin-½ fermions, a positron, e , and an electron, e , which y away from each other, and a photon, $\pi^0 \quad e \quad e \quad \gamma$. When they are very far apart, we measure the component of the spin of the electron along the x-direction, $S_{1,x}$, and find it equal to $\hbar/2$. Then we know that the component of the spin of the positron

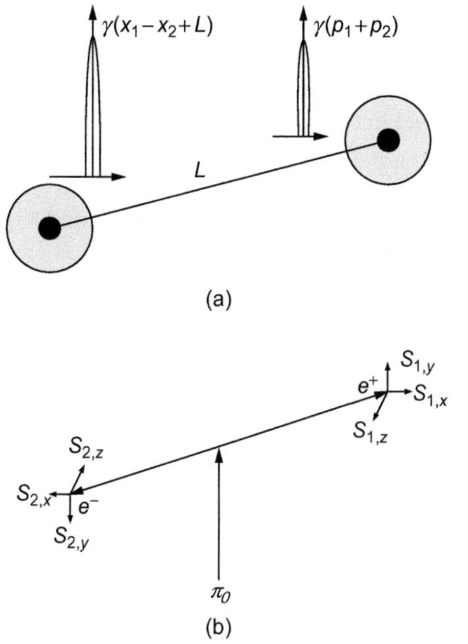

FIGURE 2.14

(a) The original EPR experiment. (b) Bohm version of the EPR experiment.

along the x-axis must have the opposite orientation, $S_{2,x} \quad \hbar/2$; moreover, it must have been in that state since it was emitted as it did not suffer any interactions. Similarly, we could have measured the component of the spin of the electron along the y-axis, $S_{1,y}$, and this measurement would have allowed us to predict with certainty the moment of the positron along the y-axis, $S_{2,y}$. This same argument applies for the measurement of the momentum along the z-axis, $S_{1,z}$ and $S_{2,z}$.

We do not claim that we actually measured the three components of the spin of the electron; all we say is *we could have* measured all of them and in that case all three components of the spin of the positron *would have* been predictable with certainty. In this case, the three components of the spin of the positron, $S_{2,x}, S_{2,y}$, and $S_{2,z}$, would have been "elements of reality," and that would have been in stark contradiction with the laws of quantum mechanics.

Peres Version of EPR

Asher Peres gives an argument in the spirit of EPR and reformulates the EPR paradox as an algebraic contradiction [325]. Given two commuting operators, **A** and **B**, for observables, A and B, quantum mechanics allows us to discover "primitive elements of reality." We can measure the observables A and B simultaneously and get as results of these measurements the eigenvalues of these operators, α and β, respectively. Let us now assume that quantum mechanics allows us to define recursively "secondary elements of reality." This means that any algebraic function $f(\mathbf{A},\mathbf{B})$ corresponds to an "element of reality," and the result of such a measurement is $f(\alpha,\beta)$.

Let us apply this argument to the spinless system in the Bohm reformulation of the EPR experiment. Call $m_{1,x}$, $m_{2,x}$, and $m_{1,y}$, $m_{2,y}$ the values of the projections of the spin on the x- and y-axis for the electron and positron, $S_{1,x}$, $S_{2,x}$ and $S_{1,y}$, $S_{2,y}$, respectively. We know that $m_{1,x} \quad m_{2,x}$ and

$m_{1,y} \quad m_{2,y}$. The spin observables, $S_{1,x}$ and $S_{2,y}$, commute, as well as $S_{1,y}$ and $S_{2,x}$. Since they correspond to elements of reality, their products, $S_{1,x}S_{2,y}$ and $S_{1,y}S_{2,x}$, also correspond to elements of reality; their numerical values are given by the products of individual numerical values, $m_{1,x}m_{2,y}$ and $m_{1,y}m_{2,x}$. These numerical products are equal in our case; thus:

$$S_{1,x}S_{2,y} \quad S_{1,y}S_{2,x} \quad m_{1,x}m_{2,y}.$$

This contradicts quantum mechanics, which asserts that these products should have opposite values because the singlet state[10], π^0, satisfies the equation:

$$(S_{1,x}S_{2,y} \quad S_{1,y}S_{2,x})\psi \quad 0.$$

In summary, the EPR is a Gedankenexperiment challenging the dogma that the description of reality given by the wave function in quantum mechanics is complete. Einstein believed that EPR forces a choice: "either (1) the quantum-mechanical description of reality given by the wave function is not complete; or (2) when the operators corresponding to two physical quantities do not commute the two quantities cannot have simultaneous reality."

Some of those questioning the nondeterminism of quantum mechanics attempted to deduce the properties of quantum systems from those of a yet unknown, but deterministic sub-quantum world and found refuge in the theory of "hidden variables." In the next section we show that the hidden variable theory is not sound.

2.17 HIDDEN VARIABLES

The EPR experiment led some physicists to the belief that the nondeterminism of quantum mechanics could be explained by the existence of "hidden variables." The proponents of this theory believed that if we knew the exact values of hidden variables, then we would have a fully deterministic view of the world.

A suggestive analogy was proposed by one of our students:

Imagine that we are behind a wall that obscures the view of the other side where several machines throw tennis balls over the wall. The trajectory of each ball depends on the setting of each machine; assuming that the tennis balls are identical and their weight and diameter are known and that there is no wind, the trajectory of each one is perfectly deterministic. Yet, to us the trajectory of a ball appears to be random. The nondeterminism is due to the lack of knowledge of the initial conditions for each trajectory, the hidden variables of this game.

We now examine more closely the hidden variable theory and show that it cannot reproduce the statistical properties of quantum mechanics evidenced by countless experiments.

The Argument of von Neumann

In 1932, von Neumann attempted to prove that quantum theory is incompatible with the existence of dispersion-free ensembles (ensembles that have dispersion equal to zero for all their observables) [441]. He considered N independent versions of a quantum system, each in a quantum state ψ_i, with $1 \quad i \quad N$. A measurement of the same observable, A, on all N versions of the system allows us to

[10]Singlet electron state: anti-symmetric state of a pair of electrons, with the *anti-parallel* spins $1/\ \overline{2}($ $)$. The electrons have different spin quantum numbers, $1/2$ and $1/2$, and the total spin of the state is zero.

compute the ensemble average of the operator, **A**. Call this ensemble average **A** :

$$\mathbf{A} \qquad \frac{1}{N}\sum_{i \;\, 1}^{N} \psi_i \, \mathbf{A} \, \psi_i \, .$$

The proof requires an ensemble of physical systems, such that for any observable A, the associated operator **A** is self-adjoint and satisfies the zero-dispersion condition:

$$\mathbf{A}^2 \qquad \mathbf{A}\;^2.$$

Yet, it is impossible to prepare an ensemble of physical systems, such that any observable A has an associated measurement operator that satisfies the condition $\mathbf{A}^2 \qquad \mathbf{A}\;^2$. The proof requires also that for any two observables, A and B, the corresponding operators, **A** and **B**, satisfy the equation:

$$\mathbf{A} \quad \mathbf{B} \qquad \mathbf{A} \qquad \mathbf{B} \; .$$

Consider the case when the operators **A** and **B** do not commute and cannot be measured simultaneously. If **A** and **B** do not commute, we need three experimental setups to measure independently the observables A, B, and $A \quad B$. The three setups could be very different, and the measurements would not justify the equality $\mathbf{A} \quad \mathbf{B} \qquad \mathbf{A} \qquad \mathbf{B}$.

John Bell s Proof of Incompatibility Between Quantum Mechanics and the Hidden Variable Theory

John Bell has devoted his scientific life to the philosophy of measurements [26], and his results are critical for understanding the subtle differences between classical and quantum information. In 1964, 32 years after von Neumann published his result, John Bell provided a rigorous proof that a nondeterministic model based on hidden variables is incompatible with quantum mechanics [25].

Consider a pair of spin-½ particles in the singlet state, $(1/\sqrt{\ }2)($, moving freely in opposite directions. Let α and β be two unit vectors with arbitrary orientations. We can measure the projections of the spins of the two particles, σ_1 and σ_2, along α using two Stern-Gerlach setups. The two measurements produce opposite results: if $\sigma_1 \;\; \alpha \qquad 1$, then $\sigma_2 \;\; \alpha \qquad 1$ and vice versa.

The question posed is, "Is there a more complete specification of the state of the two particles from the one given by the wave function?" To answer the question, we assume that we measure the spins of the particles along two independently chosen directions given by the unitary vectors, α and β, using two Stern-Gerlach setups as shown in Figure 2.15. We also assume that there is a hidden continuous random variable, λ, that would enable a more complete specification of the state; the probability density function of this random variable is $\rho(\lambda)$, and it satisfies the relationship $\int \rho(\lambda) d\lambda \quad 1$.

Let a and b be the outcomes of the measurements of the spins of the two particles, $\sigma_1 \;\; \alpha$ and $\sigma_2 \;\; \beta$, respectively. We assume that the outcome, b, of the spin measurement for the second particle does not depend on the direction, α, of the Stern-Gerlach apparatus setting for the measurement of the first particle, and that the outcome, a, of the spin measurement for the first particle does not depend on the direction, β, of the Stern-Gerlach apparatus setting for the measurement of the spin of the second particle. Thus, a and b are random variables and functions of

1. the direction we project the spin onto, and
2. the value of the hidden random variable λ:

$$a \quad a(\alpha,\lambda) \qquad 1 \;\; \text{and} \;\; b \quad b(\beta,\lambda) \qquad 1.$$

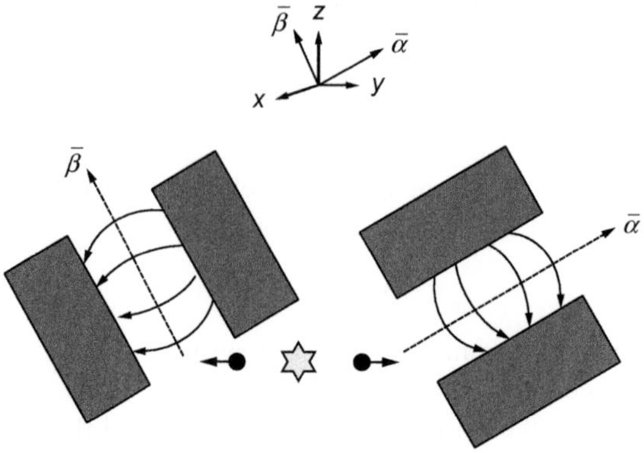

FIGURE 2.15

A pair of spin-½ particles in the singlet state $1/\ \overline{2}($ $\quad - \quad$ move freely in opposite directions. We measure the spins along two arbitrary directions, α and β, using two Stern-Gerlach experimental setups.

The expected values of a and b when λ takes on different values are functions of the two orientations α and β, respectively:

$$E(a) \quad f_1(\alpha) \quad \int a(\alpha,\lambda)\rho(\lambda)d\lambda \quad \text{and} \quad E(b) \quad f_2(\beta) \quad \int b(\beta,\lambda)\rho(\lambda)d\lambda.$$

The expected value of the product of the two random variables a and b is a function of both α and β:

$$E(ab) \quad f(\alpha,\beta) \quad \int a(\alpha,\lambda)b(\beta,\lambda)\rho(\lambda)d\lambda.$$

On the other hand, the expected value of the product of the two measurements predicted by quantum mechanics is a function of the angle between the two directions α and, respectively, β on which we project the spins of the two particles:

$$(\sigma_1 \quad \alpha)(\sigma_2 \quad \beta) \quad (\sigma_1 \quad \sigma_2)(\alpha \quad \beta) \quad -(\alpha \quad \beta).$$

In this expression, $\sigma_1 \quad \sigma_2 \quad -1$ because the spins of the two particles are opposite.

Our original question can now be formulated as, "Are the two averages, $f(\alpha,\beta)$, calculated based on a hidden variable model, $-(\alpha \quad \beta)$, predicted by quantum mechanics, equal?" John Bell proves that they are not [25, 26], and we summarize Bell's arguments in the following propositions.

Proposition. *The average of the two observables calculated based on a hidden variable model satisfies the following five conditions:*

1. Has a lower bound $f(\alpha,\beta) \quad -1$*; this follows from the definition*

$$f(\alpha,\beta) \quad \int a(\alpha,\lambda)b(\beta,\lambda)\rho(\lambda)d\lambda, \textit{ with } a(\alpha,\lambda) \quad 1, b(\beta,\lambda) \quad 1, \textit{ and } \int \rho(\lambda)d\lambda \quad 1.$$

2. *$f(\alpha,\alpha)$ $\quad$ 1 only if $a(\alpha,\lambda)$ $\quad$ $b(\alpha,\lambda)$. When the two directions coincide, α $\quad$ β and*

$$f(\alpha,\alpha) \quad \int a(\alpha,\lambda)b(\alpha,\lambda)\rho(\lambda)d\lambda.$$

If $a(\alpha,\lambda)$ $\quad$ $b(\alpha,\lambda)$, then

$$f(\alpha,\alpha) \quad \int a(\alpha,\lambda)a(\alpha,\lambda)\rho(\lambda)d\lambda.$$

But,

$$a(\alpha,\lambda) \quad \frac{1}{a(\alpha,\lambda)} \quad 1.$$

Then $a(\alpha,\lambda)a(\alpha,\lambda)$ $\quad$ 1; we also know that $\int \rho(\lambda)d\lambda$ $\quad$ 1, thus, $f(\alpha,\alpha)$ $\quad$ 1.

3. *An alternative expression for $f(\alpha,\beta)$ is*

$$f(\alpha,\beta) \quad \int a(\alpha,\lambda)a(\beta,\lambda)\rho(\lambda)d\lambda.$$

If we substitute $b(\beta,\lambda)$ with $\quad$ $a(\beta,\lambda)$ in the definition of $f(\alpha,\beta)$, then we obtain the desired result, the outcome a along direction β.

4. *Consider three arbitrary directions in space, and let α, β, and γ be three unitary vectors along these three directions. Then,*

$$1 \quad f(\beta,\gamma) \quad f(\alpha,\beta) \quad f(\alpha,\gamma) .$$

To show this, we transform the following expression:

$$f(\alpha,\beta) \quad f(\alpha,\gamma) \quad \int a(\alpha,\lambda)a(\beta,\lambda)\rho(\lambda)d\lambda \quad \int a(\alpha,\lambda)a(\gamma,\lambda)\rho(\lambda)d\lambda$$
$$\int a(\alpha,\lambda)a(\beta,\lambda)[a(\gamma,\lambda)/a(\beta,\lambda) \quad 1]\rho(\lambda)d\lambda$$
$$\int a(\alpha,\lambda)a(\beta,\lambda)[a(\gamma,\lambda)a(\beta,\lambda) \quad 1]\rho(\lambda)d\lambda.$$

The last equality is due to the fact that

$$a(\beta,\lambda) \quad 1, \quad \textit{therefore,} \quad a(\beta,\lambda) \quad \frac{1}{a(\beta,\lambda)}.$$

Considering also that $a(\alpha,\lambda)$ $\quad$ 1, we can write

$$f(\alpha,\beta) \quad f(\alpha,\gamma) \quad \int \quad [a(\beta,\lambda)a(\gamma,\lambda) \quad 1]\rho(\lambda)d\lambda$$
$$\int [1 \quad a(\beta,\lambda)a(\gamma,\lambda)]\rho(\lambda)d\lambda$$
$$1 \quad \int (a(\beta,\lambda)a(\gamma,\lambda)\rho(\lambda)d\lambda.$$

But $f(\beta,\gamma) \quad \int (a(\beta,\lambda)a(\gamma,\lambda)\rho(\lambda)d\lambda$, *which proves that*

$$1 \quad f(\beta,\gamma) \quad f(\alpha,\beta) \quad f(\alpha,\gamma)\,.$$

5. $f(\beta,\gamma)$ *cannot be stationary when it reaches the minimum value of* 1 *for* $\beta \quad \gamma$. *If* $f(\beta,\gamma)$ *is not constant and if* $\beta \quad \gamma$ *is small, then*

$$f(\alpha,\beta) \quad f(\alpha,\gamma) \quad \beta \quad \gamma\,.$$

Thus, $1 \quad f(\beta,\gamma) \quad \beta \quad \gamma$ *and* $f(\beta,\gamma) \quad 1$ *when the two directions coincide,* $\beta \quad \gamma$.

Until now we have considered expected values relative to the hidden random variable, λ, with the density function, $\rho(\lambda)$, and we have denoted the expected value of a function, $g(\theta,\lambda)$, as $E\left[g(\theta,\lambda)\right] \quad \int g(\theta,\lambda)\rho(\lambda)d\lambda$. Now we assume that the unit vectors, α and β, rotate with a small angle around their current position. We denote by $\mathcal{E}\left[g(\theta)\right]$ the expected value of a function, $g(\theta)$, when the unit vector, θ, rotates with a small angle around its current value.

Proposition. *The expected value,* $\mathcal{E}\left[f(\alpha,\beta)\right]$, *relative to* α *and* β *of the observation based on a hidden variable model and the expected value,* $\mathcal{E}\left[\alpha \quad \beta\right]$, *predicted by quantum mechanics cannot be arbitrarily close to one another.*

Proof. Let us assume that there exist two positive arbitrarily small real numbers, $\epsilon > 0$ and $\delta > 0$, such that

$$\mathcal{E}\left[f(\alpha,\beta)\right] \quad \mathcal{E}\left[\alpha \quad \beta\right] \quad \epsilon \quad \text{and} \quad \mathcal{E}\left[\alpha \quad \beta\right] \quad \alpha \quad \beta) \quad \delta.$$

First, we notice that these two conditions imply that

$$\mathcal{E}\left[f(\alpha,\beta)\right] \quad \alpha \quad \beta) \quad \epsilon \quad \delta.$$

If $\alpha \quad \beta$, then $\alpha \quad \beta \quad \beta \quad \beta \quad 1$ and

$$\mathcal{E}\left[f(\beta,\beta)\right] \quad 1 \quad \epsilon \quad \delta$$

or

$$\mathcal{E}\left[f(\beta,\beta)\right] \quad \epsilon \quad \delta \quad 1.$$

Recall that $a(\alpha,\lambda)$ and $b(\beta,\lambda)$ are outcomes of the measurements of the projections of the spins of the two particles along the two arbitrary directions, α and β, respectively; thus, $a(\alpha,\lambda) \quad 1$ and $b(\beta,\lambda) \quad 1$. It follows that

$$\mathcal{E}[a(\alpha,\lambda)] \quad 1 \quad \text{and} \quad \mathcal{E}[b(\beta,\lambda)] \quad 1.$$

Also,

$$1 \quad \mathcal{E}[a(\beta,\lambda)]\mathcal{E}[b(\gamma,\lambda)] \quad 0 \quad \text{and} \quad 1 \quad \mathcal{E}[a(\beta,\lambda)]\mathcal{E}[b(\beta,\lambda)] \quad 0.$$

Now we rewrite the expression for the expected value of $f(\alpha,\beta)$:

$$\mathcal{E}\left[f(\alpha,\beta)\right] \quad \mathcal{E}\left(\int a(\alpha,\lambda)b(\beta,\lambda)\rho(\lambda)d\lambda\right) \quad \int \rho(\lambda)d\lambda\mathcal{E}[a(\alpha,\lambda)]\mathcal{E}[b(\beta,\lambda)].$$

Then,

$$\begin{aligned}\mathcal{E}\left[f(\alpha,\beta)\right] \quad \mathcal{E}\left[f(\alpha,\gamma)\right] \quad & \int \rho(\lambda)d\lambda\left\{\mathcal{E}[a(\alpha,\lambda)]\mathcal{E}[b(\beta,\lambda)] \quad \mathcal{E}[a(\alpha,\lambda)]\mathcal{E}[b(\gamma,\lambda)]\right\}\\ & \int \rho(\lambda)d\lambda\left(\mathcal{E}[a(\alpha,\lambda)]\mathcal{E}[b(\beta,\lambda)]\left\{1 \quad \mathcal{E}[a(\beta,\lambda)]\mathcal{E}[b(\gamma,\lambda)]\right\}\right)\\ & \int \rho(\lambda)d\lambda\left(\mathcal{E}[a(\alpha,\lambda)]\mathcal{E}[b(\gamma,\lambda)]\left\{1 \quad \mathcal{E}[a(\beta,\lambda)]\mathcal{E}[b(\beta,\lambda)]\right\}\right).\end{aligned}$$

But $\mathcal{E}[a(\alpha,\lambda)]\mathcal{E}[b(\beta,\lambda)] \quad 1$ and $\mathcal{E}[a(\alpha,\lambda)]\mathcal{E}[b(\gamma,\lambda)] \quad 1$, thus,

$$\begin{aligned}\mathcal{E}\left[f(\alpha,\beta)\right] \quad \mathcal{E}\left[f(\alpha,\gamma)\right] \quad & \int \rho(\lambda)d\lambda\left(\left\{1 \quad \mathcal{E}[a(\beta,\lambda)]\mathcal{E}[b(\gamma,\lambda)]\right\}\right)\\ & \int \rho(\lambda)d\lambda\left(\left\{1 \quad \mathcal{E}[a(\beta,\lambda)]\mathcal{E}[b(\beta,\lambda)]\right\}\right).\end{aligned}$$

It follows that

$$\begin{aligned}\mathcal{E}\left[f(\alpha,\beta)\right] \quad \mathcal{E}\left[f(\alpha,\gamma)\right] \quad & \int \rho(\lambda)d\lambda\left(\left\{1 \quad \mathcal{E}[a(\beta,\lambda)]\mathcal{E}[b(\gamma,\lambda)]\right\}\right)\\ & \int \rho(\lambda)d\lambda\left(\left\{1 \quad \mathcal{E}[a(\beta,\lambda)]\mathcal{E}[b(\beta,\lambda)]\right\}\right).\end{aligned}$$

The first term on the right-hand side of this inequality can be expressed as

$$\begin{aligned}\int \rho(\lambda)d\lambda\left(\left\{1 \quad \mathcal{E}[a(\beta,\lambda)]\mathcal{E}[b(\gamma,\lambda)]\right\}\right) \quad & \int \rho(\lambda)d\lambda \quad \int \rho(\lambda)d\lambda\left\{\mathcal{E}[a(\beta,\lambda)]\mathcal{E}[b(\gamma,\lambda)]\right\}\\ & 1 \quad \mathcal{E}[f(\beta,\gamma)].\end{aligned}$$

We also notice that

$$\int \rho(\lambda)d\lambda\mathcal{E}[a(\beta,\lambda)]\mathcal{E}[b(\beta,\lambda)] \quad \mathcal{E}\left[f(\beta,\beta)\right] \quad \epsilon \quad \delta \quad 1.$$

Let us denote $\epsilon \quad \epsilon \quad \delta \quad 0$; then,

$$\int \rho(\lambda)d\lambda\left(\left\{1 \quad \mathcal{E}[a(\beta,\lambda)]\mathcal{E}[b(\beta,\lambda)]\right\}\right) \quad \epsilon .$$

The inequality becomes

$$\mathcal{E}\left[f(\alpha,\beta)\right] \quad \mathcal{E}\left[f(\alpha,\gamma)\right] \quad 1 \quad \mathcal{E}\left[f(\beta,\gamma)\right] \quad \epsilon .$$

We now evaluate the averages on both sides of this inequality. First, we notice that $\mathcal{E}\left[f(\alpha,\beta)\right]$ has an upper bound:

$$\mathcal{E}\left[f(\alpha,\beta)\right] \quad \alpha \quad \beta \quad \epsilon \qquad \mathcal{E}\left[f(\alpha,\beta)\right] \quad \epsilon \quad \alpha \quad \beta \qquad \mathcal{E}\left[f(\alpha,\beta)\right] \quad \epsilon \quad \alpha \quad \beta .$$

Similarly,

$$\mathcal{E}[f(\alpha,\gamma)] \quad \epsilon \qquad \alpha \ \gamma \quad \text{and} \quad \mathcal{E}\left[f(\beta,\gamma)\right] \quad \epsilon \qquad \beta \ \gamma \, .$$

It follows that

$$\mathcal{E}\left[f(\alpha,\beta)\right] \quad \mathcal{E}[f(\alpha,\gamma)] \qquad \alpha \ \gamma \quad \alpha \ \beta \qquad 2\epsilon \, .$$

Finally, the inequality, $\mathcal{E}\left[f(\alpha,\beta)\right] \quad \mathcal{E}[f(\alpha,\gamma)] \qquad 1 \quad \epsilon \qquad \mathcal{E}[f(\beta,\gamma)]$, becomes

$$\alpha \ \gamma \quad \alpha \ \beta \qquad 2\epsilon \qquad 1 \quad \epsilon \qquad (\epsilon \qquad \beta \ \gamma)$$

or

$$4(\epsilon \qquad \delta) \qquad \alpha \ \gamma \quad \alpha \ \beta \quad \beta \ \gamma \qquad 1.$$

This contradicts our hypothesis that given a small δ we can find an arbitrarily small ϵ. ■

Example. Assume $\alpha \ \gamma \quad 0, \alpha \ \beta \quad \beta \ \gamma \quad 1/\ \overline{2}$. Then,

$$4(\epsilon \quad \delta) \qquad 0 \quad 1/\ \overline{2} \qquad 1/\ \overline{2} \quad 1$$

or

$$\epsilon \quad \delta \qquad 1/4(\ \overline{2} \quad 1).$$

We conclude that the expected value of the observation based on a hidden variable model and the expected values predicted by quantum mechanics cannot be arbitrarily close to one another. Next, we discuss the Bell inequality, which provides deeper insights into the fallacy of the hidden variable model.

2.18 BELL AND CHSH INEQUALITIES

John Bell investigated the separability of hidden variables and derived an inequality for the outcomes reported by two observers of an experiment in a setup similar to the one used for the EPR experiments. The more general CHSH inequality, named after Clauser, Horne, Shimony, and Holt [99], is related to the Bell inequality.

Bell and CHSH inequalities establish an upper bound for the correlation of distant events based on the assumption that the principle of locality is valid. The Bell inequality is based on two common-sense assumptions:

1. *Locality:* Measurements of different physical properties of different objects, carried out by different individuals, at distinct locations, cannot in uence each other.
2. *Realism:* Physical properties are independent of observations.

The Bell Inequality

We revisit the EPR experiment and consider a pair of photons emitted in opposite directions (Figure 2.15). There are now two independent observers, each using a polarization analyzer as the measurement apparatus; the first one may choose for her analyzer one of two possible settings, the orientation, α or γ, relative to an arbitrary direction. For each orientation, the outcome is unpredictable; if she chooses the orientation to be α, then the measured observable is called σ_α, and the two possible outcome values are $a \quad 1$; if she chooses γ, then the measured observable is σ_γ, and the two possible outcome values are $c \quad 1$. A similar scenario is true for the second observer; she may choose two possible settings: a new orientation, β, or the same orientation, γ, as the first observer. If she chooses β, then the measured observable is σ_β, and the two possible outcome values are $b \quad 1$; if she chooses γ, then the measured observable is σ_γ, and the outcome values are $c \quad 1$. If both observers choose the same orientation, γ, then they obtain the same result, c. Out of the three possible measurement results, a, b, and c, only the pair (a, b) or only c can be observed.

We wish to test the following hypothesis: The outcome of the experiment for the first observer depends only on *local hidden variables* related *only* to the first photon and to the analyzer used by the first observer. This property, called the *separability of hidden variables*, is a consequence of the Einstein locality principle, which conjectures that "events occurring in a given space-time region are independent of the external parameters that may be controlled, at the same moment, by agents located in distant space-time regions." If the hidden variables have the separability property, then the outcome of the measurement performed by the first observer does not depend on the orientation of the analyzer used by the second observer.

First, we show that the three possible results, a, b, and c, satisfy the equation

$$ab \quad ac \qquad (1 \quad bc).$$

We evaluate separately the left- and the right-hand side of the equation for the two cases, correlated, $b \quad c$, and anti-correlated, $b \quad c$, photons:

$$(b \quad c) \quad b \quad c \quad 0 \quad a(b \quad c) \quad 0 \quad \text{and} \quad 1 \quad bc \quad 1 \quad (\ 1)(\ 1) \quad 1 \quad 1 \quad 0$$

$$(b \quad c) \quad b \quad c \quad\ 2 \quad a(b \quad c) \quad\ 2 \quad \text{and} \quad 1 \quad bc \quad 1 \quad (\ 1)(\ 1) \quad 1 \quad 1 \quad 2.$$

We repeat the experiment for many pairs of photons; if we accept the existence of variables we do not control, then the hidden variable for the experiment, with the i-th photon, must be different from the one for the j-th photon. From the previous equality, it follows that the results of the measurements for the i-th photon satisfy the relationship

$$a_i b_i \quad a_i c_i \qquad (1 \quad b_i c_i).$$

For correlated and anti-correlated photons, we have, respectively:

$$b_i c_i \quad 1 \quad a_i b_i \quad a_i c_i \quad (1 \quad 1) \quad 0,$$

$$b_i c_i \quad 1 \quad a_i b_i \quad a_i c_i \quad (1 \quad 1) \quad\ 2.$$

The hidden variables are different for each value of the index, i; the average of the products of pairs of hidden variables over the index, i, gives us the correlation of the hidden variables:

$$ab \quad \frac{1}{N}\sum_{i\ 1}^{N} a_i b_i, \qquad ac \quad \frac{1}{N}\sum_{i\ 1}^{N} a_i c_i, \quad \text{and} \quad bc \quad \frac{1}{N}\sum_{i\ 1}^{N} b_i c_i.$$

Then $ab \quad ac \qquad (1 \quad bc\ )$ and the well-known Bell inequality follows immediately:

$$ab \quad ac \quad 1 \quad bc\ .$$

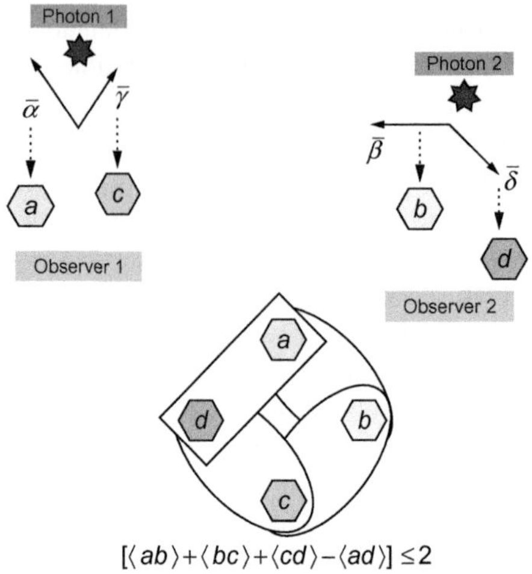

FIGURE 2.16

Two observers measure the polarization of two photons along four directions, α and γ for the first observer, and β and δ for the second one. The first observer produces the results, $a \quad 1, c \quad 1$, respectively. The second observer produces the results, $b \quad 1, d \quad 1$, respectively. The expected values of the four observables satisfy the CHSH inequality.

The Clauser, Horne, Shimony, and Holt (CHSH) Inequality

We consider a similar setup as the one for the Bell inequality, but the two observers do not share a common direction (Figure 2.16). The tests are performed for four settings: along the directions α and γ for the first observer and β and δ for the second one. The measurements produce the four pairs of outcome values, $a \quad 1, c \quad 1$, respectively, for the first observer, and $b \quad 1, d \quad 1$, respectively, for the second observer. The four random variables, a, c, b, d, have a joint probability distribution function

$$p(q,r,s,t) \quad Prob(a \quad q, c \quad r, b \quad s, d \quad t).$$

Recall that the results of the measurements performed by two distinct observers can only be $a, c, b, d \quad 1$, thus, we have two alternatives for observer 1:

$$a \quad c \quad 0 \text{ and } a-c \quad 2 \quad \text{or} \quad a \quad c \quad 2 \text{ and } a-c \quad 0.$$

The identity $(a \quad c)b-(a-c)d \quad 2$ follows from the listing of all possible combinations of values of observable a, b, c, and d in Table 2.1.

We repeat the experiment N times; for the i-th pair of photons we have

$$(a_i \quad c_i)b_i-(a_i-c_i)d_i \quad 2.$$

The average of the N measurements is

$$\frac{1}{N}\sum_{i \quad i}^{N}[(a_i \quad c_i)b_i-(a_i-c_i)d_i] \quad 2,$$

Table 2.1 Truth Table Listing Possible Combinations of Values of Observables for the CHSH Inequality

a	c	b	d	a+c	a−c	(a+c)b	(a−c)d	(a+c)b−(a−c)d
1	1	1	1	2	0	2	0	2
1	1	1	1	2	0	2	0	2
1	1	1	1	2	0	2	0	2
1	1	1	1	2	0	2	0	2
1	1	1	1	0	2	0	2	2
1	1	1	1	0	2	0	2	2
1	1	1	1	0	2	0	2	2
1	1	1	1	0	2	0	2	2
1	1	1	1	0	2	0	2	2
1	1	1	1	0	2	0	2	2
1	1	1	1	0	2	0	2	2
1	1	1	1	0	2	0	2	2
1	1	1	1	2	0	2	0	2
1	1	1	1	2	0	2	0	2
1	1	1	1	2	0	2	0	2
1	1	1	1	2	0	2	0	2

or $ab \quad bc \quad cd \quad ad \quad 2$. After taking the modulus, we obtain the CHSH inequality

$$ab \quad bc \quad cd \quad ad \quad 2.$$

We conclude that when a quantum system consists of several disjoint systems, as in the case of the EPR experiment, the hidden variables do not satisfy the *separability* requirement. Because there is an upper bound for their correlation, they do not fall in disjoint subsets.

2.19 VIOLATION OF THE BELL INEQUALITY

In the previous section we saw that if we accept two common-sense assumptions, collectively known as *local realism*, then the results of the measurements must satisfy the Bell and CHSH inequalities. We show now that *a pair of maximally entangled qubits violates the Bell inequality.* Consider a pair of entangled qubits in the singlet state $\beta_{11} \quad \psi \quad (1/ \overline{2})(01 \quad 10)$. Alice has one of the entangled particles and Bob the other. We have several one-qubit gates used to transform the state of one qubit; among them are the X gate (it transposes the components of a qubit) and the Z gate (it ips the sign of a qubit). The Pauli transformations performed by the two gates are, respectively:

$$\sigma_x \quad \begin{pmatrix} 0 & 1 \\ 1 & 0 \end{pmatrix} \quad \text{and} \quad \sigma_z \quad \begin{pmatrix} 1 & 0 \\ 0 & 1 \end{pmatrix}.$$

Alice measures two observables, a and c, by applying the operators, $\mathbf{A} \quad \sigma_z$ and $\mathbf{C} \quad \sigma_x$. Alice measures her qubit in the eigenbasis of σ_z after it has been acted on by the Z gate and in the eigenbasis

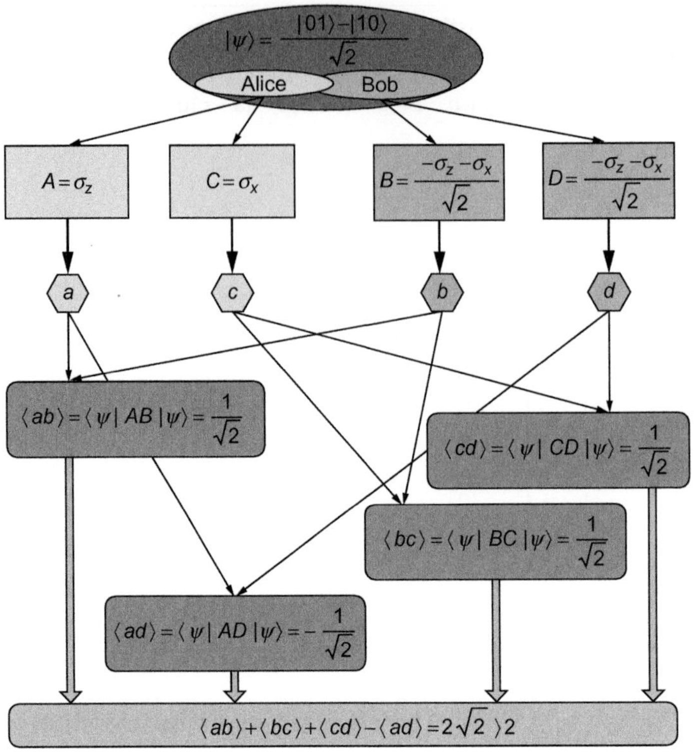

FIGURE 2.17

A pair of entangled particles, one belonging to Alice and the other to Bob, are in the state ψ . Alice measures two observables, a and c, by applying the operators, **A** σ_z and **C** σ_x. Bob measures two observables, b and d, by applying the operators, **B** $-(1/ \ 2)(\sigma_z \quad \sigma_x)$ and **D** $(1/ \ 2)(\sigma_z - \sigma_x)$. The expected values of the four observables violate the Bell inequality.

of σ_x after it has been acted on by the X gate. Similarly, Bob measures two observables, b and d, by applying the operators, **B** $-(1/ \ 2)(\sigma_z \quad \sigma_x)$ and **D** $(1/ \ 2)(\sigma_z - \sigma_x)$ (Figure 2.17).

Recall that the outcomes of measurements are eigenvalues of self-adjoint operators (projectors) and are real numbers; quantum transformations are described by unitary operators. Pauli transformations are both self-adjoint (thus, potentially physical observables) and unitary (thus, potentially gates). The eigenvectors of the transformation, σ_z, are 0 and 1 , and the corresponding eigenvalues are 1 and -1, respectively. The eigenvectors of the transformation, σ_x, are $(1/ \ 2)(0 \quad 1)$ and $(1/ \ 2)(0 - 1)$, and the corresponding eigenvalues are 1 and -1, respectively.

Proposition. *The average values of the observables are*

$$ab \quad cb \quad cd \quad \frac{1}{2} \quad \textit{and} \quad ad \quad -\frac{1}{2}.$$

Thus, $ab \quad cb \quad cd - ad \quad 2 \ 2$, *in violation of the Bell inequality.*

Proof. Indeed,

$$\sigma_z \quad \sigma_x \quad \begin{pmatrix} 1 & 1 \\ 1 & 1 \end{pmatrix} \quad \text{and} \quad \sigma_z \quad \sigma_x \quad \begin{pmatrix} 1 & 1 \\ 1 & 1 \end{pmatrix}.$$

It follows that

$$\mathbf{AB} \quad \sigma_z \quad \frac{1}{\overline{2}}(\ \sigma_z \quad \sigma_x) \quad \frac{1}{\overline{2}}\begin{pmatrix} 1 & 0 \\ 0 & 1 \end{pmatrix} \begin{pmatrix} 1 & 1 \\ 1 & 1 \end{pmatrix} \quad \frac{1}{\overline{2}}\begin{pmatrix} 1 & 1 & 0 & 0 \\ 1 & 1 & 0 & 0 \\ 0 & 0 & 1 & 1 \\ 0 & 0 & 1 & 1 \end{pmatrix},$$

$$\mathbf{CB} \quad \sigma_x \quad \frac{1}{\overline{2}}(\ \sigma_z \quad \sigma_x) \quad \frac{1}{\overline{2}}\begin{pmatrix} 0 & 1 \\ 1 & 0 \end{pmatrix} \begin{pmatrix} 1 & 1 \\ 1 & 1 \end{pmatrix} \quad \frac{1}{\overline{2}}\begin{pmatrix} 0 & 0 & 1 & 1 \\ 0 & 0 & 1 & 1 \\ 1 & 1 & 0 & 0 \\ 1 & 1 & 0 & 0 \end{pmatrix},$$

$$\mathbf{CD} \quad \sigma_x \quad \frac{1}{\overline{2}}(\sigma_z \quad \sigma_x) \quad \frac{1}{\overline{2}}\begin{pmatrix} 0 & 1 \\ 1 & 0 \end{pmatrix} \begin{pmatrix} 1 & 1 \\ 1 & 1 \end{pmatrix} \quad \frac{1}{\overline{2}}\begin{pmatrix} 0 & 0 & 1 & 1 \\ 0 & 0 & 1 & 1 \\ 1 & 1 & 0 & 0 \\ 1 & 1 & 0 & 0 \end{pmatrix},$$

and

$$\mathbf{AD} \quad \sigma_z \quad \frac{1}{\overline{2}}(\sigma_z \quad \sigma_x) \quad \frac{1}{\overline{2}}\begin{pmatrix} 1 & 0 \\ 0 & 1 \end{pmatrix} \begin{pmatrix} 1 & 1 \\ 1 & 1 \end{pmatrix} \quad \frac{1}{\overline{2}}\begin{pmatrix} 1 & 1 & 0 & 0 \\ 1 & 1 & 0 & 0 \\ 0 & 0 & 1 & 1 \\ 0 & 0 & 1 & 1 \end{pmatrix}.$$

Now we compute ab , cb , cd , and ad , the average values of the four observables ab, cb, cd, and ad, respectively. Recall that when we have n identical copies of a quantum system in the state ψ , the expected value of the observable associated, with the operator **P** applied to the systems in the state ψ , is

$$\mathbf{P} \qquad \psi\ \mathbf{P}\ \psi\ .$$

In our case, the state of the two entangled qubits is

$$\psi \quad \frac{01 \qquad 10}{\overline{2}} \qquad \psi \quad \frac{1}{\overline{2}}\begin{pmatrix} 0 \\ 1 \\ 1 \\ 0 \end{pmatrix}.$$

It follows that

$$ab \qquad \psi\ \mathbf{AB}\ \psi\ .$$

Call

$$\xi \quad \mathbf{AB}\,\psi \quad \frac{1}{\sqrt{2}}\begin{pmatrix} 1 & 1 & 0 & 0 \\ 1 & 1 & 0 & 0 \\ 0 & 0 & 1 & 1 \\ 0 & 0 & 1 & 1 \end{pmatrix} \frac{1}{\sqrt{2}}\begin{pmatrix} 0 \\ 1 \\ 1 \\ 0 \end{pmatrix} \quad \frac{1}{2}\begin{pmatrix} 1 \\ 1 \\ 1 \\ 1 \end{pmatrix}.$$

Then,

$$ab \qquad \psi\;\xi \quad \frac{1}{\sqrt{2}}\begin{pmatrix} 0 & 1 & \quad 1 & 0 \end{pmatrix}\frac{1}{2}\begin{pmatrix} 1 \\ 1 \\ 1 \\ 1 \end{pmatrix} \quad \frac{1}{\sqrt{2}}.$$

Similarly,

$$cb \qquad \psi\;\mathbf{CB}\;\psi\;.$$

Call

$$\xi \quad \mathbf{CB}\,\psi \quad \frac{1}{\sqrt{2}}\begin{pmatrix} 0 & 0 & 1 & 1 \\ 0 & 0 & 1 & 1 \\ 1 & 1 & 0 & 0 \\ 1 & 1 & 0 & 0 \end{pmatrix} \frac{1}{\sqrt{2}}\begin{pmatrix} 0 \\ 1 \\ 1 \\ 0 \end{pmatrix} \quad \frac{1}{2}\begin{pmatrix} 1 \\ 1 \\ 1 \\ 1 \end{pmatrix}.$$

Then,

$$cb \qquad \psi\;\xi \quad \frac{1}{\sqrt{2}}\begin{pmatrix} 0 & 1 & \quad 1 & 0 \end{pmatrix}\left(\;\frac{1}{2}\right)\begin{pmatrix} 1 \\ 1 \\ 1 \\ 1 \end{pmatrix} \quad \frac{1}{\sqrt{2}}.$$

For the third average we have

$$cd \qquad \psi\;\mathbf{CD}\;\psi\;.$$

Call

$$\xi \quad \mathbf{CD}\,\psi \quad \frac{1}{\sqrt{2}}\begin{pmatrix} 0 & 0 & 1 & 1 \\ 0 & 0 & 1 & 1 \\ 1 & 1 & 0 & 0 \\ 1 & 1 & 0 & 0 \end{pmatrix} \frac{1}{\sqrt{2}}\begin{pmatrix} 0 \\ 1 \\ 1 \\ 0 \end{pmatrix} \quad \frac{1}{2}\begin{pmatrix} 1 \\ 1 \\ 1 \\ 1 \end{pmatrix}.$$

Then,

$$cd \quad \psi\ \xi \quad \frac{1}{\overline{2}}\begin{pmatrix}0 & 1 & 1 & 0\end{pmatrix}\frac{1}{\overline{2}}\begin{pmatrix}1\\1\\1\\1\end{pmatrix} \quad \frac{1}{\overline{2}}.$$

Finally,

$$ad \quad \psi\ \mathbf{AD}\ \psi\ .$$

Call

$$\xi \quad \mathbf{AD}\ \psi \quad \frac{1}{\overline{2}}\begin{pmatrix}1 & 1 & 0 & 0\\1 & 1 & 0 & 0\\0 & 0 & 1 & 1\\0 & 0 & 1 & 1\end{pmatrix}\frac{1}{\overline{2}}\begin{pmatrix}0\\1\\1\\0\end{pmatrix} \quad \frac{1}{2}\begin{pmatrix}1\\1\\1\\1\end{pmatrix}.$$

Then,

$$ad \quad \psi\ \xi \quad \frac{1}{\overline{2}}\begin{pmatrix}0 & 1 & 1 & 0\end{pmatrix}\frac{1}{\overline{2}}\begin{pmatrix}1\\1\\1\\1\end{pmatrix} \quad \frac{1}{\overline{2}}.$$

It follows that

$$ab \quad bc \quad cd \quad ad \quad 2\ \overline{2}.$$

Thus, quantum mechanics predicts a value for the sum of the average values of observables in violation of the Bell inequality. ■

An interesting question is whether mixed entangled states violate the Bell inequality. There are some mixed entangled states called Werner states [453] that do not violate the Bell inequality for any number of local measurements.

2.20 ENTANGLEMENT AND HIDDEN VARIABLES

John Bell provided deeper insights into the limitations of a hidden variables theory; he showed that *it is possible to mimic the statistical properties of quantum theory by a deterministic model based on hidden variables.* He constructed such a deterministic model generating results whose averages are identical to those predicted by quantum mechanics. When the quantum system is made up of disjoint subsystems entangled like those in the EPR experiment, the hidden variables cannot be separated as disjoint subsets, local to each subsystem. Let us follow the arguments of John Bell.

Consider a spin-½ particle and an observable, A, such that the associated measurement operator is $\mathbf{A} \quad \mathbf{m}\ \boldsymbol{\sigma}$, where $\boldsymbol{\sigma}$ are the Pauli matrices and $\mathbf{m}$ are some arbitrary real numbers. The model is based on the assumption that the outcome of an experiment is determined by:

1. Two elements we can control:
   - The state ψ in which we prepare the system, and
   - The eigenvalues, $\mathbf{m}$, of the operator, $\mathbf{A}$, which can take the values, $\quad m$; they are parameters of the measuring apparatus.
2. One element we cannot control, a hidden variable uniformly distributed in the interval $1 \quad \lambda \quad 1$.

   The model is deterministic; *once we choose a value for the hidden variable, λ, the result of the measurement is certain.* The following table gives the result of a measurement and the probability of that result for a given range of the hidden variable λ:

Range of λ	**Result of Measurement**	**Probability**
$1 < \lambda < \quad \psi\ \mathbf{A}\ \psi\ /m$	m	$p(\ m) \quad (1 \quad \psi\ \mathbf{A}\ \psi\ /m)/2$
$\psi\ \mathbf{A}\ \psi\ /m < \lambda < \quad 1$	m	$p(\ m) \quad (1 \quad \psi\ \mathbf{A}\ \psi\ /m)/2$

The average value of the result of a measurement based on this deterministic model that includes a uniformly distributed hidden variable is

$$\mathbf{A} \qquad mp\ \ m \quad mp\ \ m \qquad m[(1 \quad \psi\ A\ \psi\ /m)/2] \quad m[(1 \quad \psi\ \mathbf{A}\ \psi\ /m)/2]$$
$$\psi\ \mathbf{A}\ \psi\ ,$$

in perfect agreement with quantum mechanics.

Bell showed that his model does not satisfy the separability requirement when the quantum system is in an entangled state. If the quantum system is made up of disjoint subsystems, entangled like those in the EPR experiment, then the hidden variables cannot be separated as disjoint subsets, local to each subsystem.

Is there any relationship between the local hidden variable models and shareability of entanglement? This question was answered recently [419]. A local hidden variable model assumes the existence of a random variable shared by two observers—Alice, A, and Bob, B; this variable is used to locally generate a measurement outcome. The outcome depends only on the choice of the local measurement and not on the choice of the remote observer's measurement. If each observer has a set of local measurements, $i \quad 1,2,\ldots,m_A$ for Alice, with $\lambda_A(i)$, the possible outcomes per measurement, and $k \quad 1,2,\ldots,m_B$ measurements, with $\lambda_B(k)$, the outcomes per measurement for Bob, then a local hidden variable model generates all the probability values, $P_{ij,kl}$, that Alice's i-th measurement has outcome, j, and Bob's k-th measurement has outcome, l. Each local measurement has a single outcome with probability one independently of the measurement made by the other observer. For a composite quantum system described by the density matrix, ρ, in the Hilbert space, $\mathcal{H}_A \quad \mathcal{H}_B$, the probability, $P_{ij,kl}$, is given by

$$P_{ij,kl}(\rho) \quad \mathrm{tr}((\mathbf{M}_{ij}^{A} \quad \mathbf{M}_{kl}^{B})\rho),$$

where $\mathbf{M}_{ij}^{A}$, with $\mathbf{M}_{ij}^{A}$ 0 and $\sum_j \mathbf{M}_{ij}^{A}$ I_A, the POVM elements for the i-th measurement of observer A, and where $\mathbf{M}_{kl}^{B}$, with $\mathbf{M}_{kl}^{B}$ 0 and $\sum_l \mathbf{M}_{kl}^{B}$ I_B, the POVM elements of the k-th measurement of observer B.

The Bell inequality is violated if and only if the probability value cannot be generated by a local hidden variable model. We can say that the Bell inequality quantifies how measurements[11] on entangled quantum systems can invalidate local classical models of reality. It is shown [419] that violations of Bell inequalities can be connected to the existence of a symmetric extension/quasi-extension of a quantum state. An extension of a quantum state, ρ, of a system, AB, is another quantum state defined on a system, ABC, such that when we trace over, C, we obtain the initial quantum state, ρ. The system C must be of the form

$$C \quad A^{\ (m_A \quad 1)} \quad B^{\ (m_B \quad 1)},$$

and the extension state must be invariant under all permutations of the m_A copies of A among each other and under any permutation of the m_B copies of B. Such an extension exists if the quantum state ρ of the system AB is separable—i.e.,

$$\rho \quad \sum_i p_i (\psi_i \quad \psi_i)_A \quad (\varphi_i \quad \varphi_i)_B .$$

The extension state, ρ , is obtained by copying the individual product states onto the other spaces:

$$\rho \quad \sum_i p_i (\psi_i \quad \psi_i) \ ^{m_A} \quad (\varphi_i \quad \varphi_i) \ ^{m_B} .$$

Proposition. *Consider two Hilbert spaces, $\mathcal{H}_A$ and $\mathcal{H}_{B_1}$; let ρ be a density matrix on $\mathcal{H}_A$ $\mathcal{H}_{B_1}$. If ρ has a symmetric extension, ρ , on $\mathcal{H}_A$ $\mathcal{H}_{B_1}$ $\mathcal{H}_{B_2}$ $\mathcal{H}_{B_m}$, then there exists a local hidden variable description of ρ when one of the observers performs an arbitrary number of measurements and the other observer performs m measurements.*

The intuition behind this result comes from several properties of hidden variables and measurements. When one of the two parties performs only one measurement of a quantum state, there is a local hidden variable model describing the quantum state. The second observer may carry out her m measurements, $\mathbf{M}_1, \mathbf{M}_2, \ldots, \mathbf{M}_m$, in the m Hilbert spaces, $\mathcal{H}_{B_1}, \mathcal{H}_{B_2}, \ldots, \mathcal{H}_{B_m}$, respectively. The joint probability of an outcome of a measurement $\mathbf{M}_A$ and some $\mathbf{M}_{B_i}$, 1 i m, are the same for ρ and ρ because ρ is a symmetric extension of ρ. The m measurements can be viewed as a single measurement on $\mathcal{H}_{B_1}$ $\mathcal{H}_{B_2}$ $\mathcal{H}_{B_m}$. Thus, the measurements on ρ have a local hidden variable description from which we can deduce the local hidden variable description of the original measurements on ρ.

If the quantum state ρ is a pure entangled state (nonseparable), such a symmetric extension does not exist since pure entanglement is monogamous. Therefore, a violation of the Bell inequality signals that the system is in a pure entangled state; the entanglement in this state of the composite quantum system is monogamous.

---

[11]The types and number of local measurements the observers can perform are not bounded and, correspondingly, the enumeration of Bell inequalities is computationally hard.

2.21 QUANTUM AND CLASSICAL CORRELATIONS

To answer the question of whether the correlation of quantum measurements is stronger than the correlation of classical measurements, we consider two simple systems exhibiting some degree of similarity:

1. a quantum system consisting of two spin-½ particles in a singlet state; and
2. a classical system consisting of a projectile initially at rest, which explodes and produces two fragments carrying opposite momenta.

We assume that we have N copies of each system, and we are able to measure in each case two values, a and b, of an observable, compute their correlation function, ab, and obtain ensemble averages, a , b , and ab . We show that the expected value of the correlation of the quantum measurements is stronger than the correlation of classical ones.

We examine first the quantum system; let α and β be two arbitrarily chosen unit vectors. The observables for the two particles are, respectively, $\alpha \quad \sigma_1$ and $\beta \quad \sigma_2$, with σ_1 and σ_2, the Pauli spin matrices for the two particles. The system is in a singlet state; thus:

$$\sigma_2 \; \psi \qquad \sigma_1 \; \psi \; .$$

The results of the measurements are $a, b \qquad 1$. If we repeat an experiment similar to the one suggested by Bohm, then quantum mechanics predicts that the expected value of the results is equal to zero:

$$a \quad 0 \text{ and } \quad b \quad 0.$$

The expected value of the quantum correlation is

$$ab^{\,quantum} \qquad \psi^{\dagger} \; (\alpha \quad \sigma_1)(\beta \quad \sigma_2) \; \psi \; .$$

We apply the identity [325]

$$(\alpha \quad \sigma)(\beta \quad \sigma) \quad \alpha \quad \beta \quad i(\alpha \quad \beta) \quad \sigma$$

to the system in a singlet state, $\sigma_2 \; \psi \qquad \sigma_1 \; \psi$, and conclude that

$$ab^{\,quantum} \qquad \alpha \quad \beta \qquad \cos\theta,$$

with θ, the angle between the directions of the two vectors, α, and β, freely chosen by the observer. The largest correlation occurs for $\theta \quad \pi$, then

$$\max_{\theta} \; ab^{\,quantum} \qquad \cos\pi \qquad 1.$$

Let us now discuss the classical system; the angular momentum of the projectile at rest is $\mathbf{J}$ and the angular momenta of the two fragments after the explosion are $\mathbf{J}_2 \qquad \mathbf{J}_1$. As before, α and β are two freely chosen unit vectors and θ the angle between them. The experiment is repeated N times; the observer of the first fragment measures the observable sign($\alpha \quad \mathbf{J}_1$) and obtains the results $a_i \qquad 1$, while the observer of the second fragment measures the observable sign($\beta \quad \mathbf{J}_2$) and obtains the results

$b_i \quad 1,1 \quad i \quad N$. The expected values of the outcomes of the two measurements are, respectively:

$$a \quad \frac{1}{N}\sum_{i\ 1}^{N} a_i \quad \text{and} \quad b \quad \frac{1}{N}\sum_{i\ 1}^{N} b_i.$$

These values are typically of the order $1/\ \overline{N}$, thus, very close to zero. The correlation of the two measurements is given by

$$ab^{\ classical} \quad \frac{1}{N}\sum_{i\ 1}^{N} a_i b_i,$$

an expression that does not vanish to zero. Simple mechanical considerations show that if the momenta, $\mathbf{J}_1$ and $\mathbf{J}_2$, are uniformly distributed, then

$$ab^{\ classical} \quad \frac{\theta \quad (\pi \quad \theta)}{\pi} \quad 1 \quad \frac{2\theta}{\pi}.$$

The largest correlation occurs when $\theta \quad \pi$:

$$\max_\theta\ ab^{\ classical} \quad 1 \quad \frac{2\theta}{\pi} \quad 1 \quad 2 \quad 1.$$

In the interval $\theta \quad [0,\pi]$, the two correlations are equal, with the values 0 or $\quad 1$, when θ takes the values $\frac{\pi}{2}$ or 0 and π, respectively. In the rest of the interval, the classical correlation is a linear function, while the quantum correlation is a cosine function that runs below the straight line in the negative value range for $\theta \quad (0,\frac{\pi}{2})$ and, respectively, above it in the positive value range for $\theta \quad (\frac{\pi}{2},\pi)$.

Thus, the inequality

$$ab^{\ quantum} \qquad ab^{\ classical}$$

leads us to the conclusion that *quantum mechanics, based on a nondeterministic model, allows a stronger correlation of measurements than classical mechanics, based on a deterministic model of the physical world.*

We now switch our focus from theoretical considerations regarding the measurement process to practical applications and start our analysis with a discussion of the measurements of quantum circuits.

2.22 MEASUREMENTS AND QUANTUM CIRCUITS

To report the results of any transformation produced by quantum circuits, we have to carry out a measurement; we start our discussion with an application of quantum measurements to circuits used to detect errors during a quantum error correction process and then present two principles important for the measurement of quantum circuits.

Quantum error correction is based on our ability to detect if a quantum state is in error without actually measuring the system and thus altering its state. This topic is presented at length in Chapter 5, where we discuss how to determine the *syndrome* associated with a quantum state. The stabilizer formalism allows us to establish if a system of n qubits has suffered decoherence and is in error by computing the eigenvalues of a group of operators, $\mathbf{M}_i$. We analyze next a simple circuit to measure

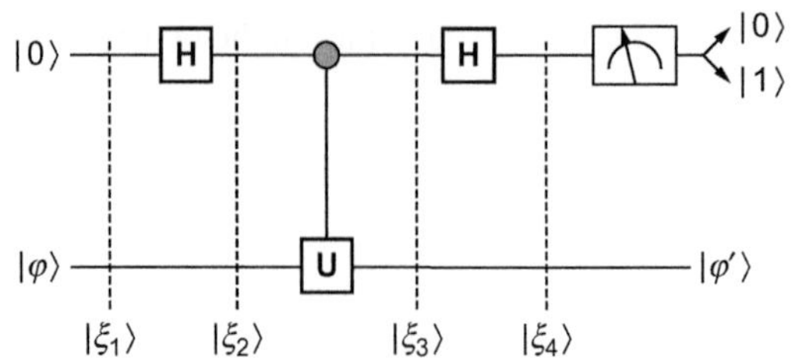

FIGURE 2.18

A circuit used to measure the eigenvalues of a single-qubit operator, **U**, with the two eigenvalues, $\lambda \quad 1$. The upper output qubit is set to 0 if the eigenvalue of **U** is $\lambda \quad 1$ and to 1 if the eigenvalue **U** is $\lambda \quad -1$. The lower output qubit is $\varphi \quad \alpha_0\ 0 \quad \alpha_1\ 1$, the corresponding eigenvector of **U**.

the eigenvalues of a single-qubit operator, **U**, with the two eigenvalues, $\lambda \quad 1$. Such an operator is unitary; thus, we can regard **U** as a one-qubit gate; **U** is also self-adjoint and we can regard it as an observable.

Example. *A circuit used to measure the eigenvalues of a single-qubit operator.* The circuit in Figure 2.18 produces an eigenvector of the operator **U**; the upper output qubit is set to 0 if the corresponding eigenvalue is $\lambda \quad 1$ and to 1 if the eigenvalue is $\lambda \quad -1$.

We want to determine the unitary transformation,

$$\mathbf{U} \quad \begin{pmatrix} u_{11} & u_{12} \\ u_{21} & u_{22} \end{pmatrix},$$

for which the computation of the eigenvalues is carried out by the circuit. Let $\varphi \quad \alpha_0\ 0 \quad \alpha_1\ 1$ be an eigenvector of **U**. The two eigenvalues, $\lambda_1 \quad 1$ and $\lambda_2 \quad -1$, identified after measurement of the first qubit are associated with the different expressions for ξ_4 in Figure 2.18:

$$\lambda_1 \quad 1 \qquad \mathbf{U}\ \varphi \quad (\ 1)\ \varphi \qquad \xi_4 \quad 0\ (\alpha_0\ 0 \quad \alpha_1\ 1\ ) \quad \alpha_0\ 00 \quad \alpha_1\ 01$$

and

$$\lambda_2 \quad -1 \qquad \mathbf{U}\ \varphi \quad (-1)\ \varphi \qquad \xi_4 \quad 1\ (\alpha_0\ 0 \quad \alpha_1\ 1\ ) \quad \alpha_0\ 10 \quad \alpha_1\ 11\ .$$

The states ξ_1 , ξ_2 , ξ_3 , and ξ_4 in Figure 2.18 are:

$$\xi_1 \quad 0 \quad \varphi \qquad \begin{pmatrix} \alpha_0 \\ \alpha_1 \\ 0 \\ 0 \end{pmatrix},$$

$$\xi_2 \quad (H \quad I)\,\xi_1 \quad \frac{1}{\overline{2}}\begin{pmatrix}1 & 0 & 1 & 0\\ 0 & 1 & 0 & 1\\ 1 & 0 & 1 & 0\\ 0 & 1 & 0 & 1\end{pmatrix}\begin{pmatrix}\alpha_0\\ \alpha_1\\ 0\\ 0\end{pmatrix} \quad \frac{1}{\overline{2}}\begin{pmatrix}\alpha_0\\ \alpha_1\\ \alpha_0\\ \alpha_1\end{pmatrix},$$

$$\xi_3 \quad G_{controlled \quad U}\,\xi_2 \quad \begin{pmatrix}1 & 0 & 0 & 0\\ 0 & 1 & 0 & 0\\ 0 & 0 & u_{11} & u_{12}\\ 0 & 0 & u_{21} & u_{22}\end{pmatrix}\frac{1}{\overline{2}}\begin{pmatrix}\alpha_0\\ \alpha_1\\ \alpha_0\\ \alpha_1\end{pmatrix} \quad \frac{1}{\overline{2}}\begin{pmatrix}\alpha_0\\ \alpha_1\\ \alpha_0 u_{11} \quad \alpha_1 u_{12}\\ \alpha_0 u_{21} \quad \alpha_1 u_{22}\end{pmatrix},$$

and

$$\xi_4 \quad \frac{1}{\overline{2}}\begin{pmatrix}1 & 0 & 1 & 0\\ 0 & 1 & 0 & 1\\ 1 & 0 & 1 & 0\\ 0 & 1 & 0 & 1\end{pmatrix}\frac{1}{\overline{2}}\begin{pmatrix}\alpha_0\\ \alpha_1\\ \alpha_0 u_{11} \quad \alpha_1 u_{12}\\ \alpha_0 u_{21} \quad \alpha_1 u_{22}\end{pmatrix} \quad \frac{1}{2}\begin{pmatrix}\alpha_0 \quad \alpha_0 u_{11} \quad \alpha_1 u_{12}\\ \alpha_1 \quad \alpha_0 u_{21} \quad \alpha_1 u_{22}\\ \alpha_0 \quad \alpha_0 u_{11} \quad \alpha_1 u_{12}\\ \alpha_1 \quad \alpha_0 u_{21} \quad \alpha_1 u_{22}\end{pmatrix}.$$

The eigenvalues, $\lambda_1 \quad 1$ and $\lambda_2 \quad 1$, require different relations among the elements of **U**; for $\lambda_1 \quad 1$, we have

$$\lambda_1 \quad 1 \quad \xi_4 \quad \begin{pmatrix}\alpha_0\\ \alpha_1\\ 0\\ 0\end{pmatrix} \quad \begin{cases}\alpha_0(1 \quad u_{11}) \quad \alpha_1 u_{12},\\ \alpha_1(1 \quad u_{22}) \quad \alpha_0 u_{21}.\end{cases}$$

This implies that $(1 \quad u_{11})(1 \quad u_{22}) \quad u_{12}u_{21}$; thus,

$$u_{11} \quad u_{22} \quad u_{11}u_{22} \quad u_{12}u_{21} \quad 1.$$

For $\lambda_2 \quad 1$, we have

$$\lambda_2 \quad 1 \quad \xi_4 \quad \begin{pmatrix}0\\ 0\\ \alpha_0\\ \alpha_1\end{pmatrix} \quad \begin{cases}\alpha_0(1 \quad u_{11}) \quad \alpha_1 u_{12},\\ \alpha_1(1 \quad u_{22}) \quad \alpha_0 u_{21}.\end{cases}$$

Then $(1 \quad u_{11})(1 \quad u_{22}) \quad u_{12}u_{21}$ and

$$u_{11} \quad u_{22} \quad u_{11}u_{22} \quad u_{12}u_{21} \quad 1.$$

It follows that $u_{22} \quad 1$, $u_{21} \quad u_{12} \quad 0$, and $u_{11} \quad 1$; thus, the transformation **U** in the eigenbasis is

$$\mathbf{U} \quad \begin{pmatrix}1 & 0\\ 0 & 1\end{pmatrix} \quad Z.$$

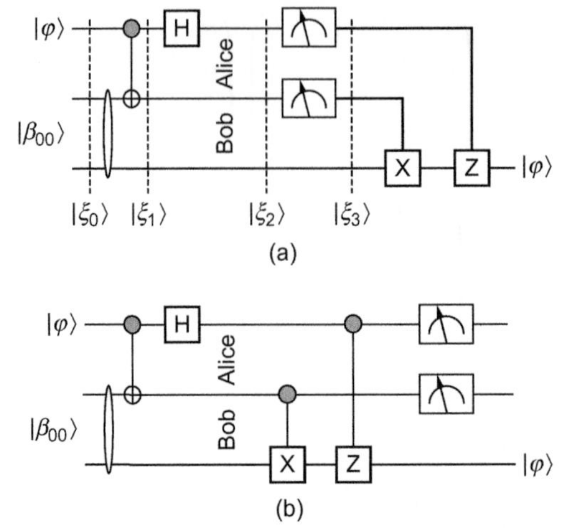

FIGURE 2.19

(a) The original circuit for teleportation. Thick lines correspond to a classical communication channel used by Alice to communicate the results of her measurements of the pairs in her possession, 00, 01, 10, or 11. (b) The circuit with deferred measurements.

The *principle of deferred measurement*: We have an algorithm that performs some quantum operations, measures, and then applies more quantum operations, depending on the outcome of the quantum measurement; *we can simulate this entire algorithm by a completely quantum algorithm with only one final measurement.* We replace classically controlled quantum operations with quantum-controlled quantum operations.

Example. Quantum teleportation. The quantum teleportation, revisited from Section 2.13, uses the circuit in Figure 2.19a. The topmost two qubits belong to Alice. The first one, whose state is to be teleported, is in the state $\varphi \quad \alpha_0\ 0 \quad \alpha_1\ 1$, and the second is entangled with Bob's qubit; the pair of qubits is in a maximally entangled state, β_{00} . The states at different stages of the quantum circuit in Figure 2.19b are

$$\xi_0 \quad \frac{1}{\sqrt{2}}[\alpha_0\ 0\ (\ 00 \quad 11\ )\ \ \alpha_1\ 1\ (\ 00 \quad 11\ )],$$

$$\xi_1 \quad \frac{1}{\sqrt{2}}[\alpha_0\ 0\ (\ 00 \quad 11\ )\ \ \alpha_1\ 1\ (\ 10 \quad 01\ )],$$

and

$$\xi_2 \quad \frac{1}{2}[\ 00\ (\alpha_0\ 0 \quad \alpha_1\ 1\ ) \quad 01\ (\alpha_0\ 1 \quad \alpha_1\ 0\ )$$
$$10\ (\alpha_0\ 0 \quad \alpha_1\ 1\ ) \quad 11\ (\alpha_0\ 1 \quad \alpha_1\ 0\ )].$$

The state, ξ_2 , is a sum of four terms, each term consisting of the tensor product of the state of two qubits in Alice's possession with the state of the qubit in Bob's possession. If Alice measures her two qubits and finds them in the state 00 , then Bob's qubit must be in the state $\alpha_0\ 0 \quad \alpha_1\ 1$, which is the original state, φ . Thus, when Bob learns that the result of the measurement is 00, he applies the identity transformation, I, to his qubit. Similarly, if Alice measures her two qubits and finds them in the state 01 , then Bob's qubit must be in the state $\alpha_0\ 1 \quad \alpha_1\ 0 \quad X\ \varphi$; thus, Bob must apply the X transformation to his qubit; if the result of the measurement is 10, then Bob's qubit must be in the state $\alpha_0\ 0 \quad \alpha_1\ 1 \quad Z\ \varphi$, and he should apply a Z transform to recover the original state. Finally, if the result of the measurement is 11, then Bob's qubit must be in the state $\alpha_0\ 1 \quad \alpha_1\ 0 \quad Y\ \varphi$, and then he should apply a $Y \quad XZ$ transform to recover the original state.

According to the principle of deferred measurement, instead of the circuit in Figure 2.19a, we can use the one in Figure 2.19b, where the measurements are moved to the end of the circuit; then, the concept of "teleportation," as discussed in this section, is altered, and no classical information is sent from Alice to Bob.

Another important concept related to the measurement of quantum circuits is provided by the *principle of implicit measurement*: *If some qubits in a computation are never used again, one can assume that they have been measured and that the result of the measurement has been ignored.* Formally, this means that the reduced density operator of the remaining qubits, the ones not measured, is the same as if the first group of qubits would have been measured and the result of the measurement ignored.

The important idea here is that a measurement is irreversible if and only if we examine its results and gather information about the quantum state. If the result of a measurement does not reveal any information about the state of the quantum system, then the measurement is reversible. It is intuitively obvious that if we totally ignore the results of a measurement, the state of the system is not affected.

For example, consider a two-qubit system. Let us perform a measurement in the computational basis of the second qubit. The following proposition is true.

Proposition. *Let ρ be the density operator of the original two-qubit system and ρ be the density operator of the system after a measurement by an observer who does not learn the result of the measurement. Then:*

$$\rho \quad P_0 \rho P_0 \quad P_1 \rho P_1 .$$

Here, P_0 and P_1 are the projectors on the basis vectors of the states of the second qubit,

$$P_0 \quad 0\ 0 \quad \text{and} \quad P_1 \quad 1\ 1 .$$

Moreover, the reduced density operator for the first qubit is not affected by the measurement:

$$tr_2(\rho) \quad tr_2(\rho\ ).$$

Proof. We rewrite the expression for ρ :

$$\begin{aligned} \rho \quad & 0\ 0\ \rho\ 0\ 0 \quad 1\ 1\ \rho\ 1\ 1 \\ & 0\ 0\ \rho\ 0\ 0 \quad (I \quad 0\ 0\ )\rho(I \quad 0\ 0\ ) \\ & I \rho I \quad \rho . \end{aligned}$$

■

A corollary of the principle of implicit measurement is that *measurement commutes with control.* Suppose we use a qubit to control a one-qubit quantum operation, U, and then we measure the qubit. We could equivalently first measure the control qubit, and then use the measurement result to classically control the application of U.

2.23 MEASUREMENTS AND ANCILLA QUBITS

A measurement of a quantum system reveals classical information. For example, if a qubit Q is in the state $|\psi_Q\rangle = |0\rangle$, then the result of the measurement is one bit of information, $q = 0$, with the probability $p = 1$. If a qubit R is in the state $|\psi_R\rangle = (1/\sqrt{2})(|0\rangle + |1\rangle)$, then the result of the measurement can be $q = 0$, with the probability $p_0 = 0.5$, or $q = 1$, with the probability $p_1 = 0.5$.

Assume that the two qubits, Q and R, are inside a black box, and we select one of them at random and perform a measurement. The original states of the two qubits are:

$$|\psi_Q\rangle = |0\rangle \quad \text{and, respectively,} \quad |\psi_R\rangle = \frac{|0\rangle + |1\rangle}{\sqrt{2}}.$$

As a result of the measurement, we obtain a classical bit of information, $q = 0$ or $q = 1$. When $q = 1$, we are certain that the original qubit was in the state $|\psi_R\rangle = (1/\sqrt{2})(|0\rangle + |1\rangle)$; thus, we have selected R. When the result is $q = 0$, we cannot be certain; there is a 50/50 chance that we have selected Q or R. It follows that we can be absolutely certain that we inferred correctly the state of the qubit selected from the black box only in 25% of the trials. Indeed, we select R in 50% of the total cases, and if we have selected R, we guess right only in 50% of those selected cases (50% of 50%).

Is there any transformation of the quantum states of the two qubits that allows us to increase our chances of guessing correctly? More precisely, we plan to transform the two qubits inside the black box, and then select and measure one of them. It is easy to see that the same one-qubit transformation applied to both qubits does not increase our probability to guess right. For example, if we apply the Walsh-Hadamard transform (use a Hadamard gate), then the qubits will end up in the following two states:

$$|\psi_Q\rangle = \frac{|0\rangle + |1\rangle}{\sqrt{2}} \quad \text{and, respectively,} \quad |\psi_R\rangle = |0\rangle.$$

If we negate the two qubits (use the X gate), then the two qubits end up in the states:

$$|\psi_Q\rangle = |1\rangle \quad \text{and, respectively,} \quad |\psi_R\rangle = \frac{|0\rangle + |1\rangle}{\sqrt{2}}.$$

In both instances we gain nothing. We are faced with either the initial problem, or we have to distinguish between the states $|1\rangle$ and $(1/\sqrt{2})(|0\rangle + |1\rangle)$. Let us now consider a different approach: We prepare an auxiliary (an ancilla) qubit, A, in a special state, $|\psi_A\rangle = |0\rangle$; then we perform a unitary transformation, U, of two systems, each one consisting of the ancilla and one of the original qubits. Finally, we measure the resulting system. Let us assume that the results of the unitary transformations are

$$|\psi_{AQ}\rangle = \alpha_0|00\rangle + \alpha_1|01\rangle \quad \text{and} \quad |\psi_{AR}\rangle = \alpha_0|00\rangle + \alpha_1|10\rangle,$$

with

$$|\alpha_0|^2 + |\alpha_1|^2 = 1.$$

A measurement could yield the result 00, 01, or 10. If the result is 01, we are certain that we have selected Q. If the result is 10, we are certain that we have selected R. If the result is 00, we cannot be sure. So the probability of identifying correctly the qubit we have selected is $\alpha_1\ {}^2$.

Now we have to construct a transformation, **U**, with the following properties:

$$\mathbf{U}(\ \psi_A \quad \psi_Q\ ) \quad \psi_{AQ} \quad \mathbf{U}\ 00 \quad \alpha_0\ 00 \quad \alpha_1\ 01\ ,$$

$$\mathbf{U}(\ \psi_A \quad \psi_R\ ) \quad \psi_{AR} \quad \mathbf{U}\frac{00 \quad 01}{\overline{2}} \quad \alpha_0\ 00 \quad \alpha_1\ 10\ ,$$

$$\mathbf{U}^{\dagger}\mathbf{U} \quad I.$$

Let us consider a transformation **U** of the form:

$$\mathbf{U} \quad \begin{pmatrix} u_{11} & u_{12} & u_{13} & u_{14} \\ u_{21} & u_{22} & u_{23} & u_{24} \\ u_{31} & u_{32} & u_{33} & u_{34} \\ u_{41} & u_{42} & u_{43} & u_{44} \end{pmatrix}.$$

The first condition can be rewritten as

$$\begin{pmatrix} u_{11} & u_{12} & u_{13} & u_{14} \\ u_{21} & u_{22} & u_{23} & u_{24} \\ u_{31} & u_{32} & u_{33} & u_{34} \\ u_{41} & u_{42} & u_{43} & u_{44} \end{pmatrix} \begin{pmatrix} 1 \\ 0 \\ 0 \\ 0 \end{pmatrix} \quad \begin{pmatrix} \alpha_0 \\ \alpha_1 \\ 0 \\ 0 \end{pmatrix}.$$

It follows that

$$u_{11} \quad \alpha_0, \quad u_{21} \quad \alpha_1, \quad u_{31} \quad 0, \quad u_{41} \quad 0.$$

The second condition can be rewritten as

$$\begin{pmatrix} \alpha_0 & u_{12} & u_{13} & u_{14} \\ \alpha_1 & u_{22} & u_{23} & u_{24} \\ 0 & u_{32} & u_{33} & u_{34} \\ 0 & u_{42} & u_{43} & u_{44} \end{pmatrix} \begin{pmatrix} 1/\ \overline{2} \\ 1/\ \overline{2} \\ 0 \\ 0 \end{pmatrix} \quad \begin{pmatrix} \alpha_0 \\ 0 \\ \alpha_1 \\ 0 \end{pmatrix}.$$

It follows that

$$u_{12} \quad \alpha_0(\ \overline{2} \quad 1), \quad u_{22} \quad \alpha_1, \quad u_{32} \quad \alpha_1\ \overline{2}, \quad u_{42} \quad 0.$$

From the third condition, we find that

$$\alpha_0^2 \quad \frac{1}{\overline{2}}, \quad \alpha_1^2 \quad 1 \quad \frac{1}{\overline{2}},$$

$$u_{13} \quad \overline{2}\alpha_1^2, \quad u_{43} \quad u_{14} \quad u_{24} \quad u_{34} \quad 0, \quad u_{44} \quad 1, \quad u_{23} \quad u_{33} \quad \frac{\alpha_1}{\alpha_0}.$$

Thus,

$$\mathbf{U} \quad \begin{pmatrix} \alpha_0 & \alpha_0(\sqrt{2} \quad 1) & \sqrt{2}\alpha_1^2 & 0 \\ \alpha_1 & \alpha_1 & \alpha_1/\alpha_0 & 0 \\ 0 & \sqrt{2}\alpha_1 & \alpha_1/\alpha_0 & 0 \\ 0 & 0 & 0 & 1 \end{pmatrix}.$$

It follows that when we use an ancilla qubit, the probability to guess correctly increases from 0.25 to $\alpha_1^2 \quad 1 \quad 1/\sqrt{2} \quad 0.2929$. An interesting question left as an exercise is whether we can further increase this probability by using more than one ancilla qubits.

In fact, we have hinted at a general principle used to measure the state of a quantum system: We let the system of interest interact with an ancilla and then we perform a von Neumann measurement on the ancilla. If ρ_Q is the density operator of the original system and ρ_A is the density operator of the ancilla system, then the composite system has a density operator that is the tensor product of the two density matrices:

$$\rho_{QA} \quad \rho_Q \quad \rho_A.$$

The interaction of the two systems is unitary:

$$\rho_{QA} \quad \mathbf{U}\rho_{QA}\mathbf{U}^{\dagger}.$$

A measurement performed on the ancilla produces a result q, with the probability

$$\text{Prob}(q) \quad \text{tr}((I \quad M_q)\mathbf{U}\rho_{QA}\mathbf{U}^{\dagger}),$$

with M_q, a set of orthogonal projector operators acting on the ancilla qubit.

Next, we discuss distinguishability of quantum states and quantum key distribution protocols.

2.24 MEASUREMENTS AND DISTINGUISHABILITY OF QUANTUM STATES

Given a set of measurement operators, $\mathbf{M}_i$, we can find a quantum measurement able to identify the state in which the quantum system has been prepared when the states, ψ_i, are orthonormal; for each possible state (each possible index, i), we can define a measurement operator

$$\mathbf{M}_i \quad \psi_i \quad \psi_i,$$

such that the probability of outcome i of the measurement when the system prepared in the state ψ_i is equal to 1:

$$p(i) \quad \psi_i \, \mathbf{M}_i \, \psi_i \quad 1,$$

If as a result of a measurement of a quantum system prepared in an unknown orthonormal state the outcome i occurs with certainty, this state is reliably distinguished as being ψ_i. We cannot reliably distinguish non-orthogonal quantum states; a rigorous proof of this statement is offered by Nielsen and Chuang [307]. This property of quantum measurements poses serious challenges for quantum

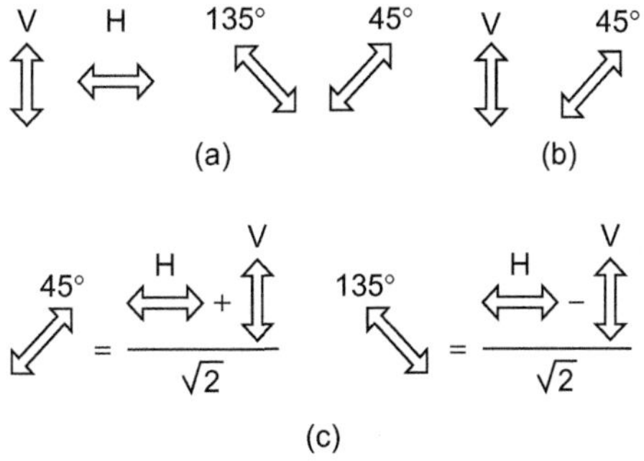

FIGURE 2.20

Distinguishability of quantum states. (a) We can reliably distinguish between orthogonal states—e.g., photons with vertical (V) versus horizontal (H) polarization, or photons with 45 versus those with 135 polarization. (b) We cannot reliably distinguish between non-orthogonal states—e.g., photons with vertical versus those with 45 polarization. (c) We can reliably distinguish between orthogonal superposition states.

computing, but, at the same time, makes quantum information very attractive for applications, such as quantum cryptography and quantum key distribution.

Figure 2.20 illustrates this fundamental property of quantum systems for polarized photons. For example, we cannot reliably distinguish if a qubit is in the state ψ_1 0 or in the state ψ_2 (0 1)/ $\overline{2}$ through a measurement in the computational basis formed by the orthonormal vectors 0 and 1 . Recall that when measuring the system in the state ψ_2 , we get the result 0 , with the probability 1/2, or 1 , with the probability 1/2. When we measure the system in the state ψ_1 , we get the result 0 , with the probability 1. Henceforth, when we get the result corresponding to 1 , it is clear that the system must have been in the state ψ_2 ; yet, when we get the result corresponding to 0 , the system could have been originally either in the state ψ_1 or in the state ψ_2 . If the states ψ_i of a quantum system are not orthonormal, we cannot find a quantum measurement able to distinguish the prepared state from other possible ones. Consider a set of measurement operators, $\mathbf{M}_i$, and assume that we have a qubit that could be in one of two non-orthogonal states, ψ_1 and ψ_2 . One of the two states, say ψ_2 , can be decomposed into two components, one (nonzero) parallel to ψ_1 , and the other, ψ_1 , orthogonal to ψ_1 :

$$\psi_2 \quad \alpha \, \psi_1 \quad \beta \, \psi_1 \; .$$

We apply the measurement operator $\mathbf{M}_i$ and based on the result of the measurement decide that the system was in the state ψ_1 . However, the previous equation shows that even when the system is in the state ψ_2 , there is a nonzero probability, α ^{2}, of deciding that the system was in the state ψ_1 . Therefore, the measurement does not reveal with certainty the state of the system prior to the measurement. The concepts discussed in this section have practical applications.

We conclude the discussion of distinguishability of quantum states with an application, the quantum key distribution. To ensure confidentiality, data, or *plaintext*, is encrypted into *ciphertext*. A *cipher* consists of algorithms transforming plaintext into ciphertext and back; a *symmetric cipher* uses the same key for encryption and decryption. Detecting eavesdropping is a major concern for key distribution protocols. Eavesdropping must be distinguished from the noise on the communication channel; to do so, "check" qubits must be interspaced randomly among the "data" qubits used to construct the encryption key. The key distribution protocols for symmetric cipher cannot guarantee that a classical system is able to detect intrusion, while quantum information theory offers an ingenious solution to the problem.

Quantum Key Distribution (QKD)

Protocols for QKD are based on the idea of transmitting non-orthogonal qubit states and then checking for the disturbance in their transmitted states, as shown in Figures 2.21a and 2.21b, respectively. The attempt of an intruder to eavesdrop increases the level of disturbance of the signal on the quantum

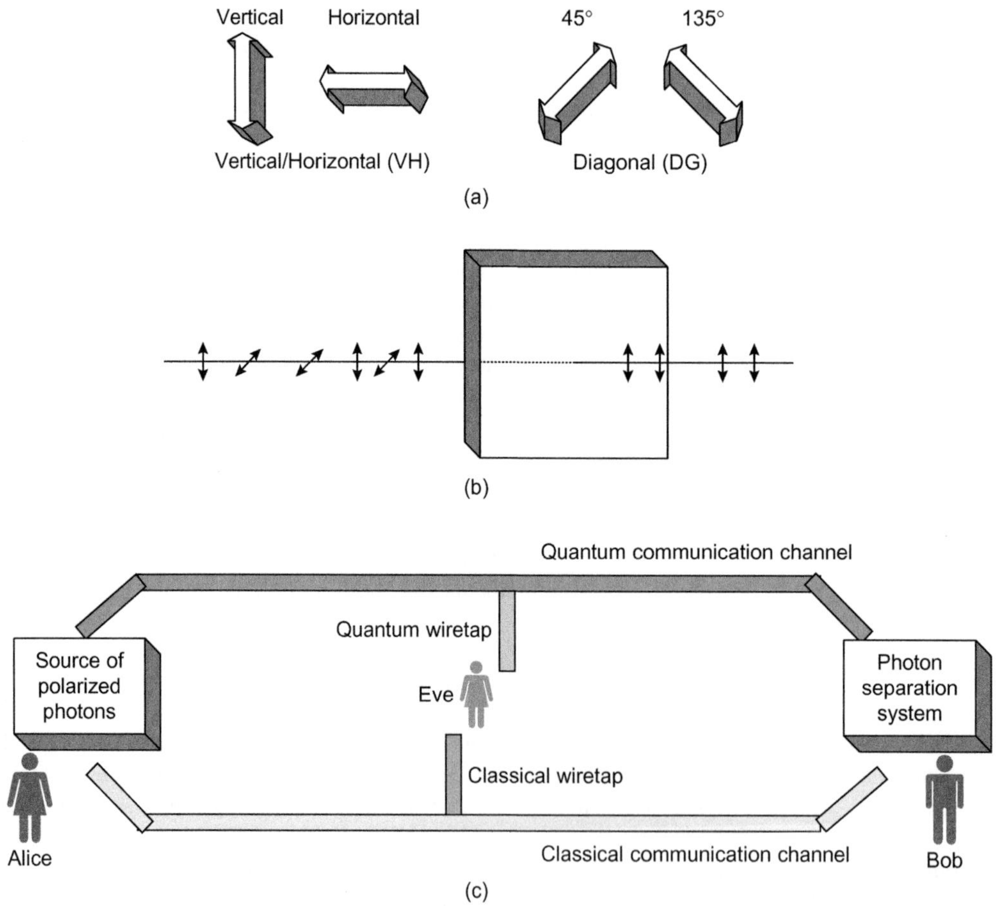

FIGURE 2.21

(a) Protocols for quantum key distribution are based on the idea of transmitting non-orthogonal qubit states—e.g., photons polarized in the X and Z bases. (b) A measurement in the wrong base alters the state. (c) The setup for quantum key distribution; Alice and Bob use quantum and classical communication channels, and Eve eavesdrops on both.

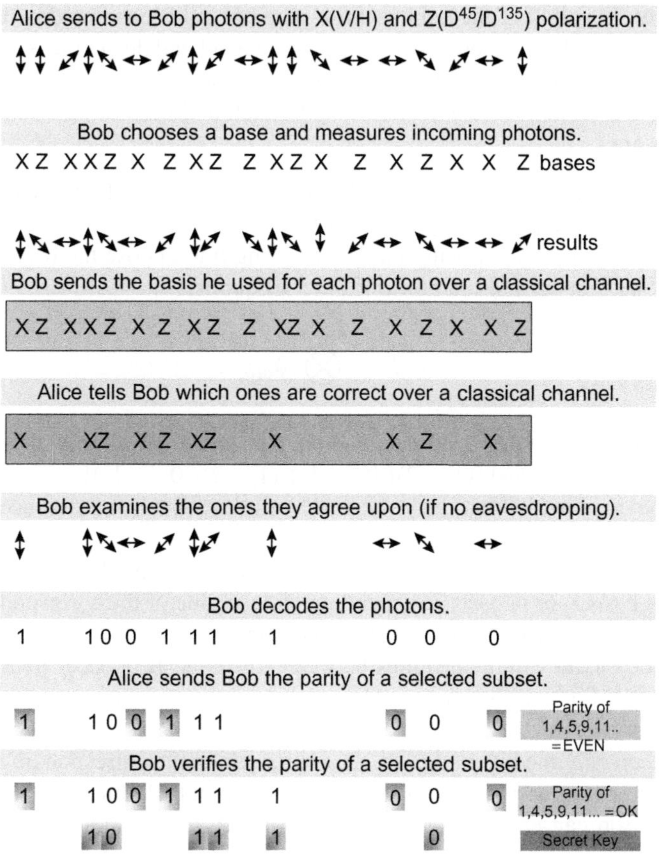

FIGURE 2.22

The main steps of the BB84 QKD protocol.

communication channel (Figure 2.21c). The two parties wishing to communicate in a secure manner establish an upper bound for the level of disturbance tolerable. They use a set of check bits to estimate the level of noise, or eavesdropping. Then they reconcile their information and distill a shared secret key.

The first QKD protocol, *BB84*, was proposed by Bennett and Brassard in 1984 [32]; a proof of the security of this protocol can be found from Shor and Preskill [384]. Informally, the protocol works as follows: Alice sends n qubits encoded randomly in one of the two bases she and Bob have initially decided on. Figure 2.22 illustrates the case when Alice wants to send the encryption key to Bob while Eve attempts to eavesdrop. Bob randomly chooses one of the two bases to decode each received qubit. Bob ends up guessing the bases correctly for about $n/2$ of the qubits. Then Alice and Bob exchange over the classical communication channel information about the base used by each one of them to encode and decode, respectively, each qubit. Alice and Bob know precisely on which qubits they have agreed and use them as the encryption key. When Eve eavesdrops and intercepts each of the n qubits, she randomly picks up one of the two bases to measure each qubit and then she resends the qubit to Bob. Approximately $n/2$ of the qubits received by Bob have their state altered by Eve. Bob guesses the correct bases for only about half of the qubits whose state has not been altered by Eve. Alice and Bob exchange information over the classical communication channel and detect the presence of an intruder when they agree on considerably fewer than $n/2$ qubits.

A formal description of the protocol is offered by Bennett and Brassard [32] and Shor and Preskill [384]; its main steps are:

1. Alice selects n, the approximate length of the desired encryption key. Then she generates two random strings of bits, a and b, of length $(4n \quad \delta)$.
2. Alice encodes the bits in string a using the bits in string b to choose the basis (either X or Z) for each qubit in a. She generates ψ , a block of $(4n \quad \delta)$ qubits,

$$\psi \qquad \bigotimes_{k \quad 1}^{4n \quad \delta} \psi_{a_k b_k} ,$$

   where a_k and b_k are the k-th bit of strings, a and b, respectively. Each qubit is in one of four pure states in two bases, $[\ 0\ ,\ 1\ ]$ and $[(1/\ \overline{2})(\ 0 \qquad 1\ ), (1/\ \overline{2})(\ 0 \qquad 1\ )]$.
3. If $\mathcal{E}$ describes the combined effect of the channel noise and Eve's interference, then the block of qubits received by Bob is $\mathcal{E}(\ \psi \quad \psi\ )$.
4. Bob constructs a random string of bits, b , of length $(4n \quad \delta)$. Bob then measures every qubit he receives either in basis X or in basis Z, depending on the value of the corresponding bit of b . As a result of his measurement, Bob constructs the binary string a .
5. From this instance on, all communications between Alice and Bob occur over the classical communication channel. First, Bob informs Alice that he now expects information about b. Alice discloses b.
6. Alice and Bob keep only the bits in the set, a, a , for which the corresponding bits of the strings, b and b , are equal. Let us assume that Alice and Bob keep only $2n$ bits. By choosing δ sufficiently large, Alice and Bob can ensure that the number of bits kept is close to $2n$, with a very high probability.
7. Alice and Bob perform several tests to determine the level of noise and eavesdropping on the channel. The set of $2n$ bits is split into two subsets of n bits each. One subset will be the *check* bits used to estimate the level of noise and eavesdropping, and the other consists of the *data* bits used for the quantum key. Alice selects n check bits at random and sends the position of the selected bits to Bob. Then Alice and Bob compare the values of the check bits. If more than, say, t bits disagree, then they abort and retry the protocol.

2.25 MEASUREMENTS AND AN AXIOMATIC QUANTUM THEORY

We conclude this chapter with a discussion of the role of the measurements in an axiomatic reformulation of quantum theory. Quantum theory is based on the five postulates discussed in Sections 1.4–1.7, and we could ask if the postulates are logical and consistent with one another. For example, how do we reconcile the fact that, according to these postulates, the state of a quantum system is subject to linear and reversible evolution transformations, while measurements are nonlinear and irreversible transformations that collapse the quantum state? We now consider the question of whether one can construct the quantum theory from a set of more logical axioms and what would be the role of measurements in the new context.

George Birkhoff and John von Neumann examined this question in their 1936 paper on the logic of quantum mechanics [56]. They start their analysis by reminding us that quantum theory shares the concept of "phase-space" with classical mechanics and with classical electrodynamics; each system, $\mathcal{S}$, is at each instance associated with a point, $p(t)$, in the phase-space, Σ, and this point represents mathematically the state of the system at time, t. The principle of "mathematical causation" is re ected by a "law of propagation," which allows us to determine the state, $p(t)$, knowing the initial state, $p(0)$. In classical mechanics, a point in the phase-space Σ corresponds to a choice of n position and n conjugate momentum coordinates, and the law of propagation is Newton's law of attraction. In electrodynamics, Σ is a function-space of infinitely many dimensions (e.g., specified by electromagnetic and electrostatic potentials), and the law of propagation is given by Maxwell equations; in quantum theory, Σ is also a function-space, the points correspond to wave-functions, and the law of propagation is given by the Schrödinger equation. The law of propagation can be viewed as causing a steady uid motion in the phase-space; in classical dynamics, this ow conserves volumes, while in quantum mechanics, it conserves distances because the equations are unitary.

Any mathematical model must be anchored in the physical reality, and this requires a mechanism to obtain information about the physical objects. We assume that the outcomes, $\sigma_1, \sigma_2, \ldots, \sigma_n$, are obtained after performing a series of measurements, $\mu_1, \mu_2, \ldots, \mu_n$, on a quantum system, $\mathcal{S}$, and the point, $(\sigma_1, \sigma_2, \ldots, \sigma_n)$ is located inside the "observation space." The formal connection between physical reality and the mathematical model in the theory proposed by Birkhoff and von Neumann is ensured by a plausible postulate, a propositional calculus of quantum mechanics, that establishes a connection between the subsets of the observation space and the subsets of the phase-space of a system $\mathcal{S}$.

More recently, Lucien Hardy proposed a set of "more reasonable axioms" [200]; he argues that a better axiomatic formulation of quantum theory is not only more aesthetically appealing, but also allows a deeper conceptual understanding of the theory.

The setup for this axiomatic derivation of quantum theory consists of three types of devices: those used to prepare a quantum state, those used to transform a state, and those used to measure the state and transform quantum information to classical information. The quantum theory is in fact a probability theory different from classical probability; measurements allow us to characterize a quantum state by a set of probabilities. To perform a measurement, we need a large number of systems that have undergone an identical preparation. The definitions of state, degrees of freedom, and dimension are:

- *The state* associated with a particular preparation is a mathematical object that can be used to determine the outcomes of any measurement of the system subject to that specific preparation.
- *The number of degrees of freedom, K,* is the minimum number of probability measurements needed to determine the state (or the cardinality of the set of real numbers required to specify the state).
- *The dimension, N,* is the maximum number of states that can be reliably distinguished from one another in a single-shot measurement.

The five axioms introduced by Hardy [200] are:

1. *Probabilities.* Relative frequencies (measured by taking the proportion of occurrences a particular outcome is observed) tend to the same value (called probability) for any case a measurement is performed on an ensemble of n systems prepared by some given preparation in the limit when n becomes infinite.
2. *Simplicity.* K is determined by a function of N, $K \; f(N)$, where, for a given N, K takes the minimum value consistent with the axioms.

3. *Subspaces.* A system whose state is constrained to belong to an M-dimensional subspace (i.e., have support on only M of a set of N possible distinguishable states) behaves like a system of dimension M.
4. *Composite systems.* A composite system consisting of subspaces, A and B, satisfies

$$N \quad N_A N_B \quad \text{and} \quad K \quad K_A K_B.$$

5. *Continuity.* There exists a continuous reversible transformation on a system between any two pure states of that system. The transformation is continuous in the sense that its probability is a continuous function.

Classical probability theory assumes that there are N distinguishable states called "basis" or "pure states." The N pure states are described by the vectors

$$p_1 \begin{pmatrix} 1 \\ 0 \\ 0 \\ \vdots \\ 0 \\ 0 \end{pmatrix}, p_2 \begin{pmatrix} 0 \\ 1 \\ 0 \\ \vdots \\ 0 \\ 0 \end{pmatrix}, p_3 \begin{pmatrix} 0 \\ 0 \\ 1 \\ \vdots \\ 0 \\ 0 \end{pmatrix}, \ldots, p_{N\ 1} \begin{pmatrix} 0 \\ 0 \\ 0 \\ \vdots \\ 1 \\ 0 \end{pmatrix}, p_N \begin{pmatrix} 0 \\ 0 \\ 0 \\ \vdots \\ 0 \\ 1 \end{pmatrix}.$$

The state of the system is described by a vector

$$p \quad \begin{pmatrix} p_1, \\ p_2, \\ p_3 \\ \vdots \\ p_N \end{pmatrix}$$

with p_i, the probability of the system being in the pure state, i. The state, p, belongs to a convex set[12]; in other words, the general state can be described as the convex sum of the pure states and the *null state*.

From this brief discussion, it follows that the state in a classical probability theory can be determined by measuring N probabilities. Thus, $K \quad N$; *the number of degrees of freedom is equal to the dimension of the space.*

In quantum theory, a state is represented by the density matrix, a self-adjoint operator ρ, with the property that $0 \quad \mathrm{tr}(\rho) \quad 1$. Measurements are represented also by self-adjoint operators, the projectors, $\mathbf{P}_k$, with the property that

$$\sum_k \mathbf{P}_k \quad I.$$

[12]A convex set, S, in a vector space over the field, $\mathbb{R}$, is one when the line segment joining any two points in S lies entirely in S.

The probability of the outcome, k, when a measurement, P_k, is performed on a system in the state ρ is

$$p(k) \quad \mathrm{tr}(\mathbf{P}_k \rho).$$

One can recast quantum theory in a similar mold; if we consider an N-dimensional Hilbert space, there are N^2 independent projectors, $P_k, 1 \quad k \quad N^2$, that span the space. Indeed, a general self-adjoint operator can be represented as a matrix, with N real numbers along the diagonal and $\binom{N}{2} \quad (N^2 \quad N)/2$ complex numbers above the diagonal; thus, a self-adjoint operator is specified by a total of N^2 real numbers. In this case, $K \quad N^2$, *the number of degrees of freedom is equal to the square of the dimension of the space.* Each projector corresponds to one degree of freedom.

While the first four axioms are consistent with classical probability, the fifth is not. For example, when $N \quad 2$, the four projectors we choose to span the space of self-adjoint operators could be

$$\begin{array}{llll} P_0 & 0 \; 0 & P_2 & (\alpha_0 \, 0 \quad \alpha_1 \, 1 \,)(\alpha_0 \, 0 \quad \alpha_1 \, 1 \,), \\ P_1 & 1 \; 1 & P_3 & (\beta_0 \, 0 \quad \beta_1 \, 1 \,)(\beta_0 \, 0 \quad \beta_1 \, 1 \,), \end{array}$$

with $\alpha_0 \,^2 \quad \alpha_1 \,^2 \quad 1$ and $\beta_0 \,^2 \quad \beta_1 \,^2 \quad 1$.

2.26 HISTORY NOTES

Philosophers have been concerned with the relationship between human perception and "reality" for centuries. *Rationalism* is a theory of knowledge based on the idea that the criterion of truth is not sensory but intellectual and deductive. The origins of this philosophy, whose main credo is self-sufficiency of reason, can be traced back to Pythagoras and Plato and was later embraced by Descartes, Spinoza, and Leibnitz who advocated the introduction of mathematical methods in philosophy. Spinoza and Leibnitz believed that in principle, all knowledge, including scientific knowledge, could be gained through the use of reason alone.

Rationalism is often contrasted to *empiricism*, a theory of knowledge whose main tenet is that scientific knowledge is related to experience and that experiments are essential for the development of scientific concepts; all hypotheses and theories must be tested against observations of the natural world. The sophists in the fifth century BC, Aristotle (384 BC–322 BC), the stoics and the epicureans a generation later, followed by Thomas Aquinas and Roger Bacon in the thirteen century, and John Locke, George Berkeley, David Hume in the twelfth and thirteenth centuries attribute sensory perception and a posteriori observation a major role in the formation of human knowledge. They believed that human knowledge of the natural world is grounded in the senses' experience. The work of Immanuel Kant bridges the gap between rationalism and empiricism. The author of the *Critique of Pure Reason*, Kant expressed the belief that we are never able to transcend the bounds of our own mind; we cannot access the "Ding an sich," German for the "thing-in-itself."

As expected from this brief discussion of rationalism and empiricism, the concept of "measurement" proved to be intensely controversial from the early days of quantum mechanics. The question of whether a numerical value of a physical quantity has any meaning until an observation has been performed was answered differently by the representatives of the metaphysical and realist schools of thought. The argument of Niels Bohr was,

If a quantity Q is measured in system S at time t, then Q has a particular value in S at time t, while Werner Heisenberg expressed his positivistic views as,

It is meaningless to assign Q a value q for S at time t, unless Q is measured to value q at time t.

Bohr argued that the later interpretation reduced the question of definability to measurability, but he and Heisenberg agreed in broad terms that experimental conditions play an important role on the description of quantum systems, an important tenet of the Copenhagen interpretation. There are some similarities between the position taken by Bohr and that of Kant in Transcendental Aesthetic:

Not only are drops of rain mere appearances, but...even the space in which they fall, are nothing in themselves but modifications of our sensible intuition.... The transcendental objects remain unknown to us.

Einstein, Podolsky, and Rosen define the concept of physical reality (see Section 2.16) as objective, independent of any measurement; a measurement re ects passively physical reality. Einstein's empiricist position has its roots in the Cartesian and Lockean notion of perception.

John von Neumann provided an axiomatic definition of a quantum measurement process consisting of a linear phase when the behavior of the system is described by the Schrödinger equation followed by a nonlinear stage when the state of the system is collapsed. The so-called "measurement problem" centers on the apparent con ict between two of the postulates of quantum mechanics, the measurements postulate and the dynamics postulate: The postulate of collapse seems to be right about what happens when we make a measurement, and the dynamics postulate seems to be bizarrely wrong about what happens when we make measurements, and yet the dynamics postulate seems to be right about what happens when we aren't making measurements [10]. Sometimes the measurements postulate is called the postulate of collapse; the dynamics of a quantum system is specified by a self-adjoint operator, **H**, called the Hamiltonian, and the time evolution of the system is described with the Schrödinger equation, according to the dynamics postulate discussed in Section 1.6.

2.27 SUMMARY AND FURTHER READINGS

Any scientific theory of physical phenomena must be anchored in reality, and this requires the means to gather information about the physical objects. Measurements are necessary to assess the consistency of the mathematical model with the physical reality. A mathematical model allows us to define the "state" of a physical system as a point in an abstract "phase-space," while measurements allow us to define the state of the physical system in an "observation space"; the consistency of the mathematical model with the physical reality requires a logical connection between corresponding subspaces of the two spaces.

Asher Peres writes [324]:

It should be clear that the interpretation of raw experimental data always necessitates the use of some theory. Concepts such as angular momentum are part of the theory, not of the experiment. Moreover correspondence rules are needed to relate the abstract notions of the theory—together with its correspondence rules—which tell us what can, or cannot, be answered. What does not exist in theory cannot be observed in any experiment to be described by that theory. Conversely, anything described by the theory is deemed to be observable, unless the theory itself prohibits to observe it. To be acceptable a theory must have predictive power about the outcomes of the experiments that it describes, so that the theory can actually fail. A good theory is one that does not fail in its domain of applicability. Today, in our present state of knowledge, quantum theory is the best available one for describing atomic, nuclear, and many other phenomena.

The concept of "measurement" and the relationship between the outcome of the measurement and our knowledge of the state of the quantum system prior, during, and after the measurement proved to be intensely controversial. The Copenhagen interpretation of quantum mechanics expounded by Niels Bohr and Werner Heisenberg was contested by many physicists, including Albert Einstein.

According to the measurements postulate, a quantum measurement *collapses the state* of a quantum system. A quantum measurement is an irreversible process that destroys quantum information and replaces it with classical information. To be reversible, a quantum measurement should not reveal any information about the quantum state.

In Section 1.14, we discussed the requirements for building a quantum computer formulated by David DiVincenzo [130]. One of the requirements states that if we have a register of n qubits and if the density operator of a qubit is

$$\rho \quad p0 \;\; 0 \quad (1 \quad p) \; 1 \;\; 1 \quad \alpha \, 0 \;\; 1 \quad \alpha \;\; 1 \;\; 0,$$

with $\alpha \quad \mathbb{C}$, then an "ideal measurement" should provide an outcome of 0, with the probability p, and an outcome of 1, with the probability $(1 \quad p)$, regardless of the value of α, regardless of the state of the neighboring qubits, and without changing the state of the quantum computer. A "nondemolition" measurement also leaves the qubit in the state 0 0 after reporting the outcome 0 and leaves it in the state 1 1 after reporting the outcome 1.

John von Neumann treats the measurement apparatus as a quantum object and argues that a measurement consists both of (1) a linear stage, when the joint state of the quantum system and the measuring instrument is a tensor product of the states of the two systems; and of (2) a nonlinear stage, when this joint state collapses to one of the tensor products of various eigenstates of the measurement operator and the state of the measuring instrument corresponding to the instantiated eigenvalue.

The outcome of a measurement of an observable $\mathcal{A}$ of a quantum system is an eigenvalue of the self-adjoint operator, $\mathbf{A}$, associated with the measurement of the observable. The quantum state immediately after the measurement is an eigenvector of the operator $\mathbf{A}$. Operator $\mathbf{A}$ is a *projector* for observable $\mathcal{A}$ of a quantum system.

Projective measurements are carried by means of self-adjoint operators, $\mathbf{M}_i$, such that $\mathbf{M}_i\mathbf{M}_j \quad \delta_{i,j}\mathbf{M}_i$. Projective measurements are restrictive and are not always possible. Sometimes we destroy a quantum particle in the process of measurement; thus, the repeatability of a projective measurement is violated. Orthogonal measurement operators commute and correspond to simultaneous observables. Measurement operators corresponding to non-orthogonal states do not commute and are therefore not simultaneously observable. The POVM measurements are concerned only with the statistics of the measurement. POVM allows us to distinguish the cases when the outcome of a measurement identifies with certainty the original state from the ones where the identification is not possible.

The characterization of the state of a quantum system by means of the density matrix allows us to distinguish between "pure" and "mixed" states and to conclude that we can acquire maximal information only about pure states. When the state of a quantum system is known exactly, the system is in a *pure state*, ψ . Then,

$$\rho \quad \psi \;\; \psi \, .$$

The trace of the density operator of a pure state satisfies the equation:

$$\text{tr}(\rho^2) \quad 1.$$

If the state of the system is not known exactly, the system is in a *mixed state*. The set $\mathcal{E} \quad p_i, \ \psi_i$ is called an *ensemble of pure states*. Given the state, ψ_i , and the probability, p_i, of the quantum system to find itself in the state ψ_i , the density operator is defined as

$$\rho \quad \sum_i p_i \ \psi_i \ \ \psi_i \quad \sum_i p_i \rho_i.$$

The trace of the density operator of a mixed state satisfies the equation

$$\text{tr}(\rho^2) < 1.$$

If a unitary operator, U, is applied to a system that is in the state ψ_i , with the probability p_i, then the resulting state is $U \ \psi_i$, with the probability p_i, and the density operator evolves as follows:

$$\rho \quad \sum_i p_i \ \psi_i \ \ \psi_i \qquad \rho^{afterU} \quad \sum_i p_i U \ \psi_i \ \ \psi_i \ U^\dagger \quad U \rho U^\dagger.$$

Mixed states are used to describe statistical mixtures of pure states as well as composite systems. Let $\mathcal{C}$ be a bipartite system consisting of systems $\mathcal{A}$ and $\mathcal{B}$; the state vectors and the density operators of the three systems are, respectively, $\xi_{\mathcal{C}}$ and $\rho_{\mathcal{C}}$, $\xi_{\mathcal{A}}$ and $\rho_{\mathcal{A}}$, and $\xi_{\mathcal{B}}$ and $\rho_{\mathcal{B}}$. The bipartite system can be either in a product state, or in an entangled state. $\mathcal{C}$ is in a product state if $\xi_{\mathcal{C}} \quad \xi_{\mathcal{A}} \quad \xi_{\mathcal{B}}$; in this case, the reduced density operators satisfy the equalities $\rho^{\mathcal{A}} \quad \rho_{\mathcal{A}}$ and $\rho^{\mathcal{B}} \quad \rho_{\mathcal{B}}$.

One of the most intriguing properties of quantum information is the shareability of quantum correlations. While classical correlations can be shared among many parties, quantum correlations cannot be shared freely; quantum correlations of pure states are monogamous. If two quantum systems, $\mathcal{A}$ and $\mathcal{B}$, are in a maximally entangled pure state, then neither of them can be correlated with any other system in the universe. The monogamy of entanglement is the deeper root of our inability to clone quantum states.

The EPR Gedankenexperiment is concerned with the measurement of entangled particles and led some to believe in the existence of hidden variables. The hidden variable theory assumes that in an EPR experiment there is a more complete specification of the state of the two entangled particles than the one given by the wave function. John Bell proved the nonexistence of local hidden variables based on the observation that any hidden variable account of quantum mechanics must have the separability property, and this is not possible. The Bell inequality is based on two common-sense assumptions:

1. *Locality:* Measurements of different physical properties of different objects, carried out by different individuals, at distinct locations, cannot in uence each other.
2. *Realism:* Physical properties are independent of observations; the existence of physical properties is independent of them being observed.

Quantum systems in pure entangled states and quantum systems in some mixed entangled states violate the Bell inequality. Quantum theory is consistent with the experimental evidence acquired along the years; thus, one of these two assumptions must be incorrect.

Bell showed that it is possible to mimic the statistical properties of quantum theory by a deterministic model based on hidden variables. He constructed such a deterministic model generating results whose averages are identical to those predicted by quantum mechanics. When the quantum system is made up of disjoint subsystems entangled like those in the EPR experiment, the hidden variables cannot be separated as disjoint subsets, local to each subsystem. The Bell inequality quantifies how measurements on quantum systems in entangled states can invalidate local classical models of reality.

John von Neumann discusses simultaneous measurability, measurements and reversibility, and the macroscopic measurements in Chapters 3 and 5 of the *Mathematical Foundations of Quantum Mechanics* [441].

Part two, "Cryptodeterminism and Quantum Inseparability," of the wonderful book by Asher Peres [325] provides insights in quantum correlations, composite systems, and the Bell theorem.

Speakable and Unspeakable in Quantum Mechanics is a collection of John Bell papers with an introduction by Alain Aspect [26]; it is recommended for its clear exposition of the EPR paradox, hidden variables, as well as the Bell theorem and other issues related to the philosophical aspects of measurements.

Roger Penrose addresses the question "What is quantum reality?" and discusses the holistic nature of the wave function in Chapter 21 of his book, *The Road to Reality: A Complete Guide to the Laws of the Universe*, and then re ects on quantum measurements in the chapter dedicated to quantum algebra [323].

The book *Science and Ultimate Reality*, dedicated to the 90th birthday of John Archibald Wheeler [23], consists of a collection of articles written by Paul Davies, David Deutsch, Freeman Dyson, Lucien Hardy, Hideo Mabuchi, Anton Zeilinger, Wojciech Zurek, and other researchers in the field of quantum computing and quantum information theory. Some of the ideas from the paper by Lucien Hardy, "Why Is Nature Described by Quantum Theory" [200], are presented in Section 2.25.

2.28 EXERCISES AND PROBLEMS

Problem 2.1. The following statement appears on page 187 of Nielsen and Chuang's *Quantum Computing and Quantum Information* [307]: "To be reversible a quantum measurement should not reveal any information about the quantum state." A measurement is expected to provide information about the system being measured; why should anyone carry out a measurement that does not reveal any information? Discuss how this statement should be understood. How is it possible to detect errors under these conditions?

Problem 2.2. Show that the trace of the product of the square matrix, $A[a_{ij}]1 \quad i,j \quad n$, and its complex conjugate, $A \quad A[a_{ij}]1 \quad i,j \quad n$, is non-negative:

$$\mathrm{tr}(A \quad A) \quad 0.$$

Problem 2.3. Prove that the set of $n \quad n$ square matrices with elements from either $\mathbb{R}$ or $\mathbb{C}$ form a vector space with the inner product defined as

$$A, B \quad \mathrm{tr}(A \quad B).$$

Problem 2.4. Consider the Bell states β_{01}, β_{10}, and β_{11}.

1. Compute the density operators, ρ_{01}, ρ_{10}, and ρ_{11}, for each of the four states, and show that each state is a pure state.
2. Compute the reduced density operators, $\rho_{01}^{(1)}, \rho_{10}^{(1)}$, and $\rho_{11}^{(1)}$, and show that one qubit of the pair is in a mixed state.
3. Consider two systems; $\mathcal{A}$ in a pure state $\psi_{\mathcal{A}}$ and $\mathcal{B}$ in a pure state $\psi_{\mathcal{B}}$. Let $\mathcal{C} \quad \mathcal{AB}$ be in the state $\psi_{\mathcal{C}} \quad \psi_{\mathcal{C}} \quad \psi_{\mathcal{B}}$. Show that $\psi_{\mathcal{C}}$ is a pure state.

Problem 2.5. Show that the density operator of an equal mixture of two orthogonal single-qubit pure states is

$$\rho \quad \frac{1}{2}\begin{pmatrix} 1 & 0 \\ 0 & 1 \end{pmatrix}.$$

Problem 2.6. Show that if the density operator of a qubit has only one eigenvalue, then the state is a pure state; show that if the density operator has more than one eigenvalue, then the state is a mixed state.

Problem 2.7. Let $a \quad \sum_i \alpha_i \; i$ and $b \quad \sum_i \beta_i \; i$ be state vectors in $\mathcal{H}_n$. Show that

$$\mathrm{tr}(\; a \quad b \;) \quad a \; b \; .$$

Problem 2.8. Show that the states at different stages of the quantum circuit in Figure 2.19a are

$$\xi_0 \quad \frac{1}{\sqrt{2}}[\alpha_0 \; 0 \; (\; 00 \quad 11 \;) \quad \alpha_1 \; 1 \; (\; 00 \quad 11 \;)],$$

$$\xi_1 \quad \frac{1}{\sqrt{2}}[\alpha_0 \; 0 \; (\; 00 \quad 11 \;) \quad \alpha_1 \; 1 \; (\; 10 \quad 01 \;)],$$

and

$$\xi_2 \quad \tfrac{1}{2}[\quad 00 \; (\alpha_0 \; 0 \quad \alpha_1 \; 1 \;) \quad 01 \; (\alpha_0 \; 1 \quad \alpha_1 \; 0 \;)$$
$$10 \; (\alpha_0 \; 0 \quad \alpha_1 \; 1 \;) \quad 11 \; (\alpha_0 \; 1 \quad \alpha_1 \; 0 \;)].$$

Problem 2.9. Show that using an ancilla system consisting of more than one qubit does not increase the probability of guessing right for the measurements discussed in Section 2.23.

CHAPTER

Classical and Quantum Information Theory

3

God exists, since mathematics is consistent, and the Devil exists, since we cannot prove it.
Attributed to Andre Weil by Paul Rosenbloom

In the late 1940s, Claude Shannon stated, "The fundamental problem of communication is that of reproducing at one point either exactly or approximately a message selected at another point" [374]. The precise nature of the agents transmitting information to one another, as well as the signaling process, can be very different; this motivated Shannon to develop abstract models of communication and establish the foundation of information theory. Signaling is an overloaded term frequently used in communication; in this context, it means the physical perturbation of the transmission media used to carry the information through the communication channel.

During the decades following the publication of Shannon's work came the realization that information theory has a broader scope. It plays an important role in understanding critical phenomena related not only to the transmission but also to the storage and the transformation of information. Indeed, information can be transmitted from one place to another; information can also be transmitted from one moment in time to another, when it is either stored or transformed with a computing device.

Today we study not only classical information, but also quantum and biological information. The three flavors of information are anchored in physical reality; the information is embodied by some property of a physical system. Classical information is transmitted, stored, and transformed using physical processes governed by the laws of classical physics, while quantum information is carried by systems governed by the laws of quantum mechanics; little is known about the properties of biological information, information manipulated at a cell level in living organisms.

The properties of classical and quantum information are very different. Classical information is carried by systems with a definite state, and it can be replicated and measured without being altered. Quantum information is encoded as a property of quantum systems (e.g., photon polarization or particle spin) and has special properties such as superposition and entanglement with no classical counterpart; quantum information cannot be cloned, and it is altered as a result of a measurement.

Quantum information theory studies: (1) transmission of classical information over quantum channels; (2) transmission of quantum information over quantum channels; (3) the effect of quantum entanglement on information transmission; and (4) the informational aspect of the quantum measurement process, the tradeoffs between the disturbance of the quantum state and the accuracy of the measurement.

The roots of quantum information theory can be traced back to John von Neumann who, in the early 1930s, introduced critical concepts such as the von Neumann entropy [441]. In 1973, Alexander Holevo found an upper bound of the accessible information in a quantum measurement [210].

Quantum communication involves a source that supplies quantum systems in a given state, a quantum channel that "transports" the quantum system, and the recipient that receives and decodes the quantum information. A quantum channel, like a classical one, can transmit quantum information *from one place to another and from one time to another.* For example, a photon transmitted through an optical fiber could carry information from site A to site B, while a storage device implemented as an ion trap could be regarded as a channel transmitting information in time. The information transmitted through a quantum channel can be affected by errors, and the means to correct such errors are critical for any practical applications of quantum information. This view of a quantum channel gives us an indication that quantum information theory and quantum error-correcting codes are important not only for the transmission, but also for the processing and storage of quantum information.

Several physical embodiments of the quantum information communication model are possible: the source could be a laser producing monochromatic photons, the channel could be an optical fiber, and the recipient a photocell; the source could also be an ion trap controlled by laser pulses, the channel an ion trap, and the receiver a photo detector reading out the state of the ions via laser-induced fluorescence [275].

The properties of quantum information analyzed in Chapters 1 and 2 have profound implications for the discussion in this chapter. We expect to reproduce, with fidelity, the input of a classical noiseless channel to its output; this is no longer true for quantum channels. A quantum channel cannot reproduce, with utmost fidelity, its input when the input consists of an ensemble of mixed non-orthogonal states; we can only distinguish orthogonal states. Thus, we expect that any measurement performed at the output of the channel will alter the states.

Rather than presenting classical and quantum information separately, we interleave the discussion of them. Concepts we are familiar with in classical information theory, such as joint and conditional entropy and mutual information, have a quantum counterpart; Fano's inequality and the data processing inequality have a quantum analog; von Neumann as well as Shannon entropy enjoy strong subadditivity, and the list could go on and on.

The organization of this chapter is summarized in Figure 3.1. We first discuss the physical support of information; next, we overview several properties of classical information based on Shannon's theory; then we concentrate on the properties of quantum information and discuss the sources and the quantum channels.

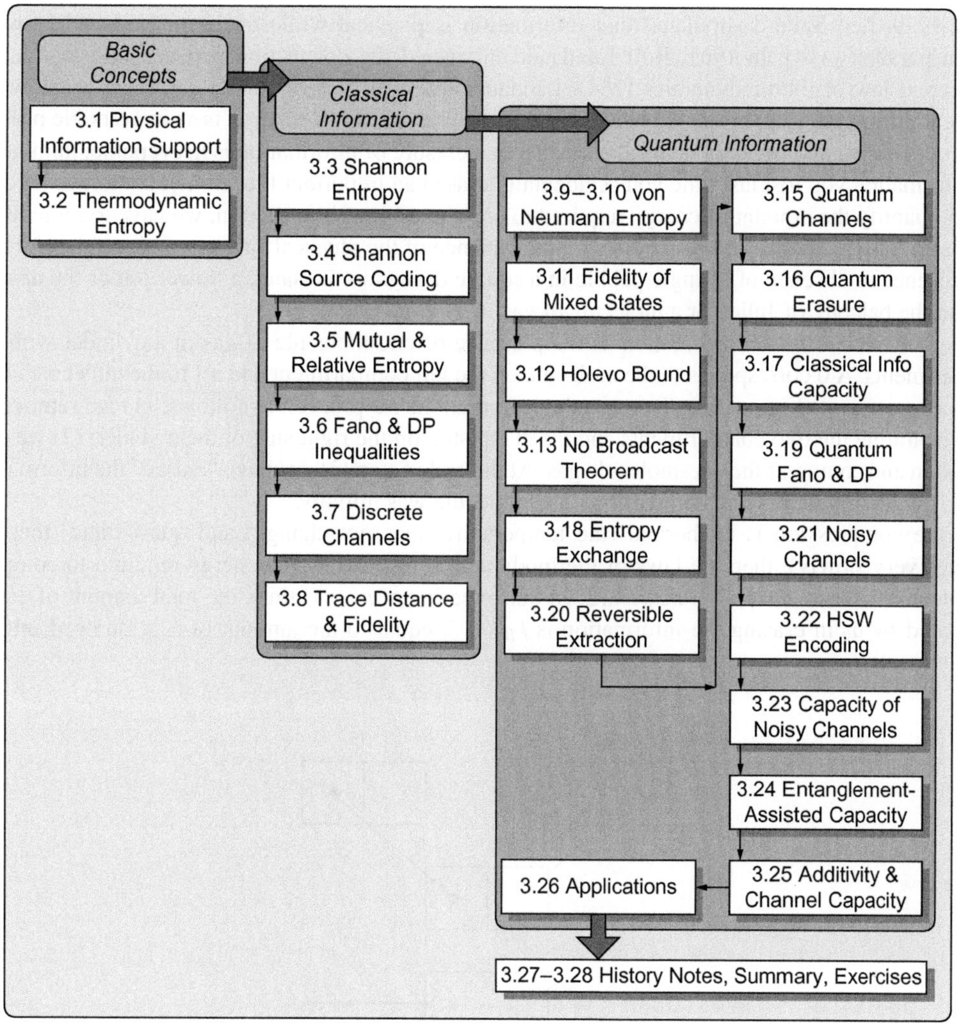

FIGURE 3.1

The organization of Chapter 3 at a glance.

3.1 THE PHYSICAL SUPPORT OF INFORMATION

Information must have a physical support. Indeed, to transmit, store, and process information, we must act on a property of the physical medium embodying the information; to extract information at the output of a channel, we have to perform a measurement of the physical media. It seems unlikely that one would dispute the statement that information is physical; nevertheless, the profound implications of this truth, the connection between the fundamental laws of physics and information, have escaped the scientific community for a very long time [329, 334].

In 1929, Leó Szilárd stipulated that information is physical while trying to explain Maxwell's demon paradox [414]. In 1961, Rolf Landauer uncovered the connection between information and the second law of thermodynamics [264]. Landauer followed a very simple argument to show that classical information can be copied reversibly, but the erasure of information is an irreversible process. We accept easily, and possibly painfully, the fact that erasure of information is irreversible; think about the information you lost last time you accidentally erased an important file from your storage media!

To quantify the amount of energy required to erase one bit of information, we have to remember a basic attribute of classical information, its independence of the physical support; a bit can be stored as the presence or absence of a single pebble in a square drawn on the sand, a flower pot or the absence of it on the balcony of Juliet, or a molecule of gas in a cylinder.

We encode one bit of information as the presence of one molecule of gas in a cylinder with two compartments. A 0 corresponds to the molecule on the left compartment and a 1 to the molecule on the right compartment (Figure 3.2). To erase the information, we proceed as follows: (1) we remove the wall separating the two compartments and insert a piston on the right side of the cylinder; (2) we push the piston and compress the one-molecule gas. At the end of step (2), we have "erased" the information. We no longer know in which compartment the molecule was initially.

If the compression is isothermal (the temperature does not change) and quasi-static (the state changes very slowly), then the laws of thermodynamics tell us that the energy required to compress m molecules of gas is equal to $mk_BT\ln 2$. In our experiment m 1; thus, the total amount of energy dissipated by us in erasing the information is $k_BT\ln 2$, equal to the amount of heat dumped into the environment.

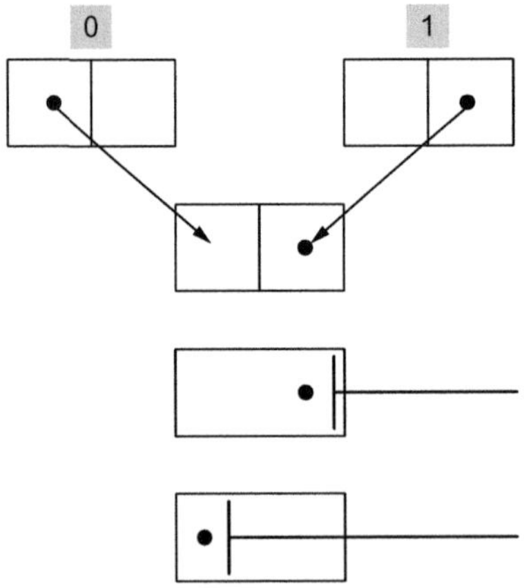

FIGURE 3.2

A Gedankenexperiment to quantify the amount of energy required to erase a bit of information. We encode one bit of information as the presence of one molecule of gas in a cylinder with two compartments. A 0 corresponds to the molecule in the left compartment, and a 1 to the molecule in the right compartment (top). To erase the information, we remove the wall separating the two compartments and insert a piston on the right side of the cylinder; we then push the piston and compress the gas. The amount of energy required to erase the information is equal to $k_BT\ln 2$.

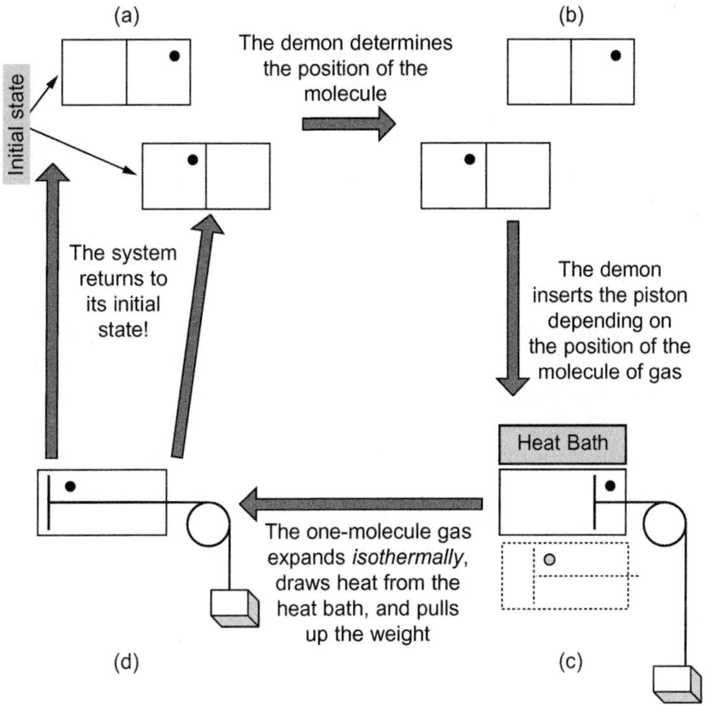

FIGURE 3.3

A cycle of Szilárd's engine. (a) One molecule of gas is located either in the left half or in the right half of a cylinder with a partition in the middle. (b) The demon measures the position of the molecule. (c) The demon removes the partition and, based on the information regarding the position of the particle, inserts a piston in the cylinder to the left of the molecule. (d) The gas expands isothermally and lifts the weight. During the expansion, the one-molecule gas is put in contact with a reservoir of heat and draws heat from the reservoir. Finally, the system returns to its initial state and a new cycle begins. The cylinder with the one-molecule gas could be returned to the initial state if and only if the demon provides the information regarding the initial position of the molecule.

Landauer's principle can be formulated as follows: *to erase one bit of information, we have to dissipate an amount of energy of at least* $k_B T \ln 2$, with k_B, the Boltzmann's constant, $k_B \;\; 1.3807 \;\; 10^{-23}$ Joules per degree Kelvin, and T, the temperature of the system.

Maxwell's Demon

Maxwell imagined a Gedankenexperiment involving the famous demon as a challenge to the second law of thermodynamics discussed in Section 3.2. Maxwell's demon is a mischievous fictional character who controls a trapdoor between two containers, A and B, filled with the same gas at equal temperatures; the containers are placed next to each other, B on the left of A. When a faster-than-average molecule from container A approaches the trapdoor, the demon opens it and allows the molecule to fly from container A to B; when a slower-than-average molecule from B approaches the trapdoor, the demon opens it, and allows the molecule to fly from B to A. The average molecular speed corresponds to temperature; therefore, the temperature will decrease in A and increase in B, in violation of the second law of thermodynamics. If you are about to invent a perpetuum mobile based on this Gedankenexperiment, read further before trying to patent your invention; if you already know what is wrong with this argument, you may skip the remainder of this section.

The Thermodynamic Engine

One of the most famous demonstrations that the second law of thermodynamics is not violated by Maxwell's demon was suggested by Leó Szilárd who considered the total entropy of the gas and demon combined. Szilárd imagined an engine powered by this demon. As before, we have one molecule of gas located either in the left half, or in the right half of a cylinder with a partition in the middle. The demon measures the position of the molecule, then removes the partition, and, based on the information regarding the position of the particle, the demon inserts a piston in the cylinder to the left of the molecule. On the right of the piston, an arm is connected to a weight hanging from a pulley. As the gas expands, the piston moves to the left and pulls up the weight. During the expansion, the one-molecule gas is put in contact with a reservoir of heat, draws heat from the reservoir, expands isothermally (the temperature, T, is maintained throughout the experiment), and lifts the weight. The work extracted by the engine in the isothermal expansion is

$$W_{isothermalExpansion} \qquad k_B T \ln 2.$$

Then the cycle repeats itself and the Szilárd engine keeps on converting the heat generated when the gas expands isothermally into work, a process forbidden by the second law of thermodynamics.

The point one could easily miss is that at the beginning of each cycle the ensemble consisting of the cylinder with its auxiliary components, *as well as the demon,* should be returned to the initial state. Figure 3.3d shows that the gas cylinder with the one-molecule gas could be returned to the initial state *if and only if the demon provides the information regarding the initial position of the molecule.*

The demon is an essential component of the engine; the contraption could not function without the demon. The demon's memory stores one bit of information acquired during the first step of the process: 0 if the molecule is located on the left partition and 1 if the molecule is on the right partition. This information should be erased at the beginning of each cycle; according to Landauer's principle, the energy dissipated in erasing one bit of information is

$$W_{erasure} \qquad k_B T \ln 2.$$

Thus, the work gained by the engine is required to erase the demon's memory; the heat generated by the erasure of the demon's memory is transferred back to the heat bath to compensate for the heat absorbed during the expansion of the one-molecule gas. No net gain is registered during a cycle of this hypothetical engine since

$$W_{cycle} \quad W_{isothermalExpansion} \quad W_{erasure} \qquad k_B T \ln 2 \quad k_B T \ln 2 \quad 0,$$

and the second law of thermodynamics is not violated.

We introduce in the next section the concept of thermodynamic entropy, a measure of the amount of chaos in a microscopic system. The information available to the demon is a part of the system state, and it is related to the entropy. One could consider a generalized definition of entropy, $\mathcal{I}$ (in bits), as the difference between the thermodynamic entropy, ΔS, and the information, I, about the system available to an external observer (in our case the demon) [334]:

$$\mathcal{I} \quad \Delta S \quad I.$$

The energy per cycle can be expressed in terms of ΔS, the difference between the entropy at the beginning and the end of one cycle, and the change in the information available to an external observer at the same instances, at temperature, T, as

$$W_{cycle} \quad W_{isothermalExpansion} \quad W_{erasure} \quad T(\Delta S \quad I) \quad T\mathcal{I}.$$

It follows that

$$W_{cycle} \quad 0 \quad \mathcal{I} \quad 0.$$

Indeed, information is physical and it contributes to define the state of a physical system.

3.2 THERMODYNAMIC ENTROPY

Entropy is one of the most important concepts in physics and in information theory. Informally, *entropy* is a measure of the amount of disorder in a physical, or a biological, system. The higher the entropy of a system, the less information we have about the system. *Hence, information is a form of negative entropy.*

We discuss first the thermodynamic entropy introduced by Ludwig Boltzmann in the 1870s; the thermodynamic entropy is proportional to the logarithm of the number of microstates of the system. After half a century, in the late 1920s, John von Neumann wanted to quantify the entropy of a quantum system in a mixed state and introduced his own definition of entropy. The von Neumann entropy is related to the number of states of a quantum system. Decades later, in the late 1940s, Claude Shannon formulated the entropy in the context of classical information theory.

According to Alfred Wehrl [445], "entropy relates macroscopic and microscopic aspects of nature and determines the behavior of macroscopic systems, i.e., real matter, in equilibrium (or close to equilibrium)." The entropy is a measure of the amount of chaos in a microscopic system. The concept of entropy was first introduced in thermodynamics. Thermodynamics is the study of energy, its ability to carry out work, and the conversion between various forms of energy, such as the internal energy of a system, heat, and work. The laws of thermodynamics are derived from statistical mechanics.

There are several equivalent formulations of each of the three laws of thermodynamics. The *first law of thermodynamics* states that energy can be neither created, nor destroyed; it can only be transformed from one form to another. An equivalent formulation is that the heat flowing into a system equals the sum of the change of the internal energy and the work done by the system.

The *second law of thermodynamics* states that it is impossible to create a process that has the unique effect of subtracting positive heat from a reservoir and converting it into positive work. An equivalent formulation is that the entropy of a closed system never decreases, whatever the processes that occur in the system: $\Delta S \quad 0$, where $\Delta S \quad 0$ refers to reversible processes and $\Delta S > 0$ refers to irreversible ones. A consequence of this law is that no heat engine can have 100% efficiency.

The *third law of thermodynamics* states that the entropy of a system at zero absolute temperature is a well-defined constant. This is due to the fact that many systems at zero temperature are in their ground states and the entropy is determined by the degeneracy of the ground state. For example, a crystal lattice has a unique ground state and has zero entropy when the temperature reaches 0 Kelvin.

Clausius defined *entropy* as a measure of energy unavailable for doing useful work. He discovered the fact that entropy can never decrease in a physical process and can only remain constant in a reversible process. This result became known as the second law of thermodynamics. There is a strong belief among astrophysicists and cosmologists that our universe started in a state of perfect order and its entropy is steadily increasing, leading some to believe in the possibility of a "heat death," a state when there is no free thermodynamic energy to sustain life.

The *thermodynamic entropy* of a gas, S, is also defined statistically, but it does not reflect a macroscopic property. *The entropy quantifies the notion that a gas is a statistical ensemble and it measures the randomness, or the degree of disorder, of the ensemble.* The entropy is larger when the vectors describing the individual movements of the molecules of gas are in a higher state of disorder than when all molecules are well organized and moving in the same direction with the same speed. Ludwig Boltzmann postulated that

$$S \quad k_B \ln(\Omega),$$

where k_B is the Boltzmann constant and Ω is the number of dynamical microstates, all of equal probability, consistent with a given macrostate; the microstates are specified by the position and the momentum of each molecule of gas.[1]

The second law of thermodynamics tells us that the entropy of an isolated system never decreases. Indeed, differentiating the previous equation, we get

$$\delta S \quad k_B \frac{\delta\Omega}{\Omega} \quad 0.$$

We wish now to gain insights into Boltzmann's definition of entropy. He wanted to provide a macroscopic characterization on an ensemble of microscopic systems; he considered, N, particles such that each particle is at one of the, m, energy levels, $E_1 \quad E_2 \qquad E_m$. The number of particles at energy levels, $E_1, E_2, \ldots, E_m$, are, $N_1, N_2, \ldots, N_m$, respectively, with $N \quad {}^{m}_{i \ 1} N_i$. The probability that a given particle is at energy level E_i is $p_i \quad N_i/N, 1 \quad i \quad m$ (Figure 3.4).

A microstate, σ_i, is characterized by an m-dimensional vector, $(p_1, p_2, \ldots, p_m)$, when N tends to infinity. The question posed by Boltzmann is how many microstates exist. To answer this question, let us assume that we have, m, boxes, and we wish to count the number of different ways the given quantitative distribution of the N particles among these boxes can be accomplished irrespective of which box a given particle is in. The answer to this question is

$$\begin{matrix} N \\ N_1, N_2, \ldots, N_m \end{matrix} \qquad \frac{N!}{N_1! N_2! \quad N_m!} \qquad \Omega.$$

---

[1]A version of the equation, $S \quad k_B \ln(\Omega)$, is engraved on Boltzmann's tombstone.

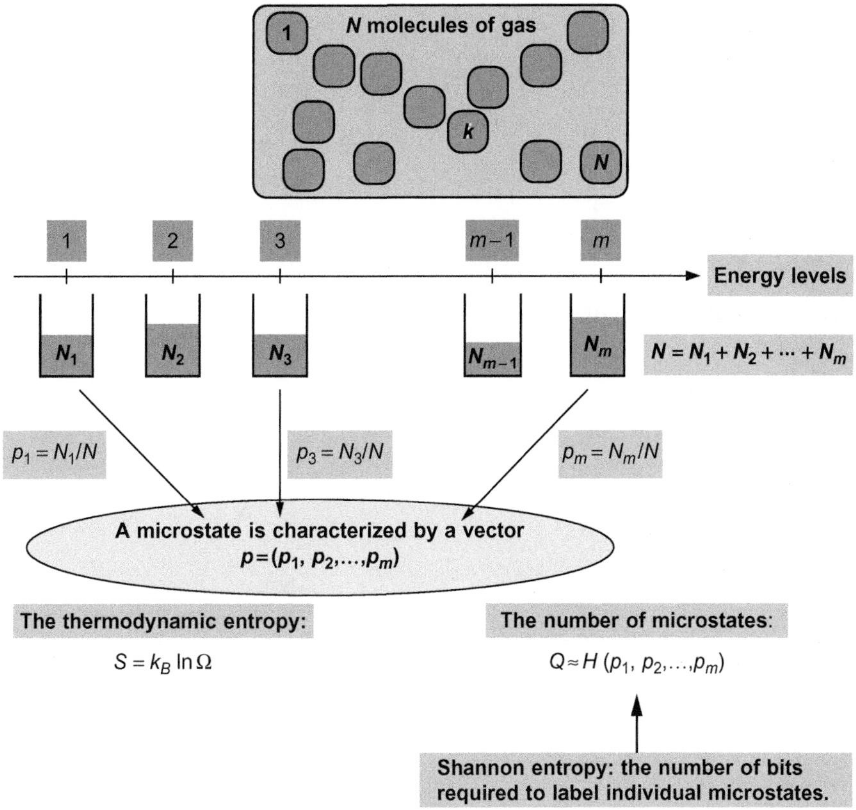

FIGURE 3.4

The relationship between thermodynamic entropy and Shannon entropy.

Call $\mathcal{Q}$ the average number of bits required to characterize the state of the system; $\mathcal{Q}$ is proportional with the logarithm of Ω, the number of microstates. More precisely,

$$\mathcal{Q} \quad \frac{1}{N} \log \Omega.$$

We will show that the number of bits, $\mathcal{Q}$, is given by

$$\mathcal{Q} \quad \frac{1}{N} \log \frac{N!}{N_1! N_2! \quad N_m!} \qquad H(p_1, p_2, \ldots, p_m) \quad \mathcal{O}(N^{-1} \log N).$$

In this expression, $H(p_1, p_2, \ldots, p_m) \quad - \quad {}_i p_i \log p_i$ denotes the *informational Shannon entropy*, a quantity we shall have much to say about later in this chapter. To prove this equality, we express $\mathcal{Q}$ as

$$\mathcal{Q} \quad \frac{1}{N} \quad \log(N!) - [\log(N_1!) \quad \log(N_2!) \qquad \log(N_m!)] \; .$$

We use Stirling's approximation, $n! \quad \overline{2\pi n}(n^n/e^n)$, and $\mathcal{Q}$ becomes

$$\mathcal{Q} \quad \frac{1}{N}[(\log \quad \overline{2\pi N} \quad N\log N \quad N\log e)$$
$$(\log \quad \overline{2\pi N_1} \quad N_1 \log N_1 \quad N_1 \log e)$$
$$(\log \quad \overline{2\pi N_2} \quad N_2 \log N_2 \quad N_2 \log e)$$
$$\vdots$$
$$(\log \quad \overline{2\pi N_m} \quad N_m \log N_m \quad N_m \log e)].$$

Now $\mathcal{Q}$ can be expressed as a sum of two terms,

$$\mathcal{Q} \quad \mathcal{Q}_1 \quad \mathcal{Q}_2,$$

with

$$\mathcal{Q}_1 \quad \log N \quad \frac{N_1}{N}\log N_1 \quad \frac{N_2}{N}\log N_2 \qquad \frac{N_m}{N}\log N_m$$

and

$$\mathcal{Q}_2 \quad \frac{1}{N}\log\frac{\overline{2\pi N}}{(2\pi N_1)(2\pi N_2) \quad (2\pi N_m)} \quad \frac{1}{N}\log \frac{1}{(2\pi)^{(m \quad 1)/2}} \quad \overline{\frac{N}{N_1 N_2 \quad N_m}}.$$

It is easy to see that

$$\log N \quad N\frac{1}{N}\log N \quad \frac{N_1 \quad N_2 \qquad N_m}{N}\log N \quad p_1 \log N \quad p_2 \log N \qquad p_m \log N.$$

Thus,

$$\mathcal{Q}_1 \quad p_1 \log N \quad p_2 \log N \qquad p_m \log N \quad [p_1 \log N_1 \quad p_2 \log N_2 \qquad p_m \log N_m].$$

But $p_i \log N \quad p_i \log N_i \qquad p_i \log(N_i/N) \qquad p_i \log p_i$. Thus,

$$\mathcal{Q}_1 \qquad\quad_{i} \; p_i \log p_i \quad H(p_1, p_2, \ldots, p_m).$$

We see that

$$\mathcal{Q}_2 \quad \mathcal{O}(N^{\;\;1} \log N).$$

We have thus shown that

$$\mathcal{Q} \quad H(p_1, p_2, \ldots, p_m) \quad \mathcal{O}(N^{\;\;1} \log N).$$

We conclude that

$$\mathcal{Q} \quad H(p_1, p_2, \ldots, p_m) \quad \text{when} \quad N \qquad \text{very large.}$$

This expression confirms our intuition; if all N particles have the same energy, then the entropy of such a perfectly ordered system is zero. If a system has an infinite number of particles, all states are equally probable and the second law of thermodynamics is inapplicable to it.

To find the most probable microstate, we have to maximize $H(p_1,p_2,\ldots,p_N)$, subject to some constraint. For example, if the constraint is $\quad {}_i p_i E_i \quad \kappa$ (the average energy is constant), then

$$p_i \quad \frac{e \quad {}^{\lambda E_i}}{{}_j e^{\lambda E_j}},$$

with λ, the solution of the equation

$$\underset{i}{} E_i \frac{e \quad {}^{\lambda E_i}}{{}_j e^{\lambda E_j}} \quad \kappa.$$

This equation has a unique solution if $E_1 \quad \kappa \quad E_m$; this solution is known as the Maxwell-Boltzmann distribution.

Now we switch our attention from thermodynamics to messages generated by a source and label each message by a binary string of length N; let $p_0 \quad N_0/N$ be the probability of a 0 in a label and $p_1 \quad N_1/N$ be the probability of a 1 in a label. N_0 is the typical number of 0s and N_1 the typical number of 1s in a label; then the number of bits required to label a message will be

$$\nu \quad \log \frac{N}{N_0,N_1} \qquad N[p_0 \log p_0 \quad p_1 \log p_1].$$

Thus, if we have an alphabet with, n, letters occurring with the probability, p_i, $1 \quad i \quad n$, the average information in bits per letter transmitted when N is large is

$$\frac{\nu}{N} \quad H(p_1,p_2,\ldots,p_n) \qquad \overset{n}{\underset{i \quad 1}{}} p_i \log p_i,$$

with $H(p_1,p_2,\ldots,p_n)$, the Shannon entropy. The connection between the thermodynamic entropy of Boltzmann and the informational entropy introduced by Shannon and discussed in depth in the next section is inescapable; Boltzmann's entropy can be written as

$$S \quad k_B \ln 2 \quad \underset{i}{} p_i \log p_i \quad \text{or} \quad S \quad k_B \ln 2 \quad H(p).$$

To convert one bit of classical information into units of thermodynamic entropy, we multiply Shannon's entropy by the constant ($k_B \ln 2$); the quantity, $k_B \ln 2$, is the amount of entropy associated with the erasure of one bit of information according to Landauer's principle.

Next, we discuss the basic concepts of the classical information theory of Shannon.

3.3 SHANNON ENTROPY

The statistical theory of communication introduced by Claude Shannon answers fundamental questions about a system consisting of agents exchanging classical information through a classical communication channel. Some of these questions are: How much information can be generated by an agent acting as a source of information? How much information can be squeezed through a classical channel? Is one agent capable of reproducing exactly, or only approximately a message generated by the source and transmitted through a classical channel? Should the information be encoded to improve the chance of exactly reproducing the message? Recall that encoding is the process of transforming information from one representation to another, or "etching" it on a physical carrier, and decoding is the process of restoring encoded information to its original form, or extracting information from a carrier.

In his endeavor to construct mathematically tractable models of communication, Shannon concentrated on stationary and ergodic[2] sources of classical information. A *stationary source* of information emits symbols with a probability that does not change over time; an *ergodic source* emits information symbols with a probability equal to the frequency of their occurrence in a long sequence. Stationary ergodic sources of information have a finite but arbitrary and potentially long correlation time.

In the late 1940s, Shannon introduced a measure of the quantity of information that a source could generate [375]. Earlier, in 1927, another scientist from Bell Labs, Ralph Hartley, had proposed to take the logarithm of the total number of possible messages as a measure of the amount of information in a message generated by a source of information, arguing that the logarithm tells us how many digits or characters are required to convey the message. Shannon recognized the relationship between thermodynamic entropy and informational entropy and, on von Neumann's advice, called the negative logarithm of probability of an event, entropy.[3]

Consider an event that happens with the probability, p; we wish to quantify the information content of a message communicating the occurrence of this event, and we impose the condition that the measure should reflect the "surprise" brought by the occurrence of this event. An initial guess for a measure of this *surprise* would be $1/p$; the lower the probability of the event, the larger the surprise. But this simplistic approach does not resist scrutiny; the surprise should be additive. If an event is composed of two independent events that occur with the probabilities, q and r, then the probability of the event should be $p \quad qr$, but we see that

$$\frac{1}{p} \quad \frac{1}{q} \quad \frac{1}{r}.$$

On the other hand, if the *surprise* is measured by the logarithm of $1/p$, then the additivity property is obeyed

$$\log\frac{1}{p} \quad \log\frac{1}{q} \quad \log\frac{1}{r}.$$

[2]A stochastic process is said to be ergodic if time averages are equal to ensemble averages—in other words, if its statistical properties such as its mean and variance can be deduced from a single, sufficiently long sample (realization) of the process.

[3]It is rumored that von Neumann told Shannon, "It is already in use under that name and, besides, it will give you a great edge in debates because nobody really knows what entropy is anyway" [77].

All logarithms are in base 2 unless stated otherwise. Given a probability distribution, $_i p_i \quad 1$, we see that the uncertainty is, in fact, equal to the average *surprise*,

$$\quad_i \; p_i \log \frac{1}{p_i}.$$

The entropy is a measure of the uncertainty of a single random variable, X, before it is observed, or the average uncertainty removed by observing it. This quantity is called entropy due to its similarity to the thermodynamic entropy.

The *entropy* of a random variable X with a probability density function, $p_X(x)$, is

$$H(X) \qquad \quad_x \; p_X(x) \log p_X(x).$$

The entropy of a random variable is a positive number. Indeed, the probability $p_X(x)$ is a positive real number between 0 and 1; therefore, $\log p_X(x) \quad 0$ and $H(X) \quad 0$.

Let X be a binary random variable and $p \quad p_X(x \quad 1)$ be the probability that X takes the value 1; then the entropy of X is

$$H(p) \qquad p \log p \quad (1 \quad p) \log(1 \quad p).$$

If the logarithm is in base 2, then the binary entropy is measured in bits. The entropy, depicted in Figure 3.5, has a maximum of 1 bit when $p \quad 0.5$ and goes to zero when $p \quad 0$ or $p \quad 1$; intuitively, we expect the entropy to be zero when the outcome is certain and reach its maximum when both outcomes are equally likely. It is easy to see that:

1. $H(X) > 0$ for $0 < p < 1$;
2. $H(X)$ is symmetric about $p \quad 0.5$;
3. $lim_p \; _0 H(X) \quad lim_p \; _1 H(X) \quad 0$;
4. $H(X)$ is increasing for $0 < p < 0.5$, decreasing for $0.5 < p < 1$, and has a maximum for $p \quad 0.5$; and
5. the binary entropy is a concave function of p, the probability of an outcome.

Before discussing this property of binary entropy, we review a few properties of convex and concave functions. A function, $f(x)$, is *convex over an interval*, (a,b), if

$$f(\gamma x_1 \quad (1 \quad \gamma)x_2) \quad \gamma f(x_1) \quad (1 \quad \gamma) f(x_2) \qquad x_1, x_2 \quad (a,b) \text{ and } 0 \quad \gamma \quad 1.$$

The function is *strictly convex* if and only if this inequality holds only for $\gamma \quad 0$ or $\gamma \quad 1$. A function $f(x)$ is *concave* only if $(\; f(x))$ is convex over the same interval.

It is easy to prove that if the second derivative of the function $f(x)$ is non-negative, then the function is convex. Call $x_0 \quad \gamma x_1 \quad (1 \quad \gamma)x_2$. From Taylor series expansion,

$$f(x) \quad f(x_0) \quad f \; (x_0)(x \quad x_0) \quad \frac{f \; (\xi)}{2}(x \quad x_0)^2, \quad x_0 \quad \xi \quad x,$$

and from the fact that the second derivative is non-negative, it follows that $f''(\xi)(x-x_0)^2 \geq 0$. If $x = x_1$, then $x_1 - x_0 = (1-\gamma)(x_1 - x_2)$ and

$$f(x_1) \geq f(x_0) + f'(x_0)(x_1 - x_0) \quad \text{or} \quad f(x_1) \geq f(x_0) + f'(x_0)(1-\gamma)(x_1 - x_2).$$

If $x = x_2$, then $x_2 - x_0 = \gamma(x_2 - x_1)$ and

$$f(x_2) \geq f(x_0) + f'(x_0)(x_2 - x_0) \quad \text{or} \quad f(x_2) \geq f(x_0) + f'(x_0)\gamma(x_2 - x_1).$$

It follows that

$$\gamma f(x_1) + (1-\gamma) f(x_2) \geq f(\gamma x_1 + (1-\gamma) x_2).$$

Convex functions enjoy a number of useful properties; for example, if X is a discrete random variable with the probability density function, $p_X(x_i)$, and $f(x)$ is a convex function, then $f(x)$ satisfies the *Jensen inequality*:

$$\sum_i p_X(x_i) f(x_i) \geq f\left(\sum_i p_X(x_i) x_i\right).$$

Example. *Given two probability density functions of the random variable X,* $p_i = p_X(x_i)$ *and* $q_i = q_X(x_i)$*, then*

$$-\sum_i p_i \log p_i \leq -\sum_i p_i \log q_i.$$

Proof. This is a consequence of the *log sum inequality*: given two sets of non-negative real numbers, $\{a_1, a_2, \ldots, a_n\}$ and $\{b_1, b_2, \ldots, b_n\}$, then

$$\sum_{i=1}^{n} a_i \log \frac{a_i}{b_i} \geq \sum_{i=1}^{n} a_i \log \frac{\sum_{i=1}^{n} a_i}{\sum_{i=1}^{n} b_i}.$$

First, we observe that $f(x) = x \log x$ is a convex function as $f''(x) > 0$ when $x > 0$. Indeed, $f(x) = x \log e \ln x$, $f'(x) = \log e(1 + \ln x)$, and $f''(x) = (1/x) \log e$.

Call $a = \sum_{i=1}^{n} a_i$ and $b = \sum_{i=1}^{n} b_i$. Then according to the Jensen inequality applied to the function $f(x) = x \log x$, we have

$$\sum_{i=1}^{n} a_i \log \frac{a_i}{b_i} = \sum_{i=1}^{n} b_i \frac{a_i}{b_i} \log \frac{a_i}{b_i} = \sum_{i=1}^{n} b_i f\left(\frac{a_i}{b_i}\right) \geq bf\left(\sum_{i=1}^{n} \frac{b_i}{b} \frac{a_i}{b_i}\right) = bf\left(\frac{a}{b}\right) = b \frac{a}{b} \log \frac{a}{b}.$$

■

Example. *Given a non-negative integer-valued random variable, X, find the probability density function,* $p_i = p_X(x_i)$*, that maximizes the entropy,* $H(X)$*, subject to the constraint,*

$$E(X) = \sum_i i p_i = A.$$

We apply the inequality, $_i p_i \log p_i \quad _i p_i \log q_i$, with $q_i \quad \alpha\beta^i$ and $0 \quad \alpha, \beta \quad 1$:

$$H(X) \quad - \quad \underset{i}{} p_i \log p_i \quad - \quad \underset{i}{} p_i \log q_i \quad - \quad \underset{i}{} p_i (\log\alpha \quad i \log\beta) \quad -\log\alpha - A\log\beta.$$

To derive this inequality, we notice that $_i p_i \quad 1$ and $_i i p_i \quad A$; we also observe that the inequality holds independently of p_i. To determine the parameters $\alpha \quad 1$ and $\beta \quad 1$, we recall that $_{i\ 0}\beta^i \quad 1/(1-\beta)$ and $_{i\ 0} i\beta^i \quad \beta/(1-\beta)^2$ and write:

$$\underset{i\ 0}{} q_i \quad \underset{i\ 0}{} \alpha\beta^i \quad 1 \quad \frac{\alpha}{1-\beta} \quad 1 \quad \text{and} \quad A \quad \underset{i\ 0}{} i q_i \quad \underset{i\ 0}{} i\alpha\beta^i \quad A \quad \frac{\beta}{1-\beta}.$$

The two parameters, α and β, the probability distribution, p_i, and the maximum entropy are:

$$\alpha \quad \frac{1}{1 \quad A}, \beta \quad \frac{A}{1 \quad A}, p_i \quad \frac{1}{1 \quad A} \quad \frac{A}{1 \quad A}^{n}, \quad \text{and} \quad H_{max}(X) \quad -\log\alpha - A\log\beta.$$

It is easy to prove that a function, $f(x)$, is concave if and only if

$$f \quad \frac{x_1 \quad x_2}{2} \quad \frac{f(x_1) \quad f(x_2)}{2}.$$

Figure 3.5 illustrates the fact that the binary entropy, $H(p)$, is a concave function of p; the function lies above any chord, in particular, above the cord connecting the points, $(p_1, H(p_1))$ and $(p_2, H(p_2))$:

$$H \quad \frac{p_1 \quad p_2}{2} \quad \frac{H(p_1) \quad H(p_2)}{2}.$$

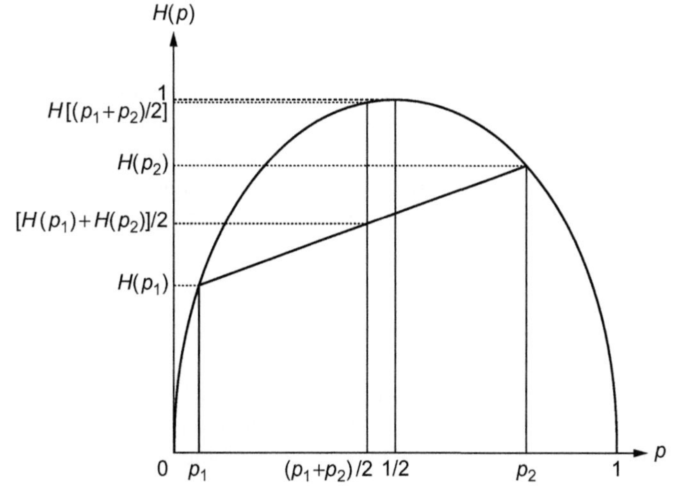

FIGURE 3.5

The entropy of a binary random variable function of the probability of an outcome.

Table 3.1 The Entropy of a Binary Random Variable for $0.0001 \quad p \quad 0.5$

p	$H(X)$	p	$H(X)$	p	$H(X)$	p	$H(X)$
0.0001	0.001	0.01	0.081	0.2	0.722	0.4	0.971
0.001	0.011	0.1	0.469	0.3	0.881	0.5	1.000

Table 3.2 The Joint Probability Distribution Matrix of Random Variables X and Y

$p_{X,Y}(x,y)$	a	b	c	d	e
a	1/10	1/20	1/40	1/80	1/80
b	1/20	1/40	1/80	1/80	1/10
c	1/40	1/80	1/80	1/10	1/20
d	1/80	1/80	1/10	1/20	1/40
e	1/80	1/10	1/20	1/40	1/80

Table 3.1 shows some values for $H(X)$ for $0.0001 \quad p \quad 0.5$.

Given two random variables, X and Y, with the probability density functions, $p_X(x)$ and $q_Y(y)$, let $p_{XY}(x,y)$ be the joint probability density function of X and Y. To quantify the uncertainty about the pair (x,y), we introduce the joint entropy of the two random variables.

The *joint entropy* of two random variables X and Y, with $p_{XY}(x,y)$, the joint probability distribution function is defined as

$$H(X,Y) \qquad \underset{x,y}{} p_{XY}(x,y)\log p_{XY}(x,y).$$

If we have acquired all the information about the random variable X, we may ask how much uncertainty is still there about the pair of the two random variables, (X,Y). To answer this question, we introduce the conditional entropy.

The *conditional entropy* of random variable, Y, given the value of X, with $p_{XY}(x,y)$, the joint probability distribution function and, $p_{Y\ X}(y\ x)$, the conditional probability distribution function, is defined as

$$H(Y\ X) \qquad \underset{x}{}\ \underset{y}{} p_{XY}(x,y)\log p_{Y\ X}(y\ x).$$

Example. *Entropy, joint entropy, and conditional entropy of X and Y.* Consider two random variables, X and Y. Each of them takes values over a five-letter alphabet consisting of the symbols, a,b,c,d,e. The joint distribution of the two random variables is given in Table 3.2.

The marginal distribution of X and Y can be computed from the relations,

$$p_X(x) \qquad \underset{y}{} p(x,y) \quad \text{and} \quad p_Y(y) \qquad \underset{x}{} p(x,y),$$

as follows:

$$p_X(a) \qquad \sum_y p_{XY}(x \quad a, y) \quad \frac{1}{10} \quad \frac{1}{20} \quad \frac{1}{40} \quad \frac{1}{80} \quad \frac{1}{80} \quad \frac{1}{5}.$$

Similarly, we obtain

$$p_X(x \quad b) \quad p_X(x \quad c) \quad p_X(x \quad d) \quad p_X(x \quad e) \quad \frac{1}{5}$$

and

$$p_Y(y \quad a) \quad p_Y(y \quad b) \quad p_Y(y \quad c) \quad p_Y(y \quad d) \quad p_Y(y \quad e) \quad \frac{1}{5}.$$

The entropy of X is thus,

$$H(X) \qquad \sum_x p_X(x) \log p_X(x) \quad 5 \quad \frac{1}{5} \log 5 \qquad \log 5 \text{ bits.}$$

Similarly, the entropy of Y is

$$H(Y) \qquad \sum_y p_Y(y) \log p_Y(y) \quad 5 \quad \frac{1}{5} \log 5 \qquad \log 5 \text{ bits.}$$

The joint probability of X and Y is

$$H(X,Y) \qquad \sum_x \sum_y p_{XY}(x,y) \log p_{XY}(x,y)$$

$$5 \quad \tfrac{1}{10} \log \tfrac{1}{10} \quad \tfrac{1}{20} \log \tfrac{1}{20} \quad \tfrac{1}{40} \log \tfrac{1}{40} \quad \tfrac{1}{80} \log \tfrac{1}{80} \quad \tfrac{1}{80} \log \tfrac{1}{80}$$

$$\log 5 \quad \tfrac{1}{2} \log \tfrac{1}{2} \quad \tfrac{1}{4} \log \tfrac{1}{4} \quad \tfrac{1}{8} \log \tfrac{1}{8} \quad \tfrac{1}{16} \log \tfrac{1}{16} \quad \tfrac{1}{16} \log \tfrac{1}{16}$$

$$\frac{15}{8} \log 5 \text{ bits.}$$

It is easy to see that

$$H(Y,X) \quad \frac{15}{8} \log 5 \text{ bits.}$$

To compute the matrix of the conditional probabilities shown in Table 3.3, we use the known relation between the joint probability density and the conditional probability, $p_{X\ Y}(x\ y) \quad p_{XY}(x,y)/p_Y(y)$. For example:

$$p_{X\ Y}(x \quad a\ y \quad b) \quad \frac{p_{XY}(x \quad a, y \quad b)}{p_Y(y \quad b)} \qquad p_{X\ Y}(x \quad a\ y \quad b) \quad \frac{1/20}{1/5} \quad \frac{1}{4}.$$

The actual value of $H(X\ Y)$ is

$$H(X\ Y) \qquad \sum_y p_Y(y) H(X\ Y \quad a,b,c,d,e) \quad 5 \frac{1}{5} H \quad \frac{1}{2}, \frac{1}{4}, \frac{1}{8}, \frac{1}{16}, \frac{1}{16} \qquad H \quad \frac{1}{2}, \frac{1}{4}, \frac{1}{8}, \frac{1}{16}, \frac{1}{16}$$

Table 3.3 The Conditional Probability Distribution Matrix of Random Variables X and Y

$p_{X\vert Y}(x\vert y)$ $p_{Y\vert X}(y\vert x)$	a	b	c	d	e
a	1/2	1/4	1/8	1/16	1/16
b	1/4	1/8	1/16	1/16	1/2
c	1/8	1/16	1/16	1/2	1/4
d	1/16	1/16	1/2	1/4	1/8
e	1/16	1/2	1/4	1/8	1/16

or

$$H(X\ Y) \quad \frac{1}{2}\log\frac{1}{2} \quad \frac{1}{4}\log\frac{1}{4} \quad \frac{1}{8}\log\frac{1}{8} \quad \frac{1}{16}\log\frac{1}{16} \quad \frac{1}{16}\log\frac{1}{16} \quad \frac{15}{8}\text{ bits.}$$

In this particular example, $H(Y\ X)$ has the same value:

$$H(Y\ X) \quad \frac{15}{8}\text{ bits.}$$

It is easy to see that $H(X), H(Y), H(X,Y)$, and $H(X\ Y)$ 0. Indeed,

$$H(X,Y) \quad H(X\ Y)H(Y) \qquad \frac{15}{8}\log 5 \quad \frac{15}{8} \quad \log 5,$$

$$H(X,Y) \quad \frac{15}{8}\log 5 \quad H(X) \quad H(Y) \quad 2\log 5,$$

$$H(X,Y) \quad H(Y,X),$$

$$H(X\ Y) \quad \frac{15}{8} \quad H(X) \quad \log 5.$$

Properties of Joint and Conditional Entropy

The joint and conditional entropy of random variables, X, Y, and Z, have several properties summarized in Table 3.4. The fact that the joint entropy is symmetric and non-negative follows immediately from its definition. The conditional entropy is non-negative because 0 $p_{Y\ X}(y\ x)$ 1 and 0 $p_{X\ Y}(x\ y)$ 1; the equality occurs only when Y is a deterministic function of X or, respectively, X is a deterministic function of Y.

From the definition of the conditional entropy and the expression of the conditional probability, $p_{X\ Y}(x\ y) \quad p_{XY}(x,y)/p_Y(y)$, it follows that

$$H(X\ Y) \qquad \underset{x\ \ y}{} p_{XY}(x,y)\log p_Y(y) \qquad \underset{x\ \ y}{} p_{XY}(x,y)\log p_{XY}(x,y)$$

$$H(Y) \quad H(X,Y).$$

Table 3.4 Properties of Joint and Conditional Entropy

$H(X,Y) \quad H(Y,X)$	Symmetry of joint entropy
$H(X,Y) \quad 0$	Non-negativity of joint entropy
$H(X\ Y) \quad 0,\ H(Y\ X) \quad 0$	Non-negativity of conditional entropy
$H(X\ Y) \quad H(X,Y) \quad H(Y)$	Conditional and joint entropy relation
$H(X,Y) \quad H(Y)$	Joint entropy versus the entropy of a single rv
$H(X,Y) \quad H(X) \quad H(Y)$	Subadditivity
$H(X,Y,Z) \quad H(Y) \quad H(X,Y) \quad H(Y,Z)$	Strong subadditivity
$H(X\ Y) \quad H(X)$	Reduction of uncertainty by conditioning
$H(X,Y,Z) \quad H(X) \quad H(Y\ X) \quad H(Z\ X,Y)$	Chain rule for joint entropy
$H(X,Y\ Z) \quad H(Y\ X,Z) \quad H(X\ Z)$	Chain rule for conditional entropy

But $H(X\ Y) \quad 0$, thus, $H(X,Y) \quad H(Y)$.

To prove *subadditivity*, we first show that

$$\ln a \quad a \quad 1 \quad \text{if } a \quad 1,$$

with equality if and only if $a \quad 1$. Indeed, if we use the substitution, $a \quad e^b$, the inequality becomes $b \quad e^b \quad 1$ for $b \quad 0$. From the Taylor series expansion of e^b, we have

$$e^b \quad 1 \quad b \quad 1 \quad \frac{b}{2!} \quad \frac{b^2}{3!} \quad \quad \quad \frac{b^k}{(k \quad 1)!} \quad \quad .$$

The parenthesis on the right-hand side of this equation is greater than 1 when $b > 0$ and is equal to 1 only when $b \quad 0$. Thus, $e^b \quad 1 \quad y$, or $a \quad 1 \quad \ln a$; the last inequality can also be written as

$$\log a \quad \frac{1}{\ln 2}(a \quad 1), \quad \text{if } a \quad 1.$$

It is easy to see that $\ln a \quad \log a \ \ln 2$; indeed, any $a \quad 1$ can be expressed as $a \quad e^{\ln a}$ but also as $a \quad 2^{\log a}$; thus, $e^{\ln a} \quad 2^{\log a}$. If we apply on both sides the natural logarithm, we get the desired expression.

Now we return to the subadditivity and express $H(X)$ and $H(Y)$ in terms of the joint probability density, $p_{XY}(x,y)$:

$$H(X) \quad \quad p_X(x) \log p_X(x) \quad \quad p_{XY}(x,y) \quad \log p_X(x)$$
$$\quad x \quad \quad x \quad y$$

and

$$H(Y) \quad \quad p_Y(y) \log p_Y(y) \quad \quad p_{XY}(x,y) \quad \log p_Y(y).$$
$$\quad x \quad \quad y \quad x$$

We wish to show that $H(X,Y) \quad H(X) \quad H(Y) \quad 0$ and compute

$$H(X,Y) \quad H(X) \quad H(Y) \qquad \underset{x \quad y}{} \quad p_{XY}(x,y)\log\frac{p_X(x)p_Y(y)}{p_{XY}(x,y)}.$$

Here, $p_X(x)p_Y(y) \quad p_{XY}(x,y)$ with equality only when X and Y are independent random variables; thus, we use the inequality, $\log a \quad \frac{1}{\ln 2}(a \quad 1)$, with $a \quad \frac{p_X(x)p_Y(y)}{p_{XY}(x,y)} \quad 1$, and rewrite the preceding expression as

$$\underset{x \quad y}{} \quad p_{XY}(x,y)\log\frac{p_X(x)p_Y(y)}{p_{XY}(x,y)} \quad \frac{1}{\ln 2} \quad \underset{x \quad y}{} \quad p_{XY}(x,y) \quad \frac{p_X(x)p_Y(y)}{p_{XY}(x,y)} \quad 1 \; .$$

Thus,

$$H(X,Y) \quad H(X) \quad H(Y) \quad \frac{1}{\ln 2} \quad \underset{x \quad y}{} \quad p_X(x)p_Y(y) \quad p_{XY}(x,y) \quad 0$$

or

$$H(X,Y) \quad H(X) \quad H(Y) \quad \frac{1}{\ln 2} \quad \underset{x \quad y}{} \quad p_X(x)p_Y(y) \quad \underset{x \quad y}{} \quad p_{XY}(x,y) \quad 0.$$

The *strong subadditivity* can be proved in a similar manner.

Finally, we show that $H(X) \quad H(X,Y)$:

$$H(X) \quad H(X,Y) \quad \underset{x \quad y}{} \quad p_{XY}(x,y)\log\frac{p_X(x)}{p_{XY}(x,y)}$$

$$\underset{x \quad y}{} \quad p_{XY}(x,y) \quad \frac{p_X(x)}{p_{XY}(x,y)} \quad 1 \; .$$

But

$$\underset{x \quad y}{} \quad [p_X(x) \quad p_{XY}(x,y)] \quad \underset{x \quad y}{} \quad [p_X(x)] \quad \underset{x \quad y}{} \quad [p_{XY}(x,y)] \quad 1 \quad 1 \quad 0.$$

We conclude that

$$H(X) \quad H(X,Y) \quad 0 \quad \text{or} \quad H(X) \quad H(X,Y),$$

with equality if and only if Y is a function of X.

The *chain rule for joint entropy* can be proved knowing the conditional and joint entropy relation, $H(X,Y) \quad H(X) \quad H(Y \; X)$. Then,

$$H(X,Y,Z) \quad H(X) \quad H(Y,Z \; X) \quad H(X) \quad H(Y \; X) \quad H(Z \; X,Y).$$

This rule can be generalized for n random variables, $X_1, X_2, \ldots, X_n$, as

$$H(X_1, X_2, \ldots, X_n) \quad \overset{n}{\underset{i \quad 1}{}} \quad H(X_i \; X_i \quad {}_1, X_i \quad {}_2, \ldots, X_2, X_1).$$

To prove the *chain rule for conditional entropy*, we start from the definition

$$H(X,Y|Z) = H(X,Y,Z) - H(Z).$$

We know that $H(X|Y) = H(X,Y) - H(Y)$; thus, the right-hand side of this equation can be written as

$$[H(X,Y,Z) - H(X,Z)] + [H(X,Z) - H(Z)] = H(Y|X,Z) + H(X|Z).$$

Thus,

$$H(X,Y|Z) = H(Y|X,Z) + H(X|Z).$$

The chain rule for the conditional entropy can be generalized for $n > 1$ random variables:

$$H(X_1,X_2,\ldots,X_n|Y) = \sum_{i=1}^{n} H(X_i|Y,X_1,\ldots X_{i-1}).$$

Last, but not least, we mention, without proof, the *grouping* property of Shannon entropy [15]:

$$\begin{aligned} H(p_1,p_2,\ldots,p_n) &= H(p_1 + p_2 + \cdots + p_\sigma, p_{\sigma+1} + \cdots + p_n) \\ &+ \left(\sum_{k=1}^{\sigma} p_k\right) H\left(p_1\Big/\sum_{k=1}^{\sigma} p_k, p_2\Big/\sum_{k=1}^{\sigma} p_k, \ldots, p_\sigma\Big/\sum_{k=1}^{\sigma} p_k\right) \\ &+ \left(\sum_{k=\sigma+1}^{n} p_k\right) H\left(p_{\sigma+1}\Big/\sum_{k=\sigma+1}^{n} p_k, p_{\sigma+2}\Big/\sum_{k=\sigma+1}^{n} p_k, \ldots, p_n\Big/\sum_{k=\sigma+1}^{n} p_k\right). \end{aligned}$$

Example. *The entropy of a function of a random variable.* If X is a discrete random variable and Y is a function of X, $Y = f(X)$, then its entropy cannot be larger than the entropy of X, $H(X) \geq H(f(X))$; when f is a one-to-one function, we have the equality $H(X) = H(f(X))$.

Proof. From the chain rule of joint entropy, it follows that

$$H(X,f(X)) = H(X) + H(f(X)|X) = H(X).$$

Indeed, $H(f(X)|X) = 0$ as f is a deterministic function; for any given value x of X, $f(x)$ is fixed. The chain rule for mutual information allows us to express $H(X,f(X))$ as $H(X,f(X)) = H(f(X)) + H(X|f(X))$. But $H(X|f(X)) \geq 0$ with equality if and only if f is a one-to-one function. We conclude that $H(f(X)) \leq H(X)$. For example, if $Y = X^{2n+1}$, then $H(Y) = H(X)$ as f is invertible; if $Z = \tan(X)$, then $H(Z) < H(X)$ because the tangent is not a one-to-one function. ■

Example. *The entropy of a sum of random variables,* $Z = X + Y$. If $Z = X + Y$, then $H(Z|X) = H(Y|Y)$. When X and Y are independent, $H(X) \leq H(Z)$ and $H(Y) \leq H(Z)$; addition of independent random variables adds uncertainty. Moreover, when X and Y are independent and the function $Z = f(X,Y) = X + Y$ is one-to-one, $H(Z) = H(X) + H(Y)$.

Proof. First, we observe that $p_{Z|X}(z|x) = p_{Y|X}(y|x)$. Indeed,

$$\text{Prob}[Z = z|X = x] = \text{Prob}[X + Y = x|X = x] = \text{Prob}[Y = z - x|X = x].$$

Then:

$$
\begin{aligned}
H(Z\ X) \quad & \ \ p_{Z\ X}(z\ x)p_X(x)\log p_{Z\ X}(z\ x) \\
& {}_{x\ \ z} \\
& p_X(x) \quad p_{Z\ X}(z\ x)\log p_{Z\ X}(z\ x) \\
& {}_{x \qquad\quad z} \\
& p_X(x) \quad p_{Y\ X}(z \quad x\ x)\log p_{Y\ X}(z \quad x\ x) \\
& {}_{x \qquad\quad z} \\
& p_X(x) \quad p_{Y\ X}(y\ x)\log p_{Y\ X}(y\ x) \\
& {}_{x \qquad\quad y} \\
& \ \ p_{Y\ X}(y\ x)p_X(x)\log p_{Y\ X}(y\ x) \\
& {}_{x\ \ y} \\
& H(Y\ X).
\end{aligned}
$$

We now show that when X and Y are independent, $H(Z) \quad H(Y)$. Indeed, in this case, $H(Y\ X) \quad H(Y)$, and then, $H(Z\ X) \quad H(Y\ X) \quad H(Y)$. But $H(Z) \quad H(Z\ X)$, so $H(Z) \quad H(Y)$; a similar argument leads to $H(Z) \quad H(X)$. Lastly, if the function f in $Z \quad f(X,Y) \quad X \quad Y$ is one-to-one, then, according to the previous example, $H(Z) \quad H(X,Y)$; if X and Y are independent, then $H(X,Y) \quad H(X) \quad HY$. Therefore, $H(Z) \quad H(X) \quad H(Y)$. ■

We conclude the discussion of entropy with the observation that in his original paper Shannon proved that a logarithmic measure for the entropy can be derived from several axioms [374]. Given a sequence of symmetric functions, $H_1(p_1), H_2(p_1,p_2), \ldots, H_n(p_1,p_2,p_3,\ldots,p_n)$, the function, H_n, must be of the form, $H_n(p_1,p_2,p_3,\ldots,p_n) \qquad {}^{n}_{i\ 1} p_i \log p_i$, when the functions satisfy three axioms:

1. *Normalization:* $H_2(1/2,1/2) \quad 1,\ldots,H_n(1/n,1/n,\ldots,1/n) \quad 1.$
2. *Continuity:* $H_2(p,1 \quad p)$ is a continuous function of p.
3. *Grouping:*
   $H_n(p_1,p_2,p_3,\ldots,p_n) \quad H_{n\ \ 1}(p_1 \quad p_2,p_3,\ldots,p_n) \quad (p_1 \quad p_2)H_2(p_1/(p_1 \quad p_2),p_2/(p_1 \quad p_2)).$

3.4 SHANNON SOURCE CODING

We know that a source of classical information, with the probability distribution, $p_X(x)$, is characterized by the Shannon entropy, $H(p_X(x))$. Source coding addresses two questions: (1) Is there an upper bound for the data compression level, and, if so, what is this upper bound? (2) Is it always possible to reach this upper bound? The answer to both questions is "yes." Classical, as well as quantum source coding discussed in Section 3.14, are based on the idea of *typical sequence of symbols* from an alphabet and *typical subspaces of a Hilbert space,* respectively.

Informally, the Shannon source encoding theorem states that a message containing, n, independent, identically distributed samples of a random variable, X, with the entropy, $H(X)$, can be compressed to a length,

$$l_X(n) \quad nH(X) \quad \mathcal{O}(n).$$

The justification of this theorem is based on the *weak law of large numbers*, which states that $\bar{x} = \sum_i x_i p_{x_i}$, the mean of a large number of independent, identically distributed random variables, x_i, approaches, $1/n \sum_i x_i$, the average, with a high probability when n is large:

$$\text{Prob}\left(\left| \frac{1}{n} \sum_{i=1}^{n} x_i - \bar{x} \right| > \delta \right) < \epsilon,$$

with δ and ϵ, two arbitrarily small positive real numbers.

Given a source of classical information, $\mathcal{A} = \{a_1, a_2, \ldots, a_m\}$, with an alphabet with m symbols, let $\{p(a_1), p(a_2), \ldots, p(a_n)\}$ be the probabilities of individual symbols; then the Shannon entropy of this source is

$$H(\mathcal{A}) = -\sum_{i=1}^{m} p(a_i) \log p(a_i).$$

A message, $a = (a_{k_1}, a_{k_2}, \ldots, a_{k_n})$, consisting of a sequence of n symbols independently selected from the input alphabet has a probability

$$P(a) = p(a_{k_1}) p(a_{k_2}) \ldots p(a_{k_n}).$$

If we define a random variable, $x_i = -\log p(a_i)$, then we can establish the following correspondence with the quantities in the expression of the weak law of large numbers:

$$\sum_{i=1}^{n} x_i = -\sum_{i=1}^{n} \log p(a_{k_i}) = -\log P(a),$$

$$\bar{x} = \sum_{i=1}^{m} x_i p(x_i) = -\sum_{i=1}^{m} p(a_i) \log p(a_i) = H(\mathcal{A}).$$

It follows that given two arbitrarily small real numbers, $(\delta, \epsilon) > 0$, then for sufficiently large n, the following inequality holds [368, 374]:

$$\text{Prob}\left(\left| -\frac{1}{n} \log P(a) - H(\mathcal{A}) \right| > \delta \right) < \epsilon.$$

This inequality partitions Λ, the set of strings of length n, into two subsets see (Figure 3.6).

1. The subset of *Typical* strings,

$$\Lambda_{Typical} = \left\{ a_{Typical} : \left| -\frac{1}{n} \log P(a_{Typical}) - H(\mathcal{A}) \right| \leq \delta \right\},$$

   that occur with high probability, $\text{Prob}(a_{Typical}) \geq (1 - \epsilon)$.

2. The subset of *Atypical* strings,

$$\Lambda_{Atypical} = \left\{ a_{Atypical} : \left| -\frac{1}{n} \log P(a_{Atipical}) - H(\mathcal{A}) \right| > \delta \right\},$$

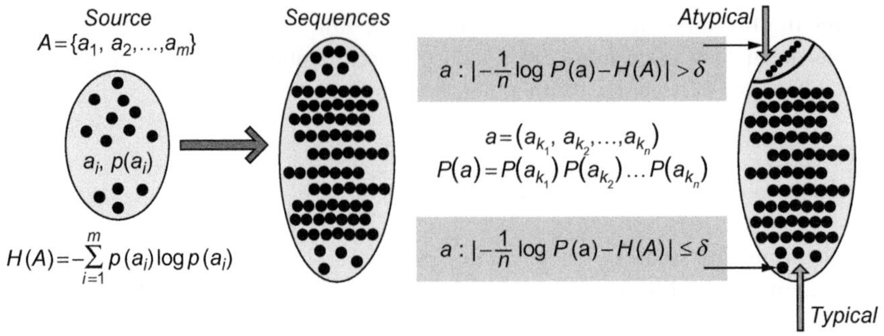

FIGURE 3.6

Source coding. The inequality, Prob $-\frac{1}{n}\log P(a) - H(\mathcal{A}) > \delta < \epsilon$, partitions the set of strings of length, n, into two classes: (1) typical strings that occur with high probability and (2) atypical strings that occur with a vanishing probability.

that occur with a vanishing probability, Prob$(a_{Atypical} < \epsilon$. The two subsets are disjoint and complementary:

$$\Lambda_{Typical} \quad \Lambda_{Atypical} \qquad \text{and} \quad \Lambda \quad \Lambda_{Typical} \bigsqcup \Lambda_{Atypical}.$$

We concentrate on typical strings generated by the source, as atypical strings occur with a vanishing probability and can be ignored. Now we shall determine $\Lambda_{Typical}$, the cardinality of the set of typical strings, $a \quad \Lambda_{Typical}$. The inequality defining the strings in this subset can be rewritten as

$$-\delta \quad -\frac{1}{n}\log P(a_{Typical}) - H(\mathcal{A}) \quad \delta \qquad 2^{-n(H(\mathcal{A}) \;\; \delta)} \quad P(a_{Typical}) \quad 2^{-n(H(\mathcal{A})-\delta)}.$$

The first inequality can be expressed in terms of the cardinality set:

$$\sum_{a \;\; \Lambda_{Typical}} P(a_{typical}) \qquad \sum_{a \;\; \Lambda_{Typical}} 2^{-n(H(\mathcal{A}) \;\; \delta)} \qquad \Lambda_{Typical}\, 2^{-n(H(\mathcal{A}) \;\; \delta)}.$$

It follows that

$$\Lambda_{Typical} \qquad 2^{n(H(\mathcal{A}) \;\; \delta)}.$$

Similarly, the second inequality can be expressed as

$$\sum_{a \;\; \Lambda_{Typical}} P(a_{typical}) \qquad \sum_{a \;\; \Lambda_{Typical}} 2^{-n(H(\mathcal{A})-\delta)} \qquad \Lambda_{Typical}\, 2^{-n(H(\mathcal{A})-\delta)}.$$

This implies that

$$\Lambda_{Typical} \qquad 2^{n(H(\mathcal{A})-\delta)}.$$

When $\delta \quad 0$, then $\Lambda_{Typical} \quad 2^{nH(\mathcal{A})}$, and the cardinality of the set of typical string converges to $2^{n(H(\mathcal{A})-\delta)}$. *There are* $2^{nH(\mathcal{A})}$ *typical strings; therefore, we need* $\log 2^{nH(\mathcal{A})} \quad nH(\mathcal{A})$ *bits to encode all possible typical strings;* this is the upper bound for the data compression provided by Shannon's classical source encoding theorem.

Example. *Message length and entropy.* Eight cars labeled, $\mathbb{C}_1, \mathbb{C}_2, \ldots, \mathbb{C}_8$, compete in several Formula I races. The probabilities of winning based on the past race history for the eight cares, $p_1, p_2, \ldots, p_8$, are:

$$p_1 \quad \frac{1}{2},\ p_2 \quad \frac{1}{4},\ p_3 \quad \frac{1}{8},\ p_4 \quad \frac{1}{16},\ p_5 \quad \frac{1}{64},\ p_6 \quad \frac{1}{64},\ p_7 \quad \frac{1}{64}, \text{ and } \ p_8 \quad \frac{1}{64}.$$

To send a binary message revealing the winner of a particular race, we could encode the identities of the winning car in several ways. For example, we can use an "obvious" encoding scheme; the identities of $\mathbb{C}_1, \mathbb{C}_2, \mathbb{C}_3, \mathbb{C}_4, \mathbb{C}_5, \mathbb{C}_6, \mathbb{C}_7, \mathbb{C}_8$ could be encoded using three bits, the binary representation of integers 0 to 7, namely, 000, 001, 010, 011, 100, 101, 110, 111, respectively. Obviously, in this case we need three bits; the average length of the string used to communicate the winner of any race is $l \quad 3$.

Let us now consider an encoding that reduces the average number of bits transmitted. The cars have different probability to win a race, and it makes sense to assign a shorter string to a car that has a higher probability to win. Thus, a better encoding of the identities of $\mathbb{C}_1, \mathbb{C}_2, \mathbb{C}_3, \mathbb{C}_4, \mathbb{C}_5, \mathbb{C}_6, \mathbb{C}_7, \mathbb{C}_8$ is

$$0, 10, 110, 1110, 111100, 111101, 111110, 111111.$$

The corresponding lengths of the strings encoding the identity of each car are:

$$l_1 \quad 1,\ l_2 \quad 2,\ l_3 \quad 3,\ l_4 \quad 4,\ l_5 \quad l_6 \quad l_7 \quad l_8 \quad 6.$$

In this case, l, the average length of the string used to communicate the winner is

$$l \qquad \overset{8}{\underset{i\ \ 1}{}}\ l_i \ p_i \quad 1 \ \frac{1}{2} \quad 2 \ \frac{1}{4} \quad 3 \ \frac{1}{8} \quad 4 \ \frac{1}{16} \quad 4 \quad 4 \ \frac{1}{64} \qquad 2 \text{ bits}.$$

Using this encoding scheme, the expected length of the string designating the winner over a large number of car races is less than the one for "obvious" encoding presented previously. Now we calculate the entropy of the random variable, W, indicating the winner:

$$H(W) \quad \frac{1}{2} \ \log \frac{1}{2} \quad \frac{1}{4} \ \log \frac{1}{4} \quad \frac{1}{8} \ \log \frac{1}{8} \quad \frac{1}{16} \ \log \frac{1}{16} \quad \frac{4}{64} \ \log \frac{1}{64} \quad 2 \text{ bits}.$$

We observe that the average length of the string identifying the outcome of a race for this particular encoding scheme is equal with the entropy, which means that this is an optimal encoding scheme. Indeed, the entropy provides the average information, or, equivalently, the average uncertainty removed by receiving a message.

While the entropy is a quantitative measure for an information source, the mutual and the conditional entropy together with the mutual information introduced in the next section provide a characterization of a communication channel, another element of Shannon's communication model.

3.5 MUTUAL INFORMATION AND RELATIVE ENTROPY

We introduce two important concepts from Shannon's information theory: the *mutual information*, which measures the reduction in uncertainty of a random variable, X, due to another random variable, Y, and the *conditional mutual information* of X, Y, and Z, which quantifies the reduction in the uncertainty of random variables X and Y given Z.

The *mutual information* of two random variables, X and Y, with the joint probability distribution function, $p_{XY}(x,y)$, is defined as

$$I(X;Y) = \sum_{x}\sum_{y} p_{XY}(x,y)\log\frac{p_{XY}(x,y)}{p_X(x)p_Y(y)}.$$

The symmetry, $I(X;Y) = I(Y;X)$, follows immediately from the definition of mutual information. To establish the relation of mutual information with entropy and conditional entropy, we use the fact that $p_{XY}(x,y) = p_{X|Y}(x|y)p_Y(y)$:

$$I(X;Y) = -\sum_{x}\sum_{y} p_{XY}(x,y)\log p_X(x) + \sum_{x}\sum_{y} p_{XY}(x,y)\log p_{X|Y}(x|y)$$
$$= H(X) - H(X|Y).$$

We also note that $H(X|X) = 0$; thus, $I(X;X) = H(X)$. We have proved earlier that $H(X) \geq H(X|Y)$; thus, $I(X,Y) \geq 0$ with equality only if X and Y are independent.

The equality, $I(X;Y) = H(X) + H(Y) - H(X,Y)$, follows immediately from the definition of $I(X;Y)$:

$$I(X;Y) = -\sum_{x}\left[\sum_{y} p_{XY}(x,y)\right]\log p_X(x) - \sum_{y}\left[\sum_{x} p_{XY}(x,y)\right]\log p_Y(y)$$
$$+ \sum_{x}\sum_{y} p_{XY}(x,y)\log p_{XY}(x,y)$$
$$= H(X) + H(Y) - H(X,Y).$$

The *conditional mutual information* of the random variables X, Y, and Z is defined as

$$I(X;Y|Z) = \sum_{x}\sum_{y}\sum_{z} p_{XYZ}(x,y,z)\log\frac{p_{XY|Z}(x,y|z)}{p_{X|Z}(x|z)p_{Y|Z}(y|z)}.$$

It follows immediately that

$$I(X;Y|Z) = H(X|Z) - H(X|Y,Z).$$

The mutual information of the random variables X, Y, and Z has several properties summarized in Table 3.5. To prove the chain rule for mutual information, we start from the definition of mutual information of the three random variables:

$$I(X,Y;Z) = H(X,Y) - H(X,Y|Z) = [H(X|Y) + H(Y)] - [H(X|Y,Z) + H(Y|Z)].$$

But,

$$H(X|Y) - H(X|Y,Z) = I(X;Z|Y) \quad \text{and} \quad H(Y) - H(Y|Z) = I(Y;Z).$$

Table 3.5 Properties of Mutual Information

$I(X;Y) \quad I(Y;X)$	Symmetry of mutual entropy
$I(X;Y) \quad H(X) \quad H(X\ Y)$	Mutual information, entropy, and conditional entropy
$I(X;Y) \quad H(Y) \quad H(Y\ X)$	Mutual information, entropy, and conditional entropy
$I(X;X) \quad H(X)$	Mutual self-information and entropy
$I(X;X) \quad 0$	Non-negativity of mutual self-information
$I(X;Y) \quad H(X) \quad H(Y) \quad H(X,Y)$	Mutual information, entropy, and joint entropy
$I(X;Y\ Z) \quad H(X\ Z) \quad H(X\ Y,Z)$	Conditional mutual information and conditional entropy
$I(X,Y;Z) \quad I(X;Z\ Y) \quad I(Y;Z)$	Chain rule for mutual information
$I(X;Y) \quad I(X;Z)$ if $X \quad Y \quad Z$	Data processing inequality

Thus,

$$I(X,Y;Z) \quad I(X;Z\ Y) \quad I(Y;Z).$$

This chain rule can be generalized to $n \quad 1$ random variables, $X_1, X_2, \ldots, X_n, Y$, as

$$I(X_1,X_2,\ldots,X_n;Y) \qquad \underset{i}{} \; I(X_i;Y\ X_1,X_2,\ldots,X_{i \quad 1}).$$

The diagrams in Figure 3.7 illustrate the relations among the mutual entropy, joint entropy, and conditional entropies of the random variables X and Y we discussed earlier:

$$I(X;Y) \quad H(X) \quad H(Y) \quad H(X,Y),$$

$$H(X,Y) \quad H(X\ Y) \quad H(Y\ X) \quad I(X;Y),$$

$$H(X\ Y) \quad H(X) \quad I(X;Y),\ H(Y\ X) \quad H(Y) \quad I(X;Y).$$

In practice, we rarely have complete information about an event; there are discrepancies between what we expect and the real probability distribution of a random variable. For example, to schedule its gate availability, an airport assumes that a particular flight arrives on time 75% of the time, while in reality the flight is on time only 61% of the time.

To measure how close two distributions are from each other, we introduce the concept of *relative entropy*. Given a random variable and two probability distributions, $p_X(x)$ and $q_X(x)$, we wish to develop an entropy-like measure of how close the two distributions are. If the real probability distribution of events is $p_X(x)$, but we erroneously assume that it is $q_X(x)$, then the surprise when an event occurs is $\log(1/q_X(x))$, and the average surprise is

$$\underset{x}{} \; p_X(x) \log \frac{1}{q_X(x)}.$$

In this expression, we use, $p_X(x)$, the correct probabilities for averaging. Yet, the correct amount of information is given by Shannon's entropy, $H(X) \qquad p_X(x) \log p_X(x)$. The relative entropy is defined as the difference between the average surprise and Shannon's entropy.

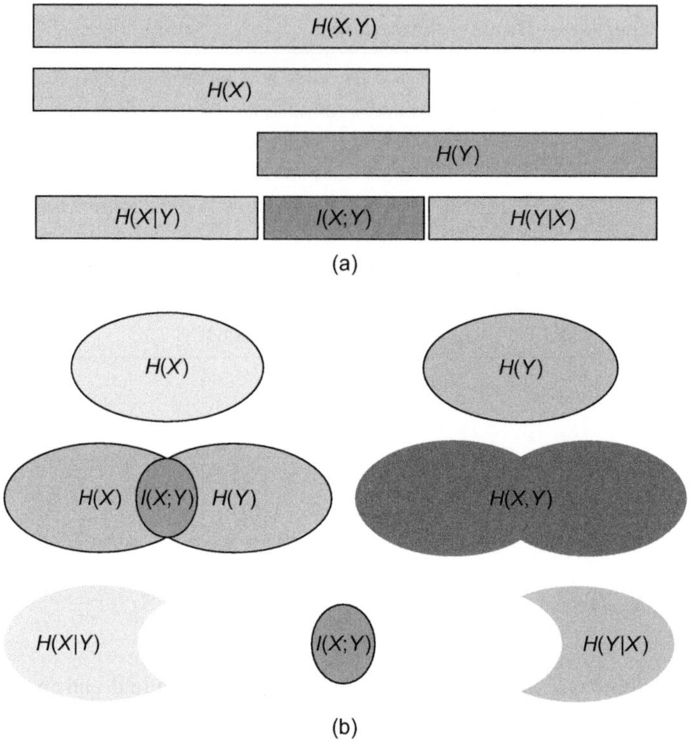

FIGURE 3.7

Graphical representation of the relationship between, $H(X)$ and $H(Y)$, the entropy of the random variables, X and Y, the mutual information, $I(X;Y)$, the joint entropy, $H(X,Y)$, and the conditional entropy, $H(X|Y)$. (a) An intuitive representation. (b) A traditional representation using Venn diagrams.

The *relative entropy* between two distributions, $p_X(x)$ and $q_X(x)$, also called the Kullback-Leibler distance, is defined as

$$H(p_X(x) \| q_X(x)) = \sum_x p_X(x) \log \frac{p_X(x)}{q_X(x)}.$$

The relative entropy is not a metrics in a mathematical sense because it is not symmetric:

$$H(p_X(x) \| q_X(x)) \neq H(q_X(x) \| p_X(x)).$$

The expression of the relative entropy can also be written as

$$H(p_X(x) \| q_X(x)) = \sum_x p_X(x) \log p_X(x) + \sum_x p_X(x) \log \frac{1}{q_X(x)} = -H(X) - \sum_x p_X(x) \log q_X(x).$$

This expression justifies our assertion that the relative entropy is equal to the difference between the average surprise and Shannon's entropy. We show now that the relative entropy is non-negative.

Thus, as pointed out by Vedral [436], we notice an "uncertainty deficit" caused by our inaccurate assumptions; our average surprise is larger than Shannon's entropy.

Proposition. *The relative entropy between two distributions, $p_X(x)$ and $q_X(x)$, of the discrete random variable, X, is non-negative; it is zero only when $p_X(x) = q_X(x)$.*

Proof. By virtue of the inequality, $\ln x \leq x - 1$ for $x > 0$, and of the fact that $\log x = \ln x / \ln 2$, it follows that

$$H(p_X(x) \| q_X(x)) = -\sum_x p_X(x) \log \frac{p_X(x)}{q_X(x)} \geq \frac{1}{\ln 2} \sum_x p_X(x)\left(1 - \frac{q_X(x)}{p_X(x)}\right).$$

But,

$$\sum_x p_X(x)\left(1 - \frac{q_X(x)}{p_X(x)}\right) = \sum_x p_X(x) - \sum_x q_X(x) = 1 - 1 = 0.$$

Thus,

$$H(p_X(x) \| q_X(x)) \geq 0$$

with equality if and only if $p_X(x) = q_X(x)$. ■

This shows that indeed relative entropy is a measure of the similarity (or distance), D, of the two distributions. For example, if X is a binary random variable and the two distributions are

$$p_X(0) = 1 - a,\ p_X(1) = a \quad \text{and} \quad q_X(0) = 1 - b,\ q_X(1) = b,\ 0 < (a,b) < 1,$$

then

$$D(p_X(x) \| q_X(x)) = (1 - a)\log \frac{1-a}{1-b} + a \log \frac{a}{b}$$

and

$$D(q_X(x) \| p_X(x)) = (1 - b)\log \frac{1-b}{1-a} + b \log \frac{b}{a}.$$

It is easy to see that $D(p_X(x) \| q_X(x)) = D(q_X(x) \| p_X(x))$ if and only if $a = b$.

As a final observation, the mutual information, $I(X;Y)$, is the relative entropy between the joint density function, $p_{XY}(x,y)$, and the product of the probability density functions, $p_X(x)$ and $p_Y(y)$,

$$I(X,Y) = \sum_x \sum_y p_{XY}(x,y) \log \frac{p_{XY}(x,y)}{p_X(x)p_Y(y)} = D(p_{XY}(x,y) \| p_X(x)p_Y(y)).$$

When X and Y are uncorrelated, $p_{XY}(x,y) = p_X(x)p_Y(y)$, and then both the mutual information and the relative entropy are equal to zero.

The next question we address is how to estimate the error when we use a random variable, Y, to guess the value of a correlated random variable, X.

3.6 FANO'S INEQUALITY AND THE DATA PROCESSING INEQUALITY

Fano's inequality is used to find a lower bound on the error probability of any decoder; it relates the average information lost in a noisy channel to the probability of the categorization error. We first introduce the concept of a Markov chain, then state Fano's inequality, and finally prove the data processing inequality, a result with many applications in classical information theory.

Random variables, X, Y, and Z, form a *Markov chain* denoted as $X \quad Y \quad Z$, if the conditional probability distribution of Z depends only on Y (it is independent of X):

$$p_{XYZ}(x,y,z) \quad p_X(x)p_{Y\,X}(y\,x)p_{Z\,Y}(z\,y) \qquad X \quad Y \quad Z.$$

Example. *Mutual information and Markov chains.* The random variables, $X_1, X_2, X_3, \ldots, X_n$, form a Markov chain in the order, $X_1 \quad X_2 \quad X_3 \quad \ldots \quad X_n$; this means that

$$p_{X_1X_2\ldots X_n} \quad p_{X_1}(x)p_{X_2\,X_1}(x_2\,x_1)p_{X_3\,X_2}(x_3\,x_2)\ldots p_{X_n\,X_{n\;1}}(x_n\,x_{n\;1}).$$

It is easy to see that

$$I(X_1 : X_2, X_3, \ldots, X_N) \quad I(X_1;X_2).$$

Indeed, we apply the chain rule for mutual information and

$$I(X_1 : X_2, X_3, \ldots, X_n) \quad I(X_1;X_2) \quad I(X_1;X_3\,X_2) \quad I(X_1;X_n\,X_2, X_3, \ldots, X_{n\;1}).$$

The random variables, $X_1, X_2, X_3, \ldots, X_n$, could represent the states of a system and, as we know, the Markov property means that the past and the future state are conditionally independent given the present state; this implies that all terms in this sum except the first are equal to zero.

Consider now a random variable, X, with the probability density function, $p_X(x)$, and let $\mathcal{X}$ denote the number of elements in the range of X. Let Y be another random variable related to X, with the conditional probability, $p_{Y\,X}(y\,x)$. Intuitively, we expect that the error when we use Y to estimate X will be small when the conditional entropy, $H(X\,Y)$, is small. Our intuition is quantified by Fano's inequality.

Proposition. *Fano's inequality: When we estimate X based on the observation of Y as $X \quad f(X)$, we make an error, $p_{err} \quad Prob(X \quad X)$, and*

$$H(p_{err}) \quad p_{err}\log(\,\mathcal{X} \quad 1) \quad H(X\,Y).$$

Weaker forms of Fano's inequality are

$$1 \quad p_{err}\log(\,\mathcal{X}\,) \quad H(X\,Y) \quad \textit{and} \quad p_{err} \quad \frac{H(X\,Y) \quad 1}{\log(\,\mathcal{X}\,)}.$$

Proof. We assume that the discrete random variable, X, has a probability distribution, $p_X(x)$, and call $\mathcal{X}$ the range of values of X. We base our estimation of X on the random variable, Y, which is related to X by the conditional density function, $p_{Y\,X}(y\,x)$; once we know Y, we estimate X as a function of

$Y, X \quad g(Y)$. Clearly, X, Y, X form a Markov chain, $X \quad Y \quad X$. Now we introduce a binary random variable, A, defined as

$$A \quad \begin{cases} 1 & \text{if} \quad X \quad X, \\ 0 & \text{if} \quad X \quad X, \end{cases}$$

We apply the chain rule for entropy and write

$$H(A,X \; Y) \quad H(X \; Y) \quad H(A \; X,Y) \quad H(X \; Y).$$

Indeed, $H(A \; X,Y) \quad 0$ as A is completely determined once we know X and Y. We apply again the chain rule for entropy to express $H(A,X \; Y)$ differently:

$$H(A,X \; Y) \quad H(A \; Y) \quad H(X \; A,Y).$$

A is a binary random variable, and Prob$(A \quad 0) \quad 1 \quad p_{err}$ and Prob$(A \quad 1) \quad p_{err}$. Thus, $H(A)$ $H(p_{err})$; also, $H(A \; Y) \quad H(A)$, since conditioning reduces the entropy. Now,

$$H(X \; A,Y) \quad \text{Prob}(A \quad 0) \; H(X \; Y,A \quad 0) \quad \text{Prob}(A \quad 1) \; H(X \; Y,A \quad 1).$$

We recognize that $H(X \; Y,A \quad 0) \quad 0$ because $X \quad X$ when $A \quad 0$; $\mathcal{X}$ is the cardinality of the target set of X; thus, $H(X \; Y,A \quad 1) \quad \log(\mathcal{X} \quad 1)$, and we conclude that

$$H(X \; A,Y) \quad H(p_{err}) \quad (1 \quad p_{err}) \; 0 \quad p_{error} \log(\mathcal{X} \quad 1) \quad H(p_{err}) \quad p_{error} \log(\mathcal{X} \quad 1).$$

Combining the two expressions of $H(X \; A,Y)$, we obtain the Fano inequality:

$$H(p_{err}) \quad p_{error} \log(\mathcal{X} \quad 1) \quad H(X \; Y).$$

But $H(p_{err}) \quad 1$ and $\mathcal{X}$, the cardinality of the set of sample values of the random variable X, is expected to be large, $\mathcal{X} \quad 1 \quad \mathcal{X}$; therefore, weaker forms of Fano's inequality are

$$1 \quad p_{error} \log(\mathcal{X}) \quad H(X \; Y) \quad \text{or} \quad p_{err} \quad \frac{H(X \; Y) \quad 1}{\log(\mathcal{X})}.$$ ■

Proposition. *Data processing inequality: If random variables, X,Y,Z, form a Markov chain, $X \quad Y \quad Z$, then*

$$I(X;Z) \quad I(X;Y).$$

Proof. The inequality follows from the chain rule, which allows us to expand the mutual information in two different ways:

$$\begin{aligned} I(X;Y,Z) &\quad I(X;Z) \quad I(X;Y \; Z) \\ &\quad I(X;Y) \quad I(X;Z \; Y). \end{aligned}$$

X, Y, and Z form a Markov chain, $X \quad Y \quad Z$, and this implies that X and Z are independent given Y and $I(X;Z \; Y) \quad 0$. Then,

$$I(X;Z) \quad I(X;Y \; Z) \quad I(X;Y) \qquad\qquad I(X;Z) \quad I(X;Y)$$

as $I(X;Y \; Z)$ is non-negative, $I(X;Y \; Z) \quad 0$.

Informally, the data processing inequality states that one cannot gather more information, Z, by processing a set of data, Y, than the information provided by the data to begin with, X; for example, no amount of signal processing could increase the information we receive from a space probe. The quantum correspondents of Fano's and data processing inequalities will be discussed in Section 3.19.

■

3.7 CLASSICAL INFORMATION TRANSMISSION THROUGH DISCRETE CHANNELS

When two agents communicate, each one can influence the physical state of the other through some physical process. The precise nature of the agents and of the signaling process can be very different, thus, it is necessary to consider an abstract model of communication. In this model, a source has the ability to select among a number of distinct physical signals and to pass the selected signal to a communication channel that then affects the physical state of the destination; finally, the destination attempts to identify precisely the specific signal that caused the change of its physical state.

We are interested in two fundamental questions regarding communication over noisy channels: (1) Is it possible to encode the information transmitted over a noisy channel to minimize the probability of errors? (2) How does the noise affect the capacity of a channel?

Intuitively, we know that whenever the level of noise in a room or on a phone line is high, we have to repeat words and sentences several times before the other party understands us. Thus, the actual rate we are able to transmit information through a communication channel is lowered by the presence of the noise. Rigorous answers to both questions, consistent with our intuition, are provided by two theorems, the *source coding theorem* discussed in Section 3.4, which provides an upper bound for the compression of a message, and the *channel coding theorem*, which states that the two agents can communicate at a rate close to the channel capacity, C, and with a probability of error, ϵ, arbitrarily small [375–377].

Encoding and Decoding

Source encoding is the process of transforming the information produced by the source into messages. The source may produce a continuous stream of symbols from the source alphabet. Then the source encoder cuts this stream into blocks of a fixed size. The source decoder performs an inverse mapping and delivers symbols from the output alphabet.

Channel encoding allows transmission of the message generated by the source through the channel. The channel encoder accepts as input a set of messages of fixed length and maps the source alphabet into a channel alphabet, then adds a set of redundancy symbols, and finally sends the message through the channel. The *channel decoder* first determines if the message is in error and takes corrective actions; then it removes the redundancy symbols, maps the channel alphabet into the source alphabet, and hands each message to the source decoder. The *source decoder* processes the message and passes it to the receiver (Figure 3.8).

Information encoding allows us to accomplish critical tasks for processing, transmitting, and storing information, such as:

1. *Error control.* An error occurs when an input symbol is distorted during transmission and interpreted by the destination as another symbol from the alphabet. The error control mechanisms

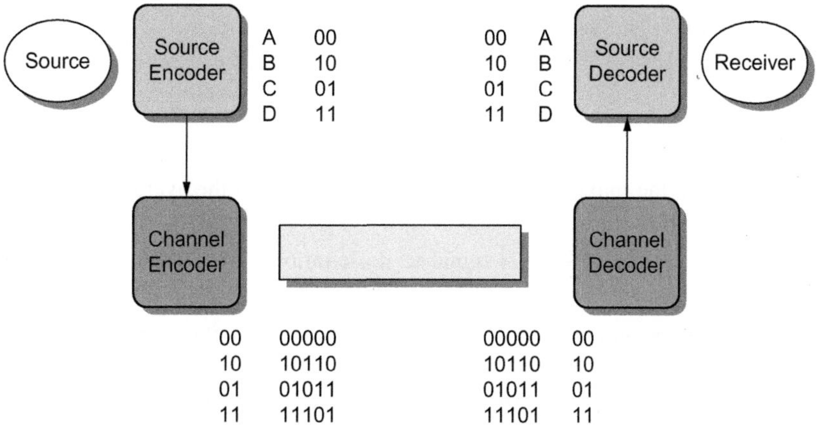

FIGURE 3.8

Source and channel encoding and decoding. The source alphabet consists of four symbols, A,B,C, and D; the source encoder maps each input symbol into a two-bit code, $A \quad 00, B \quad 10, C \quad 01$, and $D \quad 11$. The channel encoder maps a two-bit string into a five-bit string; if a one-bit or a two-bit error occurs, the channel decoder receives a string that does not map to any of the valid five-bit strings and detects an error. The source decoder performs an inverse mapping and delivers a string consisting of the four output alphabet symbols.

transform a noisy channel into a noiseless one; they are built into communication protocols to eliminate the effect of transmission errors.

2. *Data compression.* We wish to reduce the amount of data transmitted through a channel with either no or minimal effect on the ability to recognize the information produced by the source. In this case, encoding and decoding are called *compression* and *decompression*, respectively.
3. *Support confidentiality.* We want to restrict access to information to only those who have the proper authorization. The discipline covering this facet of encoding is called *cryptography*, and the processes of encoding/decoding are called *encryption/decryption*.

Channel Capacity

In this section we consider an abstraction, the *discrete memoryless channel*, defined as the triplet, $(X, Y, p_{Y\ X}(y\ x))$, with:

- The input channel alphabet, X; the selection of an input symbol, $x \quad X$, is modeled by the random variable, X, with the probability density function, $p_X(x)$.
- The output channel alphabet, Y; the selection of an output symbol, $y \quad Y$, is modeled by the random variable, Y, with the probability density function, $p_Y(y)$.
- The conditional probability density function, $p(y\ x), x \quad X, y \quad Y$, the probability of observing output symbol, y, when the input symbol is x. The channel is memoryless if the probability distribution of the output depends only on the input at that time and not on the history.

The *capacity of a discrete memoryless channel*, $(X, Y, p_{Y\ X}(y\ x))$, with $p_X(x)$, and the input distribution is defined as $C \quad \max_{p_X(x)} I(X;Y)$. In real life, various sources of noise distort transmission and lower the channel capacity; the *capacity, C, of a noisy channel* is

$$C \quad B \quad \log \quad 1 \quad \frac{Signal}{Noise} \quad ,$$

where B is the bandwidth, the maximum transmission rate, *Signal* is the average power of the signal, and *Noise* is the average noise power. The signal-to-noise ratio is usually expressed in decibels (dB) given by the formula $10 \quad \log_{10} \frac{Signal}{Noise}$. A signal-to-noise ratio of 10^3 corresponds to 30 dB, and one of 10^6 corresponds to 60 dB.

Interestingly enough, we can reach the same conclusion regarding the capacity of a noisy communication channel based on simple physical arguments and the mathematical tools used in Section 3.2 for expressing Boltzmann's entropy. Consider the following Gedankenexperiment [334]: Alice wants to send, m, bits of information, and she knows that the probability of a bit being in error is p. Alice decides to pack the m bits into a message of $n > m$ bits. When n is large, pn bits in the string received by Bob could be in error and $(1 \quad p)n$ bits may not be affected by errors. Alice is "clairvoyant" and knows ahead of time what bits will be affected by error, and she uses, k, bits to specify how the, pn, errors are distributed among the $n \quad m \quad k$ bits of the message.

To specify k, the number of ways the errors are distributed, we have to count the number of ways, pn bits, in error and, $(1 \quad p)n$ bits, that are not in error can be packed into a message of n bits. This number is $k \quad \log \, {n \atop pn}$; we have encountered this expression in Section 3.2, and we know that when n is large, the number of bits required to specify the number of ways the errors are distributed is

$$k \quad nH(p, 1 \quad p) \quad nH(p) \qquad n[p \log p \quad (1 \quad p) \log(1 \quad p)].$$

Recall that in Section 3.3 we used the notation, $H(p)$, rather than, $H(p, 1 \quad p)$, for binary information.

We conclude that in order to specify how the pn errors are distributed among the n bits of the encoded message, Alice needs k bits. Never mind that this approach is unfeasible because Alice cannot possibly know the position of the random errors when she transmits the message; there is a lesson to be learned from this experiment. When Bob receives the message of n bits and erases it, according to Landauer's principle, the amount of heat/energy he generates is $n \quad (k_B T \ln 2)$. A fraction of this heat/energy, namely, $k \quad (k_B T \ln 2)$, is generated while Bob attempts to extract the k bits of the additional information revealing where the errors are placed in the garbled string of n bits.

This additional energy is due to the fact that the communication channel is noisy and Alice had to add to the m bits she wanted to transmit the information about the errors. In other words, *a noisy channel prevents us from utilizing the full channel capacity; we have to spend an additional amount of energy to erase the additional information used to encode the message.* To locate the errors, we need at least $k \quad nH(p)$ bits. Thus, the maximum capacity of the noisy channel is the fraction of the channel capacity used to transmit useful information $m \quad n \quad k \quad n(1 \quad H(p))$.

Example. *Maximum data rate over a phone line.* A phone line allows transmission of frequencies in the range 500 Hz to 4000 Hz and has a signal-to-noise ratio of 30 dB. The maximum data rate through the phone line is

$$C \quad (4000 - 500) \quad \log(1 \quad 1000) \quad 3500 \quad \log 2^{10} \quad 35 \text{ kbps}.$$

If the signal-to-noise ratio improves to 60 dB, the maximum data rate doubles to about 70 kbps. However, improving the signal-to-noise ratio of a phone line by three orders of magnitude, from 10^3 to 10^6, is extremely difficult.

The *channel coding theorem:* Given a noisy channel with capacity, C, and given $0 < \epsilon < 1$, there is a coding scheme that allows us to transmit information through the channel at a rate arbitrarily close to channel capacity, C, with a probability of error less than ϵ. This result constitutes what mathematicians call a theorem of existence; it only states that a solution for transforming a noisy communication channel into a noiseless one exists, without giving a hint of how to achieve this result.

Simple Models of Classical Memoryless Channels

A *binary channel* is one where the input and the output alphabet consists of binary symbols, $X \quad 0, 1$ and $Y \quad 0, 1$. A *unidirectional binary communication channel* is one where the information propagates in one direction only, from the source to the destination.

A *noiseless binary channel* transmits each symbol in the input alphabet without errors, as shown in Figure 3.9a. The noiseless channel model is suitable in some cases for performance analysis, but it is not useful for reliability analysis when transmission errors have to be accounted for.

The *noisy binary symmetric channel* is a model based on the assumption that an input symbol is affected by error, with the probability $p > 0$; a 1 at the input becomes a 0 at the output, and a 0 at the input becomes a 1 at the output, as shown in Figure 3.9b.

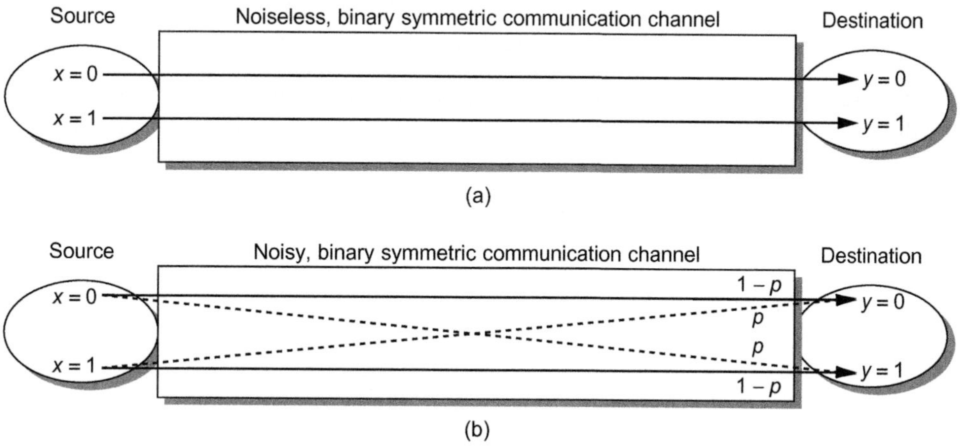

FIGURE 3.9

(a) A noiseless binary symmetric channel maps a 0 at the input into a 0 at the output, and a 1 into a 1.
(b) A noisy symmetric channel maps a 0 into a 1, and a 1 into a 0, with the probability p. An input symbol is mapped into itself, with the probability $1 - p$.

Assume that the two input symbols occur with the probabilities $\text{Prob}(X = 0) = q$ and $\text{Prob}(X = 1) = 1 - q$. In this case,

$$\begin{aligned} H(Y|X) &= \sum_{x \in X} \text{Prob}(X = x) \, H(Y|X = x) \\ &= -q \, [p \log p + (1-p) \log(1-p)] + (1-q) \, [p \log p + (1-p) \log(1-p)] \\ &= -[p \log p + (1-p) \log(1-p)]. \end{aligned}$$

Then the mutual information is

$$I(X;Y) = H(Y) - H(Y|X) = H(Y) + [p \log p + (1-p) \log(1-p)].$$

We can maximize $I(X;Y)$ over q, by making $H(Y) = 1$, and get the channel capacity per symbol in the input alphabet:

$$C_s = 1 + [p \log p + (1-p) \log(1-p)].$$

When $p = 1/2$, the channel capacity is 0 because the output is independent of the input. When $p = 0$ or $p = 1$, the capacity is 1, and we have in fact a noiseless channel.

The next model, the *binary erasure (BER) channel*, captures the following behavior: the transmitter sends a bit and the receiver receives with the probability, $1 - p_e$, the bit unaffected by error, or it receives, with the probability p_e, a message that the bit was "erased," where p_e is the "erasure probability." A binary erasure channel with the input X and output Y is characterized by the following conditional probabilities:

$$\begin{aligned} \text{Prob}(Y = 0|X = 0) &= \text{Prob}(Y = 1|X = 1) = 1 - p_e, \\ \text{Prob}(Y = e|X = 0) &= \text{Prob}(Y = e|X = 1) = p_e, \\ \text{Prob}(Y = 0|X = 1) &= \text{Prob}(Y = 1|X = 0) = 0. \end{aligned}$$

The binary erasure channel is in some sense error-free; the sender does not have to encode the information and the receiver does not have to invest any effort to decode it (Figure 3.10). C_{BEC}, the capacity of a binary erasure channel, is

$$C_{BEC} = 1 - p_e.$$

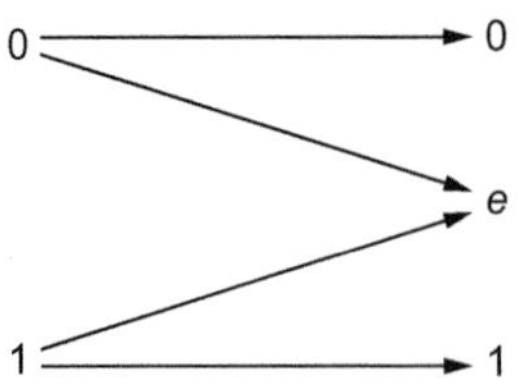

FIGURE 3.10

A binary erasure channel. The transmitter sends a bit and the receiver receives, with the probability $1 - p_e$, the bit unaffected by error, or it receives, with the probability p_e, a message e that the bit was "erased." p_e is the "erasure probability."

Indeed,

$$C_{BEC} = \max I(X;Y) = \max_{p_X(x)} (H(Y) - H(Y|X)) = \max_{p_X(x)} (H(Y) - H(p_e)).$$

If E denotes the erasure event, $Y = e$, then

$$H(Y) = H(Y,E) = H(E) + H(Y|E).$$

If the probability $\text{Prob}(X = 1) = p_1$, then

$$H(Y) = H((1 - p_1)(1 - p_e), p_e, p_1(1 - p_e)) = H(p_e) + (1 - p_e)H(p_1).$$

It follows that

$$C_{BEC} = \max_{p_X(x)} (H(Y) - H(p_e)) = \max_{p_1}((1 - p_e)H(p_1)) = (1 - p_e).$$

The following argument provides an intuitive justification for the expression of the binary erasure channel capacity: assume that the sender receives instantaneous feedback when the bit was erased and then retransmits the bit in error. This would result in an average transmission rate of $(1 - p_e)$ bits/sec. Of course, this ideal scenario is not feasible; thus, $(1 - p_e)$ is an upper bound on the channel rate.

This brings us to the role of the feedback; it can be proved that the rate, $(1 - p_e)$, is the best that can be achieved with or without feedback [106]. *It is surprising that the feedback does not increase the capacity of discrete memoryless channels.*

Another model, the *q-ary erasure channel*, assumes a channel with an alphabet of q symbols; the transmitter sends a symbol and the receiver receives, with the probability $1 - p_e$, the symbol unaffected by error, or it receives, with the probability p_e, a message that the bit was "erased." It is easy to show that the channel capacity in this case is $C_{QEC} = (1 - p_e)\log q$.

If the input symbols have the probabilities, $p_X(a_i) = p_i$, $1 \le i \le q$, then the realizations of Y could be any of the q input symbols that occur with the probabilities, $p_i(1 - p_e)$, respectively, or the erasure, E, that occurs with the probability, p_e. Then,

$$H(Y) = -\sum_{i=1}^{q} p_i(1 - q)\log p_i(1 - q) - p_e \log p_e.$$

The output entropy is maximized when $p_i = 1/q$; then,

$$\max H(Y) = -(1 - p_e)\log(1 - p_e) + (1 - p_e)\log q - p_e \log p_e.$$

As before, $H(Y|X) = H(p_e) = -p_e \log p_e - (1 - p_e)\log(1 - p_e)$; thus, $C_{QEC} = (1 - p_e)\log q$.

Lastly, the *q-ary symmetric channel* model assumes that the probability that a symbol is correctly transmitted is equal to $1 - p$, and when an error occurs, the symbol is replaced by one of the other $q - 1$ symbols with equal probability. For the following analysis, we make the assumption that *errors are introduced by the channel at random; the probability of an error in one coordinate is independent of errors in adjacent coordinates*. The probability that the received n-tuple, r is, decoded as codeword, c, at the output of a q-ary symmetric channel is

$$\text{Prob}(r,c) \quad (1 \quad p)^{n \quad d} \quad \frac{p}{q \quad 1}^{d},$$

where $d(r,c) \quad d$, d is the distance of the code, and n is the length of a codeword. Indeed, $n \quad d$ coordinate positions in, c, are not altered by the channel; the probability of this event is $(1 \quad p)^{n \quad d}$. In each of the remaining, d, coordinate positions, a symbol in c is altered and transformed into the corresponding symbol in r; the probability of this is $[p/(q \quad 1)]^d$.

The next section allows us to quantify the degree of similarity between two probability density functions, a subject important for classical as well as quantum information theory.

3.8 TRACE DISTANCE AND FIDELITY

Information theory is concerned with statistical properties of random variables and often demands an answer to the question of how similar two probability density functions are. For example, we have two information sources, S_1 and S_2, that share the same alphabet, but have different input and output probability density functions, $p_X(x)$ and $p_Y(x)$, and we ask ourselves how similar are the two sources. Another question we may pose is how well a communication channel preserves information over time. Let I be the random variable describing the input of a noisy channel and O be its output. Let I be a copy of I. Now we are interested in the relationships of the joint probability distributions, $p_{I,I}(x_i,x_i)$ and $p_{I,O}(x_i,y_i)$.

We introduce two analytical measures of the similiarity/dissimilarity of two probability density functions, $p_X(x)$ and $p_Y(x)$. Whenever there is no ambiguity, we omit the argument of the density function and write, p_X, instead of, $p_X(x)$.

The *trace distance*, $D(p_X(x),p_Y(x))$ (also called Kolmogorov, or L1 distance), of two probability density functions, $p_X(x)$ and $p_Y(x)$, is defined as

$$D(p_X(x),p_Y(x)) \quad \frac{1}{2} \quad _{x} \quad p_X(x) \quad p_Y(x) .$$

The *fidelity*, $F(p_X(x),p_Y(x))$, of two probability density functions, $p_X(x)$ and $p_Y(x)$, is defined as

$$F(p_X(x),p_Y(x)) \quad _{x} \quad \overline{p_X(x)p_Y(x)}.$$

It is easy to see that the fidelity is the inner product of unitary vectors with components equal to $\overline{p_X(x)}$ and $\overline{p_Y(x)}$. These vectors connect the center with points on a sphere of radius one; indeed, $_x \quad \overline{p_X(x)}^2 \quad 1$ and $_x \quad \overline{p_Y(x)}^2 \quad 1$. A geometric interpretation of fidelity is, thus, the cosine of θ, the angle between the two vectors,

$$F(p_X(x),p_Y(x)) \quad \cos\theta.$$

Example. *Two noisy binary symmetric channels* with the input and output alphabets, $0,1$, have probabilities of error, p and q, respectively. This means that when we transmit one of the input symbols the probability of recovering it correctly is 1 p for the first channel and 1 q for the second. The two measures of similarity between the two distributions are

$$D(p_X(x), p_Y(x)) \quad \frac{1}{2}[\,p \quad q \qquad (1 \quad p) \quad (1 \quad q)\,] \quad p \quad q$$

and

$$F(p_X(x), p_Y(x)) \qquad \overline{pq} \qquad \overline{(1 \quad p)(1 \quad q)} \qquad \overline{pq} \qquad \overline{pq \quad 1 \quad (p \quad q)}.$$

If p 0.4 and q 0.9, then

$$D(p_X(x), p_Y(x)) \quad p \quad q \quad 0.5, \quad \text{and} \quad F(p_X(x), p_Y(x)) \quad \overline{pq} \quad \overline{pq \quad 1 \quad (p \quad q)} \quad 0.6245.$$

The trace distance is a metric, while the fidelity is not. Indeed, the trace distance satisfies the properties of a metric:

$D(p_X, p_Y)$	0	non-negativity
$D(p_X, p_Y)$	0 if and only if $p_X \quad p_Y$	identity of indiscernibles
$D(p_X, p_Y)$	$D(p_Y, p_X)$	symmetry
$D(p_X, p_Z)$	$D(p_X, p_Y) \quad D(p_Y, p_Z)$	triangle inequality

The fidelity fails to satisfy one of the required properties:

$$F(p_X(x), p_X(x)) \qquad \underset{x}{} \overline{p_X(x) p_X(x)} \qquad \underset{x}{} p_X(x) \quad 1 \quad 0.$$

Next, we turn our attention to quantum information and discuss the characterization of a source of quantum information using the von Neumann entropy.

3.9 VON NEUMANN ENTROPY

Pure states of quantum systems correspond to classical states we are familiar with; they are the really physical states. In standard quantum mechanics, the pure states of a purely quantum physical system with separable complex Hilbert space, $\mathcal{H}$, are represented by the rays of this Hilbert space or by rank one operators. Normalized linear superpositions of pure states are also pure states. When the preparation process cannot be rigourously controlled, the resulting mixed state can be considered a superposition of pure states with a given probability distribution and is represented by a rank one statistical operator, the density matrix. Mixed states do not have a classical correspondent.

John von Neumann realized that quantum mechanics requires a definition of entropy [440] covering pure as well as mixed states; he associated the entropy to a statistical operator, the density matrix,[4] that he used to characterize quantum systems in a mixed state, as well as composite systems. He presented the mathematical formalism for quantum mechanics in a book published in 1932 [441].

[4] Some attribute the introduction of the density matrix to John von Neumann and believe that Lev Landau and Felix Bloch introduced it independently in 1927.

A system is in a *pure state* when its density matrix contains only one term; all entries of the diagonalized density matrix of a pure state are zero with the exception of a single term equal to one on the main diagonal. Moreover, if the state is pure, then the trace of the square of the density matrix is equal to one. In this case, there is no lack of knowledge on the preparation of the system; the preparation generates the desired state with certainty.

The *von Neumann entropy* is a function of the density matrix, ρ, of the quantum system:

$$S(\rho) = -\text{tr}[\rho \log \rho].$$

In this expression, $(\log \rho)$ is the Hermitian operator that has precisely the same eigenvectors as ρ; we simply take the logarithm base two of the corresponding eigenvalues. If we consider $0 \log 0 = 0$, then the expression, $\rho \log \rho$, may be defined for all positive semidefinite, ρ, though technically $(\rho \log \rho)$ is only defined for positive definite ρ.

The expression of von Neumann entropy requires the evaluation of a function of a matrix.[5] The function can be evaluated as an ordinary function of the eigenvalues. The eigenvalues, λ_i, are invariant under a change of basis; thus, the function itself is invariant. The eigenvalues of the density operator are probabilities, and we therefore use the notation, p_i, rather than, λ_i, for the eigenvalues of the density matrix; $\text{tr}[\rho] = 1$ reflects the fact that $\sum_i p_i = 1$.

If the states, $|e_i\rangle \in \mathcal{H}_{2^n}$, form an orthonormal basis that diagonalizes ρ, the spectral decomposition of ρ is

$$\rho = \sum_{i=1}^{2^n} p_i |e_i\rangle\langle e_i|.$$

Then [325]

$$\log(\rho) = \sum_{i=1}^{2^n} (\log p_i) |e_i\rangle\langle e_i|;$$

thus,

$$S(\rho) = -\text{tr}\left[\left(\sum_{i=1}^{2^n} p_i |e_i\rangle\langle e_i|\right)\left(\sum_{j=1}^{2^n} \log p_j |e_j\rangle\langle e_j|\right)\right] = -\sum_{i=1}^{2^n} p_i \log p_i = H(p_1, p_2, \ldots, p_n).$$

In this case, the von Neumann entropy is equal to $H(p_1, p_2, \ldots, p_n)$, the expression of the Shannon entropy of the ensemble $\{\rho_i, p_i\}$ we have encountered earlier. If we prepare the system in a *maximally mixed state*, a state when all N pure states are equally likely, then the entropy is equal to $\log N$ in agreement with Boltzmann and Shannon entropy.

If $|\varphi\rangle$ is a pure state, then its density matrix, ρ, has only one eigenvalue equal to 1, its trace is equal to 1 (then $\log \rho = 0$), and its von Neumann entropy is zero.

Let us now consider the ensemble of pure states, $|e_i\rangle$, $1 \leq i \leq N$, and prepare a mixed state where each pure state, $|e_i\rangle$, has the probability, p_i. To do so, we must first create a list and use $H(p_1, p_2, \ldots, p_N)$

---

[5]The logarithm of a matrix is another matrix, such that if B is the logarithm of a given matrix, A, then $e^B = A$. Not all matrices have a logarithm, but those that do may have more than one logarithm. A complex matrix has a logarithm if and only if it is invertible. If the matrix has only non-negative real eigenvalues, then it has a unique logarithm. The eigenvalues of this unique logarithm lie all in the domain, $\{z \in \mathbb{C} \mid -\pi < \Im(z) < \pi\}$. For any invertible matrix A, $A^{-1} = e^{-\ln A}$.

bits/letter to specify the probability of each one of the N states. After the creation of the mixed state, we have to erase this information, and, according to the Landauer principle, at temperature, T, we will generate $k_B TH(p_1, p_2, \ldots, p_N)$ units of heat per erased letter. This explains why *Shannon entropy is larger or equal to von Neumann entropy.*

Recall that we cannot distinguish among the different preparations leading to the same density matrix ρ. *If ρ is the density matrix of a mixed state, then $S(\rho)$, the von Neumann entropy, is the smallest amount of information that has to be invested to create the mixed state characterized by ρ, and also the minimum amount of classical information that we can access from the system in that state.*

We have a bipartite system consisting of two quantum registers, A and B, owned by Alice and Bob, respectively, and let φ be a pure state in the Hilbert space shared by Alice and Bob, $\varphi \quad \mathcal{H}_A \quad \mathcal{H}_B$; it is easy to see that the reduced states have the same entropy:

$$S(\mathrm{tr}_B(\ \varphi \quad \varphi\ )) \quad S(\mathrm{tr}_A(\ \varphi \quad \varphi\ )).$$

When von Neumann associated a statistical operator, ρ, to entropy [440], he first assumed that $S(\rho)$ is a continuous function of ρ. Then he established a mix-in property for orthogonal pure states:

$$S \quad \underset{i}{} p_i\ \varphi_i \quad \varphi_i \qquad \underset{i}{} p_i S(\ \varphi_i \quad \varphi_i\ ) \quad k_B \underset{i}{} p_i \log p_i.$$

Finally, he showed that $S(\ \varphi \quad \varphi\ )$ is independent of the state, φ, thus, if we normalize and set the additive constant equal to zero, we have

$$S \quad \underset{i}{} p_i\ \varphi_i \quad \varphi_i \qquad k_B \underset{i}{} p_i \log p_i.$$

The Mix-in Property

von Neumann considered a quantum system in a mixed state,

$$\psi \quad p\ \varphi_1 \quad (1 \quad p)\ \varphi_2\ .$$

He imagined a Gedankenexperiment [329] and constructed a thermodynamic model of the quantum system in a mixed state, D. In this model, N molecules of gas are mixed together, pN molecules of gas are in the state, φ_1, described by the statistical operator, D_1, and $(1 \quad p)N$ molecules are in the state, φ_2, described by D_2. The two states, D_1 and D_2, are thermodynamically disjoint in the following sense: There is a wall completely permeable for the first gas, the one in the state φ_1, but totally impermeable to the molecules of the second gas, the one in the state φ_2. This condition translates into orthogonality of the eigenvectors corresponding to the nonzero eigenvalues of the density matrices of the quantum systems modeled by the gas.

The Gedankenexperiment imagined by von Neumann is conducted as follows:

1. Initially, the gas in the mixed state, D, is in a cubic recipient, K, of volume, V. The entropy of the gas is $S_{gas}^{(1,2)}$, and it is proportional with the number of molecules, N; its density is $\frac{N}{V}$.
2. We attach a new cubic recipient, K , of an identical volume, V, to the left of K.
3. We replace the wall separating, K and K , by two walls, a and b, and add a third wall, c, to the right of b, as in Figure 3.11 (top). Wall a is impermeable for both types of molecules, wall b is permeable for molecules of type φ_1 and impermeable for molecules of type φ_2, and wall c is impermeable for molecules of type φ_1 and permeable for molecules of type φ_2.

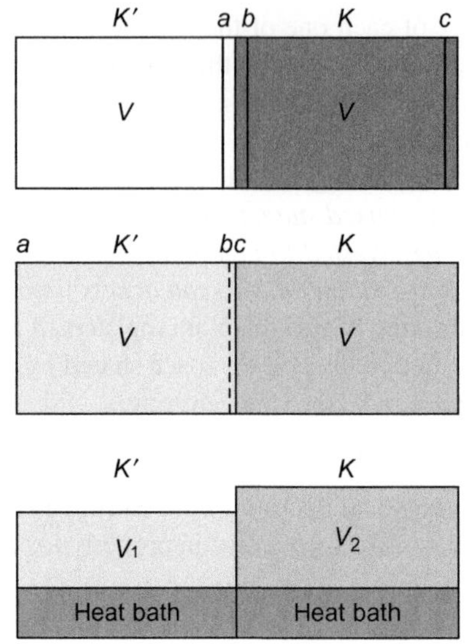

FIGURE 3.11

A Gedankenexperiment to show the mix-in property of von Neumann entropy.

4. We push walls a and c to the left, maintaining the distance between them. As a result, the φ_1 molecules are pressed through wall b into K , and φ_2 molecules are left in K, as shown in Figure 3.11 (center).
5. We replace walls b and c with impermeable walls and remove wall a. The φ_1 and φ_2 molecules have been separated without doing work, changing temperature, or producing heat. Then we compress both the gas in K from a volume, V, to a volume, $V_1 \quad pV$, and the gas in K to a volume, $V_2 \quad (1-p)V$. The densities of the two gasses are, pN/V_1 and $(1-p)N/V_2$, respectively; they are equal to the density of gas, D. The compression is isothermal; the two volumes absorb heat from two heat reservoirs, as shown in Figure 3.11 (bottom), and maintain their temperature, T. The first gas absorbs a quantity of heat equal to $k_B T(pN)\log p$, and the second gas absorbs a quantity of heat equal to $k_B T(1-p)N\log(1-p)$. The entropy of the gasses changes as follows: The entropy of the first gas decreases with the amount $S^{(1)}_{gas}(pN) \quad k_B(pN)\log p$, while the entropy of the second gas decreases with the amount $S^{(2)}_{gas}((1-p)N) \quad k_B(1-p)N\log(1-p)$.
6. We mix the two gasses of equal density and find the gas, D with N molecules, in a volume, V; we note that

$$S^{(1)}_{gas}(pN) \quad S^{(2)}_{gas}((1-p)N) \quad S^{(1,2)}_{gas}(N) \quad k_B(pN)\log p \quad k_B(1-p)N\log(1-p).$$

We assume that the entropy of a gas is proportional to the number of molecules in the gas, and if we divide the previous expression by N, we obtain

$$pS^{(1)}_{gas} \quad (1-p)S^{(2)}_{gas} \quad S^{(1,2)}_{gas} \quad k_B\,[p\log p \quad (1-p)\log(1-p)]$$

or

$$S_{gas}^{(1,2)} \quad pS_{gas}^{(1)} \quad (1 \quad p)S_{gas}^{(2)} \quad k_B[(p)\log p \quad (1 \quad p)\log(1 \quad p)].$$

The result can be extended from a two-component mixture to an infinite mixture:

$$S \quad \sum_i p_i \, \varphi_i \quad \varphi_i \qquad \sum_i p_i S(\varphi_i \quad \varphi_i) \quad k_B \quad \sum_i p_i \log p_i.$$

The proof that, $S(\varphi \quad \varphi)$, is independent of the state, φ , if the state obeys the Schrödinger equation is slightly more complex. Any two states, φ_i and φ_j , in the evolution of the system are related by a time-independent energy operator, H:

$$\varphi_j \quad e^{iHt} \quad \varphi_i \ .$$

The operator, e^{iHt}, is unitary, and the eigenvalues of the density matrix at time, t, $\rho(t)$, are the same as the eigenvalues of the density matrix at time, 0, $\rho(0)$. The density matrix at time t is obtained as

$$\rho(t) \quad e^{\ \ iHt} \rho e^{iHt}.$$

We observe that, $S[\rho(t)] \quad S[\rho]$, for a closed system with a time-independent Hamiltonian because the expression of entropy involves only the eigenvalues of the density matrix. The reality is closer to the second law of thermodynamics, which states that the entropy of a closed system never decreases; it can remain constant or increase.

John von Neumann proved that

$$S(\varphi_i \quad \varphi_i) \quad S \quad \varphi_j \quad \varphi_j$$

by performing a large number of measurements of the system in the state φ_i , such that the state after all the measurements becomes an ensemble that is arbitrarily close to $\varphi_j \quad \varphi_j$. Recall from Section 2.7 that a von Neumann projective measurement of the observable, $\mathcal{O} \qquad {}_i p_i \, e_i \quad e_i$, of a system in the state, φ , projects this state on the basis vectors, e_i , and produces the pure states, $e_i \quad e_i$, with the weights, $\varphi \ e_i$, respectively, when the eigenvalues, p_i, of the operator, $\mathcal{O}$, are distinct.

We now discuss how to erase classical information from quantum states. As we mentioned earlier, the von Neumann entropy and, thus, the information content of a pure state is equal to zero. Therefore, *to erase information from a mixed state, we have to return it to a pure state.*

Let $\rho \qquad {}_i p_i \, e_i \quad e_i$ be the density matrix of a mixed state, with e_i , the energy eigenstates of the system. We perform measurements in the energy eigenbasis; after each measurement, the system ends up in a pure state, e_i , and the corresponding eigenvalue, p_i, represents a classical record of the outcome of our measurement. To complete the erasure of the quantum information, we have to erase the classical information as well; in this process, we generate $k_B T \ln 2$ units of heat per bit. In our example, the amount of heat is $k_B T \ln 2 S(\rho) < k_B T \ln 2 H(p)$. *When the quantum measurements are conducted in the basis constructed from the eigenstates of the density matrix, ρ, the von Neumann entropy is equal to Shannon entropy, and we have an optimal erasure procedure.*

We conclude this section with the statement of two basic properties of von Neumann entropy. The proofs or references to the proofs can be found in the seminal paper of Wehrl [445].

Additivity. Let $\rho_1, \rho_2, \ldots, \rho_n$ be the density matrices of n quantum system. Then,

$$S(\rho_1 \quad \rho_2 \qquad\qquad \rho_n) \quad S(\rho_1) \quad S(\rho_2) \qquad\qquad S(\rho_n).$$

Concavity. Let $\rho_1, \rho_2, \ldots, \rho_n$ be the density matrices of n quantum systems and $p_1, p_2, \ldots, p_n$ non-negative integers, $p_i \quad 0$, such that $\;_i p_i \quad 1$. Then,

$$S \qquad \underset{i}{\;} p_i \rho_i \qquad\qquad \underset{i}{\;} p_i S(\rho_i).$$

The concavity of the entropy is the result of the concavity of the logarithm function. The concavity of von Neumann entropy indicates that when ensembles are fitted together into a composite system in a mixed state, one loses information about which ensemble a particular state arises from; thus, the entropy increases.

In summary, von Neumann used thermodynamic arguments to define the entropy of a quantum system with the density matrix, ρ, as $S(\rho) \quad \mathrm{tr}(\rho \log \rho)$; the expression, $\rho \log \rho$, is well defined as the density matrix is positive semidefinite. It is easy to see that von Neumann entropy is the Shannon entropy of the eigenvalues, p_i, of a diagonal density matrix, ρ. Indeed, if ρ is a diagonal matrix with the eigenvalues p_i, then $\mathrm{tr}(\rho) \quad \;_i p_i \quad 1$, and $\rho \log \rho$ is also diagonal with the eigenvalues $p_i \log p_i$.

3.10 JOINT, CONDITIONAL, AND RELATIVE VON NEUMANN ENTROPY

Joint, conditional, and relative von Neumann entropy are concepts similar to those defined by classical information theory. Consider a pair of quantum systems, $\mathcal{A}$ and $\mathcal{B}$, and the composite system, $(\mathcal{A}, \mathcal{B})$; let $\rho^{\mathcal{A}}$, $\rho^{\mathcal{B}}$, and $\rho^{(\mathcal{A},\mathcal{B})}$ be the density operators of $\mathcal{A}$, $\mathcal{B}$, and the composite system $(\mathcal{A}, \mathcal{B})$, respectively. The density operators, $\rho^{\mathcal{A}}$ and $\rho^{\mathcal{B}}$, are given by the partial traces of the joint density matrix:

$$\rho^{\mathcal{A}} \quad \mathrm{tr}_{\mathcal{B}}\, \rho^{(\mathcal{A},\mathcal{B})} \quad \text{and} \quad \rho^{\mathcal{B}} \quad \mathrm{tr}_{\mathcal{A}}\, \rho^{(\mathcal{A},\mathcal{B})}.$$

The *joint von Neumann entropy* of the pair of quantum systems $\mathcal{A}$ and $\mathcal{B}$ is defined as

$$S(\mathcal{A}, \mathcal{B}) \quad S\, \rho^{(\mathcal{A},\mathcal{B})}.$$

The *conditional von Neumann entropy* of system, $\mathcal{A}$, conditioned by system, $\mathcal{B}$, is defined as

$$S(\mathcal{A}\;\mathcal{B}) \quad S(\mathcal{A}, \mathcal{B}) \quad S(\mathcal{B}) \quad S\, \rho^{(\mathcal{A},\mathcal{B})} \qquad S\, \rho^{\mathcal{B}}.$$

The conditional von Neumann entropy, $S(\mathcal{A}\;\mathcal{B})$, is concave [445].

The *mutual information* of the pair of systems, ($\mathcal{A}$ and $\mathcal{B}$), is defined as

$$I(\mathcal{A}, \mathcal{B}) \quad S(\mathcal{A}) \quad S(\mathcal{B}) \quad S(\mathcal{A}, \mathcal{B}) \quad S\, \rho^{\mathcal{A}} \qquad S\, \rho^{\mathcal{B}} \qquad S\, \rho^{(\mathcal{A},\mathcal{B})}.$$

The *relative entropy*, $S(\rho_1 \; \rho_2)$, of the two states described by the density matrices, ρ_1 and ρ_2, is defined as

$$S(\rho_1 \; \rho_2) \quad \mathrm{tr}(\rho_2(\log \rho_2 \quad \log \rho_1)).$$

The two density matrices, ρ_1 and ρ_2, have the spectral decompositions

$$\rho_1 \quad \underset{j}{} q_j\ u_j\ \ u_j \quad \text{and} \quad \rho_2 \quad \underset{i}{} p_i\ e_i\ \ e_i\ ,$$

where e_i and u_j are two orthonormal bases for the same physical system.

It is easy to see that the quantum relative entropy is non-negative, $S(\rho_1\ \ \rho_2)\quad 0$, or more precisely,

$$S(\rho_1\ \ \rho_2) \quad \begin{cases} > 0 & \text{if} \quad \rho_1 \quad \rho_2, \\ 0 & \text{if} \quad \rho_1 \quad \rho_2. \end{cases}$$

Then, the relative entropy can be evaluated in the basis, where ρ_2 is diagonal:

$$\text{tr}\,(\rho_2 \log \rho_2) \quad \text{tr}\,(\rho_2 \log \rho_1) \qquad {}_i\ e_i\ \rho_2 \log(\rho_2)\ e_i \qquad {}_i\ u_i\ \rho_2 \log(\rho_1)\ u_i$$

$$\qquad {}_i p_i\ \log p_i \qquad {}_j\ \ u_j\ e_i\ \ ^2 \log q_j$$

$$\qquad \frac{1}{\ln 2}\quad {}_i\quad {}_j p_i\ \ u_j\ e_i\ \ ^2 \ln \frac{q_j}{p_i}$$

$$\qquad \frac{1}{\ln 2}\quad {}_i\quad {}_j p_i\ \ u_j\ e_i\ \ ^2\quad \frac{q_j}{p_i}\quad 1 \qquad 0.$$

This inequality is strict unless $q_j\quad p_i,\quad i,j$, satisfying $u_j\ e_i\ \ ^2 > 0$. Under this assumption, the equality occurs when

$$\underset{i}{}\ \underset{j}{}\ \ u_j\ e_i\ \ ^2 (p_i\quad q_j)^2 \quad 0.$$

This implies that $\rho_2\quad \rho_1\qquad 0$ or $\rho_2\quad \rho_1$.

The *subadditivity of von Neumann entropy* requires that

$$S\ \rho^{(\mathcal{A},\mathcal{B})} \qquad S\ \rho^{\mathcal{A}} \qquad S\ \rho^{\mathcal{B}}\ .$$

This inequality means that a pair of correlated systems contains more information than the two systems considered separately.

Given three quantum systems, $\mathcal{A}, \mathcal{B}$, and $\mathcal{C}$, the *strong subadditivity of von Neumann entropy* [445] requires that

$$S\ \rho^{(\mathcal{A},\mathcal{B},\mathcal{C})} \qquad S\ \rho^{\mathcal{A}} \qquad S\ \rho^{(\mathcal{A},\mathcal{B})} \qquad S\ \rho^{(\mathcal{A},\mathcal{C})}\ .$$

A consequence of the strong subadditivity is that when $\varphi_i^{\mathcal{A}}$ is a pure state of the system, $\mathcal{A}$, and $\rho^{(\mathcal{B},\mathcal{C})}$ describes an arbitrary joint state of the systems, $\mathcal{B}$ and $\mathcal{C}$,

$$\rho^{(\mathcal{A},\mathcal{B},\mathcal{C})} \qquad \varphi_i^{\mathcal{A}}\quad \varphi_i^{\mathcal{A}} \qquad \rho^{(\mathcal{B},\mathcal{C})}.$$

This property will be used in Section 3.12 to establish an upper bound on the mutual information between a source of classical information and the recipient of it, when information is transmitted through a quantum channel.

3.11 TRACE DISTANCE AND FIDELITY OF MIXED QUANTUM STATES

The outcome of the measurement of an observable, A, is a random variable, the eigenvalue, x_i, of the projector, **M**, applied to the system in the state, ψ . If we have, N, copies of the system and measure the same observable, then the outcomes of the measurement satisfy the equation, $\mathbf{M}\ \psi \quad x_i\ \psi$ for $i \quad 1,\ldots,N$. The expected value of the outcome of our measurements is

$$\mathbf{M} \qquad \psi\ \mathbf{M}\ \psi\ .$$

Let $p_X(x)$ be the discrete probability density function of x_i, the outcomes of the measurements. Let us assume that the quantum state, ψ , represents the state of a control register; we measure the same observable before and after the transformation performed by a quantum circuit, and we are interested in how similar are the two probability density functions, $p_X(x)$ and $p_Y(x)$, respectively, of the outcomes of the measurements. Obviously, $\ _i p_X(x_i) \quad 1$ and $\ _i p_Y(x_i) \quad 1$. We can use the two measures, the fidelity and the trace distance, to assess the effect of the transformation. The two measures introduced in this section for probability density functions can be extended to linear operators in a Hilbert space, $\mathcal{H}_2{}^n$.

Trace Distance

The concept of trace distance discussed in Section 3.8 can be extended to quantum states. Consider two quantum systems, $\mathcal{A}$ and $\mathcal{B}$, and let their states be described by density matrices, $\rho^{\mathcal{A}}$ and $\rho^{\mathcal{B}}$, respectively.

The *trace distance* between quantum systems $\mathcal{A}$ and $\mathcal{B}$ in states with the density matrices $\rho^{\mathcal{A}}$ and $\rho^{\mathcal{B}}$ is defined as

$$D\ \rho^{\mathcal{A}},\rho^{\mathcal{B}} \qquad \frac{1}{2}\mathrm{tr}\ \ \rho^{\mathcal{A}} \quad \rho^{\mathcal{B}}\ \ .$$

If $\rho^{\mathcal{A}}$ and $\rho^{\mathcal{B}}$ commute and if $\lambda_i^{\mathcal{A}}$ and $\lambda_i^{\mathcal{B}}$, $1 \quad i \quad 2^n$ are the eigenvalues of ρ^{A} and ρ^{B}, then

$$D\ \rho^{\mathcal{A}},\rho^{\mathcal{B}} \qquad \frac{1}{2}\ \ _{i\ 1}^{2^n}\ \ \lambda_i^{\mathcal{A}} \quad \lambda_i^{\mathcal{B}}\ .$$

Example. *The trace distance of two qubits is equal to the distance on the Bloch sphere between the two states.*

Consider two qubits and the corresponding vectors on a Bloch sphere, $r^{\mathcal{A}} \qquad r_x^{\mathcal{A}}\ r_y^{\mathcal{A}}\ r_z^{\mathcal{A}}$ and $r^{\mathcal{B}} \qquad r_x^{\mathcal{B}}\ r_y^{\mathcal{B}}\ r_z^{\mathcal{B}}$. If $\sigma \quad (\sigma_x\ \sigma_y\ \sigma_z)$, the two density matrices are

$$\rho^{\mathcal{A}} \quad \frac{1}{2}\ I \quad r^{\mathcal{A}} \quad \sigma \quad \text{and} \quad \rho^{\mathcal{B}} \quad \frac{1}{2}\ I \quad r^{\mathcal{B}} \quad \sigma\ .$$

Thus,

$$D\ \rho^{\mathcal{A}},\rho^{\mathcal{B}} \qquad \frac{1}{2}\mathrm{tr}\ \ \rho^{\mathcal{A}} \quad \rho^{\mathcal{B}} \qquad \frac{1}{4}\mathrm{tr}\ \ r^{\mathcal{A}} \quad r^{\mathcal{B}} \qquad \sigma\ .$$

It is easy to see that the eigenvalues of $r^{\mathcal{A}} \quad r^{\mathcal{A}} \quad \sigma$ are $\lambda_{1,2} \quad r^{\mathcal{A}} \quad r^{\mathcal{B}}$, the distance on the Bloch sphere between the two states. Let

$$R \quad r^{\mathcal{A}} \quad r^{\mathcal{B}} \quad \sigma \qquad \begin{matrix} r_z^{\mathcal{A}} \quad r_z^{\mathcal{B}} & r_x^{\mathcal{A}} \quad r_x^{\mathcal{B}} \quad i\, r_y^{\mathcal{A}} \quad r_y^{\mathcal{B}} \\ r_x^{\mathcal{A}} \quad r_x^{\mathcal{B}} \quad i\, r_y^{\mathcal{A}} \quad r_y^{\mathcal{B}} & r_z^{\mathcal{A}} \quad r_z^{\mathcal{B}} \end{matrix} .$$

The characteristic equation, $\det(R \quad \lambda I) \quad 0$, allows us to determine the eigenvalues

$$\lambda^2 \quad r_x^{\mathcal{A}} \quad r_x^{\mathcal{B}}\,^2 \quad r_y^{\mathcal{A}} \quad r_y^{\mathcal{B}}\,^2 \quad r_z^{\mathcal{A}} \quad r_z^{\mathcal{B}}\,^2.$$

Thus,

$$\lambda_{1,2} \qquad \overline{r_x^{\mathcal{A}} \quad r_x^{\mathcal{B}}\,^2 \quad r_y^{\mathcal{A}} \quad r_y^{\mathcal{B}}\,^2 \quad r_z^{\mathcal{A}} \quad r_z^{\mathcal{B}}\,^2} \qquad r^{\mathcal{A}} \quad r^{\mathcal{B}}.$$

Then

$$D\ \rho^{\mathcal{A}},\rho^{\mathcal{B}} \quad \frac{1}{4}(\lambda_1 \quad \lambda_2) \quad \frac{1}{2} \ \overline{r_x^{\mathcal{A}} \quad r_x^{\mathcal{B}}\,^2 \quad r_y^{\mathcal{A}} \quad r_y^{\mathcal{B}}\,^2 \quad r_z^{\mathcal{A}} \quad r_z^{\mathcal{B}}\,^2}.$$

The trace distance between two quantum systems $\mathcal{A}$ and $\mathcal{B}$, with the density matrices $\rho^{\mathcal{A}}$ and $\rho^{\mathcal{B}}$, respectively, is invariant to a linear transformation, $\mathbf{U}$, of the states,

$$D\ \rho^{\mathcal{A}},\rho^{\mathcal{B}} \quad D(\mathbf{U}\rho^{\mathcal{A}}\mathbf{U}^{\dagger},\mathbf{U}\rho^{\mathcal{B}}\mathbf{U}^{\dagger}).$$

The *trace distance of two operators*, **A** and **B**, is defined as

$$D(\mathbf{A},\mathbf{B}) \quad \frac{1}{2}\mathrm{tr}(\ A \quad B\ ) \quad \frac{1}{2}\mathrm{tr}\ \overline{(A \quad B)^{\dagger}(A \quad B)},$$

with A and B, the matrices corresponding to **A** and **B**, respectively.

Proposition. *If two operators,* **A** *and* **B**, *commute, then their trace distance is equal to the trace distance between the eigenvalue distribution of the two operators,* $D(\mathbf{A},\mathbf{B}) \quad D(a,b)$, *with* $a \quad a_1a_2\ldots a_{2^n}$ *and* $b \quad b_1b_2\ldots b_{2^n}$, a_i *the eigenvalues of* A *and* b_i *the eigenvalues of* B.

Proof. If $[\mathbf{A},\mathbf{B}] \quad 0$, then the corresponding matrices can be diagonalized in the same orthonormal basis formed by unitary vectors i, $1 \quad i \quad 2^n$:

$$A \qquad \underset{i}{} \ a_i\ i\ i \quad \text{and} \quad B \qquad \underset{i}{} \ b_i\ i\ i.$$

It follows that

$$D(\mathbf{A},\mathbf{B}) \quad \frac{1}{2}\mathrm{tr} \qquad \underset{i}{}\ (a_i \quad b_i)\ i\ i \qquad \frac{1}{2} \ \underset{i}{} \ a_i \quad b_i \qquad D(a,b).$$

■

Recall that a *positive operator*, **A**, is any Hermitian operator with non-negative eigenvalues; $\overline{\mathbf{A}}$ denotes its unique positive root. Then the *fidelity of positive operators* **A** and **B** is defined as

$$F(\mathbf{A},\mathbf{B}) \quad \mathrm{tr} \quad \overline{\mathbf{A}^{1/2}\mathbf{B}\mathbf{A}^{1/2}}\ .$$

Proposition. *If two operators,* **A** *and* **B**, *commute, then the fidelity,* $F(\mathbf{A},\mathbf{B})$, *is equal to the fidelity of the eigenvalue distribution,* $F(\mathbf{A},\mathbf{B}) = F(a,b)$, *with* $a = \{a_1 a_2 \ldots a_{2^n}\}$ *and* $b = \{b_1 b_2 \ldots b_{2^n}\}$, a_i *the eigenvalues of* **A** *and* b_i *the eigenvalues of* **B**.

Proof. Following the same arguments as before, we write

$$A = \sum_i a_i |i\rangle\langle i| \quad \text{and} \quad B = \sum_i b_i |i\rangle\langle i|.$$

It follows that

$$F(\mathbf{A},\mathbf{B}) = \mathrm{tr}\sqrt{\sum_i a_i b_i (|i\rangle\langle i|)^2} = \mathrm{tr}\sum_i \sqrt{a_i b_i}\,|i\rangle\langle i| = \sum_i \sqrt{a_i b_i} = F(a,b). \qquad \blacksquare$$

Next, we introduce another very important concept in quantum information theory: the fidelity of quantum states.

Fidelity of Quantum States

To quantify the properties of a quantum channel or to determine whether a quantum circuit performs its prescribed function, we correlate the information at the input with the information at the output of the quantum device.

Fidelity allows us to answer the question, "How close are two quantum states to each other?" The quantum fidelity is a measure of the distinguishability of two quantum states, $\mathcal{A}$ and $\mathcal{B}$, described by their density matrices, $\rho^{\mathcal{A}}$ and $\rho^{\mathcal{B}}$, respectively [220]. The fidelity is a real number between 0 and 1; it takes the value 0 if and only if the two states are orthogonal, and it takes the value 1 if and only if $\rho^{\mathcal{A}} = \rho^{\mathcal{B}}$.

The mathematical apparatus required for the discussion of fidelity was introduced in Section 1.3 where we defined concepts such as positive operators, the modulus, and the square root of an operator. The definition of the fidelity of two quantum states follows from the definition of the fidelity of general operators in Section 3.8; it uses the density operators of the two states rather than general operators. Recall that the density, ρ, is a Hermitian operator with non-negative eigenvalues; $\sqrt{\rho}$ denotes the unique positive square root of ρ.

Jozsa [220] defines the *fidelity of two quantum states*, with the density matrices $\rho^{\mathcal{A}}$ and $\rho^{\mathcal{B}}$, as

$$F\left(\rho^{\mathcal{A}},\rho^{\mathcal{B}}\right) = \left[\mathrm{tr}\sqrt{\left(\rho^{\mathcal{A}}\right)^{1/2}\rho^{\mathcal{B}}\left(\rho^{\mathcal{A}}\right)^{1/2}}\right]^2.$$

This quantity can be interpreted as a generalization of the transition probability for pure states. A simpler form is more convenient for some applications [307]:

$$F\left(\rho^{\mathcal{A}},\rho^{\mathcal{B}}\right) = \mathrm{tr}\sqrt{\left(\rho^{\mathcal{A}}\right)^{1/2}\rho^{\mathcal{B}}\left(\rho^{\mathcal{A}}\right)^{1/2}}.$$

The fidelity of the two pure states, $\mathcal{A}$ and $\mathcal{B}$, is *the norm of their inner product*; indeed, if

$$\rho^{\mathcal{A}} \quad \varphi^{\mathcal{A}} \ \varphi^{\mathcal{A}} \quad \text{and} \quad \rho^{\mathcal{B}} \quad \varphi^{\mathcal{B}} \ \varphi^{\mathcal{B}} ,$$

then

$$F\ \rho^{\mathcal{A}},\rho^{\mathcal{B}} \quad \left(\text{tr}\sqrt{\quad \varphi^{\mathcal{A}}\ \varphi^{\mathcal{A}}\ ^{1/2}\ \varphi^{\mathcal{B}}\ \varphi^{\mathcal{B}}\ ^{1/2}\quad \varphi^{\mathcal{B}}\ \varphi^{\mathcal{B}}\ ^{1/2}\ \varphi^{\mathcal{A}}\ \varphi^{\mathcal{A}}\ ^{1/2}}\right)^2.$$

It follows that $F\ \rho^{\mathcal{A}},\rho^{\mathcal{B}} \quad \varphi^{\mathcal{A}}\ \varphi^{\mathcal{B}}\ ^2$. Defined as such, the fidelity of two pure states represents the usual transition probability between pure states.

The fidelity of a pure state, $\mathcal{A}$, and a mixed state, $\mathcal{B}$, is *the average of the previous expression over the ensemble of pure states, with the density matrix,* $\rho^{\mathcal{B}}$; indeed, in this case, $\rho^{\mathcal{A}}$, the pure state, has the same expression as before, but now $\rho^{\mathcal{B}} \quad {}_i p_i\ \varphi_i^{\mathcal{B}}\ \ \varphi_i^{\mathcal{B}}$; therefore,

$$F\ \rho^{\mathcal{A}},\rho^{\mathcal{B}} \quad \varphi^{\mathcal{A}}\ \rho^{\mathcal{B}}\ \varphi^{\mathcal{A}} .$$

This measure of fidelity represents the probability that, $\rho^{\mathcal{B}}$, could prove to be the pure state, $\varphi^{\mathcal{A}}$, following a measurement of the observable, $\varphi^{\mathcal{A}}\ \ \varphi^{\mathcal{A}}$.

Intuitively, we expect the fidelity to be a symmetric function of its arguments, but symmetry is nontrivial to prove starting from the definition of fidelity. Before proving the symmetry of the quantum fidelity, we need to prove the Uhlmann theorem [429]. Recall that given a system in an n-dimensional Hilbert space, $\mathcal{H}_n$, with the density matrix, ρ, a purification of ρ is a pure state in an extended Hilbert space, $\varphi \quad \mathcal{H}_n \quad \mathcal{H}_n$, such that

$$\rho \quad \text{tr}_{\mathcal{H}_n}\left(\ \varphi\ \ \varphi\ \right).$$

Purification treats ρ as the reduced state of a subsystem; it allows us to perform a measurement and extract classical information without altering the state of a quantum system.

Theorem. *(Uhlmann) The fidelity of the two mixed quantum states,* $\rho^{\mathcal{A}}$ *and* $\rho^{\mathcal{B}}$, *is equal to the maximum value of the modulus of the inner product of any purifications,* $\varphi^{\mathcal{A}}$ *and* $\varphi^{\mathcal{B}}$ *, respectively, of the two density matrices,*

$$F\ \rho^{\mathcal{A}},\rho^{\mathcal{B}} \quad \max_{(\varphi^{\mathcal{A}},\varphi^{\mathcal{B}})}\ \varphi^{\mathcal{A}}\ \varphi^{\mathcal{B}}\ ^2.$$

The proof discussed in this sections is due to Richard Jozsa [220] and avoids the difficulties of the original proof of Uhlmann [429] based on the representation theory of C -algebras.

The two mixed states, $\rho^{\mathcal{A}}$ and $\rho^{\mathcal{B}}$, are defined on an n-dimensional Hilbert space, $\mathcal{H}_n$, and their purifications, $\varphi^{\mathcal{A}}$ and $\varphi^{\mathcal{B}}$, are defined on the extended space, $\mathcal{H}_n \quad \mathcal{H}_n$. Without loss of generality, we assume that $\mathcal{H}_n \quad \mathcal{H}_n$; thus, $\varphi^{\mathcal{A}},\varphi^{\mathcal{B}} \quad \mathcal{H}_n \quad \mathcal{H}_n$. We wish to express the two pure states, $\varphi^{\mathcal{A}}$ and $\varphi^{\mathcal{B}}$, in terms of the eigenvectors and eigenvalues of the corresponding density matrices, using Schmidt decomposition and

$$\varphi^{\mathcal{A}} \quad \sum_i \sqrt{p_i^{\mathcal{A}}}\ a_i \quad c_i \quad \text{and} \quad \varphi^{\mathcal{B}} \quad \sum_i \sqrt{p_i^{\mathcal{B}}}\ b_i \quad d_i .$$

In these expressions:

1. $A \quad a_1 \ , a_2 \ , \ldots, \ a_n$ is the set of orthonormal eigenvectors of $\rho^{\mathcal{A}}$ in $\mathcal{H}_n$; $p_1^{\mathcal{A}}, p_2^{\mathcal{A}}, \ldots, p_n^{\mathcal{A}}$ are the corresponding eigenvalues.
2. $B \quad b_1 \ , b_2 \ , \ldots, \ b_n$ is the set of orthonormal eigenvectors of $\rho^{\mathcal{B}}$ in $\mathcal{H}_n$; $p_1^{\mathcal{B}}, p_2^{\mathcal{B}}, \ldots, p_n^{\mathcal{B}}$ are the corresponding eigenvalues.
3. $C \quad c_1 \ , c_2 \ , \ldots, \ c_n$ and $D \quad d_1 \ , d_2 \ , \ldots, \ d_n$ are two orthonormal bases in $\mathcal{H}_n$.

There are some unitary transformations, **B**, **C**, and **D**, that allow us to express the orthonormal bases, B, C, and D, as follows:

$$b_i \quad \mathbf{B}\ a_i \ , \quad c_i \quad \mathbf{C}\ a_i \ , \quad \text{and} \quad d_i \quad \mathbf{D}\ b_i \ .$$

We also observe that

$$\overline{p_i^{\mathcal{A}}} \quad \overline{\rho^{\mathcal{A}}} \quad \text{and} \quad \overline{p_i^{\mathcal{B}}} \quad \overline{\rho^{\mathcal{B}}}.$$

Now we express the two pure states as

$$\varphi^{\mathcal{A}} \quad \underset{i}{} \ \overline{p_i^{\mathcal{A}}}\ a_i \quad c_i \quad \underset{i}{} \ \overline{\rho^{\mathcal{A}}}\ \mathbf{I} \quad a_i \quad \mathbf{C}\ a_i \quad \underset{i}{} \ \overline{\rho^{\mathcal{A}}}\mathbf{C}^{\dagger}\ a_i \quad a_i$$

and

$$\varphi^{\mathcal{B}} \quad {}_j\ \overline{p_j^{\mathcal{B}}}\ b_j \quad d_j \quad {}_j\ \overline{\rho^{\mathcal{B}}}\mathbf{D}^{\dagger}\ b_j \quad b_j \quad {}_j\ \overline{\rho^{\mathcal{B}}}\mathbf{D}^{\dagger}\mathbf{B}\ a_j \quad \mathbf{B}\ a_j$$

$$\overline{\rho^{\mathcal{B}}}\mathbf{D}^{\dagger}\,[\mathbf{B} \quad \mathbf{I}] \ {}_j\ a_j \quad \mathbf{B}\ a_j \quad {}_j\ \overline{\rho^{\mathcal{B}}}\mathbf{D}^{\dagger}\mathbf{B}\mathbf{B}^{\dagger}\ a_j \quad a_j \ .$$

Finally,

$$\varphi^{\mathcal{A}}\ \varphi^{\mathcal{B}} \quad \underset{i}{} \ a_i\ \mathbf{C}^{\dagger}\ \overline{\rho^{\mathcal{A}}}\ \overline{\rho^{\mathcal{B}}}\mathbf{D}^{\dagger}\mathbf{B}\mathbf{B}^{\dagger}\ a_i \quad \mathrm{tr} \quad \overline{\rho^{\mathcal{A}}}\ \overline{\rho^{\mathcal{B}}}\mathbf{D}^{\dagger}\mathbf{B}\mathbf{B}^{\dagger}\mathbf{C}^{\dagger} \ .$$

It follows that the maximum value of $\varphi^{\mathcal{A}}\ \varphi^{\mathcal{B}}$ is $\mathrm{tr} \quad \overline{\rho^{\mathcal{A}}}\ \overline{\rho^{\mathcal{B}}}$ as $\mathbf{B}, \mathbf{C}$, and $\mathbf{D}$ are unitary transformations. We also can write

$$\overline{\rho^{\mathcal{A}}}\ \overline{\rho^{\mathcal{B}}} \quad \overline{\rho^{\mathcal{A}}}\ \overline{\rho^{\mathcal{B}}} \quad \overline{\rho^{\mathcal{A}}}\ \overline{\rho^{\mathcal{B}}}^{\ \dagger \ 1/2} \quad \overline{\rho^{\mathcal{A}}}\ \overline{\rho^{\mathcal{B}}}\ \overline{\rho^{\mathcal{B}}}\ \overline{\rho^{\mathcal{A}}}^{\ 1/2} \quad \overline{\rho^{\mathcal{A}}}\rho^{\mathcal{B}}\ \overline{\rho^{\mathcal{A}}}^{\ 1/2} .$$

We conclude that $F \ \rho^{\mathcal{A}}, \rho^{\mathcal{B}} \quad \max_{(\varphi^{\mathcal{A}}, \varphi^{\mathcal{B}})} \ \varphi^{\mathcal{A}}\ \varphi^{\mathcal{B}} \ ^2$.

Properties of Quantum Fidelity

The first three properties follow immediately from the Uhlmann theorem, while the others can be proved with relative ease:

1. $0 \quad F \ \rho^{\mathcal{A}}, \rho^{\mathcal{B}} \quad 1$.
2. $F \ \rho^{\mathcal{A}}, \rho^{\mathcal{B}} \quad 1$ if and only if $\rho^{\mathcal{A}} \quad \rho^{\mathcal{B}}$.

3. Symmetry: $F \; \rho^{\mathcal{A}}, \rho^{\mathcal{B}} \quad F \; \rho^{\mathcal{B}}, \rho^{\mathcal{A}}$.
4. Convexity: when $\rho^{\mathcal{A}}, \rho^{\mathcal{B}} \quad 0$, and $p^{\mathcal{A}} \quad p^{\mathcal{B}} \quad 1$,

$$F \; \rho, p^{\mathcal{A}} \rho^{\mathcal{A}} \quad p^{\mathcal{B}} \rho^{\mathcal{B}} \qquad p^{\mathcal{A}} F \; \rho, \rho^{\mathcal{A}} \quad p^{\mathcal{B}} F \; \rho, \rho^{\mathcal{B}} \; .$$

5. Multiplicativity: $F \; \rho^{\mathcal{A}} \quad \rho^{\mathcal{B}}, \rho^{\mathcal{C}} \quad \rho^{\mathcal{D}} \qquad F \; \rho^{\mathcal{A}}, \rho^{\mathcal{C}} \; F \; \rho^{\mathcal{B}}, \rho^{\mathcal{D}}$.
6. $F \; \rho^{\mathcal{A}}, \rho^{\mathcal{B}}$ is invariant under unitary transformations on the state space. If $\rho^{\mathcal{A}}_{before}, \rho^{\mathcal{B}}_{before}, \rho^{\mathcal{A}}_{after}, \rho^{\mathcal{B}}_{after}$ are the density operators of systems, $\mathcal{A}$ and $\mathcal{B}$, before and after the unitary transformation, **U**, then

$$F \; \rho^{\mathcal{A}}_{after}, \rho^{\mathcal{B}}_{after} \qquad F \; \rho^{\mathcal{A}}_{before}, \rho^{\mathcal{B}}_{before} \; .$$

7. Nondecreasing: If $\rho^{\mathcal{A}}_{before}, \rho^{\mathcal{B}}_{before}, \rho^{\mathcal{A}}_{after}, \rho^{\mathcal{B}}_{after}$ are the density operators of systems, $\mathcal{A}$ and $\mathcal{B}$, before and after a measurement, **M**, then

$$F \; \rho^{\mathcal{A}}_{after}, \rho^{\mathcal{B}}_{after} \qquad F \; \rho^{\mathcal{A}}_{before}, \rho^{\mathcal{B}}_{before} \; .$$

An Alternative De nition of Quantum Fidelity [21]

Let $\mathbf{M}_i$ be a set of POVM operators, such that $\;_i \mathbf{M}_i \quad I$. The fidelity defines the minimum overlap between the probability distributions $p_{\mathcal{A}}(i)$ and $p_{\mathcal{B}}(i)$, with $p_{\mathcal{A}}(i) \quad \mathrm{tr}(\rho^{\mathcal{A}}\mathbf{M}_i)$ and $p_{\mathcal{B}}(i) \quad \mathrm{tr}(\rho^{\mathcal{B}}\mathbf{M}_i)$, for the outcomes of the POVM measurement,

$$F \; \rho^{\mathcal{A}}, \rho^{\mathcal{B}} \qquad \min_{\mathbf{M}_i} \; _i \quad \overline{\mathrm{tr} \; \rho^{\mathcal{A}}\mathbf{M}_i} \;\; \overline{\mathrm{tr} \; \rho^{\mathcal{B}}\mathbf{M}_i} \; .$$

A POVM that achieves this minimum is called an *optimal POVM*.

The proof that the two expressions of fidelity are equivalent, $F \; \rho^{\mathcal{A}}, \rho^{\mathcal{B}} \qquad F \; \rho^{\mathcal{A}}, \rho^{\mathcal{B}}$, starts with the observation that if **U** is a unitary operator, $\mathbf{U}\mathbf{U}^{\dagger} \quad \mathbf{I}$, then

$$\rho^{\mathcal{A}}\mathbf{M}_i \qquad \rho^{\mathcal{A}\;1/2}\mathbf{M}_i \; \rho^{\mathcal{A}\;1/2} \quad \mathbf{U} \; \rho^{\mathcal{A}\;1/2}\mathbf{M}_i \; \rho^{\mathcal{A}\;1/2}\mathbf{U}^{\dagger} \qquad \mathbf{U} \; \rho^{\mathcal{A}\;1/2}\mathbf{M}_i^{1/2} \quad \mathbf{U} \; \rho^{\mathcal{A}\;1/2}\mathbf{M}_i^{1/2} \;\; ^{\dagger}$$

and

$$\rho^{\mathcal{B}}\mathbf{M}_i \qquad \rho^{\mathcal{B}\;1/2}\mathbf{M}_i \; \rho^{\mathcal{B}\;1/2} \qquad \rho^{\mathcal{B}\;1/2}\mathbf{M}_i^{1/2} \qquad \rho^{\mathcal{B}\;1/2}\mathbf{M}_i^{1/2} \;\; ^{\dagger} \; .$$

The density matrices, $\rho^{\mathcal{A}}$ and $\rho^{\mathcal{B}}$, as well as the measurement operators, $\mathbf{M}_i$, are Hermitian. Then,

$$
\begin{aligned}
F\left(\rho^{\mathcal{A}},\rho^{\mathcal{B}}\right) &= \min_{\mathbf{M}_i} \sum_i \sqrt{\mathrm{tr}\left(\rho^{\mathcal{A}}\mathbf{M}_i\right)}\sqrt{\mathrm{tr}\left(\rho^{\mathcal{B}}\mathbf{M}_i\right)} \\
&= \min_{\mathbf{M}_i} \sum_i \sqrt{\mathrm{tr}\left[\left(\mathbf{U}\left(\rho^{\mathcal{A}}\right)^{1/2}\mathbf{M}_i^{1/2}\right)\left(\mathbf{U}\left(\rho^{\mathcal{A}}\right)^{1/2}\mathbf{M}_i^{1/2}\right)^{\dagger}\right]} \\
&\qquad \times \sqrt{\mathrm{tr}\left[\left(\left(\rho^{\mathcal{B}}\right)^{1/2}\mathbf{M}_i^{1/2}\right)\left(\left(\rho^{\mathcal{B}}\right)^{1/2}\mathbf{M}_i^{1/2}\right)^{\dagger}\right]} \\
&\geq \sum_i \sqrt{\mathrm{tr}\left[\left(\mathbf{U}\left(\rho^{\mathcal{A}}\right)^{1/2}\mathbf{M}_i^{1/2}\right)\left(\mathbf{U}\left(\rho^{\mathcal{A}}\right)^{1/2}\mathbf{M}_i^{1/2}\right)^{\dagger}\right]} \\
&\qquad \times \sqrt{\mathrm{tr}\left[\left(\left(\rho^{\mathcal{B}}\right)^{1/2}\mathbf{M}_i^{1/2}\right)\left(\left(\rho^{\mathcal{B}}\right)^{1/2}\mathbf{M}_i^{1/2}\right)^{\dagger}\right]} \\
&\geq \sum_i \left|\mathrm{tr}\left(\mathbf{U}\left(\rho^{\mathcal{A}}\right)^{1/2}\mathbf{M}_i^{1/2}\mathbf{M}_i^{1/2}\left(\rho^{\mathcal{B}}\right)^{1/2}\right)\right|.
\end{aligned}
$$

The last expression is based on the Schwarz inequality for the operator inner product, $\mathrm{tr}(AA^{\dagger})\mathrm{tr}(BB^{\dagger}) \geq |\mathrm{tr}(AB^{\dagger})|^2$. It is easy to see that

$$
\sum_i \left|\mathrm{tr}\left(\mathbf{U}\left(\rho^{\mathcal{A}}\right)^{1/2}\mathbf{M}_i^{1/2}\mathbf{M}_i^{1/2}\left(\rho^{\mathcal{B}}\right)^{1/2}\right)\right| \geq \left|\mathrm{tr}\left(\mathbf{U}\left(\rho^{\mathcal{A}}\right)^{1/2}\sum_i \mathbf{M}_i^{1/2}\mathbf{M}_i^{1/2}\left(\rho^{\mathcal{B}}\right)^{1/2}\right)\right|
$$

and

$$
\left|\mathrm{tr}\left(\mathbf{U}\left(\rho^{\mathcal{A}}\right)^{1/2}\sum_i \mathbf{M}_i^{1/2}\mathbf{M}_i^{1/2}\left(\rho^{\mathcal{B}}\right)^{1/2}\right)\right| = \left|\mathrm{tr}\left(\mathbf{U}\left(\rho^{\mathcal{A}}\right)^{1/2}\left(\rho^{\mathcal{B}}\right)^{1/2}\right)\right|.
$$

We address now the question of choosing the linear transformation, $\mathbf{U}$, to make this inequality as tight as possible. Given a set of linear operators, $\mathbf{U}_k$, and an operator, $\mathbf{G}$, the maximum

$$
\max_{U_k} \left|\mathrm{tr}\left(\mathbf{U}_k\mathbf{G}\right)\right| = \mathrm{tr}\left(\sqrt{\mathbf{G}^{\dagger}\mathbf{G}}\right)
$$

is obtained for a particular $\mathbf{U}_k$, with the property that $\mathbf{U}_k\mathbf{G} = \sqrt{\mathbf{G}^{\dagger}\mathbf{G}}$. It follows that when $\mathbf{G} = (\rho^{\mathcal{A}})^{1/2}(\rho^{\mathcal{B}})^{1/2}$,

$$
\mathbf{U}\left(\rho^{\mathcal{A}}\right)^{1/2}\left(\rho^{\mathcal{B}}\right)^{1/2} = \sqrt{\left(\rho^{\mathcal{B}}\right)^{1/2}\left(\rho^{\mathcal{A}}\right)^{1/2}\left(\rho^{\mathcal{A}}\right)^{1/2}\left(\rho^{\mathcal{B}}\right)^{1/2}} = \sqrt{\left(\rho^{\mathcal{B}}\right)^{1/2}\rho^{\mathcal{A}}\left(\rho^{\mathcal{B}}\right)^{1/2}} = F\left(\rho^{\mathcal{A}},\rho^{\mathcal{B}}\right).
$$

Finally, we conclude that

$$
F\left(\rho^{\mathcal{A}},\rho^{\mathcal{B}}\right) \geq \left|\mathrm{tr}\,\mathbf{U}\left(\rho^{\mathcal{A}}\right)^{1/2}\left(\rho^{\mathcal{B}}\right)^{1/2}\right| = \mathrm{tr}\sqrt{\left(\rho^{\mathcal{B}}\right)^{1/2}\rho^{\mathcal{A}}\left(\rho^{\mathcal{B}}\right)^{1/2}} = F\left(\rho^{\mathcal{A}},\rho^{\mathcal{B}}\right).
$$

The POVM is optimal when we have equality in the preceding expression for $F \;\; \rho^{\mathcal{A}}, \rho^{\mathcal{B}}$. When $\rho^{\mathcal{B}}$ is invertible, an optimal POVM is a positive operator [21] given by

$$\mathbf{M} \quad \rho^{\mathcal{B}} \quad {}^{1/2} \sqrt{\rho^{\mathcal{B}\;1/2}\,\rho^{\mathcal{A}}\;\rho^{\mathcal{B}\;1/2}} \quad \rho^{\mathcal{B}} \quad {}^{1/2}$$

and $\mathbf{M}_i \quad i \;\; i$, with $\;\; i \;\;$, an orthonormal eigenbasis of $\mathbf{M}$.

The properties of the fidelity operator will be used in Section 3.13 to show that a general mixed state of a quantum system cannot be broadcast. We are now concerned with the transmission of classical information through a quantum channel and the measurement process necessary to extract the classical information encoded as the state of a quantum system.

3.12 ACCESSIBLE INFORMATION IN A QUANTUM MEASUREMENT AND THE HOLEVO BOUND

The Holevo bound [210] was probably the first major result in quantum information theory; it provides an upper limit of the accessible information in a quantum measurement. The Holevo bound gives an upper bound on how well we can infer the value of an unknown random variable, X, based on the knowledge of another random variable, Y.

To formulate this result, we consider the following setup: A source generates letters, X, from an alphabet, with the probabilities $p_i, 1 \quad i \quad n$; Alice encodes the random variable X and transmits it through a quantum channel; Bob performs a measurement of the quantum system, and the result of this measurement is a random variable, Y. The mutual information, $I(X;Y)$, measures the amount of information about X that Bob can infer from the measurement. We know from Section 3.3 that

$$I(X;Y) \quad H(X).$$

Thus, Bob should attempt to carry out a measurement that maximizes the mutual information, $I(X;Y)$. The question posed is, "How close to $H(X)$ can Bob get?" In other words, is there an upper bound on $I(X;Y)$?

To answer this question, we assume that Alice has a choice of quantum systems in states characterized by the density matrices, $\rho_1, \rho_2, \ldots, \rho_n$; to send $X \quad k$, she chooses one of them, say the one characterized by ρ_k and the probability p_k. Recall from Chapter 2 that when Bob performs a POVM measurement, $\mathbf{M}^Y$, which produces the random variable, Y, then the probability that $Y \quad m$ when the system was prepared by Alice in the state $X \quad k$ is

$$\text{Prob}(Y \quad m\, X \quad k) \quad \text{tr} \;\; \mathbf{M}^Y \rho_k \;\; .$$

We know the distribution of X and Y; thus, we can calculate $H(X)$ and $H(Y)$. To calculate $H(X;Y)$, we only need

$$\text{Prob}(Y \quad m, X \quad k) \quad p_k \text{tr} \;\; \mathbf{M}^Y \rho_k \;\; .$$

Regardless of the measurement performed by Bob, the mutual information between the measurement outcome, Y, and the preparation, X, is limited by the *Holevo bound*:

$$I(X;Y) \quad S \quad \underset{i}{p_i\rho_i} \qquad \underset{i}{p_i S(\rho_i)}.$$

The Holevo bound is achieved if and only if the encoding states, ρ_i, commute with each other and the receiver, Bob, makes a von Neumann measurement in the basis in which ρ_i are diagonal. As we know, in a von Neumann measurement, the system is projected onto one of a complete sets of mutually orthogonal states; in Bob's case, this set of states is chosen to be the basis in which the coding states are diagonal.

Holevo Information

Consider an ensemble of mixed state with the density matrix, $\rho \qquad {}_i p_i\rho_i$, with ρ_i the density matrix of the individual mixed states of the ensemble and, p_i, the a priori probability of the mixed state, i. Then $S(\rho)$ and $S(\rho_i)$ are the von Neumann entropy of the overall ensemble of states and of the individual state i, respectively. The quantity

$$\chi \quad S \quad \underset{i}{p_i\rho_i} \qquad \underset{i}{p_i S(\rho_i)}$$

is known as the *Holevo information*, the information available in the quantum ensemble. If the density matrices of the states used as signals to transport classical information over the quantum channel are ρ_i, $1 \quad i \quad n$, with the probability, p_i, and, $\mathcal{E}$, as the mapping of the density matrices performed by the channel, then the Holevo information gives the classical capacity of a quantum channel and can be expressed as

$$\chi \quad S \quad \underset{i}{p_i\mathcal{E}(\rho_i)} \qquad \underset{i}{p_i S(\mathcal{E}(\rho_i))}.$$

The maximum classical capacity of the channel is the maximum of this expression over all input ensembles:

$$C \quad \chi_{max} \quad \max_{(p_i,\rho_i)} S \quad \underset{i}{p_i\mathcal{E}(\rho_i)} \qquad \underset{i}{p_i S(\mathcal{E}(\rho_i))} \; .$$

The von Neumann entropy is a concave function; thus, for any mixed state with the density matrix ρ, the Holevo information is non-negative:

$$\chi(\rho) \quad 0.$$

Consider now a composite quantum system consisting of the pair, $(\mathcal{X},\mathcal{Y})$, with the density matrix, $\rho^{(\mathcal{X},\mathcal{Y})}$. Let $\chi \;\rho^{\mathcal{X}}$ be the Holevo information for system, $\mathcal{X}$, and, $\chi \;\rho^{(\mathcal{X},\mathcal{Y})}$, the Holevo information for the joint system. Schumacher, Westmoreland, and Wootters [369] showed that

$$\chi \ \rho^{\mathcal{X}} \qquad \chi \ \rho^{(\mathcal{X},\mathcal{Y})} \ .$$

They observe that "since χ is an upper bound of the accessible information, it is reasonable that this bound does not increase when part of the system is discarded."

To prove that the Holevo bound, χ, cannot increase when part of a joint system is discarded, we use the relationship between the von Neumann entropy of a composite system in a product state and the Shannon entropy; consider two quantum systems, $\mathcal{A}$ and $\mathcal{B}$, in a joint state with a density matrix of the form,

$$\rho^{(\mathcal{A},\mathcal{B})} \qquad \underset{i}{\ } \ p_i \ e_i^{\mathcal{A}} \ \ e_i^{\mathcal{A}} \quad \rho_i^{\mathcal{B}},$$

with $e_i^{\mathcal{A}}$, an orthonormal basis for $\mathcal{A}$, and, $\rho_i^{\mathcal{B}}$, a set of density operators for system $\mathcal{B}$. The von Neumann entropy of the joint state is

$$S \ \rho^{(\mathcal{A},\mathcal{B})} \qquad H(p) \qquad \underset{i}{\ } \ p_i S \ \rho_i^{\mathcal{B}} \ ,$$

with $H(p)$, the Shannon entropy for the probability distribution of p_i.

Consider three systems, $\mathcal{X}, \mathcal{Y}$, and $\mathcal{Z}$, in a joint state described by a density matrix of the form

$$\rho^{(\mathcal{X},\mathcal{Y},\mathcal{Z})} \qquad \underset{k}{\ } \ p_k \rho_k^{(\mathcal{X},\mathcal{Y})} \quad k^{\mathcal{Z}} \ \ k^{\mathcal{Z}} \ ,$$

where $k^{\mathcal{Z}}$ is an orthogonal set of states for the system, $\mathcal{Z}$.

Recall from Section 3.10 that strong subadditivity requires that

$$S \ \rho^{(\mathcal{X},\mathcal{Y},\mathcal{Z})} \qquad S \ \rho^{\mathcal{X}} \qquad S \ \rho^{(\mathcal{X},\mathcal{Y})} \qquad S \ \rho^{(\mathcal{X},\mathcal{Z})} \ .$$

The relation between von Neumann and Shannon entropy allows us to express the elements of this expression as

$$S \ \rho^{(\mathcal{X},\mathcal{Y},\mathcal{Z})} \qquad H(p) \qquad \underset{k}{\ } \ p_k S \ \rho_k^{(\mathcal{X},\mathcal{Y})} \quad \text{and} \quad S \ \rho^{(\mathcal{X},\mathcal{Z})} \qquad H(p) \qquad \underset{k}{\ } \ p_k S \ \rho_k^{\mathcal{X}} \ .$$

We substitute them in the strong subadditivity expression and obtain

$$S \ \rho^{\mathcal{X}} \qquad \underset{k}{\ } \ p_k S \ \rho_k^{\mathcal{X}} \quad S \ \rho^{(\mathcal{X},\mathcal{Y})} \qquad \underset{k}{\ } \ p_k S \ \rho_k^{(\mathcal{X},\mathcal{Y})} \ ,$$

or $\chi \ \rho^{\mathcal{X}} \quad \chi \ \rho^{(\mathcal{X},\mathcal{Y})}$.

An Alternative Proof of the Holevo Bound

The proof presented by Schumacher *et al* [369] is based on a physical model of a POVM measurement process where distinct outcomes are represented by positive operators rather than projections. Three systems are considered in the measurement process: (1) $\mathcal{Q}$, the quantum system in the state, $\rho^{\mathcal{Q}}_{before}$; (2) $\mathcal{A}$, the apparatus used to perform the measurement; and (3) E, the environment.

Initially, the quantum system is in the state, $\rho^{\mathcal{Q}}_{before}$, and the additional systems, the apparatus, and the environment are in a joint state, $\rho^{(\mathcal{A},E)}_{before}$; the whole ensemble is in the joint state,

$$\rho^{(\mathcal{A},E,\mathcal{Q})}_{before} \quad \rho^{(\mathcal{A},E)}_{before} \quad \rho^{\mathcal{Q}}_{before}.$$

The measurement process proceeds in two stages: a dynamical evolution representing interactions among the three components captured by the unitary operator, $\mathbf{U}$, followed by a second stage, the discarding of the environment. The state of the system after the first stage is

$$\rho^{(\mathcal{A},E,\mathcal{Q})}_{fs} \quad \mathbf{U}\,\rho^{(\mathcal{A},E,\mathcal{Q})}_{before}\,\mathbf{U}^{\dagger}.$$

The second transformation, the decoupling from the environment, implies a partial trace over the system, E:

$$\rho^{(\mathcal{A},\mathcal{Q})}_{after} \quad \mathrm{tr}_E\ \rho^{(\mathcal{A},E,\mathcal{Q})}_{fs} \quad .$$

If $\rho^{\mathcal{Q}}_{after}$ is the state of the quantum system after the measurement, and $\varphi^{\mathcal{A}}_i$ is a set of orthogonal states independent of the input state, $\rho^{\mathcal{Q}}_{before}$, then the joint state of $\mathcal{A}$ and $\mathcal{Q}$ after the two-stage measurement should be

$$\rho^{(\mathcal{A},\mathcal{Q})}_{after} \quad {}_{i} \quad P_i\,\varphi^{\mathcal{A}}_i \quad \varphi^{\mathcal{A}}_i \quad \rho^{\mathcal{Q}}_{after}.$$

In this expression, P_i is the probability of the outcome i of the measurement. The choice of the states, $\varphi^{\mathcal{A}}_i$, is fixed for a given measurement device and the environment it is in, and it is not related to the state of the system being measured. The states $\varphi^{\mathcal{A}}_i$ are macroscopically distinguishable. A given measurement outcome does not necessarily correspond to a pure state of the apparatus system $\mathcal{A}$. If an observer does not completely resolves the state $\varphi^{\mathcal{A}}_i$, then the measurement is an average over a group of different i, and the observer acquires less information.

Now we pay closer attention to the preparation of the initial state of the quantum system, $\mathcal{Q}$. The preparer has a choice of N initial states with the density matrices, $\rho^{\mathcal{Q}}_1, \rho^{\mathcal{Q}}_2, \ldots, \rho^{\mathcal{Q}}_N$, selected with the probabilities, $p_1, p_2, \ldots, p_N$, respectively. Let us assume that the preparer chooses $\rho^{\mathcal{Q}}_{before} \quad \rho^{\mathcal{Q}}_k$. Then the corresponding state of the ensemble is

$$\rho^{(\mathcal{A},E,\mathcal{Q})}_{k,before} \quad \rho^{(\mathcal{A},\mathcal{Q})}_{before} \quad \rho^{\mathcal{Q}}_k.$$

We wish to compute $\chi^{\mathcal{Q}}$, and we observe that $\chi^{\mathcal{Q}} \quad \chi^{(\mathcal{A},E,\mathcal{Q})}$ because the joint state, $(\mathcal{A},E)$, is independent of the preparation of the system, $\mathcal{Q}$. To compute $\chi^{(\mathcal{A},E,\mathcal{Q})}$, we need $\rho^{(\mathcal{A},E,\mathcal{Q})}_{before}$, the density matrix of the joint state, an average over all possible preparation states:

$$\rho^{(\mathcal{A},E,\mathcal{Q})}_{before} \quad {}_{k} \quad p_k \rho^{(\mathcal{A},E,\mathcal{Q})}_{k,before} \ .$$

Then, after the second stage of the measurement process described earlier,

$$\rho^{(\mathcal{A},\mathcal{Q})}_{k,after} \quad \underset{i}{} \; P(i\;k)\; \varphi^{\mathcal{A}}_i \quad \varphi^{\mathcal{A}}_i \quad \rho^{\mathcal{Q}}_{ik,after},$$

with $P(i\;k)$, the conditional probability of the measurement outcome, i, given that the system was initially prepared in the state indexed by k and $\rho^{\mathcal{Q}}_{ik,after}$, the resulting states of system $\mathcal{Q}$ when the preparation, k, results in the outcome i. Again, we consider the average state of $\mathcal{A}$ and $\mathcal{Q}$ after the measurement

$$\rho^{(\mathcal{A},\mathcal{Q})}_{after} \quad \underset{k \;\; i}{} \; P(i,k)\; \varphi^{\mathcal{A}}_i \quad \varphi^{\mathcal{A}}_i \quad \rho^{\mathcal{Q}}_{ik,after},$$

where the joint probability $P(i,k) \quad p_k P(i\;k)$. Call $P(i)$ the total probability of the measurement outcome with the index i. Then,

$$P(i) \quad \underset{k}{} \; P(i,k),$$

and the final state of $\mathcal{Q}$ averaged over all preparations leading to outcome i is $\rho^{\mathcal{Q}}_{i,after} \quad {}_k P(k\;i)\rho^{\mathcal{Q}}_{ik,after}$. After summing for all k, the previous expression of the average joint state of $\mathcal{A}$ and $\mathcal{Q}$ after measurement becomes

$$\rho^{(\mathcal{A},\mathcal{Q})}_{after} \quad \underset{i}{} \; P(i)\; \varphi^{\mathcal{A}}_i \quad \varphi^{\mathcal{A}}_i \quad \rho^{\mathcal{Q}}_{i,after}.$$

The initial and the after-measurement Holevo information of the system consisting of the measuring apparatus, $\mathcal{A}$, the environment, E, and the quantum system, $\mathcal{Q}$, must be the same because the entropy is preserved under unitary transformation:

$$\chi^{(\mathcal{A},E,\mathcal{Q})}_{after} \quad \chi^{(\mathcal{A},E,\mathcal{Q})}_{before}.$$

Moreover, we proved earlier that $\chi^{(\mathcal{A},\mathcal{Q})}_{after} \quad \chi^{(\mathcal{A},E,\mathcal{Q})}_{after}$; thus, we conclude that

$$\chi^{(\mathcal{A},\mathcal{Q})}_{after} \quad \chi^{(\mathcal{Q})}_{before}.$$

The definition of Holevo information allows us to express $\chi^{(\mathcal{A},\mathcal{Q})}_{after}$ as

$$\chi^{(\mathcal{A},\mathcal{Q})}_{after} \quad S\;\rho^{(\mathcal{A},\mathcal{Q})}_{after} \quad \underset{k}{} \; p_k S\;\rho^{(\mathcal{A},\mathcal{Q})}_{k,after}\;.$$

We denote by $H(Y) \quad {}_i P(i)\log P(i)$ the Shannon entropy of the average probability distribution of $P(i)$ over the outcomes of the measurement; $H(Y\;k)$ is the Shannon entropy of the outcomes of the measurement conditioned by the initial preparation with the index, k, and $H(Y\;X) \quad {}_k p_k H(Y\;k)$ is the conditional entropy of the outcomes for all the preparations, k. Then we express the two von Neumann entropy terms in the expression of $\chi^{(\mathcal{A},\mathcal{Q})}_{after}$ as

$$S\;\rho^{(\mathcal{A},\mathcal{Q})}_{after} \quad H(Y) \quad \underset{i}{} \; P(i) S\;\rho^{\mathcal{Q}}_{i,after}$$

and

$$S\ \rho^{(\mathcal{A},Q)}_{k,after} \qquad H(Y\ k) \qquad \underset{i}{} \ P(i\ k)S\ \rho^{Q}_{ik,after}\ .$$

After substituting in the expression

$$\chi^{(Q)}_{before} \quad \chi^{(\mathcal{A},Q)}_{after},$$

it follows that

$$\chi^{Q}_{before} \quad (H(Y) \quad H(Y\ X)) \qquad \underset{i}{} P(i)\ \ S\ \rho^{Q}_{i,after} \qquad \underset{k}{} P(k\ i)S\ \rho^{Q}_{ik,after} \quad .$$

The Holevo information of the quantum system, Q, after a measurement conditioned by an outcome indexed by i is

$$\chi^{Q}_{after,i} \quad S\ \rho^{Q}_{i,after} \qquad \underset{k}{} P(k\ i)S\ \rho^{Q}_{ik,after}\ .$$

We also observe that the mutual information is $I(X;Y) \quad H(Y) \quad H(Y\ X)$; thus, we have proved that

$$I(X;Y) \quad \chi^{Q}_{after} \qquad \underset{i}{} P(i)\chi^{Q}_{after,i}.$$

This result of Schumacher, Westmoreland, and Wootters [369] is more general than the Holevo theorem; it shows that the information obtained about the preparation of a system through a measurement process is bounded by the average amount of the quantity, χ, and decreases during the measurement.

An Intuitive Representation of Von Neumann Entropy Before and After a Measurement [88]

The entities involved in the reasoning of Schumacher, Westmoreland, and Wootters [369] are:

1. The preparer, described by the random variable, X, with the probability distribution, $p_i, 1 \quad i \quad N$. The internal state of the preparer (a physical quantum system) is given by the density matrix,

$$\rho^{X}_{before} \qquad \underset{i}{} p_i\ x_i \quad x_i\ ,$$

   with x_i , a set of orthonormal preparer states.
2. The quantum channel, Q. The preparer sends through the channel a quantum system in a mixed state, ρ_i, selected out of a set of, N, mixed states according to the preparer's internal state. The joint state of the preparer and the quantum communication channel is described by the density matrix,

$$\rho^{(X,Q)}_{before} \qquad \underset{i}{} p_i\ x_i \quad x_i \qquad \rho_i.$$

This density matrix is block diagonal, the quantum analog of the "grouping" property of Shannon entropy discussed in Section 3.3. As we know, the state of the quantum channel is the partial trace of the joint state,

$$\rho^{Q}_{before} \quad \mathrm{tr}_X \ \rho^{(X,Q)} \qquad \underset{i}{} \ p_i \rho_i \quad \rho.$$

3. An ancilla, A. The measurement involves a unitary operation on the ancilla, A, and the quantum channel, Q. The information obtained as the result of the measurement is the mutual information between the preparer and the ancilla, $H_{after}(X,A)$. Initially, the ancilla is in a reference state, 0 and its density matrix is

$$\rho^{A}_{before} \quad 0 \ 0 .$$

The initial joint state of the ternary system, (X,Q,A), can be written as

$$\rho^{(X,Q,A)}_{before} \quad \rho^{(X,Q)}_{before} \quad 0 \ 0 .$$

Now we express the von Neumann entropy of the three components, as well as the joint entropy before and after the measurement:

$$S_{before}(X) \quad H(p_i), \quad S_{before}(Q) \quad S(\rho), \quad S_{before}(X,Q) \quad H(p_i) \qquad \underset{i}{} \ p_i S(\rho_i).$$

The last equality exploits the fact that $\rho^{(X,Q)}_{before}$ is block diagonal. We also observe that the von Neumann entropy of the joint state of X, Q, and A is

$$S_{before}(X,Q,A) \quad S_{before}(X,Q)$$

because A is initially in a pure state, 0 .

Then, $S_{before}(X;Q)$ (Figure 3.12a), the quantum mutual entropy, or the mutual entanglement, between the preparer and the quantum channel is equal to the Holevo bound:

$$S_{before}(X;Q) \quad S_{before}(X) \quad S_{before}(Q) \quad S_{before}(X,Q) \quad S(\rho) \qquad \underset{i}{} \ p_i S(\rho_i) \quad \chi^{(X)}.$$

The Venn diagram in Figure 3.12a summarizes this relationship between the von Neumann entropy of X and Q before the measurement, and Figure 3.12b illustrates the joint and the conditional von Neumann entropy of the three subsystems involved after the measurement.

Let us now consider the measurement, a unitary transformation, $\mathbf{U}^{(Q,A)}$, affecting only the quantum channel and the ancilla. The joint state of the three systems after the measurement is

$$\rho^{(X,Q,A)}_{after} \qquad \mathbf{I}^X \quad \mathbf{U}^{(Q,A)} \quad \rho^{(X,Q)}_{before} \quad 0 \ 0 \quad \mathbf{I}^X \quad \mathbf{U}^{(Q,A)} \ ^{\dagger} .$$

As noted earlier, we have to find an upper bound for the mutual entanglement between X and A, $S_{after}(X;A)$. To do so, we first apply the chain rule for mutual entropies:

$$S_{after}(X;(Q,A)) \quad S_{after}(X;A) \quad S_{after}(X;Q \ A).$$

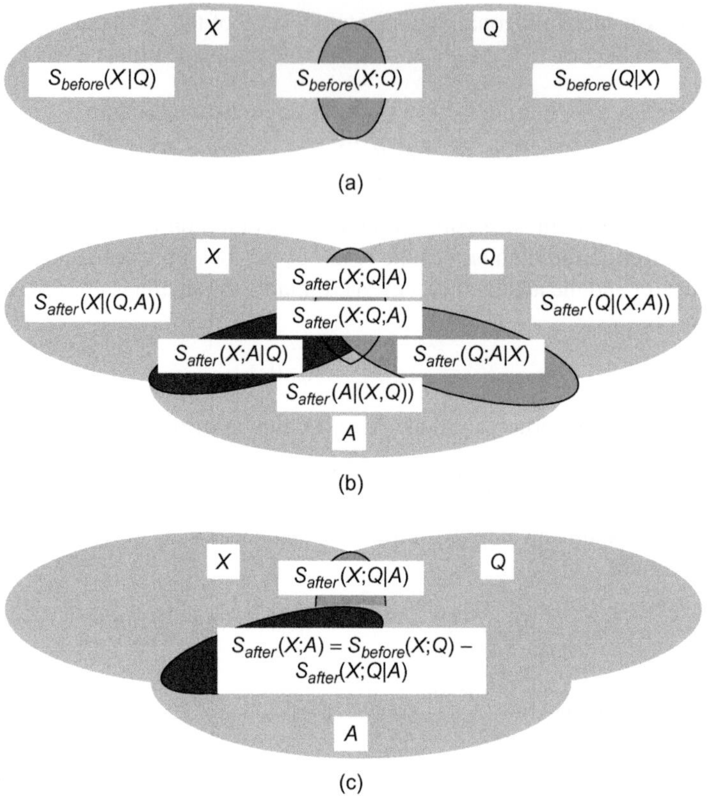

FIGURE 3.12

Venn diagrams showing the von Neumann entropy. (a) Before the measurement. (b) After the measurement. Three systems are involved: (1) the preparer, X; (2) the quantum channel, Q; and (3) the ancilla, A. (c) The mutual entropy between the ancilla and the preparer after the measurement, $S_{after}(X;A) \quad S_{before}(X;Q) - S_{after}(X;Q\ A)$, where $S_{before}(X;Q)$ is, in fact, $S_{after}(X;(Q,A))$.

Then we observe that the mutual entanglement between, X, and the joint system, (Q,A), is conserved because the measurement involves a unitary transformation of the joint system (Q,A) and does not affect the state of X. Thus,

$$S_{after}(X;(Q,A)) \quad S_{before}(X;(Q,A)) \quad S_{before}(X;Q).$$

It follows that

$$S_{after}(X;A) \quad S_{before}(X;Q) - S_{after}(X;Q\ A).$$

This equation tells us that the information obtained through the measurement is given by $S_{before}(X;Q)$ (the Holevo bound) reduced by a quantity that is the mutual entropy (after measurement) between the preparer's state X and the state of the quantum channel Q, conditional on the observed state of the ancilla A (all states existing after measurement).

Figure 3.12c illustrates the quantities involved in this expression. Strong subadditivity of von Neumann entropy implies that mutual entanglement never decreases when extending a system; it follows that $S_{after}(X;Q\ A) \quad 0$; therefore,

$$S_{after}(X;A) \quad S_{before}(X;Q).$$

The second term on the right-hand side of the expression giving the mutual information between the preparer and the ancilla is the quantum conditional mutual entropy,

$$S_{after}(X;Q\ A) \quad S_{after}(X,A) \quad S_{after}(Q,A) \quad S_{after}(A) \quad S_{after}(X,Q,A).$$

This expression is difficult to estimate.

Finally, Cerf and Adami show that a von Neumann measurement allows us to express the von Neumann entropy as a function of classical Shannon entropy [88],

$$S_{after}(X;A) \quad H(A) \quad H(A\ X) \quad I(X;A).$$

It follows that the mutual information satisfies the Holevo bound,

$$I(X;A) \quad S_{before}(X;Q) \quad S(\rho) \qquad {}_{i}\ p_i S(\rho_i).$$

This inequality shows that some information about the state before measurement of the system X might still be extractable from the state after measurement of the system Q; this is the case when the measurement is "incomplete."

An immediate consequence of the Holevo bound is that no more than n bits of information can be communicated using, n, qubits. Indeed, recall that given two correlated random variables, X and Y, the maximum amount of information about X available prior to transmission at the receiving end of a classical communication channel is $H(X)$; after the transmission, the average uncertainty is $H(X\ Y)$. Therefore, $H(X) \quad H(X\ Y)$ is the maximum amount of information that can be extracted by the recipient of information. $I(X;Y) \quad H(X) \quad H(X\ Y)$ is the mutual information between X and Y. The superdense coding scheme [33] allows the transmission of two bits of information using a single qubit, but, in addition to the quantum channel, it requires a classical channel to transmit the results of a measurement performed by the sender.

The information about the random variable X could be encoded on n qubits. The state of the n-qubit quantum register Y encoding the information about a particular value of the input random variable, $X \quad x$, is described by the density matrix, ρ, in a 2^n-dimensional Hilbert space, $\mathcal{H}_{2^n}$. Regardless of the choices of x and ρ, the POVM measurement of Y produces, at most, $\chi \quad n$ bits of information about X.

Example. *The Holevo bound for information encoded as the polarization of a photon.* Alice sends Bob information encoded as the polarization of a photon; Bob receives a photon with either vertical polarization, or a polarization at an angle θ, with the equal probabilities $p_0 \quad p_1 \quad 1/2$. We wish to determine the Holevo bound for the information Bob could extract, to study the dependence of this bound on the angle θ and to find the optimal measurement.

The possible states of the photon are $\varphi_1 \quad 0$ and $\varphi_2 \quad \cos\theta\ 0 \quad \sin\theta\ 1$. The density matrix is

$$\rho \quad p_0\rho_0 \quad p_1\rho_1 \quad \frac{1}{2} \quad \begin{matrix} 1 \\ 0 \end{matrix} \quad 1 \quad 0 \quad \begin{matrix} \cos\theta \\ \sin\theta \end{matrix} \quad \cos\theta \quad \sin\theta \quad .$$

Thus,

$$\rho \quad \frac{1}{2} \quad \begin{matrix} 1 \quad \cos^2\theta & \sin\theta\cos\theta \\ \sin\theta\cos\theta & 1 \quad \cos^2\theta \end{matrix} \quad .$$

We solve the characteristic equation, $\det(\rho \quad \lambda I) \quad 0$, to find the eigenvalues of the density matrix,

$$\det \quad \frac{1}{2} \quad \begin{matrix} 1 \quad \cos^2\theta & \sin\theta\cos\theta \\ \sin\theta\cos\theta & 1 \quad \cos^2\theta \end{matrix} \quad \begin{matrix} \lambda & 0 \\ 0 & \lambda \end{matrix} \qquad \det\frac{1}{2} \quad \begin{matrix} 1 \quad \cos^2\theta \quad 2\lambda & \sin\theta\cos\theta \\ \sin\theta\cos\theta & 1 \quad \cos^2\theta \quad 2\lambda \end{matrix} \quad .$$

Then the characteristic equation becomes

$$(1 \quad 2\lambda) \quad \cos^2\theta \quad (1 \quad 2\lambda) \quad \cos^2\theta \quad \sin^2\theta\cos^2\theta \quad 0$$

or

$$(1 \quad 2\lambda)^2 \quad \cos^2\theta.$$

The solutions are

$$\lambda_{1,2} \quad \frac{1 \quad \cos\theta}{2}.$$

We can now express the von Neumann entropy of the photon source as

$$S(\rho) \quad \lambda_1 \log\lambda_1 \quad \lambda_2 \log\lambda_2 \quad \frac{1 \quad \cos\theta}{2}\log\frac{1 \quad \cos\theta}{2} \quad \frac{1 \quad \cos\theta}{2}\log\frac{1 \quad \cos\theta}{2}.$$

Recall that $H(p) \quad p\log p \quad (1 \quad p)\log(1 \quad p)$ is the Shannon entropy of a binary random variable, with the probability p; thus,

$$S(\rho) \quad H \quad \frac{1 \quad \cos\theta}{2} \quad .$$

Figure 3.13 illustrates the polarization of the two photons, the optimal measurement, and the graph of the Holevo bound function of the angle θ. As expected, Bob extracts maximum information when $\theta \quad \pi/2$, thus, when the two states are orthogonal.

Next, we discuss the no-broadcasting theorem for mixed quantum states.

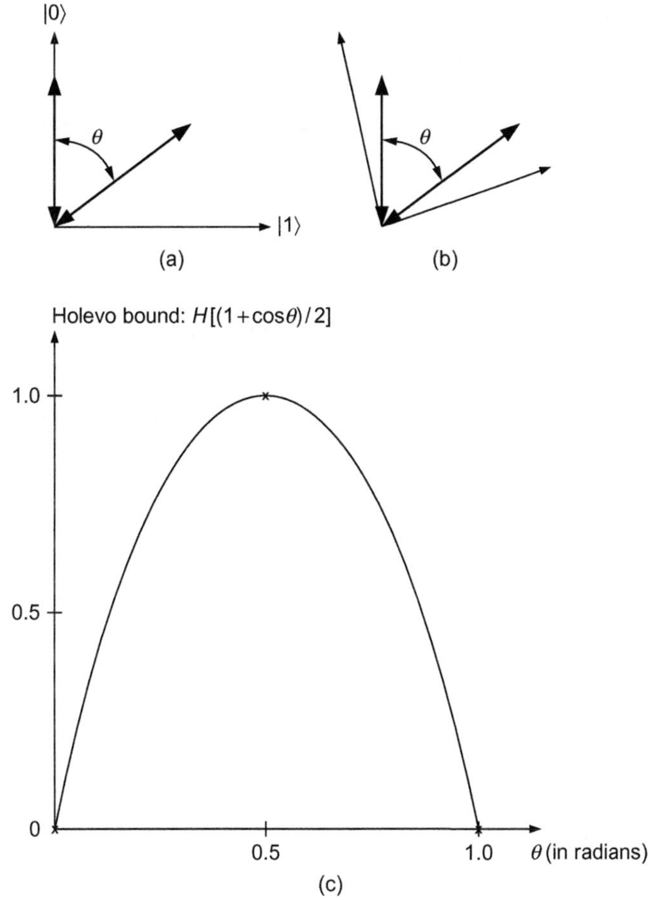

FIGURE 3.13

Information encoded as photon polarization. (a) Two non-orthogonal states. (b) An optimal measurement. (c) The plot of the Holevo bound, $H((1 \quad \cos\theta)/2)$, the function of the angle between photon polarization. Bob extracts maximum information when $\theta \quad \pi/2$, which corresponds to the orthogonal states.

3.13 NO-BROADCASTING THEOREM FOR GENERAL MIXED STATES

Broadcasting is a term used to describe the one-to-many communication where multiple recipients of information receive the same message. There is no limitation regarding broadcasting of classical information. In fact, widely used networking techniques for local area networks (e.g., Ethernet) and for radio and satellite communication use broadcasting. On the other hand, broadcasting a quantum state would allow us to replicate a mixed state, so we should not be surprised that general mixed states cannot be broadcast [21].

Table 3.6 The States of a Quantum System before and after Copy and Transposition

System	Initial State	State After *Copy*	State After *Transposition*
$\mathcal{A}$	$\varphi^{\mathcal{A}}$	$\varphi^{\mathcal{A}}$	$0^{\mathcal{A}}$
$\mathcal{B}$	$\varphi^{\mathcal{B}} \quad 0^{\mathcal{B}}$	$\varphi^{\mathcal{A}}$	$\varphi^{\mathcal{A}}$
$(\mathcal{A},\mathcal{B})$	$\varphi^{\mathcal{A}} \quad 0^{\mathcal{B}}$	$\varphi^{\mathcal{A}} \quad \varphi^{\mathcal{A}}$	$0^{\mathcal{A}} \quad \varphi^{\mathcal{A}}$

In Section 2.1, we discussed the impossibility of cloning pure or mixed non-orthogonal quantum states; cloning would allow us to replicate the states and distinguish between them with an arbitrary high reliability by measuring a large number of replicas of each state. We presented another line of reasoning for the impossibility of cloning mixed quantum states in Section 2.15; cloning an entangled state would lead to a violation of the monogamy of entanglement because the partner of an entangled state would end up being entangled with two states, the original as well as the cloned state.

Another argument for the impossibility of cloning entangled states is that neither one of two entangled systems has a definite state of its own, so it is impossible to clone a system whose state is not well defined. Now we reexamine cloning in a more general context provided by quantum information theory.

Schumacher draws a line between *copying* and *transposition*, an important issue for quantum information encoding [368]. If two quantum systems, $\mathcal{A}$, in a general state, $\varphi^{\mathcal{A}}$, and, $\mathcal{B}$, in a standard "null" state, $\varphi^{\mathcal{B}} \quad 0^{\mathcal{B}}$, are the input and, respectively, the output of a source or a channel encoder, then *copying* is described by a transformation where the original state of $\mathcal{A}$ is not disturbed and $\mathcal{B}$ is brought into a state corresponding to the original state of $\mathcal{A}$. The *transposition* does not leave a copy behind; the state of $\mathcal{A}$ is transferred to $\mathcal{B}$ and, at the same time, the original state of $\mathcal{A}$, $\varphi^{\mathcal{A}}$, is erased, (Table 3.6). Quantum information copying is possible if and only if the original states are mutually orthogonal.

We consider a slightly more general problem involving two quantum systems in an n-dimensional Hilbert space, $\mathcal{H}_n$; the two systems are $\mathcal{A}$, with two possible initial states, and $\mathcal{B}$. The states of the two systems are characterized by the density matrices, $\rho^{\mathcal{A}}_{before} \quad \rho_i, i \quad 0,1$ and $\rho^{\mathcal{B}}_{before}$. The initial state of the composite system, $(\mathcal{A},\mathcal{B})$, is

$$\rho^{(\mathcal{A},\mathcal{B})}_{before} \quad \rho^{\mathcal{A}}_{before} \quad \rho^{\mathcal{B}}_{before}.$$

The question we pose is if there is a physical process, $\mathcal{E}$, such that the state after the transformation, $\rho^{(\mathcal{A},\mathcal{B})}_{after} \quad \mathcal{E} \; \rho^{\mathcal{A}}_{before} \quad \rho^{\mathcal{B}}_{before} \quad \mathcal{H}_{n^2}$, has the following property:

$$\mathrm{tr}_{\mathcal{A}} \; \rho^{(\mathcal{A},\mathcal{B})}_{after} \quad \mathrm{tr}_{\mathcal{B}} \; \rho^{(\mathcal{A},\mathcal{B})}_{after} \quad \rho^{\mathcal{A}}_{before}.$$

If such a transformation, $\mathcal{E}$, exists for the two possible initial states, ρ_0 and ρ_1 of $\mathcal{A}$, then we say that $\mathcal{A}$ can be broadcast.

Theorem. *(No-broadcasting) $\mathcal{A}$ can be broadcast if and only if the two initial states of $\mathcal{A}$ commute,* $\rho_0 \quad \rho_1 \quad \rho_1 \quad \rho_0$.

Proof. The implication that the state can be broadcast if the two states commute is easy to prove. Indeed, if the two states commute, then there is an orthonormal basis, e_i , in which $\rho_0^{\mathcal{A}}$ and $\rho_1^{\mathcal{A}}$ can be simultaneously diagonalized:

$$\rho_0 \quad \overset{n}{\underset{i\ 1}{}} \ p_{i,0}\ e_i\ \ e_i \quad \text{and} \quad \rho_1 \quad \overset{n}{\underset{i\ 1}{}} \ p_{i,1}\ e_i\ \ e_i\ .$$

Consider a particular state of $\mathcal{B}$, namely, $\rho^{\mathcal{B}} \quad 1\ \ 1$, and let **U** be a unitary transformation of the joint system, $(\mathcal{A}, \mathcal{B})$, such that

$$\mathbf{U}(\ e_i \quad 1\ ) \quad e_i \quad e_i\ , \quad i \quad 1, n\ .$$

In this case,

$$\rho_{after}^{(\mathcal{A},\mathcal{B})} \quad \mathbf{U}\rho_{before}^{(\mathcal{A},\mathcal{B})}\mathbf{U}^{\dagger}.$$

If $\rho_{before}^{\mathcal{A}} \quad \rho_0$, then

$$\rho_{after}^{(\mathcal{A},\mathcal{B})} \quad \mathbf{U} \quad {}_i p_{i,0}\ e_i\ \ e_i \quad (\ 1\ \ 1\ )\mathbf{U}^{\dagger} \quad {}_i p_{i,0}\,[\mathbf{U}(\ e_i \quad 1\ )][\mathbf{U}(\ e_i \quad 1\ )]^{\dagger}$$

$$\quad {}_i p_{i,0}\ e_i\ \ e_i\ \ e_i\ \ e_i\ .$$

It follows that

$$\mathrm{tr}_{\mathcal{A}}\ \rho_{after}^{(\mathcal{A},\mathcal{B})} \quad \mathrm{tr}_{\mathcal{B}}\ \rho_{after}^{(\mathcal{A},\mathcal{B})} \quad \overset{n}{\underset{i\ 1}{}} \ p_{i,0}\ e_i\ \ e_i \quad \rho_0.$$

Similarly, if $\rho_{before}^{\mathcal{A}} \quad \rho_1$, then

$$\rho_{after}^{(\mathcal{A},\mathcal{B})} \quad \mathbf{U} \quad {}_i p_{i,1}\ e_i\ \ e_i \quad (\ 1\ \ 1\ )\mathbf{U}^{\dagger} \quad {}_i p_{i,1}\,[\mathbf{U}(\ e_i \quad 1\ )][\mathbf{U}(\ e_i \quad 1\ )]^{\dagger}$$

$$\quad {}_i p_{i,1}\ e_i\ \ e_i\ \ e_i\ \ e_i\ .$$

It follows that

$$\mathrm{tr}_{\mathcal{A}}\ \rho_{after}^{(\mathcal{A},\mathcal{B})} \quad \mathrm{tr}_{\mathcal{B}}\ \rho_{after}^{(\mathcal{A},\mathcal{B})} \quad \overset{n}{\underset{i\ 1}{}} \ p_{i,1}\ e_i\ \ e_i \quad \rho_1. \qquad \blacksquare$$

It is harder to prove that if $\mathcal{A}$ can be broadcast, then the two initial states of $\mathcal{A}$ commute. The proof presented by Barnum *et al* [21] is based on the following property of fidelity mentioned in Section 3.11: *fidelity does not decrease under partial trace.* We start by assuming that

$$\mathrm{tr}_{\mathcal{A}}\ \rho_{after}^{(\mathcal{A},\mathcal{B})} \quad \mathrm{tr}_{\mathcal{B}}\ \rho_{after}^{(\mathcal{A},\mathcal{B})} \quad \rho_{before}^{\mathcal{A}}.$$

Let $\mathbf{M}_i$ be an optimal POVM measurement allowing us to distinguish between the states, ρ_0 and ρ_1; the system, $\mathcal{A}$, was initially prepared in one of them. It is easy to see that

$$\mathrm{tr}\ \rho_{after}^{(\mathcal{A},\mathcal{B})}\ (\mathbf{M}_i \quad \mathbf{I}) \qquad \mathrm{tr}_{\mathcal{A}}\ \mathrm{tr}_{\mathcal{B}}\ \rho_{after}^{(\mathcal{A},\mathcal{B})}\ \ \mathbf{M}_i \qquad \mathrm{tr}_{\mathcal{A}}\ \rho_{before}^{\mathcal{A}} \mathbf{M}_i\ .$$

But $(\mathbf{M}_i \quad \mathbf{I})$ may not be an optimal POVM to distinguish between $\rho_{0,after}^{(\mathcal{A},\mathcal{B})}$ and $\rho_{1,after}^{(\mathcal{A},\mathcal{B})}$, the density matrices of the joint system after the measurement, corresponding to the system $\mathcal{A}$ being prepared in the states, ρ_0 and ρ_1, respectively. It follows that, indeed, fidelity cannot decrease under partial trace:

$$F_{\mathcal{A}}(\rho_0, \rho_1) \qquad {}_i\ \overline{\mathrm{tr}\ \rho_{0,after}^{(\mathcal{A},\mathcal{B})}\ (\mathbf{M}_i \quad \mathbf{I})}\ \ \overline{\mathrm{tr}\ \rho_{1,after}^{(\mathcal{A},\mathcal{B})}\ (\mathbf{M}_i \quad \mathbf{I})}$$

$$\min{}_{\mathbf{N}_j} \quad {}_j\ \ \overline{\mathrm{tr}\ \rho_{0,after}^{(\mathcal{A},\mathcal{B})} \mathbf{N}_j}\ \ \overline{\mathrm{tr}\ \rho_{1,after}^{(\mathcal{A},\mathcal{B})} \mathbf{N}_j}$$

$$F\ \ \rho_{0,after}^{(\mathcal{A},\mathcal{B})}, \rho_{1,after}^{(\mathcal{A},\mathcal{B})}\ \ .$$

Similarly,

$$F_{\mathcal{B}}(\rho_0, \rho_1) \qquad {}_i\ \overline{\mathrm{tr}\ \rho_{0,after}^{(\mathcal{A},\mathcal{B})}\ (\mathbf{I} \quad \mathbf{M}_i)}\ \ \overline{\mathrm{tr}\ \rho_{1,after}^{(\mathcal{A},\mathcal{B})}\ (\mathbf{I} \quad \mathbf{M}_i)}$$

$$F\ \ \rho_{0,after}^{(\mathcal{A},\mathcal{B})}, \rho_{1,after}^{(\mathcal{A},\mathcal{B})}\ \ .$$

We can also prove the opposite inequalities:

$$F_{\mathcal{A}}(\rho_0, \rho_1) \quad F\ \ \rho_{0,after}^{(\mathcal{A},\mathcal{B})}, \rho_{1,after}^{(\mathcal{A},\mathcal{B})} \qquad \text{and} \quad F_{\mathcal{B}}(\rho_0, \rho_1) \quad F\ \ \rho_{0,after}^{(\mathcal{A},\mathcal{B})}, \rho_{1,after}^{(\mathcal{A},\mathcal{B})}\ \ .$$

This means that if $\mathcal{A}$ can be broadcast, then there are density operators, $\rho_{0,after}^{(\mathcal{A},\mathcal{B})}$ and $\rho_{1,after}^{(\mathcal{A},\mathcal{B})}$, describing the joint system, $(\mathcal{A}, \mathcal{B})$, after the measurement, corresponding to ρ_0 and ρ_1, respectively, such that

$$F\ \ \rho_{0,after}^{(\mathcal{A},\mathcal{B})}, \rho_{1,after}^{(\mathcal{A},\mathcal{B})} \qquad F_{\mathcal{A}}(\rho_0, \rho_1) \quad F_{\mathcal{B}}(\rho_0, \rho_1).$$

The proof that these equalities imply that ρ_1 and ρ_2 commute can be found in Barnum, Caves, Fuchs, Jozsa, and Schumacher [21].

Cloning is a special case of broadcasting; when given a process, $\mathcal{E}$, we have

$$\mathcal{E}\ \ \rho_{before}^{\mathcal{A}} \quad \rho_{before}^{\mathcal{B}} \qquad \rho_{before}^{\mathcal{A}} \quad \rho_{before}^{\mathcal{A}}.$$

The last equality implies that $\mathcal{A}$ is clonable if and only if

$$F(\rho_0, \rho_1) \quad F(\rho_0 \quad \rho_0, \rho_1 \quad \rho_1) \quad [F(\rho_0, \rho_1)]^2 \qquad F(\rho_0, \rho_1) \quad 0 \text{ or } F(\rho_0, \rho_1) \quad 1.$$

In other words, *cloning is possible if and only if the two initial states of $\mathcal{A}$ are either identical, or orthogonal.*

We are now able to discuss the quantum equivalent of Shannon source coding, Schumacher compression.

3.14 SCHUMACHER COMPRESSION

Schumacher compression closely follows the arguments for classical source encoding from Section 3.4. The input alphabet of the source, $\mathcal{A}$, is determined by the number, D, of eigenstates of the density operator of one qubit; we assume that $D \quad 2$, though other physical implementations of qubits, with $D > 2$, may be possible.

We encode sequences of n qubits in $(\mathcal{H}_2)\ ^n$, and the input states of the encoder are

$$\alpha_k \qquad {}^{n}_{i\ 1}\ a_{i,k}\ ,\ 1 \quad k \quad K,$$

with the probability

$$p_k \qquad \overset{n}{\underset{i\ 1}{}}\ p_{i,k}.$$

The density matrix of the input to the encoder,

$$\rho^{\mathcal{A}} \qquad \overset{K}{\underset{k\ 1}{}}\ p_k\rho_k \qquad \overset{K}{\underset{k\ 1}{}}\ p_k \ \overset{n}{\underset{i\ 1}{}}\ \rho_{i,k},$$

has, d, eigenvectors, φ_i , $1 \quad i \quad d$, with the probabilities p_i, $1 \quad i \quad d$. The source is characterized by von Neumann entropy, $S(\rho) \qquad \text{tr}\ \ \rho^{\mathcal{A}} \log \rho^{\mathcal{A}}$. An eigenstate of the system is a sequence of d integers from the alphabet, $(0 \quad D)$ (in our case 0s and 1s), and the eigenvalue of this state is the probability that this particular sequence was generated by the source.

We cannot expect that the output of a quantum encoder matches exactly its input when the block length increases, n , as is the case of classical information; therefore, we have to find other ways to correlate the two. Now we use the fidelity between the input and the output states of the encoder as a measure of "goodness" for the transformation of the quantum state performed by the encoder. If ρ_{out} is the output of the encoder, then the fidelity is

$$F\ \ \rho^{\mathcal{A}}, \rho_{out} \qquad \overline{\ \rho^{\mathcal{A}}\ ^{1/2} \rho_{out}\ \ \rho^{\mathcal{A}}\ ^{1/2}\ }.$$

By analogy with a classical source, we define *typical subspaces* of the source with von Neumann entropy, $S(\rho)$, as subspaces generated by typical sequences of eigenvectors of the density matrix. Call $\Lambda_{typical}$ the subspace generated by typical sequences of eigenvectors of $\rho^{\mathcal{A}}$. Almost all the time the source produces a typical subspace.

We can now formulate Schumacher compression based on the analogy between the probabilities and the eigenvalues:

$$\dim\ \Lambda_{typical} \qquad 2^{nS\ \ \rho^{\mathcal{A}}\ \ \ \mathcal{O}(n)}.$$

Theorem. *We can compress the quantum information produced by a source,* $\mathcal{A}$, *with von Neumann entropy,* $S\ \ \rho^{\mathcal{A}}$, *up to* $\log\ 2^{nS\ \ \rho^{\mathcal{A}}\ \ \ \mathcal{O}(n)} \qquad nS\ \ \rho^{\mathcal{A}}$ *qubits; we can send,* n, *symbols using* $[nS\ \ \rho_{\mathcal{A}} \quad \mathcal{O}(n)]$ *qubits, with a fidelity approaching* 1 *as* n .

The Schumacher compression works [388] because two quantum sources, (1) one with the eigenvectors, φ_1 , φ_2 ,..., φ_d , with the probabilities, $p_i, 1 \quad i \quad d$, and (2) one with the states, α_1 , α_2 ,..., α_K , with the probabilities, $P_1, P_2, \ldots, P_K$, have the same density matrix, ρ. Therefore, the probability of the outcomes will be the same. The density matrix determines the outcomes of any measurement. It follows that the input state of the encoder is almost always close to a typical subspace of the source. More details regarding the encoding process for noiseless quantum channels are presented in Section 3.17.

Finally, we should remember that *the von Neumann entropy does not provide a complete characterization of a source of quantum information;* indeed, two different sources of quantum information, $\mathcal{A}$ and $\mathcal{B}$, may have the same density matrix:

$$\rho^{\mathcal{A}} \qquad \sum_k p_k^{\mathcal{A}} \varphi_k^{\mathcal{A}} \quad \varphi_k^{\mathcal{A}} \quad \text{and} \quad \rho^{\mathcal{B}} \qquad \sum_k p_k^{\mathcal{B}} \varphi_k^{\mathcal{B}} \quad \varphi_k^{\mathcal{B}}, \quad \rho^{\mathcal{A}} \quad \rho^{\mathcal{B}} \quad \rho.$$

The von Neumann entropy of $\mathcal{A}$ and $\mathcal{B}$ is the same, $S(\rho) \qquad \mathrm{tr}\,(\rho \log \rho)$, but the two sources, $\mathcal{A}$ and $\mathcal{B}$, are different from the quantum information theoretical point of view. The recipient of the information sent by the two sources may be able to discover the identity of the input state more accurately for one of the sources than for the other. For example, the source, $\mathcal{A}$, emits linearly polarized photons at 0 and 90 with equal probability, $p_{1,2}^{\mathcal{A}} \quad 1/2$; the density matrix of $\mathcal{A}$ is $\rho^{\mathcal{A}} \quad I/2$, with I, the identity matrix, and the entropy of $\mathcal{A}$ is one bit. The states of the two photons are orthogonal, thus, perfectly distinguishable, and the information can be measured and transmitted as classical information. The source, $\mathcal{B}$, emits linearly polarized photons at 0 ,45 ,90 , and 135 with equal probability, $p_{1,2,3,4}^{\mathcal{B}} \quad 1/4$; the density matrix is $\rho^{\mathcal{B}} \quad I/2$, but in this case the information cannot be transmitted as classical information and the recipient of information cannot identify with absolute certainty the state of the source $\mathcal{B}$ [35]. The problem of reversible extraction of classical information from quantum information is discussed in Section 3.20.

Next, we turn our attention from the source of quantum information to quantum channels.

3.15 QUANTUM CHANNELS

Quantum information theory studies transmission of classical and quantum information over quantum channels *from one point to another, or from one time to another*; whenever we talk about the "input" and the "output" of the quantum channel, we mean the state of the quantum system before and, respectively, after the transformation done by the quantum channel.

To develop a theory of transmission of quantum information, we map known concepts from the classical information theory to the quantum domain. Recall from Section 3.7 that the classical noisy channel coding theorem applies to discrete memoryless channels, channels over a finite alphabet and with a finite number of states. The memoryless property of the communication channel implies that the output of the channel is a Markov process; it is affected only by the current input and not by the history of the channel states.

A *discrete memoryless quantum channel* transforms a quantum system whose state is a vector in a finite-dimensional Hilbert space. The memoryless property has the same meaning as for the classical information; the output of the quantum channel is only determined by the current input. Figure 3.14

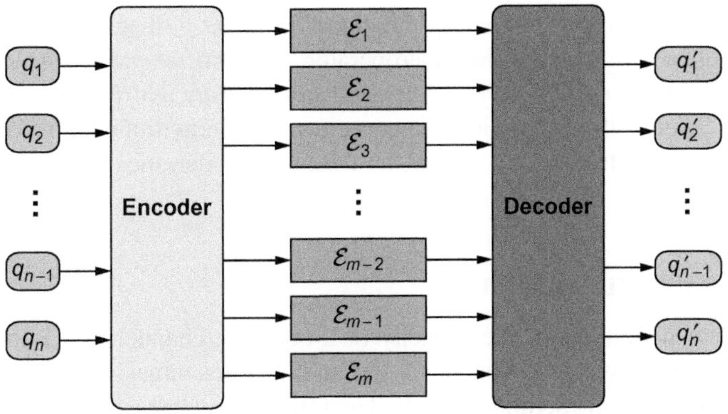

FIGURE 3.14

Communication through a quantum channel. The input is an ensemble of n qubits in a pure state, φ , with the density matrix, $\rho \quad \varphi \; \varphi$. The encoder maps the, n, input qubits into $m > n$ intermediate systems that may not be qubits and feeds them to, m, independent instances of a quantum channel characterized by the superoperator, $\mathcal{E}$; the output of each instance of the channel is generally a mixed state, with the density function, $\rho \quad \mathcal{E}(\rho)$. Finally, the decoder transforms the output of the quantum channels into a mixed state of n qubits, with the density function, ρ . The fidelity of the communication system is $F \quad \varphi \; \rho \; \varphi$.

captures the essential features of communication over a quantum channel; the input to a quantum channel is an ensemble of n qubits in a pure state. An encoder maps n input qubits into $m > n$ intermediate systems that may, or may not be qubits, and feeds them to, m, independent instances of a quantum channel characterized by the superoperator,[6] $\mathcal{E}$, describing the transformations of the input states. The output of each instance of the channel is generally a mixed state; the decoder transforms the output of the quantum channels into a mixed state of n qubits.

Quantum Channels and the Superoperator Formalism

To capture not only the essential aspects of quantum communication, but also the transformation of the quantum information during storage and processing, we consider a more general view of a quantum channel as completely positive, trace-preserving maps between spaces of operators.

The density matrix, ρ, plays an important role in quantum information theory because it captures all the information that can be obtained by an observer about a mixed state and only that; any test to identify the state of a mixed system requires the density matrix. Indeed, if the mixed state comes from an ensemble of the states, $\varphi_i : 1 \quad i \quad N$, with the probabilities $p_i : 1 \quad i \quad N$, then to determine if the system is in the state, ψ , we compute

$$\sum_{i\ 1}^{N} p_i \; \varphi_i \; \psi \;^2 \qquad \mathrm{tr}[\rho \; \psi \quad \psi \;].$$

[6]A superoperator is a linear operator, often positively defined, acting on a space of linear operators.

If a bipartite system, $\mathcal{AB}$, is characterized by the density matrix, ρ, then to determine if $\mathcal{A}$ is in the state, ψ , we compute $\mathrm{tr}_{\mathcal{B}}[\rho\ \psi\ \ \psi\ ]$. The superoperator, $\mathcal{E}$, is *a trace-preserving completely positive linear map from input-state density matrix ρ to output-state density matrix ρ , $\rho \quad \mathcal{E}(\rho)$*. The superoperator formalism captures the interaction of a system with an environment initially in a pure state and quantifies the transformation carried out by the channel encoder, the quantum channel, as well as the decoder.

The Capacity of a Quantum Channel

There are at least four definitions of the capacity of a quantum channel based on the nature of the information transmitted (classical or quantum), the auxiliary resources allowed (classical channel, entanglement), and the protocols permitted [387]. The four distinct capacities are [47] (1) C, the ordinary classical capacity for transmitting classical information; (2) Q, the ordinary quantum capacity for transmitting quantum states, typically, $Q < C$; (3) Q_2, the classically-assisted quantum capacity for transmitting quantum states with the assistance of a two-way classical channel, $Q \quad Q_2 \quad C$; and (4) C_E, the entanglement-assisted classical capacity, $Q \quad C \quad C_E$.

The *capacity Q of the quantum channel* is the largest number Q, such that for any rate, $R < Q$, $\epsilon > 0$, and block sizes, n and m, there exists an encoding procedure mapping, n, qubits in a pure state, φ , with the density matrix, $\rho \quad \varphi \ \ \varphi$, into $m > n$ intermediate systems that may not be qubits, and feeds them to, m, independent instances of a quantum channel characterized by the superoperator, $\mathcal{E}$; there should also be a decoding procedure mapping the m channel outputs to n qubits such that the original state, φ , can be recovered with a fidelity, F, at least $1 \quad \epsilon$.

Quantum Channel Fidelity

A quantum channel can be used to transport either classical or quantum information; the probability that the state at the output of the quantum channel would pass a test for being the same as the state at the input measures the reliability of the quantum channel by a quantity called *channel fidelity*. The *fidelity of quantum states, $F\ \rho^{\mathcal{A}},\rho^{\mathcal{B}}$* , introduced by Jozsa [220], provides a quantitative measure of the distinguishability of the two states, a measure of a channel fidelity. When $\varphi^{\mathcal{A}}$ and $\varphi^{\mathcal{B}}$ are pure states, the fidelity can be expressed as

$$F\ \rho^{\mathcal{A}},\rho^{\mathcal{B}} \qquad \varphi^{\mathcal{A}}\ \varphi_{\mathcal{B}} \ .$$

In the general case of mixed states, the fidelity is

$$F\ \rho^{\mathcal{A}},\rho^{\mathcal{B}} \qquad \mathrm{tr} \quad \overline{\rho^{\mathcal{A}\ 1/2}\rho^{\mathcal{B}}\ \rho^{\mathcal{A}\ 1/2}} \ .$$

The analog of a classical discrete memoryless noisy channel is a quantum system that interacts unitarily with the environment while in transit from a source to a destination. If the source is an ensemble of pure or mixed states in a Hilbert space of finite dimension, $d \quad 2^n$, with the density matrices, $\rho_1,\rho_2,\ldots,\rho_{2^n}$, with the probabilities, $p_1,p_2,\ldots,p_{2^n}$, respectively, then we can transmit, 2^n, classical messages as a sequence of n qubits. If the states, ρ_i, are orthogonal, we can obtain *complete* information about the source state by performing a measurement, and then we are faced with the classical information transmission scenario. If the source states are non-orthogonal, then no measurement can extract complete information about the state, thus, *the best we can achieve is to faithfully reproduce the input state at the receiving end.* Needless to say, in the non-orthogonal state case, the channel must let the quantum system pass through *without* learning anything about the state.

At the receiving end, we have a similar problem when we wish to assess the quality of the transmission and eventually correct the errors. Quantum error correction is possible due to the following property of quantum systems: If the bipartite system, $(\mathcal{A}, \mathcal{B})$, in the state, $\rho^{\mathcal{A}} \quad \rho^{\mathcal{B}}$, is subject to a transformation, $\mathcal{E}$, and if there is an auxiliary system, $\mathcal{C}$, in some standard state, $\rho^{\mathcal{C}}$, and some unitary operator, $\mathbf{U}$, acting on the joint system, $(\mathcal{A}, \mathcal{B}, \mathcal{C})$, and if

$$\mathcal{E} \quad \rho^{\mathcal{A}} \quad \rho^{\mathcal{B}} \qquad \mathrm{tr}_{\mathcal{C}} \; \mathbf{U} \quad \rho^{\mathcal{A}} \quad \rho^{\mathcal{B}} \quad \rho^{\mathcal{C}} \quad ,$$

then after applying the transformation, $\mathbf{U}$, we can perform a measurement on $\mathcal{C}$ that reveals how close the two states $\rho^{\mathcal{A}}$ and $\rho^{\mathcal{B}}$ are, and we then ignore the auxiliary system $\mathcal{C}$ [21]. The systems, $\mathcal{A}$ and $\mathcal{B}$, could be the input and the output of a quantum channel and, $\mathcal{E}$, the transformation of the input state density carried out by the channel; then the process described above allows us to determine if the output is in error and to pinpoint the qubit(s) in error. We defer the in-depth discussion of the solution to this problem to Chapter 5, where we analyze quantum error-correcting codes.

Mappings Between Hilbert Spaces of Different Dimensions

The analysis of communication through a quantum channel sometimes requires mappings between Hilbert spaces of different dimensions, and the superoperator formalism allows such mappings by adding dummy dimensions to the smaller spaces. Stinespring's dilation theorem [411] provides the mathematical justification for mappings between Hilbert spaces of different dimensions. A brief review of basic concepts related to such mappings follows.

If A is a unital C -algebra and $B(\mathcal{H})$ are the bounded operators on the Hilbert space $\mathcal{H}$, then for every completely positive map, $T : A \quad B(\mathcal{H})$, there exists another Hilbert space, $\mathcal{K}$, and a unital *-homomorphism[7] $L : A \quad B(\mathcal{K})$, such that

$$T(a) \quad VL(a)V \;, \quad a \quad A,$$

with $V : \mathcal{K} \quad \mathcal{H}$, an bounded operator. Then:

$$T(1) \qquad V^{\;2}.$$

This result from operator theory [411] shows that a completely positive map in a C -algebra can be represented as a composition of two completely positive maps.

An important consequence of this result for quantum information theory is that *completely positive and trace-preserving quantum operations can be considered as a unitary evolution of a larger (dilated)*

---

[7] A homomorphism, $f : A \quad B$, is algebra homomorphism compatible with involutions of A and B, i.e.,

$$f(a\;) \quad f(a) \;, \quad a \quad A.$$

A homomorphism is unital if $f(1) \quad 1$.

system of a bounded dimension, the church of larger Hilbert space according to John Smolin. The interpretation of Stinespring's dilation theorem is that any completely positive and trace-preserving map in a Hilbert space (a quantum channel) can be constructed from three operations:

1. Tensoring the input with a second Hilbert system (the ancilla system) in a specified state
2. A unitary transformation on the larger (*dilated*) space (input plus ancilla system) obtained as a result of step 1
3. Reduction to a subsystem

Theorem. *Let $\mathcal{S}_{\mathcal{H}}$ be the set of states of the Hilbert space, $\mathcal{H}$, and T be a completely positive and trace-preserving map, $T : \mathcal{S}_{\mathcal{H}} \quad \mathcal{S}_{\mathcal{H}}$. Then there exist a Hilbert space, $\mathcal{K}$, and a unitary operation, U, on $\mathcal{S}_{\mathcal{H}} \quad \mathcal{S}_{\mathcal{K}}$, with $\mathcal{S}_{\mathcal{K}}$, the set of states of the Hilbert space $\mathcal{K}$, such that*

$$T(\rho) \quad tr_{\mathcal{K}} \; U(\rho \quad 0 \;\; 0\,)U^{\dagger}$$

for all states, $\rho \quad \mathcal{S}_{\mathcal{H}}$, where $tr_{\mathcal{K}}$ denotes the partial trace on the $\mathcal{K}$ system. The dimension of the Hilbert space $\mathcal{K}$ is bounded:

$$\dim\mathcal{K} \quad \dim^2\mathcal{H}.$$

This is a version of a more general Stinespring's dilation theorem restricted to completely positive and trace-preserving maps and finite-dimensional systems; moreover, it assumes that the dimensions of the input and output spaces for the mapping, T, are the same. The triplet, $(L, V, \mathcal{K})$, is called a *Stinespring representation* of T.

In Stinespring's representation, the ancilla system has a natural interpretation as the environment of the physical system under investigation. The output of the channel, T, arises from a unitary interaction of the input state with the environment, followed by a partial trace over the degrees of freedom of the environment. Almost all information can be extracted from the output channel through a decoding operation if and only if the channel T releases almost no information to the environment. For states (i.e., channels with one-dimensional domains), dilations are called *purifications*.

In summary, a quantum channel is a completely positive trace-preserving linear map; according to Stinespring's dilation theorem, this map can be described by a unitary transformation followed by a partial trace. In the next section we discuss the quantum erasure phenomenon.

3.16 QUANTUM ERASURE

In our extensive discussion of quantum measurements in Chapter 2, we argued that the density matrix of a system characterizes all possible measurements performed on that system. The quantum erasure experiment discussed in this section raises the question of whether indeed the density matrix provides a complete description of the state of a subsystem of a composite system.

We first review the concept of interference and Young's double-slit experiment discussed in many textbooks [283]; then we introduce the coherent and the incoherent superposition states of a quantum system. The term *interference* refers to the interaction of optic, acoustic, or electromagnetic waves that are correlated, or coherent, with each other; interference occurs if the waves come from the same source, or if the waves have about the same frequency. Let us consider a transition between two quantum states, φ , ψ $\mathcal{H}_n$, and let e_i be an orthonormal basis in $\mathcal{H}_n$; then,

$$\varphi \quad \alpha_i \; e_i \quad \psi \quad \beta_i \; e_i \; .$$
$$i \qquad\qquad i$$

As we know from Section 1.7, in quantum mechanics, the probability of this transition is the square of the modulus of the inner product of the two states:

$$\text{Prob}\,(\;\varphi \quad \psi\;) \quad \varphi\;\psi\;^2 \quad \alpha_i\;\beta_i\;^2 \quad \alpha_i\;\alpha_j\beta_j\;\beta_i.$$
$$i \qquad i,j$$

The last sum can be separated into two terms; thus, the expression becomes

$$\text{Prob}\,(\;\varphi \quad \psi\;) \quad \alpha_i\;^2\;\beta_i\;^2 \quad \alpha_i\;\alpha_j\beta_j\;\beta_i.$$
$$i \qquad i,j,i\;\;j$$

The first term corresponds to a *classical interference*, while the second term reflects the *quantum interference* between i j paths of quantum evolution; this is the quantum probability rule.

A qubit could be in a superposition state, and we distinguish two types of superposition states, coherent and incoherent. A quantum system in the state, φ $_i\alpha_i\;e_i$ $\mathcal{H}_n$, is in a *coherent superposition of basis states*, e_i , if its density matrix, ρ, is not diagonal, ρ $(1/n)I_n$. If, in addition, the system is in a pure state, then it is said to be *completely coherent*.

If ρ is diagonal, ρ $(1/n)I_n$, then the system is in an *incoherent superposition of states*. In particular, a qubit is in an incoherent superposition of states if

$$\rho \quad \frac{1}{2} \quad \begin{matrix} 1 & 0 \\ 0 & 1 \end{matrix} \; .$$

In the case of an incoherent superposition, the basis states are prepared independently, and there is no definite phase relationship between the basis states; no interference is possible.

A measurement destroys coherence; in his lecture notes in physics [148], Feynman gives a clear example of the interference and how a measurement destroys coherence in a Young's experiment with electrons. When we conduct the experiment without attempting to learn which hole an electron passes through, we observe an interference pattern on the recording media positioned behind the screen with the two holes. We can identify the position of the electrons by illuminating the electrons with a laser beam; the electrons interact with the laser beam and emit a quanta of light, allowing us to know precisely the hole each electron is passing through. Now the coherence of the electrons is destroyed and we observe a Gaussian distribution on the recording media.

To discuss the quantum erasure, we consider qubits, embodied by spin-½ particles (e.g., electrons). The projections of the spin of particle, A, along the x-, y-, and z-directions can be up or down:

$$A,x \;,\quad A,x \;,\quad A,y \;,\quad A,y \;,\quad A,z \;,\quad A,z \;.$$

The orientation of the spin can be measured with a Stern-Gerlach apparatus oriented along one of the three directions. For example, when the Stern-Gerlach apparatus is oriented along the x-direction, it will measure the projection of the spin along z, the direction of the magnetic field. If we denote the basis states of a spin-½ system along a particular direction, spin up and spin down as 0 and 1 , respectively, then any superposition state can be expressed as φ $\alpha_0\;0$ $\alpha_1\;1$. The probability amplitudes, α_0 and α_1, encode the probability of an outcome in the (0 , 1) basis; the relative phases of α_0 and α_1 are also significant. The two complex numbers, α_0 and α_1, describe the three-dimensional orientation of the spin; if θ is the polar angle and ϕ the azimuthal angle, then $\alpha_0 \quad e^{-i\phi/2}\cos(\theta/2)$ and $\alpha_1 \quad e^{\ i\phi/2}\sin(\theta/2)$. The state φ can be interpreted as a spin pointing in the (θ,ϕ) direction.

If identical copies of the spin-½ system are available, then the two complex numbers α_0 and α_1 cannot be determined with a single measurement, or even with any number of measurements, *carried along only one* of the x-, y-, or z-directions. A vector in a three-dimensional space can only be determined if we know the projections along all three directions. We know by this time that unitary transformations of a qubit are rotations and there is a transformation that allows us to rotate the spin pointing in the (θ,ϕ) direction to, say, the z-direction; in that case, only one measurement will be sufficient to determine not only the modulus of the two complex numbers, but also the relative phase between them.

Consider two orthogonal superposition states when the projections of the spin of A in the x-direction are up and down, respectively, as in Figure 3.15:

$$_{A,x} \quad \frac{1}{2}(\quad _{A,z} \quad\quad _{A,z} \quad \text{and} \quad _{A,x} \quad \frac{1}{2}(\quad _{A,z} - \quad _{A,z} \ .$$

A measurement of the projection of the spin along direction x results in one of the two eigenvectors, $_{A,x}$ and $_{A,x}$, with the probability 1/2.

Now we consider the superposition state:

$$\varphi \quad \frac{1}{2}(\quad _{A,x} \quad\quad _{A,x} \ .$$

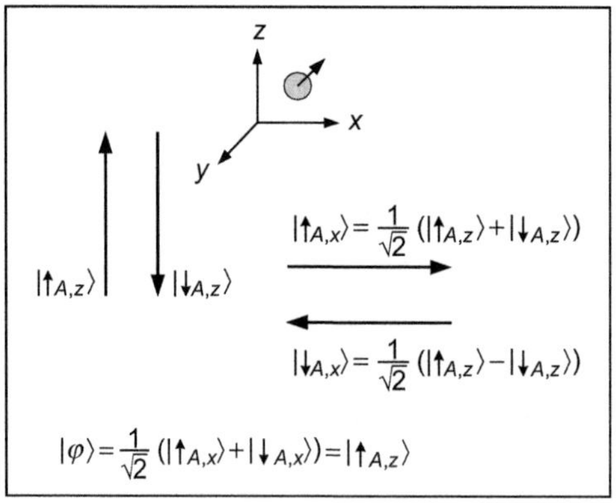

FIGURE 3.15

A superposition state, $\varphi \quad \frac{1}{2}(\quad _{A,x} \quad _{A,x} \quad _{A,z}$, of a spin-½ particle, A, where $_{A,x} \quad \frac{1}{2}(\quad _{A,z} \quad _{A,z}$ and $_{A,x} \quad \frac{1}{2}(\quad _{A,z} - \quad _{A,z}$. The projection of the spin of particle A along the x-direction is in a coherent superposition state.

When we measure the projection of the spin of the qubit in the state, φ , along the x-direction, we observe one of the two eigenvectors, $_{A,x}$ and $_{A,x}$, with the probability 1/2. If we measure the spin of the qubit in the state, φ , along the z-direction, we find $_{A,z}$, with the probability 1. A simple analysis shows that we should not be surprised at all, φ is, in fact, $_{A,z}$. Indeed,

$$\varphi \quad \frac{1}{\overline{2}} \quad {}_{A,x} \quad {}_{A,x} \quad \frac{1}{\overline{2}}\ \frac{1}{\overline{2}} \quad {}_{A,z} \quad {}_{A,z} \quad \frac{1}{\overline{2}} \quad {}_{A,z} \quad {}_{A,z} \quad {}_{A,z} \ .$$

In the case of a *coherent superposition*, we are able to distinguish the relative phases of the two spin states of A; the two states can interfere with each other. Such a coherent superposition state is

$$_{A,x},\ _{A,x} \quad \frac{1}{\overline{2}} \quad {}_{A,x} \quad {}_{A,x} \ .$$

The two pure states, $_{A,z}$ and $_{A,z}$, can interfere only if we have no means to find out if the projection of the spin of A along the x-direction is up or down.

An *incoherent superposition* state has a density matrix, $\rho \quad 1/2I$. The density matrix of this maximally mixed state can be obtained through different preparations:

$$\rho \quad \frac{1}{2} \quad {}_{A,x} \quad {}_{A,x} \quad {}_{A,x} \quad {}_{A,x} \quad \text{or} \quad \rho \quad \frac{1}{2} \quad {}_{A,z} \quad {}_{A,z} \quad {}_{A,z} \quad {}_{A,z} \ .$$

The relative phase of an incoherent superposition is totally unobservable.

Consider now two qubits, A and B, and their entangled state (Figure 3.16),

$$\varphi_{AB} \quad \frac{1}{\overline{2}} \quad {}_{A,x} \quad {}_{B,x} \quad {}_{A,x} \quad {}_{B,x} \ .$$

To determine the orientation of the spin of A, we perform a measurement on B along the direction A that is oriented. For example, Bob could measure the spin of, B, along, x, and find out that the state is $_{B,x}$; this measurement produces a decoherence of A along the x-direction; it forces it to choose a certain orientation of the spin (e.g., $_{A,x}$). The result of the measurement is sent back to Alice, the preparer of A; once this message is received, A is again in a pure state, either $_{A,x}$ or $_{A,x}$.

Now we discuss the following experimental setup (Figure 3.17): Bob feeds particle B into a Stern-Gerlach apparatus oriented to measure the projection of the spin along the x-direction and then, without recording the result of the first measurement, refocuses the two emerging beams into a second Stern-Gerlach apparatus oriented for a measurement along the z-direction. Bob then communicates to Alice if the projection of the spin of B along the z-direction is up or down; after this message reaches Alice, the coherence of the states $_{A,x}$ and $_{A,x}$ of A is restored. This phenomenon is called *quantum erasure*; the entanglement causes the loss of coherence between $_{A,z}$ and $_{A,z}$ after the first measurement, and then the second measurement along z erases this information and restores the coherence along the x-direction.

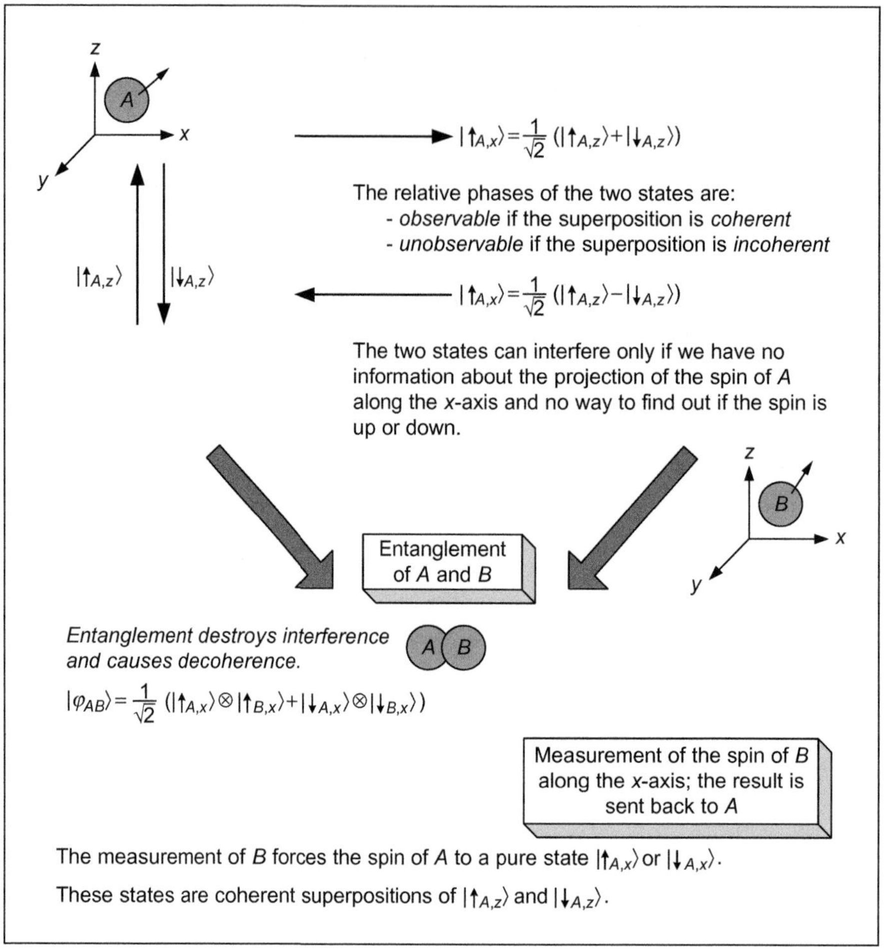

FIGURE 3.16

Entanglement destroys interference and causes decoherence. A measurement could restore a coherent superposition.

The explanation is that while Alice cannot observe the interference between $_{A,x}$ and $_{A,x}$ in the ensemble with the density matrix, ρ_A $(1/2)I$, once she gets the results of the measurement along the z-direction, this ensemble is forced into a particular preparation, while the density matrix remains the same, ρ_A $(1/2)I$. Only after receiving the information from Bob can Alice select a subensemble of her spins that are in the pure state, $_{A,z}$. The information allows Alice to distill *purity* along the z-direction from a maximally mixed state along the x-direction.

Preskill argues [342] that "the state ρ_A of system A is not the same as ρ_A accompanied by the information Alice has received from Bob." The information provided by Bob about particle,

The projections of the spin of A along the x-axis are in a coherent superposition state.

$$|\uparrow_{A,x}\rangle = \tfrac{1}{\sqrt{2}}(|\uparrow_{A,z}\rangle + |\downarrow_{A,z}\rangle)$$

$$|\uparrow_{A,x}\rangle = \tfrac{1}{\sqrt{2}}(|\uparrow_{A,z}\rangle - |\downarrow_{A,z}\rangle)$$

The entanglement of A and B causes the decoherence of the two projections of the spins of A along the x-axis.

Entanglement

Measure the spin of B along the z-axis using a Stern-Gerlach apparatus; the result is is *not* recorded.

1. Refocus the beams from the first Stern-Gerlach appartus.
2. Feed it into a second Stern-Gerlach apparatus.
3. Measure the spin of B along the x-axis.
4. Send the result to A.

$|\uparrow_{A,x}\rangle$ and $|\downarrow_{A,x}\rangle$
are now in a coherent superposition state.

The coherence of the projections of the spin of A along the x-axis is restored because the result of the first measurement of the spin of B along the z-axis was not communicated to A.

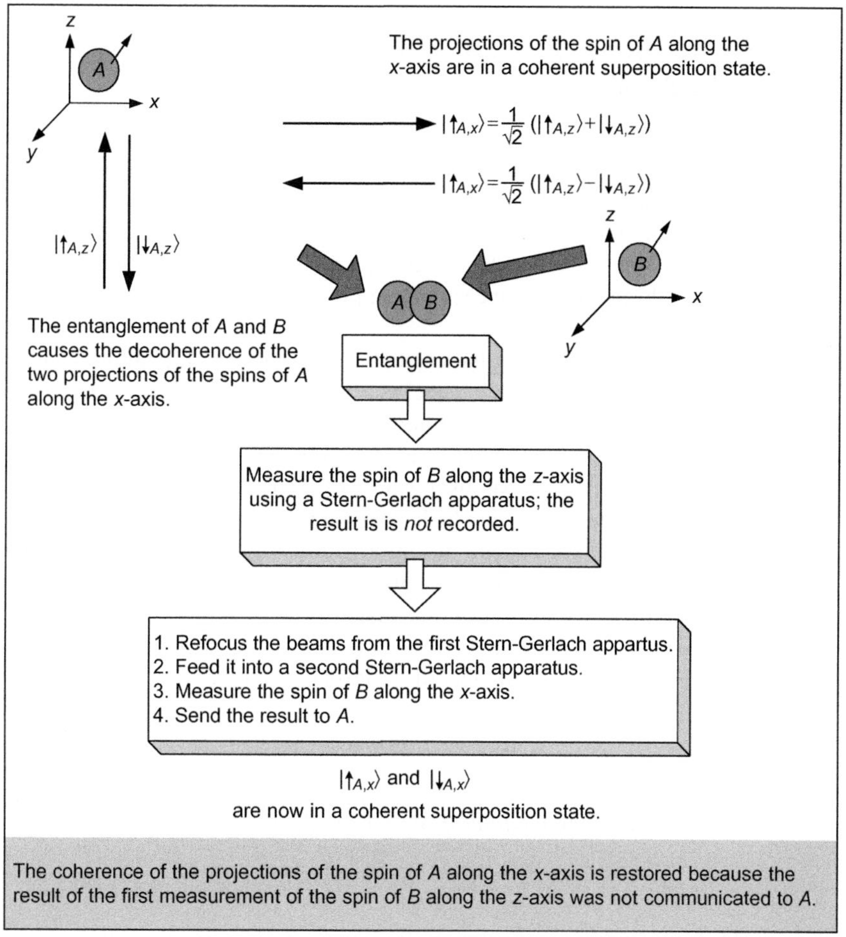

FIGURE 3.17

Quantum erasure. While Alice cannot observe the interference between, $_{A,x}$ and $_{A,x}$, in the ensemble with the density matrix, ρ_A $(1/2)I$, once she gets the results of the measurement along the z-direction, this ensemble is forced into a particular preparation, while the density matrix remains the same, ρ_A $(1/2)I$. Only after receiving the information from Bob can Alice select a subensemble of her spins that are in the pure state, $_{A,z}$. The information allows Alice to distill *purity* along the z-direction from a maximally mixed state along the x-direction.

B, changes the physical description of particle, A; the state of particle, A, is described by ρ_A $(1/2)I$ "the state of knowledge."

Quantum erasure illustrates that the behavior of particles carrying quantum information can be nonintuitive [342]. The experiments discussed in this section show that *the entanglement destroys interference and causes decoherence and that a measurement could restore a coherent superposition.*

3.17 CLASSICAL INFORMATION CAPACITY OF NOISELESS QUANTUM CHANNELS

We turn our attention to noiseless quantum channels with pure states as input and analyze their capacity to transmit classical information. Recall that the capacity of a classical channel was defined as the highest rate the information can be transmitted through the channel, the maximum of the mutual information between the input, X, and output, Y, of the channel, $C \quad \max I(X;Y)$. The mutual information depends on the fixed conditional probabilities, $p_{Y X}(y x)$, that quantify the effect of the noise on the letters transmitted by the source; the measurement process for classical information is trivial; a letter received is immediately identified.

There is a significant difference between the capacity of classical and quantum channels models discussed in this section: When the quantum channel is noiseless, the quantum states are transmitted unaltered, but, in the general case, they are non-orthogonal, and the recipient of quantum information has a choice of how to perform a measurement to identify the state transmitted by the source. The conditional probabilities arise from the probabilistic nature of quantum measurements.

We wish to determine the maximum rate at which classical information may be transmitted through a noiseless quantum channel when we encode the information into n physical qubits. We consider an ensemble of letter states, $\mathcal{A} \quad a_k$, generated with a priori probability, p_{a_k}. The communication channel is characterized by the letter states it can transmit.

Call η_k the *overall frequency* of letter, a_k , computed over the entire set of letters used (the cardinality of this set is $n \quad N$). If ν_{ik} is the number of occurrences of letter a_k in codeword, w_i , then

$$\eta_k \quad \frac{1}{n} \sum_i p_{w_i} \nu_{ik}.$$

We expect that $\eta_k \quad p_{a_k}$, and we define the *tolerance* of the code, $\mathcal{C}^k$, as

$$\tau_{\mathcal{C}^k} \quad \max_{a_k} \eta_k \quad p_{a_k} .$$

Theorem. *The noiseless quantum channel capacity. Given: (1) the source of quantum information, $\mathcal{A}$, with the density matrix, $\rho^{\mathcal{A}}$, and the von Neumann entropy, $S(\rho^{\mathcal{A}})$, and (2) an arbitrarily small number, $\delta > 0$. Let I_δ be the least upper bound on the information per letter transmissible with any code, $\mathcal{C}^k$, such that its tolerance is $\tau_{\mathcal{C}^k} \quad \delta$. Then,*

$$\lim_{\delta \quad 0} I_\delta \quad S \ \rho^{\mathcal{A}} .$$

The proof of this theorem is presented by Hausladen *et al* [204]. If we relax the restriction that letters, a_k , are generated with a fixed a priori input probability distribution, p_{a_k}, and consider this distribution to be variable, we can define the capacity of the quantum channel as the maximum von Neumann entropy:

$$C \quad \max_{p_{a_k}} S \ \rho^{\mathcal{A}} .$$

This theorem shows that C is the maximum rate at which classical information can be transmitted through a quantum channel. That is formally similar to the statement of the classical channel capacity theorem. Classical channel capacity involves maximizing over the input distribution. Quantum channel capacity implies optimizing over choice both of input distribution and of decoding observable.

We now present a formulation of the Schumacher quantum noiseless coding theorem, which resembles more closely the Shannon source coding theorem [368].

Theorem. *Let $\mathcal{A}$ be a quantum source with a letter ensemble described by the density operator, $\rho^{\mathcal{A}}$, and let (δ, ϵ) 0, some arbitrarily small numbers.*

a) *Suppose that S $\rho^{\mathcal{A}}$ δ qubits are available per $\mathcal{A}$ letter. Then for sufficiently large n, groups of n letters from the source $\mathcal{A}$ can be transposed via the available qubits, with fidelity $F > 1$ ϵ.*

b) *Suppose that S $\rho^{\mathcal{A}}$ δ qubits are available per $\mathcal{A}$ letter. Then for sufficiently large n, groups of n letters from the source $\mathcal{A}$ can be transposed via the available qubits, with fidelity $F < \epsilon$.*

Proof. The proof is similar to the one for classical information. When $\rho^{\mathcal{A}}$ is the density matrix of the source, $\mathcal{A}$, a group of n physical qubits emitted by the source is characterized by the density matrix, $\rho^{\mathcal{A}}$ ^{n}, in $(\mathcal{H}_2)$ ^{n}. The Hilbert space, $(\mathcal{H}_2)$ ^{n}, is partitioned into two orthogonal subspaces, $\Lambda_{typical}$ and $\Lambda_{typical}$ [204], such that:

1. $\Lambda_{typical}$ and $\Lambda_{typical}$ are spanned by the eigenstates of the density matrix $\rho^{\mathcal{A}}$ ^{n}.
2. If Π_{Λ} and Π_{Λ} are the projectors onto $\Lambda_{typical}$ and $\Lambda_{typical}$, then

$$\text{tr}\ \Pi_{\Lambda}\ \rho^{\mathcal{A}}\ \ ^{n}\Pi_{\Lambda}\ > 1\ \ \epsilon \qquad \text{and} \qquad \text{tr}\ \Pi_{\Lambda}\ \ \rho^{\mathcal{A}}\ \ ^{n}\Pi_{\Lambda}\ \ < \epsilon.$$

3. The eigenvalues, λ_k, of $\rho^{\mathcal{A}}$ ^{n}, corresponding to the eigenstates of $\Lambda_{typical}$, satisfy the inequality

$$2^{\ n\ S\ \rho^{\mathcal{A}}\ \ \delta}\quad \lambda_k\quad 2^{\ n\ S\ \rho^{\mathcal{A}}\ \ \delta}.$$

4. The dimension of the typical subspace is bounded:

$$(1\quad \epsilon)2^{\ n\ S\ \rho^{\mathcal{A}}\ \ \delta}\quad \dim\ \Lambda_{typical}\quad 2^{\ n\ S\ \rho^{\mathcal{A}}\ \ \delta}.$$

The typical subspace, $\Lambda_{typical}$, consists of "typical" eigenvectors of the density matrix, $\rho^{\mathcal{A}}$ ^{n}; these are eigenvectors of $\rho^{\mathcal{A}}$ ^{n}, with frequency close to the probability distribution of the eigenvalues of the density operator, $\rho^{\mathcal{A}}$ (Figure 3.18).

The recipient of information must design an optimal POVM measurement to identify the codeword, w_i , sent by the source of quantum information. Call ζ_k $\mathcal{H}_2$ ^{n} the set of vectors chosen to specify the POVM, and define the positive operator whose support is the subspace spanned by ζ_k :

$$\mathbf{M}\qquad \sum_k \zeta_k\ \ \zeta_k\ .$$

Then $\mathbf{M}^{1/2}$ exists and it is invertible in the subspace spanned by ζ_k and allows us to define the vectors

$$\mu_k\qquad \mathbf{M}\ ^{1/2}\ \zeta_k\ .$$

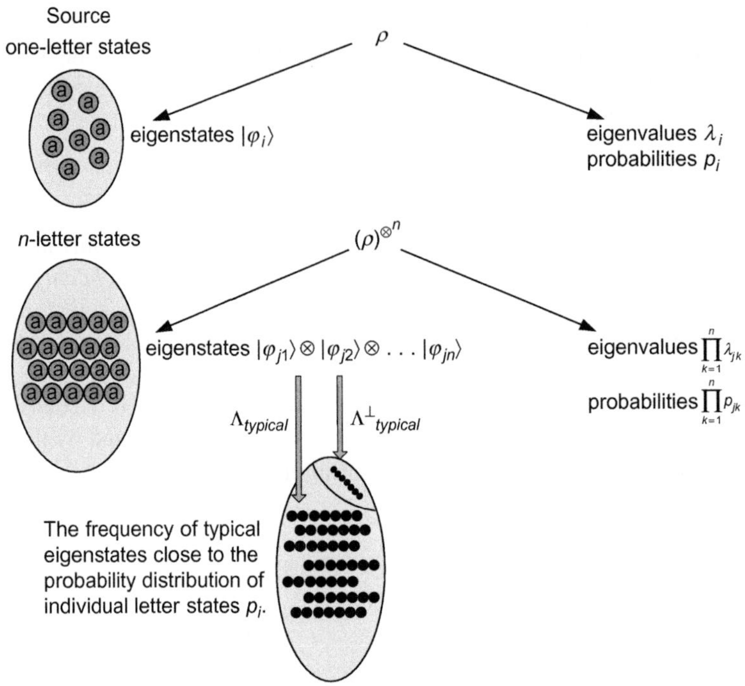

FIGURE 3.18

The construction of "typical subspaces" for Schumacher quantum noiseless encoding. The single-letter source has the density matrix, ρ, with the eigenvectors, φ_i , and, eigenvalues, λ_i, with the probability, p_i. The quantum code encodes a single logical qubit into, n, physical ones. The eigenvectors of ρ ^{n}, the density matrix of n-letter codewords, are tensor products of the eigenvectors of ρ, namely, φ_{j1} φ_{j2} φ_{jn} . The "typical subspace," $\Lambda_{typical}$, is spanned by the "typical eigenvectors" of the density matrix, ρ ^{n}. The "typical eigenvectors" of ρ ^{n} have a frequency close to the probability distribution of the eigenvalues of the density operator, ρ; indeed, each eigenvector is associated with one eigenvalue and "almost" all eigenvectors are typical.

It is easy to prove that the corresponding positive operators, μ_k μ_k , sum up to identity on this subspace:

$$ {}_{k} \ \mu_k \ \mu_k \qquad {}_{k} \ \mathbf{M}^{-1/2} \zeta_k \ \zeta_k \mathbf{M}^{-1/2} \quad \mathbf{M}^{-1/2} \ {}_{k} \ \zeta_k \ \zeta_k \quad \mathbf{M}^{-1/2} \quad \mathbf{I}. $$

These operators are the outcome operators of a POVM that is called a "*pretty good measurement*"; this POVM is used by the receiver to distinguish among the vectors, ζ_k . A variation of it will also be used in Section 3.22. If the codewords, w_i $\mathcal{C}$, are sent with an equal probability, p_i $1/N$, with N equal to the cardinality of the code $\mathcal{C}$, N $\mathcal{C}$, then the recipient of information performs the POVM measurement and obtains the outcome, μ_i , with the probability

$$ p[(\mu_i)\,(w_i)] \quad \mathrm{tr}(\mu_i \ \ \mu_i \, w_i \ \ w_i) \qquad \mu_i \, w_i \ ^2. $$

Then the expected probability of error is

$$p_{err} \quad 1 \quad \sum_i \frac{1}{N} \quad \mu_i \; w_i \quad ^2 \quad \frac{1}{N} \sum_i \quad (1 \quad \mu_i \; w_i \;)(1 \quad \mu_i \; w_i \;).$$

The modulus of the inner products between the codeword sent, w_i , and the outcome of the measurement, μ_i , is a measure of the error of the POVM measurement for w_i :

$$\mu_i \; w_i \quad \begin{array}{ll} 0 & \text{if the vectors are orthogonal,} \\ 1 & \text{if the vectors are identical,} \\ > 0 \text{ and } < 1 & \text{otherwise.} \end{array}$$

The inner product is at most equal to unity; thus, $(1 \quad \mu_i \; w_i \;) \quad 2$. It follows that

$$p_{err} \quad \frac{2}{N} \sum_i \quad (1 \quad \mu_i \; w_i \;).$$

The recipient of information employs an observable to distinguish between the signal states/codewords that, with very few exceptions, are typical sequences; therefore, instead of the signal states w_i , we consider σ_i , the projection of these states on the typical subspace $\Lambda_{typical}$:

$$\sigma_i \quad \Pi_\Lambda \; w_i \; .$$

The following diagram illustrates the relationship between the codeword, its projection, and the outcome of the POVM measurement:

w_i – codeword	μ_i – result of POVM
+	
$\sigma_i \quad \Pi_\Lambda \; w_i$ – projection of w_i on $\Lambda_{typical}$	

Let S be the complex $N \quad N$ matrix, $S \quad S_{ij}$, with positive eigenvalues; the elements of the matrix S are inner products of the N projections of w_i on $\Lambda_{typical}$, where $S_{ij} \quad \sigma_i \; \sigma_j$. The vectors, μ_k , which are contained in the subspace $\Lambda_{typical}$, are constructed, such that

$$\mu_i \; w_j \quad \mu_i \; \sigma_j \quad (\; \overline{S})_{ij}.$$

The vectors μ_k together with the projection onto the subspace perpendicular to all the vectors, σ_i , define the POVM of the receiver.

The expected probability of error can be expressed as

$$p_{err} \quad \frac{2}{N} \sum_i \quad 1 \quad (\; \overline{S})_{ii} \; .$$

The following inequality is known: when $x \quad 0$, the square root function has a parabola as a lower bound $\overline{x} \quad 1/2(3x \quad x^2)$. It follows that $\overline{S} \quad 1/2(3S \quad S^2)$ for the matrix S with non-negative eigenvalues. Call the norm of the projected codewords, $\kappa_i \quad \sigma_i \, \sigma_i \quad S_{ii}$. Then,

$$(\quad \overline{S})_{ii} \quad \frac{3}{2}\kappa_i \quad \frac{1}{2}\kappa_i^2 \quad \frac{1}{2} \quad S_{ij}S_{ji}. \quad {}_{j \quad i}$$

The expected probability of error for the code, $\mathcal{C}^k$, can now be written as

$$p_{err} \quad \frac{2}{N} \ {}_{i} \quad 1 \quad \frac{3}{2}\kappa_i \quad \frac{1}{2}\kappa_i^2 \quad \frac{1}{2} \ {}_{j \quad i} \quad S_{ij}S_{ji} \quad \frac{2}{N} \ {}_{i} \quad (1 \quad \kappa_i) \ 1 \quad \frac{\kappa_i}{2} \quad \frac{1}{2} \ {}_{j \quad i} \quad S_{ij}S_{ji}$$

or

$$p_{err} \quad \frac{2}{N} \ {}_{i} \quad (1 \quad \kappa_i) \quad \frac{1}{2} \ {}_{j \quad i} \quad S_{ij}S_{ji} \quad .$$

To prove Schumacher's noiseless coding theorem, we have to show that when N is sufficiently large so that $\log N \quad nS \ \rho^{\mathcal{A}}$, there exists a code, and the expected probability of error for this code becomes increasingly small, $p_{err} \quad 0$.

We consider a random code with n qubit codewords; each codeword w_i , $1 \quad i \quad N$, is a sequence of n letter-states generated using the a priori probabilities for the letters. The probability of a codeword, $w_i \quad a_1 \ a_2 \quad a_n$, is $P_{w_i} \quad p(a_1)p(a_2) \quad p(a_n)$. Each codeword is generated independently from the other codewords. Let $\mathcal{C} \quad \mathcal{C}^1, \mathcal{C}^2, \ldots, \mathcal{C}^k, \ldots,$ be the set of the random codes constructed by the procedure we just described. Now we compute averages over all such codes for various quantities; for example,

$$\overline{w_i \ w_i} \quad \rho^{\mathcal{A}} \quad {}^{n}.$$

The average over all random codes of the expected probability of error is then

$$\overline{p_{err}} \quad \frac{2}{N} \ {}_{i} \quad (1 \quad \overline{\kappa_i}) \quad \frac{1}{2} \ {}_{j \quad i} \quad \overline{S_{ij}S_{ji}} \quad 2(N \quad \overline{\kappa_i}) \quad (N \quad 1)\overline{S_{ij}S_{ji}} < 2(N \quad \overline{\kappa_i}) \quad N\overline{S_{ij}S_{ji}}.$$

We have used the fact that the, N, terms in the summation, with the index, i, and the, $(N \quad 1)$, terms in the summation, with the index, j, have the same value. Now we can compute the averages involved in the previous expression. First, recall that $\kappa_i \quad \sigma_i \, \sigma_i$ and $\sigma_i \quad \Pi_\Lambda \ w_i$; therefore,

$$\overline{\kappa_i} \quad \overline{\mathrm{tr}(\Pi_\Lambda \ w_i \ w_i \ \Pi_\Lambda)} \quad \mathrm{tr} \ \Pi_\Lambda \ \overline{w_i \ w_i} \ \Pi_\Lambda \quad \mathrm{tr} \ \Pi_\Lambda \ \rho^{\mathcal{A}} \ {}^{n}\Pi_\Lambda \ .$$

The projection of a codeword on the typical subspace, Π_Λ, satisfies the inequality

$$\mathrm{tr} \ \Pi_\Lambda \ \rho^{\mathcal{A}} \ {}^{n}\Pi_\Lambda \quad 1 \quad \epsilon.$$

Thus,

$$\overline{\kappa_i} \quad 1 \quad \epsilon.$$

Recall that $S_{ij} \quad w_i \, \sigma_j$. Thus,

$$S_{ij}S_{ji} \quad w_i \, \sigma_j \quad w_j \, \sigma_i \qquad w_i \, \Pi_\Lambda \, w_j \quad w_j \, \Pi_\Lambda \, w_i \; .$$

Then:

$$\overline{S_{ij}S_{ji}} \quad \overline{w_i \, \Pi_\Lambda \, w_j \quad w_j \, \Pi_\Lambda \, w_i} \quad \overline{\text{tr} \; \Pi_\Lambda \, w_i \quad w_i \quad w_j \quad w_j \, \Pi_\Lambda}$$
$$\text{tr} \; \Pi_\Lambda \quad \rho^{\mathcal{A}} \;\; {}^{n} \;\; {}^{2} \; \Pi_\Lambda \; .$$

It follows that the average over all random codes of the expected probability of error is

$$\overline{p_{err}} < 2(N \quad \overline{\kappa_i}) \quad N\overline{S_{ij}S_{ji}} < 2\epsilon \quad N\text{tr} \; \Pi_\Lambda \quad \rho^{\mathcal{A}} \;\; {}^{n} \;\; {}^{2} \; \Pi_\Lambda \; .$$

If we sum the eigenvalues of $\rho^{\mathcal{A}} \;\; {}^{n}$ only over the typical subspace, $\Lambda_{typical}$, we obtain

$$\text{tr} \; \Pi_\Lambda \quad \rho^{\mathcal{A}} \;\; {}^{n} \;\; {}^{2} \; \Pi_\Lambda \quad < \; \text{dim} \; \Lambda_{typical} \quad 2 \;\; {}^{n(S(\rho^{\mathcal{A}}) \;\; \delta)} \; {}^{2} < 2 \;\; {}^{n(S(\rho^{\mathcal{A}}) \;\; 3\delta)} .$$

We have taken into account the fact that $\text{dim} \; \Lambda_{typical} \qquad 2^{n \; S(\rho^{\mathcal{A}}) \;\; \delta}$. Finally,

$$\overline{p_{err}} < 2\epsilon \quad N2 \;\; {}^{n(S(\rho^{\mathcal{A}}) \;\; 3\delta)} .$$

From the family of codes, $\mathcal{C}$, we can then select one, $\mathcal{C}^k$, such that

$$p_{err} < 2\epsilon \quad N2 \;\; {}^{n(S(\rho^{\mathcal{A}}) \;\; 3\delta)} .$$

This concludes the proof of the quantum noiseless coding theorem. ■

The last relationship shows that if n is very large, much larger than needed to form the typical subspace, Λ, the sender can make N as large as $N \quad 2^{n(S(\rho^{\mathcal{A}}) \;\; 4\delta)}$ and still have an average probability of error $p_{err} < 3\epsilon$. In this case, the sender would use $S(\rho^{\mathcal{A}}) \quad 4\delta$ bits to encode a letter. Therefore, the existence of codes that allow the transmission at an asymptotic rate of $S(\rho^{\mathcal{A}})$ bits per letter with arbitrarily low error can be proven without actually constructing such a code.

In the proof just presented, it was assumed [204] that the sender codewords w_i are pure states and it was shown that for an appropriate choice of the sender's code and of the receiver decoding observable, for large enough n, codewords of n letters may be used to transmit $nS(\rho_{\mathcal{A}})$ bits of information (or $S(\rho_{\mathcal{A}})$ bits per letter) with arbitrarily low probability of error. The theorem was also proved for codewords that are mixed states [373]; the general procedure [204] is that of using averages over random codes and a similar prescription for the decoding POVM observables, but enforcing stronger typicality conditions on various quantities associated with the channel. Since the mixed state codewords may be interpreted as the outputs of a noisy channel, the result is applied to the case of classical information capacity of noisy channels discussed later in this chapter.

The analysis of noisy quantum channels requires the characterization of the relationship between the states at the input and the output of the channel. Next, we introduce the concepts of entropy exchange, entanglement fidelity, and coherent information necessary for the study of noisy quantum channels.

3.18 ENTROPY EXCHANGE, ENTANGLEMENT FIDELITY, AND COHERENT INFORMATION

Classical error correction uses a *syndrome* to identify the bit in error of a codeword and reverses it. In this section we discuss the theoretical underpinning of quantum error correction, namely the process to determine the error syndrome, which identifies the position of the qubits in error and the type of error. This process, analyzed in Chapter 5, consists of a measurement of a register of ancilla qubits entangled with the qubits of the codeword.

Quantum error correction requires a multi-qubit measurement that does not disturb the quantum information carried by the signaling qubits and retrieves information about the error. The outcome of the measurement tells us not only which physical qubit was affected, but also the way it was affected; the error could be either a bit-flip, or a phase-flip, or both, corresponding to the Pauli matrices σ_x, σ_z, and σ_y, respectively. The measurement of the syndrome has a projective effect; even if the error due to the noise was arbitrary, it can be expressed as a superposition of basis operations given by the Pauli matrices and the identity matrix.

We show how an environment that interacts unitarily with the quantum system, $\mathcal{Q}$, can be constructed and discuss the purification of the mixed state of the system, $\mathcal{Q}$; purification of system $\mathcal{Q}$ in a mixed state means identifying a reference system, $\mathcal{R}$, such that the composite system, $\mathcal{P} = (\mathcal{Q}, \mathcal{R})$, is in a pure state. The concepts introduced in this section allow us to quantify the interactions between the system, $\mathcal{Q}$, the environment, E, and the reference system, $\mathcal{R}$. The *entropy exchange* is a measure of the information exchanged between the system and the environment during the interaction with the environment. The *entanglement fidelity* is a measure of the overlap between the initial purification and the joint state of the system $\mathcal{Q}$ and the reference system $\mathcal{R}$ after the transmission of $\mathcal{Q}$. The *coherent information* measures the degree to which the quantum coherence is preserved by the superoperator, $\mathcal{E}$, associated with a quantum channel; it is the correspondent of the mutual information.

An important concept related to quantum information processing is that of *trace-preserving operation* introduced in Section 2.14. Informally, a trace-preserving quantum transformation takes place when a quantum system interacts with the environment E and no measurement is performed either on the quantum system or the environment. If the operation is denoted by $\mathcal{E}$ and the quantum system $\mathcal{Q}$ is in a state characterized by the density matrix, ρ, a trace-preserving operation requires that $\text{tr}(\mathcal{E}(\rho)) = 1$. The linear transformation, $\mathcal{E}(\rho)$, enjoys *complete positivity* [207] if there exist operators, $\mathbf{A}_i$, such that

$$\mathcal{E}(\rho) = \sum_i \mathbf{A}_i \rho \mathbf{A}_i^\dagger, \quad \text{with} \quad \sum_i \mathbf{A}_i^\dagger \mathbf{A}_i \leq \mathbf{I}.$$

If $\sum_i \mathbf{A}_i^\dagger \mathbf{A}_i = \mathbf{I}$, then $\text{tr}(\mathcal{E}(\rho)) = 1$ for all density operators, ρ, and the transformation is *trace-preserving*. It can be shown that the following three conditions are equivalent [256]:

1. $\mathcal{E}$ is a trace-preserving, completely positive linear map on the density operators of $\mathcal{Q}$.
2. $\mathcal{E}$ has a unitary representation.
3. $\mathcal{E}$ has a normalized operator-sum representation.

When classical information about the quantum system is made available through a measurement, the operation is *non–trace-preserving*.

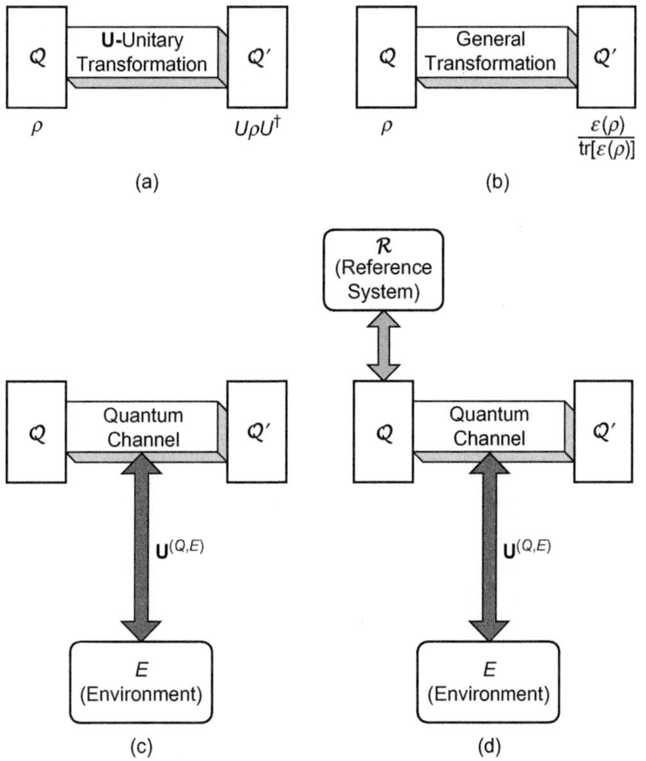

FIGURE 3.19

Quantum channel models. (a) The channel performs a unitary transformation, $\mathbf{U}$, of quantum state, $\mathcal{Q}$, described by the density matrix, ρ, into the state, $\mathcal{Q}$, described by $\mathcal{E}(\rho) \quad \mathbf{U}\rho\mathbf{U}^{\dagger}$. (b) The channel performs a general transformation, with $\frac{\mathcal{E}(\rho)}{\mathrm{tr}[\mathcal{E}(\rho)]}$, the normalized output state. (c) The *environment*, E, allows us to model a trace-preserving operation as a unitary operation based on the representation theorem for trace-preserving quantum operations. (d) A *reference system*, $\mathcal{R}$, allows purification of a mixed state of the quantum system, $\mathcal{Q}$.

Transmission of information through a quantum communication channel implies a state change described by the superoperator, $\mathcal{E}$. If ρ is the density matrix of the *input state*, then the density matrix of the *output state* is $\mathcal{E}(\rho)$. Figures 3.19a and 3.19b illustrate the transformation of the input state of the quantum channel when a closed system is subject to a unitary transformation and a general transformation, respectively. In the case of a unitary transformation, $\mathbf{U}$, the density operator of the channel output state is

$$\mathcal{E}(\rho) \quad \mathbf{U}\rho\mathbf{U}^{\dagger}.$$

In the case of a general transformation, the density operator of the normalized channel output state is

$$\frac{\mathcal{E}(\rho)}{\mathrm{tr}(\mathcal{E}(\rho))}.$$

The trace in the denominator is included to preserve the trace condition for the input state $\mathrm{tr}(\rho) \quad 1$.

The question we pose is whether we can model a trace-preserving operation as a unitary operation. The answer, provided in Barnum *et al* [22], requires the introduction of an *environment*, the quantum system unitarily interacts with, as in Figure 3.19c, and it is based on the theorem discussed next.

Theorem. *If we are given a quantum system, $\mathcal{Q}$, with the density matrix, ρ, in a d-dimensional Hilbert space, $\mathcal{H}_d$, and a trace-preserving operation, $\mathcal{E}$, it is possible to construct an environment, E, with the following properties:*

- *The environment is at most of dimensionality d^2.*
- *The environment is initially in a pure state with the density matrix, $\epsilon \quad \psi \quad \psi$.*
- *The quantum system $\mathcal{Q}$ and the environment E are initially uncorrelated.*

Then there exists a unitary transformation, **U**, *of the quantum system $\mathcal{Q}$ and the environment E, such that $\mathcal{E}(\rho) \quad tr_E \; \mathbf{U}(\rho \quad \epsilon)\mathbf{U}^\dagger$.*

In other words, *a trace-preserving quantum operation can always be modeled as a unitary evolution by constructing an environment that interacts unitarily with the quantum system;* conversely, any unitary interaction between the quantum system and an initially uncorrelated environment gives rise to a trace-preserving quantum operation.

When the input to the communication channel, the quantum system Q, is in a mixed state, it is useful to introduce in addition to the environment E a reference system, R, such that $\varphi^{(\mathcal{Q},\mathcal{R})}$, the joint state, $(\mathcal{Q} \quad \mathcal{R})$, is a pure state and also such that

$$\rho^{\mathcal{Q}}_{before} \quad \mathrm{tr}_R \quad \varphi^{(\mathcal{Q},\mathcal{R})} \quad \varphi^{(\mathcal{Q},\mathcal{R})} \quad .$$

In Section 2.4, we discussed purification of mixed states, a concept very useful to studying information transmission through noisy quantum channels. Recall that purification of a system $\mathcal{Q}$ in a mixed state means identifying another system $\mathcal{R}$ such that the composite system, $\mathcal{P} \quad (\mathcal{Q},\mathcal{R})$, is in a pure state. We now recognize that $\varphi^{(\mathcal{Q},\mathcal{R})}$ is a purification of the state $\rho^{\mathcal{Q}}_{before}$.

The analysis of noisy quantum channels requires the characterization of the relationship between the states at the input and the output of the channel.

Entropy Exchange

To model a noisy quantum channel, Schumacher [370] and Lloyd [275] introduced the entropy exchange. If $\mathcal{E}$ is the quantum transformation of the input state, ρ, the entropy exchange, S_e, is defined as the von Neumann entropy of the environment, E, after the transformation, $\mathcal{E}$, if the environment starts out in a pure state,

$$S_e(\rho,\mathcal{E}) \quad S \;\; \rho^{(\mathcal{Q}\mathcal{R})}_{after} \qquad S \;\; \rho^{E}_{after} \quad .$$

Initially, the environment E is in a pure state as a result of a purification operation; ρ^{E}_{after} is the density matrix of E after the operation. When

$$\mathcal{E}(\rho) \qquad \underset{i}{} \mathbf{A}_i \rho \mathbf{A}_i^\dagger,$$

the entropy exchange can be expressed as

$$S_e(\rho,\mathcal{E}) \quad S(w) \qquad \mathrm{tr}\,(w \log w),$$

with $w \quad [w_{ij}]$, the density operator of the environment after transformation, a matrix with elements in an orthonormal basis given by

$$w_{ij} \qquad \frac{\mathrm{tr}\ \mathbf{A}_i \rho \mathbf{A}_j^{\dagger}}{\mathrm{tr}(\mathcal{E}(\rho))},$$

where A_i are operation elements of $\mathcal{E}$. The entropy exchange is an intrinsic property of the system Q; it is a measure of the information exchanged between Q and the environment during the transformation $\mathcal{E}$. *The entropy exchange limits the amount of information that could be acquired by an eavesdropper in a quantum cryptographic protocol.*

Entanglement Fidelity

Schumacher [370] introduced the entanglement fidelity to measure how well a quantum system, $\mathcal{Q}$, with the density matrix, ρ, and its entanglement with other states are preserved by a noisy quantum channel. Recall that when φ is a pure state of a quantum system, the fidelity, F, of an arbitrary, possibly mixed state, with the density matrix ρ, is defined as $F \quad \varphi\ \rho\ \varphi$. The fidelity F is a measure of how close ρ is to $\varphi \quad \varphi$.

The entanglement fidelity is a measure of the overlap between the initial purification and the joint state of the system $\mathcal{Q}$ and the reference system $\mathcal{R}$ after the transmission of $\mathcal{Q}$:

$$F_e(\rho,\mathcal{E}) \qquad \varphi^{(\mathcal{Q},\mathcal{R})} \quad (\mathcal{I}_{\mathcal{R}} \quad \mathcal{E}) \quad \varphi^{(\mathcal{Q},\mathcal{R})} \quad \varphi^{(\mathcal{Q},\mathcal{R})} \qquad \varphi^{(\mathcal{Q},\mathcal{R})},$$

with $\varphi^{(\mathcal{Q},\mathcal{R})}$, the joint state of the quantum system $\mathcal{Q}$ and the reference system $\mathcal{R}$ before the transmission, $\mathcal{I}_{\mathcal{R}}$, the identity transformation of the reference system $\mathcal{R}$, and $\mathcal{E}$ a trace-preserving operation. When

$$\mathcal{E}(\rho) \qquad \underset{i}{\quad} \mathbf{A}_i \rho \mathbf{A}_i^{\dagger},$$

entanglement fidelity can be expressed as

$$F_e(\rho,\mathcal{E}) \qquad \frac{{}_i\ \mathrm{tr}\,(\mathbf{A}_i \rho) \quad {}^2}{\mathrm{tr}\,(\mathcal{E}(\rho))}.$$

The entanglement fidelity depends only on ρ and $\mathcal{E}$, not on the details of the purification.

Entanglement fidelity is a more stringent requirement than average pure state fidelity; as we shall see shortly, the entanglement fidelity is a lower bound on the average fidelity of pure states. Thus, *if we are able to ensure a high degree of the entanglement fidelity for an ensemble of pure states, then individual pure states will be reproduced with high fidelity at the output of the quantum channel.* A critical point is that in quantum error correction we wish to apply an error correction procedure to a subset of a quantum register even though the subset may later become entangled with other parts of the quantum computer. The properties of the entanglement fidelity [22, 370] are:

1. The entanglement fidelity is a real number between 0 and 1, $0 \quad F_e(\rho,\mathcal{E}) \quad 1$. $F_e(\rho,\mathcal{E}) \quad 1$ if and only if for all pure states, φ , in the support of, ρ, the following condition is satisfied: $\mathcal{E}(\varphi \ \varphi) \quad \varphi \ \varphi$.
2. The entanglement fidelity is a lower bound on the fidelity, $F_e(\rho,\mathcal{E}) \quad F(\rho,\mathcal{E}(\rho))$.
3. Given an ensemble of quantum systems in the states, $\varphi_i \ ,p_i$, with the density matrix, $\rho \quad {}_i p_i \ \varphi_i \ \ \varphi i$, the entanglement fidelity is a lower bound on the average fidelity of the pure states, F_p,

$$F_e(\rho,\mathcal{E}) \quad F_p \qquad \underset{i}{} \ p_i \ \varphi_i \ \mathcal{E}(\ \varphi_i \ \ \varphi i \ ) \ \varphi_i \ .$$

4. If the pure state fidelity is $F_p \quad 1 \quad \eta$, for small $\eta \quad 0$, then the entanglement fidelity is [245]

$$F_e(\rho,\mathcal{E}) \quad 1 \quad \frac{3}{2}\eta.$$

5. *Continuity.* The *continuity lemma* provides an upper bound of the entanglement fidelity when the channel input is perturbed. If $\mathcal{E}$ is a trace-preserving quantum operation, ρ is the density matrix, and Δ is a zero trace Hermitian operator, then

$$F_e(\rho \quad \Delta,\mathcal{E})) \quad F_e(\rho,\mathcal{E}) \qquad 2\mathrm{tr}(\ \Delta \ ) \quad \mathrm{tr}(\ \Delta \ )^2.$$

This lemma gives bounds on the change in entanglement fidelity when the input state is perturbed. A detailed proof of the lemma can be found in Barnum *et al* [22].

Coherent Information

The coherent information is the quantum correspondent of the mutual information in classical information theory [371] and is defined as

$$I(\rho,\mathcal{E}) \quad S \ \frac{\mathcal{E}(\rho)}{\mathrm{tr}(\mathcal{E}(\rho))} \qquad S_e(\rho,\mathcal{E}),$$

where the quantum operation, $\mathcal{E}$, takes the input state, ρ, to an output state, $(\mathcal{E}(\rho)/\mathrm{tr}[\mathcal{E}(\rho)])$.

The coherent information, $I(\rho,\mathcal{E})$, can be viewed as a measure of the degree to which the quantum coherence is preserved by the operation, $\mathcal{E}$—in other words, a measure of the degree of quantum entanglement retained by the systems, $\mathcal{Q}$ and $\mathcal{R}$, after this operation.

The role of the coherent information will be discussed in the next few sections. Now we just state several of the properties of coherent information [22]:

1. Given the quantum operator $\mathcal{E}$ and the non-negative λ,

$$I(\rho,\lambda\mathcal{E}) \quad I(\rho,\mathcal{E}).$$

2. *Convexity of coherent information*:

$$I(\rho,\mathcal{E}) \qquad \underset{i}{} p_i I(\rho,\mathcal{E}_i).$$

The coherent information, $I(\rho,\mathcal{E})$, is convex when the quantum operation, $\mathcal{E} \quad {}_i p_i \mathcal{E}_i$, is a trace-preserving convex sum of trace-preserving operations, $\mathcal{E}_i$, with $p_i \quad 0$ and ${}_i p_i \quad 1$. The proof of this property of coherent information follows immediately from the concavity of conditional von Neumann entropy discussed in Section 3.10.

3. *Generalized convexity theorem for coherent information*:

$$I \ \rho, \ \underset{i}{} \mathcal{E}_i \qquad \frac{{}_i \operatorname{tr}(\mathcal{E}_i(\rho)) I(\rho,\mathcal{E}_i)}{\operatorname{tr} \quad {}_i \mathcal{E}_i(\rho)},$$

where $\mathcal{E}_i$ are quantum operations, can be proved based on the convexity property of coherent information.

4. *Additivity of independent channels:*

$$I(\rho_1 \quad \rho_2 \qquad \rho_n, \mathcal{E}_1 \quad \mathcal{E}_2 \qquad \mathcal{E}_n) \qquad \underset{i}{} I(\rho_i,\mathcal{E}_i),$$

where $\rho_1,\rho_2,\ldots,\rho_n$ are density operators of n quantum systems (qubits) and $\mathcal{E}_1,\mathcal{E}_2,\ldots,\mathcal{E}_n$ are quantum operators that correspond to the environment acting independently on these systems. The evolution operator, $\mathcal{E}$, acting on the joint state of the n systems (qubits), $\rho \quad \rho_1 \quad \rho_2 \qquad \rho_n$, factorizes as $\mathcal{E} \quad \mathcal{E}_1 \quad \mathcal{E}_2 \qquad \mathcal{E}_n$.

The discussion in this section is very important for quantum error correction. As we shall see in Chapter 5, we separate the space of encoded states and the space of errors into orthogonal subspaces of the Hilbert space. To measure the quantum register in error, we perform a unitary transformation involving also ancilla qubits. Such a trace-preserving completely positive map, or superoperator, allows us to determine a syndrome that uniquely identifies the error. In this process, we do not disturb the encoded state, and subsequently we can correct the error using the information stored in the ancilla qubits.

From this brief presentation, it follows that we are also interested in how well subspaces of a Hilbert space can be transmitted through a quantum channel. Given a subspace with projector, **P**, the *subspace fidelity* is defined as

$$F_s(\mathbf{P},\mathcal{E}) \quad \min_{\varphi} \ \varphi \ \mathcal{E}(\ \varphi \quad \varphi \ ) \ \varphi \ ,$$

where the minimization is over all states, φ , in the subspace whose projector is **P**. The entanglement fidelity is close to 1 when the subspace fidelity is close to 1; this follows from the definition of subspace fidelity and the fourth property of the entanglement fidelity.

3.19 QUANTUM FANO AND DATA PROCESSING INEQUALITIES

In this section, we introduce the quantum equivalent of two important properties of classical information discussed in Section 3.6. Recall that Fano inequality gives us a lower bound of the error when we estimate one random variable using another one correlated with it. The data processing inequality considers a source, with the entropy, $H(X)$, and two cascaded channels. We assume that the source provides the input, X, to a first communication channel and that, Y, the output of the first channel, is the input of a second channel. X, Y, and Z, the output of the second channel, form a Markov chain. Then the mutual information between the source and the output of the second channel, $I(X,Z)$, cannot be larger than the mutual information between the source and the output of the first channel, $I(X,Y)$; in turn, $I(X,Y)$ cannot be larger than, $H(X)$, the entropy of the source. An inequality with a similar interpretation was proven by Schumacher and Nielsen [371] for quantum transformations.

We consider a quantum system characterized by the density matrix, ρ as, input to a quantum channel characterized by the transformation, $\mathcal{E}$; the output of the channel is $\mathcal{E}(\rho)$. Recall from Section 3.18 that, $S_e(\rho,\mathcal{E})$, the entropy exchange, is defined as the von Neumann entropy of the environment after the transformation of the state due to the channel; the entanglement fidelity, $F_e(\rho,\mathcal{E})$, measures how well a quantum system with the density matrix ρ and its entanglement with other states are preserved by a noisy quantum channel.

The Quantum Fano Inequality

This inequality relates the entropy exchange and the entanglement fidelity [372] as

$$S_e(\rho,\mathcal{E}) \quad h(F_e(\rho,\mathcal{E})) \quad [1 \quad F_e(\rho,\mathcal{E})]\log(d^2 \quad 1).$$

In this expression, d is the dimension of the Hilbert space of the quantum system, $\mathcal{Q}$, and $h(p)$ is the Shannon entropy of a binary random variable, $h(p) \qquad p\log p \quad (1 \quad p)\log(1 \quad p)$.

Proof. The proof is based on the grouping property of Shannon entropy:

$$H(p_1,p_2,\ldots,p_{d^2}) \quad h(p_1) \quad (1 \quad p_1)H \quad \frac{p_2}{1 \quad p_1},\frac{p_3}{1 \quad p_1},\ldots,\frac{p_{d^2}}{1 \quad p_1} \quad .$$

The proof [372] considers a set of d^2 orthonormal basis states, φ_i , of the joint system, $(\mathcal{Q},\mathcal{R})$, such that the first state in the set is $\varphi_1 \qquad \varphi^{(\mathcal{Q},\mathcal{R})}$, the entangled state, and $p_i \qquad \varphi_i\, \rho_{after}^{(\mathcal{Q},\mathcal{R})} \quad \varphi_i$. It can be shown that

$$S_e(\rho,\mathcal{E}) \quad H(p_1,p_2,\ldots,p_{d^2}),$$

where $H(p_i)$ is the Shannon information of the set p_i . With the observation that, by definition, $p_1 \quad F_e(\rho,\mathcal{E})$, and that $H \ \frac{p_2}{1 \ p_1},\frac{p_3}{1 \ p_1},\ldots,\frac{p_{d^2}}{1 \ p_1} \qquad \log(d^2 \quad 1)$, the inequality is evident.

The quantum Fano inequality shows that for a given quantum operation, if the exchange entropy is large, the entanglement fidelity must necessarily be small, indicating that the entanglement between $\mathcal{Q}$ and $\mathcal{R}$ has not been well preserved. ∎

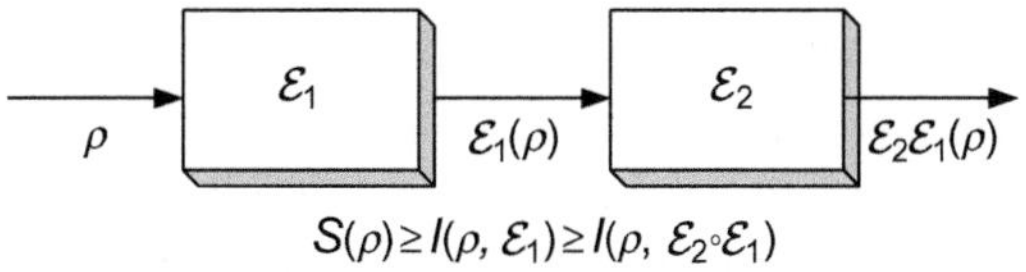

FIGURE 3.20

The quantum data processing inequality. A quantum system with the density matrix, ρ, is transmitted through two cascaded quantum channels characterized by the superoperators, $\mathcal{E}_1$ and $\mathcal{E}_2$; the output of the first channel, $\mathcal{E}_1(\rho)$, becomes the input to a second channel. The density matrix of the system after the two trace-preserving transformations is $(\mathcal{E}_2 \ \ \mathcal{E}_1)(\rho)$. The coherent information between the original input and the output of the second channel, $I(\rho, (\mathcal{E}_2 \ \ \mathcal{E}_1)(\rho))$, cannot be larger than the coherent information between the original input and the output of the first channel, $I(\rho, \mathcal{E}_1(\rho))$; in turn, $I(\rho, \mathcal{E}_1(\rho))$ cannot be larger than $S(\rho)$, the von Neumann entropy of the source. The inequality, $I(\rho, \mathcal{E}_1) \quad I(\rho, \mathcal{E}_2 \ \ \mathcal{E}_1)$, holds even when $\mathcal{E}_1$ is trace-preserving but $\mathcal{E}_2$ is not.

The Quantum Data Processing Inequality

Consider a setup like the one in Figure 3.20 where given trace-preserving quantum operations, $\mathcal{E}_1$ and $\mathcal{E}_2$, we define the quantum process

$$\rho \quad \mathcal{E}_1(\rho) \quad (\mathcal{E}_2 \ \ \mathcal{E}_1)(\rho).$$

Then $S(\rho)$, the von Neumann entropy of the source, cannot be exceeded by $I(\rho, \mathcal{E}_1)$, the coherent information between the source and the state after transformation $\mathcal{E}_1$; in turn, $I(\rho, \mathcal{E}_1)$ cannot be exceeded by $I(\rho, \mathcal{E}_2 \ \ \mathcal{E}_1)$, the coherent information between the state after transformation, $\mathcal{E}_1$, and the state after transformation $\mathcal{E}_2 \ \ \mathcal{E}_1$,

$$S(\rho) \quad I(\rho, \mathcal{E}_1) \quad I(\rho, \mathcal{E}_2 \ \ \mathcal{E}_1).$$

Schumacher and Nielsen also proved [371] that it is possible to reverse the transformation $\mathcal{E}$ if and only if no information is lost during the transformation; the entropy of the source is equal to the coherent information between the source and the output of the channel characterized by $\mathcal{E}$:

$$S(\rho) \quad I(\rho, \mathcal{E}).$$

Proof. To prove the first inequality, $S(\rho) \quad I(\rho, \mathcal{E}_1)$, we consider in addition to the quantum system, $\mathcal{Q}$, a reference system, $\mathcal{R}$, and the environment, E (see Figure 3.19d). We start from the definition of coherent information and distinguish $\rho_{before} \quad \rho$, the state of the quantum system before, and $\rho_{after\ \mathcal{E}_1}$, the state after the transformation $\mathcal{E}_1$. The coherent information is

$$I(\rho, \mathcal{E}) \quad S(\mathcal{E}_1(\rho)) - S_e(\rho, \mathcal{E}_1),$$

with $S(\mathcal{E}_1(\rho)) \quad S \ \rho^{(\mathcal{Q})}_{after\ \mathcal{E}_1}$ and $S_e(\rho, \mathcal{E}_1) \quad S \ \rho^{(E)}_{after\ \mathcal{E}_1}$. Now we observe that the state of the joint system, $(\mathcal{Q}, \mathcal{R}, E)$, was a pure state before the transformation $\mathcal{E}_1$ and remains pure after. Indeed, the joint state, $\varphi_{before}^{(Q,\mathcal{R})}$, before the transformation is a purification of ρ, the possibly mixed state of quantum system Q; the environment is initially in a pure state $\varphi_{before}^{(E)}$. The representation theorem for trace-preserving transformations tells us that a trace-preserving operation can be regarded as a unitary transformation by adding an environment initially in a pure state that the system interacts unitarily with. It follows that $S \;\; \rho_{after\ \mathcal{E}_1}^{(Q)} \qquad S \;\; \rho_{after\ \mathcal{E}_1}^{(\mathcal{R},E)}$ and that similar relations exist among all permutations of Q, $\mathcal{R}$, and E. Now by virtue of subadditivity of von Neumann entropy,

$$S \;\; \rho_{after\ \mathcal{E}_1}^{(\mathcal{R},E)} \qquad S \;\; \rho_{after\ \mathcal{E}_1}^{(\mathcal{R})} \qquad S \;\; \rho_{after\ \mathcal{E}_1}^{(E)} \;.$$

Then,

$$I(\rho,\mathcal{E}) \quad S \;\; \rho_{after\ \mathcal{E}_1}^{(Q)} \qquad S \;\; \rho_{after\ \mathcal{E}_1}^{(E)} \qquad S \;\; \rho_{after\ \mathcal{E}_1}^{(\mathcal{R})} \qquad S \;\; \rho_{after\ \mathcal{E}_1}^{(E)} \qquad S \;\; \rho_{after\ \mathcal{E}_1}^{(E)} \qquad S \;\; \rho_{after\ \mathcal{E}_1}^{(\mathcal{R})} \;.$$

Using similar arguments as before, we see that $S \;\; \rho_{after\ \mathcal{E}_1}^{(\mathcal{R})} \qquad S(\rho)$; thus, we conclude that

$$I(\rho,\mathcal{E}) \quad S(\rho).$$

To prove the second inequality, we consider the configuration in Figure 3.21 consisting of quantum systems, Q, Q, and Q, together with a reference system, $\mathcal{R}$, and two environments, E_1 and E_2. The state of quantum systems Q, Q, and Q is described by the density matrices, $\rho, \rho_{after\ \mathcal{E}_1}^{(Q)}$ and $\rho_{after\ \mathcal{E}_2}^{(Q)}$, respectively. The strong subadditivity of the von Neumann entropy applied to this system requires that

$$S \;\; \rho_{after\ \mathcal{E}_2}^{(\mathcal{R},E_1,E_2)} \qquad S \;\; \rho_{after\ \mathcal{E}_2}^{(E_1)} \qquad S \;\; \rho_{after\ \mathcal{E}_2}^{(\mathcal{R},E_1)} \qquad S \;\; \rho_{after\ \mathcal{E}_2}^{(E_1,E_2)} \;.$$

We now express the individual terms of this expression. First, we observe that the joint state, $(Q,\mathcal{R},E_1,E_2)$, of the quantum system Q, of the reference system $\mathcal{R}$, and of the two environments, E_1 and E_2, is pure at every stage in this chain of transformations; thus, it follows that

$$S \;\; \rho_{after\ \mathcal{E}_2}^{(\mathcal{R},E_1,E_2)} \qquad S \;\; \rho_{after\ \mathcal{E}_2}^{(Q)} \;.$$

From the definition of entropy exchange, it follows that

$$S \;\; \rho_{after\ \mathcal{E}_2}^{(E_1)} \qquad S \;\; \rho_{after\ \mathcal{E}_1}^{(E_1)} \qquad S_e \;\; \rho^{(Q)},\mathcal{E}_1$$

and

$$S \;\; \rho_{after\ \mathcal{E}_2}^{(E_1,E_2)} \qquad S_e \;\; \rho^{(Q)},\mathcal{E}_2 \;\; \mathcal{E}_1 \;.$$

Finally, we observe that neither the reference system $\mathcal{R}$ nor the environment E_1 is involved in the transformation $\mathcal{E}_2$. Thus,

$$\rho_{after\ \mathcal{E}_2}^{(\mathcal{R},E_1)} \qquad \rho_{after\ \mathcal{E}_1}^{(\mathcal{R},E_1)}.$$

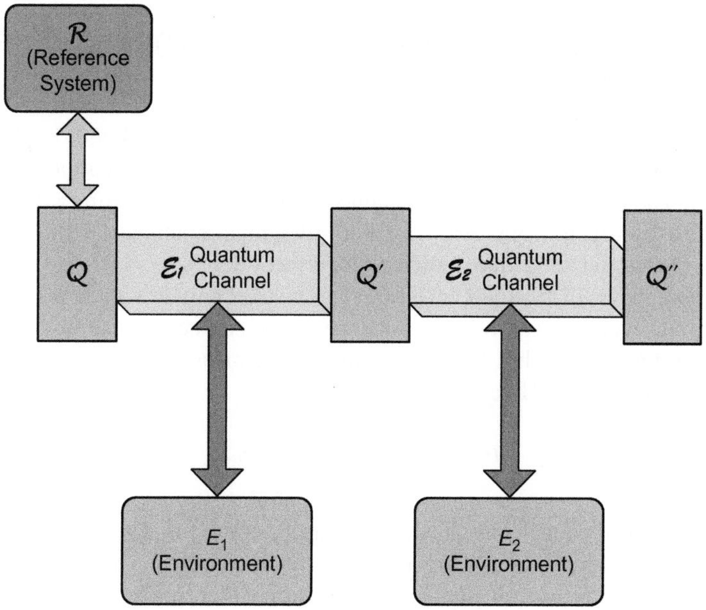

FIGURE 3.21

The configuration used to prove the second part of the quantum data processing inequality. The quantum systems, $\mathcal{Q}, \mathcal{Q}$, and $\mathcal{Q}$, together with a reference system, $\mathcal{R}$, and two environments, E_1 and E_2, are shown; the state of quantum systems $\mathcal{Q}, \mathcal{Q}$, and $\mathcal{Q}$ is described by the density matrices, $\rho, \rho^{(\mathcal{Q})}_{after\ \mathcal{E}_1}$ and $\rho^{(\mathcal{Q})}_{after\ \mathcal{E}_2}$, respectively.

Then from the purity of the joint system, $(\mathcal{Q}, \mathcal{R}, E_1)$, after the first stage, it follows that

$$S \quad \rho^{(\mathcal{R},E_1)}_{after\ \mathcal{E}_2} \qquad S \quad \rho^{(\mathcal{R},E_1)}_{after\ \mathcal{E}_1} \qquad S \quad \rho^{(\mathcal{Q})}_{after\ \mathcal{E}_1} \quad .$$

After substituting in the strong subadditivity relation, we conclude that

$$S \quad \rho^{(\mathcal{Q})}_{after\ \mathcal{E}_2} \qquad S_e \quad \rho^{(\mathcal{Q})}, \mathcal{E}_1 \qquad S \quad \rho^{(\mathcal{Q})}_{after\ \mathcal{E}_1} \qquad S_e \quad \rho^{(\mathcal{Q})}, \mathcal{E}_2 \quad \mathcal{E}_1$$

or

$$S \quad \rho^{(\mathcal{Q})}_{after\ \mathcal{E}_2} \quad - S_e \quad \rho^{(\mathcal{Q})}, \mathcal{E}_2 \quad \mathcal{E}_1 \qquad S \quad \rho^{(\mathcal{Q})}_{after\ \mathcal{E}_1} \quad - S_e \quad \rho^{(\mathcal{Q})}, \mathcal{E}_1 \quad .$$

From the definition of the coherent information, it follows that

$$I(\rho, \mathcal{E}_2 \quad \mathcal{E}_1) \qquad I(\rho, \mathcal{E}_1).$$

This concludes the proof of the quantum data processing inequality. ■

3.20 REVERSIBLE EXTRACTION OF CLASSICAL INFORMATION FROM QUANTUM INFORMATION

We saw in Section 2.13 that quantum states can be teleported, transmitted without distortion by separating quantum from classical information and using two separate communication channels, a classical and a quantum one. The classical information is uncorrelated with the arbitrary quantum input state, thus preventing an eavesdropper from observing non-orthogonal input states without disturbing them; the scheme allows input states to be transmitted with perfect fidelity.

Classical information has distinct advantages, it can be transmitted with an arbitrarily small probability of error and the recipient can recover the state of the source with almost certainty, while the recipient of quantum information has to perform an optimal measurement in order to maximize the mutual information subject to the Holevo bound

$$I(X;Y) \quad S(\rho) \qquad \underset{a}{} \; p_a \rho_a,$$

where p_a are the probabilities and ρ_a are the density matrices of the *signal states* a [324].

The example at the end of Section 3.14 illustrates the advantage of using a classical channel to transmit quantum information and, at the same time, shows that a neat separation of classical and quantum information is possible when the source states can be partitioned into mutually orthogonal subsets. Indeed, consider a three-dimensional Hilbert space with the orthonormal basis, e_1 , e_2 , e_3 ; the source, $\mathcal{A}$, transmits the state, e_1 , with the probability, p_1 $1/2$, and a random state in the subspace spanned by e_2 and e_3 , with the probability, p_2 $1/2$, and has a von Neumann entropy of 1.5 bits [35]. The information produced by $\mathcal{A}$ can be neatly separated by the following encoding scheme: (1) determine first by an incomplete measurement if the qubit is in state e_1 and transmit one bit of classical information 1 if the state is e_1 ; else, (2) transmit both a bit 0 and the qubit carrying the state in the subspace spanned by e_2 and e_3 . This encoding scheme allows the source state to be recovered with perfect fidelity.

The Schumacher compression and the quantum noiseless coding theorem tell us how to reduce the number of qubits required to encode the information produced by a quantum source viewed as an ensemble of pure, but not all orthogonal states, φ_a , with the probabilities, p_a. The question we address is whether it is advantageous to extract some classical information from the quantum information and transmit both of them as bits and qubits, respectively, through a pair of classical and quantum channels; could we improve on Schumacher compression by acting on the quantum part of the information in this scheme?

The separation of classical and quantum information and the compression of quantum information may, or may not, be always feasible and advantageous. For example, consider two sources, $\mathcal{B}$, emitting linearly polarized photons at 0 ,45 ,90 , and 135 with equal probability, and $\mathcal{C}$ emitting linearly polarized photons at 0 ,1 ,90 , and 91 with equal probability. The former has a high entropy and cannot be compressed. A classical measurement of the latter allows us to distinguish with high, though not perfect, reliability between the two groups of states 0 ,1 and 90 ,91 , and each one of these two groups has an entropy far less than one bit.

The following theorem answers the question of whether we can improve on Schumacher compression by acting on the quantum part of the information after a reversible extraction of classical information from the quantum information [35].

Improvement on Schumacher Compression

Theorem. *A quantum source consisting of the ensemble of states,* φ_a *, with the probabilities,* p_a*,* $\mathcal{A}$ $\quad \varphi_a , p_a$ *, can be encoded into, and then recovered with perfect fidelity from, an intermediate representation consisting of orthonormal distinguishable states,* X_j *, holding classical information and intermediate quantum information non-orthogonal states,* ξ_{ja} *, such that the mean entropy of the intermediate quantum states is less than that of the source ensemble,* $\mathcal{A}$*, if and only if the states of the source fall into two or more mutually orthogonal subsets of nonzero probability.*

We assume that the states of the source, φ_a $\quad \mathcal{H}_{source}$, are normalized, but not orthogonal, and that in the process of encoding we use an ancilla system initially in the state, ζ $\quad \mathcal{H}_{ancilla}$. The composite system is then in a superposition of states, X_j , that will allow us to extract classical information and normalized, but non-orthogonal states, ξ_{ja} , representing residual information not captured by the X_j states,

$$\Phi \quad \sum_j \alpha_{ja}\, \xi_{ja} \quad X_j , \quad \text{with} \quad \sum_j \alpha_{ja}{}^2 \quad 1 \quad \text{and} \quad \Phi \quad \mathcal{H}_{source} \quad \mathcal{H}_{ancilla}.$$

The encoding proceeds as follows: We first measure the "classical" subsystem; in other words, we determine if X_j is an eigenstate of the system and obtain a classical result, j, with the probability $P(j\ a) \quad \alpha_{ja}{}^2$, when the source state was φ_a . It is trivial to prove that the condition is sufficient, namely, that when the states of the source fall into two or more mutually orthogonal subsets of nonzero probability, we can recover with perfect fidelity the original states. Indeed, if the source states are separable into mutually orthogonal subsets, there is a nondemolition measurement that allows us for each original state, φ_a , to determine the subset of states it belongs to, without disturbing φ_a . The proof that this condition is necessary must show that the intermediate ensemble, $\mathcal{A}$ $\quad \xi_{ja} , p_a$, has the same entropy as the original one, $\mathcal{A}$ $\quad \varphi_a , p_a$. The proof is left as an exercise.

Thus, this theorem tells us that no compression of the quantum information is possible if the source ensemble of states is not separable into orthogonal subsets and if the decoding operation is perfectly faithful.

The question left unanswered by this theorem is what happens when instead of absolute fidelity we impose only asymptotic fidelity in decoding.

3.21 NOISY QUANTUM CHANNELS

We can think about a quantum channel as the physical medium that enables the transfer of a quantum system from a sender to a receiver. The channel is noiseless when the quantum system is undisturbed during the transfer, and it is noisy when the quantum system interacts with some other system (e.g., the environment) during the transfer.

Before we discuss noisy quantum channels, we should take a closer look at the "noise," the undesirable transformations suffered by a quantum system during the transfer from the sender to the receiver. Consider two Hilbert spaces, $\mathcal{H}_{in}$ and $\mathcal{H}_{out}$; a *quantum operation*, $\mathbf{N}$, is a map from the set of density operators in $\mathcal{H}_{in}$ to the set of density operators in $\mathcal{H}_{out}$ with the three properties:

1. If a system is characterized by the density matrix, ρ, the probability that the process represented by $\mathbf{N}$ occurs is

$$0 \quad \mathrm{tr}(\mathbf{N}(\rho)) \quad 1, \qquad \rho \quad \mathcal{H}_{in}.$$

The normalized quantum state after the mapping, **N**, is

$$\frac{\mathbf{N}(\rho)}{\mathrm{tr}(\mathbf{N}(\rho))}.$$

While $\mathrm{tr}(\rho) \quad 1$, the quantum operation **N** may not preserve the trace property of the density matrix; in the general case, $\mathrm{tr}(\mathbf{N}(\rho)) \quad 1$. To identify the outcome of the mapping **N**, we perform a measurement, and the probability of an outcome is $0 \quad \mathrm{tr}(\mathbf{N}(\rho)) \quad 1$. The result of the mapping is a valid density matrix up to a normalization factor.

2. **N** is a convex linear map

$$\mathbf{N} \qquad \underset{i}{\;} p_i\rho_i \qquad \underset{i}{\;} p_i\mathbf{N}(\rho_i).$$

The convexity of the linear map is a property relevant to ensembles of the states, ρ_i, p_i , with the density matrices, ρ_i, and the probabilities, p_i. There are two ways to look at the mapping **N**:

- It is applied to ρ, the density matrix of the entire system,

$$\mathbf{N} : \rho \qquad \frac{\mathbf{N}(\rho)}{\mathrm{tr}(\mathbf{N}(\rho))}, \quad \text{with} \quad \rho \qquad \underset{i}{\;} p_i\rho_i.$$

- It is applied to the individual states described by the density matrices ρ_i,

$$\mathbf{N} : \rho_i \qquad \frac{\mathbf{N}(\rho_i)}{\mathrm{tr}(\mathbf{N}(\rho_i))}.$$

Call $p(\mathbf{N})$ the probability that the system in the state, $\mathbf{N}(\rho)$, was initially in the state, ρ, and, $p(i \; \mathbf{N})$, the probability that the system in the state, $\mathbf{N}(\rho_i)$, was initially in the state ρ_i. According to the preceding condition 1, these two probabilities are, respectively, $p(\mathbf{N}) \quad \mathrm{tr}(\mathbf{N}(\rho))$ and $p(\mathbf{N} \; i) \quad \mathrm{tr}(\mathbf{N}(\rho_i))$. The condition that two results of applying **N** should be identical means that

$$\mathbf{N}(\rho) \quad \mathrm{tr}(\mathbf{N}(\rho)) \quad \underset{i}{\;} p(i \; \mathbf{N}) \frac{\mathbf{N}(\rho_i)}{\mathrm{tr}(\mathbf{N}(\rho_i))}.$$

But according to Bayes' rules,

$$p(i \; \mathbf{N}) p(\mathbf{N}) \quad p(\mathbf{N} \; i) p_i.$$

This implies that

$$p(i \; \mathbf{N}) \qquad \frac{p(\mathbf{N} \; i) p_i}{p(\mathbf{N})} \qquad p_i \frac{\mathrm{tr}(\mathbf{N}(\rho_i))}{\mathrm{tr}(\mathbf{N}(\rho))}.$$

It follows immediately that

$$\mathbf{N}(\rho) \quad \mathbf{N} \qquad \underset{i}{\;} p_i\rho_i \qquad \underset{i}{\;} p_i\mathbf{N}(\rho_i).$$

3. **N** is a completely positive map; if **A** is a positive operator, then **N**(**A**) is positive. Moreover, if we have an additional space, $\mathcal{H}_{aux}$, and (**I** **N**) (**A**) acts on the combined space, $\mathcal{H}_{aux}$ $\mathcal{H}_{in}$, then (**I** **N**) (**A**) is positive; in this expression, **I** is the identity transformation on $\mathcal{H}_{aux}$.

The following proposition relates the operator-sum representation discussed in Section 2.14 with the properties defining quantum operations; the proof of this proposition is presented in [307].

Proposition. *A map,* **N**, *satisfies conditions 1–3 if and only if*

$$\mathbf{N} \quad \sum_i \mathbf{E}_i \rho \mathbf{E}_i^{\dagger},$$

with $\mathbf{E}_i$, *a set of operators that map the input Hilbert space to the output Hilbert space and* $\sum_i \mathbf{E}_i \mathbf{E}_i^{\dagger} \quad \mathbf{I}$.

Example. *The partial trace is a quantum operation.* Consider two quantum systems, $\mathcal{Q}$ $\mathcal{H}_{\mathcal{Q}}$ and $\mathcal{R}$ $\mathcal{H}_{\mathcal{R}}$, and let q_i be an orthonormal basis in $\mathcal{H}_{\mathcal{Q}}$ and ρ a density operator of $\mathcal{Q}$; let j be an orthonormal basis for $\mathcal{H}_{\mathcal{R}}$ and $\rho^{\mathcal{R}}$ j j a density matrix of $\mathcal{R}$. If $(\mathcal{Q},\mathcal{R})$ $\mathcal{H}_{\mathcal{QR}}$ is a composite system of $\mathcal{Q}$ and $\mathcal{R}$, we define the linear operator, $\mathbf{E}_i : \mathcal{H}_{\mathcal{QR}} \quad \mathcal{H}_{\mathcal{Q}}$, as

$$\mathbf{E}_i \quad \sum_j \lambda_i \, q_j \quad j \quad \lambda_i \, q_i \,,$$

with λ_i complex numbers. Now we define **N** as a quantum operation, with the operation elements $\mathbf{E}_i$,

$$\mathbf{N}(\rho) \quad \sum_i \mathbf{E}_i \rho \mathbf{E}_i^{\dagger}.$$

Then,

$$\mathbf{N} \; \rho \quad \rho^{\mathcal{R}} \quad \mathbf{N} \; \rho \quad j \; j \quad \delta_{j,j} \; \rho.$$

But,

$$\rho \quad \mathrm{tr}_{\mathcal{R}} \; \rho \quad j \; j \; .$$

We conclude that

$$\mathbf{N} \; \rho \quad j \; j \quad \mathrm{tr}_{\mathcal{R}} \; \rho \quad j \; j \; .$$

Both **N** and the trace are linear operations; thus, we conclude that **N** $\mathrm{tr}_{\mathcal{R}}$.

There are several types of quantum operations carried out by noisy quantum channels: bit-flip, phase-flip, bit-phase flip, state depolarization, amplitude, and then phase damping; in all instances, some states are left invariant by the quantum operation, but the Bloch sphere is deformed.

A *bit-flip channel* transforms the qubit state, $\varphi \quad \alpha_0 \; 0 \quad \alpha_1 \; 1$, into, $\psi \quad \alpha_1 \; 0 \quad \alpha_0 \; 1$, with the probability, p. The basis states are transformed as 0 1 and 1 0 ; the operation elements are

$$E_0 \quad \sqrt{p}\,\sigma_I \quad \sqrt{p} \begin{matrix} 1 & 0 \\ 0 & 1 \end{matrix} \quad \text{and} \quad E_1 \quad \sqrt{1 \quad p}\,\sigma_x \quad \sqrt{1 \quad p} \begin{matrix} 0 & 1 \\ 1 & 0 \end{matrix} .$$

After going through a bit-flip channel, the pure states represented by points on the x-axis of a Bloch sphere are not affected, while the ones in the y–z plane are contracted by a factor $1-2p$. A *phase-flip channel* transforms the qubit state, $\varphi \quad \alpha_0\, 0 \quad \alpha_1\, 1$ into $\psi \quad \alpha_0\, 0\, -\alpha_1\, 1$, with the probability, p; the operation elements are

$$E_0 \quad \overline{p}\,\sigma_I \quad \overline{p} \begin{matrix} 1 & 0 \\ 0 & 1 \end{matrix} \quad \text{and} \quad E_1 \quad \overline{1-p}\,\sigma_z \quad \overline{1-p} \begin{matrix} 1 & 0 \\ 0 & -1 \end{matrix} .$$

After going through a phase-flip channel, the pure states represented by points on the z-axis of a Bloch sphere are not affected, while the ones in the $(x$–$y)$-plane are contracted by a factor $1-2p$. A *bit-phase flip channel* transforms the qubit state, $\varphi \quad \alpha_0\, 0 \quad \alpha_1\, 1$, into, $\psi \quad \alpha_1\, 0\, -\alpha_0\, 1$, with the probability, p; the operation elements are

$$E_0 \quad \overline{p}\,\sigma_I \quad \overline{p} \begin{matrix} 1 & 0 \\ 0 & 1 \end{matrix} \quad \text{and} \quad E_1 \quad \overline{1-p}\,\sigma_y \quad \overline{1-p} \begin{matrix} 0 & -i \\ i & 0 \end{matrix} .$$

After going through a bit-phase flip channel, the pure states represented by points on the y-axis of a Bloch sphere are not affected, while the ones in the x–z plane are contracted by a factor $1-2p$.

Consider a quantum system, $\mathcal{A} \quad \mathcal{H}_n$, with the density matrix, $\rho^{\mathcal{A}}$. A *depolarizing channel* characterized by the quantum operation, $\mathbf{N}$, replaces $\mathcal{A}$ with the completely mixed state, $\frac{1}{n}I$, with the probability, p, and leaves the system's state unchanged, with the probability, $(1-p)$:

$$\mathbf{N} \quad \rho^{\mathcal{A}} \qquad \frac{p}{n}I \quad (1-p)\rho^{\mathcal{A}}.$$

To study the effect of the depolarizing channel on a single qubit (Figure 3.22) with the density matrix ρ, we set $n \quad 2$ and note that

$$\frac{1}{2}\sigma_I \quad \frac{1}{4}(\rho \quad \sigma_x\rho\sigma_x \quad \sigma_y\rho\sigma_y \quad \sigma_z\rho\sigma_z \; ,$$

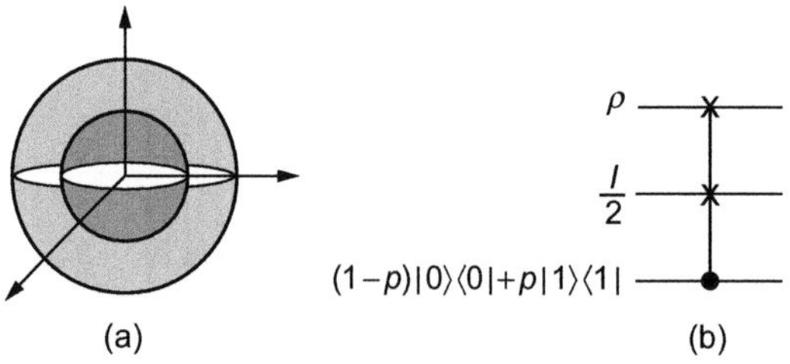

FIGURE 3.22

The effect of a depolarizing channel on one qubit. (a) The Bloch sphere representation of a depolarizing channel, with $p \quad 0.5$. The sphere contracts uniformly depending on p. (b) The quantum circuit simulating a depolarizing channel. The second and third qubit represent the environment and simulate the channel. The third qubit is in a mixture of the state, 0, with the probability, $1-p$, and of the state, 1, with the probability, p, and acts as a control qubit; it swaps the original qubit and the second qubit in a completely mixed state, $I/2$.

with $\sigma_I, \sigma_x, \sigma_y$, and σ_z, the Pauli matrices. Then,

$$\mathbf{N}(\rho) \quad 1 \quad \frac{3}{4}p \quad \rho \quad \frac{1}{4}p \quad \sigma_x\rho\sigma_x \quad \sigma_y\rho\sigma_y \quad \sigma_z\rho\sigma_z \ .$$

It follows that the operation elements are:

$$\overline{1 \quad \frac{3}{4}p}\, \sigma_I, \quad \frac{\overline{p}}{2}\sigma_x, \quad \frac{\overline{p}}{2}\sigma_y, \quad \text{and} \quad \frac{\overline{p}}{2}\sigma_z.$$

When a quantum system loses energy during its interaction with the environment, we talk about *amplitude damping*. For example, consider the interaction of a photon in a superposition state $\varphi \quad \alpha_0\ 0 \quad 1$, with a beam splitter that performs the unitary operation, $B \quad \exp\ \theta\ \alpha_0^{\dagger}\alpha_1 \quad \alpha_0\alpha_1^{\dagger}$, where α_0, α_1 and $\alpha_0^{\dagger}, \alpha_1^{\dagger}$ are the photon annihilation and creation operators, respectively, and θ is the angle of incidence of the photon. The output of the beam splitter is

$$\psi \quad \alpha_0\ 00 \quad \alpha_1 \cos(\theta)\ 01 \quad \sin(\theta)\ 10 \ .$$

The quantum operation describing the transformation of the density matrix, ρ, of the photon is

$$\mathbf{N}(\rho) \quad E_0\rho E_0 \quad E_1\rho E_1,$$

with $E_i \quad i\ B\ 0$ and $\gamma \quad \sin^2(\theta)$:

$$E_0 \quad \begin{matrix} 1 & 0 \\ 0 & \overline{1 \quad \gamma} \end{matrix} \quad \text{and} \quad E_1 \quad \begin{matrix} 0 & \overline{\gamma} \\ 0 & 0 \end{matrix} \ .$$

The, E_1, operation mimics the loss of energy to the environment; it changes 1 into 0 . The, E_0, operation reduces the amplitude of the 1 state and leaves the 0 state unchanged. In fact, energy is not lost to the environment, but the environment perceives that it is more likely that the system is in the state 0 than in the state 1 . In this case, the state 0 is left invariant.

When the quantum operation describes the energy dissipation to the environment at finite temperature, the process is called *generalized amplitude damping*; then the quantum operation describing the transformation of the density matrix due is

$$\mathbf{N}(\rho) \quad E_0\rho E_0 \quad E_1\rho E_1 \quad E_2\rho E_2 \quad E_3\rho E_3,$$

with the operation elements for a single qubit

$$E_0 \quad \overline{p} \quad \begin{matrix} 1 & 0 \\ 0 & \overline{1 \quad \gamma} \end{matrix}, \quad E_1 \quad \overline{p} \quad \begin{matrix} 0 & \overline{\gamma} \\ 0 & 0 \end{matrix}$$

and

$$E_2 \quad \overline{1 \quad p} \quad \begin{matrix} \overline{1 \quad \gamma} & 0 \\ 0 & 1 \end{matrix}, \quad E_3 \quad \overline{1 \quad p} \quad \begin{matrix} 0 & 0 \\ \overline{\gamma} & 0 \end{matrix} \ .$$

The density matrix of the resulting stationary state, $\rho_{sty} \quad \mathbf{N}(\rho)$,

$$\rho_{sty} \quad \begin{matrix} p & 0 \\ 0 & 1 \quad p \end{matrix} ,$$

satisfies the relation, $\mathbf{N} \ \rho_{sty} \quad \rho_{sty}$. The generalized amplitude damping describes, for example, the T_1 relaxation process due to the coupling of spins to a lattice system that is in thermal equilibrium at a temperature much higher than the spin temperature.

The phenomenon of *phase damping* implies a loss of information without the loss of energy, a pure quantum mechanical process unique to quantum information. When a system evolves in time, the energy eigenstates of the system do not change, but they accumulate a phase proportional to the eigenvalue of the corresponding eigenstate. This happens, for example, when photons are scattered randomly through a waveguide; it also happens when the states of an atom are perturbed due to interaction with distant charges. When a system evolves for an unknown period of time, information about the relative phase between energy eigenstates is lost. Assume that we have a qubit in the state, $\varphi \quad a_0 \ 0 \quad a_1 \ 1$, and we apply to it a rotation operation, $R_z(\theta)$, with a random angle of rotation, a so-called *phase kick* operation. The density matrix of the output state is obtained by averaging over the angle θ,

$$\rho \quad \begin{matrix} a^{\ 2} & ab \ e^{\ \ \lambda} \\ a \ be^{\ \ \lambda} & b^{\ 2} \end{matrix} ,$$

with λ proportional to the time of rotation. A characteristic of phase-damping operation is the fact that the off-diagonal terms of the density matrix decay exponentially with time. The phase damping amount increases with time for all states except the states along the z-axis, which remain invariant.

The operation elements of the phase-damping operation for two interacting oscillators—one representing the quantum system, the other the environment—and with an interaction Hamiltonian, $H \quad \chi a^{\dagger} a(b \quad b^{\dagger})$, are

$$E_0 \quad \begin{matrix} 1 & 0 \\ 0 & \overline{1 \quad \lambda} \end{matrix} \quad \text{and} \quad E_1 \quad \begin{matrix} 0 & 0 \\ 0 & \overline{\lambda} \end{matrix} ,$$

with $\lambda \quad 1 \quad \cos^2(\chi \delta t)$ for an interaction time interval, δt. These operation elements can be transformed through a unitary recombination into a new set of operation elements very similar to those of the phase-flip channel

$$E_0 \quad \overline{k} \ \begin{matrix} 1 & 0 \\ 0 & 1 \end{matrix} \quad \text{and} \quad E_1 \quad \overline{1 \quad k} \ \begin{matrix} 1 & 0 \\ 0 & 1 \end{matrix} ,$$

with $k \quad \frac{1}{2}(1 \quad \overline{1 \quad \lambda})$. This shows that the phase-damping quantum operation is exactly the same as the phase-flip channel; thus, phase errors that once were thought of as continuous can be treated discretely; that makes the quantum error-correction of phase errors possible. The phase damping is often referred to as the *spin-spin relaxation* process.

3.22 HOLEVO-SCHUMACHER-WESTMORELAND NOISY QUANTUM CHANNEL ENCODING THEOREM

When we discussed classical noisy channels, we addressed two related questions: (1) Is it possible to encode the input so that the probability of error is arbitrarily low? and (2) If such an encoding scheme exists, what is the maximum rate at which classical information can be transmitted over the channel using this encoding scheme? Similar questions, but framed differently, are discussed in this section on noisy quantum channels.

As before, we consider discrete memoryless quantum channels; the state of the quantum system is a vector in a finite-dimensional Hilbert space, and the memoryless property requires that the mean of the output of the quantum channel is only determined by the current input. This time the quantum channel is noisy; it applies a transformation, **N**, to its input. If ρ^A is the density matrix of the input, then the density matrix of the output is $\rho^{out}\quad \mathbf{N}(\rho^A)$ and the states at the input and the output of the channel are in the general case mixed states. The basic reasoning closely follows the discussion in Section 3.17; we consider product states and typical subspaces, and we are concerned with the probability of error associated with specific encoding and decoding schemes.

To give a mathematical description of the quantum channel means succinctly describing the properties of the transformation **N**. The channel maps density matrices to density matrices; thus, the mapping must be *linear, trace-preserving*, and *positive*. Linearity is required by the basic principles of quantum mechanics; the channel must preserve the trace of the density matrices and map density matrices that are positive semi-definite matrices to positive semi-definite matrices. The mapping, **N**, must be *completely positive*, and it must be positive even when tensored with the identity matrix [387].

We assume that the input to the noisy quantum channel is an ensemble of mixed states (we shall refer to them as letter states, or letters) q_i , with the density matrix, ρ_i, and the von Neumann entropy, $S(\rho_i)$; the letters are selected with a priori probability, p_i. The average density matrix and the von Neumann entropy of the ensemble are, respectively,

$$\omega \qquad \underset{i}{p_i\rho_i} \quad \text{and} \quad S(\omega) \qquad \mathrm{tr}\,(\omega \log \omega).$$

We now consider strings of n such letters:

$$a \quad q_{a,1} \quad q_{a,2} \qquad\qquad q_{a,n} \ .$$

The string, a , occurs with a priori probability,

$$P_a \quad p_{a,1}p_{a,2}\cdots p_{a,n},$$

and has the density matrix,

$$\rho_a \quad \rho_{a,1} \quad \rho_{a,2} \qquad\qquad \rho_{a,n}.$$

The density matrix, ρ_a, has a set of 1 $\;k\;$ n orthogonal eigenvectors, $e_{a,k}$, with the corresponding eigenvalues, $\gamma_{k\ a}$. Since $\mathrm{tr}(\rho)\quad 1$, these eigenvalues, $\gamma_{k\ a}$, form a probability distribution. The index, k, identifying the n-dimensional eigenvector, $e_{a,k}$, plays a significant role in quantum error correction and is called *the syndrome of the codeword a identified by the density matrix, ρ_a*. The probability of the eigenvector, $e_{a,k}$, is

$$P_{a,k} \quad P_a \, \gamma_{k\,a}.$$

This expression suggests that we can look at $P_{a,k}$ as a joint probability and at $\gamma_{k\,a}$ as the conditional probability that the eigenvector $e_{a,k}$ is observed when the string, a, was sent. Using these notations, we can express, ρ, the density matrix of the average state of the channel input, as

$$\rho \quad \sum_a P_a \rho_a \quad \omega^{\;n} \quad \sum_a \sum_k P_{a,k} \, e_{a,k} \; e_{a,k} \,,$$

with $\omega^{\;n}$, the tensor product $\omega^{\;n} \quad \omega \quad \omega \qquad \omega$.

The main result discussed in this section is known as the Holevo-Schumacher-Westmoreland noisy quantum channel encoding theorem, proven independently by Holevo [211] and by Schumacher and Westmoreland [373].

Theorem. *Consider a noisy quantum channel characterized by the trace-preserving transformation, $\mathbf{N}$; the input of the channel is an ensemble of mixed letter states with the a priori probability, p_i, the density matrix, ρ_i, and the von Neumann entropy, $S(\rho_i)$. The average density matrix of the ensemble is ω, with $\omega \quad {}_i p_i \rho_i$, and the corresponding von Neumann entropy is $S(\omega)$. Given arbitrarily small positive constants $\epsilon, \delta \quad 0$, for sufficiently large n, there exist:*

1. *a code C whose codewords are strings of n letters and*
2. *a decoding observable such that the information per letter is at least $\zeta \quad \delta$, with $\zeta \quad S(\omega) \qquad {}_j p_j S \; \rho_j$, and the probability of error is $p_{err} \quad \epsilon$.*

The capacity of the noisy quantum channel to transmit classical information is

$$C_1 \quad \max_{p_i, \rho_i} \; S(\omega) \qquad \sum_i p_i S(\rho_i) \;,$$

where the maximum is over all ensembles of possible input mixed states, p_i, ρ_i.

We use the notation, C_1, rather than C to emphasize the fact that this is the *one-shot* classical capacity, the maximum amount of classical information sent through a single use of the channel by an optimal choice of the source states and an optimal measurement. Recall from Section 3.18 that the *syndrome measurement* tells us as much as possible about the error affecting a codeword, but nothing at all about the information carried by the codeword; otherwise, the measurement would destroy any quantum superposition of the logical qubits with other qubits in the system.

Instead of considering the ensemble of mixed states, q_i, with the density matrices, ρ_i, as the input of the quantum noisy channel, we can think of them as the output of the channel, with q_i^{in} as input [373],

$$q_i \quad \mathbf{N} \; q_i^{in} \;, \quad \rho_i \quad \mathbf{N}(\rho_i^{in}), \quad \text{and} \quad \omega \quad \mathbf{N} \; \omega^{in} \;.$$

Then, the effect of the state transformation due to the noisy channel appears explicitly [307] as

$$\zeta(\mathbf{N}) \quad S \; \mathbf{N} \; \omega^{in} \qquad \sum_i p_i S \; \mathbf{N} \; \rho_i^{in}$$

and

$$C_1 \quad \max_{p_i, \rho_i} \; S \; \mathbf{N} \; \omega^{in} \qquad \underset{i}{} \; p_i S \; \mathbf{N} \; \rho_i^{in} \quad .$$

In $\zeta(\mathbf{N})$, we recognize the expression of the Holevo bound for a quantum system subjected to a trace-preserving transformation $\mathbf{N}$. The proof of this theorem, like the proof of the coding theorem for noiseless channels in Section 3.17, involves several aspects: (1) the typicality of codewords consisting of n signals from the input alphabet; (2) the question of encoding and decoding strategies; and (3) the probability of error.

Given $\epsilon, \delta > 0$, we have to find an integer n large enough such that the typicality condition on strings of letters of length n is satisfied. Call e_i and λ_i the eigenvectors and, respectively, the eigenvalues of ρ. First, we have to show that there is a typical subspace, Λ, spanned by the eigenvectors of ρ, such that

$$\mathrm{tr}(\rho\Pi) > 1 \quad \epsilon,$$

where Π is the projection onto Λ of the original states. In addition, the typical subspace Λ must satisfy the property

$$\mathrm{tr} \;\; \rho^2\Pi \;\; < 2^{\;\; n(S(\omega) \;\; 2\delta)}.$$

To bound the error, we require that the eigenvalues of ρ corresponding to eigenvectors in the typical subspace be restricted to a very narrow range of values:

$$2^{\;\; n(S(\omega) \;\; \delta)} \quad \lambda_i \quad 2^{\;\; n(S(\omega) \;\; \delta)}, \qquad e_i \quad \Lambda.$$

To provide a precise definition of typicality in this context, we introduce several new concepts. Recall that we have three probability distributions: (1) for individual letters—letter q_i occurs with probability p_i; (2) for strings of n letters—string a occurs with probability P_a; and (3) the joint probability for string-syndrome pairs—the pair $(\; a \;, k)$ occurs with probability $P_{a,k}$. Then the Shannon entropy of the individual letters of the input alphabet, of the strings of letters, and of the string-syndrome pairs are defined as

$$H(X) \qquad \underset{i}{} \; p_i \log p_i, \quad H(A) \qquad \underset{a}{} \; P_a \log P_a, \quad \text{and} \quad H(A,K) \qquad \underset{a}{} \; \underset{k}{} \; P_{a,k} \log P_{a,k}.$$

These quantities are related to one another:

$$H(A) \quad nH(X)$$

and

$$H(A,K) \quad H(A) \qquad \underset{a}{} \; P_a S(\rho_a) \quad n \;\; H(X) \qquad \underset{i}{} \; p_i S(\rho_i) \;\; .$$

Now we can provide a more precise definition of typicality. First, we say that the string, a , is a *typical string* if its probability is in a narrow range:

$$2^{\ n(H(X)\quad \delta)} < P_a < 2^{\ n(H(X)\quad \delta)}, \qquad a \quad \Lambda.$$

In addition,

$$\sum_{a \quad \Lambda} P_a > 1 \quad \epsilon, \qquad a \quad \Lambda.$$

Second, we say that, (a ,k), is a *typical string-syndrome pair* if the joint probability, $P_{a,k}$, is in a narrow range as well:

$$2^{[\ n(H(X) \qquad {}_i p_i S(\rho_i)\quad \delta)]} < P_{a,k} < 2^{[\ n(H(X) \qquad {}_i p_i S(\rho_i)\quad \delta)]}.$$

We also require that

$$\sum_{a\ ,k} P_{a,k} > 1 \quad \epsilon, \qquad \text{string-syndrome pairs } (a\ ,k).$$

Third, if a is a typical string, then the set of k, such that (a ,k) are typical string-syndrome pairs, is called the set of *relatively typical syndromes* for a . The corresponding probability, $p_{k\ a} \quad P_{a,k}/P_a$, should also be in a narrow range:

$$2^{[\ n(\ \ {}_i p_i S(\rho_i)\quad 2\delta)]} < p_{k\ a} < 2^{[\ n(\ \ {}_i p_i S(\rho_i)\quad 2\delta)]}.$$

The last inequality can be expressed as

$$2^{[\ n(S(\omega)\quad \zeta \quad 2\delta)]} < p_{k\ a} < 2^{[\ n(S(\omega)\quad \zeta \quad 2\delta)]}.$$

A code, $\mathcal{Q}$, is a subset of the set of strings of length n of letters; codewords are labeled using the Greek alphabet (e.g., α , β ,...), while we use the Roman alphabet (e.g., a , b ,...) to denote strings of n letters that are not codewords. The code $\mathcal{Q}$ consists of N codewords used with the same frequency.

The decoding procedure will be based on the "pretty good measurement" discussed in Section 3.17, but now we shall attempt to identify the codewords, as well as the syndromes. The decoding observable is a POVM described by a set of positive operators summing to unity; for each codeword-syndrome pair, (α ,k), we have a vector, $\mu_{\alpha,k}$, such that the projector, $\mu_{\alpha,k} \quad \mu_{\alpha,k}$, is an element of the decoding POVM.

The probability of errors can be derived using a logic similar to the one for noiseless quantum channels,

$$p_{err} \quad 1 \quad \frac{1}{N} \sum_{a,k} p_{k\ a} \quad \mu_{\alpha,k}\ e_{\alpha,k} \quad ^2,$$

with $e_{\alpha,k}$, the eigenvectors of the density matrix, ρ, corresponding to the codewords. To provide an upper bound for p_{err}, Schumacher and Westmoreland use arguments mirroring closely the ones in

Section 3.17. The derivation is fairly involved and the interested reader may consult, their paper [373] for the details; the final result for an upper bound for the probability of error averaged over all random codes is

$$p_{err} \quad 9\epsilon \quad N2^{\ n(\zeta \ \ 4\delta)}.$$

For sufficiently large n, one can choose N nearly as big as $2^{n\zeta}$ codewords and still have a small probability of error. The theorem shows that good codes for noisy quantum channels exist without addressing the question of how to construct such codes.

3.23 CAPACITY OF NOISY QUANTUM CHANNELS

In Section 3.15, we defined the capacity, Q, of a quantum channel, $\mathcal{N}$, characterized by the superoperator, $\mathbf{N}$, as the largest number Q such that, for any rate, $R < Q$, and for any small and positive, $\epsilon > 0$, the original state, φ , can be recovered by the receiver with a fidelity, $F \quad 1 \quad \epsilon$. While this definition resembles closely the one for classical channels, determining the capacity of quantum channels proves to be more challenging than expected due to the inherent properties of quantum information that preclude the direct mapping of classical methods to studying the quantum channel capacity. Moreover, the presence of ancillary resources, such as a classical channel and entanglement, affect in subtle ways the capacity of a quantum channel.

Quantum information cannot be cloned, and measurements generally alter the quantum state; thus, encoding and decoding are conceptually more complex and will affect the channel capacity. Non-orthogonal pure and mixed states cannot be reliably distinguished; thus, it is not possible to attain perfect fidelity, a reason to expect lower quantum channel capacity when the input states are non-orthogonal. The means to transport classical information using entangled states should be investigated "de nuovo," as entangled states do not have a classical corespondent.

As mentioned in Section 3.15, there are several definitions of the capacity of a quantum channel based on the nature of the information transmitted (classical or quantum), the ancillary resources allowed (classical channel, entanglement), and the protocols permitted [386]. The four distinct capacities as defined by Bennett *et al* [47] are (1) C, the ordinary capacity for transmitting classical information, defined as the largest asymptotic rate of reliably transmitting classical information using an encoder and a decoder characterized by the superoperators, $\mathbf{E}$ and $\mathbf{D}$, respectively; (2) Q, the quantum capacity for transmitting quantum states, defined as the largest asymptotic rate of reliably transmitting qubits under similar conditions as in the definition of C; (3) Q_2, the classically-assisted quantum capacity for transmitting qubits with the assistance of a two-way classical channel; and (4) C_E, the entanglement-assisted classical capacity.

In Section 3.22, we saw that according to the Holevo-Schumacher-Westmoreland theorem the capacity of the noisy quantum channel for transmitting classical information is

$$C_1 \quad \max_{p_i, \rho_i} \ S(\omega) \qquad \sum_i p_i S(\rho_i) \ ,$$

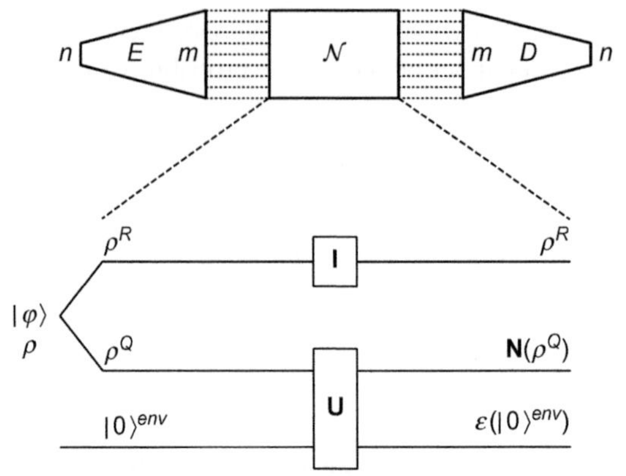

FIGURE 3.23

A quantum channel, $\mathcal{N}$, with an encoding subprotocol, **E**, and a decoding subprotocol, **D**. The encoder, E, receives an n-qubit state, φ $\mathcal{H}_n$, and produces, m n, possibly entangled states as inputs to the channel $\mathcal{N}$; the decoder, D, receives the, m, possibly entangled qubits and produces, n m, qubits as output. The noisy channel $\mathcal{N}$ could be viewed as a unitary interaction, **U**, of the quantum system, Q, with an environment initially in the state, 0 , which remains in a mixed state, $\mathcal{E}(\rho^{env})$, after the interaction. The input to $\mathcal{N}$ is a mixed state, φ , with the density matrix, ρ, which can be imagined as part of an entangled pure state of a joint system between, Q, and a purifying reference system, R. The system, R, is not physically acted upon; thus, the operation on it is **I**. The output joint state is $\mathbf{N}$ $\mathbf{I}$ ρ^Q ρ^R and has the same entropy as $\mathcal{E}(\rho^{env})$.

where the input of the channel is an ensemble of mixed letter states, with the probability, p_i, the density matrix, ρ_i, and the von Neumann entropy, $S(\rho_i)$; the average density matrix of the ensemble is ω, where ω $_i p_i \rho_i$, and the corresponding von Neumann entropy is $S(\omega)$. C_1 is maximum amount of classical information sent through a single use of the channel by an optimal choice of the source states and an optimal measurement, the *one-shot* classical capacity [373].

A common framework for various capacities of a quantum channel, $\mathcal{N}$, with an encoding subprotocol, **E**, and a decoding subprotocol, **D**, is introduced [47]. The encoder, E, receives an n-qubit state, φ $\mathcal{H}_n$, and produces, m n, possibly entangled states as inputs to the channel, N; the decoder, D, receives the, m, possibly entangled qubits and produces, n m, qubits as output (Figure 3.23). Then the generalized capacity, $C_X(\mathbf{N})$, of a quantum channel is defined asymptotically as

$$C_X(\mathcal{N}) \quad \lim_{\epsilon \quad 0} \lim_{m} \sup \quad \frac{n}{m} : \mathbf{E}\ \mathbf{D} \quad \varphi\ \mathcal{H}_n \quad F(\ \varphi\ ,\mathbf{E},\mathbf{D},\mathbf{N}) > 1 \quad \epsilon \ ,$$

with $F(\ \varphi\ ,\mathbf{E},\mathbf{D},\mathbf{N},)$, the fidelity of the output state relative to the input state, φ . The input and output von Neumann entropies of the noisy quantum channel model in Figure 3.23 are, respectively,

$$S \ \rho^Q \qquad \mathrm{tr} \ \rho^Q \log \rho^Q \qquad \text{and} \quad S \ \mathbf{N}(\rho^Q) \ .$$

The definition of the generalized capacity of a quantum channel is strikingly similar with the one for a classical binary channel, N, with the encoding and decoding subprotocols, **E** and **D**, respectively,

$$C(\mathcal{N}) \quad \lim_{\epsilon \quad 0} \lim_{m} \sup \quad \frac{n}{m} : \; \mathbf{E} \; \mathbf{D} \quad x \quad 0,1 \quad F(x, \mathbf{E}, \mathbf{D}, \mathbf{N}) > 1 \quad \epsilon \; .$$

For the classical capacity of a classical binary channel, the encoding, **E**, and decoding, **D**, protocols and the channel, **N**, are restricted to be classical stochastic maps. For the quantum capacity, the encoding and decoding protocols are completely positive trace-preserving maps from $\mathcal{H}_2{}^{n}$ to the input space of $N \;\; ^{m}$ and from the output space of $N \;\; ^{m}$ back to the space $\mathcal{H}_2{}^{n}$ of the output.

The methodology to study the capacity of a given channel, M, is sometimes based on the use of the asymptotic efficiency of a channel N, with known capacity, to simulate channel M; this method requires a comprehensive understanding of the properties of the two channels. The asymptotic capacity of channel, $\mathcal{N}$, with the superoperator, **N**, to simulate channel, $\mathcal{M}$, with the superoperator, **M**, is defined as

$$C_X(\mathcal{N}, \mathcal{M}) \quad \lim_{\epsilon \quad 0} \lim_{m} \sup \quad \frac{n}{m} : \; \mathbf{E} \; \mathbf{D} \quad \varphi \quad \mathcal{H}_{\mathbf{M}}{}^{n} \quad F \; \mathbf{M} \;\; ^{n}(\; \varphi \;), \mathbf{E}, \mathbf{D}, \mathbf{N} \; > 1 \quad \epsilon \; ,$$

where the encoder, **E**, and decoder, **D**, enable the acceptance of an input, φ , in a space, $\mathcal{H}_{\mathbf{M}}{}^{n}$, which is the tensor product of n copies of the input Hilbert space of the channel, **M**, to be simulated, and make m forward uses of the simulating channel, **N**, to produce an output state the fidelity of which is compared with that of the state that would have been generated by sending the input φ through $\mathbf{M} \;\; ^{n}$.

In Section 3.24, we shall discuss the classical channel simulation of a quantum channel. Such simulation involves some forward classical communication; the *forward classical communication cost (FCCC)* measures the capacity of the classical channel needed to simulate the quantum channel.

In some cases, exact expressions for the capacities of different types of quantum channels are known, while in many other cases only upper and lower bounds have been derived. Typically, $Q < C$ and $Q \quad C \quad C_E$. Bennett *et al* [42] made a distinction between Q_1, the capacity of a quantum channel assisted by a forward classical channel, and Q_2, the capacity of a quantum channel assisted by a two-way classical channel. It has been shown [38] that a forward classical channel does not increase the capacity of the quantum channel; thus, $Q_1(N) \quad Q(N)$. By definition, $Q \quad Q_2$, and it is conjectured that $Q_2 \quad C$.

Example. *The quantum depolarizing channel (QDC).* QDC is the quantum analog of the binary symmetric channel; QDC replaces with probability p the incoming qubit state with a random qubit state without informing the receiver of the identity of the qubit in error. The bounds for the capacity of a depolarizing channel [38, 39, 42, 137] are summarized in Figure 3.24.

Example. *The* $2/3$*-depolarizing quantum channel.* The channel transmits a qubit state intact, with the probability $1/3$, and replaces it with a random qubit state, with the probability $p \quad 2/3$. This quantum channel can be simulated by a classical channel using the following procedure: A third party selects a random orientation, R, and communicates it to the sender and the receiver; the sender measures the qubit along the direction, R, and sends through the classical channel the result of the measurement, either a 0 or a 1; thus, the FCCC of this simulation is 1 bit. The classical capacity of the $2/3$-depolarizing quantum channel can be measured when the binary 0 is encoded as the quantum state 0 and 1 is encoded as 1 in some basis and the measurement is carried out by the receiver in the same basis. The $2/3$-depolarizing quantum channel can be simulated by a classical binary symmetric

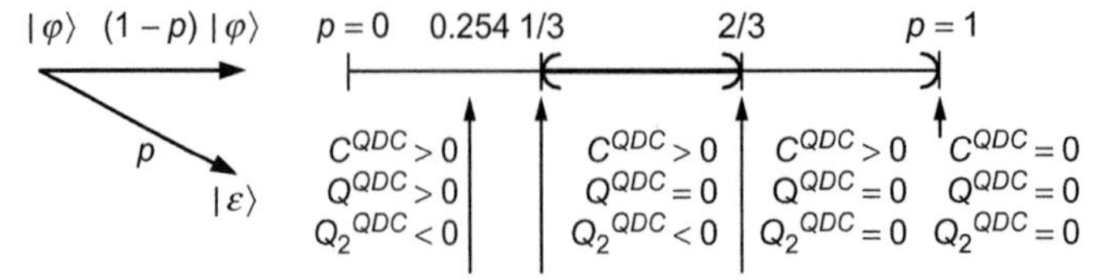

FIGURE 3.24

QDC: The state, φ , of the incoming qubit it replaced with the probability, p, with a random state, ϵ . C^{QDC}, Q^{QDC}, and Q_2^{QDC} are the capacities of the QDC for transmitting classical information, quantum states, and quantum states with the assistance of a two-way classical channel, respectively. Quantum capacity depends on the probability, p, of depolarization error: (1) if $p < 0.25408$, then $(C^{QDC}, Q^{QDC}, Q_2^{QDC}) > 0$; (2) if $1/3 < p < 2/3$, then $(C^{QDC}, Q_2^{QDC}) > 0$ but Q^{QDC} 0; (3) if $2/3$ $p < 1$, then $C^{QDC} > 0$ but Q^{QDC} Q_2^{QDC} 0; and (4) if p 1, then C^{QDC} Q^{QDC} C_2^{QDC} 0.

channel (BSC) with the error probability, p_{BSC} 1/3; the capacity of this BSC is 0.0817 bits; thus, the asymptotic classical capacity of the 2/3-depolarizing quantum channel is close to 0.0817 bits.

Example. *The quantum erasure channel (QErC).* QErC is the quantum analog of the binary erasure channel discussed in Section 3.7; the incoming qubit is replaced with the probability, p, by a qubit in a special state, the *erasure state*, a state orthogonal to 0 and 1 states, and the receiver is informed about the identity of the qubit in error. The capacities of the QErC are known: the capacity to transmit classical information is C^{QErC} $1 - p$; the one-way quantum capacity is Q^{QErC} $\max(0, 1 - 2p)$, and the two-way quantum capacity is Q_2^{QErC} $1 - p$.

It is easy to establish that C^{QErC} $1 - p$. The fraction of non-erased qubits for QErCs is $(1 - p)$, and, according to Holevo's bound, C^{QEC} cannot be larger than $(1 - p)$. Recall that the capacity of a classical erasure channel is $(1 - p)$; a classical channel emulating a QErC with the input state in the basis (0 , 1) and the output state in the basis (0 , 1 , 2) is equal to C $1 - p$.

To show that Q^{QErC} $\max(0, 1 - 2p)$, we show first that Q $1 - 2p$. If we use a one-way hash coding scheme, then the encoder adds two redundancy bits for each erased qubit and uses large block sizes; then the decoder recovers the phase and the amplitude of the erased qubits with the probability one [40]. Now we show that if p 1/2, then Q^{QErC} 0; Alice uses the following scheme to send qubits through a $(2p - 1)$ QErC channel to two receivers: she tosses a random coin and sends to Bob the information qubit and to Charles a qubit in a pure erasure state, or vice versa, depending on the outcome of the toss. This scheme is equivalent to the one when Alice is connected to Bob and to Charles via two p 1/2 QErCs. The capacity of such channels must be zero to prevent cloning. A linear interpolation between a noiseless channel and a 50% QErC shows that $(1 - 2p)$ is indeed an upper bound for Q.

To prove that the two-way quantum capacity of a QErC, Q_2^{QEC} $1 - p$, we consider a teleportation scheme like the one described in Section 2.13. Teleportation requires EPR pairs—e.g., pairs in the state β_{00} $1/\overline{\ 2}($ 00 11 $)$—and a classical channel. To simulate n instances of a QErC with the erasure probability, p, we use $n(1 - p)$ EPR pairs to teleport information from the sender to the receiver. If the capacity of the channel were larger than $(1 - p)$, the sender and the receiver would be able to deterministically increase their entanglement by local actions and communication, a proposition contrary to the laws of quantum mechanics [40]. This leads us to the next topic, the entanglement-assisted capacity of quantum channels.

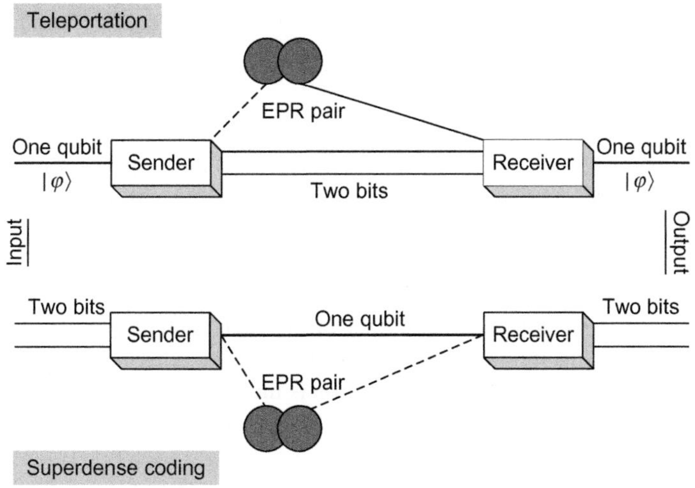

FIGURE 3.25

Superdense coding and quantum teleportation use entanglement as an ancillary resource for quantum communication. Superdense coding allows the sender to transmit one qubit instead of two bits of classical information with the aid of an EPR pair. Quantum teleportation allows the state, φ , to be transmitted with the aid of an EPR pair and classical communication; the sender sends two bits of classical information, and the receiver applies a transformation to his or her half of the EPR pair, guided by the two bits of classical information. The quantum channel used for superdense coding and the classical channel used for teleportation are noiseless.

3.24 ENTANGLEMENT-ASSISTED CAPACITY OF QUANTUM CHANNELS

The discovery of superdense coding [33] followed by quantum teleportation [34] generated a renewed interest in quantum information theory and specifically in the effect of the entanglement on the capacity of quantum channels. Superdense coding doubles the capacity of a noiseless quantum channel, $C_E \quad 2C$, and, to our surprise, the enhancement of the capacity due to entanglement increases for some channels, as the noise level increases.

Superdense coding (bottom of Figure 3.25) uses an EPR pair in the state:

$$\beta_{11} \quad \frac{1}{2}(01 \;-\; 10).$$

The sender has one of the qubits of the pair and the receiver the other qubit of the pair. To communicate one of the four possible two-bit messages, 00, 01, 10, or 11, the sender applies to her own half of the EPR pair one of four transformations given by the four Pauli matrices, $\sigma_I, \sigma_x, \sigma_y$, and σ_z, and then sends the qubit. The receiver is now in the possession of an EPR pair in one of the four Bell states:

$$\beta_{11} \quad \frac{1}{2}(01 \;-\; 10), \quad \beta_{10} \quad \frac{1}{2}(00 \;-\; 11),$$

$$\beta_{00} \quad \frac{i}{2}(00 \quad 11), \quad \beta_{01} \quad \frac{1}{2}(01 \quad 10).$$

The receiver decides which one of the four messages has been sent after she identifies the Bell state:

$$\beta_{11} \quad 00, \quad \beta_{10} \quad 01, \quad \beta_{00} \quad 10, \quad \beta_{01} \quad 11.$$

As we can see, the capacity of a noiseless quantum channel is doubled by the presence of entanglement.

Quantum teleportation (top of Figure 3.25) is in some sense the reciprocal of superdense coding. Quantum teleportation allows the state φ to be transmitted with the aid of an EPR pair and classical communication. The sender sends two bits of classical information and the receiver applies a transformation to his or her half of the EPR pair, guided by the two bits of classical information, as we saw in Section 2.13.

Superdense coding and teleportation show that prior entanglement between the sender and the receiver does increase the capacity of a quantum channel. On the other hand, entanglement does not increase the capacity of a classical channel. The reason is the so-called *constraint on causality*, which does not allow us to increase the expectation of any local variable of the other system; its violation would allow us to send messages to the past.

In this section, we consider a strategy called *measure/re-prepare* to simulate a quantum channel with the aid of a forward classical communication channel: (a) the sender and the receiver use the same orthonormal basis, and the sender prepares the quantum channel input in a pure state in this basis and sends the information regarding the state through a classical channel; (b) the receiver then prepares a quantum state in the same orthonormal basis using the classical information she or he got. The amount of classical information sent through the classical channel in this case is called *forward classical communication cost*, $FCCC_{MR}$.

We begin our discussion of the entanglement-assisted capacity of quantum channels with an example when the exact expression of C_E is known. The *d-dimensional depolarizing channel with depolarization probability p*, $D_p^{(d)}$, transmits a d-state of a quantum system unaltered, with the probability, $(1 \quad p)$, and randomizes the input state, with the probability, p. We show that asymptotically, when $p \quad 1$, the entanglement-assisted capacity of the channel, $D_p^{(d)}$, is $C_E \quad (d \quad 1)C_1(D_p^{(d)})$ [45].

First, we calculate, $C_E(D_p^{(d)})$, the entanglement-assisted capacity of $D_p^{(d)}$. In this case, as in the case of the erasure channel, we use the following methodology to obtain an exact value for C_E: we consider two quantum communication protocols that allow us to establish lower and upper bounds for the entanglement-assisted capacity of a specific type of quantum channel, $\mathcal{N}$, characterized by the superoperator, **N**. If the lower and the upper bounds coincide, then we obtain exact expressions for the capacity of the quantum channel $\mathcal{N}$.

The choices for the communication protocols used to establish the lower and the upper bounds for the entanglement-assisted capacity of the quantum channel, N, are superdense coding, Sd, and teleportation, Tp, depicted in Figure 3.25. To establish a lower bound for C_E, we consider a modified setup for the superdense coding in Figure 3.26 (bottom); we substitute the noiseless quantum channel used to transport the sender's half of the EPR pair by a noisy quantum channel, $\mathcal{N}$. We denote by $C_{Sd}(\mathcal{N})$ the lower bound on the entanglement-assisted capacity estimated for the superdense coding protocol for channel $\mathcal{N}$.

For the upper bound, we consider a teleportation setup (top of Figure 3.26), where the noiseless classical channel between the sender and the receiver is substituted by a noisy one over a d^2-letter alphabet. Due to causality considerations, $FCCC_{Tp}$, the classical communication cost of simulating a quantum channel with teleportation cannot be smaller than the classical capacity of the quantum channel. This

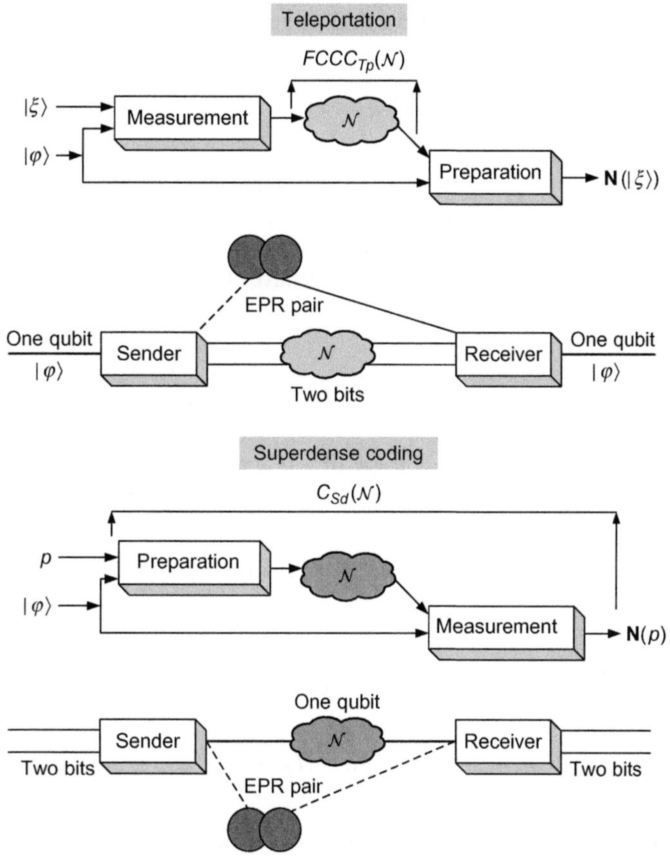

FIGURE 3.26

Superdense coding and quantum teleportation used to simulate the d-dimensional depolarizing channel, with the depolarization probability, p. Superdense coding allows us to establish a lower bound for C_E; the noiseless quantum channel used to transport the sender's half of the EPR pair is substituted by a noisy quantum channel, $\mathcal{N}$. $C_{Sd}(\mathcal{N})$ is the lower bound on the entanglement-assisted capacity using the superdense coding protocol, with the channel, $\mathcal{N}$. Quantum teleportation allows us to establish an upper bound for C_E; the noiseless classical channel between the sender and the receiver is substituted by a noisy one over a d^2-letter alphabet. $FCCC_{Tp}$, the classical communication cost of simulating a quantum channel with teleportation, cannot be smaller than the classical capacity of the quantum channel.

argument justifies the use of the teleportation protocol for the upper bound of the entanglement-assisted capacity of $D_p^{(d)}$.

It can be shown [45] that there are one-to-one mappings between a d-dimensional quantum channel with the depolarization probability, p, and, d^2-ary, the symmetric channel with the randomizing probability, p; the two mappings are done via superdense coding and teleportation. Moreover, the upper and lower bounds for the entanglement-assisted capacity of the quantum channel, $D_p^{(d)}$, obtained through the simulation of $D_p^{(d)}$, with superdense coding and teleportation, coincide:

$$C_E \; D_p^{(d)} \quad C_{sd} \quad FCCC_{Tp} \quad 2\log d \quad H_{d^2} \; 1 \quad p\frac{d^2 \quad 1}{d^2} \; .$$

To compute $C_1(D_p^{(d)})$, we use the measure-reprepare strategy and consider a quantum channel with an input consisting of an arbitrary orthonormal basis, $\; 1 \; , \; 2 \; , \ldots , \; d \quad \mathcal{H}_d$. We assign an equal probability, $1/d$, to each basis vector at the input, and we perform a von Neumann measurement of the output of the channel in the same basis. This quantum channel behaves like a d-ary symmetric classical channel, with the probability of error, p; thus,

$$C_1 \; D_p^{(d)} \quad \log d \quad H_d \; 1 \quad p\frac{d \quad 1}{d} \; .$$

In this expression, $H_d(q)$ is the Shannon entropy of a d-ary distribution consisting of an element of probability, q, and $(d \quad 1)$ elements of probability, $(1 \quad p)/(d \quad 1)$, each:

$$H_d(q) \quad q\log q \quad (1 \quad q)\log \frac{1 \quad q}{d \quad 1} \; .$$

Clearly, when $p \quad 1$, then $C_1(D_p^{(d)}) \quad \log d$. When $p \quad d/(d \quad 1)$, the d-dimensional depolarizing channel can be simulated classically; the amount of classical forward communication required is

$$FCCC_{MR} \; D_p^{(d)} \quad \log d \quad H_d \; d \quad p \quad d \; \frac{1}{d} \; .$$

We continue our discussion of the entanglement-assisted capacity of a quantum channel with the presentation of the general case. Consider a quantum channel, N, mapping the input Hilbert space, $\mathcal{H}_{in}$, to $\mathcal{H}_{out}$, and assume that the sender, Alice, and the receiver, Bob, have an unlimited supply of EPR pairs. If C_E denotes the entanglement-assisted classical capacity of N, then C_E is the maximal quantum mutual information:

$$C_E \quad \max_{\rho \;\; \mathcal{H}_{in}} \; S(\rho) \quad S(\mathbf{N}(\rho)) \quad S \; (\mathbf{N} \quad \mathbf{I}) \; \psi_{pur} \; .$$

In this expression, which resembles the expression of capacity of a classical channel, $S(\rho) \quad \mathcal{H}_{in}$, and $S(\mathbf{N}(\rho))$ are the von Neumann entropy of the input and the output, respectively, of the quantum channel; $S \; (\mathbf{N} \quad \mathbf{I}) \; \psi_{pur}$ is the von Neumann entropy of a purification, ψ_{pur}, of the input density matrix, ρ, a partial trace over a reference system R, $\mathrm{tr}_R \; \psi_{pur} \; \psi_{pur} \quad \rho$. In this model, half of the input, the one corresponding to the entangled state Bob holds at the beginning of the exchange, is sent through the reference system represented by the identity channel, I, and the other half, Alice's half, is sent through the noisy channel, N.

The analysis of the entanglement-assisted capacity of a quantum channel [47] proves first a lower bound for C_E:

$$C_E \quad \max_{\rho \;\; \mathcal{H}_{in}} \; S(\rho) \quad S(\mathbf{N}(\rho)) \quad S \; (\mathbf{N} \quad \mathbf{I}) \; \psi_{pur} \; .$$

The proof uses frequency-typical subspaces, a technique common in classical information theory. Then it proves an upper bound:

$$C_E \quad \max_{\rho \quad \mathcal{H}_{in}} \; S(\rho) \quad S(\mathbf{N}(\rho)) \quad S\;(\mathbf{N} \quad \mathbf{I})\;\psi_{pur} \quad .$$

The proofs of Bennett *et al* [47] are quite intricate and will not be presented here. In conclusion, the entanglement-assisted capacity, C_E, of a quantum channel is given by an expression similar to that for a purely classical channel as a maximum, over the channel input states, ρ, of the entropy of the channel input plus the entropy of the channel output minus their joint entropy. Here, the joint entropy is defined as the entropy of an entangled purification of ρ after half of it has passed through the channel.

3.25 ADDITIVITY AND QUANTUM CHANNEL CAPACITY

The channel capacity quantifies the amount of classical or quantum information that can be protected through error correction, a topic mentioned in Section 3.26 and discussed in depth in Chapters 4 and 5. While a classical channel has a definite capacity to carry classical information, it may not be possible to find a simple expression for the capacity of a quantum channel [390].

In 2008, Smith and Yard showed that *communication is possible over zero-capacity quantum channels* [398]. This result raised new questions regarding quantum channel capacity as a complete description of a quantum channel's ability to transmit quantum information. In 2009, Matthew Hastings provided an answer to one of the most important open questions in quantum information theory showing that the minimum entropy output of a communication channel is not additive [203]. We now discuss briefly these two results.

There are several classes of *zero-capacity channels* [398] useless for quantum information transmission when used alone; one class consists of quantum channels when the joint state of the output and the environment is symmetric under interchange and information transmission would violate the no-cloning theorem. The other class consists of the so-called *entanglement-binding* channels that produce only weakly entangled states. The amazing result is that when we combine one channel from each of these classes with a channel from the other, the resulting channel capacity is positive under some conditions. This means that when two parties attempt to communicate on channels, $\mathcal{N}_1$ and $\mathcal{N}_2$, using separate encoders, E_1 and E_2, as well as separate decoders, D_1 and D_2, the communication fails; when the two channels are used in parallel, the communication is possible, as shown in Figure 3.27.

This result requires two new concepts introduced by Smith *et al* [397], $\mathcal{P}(\mathcal{N})$, the *private capacity* of a quantum channel, $\mathcal{N}$, and, $\mathcal{Q}_{\mathcal{A}}(\mathcal{N})$, the *assisted capacity* of $\mathcal{N}$. The private capacity is the rate at which the channel $\mathcal{N}$ can be used to send classical data secure against eavesdropping for QKD protocols [32]; the assisted capacity reflects the free use of arbitrary symmetric channels to assist quantum communication over $\mathcal{N}$. The assisted capacity is at least as large as half of the channel's private capacity [398].

The additivity conjecture we discuss next would make it possible to compute the classical capacity of a quantum channel. Recall from Section 3.12 that the classical capacity of a quantum channel $\mathcal{N}$ characterized by the mapping, $\mathcal{E}$, of the density matrices ρ_i, of an ensemble of mixed states, with p_i the a priori probability of mixed state i, is given by the Holevo expression

$$\chi(\mathcal{N}) \quad S \quad \underset{i}{\quad} p_i\mathcal{E}(\rho_i) \qquad \underset{i}{\quad} p_i S(\mathcal{E}(\rho_i)),$$

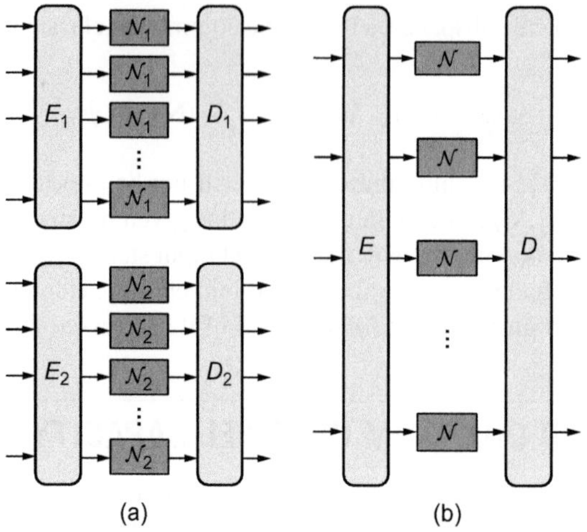

FIGURE 3.27

Communication over zero-capacity quantum channels. (a) Two parties attempt to communicate over zero-capacity channels, $\mathcal{N}_1$ and $\mathcal{N}_2$, using separate encoders, E_1 and E_2, as well as separate decoders, D_1 and D_2. (b) The sender's encoder, E, has access to the inputs of all channels, and the receiver's decoder, D, acts jointly on the output of all channels; communication is possible [398].

with $S(\rho)$ and $S(\rho_i)$, the von Neumann entropy of the overall ensemble of states and of the individual state i, respectively.

Given two quantum channels, $\mathcal{N}_1$ and $\mathcal{N}_2$, the *additivity of the classical information capacity* requires an answer to the question whether

$$\max[\chi(\mathcal{N}_1) \quad \chi(\mathcal{N}_2)] \quad \max[\chi(\mathcal{N}_1)] \quad \max[\chi(\mathcal{N}_2)].$$

Shor has shown [389] that there are three additivity problems equivalent with the additivity of the Holevo channel capacity. The first, the *additivity of the minimum output entropy of a quantum channel*, reflects the desire to minimize the second term in the expression of the Holevo capacity and is formulated as

$$\min_{\rho}[S(\mathcal{N}_1 \quad \mathcal{N}_2)(\rho)] \quad \min_{\rho} S(\mathcal{N}_1(\rho)) \quad \min_{\rho} S(\mathcal{N}_1(\rho)).$$

The other two problems are the *additivity of the entanglement of formation* and *strong superadditivity of the entanglement of formation* [389].

Hastings constructs a counter-example for the additivity of the minimum output entropy and shows that the additivity conjecture is false [203]. The basic idea is to model a situation when the unitary evolution of a system is determined by an unknown state of the environment. This is done with a pair of channels, $\mathcal{N}_1$, and its complex conjugate, $\mathcal{N}_2$ (Figure 3.28); each channel randomly chooses a unitary transformation, $\mathbf{U}_i$, $1 \quad i \quad D$, from a small set and applies that transformation to the input density, ρ:

$$\mathcal{E}(\rho) \quad \sum_{i \; 1}^{D} p_i U_i^{\dagger} \rho U_i \quad \text{and} \quad \mathcal{E}(\rho) \quad \sum_{i \; 1}^{D} p_i U_i^{\dagger} \rho U_i.$$

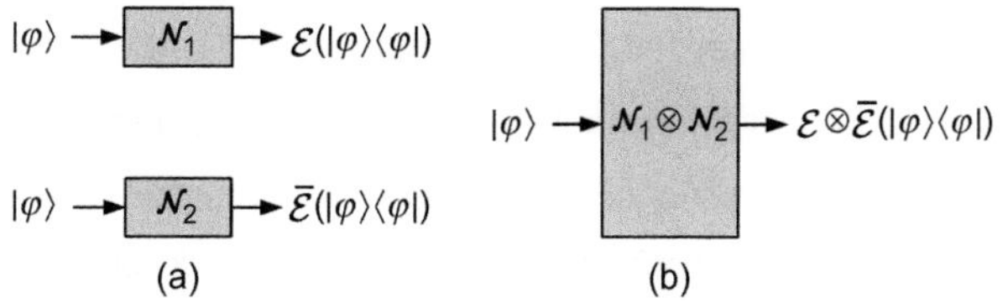

FIGURE 3.28

Superadditivity of the minimum output entropy. (a) Minimization of the output entropy, $S(\mathcal{E}(\rho))$ and $S(\mathcal{E}(\rho))$, of the two channels, $\mathcal{N}_1$, and its complex conjugate, $\mathcal{N}_2$, with a pure state, φ , as input and possibly a mixed state, with the density, ρ, at the output. (b) The entangled input state, φ , is transformed according to the mapping, $\mathcal{E}\quad\mathcal{E}$, and the entropy of the output state is minimized [203].

In this expression, U_i are $N \quad N$ unitary matrices chosen at random from the Haar measure, the probabilities, p_i are also randomly chosen and roughly equal, and $1 << D << N$. The main result in [203] is that *when D and N are sufficiently large, there is a nonzero probability that the random choice of U_i and p_i will lead to a channel, with a mapping $\mathcal{E}\quad\mathcal{E}$, such that*

$$S^{min}(\mathcal{E}\quad\mathcal{E})(\rho) < S^{min}(\mathcal{E}(\rho))\quad S^{min}(\mathcal{E}(\rho))\quad 2S^{min}(\mathcal{E}(\rho)).$$

The superadditivity of the minimum output entropy and the equivalence of the four additivity problems indicate that the problem of quantum channel capacity remains open.

We conclude the chapter with an overview of applications of information theory.

3.26 APPLICATIONS OF INFORMATION THEORY

Classical information theory is critical for the development of modern computing and communication systems. Error control mechanisms enable reliable transport of information on the Internet and allow us to build reliable computing and communication systems. Other important applications of classical information theory are in the area of secure communications and information storage.

Since the early 1990s, quantum information theory has produced significant results in areas such as quantum error-correcting codes, quantum data compression, superdense coding, quantum teleportation, and entanglement concentration; these results give us hope that we will be able to compute and communicate more reliably and securely in the future. Table 3.7 summarizes the applications of classical and quantum information [46].

Error correction is one of the most important applications of information theory and, in turn, a critical element for fault-tolerance. A preview of classical and quantum error correction, topics discussed in depth in the next two chapters, gives us some idea of the similarities as well as the dissimilarities of classical and quantum information processing and of the challenges posed by practical applications of quantum information theory.

Quantum error correction is possible; indeed, consider systems, $\mathcal{A}$ and $\mathcal{B}$, the input and the output of a quantum channel, and, $\mathcal{E}$, the transformation of the input state density, $\rho^{\mathcal{A}}$, carried out by the channel. If the bipartite system, $(\mathcal{A}, \mathcal{B})$, in the product state, $\rho^{\mathcal{A}}\quad\rho^{\mathcal{B}}$, is subject to a transformation $\mathcal{E}$, if there is an auxiliary system, $\mathcal{C}$, in some standard state, $\rho^{\mathcal{C}}$, and some unitary operator, **U**, acting on

Table 3.7 Application of Classical and Quantum Information Theory

Application	Classical Information	Quantum Information
Error correction	Classical error-correcting codes.	Quantum error-correcting codes.
Fault-tolerance	von Neumann's analysis: error probability per gate smaller than a threshold.	The decoherence time per gate below a threshold estimated to be $10^{\ 6}$ to $10^{\ 2}$ seconds.
Communication	Bits over classical channels.	Qubits over quantum channels.
Entanglement-assisted communication	Not possible.	Superdense coding; quantum teleportation.
Cryptographic key distribution	Classical protocols are insecure.	Quantum protocols are secure.
Cryptographic protocols	Insecure against attacks by quantum computers.	Secure against attacks by quantum computers.
Two-party bit commitment	Known protocols insecure against attacks by quantum computers.	Known protocols insecure against attacks by quantum computers.

the joint system, $(\mathcal{A}, \mathcal{B}, \mathcal{C})$, and if

$$\mathcal{E}\ \ \rho^{\mathcal{A}}\ \ \ \rho^{\mathcal{B}}\ \ \ \ \mathrm{tr}_{\mathcal{C}}\ \mathbf{U}\ \ \rho^{\mathcal{A}}\ \ \ \rho^{\mathcal{B}}\ \ \ \rho^{\mathcal{C}}\ \ ,$$

then, after applying the transformation $\mathbf{U}$, we can perform a measurement on $\mathcal{C}$ that reveals how close the two states, $\rho^{\mathcal{A}}$ and $\rho^{\mathcal{B}}$, are, and then ignore the auxiliary system $\mathcal{C}$.

As we can see in Figure 3.29a, classical information is encoded into codewords sparsely distributed in the input space of the communication channel. The codewords are affected by errors during the transmission through the channel; each codeword is surrounded now by a Hamming sphere of possible outputs, and, as long as these spheres are disjoint, the decoding process is able to reconstruct the original codeword. The number of distinct messages is the ratio of the volume of the output space to the volume of a sphere, and the effective channel capacity is the logarithm of this ratio.

Inspired by classical error correction, quantum error correction separates the space of encoded states into *orthogonal subspaces of the Hilbert space* and then performs a measurement of the error state without disturbing the encoded state. Any transformation of quantum information can be regarded as a unitary interaction of the quantum system encoding the information with ancilla qubits initially in a null state (0).

There is a conceptual difference between coding classical and quantum information; *the objective of classical coding is to increase the redundancy and allow each codeword to "represent" an equivalence class of n-tuples that will be decoded as the codeword, while the objective of quantum coding is to increase the distinguishability of the codeword states.*

When we transmit quantum information, we encode every basis vector in the quantum message space as a quantum codeword in a Hilbert space (Figure 3.29b). Quantum noise and decoherence affect the quantum state. As a result, the output for a given input lies with a high probability in some subspace of the original Hilbert space. As long as the overlap of these subspaces is either null, or minimal, we can

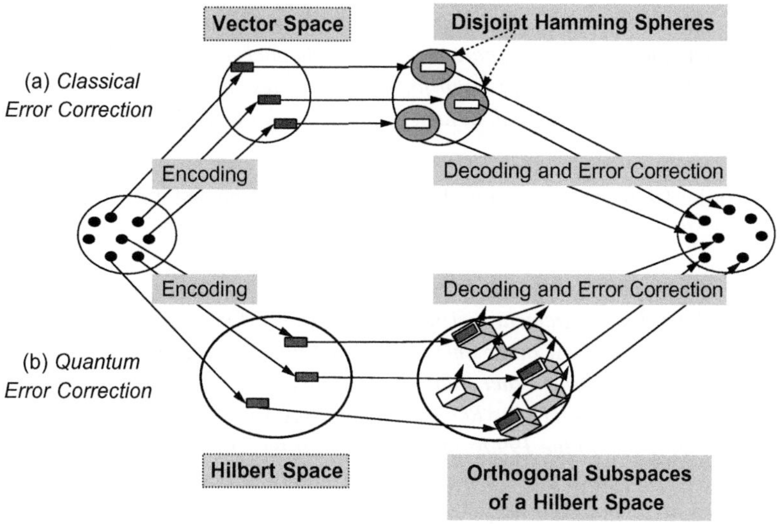

FIGURE 3.29

(a) Classical error correction: Messages are encoded as codewords in a sparsely populated n-dimensional space; codewords are surrounded by disjoint Hamming spheres of n-tuples. Error correction implies that all n-tuples within a Hamming sphere are decoded as the codeword at the center of the sphere. (b) Quantum error correction: A logical qubit is encoded as several physical qubits. The states of the physical qubits used to encode distinct logical qubits span orthogonal subspaces; the transformations applied to these subspaces to determine if an error has occurred and to identify the error should preserve the relationships between the physical qubits used to encode a single logical qubit.

reliably decode each quantum codeword. We note that quantum encoding must preserve the phases and the amplitudes of all qubits of the quantum codeword. The vectors describing quantum codewords in each of these subspaces are perturbed differently. The decoding process must preserve superpositions of the basis vectors; therefore, the amplitudes and the phases of the perturbations must be the same as the initial ones in order to recreate the original states. Questions such as, "How do we carry out a measurement to determine if the quantum codeword is in error without altering the state of the quantum register?" or "How do we distinguish non-orthogonal states?" do not have trivial answers.

3.27 HISTORY NOTES

In his quest to provide a rigorous mathematical foundation for communication, Claude Shannon created, in the late 1940s, the discipline of information theory. The first "golden age" of information theory research can be traced back to the work carried out at Bell Labs starting in the late 1940s and continued at MIT until the end of the 1970s by the group around Claude Shannon.

In his 1995 Shannon Lecture [154], David Forney writes:

Our field has a rather remarkable history, in that it sprang almost fully fledged from the brow of one individual. In his original paper [374] Claude Shannon both posted the fundamental problems of information theory and also, to a large extent, answered them. Shannon's good news was this: for every memoryless channel there exists a parameter C, the channel capacity, such that the probability of error Pr(E) can be made arbitrarily small for any rate R less than C by use of an appropriate code and decoder. Good codes exist and, in fact, a randomly chosen code will with high probability turn out to be good.

The article "Profile: Claude E. Shannon" by John Hogan in the January 1990 issue of *Scientific American* notes that:

Shannon fits the stereotype of an eccentric genius to a T. At Bell Labs (and later at MIT where he returned in 1958 until his retirement in 1978) he was known for riding in the halls on a unicycle sometimes juggling as well. At other times he hopped along the hallways on a pogo stick. He was a lover of gadgets and among other things built a robotic mouse that solved mazes and also a computer called the Throbac—THrifty ROman-numeral BAckward-looking Computer—that computed in roman numerals.... In the 1990s, in one of life's tragic ironies, Shannon came down with Alzheimer's disease, which could be described as the insidious loss of information in the brain. The communication channel to one's memory—one's past and one's very personality—is progressively degraded until every effort at error correction is overwhelmed and no meaningful signal can pass through. The bandwidth falls to zero. The extraordinary pattern of information processing that was Claude Shannon finally succumbed to the depredation of thermodynamic entropy in February 2001. But some of the signal generated by Shannon lives on, expressed in the information technology in which our lives are immersed.

The group of Thomas Cover at Stanford continues to this day the tradition of Claude Shannon. Many distinguished researchers in the field of information theory are either disciples or collaborators of this group. Papers describing significant research results as well as textbooks have been published by the members of the group [106].

The roots of quantum information theory can be traced back to John von Neumann, who introduced the von Neumann entropy in the early 1930s [441]. The relation between information and thermodynamics was captured in 1961 by Rolf Landauer [264]. During the next two decades, Charles Bennett provided deeper insights into the logical reversibility [30] and the thermodynamics of computations [31]; he was able to reconcile the Maxwell demon paradox with the second law of thermodynamics.

In 1973, Alexander Holevo found an upper bound of the accessible information in a quantum measurement [210]; the Holevo bound was probably the first major result in quantum information theory.

The last decade of the twentieth century witnessed a spectacular development of quantum information theory: in the late 1980s, Charles Bennett and Gilles Brassard developed a practical system of quantum cryptography, then, in 1992, Bennett and Wiesener presented a scheme for superdense coding using EPR pairs. In 1993, Bennett and Brassard, in collaboration with Claude Crépeau, Richard Jozsa, Asher Peres, and William Wootters, discovered quantum teleportation. In the mid-1990s, Bennett, Smolin, Wootters, DiVincenzo, and others developed a quantitative theory of entanglement and introduced several techniques for faithful transmission of classical and quantum information through noisy channels; during the same period, Benjamin Schumacher, Richard Jozsa, Seth Lloyd, and others had major contributions to the study of noisy memoryless quantum channels.

3.28 SUMMARY AND FURTHER READINGS

The Landauer principle [264] relates information with thermodynamic entropy: *When a bit of information is erased, the thermodynamic entropy of the environment increases by at least $k_B \ln 2$, with k_B being Boltzmann's constant.* Once the thermodynamic effect of information erasure was established, the physical nature of information could be quantified.

It seems so obvious now that *the amount of information that can be stored or transmitted by a physical system is related to the number of states the system can find itself in* that one may wonder why it took so long to fully comprehend the physical nature of information and the role of entropy. The entropy, an important concept in thermodynamics, Shannon's information theory, as well as quantum information theory, is a function of the logarithm of the number of states of a system; this logarithm is equal to the number of bits required to uniquely identify the state of the system—in other words, to "label" a state. The label becomes an element of the state and any state transformation will affect the label; to prepare a system in a certain state, we have to erase the label of the previous state.

Shannon entropy is used to characterize a source of classical information as well as the properties of a classical communication channel. The properties of joint and conditional entropy are summarized in Table 3.4. The mutual information between the two random variables is $I(X;Y) \quad H(X) \quad H(Y) \quad H(X,Y)$. If the inputs and the outputs of the channel are the random variables X and Y, respectively, then the channel capacity, C, defined as the maximum rate of information transmission over the channel, is $C \quad \max_{p(X)} I(X;Y)$. Two important theorems are the cornerstones of Shannon's information theory, the *source coding theorem* and the *channel coding theorem.*

The original papers of Claude Shannon [374–377] and Robert Fano [144] as well as highly regarded textbooks on classical information theory by Robert Ash [15], Robert Gallagher [167], and Thomas Cover and Joy Thomas [106] published in 1965, 1968, and 1991, respectively, provide the necessary material to understand the basic concepts of classical information theory.

Despite the significant progress made during the past half century, profound questions remain to be answered unequivocally by a unified information theory. Some believe that information has three aspects: (1) a syntactic aspect, the relationship between the symbols of the alphabet used to construct a message, (2) a semantic aspect, the meaning of the message, and (3) a pragmatic aspect, the actions taken by the parties involved in the exchange. Shannon's theory ignores the semantic aspect of information and cannot describe quantum and biological information [261], which most certainly will play a critical role in the third millennium.

Nowadays, some believe that the focus of information theory should migrate from the communication channel to the recipient of information and from the process of communication to the consequences of receiving information. Indeed, in many instances the timeliness of information is important; the context of the information and the goals of the recipient of information (whether a human or a machine) affect the usefulness of information.

John von Neumann used thermodynamic arguments to define the entropy of a quantum system with the density matrix, ρ as $S(\rho) \quad \mathrm{tr}(\rho \log \rho)$; the expression, $\rho \log \rho$, is well defined, as the density matrix is positive semidefinite. It is easy to see that von Neumann entropy is the Shannon entropy of the eigenvalues, p_i, of a diagonal density matrix, ρ. Indeed, if ρ is a diagonal matrix with the eigenvalues, p_i, then $\mathrm{tr}(\rho) \quad {}_i p_i \quad 1$ and $\rho \log \rho$ is also diagonal with the eigenvalues, $p_i \log p_i$. A source of quantum information is not fully characterized by the von Neumann entropy; two different sources of quantum information may have the same density matrix. Thus, they may have the same von Neumann entropy, but be very different from the quantum information theoretical point of view.

A quantum channel is a completely positive, trace-preserving linear map from density matrices to density matrices, $\rho \quad {}_i A_i \rho A_i^{\dagger}$ with ${}_i A_i^{\dagger} A_i \quad I$. An alternative definition of a quantum channel is a partial trace of a unitary transformation on a larger Hilbert space.

A quantum channel has four distinct capacities [47]:

1. C, the ordinary capacity for transmitting classical information: the maximum rate at which classical bits can be transmitted reliably through a quantum channel encoded as the state of a quantum system and decoded after a measurement of the quantum state
2. $Q \quad C$, the ordinary capacity for transmitting quantum states: the maximum asymptotic rate at which qubits can be transmitted reliably through a quantum channel
3. $Q_2 \quad Q$, the classically assisted quantum capacity: the asymptotic rate for reliable qubit transmission with the assistance of a two-way classical channel
4. $C_E \quad C \quad Q$, the entanglement-assisted classical capacity: the maximum asymptotic rate of reliable bit transmission with the assistance of unlimited prior entanglement between the sender and receiver

Table 3.8 summarizes the properties of classical and quantum information and the main results in classical and quantum information theory.

We use the term "ebit" to describe a source of a shared pair of maximally entangled particles. Table 3.9 summarizes the amount of resources necessary to transmit one unit of information [37].

The papers co-authored by Charles Bennett [33–35, 37–40, 42, 44, 46–48] introduce many basic concepts of quantum information theory. Major contributions are also due to Benjamin Schumacher [368–371, 373], Richard Jozsa [220, 222, 226], Seth Lloyd [273, 275], and Peter Shor [384].

A number of dissertations on quantum information theory were written, by Benjamin Schumacher (under Archibald Wheeler) in 1990, and by Howard Barnum and Michael Nielsen [306] in 1999 and 2000, respectively. The textbook by Michael Nielsen and Isaac Chuang [307] covers many aspects of quantum information theory.

The last decade of the twentieth century witnessed a very spectacular development of the field. In 1992, Charles Bennett and Steven Wiesener presented a scheme for superdense coding using EPR pairs. An influential paper on teleportation of quantum states was published in 1993 by Charles Bennett, Gilles Brassard, Claude Crépeau, Richard Jozsa, Asher Peres, and William Wootters [34]. Quantum error-correcting codes were introduced during 1994–1998 by Richard Calderbank, Peter Shor, Andrew Steane, Emmanuel Knill, Raymond Laflamme, Daniel Gottesman, and others [80, 173, 244, 380, 382].

Benjamin Schumacher, Richard Jozsa, Seth Lloyd, Howard Barnum, and Michael Nielsen contributed to the study of noisy memoryless quantum channels [22, 275, 368–370]. In 1998, Charles Bennett and Peter Shor published a seminal paper on quantum information theory [44], and in 2002, Charles Bennett, Peter Shor, John Smolin, and Ashish Thapliyal discussed the entanglement-assisted capacity of a quantum channel [47].

In 2009, Matthew Hastings provided an answer to one of the most important open questions in quantum information theory, showing that the minimum entropy output of a communication channel is not additive [203]. As Shor notes [390],

> *"It therefore remains an open question whether it is possible to find a simple expression for the information capacity of a quantum channel—similar to Shannon's formula for classical channels—that would make practical calculation of the quantum channel capacity possible."*

Table 3.8 Side-by-Side Properties of Classical and Quantum Information and the Main Results in Classical and Quantum Information Theory

Property	Classical Information	Quantum Information
Redundancy	Classical bits can be cloned.	Qubits cannot be cloned.
Distinguishability	Classical states are always distinguishable.	Only orthogonal states can be reliably distinguished.
Source of information	Alphabet $\mathcal{A}$ $a_1, a_2, \ldots, a_n$ probability distribution p_i	Pure/mixed quantum states density matrix ρ
Entropy of a source	Shannon entropy $H \quad\quad p_i \log p_i$	von Neumann entropy $S \quad \mathrm{tr}[\rho \log \rho]$
Measurements	Do not alter the state	Alter the state; project the state onto a subspace
Communication channel	X, Y input/output RVs $X \quad Y$	Input/output states ρ, $\mathcal{E}(\rho)$ $\rho \quad \mathcal{E}(\rho)$
Trace distance input–output	$1/2 \quad _x (p_X(x) \quad p_Y(x))$	$1/2\,\mathrm{tr}\;\rho \quad \mathcal{E}(\rho)$
Fidelity of the channel output	$F(p_X(x), p_Y(y)) \quad\quad _x \overline{p_X(x) p_Y(y)}$	$F(\rho, \sigma) \quad\quad \mathrm{tr} \quad \overline{\rho^{1/2} \sigma \rho^{1/2}} \quad ^2$
Correlation of channel	Mutual information $I(X;Y)$	Coherent information $I(\rho, \mathcal{E}) \quad S(\mathcal{E}(\rho)) \quad S(\rho, \mathcal{E})$
input–output		Entropy exchange $S_e(\rho, \mathcal{E}) \quad S(\rho^{(E)}_{after})$ Entanglement fidelity $F_e(\rho, \mathcal{E}) \quad \varphi^{(Q,R)} \; (\mathcal{I}_R \quad \mathcal{E}) \; \varphi^{(Q,R)} \quad \varphi^{(Q,R)} \quad\quad \varphi^{(Q,R)}$
Information from a measurement of channel output	$I(X;Y) \quad H(X) \quad H(X \; Y)$	Holevo bound $I(X;Y) \quad S \quad\quad _i p_i \rho_i \quad\quad _i p_i S(\rho_i)$
Fano inequality	$H(X \; Y) \quad H(p_{err}) \quad p_{err} \log(\; \mathcal{X} \;)$	$S_e(\rho, \mathcal{E}) \quad h(p) F_s(\rho, \mathcal{E}) \quad [1 \quad F_s(\rho, \mathcal{E})] \log(d^2 \quad 1)$
Data processing inequality	$X \quad Y \quad Z$ $H(X) \quad I(X;Y) \quad I(X;Z)$	$\rho \quad \mathcal{E}_1(\rho) \quad\quad \mathcal{E}_2 \;\; \mathcal{E}_1(\rho)$ $S(\rho) \quad I(\rho, \mathcal{E}_1) \quad I(\rho, \mathcal{E}_2 \;\; \mathcal{E}_1)$
Noiseless channel coding	Source entropy $H(X)$ $n_{bits} \quad H(X)$	Pure states ρ_i, p_i $n_{qubits} \quad S \quad\quad _i p_i \rho_i$
Noisy channel capacity	$C \quad \max I(X;Y)$	Classical capacity, $C(\mathcal{E})$ $C(\mathcal{E}) \quad \max_{p_i, \rho_i} \; S(\sigma) \quad\quad _i p_i S(\sigma_i)$ $\sigma_i \quad \mathcal{E}(\rho_i), \sigma \quad\quad _i p_i \sigma_i$ Quantum capacity, $Q \quad C$ Classically, assisted quantum capacity, Q_2, $Q \quad Q_2 \quad C$ Entanglement-assisted classical capacity, C_E, $Q \quad C \quad C_E$

Table 3.9 Resources Necessary to Transmit one Unit of Information

1 bit	1 qubit		To transmit one bit, we need at least one qubit.
1 ebit	1 qubit		To transmit one ebit, we need at least one qubit.
1 qubit	1 ebit	2 bits	To transmit one qubit, we need at least one ebit and two bits.
2 bits	1 ebit	1 qubit	To transmit two bits, we need at least one ebit and one qubit.

3.29 EXERCISES AND PROBLEMS

Problem 3.1 Prove that binary entropy is a concave function of p.

Problem 3.2 Let $p(x)$ be the distribution of a discrete random variable, X, and, $p_i, 1 \quad i \quad n$, the probability of individual outcomes, x_i, of X. Show that

$$H(X) \quad \log n$$

with equality if and only if we have a uniform distribution.

Problem 3.3 Prove the chain rule for the entropy of n random variables $X_1, X_2, \ldots, X_n$:

$$H(X_1, X_2, \ldots, X_n) \quad \overset{n}{\underset{i\ 1}{\quad}} H(X_i \ X_{i\ 1}, \ldots, X_1).$$

Problem 3.4 Let X and Y be independent binary variables, with the probability density functions, $p_X(x)$ and $p_Y(y)$. Z is also a random variable and $z \quad x \quad y$ mod 2. Depict the relationship among $H(X), H(Y), H(Z)$, the conditional entropy, the joint entropy, and the mutual information using the representations in Figures 3.7(a) and Figure 3.7(b). Discuss the accuracy of each representation.

Problem 3.5 Consider a classical binary symmetric channel, with the error probability $p \quad 1/3$; show that the capacity of this BSC is 0.0817 bits.

Problem 3.6 Prove Jensen's inequality:

$$\underset{i}{\quad} p_X(x_i) f(x_i) \quad f \quad \underset{i}{\quad} p_X(x_i) x_i \quad ,$$

where X is a random variable and $f(x)$ is a convex function.

Problem 3.7 Prove Fano's inequality:

$$H(p_{err}) \quad p_{err} \log(\mathcal{X}) \quad H(X \ Y),$$

with X, a random variable with the probability density function, $p_X(x)$, and, $\mathcal{X}$, the number of elements in the range of X. Y is another random variable related to X, with the conditional probability $p_{Y\ X}(y\ x)$. $p_{err} \quad \text{Prob}(X \quad X)$ is the error when we estimate X based on the observation of Y as $X \quad f(X)$.

Problem 3.8 Let $\alpha_1,\alpha_2,\ldots,\alpha_n$ and $\beta_1,\beta_2,\ldots,\beta_n$ be non-negative real numbers. Show that

$$\overset{n}{\underset{i\ \ 1}{}} \alpha_i \log \frac{\alpha_i}{\beta_i} \qquad \overset{n}{\underset{i\ \ 1}{}} \alpha_i \quad \log \frac{\overset{n}{\underset{i\ \ 1}{}}\alpha_i}{\overset{n}{\underset{i\ \ 1}{}}\beta_i}.$$

Problem 3.9 Prove the grouping property of Shannon entropy:

$$H(p_1,p_2,\ldots,p_n) \quad h(p_1) \quad (1 \quad p_1)H \quad \frac{p_2}{1 \quad p_1},\frac{p_3}{1 \quad p_1},\ldots,\frac{p_n}{1 \quad p_1} \quad ,$$

with $h(p) \qquad p\log p \quad (1 \quad p)\log(1 \quad p)$ and $H(p_1,p_2,\ldots,p_n)$, the Shannon entropy. Show also that

$$H \quad \frac{p_2}{1 \quad p_1},\frac{p_3}{1 \quad p_1},\ldots,\frac{p_n}{1 \quad p_1} \qquad \log(n \quad 1).$$

Problem 3.10 A multiple access channel can be modeled as a channel with n sets of inputs and one output. For example, the input is $x_1,x_2,\ldots,x_n$, with $x_i \quad 0,1$, and the output, $y \quad 0,1,\ldots,n$. What is the capacity of the channel? When $n \quad 2$ and the two sources transmit with rates, R_1 and R_2, respectively, plot the achievable rate.

A broadcast channel consists of a transmitter and n receivers. Such a channel can be modeled as, n, binary symmetric channels, $C_1,C_2,\ldots,C_n$, with capacities, $C_1 > C_2 > \quad > C_n$, and the probability of error, $p_1 < p_2 < \quad < p_n < 1/2$, respectively. Call R the rate of reliable communication through the channel; plot the, R, function of the probability of error, p.

Problem 3.11 Let $X \quad X_1,X_2,\ldots,X_i,\ldots$ be a stationary Markov chain. Show that $H(X_i\, X_1)$ is monotonously increasing even though $H(X_i)$ does not change with i by stationarity of the Markov chain. Discuss how this result is related to the second law of thermodynamics.

Problem 3.12 Prove the quantum Fano inequality:

$$S_e(\rho,\mathcal{E}) \quad h(p)F_s(\rho,\mathcal{E}) \quad [1 \quad F_s(\rho,\mathcal{E})]\log(d^2 \quad 1).$$

In this expression, d is the dimension of the Hilbert space of the quantum system, Q, and $h(p)$ is the binary entropy associated with the probability, p, $h(p) \qquad p\log p \quad (1 \quad p)\log(1 \quad p)$.

Problem 3.13 Prove the upper bound of the entanglement fidelity when the channel input is perturbed. If $\mathcal{E}$ is a trace-preserving quantum operation, ρ is the density matrix, and Δ is a zero trace Hermitian operator, then:

$$F_e(\rho \quad \Delta,\mathcal{E})) \quad F_e(\rho,\mathcal{E}) \qquad 2\mathrm{tr}(\ \Delta\ ) \quad \mathrm{tr}(\ \Delta\ )^2.$$

Problem 3.14 Let ρ be the density matrix of system, Q, the input to a quantum channel and, $\mathcal{E}$, the transformation performed by the channel to its input. Show that the following three conditions are equivalent:

1. $\mathcal{E}$ has a unitary representation.
2. $\mathcal{E}$ has a normalized operator-sum representation.
3. $\mathcal{E}$ is a trace-preserving, completely positive linear map on the density operators of Q.

Hint: See reference [256].

Problem 3.15 Show that the second part of the quantum data processing inequality,

$$I(\rho,\mathcal{E}_1) \quad I(\rho,\mathcal{E}_2 \quad \mathcal{E}_1),$$

holds even when $\mathcal{E}_1$ is trace-preserving but $\mathcal{E}_2$ is not.

Problem 3.16 Prove properties 4–7 of the quantum fidelity discussed in Section 3.11.

Problem 3.17 Show that ρ_1 and ρ_2 in Section 3.13 commute if

$$F(\rho_0,\rho_1) \quad F_{\mathcal{A}}(\rho_0,\rho_1) \quad F_{\mathcal{B}}(\rho_0,\rho_1).$$

Problem 3.18 Show that the intermediate ensemble, $\mathcal{A} \quad \xi_{ja}, p_a$, in Section 3.20, has the same entropy as the original one, $\mathcal{A} \quad \varphi_a, p_a$. *Hint:* See reference [35].

Problem 3.19 Show that the following equality holds:

$$\quad_k \mu_k \quad 1,$$

where μ_k are defined in Section 3.17.

Problem 3.20 Prove Schumacher's noiseless coding theorem discussed in Section 3.17. Show that when N is sufficiently large so that $\log N \quad nS(\rho^{\mathcal{A}})$, the expected probability of error becomes increasingly small, $p_{err} \quad 0$.

CHAPTER

Classical Error-Correcting Codes

4

Machines should work. People should think.
Richard Hamming

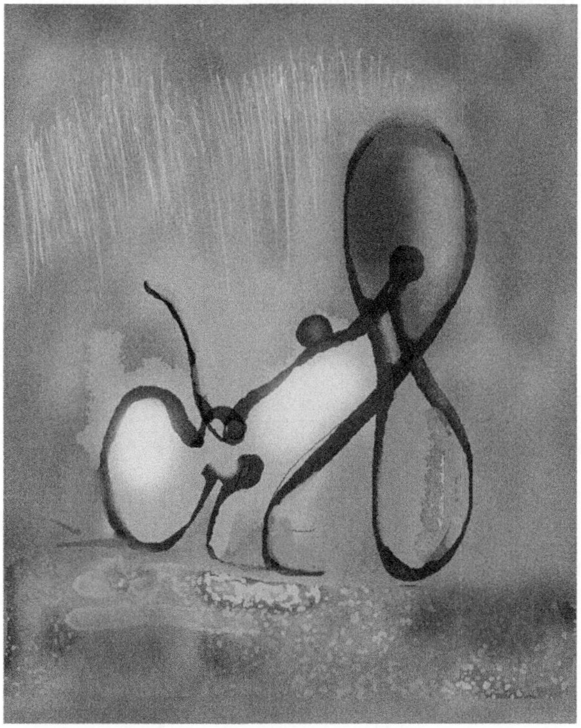

Coding theory is an application of information theory critical for reliable communication and fault-tolerant information storage and processing; indeed, the Shannon channel coding theorem tells us that we can transmit information on a noisy channel with an arbitrarily low probability of error. A code is designed based on a well-defined set of specifications and *protects the information* only *for the type and number of errors prescribed in its design.* A good code has the following characteristics:

1. It adds a minimum amount of redundancy to the original message.
2. Efficient encoding and decoding schemes for the code exist; this means that information is easily mapped to and extracted from a codeword.

Reliable communication and fault-tolerant computing are intimately related to each other; the concept of a communication channel, an abstraction for a physical system used to transmit information from one place to another and/or from one time to another, is at the core of communication, as well as information storage and processing. It should, however, be clear that the existence of error-correcting codes does not guarantee that logic operations can be implemented using noisy gates and circuits. The strategies to build reliable computing systems using unreliable components are based on John von Neumann's studies of error detection and error correction techniques for information storage and processing [442].

This chapter covers topics from the theory of classical error detection and error correction and introduces concepts useful for understanding quantum error correction, the subject of Chapter 5. For simplicity, we often restrict our discussion to the binary case. The organization of Chapter 4 is summarized in Figure 4.1. After introducing the basic concepts, we discuss linear, polynomial, and several other classes of codes.

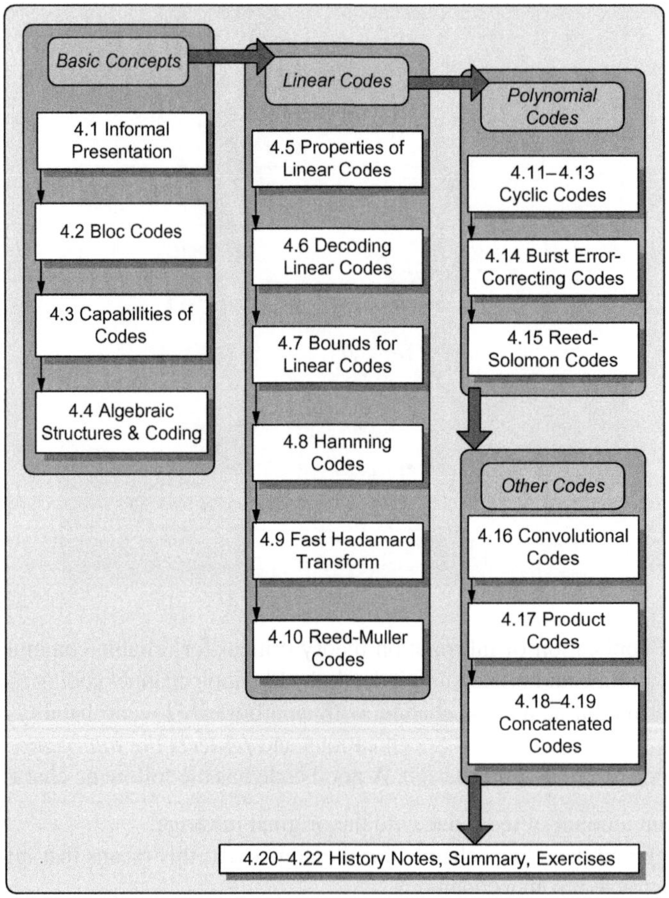

FIGURE 4.1

The organization of Chapter 4 at a glance.

4.1 INFORMAL INTRODUCTION TO ERROR DETECTION AND ERROR CORRECTION

Error detection is the process of determining if a message is in error; *error correction* is the process of restoring a message in error to its original content. Information is packaged into codewords, and the intuition behind error detection and error correction is to artificially increase some "distance" between codewords so that the errors cannot possibly transform one valid codeword into another valid codeword.

Error detection and error correction are based on schemes to increase the redundancy of a message. A crude analogy is to bubble wrap a fragile item and place it into a box to reduce the chance that the item will be damaged during transport. Redundant information plays the role of the packing materials; it increases the amount of data transmitted, but it also increases the chance that we will be able to restore the original contents of a message distorted during communication. Coding corresponds to the selection of both the packing materials and the strategy to optimally pack the fragile item subject to the obvious constraints: use the least amount of packing materials and the least amount of effort to pack and unpack.

Example. *A trivial example of an error detection scheme is the addition of a parity check bit to a word of a given length.*

This is a simple scheme but very powerful; it allows us to detect an odd number of errors, but fails if an even number of errors occur. For example, consider a system that enforces even parity for an eight-bit word. Given the string (10111011), we add one more bit, in this case a 0, to ensure that the total number of 1s is even (in this case a 0), and we transmit the nine-bit string (101110110). The error detection procedure is to count the number of 1s; we decide that the string is in error if this number is odd. When the information symbols are embedded into the codeword, as in this example, the code is called a *systematic code*; the information symbols are scattered throughout the codeword of a nonsystematic code.

This example also hints at the limitations of error detection mechanisms. A code is designed with certain error detection or error correction capabilities and fails to detect or to correct error patterns not covered by the original design of the code. In the previous example, we transmit (101110110), and when two errors occur, in the fourth and the seventh bits, we receive (101010010). This tuple has even parity (an even number of 1s) and our scheme for error detection fails.

Error detection and error correction have practical applications in our daily life. Error detection is used in cases when an item needs to be uniquely identified by a string of digits and there is the possibility of misidentification. Bar codes, also called universal product codes (UPCs), and the codes used by the postal service, the American banking system, or the airlines for paper tickets, all have some error detection capability. For example, in the case of ordinary bank checks and travelers' checks, there are means to detect if the check numbers are entered correctly during various financial transactions.

Example. *The international standard book number (ISBN) coding scheme.* The ISBN code is designed to detect any single digit in error and any transposition of adjacent digits. The ISBN number consists of 10 digits:

$$ISBN \quad d_1\, d_2\, d_3\, d_4\, d_5\, d_6\, d_7\, d_8\, d_9\, d_{10}.$$

Here, d_{10} is computed as follows:

$$A \quad w_1 d_1 \quad w_2 d_2 \quad w_3 d_3 \quad w_4 d_4 \quad w_5 d_5 \quad w_6 d_6 \quad w_7 d_7 \quad w_8 d_8 \quad w_9 d_9 \quad \text{mod } 11,$$

with $w_1 \quad 10$, $w_2 \quad 9$, $w_3 \quad 8$, $w_4 \quad 7$, $w_5 \quad 6$, $w_6 \quad 5$, $w_7 \quad 4$, $w_8 \quad 3$, $w_9 \quad 2$. Then:

$$d_{10} \quad \begin{cases} 11 \quad A & \text{if } 2 \quad A \quad 10, \\ X & \text{if } A \quad 1. \end{cases}$$

For example, consider the book with ISBN 0-471-43962-2. In this case:

$$A \quad 10 \quad 0 \quad 9 \quad 4 \quad 8 \quad 7 \quad 7 \quad 1 \quad 6 \quad 4 \quad 5 \quad 3 \quad 4 \quad 9 \quad 3 \quad 6 \quad 2 \quad 2 \quad \text{mod } 11,$$

$$A \quad 36 \quad 56 \quad 7 \quad 24 \quad 15 \quad 36 \quad 18 \quad 4 \quad \text{mod } 11 \quad 196 \quad \text{mod } 11 \quad 9,$$

$$d_{10} \quad 11 \quad A \quad 2.$$

The code presented here detects any single errors because the weights are relatively prime with 11. Indeed, if digit d_i is affected by an error and becomes $d_i \quad e_i$, then

$$e_i, \quad w_i(d_i \quad e_i) \quad \text{mod } 11 \quad w_i d_i \quad \text{mod } 11,$$

because $w_i e_i \quad 0$ mod 11, w_i is relatively prime to 11. The code also detects any transpositions of adjacent digits. Indeed, the difference of two adjacent weights, $w_i \quad w_{i \quad 1} \quad 1$ and $d_i \quad d_{i \quad 1} < 11$; thus, it cannot be a multiple of 11. We conclude that the ISBN code detects any single digit in error and any transposition of adjacent digits.

Assume that we wish to transmit through a binary channel a text from a source, A, with an alphabet with 32 letters. We can encode each one of the $32 \quad 2^5$ letters of the alphabet as a five-bit binary string, and we can rewrite the entire text in the binary alphabet accepted by the channel. This strategy does not support either detection or correction of transmission errors because all possible five-bit combinations are exhausted to encode the 32 letters of the alphabet; any single-bit error will transform a valid letter into another valid letter. However, if we use more than five bits to represent each letter of the alphabet, we have a chance to correct a number of single-bit errors.

Adding redundancy to a message increases the amount of information transmitted through a communication channel. The ratio useful information versus total information packed in each codeword defines the rate of a code (see Section 4.2).

Example. *We now consider a very simple error-correcting code, a "repetitive" code.* Instead of transmitting a single binary symbol, 0 or 1, we map each bit into a string of length, n, of binary symbols:

$$0 \quad (00\ldots0) \quad \text{and} \quad 1 \quad (11\ldots1).$$

If $n \quad 2k \quad 1$, we use a *majority voting scheme* to decode a codeword of n bits. We count the number of 0s; if this number is larger than k, we decode the codeword as 0, else we decode it as 1.

Consider a binary symmetric channel with random errors occurring with the probability, p, and call p^n_{err} the probability that an n-bit codeword of this repetitive code is in error; then,

$$p^n_{err} \quad 1 \quad (1 \quad p)^{k \quad 1} \quad (1 \quad p)^k \quad \genfrac{}{}{0pt}{}{2k \quad 1}{1} \quad p(1 \quad p)^{k \quad 1} \qquad \genfrac{}{}{0pt}{}{2k \quad 1}{k} \quad p^k(1 \quad p)^0 \ .$$

Indeed, the probability of error is the probability that $(k \quad 1)$ or more bits are in error, which in turn is equal to 1 minus the probability that at most, k, bits are in error.

As we can see, the larger k is, the smaller the probability of error; this encoding scheme is better than sending a single bit as long as $p^n_{err} < p$. For example, when $n \quad 3$ and $k \quad 1$, we encode a single bit as follows: 0 (000) and 1 (111). Then the probability of error is

$$p^3_{err} \quad 1 \quad (1 \quad p)^2[(1 \quad p)^1 \quad \genfrac{}{}{0pt}{}{3}{1} \quad p(1 \quad p)^0] \quad 3p^2 \quad 2p^3.$$

The encoding scheme increases the reliability of the transmission when $p < 1/2$.

4.2 BLOCK CODES, DECODING POLICIES

In this section we discuss information transmission through a classical channel using codewords of fixed length. The *channel alphabet*, $\mathcal{A}$, is a set of q symbols that could be transmitted through the communication channel; when $q \quad 2$ we have the *binary alphabet*, $\mathcal{A} \quad 0,1$. The source encoder maps a message into blocks of, k, symbols from the code alphabet; then the channel encoder adds, r, symbols to increase redundancy and transmits n-tuples, or codewords, of length, $n \quad k \quad r$, as shown in Figure 3.8.

At the destination, the channel decoder maps n-tuples into k-tuples, and the decoder reconstructs the message. The quantity, $r \quad n \quad k > 0$, is called the *redundancy* of the code. The channel encoder adds redundancy, and this leads to the expansion of a message. While the added redundancy is desirable from the point of view of error control, it decreases the efficiency of the communication channel by reducing its effective capacity. The ratio, $\frac{k}{n}$, measures the *ef ciency* of a code. First, we give several definitions related to block codes and then discuss decoding policies.

A *block code*, $\mathcal{C}$, of length, n, over the alphabet, $\mathcal{A}$, with q symbols, is a set of M n-tuples, where each n-tuple (codeword) uses symbols from $\mathcal{A}$. Such a block code is also called an $[n,M]$ code over $\mathcal{A}$. The codewords of an $[n,M]$ code, $\mathcal{C}$, over $\mathcal{A}$ are a subset, M, of all n-tuples over $\mathcal{A}$, with $M < q^n$. Given an $[n,M]$ code, $\mathcal{C}$, that encodes k-tuples into n-tuples, the *rate of the code* is $R \quad k/n$.

We now introduce several metrics necessary to characterize the error-detecting and the error-correcting properties of a code, $\mathcal{C}$, with symbols from the alphabet, $\mathcal{A}$.

Given two n-tuples, v_i and v_j, the *Hamming distance*, $d(v_i, v_j)$, is the number of coordinate positions in which the two n-tuples differ. The Hamming distance is a metric; indeed, $v_i, v_j, v_k \quad \mathcal{C}$, and we have:

1. $d(v_i, v_j) \quad 0$, with equality if and only if $v_i \quad v_j$.
2. $d(v_i, v_j) \quad d(v_j, v_i)$.
3. $d(v_i, v_j) \quad d(v_j, v_k) \quad d(v_i, v_k)$ (triangle inequality).

The proof of this proposition is left as an exercise.

The *Hamming distance of a code* is the minimum distance between any pairs of codewords:

$$d = \min\{d(c_i, c_j) : \forall c_i, c_j \in \mathcal{C}, c_i \neq c_j\}.$$

To compute the Hamming distance for an $[n, M]$ code, $\mathcal{C}$, it is necessary to compute the distance between, $\binom{M}{2}$, pairs of codewords and then to find the pair with the minimum distance. A block code with the block length, n, with k information symbols and with distance d, will be denoted as $[n, k, d]$; if the alphabet, $\mathcal{A}$, has q symbols, then $M = q^k$.

Given a codeword $c \in \mathcal{C}$, the *Hamming weight*, $w(c)$, is the number of nonzero coordinates of the codeword c. Let c_i, $1 \leq i \leq M$ be the codewords of $\mathcal{C}$, an $[n, M]$ code, and let S be the set of all n-tuples over the alphabet of $\mathcal{C}$; the *Hamming sphere of radius, d,* around the codeword, c_i, consists of all n-tuples, c, within distance, d, of the codeword, c_i, the *center* of the Hamming sphere:

$$S_{c_i} = \{c \in S : d(c, c_i) \leq d\}.$$

Figure 4.2 shows the two Hamming spheres of radius 1 around the 3-tuples 000 and 111.

Example. *The Hamming distance of two codewords.* Consider the binary alphabet $\{0, 1\}$, and let the two codewords be $v_i = (010110)$ and $v_j = (011011)$. The Hamming distance between the two codewords is $d(v_i, v_j) = 3$. Indeed, if we number the bit position in each n-tuple from left to right as 1 to 6, the two n-tuples differ in bit positions 3, 4, and 6.

Example. *The Hamming distance of a code.* Consider the code $\mathcal{C} = \{c_0, c_1, c_2, c_3\}$, where $c_0 = (000000)$, $c_1 = (101101)$, $c_2 = (010110)$, $c_3 = (111011)$. This code has distance $d = 3$. Indeed, $d(c_0, c_1) = 4$, $d(c_0, c_2) = 3$, $d(c_0, c_3) = 5$, $d(c_1, c_2) = 5$, $d(c_1, c_3) = 3$, and $d(c_2, c_3) = 4$.

Example. *The Hamming sphere of radius, $d = 1$, around the codeword,* (000000), consists of the center, (0000000), and all binary 6-tuples that differ from (000000) in exactly one bit position, {(000000), (100000), (010000), (001000), (000100), (000010), (000001)}. In this case, one can construct a diagram similar to the one in Figure 4.2 where the 6-tuples are the vertices of a hypercube in a space with six dimensions.

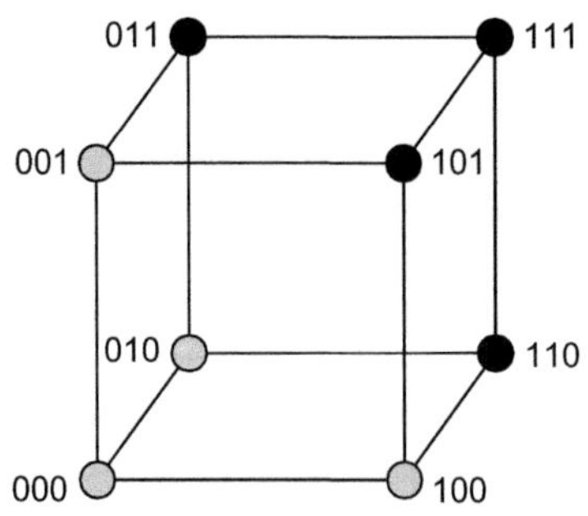

FIGURE 4.2

The Hamming sphere of radius 1 around 000 contains {000, 100, 010, 001}, and the Hamming sphere of radius 1 around 111 contains {111, 011, 101, 110}.

Decoding Policy

Given an $[n, M]$ code $\mathcal{C}$, with distance, d, the decoding policy tells the channel decoder what actions to take when it receives an n-tuple, v. This policy consists of two phases:

1. The recognition phase when the received n-tuple v is compared with all the codewords of the code until either a match is found or we decide that v is not a codeword
2. The error correction phase, when the received n-tuple v is mapped into a codeword

The actions taken by the channel decoder are:

1. If v is a codeword, conclude that no errors have occurred and decide that the codeword sent was $c = v$.
2. If v is not a codeword, conclude that errors have occurred and either correct v to a codeword c, or declare that correction is not possible.

We observe that any decoding policy fails when $c \in \mathcal{C}$, the codeword sent, is affected by errors and transformed into another valid codeword, $c' \in \mathcal{C}$. This is a fundamental problem with error detection, and we shall return to it later. Once we accept the possibility that the channel decoder may fail to properly decode in some cases, our goal is to take the course of action with the highest probability of being correct. Here we discuss three decoding policies.

(A) *Minimum distance or nearest neighbor decoding.* If an n-tuple v is received, and there is a unique codeword, $c \in \mathcal{C}$, such that $d(v, c)$ is the minimum over all $c \in \mathcal{C}$, then correct v as the codeword c. If no such c exists, report that errors have been detected, but no correction is possible. If multiple codewords are at the same minimum distance from the received codeword, select at random one of them and decode v as that codeword.

(B) *Bounded distance decoding.* This decoding policy requires that all patterns of at most, e, errors and no other errors are corrected. When bounded distance decoding is used and more than, e, errors occur, an n-tuple, v, is either not decoded or is decoded to a wrong codeword; the first case will be called a *decoding failure* and the second one a *decoding error*. It is easy to see that if

$$e = \left\lfloor \frac{d-1}{2} \right\rfloor,$$

then the bounded distance decoding is identical to minimum distance decoding.

Recall from Section 3.7 that in the case of a binary symmetric channel (BSC) the errors occur with the probability, p, and are mutually independent. Given an $[n, k]$ linear code operating on a BSC the probabilities of decoding failure, P_{fail}, and of decoding error, P_{err}, when at most, e, errors occur are

$$P_{fail} = 1 - \sum_{i=0}^{e} \binom{n}{i} p^i (1-p)^{n-i} - \sum_{i=e+1}^{n} \binom{n}{i} p^i (1-p)^{n-i}$$

and

$$P_{err} = \sum_{w>0} \sum_{i=w-e}^{w+e} \sum_{j=0}^{e} A_w T(i,j,w) p^i (1-p)^{n-i}.$$

In this expression, A_w is the number of codewords of code, $\mathcal{C}$, of weight equal to w, and $T(i,j,w)$ is the number of n-tuples at distance, i, from a codeword with weight, w, and at distance, j, from the decoded codeword; e is the number of errors.

It is easy to prove that given a binary $[n,k]$ linear code, $\mathcal{C}$, $T(i,j,w)$ is given by the following expression:

$$T(i,j,w) = \binom{w}{i-k}\binom{n-w}{k},$$

with $k = 1/2(i+j-w)$ when $(i+k-w)$ is even and when $w-j \le i \le w+j$; otherwise, $T = 0$.

(C) *Maximum likelihood decoding.* Under this decoding policy, of all possible codewords, $c \in \mathcal{C}$, the n-tuple, v, is decoded to that codeword, c, which maximizes the probability, $P(v,c)$, that v is received, given that c is sent.

Proposition. *If v is received at the output of a q-ary symmetric channel, and if $d_1 \neq d_2$, with $d_1 = d(v,c_1)$ and $d_2 = d(v,c_2)$, then*

$$P(v,c_1) \geq P(v,c_2) \quad \text{if and only if} \quad p < \frac{(q-1)}{q}.$$

In other words, if $p < (q-1)/q$, the maximum likelihood decoding policy is equivalent to the nearest neighbor decoding; a received vector is decoded to the codeword "closest" to it, in terms of Hamming distance.

Proof. Without loss of generality, we assume that $d_1 \leq d_2$ and that

$$P(v,c_1) > P(v,c_2).$$

It follows that

$$(1-p)^{n-d_1}\left(\frac{p}{q-1}\right)^{d_1} > (1-p)^{n-d_2}\left(\frac{p}{q-1}\right)^{d_2},$$

$$(1-p)^{d_2-d_1} > \left(\frac{p}{q-1}\right)^{d_2-d_1},$$

and

$$\left(\frac{p}{(1-p)(q-1)}\right)^{d_2-d_1} < 1.$$

If $d_1 = d_2$, this is false, and in fact, $P(v,c_1) = P(v,c_2)$. Otherwise, $d_2 - d_1 \geq 1$, and the inequality is true if and only if

$$\frac{p}{(1-p)(q-1)} < 1, \quad \text{i.e.,} \quad p < \frac{(q-1)}{q}.$$

In conclusion, when the probability of error is $p < (q-1)/q$ and we receive the n-tuple v, we decide that codeword c at the minimum distance from v is the one sent by the source. ∎

Example. *Maximum likelihood decoding.* Assume that the probability of error is $p \quad 0.15$. Let the binary code, $\mathcal{C}$, be $\mathcal{C} \quad (000000),(101100),(010111),(111011)$. When we receive the 6-tuple $v \quad (111111)$, we decode it as (111011) because the probability, $P((v),(111011))$, is the largest. Indeed,

$$P(v,(000000)) \quad (0.15)^6 \quad 0.000011,$$

$$P(v,(101100)) \quad (0.15)^3 \quad (0.85)^3 \quad 0.002076,$$

$$P(v,(010011)) \quad (0.15)^3 \quad (0.85)^3 \quad 0.002076,$$

$$P(v,(111011)) \quad (0.15)^1 \quad (0.85)^5 \quad 0.066555.$$

A code is capable of *correcting, e, errors* if the channel decoder is capable of correcting any pattern of e or fewer errors, using the preceding algorithm.

4.3 ERROR CORRECTING AND DETECTING CAPABILITIES OF A BLOCK CODE

The intuition behind error detection is to use a small subset of the set of n-tuples over an alphabet, $\mathcal{A}$, as codewords of a code, $\mathcal{C}$, and detect an error whenever the n-tuple received is not a codeword. To detect errors, we choose the codewords of $\mathcal{C}$ so that the Hamming distance between any pair of codewords is large enough and one valid codeword cannot be transformed by errors into another valid codeword. The Hamming distance of a code allows us to accurately express the capabilities of a code to detect and to correct errors. Several propositions quantify the error detection and error correction properties of a code.

Proposition. *Let $\mathcal{C}$ be an $[n,k,d]$ code with an odd distance, $d \quad 2e \quad 1$. Then $\mathcal{C}$ can correct, e, errors and can detect, $2e$, errors.*

Proof. We first show that, $S_{c_i}^{(e)}$ and $S_{c_j}^{(e)}$, the Hamming spheres of radius, e, around two distinct codewords, $c_i \quad c_j \quad \mathcal{C}$, are disjoint, $S_{c_i}^{(e)} \quad S_{c_j}^{(e)} \quad \emptyset$. Assume by contradiction that there exists a codeword, $v \quad \mathcal{C}$, with the property that $v \quad S_{c_i}^{(e)} \quad S_{c_j}^{(e)}$ and that $d(v,c_i) \quad e$ and $d(v,c_j) \quad e$. The triangle inequality requires that

$$d(c_i,v) \quad d(v,c_j) \quad d(c_i,c_j) \qquad\qquad 2e \quad d(c_i,c_j).$$

But the distance of the code is $d \quad 2e \quad 1$ so $d(c_i,c_j) \quad 2e \quad 1$. This is a contradiction, and we conclude that $S_{c_i}^{(e)} \quad S_{c_j}^{(e)} \quad \emptyset$.

Therefore, when the codeword, c_i, is transmitted and $t \quad e$ errors are introduced, the received n-tuple, v, is inside the Hamming sphere, $S_{c_i}^{(e)}$, and c_i is the unique codeword closest to v. The decoder can always correct any error pattern of this type.

If we use the code, $\mathcal{C}$, for error detection, then at least $2e \quad 1$ errors must occur in any codeword, $c \quad \mathcal{C}$, to transform it into another codeword, $c \quad \mathcal{C}$. If at least 1 and at most, $2e$, errors are introduced, the received n-tuple will never be a codeword, and error detection is always possible. ■

The distance of a code may be even or odd. The case for even distance is proved in a similar manner. Throughout this chapter, we use the traditional notation, a , for the largest integer smaller than or equal to a.

Proposition. *Let C be an $[n,k,d]$ code. Then C can correct,* $\frac{(d\ \ 1)}{2}$, *errors and can detect d 1 errors.*

A geometric interpretation of this proposition is illustrated in Figure 4.3. The $[n,k,d]$ code, C $c_1, c_2, c_3, c_4, \ldots$, with distance, d $2e$ 1, is used for error correction as well as for error detection. The Hamming spheres of radius, e, around all codewords do not intersect because the minimum distance between any pair of codewords is $2e$ 1. According to the channel decoding rules, any n-tuple, t, in the Hamming sphere of radius, e, around codeword, c_1, is decoded as the center of the Hamming sphere, c_1. If at most, $2e$, errors occur when the codeword, c_i, is transmitted, the received n-tuple, v, cannot be masquerading as a valid codeword, c_j, since the distance between c_i and c_j is at least $2e$ 1.

From Figure 4.3, we see that there are n-tuples that are not contained in any of the Hamming spheres around the codewords of a code C. If an n-tuple v is received and the decoder cannot place it in any of the Hamming spheres, then the decoder knows that at least e 1 errors have occurred and no correction is possible.

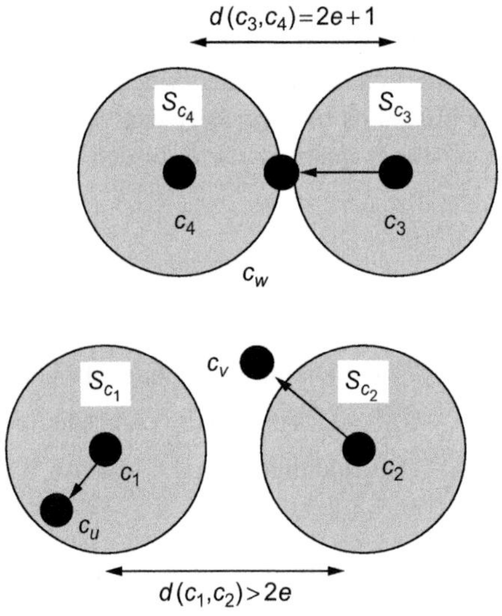

FIGURE 4.3

A code, C $c_1, c_2, c_3, c_4, \ldots$, with minimum distance, d $2e$ 1, is used for error detection as well as for error correction. The Hamming spheres, $S_{c_1}, S_{c_2}, S_{c_3}$, and S_{c_4}, are centered around the codewords, c_1, c_2, c_3, and c_4, respectively. The decoder identifies c_u, c_v, and c_w as errors; one of them is corrected, and the other two are not. c_u, received when the c_1 was sent, is correctly decoded as c_1 because it is located inside the Hamming sphere around, codeword, c_1, $d(c_1, c_u) < e$. The decoder fails to decode c_v, received when c_2 was sent, because $d(c_2, c_v) > e$, and c_v is not inside any of the Hamming spheres. The decoder incorrectly decodes c_w, received when c_3 was sent, as c_4 because $d(c_3, c_w)$ e 1 and $d(c_4, c_w)$ e.

If a code, C, with distance $2e \quad 1$, is used *for error detection as well as error correction*, some patterns of less than $2e$ errors could escape detection. For example, patterns of $e \quad 1$ errors that transform a codeword, c_3, into an n-tuple, c_w, inside the Hamming sphere, S_{c_4}, force the decoder to correct the received n-tuple to c_4. In this case, the $(e \quad 1)$-error pattern causes a faulty decoding.

In conclusion, when $d \quad 2e \quad 1$, the code, C, can correct, e, errors, in general, but it is unable to detect additional errors at the same time. If the distance of the code is an even number, the situation changes slightly according to the following proposition.

Proposition. *Let C be an $[n,k,d]$ code, with $d \quad 2e$. Then C can correct, $e \quad 1$, errors and, simultaneously, detect, e, errors.*

Proof. According to a previous proposition, the code, C, can correct up to

$$\frac{d \quad 1}{2} \qquad \frac{2e \quad 1}{2} \qquad e \quad \frac{1}{2} \qquad e \quad 1$$

errors. Since the Hamming spheres around codewords have radius, $e \quad 1$, any pattern of e errors cannot take a codeword into an n-tuple contained in some Hamming sphere around another codeword. Otherwise, the codewords at the centers of these two Hamming spheres would have distance at most, $e \quad e \quad 1 \quad 2e \quad 1$, which is impossible since $d \quad 2e$. Hence, a received word obtained from a codeword by introducing, k, errors cannot lie in any codeword Hamming sphere and the decoder can detect the occurrence of errors. The decoder cannot detect, $e \quad 1$, errors, in general, since a vector at distance $e \quad 1$ from a given codeword may be in the Hamming sphere of another codeword, and the decoder would erroneously correct such a vector to the codeword at the center of the second Hamming sphere. ■

To evaluate the reliability of the nearest neighbor decoding strategy, we have to compute the probability that a codeword, c, sent over the channel is decoded correctly. Let us consider an $[n,k,d]$ code C over an alphabet, $\mathcal{A}$, with the q symbols, $A \quad q$. Assume a probability of error, p, such that $0 \quad p \quad (q \quad 1)/q$.

When codeword, c, is sent, the received binary n-tuple, v, will be decoded correctly if it is in the Hamming sphere, S_c, of radius, $e \quad (d \quad 1)/2$, about c. The probability of this event is

$$\underset{v \; S_c}{P(v,c)} \qquad \overset{e}{\underset{i \; 0}{}} \; \binom{n}{i} p^i(1 \quad p)^{n \quad i}.$$

Indeed, the probability of receiving an n-tuple, with, i, positions in error, is $\binom{n}{i} p^i(1 \quad p)^{n \quad i}$ and i takes values in the range from 0, no errors, to a maximum of e errors. This expression gives a lower bound on the probability that a transmitted codeword is correctly decoded.

Example. *Decoding rule*. Consider the code, $C \quad c_1, c_2, c_3$, where

$$c_1 \quad (101100) \quad c_2 \quad (010111) \quad c_3 \quad (111011).$$

The code, C, has distance, $d \quad 3$, and hence can correct one error. The set, S, of all possible words over the alphabet, $0,1$, consists of all possible binary 6-tuples; hence, $S \quad 64$. Let us now construct the four Hamming spheres of radius 1 about each codeword:

$$S_{c_1} \quad (101100),(001100),(111100),(100100),(101000),(101110),(101101) \ ,$$

$$S_{c_2} \quad (010111),(110111),(000111),(011111),(010011),(010101),(010110) \ ,$$

$$S_{c_3} \quad (111011),(011011),(101011),(110011),(111111),(111001),(111010) \ .$$

The Hamming spheres cover 21 of the 64 possible 6-tuples in S. Let S be the set of 6-tuples that are not included in any Hamming sphere; then $S \quad 64 \quad 21 \quad 43$. When the decoder receives the binary 6-tuple, $v \quad (000111)$, the distance to each codeword is computed as follows:

$$d(c_1,v) \quad 4, \quad d(c_2,v) \quad 1, \quad d(c_3,v) \quad 4.$$

According to the minimum distance rule, v is decoded as c_2; v lies in the Hamming sphere, S_{c_2}.

4.4 ALGEBRAIC STRUCTURES AND CODING THEORY

A brief excursion into algebraic structures used by coding theory will help us follow the presentation of the codes discussed in this chapter, as well as the quantum error-correcting codes presented in Chapter 5. First, we review basic concepts regarding the algebraic structures used in this chapter and in the next one.

A *group* consists of a set, G, equipped with a map, $:(G \quad G) \quad G$ and $^{1}: G \quad G$, and an identity element, $e \quad G$, so that the following axioms hold:

1. $(g_i, g_j) \quad G \quad (g_i \quad g_j) \quad G.$
2. $(g_i, g_j, g_k) \quad G \quad (g_i \quad g_j) \quad g_k \quad g_i \quad (g_j \quad g_k).$
3. $g_i \quad G \quad g_i \quad e \quad g_i.$
4. $g_i \quad G \quad g_i^{\ 1}$, such that $g_i \quad g_i^{\ 1} \quad e.$

G is an *Abelian or commutative group* if an additional axiom is satisfied: $(g_i, g_j) \quad F \quad g_i \quad g_j \quad g_j \quad g_i$. The group is either an additive group when the operation, ( ), is addition, denoted as ( ), or a multiplicative group when the operation is multiplication, denoted as ( ). A subset, G , is called a *subgroup* of G if $g_1, g_2 \quad G$ implies $g_1 \quad g_2^{\ 1} \quad G$. The number of distinct elements of a group is called the *order* of the group.

Example. *The set of rotational symmetries of an ordinary sphere.* This symmetry group has an infinite number of elements since we can rotate a sphere through any angle about any direction in the three-dimensional space. The group is called $SO(3)$ [323].

A subgroup, $G \quad G$, is a *normal subgroup* of G if each element of the group, G, commutes with the normal subgroup, G :

$$G \quad g \quad g \quad G , \quad g \quad G.$$

A set, R, with two binary operations, addition ( ) and multiplication ( ), and a distinct element, $0 \quad G$, is a *commutative ring with unity*, or simply a ring, if the following axioms are satisfied:

1. $(R, \ )$ is an Abelian group with the identity element 0.
2. Multiplication is commutative: $(a,b) \quad R \quad a \ b \quad b \ a.$
3. The identity/unity element for multiplication is $1 \quad R$, such that $a \quad R \ \ 1 \ a \quad a \ 1 \quad a.$
4. Multiplication is associative: $(a,b,c) \quad R \quad (a \ b) \ c \quad a \ (b \ c).$
5. Multiplication is distributive: $(a,b,c) \quad R \quad (a \quad b) \ c \quad a \ c \quad b \ c.$

Note that there is no multiplicative inverse of the elements of a ring.

Example. *The set of integers with integer addition and integer multiplication,* $(Z, \ , \ )$. This set forms a ring, denoted as $\mathbb{Z}$, with an infinite number of elements.

An algebraic structure, $(F, \ , \ )$, consisting of a set, F; two binary operations, addition ( ) and multiplication ( ); and two distinct elements, 0 and 1, is a *eld* if the following axioms are satisfied:

1. $(a,b) \quad F \quad a \quad b \quad b \quad a.$
2. $(a,b,c) \quad F \quad (a \quad b) \quad c \quad a \quad (b \quad c).$
3. $a \quad F \quad a \quad 0 \quad a.$
4. $a \quad F \quad (\ a) \quad a \quad (\ a) \quad 0.$
5. $(a,b) \quad F \quad a \ b \quad b \ a.$
6. $(a,b,c) \quad F \quad (a \ b) \ c \quad a \ (b \ c).$
7. $a \quad F \quad a \ 1 \quad a.$
8. $a \quad F \quad a^{\ 1} \quad a \ a^{\ 1} \quad 1.$
9. $(a,b,c) \quad F \quad a \ (b \quad c) \quad a \ b \quad a \ c.$

A field, F, is at the same time an Abelian additive group, $(F, \ )$, and an Abelian multiplicative group, $(F, \ )$.

A field, F, with a finite number of elements, q, is called a *nite or Galois eld*, denoted as $GF(q)$. It is also named after the French mathematician Évariste Galois.

We now discuss the congruence relationship and show that it allows us to construct finite algebraic structures consisting of equivalence classes. Given a positive integer, $p \quad Z$, the *congruence* relation among integers is defined as

$$(m,n) \quad Z \quad m \quad n \mod p \qquad (m \quad m/p \quad p) \quad (n \quad n/p \quad p).$$

The congruence is an *equivalence relation*, $\mathcal{R}$: it is re exive, $a\mathcal{R}a$; symmetric, $a\mathcal{R}b \qquad b\mathcal{R}a$; and transitive, $[a\mathcal{R}b$ and $b\mathcal{R}c] \qquad a\mathcal{R}c$. It partitions a set into equivalence classes. For example, the set of integers modulo, 3, is partitioned into three equivalence classes:

[0] $0,3,6,9\ldots$, the integers divisible by 3;
[1] $1,4,7,10\ldots$, the integers with a remainder of 1 when divided by 3; and
[2] $2,5,8,11\ldots$, the integers with a remainder of 2 when divided by 3.

Proposition. *The set of integers modulo p,* $\mathbb{Z}_p$*, with the standard integer addition and integer multiplication is a nite eld if and only if p is a prime number.*

Proof. To prove that the condition is necessary, let us assume that p is prime. First, we observe that, $\mathbb{Z}_p$, has a finite number of elements, $0, 1, 2, \ldots, (p \quad 1)$. Then we verify all the properties or axioms of a field. Here, we only show that given, $0 < a < p$, the multiplicative inverse, $a^{\ 1}$ exists. The $\gcd(p, a) \quad 1$ because p is prime; then according to Euclid's algorithm, there exist integers, s and t, such that $s \ p \quad t \ a \quad 1$. But for all integers, s, we have $s \ p \quad 0 \bmod p$; this implies that $t \ a \quad 1 \bmod p$ or $t \quad a^{\ 1} \bmod p$. Thus, if p is prime, then $\mathbb{Z}_p$ is a finite field.

We start with the observation that if p is not a prime, then it is a product of two terms, $p \quad a \ b$, with $a, b \quad 0 \bmod p$. To prove that the condition is sufficient, we have to show that b has no multiplicative inverse in $\mathbb{Z}_p$. In other words, we must show that there is no c, such that $b \quad c^{\ 1}$ or $b \ c \quad 1 \bmod p$. If such c exists, then

$$p \ c \quad a \ b \ c \quad a \ (b \ c) \quad a \ 1 \quad a \bmod p \quad 0 \bmod p.$$

But, $a \quad 0 \bmod p$ contradicts the assumption that $a \quad 0 \bmod p$. ■

Example. *The nite eld of integers, modulo 5, denoted as $\mathbb{Z}_5$ or $GF(5)$:* $\mathbb{Z}_5$, consists of five equivalence classes of integers modulo 5, namely, [0], [1], [2], [3], and [4] corresponding to the remainders $0, 1, 2, 3$, and 4, respectively:

[0]	0	5	10	15	20	25	30	...,
[1]	1	6	11	16	21	26	31	...,
[2]	2	7	12	17	22	27	32	...,
[3]	3	8	13	18	23	28	33	...,
[4]	4	9	14	19	24	29	34	... .

The addition and multiplication tables of $\mathbb{Z}_5$ are:

	[0]	[1]	[2]	[3]	[4]
[0]	[0]	[1]	[2]	[3]	[4]
[1]	[1]	[2]	[3]	[4]	[0]
[2]	[2]	[3]	[4]	[0]	[1]
[3]	[3]	[4]	[0]	[1]	[2]
[4]	[4]	[0]	[1]	[2]	[3]

and

	[0]	[1]	[2]	[3]	[4]
[0]	[0]	[0]	[0]	[0]	[0]
[1]	[0]	[1]	[2]	[3]	[4]
[2]	[0]	[2]	[4]	[1]	[3]
[3]	[0]	[3]	[1]	[4]	[2]
[4]	[0]	[4]	[3]	[2]	[1]

.

From these tables we see that [3] is the multiplicative inverse of [2] (indeed, [3] [2] [1]) and that [1] and [4] are their own multiplicative inverses ([1] [1] [1] and [4] [4] [1]).

$\mathbb{Z}_5, \mathbb{Z}_7, \mathbb{Z}_{11}$, and $\mathbb{Z}_{13}$ are finite fields, but $\mathbb{Z}_9$ and $\mathbb{Z}_{15}$ are not because 9 and 15 are not prime numbers. Indeed, the set, $\mathbb{Z}_9$, is finite and consists of the following elements: $[0], [1], [2], [3], [4], [5], [6], [7],$

and [8]. The multiplication table of $\mathbb{Z}_9$ is:

	[0]	[1]	[2]	[3]	[4]	[5]	[6]	[7]	[8]
[0]	[0]	[0]	[0]	[0]	[0]	[0]	[0]	[0]	[0]
[1]	[0]	[1]	[2]	[3]	[4]	[5]	[6]	[7]	[8]
[2]	[0]	[2]	[4]	[6]	[8]	[1]	[3]	[5]	[7]
[3]	[0]	[3]	[6]	[0]	[3]	[6]	[0]	[3]	[6]
[4]	[0]	[4]	[8]	[3]	[7]	[2]	[6]	[1]	[5]
[5]	[0]	[5]	[1]	[6]	[2]	[7]	[3]	[8]	[4]
[6]	[0]	[6]	[3]	[0]	[6]	[3]	[0]	[6]	[3]
[7]	[0]	[7]	[5]	[3]	[1]	[8]	[6]	[4]	[2]
[8]	[0]	[8]	[7]	[6]	[5]	[4]	[3]	[2]	[1]

We see immediately that $\mathbb{Z}_9$ is not a field because some nonzero elements (e.g., [3] and [6]) do not have a multiplicative inverse. The multiplicative inverse of [2] is [5] (indeed, [2] [5] [1]), the multiplicative inverse of [7] is [4] (indeed, [7] [4] [1]), and the multiplicative inverse of [1] is [1].

It is easy to prove that a finite field, $GF(q)$, with $q \quad p^n$ and p a prime number, has, p^n, elements. The finite field $GF(q)$ can be considered a vector space, V, over $\mathbb{Z}_p$. $GF(q)$ is a finite field; thus, the vector space, V, is finite dimensional. If n represents the number of dimensions of V, there are n elements, $v_1, v_2, \ldots, v_n \quad V$, that form a basis for V:

$$V \qquad \overset{n}{\underset{i\ \ 1}{}} \lambda_i v_i \quad \lambda_i \quad Z_p.$$

Example. *Consider p 2, the nite eld of binary numbers,* $\mathbb{Z}_2$. For $n \quad 2$, we construct the set of all 2^n polynomials with binary coefficients:

$$\mathbb{Z}_2[x] \quad 0, 1, x, 1 \quad x, 1 \quad x^2, x^2, x \quad x^2, 1 \quad x \quad x^2, x^3, 1 \quad x^3, 1 \quad x \quad x^3, 1 \quad x^2 \quad x^3, 1 \quad x \quad x^2 \quad x^3,$$
$$x \quad x^3, x^2 \quad x^3, \ldots .$$

Given a finite field with q elements, there is an element, $\beta \quad GF(q)$, called the *primitive element*, or characteristic element, of the finite field, such that the set of nonzero elements of $GF(q)$ is

$$GF(q) \quad 0 \qquad \beta^1, \beta^2, \ldots, \beta^{q \ \ 2}, \beta^{q \ \ 1} \ , \quad \text{with} \quad \beta^{q \ \ 1} \quad 1.$$

Multiplication in a finite field with q elements, $GF(q)$, can be carried out as addition modulo, $(q \quad 1)$. Indeed, if α is the primitive element of $GF(q)$, then any two elements can be expressed as powers of β, $a \quad \beta^i$ and $b \quad \beta^j$. Then,

$$a \ \ b \quad \beta^i \ \ \beta^j \quad \beta^{(i \ \ j) \ \bmod (q \ \ 1)}.$$

Example. *Table 4.1 displays the four nonzero elements of GF(5) as powers of the primitive element* $\beta \quad [3]$.

Table 4.1 The Nonzero Elements of $GF(5)$ Expressed as a Power of β [3]

Element of $GF(5)$	Element of $GF(5)$ as a Power of β	Justification
[3]	β^1	$3 \quad \text{mod } 5 \quad 3$
[4]	β^2	$3^2 \text{ mod } 5 \quad 9 \quad \text{mod } 5 \quad 4$
[2]	β^3	$3^3 \text{ mod } 5 \quad 27 \text{ mod } 5 \quad 2$
[1]	β^4	$3^4 \text{ mod } 5 \quad 81 \text{ mod } 5 \quad 1$

The product of two elements of $GF(5)$ can be expressed as powers of the primitive element. For example, $[2] \quad [4] \quad [3]^3 \quad [3]^2 \quad [3]^{5 \text{ mod } 4} \quad [3]$; indeed, $(2 \quad 4 \text{ mod } 5) \quad (8 \text{ mod } 5) \quad 3$.

Let $GF(q)$ be a finite field and $a \quad GF(q)$; the *order of element a*, ord(a), is the smallest integer, s, such that $a^s \quad 1$. The set, $\beta^1, \beta^2, \ldots$, is finite; thus, there are two positive integers, i and j, such that $\beta^i \quad \beta^j$. This implies that $\beta^{i \quad j} \quad 1$; in turn, this means that there is a smallest s, such that $a^s \quad 1$, so the order is well defined.

By analogy with the ring of integers modulo a positive integer p, we can use the congruence relation to construct equivalence classes of polynomials. The set of polynomials, $q(x)$, in x and with coefficients from a field, F, form a ring with an infinite number of elements denoted as $F[x]$. If $f(x) \quad F[x]$ is a nonzero polynomial, we define the equivalence class, $[g(x)]$, containing the polynomial, $g(x) \quad F[x]$, as the set of polynomials that produce the same remainder as $g(x)$ when divided by $f(x)$.

Let $\mathbb{F}[x]$ denote the set of polynomials in variable, x, with coefficients from a finite field, F, and let $f(x), g(x) \quad \mathbb{F}[x]$ be nonzero polynomials; the *equivalence class of polynomials*, $[g(x)]$, relative to, $f(x)$, is defined as

$$[g(x)] \qquad q(x) : q(x) \qquad g(x) \quad \text{mod} f(x) \ .$$

Addition and multiplication of equivalence classes of polynomials are defined as

$$[g(x)] \quad [h(x)] \quad [g(x) \quad h(x)] \quad \text{and} \quad [g(x)] \; [h(x)] \quad [g(x) \;\; h(x)].$$

We can use the congruence relation to construct a finite ring of polynomials over F modulo the nonzero polynomial $f(x)$. A *nite ring of polynomials* over F modulo the nonzero polynomial $f(x)$, denoted by $R \quad \mathbb{F}[x]/(f(x))$, is

$$R \quad \mathbb{F}[x]/(f(x)) \qquad [g(x)] : g(x) \quad F[x] \ .$$

We are particulary interested in the special case, $f(x) \quad x^n \quad 1$, when

$$x^n \quad 1 \quad 0 \quad \text{mod} f(x) \qquad\qquad x^n \quad 1 \quad \text{mod} f(x).$$

Example. *Addition and multiplication in* $\mathbb{Z}_2[x]/(x^4 \quad 1)$:

$$[1 \quad x^2 \quad x^3] \quad [1 \quad x \quad x^2 \quad x^3] \quad [1 \quad x^2 \quad x^3 \quad 1 \quad x \quad x^2 \quad x^3] \quad [x]$$

and

$$[1 \quad x] \; [1 \quad x \quad x^2 \quad x^3] \quad [1 \quad x \quad x^2 \quad x^3 \quad x \quad x^2 \quad x^3 \quad x^4] \quad [1 \quad x^4] \quad [0].$$

An *ideal, I, of a ring,* $(R, \;\; , \;)$, is a non-empty subset of R, $I \quad R, I \qquad$, with the two properties:

1. $(I, \;\;)$ is a group.
2. $i \; r \quad I \qquad i \quad I$ and $\; r \quad R$.

The *ideal generated by the polynomial,* $g(x) \quad R$, is the set of all polynomials multiple of $g(x)$:

$$I \qquad g(x) \; p(x), \qquad p(x) \quad R \; .$$

A *principal ideal ring,* $(R, \;\; , \;)$, is one when $\; I \quad R$ and there exists $g(x) \quad I$, such that I is the ideal generated by $g(x)$.

It is easy to prove that $\mathbb{F}[x]$, the set of polynomials $f(x)$, with coefficients from a field F, and $\mathbb{F}(x)/q(x)$, the finite ring of polynomials over the field F modulo the nonzero polynomial $q(x)$, are principal ideal rings.

In summary, the congruence relationship allows us to construct finite algebraic structures consisting of equivalence classes; this process is illustrated in Figure 4.4. Given the set of integers, we can construct the ring, $\mathbb{Z}_p$, of integers modulo p. Similarly, given a nonzero polynomial, with coefficients in $\mathbb{Z}_2$, we construct 16 equivalence classes of polynomials modulo the polynomial, $f(x) \quad x^4 \quad 1$:

$$
\begin{array}{rl}
[0] & \text{the set of polynomials with remainder } 0 \text{ when divided by } f(x) \\
[1] & \text{the set of polynomials with remainder } 1 \text{ when divided by } f(x) \\
[x] & \text{the set of polynomials with remainder } x \text{ when divided by } f(x) \\
[1 \quad x] & \text{the set of polynomials with remainder } 1 \quad x \text{ when divided by } f(x) \\
 & \vdots \\
[x^2 \quad x^3] & \text{the set of polynomials with remainder } x^2 \quad x^3 \text{ when divided by } f(x) \\
[x \quad x^2 \quad x^3] & \text{the set of polynomials with remainder } x \quad x^2 \quad x^3 \text{ when divided by } f(x).
\end{array}
$$

We now brie y discuss *extension Galois elds,* $GF(q)$, with $q \quad p^m$, and p a prime number. The extension field, $GF(p^m)$, consists of p^m vectors, with m components; each component is an element of $GF(p)$. The vector, $a \quad (a_1, a_2, \ldots, a_{m \; 2}, a_m)$, with $a_i \quad GF(p)$, can also be represented as the polynomial $a(x) \quad a_{m \; 1}x^{m \; 1} \quad a_{m \; 2}x^{m \; 2} \qquad a_1 x \quad a_0$.

The addition operation in $GF(p^m)$ is defined componentwise. The multiplication is more intricate: if $f(x) \quad GF(p^m)$ is an irreducible polynomial of degree, m, with coefficients in $GF(p)$, then multiplication is defined as $a(x) \; b(x) \; \bmod f(x)$; thus, if $a(x) \; b(x) \quad q(x) \; f(x) \quad r(x)$, then $r(x) \quad r_{m \; 1}x^{m \; 1} \quad r_{m \; 2}x^{m \; 2} \qquad r_1 x \quad r_0$ is a polynomial of degree at most, $m \quad 1$. The identity element is $I(x) \quad 1$ or the m-tuple $(10\ldots00)$. It is left as an exercise to prove that the axioms for a finite field are satisfied. For example, the 2^m elements of $GF(2^m)$ are binary m-tuples. The addition is componentwise, modulo 2. If $a, b \quad GF(2^m)$ and

$$a \quad (a_0, a_1, \ldots, a_{m \; 2}, a_{m \; 1}), \quad b \quad (b_0, b_1, \ldots, b_{m \; 2}, b_{m \; 1}), \quad a_i, b_i \quad GF(2), \quad 0 \quad i \quad m \quad 1,$$

then

$$a \quad b \quad [(a_0 \quad b_0) \bmod 2, (a_1 \quad b_1) \bmod 2, \ldots, (a_{m \; 2} \quad b_{m \; 2}) \bmod 2, (a_{m \; 1} \quad b_{m \; 1}) \bmod 2].$$

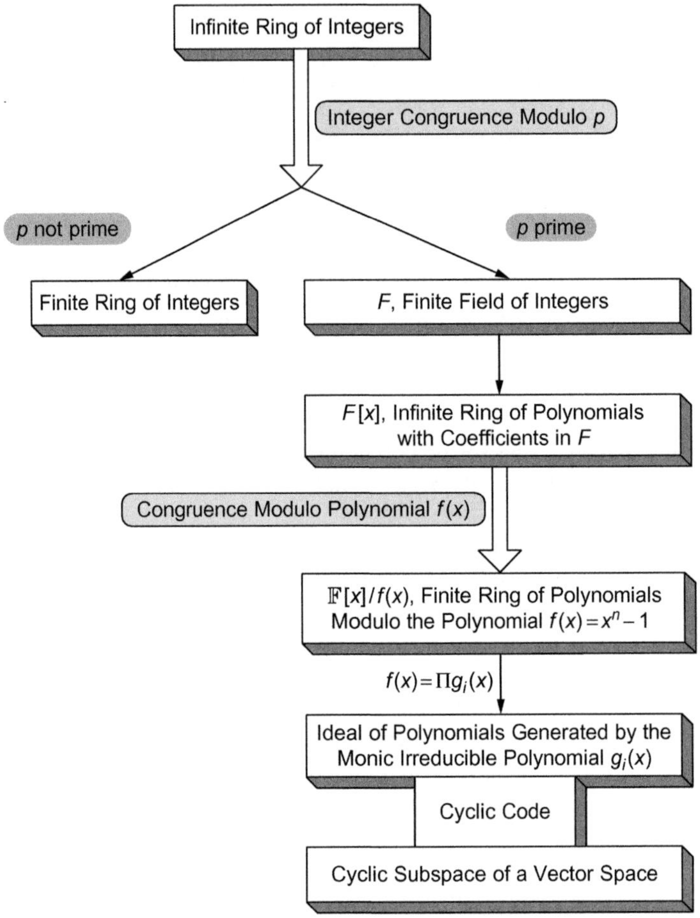

FIGURE 4.4

The congruence relationship allows us to construct finite algebraic structures. The integer congruence modulo p leads to a finite ring of integers when p is not a prime number, or to a finite field, F, when p is prime. The infinite ring of polynomials over the finite field, F, and the congruence modulo the polynomial, $f(x)$, allow us to construct a finite ring of polynomials, $\mathbb{F}[x]/f(x)$. When $f(x) \quad x^n - 1 \quad \Pi_i^q {}_1 g_i(x)$, we can construct ideals of polynomials generated by $g_i(x)$. Such ideals of polynomials are equivalent to cyclic subspaces of a vector space over $F, V_n(F)$, or cyclic codes as discussed in Section 4.11.

An important property of an extensions field $GF(q^m)$: a polynomial, $g(x)$, irreducible over $GF(q)$ could be factored in $GF(q^m)$. For example, the polynomial, $g(x) \quad x^2 \quad x \quad 1$, is irreducible over $GF(2)$ but can be factored over $GF(2^2)$:

$$g(x) \quad (x \quad \alpha)(x \quad \alpha^2),$$

with α, a primitive element of $GF(2^2) \quad 0, \alpha^0, \alpha^1, \alpha^2$ and $g(\alpha) \quad 0$.

Example. *Construct $GF(2^4)$ using the irreducible polynomial $f(x)\ \ x^4\ \ x\ \ 1$.*

First, we identify the primitive element, α, which must satisfy the equation, $\alpha^{q\ \ 1}\ \ 1$. In this case, $q\ \ 2^4\ \ 16$ and $\alpha\ \ x$ and $\alpha^{15}\ \ x^{15}\ \ (x^4\ \ x\ \ 1)(x^{11}\ \ x^8\ \ x^7\ \ x^5\ \ x^3\ \ x^2\ \ x)\ \ 1$; thus, $\alpha^{15}\ \ 1 \bmod (x^4\ \ x\ \ 1)$.

We construct all powers of $\alpha\ \ x$, or the binary 4-tuple $(0,0,1,0)$. The first elements are $\alpha^2\ \ x^2$, $\alpha^3\ \ x^3$. Then $\alpha^4 \bmod (x^4\ \ x\ \ 1)\ \ x\ \ 1$. Indeed, the remainder when dividing x^4 by $(x^4\ \ x\ \ 1)$ is $(x\ \ 1)$ because $x^4\ \ 1\ \ (x^4\ \ x\ \ 1)\ \ (x\ \ 1)$. The next two elements are $\alpha^5\ \ x\ x^4\ \ x(x\ \ 1)\ \ x^2\ \ x$ and $\alpha^6\ \ x\ x^5\ \ x\ (x^2\ \ x)\ \ x^3\ \ x^2$.

The other elements are $\alpha^7\ \ x^3\ \ x\ \ 1$, $\alpha^8\ \ x^2\ \ 1$, $\alpha^9\ \ x^3\ \ x$, $\alpha^{10}\ \ x^2\ \ x\ \ 1$, $\alpha^{11}\ \ x^3\ \ x^2\ \ x$, $\alpha^{12}\ \ x^3\ \ x^2\ \ x\ \ 1$, $\alpha^{13}\ \ x^3\ \ x^2\ \ 1$, $\alpha^{14}\ \ x^3\ \ 1$, and $\alpha^{15}\ \ 1$.

Example. *Construct $GF(2^4)$ using the irreducible polynomial $h(x)\ \ x^4\ \ x^3\ \ x^2\ \ x\ \ 1$.*

The primitive element is $\beta\ \ x\ \ 1$. To show that $\beta^{15}\ \ 1$, we carry out a binomial expansion and a polynomial division and conclude that $(x\ \ 1)^{15}\ \ 1 \bmod (x^4\ \ x^3\ \ x^2\ \ x\ \ 1)$.

Now $\beta^2\ \ (x\ \ 1)\ (x\ \ 1)\ \ x^2\ \ 1$. Then $\beta^{k\ \ 1}\ \ \beta^k(x\ \ 1)$. For example, after computing $\beta^5\ \ x^3\ \ x^2\ \ 1$, we can compute $\beta^6\ \ (x^3\ \ x^2\ \ 1)\ (x\ \ 1)$; as $x^4\ \ (x^3\ \ x^2\ \ x\ \ 1)$, it follows that $\beta^6\ \ x^3$.

Table 4.2 shows the correspondence between the 4-tuples, the powers of the primitive element, and polynomials for the two examples.

Table 4.2 $GF(2^4)$ Construction When the Irreducible Polynomial and the Primitive Elements are (left) $f(x)\ \ x^4\ \ x\ \ 1$ and $\alpha\ \ x$; (right) $h(x)\ \ x^4\ \ x^3\ \ x^2\ \ x\ \ 1$ and $\beta\ \ x\ \ 1$

Vector	Primitive Element	Polynomial	Vector	Primitive Element	Polynomial
0000		0	0000		0
0010	α^1	x	0011	β^1	$x\ \ 1$
0100	α^2	x^2	0101	β^2	$x^2\ \ 1$
1000	α^3	x^3	1111	β^3	$x^3\ \ x^2\ \ x\ \ 1$
0011	α^4	$x\ \ 1$	1110	β^4	$x^3\ \ x^2\ \ x$
0110	α^5	$x^2\ \ x$	1101	β^5	$x^3\ \ x^2\ \ 1$
1100	α^6	$x^3\ \ x^2$	1000	β^6	x^3
1011	α^7	$x^3\ \ x\ \ 1$	0111	β^7	$x^2\ \ x\ \ 1$
0101	α^8	$x^2\ \ 1$	1001	β^8	$x^3\ \ 1$
1010	α^9	$x^3\ \ x$	0100	β^9	x^2
0111	α^{10}	$x^2\ \ x\ \ 1$	1100	β^{10}	$x^3\ \ x^2$
1110	α^{11}	$x^3\ \ x^2\ \ x$	1011	β^{11}	$x^3\ \ x\ \ 1$
1111	α^{12}	$x^3\ \ x^2\ \ x\ \ 1$	0010	β^{12}	x
1101	α^{13}	$x^3\ \ x^2\ \ 1$	0110	β^{13}	$x^2\ \ x$
1001	α^{14}	$x^3\ \ 1$	1010	β^{14}	$x^3\ \ x$
0001	α^{15}	1	0001	β^{15}	1

Example. *Identify the primitive element of $GF(3^2)$ constructed as a class of remainders of the irreducible polynomial, $f(x) \quad x^2 \quad 1$.* This is another example when the primitive element is not x but $\alpha \quad x \quad 1$. Indeed, the nonzero elements of the field are:

$$1 \quad \alpha^0 \qquad 2 \quad \alpha^4 \qquad x \quad \alpha^6 \qquad 2x \quad \alpha^2$$
$$1 \quad x \quad \alpha^1 \quad 2 \quad x \quad \alpha^7 \quad 1 \quad 2x \quad \alpha^3 \quad 2 \quad 2x \quad \alpha^5$$

The order of each of the eight elements of $GF(3^2)$ is given by the power of the primitive element. For example, the order of $2 \quad x$ is 7, and the order of x is 6.

It is easy to see that if $\beta \quad GF(q) \quad 0$, then $\beta^{q \quad 1} \quad 1$. Observe that the multiplicative inverse of β is in $GF(q)$ and let the $(q \quad 1)$ distinct nonzero elements $GF(q)$ be $a_1, a_2, \ldots, a_{q \quad 1}$. Then all the products, $\beta \quad a_1, \beta \quad a_2, \ldots, \beta \quad a_{q \quad 1}$, must be distinct. Indeed, if $\beta \quad a_i \quad \beta \quad a_j$, $i \quad j$, then $(\beta^{\quad 1} \quad \beta) \quad a_i \quad (\beta^{\quad 1} \quad \beta) \quad a_j$, $i \quad j$, or $a_i \quad a_j$, which contradicts the assumption that a_i and a_j are distinct. It follows immediately that the product of $(q \quad 1)$ distinct nonzero elements of $GF(q)$ must be the same:

$$(\beta \quad a_1) \quad (\beta \quad a_2) \quad (\beta \quad a_{q \quad 1}) \quad a_1 \quad a_2 \quad a_{q \quad 1} \quad \text{or} \quad \beta^{q \quad 1} \quad (a_1 \quad a_2 \quad a_{q \quad 1}) \quad a_1 \quad a_2 \quad a_{q \quad 1},$$

and this implies that $\beta^{q \quad 1} \quad 1$. Another way of reading this equation is that any element, $\beta \quad GF(q)$, satisfies the equation, $\beta^q \quad \beta \quad 0$; this leads us to the conclusion that any element of $GF(q)$ must be a root of the equation $x^q \quad x \quad 0$.

It is also easy to show that any element, $\beta \quad GF(q^m)$, is in $GF(q)$ if and only if $\beta^q \quad \beta$. A polynomial of degree, n, in x with coefficients in $GF(q)$ has at most, n, roots in $GF(q)$. The condition, $\beta^q \quad \beta$, shows that β is a root of the polynomial equation $x^q \quad x \quad 0$, which has as roots all the elements of $GF(q)$. We conclude that β is in $GF(q)$ if and only if $\beta^q \quad \beta$.

We now show that the orders of all nonzero elements of $GF(q)$ divide $(q \quad 1)$. Consider an element $\beta \quad GF(q)$ of order $a \quad \text{ord}(\beta)$; thus, $\beta^a \quad 1$. Let us assume that, and a, does not divide $(q \quad 1)$ so $q \quad 1 \quad a \quad b \quad c$, with $1 \quad c < a$. We showed earlier that $\beta \quad GF(q)$, and we have the equality $\beta^{q \quad 1} \quad 1$. Thus,

$$\beta^{q \quad 1} \quad \beta^{a \quad b \quad c} \quad \beta^{a \quad b} \quad \beta^c \quad (\beta^a)^b \quad \beta^c \quad 1.$$

But $(\beta^a)^b \quad 1$; thus, $\beta^c \quad 1$. This is possible only if $c \quad 0$; thus, $q \quad 1 \quad a \quad b$ and $a \quad \text{ord}(\beta)$ divides $q \quad 1$.

Finally, we show that there exists a primitive element in any finite field $GF(q)$. Assume that β is the element of the largest order, b, in $GF(q)$, $\beta^b \quad 1$. In the trivial case, $b \quad q \quad 1$, and we have concluded the proof; thus, β is the primitive element of $GF(q)$.

Let us now concentrate on the case when $b < q \quad 1$. Assume that the orders of all elements in $GF(q)$ divide b; this implies that every nonzero element of $GF(q)$ satisfies the polynomial equation, $y^b \quad 1 \quad 0$. This equation has at most, b, roots and can generate at most, b, elements of $GF(q)$; we have reached a contradiction with the assumption $b < q \quad 1$.

Let γ be the element of $GF(q)$ whose order, c, does not divide b; thus,

$$\gcd(b,c) \quad d > 1, \qquad c \quad d \quad g, \quad \text{with} \quad \gcd(g,b) \quad 1.$$

But $\gamma^c \quad 1$ implies that there exists an element, δ, of order d, in $GF(q)$; indeed, $\gamma^c \quad \gamma^{c \quad g} \quad (\gamma^g)^d \quad 1$. Thus, $\delta \quad \gamma^g \quad GF(q)$.

Let us now consider another element of $GF(q)$ constructed as the product of two elements, $\beta \;\; \delta$; it is easy to see that this is an element of order, $b \;\; d$, because $\beta^b \;\; \delta^d \;\;\; 1$. We have thus found an element in $GF(q)$ of order $b \;\; d$, and this contradicts the assumption that b is the order of the largest element.

The following proposition summarizes some of the results just discussed.

Proposition. *Consider a nite eld, with q elements, $GF(q)$, and an element, $\alpha \;\; GF(q)$; the following properties are true:*

1. $\alpha^j \;\; 1 \qquad \text{ord}(\alpha) \;\; j$.
2. $\text{ord}(\alpha) \;\; s \qquad \alpha^1 \;\; \alpha^2 \qquad \alpha^s$.
3. $\text{ord}(\alpha) \;\; s$, $\text{ord}(\beta) \;\; r$, $\text{gcd}(s,r) \;\; 1 \qquad \text{ord}(\alpha \;\; \beta) \;\; s \;\; r$.
4. $\text{ord}(\alpha^j) \;\; \frac{\text{ord}(\alpha)}{\text{gcd}(\text{ord}(\alpha), j)}$.
5. $GF(q)$ has an element of order $(q \;\; 1)$.
6. The order of any element in $GF(q)$ divides $(q \;\; 1)$.
7. $\gamma^q \;\; \gamma \;\; 0, \;\; \gamma \;\; GF(q)$.

After this brief review of algebraic structures used in coding theory, we start our discussion of error-correcting codes with an informal introduction.

4.5 LINEAR CODES

Linear algebra allows an alternative description of a class of block codes called *linear codes*. In this section, we show that an $[n,k]$ linear code, $\mathcal{C}$, is a k-dimensional subspace, $V_n^{(k)}$, of an n-dimensional vector space, V_n. Once we select a set of, k, linearly independent vectors in V_n, we can construct a generator matrix, G, for $\mathcal{C}$. Then the code $\mathcal{C}$ can be constructed by multiplying the generator matrix G with the message vectors in V_k. Several classes of linear codes, codes defined as subspaces of a vector space, are known, including the cyclic codes and the Bose-Chaudhuri-Hocquenghem codes discussed in Section 4.11. Throughout the presentation of linear codes, we use the term *n-tuple* instead of *vector in an n-dimensional vector space*.

Recall that an $[n,M]$ block code, $\mathcal{C}$, over the alphabet, $\mathcal{A}$, with q symbols, encodes $k \;\; \log_q M$ information symbols into codewords, with $n \;\; k \;\; r$ symbols. Two alternative descriptions of a linear code $\mathcal{C}$ are used: $[n,k]$ code and $[n,k,d]$ code with, d, the Hamming distance of the code. Decoding is the reverse process, extracting the message from a codeword. Both processes can be simplified when the code has an algebraic structure.

We consider an alphabet with q symbols from a finite field, $F \;\; GF(q)$, and vectors with n or k components also called n-tuples and, respectively, k-tuples; the corresponding vector spaces are denoted as $V_n(F)$ and $V_k(F)$. When there is no confusion about the alphabet used instead of $V_n(F)$ and $V_k(F)$, we shall use the notation V_n and V_k, respectively. These two vector spaces can also be regarded as the finite fields, $GF(p^n)$ and $GF(p^k)$, respectively. A subspace of dimension, k, of the vector space, V_n, is denoted as $V_n^{(k)}$. Most of our examples cover binary codes, the case when $q \;\; 2$.

We start this section with formal definitions and then discuss relevant properties of linear codes. In our presentation, we refer to the linear code, $\mathcal{C}$, as an $[n,k]$ code.

A *linear* $[n,k]$ *code* over F is a k-dimensional subspace of the vector space, V_n. A linear code is a one-to-one mapping, f, of k-tuples from the *message space* to n-tuples,

$$f : V_k \quad V_n,$$

with $n > k$. The n-tuples selected as codewords form a *subspace* of $V_n^{(k)} \quad V_n$ spanned by the k linearly independent vectors. Given a set of messages,

$$M \quad m_1, m_2, \ldots, m_{q^k}\,, \quad \text{where} \quad m_i \quad (m_{i,1}, m_{i,2}, \ldots, m_{i,n}) \quad V_k \quad \text{and} \quad m_{i,j} \quad F,$$

and a basis, B_k, for the k-dimensional subspace, $V_n^{(k)} \quad V_n$, $B_k \quad v_1, v_2, \ldots, v_k$, with $v_i \quad V_n$, *encoding* can be defined as the product of vectors in, M, with vectors in, B_k; the code, $\mathcal{C}$, is then

$$\mathcal{C} \quad c_i \quad \sum_{j\ 1}^{k} m_{i,j}\ v_j \quad m_{i,j} \quad M \text{ with } v_j \quad B_k\ .$$

Example. Consider a three-dimensional subspace of a vector space over the finite field, with $q \quad 2$ and $n \quad 6$. We choose the following three vectors as a basis:

$$B_k \quad v_1 \quad (111000) \quad v_2 \quad (100110) \quad v_3 \quad (110011)\ .$$

The message space consists of all 3-tuples:

$$M \quad m_1 \quad (000),\ m_2 \quad (001),\ m_3 \quad (010),\ m_4 \quad (011),$$
$$m_5 \quad (100),\ m_6 \quad (101),\ m_7 \quad (110),\ m_8 \quad (111)\ .$$

Each codeword of the code, $\mathcal{C}$, is a 6-tuple obtained as a mapping of the products of individual message bits, with the basis vectors:

$$m_1 \quad (000) \qquad c_1 \quad 0\ v_1 \quad 0\ v_2 \quad 0\ v_3 \quad (000000),$$
$$m_2 \quad (001) \qquad c_2 \quad 0\ v_1 \quad 0\ v_2 \quad 1\ v_3 \quad (110011),$$
$$m_3 \quad (010) \qquad c_3 \quad 0\ v_1 \quad 1\ v_2 \quad 0\ v_3 \quad (100110),$$
$$m_4 \quad (011) \qquad c_4 \quad 0\ v_1 \quad 1\ v_2 \quad 1\ v_3 \quad (010101),$$
$$m_5 \quad (100) \qquad c_5 \quad 1\ v_1 \quad 0\ v_2 \quad 0\ v_3 \quad (111000),$$
$$m_6 \quad (101) \qquad c_6 \quad 1\ v_1 \quad 0\ v_2 \quad 1\ v_3 \quad (001011),$$
$$m_7 \quad (110) \qquad c_7 \quad 1\ v_1 \quad 1\ v_2 \quad 0\ v_3 \quad (011110),$$
$$m_8 \quad (111) \qquad c_8 \quad 1\ v_1 \quad 1\ v_2 \quad 1\ v_3 \quad (101101).$$

An $[n,k]$ linear code, $\mathcal{C}$, is characterized by a $k \quad n$ matrix whose rows form a basis in V_n; this matrix is called the *generator matrix* of the linear code $\mathcal{C}$.

Example. *The generator matrix of the code $\mathcal{C}$, with $q \quad 2, n \quad 6, k \quad 3$, is*

$$\mathrm{G} \quad \begin{matrix} v_1 \\ v_2 \\ v_3 \end{matrix} \qquad \begin{matrix} 1 & 1 & 1 & 0 & 0 & 0 \\ 1 & 0 & 0 & 1 & 1 & 0 \\ 1 & 1 & 0 & 0 & 1 & 1 \end{matrix} .$$

The code, $\mathcal{C} \quad c_1, c_2, \ldots, c_8$, generated by matrix, G, is obtained as products of the vectors in the message space, M, with G. For example:

$$c_3 \quad m_3 \; \mathrm{G} \quad (010) \begin{matrix} 1 & 1 & 1 & 0 & 0 & 0 \\ 1 & 0 & 0 & 1 & 1 & 0 \\ 1 & 1 & 0 & 0 & 1 & 1 \end{matrix} \quad (100110),$$

$$c_7 \quad m_7 \; \mathrm{G} \quad (110) \begin{matrix} 1 & 1 & 1 & 0 & 0 & 0 \\ 1 & 0 & 0 & 1 & 1 & 0 \\ 1 & 1 & 0 & 0 & 1 & 1 \end{matrix} \quad (011110).$$

Given two $[n,k]$ linear codes, $\mathcal{C}_1$ and $\mathcal{C}_2$, over the filed, F, with generator matrices, G_1 and G_2, respectively, we say that they are *equivalent codes* if

$$\mathrm{G}_2 \quad \mathrm{G}_1 \; P,$$

with P, a permutation matrix, and ( ) denoting matrix multiplication.

Example. *Consider the generator matrix, G, of the code, $\mathcal{C}$,*

$$\mathrm{G} \quad \begin{matrix} 1 & 1 & 1 & 0 & 0 & 0 \\ 1 & 0 & 0 & 1 & 1 & 0 \\ 1 & 1 & 0 & 0 & 1 & 1 \end{matrix} ,$$

and a permutation matrix, P, given by

$$P \quad \begin{matrix} 0 & 0 & 1 & 0 & 0 & 0 \\ 1 & 0 & 0 & 0 & 0 & 0 \\ 0 & 0 & 0 & 1 & 0 & 0 \\ 0 & 0 & 0 & 0 & 1 & 0 \\ 0 & 1 & 0 & 0 & 0 & 0 \\ 0 & 0 & 0 & 0 & 0 & 1 \end{matrix} .$$

We construct the generator matrix, $G_p \quad GP$, as

$$G_p \quad \begin{pmatrix} 1 & 1 & 1 & 0 & 0 & 0 \\ 1 & 0 & 0 & 1 & 1 & 0 \\ 1 & 1 & 0 & 0 & 1 & 1 \end{pmatrix} \begin{pmatrix} 0 & 0 & 1 & 0 & 0 & 0 \\ 1 & 0 & 0 & 0 & 0 & 0 \\ 0 & 0 & 0 & 1 & 0 & 0 \\ 0 & 0 & 0 & 0 & 1 & 0 \\ 0 & 1 & 0 & 0 & 0 & 0 \\ 0 & 0 & 0 & 0 & 0 & 1 \end{pmatrix} \begin{pmatrix} 1 & 0 & 1 & 1 & 0 & 0 \\ 0 & 1 & 1 & 0 & 1 & 0 \\ 1 & 1 & 1 & 0 & 0 & 1 \end{pmatrix}.$$

Recall that the code $\mathcal{C}$ is

$$\mathcal{C} \quad (000000)\ (110011)\ (100110)\ (010101)\ (111000)\ (001011)\ (011110)\ (101101)\ .$$

It is easy to see that, $\mathcal{C}_p$, the code generated by G_p, is

$$\mathcal{C}_p \quad (000000)\ (111001)\ (011010)\ (100011)\ (101100)\ (010101)\ (110110)\ (001111)\ .$$

The permutation matrix P maps columns of matrix G to columns of G_p as follows: $2 \quad 1, 5 \quad 2, 1 \quad 3, 3 \quad 4, 4 \quad 5$, and $6 \quad 6$. To map column, i, of the original matrix G into column, j, of matrix G_p, the element, p_{ij}, of the permutation matrix must be $p_{ij} \quad 1$.

This example shows that being equivalent does not mean that the two codes are identical. Since a k-dimensional subspace of a vector space is not unique, the generator matrix of a code is not unique but one of them may prove to be more useful than the others. We shall see shortly that among the set of equivalent codes there is one whose generator matrix is of the form $G \quad [I_k A]$. Indeed, given the $k \quad n$ generator matrix G of code $\mathcal{C}$, with a subset of columns being the binary representation of the integers 1 to k, we can easily construct the permutation matrix that exchanges the columns of G to obtain the generator matrix, $G_I \quad [I_k A]$, of the equivalent code, $\mathcal{C}_I$.

Given an $[n, k]$-linear code $\mathcal{C}$ over the field F, the orthogonal complement, or the *dual of the code* $\mathcal{C}$ denoted as $\mathcal{C}$, consists of vectors (or n-tuples) orthogonal to every vector in $\mathcal{C}$:

$$\mathcal{C} \quad c \quad V_n : \quad c\ w \quad 0, \quad w \quad \mathcal{C}\ .$$

Given an $[n, k]$-linear code $\mathcal{C}$ over the field F, let H be the generator matrix of the dual code $\mathcal{C}$. Then $c \quad V_n$, $Hc^T \quad 0 \quad c \quad \mathcal{C}$; the matrix, H, is called the *parity-check matrix of* $\mathcal{C}$.

Given an $[n, k]$-linear code $\mathcal{C}$ over the field F, with the parity-check matrix H, the *error syndrome*, s, of the n-tuple v is defined as $s \quad Hv^T$. There is a one-to-one correspondence between the error syndrome and the bits in error; thus, the syndrome is used to determine if an error has occurred and to identify the bit in error. We wish to have an efficient algorithm to identify the bit(s) in error once the syndrome is calculated, a topic addressed in Section 4.6. A linear code $\mathcal{C}$, with the property that $\mathcal{C} \quad \mathcal{C}$, is called *weakly self-dual*. When $\mathcal{C} \quad \mathcal{C}$ the code is called *strictly self-dual*.

Some Properties of Linear Codes

We discuss several properties of linear codes over a finite field. There are many k-dimensional subspaces of the vector space V_n, thus, many $[n, k]$ codes; their exact number is given by the following proposition.

Proposition. *The number of k-dimensional subspaces, $V_n^{(k)}$, of the vector space, V_n, over $GF(q)$, with $n > k$, is*

$$s_{n,k} \quad \frac{(q^n \quad 1)(q^{n \quad 1} \quad 1)\ldots(q^{n \quad k \quad 1} \quad 1)}{(q^k \quad 1)(q^{k \quad 1} \quad 1)\ldots(q \quad 1)}.$$

Proof. First, we show that there are

$$(q^n \quad 1)(q^n \quad q)\ldots(q^n \quad q^{k \quad 1})$$

ordered sets of k linearly independent vectors in V_n. The process of selecting one by one the k linearly independent vectors is now described. To select the first one, we observe that there are $q^n \quad 1$ nonzero vectors in V_n. This explains the first term. There are q vectors that are linearly dependent on a given vector. Thus, for every vector in the first group, there are, $(q^n \quad q)$, linearly independent vectors. Therefore, there are $(q^n \quad 1)(q^n \quad q)$ different choices for the first two vectors. The process continues until we have selected all k linearly independent vectors of the k-dimensional subspace.

Similar reasoning shows that a k-dimensional subspace consists of

$$(q^k \quad 1)(q^k \quad q)\ldots(q^k \quad q^{k \quad 1})$$

ordered sets of k linearly independent vectors. This is the number of the ordered sets of k linearly independent vectors generating the same k-dimensional subspace. To determine the number of distinct k-dimensional subspaces, we divide the two numbers. ■

Example. *If $q \quad 2$, $n \quad 6$, and $k \quad 3$, the number of three-dimensional subspaces of V_6 is*

$$s_{n,k} \quad \frac{(2^6 \quad 1)(2^5 \quad 1)(2^4 \quad 1)}{(2^3 \quad 1)(2^2 \quad 1)(2^1 \quad 1)} \quad \frac{63 \quad 31 \quad 15}{7 \quad 3 \quad 1} \quad 1395.$$

The weight of the minimum weight nonzero vector, or n-tuple, $w \quad C$, is called the *Hamming weight of the code* C, $w(C) \quad \min\ w(c) : \ c \quad C, c \quad 0$.

Proposition. *The Hamming distance of an $[n,k]$ linear code, C, over $GF(2)$, is equal to the Hamming weight of the code,*

$$d(C) \quad w(C).$$

Proof. $c_i, c_j \quad C, c_i \quad c_j, c_i \quad c_j \quad c_k \quad C$ and $d(c_i,c_j) \quad w(c_i \quad c_j)$, with , the bitwise addition modulo 2 in $GF(2)$. Therefore, $d(C) \quad \min\ w(c_k) : \ c_k \quad C, c_k \quad 0 \quad w(C)$. ■

Proposition. *Given an $[n,k]$ linear code, C, over the eld, F, among the set of generator matrices of C, or of codes equivalent to C, there is one, where the information symbols appear in the rst, k, positions of a codeword. Thus, the generator matrix is of the form*

$$G \quad [I_k A].$$

Here, I_k is the $k \quad k$ identity matrix and A is a $k \quad (n \quad k)$ matrix.

Proposition. *Given an [n,k] linear code, C, over the eld, F, its dual, C , is an [n,n-k] linear code over the same eld, F. If* G $[I_kA]$ *is a generator matrix for C, or for a code equivalent to C, then* H $[\ A^T I_{n\ k}]$ *is a generator matrix for its dual code, C .*

Proof. It is relatively easy to prove that C is a subspace of V_n. From the definition of H,

$$\mathrm{H} \quad [\ A^T I_{n\ k}] \quad \begin{matrix} a_{1,1} & a_{2,1} & a_{3,1} & \cdots & a_{n,1} & 1 & 0 & 0 & \ldots & 0 \\ a_{1,2} & a_{2,2} & a_{3,3} & \cdots & a_{n,2} & 0 & 1 & 0 & \ldots & 0 \\ a_{1,3} & a_{2,3} & a_{3,3} & \cdots & a_{n,3} & 0 & 0 & 1 & \ldots & 0 \\ \cdots & & & & & & & & & \\ a_{1,n\ k} & a_{2,n\ k} & a_{3,n\ k} & \cdots & a_{n,n\ k} & 0 & 0 & 0 & \ldots & 1 \end{matrix},$$

and it follows that

$$\mathrm{H}^T \quad \begin{matrix} a_{1,1} & a_{1,2} & a_{1,3} & \cdots & a_{1,n\ k} \\ a_{2,1} & a_{2,2} & a_{2,3} & \cdots & a_{2,n\ k} \\ a_{3,1} & a_{3,2} & a_{3,3} & \cdots & a_{3,n\ k} \\ \cdots & & & & \\ a_{n,1} & a_{n,2} & a_{n,3} & \cdots & a_{n,n\ k} \\ 1 & 0 & 0 & \ldots & 0 \\ 0 & 1 & 0 & \ldots & 0 \\ 0 & 0 & 1 & \ldots & 0 \\ \cdots & & & & \\ 0 & 0 & 0 & \ldots & 1 \end{matrix} \quad \begin{matrix} A \\ I_{n\ k} \end{matrix}.$$

Then,

$$\mathrm{GH}^T \quad [I_kA] \begin{matrix} A \\ I_{n\ k} \end{matrix} \quad 0.$$

It is easy to see that the element, c_{ij} $1 \quad i \quad k$, $1 \quad j \quad n \quad k$, of the product is $c_{ij} \quad a_{ij} \quad a_{ij} \quad 0$. Note also that in $GF(q)$ the additive inverse of a_{ij} is an element, such that $a_{ij} \quad (\ a_{ij}) \quad 0$; for example, $(\ 1) \quad 1$ in $GF(2)$ and $(\ 1) \quad 2$ in $GF(3)$.

Call span(H) the subspace generated by the rows of H. The rows of H are linearly independent so the row space of H spans an $(n \quad k)$ subspace of V_n. If the rows of H are orthogonal to the rows of G, it follows that

$$\mathrm{span(H)} \quad C\ .$$

To complete the proof, we have to show that C span(H). We leave this as an exercise for the reader. Thus, C span(H) and dim(C) $n \quad k$. ■

Example. *The previous proposition suggests an algorithm to construct the generator matrix of the dual code, C . Let the generator matrix of the code*

$$C \quad (000000)\ (101001)\ (011010)\ (110011)\ (101100)\ (000101)\ (110110)\ (011111)$$

be

$$G \quad \begin{matrix} 1 & 0 & 1 & 1 & 0 & 0 \\ 0 & 1 & 1 & 0 & 1 & 0 \\ 1 & 1 & 1 & 0 & 0 & 1 \end{matrix} .$$

This matrix can be transformed by permuting the columns of G. We map columns 1 4, 2 5, 3 6, 4 1, 5 2, and 6 3 to construct G_I:

$$G_I \quad GP \quad \begin{matrix} 1 & 0 & 1 & 1 & 0 & 0 \\ 0 & 1 & 1 & 0 & 1 & 0 \\ 1 & 1 & 1 & 0 & 0 & 1 \end{matrix} \quad \begin{matrix} 0 & 0 & 0 & 1 & 0 & 0 \\ 0 & 0 & 0 & 0 & 1 & 0 \\ 0 & 0 & 0 & 0 & 0 & 1 \\ 1 & 0 & 0 & 0 & 0 & 0 \\ 0 & 1 & 0 & 0 & 0 & 0 \\ 0 & 0 & 1 & 0 & 0 & 0 \end{matrix} \quad \begin{matrix} 1 & 0 & 0 & 1 & 0 & 1 \\ 0 & 1 & 0 & 0 & 1 & 1 \\ 0 & 0 & 1 & 1 & 1 & 1 \end{matrix}$$

or

$$G_I \quad [I_3 A], \quad \text{with} \quad A \quad \begin{matrix} 1 & 0 & 1 \\ 0 & 1 & 1 \\ 1 & 1 & 1 \end{matrix} .$$

It is easy to observe that in this case, A^T A and H $[\ A\, I_3]$, or

$$H \quad \begin{matrix} 1 & 0 & 1 & 1 & 0 & 0 \\ 0 & 1 & 1 & 0 & 1 & 0 \\ 1 & 1 & 1 & 0 & 0 & 1 \end{matrix} \quad G.$$

Proposition. *The necessary and suf cient condition for the linear code, C, with generator matrix,* G, *to be weakly self-dual is that*

$$G^T G \quad 0.$$

The proof of this proposition is left as an exercise for the reader.

Proposition. *Given an $[n,k]$ linear code, C, over the eld, F, with the parity-check matrix,* H, *every set of $(d \quad 1)$ columns of* H *are linearly independent if and only if C has a distance at least equal to d.*

Example. *The parity-check matrix of a code capable of correcting a single error.* A single-error–correcting code must have a distance d 3. According to the previous proposition, we need a matrix H such that no two or fewer than two columns are linearly dependent. Let us first examine the set of single columns. If we eliminate the all-zero column, then no single column can be linearly dependent. Two columns, h_i and h_j, are linearly dependent if there are nonzero scalars, λ_i, λ_j, such that

$$\lambda_i h_i \quad \lambda_j h_j \quad 0.$$

This implies that the two columns are scalar multiples of each other. Indeed, the previous equation allows us to express $h_i \quad (\lambda_j/\lambda_i)h_j$.

In summary, if we want a code capable of correcting any single error we have to construct a matrix, H, to be used as the parity-check matrix of the code such that:

1. The matrix H does not contain an all-zero column.
2. No two columns of H are scalar multiples of each other.

Example. *Generate the code with the parity-check matrix,* H.

First, we transform the parity-check matrix by column operations as follows:

$$\mathrm{H} \qquad \mathrm{H} \quad [I_k A].$$

Then we construct the generator matrix of the code, G $\quad [\ A^T I_{n\ k}]$. Finally, we multiply all vectors in the message space with the generator matrix and obtain the set of codewords.

Example. *Correct a single error in the n-tuple, v, using the parity-check matrix,* H, *of the* $[n,k]$ *linear code,* $\mathcal{C}$. Apply the following algorithm:

1. Compute $s^T \quad \mathrm{H}v^T$. If $s^T \quad 0$, then $v \quad \mathcal{C}$.
2. If $s^T \quad 0$, compare s^T with $h_1, h_2, \ldots, h_n$, the columns of H.
   - If there is an index, j, such that $s^T \quad h_j$, then the error is an n-tuple, e, with a 1 in the j-th bit position and 0s elsewhere. The codeword is then $c \quad v \quad e$.
   - Else, there are two or more errors.

For example, assume that the codeword, $c \quad (0111100)$, of the code, $\mathcal{C}$, with the parity-check matrix, H, is affected by an error in the last bit position; the n-tuple received is $v \quad (0111101)$. Then,

$$s^T \quad \mathrm{H}v^T \quad \begin{matrix} 1 & 0 & 1 & 0 & 1 & 0 & 1 \\ 0 & 1 & 1 & 0 & 0 & 1 & 1 \\ 0 & 0 & 0 & 1 & 1 & 1 & 1 \end{matrix} \quad \begin{matrix} 0 \\ 1 \\ 1 \\ 1 \\ 1 \\ 0 \\ 1 \end{matrix} \quad \begin{matrix} 1 \\ 1 \\ 1 \end{matrix}.$$

The vector, s^T, is identical to the last column of H; thus, $e \quad (0000001)$, and the codeword is $c \quad v \quad e \quad (0111101) \quad (0000001) \quad (0111100)$. In this example, H is the parity-check matrix of a Hamming code; Hamming codes are discussed in Section 4.8.

Proposition. *If* $\mathcal{C}$ *is an* $[n,k]$ *code with a generator matrix,* G, *whose columns are the* $2^k \quad 1$ *nonzero binary k-tuples, then all nonzero codewords have weight* $2^{k \quad 1}$.

Hint: About half of the $(2^k \quad 1)$ k-tuples that are columns of G have an even weight and half have an odd weight. More precisely, $2^{k \quad 1}$ have an odd weight and $2^{k \quad 1} \quad 1$ have an even weight. It follows that half of the components of a codeword obtained by linear combinations of the rows of G are 1s.

The *weight distribution* of a code, C, is a vector, $A_C \quad (A_0, A_1, \ldots, A_n)$, with A_w, the number of codewords of weight, w; the *weight enumerator* of C is the polynomial

$$A(z) \qquad \overset{n}{\underset{w \quad 0}{}} A_w z^w.$$

If C is an $[n,k,d]$ linear code, then A_w represents also the number of codewords at distance, $d \quad w$, from a given codeword.

Proposition. *If $A(z)$ is the weight enumerator of the $[n,k,d]$ linear code, C, then the weight enumerator of its dual, $C \quad$, is*

$$B(z) \quad 2^k (1 \quad z)^n A \; \frac{1 \quad z}{1 \quad z} \, .$$

In the next section, we discuss algorithms for efficient decoding of linear codes.

4.6 SYNDROME AND STANDARD ARRAY DECODING OF LINEAR CODES

Now we discuss the partitioning of a vector space, V_n, over the finite field, F, into equivalence classes such that all n-tuples in the same class have the same syndrome; then we present a decoding procedure that exploits this partitioning. We start with a few definitions of algebraic structures.

Let G be an additive Abelian group and S be a subgroup of G, $S \quad G$. Two elements, $g_1, g_2 \quad G$, are *congruent modulo the subgroup, S,* if $g_1 \quad g_2 \quad S$,[1] or

$$g_1 \quad g_2 \bmod S \qquad\qquad g_1 \quad g_2 \quad S.$$

Let G be an additive Abelian group and S be a subgroup of G, $S \quad G$; the equivalence classes for congruence modulo S are called the *cosets of the subgroup S.*[2]

Proposition. *The order, m, of a subgroup, $S \quad G$, divides, n, the order of the group (Lagrange theorem).*

Proof. The elements of the subgroup are $S \quad s_1, s_2, \ldots, s_m$. Coset i consists of the elements congruent with s_i and is denoted as Ss_i, and coset j consists of the elements congruent with s_j and is denoted as Ss_j. Let f be the mapping

$$f : Ss_i \qquad Ss_j \qquad\qquad f(g_k s_i) \quad g_k s_j.$$

The function, f, maps distinct elements of the coset, Ss_i, into distinct elements of the coset, Ss_j. All elements of Ss_j are mapped onto. Thus, f is a one-to-one map. This implies that the two cosets, and

---

[1]In the general case, the congruence relationship implies that $g_1 \quad g_2^{\;\;1} \quad S$, where " " is the group operation and $g_2^{\;\;1}$ is the inverse of g_2 under " ". In other words, $g_2 \quad g_2^{\;\;1} \quad e$, with e, the neutral element of G under the group operation. In the case of an additive group, the operation is , "addition," and the "inverse" $g^{\;\;1} \quad g$.

[2]In the general case, the elements defined here are called the *right cosets*, but for an Abelian group they are simply called cosets.

then obviously all cosets, have the same cardinality, m. But congruence does partition the set, G, of n elements into a number of disjoint cosets and $n \quad (\textit{number of cosets of } S) \quad m$. ■

We now present an alternative definition of the cosets useful for syndrome calculation and decoding of linear codes. We discuss only binary codes, codes over $GF(2^n)$, but similar definitions and properties apply to codes over $GF(q^n)$. Throughout this section, V_n is $GF(2^n)$, and we consider binary n-tuples; the code, $\mathcal{C}$, is a subset of $GF(2^n)$.

Let $\mathcal{C}$ be an $[n,k]$ code, $\mathcal{C} \quad GF(2^n)$. The *coset of the code*, $\mathcal{C}$, containing $v \quad GF(2^n)$ is the set $v \quad \mathcal{C} \quad v \quad c,\ c \quad \mathcal{C}$. The binary n-tuple of minimum weight in $\mathcal{C}_i$ is called the *leader of the coset* $\mathcal{C}_i$. The decoding procedure we present later is based on the following two propositions.

Proposition. *Two binary n-tuples are in the same coset if and only if they have the same syndrome.*

Proof. Recall that if H is the parity-check matrix of the code, $\mathcal{C}$, and $c \quad \mathcal{C}$, then $\mathrm{H}c^T \quad 0$. The syndrome corresponding to an n-tuple, $v \quad GF(2^n)$, is $\sigma_v \quad \mathrm{H}v^T$.

If the two binary n-tuples are in the same coset, they can be written as $v_1 \quad c_1 \quad e$ and $v_2 \quad c_2 \quad e$, with $c_1, c_2 \quad \mathcal{C},\ e \quad GF(2^n)$. The two syndromes are equal:

$$\sigma_{v_1} \quad \mathrm{H}v_1^T \quad \mathrm{H}(c_1 \quad e)^T \quad \mathrm{H}e^T \quad \text{and} \quad \sigma_{v_2} \quad \mathrm{H}v_2^T \quad \mathrm{H}(c_2 \quad e)^T \quad \mathrm{H}e^T \qquad \sigma_{v_1} \quad \sigma_{v_2}.$$

If the two n-binary tuples, v_1 and v_2, have the same syndrome, then

$$\mathrm{H}v_1^T \quad \mathrm{H}v_2^T \qquad \mathrm{H}(v_1 \quad v_2)^T \quad 0 \qquad (v_1 \quad v_2) \quad \mathcal{C}.$$

Because $(v_1 \quad v_2)$ is a codeword, v_1 and v_2 must be in the same coset. ■

It is easy to see that there is a one-to-one correspondence between the coset leaders and the syndromes.

Proposition. *Let $\mathcal{C}$ be an $[n,k]$ binary linear code. $\mathcal{C}$ can correct t errors if and only if all binary n-tuples of weight, t, or less, are coset leaders.*

Proof. To prove that this condition is necessary, assume that $\mathcal{C}$ has the distance, $d \quad 2t \quad 1$; thus, it can correct, t, errors. Let $v_1, v_2 \quad GF(2^n)$; if v_1 and v_2 belong to the same coset, then the binary n-tuple, $(v_1 \quad v_2)$, is a codeword, $(v_1 \quad v_2) \quad \mathcal{C}$, according to the definition of the coset and to the previous proposition. If their weight is at most, t, $w(v_1), w(v_2) \quad t$, then the weight of the codeword, $(v_1 \quad v_2)$, is $w(v_1 \quad v_2) \quad 2t$, which contradicts the fact that the distance of the code is $d \quad 2t \quad 1$. Thus, if the code can correct t errors, all binary n-tuples of weight less than or equal to t must be in distinct cosets and each one can be elected as coset leader.

To prove that this condition is sufficient, assume that all binary n-tuples of weight less than or equal to t are coset leaders but there are two codewords, c_1 and c_2, such that $d(c_1, c_2) \quad 2t$. There exist two coset leaders, v_1 and v_2, such that $w(v_1), w(v_2) \quad t$, and $c_1 \quad v_1 \quad c_2 \quad v_2$. Indeed, we could choose them such that $c_1 \quad c_2 \quad v_2 \quad v_1$. This is possible because $d(c_1, c_2) \quad 2t$ and also $d(v_1, v_2) \quad 2t$ as each coset leader has a weight of at most, t. But $c_1 \quad c_2 \quad v_2 \quad v_1$ implies that the two syndromes are equal, $\sigma_{c_1 \quad v_1} \quad \sigma_{c_2 \quad v_2}$. This is a contradiction because we assumed that two coset leaders, v_1 and v_2, are in different cosets. Thus, the assumption that $d(c_1, c_2) \quad 2t$ is incorrect and the code is capable of correcting t errors. ■

Table 4.3 The Correspondence Between the Coset Leaders and the Syndromes

Coset Leader	Syndrome
0000000	$(000)^T$
0000001	$(111)^T$
0000010	$(011)^T$
0000100	$(110)^T$
0001000	$(101)^T$
0010000	$(001)^T$
0100000	$(010)^T$
1000000	$(100)^T$

Example. *Let C be a $[7,4,3]$ code, with the parity-check matrix*

$$H \quad \begin{matrix} 1 & 0 & 0 & 1 & 1 & 0 & 1 \\ 0 & 1 & 0 & 0 & 1 & 1 & 1 \\ 0 & 0 & 1 & 1 & 0 & 1 & 1 \end{matrix} .$$

Table 4.3 shows the correspondence between the coset leaders and the syndromes for code C.

Next, we discuss an efficient decoding procedure for a linear code C.

Standard Array Decoding

Consider an $[n,k]$ code, $C \quad c_1, c_2, \ldots, c_{2^k}$, over $GF(2)$. Let V_{2^n} be the vector space of all n-tuples over $GF(2)$. C is a subgroup of order 2^k of V_{2^n}, with 2^n elements. According to the Lagrange theorem, there are $2^n/2^k \quad 2^{n \ k}$ cosets of C denoted as $\mathcal{C}_0, \mathcal{C}_1, \ldots, \mathcal{C}_i, \ldots, \mathcal{C}_{2^{n \ k \ 1}}$.

We start with the coset leader, $\mathcal{L}_0 \quad 00\ldots0$, the all-zero n-tuple, and then construct the coset $\mathcal{C}_0 \quad \mathcal{L}_0, c_1, c_2, \ldots, c_{2^k}$. Then we select as coset leaders n-tuples of weight 1. After we select all n-tuples of weight 1 as coset leaders, we move to select those n-tuples of weight 2 that do not appear in any previously constructed coset and continue the process until all $2^{n \ k}$ cosets have been constructed. Once we select the leader of a coset j, $\mathcal{L}_j$, the i-th element of the coset is $\mathcal{L}_j \quad c_i$.

The cosets could be arranged in order as rows of a table, with $2^{n \ k}$ rows and with 2^k columns. This table is called a *standard array* for code C.

Example. *Consider the binary $[6,3]$ code, with distance 3 and the generator matrix*

$$G \quad \begin{matrix} 0 & 0 & 1 & 0 & 1 & 1 \\ 0 & 1 & 0 & 1 & 1 & 0 \\ 1 & 0 & 0 & 1 & 0 & 1 \end{matrix} \qquad \begin{matrix} c_1 \\ c_2 \\ c_3 \end{matrix} .$$

The standard array has in the first row the codewords, $c_1, c_2, c_3, c_1 \quad c_2, c_1 \quad c_3, c_2 \quad c_3, c_1 \quad c_2 \quad c_3$, and in the first column, the coset leaders,

000000	001011	010110	100101	011101	101100	110011	111000
000001	001010	010111	100100	011100	101101	110010	111001
000010	001001	010100	100111	011111	101110	110001	111010
000100	001111	010010	100001	011001	101000	110111	111100
001000	000011	011110	101101	010101	100100	111011	110000
010000	011011	000110	110101	001101	111100	100011	101000
100000	101011	110110	000101	111101	001100	010011	011000.

Proposition. *Consider an* $[n,k]$ *code,* $\mathcal{C} \quad c_1, c_2, \ldots, c_{q^k \quad 1}$ *, over* $GF(2)$*. Let SA be a standard array, with cosets* $\mathcal{C}_i : i \quad 1, 2^{n \quad k}$ *and coset leaders* $\mathcal{L}_i : i \quad 1, 2^{n \quad k}$ *. The* $(i \quad 1, j \quad 1)$ *entry is* $\mathcal{L}_i \quad c_j$*, as described earlier. Then* $h \quad 0, 2^k$ *:*

$$d(\mathcal{L}_i \quad c_j, c_j) \quad d(\mathcal{L}_i \quad c_j, c_h).$$

In other words, the $j \quad 1$ *entry in column, i, is closer to the codeword at the head of its column,* c_j*, than to any other codeword of* $\mathcal{C}$*.*

Proof. Recall that $w(c_i)$ is the weight of the codeword, c_i. Then from the definition of the entries of the standard table,

$$d(\mathcal{L}_i \quad c_j, c_j) \quad w(\mathcal{L}_i),$$
$$d(\mathcal{L}_i \quad c_j, c_h) \quad w(\mathcal{L}_i \quad c_j \quad c_h).$$

Coset $\mathcal{C}_i$ is obtained by adding the coset leader, $\mathcal{L}_i$, to the codewords of $\mathcal{C}$, $c_i \quad \mathcal{C}$. Clearly, the n-tuple, $(\mathcal{L}_i \quad c_j \quad c_h)$, is a member of the same coset, $\mathcal{C}_i$ because $c_j \quad c_h \quad \mathcal{C}$. Yet, by construction, the coset leader, $\mathcal{L}_i$, is the n-tuple with the minimum weight in $\mathcal{C}_i$:

$$w(\mathcal{L}_i) \quad w(\mathcal{L}_i \quad c_j \quad c_h).$$

■

Proposition. *The following procedure based on a standard table can be used to decode a received n-tuple, v, of a linear code:*

1. *Locate v in the standard table.*
2. *Correct v as the codeword at the top of its column.*

Example. *If we receive v* 111101 *we identify it at the fth element of the last row of the standard table constructed for the previous example and decode it as the codeword at the top of the fth column, c* 011101.

In the next section, we discuss the limitations imposed on our ability to construct "good" linear codes.

4.7 HAMMING, SINGLETON, GILBERT-VARSHAMOV, AND PLOTKIN BOUNDS

Ideally, we want codes with a large number of codewords (thus, large value of k) that can correct as large a number of errors, e, as possible, and with the lowest possible number of parity-check symbols, $r \quad n \quad k$. These contradictory requirements cannot be satisfied concurrently. To illustrate these limitations, Table 4.4[3] lists the number of binary codewords for the case when the length, n, of a codeword is in the range, $5 \quad n \quad 15$, and the number of errors the code is able to correct is in the range, $1 \quad e \quad 3$.

Several bounds on the parameters of a linear code have been established, and in this section we discuss Hamming, Singleton, Gilbert-Varshamov, and Plotkin bounds. When n and k are given, the Hamming bound shows the limit of the efficiency of an $[n,k]$ code, it specifies the minimum number of redundancy symbols for the code, while the Singleton bound provides an upper limit to the distance d of an $[n,k,d]$ code. The Gilbert-Varshamov and the Plotkin bounds limit the number of codewords on an $[n,k,d]$ code when n and d are specified.

We consider $[n,k,d]$ linear codes, $\mathcal{C}$, over an alphabet, with q symbols; recall that an $[n,k,d]$ code encodes, k, information symbols into codewords consisting of, n, symbols and guarantees a minimum distance, d, between any pair of codewords. The codes are vector spaces over the field, $GF(q)$, and the codewords, $c \quad \mathcal{C}$, are vectors in $GF(q^n)$.

Table 4.4 The Number of Binary Codewords for $5 \quad n \quad 15$ and $e \quad 1,2,3$

n	$e=1$	$e=2$	$e=3$
5	4	2	
6	8	2	
7	16	2	2
8	20	4	2
9	40	6	2
10	72	12	2
11	144	24	4
12	256	32	4
13	512	64	8
14	1024	128	16
15	2048	256	32

[3] The table is from http://www.ams.org.

Hamming Bound

Let C be an $[n,k]$ code over, $GF(q)$, capable of correcting, e, errors; then C satisfies the Hamming bound

$$q^n \geq q^k \sum_{i=0}^{e} \binom{n}{i} (q-1)^i.$$

Proof. The number of n-tuples within a Hamming sphere of radius, e, around a codeword is

$$\sum_{i=0}^{e} \binom{n}{i} (q-1)^i.$$

There are q^k such Hamming spheres, and the total number of n-tuples is q^n. ■

Example. *Determine the minimum number, r, of redundancy bits required for a binary code able to correct any single error.* Our analysis is based on the following observations:

1. To construct an $[n,M]$ block code, we select a subset of cardinality, $M = 2^k$, as codewords out of the set of all 2^n n-tuples.
2. To correct all one-bit errors, the Hamming spheres of radius 1 centered at all, M, codewords must be disjoint.
3. A Hamming sphere of radius 1 centered around an n-tuple consists of $(n+1)$ n-tuples; indeed, the number of n-tuples at distance 1 from the center of the Hamming sphere is equal to n, one for each bit position, and we should add one for the center.

It is easy to see that $M \times (n+1) \leq 2^n$. Indeed, the number of n-tuples of all the M Hamming spheres of radius 1 cannot exceed the total number of n-tuples. But $n = k + r$; thus, this inequality becomes

$$2^k \times (k + r + 1) \leq 2^{(k+r)} \quad \text{or} \quad 2^r \geq r + k + 1.$$

We conclude that the minimum number of redundancy bits has a lower bound:

$$2^r \geq n + 1.$$

For example, if $k = 15$, the minimum value of r is 5.

A binary code is able to correct at most, e, errors if

$$2^n \geq 2^k \sum_{i=0}^{e} \binom{n}{i}$$

or, using the fact that $d = 2e + 1$ and $r = n - k$,

$$2^r \geq \sum_{i=0}^{(d-1)/2} \binom{n}{i}.$$

When $d = 3$, this becomes $2^r \geq n + 1$ as before.

Singleton Bound

Every $[n,k,d]$ linear code, $\mathcal{C}$, over the field, $GF(q)$, satisfies the inequality

$$k \quad d \quad n \quad 1.$$

Proof. Let $e_1, e_2, \ldots$ be a basis in the vector space, $GF(q^n)$. Express every $c \quad \mathcal{C}$ in this basis; then construct a new code, $\mathcal{C}_s$, consisting of codewords of length, $n \quad d \quad 1$, obtained by deleting the first $d \quad 1$ symbols of every $c \quad \mathcal{C}$. The codewords of $\mathcal{C}$ are at a distance at least, d, from one another. When we remove the first $d \quad 1$ symbols from every $c \quad \mathcal{C}$, the remaining $(n \quad d \quad 1)$-tuples, $c_s \quad \mathcal{C}_s$, are at least at distance 1; thus, they are distinct. The original code, $\mathcal{C}$, consists of $\mathcal{C} \quad q^k$ distinct codewords; thus, the cardinality of the new code is

$$\mathcal{C}_s \quad q^k.$$

There cannot be more than, $q^{n \quad d \quad 1}$, tuples of length, $(n \quad d \quad 1)$, in $GF(q^{n \quad d \quad 1})$. Thus,

$$q^k \quad q^{n \quad k \quad 1} \qquad\qquad k \quad n \quad d \quad 1$$

or

$$k \quad d \quad n \quad 1.$$

■

A *maximum distance separable (MDS)*, $[n,k,d]$, linear code is a code, $\mathcal{C}$, with the property

$$k \quad d \quad n \quad 1.$$

MDS codes are optimal in the following sense: for a given length of a codeword, n, and a given number of information symbols, k, the distance, d, of the code reaches its maximum.

Corollary. *Consider an in nite sequence of linear codes, $[n_i, k_i, d_i]$, such that when $i \qquad$, then $n_i \qquad$ and:*

$$R \quad \lim_{i} \frac{k_i}{n_i}, \quad \delta \quad \lim_{i} \frac{d_i}{n_i}.$$

Then the Singleton bound requires that

$$\frac{k_i}{n_i} \quad \frac{d_i}{n_i} \quad 1 \quad \frac{1}{n_i}.$$

When $i \qquad$, then

$$R \quad \delta \quad 1.$$

Figure 4.5(a) shows that in the limit, any sequence of codes must have the rate, (R), and the relative minimum distance, $(\delta \quad d/n)$, within the triangle, $R \quad \delta \quad 1$.

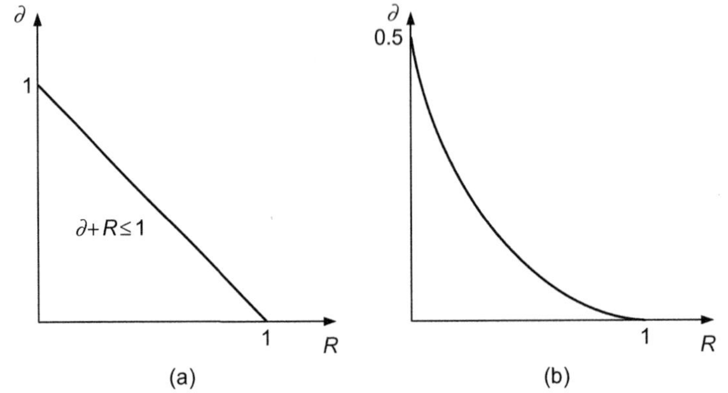

FIGURE 4.5

(a) The Singleton bound. (b) The Gilbert-Varshamov bound.

Gilbert-Varshamov Bound

If $\mathcal{C}$ is an $[n,k,d]$ linear code over $GF(q)$ and if $M(n,d) \quad \mathcal{C}$ is the largest possible number of codewords for given n and d values, then the code satisfies the Gilbert-Varshamov bound:

$$M(n,d) \quad \frac{q^n}{\,{}^{d-1}_{i\ 0} \binom{n}{i} (q-1)^i}.$$

Proof. Let $c \quad \mathcal{C}$. Then there is an n-tuple, $v \quad GF(q^n)$, such that, $d(c,v)$, the distance between c and v, satisfies the inequality

$$d(c,v) \quad d-1.$$

The existence of such an n-tuple is guaranteed because otherwise v could be added to the code $\mathcal{C}$, while maintaining the minimum distance of the code as d. Recall that $M(n,d)$ is the largest possible number of codewords for given n and d values. A Hamming sphere of radius, $(d-1)$, about a codeword, $c \quad \mathcal{C}$, contains

$$\,{}^{d-1}_{i\ 0} \quad {}^{n}_{i} \ (q-1)^i$$

n-tuples. We assume that there are $M(n,d)$ codewords in $\mathcal{C}$ and therefore an equal number of Hamming spheres of radius $(d-1)$, one about each codeword, S_{c_i}, $c_i \quad \mathcal{C}$. Thus, the entire space, $GF(q^n)$, is contained in the union of all Hamming spheres of radius, $d-1$, around individual codewords:

$$GF(q^n) \quad {}_{c_j \ \mathcal{C}} S_{c_j}(d-1)$$

or

$$|GF(q^n)| \le M(n,d) \sum_{i=0}^{d-1} \binom{n}{i} (q-1)^i.$$

It follows that

$$q^n \le M(n,d) \sum_{i=0}^{d-1} \binom{n}{i} (q-1)^i$$

or

$$M(n,d) \ge \frac{q^n}{\sum_{i=0}^{d-1} \binom{n}{i} (q-1)^i}.$$

■

The Gilbert-Varshamov bound applies not only to linear codes but also to any code of length, n, and distance, d. If the code has distance d, the Hamming spheres of radius, $\lfloor (d-1)/2 \rfloor$, are disjoint but those of radius, d, are not disjoint. Two spheres of radius d have many n-tuples in common, thus, the sum of the volumes of all these spheres has q^n as a lower bound, as we are guaranteed to include all n-tuple and some of them are counted multiple times.

Plotkin Bound

If C is an $[n,k,d]$ linear code and if the condition, $(2d > n)$, is satisfied, then the code satisfies the Plotkin bound,

$$M \le 2 \left\lfloor \frac{d}{2d-n} \right\rfloor,$$

with $M = |C|$.

Proof. Let $c_x, c_y \in C$. We find first a lower bound for $\sum_{c_x,c_y \in C} d(c_x,c_y)$; we observe that

$$\sum_{c_x,c_y \in C} d(c_x,c_y) \ge dM(M-1).$$

Indeed, there are $M(M-1)$ pairs of codewords in C and d is the minimum distance between any pair of codewords.

We can arrange the codewords of C in a table, with M rows and n columns. Consider column, i, of this table and call z_i the number of 0s and $(M-z_i)$ the number of 1s in that column. Then,

$$\sum_{c_x,c_y \in C} d(c_x,c_y) = \sum_{i=1}^{n} 2z_i(M-z_i)$$

because each choice of a 0 and a 1 in the same column of the table contributes 2 to the sum of distances. Now we consider two cases:

1. M is an even integer. In that case the sum on the right-hand side of the equality is maximized when $z_i \quad M/2$. Thus, we obtain an upper bound for the sum of distances:

$$\sum_{c_x, c_y \quad C} d(c_x, c_y) \quad \frac{1}{2} n M^2.$$

   Now we combine the lower and the upper bound to get

$$M(M \quad 1)d \quad \frac{1}{2} n M^2$$

   or

$$M \quad \frac{2d}{2d \quad n}$$

   because we assumed that $2d \quad n > 0$. Since M is an even integer, it follows that

$$M \quad 2 \quad \frac{d}{2d \quad n} \quad .$$

2. M is an odd integer. In that case, the sum on the right-hand side of the equality is maximized when $z_i \quad (M \quad 1)/2$. Thus, we obtain an upper bound for the sum of distances:

$$\sum_{c_x, c_y \quad C} d(c_x, c_y) \quad \frac{1}{2} n (M^2 \quad 1).$$

   Now,

$$M(M \quad 1)d \quad \frac{1}{2} n (M^2 \quad 1)$$

   or

$$M \quad \frac{2d}{2d \quad n} \quad 1.$$

   Again, we use the fact that M is an integer:

$$M \quad \frac{2d}{2d \quad n} \quad 1 \quad \frac{2d}{2d \quad n} \quad 1 \quad 2 \quad \frac{d}{2d \quad n} \quad .$$

■

It is now time to present several families of codes, and, not unexpectedly, we start our journey with the codes discovered by Richard Hamming.

4.8 HAMMING CODES

Hamming codes, introduced in 1950 by Richard Hamming [199], are the first error-correcting codes ever invented. Hamming codes are linear codes; they can detect up to two simultaneous bit errors, and correct single-bit errors. At the time of their discovery, Hamming codes were considered a great improvement over simple parity codes, which cannot correct errors and can only detect an odd number of errors. Throughout this section, we refer to the linear code, $\mathcal{C}$, consisting of n-tuples, with, k, information symbols in each codeword and with distance, d, as an $[n,k,d]$ code.

A *q-ary Hamming code of order, r,* over the field, $GF(q)$, is an $[n,k,3]$ linear code, where

$$n = \frac{q^r - 1}{q - 1} \quad \text{and} \quad k = n - r.$$

The parity-check matrix of the code, H_r, is an $r \times n$ matrix, such that it contains no all-zero columns and no two columns are scalar multiples of each other.

A code, $\mathcal{C}$, such that the Hamming spheres of radius, e, about each codeword are disjoint and exhaust the entire space of n-tuples is called a *perfect code* capable of correcting, e, errors.

The last statement means that every n-tuple is contained in the Hamming sphere of radius e about some codeword.

Proposition. *The Hamming code of order, r, over $GF(q)$ is a perfect code, $\mathcal{C}_r$.*

Proof. If $c \in \mathcal{C}_r$, then the Hamming sphere of radius, e, about c, contains one n-tuple at distance 0 from c (c itself) and $n(q-1)$ n-tuples at distance 1 from c (there are, n, symbols in an n-tuple and each symbol can be altered to any of the other, $q-1$, letters of the alphabet). Thus, the total number of n-tuples in a Hamming sphere about a codeword is $(1 + n(q-1))$.

There are, k, information symbols; thus, the total number of messages is q^k and the total number of such Hamming spheres is also q^k. The total number of n-tuples in all the Hamming spheres about codewords is equal to the number of Hamming spheres times the number of n-tuples in each Hamming sphere:

$$q^k(1 + n(q-1)) = q^{n-r}\left(1 + \frac{(q^r - 1)}{(q-1)}(q-1)\right) = q^{n-r}(1 + (q^r - 1)) = q^n.$$

But q^n is precisely the total number of n-tuples with elements from an alphabet with q symbols. This proves that the Hamming spheres of radius, $e = 1$, about all codewords contain all possible n-tuples. The spheres of radius 1 about codewords are disjoint because the minimum distance of the code is 3. ■

Assume that we transmit a codeword, $c \in \mathcal{C}$, where $\mathcal{C}$ is a linear code, with distance, $d \geq 3$, and parity-check matrix, H. A single error occurs and the error vector, e, has a Hamming weight of 1; what we receive is an n-tuple, $v = c + e$. From the definition of the parity-check matrix, it follows that the syndrome associated with the n-tuple, v, is

$$\mathrm{H}v^T = \mathrm{H}(c + e)^T = \mathrm{H}e^T.$$

Since e has only one nonzero coordinate, say coordinate, j, which has value, β; then $\mathrm{H}e^T = \beta h_j$, where h_j is the j-th column of the parity-check matrix H. This justifies the following proposition.

Proposition. *The following procedure can be used to decode any single-error linear code, and in particular, Hamming codes:*

1. *Compute the syndrome,* $\mathrm{H}v^T$.
2. *If the syndrome is zero, no error has occurred.*
3. *If* $\mathrm{H}e^T \quad s^T \quad 0$, *then compare the syndrome with the columns of the parity-check matrix,* H.
4. *If there is some integer, j, such that* $s^T \quad \beta h_j$, *then the error vector, e, is an n-tuple, with* β *in position j and zero elsewhere. Then* $c \quad v \quad e$.
5. *If there is no integer, j, such that* $s^T \quad \beta h_j$, *then more than one error has occurred.*

An $[n, n \quad r, d]$ code whose parity-check matrix has all nonzero binary vectors of length, r, as columns is a *binary Hamming code.*

There are $2^r \quad 1$ nonzero vectors of length, r; thus, $n \quad 2^r \quad 1$. The number of information symbols is $k \quad n \quad r \quad 2^r \quad 1 \quad r$. Recall that the minimum distance of the code is equal to the minimum number of linearly dependent columns of the parity-check matrix. By definition, any two rows are linearly independent; thus, $d \quad 3$. It follows that a binary Hamming code is a $[(2^r \quad 1), (2^r \quad 1 \quad r), 3]$ linear code.

Example. *The binary Hamming code of order,* $r \quad 3$, *is a* $[7,4,3]$ *linear code.*

Indeed, $(2^3 \quad 1)/(2 \quad 1) \quad 7$; thus, $n \quad 7$ and $k \quad n \quad r \quad 7 \quad 3 \quad 4$. The parity-check matrix of the code is

$$\mathrm{H}_3 \quad \begin{array}{ccccccc} 1 & 0 & 0 & 1 & 1 & 0 & 1 \\ 0 & 1 & 0 & 0 & 1 & 1 & 1 \\ 0 & 0 & 1 & 1 & 0 & 1 & 1 \end{array}.$$

A code obtained by adding one all-0s column and one all-1s row to the generator matrix of a binary Hamming code is called an *extended binary Hamming code.* An extended binary Hamming code is a $[2^r, (2^r \quad r \quad 1), 4]$ linear code.

Example. *The binary Hamming code and the extended binary Hamming code of order* $r \quad 4$. They are $[15,11,3]$ and, respectively, $[16,11,4]$ linear codes with parity-check matrices, H_4 and HE_4, given by:

$$H_4 \quad \begin{array}{ccccccccccccccc} 1 & 0 & 1 & 0 & 1 & 0 & 1 & 0 & 1 & 0 & 1 & 0 & 1 & 0 & 1 \\ 0 & 1 & 1 & 0 & 0 & 1 & 1 & 0 & 0 & 1 & 1 & 0 & 0 & 1 & 1 \\ 0 & 0 & 0 & 1 & 1 & 1 & 1 & 0 & 0 & 0 & 0 & 1 & 1 & 1 & 1 \\ 0 & 0 & 0 & 0 & 0 & 0 & 0 & 1 & 1 & 1 & 1 & 1 & 1 & 1 & 1 \end{array}$$

and

$$HE_4 \quad \begin{array}{cccccccccccccccc} 1 & 0 & 1 & 0 & 1 & 0 & 1 & 0 & 1 & 0 & 1 & 0 & 1 & 0 & 1 & 0 \\ 0 & 1 & 1 & 0 & 0 & 1 & 1 & 0 & 0 & 1 & 1 & 0 & 0 & 1 & 1 & 0 \\ 0 & 0 & 0 & 1 & 1 & 1 & 1 & 0 & 0 & 0 & 0 & 1 & 1 & 1 & 1 & 0 \\ 0 & 0 & 0 & 0 & 0 & 0 & 0 & 1 & 1 & 1 & 1 & 1 & 1 & 1 & 1 & 0 \\ 1 & 1 & 1 & 1 & 1 & 1 & 1 & 1 & 1 & 1 & 1 & 1 & 1 & 1 & 1 & 1 \end{array}.$$

4.9 PROPER ORDERING AND THE FAST WALSH-HADAMARD TRANSFORM

In Section 1.21, we introduced Hadamard matrices and the Walsh-Hadamard transform. In this section, we continue our discussion, provide an alternative definition of the Walsh-Hadamard transform, and introduce the fast Walsh-Hadamard transform.

Consider all binary q-tuples, $b_1, b_2, \ldots, b_{2^q}$. The *proper ordering*, π_q, of binary q-tuples is defined recursively:

$$\pi_1 \quad [0,1],$$
$$\vdots$$
$$\pi_i \quad [b_1, b_2, \ldots, b_{2^i}],$$
$$\pi_{i \ \ 1} \quad [b_1 0, b_2 0, \ldots, b_{2^i} 0, b_1 1, b_2 1, \ldots, b_{2^i} 1] \quad i \quad 2, q \quad 1 .$$

Example. *Given that* $\pi_1 \quad [0,1]$, *it follows that*

$$\pi_2 \quad [00, 10, 01, 11],$$
$$\pi_3 \quad [000, 100, 010, 110, 001, 101, 011, 111],$$
$$\pi_4 \quad [0000, 1000, 0100, 1100, 0010, 1010, 0110, 1110, 0001, 1001, 0101, 1101, 0011, 1011, 0111, 1111].$$

Let $n \quad 2^q$ and let the n q-tuples under the proper order be

$$\pi_q \quad [u_0, u_1, \ldots, u_{n \ \ 1}].$$

Let us enforce the convention that the leftmost bit of u_i is the least significant bit of the integer represented by u_i as a q-tuple. For example, consider the following mappings of 4-tuples to integers:

$$0000 \quad 0,\ 1000 \quad 1,\ 0100 \quad 2,\ 1100 \quad 3,\ 0010 \quad 4,\ 1010 \quad 5, \ldots, 0111 \quad 14,\ 1111 \quad 15.$$

Proposition. *The matrix,* $H \quad [h_{ij}]$, *with*

$$h_{ij} \quad (\ \ 1)^{u_i \ \ u_j} \quad (i,j) \quad 0, n \quad 1 ,$$

where u_i *and* u_j *are members of*

$$\pi_q \quad [u_0, u_1, \ldots, u_{n \ \ 1}],$$

the Hadamard matrix of order $n \quad 2^q$. *Observe that the rows and the columns of the matrix,* H, *are numbered from* 0 *to* $n \quad 1$.

As an example, consider the case $q \quad 3$. It is easy to see that the matrix elements of the first row, h_{0i}, and the first column, h_{i0}, are all $\ 1$ because $u_0 \quad u_i \quad u_i \quad u_0 \quad 0,\ i \quad 1,7$, and $(\ \ 1)^0 \quad 1$. It is also easy to see that the diagonal element, $h_{ii} \quad 1$, when the weight of i (the number of 1s in the binary representation of integer i) is odd, and, $h_{ii} \quad 1$, when the weight of i is even. Individual calculations of $h_{ij},\ i \quad j$ are also trivial. For example, $h_{32} \quad (\ \ 1)^{(110) \ (010)} \quad (\ \ 1)^1 \quad 1$. Indeed, $(110) \ (010) \quad 1 \ 0 \quad 1 \ 1 \quad 0 \ 0 \quad 1$. This shows that $H \quad H_3$.

Let $c \quad (c_0 c_1 c_2 \ldots c_i \ldots c_{2^q \quad 1})$ be a binary 2^q-tuple, $c_i \quad\quad 0, 1$, and let B_q be a $q \quad 2^q$ matrix whose columns are all 2^q possible q-tuples b_i. We define $c(b)$ to be the component of, c, selected by the q-vector, b, according to the matrix, B. The binary value, $c(b)$, can be either 0 or 1, depending on the value of the corresponding element of the binary 2^q-tuple, c.

For example, let $q \quad 3$ and B_3 be given by

$$B_3 \quad \begin{matrix} 1 & 0 & 0 & 1 & 0 & 1 & 1 & 0 \\ 0 & 1 & 0 & 1 & 1 & 0 & 1 & 0 \\ 0 & 0 & 1 & 0 & 1 & 1 & 1 & 0 \end{matrix} .$$

Let $c \quad (0\ 1\ 1\ 1\ 1\ 0\ 0\ 1)$. Then $c(1\ 1\ 0) \quad 1$ because $(1\ 1\ 0)$ is the fourth column of B and the fourth element of c is 1. Similarly, $c(1\ 1\ 1) \quad 0$ because $(1\ 1\ 1)$ is the seventh column of B and the seventh element of c is 0.

We can define a vector, $R(c)$, whose components, $R_i(c)$, are $(\quad 1)^{c(b)}$, and their value can be either 1 or 1. In our example:

$$R(c) \quad (\ 1, \ 1, \ 1, \ 1, \ 1, \ 1, \ 1, \ 1).$$

Indeed,

$$\begin{aligned}
R_1(c) &\quad (\ 1)^{c(100)} \quad (\ 1)^0 \quad 1,\\
R_2(c) &\quad (\ 1)^{c(010)} \quad (\ 1)^1 \quad 1,\\
R_3(c) &\quad (\ 1)^{c(001)} \quad (\ 1)^1 \quad 1,\\
R_4(c) &\quad (\ 1)^{c(110)} \quad (\ 1)^1 \quad 1,\\
R_5(c) &\quad (\ 1)^{c(011)} \quad (\ 1)^1 \quad 1,\\
R_6(c) &\quad (\ 1)^{c(101)} \quad (\ 1)^0 \quad 1,\\
R_7(c) &\quad (\ 1)^{c(111)} \quad (\ 1)^0 \quad 1,\\
R_8(c) &\quad (\ 1)^{c(000)} \quad (\ 1)^1 \quad 1.
\end{aligned}$$

Let d be a binary q-tuple and let c be a binary 2^q-tuple. Let $R(c) \quad (\quad 1)^{c(b)}$ be a 2^q-tuple, with entries either 1 or 1, as defined earlier; then the *Walsh-Hadamard transform of* $R(c)$ is defined as

$$R(d) \quad \underset{b \ B_q}{\quad} (\ 1)^{d \ b} R(c)$$

or

$$R(d) \quad \underset{b \ B_q}{\quad} (\ 1)^{d \ b \quad c(b)}.$$

Example. *Let* $q\ \ 3$, $d\ \ (1\,1\,1)^T$, *and* $c\ \ (0\,1\,1\,1\,1\,0\,0\,1)$. *Then,*

$$R(1\,1\,1)\qquad \underset{b\ B_3}{}(\ 1)^{(1\,1\,1)\ b\ \ c(b)}$$

$$(\ 1)^{(1\,1\,1)\ (1\,0\,0)\ \ c(1\,0\,0)}\quad (\ 1)^{(1\,1\,1)\ (0\,1\,0)\ \ c(0\,1\,0)}\quad (\ 1)^{(1\,1\,1)\ (0\,0\,1)\ \ c(0\,0\,1)}$$

$$(\ 1)^{(1\,1\,1)\ (1\,1\,0)\ \ c(1\,1\,0)}\quad (\ 1)^{(1\,1\,1)\ (0\,1\,1)\ \ c(0\,1\,1)}\quad (\ 1)^{(1\,1\,1)\ (1\,0\,1)\ \ c(1\,0\,1)}$$

$$(\ 1)^{(1\,1\,1)\ (1\,1\,1)\ \ c(1\,1\,1)}\quad (\ 1)^{(1\,1\,1)\ (0\,0\,0)\ \ c(0\,0\,0)}$$

$$(\ 1)^{1\ \ 0}\quad (\ 1)^{1\ \ 1}\quad (\ 1)^{1\ \ 1}\quad (\ 1)^{0\ \ 1}\quad (\ 1)^{0\ \ 1}\quad (\ 1)^{0\ \ 0}\quad (\ 1)^{1\ \ 0}\quad (\ 1)^{0\ \ 1}$$

$$2.$$

Proposition. *Given the binary vector*

$$t\ \ c\ \ \ \sum_{i\ 1}^{q} d_i v_i,$$

with $d\ \ (d_1 d_2 \ldots d_q)^T$ *a binary q-tuple and* v_i *the i-th row of* B_q, *then* $R(d)$ *is the number of* 0*s minus the number of* 1*s in t. This proposition is the basis of a decoding scheme for the rst-order Reed-Muller codes,* $R(1,r)$, *discussed in Section 4.10.*

There is a faster method to compute the Walsh-Hadamard transform by using matrices $M_{2^q}^{(i)}\ \ I_{2^{q\ i}}\ \ H(2)\ \ I_{2^{i\ 1}}$. If q is a positive integer the *fast Walsh-Hadamard transform* allows a speedup of $(2^{q\ 1}\ \ 1)/3q$ over the classical Walsh-Hadamard transform.

Proposition. *If q is a positive integer and* $M_{2^q}^{(i)}\ \ I_{2^{q\ i}}\ \ H(2)\ \ I_{2^{i\ 1}}$, *with* $H(2)$ *the Hadamard matrix,* $I_{2^{q\ i}}$ *the identity matrix of size* $2^{q\ i}\ \ 2^{q\ i}$ *and* $I_{2^{i\ 1}}$ *the identity matrix of size* $2^{i\ 1}\ \ 2^{i\ 1}$, *then*

$$H(2^q)\ \ M_{2^q}^{(1)} M_{2^q}^{(2)} \ldots M_{2^q}^{(q)}.$$

We can prove this equality by induction. For $q\ \ 1$ we need to prove that $H(2)\ \ M_2^{(1)}$. We denote by $I_{n\ \ n}$ or simply, I_n, the $n\ \ n$ identity matrix. By definition:

$$M_2^{(1)}\ \ I_{2^{1\ 1}}\ \ H(2)\ \ I_{2^{1\ 1}}\ \ H(2).$$

Assume that this is true for $q\ \ k$ and consider the case, $q\ \ k\ \ 1$. For $q\ \ k$ we have:

$$H(2^k)\ \ M_{2^k}^{(1)} M_{2^k}^{(2)} \ldots M_{2^k}^{(k)}.$$

Now for $q\ \ k\ \ 1$ we have to prove that

$$H(2^{k\ 1})\ \ M_{2^{k\ 1}}^{(1)} M_{2^{k\ 1}}^{(2)} \ldots M_{2^{k\ 1}}^{(k)} M_{2^{k\ 1}}^{(k\ 1)}.$$

Then,

$$M_{2^{k\ 1}}^{(i)}\ \ I_{2^{k\ 1\ i}}\ \ H(2)\ \ I_{2^{i\ 1}}\ \ I_2\ \ I_{2^{k\ i}}\ \ H(2)\ \ I_{2^{i\ 1}}\ \ I_2\ \ M_{2^k}^{(i)}.$$

Thus,

$$H(2^{k\ \ 1})\quad (I_2\quad M_{2^k}^{(1)})(I_2\quad M_{2^k}^{(2)})\dots(I_2\quad M_{2^k}^{(k)})M_{2^{k}\ 1}^{(k\ \ 1)}.$$

We know that the tensor product of square matrices, V,W,X,Y, has the following property: $(V\quad W)(X\quad Y)\quad VX\quad WY$. Applying this property repeatedly after substituting $H(2^k)$ for $M_{2^k}^{(1)}M_{2^k}^{(2)}\dots M_{2^k}^{(k)}$, we get

$$H(2^{k\ \ 1})\quad (I_2\quad M_{2^k}^{(1)}M_{2^k}^{(2)}\dots M_{2^k}^{(k)})(M_{2^{k}\ 1}^{(k\ \ 1)})\quad (I_2\quad H(2^k))(M_{2^{k}\ 1}^{(k\ \ 1)}).$$

But from the definition of $M_{2^{k}\ 1}^{(k\ \ 1)}$, we see that

$$M_{2^{k}\ 1}^{(k\ \ 1)}\quad H(2)\quad I_{2^k}.$$

Thus,

$$H(2^{k\ \ 1})\quad (I_2\quad H(2^k))(H(2)\quad I_{2^k})\quad (I_2H(2))\quad (H(2^k)I_{2^k})\quad H(2)\quad H(2^k)\quad H(2^{k\ \ 1}).$$

Compared with the traditional Walsh-Hadamard transform, the fast Walsh-Hadamard transform allows a speedup, $S_{FHT/HT}$, given by:

$$S_{FHT/HT}\quad \frac{2^{q\ \ 1}\quad 1}{3q}.$$

For example, for $q\quad 16$, we have $S_{FHT/HT}\quad 2^{17}/48\quad 128\,000/48\quad 4200$.

Example. *Given* $R(c)\quad (\ 1,\ \ 1,\ \ 1,\ \ 1,\ \ 1,\ \ 1,\ \ 1,\ \ 1)$, *let us compute* $R\quad RH$ *with* $H\quad H(2^3)\quad M_8^1M_8^2M_8^3$. First, we calculate M_8^1, M_8^2, and M_8^3 as follows:

$$M_8^1\quad I_4\quad H(2)\quad I_1\quad \begin{matrix}1&0&0&0\\0&1&0&0\\0&0&1&0\\0&0&0&1\end{matrix}\quad \begin{matrix}1&1\\1&1\end{matrix}.$$

Thus,

$$M_8^1\quad \begin{matrix}1&1&0&0&0&0&0&0\\1&1&0&0&0&0&0&0\\0&0&1&1&0&0&0&0\\0&0&1&1&0&0&0&0\\0&0&0&0&1&1&0&0\\0&0&0&0&1&1&0&0\\0&0&0&0&0&0&1&1\\0&0&0&0&0&0&1&1\end{matrix}.$$

Then,

$$M_8^2 \quad I_2 \quad H(2) \quad I_2 \qquad \begin{matrix} 1 & 0 \\ 0 & 1 \end{matrix} \qquad \begin{matrix} 1 & 1 \\ 1 & 1 \end{matrix} \qquad \begin{matrix} 1 & 0 \\ 0 & 1 \end{matrix} \qquad \begin{matrix} 1 & 1 & 0 & 0 \\ 1 & 1 & 0 & 0 \\ 0 & 0 & 1 & 1 \\ 0 & 0 & 1 & 1 \end{matrix} \qquad \begin{matrix} 1 & 0 \\ 0 & 1 \end{matrix}.$$

Thus,

$$M_8^2 \qquad \begin{matrix} 1 & 0 & 1 & 0 & 0 & 0 & 0 & 0 \\ 0 & 1 & 0 & 1 & 0 & 0 & 0 & 0 \\ 1 & 0 & 1 & 0 & 0 & 0 & 0 & 0 \\ 0 & 1 & 0 & 1 & 0 & 0 & 0 & 0 \\ 0 & 0 & 0 & 0 & 1 & 0 & 1 & 0 \\ 0 & 0 & 0 & 0 & 0 & 1 & 0 & 1 \\ 0 & 0 & 0 & 0 & 1 & 0 & 1 & 0 \\ 0 & 0 & 0 & 0 & 0 & 1 & 0 & 1 \end{matrix}.$$

Finally,

$$M_8^3 \quad I_1 \quad H(2) \quad I_4 \qquad \begin{matrix} 1 & 1 \\ 1 & 1 \end{matrix} \qquad \begin{matrix} 1 & 0 & 0 & 0 \\ 0 & 1 & 0 & 0 \\ 0 & 0 & 1 & 0 \\ 0 & 0 & 0 & 1 \end{matrix}.$$

Thus,

$$M_8^3 \qquad \begin{matrix} 1 & 0 & 0 & 0 & 1 & 0 & 0 & 0 \\ 0 & 1 & 0 & 0 & 0 & 1 & 0 & 0 \\ 0 & 0 & 1 & 0 & 0 & 0 & 1 & 0 \\ 0 & 0 & 0 & 1 & 0 & 0 & 0 & 1 \\ 1 & 0 & 0 & 0 & 1 & 0 & 0 & 0 \\ 0 & 1 & 0 & 0 & 0 & 1 & 0 & 0 \\ 0 & 0 & 1 & 0 & 0 & 0 & 1 & 0 \\ 0 & 0 & 0 & 1 & 0 & 0 & 0 & 1 \end{matrix}.$$

Then,

$$RM_8^1 \quad (\ 1,\ 1,\ 1,\ 1,\ 1,\ 1,\ 1,\ 1) \begin{matrix} 1 & 1 & 0 & 0 & 0 & 0 & 0 & 0 \\ 1 & 1 & 0 & 0 & 0 & 0 & 0 & 0 \\ 0 & 0 & 1 & 1 & 0 & 0 & 0 & 0 \\ 0 & 0 & 1 & 1 & 0 & 0 & 0 & 0 \\ 0 & 0 & 0 & 0 & 1 & 1 & 0 & 0 \\ 0 & 0 & 0 & 0 & 1 & 1 & 0 & 0 \\ 0 & 0 & 0 & 0 & 0 & 0 & 1 & 1 \\ 0 & 0 & 0 & 0 & 0 & 0 & 1 & 1 \end{matrix}$$

or

$$RM_8^1 \quad (0,\ 2,\ 2,0,0,\ 2,0,\ 2).$$

Next,

$$(RM_8^1)M_8^2 \quad (0,\ 2,\ 2,0,0,\ 2,0,\ 2) \begin{matrix} 1 & 0 & 1 & 0 & 0 & 0 & 0 & 0 \\ 0 & 1 & 0 & 1 & 0 & 0 & 0 & 0 \\ 1 & 0 & 1 & 0 & 0 & 0 & 0 & 0 \\ 0 & 1 & 0 & 1 & 0 & 0 & 0 & 0 \\ 0 & 0 & 0 & 0 & 1 & 0 & 1 & 0 \\ 0 & 0 & 0 & 0 & 0 & 1 & 0 & 1 \\ 0 & 0 & 0 & 0 & 1 & 0 & 1 & 0 \\ 0 & 0 & 0 & 0 & 0 & 1 & 0 & 1 \end{matrix}$$

or

$$(RM_8^1)M_8^1 \quad (\ 2,\ 2,\ 2,\ 2,0,0,0,\ 4).$$

Finally,

$$(RM_8^1M_8^2)M_8^3 \quad (\ 2,\ 2,\ 2,\ 2,0,0,0,\ 4) \begin{matrix} 1 & 0 & 0 & 0 & 1 & 0 & 0 & 0 \\ 0 & 1 & 0 & 0 & 0 & 1 & 0 & 0 \\ 0 & 0 & 1 & 0 & 0 & 0 & 1 & 0 \\ 0 & 0 & 0 & 1 & 0 & 0 & 0 & 1 \\ 1 & 0 & 0 & 0 & 1 & 0 & 0 & 0 \\ 0 & 1 & 0 & 0 & 0 & 1 & 0 & 0 \\ 0 & 0 & 1 & 0 & 0 & 0 & 1 & 0 \\ 0 & 0 & 0 & 1 & 0 & 0 & 0 & 1 \end{matrix}$$

or

$$R \quad (RM_8^1M_8^2)M_8^3 \quad (\ 2,\ 2,\ 2,\ 2,\ 2,\ 2,\ 2,\ 6).$$

4.10 REED-MULLER CODES

Reed-Muller codes are linear codes with numerous applications in communication; they were introduced in 1954 by Irving S. Reed and D. E. Muller. Muller invented the codes [302] and then Reed proposed the majority logic decoding scheme [352]. In Section 4.8, we defined, H_r, the parity-check matrix of a binary, $(2^r \;\; 1, 2^r \;\; 1 \;\; r)$, Hamming code. H_r has $2^r \;\; 1$ columns, consisting of all nonzero binary r-tuples.

The *rst-order Reed-Muller code*, $R(1,r)$, is a binary code, with the generator matrix

$$\mathrm{G} \quad \begin{array}{cc} \mathbf{1} & 1 \\ \mathrm{H}_r & 0 \end{array} \qquad \begin{array}{c} \mathbf{1} \\ B_r \end{array} ,$$

where $\mathbf{1}$ is a $(2^r \;\; 1)$-tuple of 1s. The generator matrix, G, of the code is constructed by augmenting H_r with a zero column to obtain a matrix, B_r, and then adding as a first row a 2^r-tuple of 1s.

Example. *For $r \;\; 3$, given the parity-check matrix*

$$\mathrm{H}_3 \quad \begin{array}{ccccccc} 1 & 0 & 0 & 1 & 1 & 0 & 1 \\ 0 & 1 & 0 & 1 & 1 & 1 & 0 \\ 0 & 0 & 1 & 1 & 0 & 1 & 1 \end{array} ,$$

we construct first

$$B_3 \quad \begin{array}{cccccccc} 1 & 0 & 0 & 1 & 1 & 0 & 1 & 0 \\ 0 & 1 & 0 & 1 & 1 & 1 & 0 & 0 \\ 0 & 0 & 1 & 1 & 0 & 1 & 1 & 0 \end{array}$$

and then

$$\mathrm{G} \quad \begin{array}{cccccccc} 1 & 1 & 1 & 1 & 1 & 1 & 1 & 1 \\ 1 & 0 & 0 & 1 & 1 & 0 & 1 & 0 \\ 0 & 1 & 0 & 1 & 1 & 1 & 0 & 0 \\ 0 & 0 & 1 & 1 & 0 & 1 & 1 & 0 \end{array} .$$

Proposition. *The code with the generator matrix, G, a rst-order Reed-Muller code, has codewords of length, $n \;\; 2^r$, and the number of information symbols is $k \;\; r \;\; 1$. The code has distance $d \;\; 2^{r \;\; 1}$.*

Proof. Recall that H_r is an $r \;\; n$ matrix, with $n \;\; 2^r \;\; 1$. Then B_r is an $r \;\; n$ matrix, with $n \;\; n \;\; 1 \;\; 2^r$; its first $n \;\; 1$ columns, b_i, $1 \;\; i \;\; n \;\; 1$, are all nonzero binary r-tuples, and its last column is the all-zero r-tuple, $B_r \;\; (b_1 b_2 \ldots b_{n \;\; 1} 0)$. It follows that half of the $n \;\; 2^r$ columns of B_r have even parity and half have odd parity.

The generator matrix, G, of the code, $\mathcal{C}$, is an $(r \;\; 1) \;\; n$ matrix, with the first row an n-tuple of all 1s and the other r rows the rows of B_r. Half of the columns of G have an even parity and half have an odd parity as every column of G adds a 1 as its first element to a column of B_r. Every codeword, $c \quad \mathcal{C}$, is obtained as a product of a message, $(r \quad 1)$-tuple, with the generator matrix G; thus, half of each codeword's 2^r components will be 1s and half will be 0s. Thus, the minimum weight of any codeword is exactly $2^{r \quad 1}$. ■

Proposition. *Assume that the columns of B_r are in the proper order, π_r, and that H is the Hadamard matrix, $H \quad H(2^r)$. Let w be a received n-tuple when a rst-order Reed-Muller code is used for transmission. The following procedure is used to decode w:*

1. *Compute R and R.*
2. *Find a component, $R(u)$, whose magnitude is maximum. Let $u \quad (u_1, u_2, \ldots, u_r)^T$ be a binary r-tuple, and let b_i be the i-th row of B_r.*
3. *If $R(u) > 0$, then decode w as $\quad_{i \quad 1}^{r} u_i b_i$.*
4. *If $R(u) \quad 0$, then decode w as $I \quad \quad_{i \quad 1}^{r} u_i b_i$, with I, an n-tuple of 1s.*

A justification of this decoding procedure can be found in most standard texts on error-correcting codes such as [435].

Example. Consider the B_3 matrix whose columns are π_3; in other words, they are the binary expression of all 3-tuples in proper order. For example, the fourth column is (0 1 1), the binary expression of integer 3; the seventh column is (1 1 0), the binary expression of integer 6; and so on. We assume that the leftmost bit is the least significant one:

$$B_3 \quad \begin{matrix} 0 & 1 & 0 & 1 & 0 & 1 & 0 & 1 \\ 0 & 0 & 1 & 1 & 0 & 0 & 1 & 1 \\ 0 & 0 & 0 & 0 & 1 & 1 & 1 & 1 \end{matrix} \qquad \begin{matrix} b_1 \\ b_2 \\ b_3 \end{matrix}.$$

The corresponding generator matrix of the Reed-Muller code is

$$G \quad \begin{matrix} 1 & 1 & 1 & 1 & 1 & 1 & 1 & 1 \\ 0 & 1 & 0 & 1 & 0 & 1 & 0 & 1 \\ 0 & 0 & 1 & 1 & 0 & 0 & 1 & 1 \\ 0 & 0 & 0 & 0 & 1 & 1 & 1 & 1 \end{matrix}.$$

Assume that we receive $w \quad (0\,1\,1\,1\,1\,0\,0\,1)$. Then,

$$R \quad (\ 1,\ \ 1,\ \ 1,\ \ 1,\ \ 1,\ \ 1,\ \ 1,\ \ 1).$$

The Hadamard transform of R is (see, for example, [283])

$$R \quad RH_8 \quad (\ 2,\ \ 2,\ \ 2,\ \ 2,\ \ 2,\ \ 2,\ \ 2,\ \ 4),$$

with

$$H_8 \quad \begin{matrix} 1 & 1 & 1 & 1 & 1 & 1 & 1 & 1 \\ 1 & 1 & 1 & 1 & 1 & 1 & 1 & 1 \\ 1 & 1 & 1 & 1 & 1 & 1 & 1 & 1 \\ 1 & 1 & 1 & 1 & 1 & 1 & 1 & 1 \\ 1 & 1 & 1 & 1 & 1 & 1 & 1 & 1 \\ 1 & 1 & 1 & 1 & 1 & 1 & 1 & 1 \\ 1 & 1 & 1 & 1 & 1 & 1 & 1 & 1 \\ 1 & 1 & 1 & 1 & 1 & 1 & 1 & 1 \end{matrix} .$$

We see that the largest component of R occurs in position 8, so we set, u, equal to the last column of B_r, $u \quad (1\ 1\ 1)^T$. Since the largest component is negative (it is equal to 4), we decode, w, as

$$v \quad I \quad \sum_{i\ 1}^{3} u_i b_i \quad (1\,1\,1\,1\,1\,1\,1\,1) \quad 1\ b_1 \quad 1\ b_2 \quad 1\ b_3$$

$$v \quad (1\,1\,1\,1\,1\,1\,1\,1) \quad 1\ (0\,1\,0\,1\,0\,1\,0\,1) \quad 1\ (0\,0\,1\,1\,0\,0\,1\,1) \quad 1\ (0\,0\,0\,0\,1\,1\,1\,1)$$
$$(1\,1\,1\,1\,1\,1\,1\,1) \quad (0\,1\,1\,0\,1\,0\,0\,1) \quad (1\,0\,0\,1\,0\,1\,1\,0).$$

An alternative description of Reed-Muller codes is based on manipulation of Boolean monomials. A *monomial* is a product of powers of variables. For example, given the variables, x, y, and z, a monomial is of the form, $x^a y^b z^c$, with a, b, and c non-negative integers, $a, b, c \quad Z$. A *Boolean monomial* in variables, $x_1, x_2, \ldots, x_n$, is an expression of the form

$$p \quad x_1^{r_1} x_2^{r_2} \ldots x_n^{r_n}, \quad \text{with} \quad r_i \quad Z \ , \quad 1 \quad i \quad n.$$

The *reduced form* of p is the result of applying two rules: (1) $x_i x_j \quad x_j x_i$, and (2) $x_i^2 \quad x_i$.

A *Boolean polynomial* is a linear combination of Boolean monomials, where x_i are binary vectors of length, 2^m, and the exponents, r_i, are binary. The degree 0 monomial is a vector, **1**, of 2^m ones; for example, when $m \quad 3$, this monomial is (11111111). The degree 1 monomial, x_1, is a vector of $2^{m \ 1}$ ones followed by $2^{m \ 1}$ zeros. The monomial associated with x_2 is a vector of $2^{m \ 2}$ ones followed by $2^{m \ 2}$ zeros followed by $2^{m \ 2}$ ones followed by $2^{m \ 2}$ zeros. In general, the vector associated with the monomial, x_i, consists of $2^{m \ i}$ ones followed by $2^{m \ i}$ zeros repeated until a vector of length, 2^m, is obtained. For example, when $m \quad 3$, the vector associated with the monomial, $x_1 x_2 x_3$, is

$$(11110000) \quad (11001100) \quad (10101010) \quad (10000000),$$

with " ", the regular bitwise XOR operation. The scalar product of two binary vectors is defined as the sum of the product of individual components; for example,

$$(11110000)\ (11001100) \quad 1 \quad 1 \quad 0 \quad 0 \quad 0 \quad 0 \quad 0 \quad 0 \quad 0.$$

An r-th order Reed-Muller code, $R(r, m)$, is the set of all binary strings of length, $n \quad 2^m$, associated with the Boolean polynomials, $p \quad x_1^{r_1} x_2^{r_2} \ldots x_n^{r_n}$, of degree at most, r. For example, the 0-th order code,

$R(0,m)$, is a repetition of strings of zeros or ones of length, 2^m; the m-th order code, $R(m,m)$, consists of all binary strings of length, 2^m.

The generator matrix of an $R(r,m)$, with $r > 1$, code is formed by adding, $\binom{m}{r}$, rows to the generator matrix of an $R(r-1,m)$ code. For example, the generator matrix of an $R(1,3)$ code is

$$G_{R(1,3)} \quad \begin{array}{c} \mathbf{1} \\ x_1 \\ x_2 \\ x_3 \end{array} \quad \begin{array}{cccccccc} 1 & 1 & 1 & 1 & 1 & 1 & 1 & 1 \\ 1 & 1 & 1 & 1 & 0 & 0 & 0 & 0 \\ 1 & 1 & 0 & 0 & 1 & 1 & 0 & 0 \\ 1 & 0 & 1 & 0 & 1 & 0 & 1 & 0 \end{array}.$$

We add the rows, $x_1 \wedge x_2$, $x_1 \wedge x_3$ and $x_2 \wedge x_3$, to, $G_{R(1,3)}$, to obtain the generator matrix of an $R(2,3)$ code,

$$G_{R(2,3)} \quad \begin{array}{c} \mathbf{1} \\ x_1 \\ x_2 \\ x_3 \\ x_1 \wedge x_2 \\ x_1 \wedge x_3 \\ x_2 \wedge x_3 \end{array} \quad \begin{array}{cccccccc} 1 & 1 & 1 & 1 & 1 & 1 & 1 & 1 \\ 1 & 1 & 1 & 1 & 0 & 0 & 0 & 0 \\ 1 & 1 & 0 & 0 & 1 & 1 & 0 & 0 \\ 1 & 0 & 1 & 0 & 1 & 0 & 1 & 0 \\ 1 & 1 & 0 & 0 & 0 & 0 & 0 & 0 \\ 1 & 0 & 1 & 0 & 0 & 0 & 0 & 0 \\ 1 & 0 & 0 & 0 & 1 & 0 & 0 & 0 \end{array}.$$

The generator matrix of an $R(2,4)$ code is

$$G_{R(2,4)} \quad \begin{array}{c} \mathbf{1} \\ x_1 \\ x_2 \\ x_3 \\ x_4 \\ x_1 \wedge x_2 \\ x_1 \wedge x_3 \\ x_1 \wedge x_4 \\ x_2 \wedge x_3 \\ x_2 \wedge x_4 \\ x_3 \wedge x_4 \end{array} \quad \begin{array}{cccccccccccccccc} 1 & 1 & 1 & 1 & 1 & 1 & 1 & 1 & 1 & 1 & 1 & 1 & 1 & 1 & 1 & 1 \\ 1 & 1 & 1 & 1 & 1 & 1 & 1 & 1 & 0 & 0 & 0 & 0 & 0 & 0 & 0 & 0 \\ 1 & 1 & 1 & 1 & 0 & 0 & 0 & 0 & 1 & 1 & 1 & 1 & 0 & 0 & 0 & 0 \\ 1 & 1 & 0 & 0 & 1 & 1 & 0 & 0 & 1 & 1 & 0 & 0 & 1 & 1 & 0 & 0 \\ 1 & 0 & 1 & 0 & 1 & 0 & 1 & 0 & 1 & 0 & 1 & 0 & 1 & 0 & 1 & 0 \\ 1 & 1 & 1 & 1 & 0 & 0 & 0 & 0 & 0 & 0 & 0 & 0 & 0 & 0 & 0 & 0 \\ 1 & 1 & 0 & 0 & 1 & 1 & 0 & 0 & 0 & 0 & 0 & 0 & 0 & 0 & 0 & 0 \\ 1 & 0 & 1 & 0 & 1 & 0 & 1 & 0 & 0 & 0 & 0 & 0 & 0 & 0 & 0 & 0 \\ 1 & 1 & 0 & 0 & 0 & 0 & 0 & 0 & 1 & 1 & 0 & 0 & 0 & 0 & 0 & 0 \\ 1 & 0 & 1 & 0 & 0 & 0 & 0 & 0 & 1 & 0 & 1 & 0 & 0 & 0 & 0 & 0 \\ 1 & 0 & 0 & 0 & 1 & 0 & 0 & 0 & 1 & 0 & 0 & 0 & 1 & 0 & 0 & 0 \end{array}.$$

The *set of characteristic vectors* of any row of the generator matrix is the set of all monomials in x_i and $\bar{x}_i$, the bitwise reverse of x_i, that are not in the monomial associated with that row. For example, the characteristic vectors of the row before the last of the generator matrix of the $R(2,4)$ code correspond to the vector, $x_2 \wedge x_4$, and are, $x_1 \wedge x_3$, $\bar{x}_1 \wedge x_3$, $x_1 \wedge \bar{x}_3$, and $\bar{x}_1 \wedge \bar{x}_3$. The scalar product of the vectors in this set with all the rows of the generator matrix of the code, except the row they belong to, is equal to zero.

The decoding algorithm for Reed-Muller codes described next is not very efficient, but it is straightforward; it allows us to extract the message, μ, from the n-tuple received, v, and to determine, c, the original codeword sent.

- Step 1. Compute the, $2^{r\quad m}$, characteristic vectors, κ_i^j, for each row of the generator matrix, $G_{R(m,r)}$, except the first row.
- Step 2. Compute the scalar products of v, with all the characteristic vectors, κ_i^j, for rows, $g_i, i > 1$, of the generator matrix. Determine the binary indicator, f_i, for each row, starting with the bottom row; if the majority of the scalar products, $v \quad \kappa_i^j$, are equal to zero, then $f_i \quad 0$, else $f_i \quad 1$. For example, the rows for the $R(1,3)$ code are $g_1 \quad x_1, g_2 \quad x_2$, and $g_3 \quad x_3$.
- Step 3. Compute the binary vector, v, as follows: multiply each, f_i, with the corresponding row, g_i, of the generator matrix and then add them together, $v \qquad {}_i f_i g_i$. For example, $v \quad f_3x_3 \quad f_2x_2 \quad f_1x_1$ for the $R(1,3)$ code.
- Step 4. Add the result of step 3 to the received n-tuple; if $v \quad v$ has a majority of zeros, then assign a zero to the binary indicator of the first row of the generator matrix, $f_0 \quad 0$, else $f_0 \quad 1$. The message sent is $\mu \quad (f_0f_1f_2\ldots)$. For example, $\mu \quad f_0f_1f_2f_3$ for the $R(1,3)$ code.

Example. *Correction of one error for the $R(1,3)$ Reed-Muller code.* The distance of this code is $d \quad 2^{r\quad m} \quad 2^{3\quad 1} \quad 4$; since $d \quad 3$, the code can correct a single error in v. The message sent is $\mu \quad (1110)$; it is encoded as the codeword, $c \quad R(1,3)$, by multiplying the message vector with the generator matrix of the code:

$$c \quad \mu G_{R(1,3)} \quad (1\,1\,1\,0) \begin{matrix} 1 & 1 & 1 & 1 & 1 & 1 & 1 & 1 \\ 1 & 1 & 1 & 1 & 0 & 0 & 0 & 0 \\ 1 & 1 & 0 & 0 & 1 & 1 & 0 & 0 \\ 1 & 0 & 1 & 0 & 1 & 0 & 1 & 0 \end{matrix} \quad (1\,1\,0\,0\,0\,0\,1\,1).$$

The message is affected by an error in the second bit, and the 8-tuple received is $w \quad 10000011$.

Let us now apply the algorithm outlined above. We first compute the characteristic set of the vector, $x_3 \quad 10101010$, corresponding to the last row of the generator matrix; this set includes $x_1 \quad x_2$, $x_1 \quad x_2$, $x_1 \quad x_2$, and $x_1 \quad x_2$. Recall that $x_1 \quad (11110000)$, thus, $x_1 \quad (00001111)$; similarly, $x_2 \quad (11001100)$, so $x_2 \quad (00110011)$. Therefore, $x_1 \quad x_2 \quad (11000000)$, $x1 \quad x_2 \quad (00110000)$, $x_1 \quad x_2 \quad (00001100)$, and $x_1 \quad x_2 \quad (00000011)$. The scalar products of these vectors, with v, are

$$(11000000)\ (10000011) \quad 1;\ (00110000)\ (10000011) \quad 0;$$
$$(00001100)\ (10000011) \quad 0;\ (00000011)\ (10000011) \quad 0.$$

The majority logic leads us to conclude that $f_3 \quad 0$. We continue with the third row of the matrix, $x_2 \quad (11001100)$. The characteristic vectors are $x_1 \quad x_3$, $x_1 \quad x_3$, $x_1 \quad x_3$, and $x_1 \quad x_3$ These vectors are, (10100000), (01010000), (00001010), and (00000101), respectively. The scalar products of these vectors, with v, are

$$(10100000)\ (10000011)\quad 1;\ (01010000)\ (10000011)\quad 0;$$

$$(00001010)\ (10000011)\quad 1;\ (00000101)\ (10000011)\quad 1.$$

We conclude that $f_2 \quad 1$. The second row of the matrix is $x_1 \quad (11110000)$. The characteristic vectors are $x_2 \quad x_3, x_2 \quad x_3, x_2 \quad x_3$, and $x_2 \quad x_3$. These vectors are, (10001000), (00100010), (00001010), and (00010001), respectively. The scalar products of these vectors, with v, are

$$(10001000)\ (10000011)\quad 1;\ (00100010)\ (10000011)\quad 1;$$

$$(01000100)\ (10000011)\quad 0;\ (00010001)\ (10000011)\quad 1.$$

We conclude that $f_1 \quad 1$. Now we construct

$$v \quad f_3x_3 \quad f_2x_2 \quad f_1x_1 \quad 0\ (10101010) \quad 1\ (11001100) \quad 1\ (11110000) \quad (00111100)$$

and then compute

$$v \quad w \quad (00111100) \quad (10000011) \quad (10111111).$$

This string has more ones than zeros; this implies that the coefficient of the first row, f_0, of the generator matrix is 1; thus, the message is

$$\mu \quad (f_0 f_1 f_2 f_3) \quad (1110).$$

The zero in the second position of ($v \quad w$) indicates an error; namely, a zero has been received in error instead of a one.

We note that a Reed-Muller code, $R(r,m)$, exists for any integers, $m \quad 0$ and $0 \quad r \quad m$; the code $R(r,m)$ is a binary $[n,k,d]$ code, with codewords of length $n \quad 2^m$, distance $d \quad 2^{r \quad m}$, and $k(r,m) \quad k(r,m \quad 1) \quad k(r \quad 1,m \quad 1)$. For example, $R(3,4)$ and $R(3,5)$ are $[16,15,2]$ and $[32,26,4]$ Hamming codes, respectively, while $R(0,1)$, $R(1,3)$, and $R(2,5)$ are $[2,1,2]$, $[8,4,4]$, and $[32,16,8]$ self-dual codes, respectively. The dual code of $R(r,m)$ is the code $R(m \quad r \quad 1,m)$. The majority logic decoding used for Reed-Muller codes computes several checksums for each received codeword element. To decode an r-th order code, we have to decode iteratively $r \quad 1$ times before we identify the original codeword.

We continue our discussion with one of the most important classes of linear codes, cyclic codes.

4.11 CYCLIC CODES

There is a one-to-one mapping between vectors from, V_n, an n-dimensional vector space over the finite field, F, and polynomials of degree $(n \quad 1)$, with coefficients from the same field, F. This observation is the basis of another formalism to describe a class of linear codes, the cyclic codes, a formalism based on algebras of polynomials over finite fields.

A cyclic code is characterized by a generator polynomial; encoding and decoding of cyclic codes reduce to algebraic manipulation of polynomials. We start our discussion of cyclic codes with several definitions and then discuss some properties of polynomials with coefficients from a finite field.

Given the vector, $c \quad (c_0 c_1 \ldots c_{n\ 2} c_{n\ 1}) \quad V_n$, a *cyclic shift* is a transformation, $c \quad c$, with $c \quad (c_{n\ 1} c_0 c_1 \ldots c_{n\ 2}) \quad V_n$.

We wish to establish a correspondence between vector spaces and sets of polynomials. Such a correspondence will facilitate the construction of codes with an algebraic structure and will allow us to express properties of linear codes, as well as algorithms for error detection and error correction in terms of the algebra of polynomials. We associate a codeword, a vector, $c \quad V_n$, with a polynomial of degree, $(n \quad 1)$, as follows: the coefficients of the polynomial terms, from low to high order, are the vector elements,

$$c \quad (c_0 c_1 \ldots c_{n\ 1}) \quad V_n \qquad\qquad c(x) \quad c_0 \quad c_1 x \quad c_2 x^2 \qquad\qquad c_{n\ 1} x^{n\ 1}.$$

Let us now introduce cyclic codes. First, we define the concept of a *cyclic subspace of a vector space* as a subspace closed to cyclic shifts.

A *cyclic subspace*, $\mathcal{C}$, of an n-dimensional vector space, V_n, over the field, F, is a set of vectors, $c \quad \mathcal{C}$, with the property

$$c \quad (c_0 c_1 c_2 \ldots c_{n\ 2} c_{n\ 1}) \quad \mathcal{C} \qquad\qquad c \quad (c_{n\ 1} c_0 c_1 \ldots c_{n\ 2}) \quad \mathcal{C}.$$

A linear code, $\mathcal{C}$, is a *cyclic code* if $\mathcal{C}$ is a cyclic subspace of an n-dimensional vector space, V_n, over the field, F.

Consider again the case, $f(x) \quad x^n \quad 1$, and the finite ring of polynomials modulo $f(x)$, $\mathbb{F}[x]/f(x)$. It is easy to prove the following proposition.

Proposition. *Multiplication by x of a polynomial, $c(x)$, in the nite ring of polynomials modulo $f(x)$, $\mathbb{F}[x]/f(x)$, is equivalent with a cyclic shift of the corresponding vector, $c \quad (c_0, c_1, c_2, \ldots, c_{n\ 2} c_{n\ 1})$, into, $c \quad (c_{n\ 1}, c_0, c_1, \ldots, c_{n\ 2})$, with $c, c \quad V_n$.*

Proof. Let $c(x) \quad c_0 \quad c_1 x \quad c_2 x^2 \qquad c_{n\ 2} x^{n\ 2} \quad c_{n\ 1} x^{n\ 1} \bmod f(x)$. Then,

$$c\ (x) \quad x\ c(x) \quad c_0 x \quad c_1 x^2 \quad c_2 x^3 \qquad c_{n\ 2} x^{n\ 1} \quad c_{n\ 1} x^n \bmod f(x).$$

But $x^n \quad 1 \bmod f(x)$; thus,

$$c\ (x) \quad c_{n\ 1} \quad c_0 x \quad c_1 x^2 \quad c_2 x^3 \qquad c_{n\ 2} x^{n\ 1} \bmod f(x).$$

■

Example. *Let $\mathbb{Z}_2$ be the binary eld and $f(x) \quad 1 \quad x^4$. Then*

$$\mathbb{Z}_2[x]/(x^4 \quad 1) \quad [0], [1], [x], [1 \quad x], [x^2], [1 \quad x^2], [x \quad x^2], [1 \quad x \quad x^2], [x^3], [1 \quad x^3],$$
$$[1 \quad x \quad x^3], [1 \quad x^2 \quad x^3], [1 \quad x \quad x^2 \quad x^3], [x \quad x^3], [x^2 \quad x^3], [x \quad x^2 \quad x^3] .$$

The corresponding vectors in $V_4(GF(2))$ are

$$\mathcal{C} \quad 0000,\ 1000,\ 0100,\ 1100,\ 0010,\ 1010,\ 0110,\ 1110,$$
$$0001,\ 1001,\ 1101,\ 1011,\ 1111,\ 0101,\ 0011,\ 0111 .$$

C is a cyclic subspace of $V_4(GF(2))$, and it is closed to cyclic shifts. Indeed, if we apply a cyclic shift we get:

0000	0000,	1000	0100,	0100	0010,	1100	0110,	0010	0001,	1010	0101,
0110	0011,	1110	0111,	0001	1000,	1001	1100,	1101	1110,	1011	1101,
1111	1111,	0101	1010,	0011	1001,	0111	1011.				

We notice that

$$x \quad [0] \quad [0], \quad x \quad [1] \quad [x], \ldots, x \quad [x^2 \quad x^3] \quad [1 \quad x^3], \quad x \quad [x \quad x^2 \quad x^3] \quad [1 \quad x^2 \quad x^3].$$

Therefore, multiplication with x of the polynomials in $\mathbb{Z}_2[x]/(x^4 \quad 1)$ leads to the set

$$[0],[1],[x],[1 \quad x],[x^2],[1 \quad x^2],[x \quad x^2],[1 \quad x \quad x^2],[x^3],[1 \quad x^3],[1 \quad x \quad x^3],$$
$$[1 \quad x^2 \quad x^3],[1 \quad x \quad x^2 \quad x^3],[x \quad x^3],[x^2 \quad x^3],[x \quad x^2 \quad x^3] \; .$$

We can easily verify that the correspondence between the vectors subject to a cyclic shift and polynomials multiplied by x is preserved.

Now we discuss a very important proposition that relates cyclic subspaces of an n-dimensional vector space with ideals of a ring of polynomials. We show that there is a one-to-one correspondence between cyclic subspaces of the vector space, V_n, and ideals of polynomials generated by a polynomial, $g(x)$, in a finite ring of polynomials over, F, modulo the nonzero polynomial, $f(x) \quad x^n \quad 1$, when $g(x)$ divides $f(x)$.

Proposition. *There is a one-to-one correspondence between the cyclic subspaces of an n-dimensional vector space, V_n, over the eld, $F \quad GF(q)$, and the monic polynomials*[4]*, $g(x) \quad \mathbb{F}[x]$, that divide $f(x) \quad x^n \quad 1$, with $\mathbb{F}[x]$, a ring of polynomials. If*

$$f(x) \qquad \overset{q}{\underset{i \quad 1}{}} \; g_i^{a_i}(x),$$

with a_i positive integers and $g_i^{a_i}(x)$, $1 \quad i \quad q$, distinct irreducible monic polynomials, then V_n contains

$$Q \qquad \overset{q}{\underset{i \quad 1}{}} \; (a_i \quad 1)$$

cyclic subspaces. If $g(x)$ is a monic polynomial of degree $(n \quad k)$ and it divides $f(x) \quad x^n \quad 1$, then $g(x)$ is the generator polynomial of a cyclic subspace of V_n. The dimension of this cyclic subspace is equal to k.

This proposition follows from several propositions whose proofs are left to the reader.[5] First, let S be a subspace of an n-dimensional vector space, V_n, over the field F, $S \quad V_n$. Let R be the ring of

---

[4]The coefficient of the highest order term of a monic polynomial is 1: $f(x) \quad x^n \quad a_{n \quad 1}x^{n \quad 1} \qquad a_1 x \quad 1$.

[5]See the exercises at the end of this chapter.

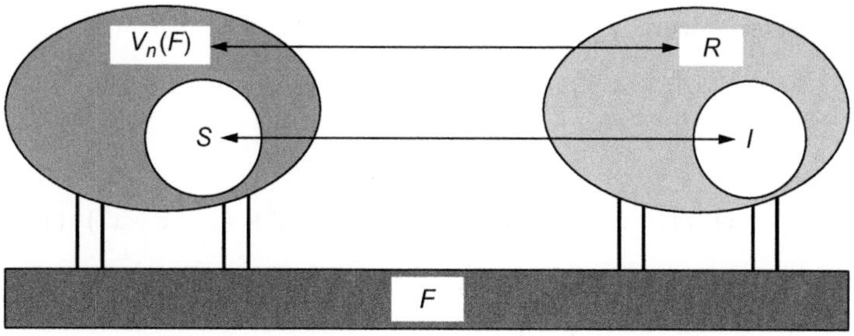

FIGURE 4.6

There is a one-to-one correspondence between the vector space, $V_n(F)$, a q^n-dimensional vector space over the finite field, F, and, R, the ring of polynomials with coefficients from the finite field, F, modulo the polynomial, $f(x) \quad x^n - 1$. There is also a one-to-one correspondence between, S, a cyclic subspace of V_n, and, I, an ideal of the ring, R, generated by the polynomial, $g(x) \quad R$. Moreover, $f(x)$ is a multiple of $g(x), f(x) \quad g(x) \; h(x)$.

polynomials associated with V_n, and let $I \quad R$ be the set of polynomials corresponding to S. We can show that S is a cyclic subspace of V_n if and only if I is an ideal in R (Figure 4.6).

If $f(x) \quad x^n - 1$, R is the ring of equivalence classes of polynomials modulo $f(x)$, with coefficients in the field F, $R \quad \mathbb{F}[x]/(f(x))$, and $g(x)$ is a monic polynomial that divides $f(x)$; then we can show that $g(x)$ is the generator of the ideal of polynomials $I \quad m(x) \; g(x), \; m(x) \quad R$. We can also show that there is a one-to-one correspondence between the cyclic subspaces of V_n and the monic polynomials, $g(x)$, that divide $f(x) \quad x^n - 1$.

Example. We consider several values of n, namely, $5, 10, 15, 20,$ and 25. Show the generator polynomials for the cyclic subspaces of $V_n(GF(2))$ and calculate their dimension.

- $n \quad 5$. Then:

  $$x^5 - 1 \quad (1 \quad x) \; (1 \quad x \quad x^2 \quad x^3 \quad x^4); \quad g_1(x) \quad 1 \quad x \quad \text{and} \quad g_2(x) \quad 1 \quad x \quad x^2 \quad x^3 \quad x^4.$$

  The number of cyclic subspaces is $Q \quad 2 \quad 2 \quad 4$.
  The cyclic subspace generated by g_1 has dimension 4 as

  $$n - k_1 \quad 1 \qquad k_1 \quad n - 1 \quad 4.$$

  The cyclic subspace generated by g_2 has dimension 1 as

  $$n - k_2 \quad 4 \qquad k_2 \quad n - 4 \quad 1.$$

- $n \quad 10$. Then:

  $$x^{10} - 1 \quad (1 \quad x)^2 \; (1 \quad x \quad x^2 \quad x^3 \quad x^4)^2; \quad g_1(x) \quad 1 \quad x \quad \text{and} \quad g_2(x) \quad 1 \quad x \quad x^2 \quad x^3 \quad x^4.$$

  The number of cyclic subspaces is $Q \quad 3 \; 3 \quad 9$.
  The cyclic subspace generated by g_1 has dimension 9 as

  $$n - k_1 \quad 1 \qquad k_1 \quad n - 1 \quad 9.$$

The cyclic subspace generated by g_2 has dimension 6 as

$$n \quad k_2 \quad 4 \qquad k_2 \quad n \quad 4 \quad 6.$$

- n 15. Then:

$$x^{15} \quad 1 \quad (1 \quad x) \; (1 \quad x \quad x^2) \; (1 \quad x \quad x^2 \quad x^3 \quad x^4) \; (1 \quad x \quad x^4) \; (1 \quad x^3 \quad x^4);$$

$$g_1(x) \quad 1 \quad x, \quad g_2(x) \quad 1 \quad x \quad x^2, \quad g_3(x) \quad 1 \quad x \quad x^2 \quad x^3 \quad x^4,$$

$$g_4(x) \quad 1 \quad x \quad x^4, \quad g_5(x) \quad 1 \quad x^3 \quad x^4.$$

The number of cyclic subspaces is $Q \quad 2 \quad 2 \quad 2 \quad 2 \quad 2 \quad 32.$
The cyclic subspace generated by g_1 has dimension 14 as

$$n \quad k_1 \quad 1 \qquad k_1 \quad n \quad 1 \quad 14.$$

The cyclic subspace generated by g_2 has dimension 13 as

$$n \quad k_2 \quad 2 \qquad k_2 \quad n \quad 2 \quad 13.$$

The cyclic subspaces generated by g_3, g_4, and g_5 have dimension 11 as

$$n \quad k_3 \quad 4 \qquad k_3 \quad n \quad 4 \quad 11.$$

- n 20. Then:

$$x^{20} \quad 1 \quad (1 \quad x)^4 \; (1 \quad x \quad x^2 \quad x^3 \quad x^4)^4; \quad g_1(x) \quad 1 \quad x \quad \text{and} \quad g_2(x) \quad 1 \quad x \quad x^2 \quad x^3 \quad x^4.$$

The number of cyclic subspaces is $Q \quad 5 \quad 5 \quad 25.$
The cyclic subspace generated by g_1 has dimension 19 as

$$n \quad k_1 \quad 1 \qquad k_1 \quad n \quad 1 \quad 19.$$

The cyclic subspace generated by g_2 has dimension 16 as

$$n \quad k_2 \quad 4 \qquad k_2 \quad n \quad 4 \quad 16.$$

- n 25. Then:

$$x^{25} \quad 1 \quad (1 \quad x) \; (1 \quad x \quad x^2 \quad x^3 \quad x^4) \; (1 \quad x^5 \quad x^{10} \quad x^{15} \quad x^{20});$$

$$g_1(x) \quad 1 \quad x, \quad g_2(x) \quad 1 \quad x \quad x^2 \quad x^3 \quad x^4, \quad \text{and} \quad g_3(x) \quad 1 \quad x^5 \quad x^{10} \quad x^{15} \quad x^{20}.$$

The number of cyclic subspaces is $Q \quad 2 \quad 2 \quad 2 \quad 8.$
The cyclic subspace generated by g_1 has dimension 24 as

$$n \quad k_1 \quad 1 \qquad k_1 \quad n \quad 1 \quad 24.$$

The cyclic subspace generated by g_2 has dimension 21 as

$$n \quad k_2 \quad 4 \qquad k_2 \quad n \quad 4 \quad 21.$$

The cyclic subspace generated by g_3 has dimension 5 as

$$n \quad k_3 \quad 20 \qquad\qquad k_2 \quad n \quad 20 \quad 5.$$

In summary, a cyclic code is an ideal, I, in the ring, R, of polynomials modulo, $f(x) \quad x^n \quad 1$, over a finite field, $GF(q)$; $f(x)$ is a product of irreducible polynomials, $f(x) \quad \Pi_i(g_i(x))$. An $[n,k]$ cyclic code is generated by a polynomial, $g(x) \quad I$, of degree, $(n \quad k)$; all codewords of C_g are multiples of the generator polynomial, $g(x)$. The generator polynomial of a code, C_g, may be irreducible, $g(x) \quad g_i(x)$, or a product of several irreducible polynomials, $g_i(x)$.

4.12 ENCODING AND DECODING CYCLIC CODES

In this section, we recast familiar concepts from the theory of linear codes, such as generator and parity-check matrices and syndrome in terms of algebras of polynomials and discuss encoding and decoding cyclic codes. Without loss of generality, we only consider vector spaces over the finite field, $GF(2)$, a fancy way of expressing the fact that the coefficients of polynomials are only zeros and ones. The $[n,k]$ linear code, C, is characterized by the generator polynomial, $g(x)$, of degree, $(n \quad k)$:

$$g(x) \quad g_0 \quad g_1x \quad g_2x^2 \qquad\qquad g_{n \ k \ 1}x^{n \ k \ 1} \quad g_{n \ k}x^{n \ k}, \quad g_i \quad 0,1 \ .$$

All operations are performed in an algebra modulo the polynomial, $f(x)$, with

$$f(x) \quad x^n \quad 1,$$

subject to the condition that $g(x)$ divides $f(x)$:

$$f(x) \quad x^n \quad 1 \quad g(x) \ h(x).$$

A message of k bits is now represented by $m(x)$, a polynomial of degree at most, $(k \quad 1)$:

$$m(x) \quad m_0 \quad m_1x \qquad\qquad m_{k \ 2}x^{k \ 2} \quad m_{k \ 1}x^{k \ 1}, \quad m_i \quad 0,1 \ .$$

The message, $m(x)$, is encoded as the polynomial, $c(x)$:

$$c(x) \quad g(x) \ m(x).$$

The degree of the polynomial, $c(x)$, cannot be larger than $(n \quad 1)$:

$$\deg[c(x)] \quad \deg[g(x)] \quad \deg[m(x)] \quad (n \quad k) \quad (k \quad 1) \quad n \quad 1.$$

Thus, when we encode the message polynomial, $m(x)$, into the codeword polynomial, $c(x)$, we do not need to perform a reduction modulo $f(x)$. We can anticipate a very important property of cyclic codes, a direct consequence of the way we encode a message into a polynomial: All codewords are represented by polynomials multiple of the generator polynomial of the code. Thus, *to decide if a received polynomial, $r(x)$, belongs to the code, C, we divide it by the generator polynomial of the code, $g(x)$.*

We now revisit concepts, such as the generator and the parity matrix of a code, dual code, and encoding/decoding algorithms using polynomial representation of codewords and the properties of cyclic subspaces of a vector space.

A *generator matrix*, G_g, of the code, $C[n,k]$, is a $k \quad n$ matrix having as elements the coefficients of the generator polynomial, $g(x)$,

$$G_g \quad \begin{matrix} g(x) \\ x \; g(x) \\ x^2 \; g(x) \\ \vdots \\ x^{k \;\; 1} \; g(x) \end{matrix} \quad \begin{matrix} g_0 & g_1 & \cdots & g_{n \; k \; 1} & g_{n \; k} & 0 & \cdots & 0 \\ 0 & g_0 & g_1 & \cdots & g_{n \; k \; 1} & g_{n \; k} & \cdots & 0 \\ 0 & 0 & g_0 & g_1 & \cdots & g_{n \; k \; 1} & \cdots & 0 \\ \vdots & \vdots & \vdots & \cdots & \vdots & \vdots & \vdots & \\ 0 & 0 & 0 & \cdots & g_0 & \cdots & g_{n \; k \; 1} & g_{n \; k} \end{matrix} \; .$$

Example. *Consider the case, $n \quad 10$, and express*

$$x^{10} \quad 1 \quad (1 \quad x)^2 \; (1 \quad x \quad x^2 \quad x^3 \quad x^4)^2.$$

If we choose $g(x) \quad 1 \quad x \quad x^2 \quad x^3 \quad x^4$ as the generator polynomial of the code, it follows that $\deg[g(x)] \quad n \quad k \quad 4$, and $k \quad 10 \quad 4 \quad 6$. A generator matrix of the code is

$$G \quad \begin{matrix} 1 & 1 & 1 & 1 & 1 & 0 & 0 & 0 & 0 & 0 \\ 0 & 1 & 1 & 1 & 1 & 1 & 0 & 0 & 0 & 0 \\ 0 & 0 & 1 & 1 & 1 & 1 & 1 & 0 & 0 & 0 \\ 0 & 0 & 0 & 1 & 1 & 1 & 1 & 1 & 0 & 0 \\ 0 & 0 & 0 & 0 & 1 & 1 & 1 & 1 & 1 & 0 \\ 0 & 0 & 0 & 0 & 0 & 1 & 1 & 1 & 1 & 1 \end{matrix} \; .$$

The message, $m(x) \quad 1 \quad x \quad x^5$, corresponding to the 6-tuple, (110001), is encoded as

$$\begin{aligned} c(x) \quad & m(x) \;\; g(x) \\ & (1 \quad x \quad x^5) \; (1 \quad x \quad x^2 \quad x^3 \quad x^4) \\ & (1 \quad x \quad x^2 \quad x^3 \quad x^4) \quad (x \quad x^2 \quad x^3 \quad x^4 \quad x^5) \quad (x^5 \quad x^6 \quad x^7 \quad x^8 \quad x^9) \\ & 1 \quad x^6 \quad x^7 \quad x^8 \quad x^9. \end{aligned}$$

Recall that the codeword, $c(x)$, corresponds to the 10-tuple, (1000001111), the vector obtained by multiplying the message vector with the generator matrix:

$$(1\,1\,0\,0\,0\,1) \quad \begin{matrix} 1 & 1 & 1 & 1 & 1 & 0 & 0 & 0 & 0 & 0 \\ 0 & 1 & 1 & 1 & 1 & 1 & 0 & 0 & 0 & 0 \\ 0 & 0 & 1 & 1 & 1 & 1 & 1 & 0 & 0 & 0 \\ 0 & 0 & 0 & 1 & 1 & 1 & 1 & 1 & 0 & 0 \\ 0 & 0 & 0 & 0 & 1 & 1 & 1 & 1 & 1 & 0 \\ 0 & 0 & 0 & 0 & 0 & 1 & 1 & 1 & 1 & 1 \end{matrix} \quad (1\,0\,0\,0\,0\,0\,1\,1\,1\,1).$$

An Alternative Method to Construct the Generator Matrix

Consider a cyclic code with the generator polynomial, $g(x)$; when we divide $x^{n \ k \ i}$, $0 \ i \ k \ 1$ by the generator polynomial, we obtain a quotient, $q^{(i)}(x)$, and a remainder, $r^{(i)}(x)$:

$$x^{n \ k \ i} \quad g(x) \ q^{(i)}(x) \quad r^{(i)}(x), \quad 0 \ i \ k \ 1.$$

The degree of the remainder satisfies the following condition:

$$\deg[r^{(i)}(x)] < \deg[g(x)] \quad n \ k, \quad 0 \ i \ k \ 1.$$

Clearly,

$$x^{n \ k \ i} \quad r^{(i)}(x) \quad g(x) \ q^{(i)}(x), \quad 0 \ i \ k \ 1.$$

Consider a $k \ (n \ k)$ matrix, R_g, with k rows corresponding to the polynomials $[\ r^{(i)}(x)]$, $0 \ i \ k \ 1$,

$$R_g \quad \begin{matrix} r^0(x) \\ r^1(x) \\ \vdots \\ r^{k \ 1}(x) \end{matrix} .$$

Then a generator matrix of the code is $G_g \ [R_g \ I_k]$. We will show that a parity-check matrix of the code is $H_g \ [I_{n \ k} \ R_g^T]$, with I_k being a $k \ k$ identity matrix.

Example. *Alternative construction of the generator matrix of a cyclic code when n 10, k 6, and $g(x)$ 1 x x^2 x^3 x^4.* The following table gives the polynomials, $r^{(i)}(x)$, necessary for the construction of the matrix, R:

i	x^{n-k+i}	$r^{(i)}(x)$
0	$x^4 \ [(1) \ (1 \ x \ x^2 \ x^3 \ x^4)] \ 1 \ x \ x^2 \ x^3$	$1 \ x \ x^2 \ x^3$
1	$x^5 \ [(x \ 1) \ (1 \ x \ x^2 \ x^3 \ x^4)] \ 1$	1
2	$x^6 \ [(x^2 \ x) \ (1 \ x \ x^2 \ x^3 \ x^4)] \ x$	x
3	$x^7 \ [(x^3 \ x^2) \ (1 \ x \ x^2 \ x^3 \ x^4)] \ x^2 \ x^3$	$x^2 \ x^3$
4	$x^8 \ [(x^4 \ x^3) \ (1 \ x \ x^2 \ x^3 \ x^4)] \ x^3$	x^3
5	$x^9 \ [(x^5 \ x^4 \ 1) \ (1 \ x \ x^2 \ x^3 \ x^4)] \ 1 \ x \ x^2 \ x^3$	$1 \ x \ x^2 \ x^3$

Thus,

$$R = \begin{pmatrix} 1 & 1 & 1 & 1 \\ 1 & 0 & 0 & 0 \\ 0 & 1 & 0 & 0 \\ 0 & 0 & 1 & 1 \\ 0 & 0 & 0 & 1 \\ 1 & 1 & 1 & 1 \end{pmatrix}.$$

A generator matrix of the code is $G = [R I_k]$:

$$G = \begin{pmatrix} 1 & 1 & 1 & 1 & 1 & 0 & 0 & 0 & 0 & 0 \\ 1 & 0 & 0 & 0 & 0 & 1 & 0 & 0 & 0 & 0 \\ 0 & 1 & 0 & 0 & 0 & 0 & 1 & 0 & 0 & 0 \\ 0 & 0 & 1 & 1 & 0 & 0 & 0 & 1 & 0 & 0 \\ 0 & 0 & 0 & 1 & 0 & 0 & 0 & 0 & 1 & 0 \\ 1 & 1 & 1 & 1 & 0 & 0 & 0 & 0 & 0 & 1 \end{pmatrix}.$$

The message, $m(x) = 1 + x + x^5$, that corresponds to the 6-tuple, (110001), is encoded as

$$(1\,1\,0\,0\,0\,1) \begin{pmatrix} 1 & 1 & 1 & 1 & 1 & 0 & 0 & 0 & 0 & 0 \\ 1 & 0 & 0 & 0 & 0 & 1 & 0 & 0 & 0 & 0 \\ 0 & 1 & 0 & 0 & 0 & 0 & 1 & 0 & 0 & 0 \\ 0 & 0 & 1 & 1 & 0 & 0 & 0 & 1 & 0 & 0 \\ 0 & 0 & 0 & 1 & 0 & 0 & 0 & 0 & 1 & 0 \\ 1 & 1 & 1 & 1 & 0 & 0 & 0 & 0 & 0 & 1 \end{pmatrix} = (1\,0\,0\,0\,1\,1\,0\,0\,0\,1).$$

Note that the information symbols appear as the last, k, bits of the codeword; this is a general property of this method of constructing the generator matrix of a cyclic code.

Given a polynomial of degree m, $h(x) = \sum_{k=0}^{m} h_k x^k$, $h_m \neq 0$, its *reciprocal*, $\bar{h}(x)$, is the polynomial defined by

$$\bar{h}(x) = \sum_{k=0}^{m} h_{m-k} x^k.$$

The vectors corresponding to the two polynomials, $h(x)$ and $\bar{h}(x)$, are, respectively,

$$h = (h_0 h_1 \ldots h_{m-1} h_m) \quad \text{and} \quad \bar{h} = (h_m h_{m-1} \ldots h_1 h_0).$$

For example, when $m \quad 4$:

$$h(x) \quad 1 \quad x \quad x^4 \qquad h(x) \quad 1 \quad x^3 \quad x^4 \quad \text{or} \quad (11001) \qquad (10011),$$

$$h(x) \quad x^2 \quad x^3 \quad x^4 \qquad h(x) \quad 1 \quad x \quad x^2 \quad \text{or} \quad (00111) \qquad (11100).$$

The Equivalent and the Dual of a Cyclic Code, with the Generator Polynomial $g(x)$

Recall that two linear codes, C, with generator matrix, G, and, C , with generator matrix, G , are said to be equivalent if G G P, with P, a permutation matrix. Recall also that a linear code and its dual are orthogonal subspaces of a vector space and the generator matrix of a code is the parity-check matrix of the dual code.

The polynomial, $h(x)$, defined by the expression, $f(x) \quad x^n \quad 1 \quad g(x) \; h(x)$, is called the *parity-check polynomial* of C and has the following properties: (1) it is a monic polynomial of degree, k, and (2) it is the generator of a cyclic code, C_h. Indeed, any monic polynomial that divides $f(x)$ generates an ideal of polynomials, thus, a cyclic code. The codewords of the two cyclic codes generated by $g(x)$ and $h(x)$, C_g and C_h, respectively, have the following property:

$$c_g(x) \; c_h(x) \quad 0 \quad \text{mod} f(x) \qquad c_g(x) \quad C_g, \qquad c_h(x) \quad C_h.$$

A codeword of a cyclic code is represented by a polynomial multiple of the generator polynomial of the code. Thus:

$$c_g(x) \quad g(x) \; q_g(x) \quad \text{mod} f(x) \quad \text{and} \quad c_h(x) \quad h(x) \; q_h(x) \quad \text{mod} f(x).$$

Thus:

$$c_g(x) \; c_h(x) \quad [g(x) \; q_g(x)] \; [h(x) \; q_h(x)] \quad [g(x) \; h(x)] \; [q_g(x) \; q_h(x)] \quad 0 \quad \text{mod} f(x)$$

as $g(x) \; h(x) \quad 0 \; \text{mod} f(x)$. The equation, $c_g(x) \; c_h(x) \quad 0 \; \text{mod} f(x)$, does not imply that the two codes are orthogonal. Indeed, $C_h \quad C_g$, i.e., the two codes, C_h and C_g , are not each other's dual.

Given that $f(x) \quad x^n \quad 1 \quad g(x) \; h(x)$, we define the codes, C_g, C_h, C_g , and C_h, as follows:

- C_g is the cyclic code generated by the monic polynomial, $g(x)$, of degree k; G_g, the generator matrix of C_g, is a $(k \quad n)$ matrix; and, H_g, the parity-check matrix of C_g, is a $(n \quad k) \quad n$ matrix.
- C_h is the cyclic code generated by the monic polynomial, $h(x)$, of degree k; the generator matrix of C_h is an $(n \quad k) \quad n$ matrix denoted by G_h.
- C_g is the dual of C_g; G_g , the generator matrix of C_g , is an $(n \quad k) \quad n$ matrix.
- C_h is the code generated by, $h(x)$, the reciprocal of the polynomial, $h(x)$; the generator matrix of C_h is an $(n \quad k) \quad n$ matrix denoted by G_h.

Proposition. *The codes, C_g, C_h, C_g , and C_h, have the following properties:*

1. *C_g and C_h are equivalent codes; the generator matrices of the two codes satisfy the relation,*

$$G_h \quad G_g \quad P.$$

2. *The dual of the code, $\mathcal{C}_g$, is the code generated by $\bar{h}(x)$, the reciprocal of $h(x)$: $\mathcal{C}_g^{\perp} \;\; \mathcal{C}_{\bar{h}}$; the parity-check matrix of $\mathcal{C}_g$ is the generator matrix of $\mathcal{C}_{\bar{h}}$:*

$$\mathrm{H}_g \quad \mathrm{G}_{\bar{h}}.$$

The generator matrices of the cyclic codes, $\mathcal{C}_h$ and $\mathcal{C}_{\bar{h}}$, are

$$\mathrm{G}_h \quad \begin{matrix} h(x) \\ x \;\; h(x) \\ x^2 \;\; h(x) \\ \vdots \\ x^{n \;\; k \;\; 1} \;\; h(x) \end{matrix} \quad \text{and} \quad \mathrm{G}_{\bar{h}} \quad \begin{matrix} \bar{h}(x) \\ x \;\; \bar{h}(x) \\ x^2 \;\; \bar{h}(x) \\ \vdots \\ x^{n \;\; k \;\; 1} \;\; \bar{h}(x) \end{matrix} .$$

It is easy to see that the parity-check matrix of the code, $\mathcal{C}_g$, is $\mathrm{H}_g \;\; \mathrm{G}_{\bar{h}} \;\; [I_{n \;\; k} \;\; R_g^T]$, with R_g^T, an $(n \;\; k) \;\; k$ matrix with the rows corresponding to $[\;\; r^{(i)}(x)]$, the remainder when we divide $x^{n \;\; k \;\; i}$ by $g(x)$.

Let two arbitrary codewords, $c_g(x) \;\; \mathcal{C}_g$ and $c_h(x) \;\; \mathcal{C}_h$, and the corresponding vectors be

$$c_g(x) \quad a_0 \quad a_1 x \quad \quad a_{n \;\; 1} x^{n \;\; 1} \quad \quad c_g \quad (a_0 a_1 \ldots a_{n \;\; 1}),$$

$$c_h(x) \quad b_0 \quad b_1 x \quad \quad b_{n \;\; 1} x^{n \;\; 1} \quad \quad c_h \quad (b_0 b_1 \ldots b_{n \;\; 1}).$$

Then,

$$c_g(x) \; c_h(x) \quad \sum_{i \;\; 0}^{n \;\; 1} a_i x^i \quad \sum_{i \;\; 0}^{n \;\; 1} b_i x^i \quad \sum_{i \;\; 0}^{n \;\; 1} d_i x^i \mod (x^n \quad 1).$$

Call $\bar{c}_h \quad (b_{n \;\; 1} b_{n \;\; 2} \ldots b_1 b_0)$ the reciprocal of the vector, $c_h \quad (b_0 b_1 \ldots b_{n \;\; 2} b_{n \;\; 1})$. Then the coefficient of x^0 can be expressed as the inner product, $d_0 \quad c_g \;\; \bar{c}_h$:

$$d_0 \quad c_g \;\; \bar{c}_h \quad (a_0 a_1 \ldots a_{n \;\; 1}) \;\; (b_{n \;\; 1} b_{n \;\; 2} \ldots b_1 b_0) \quad a_0 b_{n \;\; 1} \quad a_1 b_{n \;\; 2} \quad a_2 b_{n \;\; 3} \quad \quad a_{n \;\; 1} b_0.$$

Now compute d_k, $1 \;\; k \;\; n \;\; 1$, and observe that d_k becomes the constant term if we multiply the product, $\;\; {}^{n \;\; 1}_{i \;\; 0} d_i x^i$ by $x^{n \;\; k}$. The multiplication corresponds to the following product of polynomials:

$$c_g(x) \quad x^{n \;\; k} \quad c_h(x) \;\; .$$

Let $c_h^{(k)}$ be the vector obtained by cyclically shifting, $c_h(x)$, to the left by $k \;\; 1$ positions and then switching the last component of the vector with the first one. Then,

$$d_k \quad c_g \;\; c_h^{(k)}.$$

Example. *The procedure to compute the coef cients* c_k, $0 \quad k \quad n$. Consider the case $n \quad 5$ and $f(x) \quad x^5 \quad 1$. Let

$$c_g(x) \quad a_0 \quad a_1x \quad a_2x^2 \quad a_3x^3 \quad a_4x^4 \quad \text{and} \quad c_h(x) \quad b_0 \quad b_1x \quad b_2x^2 \quad b_3x^3 \quad b_4x^4.$$

The corresponding vectors are

$$c_g \quad (a_0a_1a_2a_3a_4) \quad \text{and} \quad c_h \quad (b_0b_1b_2b_3b_4).$$

Then,

$$\begin{aligned} c_g(x) \; c_h(x) \quad & (a_0 \quad a_1x \quad a_2x^2 \quad a_3x^3 \quad a_4x^4) \; (b_0 \quad b_1x \quad b_2x^2 \quad b_3x^3 \quad b_4x^4) \\ & (a_0b_0 \quad a_1b_4 \quad a_2b_3 \quad a_3b_2 \quad a_4b_1) \quad (a_0b_1 \quad a_1b_0 \quad a_2b_4 \quad a_3b_3 \quad a_4b_2)x \\ & (a_0b_2 \quad a_1b_1 \quad a_2b_0 \quad a_3b_4 \quad a_4b_3)x^2 \quad (a_0b_3 \quad a_1b_2 \quad a_2b_1 \quad a_3b_0 \quad a_4b_4)x^3 \\ & (a_0b_4 \quad a_1b_3 \quad a_2b_2 \quad a_3b_1 \quad a_4b_0)x^4 \quad \text{mod} \, f(x) \quad x^5 \quad 1. \end{aligned}$$

The coefficient of the constant term in this expression is given by the inner product of the vectors, c_g and $c_h^{(1)}$:

$$c_g \; c_h^{(1)} \quad (a_0a_1a_2a_3a_4) \; (b_0b_2b_3b_4b_1) \quad a_0b_0 \quad a_1b_2 \quad a_2b_3 \quad a_3b_4 \quad a_4b_1.$$

Indeed, $c_h^{(1)}$ is obtained by a cyclic shift of $c_h \quad (b_0b_1b_2b_3b_4)$ one position to the left to obtain $(b_1b_2b_3b_4b_0)$ and then by switching the last with the first components, leading to $(b_0b_2b_3b_4b_1)$.

Similarly, the coefficient of x^2 in this expression is the inner product of c_g and $c_h^{(3)}$:

$$c_g \; c_h^{(3)} \quad (a_0a_1a_2a_3a_4) \; (b_2b_1b_0b_4b_3) \quad a_0b_2 \quad a_1b_1 \quad a_2b_0 \quad a_3b_4 \quad a_4b_3.$$

Now, $c_h^{(3)}$ is obtained by a cyclic shift of c_h three positions to the left to obtain, $(b_3b_4b_0b_1b_2)$, and then by switching the last with the first components to obtain, $(b_2b_1b_0b_4b_3)$.

Now we consider the general case:

$$c_g(x) \; c_h(x) \quad 0 \quad \text{mod} \; (x^n \quad 1).$$

The coefficients of every power of x in the product of polynomials must be zero. But these coefficients are precisely the scalars, d_k, computed above. Now consider, d, to be any cyclic shift of the vector obtained from, c_h, by reversing its components. Then:

$$c_g \; d \quad 0.$$

Thus, the code, C_h, obtained by such cyclic shifts of c_h followed by reversal of the components is the dual of C_g. It is left as an exercise to prove that C_h and C_h are equivalent codes.

Example. *Consider the polynomial,* $f(x) \quad x^9 \quad 1 \quad (1 \quad x) \; (1 \quad x \quad x^2) \; (1 \quad x^3 \quad x^6)$, *and choose* $g(x) \quad 1 \quad x^3 \quad x^6$. In this case, $n \quad 9$ and $\deg[g(x)] \quad n \quad k \quad 6$; thus, $k \quad 3$. It follows that

$h(x) \quad (1 \quad x)(1 \quad x \quad x^2) \quad (1 \quad x \quad x^2) \quad (x \quad x^2 \quad x^3) \quad 1 \quad x^3$. The vectors corresponding to $g(x)$ and $h(x)$ are $g(x) \quad 1 \quad x^3 \quad x^6 \quad (100100100)$ and $h(x) \quad 1 \quad x^3 \quad (100100000)$. The generator and the parity matrices of the code, C_g, are $k \quad n$ and $(n \quad k) \quad n$ matrices:

$$G_g \quad \begin{array}{r} g(x) \\ x\ g(x) \\ x^2\ g(x) \end{array} \quad \begin{array}{ccccccccc} 1 & 0 & 0 & 1 & 0 & 0 & 1 & 0 & 0 \\ 0 & 1 & 0 & 0 & 1 & 0 & 0 & 1 & 0 \\ 0 & 0 & 1 & 0 & 0 & 1 & 0 & 0 & 1 \end{array} \quad [I_3 A],$$

$$H_g \quad [\ A^T I_{n\ k}] \quad \begin{array}{ccccccccc} 1 & 0 & 0 & 1 & 0 & 0 & 0 & 0 & 0 \\ 0 & 1 & 0 & 0 & 1 & 0 & 0 & 0 & 0 \\ 0 & 0 & 1 & 0 & 0 & 1 & 0 & 0 & 0 \\ 1 & 0 & 0 & 0 & 0 & 0 & 1 & 0 & 0 \\ 0 & 1 & 0 & 0 & 0 & 0 & 0 & 1 & 0 \\ 0 & 0 & 1 & 0 & 0 & 0 & 0 & 0 & 1 \end{array} \quad G_g .$$

The generator matrix of the code, C_h, is an $(n \quad k) \quad n \quad 6 \quad 9$ matrix:

$$G_h \quad \begin{array}{r} h(x) \\ x\ h(x) \\ x^2\ h(x) \\ x^3\ h(x) \\ x^4\ h(x) \\ x^5\ h(x) \end{array} \quad \begin{array}{ccccccccc} 1 & 0 & 0 & 1 & 0 & 0 & 0 & 0 & 0 \\ 0 & 1 & 0 & 0 & 1 & 0 & 0 & 0 & 0 \\ 0 & 0 & 1 & 0 & 0 & 1 & 0 & 0 & 0 \\ 0 & 0 & 0 & 1 & 0 & 0 & 1 & 0 & 0 \\ 0 & 0 & 0 & 0 & 1 & 0 & 0 & 1 & 0 \\ 0 & 0 & 0 & 0 & 0 & 1 & 0 & 0 & 1 \end{array} .$$

As we know, the parity-check matrix, G_h, is the generator matrix of the dual code, C . The vector $h \quad (100100000)$; thus, $h \quad (000001001)$ and G_h is

$$G_h \quad \begin{array}{r} h(x) \\ x\ h(x) \\ x^2\ h(x) \\ x^3\ h(x) \\ x^4\ h(x) \\ x^5\ h(x) \end{array} \quad \begin{array}{ccccccccc} 0 & 0 & 0 & 0 & 0 & 1 & 0 & 0 & 1 \\ 1 & 0 & 0 & 0 & 0 & 0 & 1 & 0 & 0 \\ 0 & 1 & 0 & 0 & 0 & 0 & 0 & 1 & 0 \\ 0 & 0 & 1 & 0 & 0 & 0 & 0 & 0 & 1 \\ 1 & 0 & 0 & 1 & 0 & 0 & 0 & 0 & 0 \\ 0 & 1 & 0 & 0 & 1 & 0 & 0 & 0 & 0 \end{array} .$$

Example. *Consider the* [7,4] *cyclic code, with generator polynomial,* $g(x) \quad 1 \quad x \quad x^3$. In this case, $n \quad 7$ and $k \quad 3$, and $x^n \quad 1 \quad x^7 \quad 1 \quad g(x)\ h(x)$, with $h(x) \quad 1 \quad x \quad x \quad x^4$. A generator matrix of the code is

$$G^a \quad \begin{array}{r} g(x) \\ x\ g(x) \\ x^2\ g(x) \\ x^2\ g(x) \end{array} \quad \begin{array}{ccccccc} 1 & 1 & 0 & 1 & 0 & 0 & 0 \\ 0 & 1 & 1 & 0 & 1 & 0 & 0 \\ 0 & 0 & 1 & 1 & 0 & 1 & 0 \\ 0 & 0 & 0 & 1 & 1 & 0 & 1 \end{array} .$$

Now we compute, $r^{(i)}(x) \quad r_0^i \quad r_1^i x \quad r_2^i x^2$, for $0 \quad i \quad k$, the remainders, when we divide $x^{n \;\; k \;\; i}$ by $g(x)$:

$$\begin{array}{llll} x^3 & 1\ (1 \quad x \quad x^3) \quad (1 \quad x) & & r^{(0)}(x) \quad 1 \quad x, \\ x^4 & (x)\ (1 \quad x \quad x^3) \quad (x \quad x^2) & & r^{(1)}(x) \quad x \quad x^2, \\ x^5 & (1 \quad x^2)\ (1 \quad x \quad x^3) \quad (1 \quad x \quad x^2) & & r^{(2)}(x) \quad 1 \quad x \quad x^2, \\ x^6 & (1 \quad x \quad x^3)\ (1 \quad x \quad x^3) \quad (1 \quad x^2) & & r^{(3)}(x) \quad 1 \quad x^2, \end{array}$$

and construct another generator matrix, $G_g^b \quad [RI_4]$:

$$G_g^b \quad \begin{array}{ccccccc} r_0^{(0)} & r_1^{(0)} & r_2^{(0)} & 1 & 0 & 0 & 0 \\ r_0^{(1)} & r_1^{(1)} & r_2^{(1)} & 0 & 1 & 0 & 0 \\ r_0^{(2)} & r_1^{(2)} & r_2^{(2)} & 0 & 0 & 1 & 0 \\ r_0^{(3)} & r_1^{(3)} & r_2^{(3)} & 0 & 0 & 0 & 1 \end{array} \qquad \begin{array}{ccccccc} 1 & 1 & 0 & 1 & 0 & 0 & 0 \\ 0 & 1 & 1 & 0 & 1 & 0 & 0 \\ 1 & 1 & 1 & 0 & 0 & 1 & 0 \\ 1 & 0 & 1 & 0 & 0 & 0 & 1 \end{array},$$

To construct the parity-check matrix, $H_g \quad [\ \ A^T I_{n \;\; k}]$, of the code, we transform G_g^b to the canonical form $G_g \quad [I_k A]$. The permutation matrix, P, transforms the columns of G_g^b as follows: $1 \quad 5, 2 \quad 6, 3 \quad 7, 4 \quad 1, 5 \quad 2, 6 \quad 3$, and $7 \quad 4$. Then $G_g \quad G_g^b P$:

$$G_g \quad \begin{array}{ccccccc} 1 & 1 & 0 & 1 & 0 & 0 & 0 \\ 0 & 1 & 1 & 0 & 1 & 0 & 0 \\ 1 & 1 & 1 & 0 & 0 & 1 & 0 \\ 1 & 0 & 1 & 0 & 0 & 0 & 1 \end{array} \qquad \begin{array}{ccccccc} 0 & 0 & 0 & 0 & 1 & 0 & 0 \\ 0 & 0 & 0 & 0 & 0 & 1 & 0 \\ 0 & 0 & 0 & 0 & 0 & 0 & 1 \\ 1 & 0 & 0 & 0 & 0 & 0 & 0 \\ 0 & 1 & 0 & 0 & 0 & 0 & 0 \\ 0 & 0 & 1 & 0 & 0 & 0 & 0 \\ 0 & 0 & 0 & 1 & 0 & 0 & 0 \end{array} \qquad \begin{array}{ccccccc} 1 & 0 & 0 & 0 & 1 & 1 & 0 \\ 0 & 1 & 0 & 0 & 0 & 1 & 1 \\ 0 & 0 & 1 & 0 & 1 & 1 & 1 \\ 0 & 0 & 0 & 1 & 1 & 0 & 1 \end{array}.$$

Then,

$$G_g \quad H_g \quad [\ \ A^T I_{n \;\; k}] \qquad \begin{array}{ccccccc} 1 & 0 & 1 & 1 & 1 & 0 & 0 \\ 1 & 1 & 1 & 0 & 0 & 1 & 0 \\ 0 & 1 & 1 & 1 & 0 & 0 & 1 \end{array}.$$

Recall that $h(x) \quad h(x) \quad 1 \quad x \quad x \quad x^4$ and G_h is

$$G_h \quad \begin{array}{c} h(x) \\ x \;\; h(x) \\ x^2 \;\; h(x) \end{array} \qquad \begin{array}{ccccccc} 1 & 1 & 1 & 0 & 1 & 0 & 0 \\ 0 & 1 & 1 & 1 & 0 & 1 & 0 \\ 0 & 0 & 1 & 1 & 1 & 0 & 1 \end{array}.$$

It is easy to see that $G_h \quad G_g P_1$, with P_1, the permutation matrix exchanging the columns of G_g as follows: 1 2, 2 4, 3 3, 4 5, 5 1, 6 6, and 7 7:

$$G_h \quad \begin{pmatrix} 1 & 0 & 1 & 1 & 1 & 0 & 0 \\ 1 & 1 & 1 & 0 & 0 & 1 & 0 \\ 0 & 1 & 1 & 1 & 0 & 0 & 1 \end{pmatrix} \begin{pmatrix} 0 & 1 & 0 & 0 & 0 & 0 & 0 \\ 0 & 0 & 0 & 1 & 0 & 0 & 0 \\ 0 & 0 & 1 & 0 & 0 & 0 & 0 \\ 0 & 0 & 0 & 0 & 1 & 0 & 0 \\ 1 & 0 & 0 & 0 & 0 & 0 & 0 \\ 0 & 0 & 0 & 0 & 0 & 1 & 0 \\ 0 & 0 & 0 & 0 & 0 & 0 & 1 \end{pmatrix} \quad \begin{pmatrix} 1 & 1 & 1 & 0 & 1 & 0 & 0 \\ 0 & 1 & 1 & 1 & 0 & 1 & 0 \\ 0 & 0 & 1 & 1 & 1 & 0 & 1 \end{pmatrix}.$$

Decoding Cyclic Codes

We recast linear codes decoding strategies in terms of algebras of polynomials. Recall that the parity-check matrix, H, of an $[n,k]$ linear code, $\mathcal{C}$, is used to compute the error syndrome, s, given the n-tuple, v: $s^T \quad Hv^T$. Once we compute the syndrome, we are able to determine the error pattern, e; there is a one-to-one correspondence between the two. Indeed, if the codeword, $c \quad \mathcal{C}$, is affected by the error, e, and received as, v, then

$$v \quad c \quad e \qquad s^T \quad Hv^T \quad H(c \quad e)^T \quad He^T \quad \text{as} \quad Hc^T \quad 0, \quad c \quad \mathcal{C}.$$

The following proposition gives a simple algorithm for computing the syndrome polynomial. This algorithm requires the determination of the remainder at the division of polynomials and can be implemented very efficiently using a family of linear circuits called *feedback shift registers*.

Proposition. *The syndrome corresponding to the polynomial,* $v(x) \quad v_0 \quad v_1 x \qquad v_{n \ 1} x^{n \ 1}$, *is the polynomial,* $s(x)$, *satisfying the condition,* $\deg[s(x)] \quad n \quad k \quad 1$, *obtained as the remainder when* $v(x)$ *is divided by* $g(x)$*:*

$$v(x) \quad q(x)g(x) \quad s(x).$$

Proof. We have shown that a parity-check matrix, H_g, of the cyclic code, $\mathcal{C}_g$, with generator polynomial, $g(x)$, is $H_g \quad [I_{n \ k} \quad R_g^T]$; it follows that the columns of H_g correspond to the following polynomials:

$$h_i(x) \quad \begin{cases} x^i & \text{if } 0 \quad i \quad n \quad k \quad 1, \\ r^{(i \ n \ k)}(x) & \text{if } n \quad k \quad i \quad n \quad 1, \end{cases}$$

with $r^{(i \quad n \quad k)}(x)$ defined by $x^{n \quad k \quad i} \quad g(x) \; q^{(i)}(x) \quad r^{(i)}(x)$, $0 \quad i \quad k \quad 1$; the degree of the polynomial, $r^{(i)}(x)$, is less than $n \quad k \quad 1$. It follows that

$$
\begin{aligned}
s(x) \quad & [v_0 \quad v_1 x \qquad v_{n \; k \; 1} x^{n \; k \; 1}] \quad [v_{n \; k} r^0(x) \qquad v_{n \; 1} r^{k \; 1}(x)] \\
& [v_0 \quad v_1 x \qquad v_{n \; k \; 1} x^{n \; k \; 1}] \\
& \quad [v_{n \; k}(x^{n \; k} \quad q^{(0)}(x) g(x)) \qquad v_{n \; 1}(x^{n \; 1} \quad q^{(k \; 1)}(x) g(x))] \\
& [v_0 \quad v_1 x \qquad v_{n \; k \; 1} x^{n \; k \; 1} \quad v_{n \; k} x^{n \; k} \qquad v_{n \; 1} x^{n \; 1}] \\
& \quad [v_{n \; k} q^{(0)}(x) g(x) \qquad v_{n \; 1} q^{(k \; 1)}(x) g(x)] \\
& v(x) \quad [v_{n \; k} q^{(0)}(x) \qquad v_{n \; 1} q^{(k \; 1)}(x)] g(x) \\
& v(x) \quad h(x) g(x).
\end{aligned}
$$

Thus, $v(x) \quad h(x)g(x) \quad s(x)$, with $h(x) \quad v_{n \; k} q^{(0)}(x) \qquad v_{n \; 1} q^{(k \; 1)}(x)$. The uniqueness of the quotient and the remainder of polynomial division guarantees that the syndrome polynomial, $s(x)$, constructed following the definition of the syndrome for linear codes can be used to associate an error polynomial, $e(x)$, with a polynomial, $v(x)$, that may or may not be a codeword; if $s(x) \quad 0$, then $e(x) \quad 0$ and $v(x) \quad c(x)$, as all codewords, $c(x) \quad C_g$, are multiples of the generator polynomial, $g(x)$. The degree of $s(x)$ is $\deg[s(x)] \quad (n \quad k \quad 1)$. ■

An algorithm for decoding binary cyclic codes is discussed next. This algorithm corrects individual errors, as well as bursts of errors, a topic discussed in the next section. Before presenting the algorithm, we introduce the concept of *cyclic run of length, $m \quad n$*, a succession of, m, cyclically consecutive components of an n-tuple. For example, $v \quad (011100101)$ has a cyclic run of two 0s and of three 1s.

Algorithm. *Error trapping*: Given an $[n,k,d]$ cyclic code, C_g, with generator polynomial $g(x)$ and $d \quad 2t \quad 1$, let $s^{(i)}(x) \quad \sum_{j \; 0}^{n \; k \; 1} s_j x^j$ be the syndrome polynomial for $x^i v(x)$, with $v(x) \quad c(x) \quad e(x)$. The error polynomial, $e(x)$, satisfies two conditions: (1) it corresponds to an error pattern of at most, t, errors, $w[e(x)]) \quad t$; and (2) it contains a cyclic run of at least, k, zeros. To decode $v(x)$, we carry out the following steps:

1. Set $i \quad 0$; compute syndrome, $s^{(0)}(x)$, as the remainder at the division of $v(x)$ by $g(x)$: $v(x) \quad q(x) \; g(x) \quad s^{(0)}(x)$.
2. If the Hamming weight of $s^{(i)}(x)$ satisfies the condition, $w[s^{(i)}(x)] \quad t$, then the error polynomial is $e(x) \quad x^{n \; i} s^{(i)}(x) \mod (x^n \quad 1)$ and terminate.
3. Increment i:
   - If $i \quad n$, stop and report that the error is not trappable.
   - Else:

$$
s^{(i)}(x) \quad \begin{cases} x s^{(i \; 1)}(x) & \text{if } \deg[s^{(i \; 1)}(x)] < n \quad k \quad 1, \\ x s^{(i \; 1)}(x) \quad g(x) & \text{if } \deg[s^{(i \; 1)}(x)] \quad n \quad k \quad 1. \end{cases}
$$

4. Go to step 2.

To justify this algorithm, we first show that the syndrome of $xv(x)$ is

$$s^{(1)}(x) = \begin{cases} xs(x) & \text{if} \quad \deg[s(x)] < n-k-1, \\ xs(x) - g(x) & \text{if} \quad \deg[s(x)] = n-k-1. \end{cases}$$

According to the previous proposition, if $s(x)$ is the syndrome corresponding to a polynomial $v(x)$, then the following condition must be satisfied: $\deg[s(x)] \le n-k-1$. It follows that $xs(x)$ is the syndrome of $xv(x)$ if $\deg[s(x)] < n-k-1$; then, $\deg[xs(x)] \le n-k-1$, and the condition is satisfied. If $\deg[s(x)] = n-k-1$, then we express $s(x)$, $g(x)$, and $xs(x)$ as

$$s(x) = \sum_{i=0}^{n-k-1} s_i x^i = a(x) + s_{n-k-1}x^{n-k-1}, \quad \text{with} \quad a(x) = \sum_{i=0}^{n-k-2} s_i x^i, \quad \deg[a(x)] \le n-k-2,$$

$$g(x) = \sum_{i=0}^{n-k} g_i x^i = b(x) + x^{n-k}, \quad \text{with} \quad b(x) = \sum_{i=0}^{n-k-1} g_i x^i, \quad \deg[b(x)] \le n-k-1,$$

$$\begin{aligned} xs(x) &= xa(x) + xs_{n-k-1}x^{n-k-1} = xa(x) + s_{n-k-1}x^{n-k} \\ &= xa(x) + s_{n-k-1}[g(x) - b(x)] = s_{n-k-1}g(x) + [xa(x) - s_{n-k-1}b(x)]. \end{aligned}$$

We see that $d(x) = [xa(x) - s_{n-k-1}b(x)]$ is the remainder when $xs(x)$ is divided by $g(x)$; $\deg[d(x)] < \deg[g(x)] = n-k-1$, thus, $d(x)$ satisfies the condition to be the syndrome associated with $xs(x)$.

Let us now return to the justification of the error trapping algorithm and recall that the error pattern has weight at most, t, and has a run of at least, k, 0s. Then there exists an integer, i, with $0 \le i \le n-1$, such that a cyclic shift of i positions would bring all nonzero components of $e(x)$ within the first, $n-k$, components. This new polynomial, $x^i e(x) \mod (x^n - 1)$, has the syndrome, $s^{(i)}(x)$, and its Hamming weight is $w[s^{(i)}(x)] \le t$; this syndrome is the remainder when we divide the polynomial, $x^i e(x) \mod (x^n - 1)$, by $g(x)$. Indeed, if the parity-check matrix of a linear, $[n,k,d]$, code is $H = [I_{n-k} A_{n-k,k}]$ and the error pattern has the last $n-k$ bits equal to zero, $e = (e_1\, e_2 \ldots e_{n-k}\, 0\, 0 \ldots 0)$, then the syndrome is

$$s^T = He^T = [I_{n-k}A_{n-k,k}]e^T = (e_1\, e_2 \cdots e_{n-k})^T.$$

In this case, the syndrome is the same as the error polynomial.

Example. *Let C_g be the $[7,4,3]$ cyclic code, with the generator polynomial, $g(x) = 1 + x^2 + x^3$.* In this case, $n = 7$, $k = 4$; the code can correct one error as $d = 2t + 1 = 3$, and $t = 1$. We see also that $x^7 - 1 = (x^3 + x^2 + 1)(x^4 + x^3 + x^2 + 1)$.

We apply the error trapping algorithm for the case when we transmit the code polynomial, $c(x) = (1 + x) \cdot g(x)$, and $c(x)$ is affected by the error polynomial, $e(x) = x^3$, and we receive $v(x) = c(x) + e(x)$. Then the codeword is $c(x) = 1 + x^2 + x^3 + x + x^3 + x^4 = 1 + x + x^2 + x^4$, and the received 7-tuple is $v(x) = c(x) + e(x) = 1 + x + x^2 + x^3 + x^4$.

Set $i = 0$ and compute the remainder when we divide $v(x)$ by $g(x)$:

$$v(x) = q(x)g(x) + r^0(x) \Rightarrow x^4 + x^3 + x^2 + x + 1 = x(x^3 + x^2 + 1) + (x^2 + 1).$$

Then $s^{(0)}(x) \quad 1 \quad x^2$ and the Hamming weight, $w[s^{(0)}(x)] \quad 2 > t$; thus, we set $i \quad 1$ and evaluate $\deg[s^{(0)}(x)] \quad 2$, and $n \quad k \quad 1 \quad 7 \quad 4 \quad 1 \quad 2$. We see that $\deg[s^{(0)}(x)] \quad 2$; thus, we compute $s^{(1)}(x)$ as

$$s^{(1)}(x) \quad xs^{(0)}(x) \quad g(x) \quad x(1 \quad x^2) \quad (1 \quad x^2 \quad x^3) \quad 1 \quad x \quad x^2.$$

We see that $w[s^{(1)}(x)] \quad 3 > t$; we have to go to the next iteration, $i \quad 2$. Then $\deg[s^1(x)] \quad 2$; thus, we compute $s^{(2)}(x)$ as

$$s^{(2)}(x) \quad xs^{(1)}(x) \quad g(x) \quad x(1 \quad x \quad x^2) \quad (1 \quad x^2 \quad x^3) \quad 1 \quad x.$$

Once again, $w[s^{(2)}(x)] \quad 2 > t$; we have to go to the next iteration, $i \quad 3$. We evaluate $\deg[s^{(2)}(x)] \quad 1$, and the rule to compute, $s^{(3)}(x)$, is

$$s^{(3)}(x) \quad xs^{(2)}(x) \quad x(1 \quad x) \quad x \quad x^2.$$

Again, $w[s^3(x)] \quad 2 > t$, and we set $i \quad 4$. We evaluate $\deg[s^2(x)] \quad 2$; thus, the rule to compute $s^{(4)}(x)$ is

$$s^{(4)}(x) \quad xs^{(3)}(x) \quad g(x) \quad x(x \quad x^2) \quad (1 \quad x^2 \quad x^3) \quad 1.$$

We have now reached the value of the Hamming weight, which allows us to identify the error pattern: $w[s^{(4)}(x)] \quad 1$. Thus, $e(x) \quad x^{n \quad i} s^{(4)}(x) \mod (x^n \quad 1)$ and

$$e(x) \quad x^{7 \quad 4} \mod (x^7 \quad 1) \quad x^3.$$

Example. *Cyclic redundancy check.* Cyclic codes are used for error detection by link layer protocols where the data transport unit is called a frame. It is easy to show that *given $p(x)$, a primitive polynomial of degree m, the cyclic binary code with the generator polynomial, $g(x) \quad (x \quad 1)p(x)$, is an $[n,k,d]$ code, with*

$$n \quad 2^m \quad 1, \quad k \quad 2^m \quad m \quad 2, \quad \text{and} \quad d \quad 4.$$

The $n \quad k \quad m \quad 1$ parity-check bits are called *cyclic redundancy check (CRC)* bits and are included in the trailer header of the frame. A cyclic code with $g(x)$ of this form can correct one error. The probability that two or more errors were undetected and the receiver accepted the frame, P_{underr}, is equal to the probability that the, $(m \quad 1)$, parity-check bits are all zero when $e \quad 2$ (i.e., two or more errors occur):

$$P_{underr} \quad \text{Prob}(e \quad 2) \quad 2^{\quad (m \quad 1)}.$$

When m is large, undetected errors have a vanishing probability.

Two standard CRC polynomials, CRC-ANSI and CRC-ITU-T, are used [227]:

$$x^{16} \quad x^{15} \quad x^2 \quad 1 \quad (x \quad 1)(x^{15} \quad x \quad 1)$$

and

$$x^{16} \quad x^{12} \quad x^5 \quad 1 \quad (x \quad 1)(x^{15} \quad x^{14} \quad x^{13} \quad x^{12} \quad x^4 \quad x^3 \quad x^2 \quad x \quad 1).$$

In this case, $2^{\ 17} < 10^{\ 5}$ and the probability of undetected errors is less than $10^{\ 3}$ if the probability of two or more errors is less than $10^{\ 2}$.

The question we address next is if there are bounds restricting our ability to design "good" cyclic codes.

4.13 THE MINIMUM DISTANCE OF A CYCLIC CODE AND THE BCH BOUND

We wish to establish a bound on the distance of a cyclic code, a question answered by the so-called Bose-Chaudhuri-Hocquenghem (BCH) bound. We recall a theorem discussed in Section 4.6 that states that every set of $(d \quad 1)$ columns of H, the parity-check matrix of a linear code, are linearly independent if and only if the code has a distance of at least d. Thus, we should concentrate on the properties of the parity-check matrix of a cyclic code. First, we introduce Vandermonde matrices and then review some properties of Vandermonde matrices, with elements from a finite field, $GF(q)$ [227].

A *Vandermonde matrix of order, n,* with elements, $x_1, x_2, \ldots, x_n \quad GF(q)$, is of the form

$$A_n \quad \begin{array}{cccccc} 1 & x_1 & (x_1)^2 & (x_1)^3 & \ldots & (x_1)^{n \quad 1} \\ 1 & x_2 & (x_2)^2 & (x_2)^3 & \ldots & (x_2)^{n \quad 1} \\ \vdots & \vdots & \vdots & \vdots & & \vdots \\ 1 & x_n & (x_n)^2 & (x_n)^3 & \ldots & (x_n)^{n \quad 1} \end{array} .$$

Proposition. *Let β be a primitive element of order, n, of the nite eld, $GF(q)$; thus, $\beta^n \quad 1$ and $\beta^i \quad 1, 0 \quad i < n$. Let $x_j \quad \beta^{j \quad 1}$. Then,*

$$C \quad BA^T \quad 0 \ \textit{and} \ D \quad AB^T \quad 0,$$

with

$$A \quad \begin{array}{cccccc} 1 & 1 & 1 & \ldots & 1 & 1 \\ x_1 & x_2 & x_3 & \ldots & x_{n \quad 1} & x_n \\ (x_1)^2 & (x_2)^2 & (x_3)^2 & \ldots & (x_{n \quad 1})^2 & (x_n)^2 \\ \vdots & \vdots & \vdots & & \vdots & \vdots \\ (x_1)^{\alpha \quad 1} & (x_2)^{\alpha \quad 1} & (x_3)^{\alpha \quad 1} & \ldots & (x_{n \quad 1})^{\alpha \quad 1} & (x_n)^{\alpha \quad 1} \\ (x_1)^{\alpha} & (x_2)^{\alpha} & (x_3)^{\alpha} & \ldots & (x_{n \quad 1})^{\alpha} & (x_n)^{\alpha} \end{array} ,$$

$$B \quad \begin{array}{cccccc} x_1 & x_2 & x_3 & \ldots & x_{n \quad 1} & x_n \\ (x_1)^2 & (x_2)^2 & (x_3)^2 & \ldots & (x_{n \quad 1})^2 & (x_n)^2 \\ (x_1)^3 & (x_2)^3 & (x_3)^3 & \ldots & (x_{n \quad 1})^3 & (x_n)^3 \\ \vdots & \vdots & \vdots & & \vdots & \vdots \\ (x_1)^{s \quad 1} & (x_2)^{s \quad 1} & (x_3)^{s \quad 1} & \ldots & (x_{n \quad 1})^{s \quad 1} & (x_n)^{s \quad 1} \\ (x_1)^{s} & (x_2)^{s} & (x_3)^{s} & \ldots & (x_{n \quad 1})^{s} & (x_n)^{s} \end{array} ,$$

and

$$s \quad \alpha \quad 1 \quad n.$$

Proof. If $C = [c_{ik}]$, $A = [a_{kj}]$, and $B = [b_{ij}]$, with $a_{kj} = (x_j)^{k-1}$ and $b_{ij} = (x_j)^i$, then

$$c_{ik} = \sum_{j=1}^{n} b_{ij}a_{jk} = \sum_{j=1}^{n} (x_j)^i (x_j)^{k-1} = \sum_{j=1}^{n} (x_j)^{i+k-1} = \sum_{j=1}^{n} (\beta^{i+k-1})^{j-1}.$$

But $\beta^{i+k-1} \neq 1$ and

$$i+k-1 \leq s+a \leq n-1 \Rightarrow \beta^{i+k-1} \neq 1.$$

It follows immediately that the sum of the geometric series is

$$c_{ik} = \frac{(\beta^{i+k-1})^n - 1}{(\beta^{i+k-1}) - 1} = \frac{(\beta^n)^{i+k-1} - 1}{(\beta^{i+k-1}) - 1} = 0$$

because

$$(\beta^n)^{i+k-1} = 1.$$

The proof that $D = AB^T = 0$ follows a similar path. ■

Proposition. *Call* $D_n = \det[A_n]$. *Then:*

$$D_n = \prod_{i>j}^{n} (x_i - x_j)$$
$$= (x_2 - x_1)(x_3 - x_2)(x_3 - x_1)\ldots(x_n - x_{n-1})(x_n - x_{n-2})\ldots(x_n - x_1).$$

Proof. We prove this equality by induction:

1. It is easy to see that $D_2 = x_2 - x_1$.
2. Assume that

$$D_{n-1} = \prod_{i>j}^{n-1} (x_i - x_j)$$
$$= (x_2 - x_1)(x_3 - x_2)(x_3 - x_1)\ldots(x_{n-1} - x_{n-3})(x_{n-1} - x_{n-2})\ldots(x_{n-1} - x_1).$$

We can consider, D_n, a polynomial of degree, $(n-1)$, in x_n, with $x_1, x_2, \ldots, x_{n-1}$ as zeros. The coefficient of $(x_n)^{n-1}$ is D_{n-1}. Thus:

$$D_n = \prod_{i>j}^{n} (x_i - x_j) = \prod_{i=1}^{n-1} (x_n - x_i) D_{n-1}.$$

■

Proposition. *The BCH bound. Let $g(x)$ be the generator polynomial of the $[n,k]$ cyclic code, C, over $GF(q)$. Let $\beta \in GF(q)$ be a primitive element of order n. If $g(x)$ has among its zeros, $(d-1)$, elements of $GF(q)$, $\beta^{\alpha}, \beta^{\alpha+1}, \ldots, \beta^{\alpha+d-2}$, then d is the distance of the cyclic code.*

Proof. Recall that the minimum distance between any pair of codewords called the distance of a linear code is equal to the number of independent column vectors of, H, the parity-check matrix of the code. The parity-check matrix of code, C, is

$$\mathrm{H} \quad \begin{matrix} 1 & (\beta)^{\alpha} & (\beta)^{2\alpha} & (\beta)^{3\alpha} & \cdots & (\beta)^{(n\ \ 1)\alpha} \\ 1 & (\beta)^{\alpha\ \ 1} & (\beta)^{2(\alpha\ \ 1)} & (\beta)^{3(\alpha\ \ 1)} & \cdots & (\beta)^{(n\ \ 1)(\alpha\ \ 1)} \\ \vdots & \vdots & \vdots & \vdots & & \vdots \\ 1 & (\beta)^{(\alpha\ \ d\ \ 2)} & (\beta)^{2(\alpha\ \ d\ \ 2)} & (\beta)^{3(\alpha\ \ d\ \ 2)} & \cdots & (\beta)^{(n\ \ 1)(\alpha\ \ d\ \ 2)} \end{matrix} .$$

The determinant of, $(d\ \ 1)$, columns of the parity-check matrix, H, is

$$\det[\mathrm{H}] \quad \begin{vmatrix} (\beta)^{\alpha i_1} & (\beta)^{\alpha i_2} & (\beta)^{\alpha i_3} & \cdots & (\beta)^{\alpha i_{d\ 1}} \\ (\beta)^{(\alpha\ \ 1)i_1} & (\beta)^{(\alpha\ \ 1)i_2} & (\beta)^{(\alpha\ \ 1)i_3} & \cdots & (\beta)^{(\alpha\ \ 1)i_{d\ 1}} \\ \vdots & \vdots & \vdots & & \vdots \\ (\beta)^{(\alpha\ \ d\ \ 2)i_1} & (\beta)^{(\alpha\ \ d\ \ 2)i_2} & (\beta)^{(\alpha\ \ d\ \ 2)i_3} & \cdots & (\beta)^{(\alpha\ \ d\ \ 2)i_{d\ 1}} \end{vmatrix} .$$

But $x_j \quad \beta^{i_j}$; thus,

$$\det[\mathrm{H}] \quad \begin{vmatrix} (x_1)^{\alpha} & (x_2)^{\alpha} & (x_3)^{\alpha} & \cdots & (x_{d\ \ 1})^{\alpha} \\ (x_1)^{(\alpha\ \ 1)} & (x_2)^{(\alpha\ \ 1)} & (x_3)^{(\alpha\ \ 1)} & \cdots & (x_{d\ \ 1})^{(\alpha\ \ 1)} \\ \vdots & \vdots & \vdots & & \vdots \\ (x_1)^{(\alpha\ \ d\ \ 2)} & (x_2)^{(\alpha\ \ d\ \ 2)} & (x_3)^{(\alpha\ \ d\ \ 2)} & \cdots & (x_{d\ \ 1})^{(\alpha\ \ d\ \ 2)} \end{vmatrix}$$

or

$$\det[\mathrm{H}] \quad x_1^{\alpha} x_2^{\alpha} x_3^{\alpha} \ldots x_{d\ \ 1}^{\alpha} \begin{vmatrix} 1 & 1 & 1 & \cdots & 1 \\ x_1 & x_2 & x_3 & \cdots & x_{d\ \ 1} \\ \vdots & \vdots & \vdots & & \vdots \\ (x_1)^{(d\ \ 2)} & (x_2)^{(d\ \ 2)} & (x_3)^{(d\ \ 2)} & \cdots & (x_{d\ \ 1})^{(d\ \ 2)} \end{vmatrix} .$$

Finally, if $x_i \quad x_j$

$$\det[\mathrm{H}] \quad x_1^{\alpha} x_2^{\alpha} x_3^{\alpha} \ldots x_{d\ \ 1}^{\alpha} \quad \overset{d\ \ 1}{\underset{i>j}{}} \ (x_i \quad x_j) \quad 0.$$

This proves that $(d\ \ 1)$ columns of the parity-check matrix are linearly independent; thus, the minimum distance of the code is d. ■

4.14 BURST ERRORS AND INTERLEAVING

So far, we have assumed that errors are uncorrelated, and we have only been concerned with the construction of error-correcting codes able to handle errors with a random distribution. Yet, physical systems may be affected by errors concentrated in a short time interval, or affecting a number of adjacent bits; communication channels, as well as memory and information processing systems, may experience correlated errors. A burst of noise triggered by an electrostatic phenomenon, a speck of dust on a recording medium, a slight variation of the voltage of a power supply, may affect a contiguous stream of bits transmitted, stored, or being processed. Codes designed to handle *bursts* of errors are called *burst error correcting (BEC)* codes. We start this section with a few definitions, then present several properties and discuss several classes of BEC codes.

An *error burst* of length, l, is an n-tuple whose nonzero symbols are confined to a span of no fewer than, l, symbols. A *wraparound burst* of length l is a cyclic shift of a burst of length l. A burst is specified by its *location*, the least significant digit of the burst, and the *burst pattern*. If i is the location and $b(x)$ is the pattern, $(b_0 \quad 0)$, then the error burst, $e(x)$, and the wraparound burst are

$$e(x) \quad x^i\, b(x), \quad e(x) \quad x^i\, b(x) \quad \bmod\,(x^n \quad 1).$$

Example. *Assume* $n \quad 10$. Then 0001101000 is a burst of length 4, the location is 3, and the pattern is $1 \quad x \quad x^3$; 0100000011 is a wraparound burst of length 4, location 8, and the same pattern, $1 \quad x \quad x^3$.

Given an $[n,k]$ linear code capable of correcting error bursts of length, l, its *burst error correcting ef ciency* is defined as

$$\eta \quad \frac{2l}{(n \quad k)}.$$

The following two propositions show that cyclic codes and block codes can be used for burst error detection and correction.

Proposition. *A cyclic code with generator polynomial,* $g(x)$, *of degree,* $r \quad n \quad k$, *can detect all burst errors of length* $l \quad r$.

Proof. Call $c(x)$ the polynomial representing the codeword sent and $e(x)$ the error polynomial. We are able to detect an error when $v(x) \quad c(x) \quad e(x)$ is not a codeword. This condition requires that $e(x)$ is not a codeword; thus, it is not a multiple of the generator polynomial, $g(x) \quad g_0 \quad g_1 x \quad g_2 x^2 \quad x^r$. Assume that the error burst starts at position, i, and affects, $l \quad r$, consecutive bits; thus, $e(x) \quad x^i\, b(x)$. We see immediately that $g(x)$ and x^i are relatively prime because $g_0 \quad 0$. The length of the burst is $l \quad r$, and this implies that $\deg[b(x)] \quad r \quad 1 < \deg[g(x)] \quad r$; thus, $b(x)$ is not a multiple of $g(x)$. It follows immediately that $e(x) \quad x^i\, b(x)$ is not a multiple of $g(x)$, and it is not a codeword.

A cyclic code will detect most burst errors of length, $l > r$; indeed, the probability that a binary cyclic code with generator polynomial of degree, r, will detect error bursts of length, $l > r$, is equal to $1 \quad 2^{\ r}$, and the probability of detecting error bursts of length l is $1 \quad 2^{\ r \ 1}$. ■

Proposition. *A block code can correct all bursts of length,* $\quad l$, *if and only if no two codewords differ by the sum of two bursts of length* $\quad l$.

An equivalent statement is that a linear block code can correct all bursts of length, l, if and only if no codeword is the sum of two bursts of length l.

Proposition. *The Rieger bound: The BEC ability, l, of an $[n,k]$ linear block code satis es the condition*

$$l \quad \frac{n \quad k}{2}.$$

This inequality is an immediate consequence of the fact that all polynomials of degree smaller than l must have distinct syndromes; otherwise, the difference of two polynomials would be a codeword and, at the same time, the sum of two bursts of length l.

We now examine several classes of BEC codes and their BEC efficiency. The Rieger bound imposes that linear codes have error-correcting efficiency, $\eta \quad 1$. On the other hand, the Reed-Solomon codes discussed in Section 4.15 have $\eta \quad 1$. Reed-Solomon codes use, $2t$, check symbols and can correct, t, random errors and, thus, bursts of length t. The Fire codes have an error correcting efficiency of $2/3$.

Cyclic codes have some desirable properties for burst error correction. Indeed, if symbol, j, of cyclic shift, $r^{(i)}(x)$, can be corrected, then the same error magnitude can be used to correct the symbol, $(j \quad i)$ mod n of $r(x)$. To decode a cyclic code, we use a feedback shift register implementing the generator polynomial of the code, $g(x)$. If $r(x)$ is the received polynomial, we use the following algorithm: (1) compute $s(x) \quad r(x) \mod g(x)$; (2) shift syndrome register with feedback until $s^{[i]}(x)$ contains a syndrome with a known error, $e(x)$; and (3) decode $r(x)$ using $e^{(\quad i)}(x)$. For example, $g(x) \quad x^4 \quad x \quad 1$ is a primitive polynomial over $GF(2)$, which generates a $(15,11)$ cyclic Hamming code. The code with $g(x) \quad (x \quad 1)(x^4 \quad x^3 \quad 1)$ can correct any bursts of length 2.

Fire codes are cyclic codes over $GF(q)$, with a generator polynomial, $g(x) \quad (x^{2t \quad 1} \quad 1)p(x)$, such that (1) $p(x)$ is a prime polynomial of degree, m, and order, e, over $GF(q)$, (2) $m \quad t$, and (3) $p(x)$ does not divide $(x^{2t \quad 1} \quad 1)$. The block length of a Fire code, n, is the order of the generator polynomial, $g(x)$; $n \quad \mathrm{lcm}(e, 2t \quad 1)$.

Interleaving is a simple technique to ensure burst error correction capabilities of a code originally designed for random error correction, or to enhance the capabilities of an existing BEC code. The basic idea of interleaving is to alternate the transmission of the individual bits of a set of codewords (Figure 4.7). If we have an original $[n,k]$ code, $C \quad c_1, c_2, \ldots, c_m$, with, m, codewords capable of correcting a burst of length, $b \quad 1$, errors or less, then interleaving allows us to correct a burst of length, mb, or less. We construct a table, and every row of this table is a codeword of the original code. When transmitted or stored, the symbols are interleaved; we send the symbols column-wise, the first symbol of c_1, then the first symbol of c_2, and so on. A burst of length, mb, or less can have at most, b, symbols in any row affected by the error. But each row is a codeword of the original code, and it is able to correct a burst of length, b, or less. If $b \quad 1$, we start with a code with no burst error correction capabilities.

Let us now give an example of interleaving. We start with the Reed-Muller code with the generator matrix:

$$G \quad \begin{matrix} 1 & 1 & 1 & 1 & 1 & 1 & 1 & 1 \\ 1 & 0 & 0 & 1 & 1 & 0 & 1 & 0 \\ 0 & 1 & 0 & 1 & 1 & 1 & 1 & 0 \\ 0 & 0 & 1 & 1 & 0 & 1 & 1 & 0 \end{matrix}.$$

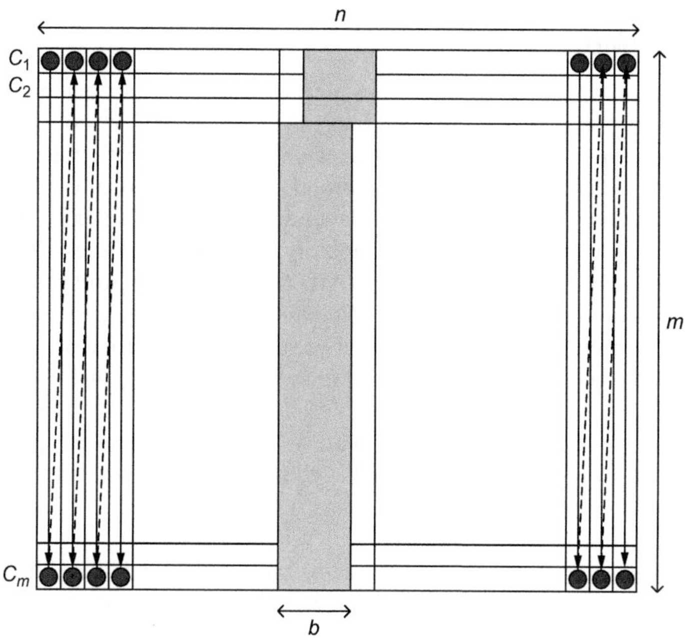

FIGURE 4.7

Schematic representation of interleaving. We wish to transmit, m, codewords and construct a table with, m, rows and, n, columns; every row of this table is a codeword of length, n, of the original code. We send the table column-wise, thus, the symbols are interleaved. Now the code is capable of correcting a burst of length, mb, or less because in any row at most, b, symbols could be affected by the error and the original code can correct a burst of this length.

This code is able to correct a single error; its distance is $d = 3$, and the length of a codeword is $n = 8$. Consider the set of five messages, $\mathcal{M} = \{0001, 0010, 0100, 0011, 0111\}$, encoded as the codewords, $\mathcal{C} = \{00110110, 01011110, 10011010, 01101000, 11110000\}$, respectively. A first option is to send the five codewords in sequence:

$$00110110\ 01011110\ 10011010\ 01101000\ 11110000.$$

In this case, a burst of errors of length, $b = 2$, would go undetected. Note that the vertical bars are not part of the string transmitted; they mark the termination of each one of the five codewords. If we interleave the symbols and transmit the first bits of the five codewords followed by the second bits, and so on, then the string of bits transmitted is

$$00101\ 01011\ 10011\ 01001\ 01110\ 11000\ 11100\ 00000.$$

The interleaving allows us to correct any burst of errors of length, $b = 4$. Now there are eight groups, one for each symbol of a codeword, and the vertical bars mark the end of a group.

4.15 REED-SOLOMON CODES

Many of the codes we have examined so far are binary codes. We start our presentation of nonbinary codes with a brief overview and then introduce a class of nonbinary codes, the Reed-Solomon (R-S) codes. R-S codes are nonbinary cyclic codes widely used for burst error correction; they were introduced in 1960 by Irving Reed and Gustav Solomon [353]. Reed-Solomon codes are used for data storage on CDs, on DVDs and Blu-ray Discs, in data transmission technologies such as DSL (digital subscriber line) and WiMAX (worldwide interoperability for microwave access), in broadcast systems such as DVB (digital video broadcasting), and in RAID 6 computer disk storage systems.

The *density of codewords* of the linear code, $\mathcal{C}$, over the alphabet, $\mathcal{A}$, is the ratio of the number of codewords to the number of n-tuples constructed with the symbols from the alphabet $\mathcal{A}$. Table 4.5 shows the density of the codewords, $\mathcal{C} \ / \ V_n$, and the rate, R, of binary and nonbinary, $[n,k]$, codes over $GF(2^m)$.

We notice that the rates of a binary, $[n,k]$, and of a, $q \quad 2^m$-ary $[n,k]$, code are equal. We also see that nonbinary codes have a lower density of codewords—thus, a larger distance between codewords—and this implies increased error correction capability.

Example. *A binary Hamming code versus a code using three-bit symbols.* The binary, $[7,4,3]$, code is a subspace of cardinality, $\mathcal{C} \quad 2^4 \quad 16$, of a vector space of cardinality, $V_n \quad 2^7 \quad 128$. The density of the codewords is

$$\frac{16}{128} \quad \frac{1}{8}.$$

If the code uses an alphabet with eight symbols, each symbol requiring $m \quad 3$ bits, then the total number of codewords becomes $2^{4 \ 3} \quad 2^{12} \quad 4096$, while the number of 7-tuples is $2^{7 \ 3} \quad 2^{21} \quad 2097152$. Now the density of the codewords is much lower:

$$\frac{4096}{2097152} \quad \frac{1}{2^9} \quad \frac{1}{512}.$$

Let $\mathbb{P}_{k \ 1}[x]$ be the set of polynomials, $p_i(x)$, of degree, at most, $(k \quad 1)$, over the finite field, $GF(q)$. Given $n > k$, an $[n,k]$ *Reed-Solomon (R-S) code* over $GF(q)$ consists of all codewords, c_i, generated by polynomials, $p_i \quad \mathbb{P}_{k \ 1}[x]$:

$$c_i \quad (p_i(x_1), p_i(x_2), p_i(x_3), \ldots, p_i(x_{n \ 1}), p_i(x_n)),$$

Table 4.5 The Cardinality of the n-Dimensional Vector Space, V_n, the Cardinality of an $[n,k]$ Code, $\mathcal{C}$, the Density of the Codewords, and the Rates of a Binary and of a Nonbinary Code Over $GF(2^m)$

Code	$\lvert V_n \rvert$	$\lvert \mathcal{C} \rvert$	$\lvert \mathcal{C} \rvert / \lvert V_n \rvert$	R
Binary	2^n	2^k	$2^{\ (n \ k)}$	k/n
Nonbinary	$q^n \quad 2^{mn}$	$q^k \quad 2^{mk}$	$2^{\ (n \ k)m}$	$k2^m/n2^m \quad k/n$

where $x_1, x_2, \ldots, x_n$ are n different elements of $GF(q)$; $x_i \quad \beta^{i \quad 1}, 1 \quad i \quad n$, with β, a primitive element of $GF(q)$. The polynomial, $p_i(x)$, is called the generator of the codeword, c_i.

Reed-Solomon codes have several desirable properties; for example, an $[n,k]$ R-S code can achieve the Singleton bound, the largest possible code distance (minimum distance between any pair of codewords) of any linear code: $d \quad n \quad k \quad 1$. This means that the code can correct at most, t, errors, with

$$t \quad \frac{d \quad 1}{2} \quad \frac{n \quad k}{2} \quad .$$

Recall that a cyclic code, C_g, has a generator polynomial, $g(x)$, and all codewords, $c_i(x)$, are multiple of the generator polynomial, $c_i(x) \quad g(x) \quad p_i(x), \quad c_i \quad C$. The Reed-Solomon codes are constructed differently. First we identify, $p_i(x)$, the generator polynomial of each codeword, $c_i(x)$, and then evaluate this polynomial for the elements, $(x_1, x_2, \ldots, x_n)$, of the finite field, $GF(q)$, to identify the, n, coordinates, $c_{i,j}$, of the codeword, $c_i \quad {}_j^n \, c_{i,j} x^j$. We shall later see that an R-S code is a cyclic code, and we can define a generator polynomial. The difference between R-S codes and the BCH codes discussed at the end of this section is that for R-S codes, β is a primitive element of $GF(q)$, while in the case of BCH codes, β is an element of an extension Galois field.

As noted earlier, $\mathbb{P}_{k \quad 1}[x] \quad GF(q)[x]$ is a vector space of dimension $(k \quad 1)$. For example, when $k \quad 3$, the polynomials of degree at most, two, over $GF(2)$, are

$$\mathbb{P}_2[x] \quad p_0(x) \quad 1, \; p_1(x) \quad x, \; p_2(x) \quad x \quad 1, \; p_3(x) \quad x^2, \; p_4(x) \quad x^2 \quad 1, \; p_5(x) \quad x^2 \quad x,$$
$$p_6(x) \quad x^2 \quad x \quad 1 \; .$$

It is easy to see that an R-S code is a linear code. Let two codewords of an R-S code be:

$$c_i \quad (p_i(x_1), p_i(x_2), \ldots, p_i(x_n)) \quad \text{and} \quad c_j \quad p_j(x_1), p_j(x_2), \ldots, p_j(x_n) \; .$$

Then $\alpha c_i \quad \beta c_j \quad (g(x_1), g(x_2), \ldots, g(x_n))$, with $\alpha, \beta \quad GF(q)$ and $g(x) \quad \alpha p_i(x) \quad \beta p_j(x)$.

The minimum distance of an $[n,k]$ *Reed-Solomon code is* $(n \quad k \quad 1)$. Indeed, each codeword consists of n symbols from $GF(q)$, and it is generated by a polynomial, $p_i(x) \quad \mathbb{P}_{k \quad 1}$, of degree at most, $(k \quad 1)$. Such a polynomial may have at most, $(k \quad 1)$, zeros. It follows that the minimum weight of a codeword is $n \quad (k \quad 1) \quad n \quad k \quad 1$; thus, the minimum distance the code is $d_{min} \quad n \quad k \quad 1$.

Encoding Reed-Solomon Codes

We can easily construct a generator and a parity-check matrix of an $[n,k]$ R-S code when the, n, elements of the finite field, $x_1, x_2, \ldots, x_n \quad GF(q)$, can be expressed as powers of, β, the characteristic element of the finite field. Each row of the generator matrix, G, consists of one polynomial in $\mathbb{P}_k[x]$ evaluated for all elements, $x_i \quad GF(q)$. For example, the rows 1 to k may consist of the polynomials,

$p_0(x) = x^0, p_1(x) = x^1, \ldots, p_{k-1}(x) = x^{k-1}$; thus, a generator matrix of an $[n,k]$ R-S code is [227]

$$G = \begin{pmatrix} 1 & 1 & 1 & \cdots & 1 & 1 \\ x_1 & x_2 & x_3 & \cdots & x_{n-1} & x_n \\ (x_1)^2 & (x_2)^2 & (x_3)^2 & \cdots & (x_{n-1})^2 & (x_n)^2 \\ \vdots & \vdots & \vdots & & \vdots & \vdots \\ (x_1)^{k-1} & (x_2)^{k-1} & (x_3)^{k-1} & \cdots & (x_{n-1})^{k-1} & (x_n)^{k-1} \end{pmatrix}.$$

To construct the parity-check matrix, H, of the R-S code with the generator matrix, G, we use a property discussed in Section 4.13: If G is a $k \times n$ matrix, then there exists an $(n-k) \times n$ matrix, H, such that $HG^T = GH^T = 0$. It follows immediately that the parity-check matrix, H, of the code is

$$H = \begin{pmatrix} x_1 & x_2 & x_3 & \cdots & x_{n-1} & x_n \\ (x_1)^2 & (x_2)^2 & (x_3)^2 & \cdots & (x_{n-1})^2 & (x_n)^2 \\ (x_1)^3 & (x_2)^3 & (x_3)^3 & \cdots & (x_{n-1})^3 & (x_n)^3 \\ \vdots & \vdots & \vdots & & \vdots & \vdots \\ (x_1)^{n-k} & (x_2)^{n-k} & (x_3)^{n-k} & \cdots & (x_{n-1})^{n-k} & (x_n)^{n-k} \end{pmatrix}.$$

The two matrices can be expressed in terms of the primitive element, β, with the properties $\beta^n = 1$ and $\beta^i \neq 1, 0 < i < n$. Then:

$$x_j = \beta^{j-1}\ 1 \leq j \leq q-1, \quad \text{such that} \quad x_j^n = \beta^{j-1}, 1 \leq j \leq n, \quad \text{and} \quad n = q-1.$$

Now the generator and the parity-check matrices are

$$G = \begin{pmatrix} 1 & 1 & 1 & \cdots & 1 & 1 \\ 1 & (\beta)^1 & (\beta)^2 & \cdots & (\beta)^{n-2} & (\beta)^{n-1} \\ 1 & (\beta^2)^1 & (\beta^2)^2 & \cdots & (\beta^2)^{n-2} & (\beta^2)^{n-1} \\ \vdots & \vdots & \vdots & & \vdots & \vdots \\ 1 & (\beta^{k-1})^1 & (\beta^{k-1})^2 & \cdots & (\beta^{k-1})^{n-2} & (\beta^{k-1})^{n-1} \end{pmatrix},$$

and

$$H = \begin{pmatrix} 1 & (\beta)^1 & (\beta)^2 & \cdots & (\beta)^{n-2} & (\beta)^{n-1} \\ 1 & (\beta^2)^1 & (\beta^2)^2 & \cdots & (\beta^2)^{n-2} & (\beta^2)^{n-1} \\ 1 & (\beta^3)^1 & (\beta^3)^2 & \cdots & (\beta^3)^{n-2} & (\beta^3)^{n-1} \\ \vdots & \vdots & \vdots & & \vdots & \vdots \\ 1 & (\beta^{n-k})^1 & (\beta^{n-k})^2 & \cdots & (\beta^{n-k})^{n-2} & (\beta^{n-k})^{n-1} \end{pmatrix}.$$

Algorithm. To determine the codeword, c_i, of an $[n,k]$ Reed-Solomon code generated by the information polynomial, $p_i(x)\quad p_{i,0}\quad p_{i,1}x^1 \qquad\qquad p_{i,n}x^n$, with $p_i(x)\quad \mathbb{P}_{k\ 1}$, carry out the following three steps:

1. Choose k, q, and n, with $k < n < q$.
2. Select n elements, $x_j\quad GF(q), 1\quad j\quad n$.
3. Evaluate the polynomial, $p_i(x)$ for $x\quad x_j, 1\quad j\quad n$, and then construct:

$$c_i \quad (p_i(x_1), p_i(x_2), \ldots, p_i(x_n)).$$

Example. *Encoding for a* $[10,5]$ *Reed-Solomon code over* $GF(11)$. In this case, $n\quad 10$, $k\quad 5$, and $q\quad 11$. First, we have to find a primitive element, $\beta^n\quad 1$; we notice that 2^{10} mod 11 $\quad 1$; then the, $q\quad 1\quad 10$, nonzero elements of $GF(11)$ can be expressed as $x_j\quad 2^{j\ 1}$, with $1\quad j\quad 10$:

$x_1\quad 2^0$ mod 11	1	$x_5\quad 2^4$ mod 11	5	$x_9\quad 2^8$ mod 11	3
$x_2\quad 2^1$ mod 11	2	$x_6\quad 2^5$ mod 11	10	$x_{10}\quad 2^9$ mod 11	6.
$x_3\quad 2^2$ mod 11	4	$x_7\quad 2^6$ mod 11	9		
$x_4\quad 2^3$ mod 11	8	$x_8\quad 2^7$ mod 11	7		

The information vectors are denoted as $(i_0, i_1, i_2, i_3, i_4)$; the 31 polynomials of degree at most, $k\quad 4$, with coefficients from $GF(2)$ corresponding to these vectors and the codewords they generate are shown in Table 4.6.

Table 4.6 The Information Vector, the Generator Polynomials, and the Codewords for the [10,4] R-S Code Over *GF*(11)

Vector	Generator Polynomial	Codeword
(1,0,0,0,0)	$p_1(x)\quad 1$	(1,1,1,1,1,1,1,1,1,1)
(0,1,0,0,0)	$p_2(x)\quad x$	(1,2,4,8,5,10,9,7,3,6)
(0,0,1,0,0)	$p_3(x)\quad x^2$	(1,4,5,9,3,1,4,5,9,3)
(0,0,0,1,0)	$p_4(x)\quad x^3$	(1,8,9,6,4,10,3,2,5,7)
(0,0,0,0,1)	$p_5(x)\quad x^4$	(1,5,3,4,9,1,5,3,4,9)
(1,1,0,0,0)	$p_6(x)\quad 1\quad x$	(2,3,5,9,6,0,10,8,4,7)
(1,0,1,0,0)	$p_7(x)\quad 1\quad x^2$	(2,5,6,10,4,2,5,6,10,4)
(1,0,0,1,0)	$p_8(x)\quad 1\quad x^3$	(2,9,10,7,5,0,4,3,6,8)
(1,0,0,0,1)	$p_9(x)\quad 1\quad x^4$	(2,6,4,5,10,2,6,4,5,10)
⋮	⋮	⋮
(1,1,1,1,1)	$p_{31}(x)\quad x^4\quad x^3\quad x^2\quad x^1\quad 1$	(5,9,0,6,0,1,5,0,0,4)

For example, when the information vector is $(0,0,0,0,1)$, the codeword is generated by the polynomial, $p_5(x) \quad x^4$, then $c_{(0,0,0,0,1)} \quad (1,5,3,4,9,1,5,3,4,9)$. Indeed,

$p_5(x_1)$	$(x_1)^4$	1^4	mod 11	1	$p_5(x_6)$	$(x_6)^4$	10^4	mod 11	1
$p_5(x_2)$	$(x_2)^4$	2^4	mod 11	5	$p_5(x_7)$	$(x_7)^4$	9^4	mod 11	5
$p_5(x_3)$	$(x_3)^4$	4^4	mod 11	3	$p_5(x_8)$	$(x_8)^4$	7^4	mod 11	3
$p_5(x_4)$	$(x_4)^4$	8^4	mod 11	4	$p_5(x_9)$	$(x_9)^4$	3^4	mod 11	4
$p_5(x_5)$	$(x_5)^4$	5^4	mod 11	9	$p_5(x_{10})$	$(x_{10})^4$	6^4	mod 11	9

A generator matrix of the code is

$$G \quad \begin{array}{cccccccccc} 1 & 1 & 1 & 1 & 1 & 1 & 1 & 1 & 1 & 1 \\ 1 & 2 & 4 & 8 & 5 & 10 & 9 & 7 & 3 & 6 \\ 1 & 4 & 5 & 9 & 3 & 1 & 4 & 5 & 9 & 3 \\ 1 & 8 & 9 & 6 & 4 & 10 & 3 & 2 & 5 & 7 \\ 1 & 5 & 3 & 4 & 9 & 1 & 5 & 3 & 4 & 9 \end{array} .$$

Decoding Reed-Solomon Codes

Assume the codeword, c_i, of an $[n,k]$ Reed-Solomon code is generated by the polynomial, $p_i(x)$; an error, e_i, occurs when we send $c_i \quad (p_i(x_1), p_i(x_2), \ldots, p_i(x_{n\ \ 1}), p_i(x_n))$, and we receive the n-tuple, v, affected by at most, t, errors:

$$v \quad c_i \quad e, \quad \text{with} \quad w(e) \quad t \quad \frac{n \quad k}{2} .$$

Before presenting the decoding algorithm for an R-S codes, we introduce two new concepts, the interpolating polynomial and the error locator polynomial; then we prove a theorem of existence for the interpolation polynomial and show that the polynomial used to generate the codeword, c_j, is a ratio of the interpolating and error-locating polynomials. The bivariate *interpolating polynomial* is defined as

$$Q(x,y) \quad Q_0(x) \quad yQ_1(x).$$

$Q(x,y)$ is a polynomial in x and y, with $Q_0(x)$ and $Q_1(x)$ polynomials of degrees m_0 and m_1, respectively,

$$Q_0(x) \quad Q_{0,0} \quad Q_{0,1}x \qquad Q_{0,j}x^j \qquad Q_{0,m_0}x^{m_0}$$

and

$$Q_1(x) \quad Q_{1,0} \quad Q_{1,1}x \qquad Q_{1,j}x^j \qquad Q_{0,m_1}x^{m_1},$$

the so-called *error locator polynomial.*

Proposition. *When we receive an n-tuple,* $v \quad (v_0, v_2, \ldots, v_n \;\; 1)$, *an interpolation polynomial for the* $[n,k]$ *Reed-Solomon code, with* $t \quad (n \quad k)/2$, *exists if the following three conditions are satis ed:*

1. $Q(x_i, v_i) \quad Q_0(x_i) \quad v_i Q_1(x_i) \quad 0, \quad i \quad 1,2,\ldots,(n \quad 1),n, \quad x_i, v_i \quad GF(q).$
2. $\deg[Q_0(x)] \quad m_0 \quad n \quad 1 \quad t.$
3. $\deg[Q_1(x)] \quad m_1 \quad n \quad 1 \quad t \quad (k \quad 1).$

Proof. To prove the existence of $Q(x_i, v_i)$, we have to show that we can determine the $m_0 \quad 1$ coefficients, $Q_{0,j}$, and the $m_1 \quad 1$ coefficients, $Q_{1,j}$, given the n equations, $Q(x_i, v_i) \quad 0, \; 1 \quad i \quad n$, and that not all of them are equal to zero. The total number of coefficients of the bivariate polynomial is $(m_0 \quad 1) \quad (m_1 \quad 1) \quad (n \quad t) \quad (n \quad t \quad (k \quad 1)) \quad 2n \quad k \quad 1 \quad 2t$. If we substitute $2t \quad d \quad n \quad k \quad 1$, we see that the total number of coefficients is $2n \quad k \quad 1 \quad n \quad k \quad 1 \quad n$. The system of, n, equations, with, n, unknowns, has a nonzero solution if the determinant of the system is nonzero, and in this case, a polynomial, $Q(x_i, v_i)$, that satisfies the three conditions exists. ∎

Proposition. *Let C be a Reed-Solomon code, with* $t \quad \frac{n \;\; k}{2}$. *If we transmit the codeword,* $c_i(x)$, *generated by the polynomial,* $p_i(x) \quad \mathbb{P}_{k \;\; 1}[x]$, *and we receive* $v(x) \qquad {}_j v_j x^i \quad c_i(x) \quad e(x)$, *and if the number of errors is* $w[e(x)] \quad t$, *then the codeword,* $c_i(x)$, *is generated by the polynomial,* $p_i(x) \quad \mathbb{P}_{k \;\; 1}[x]$, *given by*

$$p_i(x) \qquad \frac{Q_0(x)}{Q_1(x)}.$$

Proof. If $v \quad (v_0, v_2, \ldots, v_n \;\; 1)$ is the received n-tuple and $e(x) \quad (e_0, e_1, \ldots, e_n \;\; 1)$ is the error polynomial, then at most, t, components of, e, are nonzero, $e_l \quad r_l$; thus, at least, $n \quad t$, components, e_j, of e are zero. An interpolating polynomial, $Q(x,v) \quad Q(x,(p_j(x) \quad e(x)))$, exists and has at least, $n \quad t$, zeros, namely, those values, $x \quad x_j$, when $p_i(x_j) \quad v_j \quad e_j$, with $e_j \quad 0$. But the degree of $Q(x, p_i(x))$ is $m_0 \quad n \quad t \quad 1$. It follows that $Q(x, p_i(x)) \quad 0$ for $x \quad x_j, j \quad 1,2,\ldots,n$. But $Q(x, p_i(x)) \quad 0$ implies that

$$Q_0(x) \quad p_i(x) Q_2(x) \quad 0, \quad \text{thus,} \quad p_i(x) \qquad \frac{Q_0(x)}{Q_1(x)}.$$

∎

Algorithm. Let C be an $[n,k]$ Reed-Solomon code able to correct at most, $t \quad (n \quad k)/2$, errors. We send $c_i \quad (p_i(x_1), p_i(x_2), \ldots, p_i(x_n))$ generated by the polynomial, $p_i(x)$, and we receive $v \quad c_i \quad e$, with $v \quad (v_0, v_2, \ldots, v_n \;\; 1)$. To determine c_i, we first compute the generator polynomial, $p_i(x)$, and then evaluate it for $x \quad x_j, 1 \quad j \quad n$. The decoding algorithm consists of the following steps:

Step 1. Compute $m_0 \quad n \quad 1 \quad t$ and $m_1 \quad n \quad 1 \quad t \quad (k \quad 1)$. Solve the system of linear equations:

$$\begin{pmatrix} 1 & x_1 & x_1^2 & \ldots x_1^{m_0} & v_1 & v_1x_1 & \ldots v_1x_1^{m_1} \\ 1 & x_2 & x_2^2 & \ldots x_2^{m_0} & v_2 & v_2x_2 & \ldots v_2x_2^{m_1} \\ 1 & x_3 & x_3^2 & \ldots x_3^{m_0} & v_3 & v_3x_1 & \ldots v_3x_3^{m_1} \\ \vdots & & & & & & \\ 1 & x_n & x_n^2 & \ldots x_n^{m_0} & v_n & v_nx_n & \ldots v_nx_n^{m_1} \end{pmatrix} \begin{pmatrix} Q_{0,0} \\ Q_{0,1} \\ Q_{0,2} \\ \vdots \\ Q_{0,m_0} \\ Q_{1,0} \\ Q_{1,1} \\ Q_{1,2} \\ \vdots \\ Q_{1,m_1} \end{pmatrix} \quad \begin{pmatrix} 0 \\ 0 \\ 0 \\ \vdots \\ 0 \\ 0 \\ 0 \\ 0 \\ \vdots \\ 0 \end{pmatrix}.$$

Step 2. Construct the following polynomials:

$$Q_0(x) \quad \sum_{j\ 0}^{m_0} Q_{0,j}x^j, \quad Q_1(x) \quad \sum_{j\ 1}^{m_1} Q_{1,j}x^j, \quad \text{and} \quad p(x) \quad \frac{Q_0(x)}{Q_1(x)}.$$

Step 3. If $p(x) \quad \mathbb{P}_{k\ 1}$, then the algorithm succeeds and produces $(p(x_1), p(x_2), \ldots, p(x_n))$.

An Alternative De nition of Reed-Solomon Codes

We discuss now several properties of extension finite fields, $GF(2^m)$, and show that the polynomial, $x^n \quad 1$, can be factored as a product of irreducible monic polynomial over $GF(2^m)$. Recall that the set of polynomials, $p(x) \quad \sum_{i\ 0}^{k} \alpha_i x^i$, of degree at most, k, and with $\alpha_i \quad GF(2^m)$, is denoted as $\mathbb{P}_{2^m}[x]$.

Proposition. *If the polynomial* $p(x) \quad \mathbb{P}_{2^m}[x]$, *then* $p(x^2) \quad [p(x)]^2$.

Proof. If $\alpha, \beta \quad GF(2^m)$, then $(\alpha \quad \beta)^2 \quad \alpha^2 \quad \beta^2$. Indeed, the term $2\alpha \quad \beta \quad 0$ as $\alpha, \beta \quad GF(2^m)$. If $p(x) \quad \alpha_k x^k \quad \alpha_{k\ 1} x^{k\ 1} \quad \quad \alpha_1 x \quad \alpha_0 x^0$, with $\alpha_0, \alpha_1, \ldots, \alpha_k \quad GF(2^m)$, then

$$[p(x)]^2 \quad [\alpha_k x^k \quad \alpha_{k\ 1} x^{k\ 1} \quad \quad \alpha_1 x \quad \alpha_0 x^0]^2 \quad \alpha_k^2 x^{2k} \quad \alpha_{k\ 1}^2 x^{2(k\ 1)} \quad \quad \alpha_1^2 x^2 \quad \alpha_0^2,$$

$$p(x^2) \quad \alpha_k x^{2k} \quad \alpha_{k\ 1} x^{2(k\ 1)} \quad \quad \alpha_1 x^2 \quad \alpha_0.$$

Thus, $p(x^2) \quad [p(x)]^2$ if and only if $\alpha_i \quad (\alpha_i)^2$, $1 \quad i \quad k$. But this is true if and only if $\alpha_i \quad GF(2^m)$. It follows immediately that if β is a zero of $p(x)$, then so is β^2:

$$p(\beta) \quad 0 \qquad p(\beta^2) \quad 0.$$

■

Proposition. *The extension Galois eld,* $GF(2^m)$, *is a sub eld of* $GF(2^n)$ *if and only if* $m\ n$ *(m divides n).*

Proof. It is easy to see that $a^q \quad a$ if and only if $a \quad GF(q)$. If $a \quad GF(q)$, then it can be expressed as a power of the primitive element, β, of $GF(q)$ as $a \quad \beta^i$. Then,

$$a^{q \quad 1} \quad (\beta^i)^{(q \quad 1)} \quad (\beta^{q \quad 1})^i \quad 1^i \quad 1,$$

as $\beta^{q \quad 1} \quad 1$. If $a^{q \quad 1} \quad 1$, then $a^q \quad a$, and a is the primitive element of a finite field, $GF(q)$.

We can thus express $GF(2^n) \quad x\, x^{2^n} \quad x$ and $GF(2^m) \quad x\, x^{2^m} \quad x$. In preparation for our proof we show that

$$(x^m \quad 1)\ (x^n \quad 1) \qquad\qquad m\ n.$$

Given a non-negative integer, k, we have the identity

$$(x^n \quad 1) \quad (x^m \quad 1)(x^{n \quad m} \quad x^{n \quad 2m} \qquad\qquad x^{n \quad km}) \quad x^{n \quad km} \quad 1$$

or

$$\frac{x^n \quad 1}{x^m \quad 1} \quad (x^{n \quad m} \quad x^{n \quad 2m} \qquad\qquad x^{n \quad km}) \quad \frac{x^{n \quad km} \quad 1}{x^m \quad 1}.$$

If $n \quad km \quad m$, then

$$\frac{x^n \quad 1}{x^m \quad 1} \quad (x^{n \quad m} \quad x^{n \quad 2m} \qquad\qquad x^{n \quad km}) \quad 1.$$

In this case, $n \quad m(k \quad 1)$ so that when $m\ n$, then $(x^m \quad 1)\ (x^n \quad 1)$.

Now we show that $m\ n$ is a necessary and sufficient condition for $GF(2^m)$ to be a subfield of $GF(2^n)$. The condition is necessary: If $GF(2^m)$ is a subfield of $GF(2^n)$, then $m\ n$. Indeed, $GF(2^m)$ contains an element of order $2^m \quad 1$. If $GF(2^m)$ is a subfield of $GF(2^n)$, then $GF(2^n)$ contains an element of order $2^m \quad 1$. Thus, $(x^m \quad 1)\ (x^n \quad 1)$, and according to the previous statement, $m\ n$. The condition is sufficient: If $m\ n$, then $GF(2^m)$ is a subfield of $GF(2^n)$. Indeed, $m\ n \qquad (2^m \quad 1)\ (2^n \quad 1)$, and this implies that $(x^{2^m} \quad 1)\ (x^{2^n} \quad 1)$, in other words, that $GF(2^m)$ is a subfield of $GF(2^n)$. ■

The *minimal polynomial*, $m_\gamma(x)$, is the lowest degree polynomial, with binary coefficients, that has $\gamma \quad GF(2^m)$ as a zero.

Proposition. *The minimal polynomial in $GF(2^m)$ has the following properties:*

1. *$m_\gamma(x)$ exists and is unique.*
2. *$m_\gamma(x)$ is irreducible.*
3. *If $p(x)$ is a polynomial with binary coef cients and if $p(\gamma) \quad 0$, then minimal polynomial, $m_\gamma(x)$, divides $p(x)$.*
4. *The polynomial, $g(x) \quad x^{2^m} \quad x$, is the product of different minimal polynomials, $m_{\gamma_i}(x)$, with $\gamma_i \quad GF(2^m)$.*
5. $\deg m_\gamma(x) \quad m$. *If γ is a primitive element of $GF(2^m)$, then* $\deg m_\gamma(x) \quad m$.

Proof. If γ is a primitive element of $GF(2^m)$, then $\gamma^{2^m \quad 1} \quad 1$ and $\gamma^{2^m} \quad \gamma \quad 0$. There exists a polynomial, such that $p(\gamma) \quad 0$; the polynomial, $p(x) \quad x^{2^m} \quad x$, satisfies the condition, $p(\gamma) \quad 0$. This polynomial must be unique; otherwise, if two polynomials, $f_1(x)$ and $f_2(x)$, of the same degree existed, then their difference, $g(x) = f_1(x) - f_2(x)$, would have (1) γ as a zero and (2) a lower degree, $\deg g(x) < \deg f_1(x)$. This contradicts the assumption that $f_1(x)$ and $f_2(x)$ are minimal polynomials.

To show that $m_\gamma(x)$ is irreducible, we assume that $m_\gamma(x) = p(x)q(x)$. Then $\deg[p(x)], \deg[q(x)] < \deg[m_\gamma(x)]$ and at least one of the two polynomials, $p(x)$ and $q(x)$, must have γ as a root. This contradicts again the assumption that $m_\gamma(x)$ is a minimal polynomial.

It is easy to see that the minimal polynomial, $m_\gamma(x)$, divides any polynomial, $p(x)$, with binary coefficients if $p(\gamma) = 0$. Indeed, we can express $p(x) = m_\gamma(x)q(x) + r(x)$. But $p(\gamma) = 0$. Therefore, $r(\gamma) = 0$ and $r(x) = 0$.

To prove that the polynomial, $g(x) = x^{2^m} - x$, is the product of minimal polynomials, $m_{\gamma_i}(x)$, with $\gamma_i \in GF(2^m)$, we observe that all elements, $a \in GF(2^m)$, are zeros of $g(x)$ as they satisfy the equation $a^{2^m - 1} - a = 0$. Then the minimal polynomial, $m_a(x)$, exists, and it is unique; according to the fourth property, $m_a(x)$ divides the polynomial $g(x)$.

Finally, to show that $\deg[m_\gamma(x)] \le m$, we recall that $GF(2^m)$ can be viewed as an m-dimensional vector space over $GF(2)$; thus, the elements, $1, \gamma^1, \gamma^2, \ldots, \gamma^{m-1}, \gamma^m$, are linearly dependent. This means that there are $m + 1$ coefficients, $g_i \in GF(2)$, $0 \le i \le m$, such that

$$g_m\gamma^m + g_{m-1}\gamma^{m-1} + \cdots + g_1\gamma^1 + g_0\gamma^0 = 0.$$

Thus, $\deg[m_\gamma(x)]$ can be at most m. If γ is a primitive element of $GF(2^m)$, then $GF(2^m)$ must consist of $2^m - 1$ different elements; thus, $1, \gamma^1, \gamma^2, \ldots, \gamma^{m-1}$ must be linearly independent. This shows that when γ is a primitive element of $GF(2^m)$, then $\deg[m_\gamma(x)] = m$. ■

The following proposition allows us to introduce the generator matrix of a Reed-Solomon code.

Proposition. *If C_g is a cyclic code with generator polynomial, $g(x)$ and $h(x) = (x^n - 1)/g(x)$, then the dual code, $C_g^\perp$, has a generator polynomial, $g^\perp(x) = x^{k-1}h(1/x)$, with $h(x) = \sum_{i=0}^{k-1} h_i x^i$. Thus, $g^\perp(x) = h^0x^{k-1} + h_1x^{k-1} + \cdots + h_{k-1}$.*

Let α be a primitive element of $GF(q)$. If n is a divisor of $(q-1)$ and $\beta = \alpha^{(q-1)/n}$, then $\beta^n = 1$. Let C be the $[n,k,d]$ Reed-Solomon code obtained by evaluating polynomials of degree at most, $(k-1)$, $p(x) \in \mathbb{P}_{k-1}[x]$ over $GF(2^m)$ at $x_i = \beta^{i-1}$, $1 \le i \le n$. Then C is a cyclic code over $GF(q)$, with generator polynomial

$$g(x) = (x-\beta)(x-\beta^2)\ldots(x-\beta^{n-k}).$$

Proof. If $c \in C$, then $c = (p(\beta^0), p(\beta^1), p(\beta^2), \ldots, p(\beta^{n-2}), p(\beta^{n-1}))$. A cyclic shift of the codeword, c, leads to the n-tuple

$$c' = (p(\beta^{n-1}), p(\beta^0), p(\beta^1), \ldots, p(\beta^{n-3}), p(\beta^{n-2})).$$

Recall that a cyclic code is a cyclic subspace of a vector space and from the fact that, $c' \in C$, we conclude that C is a cyclic code. We now construct the parity-check matrix of the code and then show that this parity-check matrix corresponds to a cyclic code with a generator polynomial having $\beta, \beta^2, \ldots, \beta^{n-k}$ as roots. First, we express, c, as

$$c = (p_1(\beta^0), p_1(\beta^1), p_1(\beta^2), \ldots, p_1(\beta^{n-2}), p_1(\beta^{n-1})),$$

with $p_1(x) \quad p(\beta^{\ 1})$. According to the previous proposition, the generator matrix of the dual code is an $(n \quad k) \quad n$ matrix, H $\quad [h_{ij}]$, with $h_{ij} \quad p(1/(x_j)^i)$, $1 \quad i \quad n \quad k, 1 \quad j \quad n$; when we substitute $x_j \quad \beta^{j \ 1}$, the generator matrix of the dual code, which is at the same time the parity-check matrix of the Reed-Solomon code, can be expressed as

$$\mathrm{H} \quad \begin{matrix} 1 & (\beta)^1 & (\beta)^2 & \dots & (\beta)^{n \ 2} & (\beta)^{n \ 1} \\ 1 & (\beta^2)^1 & (\beta^2)^2 & \dots & (\beta^2)^{n \ 2} & (\beta^2)^{n \ 1} \\ 1 & (\beta^3)^1 & (\beta^3)^2 & \dots & (\beta^3)^{n \ 2} & (\beta^3)^{n \ 1} \\ \vdots & \vdots & \vdots & & \vdots & \vdots \\ 1 & (\beta^{n \ k})^1 & (\beta^{n \ k})^2 & \dots & (\beta^{n \ k})^{n \ 2} & (\beta^{n \ k})^{n \ 1} \end{matrix} .$$

Recall from Section 4.13 that when $g(x)$ is the generator polynomial of the cyclic $[n,k]$ code, C_g, over $GF(q)$ and when $g(x)$ has among its zeros, $(d \quad 1)$, elements of $GF(q)$, $\beta^\alpha, \beta^{\alpha \ 1}, \dots, \beta^{\alpha \ d \ 2}$, with $\beta \quad GF(q)$, the parity-check matrix of cyclic code, C_g, is

$$\mathrm{H}_g \quad \begin{matrix} 1 & (\beta)^\alpha & (\beta)^{2\alpha} & (\beta)^{3\alpha} & \dots & (\beta)^{(n \ 1)\alpha} \\ 1 & (\beta)^{\alpha \ 1} & (\beta)^{2(\alpha \ 1)} & (\beta)^{3(\alpha \ 1)} & \dots & (\beta)^{(n \ 1)(\alpha \ 1)} \\ \vdots & \vdots & \vdots & \vdots & & \vdots \\ 1 & (\beta)^{(\alpha \ d \ 2)} & (\beta)^{2(\alpha \ d \ 2)} & (\beta)^{3(\alpha \ d \ 2)} & \dots & (\beta)^{(n \ 1)(\alpha \ d \ 2)} \end{matrix} .$$

The similarity of H and H_g allows us to conclude that, $\beta, \beta^2, \dots, \beta^{n \ k}$, are the zeros of the generator matrix of the R-S code; thus,

$$g(x) \quad (x \quad \beta)(x \quad \beta^2)\dots(x \quad \beta^{n \ k}).$$

■

Example. Construct the generator polynomial of the [7,3] *Reed-Solomon code, C, over* $GF(2^3)$, with the primitive element, $\beta \quad GF(2^3)$, satisfying the condition, $\beta^3 \quad \beta \quad 1 \quad 0$.

The distance of the code is $d \quad n \quad k \quad 1 \quad 5$. We can describe $GF(8)$ by binary 3-tuples, $000, 001, \dots, 111$, by listing its elements expressed as powers of the primitive element, β, or by polynomials, $p(x)$, of degree at most, $k \quad 1 \quad 2$. For example, the tuple 011 corresponds to β^3 and to the polynomial $p(x) \quad x \quad 1$.

Table 4.7 shows the correspondence between the 3-tuples, the powers of the primitive element, β, and the polynomials. Indeed, we observe that the polynomial, $x^2 \quad x \quad 1$, is irreducible and thus can be used to construct $GF(8)$. In $GF(8)$:

$$p_0(x) \quad x^0 \quad 1, \quad p_1(x) \quad x^1 \quad x, \quad p_2(x) \quad x^2 \quad x^2,$$

$$p_3(x) \quad x^3 \quad x \quad 1,$$

$$p_4(x) \quad x^4 \quad x(x^3) \quad x(x \quad 1) \quad x^2 \quad x,$$

$$p_5(x) \quad x^5 \quad x(x^4) \quad x(x^2 \quad x) \quad x^3 \quad x^2 \quad x^2 \quad x \quad 1,$$

$$p_6(x) \quad x^6 \quad x(x^5) \quad x(x^2 \quad x \quad 1) \quad x^3 \quad x^2 \quad x \quad x \quad 1 \quad x^2 \quad x \quad x^2 \quad 1.$$

Table 4.7 The Correspondence Between the 3-Tuples, the Powers of β, and the Polynomials

Binary 3-Tuple	Power of Primitive Element	Polynomial
000		0
001	β^0	1
010	β^1	x
100	β^2	x^2
011	β^3	$x \quad 1$
101	β^4	$x^2 \quad x$
110	β^5	$x^2 \quad x \quad 1$
111	β^6	$x^2 \quad 1$

The generator polynomial of, C, and of its dual, C , when $d \quad 5$ are, respectively,

$$g(x) \quad (x \quad \beta^0)(x \quad \beta^1)(x \quad \beta^2)(x \quad \beta^3) \quad (x \quad 1)(x \quad \beta)(x \quad \beta^2)(x \quad \beta^3)$$

and

$$g \ (x) \quad (x \quad \beta^{\ 4})(x \quad \beta^{\ 5})(x \quad \beta^{\ 6}) \quad (x \quad \beta^3)(x \quad \beta^2)(x \quad \beta^1) \quad (x \quad \beta)(x \quad \beta^2)(x \quad \beta^3).$$

We notice that $g(x)/g \ (x) \quad x \quad 1$; thus, the code C is self-dual.

Burst-Error Correction and Reed-Solomon Codes

Reed-Solomon codes are well suited for burst-error correction. The following proposition quantifies the burst error correcting capability of a Reed-Solomon code, with symbols from $GF(2^m)$.

Proposition. *If C is an $[n,k]$ Reed-Solomon code over $GF(2^m)$, with $t \quad (n \quad k)/2$, then the corresponding, $[n \quad m, k \quad m]$, binary code is able to correct burst errors of length, $(m \quad t)$, bits.*

An R-S code using an alphabet, with m-bit symbols, could correct any burst of $[(t \quad 1) \quad m \quad 1]$ bits errors with, t, the number of symbols in error. The code can correct a burst of $(m \quad t)$ bit errors, provided that the errors occur in consecutive symbols and start at a symbol boundary; the code could not correct any arbitrary pattern of $(m \quad t)$ bit errors because such a pattern could be spread over $(m \quad t)$ symbols, each with one bit in error.

Example. *The* $[255,246]$ *Reed-Solomon code.* In this case, a symbol is an eight-bit byte, $m \quad 8$, $n \quad 255$, and $k \quad 246$ (Figure 4.8). We transmit bytes of information in blocks of 255 bits, with 246 information bits in each block. The code has a minimum distance of $d_{min} \quad n \quad k \quad 1 \quad 10$ and can correct, t, symbols in error:

$$t \quad \frac{n \quad k}{2} \quad \frac{255 \quad 246}{2} \quad \frac{9}{2} \quad 4.$$

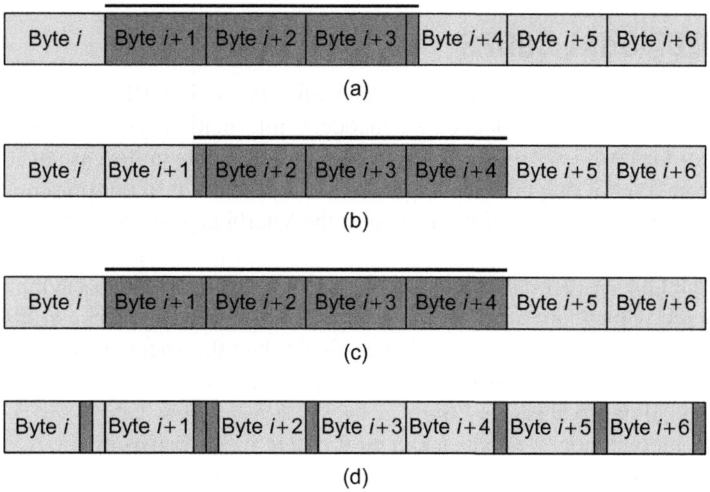

FIGURE 4.8

Correction of burst errors with the [255, 246] Reed-Solomon code. The upper bar marks consecutive bits affected by error. The code can correct: (a) and (b) any burst of 25 consecutive bit errors; (c) some bursts of 32 consecutive bit errors. The code cannot correct any arbitrary pattern of 32 bit errors (d).

The R-S code could correct all bursts of at most, $(t-1) \times m + 1 = 3 \times 8 + 1 = 25$, consecutive bits in error (see Figures 4.8a and b), and some bursts of $t \times m = 4 \times 8 = 32$ bits in error (Figure 4.8c). It cannot correct any arbitrary pattern of 32 errors (Figure 4.8d).

Bose-Chaudhuri-Hocquenghem (BCH) Codes

Given the prime power, q, and integers, m and d, the *BCH code*, $BCH_{q;m;d}$, is obtained as follows: let $n = q^m$ and $GF(q^m)$ be an extension of $GF(q)$ and let C be the (extended) $[n; n-(d-1); d]_{q^m}$ Reed-Solomon code obtained by evaluating polynomials of degree at most, $(n-d)$, over $GF(q^m)$, at all the points of $GF(q^m)$. Then the code, $BCH_{q;m;d}$, is the $GF(q)$-subfield subcode of C.

A simple argument shows that a BCH code has dimension at least, $n - m(d-1)$ [412]. Indeed, every function, from $GF(q^m)$ to $GF(q)$, is a polynomial over $GF(q)$ of degree at most, $(n-1)$. The space of polynomials, from $GF(q^m)$ to $GF(q)$, is a linear space of dimension, n. Now we ask what is the dimension of the subspace of polynomials of degree at most, $(n-d)$. The restriction that a polynomial, $p(x) = \sum_{i=0}^{n-1} c_i x^i$, has degree at most, $(n-d)$, implies that, $c_i = 0, i = n-(d-1), \ldots, (n-1)$. Each such condition is a linear constraint over $GF(q^m)$, and this translates into a block of, m, linear constraints over $GF(q)$. Since we have $(d-1)$ such blocks, the restriction that the functions have degree at most, $(n-d)$, translates into at most, $m(d-1)$, linear constraints. Thus, the resulting space has dimension at least $n - m(d-1)$.

BCH codes were independently discovered by Hocquenghem [208] and Bose and Ray-Chaudhuri [67]; the extension to the general q-ary case is due to Gorenstein and Zierler [172].

Next, we examine codes used in radio and satellite communication to encode streams of data.

4.16 CONVOLUTIONAL CODES

Binary convolutional codes were introduced by Peter Elias in 1955 [140] as a generalization of block codes. Recall that in the case of a block code we encode, k, information symbols into a block of, n, symbols, and the rate of the code is $R \quad k/n$. In some cases, for example in the case of R-S codes, we use one polynomial from a set of polynomials to generate one codeword. In our presentation we only discuss the encoding algorithm for convolutional codes; the Viterbi algorithm for decoding convolutional codes [439] is not analyzed here.

We consider an infinite sequence of blocks, $\ldots b \ \ _2, b \ \ _1, b_0, b_1, b_2, \ldots$, with k symbols per block. The input symbols are fed sequentially into an encoder that first creates a sequence of *information blocks*, $\ldots In \ \ _2, In \ \ _1, In_0, In_1, In_2, \ldots$, with, n, symbols per block; then the encoder cyclically convolutes the information blocks with a *generating vector* $g \quad (g_0 g_1 \ldots g_m)$, with $g_0 \quad g_m \quad 1$, and produces the output, Out_j; the rate of the code is $R \quad k/n$.

To facilitate the presentation, we consider an *information vector* consisting of, N, bits or, NR, information blocks obtained from, K, input blocks; the output of the encoder is a *codeword* of NR blocks or N bits. The rate of the code is $R \quad K/N \quad k/n$. For example, in the binary case, with $k \quad 1$ and $n \quad 2$, the rate is $R \quad 1/2$. If we choose $K \quad 8$, then the information vector and an output vector consist of $N \quad KR \quad 16$ bits. A rule to construct the information vector could be

$$In \quad InBit_0, InBit_1, \ldots, InBit_{N \ \ 1}, \quad \text{with} \quad InBit_j \quad \begin{cases} b_i & \text{if } j \quad iR, \\ 0 & \text{otherwise.} \end{cases}$$

In this case, each input block consists of one bit, and each information block consists of two bits, the first equal to the corresponding bit of the information block and the second bit being a 0.

Given N, R, and $g \quad (g_0 g_1 \ldots g_m)$, with $m < N$, the *convolutional code*, $\mathcal{C}$, encodes the information vector, $In \quad In_0, In_1, \ldots, In_{N \ \ 1}$, as the N-vector, $Out \quad Out_0, Out_1, \ldots Out_{N \ \ 1}$, with

$$Out_j \qquad \sum_{i \ 0}^{m} g_i In_{j \ \ i}, \quad 0 \quad j \quad N \quad 1.$$

The sum is modulo 2 and all indices are modulo N. We recognize that an N-vector (codeword) results from the *cyclic convolution* of an information vector with the generating vector, hence, the name convolutional code. The main difference between a convolutional encoder and a block code encoder is that the j-th code block, Out_j, depends not only on the current, In_j, but also on the earlier, m, information blocks, $In_{j \ \ 1}, In_{j \ \ 2}, \ldots, In_{j \ \ m}$.

Example. *Encode the frame, $Fr \quad (11101010)$, from the input stream when $N \quad 16$, $R \quad 1/2$, and $g \quad (11.10.11)$.* To facilitate reading, in our examples the blocks of a codeword are separated by periods.

The process of codeword construction is illustrated in Figure 4.9. Each bit of the frame generates two bits of the information vector; Table 4.8 gives the expression of the individual bits of the codeword as the convolution of the generating vector and the information vector. Note that all indices are modulo $N \quad 16$; thus, $In \ \ _1 \quad In_{15}, In \ \ _2 \quad In_{14}, \ldots, In \ \ _5 \quad In_{11}$.

Given N, the length of a block, and R, the rate, the convolutional encoding algorithm discussed in this section generates a linear, $[N, K]$, code if $K \quad R \quad N$. In this case, there is a $K \quad N$ generating

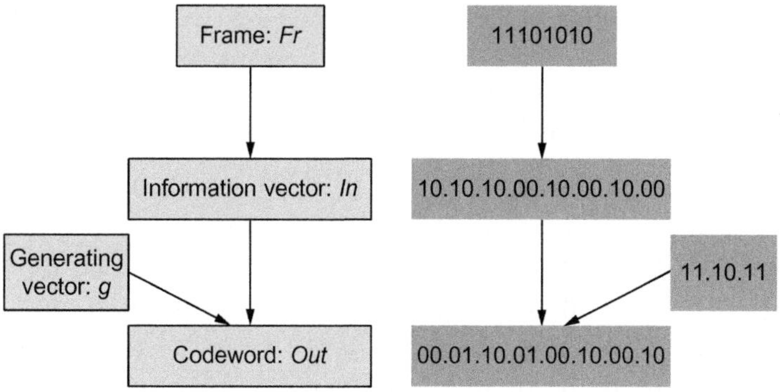

FIGURE 4.9

The construction of the codeword when R $1/2$, the frame size is $N/2$ 8, and the generating polynomial of degree, m 5, is g (11.10.11).

Table 4.8 The Expression of the Individual Bits of the Codeword When Fr (11101010), N 16, R 1/2, and g (11.10.11)

Out_0	$g_0\ In_0$	$g_1\ In_{15}$	$g_2\ In_{14}$	$g_3\ In_{13}$	$g_4\ In_{12}$	$g_5\ In_{11}$	0
Out_1	$g_0\ In_1$	$g_1\ In_0$	$g_2\ In_{15}$	$g_3\ In_{14}$	$g_4\ In_{13}$	$g_5\ In_{12}$	0
Out_2	$g_0\ In_2$	$g_1\ In_1$	$g_2\ In_0$	$g_3\ In_{15}$	$g_4\ In_{14}$	$g_5\ In_{13}$	0
Out_3	$g_0\ In_3$	$g_1\ In_2$	$g_2\ In_1$	$g_3\ In_0$	$g_4\ In_{15}$	$g_5\ In_{14}$	1
Out_4	$g_0\ In_4$	$g_1\ In_3$	$g_2\ In_2$	$g_3\ In_1$	$g_4\ In_0$	$g_5\ In_{15}$	1
Out_5	$g_0\ In_5$	$g_1\ In_4$	$g_2\ In_3$	$g_3\ In_2$	$g_4\ In_1$	$g_5\ In_0$	0
Out_6	$g_0\ In_6$	$g_1\ In_5$	$g_2\ In_4$	$g_3\ In_3$	$g_4\ In_2$	$g_5\ In_1$	0
Out_7	$g_0\ In_7$	$g_1\ In_6$	$g_2\ In_5$	$g_3\ In_4$	$g_4\ In_3$	$g_5\ In_2$	1
Out_8	$g_0\ In_8$	$g_1\ In_7$	$g_2\ In_6$	$g_3\ In_5$	$g_4\ In_4$	$g_5\ In_3$	0
Out_9	$g_0\ In_9$	$g_1\ In_8$	$g_2\ In_7$	$g_3\ In_6$	$g_4\ In_5$	$g_5\ In_4$	0
Out_{10}	$g_0\ In_{10}$	$g_1\ In_9$	$g_2\ In_8$	$g_3\ In_7$	$g_4\ In_6$	$g_5\ In_5$	1
Out_{11}	$g_0\ In_{11}$	$g_1\ In_{10}$	$g_2\ In_9$	$g_3\ In_8$	$g_4\ In_7$	$g_5\ In_6$	0
Out_{12}	$g_0\ In_{12}$	$g_1\ In_{11}$	$g_2\ In_{10}$	$g_3\ In_9$	$g_4\ In_8$	$g_5\ In_7$	0
Out_{13}	$g_0\ In_{13}$	$g_1\ In_{12}$	$g_2\ In_{11}$	$g_3\ In_{10}$	$g_4\ In_9$	$g_5\ In_8$	0
Out_{14}	$g_0\ In_{14}$	$g_1\ In_{13}$	$g_2\ In_{12}$	$g_3\ In_{11}$	$g_4\ In_{10}$	$g_5\ In_9$	1
Out_{15}	$g_0\ In_{15}$	$g_1\ In_{14}$	$g_2\ In_{13}$	$g_3\ In_{12}$	$g_4\ In_{11}$	$g_5\ In_{10}$	0

matrix, G, of the code [227]. Then:

$$Out \quad Fr\ \mathrm{G}.$$

The K rows of the generator matrix are as follows: the first row consists of the m 1 bits of the generating vector followed by $N - (m$ 1) zeros; rows 2 to K are cyclic shifts of the previous row $1/R$ positions to the right.

Example. *The generator matrix of the convolutional code, with $N \quad 16$, $R \quad 1/2$, and $g \quad (11.10.11)$,* is an $8 \quad 16$ matrix. The first $m \quad 1 \quad 6$ elements of the first row are the elements of g, (111011), followed by $N \quad (m \quad 1) \quad 16 \quad 6 \quad 10$ zeros; the next $K \quad 1 \quad 7$ rows are cyclic shifts to the right of the previous row, with two positions as $1/R \quad 2$.

$$G \quad \begin{matrix} 1 & 1 & 1 & 0 & 1 & 1 & 0 & 0 & 0 & 0 & 0 & 0 & 0 & 0 & 0 & 0 \\ 0 & 0 & 1 & 1 & 1 & 0 & 1 & 1 & 0 & 0 & 0 & 0 & 0 & 0 & 0 & 0 \\ 0 & 0 & 0 & 0 & 1 & 1 & 1 & 0 & 1 & 1 & 0 & 0 & 0 & 0 & 0 & 0 \\ 0 & 0 & 0 & 0 & 0 & 0 & 1 & 1 & 1 & 0 & 1 & 1 & 0 & 0 & 0 & 0 \\ 0 & 0 & 0 & 0 & 0 & 0 & 0 & 0 & 1 & 1 & 1 & 0 & 1 & 1 & 0 & 0 \\ 0 & 0 & 0 & 0 & 0 & 0 & 0 & 0 & 0 & 0 & 1 & 1 & 1 & 0 & 1 & 1 \\ 1 & 1 & 0 & 0 & 0 & 0 & 0 & 0 & 0 & 0 & 0 & 0 & 1 & 1 & 1 & 0 \\ 1 & 0 & 1 & 1 & 0 & 0 & 0 & 0 & 0 & 0 & 0 & 0 & 0 & 0 & 1 & 1 \end{matrix} .$$

When $Fr \quad (11101010)$, it is easy to check that $Out \quad (00.01.10.01.00.10.00.10)$, a result consistent with the previous example. Indeed, $(11101010) \quad G \quad (00.01.10.01.00.10.00.10)$.

Proposition. *Every symbol encoded by a convolutional code depends on at most, $M \quad 1$, previous information symbols, where the integer, $M \quad R \quad (m \quad 1) \quad 1$, is called the* memory *of the convolutional code.*

Proof. Recall that $Out_j \quad {}^{m}_{i \; 0} g_i In_{j \; i}$; thus, each symbol of the resulting N-vector is the result of the cyclic convolution of $(m \quad 1)$ symbols of the information vector. Let us concentrate on the first of these $(m \quad 1)$ symbols of the information vector, the symbol, In_j, inherited from the i-th symbol of the frame, Fr_i, with $j \quad i/R$. Similarly, the information symbol, $In_{j_{out}}$, is inherited from the symbol, $Fr_{M \; i \; 1}$, with $j_{out} \quad (M \quad i \quad 1)/R$. It is easy to see that $Fr_{M \; i \; 1}$ is outside of the range of the summation required by convolution. To show this, we recall that when $a \quad b \quad r$ and $r < 1$, then $a/r \quad b$. In our case, $M \quad 1 \quad R \quad (m \quad 1)$, with $R < 1$, thus, $M \quad 1/R \quad M \quad 1$. Recall that $j \quad i/R$, so we conclude that

$$\frac{M \quad i \quad 1}{R} \quad \frac{i}{R} \quad \frac{M \quad 1}{R} \quad j \quad M \quad 1.$$

It follows that Out_j may only depend on at most, $(M \quad 1)$, symbols of the original frame. ■

The cyclic convolution rule formulated at the beginning of this section when used with relatively short frames justifies the name *tail-biting code* given to this class of codes. The encoding rule we discussed so far is too restrictive. Instead of a generating vector, we can consider a periodic function, g, alternating between, k, generating vectors or polynomials; in this case the generating matrix consists of k different types of rows.

An (n,k,M) *convolutional code* is a sequence of block codes of length, $N \quad q \quad n$, encoded using the cyclic convolution rule; q is an integer greater than 1. A *terminated, (n,k,M), convolutional code* is an (n,k,M) convolutional code, with the last, M, input bits set to 0.

Example. *An encoder for a convolutional code, with multiple generator polynomials.* The encoder consists of, m, registers each holding one input bit and n modulo 2 adders. Each adder has a generator polynomial.

The encoded sequence of, say, q input bits is of length, $4q$, and represents a path in the state transition diagram. The diagram in Figure 4.11 shows that not all transitions are possible and gives us a hint at the decoding procedure for convolutional codes.

A sequence of symbols is valid only if it fits the state transition diagram of the encoder. If a sequence does not fit the state transition diagram of the encoder, we have to identify the nearest sequence that fits the state transition diagram.

The encoder has, 2^n, states and operates as follows:

- Initially, the, m, input registers are all set to 0.
- An input bit is fed to the m-bit register.
- The encoder outputs, n, bits, each computed by one of the, n, adders.
- The next input bit is fed and the register shifts one bit to the right.
- The next group of n bits is calculated.

We present an encoder, with m 1, n 4, and k 1, in Figure 4.10; the encoder is a finite state machine and its transition diagram is shown in Figure 4.11.

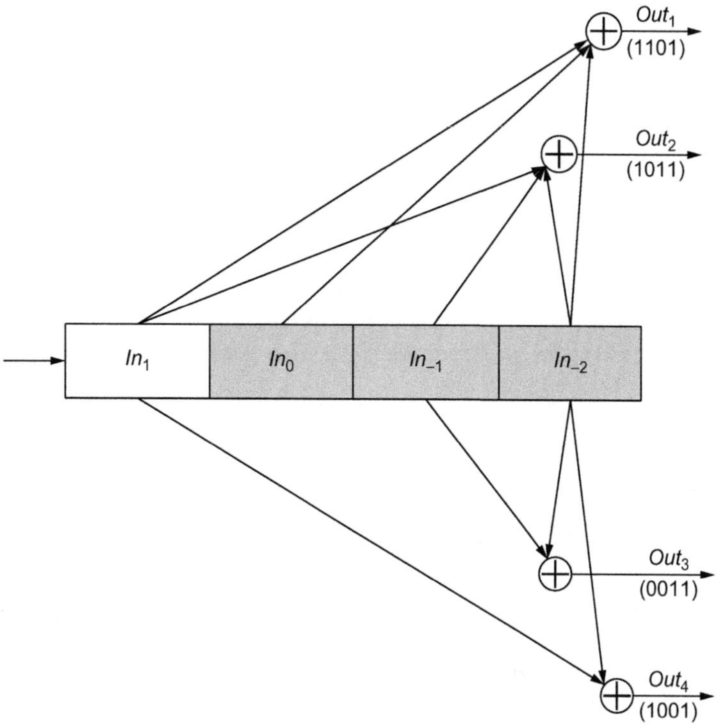

FIGURE 4.10

A convolutional nonrecursive, nonsystematic encoder, with k 1 and n 4, and generator polynomials, $g_1(x)$ x^3 x^2 1, $g_2(x)$ x^3 x 1, $g_3(x)$ x 1, and $g_4(x)$ x^3 1.

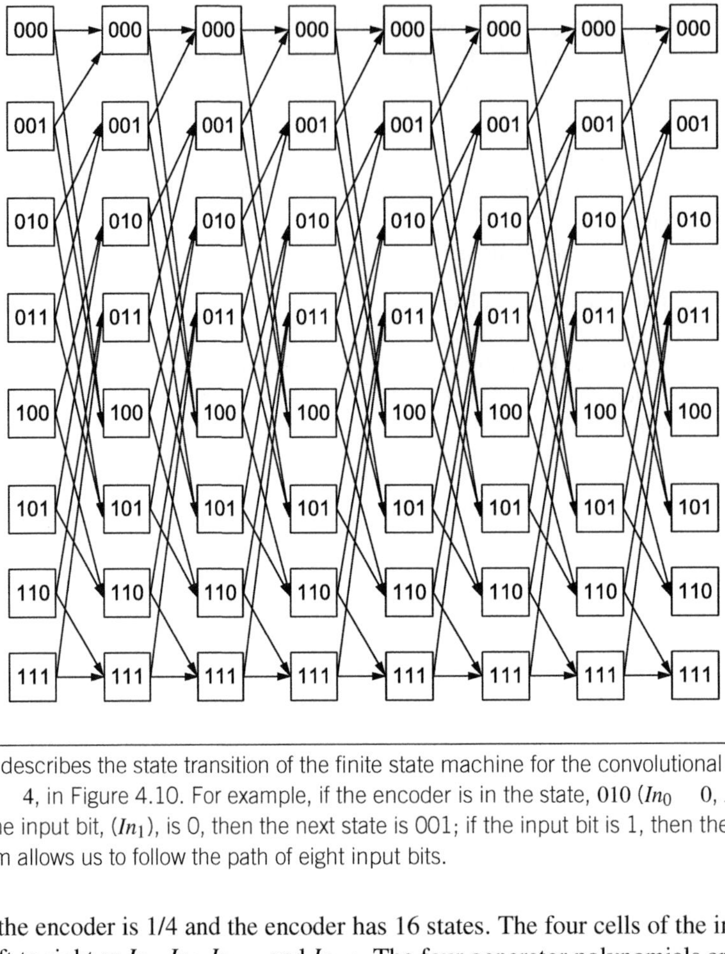

FIGURE 4.11

A trellis diagram describes the state transition of the finite state machine for the convolutional code encoder, with $k\ \ 1$ and $n\ \ 4$, in Figure 4.10. For example, if the encoder is in the state, 010 ($In_0\ \ 0$, $In_{-1}\ \ 1$, and $In_{-2}\ \ 0$), and the input bit, (In_1), is 0, then the next state is 001; if the input bit is 1, then the next state is 101. The diagram allows us to follow the path of eight input bits.

The rate of the encoder is 1/4 and the encoder has 16 states. The four cells of the input register are labeled from left to right as In_1, In_0, In_{-1}, and In_{-2}. The four generator polynomials are

$$g_1(x)\ \ x^3\ \ x^2\ \ 1,\quad g_2(x)\ \ x^3\ \ x\ \ 1,\quad g_3(x)\ \ x\ \ 1,\quad \text{and}\quad g_4(x)\ \ x^3\ \ 1.$$

Thus, the four output symbols are calculated as follows:

$$Out_1\ \ In_1\ \ In_0\ \ In_{-2} \bmod 2 \quad\Big\|\quad Out_2\ \ In_1\ \ In_{-1}\ \ In_{-2} \bmod 2$$
$$Out_3\ \ In_{-1}\ \ In_{-2} \bmod 2 \quad\Big\|\quad Out_4\ \ In_1\ \ In_{-2} \bmod 2.$$

Table 4.9 summarizes the state transitions. The state is given by the contents of memory cells, In_0, In_{-1}, In_{-2}; the input is In_1. If at time, t, the system is in the state 010 ($In_0\ \ 0, In_{-1}\ \ 1$, and $In_{-2}\ \ 0$) and

- the input is $In_1\ \ 0$, then the next state of the encoder is 001 and the four output bits produced by the encoder are 0110 as

$$Out_1\ \ 0\ \ 0\ \ 0\ \ 0,\quad Out_2\ \ 0\ \ 1\ \ 0\ \ 1,\quad Out_3\ \ 1\ \ 0\ \ 1,\quad Out_4\ \ 0\ \ 0\ \ 0;$$

Table 4.9 The State Transitions for the Convolutional Encoder

Time	Current State	Input	Input Register	Output Register	Next State
t	010	0	0010	0110	001
		1	1010	1011	101
t 1	001	0	0001	1111	000
		1	1001	0010	100
	101	0	0101	0111	010
		1	1101	1010	110

- the input is In_1 1, then the next state of the encoder is 101 and the four output bits produced by the encoder are 1011 as

$$Out_1 \quad 1 \quad 0 \quad 0 \quad 1, \quad Out_2 \quad 1 \quad 1 \quad 0 \quad 0, \quad Out_3 \quad 1 \quad 0 \quad 1, \quad Out_4 \quad 1 \quad 0 \quad 1.$$

If at time, t 1, the system is in the state 001 and

- the input is In_1 0, then the next state of the encoder is 000 and the four output bits produced by the encoder are 1111 as

$$Out_1 \quad 0 \quad 0 \quad 1 \quad 1, \quad Out_2 \quad 0 \quad 0 \quad 1 \quad 1, \quad Out_3 \quad 0 \quad 1 \quad 1, \quad Out_4 \quad 0 \quad 1 \quad 1;$$

- the input is In_1 1, then the next state of the encoder is 100 and the four output bits produced by the encoder are 0010 as

$$Out_1 \quad 1 \quad 0 \quad 1 \quad 0, \quad Out_2 \quad 1 \quad 0 \quad 1 \quad 0, \quad Out_3 \quad 0 \quad 1 \quad 1, \quad Out_4 \quad 1 \quad 1 \quad 0.$$

Figure 4.12 shows the state transitions for a sequence of two input bits, the contents of the memory register, and the four output bits.

Convolutional codes are widely used in applications such as digital video, radio, mobile communication, and satellite communication. We now focus our attention on other families of codes able to encode large data frames and capable of approaching the Shannon limit.

4.17 PRODUCT CODES

We first discuss the encoding of large blocks of data and methods to construct composite codes from other codes. Then we survey some of the properties of product codes.

The question we address is how to encode large frames and decode them efficiently. The Shannon channel encoding theorem tells us that we are able to transmit reliably at rates close to channel capacity, even though the number of errors per frame, e p N, increases with the frame size, N, and with the probability of a symbol error, p. We also know that decoding complexity increases exponentially with N; indeed, after receiving a frame encoded as a codeword, we check if errors have occurred and, if so, compute the likelihood of every possible codeword to correct the error(s).

A trivial solution is to split the frame into N/n blocks of size, n, and use short codes with small n. We distinguish *decoder errors,* the cases when an n-tuple in error is decoded as a valid codeword, from

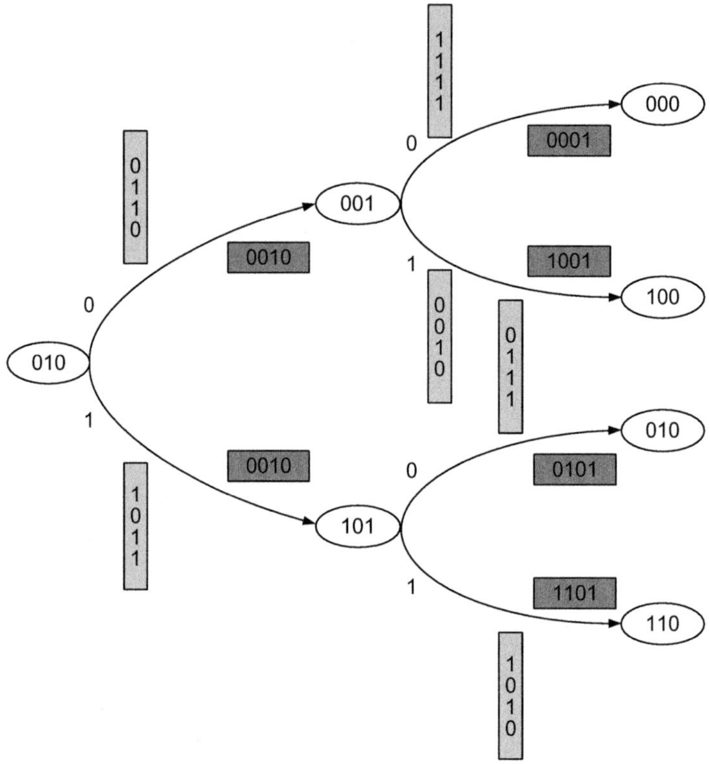

FIGURE 4.12

The state transition of the encoder in Figure 4.10 for a sequence of two input bits. The state is given by the contents of memory cells, In_0, In_{-1}, In_{-2}; the input is In_1. The vertical box next to a transition shows the output, $Out_1, Out_2, Out_3, Out_4$, generated by the encoder, and the horizontal box shows the contents of the four cells, $In_1, In_0, In_{-1}, In_{-2}$.

decoding failures, when the decoder fails to correct the errors; if we denote by p_{err} the probability of a decoding error and by p_{fail} the probability of a decoding failure, then P_{cF}, the probability of a correct frame, and P_{eF}, the probability of a frame error, are

$$P_{cF} \quad (1-p_{err})^{\frac{N}{n}} \quad \text{and} \quad P_{eF} \quad 1-(1-p_{fail})^{\frac{N}{n}}.$$

The undetected errors have a small probability only if the error rate and the block size, n, are small; but a small n leads to a large P_{eF}. Thus, we have to search for other means to encode large frames. A possible solution is to construct codes from other codes using one of the following strategies:

- *Parity check.* Add a parity bit to an $[n,k,2t-1]$ code to obtain an $[n \quad 1,k,2t]$ code.
- *Punctuating.* Delete a coordinate of an $[n,k,d]$ code to obtain an $[n-1,k,d-1]$ code.
- *Restricting.* Take the subcode corresponding to the first coordinate being "the most common element" and then delete the first coordinate: $[n,k,d] \quad [n-1,k-1,d]$.
- *Direct product.* Use two or more codes to encode a block of data.

The methods to construct codes from other codes weaken codes asymptotically, with one exception, the product codes discussed next.

A *product code* is a vector space of n_1 n_2 arrays. Each row of this array is a codeword of a linear, $[n_1, k_1, d_1]$, code, and each column is a codeword of the linear code, $[n_2, k_2, d_2]$.

Proposition. *C, the product code of C_1 and C_2, $[n_1, k_1, d_1]$ and $[n_2, k_2, d_2]$ linear codes, respectively, is an $[n_1$ n_2, k_1 k_2, d_1 $d_2]$ linear code.*

Proof. We arrange the information symbols into an k_1 k_2 array and compute the parity-check symbols on rows and columns independently, using C_1 for the rows and C_2 for the columns; the parity checks on the parity checks computed either using C_1 or C_2 are the same due to linearity (Figure 4.13). It follows that C is an $[n_1$ n_2, k_1 $k_2]$ linear code.

The minimum distance codewords of C are those resulting from repeating a codeword of minimum weight equal to d_1 of C_1 in at most, d_2, rows; the weight of these codewords is d_1 d_2, and this is the distance of C. ■

Proposition. *If a_1 and a_2 are the numbers of minimum weight codewords in C_1 and C_2, respectively, then the product code, C, has a_1 a_2 codewords of minimum weight d_1 d_2.*

Proof. The minimum weight of any row is d_1, and there can be at most, a_1, such rows; the minimum weight of any column is d_2, and there can be at most, a_2, such columns. A minimum weight codeword of C is one, corresponding to a choice of a row and of a column of minimum weight; there are only a_1 a_2 such choices of weight d_1 d_2. ■

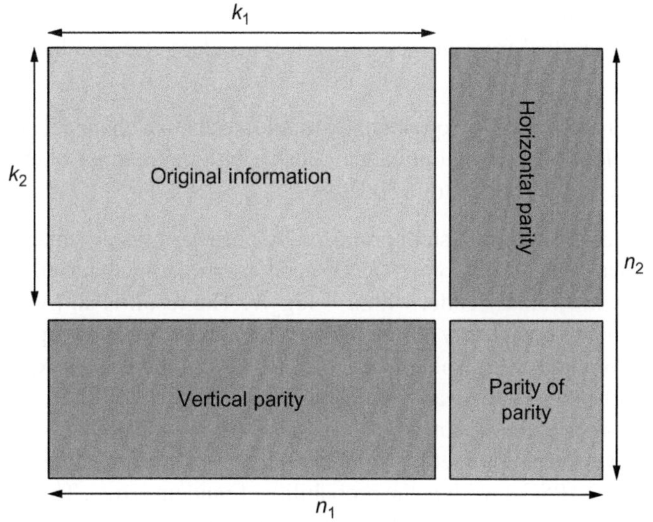

FIGURE 4.13

Schematic representation of product codes. Horizontal, vertical, and parity of parity symbols are added to the original information.

Serial decoding of a product code $\mathcal{C}$: Each column is decoded separately using the decoding algorithms for $\mathcal{C}_2$; then errors are corrected and the decoding algorithm for $\mathcal{C}_1$ is applied to each row.

Proposition. *When both $\mathcal{C}_1$ and $\mathcal{C}_2$ are $[n,k,d]$ codes, the resulting product code, $\mathcal{C}$, is an $[n^2,k^2,d^2]$ linear code and can correct any error pattern of at most, $e\quad(d^2\quad 1)/4$, errors when serial decoding is used.*

Proof. $e\quad(d^2\quad 1)/4\quad[(d\quad 1)/2][(d\quad 1)/2]$; if $d\quad 2t\quad 1$, then $e\quad t(t\quad 1)$ and several columns affected by $t\quad 1$ errors will be decoded incorrectly; indeed, $\mathcal{C}_2$, the code used for encoding the columns, is an $[n,k,d]$ code and can correct at most, t, errors. But, as the total number of errors is limited by $e\quad t(t\quad 1)$, we may have at most, t, such columns affected by errors; thus, the maximum number of errors left after the first decoding step in any row is t. These errors can be corrected by $\mathcal{C}_1$, the $[n,k,d]$ row code. ■

In the next section, we introduce concatenated codes; we show that as the length of a codeword increases, concatenated codes can be decoded more efficiently than other classes of composite codes.

4.18 SERIALLY CONCATENATED CODES AND DECODING COMPLEXITY

The concatenated codes were discovered by David Forney [153] at a time when the decoding problem was considered solved through (1) long, randomly chosen convolutional codes with a dynamically branching tree structure and (2) sequential decoding, namely, exclusive tree search techniques characterized by a probabilistic distribution of the decoding computations. The balance between code performance and decoding complexity was tilted in favor of code performance, measured by the number of errors corrected by the code.

Concatenated codes resulted from Forney's search for a class of codes having two simultaneous properties:

1. The probability of error decreases exponentially at all rates below channel capacity.
2. The decoding complexity increases only algebraically with the number of errors corrected by the code and, implicitly, with the length of a codeword.

Figure 4.14 illustrates the coding system proposed by Forney in his dissertation; a message is first encoded using an "outer" Reed-Solomon $[N,K,D]$ code over a large alphabet, with N symbols, and then with an "inner" $[n,k,d]$ binary code, with $k\quad \log_2 N$. The result is an $[Nn,Kk,Dd]$ binary code (Figure 4.15). The decoding complexity of the code is dominated by the complexity of the algebraic decoder for the Reed-Solomon code, while the probability of error decreases exponentially with n, the code length, at all rates below channel capacity.

Information transmission at the Shannon rate is achievable using an efficient encoding and decoding scheme, and Forney's concatenation method provides such an efficient decoding scheme. Consider a binary symmetric channel, (BSC_p), with p, the probability of random errors. The following formulation of Shannon channel coding theorem [412] states that reliable information transmission through a binary symmetric channel is feasible at a rate

$$R\quad 1\quad H(p)\quad \epsilon,\quad \text{with}\quad \epsilon > 0.$$

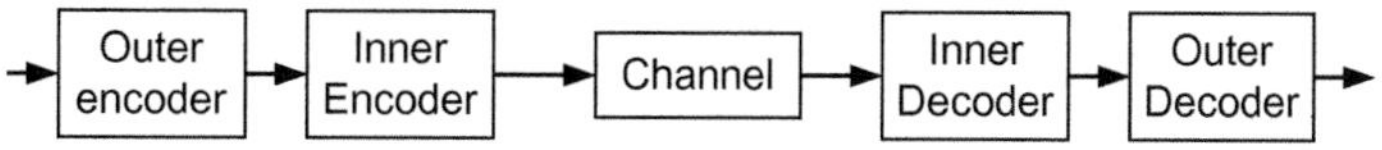

FIGURE 4.14

The original concatenated code was based on a relatively short random "inner" code and used maximum likelihood decoding to achieve a modest error probability, Prob(*error*) $\approx 10^{-2}$, at a near capacity rate. A long, high-rate Reed-Solomon code and a generalized-minimum-distance decoding scheme were used for the "outer" level.

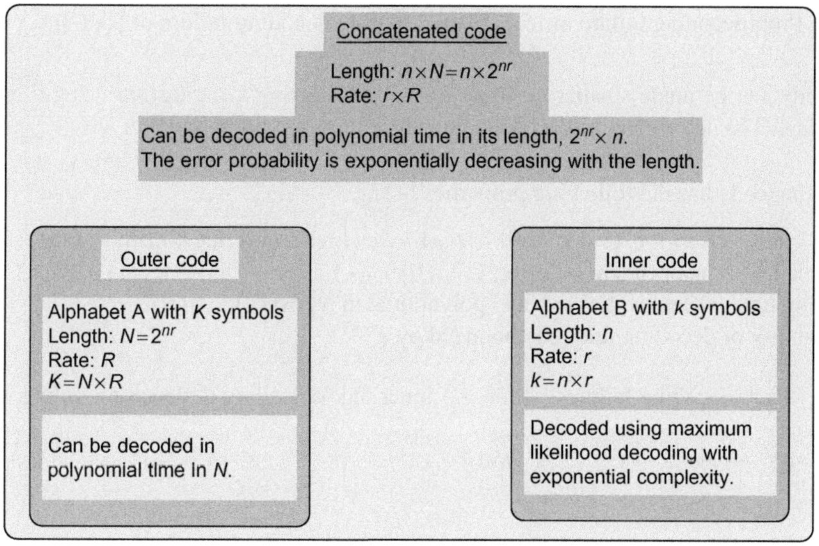

FIGURE 4.15

The "inner" and the "outer" codes of a concatenated code.

Theorem. *There exists a code, C, with an encoding algorithm, $\mathcal{E}: \{0,1\}^k \to \{0,1\}^n$, and a decoding algorithm, $\mathcal{D}: \{0,1\}^n \to \{0,1\}^k$, with $k \geq (1 - H(p) - \epsilon) \cdot n$, such that*

$$\text{Prob(decoding error for } BSC_p) < e^{-n}.$$

A corollary of this theorem, discussed next, shows that given a binary symmetric channel, if we select an "outer" Reed-Solomon, $(N, K, \epsilon N)$, code, with $K \geq (1 - \epsilon)N$, and an "inner" binary, (n, k, d), code, with $k \geq (1 - H(p) - \epsilon) \cdot n$, then we should be able to correct random errors occurring with the probability, p, and in this process, limit the decoding failures by e^{-n} when $k \approx log_2 N$.

Corollary. *There exists a code, C, with an encoding algorithm, $\mathcal{E} : \{0,1\}^K \to \{0,1\}^N$, and a decoding algorithm, $\mathcal{D} : \{0,1\}^N \to \{0,1\}^K$, such that*

$$\text{Prob(decoding error)} < \frac{1}{N},$$

where $\mathcal{E}$ and $\mathcal{D}$ are polynomial time algorithms.

Proof. We divide a message, $m \quad 0,1^{\;K}$, into K/k blocks of size, $k \quad \log K$. Then we use Shannon's encoding function, $\mathcal{E}$, to encode each of these blocks into words of size, n, and concatenate the results, such that $N \quad (K/k)n$. The rate is preserved:

$$\frac{K}{N} \quad \frac{k}{n}.$$

It follows that

$$\text{Prob(decoding failure of } [\mathcal{E}\;, \mathcal{D}\;]_K) \quad \frac{K}{k}\text{Prob(decoding failure of } [\mathcal{E}, \mathcal{D}]_k).$$

This probability can be made smaller or equal to $\frac{1}{KN}$ by selecting $k \quad \textit{constant} \quad \log K$, with a large enough constant. The new encoding and decoding algorithms are polynomial in N. ■

The resulting code has the following properties [412]:

1. A rate, $R \quad (1 \quad H(p) \quad \epsilon)(1 \quad \epsilon) \quad 1 \quad H(p) \quad 2\epsilon$, lower than the Shannon rate for a random error channel with independent bit error, $(1 \quad H(p) \quad \epsilon)$.
2. The running time is exponential in k and polynomial in N, $\exp(k)\text{poly}(N)$.
3. The probability of decoding failure is bounded by $e^{\quad nN}$:

$$\begin{aligned}\text{Prob(decoding failure)} \quad & \text{Prob} \; \tfrac{\epsilon N}{2} \text{ inner blocks leading to decoding failure} \\ & \binom{N}{\epsilon N/2} e^{\quad n\epsilon N/2} \\ & e^{\quad nN}.\end{aligned}$$

The decoding effort can be broken into the time to decode the R-S code, $\mathcal{O}(n^2)$, and the time to decode the "inner code," n, times (once for each "inner" codeword for $n \log n$ bits). Thus, the decoding complexity for the concatenated code is $\mathcal{O}(n^2 \log n)$.

Concatenated codes with an interleaver placed between the outer and the inner encoder to spread out bursts of errors and with a deinterleaver between the inner and outer decoder have been used for space exploration since the late 1970s. The serially concatenated codes have a very large minimum distance; they could be improved if the inner decoder could provide information to the outer decoder, but such a scheme would introduce an asymmetry between the two. The turbo codes discussed next address these issues.

4.19 PARALLEL CONCATENATED CODES: TURBO CODES

Turbo codes were introduced by Claude Berrou and his coworkers in 1993 as a refinement of concatenated codes [51]. The encoding structure of concatenated codes is complemented by an iterative algorithm for decoding; by analogy with a turbo engine, the decoder uses the processed output values as a posteriori input for the next iteration, hence, the name "turbo," as shown in Figure 4.16. Turbo codes achieve small bit error rates at information rates much closer to the capacity of the communication

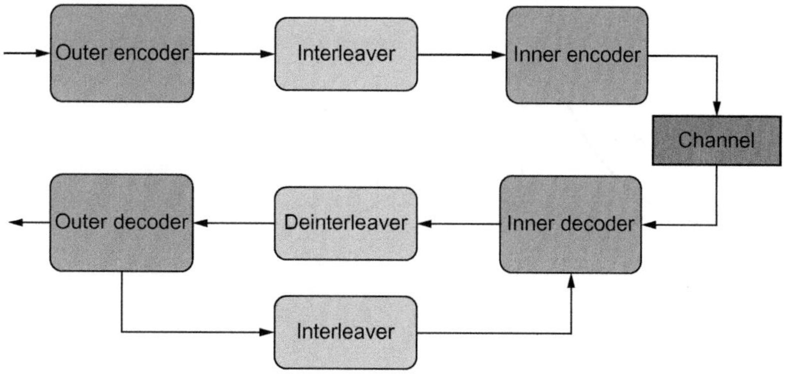

FIGURE 4.16

The feedback from the outer to the inner decoder of a turbo code.

channel than previously possible. If the rates of the inner and outer codes are, r and R, respectively, then the rate of a serially concatenated code is $R_s = r \times R$, while the rate of a parallel concatenated code is

$$R_p = \frac{r \times R}{1 - (1 - r)(1 - R)}.$$

We start the discussion of turbo codes with a review of *likelihood functions*. Consider a random variable, X, with the probability density function, $p_X(x)$; the result of an observation of the random variable X should be classified in one of, M, classes. Bayes's theorem expresses the "a posteriori probability (APP)" of a decision, d, to classify the observation of X in a class, i, $d = i$, $1 \leq i \leq M$, conditioned by an observation, x, as

$$\text{Prob}(d = i \mid x) = \frac{p_X[x \mid (d = i)]\text{Prob}(d = i)}{p_X(x)}, \quad 1 \leq i \leq M,$$

where

$$p_X(x) = \sum_{i=1}^{M} p_X[x \mid (d = i)]\text{Prob}(d = i).$$

Two decisions rules are widely used, *ML, maximum likelihood* (Figure 4.17) and, *MAP, maximum a posteriori*.

When d is a binary decision, $d \in \{0, 1\}$,

$$p_X(x) = p_X[x \mid (d = 0)]\,[\text{Prob}(d = 0) + p_X[x \mid (d = 1)]\,]\text{Prob}(d = 1),$$

and the *maximum a posteriori decision* is a minimum probability error rule formulated as:

$\text{Prob}(d = 0 \mid x) > \text{Prob}(d = 1 \mid x) \Rightarrow d = 0$ and $\text{Prob}(d = 1 \mid x) > \text{Prob}(d = 0 \mid x) \Rightarrow d = 1.$

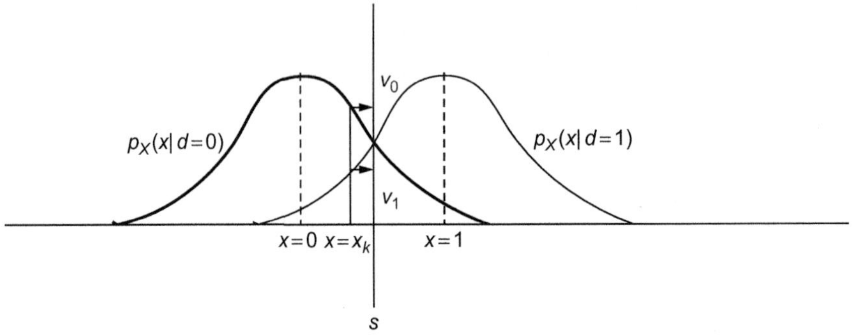

FIGURE 4.17

The probability distribution functions, $p_X[x\ (d\quad 0)]$ and $p_X[x\ (d\quad 1)]$, for a continuous random variable, X, and a binary decision. The decision line is s. In the case of maximum likelihood, we decide that x_k, the value of the random variable, X, observed at time, k, is 0 $(d\quad 0)$ because $v_0 > v_1$; x_k falls on the left of the decision line.

These conditions can be expressed as

$$\frac{p_X[x\ (d\quad 1)]}{p_X[x\ (d\quad 0)]} < \frac{\text{Prob}(d\quad 0\ x)}{\text{Prob}(d\quad 1\ x)} \qquad d\quad 0 \quad \text{and} \quad \frac{p_X[x\ (d\quad 1)]}{p_X[x\ (d\quad 0)]} > \frac{\text{Prob}(d\quad 0\ x)}{\text{Prob}(d\quad 1\ x)} \qquad d\quad 1.$$

Next, we introduce the *log-likelihood ratios, LLR*, real numbers used by a decoder; the absolute value of LLR quantifies the reliability of the decision and represents the "soft" component of the decision, while the sign selects either the value 0 or the value 1 and represents the "hard" component of the decision. The LLR, the logarithm of the ratio of probabilities required by the MAP rule, is

$$L(d\ x)\quad \log \frac{\text{Prob}(d\quad 1\ x)}{\text{Prob}(d\quad 0\ x)} \qquad \log \frac{p_X[x\ (d\quad 1)]\text{Prob}(d\quad 1)}{p_X[x\ (d\quad 0)]\text{Prob}(d\quad 0)}$$

or

$$L(d\ x)\quad \log \frac{p_X[x\ (d\quad 1)]}{p_X[x\ (d\quad 1)]} \qquad \log \frac{\text{Prob}(d\quad 1)}{\text{Prob}(d\quad 0)}.$$

We rewrite the previous equality as

$$L_X(d)\quad L_{channel}(x)\quad L_X(d).$$

In this expression, $L_{channel}(x)$ is the result of a measurement performed on the output of the channel, and $L_X(d)$ is based on the a priori knowledge of the input data.

A turbo decoder is at the heart of the scheme discussed in this section. As we can see in Figure 4.18, the decoder needs as input:

- a channel measurement,
- the a priori knowledge about the data transmitted, and
- feedback from the previous iteration, called "extrinsic" information.

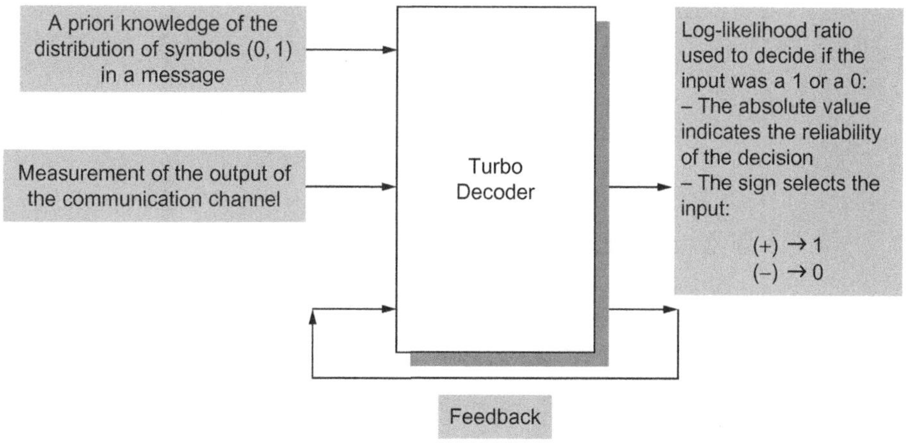

FIGURE 4.18

The decoder of a turbo code uses feedback to complement a priori information and the result of output channel measurement.

The soft output of the decoder, denoted as $L(d)$, will consist of two components, $L_X(d)$, the LLR representing the soft decision of the decoder, and $L_{extrn}(d)$, the feedback re ecting the knowledge from the previous iteration of the decoding process:

$$L(d) \quad L_X(d) \quad L_{extrn}(d) \quad L_{channel}(x) \quad L_X(d) \quad L_{extrn}(d).$$

We illustrate these ideas with an example for a two-dimensional code [393]. An $[n_1, k_1]$ code is used for horizontal encoding and an $[n_2, k_2]$ code is used for vertical encoding (Figure 4.19).

At each step, we compute also the horizontal and vertical extrinsic LLRs, $L_{extrnHoriz}$ and $L_{extrnVert}$, respectively. The decoding algorithm consists of the following steps:

1. If a priori probability is available, set $L(d)$ to that value; if not, set $L(d) \quad 0$.
2. Decode horizontally and set $L_{extrnHoriz}(d) \quad L(d) \quad L_{chan}(x) \quad L(d)$.
3. Set $L(d) \quad L_{extrnHoriz}(d)$.
4. Decode vertically and set $L_{extrnVert}(d) \quad L(d) \quad L_{chan}(x) \quad L(d)$.
5. Set $L(d) \quad L_{extrnVert}(d)$.
6. Repeat steps 2–5 until a reliable decision can be made, then provide the soft output $L(d) \quad L_{chan}(x) \quad L_{extrnHoriz}(d) \quad L_{extrnVert}(d)$.

To verify that indeed the reliability of a decision made using two independent random variables is the minimum of the absolute value of individual LLRs and the "hard" component of the decision is given by the signs of the two LLRs, we consider symbol-by-symbol MAP estimation of a set of symbols, c_k, of a transmitted vector of symbols, c, encoded, transmitted through a communication channel

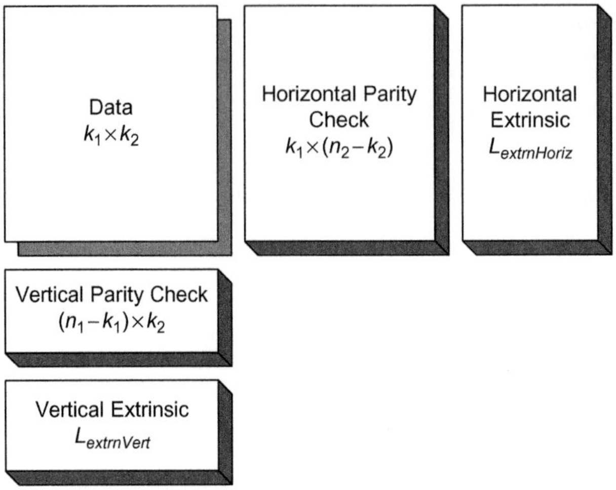

FIGURE 4.19

Schematic representation of a two-dimensional turbo code constructed from a horizontal, $[n_1, k_1]$, code and a vertical, $[n_2, k_2]$, code. The data are represented by a, $k_1 \quad k_2$, rectangle, the horizontal parity-check bits by a, $k_1 \quad (n_2 \quad k_2)$, v rectangle, and the vertical parity-check bits by a, $k_2 \quad (n_1 \quad k_1)$, rectangle. The horizontal extrinsic and the vertical extrinsic LLR bits computed at each step are shown.

subject to a Gauss-Markov process[6] according to the distribution, $p_{X\ Y}(r\ c)$, and finally received as a vector, $v \quad (v_1, v_2, \ldots, v_k, \ldots)$.

To distinguish between two distinct random variables, instead of $L(d)$ we use the notation, $L_X(d)$, and assume that

$$P_X(x \quad 0) \quad \frac{e^{\ \ L_X(d)}}{1 \quad e^{\ \ L_X(d)}} \quad \text{and} \quad P_X(x \quad 1) \quad \frac{e^{L_X(d)}}{1 \quad e^{L_X(d)}}.$$

If X_1 and X_2 are independent binary random variables, then

$$L_{X_1, X_2}(d_1 \quad d_2) \quad \log \frac{1 \quad e^{L_{X_1}(d_1)} e^{L_{X_2}(d_2)}}{e^{L_{X_1}(d_1)} \quad e^{L_{X_2}(d_2)}},$$

with , the addition modulo 2. This expression becomes

$$L_{X_1, X_2}(d_1 \quad d_2) \quad \text{sign}[L_{X_1}(d_1)] \ \text{sign}[L_{X_2}(d_2)] \ \min[\ L_{X_1}(d_1)\ ,\ L_{X_2}(d_2)\ ].$$

This brief introduction of turbo codes concludes our presentation of classical error-correcting codes.

---

[6]A Gauss-Markov process is a stochastic process, $X(t)$, with three properties: (1) If $h(t)$ is a nonzero scalar function of, t, then $Z(t) \quad h(t)X(t)$ is also a Gauss-Markov process. (2) If $f(t)$ is a nondecreasing scalar function of, t, then $Z(t) \quad X(f(t))$ is also a Gauss-Markov process. (3) There exists a nonzero scalar function, $h(t)$, and a nondecreasing scalar function, $f(t)$, such that $X(t) \quad h(t)W(f(t))$, where $W(t)$ is the standard Wiener process.

4.20 HISTORY NOTES

While Claude Shannon was developing the information theory, Richard Hamming, a colleague of Shannon's at Bell Labs, understood that a more sophisticated method than the parity checking used in relay-based computers was necessary. Hamming realized the need for error correction, and in the early 1950s, he discovered the single-error–correcting binary Hamming codes and the single-error–correcting, double-error–detecting extended binary Hamming codes. Hamming's work marked the beginning of coding theory; he introduced fundamental concepts of coding theory, such as Hamming distance, Hamming weight, and Hamming bound.

In 1955, Peter Elias introduced probabilistic convolutional codes with a dynamic structure pictured at that time as a branching tree [140], and in 1961, John Wozencraft and Barney Reiffen proposed *sequential decoding* based on exhaustive tree search techniques for long convolutional codes [462]. The probability that the number of operations required for the sequential decoding exceeds a threshold, D, for an optimal choice of algorithm parameters is a Pareto (algebraic) distribution:

$$\text{Prob}(\text{Number of Operations} > D) \quad D^{\alpha(R)},$$

with $\alpha(R) > 1$, and, R, the transmission rate; this implies that the number of operations is bounded when $R < R_0$ and R_0 was regarded as the "practical capacity" of a memoryless channel [154]. In 1963, Robert Fano developed an efficient algorithm for sequential decoding [144].

Work on algebraic codes started in the late 1950s with codes over Galois fields [67] and the Reed-Solomon codes [353]. This line of research continued throughout the 1960s with the development of cyclic codes, including the Bose-Chaudhuri-Hocquenghem codes, and the publication of the seminal book of Elwyn Berlekamp [49].

In the early 1960s, the center of gravity in the field of error-correcting codes moved to MIT. The information theory group at MIT was focused on two problems: (a) how to reduce the probability of error of block codes, and (b) how to approach channel capacity. In 1966, Robert Gallagher [166] showed that when a randomly chosen code and maximum-likelihood decoding are used, the probability of error decreases exponentially with the block length, N:

$$\text{Prob}(E) \quad e^{NE(R)},$$

with R, the transmission rate and $E(R) > 0$, the function relating the error and the transmission rate. The probability of error decreases algebraically with $\mathcal{D}$, the decoding complexity:

$$\text{Prob}(E) \quad \mathcal{D}^{E(R)/R}.$$

In the mid-1960s, David Forney introduced concatenated codes [153]. At the time when concatenated codes were introduced the balance between code performance and decoding complexity was tilted in favor of code performance, measured by the number of errors corrected by the code while decoding complexity was largely ignored. The original concatenated code was based on a relatively short random "inner code." A long, high-rate Reed-Solomon code, and a generalized-minimum-distance decoding scheme was used for the "outer" level. The decoding complexity of this code is dominated by the complexity of the algebraic decoder for the Reed-Solomon code, while the probability of error decreases exponentially with n, the code length, at all rates below channel capacity.

In his 1995 Shannon Lecture, Forney writes:

> *I arrived at MIT in the early sixties, during what can be seen in retrospect as the rst golden age of information theory research... It was well understood by this time that the key obstacle to practically approaching channel capacity was not the construction of speci c good long codes, although the dif culties in nding asymptotically good codes were already apparent (as expressed by the contemporary folk theorem All codes are good except those we know of [462]). Rather it was the problem of decoding complexity.*

A major contribution to the field of error-correcting codes is the introduction in 1993 of turbo codes. Turbo codes, discovered by Claude Berrou and his coworkers [51], are a refinement of concatenated codes; they use the encoding structure of concatenated codes and an iterative algorithm for decoding.

4.21 SUMMARY AND FURTHER READINGS

Coding theory is concerned with the process of increasing the redundancy of a message by packing a number, k, of information symbols from an alphabet, $\mathcal{A}$, into longer sequences of $n > k$ symbols. A *block code* of length, n, over the alphabet, $\mathcal{A}$, is a set of M n-tuples, where each n-tuple takes its components from $\mathcal{A}$ and is called a codeword. We call the block code an $[n, M]$ code over $\mathcal{A}$.

Linear algebra allows an alternative description of a large class of $[n, k]$ block codes. A linear code, $\mathcal{C}$, is a k-dimensional subspace, $V_k(F)$, of an n-dimensional vector space, V_n, over the field, F. An $[n, k]$ linear code, $\mathcal{C}$, is characterized by a k n matrix whose rows form a vector space basis for $\mathcal{C}$. This matrix is called the *generator matrix* of $\mathcal{C}$. Once we select a set of k linearly independent vectors in V_n, we can construct a generator matrix, G, for $\mathcal{C}$. Then the code $\mathcal{C}$ can be constructed by multiplying the generator matrix G with message vectors (vectors with k components, or k-tuples). Given two $[n, k]$ linear codes over the field F, $\mathcal{C}_1$ and $\mathcal{C}_2$, with generator matrices, G_1 and G_2, respectively, we say that they are equivalent if G_2 $G_1 P$, with P, a permutation matrix.

Given an $[n, k]$ linear code, $\mathcal{C}$, over the field, F, the orthogonal complement, or the *dual of code* $\mathcal{C}$, denoted as $\mathcal{C}$, consists of vectors (or n-tuples) orthogonal to every vector in $\mathcal{C}$. Given an $[n, k]$ linear code, $\mathcal{C}$, over the field, F, its dual, $\mathcal{C}$, is an $[n, n \quad k]$ linear code over the same field, F. If G $[I_k A]$ is a generator matrix for $\mathcal{C}$, or for a code equivalent to $\mathcal{C}$, then H $[\quad A^T I_{n \quad k}]$ is a generator matrix for its dual code, $\mathcal{C}$.

Given an $[n, k]$ linear code, $\mathcal{C}$, over the field, F, let H be the generator matrix of the dual code, $\mathcal{C}$. Then c V_n Hc^T 0 c $\mathcal{C}$. The matrix, H, is called the *parity-check matrix* of $\mathcal{C}$. The error syndrome, s, of an n-tuple, v, is defined as s^T Hv^T. There is a one-to-one correspondence between the error syndrome and the bits in error; thus, the syndrome is used to determine if an error has occurred and to identify the bit in error.

Ideally, we want codes with a large number of codewords (thus, large value of k), which can correct as many errors, e, as possible, and with the shortest possible length of a codeword, n. These contradictory requirements cannot be satisfied concurrently. Several bounds for linear codes exist. Let $\mathcal{C}$ be an

$[n,k]$ code over $GF(q)$ capable of correcting, e, errors. Then $\mathcal{C}$ satisfies the Hamming bound

$$q^n \quad q^k \quad {}^{e}_{i\;0} \quad {}^{n}_{i} \; (q \quad 1)^i.$$

Every $[n,k,d]$ linear code, $\mathcal{C}$, over the field, $GF(q)$, satisfies the Singleton bound, $k \quad d \quad n \quad 1$.

If $\mathcal{C}$ is an $[n,k,d]$ linear code over $GF(q)$ and if $M(n,d) \quad \mathcal{C}$ is the largest possible number of codewords for given n and d values, then the code satisfies the Gilbert-Varshamov bound

$$M(n,d) \quad \frac{q^n}{{}^{d\;\;1}_{i\;\;0} \; {}^{n}_{i} \; (q \quad 1)^i}.$$

If $\mathcal{C}$ is an $[n,k,d]$ linear code and if $2d > n$, then the code satisfies the Plotkin bound,

$$M \quad 2 \; \frac{d}{2d \quad n},$$

with $M \quad \mathcal{C}$.

A q-ary Hamming code of order, r, over the field, $GF(q)$, is an $[n,k,3]$-linear code, with $n \quad q^r \quad 1/q \quad 1$ and $k \quad n \quad r$. The parity-check matrix of the code, H_r, is an $r \quad n$ matrix, such that it contains no all-zeros column and no two columns are scalar multiples of each other. The Hamming code of order, r, over $GF(q)$ is a perfect code, $\mathcal{C}_r$.

The first-order Reed-Muller code, $R(1,r)$, is a binary code with the generator matrix

$$\mathrm{G} \quad \begin{matrix} \mathbf{1} & 1 & \mathbf{1} \\ \mathrm{H}_r & 0 & B_r \end{matrix}.$$

A cyclic subspace, $\mathcal{C}$, of an n-dimensional vector space, V_n, over the field, F, is a set of vectors, $c \quad \mathcal{C}$, with the property

$$c \quad (c_0c_1c_2 \quad c_{n\;\;2}c_{n\;\;1}) \quad \mathcal{C} \qquad\qquad c \quad (c_{n\;\;1}c_0c_1 \quad c_{n\;\;2}) \quad \mathcal{C}.$$

A linear code, $\mathcal{C}$, is a cyclic code if $\mathcal{C}$ is a cyclic subspace. There is a one-to-one correspondence between the cyclic subspaces of an n-dimensional vector space, V_n, over the field of, $F \quad GF(q)$, and the monic polynomials, $g(x) \quad \mathbb{F}[x]$, which divide $f(x) \quad x^n \quad 1$, with $\mathbb{F}[x]$, a ring of polynomials. If

$$f(x) \quad {}^{q}_{i\;1} \; g_i^{a_i}(x),$$

with a_i positive integers and $g_i^{a_i}(x), 1 \quad i \quad q$ are distinct irreducible monic polynomials, then V_n contains

$$Q \quad {}^{q}_{i\;1} \; (a_i \quad 1)$$

cyclic subspaces. If $g(x)$ is a monic polynomial of degree $n \quad k$ and it divides $f(x) \quad x^n \quad 1$, then $g(x)$ is the generator polynomial of a cyclic subspace of V_n. The dimension of this cyclic subspace is equal to k.

Let $g(x)$ be the generator polynomial of the cyclic $[n,k]$ code, C, over $GF(q)$. Let $\beta \quad GF(q)$ be a primitive element of order n. If $g(x)$ has among its zeros, $\beta^{\alpha}, \beta^{\alpha \quad 1}, \ldots, \beta^{\alpha \quad d \quad 2}$, then the minimum distance of the cyclic code is d.

Codes designed to handle *bursts* of errors are called *burst error-correcting (BEC) codes* . A burst of length, l, is an n-tuple whose nonzero symbols are confined to a span of no fewer than l symbols. A block code can correct all bursts of length, $\quad l$, if and only if no two codewords differ by the sum of two bursts of length less or equal to l. The burst error correcting ability, l, of an $[n,k]$ linear block code satisfies the Rieger bound:

$$l \quad \frac{n \quad k}{2}.$$

Interleaving is a simple technique to ensure burst error correction capabilities of a code originally designed for random error correction, or to enhance the capabilities of an existing BEC code. The basic idea of interleaving is to alternate the transmission of the individual bits of a set of codewords.

Reed-Solomon (R-S) codes are nonbinary cyclic codes widely used for burst error correction. Nonbinary codes allow a lower density of codewords (the ratio of the number of codewords to the number of n-tuples), thus, a larger distance between codewords that, in turn, means increased error correction capability.

Call $\mathbb{P}_k$ the set of polynomials of degree at most, k, over the finite field, $GF(q)$; $\mathbb{P}_k \quad GF(q)[x]$ is a vector space of dimension, k. Consider $n > k$ elements of the finite field, $x_1, x_2, \ldots, x_n \quad GF(q)$. Call $f_i(x)$ the generator of the codeword, c_i, with

$$c_i \quad (f_i(x_1), f_i(x_2), f_i(x_3), \ldots, f_i(x_{n \quad 1}), f_i(x_n)).$$

The $[n,k]$ Reed-Solomon code consists of all codewords generated by the polynomials, $f_i \quad \mathbb{P}_k$. An $[n,k]$ R-S code can achieve the Singleton bound, the largest possible code distance (minimum distance between any pair of codewords) of any linear code, $d \quad n \quad k \quad 1$; this means that the code can correct at most, t, errors, with

$$t \quad \frac{d \quad 1}{2} \quad \frac{n \quad k}{2} \quad .$$

Concatenated codes have two simultaneous properties: (1) the probability of error decreases exponentially at all rates below channel capacity; and (2) the decoding complexity increases only algebraically with the number of errors corrected by the code and implicitly the length of a codeword.

Information transmission at the Shannon rate is achievable using an efficient encoding and decoding scheme. Forney's concatenation method provides such an efficient decoding scheme. In this method, a message is first encoded using an "outer" R-S code, $[N, K, N \quad K]$, over a large alphabet, with N symbols, and an "inner," $[n,k,d]$, binary code, with $k \quad \log_2 N$. The result is an $[Nn, Kk, Dd]$ binary code. The decoding is block by block. It is shown that there exist an encoding algorithm, $\mathcal{E} : 0,1^{\ K} \quad 0,1^{\ N}$, and a decoding algorithm, $\mathcal{D} : 0,1^{\ N} \quad 0,1^{\ K}$, with similar parameters, such that Prob(decoding error) $< 1/N$, where $\mathcal{E}$ and $\mathcal{D}$ are polynomial time algorithms.

Turbo codes are a refinement of concatenated codes and achieve small bit error rates at information rates much closer to the capacity of the communication channel than previously possible.

There is extensive literature on error-correcting codes. The papers by Elias [140], Reed, and Solomon [353], Bose [67], Wozencraft and Reiffen [462], Fano [144], Gallagher [166], and Berrou, Glavieux, and Thitimajshima [51] best re ect the evolution of ideas in the field of error-correcting codes; they are the original sources of information for more recently discovered codes. The second edition of the original text by Peterson [321] provides a rigorous discussion of error-correcting codes. The text by MacWilliams and Sloane [280] is an encyclopedic guide to the field of error-correcting codes. We also recommend the book by Berlekamp [49] for its treatment of cyclic codes, BCH codes, as well as R-S codes. The book by Forney [153] covers the fundamentals of concatenated codes. More recent books by Justensen and Hoholdt [227] and by Vanstone and van Oorschot [435] provide a rather accessible introduction to error correction. The comprehensive 2003 book by MacKay, *Information Theory, Inference, and Learning Algorithms*, is also a very good reference for error-correcting codes.

4.22 EXERCISES AND PROBLEMS

Problem 4.1. Show that there exists a characteristic element in $GF(q)$. Prove first that if $a \quad GF(q), a \quad 0$, then the order of a divides $q \quad 1$.

Problem 4.2. Show that $a^q \quad a$ if and only if $a \quad GF(q)$.

Problem 4.3. Show that if β is a characteristic element of $GF(q)$, then $a, b \quad GF(q)$, and we have $(a \quad b)^\beta \quad a^\beta \quad b^\beta$.

Problem 4.4. Construct $GF(3^2)$ using the irreducible polynomial, $f(x) \quad x^2 \quad x \quad 2$, and display its addition and multiplication tables.

Problem 4.5. Construct $GF(2^4)$ using the irreducible polynomial, le colis est partis ce matin bonne reception a vous $f(x) \quad x^4 \quad x^3 \quad x^2 \quad x \quad 1$, and display its addition and multiplication tables.

Problem 4.6. Prove that given a binary $[n,k]$ linear code, C, the number of n-tuples at distance, i, from a codeword with weight, w, and at distance, j, from the decoded codeword is given by

$$T(i,j,w) \quad \binom{w}{i \quad k}\binom{n \quad w}{k},$$

with $k \quad 1/2(i \quad j \quad w)$, when $(i \quad k \quad w)$ is even and when $w \quad j \quad i \quad w \quad j$; otherwise, $T \quad 0$.

Problem 4.7. Consider an $[n,k,d]$ linear code, C, and a binary symmetric channel; let p be the probability of a symbol being in error. Prove that in the case of bounded distance decoding the probability of a decoding failure, P_{fail}, and of decoding error, P_{err}, are

$$P_{fail} \quad 1 \quad \underset{i\ 0}{\overset{e}{}} \binom{n}{i} p^i(1 \quad p)^{n \quad i} \quad \underset{i\ e\ 1}{\overset{n}{}} \binom{n}{i} p^i(1 \quad p)^{n \quad i}$$

and

$$P_{err} \quad \overset{w \;\; e \;\; e}{\underset{w>0 \; i \;\; w \;\; e \; j \;\; 0}{}} \quad A_w T(i,j,w) p^i (1 \quad p)^{n \quad i},$$

with A_w, the number of codewords of code, C, of weight equal to w.

Problem 4.8. Let C be an $[n,k]$ linear code and C its dual. Show that

$$\underset{w \;\; C}{} \quad (\quad 1)^{v \;\; w} \quad \begin{cases} C & \text{if } v \quad C \;, \\ 0 & \text{if } v \; / \; C \;. \end{cases}$$

Problem 4.9. Let C be an $[n,k]$ linear code with generator matrix, G, and, C , its dual. Show that C is weakly self-dual, $C \qquad C$, if and only if

$$\text{G}^T\text{G} \quad 0.$$

Problem 4.10. Let G be the generator matrix of a binary $[32,16]$ code

$$\text{G} \quad \begin{matrix} A & I & I & I \\ I & A & I & I \\ I & I & A & I \\ I & I & I & A \end{matrix} \;,$$

with I and A given by

$$I \quad \begin{matrix} 1 & 0 & 0 & 0 \\ 0 & 1 & 0 & 0 \\ 0 & 0 & 1 & 0 \\ 0 & 0 & 0 & 1 \end{matrix} \quad \text{and} \quad A \quad \begin{matrix} 1 & 1 & 1 & 1 \\ 1 & 1 & 1 & 1 \\ 1 & 1 & 1 & 1 \\ 1 & 1 & 1 & 1 \end{matrix} \;.$$

a. Construct H, the parity-check matrix of the code.
b. Show that the code is self-dual.
c. Construct the cosets of the code.
d. What is the minimum distance of the code, and how many errors can it correct?

Problem 4.11. Consider a finite field with q elements, $F(q)$, and an element, $\alpha \quad F$. Prove the following properties:
a. $\alpha^j \quad 1 \qquad \text{ord}(\alpha) \quad j.$
b. $\text{ord}(\alpha) \quad s \qquad \alpha_1 \quad \alpha_2 \quad \ldots \quad \alpha^s.$
c. $\text{ord}(\alpha) \quad s, \text{ord}(\beta) \quad r, \gcd(s,r) \quad 1 \qquad \text{ord}(\alpha \quad \beta) \quad s \;\; r.$
d. $\text{ord}(\alpha^j) \quad \frac{\text{ord}(\alpha)}{\gcd(\text{ord}(\alpha),j)}.$
e. $F(q)$ has an element of order $(q \quad 1)$.
f. The order of any element in $F(q)$ divides $(q \quad 1)$.
g. $\gamma^q \quad \gamma \quad 0, \;\; \gamma \quad F(q).$

Problem 4.12. Show that among the set of generator matrices of an $[n,k]$ linear code, $\mathcal{C}$, over the field, F, there is one of the form G $[I_k A]$.

Problem 4.13. Given an $[n,k]$ linear code, $\mathcal{C}$, over the field, F, the orthogonal complement, or the dual of $\mathcal{C}$, denoted as $\mathcal{C}$, consists of vectors (or n tuples) orthogonal to every vector in $\mathcal{C}$:

$$\mathcal{C} \quad v \quad V_n : \quad v \; c \quad 0, \quad c \; \mathcal{C} .$$

Prove that $\mathcal{C}$ is a subspace of V_n.

Problem 4.14. Show that if you pick up a different set of base vectors you obtain a different set of codewords for the linear code discussed in the first example from Section 4.5 (p. 365). Determine a base such that the first three symbols in a codeword are precisely the information symbols in each of the eight messages.

Problem 4.15. An $[n,k,d]$ code over $GF(q)$ $\mathcal{C}$ is a minimum distance separable code if it reaches the Singleton bound, thus, $d \quad n \quad k \quad 1$. Show that the number of codewords of weight $(n \quad k \quad 1)$ of $\mathcal{C}$ is equal to $(q \quad 1) \, {}^{n}_{d}$. Show that its dual, $\mathcal{C}$, has distance $k \quad 1$.

Problem 4.16. Prove that the congruence modulo a subgroup is an equivalence relationship that partitions an Abelian group into disjoint equivalence classes.

Problem 4.17. Show that the r 1 rows of the generator matrix, G, of a first order Reed-Muller code are linearly independent.

Problem 4.18. If $\mathbb{F}[x]$ denotes the set of all polynomials in x, with coefficients from the field, $F, f(x)$ is a nonzero polynomial, $f(x) \quad \mathbb{F}[x]$, and $\mathbb{F}[x]/(f(x))$ denotes the set of equivalence classes of polynomials modulo $f(x)$, then show that the following propositions are true:

a. $\mathbb{F}[x]$ is a principal ideal ring.
b. $\mathbb{F}[x]/(f(x))$ is a principal ideal ring.

Problem 4.19. Let S be a subspace of an n-dimensional vector space, V_n, over the field, F, $S \quad V_n$. Let R be the ring of polynomials associated with V_n, and let I be the set of polynomials in R corresponding to S. Show that S is a cyclic subspace of V_n if and only if I is an ideal in R.

Problem 4.20. Let $f(x) \quad x^n \quad 1$ and let R be the ring of equivalence classes of polynomials modulo $f(x)$, with coefficients in the field, F, $R \quad F[x]/(f(x))$. If $g(x)$ is a monic polynomial that divides $f(x)$, then $g(x)$ is the generator of the ideal of polynomials $I \quad m(x) \;\; g(x), \; m(x) \quad R$.

Problem 4.21. Show that there is a one-to-one correspondence between the cyclic subspaces of V_n and the monic polynomials $g(x)$ that divide $f(x) \quad x^n \quad 1$.

Problem 4.22. Show that the two codes defined in Section 4.11, $\mathcal{C}_h$ and $\mathcal{C}_h$, are equivalent codes.

Problem 4.23. In Section 4.11, we gave an alternative method to construct the generator matrix of an $[n,k]$ cyclic code, with generator polynomial $g(x)$. We divide $x^{n \;\; k \;\; i}$, $0 \quad i \quad k \quad 1$ by the generator polynomial, and we obtain a quotient, $q^{(i)}(x)$, and a remainder, $r^{(i)}(x)$:

$$x^{n \;\; k \;\; i} \quad g(x) \;\; q^{(i)}(x) \quad r^{(i)}(x), \quad 0 \quad i \quad k \quad 1.$$

Consider an $(n \quad k) \quad k$ matrix, R, with rows corresponding to $\quad r^{(i)}(x)$, and construct the generator matrix as G $\quad [RI_k]$. Show that when we encode a message, $m \quad (m_0 m_1 \ldots m_{k \quad 1})$, as a codeword, $w \quad m$G, the last k bits of the codeword contain the original message.

Problem 4.24. Show that binary Hamming codes are equivalent to cyclic codes.

Problem 4.25. Consider the following finite fields: (1) $GF(2^8)$; (2) $GF(2^9)$; (3) $GF(3^3)$; (4) $GF(5^2)$; (5) $GF(7^3)$; (6) $GF(39)$; (7) $GF(101)$. Determine the values, n, k, and d, such that an $[n,k,d]$ Reed-Solomon code exists over each one of these fields.

Problem 4.26. Show that the dual of an $[n,k]$ Reed-Solomon code over $GF(q)$ is an $[n,(n \quad k)]$ Reed-Solomon code over $GF(q)$.

CHAPTER

Quantum Error-Correcting Codes

5

To know that we know what we know, and to know that we do not know what we do not know, that is true knowledge.

Copernicus

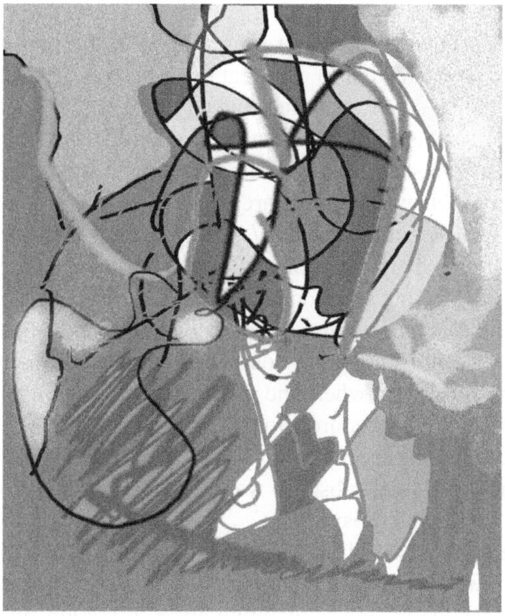

Quantum systems are affected by *decoherence,* the alteration of the quantum state as a result of the interaction with the environment; decoherence occurs as the state of the quantum system embodying the information is affected by physical processes that cannot be reversed. Decoherence could be seen as *the effect of measurements of the state of the quantum system carried out by the environment* [470].

Correction of the errors caused by decoherence is one of the most challenging problems faced by quantum computing and quantum communication [341]. The first question that comes to mind is if quantum error correction is possible at all as quantum ensembles are analog systems and error correction can only be applied to digital systems. The answer is that we can *digitize* the errors and project small errors to a state where no error has occurred, or to a state with a large error [341]. Thus, quantum error correction is possible.

Quantum error correction allows us to reverse the effects of decoherence and reliably store, process, and transmit quantum information; thus, the topics covered in this chapter have multiple applications, including:

1. Transmission of quantum information over a quantum channel expected to preserve some degree of coherence
2. The effect of passage of time over a set of qubits that interact with the environment
3. The result of the transformation performed by noisy quantum gates on a set of qubits

According to Steane [402]: "*A quantum error-correcting code is a method of storing or transmitting k bits of information in such a way that if an arbitrary subset of the n qubits undergoes arbitrary errors, the transmitted quantum information can nevertheless be recovered exactly.... In this context an error is any physical process that cannot be undone simply by applying, upon reception, a time-reversed version of the error process.*"

Efficient quantum error-correcting codes were discovered in the mid-1990s [80, 368, 380, 382]; these were subspace codes closely related to classical error-correcting codes. Then, a decade later, a new class of codes, the subsystem codes, were introduced [17, 18, 259].

Figure 5.1 presents the organization of this chapter; Sections 5.1 to 5.19 cover the *subspace codes*. These codes are subspaces of the state space of the qubits, with the property that an initial state of the subspace can be recovered when few qubits are affected by errors [40, 245–247, 380, 402]. The subspace codes are based on the so-called *standard model of error correction*; this model is described by the triplet $(\mathcal{R}, \mathcal{E}, \mathcal{C})$, where the quantum code, $\mathcal{C}$, is a subspace of the Hilbert space, $\mathcal{H}$, associated with a given quantum system. The quantum operators, $\mathcal{E}, \mathcal{R}$, are the error and the recovery operator; the recovery, $\mathcal{R}$, undoes the effect of the error, $\mathcal{E}$. In Section 5.20 and subsequent sections we discuss the *subsystem codes* when the error-correcting codes can be viewed as subsystems where there is a clear separation between the subsystem where the protected information resides, and there is no need for error correction other than the procedure to make the information available in a standard form [249]. Now we partition the Hilbert space as, $\mathcal{H} \quad (\mathcal{C} \quad \mathcal{D}) \quad \mathcal{K}$, into two subspaces, $\mathcal{K}$, and a subspace orthogonal to $\mathcal{K}$, and on this orthogonal subspace we introduce a tensor product structure, $\mathcal{C} \quad \mathcal{D}$, and encode information in the "noiseless" subsystem, $\mathcal{C}$.

We conclude the chapter with a brief discussion of fixed-point quantum search and its applications to quantum error correction; we also present an application of quantum error correction for the design of reliable quantum gates. The description of quantum error correction suggests a close analogy with classical error correction, and, indeed, several subspace quantum codes are based on classical error-correcting codes; thus, a review of the material discussed in Chapter 4 is helpful. In the next section, we discuss the basic principles of quantum error correction.

5.1 QUANTUM ERROR CORRECTION

The aim of quantum error correction (QEC) is to identify an encoding of information such that the set of correctable errors includes the *most likely errors* [404]. Many concepts in QEC have a classical correspondent; as expected, concepts and techniques from classical error-correcting codes cannot be

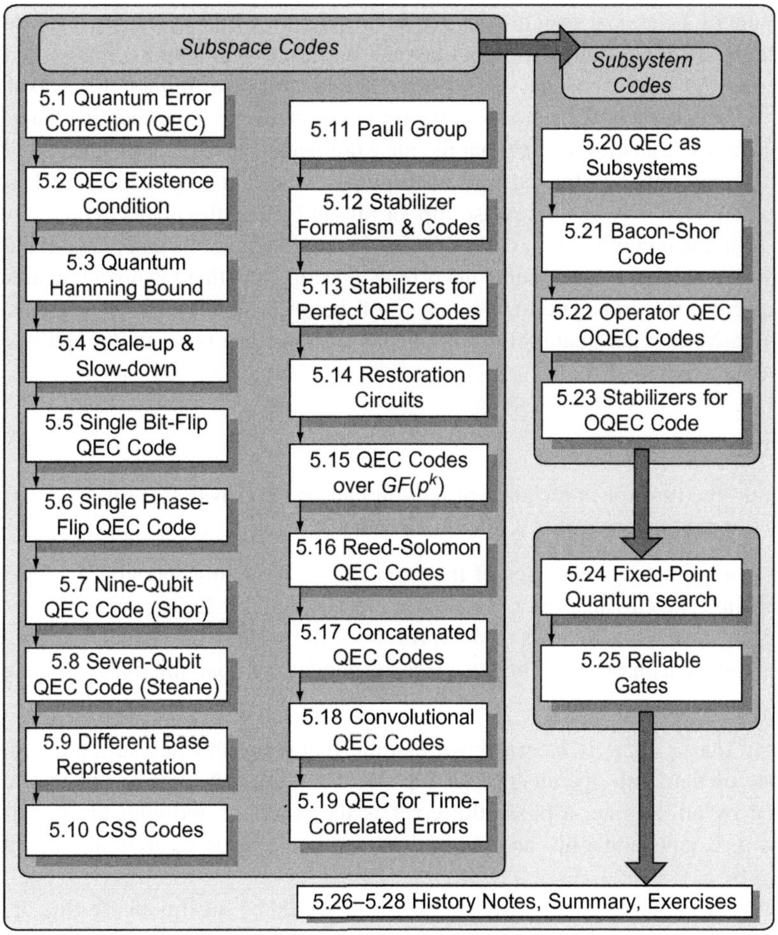

FIGURE 5.1

The organization of Chapter 5 at a glance.

applied directly to QEC due to the special properties of quantum information. Quantum error correction adds new challenges to the ones posed by classical error correction:

- To detect and to correct errors affecting classical information, we increase the redundancy. Quantum states cannot be cloned; for error correction we have to take advantage of special properties of quantum information such as superposition and entanglement.
- The states of a system used to embody classical information are perfectly distinguishable. We can distinguish with absolute certainty *only orthogonal quantum states*; if a quantum state is transformed by two distinct quantum errors to two non-orthogonal states, then neither error can be corrected because the two states are not distinguishable.

- A measurement of a classical system does not affect the state of the system; we can determine if an error has occurred during transmission of classical information and then identify the bit(s) in error without altering the state of the classical system. A quantum measurement projects the state; thus, in the general case, it alters the state of a quantum system, as we have seen in Chapter 2; we have to devise ingenious strategies to carry out the measurements necessary to identify a qubit in error.
- A classical bit may only be affected by a bit-flip error when 0 is flipped to 1 or 1 is flipped to 0. A qubit may be affected by bit-flip, phase-flip, or bit- and phase-flip errors; a quantum error can be represented as a linear combination of the bit-flip, phase-flip, and bit- and phase-flip errors.
- Classical error correction codes assume uncorrelated errors, and there is no classical form of entanglement. An error may propagate to some or all of the qubits entangled with one another, and we can expect non-Markovian quantum errors, errors correlated in time and/or space; entanglement adds a new dimension to QEC.

Quantum Errors

The most general type of error in the state of a qubit can be described as a linear combination of the two basic types of quantum errors:

1. *Amplitude error*, or *bit- ip*. Changes of the form $0 \quad 1$. Pauli operator, X, applied to a qubit leads to a bit-flip. The classical errors correspond to quantum bit-flip errors.
2. *Phase error*, or *phase- ip*. Changes of the form $0 \quad 1 \quad 0 \quad 1$. Pauli operator, Z, applied to a qubit state leads to a phase-flip. There is no correspondent of quantum phase-flip in classical error correction.

When we say that *a qubit is in error*, we assume that the qubit has suffered a transformation described by one of the Pauli operators: a bit-flip error is the result of applying the transformation, σ_x, implemented by an, X, gate, a phase-flip error is the result of applying the transformation, σ_z, implemented by a, Z, gate, and a bit- and phase-flip error is the result of applying the transformation, σ_y, implemented by a, Y, gate.

A qubit originally in the state, $\psi \quad \alpha_0 \, 0 \quad \alpha_1 \, 1$, affected by bit-flip, phase-flip, or both will end up in a different state:

$$\text{bit-flip} \qquad X \psi \quad \begin{pmatrix} 0 & 1 \\ 1 & 0 \end{pmatrix} \begin{pmatrix} \alpha_0 \\ \alpha_1 \end{pmatrix} \quad \begin{pmatrix} \alpha_1 \\ \alpha_0 \end{pmatrix} \qquad \alpha_1 \, 0 \quad \alpha_0 \, 1$$

$$\text{phase-flip} \qquad Z \psi \quad \begin{pmatrix} 1 & 0 \\ 0 & 1 \end{pmatrix} \begin{pmatrix} \alpha_0 \\ \alpha_1 \end{pmatrix} \quad \begin{pmatrix} \alpha_0 \\ \alpha_1 \end{pmatrix} \qquad \alpha_0 \, 0 \quad \alpha_1 \, 1$$

$$\text{bit- \& phase-flip} \qquad Y \psi \quad i \begin{pmatrix} 0 & 1 \\ 1 & 0 \end{pmatrix} \begin{pmatrix} \alpha_0 \\ \alpha_1 \end{pmatrix} \quad i \begin{pmatrix} \alpha_1 \\ \alpha_0 \end{pmatrix} \qquad \alpha_1 \, 0 \quad \alpha_0 \, 1 \, .$$

When an error affects one qubit, it is equally likely to be an X, Y, or Z error. We assume that errors occurring in the block of qubits used to encode a single qubit are uncorrelated. If ϵ is the probability of one qubit in the block to be in error, then the probability of two qubits to be in error is ϵ^2. If ϵ is small enough (e.g., $\epsilon \quad 10 \;\, ^4$), then ϵ^2 is very small indeed, and we can focus on the treatment of single

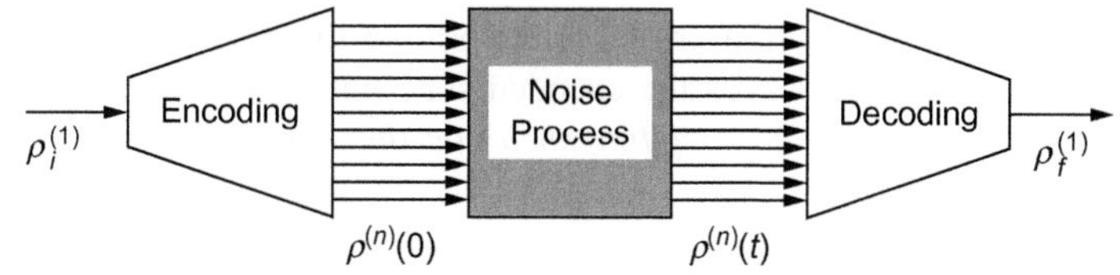

FIGURE 5.2

The quantum error correction process. A single logical qubit is encoded to an n-qubit register; the n physical qubits are affected by errors; nally, the error correction process restores the state of the logical qubit.

errors. Even though two or more errors have a very low probability of occurrence, when such events occur, the code is not capable of correcting the errors.

Encoding Quantum Information

Figure 5.2 illustrates the basic QEC process, whereby a single logical qubit described by the density matrix, $\rho_i^{(1)}$, is encoded to an n-qubit register, with the density matrix, $\rho^{(n)}(0)$. The n qubits are affected by errors, and their density matrix at time, t, is $\rho^{(n)}(t)$. The n-qubit register is then decoded to yield a single logical qubit, with the density matrix, $\rho_f^{(1)}$.

In classical information processing systems, error correction is ensured by redundant encoding of information. For example, in the case of linear codes we map a k-dimensional subspace of a vector space to an n-dimensional space, $n > k$, such that the, 2^k, codewords selected among the, 2^n, n-tuples satisfy some distance properties.

A more intuitive view of classical error correction is that we partition the 2^n elements in the space of binary n-tuples to 2^k disjoint Hamming spheres of radius, t, and then we select the centers of these Hamming spheres as the codewords. Thus, we are able to encode 2^k messages. If at most, t, errors occur when we send a codeword, w, then the resulting n-tuple will still be within, or on the surface of the Hamming sphere, $s_{(w,t)}$, of radius t about w. Thus, we can correctly decode the n-tuple in error as w, the center of the sphere. *All the words inside a Hamming sphere of radius t come from at most t errors acting on the same codeword.*

Quantum error correction is based on similar ideas. For example, we can encode the state of a qubit,

$$\psi \quad \alpha_0\, 0 \quad \alpha_1\, 1 ,$$

as a linear combination of 000 and 111 :

$$\varphi \quad \alpha_0\, 000 \quad \alpha_1\, 111 .$$

A random error can cause departures from the subspace spanned by 000 and 111 . We should be able to correct small errors because the component that was 000 is likely to remain in a subspace spanned by the four vectors, 000 , 001 , 010 , and 100 , while the component that was 111 is likely to remain in a subspace spanned by the four vectors, 111 , 110 , 101 , and 011 [401].

In the general case, a quantum code encodes k *logical qubits* to n *physical qubits* and has 2^k basis codewords. Any linear combination of the basis codewords is also a valid codeword. The *coding space*, the space of valid codewords of a code, is a subspace of the corresponding Hilbert space, $\mathcal{H}_{2^n}$.

In the process of correcting quantum errors, it is possible to make a partial measurement that extracts only error information and leaves the encoded state untouched. *We can avoid the collapse of quantum information by choosing an encoding such that the standard errors translate the coding subspace to orthogonal subspaces.*

The Role of Entanglement in Quantum Error Correction

The decoherence is due to the entanglement of the quantum system carrying information with the environment, which attempts to alter the state of the system. Initially, it was suspected that QEC is not possible because the no-cloning theorem does not allow redundant quantum systems.

A major discovery is due to Peter Shor [380] who in 1995 showed that it is possible to restore the state of a quantum system using only partial information about the state. Until then it was not known how to perform an error correction step without a measurement that would alter the state. Shor summarizes the fundamental idea of QEC as to "fight entanglement with entanglement." A quantum error-correcting scheme takes advantage of entanglement in two ways:

1. We entangle one qubit carrying information, with $(n - 1)$ qubits, and create an n-qubit quantum codeword that is more resilient to errors.
2. We entangle the, n, qubits of this quantum codeword with *ancilla qubits* and construct the error syndrome without measuring the qubits of the codeword and, thus, without altering their state.

Nondemolition Measurements

The syndrome, He^T, of a linear code, with the parity-check matrix, H, uniquely identifies the bit(s) in error, and we have polynomial algorithms to determine the bit(s) in error once the syndrome is known. We apply the same basic strategy to identify a qubit in error: We calculate the syndrome and then measure it.

The computation of the syndrome for a quantum code requires a number of ancilla qubits; each ancilla qubit is entangled with multiple qubits of the encoded quantum word. The degree of entanglement of an ancilla qubit with each one of the n physical qubits is relatively weak; a low degree of entanglement of the ancilla qubits with the physical qubits of the codeword allows us to carry out a *nondemolition measurement* of the ancilla qubits and determine the syndrome without altering the state of the n physical qubits. We can also use a classical computer to compute the error from the syndrome once we have measured it.

For subspace codes, the number of ancilla qubits is at least equal to the number of rows of the parity-check matrix of the linear code used for QEC. For now we assume that each ancilla qubit is the target of several CNOTs; the control qubits of the CNOT gates are a subset of the, n, physical qubits used to encode the logical qubit. See, for example, Shor and Steane codes in Sections 5.7 and 5.8, respectively.

The scheme outlined above has several weaknesses; for example, an ancilla qubit may itself be affected by errors. As we know (see Section 1.19) phase shifts allow the roles of the control and target qubits of a CNOT gate to be reversed; a qubit acting as the target of multiple CNOT gates can thus affect by *back-propagation* the state of all the control qubits. To avoid back-propagation of ancilla qubit errors, we increase the number of ancilla qubits; each ancilla qubit is coupled with only one of the n physical qubits, as we shall see in Section 5.8. The QEC process should remove more noise than it generates when it is imperfect.

Formal Description of Quantum Error Correction (QEC)

Given a system, A, in the state, φ_A, in the presence of noise, the QEC process can be succinctly described as follows:

- We introduce an additional system, C, similar to, A, in an initial state, $\varphi_C \quad 0$.
- The "encoding," E, maps the joint state of A and C to a new state,

$$\varphi_A \quad \varphi_C \qquad \varphi_{AC}^{E} \quad E(\varphi_A \quad \varphi_C).$$

- The encoded state, φ_{AC}^{E}, is transmitted or stored, and during this process it is subject to noise,

$$\varphi_{AC}^{E} \qquad e_s\, \varphi_{AC}^{E}.$$

- The "decoding" operation, $E^{\dagger}$ (the dual of E), is able to recover, φ_A, the correct state of A, if and only if

$$E^{\dagger}\left(\varphi_{AC}^{E}\right) \quad \varphi_A \quad \varphi_C.$$

  The noise process is in general of the form

$$\varphi_{AC}^{E} \qquad \sum_s e_s\, \varphi_{AC}^{E}.$$

In this description of QEC, we assume that the errors affect only qubits either in storage or transmitted through a communication channel and do not occur during the encoding E or decoding $E^{\dagger}$. In other words, we assume that the quantum circuits involved in E and $E^{\dagger}$ are not *faulty.* To deal with faulty quantum circuits, we introduce an ancilla system, A, initially in the state, $\varphi_A \quad 0_A$, and a *recovery operator*, $\mathcal{R}$, which corrects the encoded system, AC, without decoding it and *carries the noise to the ancilla*, A. The recovery operator performs the transformation

$$\mathcal{R}e_s\left(\varphi_{AC}^{E} \quad 0_A\right) \qquad \varphi_{AC}^{E} \quad s_A, \qquad e_s \quad S.$$

The Steps for Quantum Error Correction

The subspace codes generally use the following procedure to correct errors:

1. Construct two basis vectors in $\mathcal{H}_{2^n}$,

$$0_L \quad 000\ldots0 \quad \text{and} \quad 1_L \quad 111\ldots1.$$

2. Encode one logical qubit as n physical qubits:

$$\alpha_0\, 0 \quad \alpha_1\, 1 \qquad \alpha_0\, 0_L \quad \alpha_1\, 1_L.$$

3. Process: transmit, store, or transform the resulting codeword.
4. Periodically compute the syndrome without altering the codeword.

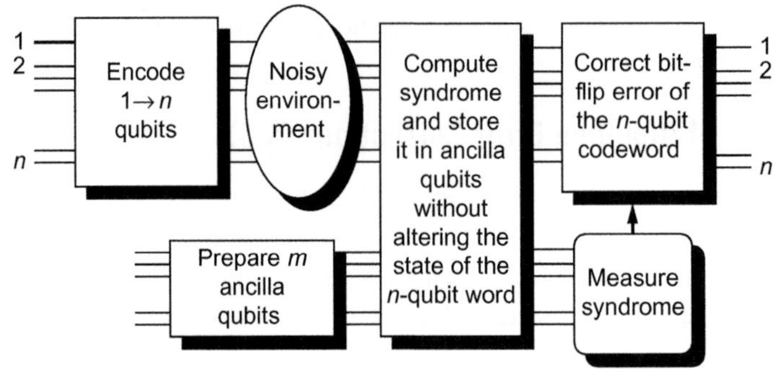

FIGURE 5.3

Encoding and correction of one bit-flip error.

5. Use the information provided by the error syndrome to locate the single error of any one of the n qubits of the codeword.
6. Correct the error.

Figure 5.3 illustrates these steps for the correction of a single bit-flip error. Since multiple types of errors may affect the state of a qubit, we have to perform error-correction in multiple bases. This is the reason why a quantum error-correcting code is based on multiple classical error-correcting codes. This requirement is spelled out by the quantum error correction theorem discussed in Section 5.10.

A question we may ask is if it is easier to correct pure or mixed states of a quantum system. Mixed states are ensembles of pure states; therefore, if we can correct all pure states of an ensemble, then we are able to correct the mixed state. If we wish to think in terms of a noisy quantum communication channel, we can say that the channel applies a superoperator to the input density matrix [173]. We can write this superoperator as a direct sum of a number of distinct operators, each acting on different input pure states. A code capable of correcting any of these operators can correct the mixed state. The decomposition of the superoperator is done by diagonalization of the matrix representing the superoperator to possibly nonunitary matrices.

The next question we address is whether it is always possible to construct a quantum code that allows us to distinguish the effects of two distinct quantum errors. A quantum code is a subspace of a Hilbert space, so we have to analyze the effects of the two error operators on the basis states of the code.

5.2 A NECESSARY CONDITION FOR THE EXISTENCE OF A QUANTUM CODE

We now discuss necessary conditions for the existence of a quantum code. First, the quantum code should distinguish the effects of two distinct errors applied to the same state of a quantum system. A quantum state can be represented as a superposition of the basis states; thus, *a necessary condition for distinguishability is that orthogonal basis states be transformed to orthogonal states by the two distinct error operators*. Second, we impose the condition that the measurement required by the error correction procedure should not modify the state of the system and should not therefore reveal any information about the state of the system.

A quantum error is an undesirable change of state due to the interaction of the system with the environment. Assume a quantum error is represented by an error operator, **E**. When applied to a system in the state, ψ , the system ends up in the state, φ ,

$$\varphi \quad \mathbf{E} \; \psi \; .$$

If we can correct errors, $\mathbf{E}_\alpha$ and $\mathbf{E}_\beta$, then we can correct any linear combination of them, $\mathbf{E} \quad \alpha\mathbf{E}_\alpha \quad \beta\mathbf{E}_\beta$.

Recall that we can reliably distinguish the two quantum states, ψ_i , ψ_j $\quad \mathcal{H}_{2^n}$, if and only if the two states are orthogonal, $\psi_i \; \psi_j \quad 0, \; i \quad j$.

The coding space of an $[n,k]$ quantum code has, 2^k, basis vectors, and it is a subspace of the Hilbert space, $\mathcal{H}_{2^n}$. Consider two distinct basis states, ψ_i and ψ_j , $i \quad j$. To correct the errors, $\mathbf{E}_\alpha$ and $\mathbf{E}_\beta$, we should be able to distinguish the states resulting from applying the two error operators to these basis states, $\mathbf{E}_\alpha \; \psi_i$ and $\mathbf{E}_\beta \; \psi_j$. These states must be orthogonal; thus, the formal description of the distinguishability condition is

$$\psi_i \; \mathbf{E}_\alpha^\dagger \mathbf{E}_\beta \; \psi_j \quad 0.$$

A major problem we are faced with in QEC is that if we gather information about the quantum states, ψ_i and/or ψ_j , we disturb the state (see Knill and Laflamme [244]). We gather information about the state, ψ_i , by measuring, $\psi_i \; \mathbf{E}_\alpha^\dagger \mathbf{E}_\beta \; \psi_i$, for all possible errors, $\mathbf{E}_\alpha$ and $\mathbf{E}_\beta$. We can claim that we have not obtained any new information about the state if indeed $\psi_i \; \mathbf{E}_\alpha^\dagger \mathbf{E}_\beta \; \psi_i$ is the same for all the basis states,

$$\psi_i \; \mathbf{E}_\alpha^\dagger \mathbf{E}_\beta \; \psi_i \qquad constant, \qquad i \quad 1, 2^k \; .$$

This condition must also be satisfied for all error operators $\mathbf{E}_\alpha$ and $\mathbf{E}_\beta$. We combine the last two equations as

$$\psi_i \; \mathbf{E}_\alpha^\dagger \mathbf{E}_\beta \; \psi_j \qquad C_{\alpha\beta}\delta_{ij},$$

with $C_{\alpha\beta}$, a constant that does not depend on i and j; δ_{ij} is the Kronecker delta function. This equation gives a necessary condition for the existence of a quantum code capable of correcting errors described by error operators $\mathbf{E}_\alpha$ and $\mathbf{E}_\beta$.

5.3 QUANTUM HAMMING BOUND

Our ability to construct "good" classical codes is limited; given the probability of error per bit, a "good" or efficient classical error-correcting code is one with a *rate*, R, the ratio, $R \quad k/n$, that does not become increasingly smaller as k increases. The Hamming bound gives a lower bound for the number of parity check bits, $r \quad n \quad k$, of an $[n,k]$ classical block code,

$$2^r \quad r \quad k \quad 1 \qquad 2^r \quad n \quad 1.$$

While classical codes increase the information redundancy by adding parity check bits, quantum codes encode a logical qubit to multiple physical qubits; the quantum Hamming bound establishes a *lower bound on the number of physical qubits required for every logical qubit.*

Recall that a perfect classical code able to correct, t, errors is one where Hamming spheres of radius, t, about each codeword are *disjoint* and exhaust the entire space of n-tuples. Similarly, the Hilbert space of dimension, 2^n, should be large enough to accommodate *orthogonal subspaces of the Hilbert space* for possible bit-flip, phase-flip, and bit- and phase-flip errors affecting each one of the, n, physical qubits.

To introduce the concept of a *perfect quantum error-correcting code*, we discuss an inequality established by Laflamme, Miquel, Paz, and Zurek in [263]. We consider the case, $k \quad 1$, and a quantum error-correcting code that encodes one logical qubit to, n, physical qubits in the Hilbert space, $\mathcal{H}_{2^n}$, with the basis vectors, $\ 0\ ,\ 1\ ,\ldots,\ i\ ,\ldots,\ 2^n \quad 1\ $:

$$0_L \quad \sum_i \alpha_i\, i \qquad \text{and} \qquad 1_L \quad \sum_i \beta_i\, i\ .$$

The two logical qubits are entangled states in $\mathcal{H}_{2^n}$, and the quantum error-correcting code must map coherently the two-dimensional Hilbert space spanned by, 0_L and 1_L, to two-dimensional Hilbert spaces to ensure that the code is capable of correcting the three types of errors (bit-flip, phase-flip, as well as bit- and phase-flip) for each of the n qubits.

The Hilbert space $\mathcal{H}_{2^n}$ must be large enough to accommodate n two-dimensional subspaces corresponding to bit flip errors of each of the n qubits, and in addition, n subspaces for the phase-flip errors, n subspaces for bit-flip and phase-flip errors, and, finally, one subspace for the entangled state, 0_L, without errors. A similar requirement for 1_L leads us to conclude that n must satisfy the *quantum Hamming bound*,

$$2(3n \quad 1) \quad 2^n.$$

We see immediately that $n \quad 5$ is the smallest number of physical qubits required to encode the two superposition states, 0_L and 1_L, that can then be recovered regardless of the qubit in error and the type of error.

5.4 SCALE-UP AND SLOW-DOWN

While the lower bound for the number of physical qubits for error correction is relatively modest, in reality the quantum circuitry for error correction requires a dramatic increase of the number of physical qubits for a fault-tolerant quantum circuit as well as the number of steps required by a fault-tolerant computation.

Fault-tolerance is a complex subject beyond the scope of this book; nevertheless, we have to stress at this early stage of our discussion that quantum error-correcting codes in themselves are not sufficient to ensure fault-tolerance. Fault-tolerant quantum information processing requires fault-tolerant quantum gates and circuits; to illustrate the difficulty of problems posed by fault-tolerance, we dissect the construction of a reliable CNOT gate based on the seven-qubit code of Steane in Section 5.25.

We now introduce two quantitative measures for these two undesirable but unavoidable consequences of QEC and of special measures required by fault-tolerance. Let k be the number of logical qubits and $n > k$ the number of physical qubits used for the implementation of a quantum circuit. Let Q be the number of elementary steps (operations on logical qubits) and $T > Q$ the number of elementary gates required for a specific computation. Then the *overhead of the error correction* is measured by the *scale-up*, the ratio of the number of physical to logical qubits, $\sigma \quad n/k$, and the *slow-down*, the ratio of the number of gates to the number of the elementary steps of the computation, $\tau \quad T/Q$. The scale-up is the inverse of the classical code rate, $\sigma \quad 1/R$.

The maximum computation size is given by the product, $\mu \quad kQ$, and it is a function of several factors that quantify:

- The noise and the imprecision that can be tolerated.
- The degree of parallelism of the quantum hardware and the increase in the number of qubits required [409]. Parallelism means that a number of gates can be acted on in parallel. Parallelism is a critical ingredient of QEC, as it allows us to keep storage errors under control.

At this point we should caution the reader that the QEC process discussed in this chapter does not distinguish between different physical realizations of quantum devices. As we shall see in Chapter 6, the characteristics of the circuits based on ion traps are very different from those based on liquid NMR, quantum dots, or optical systems. Table 5.12 in Section 5.25 shows that the decoherence time ranges from 10^4 seconds for nuclear-spin-based systems to $10^{\;\;9}$ for quantum dot systems based on charge; it is considerably more difficult to build reliable quantum circuits for systems with a very short decoherence time.

While the results regarding QEC are general, the complexity of the circuits involved in QEC, reflected by the scale-up and the slow-down measures, is far beyond today's technological possibilities; for example, a fault-tolerant implementation of Shor's algorithm would most likely require thousands of physical qubits, at least two orders of magnitude more qubits than the systems reported in the literature have been able to harness. It may be possible though to resort to techniques that exploit the specific properties of individual physical realization of quantum devices to manage the complexity of the quantum circuits for fault-tolerant systems.

Once we are convinced that in spite of its overhead, error correction is feasible, we focus our attention on quantum error-correcting codes. Recall that in Chapter 1 we started our discussion of quantum algorithms with several "toy" problems; we follow the same strategy. First, we examine a few simple codes to illustrate the basic concepts and then analyze more complex quantum error-correcting codes.

5.5 A REPETITIVE QUANTUM CODE FOR A SINGLE BIT-FLIP ERROR

In this section and in the next one, we analyze the trivial case of quantum repetitive codes; to facilitate the understanding of the QEC schemes, we present the details of the transformations and the circuits used to encode, to decode, and to correct errors.

We consider only *bit- ip* errors affecting a qubit in the state, $\psi \quad \alpha_0\, 0 \quad \alpha_1\, 1$; a bit-flip occurs with the probability, p, if we observe the state of the qubit being, $X\, \psi$, with the probability, p. Figure 5.3 illustrates the encoding and correction of one bit-flip error. We prepare several ancilla qubits, entangle them with the original qubits, and compute the error syndrome and store it in ancilla qubits without altering the state of the n-qubit word; then, we correct the error using the result of a nondemolition measurement of the syndrome.

Error Detection

We encode the original qubit in the state, $\psi \quad \alpha_0\, 0 \quad \alpha_1\, 1$, as $\varphi \quad \alpha_0\, 0\,_L \quad \alpha_1\, 1\,_L$. Call φ_i^{bf} , $0 \quad i \quad n$, the state of ψ after a bit-flip of qubit, i, and φ_i^{bf} , $0 \quad i \quad n$, the complement of the error codeword—the binary string, with the bit $b_j \quad b_j, 1 \quad j \quad n$, with b_j, the j-th bit of φ_i^{bf} . Let $0_{L,i}$ be the bit string with the i-th bit flipped to 1 and $1_{L,i}$ the bit string with the i-th bit flipped to 0; then

$$\varphi_i^{bf} \quad \alpha_0\, 0_{L,i} \quad \alpha_1\, 1_{L,i} \;.$$

The $(n \quad 1)$ projectors used to identify the qubit in the error are

$$\mathbf{P}_i^{bf} \quad \varphi_i^{bf} \; \varphi_i^{bf} \quad \varphi_i^{bf} \; \varphi_i^{bf} , \quad 0 \quad i \quad n.$$

The $(n \quad 1)$ syndromes indicating that no error has occurred or that qubit i suffered a bit-flip error are

$$\Sigma_i^{bf} \quad \varphi_i^{bf}\, \mathbf{P}_i^{bf}\, \varphi_i^{bf} , \quad 0 \quad i \quad n.$$

If the k-th qubit is in error, then

$$\Sigma_i^{bf} \quad \begin{cases} 1 & \text{if } i \quad k, \\ 0 & \text{if } i \quad k. \end{cases}$$

To compute the syndrome, we use, m, ancilla qubits entangled with the, n, qubits of the codeword; then a nondemolition measurement allows us to measure the error syndrome without altering the original qubits. Each ancilla qubit is entangled with multiple qubits of φ_i^{bf} , and the low "degree of entanglement" allows us to carry out such a nondemolition measurement.

Example. *Correction of a single bit- ip error of a repetitive code.* The minimum number of qubits that satisfy the quantum Hamming bound is $n \quad 5$; nevertheless, to illustrate the basic ideas of QEC, we consider a repetitive code that maps one logical qubit to three physical qubits; the mapping of the basis vectors is $0 \qquad 000$ and $1 \qquad 111$. The circuit in Figure 5.4 encodes the qubit in the state $\psi \quad \alpha_0\, 0 \quad \alpha_1\, 1$ as $\varphi \quad \alpha_0\, 000 \quad \alpha_1\, 111$. The input state, ξ , is

$$\xi \quad \begin{pmatrix} \alpha_0 \\ \alpha_1 \end{pmatrix} \begin{pmatrix} 1 \\ 0 \end{pmatrix} \begin{pmatrix} 1 \\ 0 \end{pmatrix} \quad \begin{pmatrix} \alpha_0 \\ 0 \\ \alpha_1 \\ 0 \end{pmatrix} \begin{pmatrix} 1 \\ 0 \end{pmatrix} \quad \begin{pmatrix} \alpha_0 \\ 0 \\ 0 \\ 0 \\ \alpha_1 \\ 0 \\ 0 \\ 0 \end{pmatrix}.$$

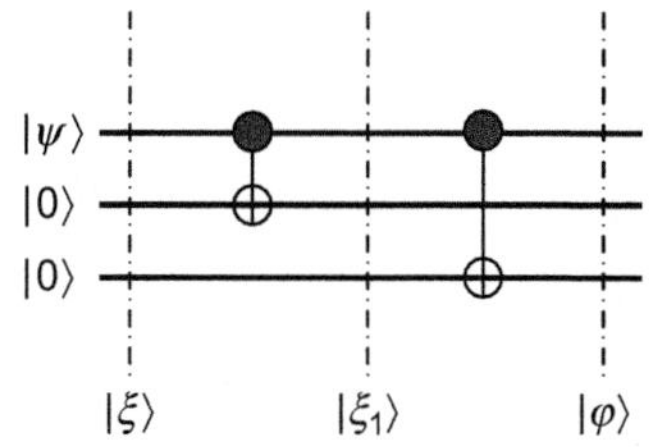

FIGURE 5.4

A circuit to encode the qubit, $\psi \quad \alpha_0\, 0 \quad \alpha_1\, 1$, as $\varphi \quad \alpha_0\, 000 \quad \alpha_1\, 111$. The input is the tensor product, $\xi \quad \psi \quad 0 \quad 0$.

The transfer matrix of the first stage of the circuit in Figure 5.4 is

$$G_1 \quad G_{CNOT} \quad \mathrm{I} \quad \begin{pmatrix} 1 & 0 & 0 & 0 \\ 0 & 1 & 0 & 0 \\ 0 & 0 & 0 & 1 \\ 0 & 0 & 1 & 0 \end{pmatrix} \quad \begin{pmatrix} 1 & 0 \\ 0 & 1 \end{pmatrix} \quad \begin{pmatrix} 1 & 0 & 0 & 0 & 0 & 0 & 0 & 0 \\ 0 & 1 & 0 & 0 & 0 & 0 & 0 & 0 \\ 0 & 0 & 1 & 0 & 0 & 0 & 0 & 0 \\ 0 & 0 & 0 & 1 & 0 & 0 & 0 & 0 \\ 0 & 0 & 0 & 0 & 0 & 0 & 1 & 0 \\ 0 & 0 & 0 & 0 & 0 & 0 & 0 & 1 \\ 0 & 0 & 0 & 0 & 1 & 0 & 0 & 0 \\ 0 & 0 & 0 & 0 & 0 & 1 & 0 & 0 \end{pmatrix}.$$

The state after the first stage is

$$\xi_1 \quad G_1\, \xi \quad \begin{pmatrix} 1 & 0 & 0 & 0 & 0 & 0 & 0 & 0 \\ 0 & 1 & 0 & 0 & 0 & 0 & 0 & 0 \\ 0 & 0 & 1 & 0 & 0 & 0 & 0 & 0 \\ 0 & 0 & 0 & 1 & 0 & 0 & 0 & 0 \\ 0 & 0 & 0 & 0 & 0 & 0 & 1 & 0 \\ 0 & 0 & 0 & 0 & 0 & 0 & 0 & 1 \\ 0 & 0 & 0 & 0 & 1 & 0 & 0 & 0 \\ 0 & 0 & 0 & 0 & 0 & 1 & 0 & 0 \end{pmatrix} \begin{pmatrix} \alpha_0 \\ 0 \\ 0 \\ 0 \\ \alpha_1 \\ 0 \\ 0 \\ 0 \end{pmatrix} \quad \begin{pmatrix} \alpha_0 \\ 0 \\ 0 \\ 0 \\ 0 \\ 0 \\ \alpha_1 \\ 0 \end{pmatrix}.$$

The transfer matrix of the second stage was derived [283] as

$$G_Q \quad \begin{pmatrix} 1 & 0 & 0 & 0 & 0 & 0 & 0 & 0 \\ 0 & 1 & 0 & 0 & 0 & 0 & 0 & 0 \\ 0 & 0 & 1 & 0 & 0 & 0 & 0 & 0 \\ 0 & 0 & 0 & 1 & 0 & 0 & 0 & 0 \\ 0 & 0 & 0 & 0 & 0 & 1 & 0 & 0 \\ 0 & 0 & 0 & 0 & 1 & 0 & 0 & 0 \\ 0 & 0 & 0 & 0 & 0 & 0 & 0 & 1 \\ 0 & 0 & 0 & 0 & 0 & 0 & 1 & 0 \end{pmatrix}.$$

The output of the encoding circuit is

$$\varphi \qquad G_Q\,\xi_1 \quad \begin{pmatrix} 1 & 0 & 0 & 0 & 0 & 0 & 0 & 0 \\ 0 & 1 & 0 & 0 & 0 & 0 & 0 & 0 \\ 0 & 0 & 1 & 0 & 0 & 0 & 0 & 0 \\ 0 & 0 & 0 & 1 & 0 & 0 & 0 & 0 \\ 0 & 0 & 0 & 0 & 0 & 1 & 0 & 0 \\ 0 & 0 & 0 & 0 & 1 & 0 & 0 & 0 \\ 0 & 0 & 0 & 0 & 0 & 0 & 0 & 1 \\ 0 & 0 & 0 & 0 & 0 & 0 & 1 & 0 \end{pmatrix} \begin{pmatrix} \alpha_0 \\ 0 \\ 0 \\ 0 \\ 0 \\ 0 \\ \alpha_1 \\ 0 \end{pmatrix} \quad \begin{pmatrix} \alpha_0 \\ 0 \\ 0 \\ 0 \\ 0 \\ 0 \\ 0 \\ \alpha_1 \end{pmatrix}.$$

Thus, we have

$$\varphi \qquad \alpha_0\, 000 \qquad \alpha_1\, 111 \,.$$

We turn our attention from encoding to error detection and error correction of the quantum codeword, φ , for a bit-flipping channel; there are four projection operators, $\mathbf{P}_i^{bf}$, corresponding to the four possible cases: either there is no error or the first qubit, the second qubit, or the third qubit is bit-flipped:

$$\text{No error} \quad (i \quad 0) \qquad \mathbf{P}_0^{bf} \quad 000\ \ 000 \qquad 111\ \ 111\,,$$

$$\text{Qubit 1 in error} \quad (i \quad 1) \qquad \mathbf{P}_1^{bf} \quad 100\ \ 100 \qquad 011\ \ 011\,,$$

$$\text{Qubit 2 in error} \quad (i \quad 2) \qquad \mathbf{P}_2^{bf} \quad 010\ \ 010 \qquad 101\ \ 101\,,$$

$$\text{Qubit 3 in error} \quad (i \quad 3) \qquad \mathbf{P}_3^{bf} \quad 001\ \ 001 \qquad 110\ \ 110\,,$$

with

$$\mathbf{P}_0^{bf} \begin{pmatrix} 1 & 0 & 0 & 0 & 0 & 0 & 0 & 0 \\ 0 & 0 & 0 & 0 & 0 & 0 & 0 & 0 \\ 0 & 0 & 0 & 0 & 0 & 0 & 0 & 0 \\ 0 & 0 & 0 & 0 & 0 & 0 & 0 & 0 \\ 0 & 0 & 0 & 0 & 0 & 0 & 0 & 0 \\ 0 & 0 & 0 & 0 & 0 & 0 & 0 & 0 \\ 0 & 0 & 0 & 0 & 0 & 0 & 0 & 0 \\ 0 & 0 & 0 & 0 & 0 & 0 & 0 & 1 \end{pmatrix}, \quad \mathbf{P}_1^{bf} \begin{pmatrix} 0 & 0 & 0 & 0 & 0 & 0 & 0 & 0 \\ 0 & 0 & 0 & 0 & 0 & 0 & 0 & 0 \\ 0 & 0 & 0 & 0 & 0 & 0 & 0 & 0 \\ 0 & 0 & 0 & 1 & 0 & 0 & 0 & 0 \\ 0 & 0 & 0 & 0 & 1 & 0 & 0 & 0 \\ 0 & 0 & 0 & 0 & 0 & 0 & 0 & 0 \\ 0 & 0 & 0 & 0 & 0 & 0 & 0 & 0 \\ 0 & 0 & 0 & 0 & 0 & 0 & 0 & 0 \end{pmatrix},$$

$$\mathbf{P}_2^{bf} \begin{pmatrix} 0 & 0 & 0 & 0 & 0 & 0 & 0 & 0 \\ 0 & 0 & 0 & 0 & 0 & 0 & 0 & 0 \\ 0 & 0 & 1 & 0 & 0 & 0 & 0 & 0 \\ 0 & 0 & 0 & 0 & 0 & 0 & 0 & 0 \\ 0 & 0 & 0 & 0 & 0 & 0 & 0 & 0 \\ 0 & 0 & 0 & 0 & 0 & 1 & 0 & 0 \\ 0 & 0 & 0 & 0 & 0 & 0 & 0 & 0 \\ 0 & 0 & 0 & 0 & 0 & 0 & 0 & 0 \end{pmatrix}, \quad \mathbf{P}_3^{bf} \begin{pmatrix} 0 & 0 & 0 & 0 & 0 & 0 & 0 & 0 \\ 0 & 1 & 0 & 0 & 0 & 0 & 0 & 0 \\ 0 & 0 & 0 & 0 & 0 & 0 & 0 & 0 \\ 0 & 0 & 0 & 0 & 0 & 0 & 0 & 0 \\ 0 & 0 & 0 & 0 & 0 & 0 & 0 & 0 \\ 0 & 0 & 0 & 0 & 0 & 0 & 0 & 0 \\ 0 & 0 & 0 & 0 & 0 & 0 & 1 & 0 \\ 0 & 0 & 0 & 0 & 0 & 0 & 0 & 0 \end{pmatrix}.$$

The error-correction procedure is based on a measurement leading to the calculation of the *error syndrome*. Call φ_i^{bf} ,0 i 3, the error codeword, with i 0 corresponding to no error and with i 1,2,3 corresponding to bit-flip of the first, second, or third qubit of the quantum codeword, φ α_0 000 α_1 111 , respectively. Then the error syndromes corresponding to the four possible cases are, respectively:

$$\varphi_i^{bf}\, \mathbf{P}_0^{bf}\, \varphi_i^{bf}\,, \quad 0 \;\; i \;\; 3 \qquad \varphi_i^{bf}\, \mathbf{P}_1^{bf}\, \varphi_i^{bf}\,, \quad 0 \;\; i \;\; 3$$
$$\varphi_i^{bf}\, \mathbf{P}_2^{bf}\, \varphi_i^{bf}\,, \quad 0 \;\; i \;\; 3 \qquad \varphi_i^{bf}\, \mathbf{P}_3^{bf}\, \varphi_i^{bf}\,, \quad 0 \;\; i \;\; 3.$$

When one bit-flip error occurs, there is a one-to-one correspondence between the error syndrome and the qubit in error; consider first the case of an error in the first qubit:

$$\varphi_1^{bf} \quad \alpha_0\, 100 \quad \alpha_1\, 011 \quad \begin{pmatrix} 0 \\ 0 \\ 0 \\ \alpha_1 \\ \alpha_0 \\ 0 \\ 0 \\ 0 \end{pmatrix} \quad \text{or} \quad \varphi_1^{bf} \qquad (0\,0\,0\,\alpha_1\,\alpha_0\,0\,0\,0).$$

It is easy to see that $\varphi_1^{bf}\ \mathbf{P}_1^{bf}\ \varphi_1^{bf} \quad 1$, while

$$\varphi_1^{bf}\ \mathbf{P}_0^{bf}\ \varphi_1^{bf} \quad 0, \quad \varphi_1^{bf}\ \mathbf{P}_2^{bf}\ \varphi_1^{bf} \quad 0, \quad \varphi_1^{bf}\ \mathbf{P}_3^{bf}\ \varphi_1^{bf} \quad 0.$$

Indeed,

$$\varphi_1^{bf}\ \mathbf{P}_1^{bf} \quad (0\,0\,0\,\alpha_1\ \alpha_0\,0\,0\,0)\begin{pmatrix} 0 & 0 & 0 & 0 & 0 & 0 & 0 & 0 \\ 0 & 0 & 0 & 0 & 0 & 0 & 0 & 0 \\ 0 & 0 & 0 & 0 & 0 & 0 & 0 & 0 \\ 0 & 0 & 0 & 1 & 0 & 0 & 0 & 0 \\ 0 & 0 & 0 & 0 & 1 & 0 & 0 & 0 \\ 0 & 0 & 0 & 0 & 0 & 0 & 0 & 0 \\ 0 & 0 & 0 & 0 & 0 & 0 & 0 & 0 \\ 0 & 0 & 0 & 0 & 0 & 0 & 0 & 0 \end{pmatrix} \quad (0\,0\,0\,\alpha_1\ \alpha_0\,0\,0\,0),$$

$$\varphi_1^{bf}\ \mathbf{P}_1^{bf}\ \varphi_1^{bf} \quad (0\,0\,0\,\alpha_1\ \alpha_0\,0\,0\,0)\begin{pmatrix} 0 \\ 0 \\ 0 \\ \alpha_1 \\ \alpha_0 \\ 0 \\ 0 \\ 0 \end{pmatrix} \quad \alpha_0\ ^2 \quad \alpha_1\ ^2 \quad 1.$$

We only show that, $\varphi_1^{bf}\ \mathbf{P}_0^{bf}\ \varphi_1^{bf} \quad 0$:

$$\varphi_1^{bf}\ \mathbf{P}_0^{bf} \quad (0\,0\,0\,\alpha_1\ \alpha_0\,0\,0\,0)\begin{pmatrix} 1 & 0 & 0 & 0 & 0 & 0 & 0 & 0 \\ 0 & 0 & 0 & 0 & 0 & 0 & 0 & 0 \\ 0 & 0 & 0 & 0 & 0 & 0 & 0 & 0 \\ 0 & 0 & 0 & 0 & 0 & 0 & 0 & 0 \\ 0 & 0 & 0 & 0 & 0 & 0 & 0 & 0 \\ 0 & 0 & 0 & 0 & 0 & 0 & 0 & 0 \\ 0 & 0 & 0 & 0 & 0 & 0 & 0 & 0 \\ 0 & 0 & 0 & 0 & 0 & 0 & 0 & 1 \end{pmatrix} \quad (0\,0\,0\,0\,0\,0\,0\,0),$$

$$\varphi_1^{bf} \; \mathbf{P}_0^{bf} \; \varphi_1^{bf} \qquad (0\,0\,0\,0\,0\,0\,0\,0) \begin{pmatrix} 0 \\ 0 \\ 0 \\ \alpha_1 \\ \alpha_0 \\ 0 \\ 0 \\ 0 \end{pmatrix} \qquad 0.$$

The case of a bit-flip of the second or third qubit is treated similarly.

Once we have calculated the syndrome, the error correction is straightforward. We flip back the corresponding qubit of the quantum codeword if an error is detected. We apply a transformation given by a unitary operator, $\mathbf{C}_0, \mathbf{C}_1, \mathbf{C}_2$, or $\mathbf{C}_3$, to the codeword:

$$\begin{array}{rl} \text{No error} & \mathbf{C}_0 \quad \mathrm{I} \quad \mathrm{I}_2 \quad \mathrm{I}, \\ \text{Bit-flip error of first qubit} & \mathbf{C}_1 \quad X \quad \mathrm{I} \quad \mathrm{I}_2, \\ \text{Bit-flip error of second qubit} & \mathbf{C}_2 \quad \mathrm{I} \quad X \quad \mathrm{I}, \\ \text{Bit-flip error of third qubit} & \mathbf{C}_3 \quad \mathrm{I} \quad \mathrm{I} \quad X. \end{array}$$

The syndrome is computed as $\mathrm{H}e_i^T$; the parity-check matrix of the code discussed in this section is

$$\mathrm{H} \quad \begin{pmatrix} 0 & 1 & 1 \\ 1 & 0 & 1 \end{pmatrix}.$$

The circuit to calculate the syndrome requires a number of ancillas larger or equal to the number of rows of the parity-check matrix; in the case of equality, *each ancilla is entangled with the physical qubits selected by the corresponding row of the parity-check matrix.*

A circuit for encoding and syndrome calculation for this code is presented in Figure 5.5; the three qubits of the encoded state, $\varphi \quad \alpha_0\, 000 \quad \alpha_1\, 111$, act as control qubits of the CNOT gates used to set the ancilla qubits. Each ancilla qubit is entangled with multiple qubits of the original codeword, and the resulting degree of entanglement allows for nondemolition measurements. The first ancilla qubit is set to 1 if and only if only one of the second or third qubit is flipped; the second ancilla qubit is set to 1 if and only if only one of the first or third qubit is flipped. So the pair of ancilla qubits is set to 00 if there is no error, 01 if the topmost qubit is bit-flipped, 10 if the middle qubit is bit-flipped, and 11 if the third qubit is bit-flipped. If two qubits are bit-flipped, the error cannot be corrected. The syndrome does not contain any information regarding the values of α_0 or α_1.

An alternative analysis of the error-correction process considers an observable q_{12}, which provides information about whether the first or the second qubit has been flipped and returns the value 1 if none of them has been flipped and 1 if one has been flipped. The projective measurement uses two projectors, $(\,00 \;\; 00 \quad 11 \;\; 11\,) \;\; \mathrm{I}$ and $(\,01 \;\; 01 \quad 10 \;\; 10\,) \;\; \mathrm{I}$, and has the spectral decomposition

$$q_{12} \quad (\,00 \;\; 00 \quad 11 \;\; 11\,) \;\; \mathrm{I} \quad (\,01 \;\; 01 \quad 10 \;\; 10\,) \;\; \mathrm{I}.$$

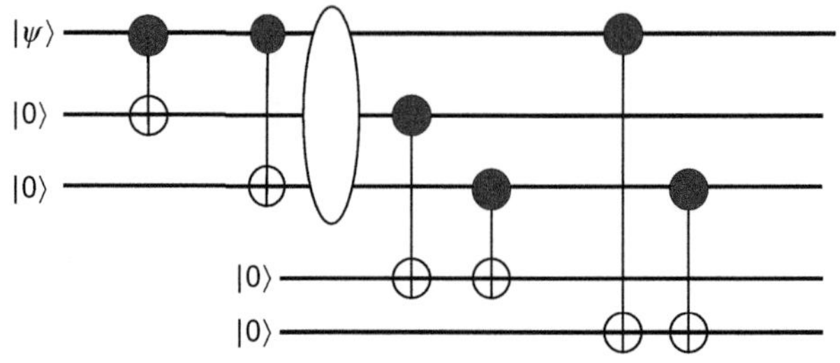

FIGURE 5.5

A circuit for encoding the qubit, $\psi \quad \alpha_0\, 0 \quad \alpha_1 1$, as $\alpha_0\, 000 \quad \alpha_1 111$, and for syndrome calculation; the qubits are numbered top to bottom. To calculate the syndrome, we need two ancilla qubits; the rst ancilla is the target qubit of CNOT gates selected by the rst row of the parity-check matrix, 011, and the second is the target selected by the second row of the parity-check matrix, 101.

A second observable, q_{23}, provides information about whether the second or the third qubit has been flipped and returns the value 1 if none of them has been flipped and 1 if one of them has been flipped. The projectors are

$$\mathrm{I} \quad (00 \quad 00 \quad\quad 11 \quad 11) \quad \text{and} \quad \mathrm{I} \quad (01 \quad 01 \quad\quad 10 \quad 10).$$

The projective measurement has the spectral decomposition

$$q_{23} \quad \mathrm{I} \quad (00 \quad 00 \quad\quad 11 \quad 11) \quad \mathrm{I} \quad (01 \quad 01 \quad\quad 10 \quad 10).$$

By combining the two measurements, we can determine if any of the three qubits of the quantum word has been affected by an error and if so, which one. Our next challenge is to correct phase-flip errors, the topic of the following section.

5.6 A REPETITIVE QUANTUM CODE FOR A SINGLE PHASE-FLIP ERROR

A phase-flip error is described as the effect of applying the, Z, operator to a qubit in the state, ψ :

$$Z\,\psi \quad Z(\alpha_0\, 0 \quad \alpha_1\, 1) \quad \alpha_0\, 0 \quad \alpha_1\, 1 \qquad \begin{pmatrix} 1 & 0 \\ 0 & 1 \end{pmatrix}\begin{pmatrix} \alpha_0 \\ \alpha_1 \end{pmatrix} \quad \begin{pmatrix} \alpha_0 \\ \alpha_1 \end{pmatrix}.$$

The phase-flip error has no classical equivalent and the solution is to convert the phase-flip to a bit-flip error by a change of basis carried out by the Hadamard transform, H:

$$0\,,\,1 \quad \overset{H}{} \quad , \quad \overset{H}{} \quad 0\,,\,1\,,$$

with HH I; the new basis state vectors, , , are

$$H\,0 \quad \frac{0 \quad 1}{\overline{2}} \qquad \frac{1}{\overline{2}}\begin{pmatrix} 1 & 1 \\ 1 & 1 \end{pmatrix}\begin{pmatrix} 1 \\ 0 \end{pmatrix} \quad \frac{1}{\overline{2}}\begin{pmatrix} 1 \\ 1 \end{pmatrix}$$

and

$$H\,1 \quad \frac{0 \quad 1}{\overline{2}} \qquad \frac{1}{\overline{2}}\begin{pmatrix} 1 & 1 \\ 1 & 1 \end{pmatrix}\begin{pmatrix} 0 \\ 1 \end{pmatrix} \quad \frac{1}{\overline{2}}\begin{pmatrix} 1 \\ 1 \end{pmatrix}.$$

In the rotated basis, Z becomes a bit-flip operator

$$Z \qquad \begin{pmatrix} 1 & 0 \\ 0 & 1 \end{pmatrix}\frac{1}{\overline{2}}\begin{pmatrix} 1 \\ 1 \end{pmatrix} \quad \frac{1}{\overline{2}}\begin{pmatrix} 1 \\ 1 \end{pmatrix}$$

and

$$Z \qquad \begin{pmatrix} 1 & 0 \\ 0 & 1 \end{pmatrix}\frac{1}{\overline{2}}\begin{pmatrix} 1 \\ 1 \end{pmatrix} \quad \frac{1}{\overline{2}}\begin{pmatrix} 1 \\ 1 \end{pmatrix}.$$

The error-correction process closely follows the one presented in Section 5.5. Call φ_i^{phf} the state ψ after a phase-flip of qubit i, with 0 i n; the (n 1) projectors used to identify the qubit in the error are

$$\mathbf{P}_i^{phf} \quad \varphi_i^{phf} \quad \varphi_i^{phf} \qquad \varphi_i^{phf} \quad \varphi_i^{phf}, \quad 0 \quad i \quad n.$$

The (n 1) syndromes indicating that either no error has occurred or that qubit i suffered a phase-flip error are

$$\Sigma_i^{phf} \qquad \varphi_i^{bf} \; \mathbf{P}_i^{bf} \; \varphi_i^{bf}, \quad 0 \quad i \quad n.$$

If the k-th qubit is in error, then

$$\Sigma_i^{phf} \qquad \begin{cases} 1 & \text{if } i \quad k, \\ 0 & \text{if } i \quad k. \end{cases}$$

Example. *Consider the three-qubit repetitive code discussed in Section 5.5 and apply a rotation of the basis vectors to correct phase- ip errors.*

The scheme for error correction follows closely the one presented in Figure 5.3; the states, 0 and 1 , are encoded as

$$0 \qquad\qquad \text{and} \quad 1 \qquad\qquad .$$

The encoding, the error detection, and the recovery are performed with respect to the logical states, and ; the circuit in Figure 5.6 encodes a qubit state, ψ α_0 0 α_1 1 to ζ α_0 α_1 .

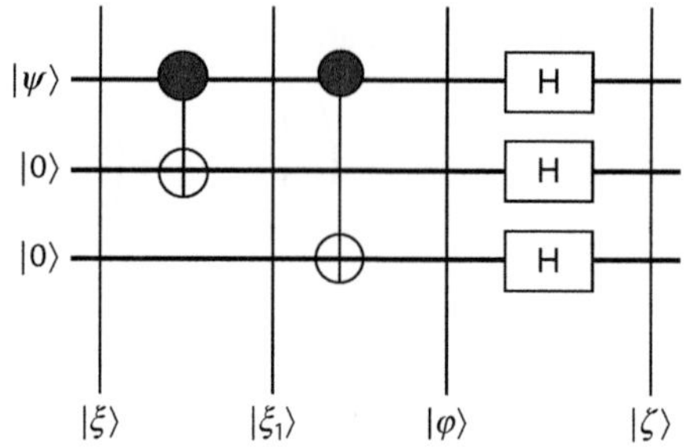

FIGURE 5.6

An encoding circuit to correct a phase-flip error. The state, $\psi \quad \alpha_0\, 0 \quad \alpha_1\, 1$, of the qubit is encoded rst to $\varphi \quad \alpha_0\, 000 \quad \alpha_1\, 111$, as in the case of bit-flip error correction; then φ is encoded as $\zeta \quad \alpha_0 \qquad \alpha_1$.

The first two stages of this circuit are similar to the circuit in Figure 5.4 used for encoding in the basis, 000 , 111 , and the third stage consists of Hadamard gates, one for each qubit. If the qubit to be encoded is in the state, $\psi \quad \alpha_0\, 0 \quad \alpha_1\, 1$, the input state for the third stage is φ and the transfer matrix of the third stage is

$$H^{\ 3} \quad H \; H \; H \quad \frac{1}{2\ \overline{2}}\begin{pmatrix} 1 & 1 \\ 1 & 1 \end{pmatrix} \begin{pmatrix} 1 & 1 \\ 1 & 1 \end{pmatrix} \begin{pmatrix} 1 & 1 \\ 1 & 1 \end{pmatrix}.$$

The state of the three-qubit system at the output of the encoding circuit is

$$\zeta \quad H^{\ 3}\,\varphi \quad \frac{1}{2\ \overline{2}}\begin{pmatrix} 1 & 1 & 1 & 1 & 1 & 1 & 1 & 1 \\ 1 & 1 & 1 & 1 & 1 & 1 & 1 & 1 \\ 1 & 1 & 1 & 1 & 1 & 1 & 1 & 1 \\ 1 & 1 & 1 & 1 & 1 & 1 & 1 & 1 \\ 1 & 1 & 1 & 1 & 1 & 1 & 1 & 1 \\ 1 & 1 & 1 & 1 & 1 & 1 & 1 & 1 \\ 1 & 1 & 1 & 1 & 1 & 1 & 1 & 1 \\ 1 & 1 & 1 & 1 & 1 & 1 & 1 & 1 \end{pmatrix}\begin{pmatrix} \alpha_0 \\ 0 \\ 0 \\ 0 \\ 0 \\ 0 \\ 0 \\ \alpha_1 \end{pmatrix} \quad \frac{1}{2\ \overline{2}}\begin{pmatrix} \alpha_0 \quad \alpha_1 \\ \alpha_0 \quad \alpha_1 \\ \alpha_0 \quad \alpha_1 \\ \alpha_0 \quad \alpha_1 \\ \alpha_0 \quad \alpha_1 \\ \alpha_0 \quad \alpha_1 \\ \alpha_0 \quad \alpha_1 \\ \alpha_0 \quad \alpha_1 \end{pmatrix}.$$

The detailed calculations related to the first two stages of the circuit in Figure 5.6 are presented in Section 5.5. In Section 5.9, we show that

$$\frac{1}{2\ \overline{2}}(\ 000 \quad 001 \quad 010 \quad 011 \quad 100 \quad 101 \quad 110 \quad 111\ )$$

and

$$\frac{1}{2\ \overline{2}}(\ 000 \quad 001 \quad 010 \quad 011 \quad 100 \quad 101 \quad 110 \quad 111\ ).$$

Thus, our circuit performs the desired transformation:

$$\psi \quad \alpha_0\, 0 \quad \alpha_1\, 1 \qquad \zeta \quad \alpha_0 \qquad \alpha_1 \quad .$$

We now discuss the projector operators for phase-flip errors; there are four projectors, $\mathbf{P}_i^{phf}$, corresponding to the four possible cases: either there is no error or the first qubit, the second qubit, or the third qubit is phase-flipped:

No error	$(i \quad 0)$	$\mathbf{P}_0^{phf}$	,
Qubit 1 in error	$(i \quad 1)$	$\mathbf{P}_1^{phf}$	,
Qubit 2 in error	$(i \quad 2)$	$\mathbf{P}_2^{phf}$	,
Qubit 3 in error	$(i \quad 3)$	$\mathbf{P}_3^{phf}$	.

The phase-flip projectors are related to the ones for bit-flip errors by the following relation:

$$\mathbf{P}_i^{phf} \quad H^{\ 3}\mathbf{P}_i^{bf} H^{\ 3}, \quad 0 \quad i \quad 3.$$

Thus,

$$\mathbf{P}_0^{phf} \quad H^{\ 3}\mathbf{P}_0^{bf} H^{\ 3} \quad \frac{1}{4}\begin{pmatrix} 1 & 0 & 0 & 1 & 0 & 1 & 1 & 0 \\ 0 & 1 & 1 & 0 & 1 & 0 & 0 & 1 \\ 0 & 1 & 1 & 0 & 1 & 0 & 0 & 1 \\ 1 & 0 & 0 & 1 & 0 & 1 & 1 & 0 \\ 0 & 1 & 1 & 0 & 1 & 0 & 0 & 1 \\ 1 & 0 & 0 & 1 & 0 & 1 & 1 & 0 \\ 1 & 0 & 0 & 1 & 0 & 1 & 1 & 0 \\ 0 & 1 & 1 & 0 & 1 & 0 & 0 & 1 \end{pmatrix},$$

$$\mathbf{P}_1^{phf} \quad H^{\ 3}\mathbf{P}_1^{bf} H^{\ 3} \quad \frac{1}{4}\begin{pmatrix} 1 & 0 & 0 & 1 & 0 & 1 & 1 & 0 \\ 0 & 1 & 1 & 0 & 1 & 0 & 0 & 1 \\ 0 & 1 & 1 & 0 & 1 & 0 & 0 & 1 \\ 1 & 0 & 0 & 1 & 0 & 1 & 1 & 0 \\ 0 & 1 & 1 & 0 & 1 & 0 & 0 & 1 \\ 1 & 0 & 0 & 1 & 0 & 1 & 1 & 0 \\ 1 & 0 & 0 & 1 & 0 & 1 & 1 & 0 \\ 0 & 1 & 1 & 0 & 1 & 0 & 0 & 1 \end{pmatrix},$$

$$\mathbf{P}_2^{phf} \quad H^{\;3}\mathbf{P}_2^{bf}H^{\;3} \quad \frac{1}{4}\begin{pmatrix} 1 & 0 & 0 & 1 & 0 & 1 & 1 & 0 \\ 0 & 1 & 1 & 0 & 1 & 0 & 0 & 1 \\ 0 & 1 & 1 & 0 & 1 & 0 & 0 & 1 \\ 1 & 0 & 0 & 1 & 0 & 1 & 1 & 0 \\ 0 & 1 & 1 & 0 & 1 & 0 & 0 & 1 \\ 1 & 0 & 0 & 1 & 0 & 1 & 1 & 0 \\ 1 & 0 & 0 & 1 & 0 & 1 & 1 & 0 \\ 0 & 1 & 1 & 0 & 1 & 0 & 0 & 1 \end{pmatrix},$$

and

$$\mathbf{P}_3^{phf} \quad H^{\;3}\mathbf{P}_3^{bf}H^{\;3} \quad \frac{1}{4}\begin{pmatrix} 1 & 0 & 0 & 1 & 0 & 1 & 1 & 0 \\ 0 & 1 & 1 & 0 & 1 & 0 & 0 & 1 \\ 0 & 1 & 1 & 0 & 1 & 0 & 0 & 1 \\ 1 & 0 & 0 & 1 & 0 & 1 & 1 & 0 \\ 0 & 1 & 1 & 0 & 1 & 0 & 0 & 1 \\ 1 & 0 & 0 & 1 & 0 & 1 & 1 & 0 \\ 1 & 0 & 0 & 1 & 0 & 1 & 1 & 0 \\ 0 & 1 & 1 & 0 & 1 & 0 & 0 & 1 \end{pmatrix}.$$

Call ζ_i^{phf} the codeword, with $i \quad 1,2,3$ corresponding to the phase-flip of the first, second, or third qubit of the quantum codeword, $\zeta \quad \alpha_0 \qquad\qquad \alpha_1$, respectively, and $i \quad 0$ corresponding to no error. The error syndromes for the four possible cases are, respectively:

$$\zeta_i^{phf}\,\mathbf{P}_0^{phf}\,\zeta_i^{phf}\,, \quad 0 \quad i \quad 3 \qquad \zeta_i^{phf}\,\mathbf{P}_1^{phf}\,\zeta_i^{phf}\,, \quad 0 \quad i \quad 3$$

$$\zeta_i^{phf}\,\mathbf{P}_2^{phf}\,\zeta_i^{phf}\,, \quad 0 \quad i \quad 3 \qquad \zeta_i^{phf}\,\mathbf{P}_3^{phf}\,\zeta_i^{phf}\,, \quad 0 \quad i \quad 3.$$

There is a one-to-one correspondence between the error syndrome and the qubit in error, if one error is present; we discuss only the phase-flip of the third qubit and show that

$$\zeta_0^{phf}\,\mathbf{P}_3^{phf}\,\zeta_0^{phf} \qquad \zeta_1^{phf}\,\mathbf{P}_3^{phf}\,\zeta_1^{phf} \qquad \zeta_2^{phf}\,\mathbf{P}_3^{phf}\,\zeta_2^{phf} \qquad 0$$

and

$$\zeta_3^{phf}\ \mathbf{P}_3^{phf}\ \zeta_3^{phf} \qquad 1,$$

where

$$\zeta_3^{phf} \qquad \alpha_0 \qquad\qquad \alpha_1 \qquad\qquad \frac{1}{2\ \overline{2}}\begin{pmatrix} \alpha_0 & \alpha_1 \\ \alpha_0 & \alpha_1 \\ \alpha_0 & \alpha_1 \\ \alpha_0 & \alpha_1 \\ \alpha_0 & \alpha_1 \\ \alpha_0 & \alpha_1 \\ \alpha_0 & \alpha_1 \\ \alpha_0 & \alpha_1 \end{pmatrix}.$$

Now

$$\mathbf{P}_3^{phf}\ \zeta_3^{phf} \qquad \frac{1}{4}\begin{pmatrix} 1 & 0 & 0 & 1 & 0 & 1 & 1 & 0 \\ 0 & 1 & 1 & 0 & 1 & 0 & 0 & 1 \\ 0 & 1 & 1 & 0 & 1 & 0 & 0 & 1 \\ 1 & 0 & 0 & 1 & 0 & 1 & 1 & 0 \\ 0 & 1 & 1 & 0 & 1 & 0 & 0 & 1 \\ 1 & 0 & 0 & 1 & 0 & 1 & 1 & 0 \\ 1 & 0 & 0 & 1 & 0 & 1 & 1 & 0 \\ 0 & 1 & 1 & 0 & 1 & 0 & 0 & 1 \end{pmatrix} \frac{1}{2\ \overline{2}}\begin{pmatrix} \alpha_0 & \alpha_1 \\ \alpha_0 & \alpha_1 \\ \alpha_0 & \alpha_1 \\ \alpha_0 & \alpha_1 \\ \alpha_0 & \alpha_1 \\ \alpha_0 & \alpha_1 \\ \alpha_0 & \alpha_1 \\ \alpha_0 & \alpha_1 \end{pmatrix}$$

$$\frac{1}{8\ \overline{2}}\begin{pmatrix} 4\alpha_0 & 4\alpha_1 \\ 4\alpha_0 & 4\alpha_1 \\ 4\alpha_0 & 4\alpha_1 \\ 4\alpha_0 & 4\alpha_1 \\ 4\alpha_0 & 4\alpha_1 \\ 4\alpha_0 & 4\alpha_1 \\ 4\alpha_0 & 4\alpha_1 \\ 4\alpha_0 & 4\alpha_1 \end{pmatrix} \qquad \frac{1}{2\ \overline{2}}\begin{pmatrix} \alpha_0 & \alpha_1 \\ \alpha_0 & \alpha_1 \\ \alpha_0 & \alpha_1 \\ \alpha_0 & \alpha_1 \\ \alpha_0 & \alpha_1 \\ \alpha_0 & \alpha_1 \\ \alpha_0 & \alpha_1 \\ \alpha_0 & \alpha_1 \end{pmatrix} \qquad \zeta_3^{phf}$$

and

$$\zeta_3^{phf}\ \zeta_3^{phf} \qquad 1.$$

In the case of no error,

$$\zeta_0^{phf} \quad \alpha_0 \qquad \alpha_1 \qquad \frac{1}{2\sqrt{2}}\begin{pmatrix} \alpha_0 & \alpha_1 \\ \alpha_0 & \alpha_1 \\ \alpha_0 & \alpha_1 \\ \alpha_0 & \alpha_1 \\ \alpha_0 & \alpha_1 \\ \alpha_0 & \alpha_1 \\ \alpha_0 & \alpha_1 \\ \alpha_0 & \alpha_1 \end{pmatrix}$$

and

$$\mathbf{P}_3^{phf} \ \zeta_0^{phf} \quad \frac{1}{4}\begin{pmatrix} 1 & 0 & 0 & 1 & 0 & 1 & 1 & 0 \\ 0 & 1 & 1 & 0 & 1 & 0 & 0 & 1 \\ 0 & 1 & 1 & 0 & 1 & 0 & 0 & 1 \\ 1 & 0 & 0 & 1 & 0 & 1 & 1 & 0 \\ 0 & 1 & 1 & 0 & 1 & 0 & 0 & 1 \\ 1 & 0 & 0 & 1 & 0 & 1 & 1 & 0 \\ 1 & 0 & 0 & 1 & 0 & 1 & 1 & 0 \\ 0 & 1 & 1 & 0 & 1 & 0 & 0 & 1 \end{pmatrix} \frac{1}{2\sqrt{2}}\begin{pmatrix} \alpha_0 & \alpha_1 \\ \alpha_0 & \alpha_1 \\ \alpha_0 & \alpha_1 \\ \alpha_0 & \alpha_1 \\ \alpha_0 & \alpha_1 \\ \alpha_0 & \alpha_1 \\ \alpha_0 & \alpha_1 \\ \alpha_0 & \alpha_1 \end{pmatrix}$$

$$\frac{1}{8\sqrt{2}}\begin{pmatrix} 0 \\ 0 \\ 0 \\ 0 \\ 0 \\ 0 \\ 0 \\ 0 \end{pmatrix}.$$

Thus, $\zeta_0^{phf} \ \mathbf{P}_3^{phf} \ \zeta_0^{phf} \quad 0$. In summary, when we apply a Hadamard transform to each qubit we transform phase-flip errors to bit-flip-errors; then we correct the bit-flip errors, and finally apply another Hadamard transform to each qubit of the register to return to the original basis.

5.7 THE NINE-QUBIT ERROR-CORRECTING CODE OF SHOR

The repetitive quantum code presented in Section 5.5 can only correct a single bit-flip error, and the one in Section 5.6 can only correct a single phase-flip error. Shor [380] created a nine-qubit quantum error-correcting code able to correct a single bit-flip error, a single phase-flip error, or a single bit- and phase-flip error; the distance of the code is 3.

A qubit is first encoded using the phase-flip error correction code in Section 5.6, then each of the three qubits is encoded using the circuit for bit-flip error correction discussed in Section 5.5. The quantum circuits for syndrome calculation for each group of three qubits are shown in Figure 5.7; the

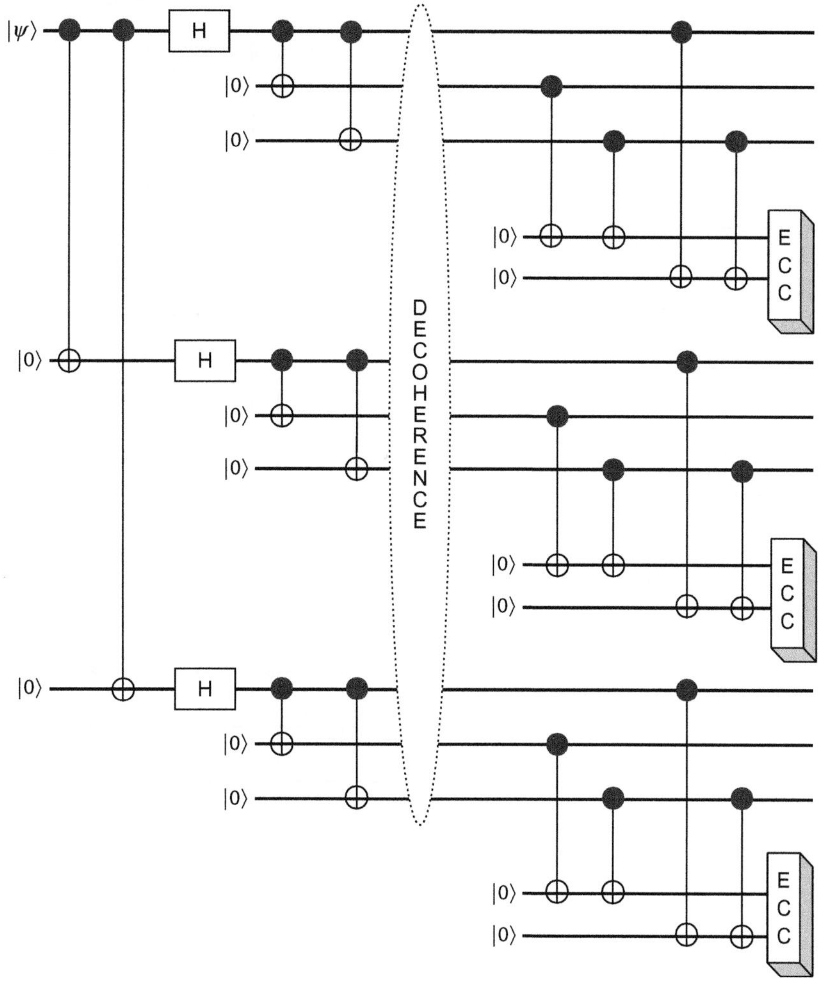

FIGURE 5.7

A circuit for Shor s algorithm.

qubits are counted left to right starting with qubit 0. The circuit uses six ancilla qubits, two for each group of three qubits, though we only need four ancilla qubits for that computation of the syndrome as seen in Figure 5.8.

First, a qubit is encoded for phase-flip error correction, then each of the resulting three qubits is encoded for bit-flip error correction. The two basis vectors in $\mathcal{H}_2$, 0 , 1 , are transformed in

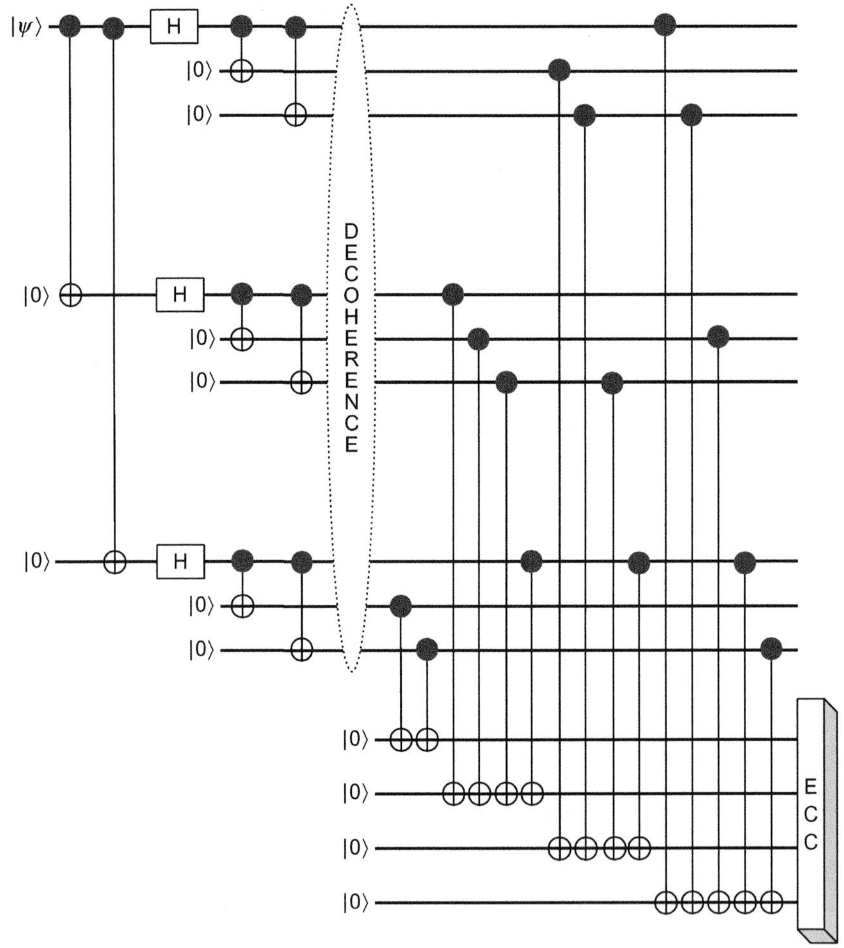

FIGURE 5.8

An alternative circuit for syndrome decoding for Shor s algorithm that uses only four ancilla qubits, while the one in Figure 5.7 needs six. Each one of the four ancilla qubits reflects the transformations expressed by one row of the parity-check matrix of the code.

0_L, 1_L in $\mathcal{H}_{2^9}$; the encoding is

	Phase- ip		*Bit- ip*
0	$\frac{0 \quad 1}{2} \; \frac{0 \quad 1}{2} \; \frac{0 \quad 1}{2}$	0_L	$\frac{000 \quad 111}{2} \; \frac{000 \quad 111}{2} \; \frac{000 \quad 111}{2}$
1	$\frac{0 \quad 1}{2} \; \frac{0 \quad 1}{2} \; \frac{0 \quad 1}{2}$	1_L	$\frac{000 \quad 111}{2} \; \frac{000 \quad 111}{2} \; \frac{000 \quad 111}{2}$.

The parity-check matrix for this code is

$$H \quad \begin{pmatrix} 0 & 0 & 0 & 0 & 0 & 0 & 0 & 1 & 1 \\ 0 & 0 & 0 & 1 & 1 & 1 & 1 & 0 & 0 \\ 0 & 1 & 1 & 0 & 0 & 1 & 1 & 0 & 0 \\ 1 & 0 & 1 & 0 & 1 & 0 & 1 & 0 & 1 \end{pmatrix}.$$

The error syndrome identifies the qubit in error; for example, when the fifth qubit is in error, the error vector is e_5 (000010000) and the syndrome is He_5^T (0101), where $(0101)_2$ 5.

5.8 THE SEVEN-QUBIT ERROR-CORRECTING CODE OF STEANE

Now we turn our attention to a quantum system where we encode a single logical qubit as a block of seven physical qubits and correct one bit-flip or one phase-flip error. Steane's approach [401] is to encode an arbitrary state in a two-dimensional Hilbert space, $\mathcal{H}_2$, by embedding $\mathcal{H}_2$ into $\mathcal{H}_{2^7}$. Steane's code is based on a (7,4,3) Hamming code, $\mathcal{C}$, with the generator matrix, G, given by

$$G \quad \begin{pmatrix} 1 & 0 & 0 & 0 & 0 & 1 & 1 \\ 0 & 1 & 0 & 0 & 1 & 0 & 1 \\ 0 & 0 & 1 & 0 & 1 & 1 & 0 \\ 0 & 0 & 0 & 1 & 1 & 1 & 1 \end{pmatrix} \quad [I_4A], \quad \text{with} \quad A \quad \begin{pmatrix} 0 & 1 & 1 \\ 1 & 0 & 1 \\ 1 & 1 & 0 \\ 1 & 1 & 1 \end{pmatrix}.$$

$\mathcal{C}$ is a code that maps, k 4, information symbols to codewords of length, n 7; the 2^4 16 codewords of $\mathcal{C}$ are generated by multiplying all 4-tuples, v_i, 1 i 16, with the generator matrix, G, of the code, $\mathcal{C}$ v_iG, 1 i 16 , as shown in Table 5.1.

The four information symbols are the prefix of each codeword; the subset of codewords of $\mathcal{C}$ marked, , are also the codewords of the dual code, $\mathcal{C}$. The distance of code $\mathcal{C}$ is equal to the weight of the minimum weight codeword, d 3; the number of linearly independent columns of G is equal to 4.

To construct the dual code, $\mathcal{C}$, we first derive, H, the parity-check matrix of $\mathcal{C}$, which is the generator matrix of the dual code, $\mathcal{C}$; from Section 4.5 we know that

$$H \quad [A^T I_{n\ k}] \quad [A^T I_3] \quad \begin{pmatrix} 0 & 1 & 1 & 1 & 1 & 0 & 0 \\ 1 & 0 & 1 & 1 & 0 & 1 & 0 \\ 1 & 1 & 0 & 1 & 0 & 0 & 1 \end{pmatrix}.$$

Table 5.1 Mapping of Information Symbols to the Codewords for the Code, $\mathcal{C}$, Used by the Seven-qubit Quantum Code of Steane

i	v_i		$\mathcal{C}$
1	0000		**0000**000
2	0001		**0001**111
3	0010		**0010**110
4	0011		**0011**001
5	0100		**0100**101
6	0101	$\begin{pmatrix} 1 & 0 & 0 & 0 & 0 & 1 & 1 \\ 0 & 1 & 0 & 0 & 1 & 0 & 1 \\ 0 & 0 & 1 & 0 & 1 & 1 & 0 \\ 0 & 0 & 0 & 1 & 1 & 1 & 1 \end{pmatrix}$	**0101**010
7	0110		**0110**011
8	0111		**0111**100
9	1000		**1000**011
10	1001		**1001**100
11	1010		**1010**101
12	1011		**1011**010
13	1100		**1100**110
14	1101		**1101**001
15	1110		**1110**000
16	1111		**1111**111

Table 5.2 Mapping Information Symbols to the Codewords for the Dual Code, $\mathcal{C}$, Used by the Seven-qubit Quantum Code of Steane

v'		$\mathcal{C}^{\perp}$
000		0000**000**
001		1101**001**
010	$\begin{pmatrix} 0 & 1 & 1 & 1 & 1 & 0 & 0 \\ 1 & 0 & 1 & 1 & 0 & 1 & 0 \\ 1 & 1 & 0 & 1 & 0 & 0 & 1 \end{pmatrix}$	1011**010**
011		0110**011**
100		0111**100**
101		1010**101**
110		1100**110**
111		0001**111**

The codewords of $\mathcal{C}$ are constructed using H as the generator matrix as shown in Table 5.2. We observe that, $\mathcal{C}$ $\mathcal{C}$ as $\mathcal{C}$, consists of all the even codewords of $\mathcal{C}$.

Code $\mathcal{C}$ can be used to encode one bit of information to seven bits. If the i-th bit of a codeword, c $\mathcal{C}$, is affected by an error, instead of c we receive a different 7-tuple, $(c \quad e_i)$.

The error 7-tuple, e_i, has only one 1 in the position, i, in error. There is a one-to-one correspondence between an error e_i and the syndrome Σ_i:

$$\Sigma_i \quad H(c \quad e_i)^T \quad He_i^T \quad \text{as} \quad Hc^T \quad 0 \quad \text{when} \quad c \quad \mathcal{C}.$$

Encoding for Steane s Code

The *logical zero* qubit is the superposition of all codewords with an even number of 1s (even codewords) of the [3, 4, 7] Hamming code

$$0_L \quad \frac{1}{8}(0000000 \quad 0011101 \quad 0100111 \quad 0111010 \quad 1001110 \quad 1010011$$
$$1101001 \quad 1110100).$$

The *logical one* qubit is the superposition of all codewords with an odd number of 1s (odd codewords) of the Hamming code

$$1_L \quad \frac{1}{8}(1111111 \quad 1100010 \quad 1011000 \quad 1000101 \quad 0110001 \quad 0101100$$
$$0010110 \quad 0001011).$$

The quantum circuit in Figure 5.9 encodes a qubit, $\psi \quad \alpha_0\, 0 \quad \alpha_1\, 1$, as a block of seven qubits, $\varphi \quad \alpha_0\, 0_L \quad \alpha_1\, 1_L$; two CNOT gates encode the state, $\alpha_0\, 0 \quad \alpha_1\, 1$, as the state $\alpha_0\, 000 \quad \alpha_1\, 111$,

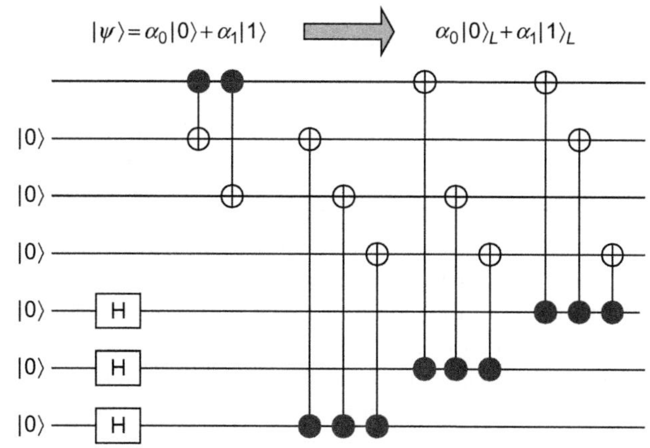

FIGURE 5.9

A circuit for encoding the qubit, $\psi \quad \alpha_0\, 0 \quad \alpha_1\, 1$, as a block of seven qubits, $\varphi \quad \alpha_0\, 0_L \quad \alpha_1\, 1_L$, for Steane s code. The two CNOT gates to encode the state, $\alpha_0\, 0 \quad \alpha_1\, 1$, as the state $\alpha_0\, 000 \quad \alpha_1\, 111$. The three Hadamard gates prepare an equal superposition of all eight possible values of the last three qubits. The three groups of CNOTs gates switch on the parity bits dictated by H; the rst group of CNOTs corresponds to the pre x, (0 1 1 1), of the rst row, the second group to the pre x, (1 0 1 1), of the second row, and the third group to the pre x, (1 1 0 1), of the third row.

a superposition of even and odd Hamming codewords. The three Hadamard gates prepare an equal superposition of all eight possible values of the last three qubits. The state of the system at this stage is

$$\xi \quad (\alpha_0\ 000 \quad \alpha_1\ 111\ ) \quad (\ 0\ ) \quad \frac{0 \quad 1}{\overline{2}} \quad \frac{0 \quad 1}{\overline{2}} \quad \frac{0 \quad 1}{\overline{2}}$$

or

$$\begin{aligned} \xi \quad \tfrac{1}{\overline{8}}[\alpha_0(\quad & 0000000 \quad 0000001 \quad 0000010 \quad 0000011 \\ & 0000100 \quad 0000101 \quad 0000110 \quad 0000111\ ) \\ \alpha_1(\ & 1110000 \quad 1110001 \quad 1110010 \quad 1110011 \\ & 1110100 \quad 1110101 \quad 1110110 \quad 1110111\ )]. \end{aligned}$$

The three groups of CNOT gates act on, ξ , and switch on the parity bits dictated by the parity-check matrix, H; the first group of CNOTs corresponds to the prefix, (0 1 1 1), of the first row, the second group to the prefix, (1 0 1 1), of the second row, and the third group to the prefix, (1 1 0 1), of the third row; the first four bits of the Hamming codeword are identical to the four information bits.

Error Correction

Let $\psi \quad \alpha_0\ 0 \quad \alpha_1\ 1$ and $A\ _{env}$ be the states of the qubit and of the environment, respectively. The decoherence is caused by the entanglement of the qubit with the environment; thus, the effect of an error is captured by a linear combination of the four states corresponding to no error, bit-flip, phase-flip, and both bit- and phase-flip [341]:

$$\begin{aligned} \alpha_0\ 0 \quad \alpha_1\ 1 \qquad\qquad & (\alpha_0\ 0 \quad \alpha_1\ 1\ ) \quad A_{no\ error\ \ env} \\ & (\alpha_0\ 1 \quad \alpha_1\ 0\ ) \quad A_{bit\text{-}\ ip\ \ env} \\ & (\alpha_0\ 0 \quad \alpha_1\ 1\ ) \quad A_{phase\text{-}\ ip\ \ env} \\ & (\alpha_0\ 1 \quad \alpha_1\ 0\ ) \quad A_{bit\ \ phase\text{-}\ ip\ \ env}. \end{aligned}$$

The system "qubit+environment" evolves according to a unitary transformation, but we make no assumption regarding the orthogonality and normalization of the states, $A_{x...x\ \ env}$.

We perform a measurement that projects the current state of the qubit onto the basis formed by the four states and determine the effect of the environment on the qubit: no error, bit-flip, phase-flip, bit- and phase-flip. To correct a bit-flip, phase-flip, or bit- and phase-flip error, we apply to the current state of the qubit the X, Z, Y transformations, respectively. In this manner we restore the qubit to its original state and disentangle the qubit from its environment. The correction circuit consists of the I, X, Z, Y gates activated by the result of the measurement of the syndrome.

The circuit in Figure 5.10 is used for calculating the bit-flip syndrome for the seven-qubit Steane code, with the parity-check matrix, H. Each ancilla qubit is determined by the corresponding row of the parity-check matrix: the first is determined by (0 1 1 1 1 0 0), where the control qubits of the CNOTs are the second, third, fourth, and fifth qubits of the codeword; the second is determined by (1 0 1 1 0 1 0); and the third is determined by (1 0 1 0 1 0 1).

We need at least three ancilla qubits to identify the individual qubits of the block of seven qubits. If ξ is the state of the ensemble consisting of the codeword initially in the state, ξ , and the

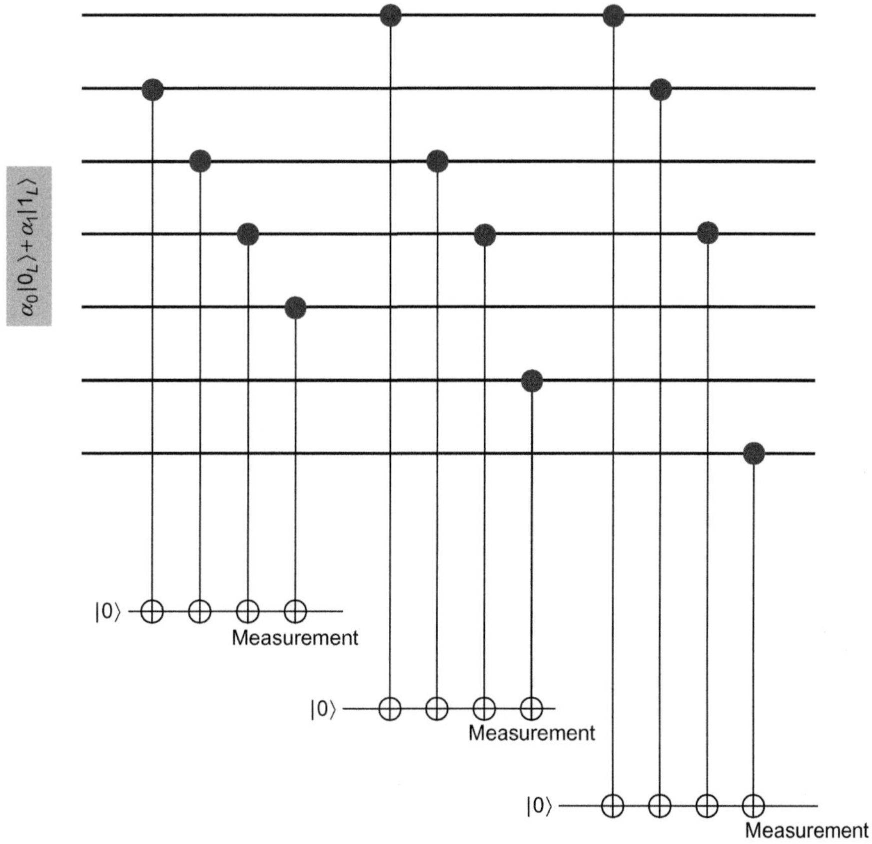

FIGURE 5.10

A circuit for calculating the syndrome for bit-flip errors for the seven-qubit Steane code. The input is the encoded information, the codeword, $\varphi \quad \alpha_0\ 0_L \quad \alpha_1\ 1_L$. The seven qubits of the codeword act as the control qubits of the 12 CNOT gates that set the ancilla qubits; then we perform a nondemolition measurement of the syndrome. If a phase error occurs in one of the ancilla qubits used as the target of a CNOT gate, then the error can "back-propagate" to the control qubits and alter the state of the subset of the seven physical qubits that act as control qubits for the ancilla.

ancilla qubits, initially in the state, 000 , then the ensemble codeword ancilla qubits is transformed as follows:

$$\xi \quad 0_{ancilla} \quad \xi \quad H\xi_{ancilla},$$

where $H\xi_{aniclla}$ is the final state of the three ancilla qubits used to compute the syndrome. If only one of the seven qubits of a block has suffered a bit-flip, the measurement of the three ancilla qubits identifies which one of the qubits of the block is in error, without revealing anything about the quantum information, the state ψ , encoded to the block.

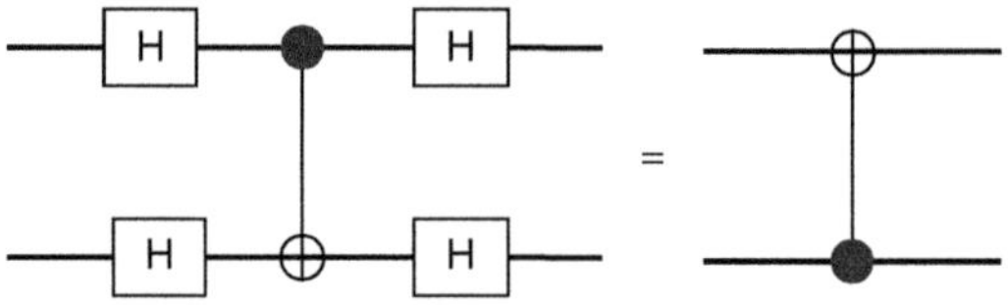

FIGURE 5.11

Back error propagation. We can reverse the role of a target and control qubit of a CNOT gate with Hadamard gates. The rotation of the basis performed by the Hadamard gate transforms a bit-flip error to a phase-flip error. When a phase error occurs in one qubit used as the target of a CNOT gate, the error can propagate "backward" to the control qubit.

The measurement should be nondestructive and should preserve the code subspace. To ensure a nondemolition measurement, we have to prepare the ancilla qubits in a special state. Then we "write" the parity qubits to the ancilla and then we measure the ancilla. Clearly, the parity bits of a Hamming code are either 0 or 1; thus, we can "write" the parity qubits to the ancilla. To diagnose phase-flip errors, we have to repeat the process in a rotated base.

It is now a good time to observe that we need more than three ancilla qubits, even though three bits are sufficient to express integers in the range 1 to 7. The circuit to compute the ancilla qubits for Steane's code (Figure 5.10) uses one ancilla qubit as the target of multiple CNOT gates. Such a circuit is not fault-tolerant.

If the control qubit of a CNOT gate is in error, then it can propagate "forward" to the target qubit. Recall that we can reverse the role of a target and control qubit of a CNOT gate with Hadamard gates (see Figure 5.11) [283]. We also know that the rotation of the basis performed by the Hadamard gate transforms a bit-flip error to a phase-flip error. Thus, if a phase error occurs in one qubit used as the target of a CNOT gate, then the error can propagate "backward" to the control qubit. The result of the CNOT transformation corresponds to a virtual bit-flip error of the control qubit. When a single qubit is the target of multiple CNOT gates, *an error of the target qubit will affect all the control qubits.* The fault-tolerant version of the circuit for calculating the bit-flip syndrome for the seven-bit Steane code in Figure 5.12 has 12 rather than 3 ancilla qubits.

In summary, Steane's code uses the logical, 0_L and 1_L, qubits that are superpositions of Hamming codewords. In the original and the rotated basis, we can thus perform the parity check, which allows us to diagnose and then correct a single-qubit error.

5.9 AN INEQUALITY FOR REPRESENTATIONS IN DIFFERENT BASES

Classical error-correcting codes satisfy the bounds discussed in Section 4.7; these bounds relate the number of codewords, C, of a code, C, the length, n, of a codeword, and the distance, d, of a code. From the discussion of Shor and Steane codes, we know that a quantum error-correcting code requires two classical error-correcting codes, one to correct bit-flip and the other to correct phase-flip errors. In this section, we discuss the "entropic uncertainty relation," (introduced in [55, 114]), that provides an upper bound on the largest distance achievable in the two bases required to correct both bit-flip, (0   1), and phase-flip, (0   1   0   1), errors.

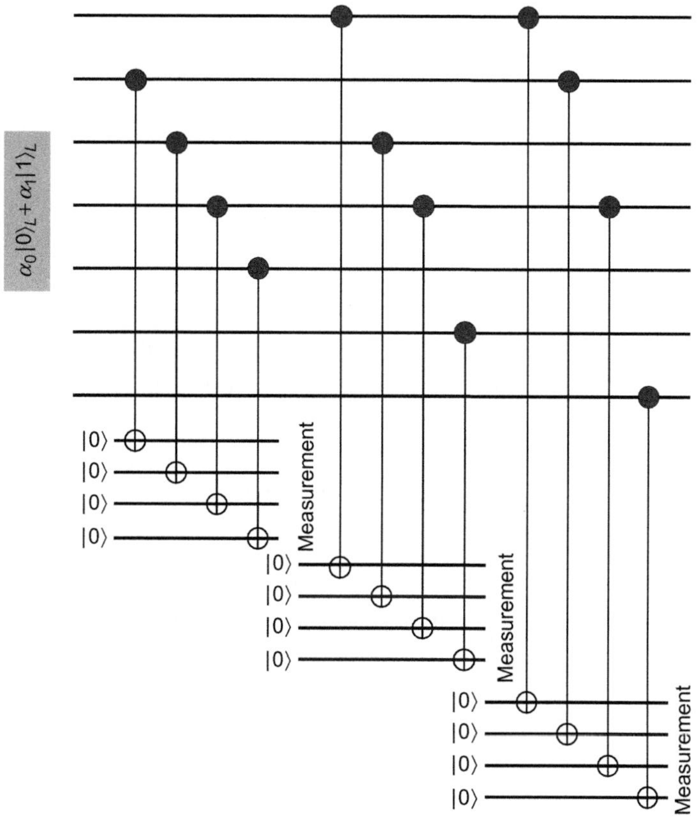

FIGURE 5.12

The fault-tolerant version of the circuit for calculating the bit-flip syndrome for the seven-bit Steane code in Figure 5.10. Each one of the 12 ancilla qubits is the target of a single CNOT gate. The circuit projects the state, $(\alpha_0\ 0\ _L \quad \alpha_1\ 1\ _L)$, onto one of the basis states, $0\ _L$ and $1\ _L$.

The transformation performed by a Hadamard gate, H, maps the basis, $B_1 \quad 0\ , 1$ to $B_2 \quad\quad , - \ $, with

$$H\ 0 \quad \frac{0 \quad 1}{\overline{2}} \quad \text{and} \quad - \quad H\ 1 \quad \frac{0 - 1}{\overline{2}}.$$

This rotation of basis transforms phase-flip errors to bit-flip errors and vice versa:

$$(\ 0 \quad 1\ )_{B_1} \quad 0\ _{B_2} \quad \text{and} \quad (\ 0 - 1\ )_{B_1} \quad 1\ _{B_2}.$$

In $\mathcal{H}_{2^n}$ we consider the basis, B_1, with the basis vectors, $e_0\ , e_1\ , \ldots, e_{2^n-1}$, and, B_2, with the basis vectors, $e_0\ , e_1\ , \ldots, e_{2^n-1}$. Table 5.3 shows the states, $e_0 e_1, \ldots, e_{2^n-1}\ _{B_2}$, represented as a superposition of the basis vectors, $e_0\ , e_1\ , \ldots, e_{2^n-1}$.

Table 5.3 Changes of Basis in $\mathcal{H}_{2^n}$ from B_1 e_0 , e_1 ... e_{2^n} 1 to B_2 e_0 , e_1 ... e_{2^n} 1 ; the States, $b_0 b_1 \ldots b_{2^n}$ 1 $_{B_2}$, Represented as a Superposition of the Basis Vectors, e_0 , e_1 ... e_{2^n} 1 , in B_1 for n 1,2,3

Base B_2	***Base B_1***
n 1	
	$\frac{1}{2}$(0 1)
	$\frac{1}{2}$(0 1)
n 2	
	$\frac{1}{2}$(0 1) $\frac{1}{2}$(0 1) $\frac{1}{2}$(00 01 10 11)
	$\frac{1}{2}$(0 1) $\frac{1}{2}$(0 1) $\frac{1}{2}$(00 01 10 11)
	00 11
	01 10
	00 11
n 3	
	$\frac{1}{2}$(0 1) $\frac{1}{2}$(00 01 10 11)
	$\frac{1}{2\ 2}$(000 001 010 011 100 101 110 111)
	$\frac{1}{2}$(0 1) $\frac{1}{2}$(00 01 10 11)
	$\frac{1}{2\ 2}$(000 001 010 011 100 101 110 111)
	$\frac{1}{2}$(000 011 101 110)
	$\frac{1}{2}$(001 010 100 111)
	$\frac{1}{2}$(0 1) $\frac{1}{2}$(0 1) $\frac{1}{2}$(0 1)
	$\frac{1}{2}$(0 1) $\frac{1}{2}$(0 1) $\frac{1}{2}$(0 1)
	$\frac{1}{2}$(000 011 110 101)
	$\frac{1}{2}$(0 1) $\frac{1}{2}$(0 1) $\frac{1}{2}$(0 1)
	$\frac{1}{2}$(0 1) $\frac{1}{2}$(0 1) $\frac{1}{2}$(0 1)
	$\frac{1}{2}$(000 011 110 110)

We say that a basis vector, u_j , has an "odd" parity if the number of 1s in the binary representation of integer, u_j, is an odd number, and u_j has an "even" parity if the number of 1s in the binary representation of integer, u_j, is an even number. For example, 001 has odd parity and 101 has even parity. It is easy to infer and then verify the following statements:

1. The 00...0 $_{B_2}$ state is a superposition of all basis vectors in B_1, with equal and positive coefficients.
2. The 11...1 $_{B_2}$ state is a superposition of all basis vectors in B_1, with equal coefficients of positive (even parity) and negative (odd parity) sign.

3. The $(00\ldots0 \quad 11\ldots1)_{B_2}$ state is a superposition of basis vectors in B_1, with even parity. All coefficients are equal and positive.
4. The $(00\ldots0 \quad 11\ldots1)_{B_2}$ state is a superposition of basis vectors in B_1, with odd parity. All coefficients are equal and positive.
5. If the i-th bit in the state, $(00\ldots0 \quad 11\ldots1)_{B_2}$, is flipped, then the mapping of this state is similar with the original superposition, but the coefficients of the basis states when this bit is equal to 1 are negative. For example, when the first bit is flipped, the state becomes $100 \quad 011$ and then, instead of the superposition, $\frac{1}{\overline{2}}(000 \quad 011) \quad 101 \quad 110)$, we have $\frac{1}{\overline{2}}(000 \quad 011) \quad 101 \quad 110)$.

Let $\psi \quad \alpha_0\, e_0 \quad \alpha_1\, e_1 \qquad \alpha_{2^n \; 1}\, e_{2^n \; 1}$ be a vector in a Hilbert space, $\mathcal{H}_{2^n}$. The basis vectors are $e_i \quad 2^i \quad 1\,, 0 \quad i \quad 2^n \quad 1$. Call $w(\psi)_i$ the *weight* of the state vector in the basis B_i, $i \quad 1, 2$. The weight is equal to the number of nonzero complex numbers α_i, $0 \quad i \quad 2^n \quad 1$, or the number of basis vectors used to express ψ in B_i, $i \quad 1, 2$.

Proposition. *The state of an n-qubit system can be expressed as a superposition of w_1 basis vectors in base, B_1, and as a superposition of w_2 basis vectors in a rotated base, B_2. The weights, w_1 and w_2, satisfy the following inequality:*

$$w_1 \; w_2 \quad 2^n.$$

Proof. Recall that the Hadamard gate performs a Walsh-Hadamard transform of one qubit,

$$H \quad \frac{1}{\overline{2}} \begin{pmatrix} 1 & 1 \\ 1 & 1 \end{pmatrix},$$

and that $H^{\;\;n}$ denotes the Walsh-Hadamard transform applied to, n, qubits. Consider the state,

$$\psi \qquad \alpha_0\, e_0 \quad \alpha_1\, e_1 \qquad \alpha_{2^n \; 1}\, e_{2^n \; 1} \, ,$$

expressed in B_1, with the basis vectors, $e_0\,,\ e_1\,, \ldots,\ e_{2^n \; 1}$. The same state is expressed in base, B_2, as

$$\varphi \quad H^{\;\;n}\, \psi \, .$$

The respective weights are

$$w_1 \quad w(\psi) \;\text{ and }\; w_2 \quad w(\varphi).$$

Our proof is by induction. For $n \quad 1$, the vectors in B_1 and B_2 are, respectively,

$$0 \quad \begin{pmatrix} 1 \\ 0 \end{pmatrix}, \; 1 \quad \begin{pmatrix} 1 \\ 0 \end{pmatrix} \quad \text{and} \quad \frac{0 \quad 1}{\overline{2}} \quad \begin{pmatrix} 1 \\ 1 \end{pmatrix}, \frac{0 \quad 1}{\overline{2}} \quad \begin{pmatrix} 1 \\ 1 \end{pmatrix}.$$

Thus, for $n \quad 1$ the relationship is true: $w_1 \quad 1$, $w_2 \quad 2$, and, $w_1 \; w_2 \quad 1 \; 2 \quad 2 \quad 2$.

Now assume that this is true for $n \quad k$ and attempt to prove it for $n \quad k \quad 1$. We represent the integer, k, by a binary string of length, n, denoted as k_n. For example, for $n \quad 2$ we write the basis vectors as follows:

$$\psi \quad \alpha_0\, 00 \quad \alpha_1\, 01 \quad \alpha_2\, 10 \quad \alpha_3\, 11 \qquad \psi \quad \alpha_0\, 0_2 \quad \alpha_1\, 1_2 \quad \alpha_2\, 2_2 \quad \alpha_3\, 3_2 \; .$$

We observe that

$$\begin{aligned} \psi \quad & \alpha_0\, 0_{n\ 1} \quad \alpha_1\, 1_{n\ 1} \qquad \alpha_{2^{n\ 1}\ 1}\, (2^{n\ 1} \quad 1)_{n\ 1} \\ & 0_1 \quad [\alpha_0\, 0_n \quad \alpha_1\, 1_n \qquad \alpha_{2^n\ 1}\, (2^n \quad 1)_n\,] \\ & 1_1 \quad [\alpha_{2^n}\, 0_n \quad \alpha_{2^n\ 1}\, 1_n \qquad \alpha_{2^{n\ 1}\ 1}\, (2^n \quad 1)_n\,]. \end{aligned}$$

We use the following notations:

$$\psi_F \quad \alpha_0\, 0_n \quad \alpha_1\, 1_n \qquad \alpha_{2^n\ 1}\, (2^n \quad 1)_n$$

and

$$\psi_L \quad \alpha_{2^n}\, 0_n \quad \alpha_{2^n\ 1}\, 1_n \qquad \alpha_{2^{n\ 1}\ 1}\, (2^n \quad 1)_n \; .$$

Thus,

$$\psi \quad 0_1 \quad \psi_F \quad 1_1 \quad \psi_L \; .$$

Now

$$\begin{aligned} \varphi \quad & H^{\ k}\, \psi \\ & (H \quad H^{\ k})[\, 0_1 \quad \psi_F \quad 1_1 \quad \psi_L\,] \\ & (\, 0_1 \quad 1_1\,)H^{\ k}\, \psi_F \quad (\, 0_1 \quad 1_1\,)H^{\ k}\, \psi_L \\ & 0_1\, (H^{\ k}\, \psi_F \quad H^{\ k}\, \psi_L) \quad 1_1(\, H^{\ k}\, \psi_F \quad H^{\ k}\, \psi_L\,). \end{aligned}$$

Assume that $\psi_F \quad 0$ and then, by inductive hypothesis, $\psi_L \quad 0$:

$$w(\, \psi\,)\;\; w(\, \varphi\,) \quad 2\;\; w(\, \psi_L\,)\;\; w(H^{\ k}\, \psi_L\,) \quad 2\;\; 2^k \quad 2^{k\ 1}.$$

Assume that $\psi_L \quad 0$ and $\psi_F \quad 0$. A similar derivation allows us to prove that in this case, $w_1 \;\; w_2 \quad 2^{k\ 1}$.

Now consider the case, $\psi_F \quad 0$ and $\psi_L \quad 0$. Then

$$w(\, \psi_F\,)\;\; w(H^{\ k}\, \psi_F\,) \quad 2^k \quad \text{and} \quad w(\, \psi_L\,)\;\; w(H^{\ k}\, \psi_L\,) \quad 2^k.$$

Also, if $\alpha_i \quad 0$, then $\alpha_i \quad \alpha_{i\ 2^k} \quad 0$ or $\alpha_i \quad \alpha_{i\ 2^k} \quad 0$, where we have addition modulo, $2^{k\ 1}$, for the subscripts. Thus,

$$w(\, \varphi\,) \quad w(H^{\ k}\, \psi_F\,) \quad \text{and} \quad w(\, \varphi\,) \quad w(H^{\ k}\, \psi_L\,).$$

Finally, in this case

$$
\begin{aligned}
w(\ \psi\ )w(\ \varphi\ ) \quad & [w(\ \psi_F\ ) \quad w(\ \psi_L\ )]\ w(\ \varphi\ ) \\
& w(\ \psi_F\ )\ w(\ \varphi \quad w(\ \psi_L\ )\ w(\ \varphi\ ) \\
& w(\ \psi_F\ )\ w(H\ ^{k})\ \psi_F\ ) \quad w(\ \psi_L\ )\ w(H\ ^{k}\ \psi_L\ ) \\
& 2^k \quad 2^k \\
& 2^{k\quad 1}.
\end{aligned}
$$

This concludes the proof that the weights, w_1 and w_2, satisfy the inequality, $w_1 \quad w_2 \quad 2^n$. ■

5.10 CALDERBANK-SHOR-STEANE (CSS) CODES

The Calderbank-Shor-Steane (CSS) codes were discovered independently by Calderbank and Shor [80] and Steane [401]. CSS codes are a subclass of the larger class of stabilizer codes discussed in Section 5.12.

First, we review basic concepts from linear coding theory discussed in Section 4.5. Let $\mathcal{C}$ be a linear $[n,k,d]$ code over the binary field, $\mathcal{F}_2^n$, capable of correcting, $t \quad \frac{d\ 1}{2}$, errors; $\mathcal{C}$, the dual of the code $\mathcal{C}$, is the set of all binary n-tuples orthogonal to all codewords of $\mathcal{C}$:

$$\mathcal{C} \quad v \quad F_2^n : v \ c \quad 0 \quad c \quad \mathcal{C}\ .$$

The generator matrix, G, of, $\mathcal{C}$, is a $k \quad n$ matrix whose row space consists of independent n-tuples spanning the code, $\mathcal{C}$; the parity-check matrix, H, of code, $\mathcal{C}$, is an $(n \quad k) \quad n$ matrix, such that $\mathrm{H}c^T \quad 0, \ c \quad \mathcal{C}$. Code $\mathcal{C}$ is spanned by, k, independent n-tuples, $\dim(\mathcal{C}) \quad k$, and $\mathcal{C}$ is spanned by, $n \quad k$, independent n-tuples, $\dim(\mathcal{C}\ ) \quad n \quad k$. If G and H are the generator and, respectively, the parity-check matrix of $\mathcal{C}$, then H and G are the generator and, respectively, the parity-check matrix of $\mathcal{C}$. The code $\mathcal{C}$ is weakly self-dual if $\mathcal{C} \quad \mathcal{C}$.

Calderbank and Shor proved the quantum error-correction theorem relating classical and quantum error-correcting codes [80], discussed next.

Theorem. *If $\mathcal{C}_1$ and $\mathcal{C}_2$ are both $[n,k,d]$ codes, with $0 \quad \mathcal{C}_2 \quad \mathcal{C}_1 \quad \mathcal{F}_2^n$, then the quantum code, $\mathcal{Q}_{\mathcal{C}_1,\mathcal{C}_2}$, is a t–error–correcting code, with $t \quad (d \quad 1)/2$.*

Steane reformulated this theorem based on several properties of $\mathcal{C}_2$: (1) the dimension of $\mathcal{C}_2$ is $(n \quad k)$; (2) $\mathcal{C}_2 \quad \mathcal{C}_1$; and (3) $\mathcal{C}_2$ is a K-th order subcode of $\mathcal{C}_1$, with $K \quad (2k \quad n)$.

Theorem. *To encode K qubits, with minimum distance, d_1, in one basis and, d_2, in another, it is suf cient to nd a linear code of minimum distance d_1, whose K-th order subcode is the dual of a distance d_2 code.*

A quantum code, $\mathcal{Q}$, is denoted by $[n,k,d_1,d_2]$, with n, the number of qubits used to store or transmit, k, bits of information; $\mathcal{Q}$ allows correction of up to $(d_1 \quad 1)/2$ bit-flip and simultaneously up to $(d_2 \quad 1)/2$ phase-flip errors.

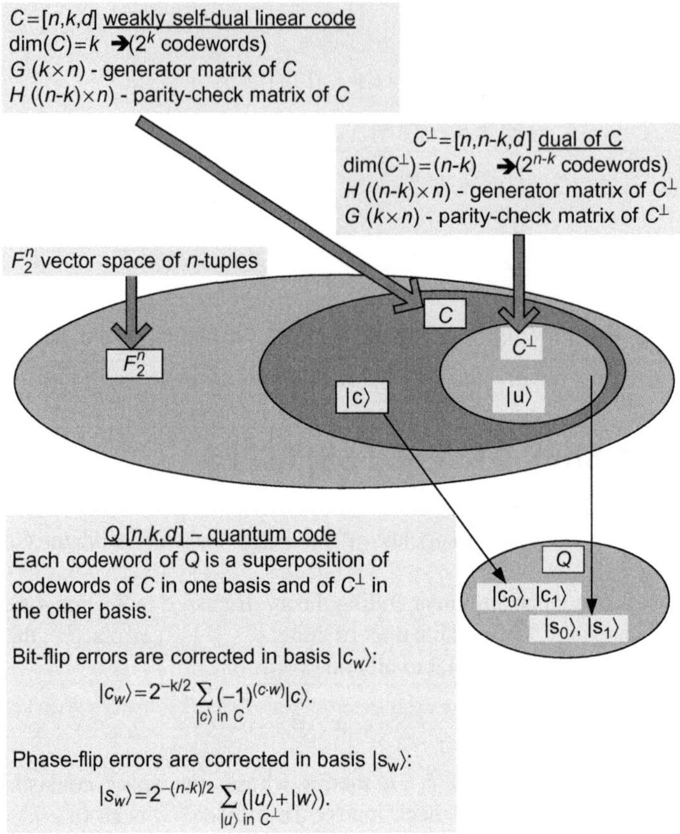

FIGURE 5.13

The construction of a CSS code, $\mathcal{Q}$, based on a self-dual linear code, $\mathcal{C}$.

Construction of a Quantum Code, $\mathcal{Q}$, Based on a Classical Weakly Self-Dual Code

The codewords of $\mathcal{Q}$ (see Figure 5.13) are superpositions of the codewords of $\mathcal{C}$ assembled according to the following rules [80]:

1. Construct first a c-basis. If $v \quad \mathcal{C}$, then a quantum state, c_w , is defined as

$$c_w \quad 2^{-\frac{\dim(\mathcal{C})}{2}} \sum_{v \quad \mathcal{C}} (-1)^{v \quad w} \; v \; , \quad w \quad F_2^n,$$

with "$v \quad w$", the bitwise addition modulo 2 of two n-tuples, v and w, $\sum_{i \quad 1}^{n} v_i w_i \mod 2$.

2. Construct an s-basis. According to rule 1, the codewords of the quantum code, $\mathcal{Q}$, are the set of c_w , w $\mathcal{C}$. We rotate each of the, n, qubits of a codeword:

$$0 \quad \frac{0 \quad 1}{\overline{2}} \quad \text{and} \quad 1 \quad \frac{0 \quad 1}{\overline{2}}.$$

Phase errors in the original basis become bit errors in the new basis, 0 0 and 1 1 . After the change of basis applied to a qubit in the state, c_w , we obtain the state,

$$s_w \quad 2^{\frac{\dim(\mathcal{C}) \quad n}{2}} \sum_{u \quad \mathcal{C}} u \quad w \;, \quad w \quad F_2^n.$$

It is left as an exercise for the reader to prove that given (v , u) $\mathcal{C}$, two basis vectors are identical, s_v s_u and c_v c_u , if and only if v u . The dimension of the quantum code $\mathcal{Q}$ is

$$\dim(\mathcal{Q}) \quad \dim(\mathcal{C}) \quad \dim(\mathcal{C} \;) \quad k \quad (n \quad k) \quad 2k \quad n.$$

Codes, $\mathcal{C}$ and $\mathcal{C}$, have, 2^k and $2^{n \quad k}$, codewords, respectively; thus, the quantum code, $\mathcal{Q}$, has $2^{2k \quad n}$ codewords.

We first correct the bit-flip errors in the c-basis, then perform the rotation of basis and correct the bit-flip errors in the s-basis; the phase-flip errors in the c-basis, if any, become bit-flip errors in the s-basis. We can correct at most, t, bit-flip and phase-flip errors simply because the classical code, $\mathcal{C}$, can correct up to, t, errors. Shor and Calderbank show that the two error correction steps do not interfere with each other [80]. The following example from their work [80] illustrates the construction of a quantum code.

Example. *Construct quantum code, $\mathcal{Q}$, based on $\mathcal{C}$ and $\mathcal{C}$, with $\mathcal{C}$, the* [7,4,3] *Hamming code*

$$\mathcal{C} \quad 0000000 \;\; 0001011 \;\; 0010110 \;\; 0011101 \;\; 0100111 \;\; 0101100 \;\; 0110001 \;\; 0111010$$
$$1000101 \;\; 1001110 \;\; 1010011 \;\; 1011000 \;\; 1100010 \;\; 1101001 \;\; 1110100 \;\; 1111111 \;.$$

The dimension of this code is $\dim(\mathcal{C})$ 4; the dual code, $\mathcal{C}$, is

$$\mathcal{C} \quad 0000000 \;\; 0011101 \;\; 0100111 \;\; 0111010 \;\; 1001110 \;\; 1010011 \;\; 1101001 \;\; 1110100 \;.$$

$\mathcal{C}$ $\mathcal{C}$; thus, $\mathcal{C}$ is a weakly self-dual code. The dimension of $\mathcal{C}$ is

$$\dim(\mathcal{C} \;) \quad n \quad \dim(\mathcal{C}) \quad 7 \quad 4 \quad 3.$$

We see that $2^{\dim(\mathcal{C}) \quad \dim(\mathcal{C} \;)}$ 2; thus, $\mathcal{Q}$ has two orthogonal vectors, c_0 and c_1 , constructed according to the rule

$$c_w \quad 2^{\;\frac{\dim(\mathcal{C})}{2}} \sum_{c \quad \mathcal{C}} (\quad 1)^{c \quad w} \; c \;, \quad w \quad 0, 1.$$

If w 0, then c C, c w c (0000000) 0; thus, all the codewords in the sum have positive amplitudes

$$c_0 \quad \frac{1}{4}(\ 0000000 \quad 0001011 \quad 0010110 \quad 0011101$$
$$0100111 \quad 0101100 \quad 0110001 \quad 0111010$$
$$1000101 \quad 1001110 \quad 1010011 \quad 1011000$$
$$1100010 \quad 1101001 \quad 1110100 \quad 1111111\ ).$$

When w 1,

$$c \ w \quad c \ (1111111) \quad \begin{cases} 1 & \text{if the weight of c is odd,} \\ 0 & \text{if the weight of c is even.} \end{cases}$$

The codewords with an even weight have positive amplitudes, while those with an odd weight have negative amplitudes:

$$c_1 \quad \frac{1}{4}(\ 0000000 \quad 0001011 \quad 0010110 \quad 0011101$$
$$0100111 \quad 0101100 \quad 0110001 \quad 0111010$$
$$1000101 \quad 1001110 \quad 1010011 \quad 1011000$$
$$1100010 \quad 1101001 \quad 1110100 \quad 1111111\ ).$$

We use a similar strategy to construct s_0 and s_1 :

$$s_w \quad 2^{\frac{\dim(C\ )\ n}{2}} \sum_{u\ C} u \quad w\ , w \quad 0, 1.$$

If w 0, then u w u, u C and

$$s_0 \quad \frac{1}{2\ 2}(\ 0000000 \quad 0011101 \quad 0100111 \quad 0111010$$
$$1001110 \quad 1010011 \quad 1101001 \quad 1110100\ ).$$

If w 1, then u w u (1111111) u u C ; here, u is the binary n-tuple with every bit complemented and

$$s_1 \quad \frac{1}{2\ 2}(\ 1111111 \quad 1100010 \quad 1011000 \quad 1000101$$
$$0110001 \quad 0101100 \quad 0010110 \quad 0001011\ ).$$

Steane s Version

We construct a quantum code, $\mathcal{Q}(C_1, C_2)$, based on two classical codes, C_1 and C_2 C_1 [401]. Let C_2 be the dual code of C_2. Both C_1 and C_2 are capable of correcting, t $\frac{d\ 1}{2}$, errors. C_1 is an $[n, k_1, d]$ code, C_2 is an $[n, k_2, d\ ]$ code, and C_2 is an $[n, n \quad k_2, d]$ code. Call H_1 and H_2 the parity-check matrices of C_1 and C_2 , respectively.

The quantum code, $\mathcal{Q}(C_1,C_2)$, should be capable of correcting, t, or fewer bit-flip and phase-flip errors. We construct first an s-basis. If $u \quad C_1$, then we define the state,

$$s_{u \quad w} \quad \frac{1}{2^{\frac{k_2}{2}}} \sum_{u \quad C_1, w \quad C_1/C_2} u \quad w \,,$$

with w chosen from the cosets $\frac{C_1}{C_2}$. We can compute the cosets of C_2 in C_1 because C_2 is an additive subgroup of C_1, $C_2 \quad C_1$. There are 2^{k_2} elements in C_2 and therefore as many coset leaders. It follows that each coset consists of

$$\frac{C_1}{C_2} \quad \frac{2^{k_1}}{2^{k_2}} \quad 2^{k_1-k_2}$$

n-tuples. The corresponding table has 2^{k_1} elements, 2^{k_2} rows, and $2^{k_1-k_2}$ columns. It is easy to see that the state, $s_{u \quad w}$, depends only on the coset of C_1/C_2. Indeed, $s_{u \quad w} \quad s_{u \quad w}$ if and only if $u - u \quad C_2$. Thus, the number of codewords in $\mathcal{Q}(C_1,C_2)$ is equal to $2^{k_1-k_2}$. Note that Steane's codewords, s_w, generate the same subspace in $\mathcal{H}_2{}^n$ as the codewords, c_w, of Calderbank and Shor.

The Error Correction

The decoding scheme for CSS codes based on weakly self-dual codes is depicted in Figure 5.14. Assume that an error is represented by an n-tuple, e, such that:

1. Both bit-flip and phase-flip errors could be present, $e \quad e_{bf} \quad e_{pf}$. The n-tuples, e_{bf} and e_{pf}, have a 1 in a position corresponding to a bit-flip and phase-flip error, respectively.

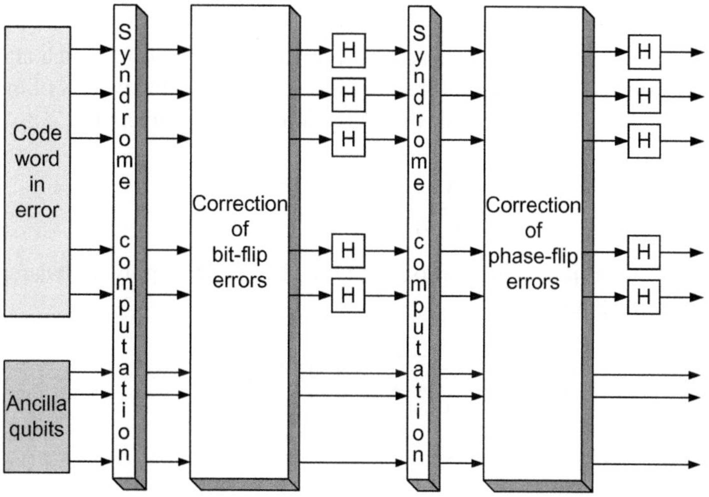

FIGURE 5.14

The decoding scheme for CSS codes constructed from weakly self-dual codes.

2. There are at most t errors, $w(e)\ \ t$, the weight of the error vector is less than or equal to t; thus, $w(e_{bf})\ \ w(e_{pf})\ \ t$.

If the original state, $s_{u\ \ w}$, is affected by e_{bf} bit-flip and e_{pf} phase-flip errors, then the corrupted state is

$$s_{(u\ \ w),error}\ \ \frac{1}{2^{\frac{k_2}{2}}}\sum_{u\ \ C_1, w\ \ C_1/C_2} (\ \ 1)^{(u\ \ w)\ \ e_{pf}}\ u\ \ w\ \ e_{bf}\ .$$

We use a sufficient number of ancilla qubits initially in the all-zero state, $00\ldots0$, to store the syndrome for the code, C_1. When we apply the parity-check matrix, H_1, to the corrupted state, we produce the state,

$$\frac{1}{2^{\frac{k_2}{2}}}\sum_{u\ \ C_1, w\ \ C_1/C_2} (\ \ 1)^{(u\ \ w)\ \ e_{pf}}\ u\ \ w\ \ e_{bf}\ \ \ H_1(e_{bf})\ .$$

We write explicitly, ab as $a\ \ \ b$, to emphasize the action of the parity-check matrix, which calculates the syndrome in the ancilla qubits. The application of H_1 transforms the joint state of an original element of the superposition and the ancilla qubits as follows:

$$u\ \ w\ \ e_{bf}\ \ 0\ \ \overset{H_1}{}\ \ u\ \ w\ \ e_{bf}\ \ \ H_1(u\ \ w\ \ e_{bf})\ \ \ u\ \ w\ \ e_{bf}\ \ \ H_1(e_{bf})\ .$$

The computation of the syndrome does not affect the state and that for $u\ \ w\ \ C_1$, $H_1(u\ \ w\ \ e_{bf})\ \ H_1(e_{bf})$ because $H_1(u\ \ w)\ \ 0$; after we discard the ancilla qubits, the state is

$$\frac{1}{2^{\frac{k_2}{2}}}\sum_{u\ \ C_1, w\ \ C_1/C_2} (\ \ 1)^{(u\ \ w)\ e_{pf}}\ u\ \ w\ \ e_{bf}\ .$$

The ancilla qubits contain a portion of the state in error, namely, $H_1(e_{bf})$, which allows us to correct the, e_{bf}, bit-flip errors since code, C_1, can correct at most, t, errors and the weight of the phase-flip error vector is at most t, $w(e_{bf})\ \ t$; the state after the correction of bit-flip errors is

$$\frac{1}{2^{\frac{k_2}{2}}}\sum_{u\ \ C_1, w\ \ C_1/C_2} (\ \ 1)^{(u\ \ w)\ e_{pf}}\ u\ \ w\ .$$

Now we correct the residual phase-flip errors, if any; as before, we apply a Hadamard Transform to each qubit and the resulting state is

$$\frac{1}{2^{\frac{n\ \ k_2}{2}}}\sum_{z}\ \sum_{u\ \ C_1, w\ \ C_1/C_2} (\ \ 1)^{(u\ \ w)\ (e_{pf}\ \ z)}\ z\ ,$$

with z, an n-tuple. This can be rewritten as

$$\frac{1}{2^{\frac{n\ \ k_2}{2}}}\sum_{z}\ \sum_{u\ \ C_1, w\ \ C_1/C_2} (\ \ 1)^{(u\ \ w)\ z}\ z\ \ \ e_{pf}\ ,$$

with $z \quad z \quad e_{pf}$. If $z \quad C_2$, then $\sum_{w \ C_2}(\ 1)^{w\ z} \quad C_2$. We can take advantage of the fact that if C_2 is a linear code and C_2 its dual, then

$$\sum_{w \ C_2} (\ 1)^{w\ z} \quad \begin{cases} C_2 \quad 2^{k_2} & \text{if } z \quad C_2\,, \\ 0 & \text{if } z \ / C_2\,. \end{cases}$$

Indeed, if $z \quad C_2$, then the two vectors are orthogonal, $w \ z \quad 0$, and the sum equals C_2; the state after the Hadamard transform can be expressed as

$$\frac{1}{2^{\frac{n \ k_2}{2}}} \sum_{z \ C_2} (\ 1)^{u\ z} \ z \quad e_{pf} \ .$$

We notice that this state resembles a state affected by a bit-flip error vector, but in this case the error vector is e_{pf}. We use the parity-check matrix, H_2, to correct the errors described by e_{pf} as before; we first compute the syndrome, then correct the errors and reach the state

$$\frac{1}{2^{\frac{n \ k_2}{2}}} \sum_{z \ C_2} (\ 1)^{u\ z} \ z \ .$$

Finally, we apply a Hadamard Transform to each qubit and recover the original state:

$$\frac{1}{2^{\frac{k_2}{2}}} \sum_{w \ C_2} u \quad w \ .$$

In preparation for the discussion of stabilizer codes, we present in the next section a group theoretic view of QEC.

5.11 THE PAULI GROUP

The four Pauli operators, I, X, Z, Y, allow us to express the four possible effects of the environment on a qubit in the state, $\psi \quad \alpha_0\ 0 \quad \alpha_1\ 1$: no error (the qubit is unchanged), bit-flip, phase-flip, and bit- and phase-flip:

No error	I	$\begin{pmatrix} 1 & 0 \\ 0 & 1 \end{pmatrix}$	$\varphi \quad \mathrm{I}\ \psi \quad \alpha_0\ 0 \quad \alpha_1\ 1\,,$
Bit-flip	X	$\begin{pmatrix} 0 & 1 \\ 1 & 0 \end{pmatrix}$	$\varphi \quad X\ \psi \quad \alpha_1\ 0 \quad \alpha_0\ 1\,,$
Phase-flip	Z	$\begin{pmatrix} 1 & 0 \\ 0 & 1 \end{pmatrix}$	$\varphi \quad Z\ \psi \quad \alpha_0\ 0 \quad \alpha_1\ 1\,,$
Bit- & phase-flip	$Y \quad iXZ$	$\begin{pmatrix} 0 & i \\ i & 0 \end{pmatrix}$	$\varphi \quad Y\ \psi \quad i(\alpha_1\ 0 \quad \alpha_0\ 1\ .$

Pauli operators, I, X, Y, and Z, form a group and have several nice properties:

1. They anticommute:

$$
\begin{array}{llll}
X,Z & 0 & XZ & ZX, \\
X,Y & 0 & XY & YX, \\
Y,Z & 0 & YZ & ZY.
\end{array}
$$

2. The square of each is equal to unity; this means that the transformations they perform on a qubit are reversible.
3. They span the space of 2 2 matrices; any operator describing the transformation of a single qubit is a rotation, $\mathcal{R}_\theta$, with an angle, θ, and any rotation operator can be expressed as a linear combination of Pauli operators,

$$\mathcal{R}_\theta \quad e^{\ i\frac{\theta}{2} n \ \sigma} \quad \cos\frac{\theta}{2}\mathrm{I} \quad i \sin\frac{\theta}{2}(n \ \sigma),$$

where n (x,y,z) is a vector about which the rotation is performed and $\boldsymbol{\sigma}$ (X,Y,Z) is a vector of Pauli operators.

Two Pauli matrices, σ_i and σ_j, are *equivalent* if σ_j c (σ_i), with the constant c, equal to 1 or i, where i $\overline{\quad 1}$:

$$\sigma_i \quad \sigma_j \qquad \sigma_j \quad \sigma_i \ \text{ or } \ \sigma_j \quad i\sigma_i.$$

Thus, we can define the set of equivalence classes of Pauli operators, $[\mathcal{P}]$. The set of Pauli operators, $\mathcal{P}$, is not an Abelian group,[1] but the set Π of equivalence classes, $[\mathcal{P}]$, of Pauli operators, also called the *projective Pauli group* [155], forms an Abelian group. The multiplication tables of the two groups, $\mathcal{P}$ and $[\mathcal{P}]$, are, respectively,

	I	X	Y	Z
I	I	X	Y	Z
X	X	I	iZ	iY
Y	Y	iZ	I	iX
Z	Z	iY	iX	I

and

	[I]	$[X]$	$[Y]$	$[Z]$
[I]	[I]	$[X]$	$[Y]$	$[Z]$
$[X]$	$[X]$	[I]	$[\sigma_z]$	$[Y]$
$[Y]$	$[Y]$	$[Z]$	[I]	$[X]$
$[Z]$	$[Z]$	$[Y]$	$[X]$	[I]

Indeed, from the multiplication table of $\mathcal{P}$, we see that

$$XZ \quad iY \ \text{ and } \ ZX \quad iY,$$

while from the multiplication table of $[\mathcal{P}]$ we see that

$$[X][Z] \quad [Z][X] \quad [Y].$$

[1]A multiplicative group, G, is Abelian if and only if $a,b \quad G \ a \quad b \quad b \quad a$.

The *1-qubit Pauli group*, $\mathcal{P}_1$, consists of the Pauli operators, I, X, Y, and Z, together with the multiplicative factors, 1 and i:

$$\mathcal{P}_1 \quad \mathrm{I}, \ i\mathrm{I}, \ X, \ iX, \ Y, \ iY, \ Z, \ iZ \ .$$

The cardinality of $\mathcal{P}_1$ is $\mathcal{P}_1 \quad 16 \quad 2^4$.

Proposition. *The members of the* 1*-qubit Pauli group,* $\mathcal{P}_1$,

1. *are unitary;*
2. *either commute or anticommute; and*
3. *are either Hermitian or anti-Hermitian.*

Proof. Clearly, all members are unitary because the four Pauli operators are multiplied, with 1 or i. X, Y, Z anticommute, but I commutes with each of them. I, X, Y, and Z are Hermitian, but

$$(i\mathrm{I})^\dagger \quad i\mathrm{I}.$$

■

We introduce now the generator of a group, a concept that allows us to describe succinctly the group, G; instead of listing a possibly very large number of elements of a group, we only specify the much smaller number of generators of the group.

A set of μ elements, $g_1, g_2, \ldots, g_\mu$, is called the *generator of the group G* if the element, g_i, can be written as a product of (possibly repeated) elements from the list, $g_1, g_2, \ldots, g_\mu$, $g_i \quad G$; we write $G \quad g_1, g_2, \ldots, g_\mu$.

Proposition. *The number,* μ, *of generators of the group, G, satis es the inequality*

$$\mu \quad \log_2(\ G \ ).$$

Proof. Consider two elements, $f, g \quad G$, such that, $f \quad g_1, g_2, \ldots, g_\mu$ and $g \quad g_1, g_2, \ldots, g_\mu$; then $fg \quad g_1, g_2, \ldots, g_\mu$. If we assume

$$fg \quad g_1, g_2, \ldots, g_\mu \ , \quad \text{then } f \ ^1 fg \quad g_1, g_2, \ldots, g_\mu \qquad g \quad g_1, g_2, \ldots, g_\mu \ ,$$

which contradicts our assumption that $g \quad g_1, g_2, \ldots, g_\mu$; thus, $f \quad g_1, g_2, \ldots, g_\mu$, and we have

$$fg \quad g_1, g_2, \ldots, g_\mu, g \quad \text{and} \quad fg \quad g_1, g_2, \ldots, g_\mu \ .$$

It follows that by adding the element, g, to the set of generators, we double the number of elements of the group described by the generator; we conclude that the cardinality of the set of generators is at most the logarithm base 2 of the cardinality of the group. ■

Example. *The generators of* $\mathcal{P}_1$ *are*

$$\mathcal{G}_1 \quad X, Z, i\mathrm{I} \ .$$

Indeed, i^2 1, i^3 i, and i^4 1; every element of $\mathcal{P}_1$ can be expressed as a product of a finite number of generators:

$$\mathrm{I} \quad i\mathrm{I}i\mathrm{I}i\mathrm{I}i\mathrm{I}, \qquad \mathrm{I} \quad i\mathrm{I}i\mathrm{I}, \qquad i\mathrm{I} \quad i\mathrm{I}i\mathrm{I}i\mathrm{I};$$

while

$$X \quad i\mathrm{I}i\mathrm{I}X, \qquad iX \quad i\mathrm{I}X, \qquad iX \quad i\mathrm{I}i\mathrm{I}i\mathrm{I}X$$

and

$$Y \quad i\mathrm{I}XZ, \qquad Y \quad i\mathrm{I}i\mathrm{I}i\mathrm{I}XZ,$$

$$iY \quad i\mathrm{I}i\mathrm{I}XZ, \qquad iY \quad i\mathrm{I}i\mathrm{I}XZ,$$

and, finally,

$$Z \quad i\mathrm{I}i\mathrm{I}Z, \qquad iZ \quad i\mathrm{I}Z, \qquad iZ \quad i\mathrm{I}i\mathrm{I}i\mathrm{I}Z.$$

We conclude that the three generators, $X, Z, i\mathrm{I}$, allow us to express any of the, 16 2^4, operators in $\mathcal{P}_1$. The cardinality of the set of generators satisfies the condition

$$X, Z, i\mathrm{I} \qquad \log_2 \mathcal{P}_1 , \quad \text{or} \quad 3 < 4.$$

The *n-qubit Pauli group*, $\mathcal{P}_n$, consists of the 4^n tensor products of I, X, Y, and Z and an overall phase of 1 or i; the group has $4^{n \ \ 1}$ elements.

One element of the group is an n-tuple, the tensor product of, n, one-qubit Pauli operators, and can be used to describe the error operator applied to an n-qubit register; the *weight* of such an operator in $\mathcal{P}_n$ is equal to the number of tensor factors that are not equal to I.

Example. *Let n 5; the ve-qubit Pauli group consists of the tensor products of the form:*

$$\mathbf{E}^{(1)} \quad \mathbf{E}^{(2)} \quad \mathbf{E}^{(3)} \quad \mathbf{E}^{(4)} \quad \mathbf{E}^{(5)}, \quad \text{with} \quad \mathbf{E}^{(i)} \quad \mathcal{P}_1, 1 \quad i \quad 5.$$

The operator, $\mathbf{E}_\alpha$ I X I Z I, has a weight of 2 and represents no error for qubits 1, 3, and 5; a bit-flip of qubit 2; and a phase-flip of qubit 4. The qubits are numbered from left to right.

5.12 STABILIZER CODES

Classical codes use simple tests to determine if an n-tuple is a codeword:

- If H is the parity-check matrix of the linear code C, v C is a codeword, and w v e is an n-tuple in error, then Hv^T 0 and Σ Hw^T He^T 0; we use the syndrome, Σ, to locate the error.
- A cyclic code, C, is an ideal of polynomials with the generator polynomial, $g(x)$, and we conclude that the received polynomial is a codeword, $r(x)$ C, if $g(x)$ divides $r(x)$; otherwise, we use the remainder of the division of $r(x)$ by $g(x)$ to locate the error(s).

We wish to construct some tests for quantum codes that resemble the effect of the parity-check matrix or of the generator polynomials and allow us to detect if errors have occurred, and if so, to correct them. Assume the, m, codewords of a code are described by the vectors, ψ_i , and that we have identified a group of, q, operators, $\mathbf{M}_j$, that allow us to detect the errors that may affect any codeword. In this context, the word "detect" means that we carry out a measurement that does not reveal any information about the actual state, ψ_i , but indicates if the codeword is affected by error(s). This line of thought leads us to an alternative way of specifying a quantum code as *the set of quantum operators that correct all possible errors affecting all codewords.*

The stabilizer formalism is a succinct manner of describing a quantum error-correcting code by a set of quantum operators. In this section we first describe the basic idea and illustrate its application to the nine-qubit code of Shor, then we discuss the general stabilizer formalism.

Basic Concepts

Consider a set of n-qubit registers in the state, ψ_i ,1 i m, each one representing a word of a quantum code, $\mathcal{Q}$. Informally, the stabilizer, $\mathcal{S}$, of $\mathcal{Q}$ is a subgroup of the n-qubit Pauli group,

$$\mathcal{S} \quad \mathcal{P}_n.$$

The generators of the subgroup, $\mathcal{S}$, are

$$M \quad \mathbf{M}_1, \mathbf{M}_2, \ldots, \mathbf{M}_q \ .$$

The *eigenvectors of the generators,* $\mathbf{M}_1, \mathbf{M}_2, \ldots, \mathbf{M}_q$, have special properties: Those corresponding to eigenvalues of   1 are the codewords of $\mathcal{Q}$, and those corresponding to eigenvalues of   1 are codewords affected by errors. If $\mathbf{M}_j \quad M$ and if

$$\mathbf{M}_j \ \psi_i \quad (\ 1) \ \psi_i \ ,$$

then ψ_i is a codeword, $\psi_i \quad \mathcal{Q}$; This justifies the name "stabilizer" given to the set, $\mathcal{S}$; any operator in $\mathcal{S}$ leaves the state of a word unchanged. On the other hand, if

$$\mathbf{M}_j \ \varphi_k \quad (\ 1) \ \varphi_k \ ,$$

then $\varphi_k \quad \mathbf{E}_\alpha \ \psi_k$, and the state, φ_k , is the result of an error affecting the codeword, $\psi_k \quad \mathcal{Q}$.

The errors, $E \quad \mathbf{E}_1, \mathbf{E}_2, \ldots$, affecting a codeword in $\mathcal{Q}$ are also a subgroup of the n-qubit Pauli group, $E \quad \mathcal{P}_n$; each error operator, $\mathbf{E}_i$, is a tensor product of, n, Pauli matrices. The *weight of an error operator* is equal to the number of errors affecting a quantum word, thus, the number of Pauli operators other than, I, in this n-dimensional tensor product.

We show that the error operators anticommute with the generators of the stabilizer group, $\mathcal{S}$. Therefore, to detect errors we have to compute the eigenvectors of the generators and identify those with an eigenvalue of   1.

Example. *A pair of entangled qubits.* Their state is

$$\beta_{00} \quad \frac{00 \quad 11}{\overline{2}} \quad \text{or} \quad \beta_{00} \quad \begin{pmatrix} 1 \\ 0 \\ 0 \\ 1 \end{pmatrix}.$$

The following transformations are applied to the qubits:

$$\mathbf{M}_x \quad X \quad X \quad \begin{pmatrix} 0 & 0 & 0 & 1 \\ 0 & 0 & 1 & 0 \\ 0 & 1 & 0 & 0 \\ 1 & 0 & 0 & 0 \end{pmatrix} \quad \text{and} \quad \mathbf{M}_z \quad Z \quad Z \quad \begin{pmatrix} 1 & 0 & 0 & 0 \\ 0 & 1 & 0 & 0 \\ 0 & 0 & 1 & 0 \\ 0 & 0 & 0 & 1 \end{pmatrix}.$$

We see that

$$\mathbf{M}_x \ \beta_{00} \quad \begin{pmatrix} 0 & 0 & 0 & 1 \\ 0 & 0 & 1 & 0 \\ 0 & 1 & 0 & 0 \\ 1 & 0 & 0 & 0 \end{pmatrix} \begin{pmatrix} 1 \\ 0 \\ 0 \\ 1 \end{pmatrix} \quad \begin{pmatrix} 1 \\ 0 \\ 0 \\ 1 \end{pmatrix}.$$

Thus,

$$\mathbf{M}_x \ \beta_{00} \quad \beta_{00} \quad \text{or} \quad \mathbf{M}_x \ \beta_{00} \quad (\quad 1) \ \beta_{00} \ .$$

Similarly,

$$\mathbf{M}_z \ \beta_{00} \quad \beta_{00} \quad \text{or} \quad \mathbf{M}_z \ \beta_{00} \quad (\quad 1) \ \beta_{00} \ .$$

These two equalities show that, β_{00} , is an eigenvector of the operators, $\mathbf{M}_x$ and $\mathbf{M}_z$, respectively, and that the corresponding eigenvalues are equal to 1. The generators commute, thus, $\mathcal{S}$ is an Abelian group.

Example. *The nine-qubit error-correcting code of Shor.* The logical qubits are

$$0_L \quad \frac{000 \quad 111}{\overline{2}} \quad \frac{000 \quad 111}{\overline{2}} \quad \frac{000 \quad 111}{\overline{2}},$$

$$1_L \quad \frac{000 \quad 111}{\overline{2}} \quad \frac{000 \quad 111}{\overline{2}} \quad \frac{000 \quad 111}{\overline{2}}.$$

Consider the following set of operators acting on nine-qubit codewords:

Operator/Qubit	1	2	3	4	5	6	7	8	9
$\mathbf{M}_1$	Z	Z	I	I	I	I	I	I	I
$\mathbf{M}_2$	Z	I	Z	I	I	I	I	I	I
$\mathbf{M}_3$	I	I	I	Z	Z	I	I	I	I
$\mathbf{M}_4$	I	I	I	Z	I	Z	I	I	I
$\mathbf{M}_5$	I	I	I	I	I	I	Z	Z	I
$\mathbf{M}_6$	I	I	I	I	I	I	Z	I	Z
$\mathbf{M}_7$	X	X	X	X	X	X	I	I	I
$\mathbf{M}_8$	X	X	X	I	I	I	X	X	X

The eight operators, $\mathbf{M}_1, \mathbf{M}_2, \ldots, \mathbf{M}_8$, are the generators of the group, $\mathcal{P}$, of operators that stabilize the nine-qubit error-correcting code. The logical qubits, 0_L and 1_L, are eigenvectors of the eight operators, and the corresponding eigenvalues are equal to 1:

$$\mathbf{M}_j\, 0_L \quad (\;1)\, 0_L \quad \text{and} \quad \mathbf{M}_j\, 1_L \quad (\;1)\, 1_L, \; 1 \quad j \quad 8.$$

For example, 0_L is an eigenvector of $\mathbf{M}_7$ corresponding to an eigenvalue λ 1,

$$\mathbf{M}_7\, 0_L \quad (\;1)\, 0_L,$$

where

$$0_L \qquad 000000000 \quad 000000111 \quad 000111000 \quad 000111111$$
$$111000000 \quad 111000111 \quad 111111000 \quad 111111111 \;.$$

The tensor product of two operators, $\sigma_a^{(i)}$ and $\sigma_b^{(j)}$, acting on qubits, i and j, of an n-qubit word is $\sigma_a^{(i)} \quad \sigma_b^{(j)}$. According to its definition,

$$\mathbf{M}_7 \quad X^{(1)} \quad X^{(2)} \quad X^{(3)} \quad X^{(4)} \quad X^{(5)} \quad X^{(6)} \quad \mathrm{I}^{(7)} \quad \mathrm{I}^{(8)} \quad \mathrm{I}^{(9)},$$

the operator, $\mathbf{M}_7$, flips the first six qubits of a nine-qubit word and leaves the remaining three unchanged; thus,

$$\mathbf{M}_7\, 0_L \qquad 111111000 \quad 111111111 \quad 111000000 \quad 111000111$$
$$000111000 \quad 000111111 \quad 000000000 \quad 000000111 \;.$$

Let $\mathbf{E}_x^{(1)}$ and $\mathbf{E}_x^{(2)}$ be the two operators describing a single bit-flip error of the first and the second qubit of a nine-qubit codeword, respectively:

$$\mathbf{E}_x^{(1)} \quad X^{(1)} \quad \mathrm{I}^{(2)} \quad \mathrm{I}^{(3)} \quad \mathrm{I}^{(4)} \quad \mathrm{I}^{(5)} \quad \mathrm{I}^{(6)} \quad \mathrm{I}^{(7)} \quad \mathrm{I}^{(8)} \quad \mathrm{I}^{(9)},$$
$$\mathbf{E}_x^{(2)} \quad \mathrm{I}^{(1)} \quad X^{(2)} \quad \mathrm{I}^{(3)} \quad \mathrm{I}^{(4)} \quad \mathrm{I}^{(5)} \quad \mathrm{I}^{(6)} \quad \mathrm{I}^{(7)} \quad \mathrm{I}^{(8)} \quad \mathrm{I}^{(9)}.$$

Both operators anticommute with $\mathbf{M}_1$:

$$\mathbf{M}_1 \quad Z^{(1)} \quad Z^{(2)} \quad I^{(3)} \quad I^{(4)} \quad I^{(5)} \quad I^{(6)} \quad I^{(7)} \quad I^{(8)} \quad I^{(9)}.$$

For example,

$$\begin{aligned}\mathbf{M}_1\mathbf{E}_x^{(1)} \quad & (Z^{(1)} \quad Z^{(2)} \quad I^{(3)} \quad I^{(4)} \quad I^{(5)} \quad I^{(6)} \quad I^{(7)} \quad I^{(8)} \quad I^{(9)}) \\ & (X^{(1)} \quad I^{(2)} \quad I^{(3)} \quad I^{(4)} \quad I^{(5)} \quad I^{(6)} \quad I^{(7)} \quad I^{(8)} \quad I^{(9)}) \\ & (Z^{(1)}X^{(1)}) \quad (Z^{(2)}I^{(2)}) \quad (I^{(3)}I^{(3)}) \quad (I^{(4)}I^{(4)}) \\ & \quad (I^{(5)}I^{(5)}) \quad (I^{(6)}I^{(6)}) \quad (I^{(7)}I^{(7)}) \quad (I^{(8)}I^{(8)}) \quad (I^{(9)}I^{(9)}) \\ & (X^{(1)}Z^{(1)}) \quad Z^{(2)} \quad I^{(3)} \quad I^{(4)} \quad I^{(5)} \quad I^{(6)} \quad I^{(7)} \quad I^{(8)} \quad I^{(9)} \\ & \mathbf{E}_x^{(1)}\mathbf{M}_1.\end{aligned}$$

From our previous derivation, it follows that

$$\mathbf{M}_1\mathbf{E}_x^{(1)}\ 0_L \qquad \mathbf{E}_x^{(1)}\mathbf{M}_1\ 0_L \qquad \mathbf{E}_x^{(1)}[\mathbf{M}_1\ 0_L] \quad (\ \ 1)\mathbf{E}_x^{(1)}\ 0_L.$$

This means that, $\mathbf{E}_x^{(1)}\ 0_L$, is an eigenvector of the stabilizer, $\mathbf{M}_1$, with an eigenvalue of $\lambda \quad 1$. Similarly, $\mathbf{E}_x^{(1)}\ 1_L$, $\mathbf{E}_x^{(2)}\ 0_L$, and $\mathbf{E}_x^{(2)}\ 1_L$, are eigenvectors of the stabilizer $\mathbf{M}_1$, with eigenvalues equal to 1. Thus, $\mathbf{M}_1$ can be used to detect a single bit-flip error of qubits, 1 and 2, of the a nine-qubit word of Shor's code. The operators corresponding to a single bit-flip error of qubits, 3 to 9, commute with $\mathbf{M}_1$; thus, $\mathbf{M}_1$ cannot be used to detect a single bit-flip error of qubits 3 to 9. Indeed,

$$\mathbf{M}_1\mathbf{E}_x^{(i)}\ 0_L \quad \mathbf{E}_x^{(i)}\mathbf{M}_1\ 0_L \quad \text{and} \quad \mathbf{M}_1\mathbf{E}_x^{(i)}\ 1_L \quad \mathbf{E}_x^{(i)}\mathbf{M}_1\ 1_L \quad 3 \quad i \quad 9.$$

It is left as an exercise to show that any error operator, $\mathbf{E}^i_{x\ y\ z}, 1 \quad i \quad 9$, anticommutes with at least one of the eight operators $\mathbf{M}_j, 1 \quad j \quad 8$; thus, any single-qubit error can be detected.

This example shows that it is sufficient only to measure the eigenvalues of the stabilizers, $\mathbf{M}_i$, to detect errors; an eigenvalue of 1 shows that the corresponding codeword was affected by the error, E. We can detect the occurrence of an error if the corresponding error operator anticommutes with the generators of a code regarded as a subgroup.

The General Stabilizer Formalism

The stabilizer formalism provides an alternative description of an $[n,k,d]$ quantum code; we use the following notations:

- ψ_i is a codeword.
- $\mathbf{M}$ is a member of the group of operators that stabilize codewords, $\mathbf{M}\ \psi_i \quad (\ \ 1)\ \psi_i$.
- $\mathbf{E}$ is an error operator. Sometimes we denote the error operator by $\mathbf{E}^k_{x\ y\ z}$ to specify a bit-flip error, X, a phase-flip error, Z, or a bit- and phase-flip error, Y, of qubit k.

We show that if the two operators, $\mathbf{M}$ and $\mathbf{E}$, anticommute, $\mathbf{M},\mathbf{E} \quad 0$, then

$$\mathbf{M}(\mathbf{E}\ \psi_i\ ) \quad (\ \ 1)\ (\mathbf{E}\ \psi_i\ ).$$

This means that, $\mathbf{E}\ \psi_i$, is an eigenvector of $\mathbf{M}$ corresponding to the eigenvalue 1. We detect the error, $\mathbf{E}$, by measuring $\mathbf{M}$ on the original codeword, ψ_i , after error $\mathbf{E}$ forced it to the state, $\mathbf{E}\ \psi_i$.

We review first a few properties of the Pauli group. All elements of $\mathcal{P}$ are unitary. I, X, Y, and Z are Hermitian, but $(iI)^\dagger \quad iI$; thus, all elements of $\mathcal{P}$ are either Hermitian or anti-Hermitian. All elements of $\mathcal{P}$ either commute or anticommute. Recall that the *commutator of two operators* is

$$[\mathbf{M}_\alpha, \mathbf{M}_\beta] \quad \mathbf{M}_\alpha \mathbf{M}_\beta \quad \mathbf{M}_\beta \mathbf{M}_\alpha.$$

If the two operators commute, $\mathbf{M}_\alpha \mathbf{M}_\beta \quad \mathbf{M}_\beta \mathbf{M}_\alpha$, then $[\mathbf{M}_\alpha, \mathbf{M}_\beta] \quad 0$. The *anticommutator of two operators* is

$$\mathbf{M}_\alpha, \mathbf{M}_\beta \quad \mathbf{M}_\alpha \mathbf{M}_\beta \quad \mathbf{M}_\beta \mathbf{M}_\alpha.$$

If the two operators anticommute, $\mathbf{M}_\alpha \mathbf{M}_\beta \quad \mathbf{M}_\beta \mathbf{M}_\alpha$, then $\mathbf{M}_\alpha, \mathbf{M}_\beta \quad 0$.

We continue this discussion with a few definitions.

The *stabilizer*, $\mathcal{S}$, of an $[n,k,d]$ code is an Abelian subgroup of the Pauli group, $\mathcal{P}$. The generators of $\mathcal{S}$ are $\mathbf{M}_1, \mathbf{M}_2, \ldots, \mathbf{M}_{n\ k}$. The *weight* of, $\mathbf{M}_i \quad \mathcal{S}$, is the number of elements of $\mathbf{M}_i$ different from I.

A *nondegenerate stabilizer code*, $[n,k,d]$, is one where none of the elements of the stabilizer $\mathcal{S}$ (except the identity) has weight less than d; in the case of a *degenerate stabilizer code*, at least one element of the stabilizer $\mathcal{S}$ other than the identity has a weight less than d.

Example. *Shor s code.* The weight of $\mathbf{M}_1$, the first generator for Shor's code, is equal to 2; thus, the nine-qubit code is degenerate.

The *coding space*,

$$T \quad \psi \text{ , such that } \mathbf{M}\ \psi \quad (\ 1)\ \psi \quad \mathbf{M} \quad \mathcal{S}\ ,$$

is the space of all the vectors, ψ , fixed by $\mathcal{S}$.

Recall that *two commuting matrices can be simultaneously diagonalized; thus, we can measure the eigenvalue of one of them without disturbing the eigenvector of the other. Thus, $\mathcal{S}$ must be Abelian* because only commuting operators may have simultaneous eigenvectors and the code consists of eigenvectors of the generators of $\mathcal{S}$. Because $\mathcal{S}$ is Abelian and because neither $\ 1$ nor $\ i$ is in $\mathcal{S}$, it follows that the coding space, T, has dimension 2^k. Indeed, an $[n,k,d]$ code should be able to encode, 2^k, messages and must therefore have, 2^k, codewords.

Proposition. *$\mathcal{S}$ is a stabilizer of a nontrivial Hilbert subspace, $V_{2^n} \quad \mathcal{H}_{2^n}$, if and only if:*

1. $\mathcal{S} \quad \mathbf{S}_1, \mathbf{S}_2, \ldots$ *is an Abelian group:*

$$\mathbf{S}_i \mathbf{S}_j \quad \mathbf{S}_j \mathbf{S}_i, \qquad \mathbf{S}_i, \mathbf{S}_j \quad S,\ i \quad j.$$

2. *The identity matrix multiplied by $\ 1$ is not in $\mathcal{S}$:*

$$\mathrm{I}\ ^n \ / \ \mathcal{S}.$$

Proof. We show only that the two conditions are necessary. First, we note that $\mathbf{S}_i, \mathbf{S}_j \quad S$ are tensor products of Pauli operators; thus, they either commute or anticommute. Then

$$\mathbf{S}_i \mathbf{S}_j\ \psi \qquad \psi\ , \qquad \psi \quad V_{2^n}.$$

Indeed, $\mathbf{S}_i$ and $\mathbf{S}_j$ are both stabilizers, $\mathbf{S}_i\ \psi \quad \psi$ and $\mathbf{S}_j\ \psi \quad \psi$; thus,

$$\mathbf{S}_i\mathbf{S}_j\ \psi \quad \mathbf{S}_i(\mathbf{S}_j\ \psi\ ) \quad \mathbf{S}_i\ \psi \quad \psi\ .$$

If $\mathbf{S}_i$ and $\mathbf{S}_j$ anticommute, then

$$\mathbf{S}_j\mathbf{S}_i\ \psi \quad \psi\ , \quad \psi \quad V_{2^n}.$$

We see that

$$\mathbf{S}_j\mathbf{S}_i\ \psi \quad \mathbf{S}_j\,(\mathbf{S}_i\ \psi\ ) \quad \mathbf{S}_j\ \psi \quad \psi\ .$$

We assumed that the two operators anticommute, $\mathbf{S}_i\mathbf{S}_j \quad \mathbf{S}_j\mathbf{S}_i$; thus,

$$\mathbf{S}_i\mathbf{S}_j\ \psi \quad \mathbf{S}_j\mathbf{S}_i\ \psi \quad \psi \quad \psi \quad \psi \quad V_{2^n},$$

which is a contradiction, since we assumed that, V_{2^n}, is a nontrivial Hilbert subspace. Therefore, we conclude that all members of the stabilizer group commute and $\mathcal{S}$ is an Abelian group.

Now, assume that $\mathrm{I}\ ^n \quad \mathcal{S}$; then

$$\mathrm{I}\ ^n\ \psi \quad \psi \quad \text{or} \quad \mathrm{I}\ ^n\ \psi \quad \psi\ , \quad \psi \quad V_{2^n}.$$

This leads to a contradiction because $\mathrm{I}\ ^n\ \psi \quad \psi$; we conclude that $\mathrm{I}\ ^n\ /\ \mathcal{S}$. ■

Proposition. *Let* **E** *be an error operator. If* **M** $\ \mathcal{S}$ *and* **M** *and* **E** *anticommute, then*

$$\mathbf{ME}\ \psi_i \quad \mathbf{E}\ \psi_i \quad \psi_i \quad T.$$

We can detect the error, **E**, by measuring **M**; indeed, $\mathbf{E}\ \psi_i$ is an eigenvector of the generator, **M**, corresponding to the eigenvalue 1.

Proof. We know that **M** and **E** anticommute, **ME** **EM**, and that **M** is a stabilizer of the code, $\mathbf{M}\ \psi_i \quad \psi_i \quad \psi_i \quad T$. Thus:

$$\mathbf{M}(\mathbf{E}\ \psi_i\ ) \quad \mathbf{ME}\ \psi_i \quad \mathbf{EM}\ \psi_i \quad \mathbf{E}(\mathbf{M}\ \psi_i\ ) \quad (\ \ 1)\,\mathbf{E}\ \psi_i\ .$$

We also observe that given any codewords, $\psi_i\ ,\ \psi_j \quad T$:

$$\psi_i\ \mathbf{E}\ \psi_j \quad \psi_i\ \mathbf{EM}\ \psi_j \quad \psi_i\ \mathbf{ME}\ \psi_j \quad \psi_i\ \mathbf{E}\ \psi_j \quad 0.$$

Thus, the code satisfies the necessary condition for the existence of a quantum code discussed in Section 5.1. ■

Proposition. *Given an* $[n,k,d]$ *stabilizer code (with n the length of a codeword, k the number of information symbols, and d the distance of the code), the cardinality of the stabilizer,* $\mathcal{S}$, *and the cardinality of its generator, M, are, respectively:*

$$\mathcal{S} \quad 2^{n\ \ k}, \quad M \quad n\ \ k.$$

The *normalizer*, $N(\mathcal{S})$, $\mathcal{S} \quad \mathcal{P}$, is the set of elements in, $\mathcal{P}$, that fix the stabilizer, $\mathcal{S}$, under conjugation. It is left as an exercise to show that $\mathcal{S}$ is a normal subgroup of $N(\mathcal{S})$:

$$\mathcal{S} \quad N(\mathcal{S}).$$

Since the elements of the normalizer, $N(\mathcal{S})$, move codewords around in the coding space, T, they are in fact the *encoding operators*. Only the elements, $\mathbf{A} \quad N(\mathcal{S})/\mathcal{S}$, act on the codewords of T nontrivially; if $\mathbf{B} \quad \mathcal{S}$ and $\psi_i \quad T$, then $\mathbf{B}\ \psi_i \qquad \psi_i$.

The *centralizer*, $C(\mathcal{S})$, $\mathcal{S} \quad \mathcal{P}$, is the set of elements in $\mathcal{P}$ that commute with all the elements of the stabilizer $\mathcal{S}$. Figure 5.15 shows the relationship between groups of operators, error operators, and the coding space for stabilizer codes.

Proposition. *The normalizer is equal to the centralizer,* $N(\mathcal{S}) \quad C(\mathcal{S})$, *and they contain,* $4 \quad 2^{n \quad k}$, *elements.*

$\mathcal{S}$ is an Abelian subgroup of $\mathcal{P}$ thus, $\mathbf{A} \quad \mathcal{P}$ and $\mathbf{M} \quad \mathcal{S}$, $\mathbf{A}$ and $\mathbf{M}$ either commute or anti-commute, $\mathbf{MA} \qquad \mathbf{AM}$; moreover, $\mathbf{M}$ is an element of the stabilizer, $\mathbf{M} \quad \mathcal{S}$, and this implies that $\mathbf{AM} \quad \mathbf{M}$; thus,

$$\mathbf{A}^{\dagger}\mathbf{MA} \qquad \mathbf{A}^{\dagger}\mathbf{AM} \qquad \mathbf{M}.$$

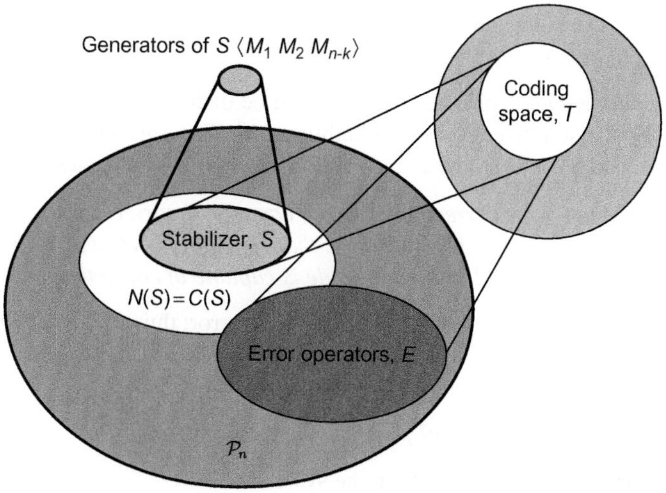

FIGURE 5.15

A quantum code with stabilizer, $\mathcal{S}$, and normalizer, $N(\mathcal{S})$, detects all errors, $\mathbf{E}$, that are either in $\mathcal{S}$ or anticommute with some element of $\mathcal{S}$. The stabilizer, $\mathcal{S}$, of an $[n,k,d]$ code has $2^{n \quad k}$ elements; the coding space, T, has 2^k elements and is a subspace of the larger, 2^n, space. The code has distance, d, if and only if $N(S) \quad S$ contains no elements with weight less than d. $\mathbf{M}_1, \mathbf{M}_2, \ldots, \mathbf{M}_{n \quad k}$ are the $n \quad k$ generators of $\mathcal{S}$. $N(S) \quad C(S)$ and E / S. The group, $N(S)$, has $4 \quad 2^{n \quad k}$ elements. If $\mathbf{E}$ / $N(S) \quad S$, then $\mathbf{E}$ rearranges elements of T; it does not take codewords outside T. $\mathcal{P}_n$ is the n-qubit Pauli group.

We conclude that $N(\mathcal{S}) \quad C(\mathcal{S})$ because $\quad 1 \,/\, \mathcal{S}$, and $\mathbf{A} \quad N(\mathcal{S})$ if and only if $\mathbf{A} \quad C(\mathcal{S})$. It is left as an exercise to prove that $N(\mathcal{S}) \qquad 4 \;\; 2^{n \quad k}$ using the fact that the Pauli group, $\mathcal{P}_n$, has $4^{n \quad 1}$ elements including an overall phase of $\quad 1$ and $\quad i$.

Proposition. *A code with stabilizer, S, and normalizer, $N(\mathcal{S})$, will have distance, d, if and only if $N(\mathcal{S}) \quad \mathcal{S}$ contains no elements with weight less than d.*

The *error syndrome* corresponding to stabilizer, $\mathbf{M}$, is a function of, $\mathbf{E}$, the error operator, $f_{\mathbf{M}}(\mathbf{E})\colon \mathcal{P} \quad \mathbb{Z}_2$ is defined as

$$f_{\mathbf{M}}(\mathbf{E}) \quad \begin{cases} 0 & \text{if } [\mathbf{M},\mathbf{E}] \quad 0, \\ 1 & \text{if } \;\mathbf{M},\mathbf{E} \quad 0. \end{cases}$$

Let $f(\mathbf{E})$ be the $(n \quad k)$-bit integer given by the binary vector

$$f(\mathbf{E}) \quad (f_{\mathbf{M}_1}(\mathbf{E}) f_{\mathbf{M}_2}(\mathbf{E}) \ldots f_{\mathbf{M}_{n \quad k}}(\mathbf{E})).$$

If $\mathbf{E} \quad N(\mathcal{S})$, then $f(\mathbf{E}) \quad 0$.

To "detect" an error for a stabilizer code, we have to measure the eigenvalues of each generator of the stabilizer of the code. The eigenvalue of, $\mathbf{M}_i$, is equal to $(\quad 1)^{f_{\mathbf{M}_i}(\mathbf{E})}$ and gives us the error syndrome. For a nondegenerate code, the error syndrome reveals position and type of error, and for a degenerate code the error syndrome reveals the set of degenerate errors.

Proposition. *The error syndrome uniquely identi es the qubit(s) in error if and only if the subsets of the stabilizer group that anticommute with the error operators are distinct.*

An error can be identified and corrected only if it can be distinguished from any other error in the error set. Let $Q(\mathcal{S})$ be the stabilizer code with stabilizer, $\mathcal{S}$. The *correctable set of errors* for $Q(\mathcal{S})$ includes all errors that can be detected by $\mathcal{S}$ and have distinct error syndromes.

Corollary. *Given a quantum error correcting code, Q, capable of correcting, e_u, errors, the syndrome does not allow us to distinguish the case when more than, e_u, qubits are in error. When we have exact prior knowledge about, e_c, correlated errors, the code is capable of correcting these $e_u \quad e_c$ errors.*

Proof. Assume that F_1, F_2 cause at most, e_u, qubits to be in error; thus, F_1, F_2 are included in the *correctable set of errors* of Q. Since errors F_1 and F_2 are distinguishable, there must exist some operator, $M \quad S$, that commutes with one of them and anticommutes with the other:

$$F_1^T F_2 M \qquad M F_1^T F_2.$$

If we know the exact correlated errors, E, in the system, then

$$(E^T F_1)^T (E^T F_2) M \quad (F_1^T E E^T F_2) M \quad F_1^T F_2 M \qquad M F_1^T F_2 \qquad M (E^T F_1)^T (E^T F_2),$$

which means that the stabilizer, M, commutes with one of the two errors, $E^T F_1$, $E^T F_2$, and anticommutes with the other. So error, $E^T F_1$, is distinguishable from error, $E^T F_2$. Therefore, if we know the exact prior errors, E, we can identify and correct any, $E^T F_i$, errors with the weight of, F_i, equal to or less than, e_u. ■

Example. *Steane s code.* The generators of the stabilizer group for Steane's code are:

$$\mathbf{M}_1 \quad I \ I \ I \ X \ X \ X \ X, \quad \mathbf{M}_2 \quad I \ X \ X \ I \ I \ X \ X,$$
$$\mathbf{M}_3 \quad X \ I \ X \ I \ X \ I \ X, \quad \mathbf{M}_4 \quad I \ I \ I \ Z \ Z \ Z \ Z,$$
$$\mathbf{M}_5 \quad I \ Z \ Z \ I \ I \ Z \ Z, \quad \mathbf{M}_6 \quad Z \ I \ Z \ I \ Z \ I \ Z.$$

The weight of all stabilizers is four, $w(\mathbf{M}_i) \quad 4, 1 \quad i \quad 6$, and the distance is, $d \quad 3$; thus, the code is a nondegenerate stabilizer code.

Table 5.4 summarizes the single-error operators and generators that anticommute with each operator for Steane's code; Table 5.5 presents the single errors and the corresponding syndromes for the same code. Subscripts indicate positions of errors; one bit-flip error is denoted as $X_{1 \quad 7}$; one phase-flip error as $Z_{1 \quad 7}$; the combination of one bit-flip and one phase-flip as $Y_{1 \quad 7}$.

Next, we discuss quantum codes that use the smallest number of qubits for encoding the quantum information.

Table 5.4 Single–error Operators for the Steane Seven-qubit Code and the Generator(s) that Anticommute with Each Operator

Error	Generator(s)	Error	Generator(s)	Error	Generator(s)
$\mathbf{X}_1$	$\mathbf{M}_6$	$\mathbf{Z}_1$	$\mathbf{M}_3$	$\mathbf{Y}_1$	$\mathbf{M}_3, \mathbf{M}_6$
$\mathbf{X}_2$	$\mathbf{M}_5$	$\mathbf{Z}_2$	$\mathbf{M}_2$	$\mathbf{Y}_2$	$\mathbf{M}_2, \mathbf{M}_5$
$\mathbf{X}_3$	$\mathbf{M}_5, \mathbf{M}_6$	$\mathbf{Z}_3$	$\mathbf{M}_2, \mathbf{M}_3$	$\mathbf{Y}_3$	$\mathbf{M}_2, \mathbf{M}_3, \mathbf{M}_5, \mathbf{M}_6$
$\mathbf{X}_4$	$\mathbf{M}_4$	$\mathbf{Z}_4$	$\mathbf{M}_1$	$\mathbf{Y}_4$	$\mathbf{M}_1, \mathbf{M}_4$
$\mathbf{X}_5$	$\mathbf{M}_4, \mathbf{M}_6$	$\mathbf{Z}_5$	$\mathbf{M}_1, \mathbf{M}_3$	$\mathbf{Y}_5$	$\mathbf{M}_1, \mathbf{M}_3, \mathbf{M}_5, \mathbf{M}_6$
$\mathbf{X}_6$	$\mathbf{M}_4, \mathbf{M}_5$	$\mathbf{Z}_6$	$\mathbf{M}_1, \mathbf{M}_2$	$\mathbf{Y}_6$	$\mathbf{M}_1, \mathbf{M}_2, \mathbf{M}_4, \mathbf{M}_5$
$\mathbf{X}_7$	$\mathbf{M}_4, \mathbf{M}_5, \mathbf{M}_6$	$\mathbf{Z}_7$	$\mathbf{M}_1, \mathbf{M}_2, \mathbf{M}_3$	$\mathbf{Y}_7$	$\mathbf{M}_1, \mathbf{M}_2, \mathbf{M}_3, \mathbf{M}_4, \mathbf{M}_5, \mathbf{M}_6$

Table 5.5 The Single Errors and the Corresponding Syndromes for the Steane Seven-qubit Code

Error	Syndrome a_{1-6}	Error	Syndrome a_{1-6}	Error	Syndrome a_{1-6}
X_1	000001	Z_1	001000	Y_1	001001
X_2	000010	Z_2	010000	Y_2	010010
X_3	000011	Z_3	011000	Y_3	011011
X_4	000100	Z_4	100000	Y_4	100100
X_5	000101	Z_5	101000	Y_5	101101
X_6	000110	Z_6	110000	Y_6	110110
X_7	000111	Z_7	111000	Y_7	111111

5.13 STABILIZERS FOR PERFECT QUANTUM CODES

Recall from the discussion of the quantum Hamming bound in Section 5.3 that we need at least five physical qubits to encode one logical qubit. A *perfect quantum error correcting code* is one that uses the smallest dimension Hilbert space to encode, 0_L and 1_L, as entangled states in $\mathcal{H}_{2^5}$.

A perfect quantum error-correcting code is described by Laflamme *et al* [263]. This code can be described in different ways all related by a change of base of one of the five physical qubits. We start with the definition of the code in [263], then the one in [307], and in the next section, we use a third definition from DiVincenzo and Shor [128].

To introduce the two logical qubits, 0_L and 1_L, according to [263], we first define

$$\varphi_1 \quad 000 \quad 111, \quad \varphi_3 \quad 100 \quad 011, \quad \varphi_5 \quad 010 \quad 101, \quad \varphi_7 \quad 110 \quad 001,$$
$$\varphi_2 \quad 000 \quad 111, \quad \varphi_4 \quad 100 \quad 011, \quad \varphi_6 \quad 010 \quad 101, \quad \varphi_8 \quad 110 \quad 001.$$

The two logical qubits for one of the five-qubit codes are

$$0_L \quad \varphi_1\ 00 \quad \varphi_3\ 11 \quad \varphi_5\ 01 \quad \varphi_7\ 10,$$
$$1_L \quad \varphi_2\ 11 \quad \varphi_4\ 00 \quad \varphi_6\ 10 \quad \varphi_8\ 01,$$

or

$$0_L \quad 00000 \quad 11100 \quad 10011 \quad 01111 \quad 01001 \quad 10101 \quad 11010 \quad 00110,$$
$$1_L \quad 00011 \quad 11111 \quad 10000 \quad 01100 \quad 01010 \quad 10110 \quad 11001 \quad 00101.$$

The encoding circuit for this code is shown in Figure 5.16. The input state, ψ, is identical with the output state, φ, when there is no error. The circuit uses single-qubit Hadamard gates, two-qubit CNOT gates, as well as two- and three-qubit controlled, π, gates that rotate the phase of a qubit by π; the leftmost π gate rotates the phase of d when all control qubits are 1; the second to the right π gate rotates the phase of d when b and c are 0 and ψ is 1. The error operators and the syndromes are summarized in Table 5.6; the resulting state, φ, after an error occurs in the original state, $\psi \quad \alpha_0\, 0 \quad \alpha_1\, 1$, is shown in Table 5.7. As before, subscripts indicate positions of errors: one bit-flip error is denoted as $X_{1\ \ 5}$; one phase-flip error as $Z_{1\ \ 5}$; and the combination of one bit-flip and one phase-flip as $Y_{1\ \ 5}$.

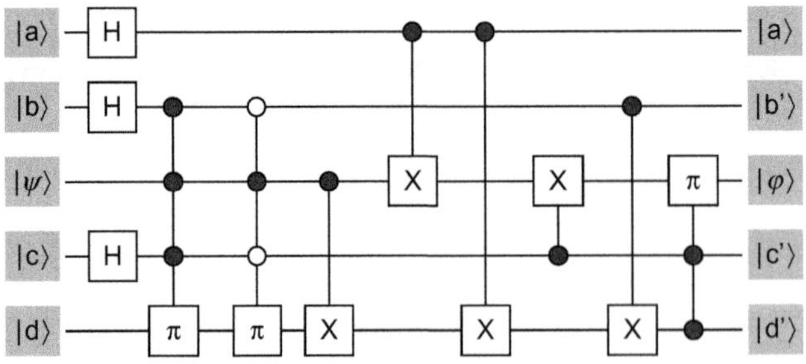

FIGURE 5.16

The encoder for the perfect quantum code of Laflamme *et al* [263].

Table 5.6 The Single Errors and the Corresponding Syndromes for the Five-qubit Code

Error Operator	Syndrome $a_1a_2a_3a_4$	Error Operator	Syndrome $a_1a_2a_3a_4$	Error Operator	Syndrome $a_1a_2a_3a_4$
$\mathbf{X}_1$	0110	$\mathbf{Z}_1$	1000	$\mathbf{Y}_1$	1110
$\mathbf{X}_2$	0001	$\mathbf{Z}_2$	0100	$\mathbf{Y}_2$	0101
$\mathbf{X}_3$	0111	$\mathbf{Z}_3$	1010	$\mathbf{Y}_3$	1101
$\mathbf{Y}_4$	1011	$\mathbf{Z}_4$	0010	$\mathbf{Y}_4$	1001
$\mathbf{X}_5$	0011	$\mathbf{Z}_5$	1100	$\mathbf{Y}_5$	1111

Table 5.7 The Resulting State, φ , After an Error

Error(s)	$\|\varphi\rangle$
$\mathbf{Y}_3$	α_0 1 α_1 0
$\mathbf{Y}_5$	α_0 0 α_1 1
$\mathbf{X}_2, \mathbf{Z}_3, \mathbf{Z}_5, \mathbf{Y}_2$	α_0 0 α_1 1
$\mathbf{X}_5, \mathbf{Z}_1, \mathbf{Z}_2, \mathbf{Z}_4$	α_0 0 α_1 1
$\mathbf{X}_1, \mathbf{X}_3, \mathbf{X}_4, \mathbf{Y}_1, \mathbf{Y}_4$	α_0 1 α_1 0

Another version of the perfect code [307] encodes the logical qubits as

$$0_L \quad \tfrac{1}{4}[\ 00000 \quad 10010 \quad 01001 \quad 10100 \quad 01010 \quad 11011 \quad 00110 \quad 11000$$
$$11101 \quad 00011 \quad 11110 \quad 01111 \quad 10001 \quad 01100 \quad 10111 \quad 01101\ ],$$

$$1_L \quad \tfrac{1}{4}[\ 11111 \quad 01101 \quad 10110 \quad 01011 \quad 10101 \quad 00100 \quad 11001 \quad 00111$$
$$00010 \quad 11100 \quad 00001 \quad 10000 \quad 01110 \quad 10011 \quad 01000 \quad 11010\ ].$$

The stabilizer, $\mathcal{S}$, of this code is described by a group, M, of four generators:

$$M \quad \mathbf{M}_1, \mathbf{M}_2, \mathbf{M}_3, \mathbf{M}_4$$
$$\mathbf{M}_1 \quad X \ Z \ Z \ X \ \mathrm{I}, \quad \mathbf{M}_2 \quad \mathrm{I} \ X \ Z \ Z \ X,$$
$$\mathbf{M}_3 \quad X \ \mathrm{I} \ X \ Z \ Z, \quad \mathbf{M}_4 \quad Z \ X \ \mathrm{I} \ X \ Z.$$

The codewords, 0_L and 1_L, are the eigenvectors, with an eigenvalue of (1) of the stabilizers,

$$\mathbf{M}_j\, 0_L \quad (\ \ 1)\, 0_L \quad \text{and} \quad \mathbf{M}_j\, 1_L \quad (\ \ 1)\, 1_L, \quad 1 \quad j \quad 4.$$

M is an Abelian subgroup; each generator commutes with all the others. For example,

$$\mathbf{M}_1\mathbf{M}_3 \quad (X \quad Z \quad Z \quad X \quad \mathrm{I})(X \quad \mathrm{I} \quad X \quad Z \quad Z) \quad \mathrm{I} \quad Z \quad (ZX) \quad (XZ) \quad Z,$$

$$\mathbf{M}_3\mathbf{M}_1 \quad (X \quad \mathrm{I} \quad X \quad Z \quad Z)(X \quad Z \quad Z \quad X \quad \mathrm{I}) \quad \mathrm{I} \quad Z \quad (XZ) \quad (ZX) \quad Z.$$

The Pauli operators anticommute, $XZ \quad ZX$ but $\mathbf{M}_1\mathbf{M}_3 \quad \mathbf{M}_3\mathbf{M}_1$.

The circuit for computing the syndrome is shown in Figure 5.17. The syndrome, Σ, consists of four ancilla qubits, $a_1a_2a_3a_4$, each one measured using one of the four generators, $\mathbf{M}_1, \mathbf{M}_2, \mathbf{M}_3, \mathbf{M}_4$ (see Table 5.8. Table 5.9 lists single-error operators and the generator(s) that anticommute with each operator. For example, $\mathbf{X}_1$ (where $\mathbf{X}_1 \quad X \quad \mathrm{I} \quad \mathrm{I} \quad \mathrm{I} \quad \mathrm{I}$) anticommutes with $\mathbf{M}_4$; thus, a bit-flip on the first qubit can be detected. $\mathbf{Z}_1$ anticommutes with $\mathbf{M}_1$ and $\mathbf{M}_3$; thus, a phase-flip of the first qubit can also be detected. Since each of these 15 errors anticommute with distinct subsets of $\mathcal{S}$, we are able to distinguish individual errors and then correct them.

The code cannot detect two-qubit errors; for example, the two bit-flip errors, $\mathbf{X}_1\mathbf{X}_2 \quad X \quad X \quad \mathrm{I} \quad \mathrm{I} \quad \mathrm{I}$, are indistinguishable from $\mathbf{Z}_4 \quad \mathrm{I} \quad \mathrm{I} \quad \mathrm{I} \quad Z \quad \mathrm{I}$ because both $\mathbf{X}_1\mathbf{X}_2$ and $\mathbf{Z}_4$ anticommute with the same subset of stabilizers, $\mathbf{M}_1, \mathbf{M}_4$, and give the same error syndrome. We conclude that *the ve-qubit code can correct any single-qubit error, but cannot correct two-qubit errors*.

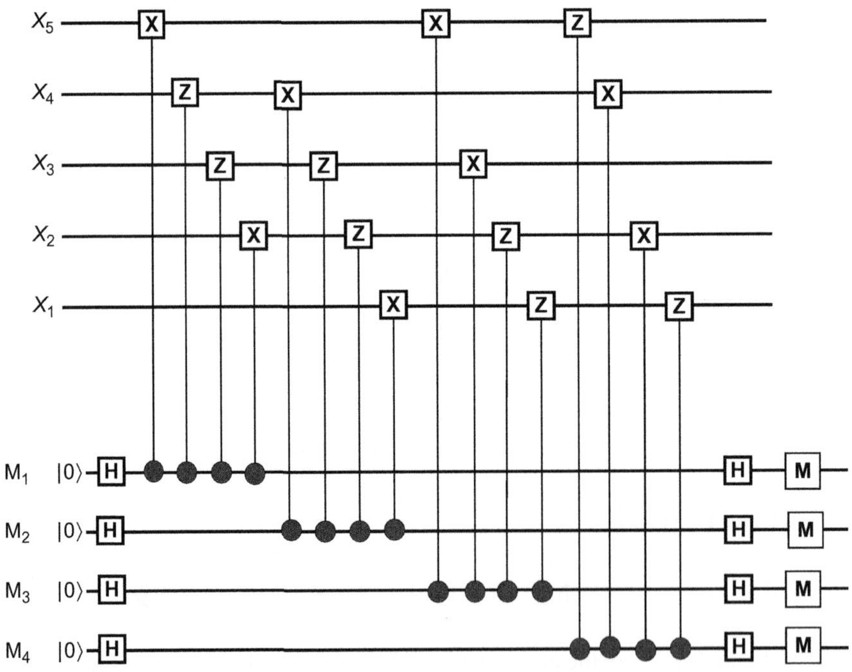

FIGURE 5.17

The quantum circuit for the syndrome measurement of the perfect quantum code de ned in [307]. Ancilla qubit $i, 1 \quad i \quad 4$, acts as the control qubit for the four gates used by the generators $\mathbf{M}_i, 1 \quad i \quad 4$.

Table 5.8 The Single Errors and the Corresponding Syndromes for the Second Definition of the Perfect Quantum Code

Error Operator	Syndrome $a_1a_2a_3a_4$	Error Operator	Syndrome $a_1a_2a_3a_4$	Error Operator	Syndrome $a_1a_2a_3a_4$
$\mathbf{X}_1$	0001	$\mathbf{Z}_1$	1010	$\mathbf{Y}_1$	1011
$\mathbf{X}_2$	1000	$\mathbf{Z}_2$	0101	$\mathbf{Y}_2$	1101
$\mathbf{X}_3$	1100	$\mathbf{Z}_3$	0010	$\mathbf{Y}_3$	1110
$\mathbf{Y}_4$	0110	$\mathbf{Z}_4$	1001	$\mathbf{Y}_4$	1111
$\mathbf{X}_5$	0011	$\mathbf{Z}_5$	0100	$\mathbf{Y}_5$	0111

Table 5.9 Single Error Operators for the Second Definition of the Perfect Quantum Code and the Generator(s) that Anticommute with Each Operator

Error Operator	Generator(s)	Error Operator	Generator(s)	Error Operator	Generator(s)
$\mathbf{X}_1$	$\mathbf{M}_4$	$\mathbf{Z}_1$	$\mathbf{M}_1, \mathbf{M}_3$	$\mathbf{Y}_1$	$\mathbf{M}_1, \mathbf{M}_3, \mathbf{M}_4$
$\mathbf{X}_2$	$\mathbf{M}_1$	$\mathbf{Z}_2$	$\mathbf{M}_2, \mathbf{M}_4$	$\mathbf{Y}_2$	$\mathbf{M}_1, \mathbf{M}_2, \mathbf{M}_4$
$\mathbf{X}_3$	$\mathbf{M}_1, \mathbf{M}_2$	$\mathbf{Z}_3$	$\mathbf{M}_3$	$\mathbf{Y}_3$	$\mathbf{M}_1, \mathbf{M}_2, \mathbf{M}_3$
$\mathbf{X}_4$	$\mathbf{M}_2, \mathbf{M}_3$	$\mathbf{Z}_4$	$\mathbf{M}_1, \mathbf{M}_4$	$\mathbf{Y}_4$	$\mathbf{M}_1, \mathbf{M}_2, \mathbf{M}_3, \mathbf{M}_4$
$\mathbf{X}_5$	$\mathbf{M}_3, \mathbf{M}_4$	$\mathbf{Z}_5$	$\mathbf{M}_2$	$\mathbf{Y}_5$	$\mathbf{M}_2, \mathbf{M}_3, \mathbf{M}_4$

5.14 QUANTUM RESTORATION CIRCUITS

The *parity of a qubit* is given by the eigenvalue of the Pauli operator, Z, applied to the qubit; the *parity of an n-qubit codeword* is given by the eigenvalue of the product operator, $\sigma_{z,1}\sigma_{z,2},\ldots,\sigma_{z,n}$; an eigenvalue of 1 corresponds to even parity and 1 to odd parity of the qubit or of the codeword.

Now we discuss a family of circuits DiVincenzo calls *quantum restoration circuits* [129]. We examine a circuit described in [128] that restores a five-qubit codeword to its original state after determining the error syndrome that identifies the location and the type of a single-qubit error, Figure 5.18. When both the control and the target qubit of a CNOT gate are in a rotated base, then the roles of the control and target are reversed (see Figure 1.10(d)).

The encoding of the logical basis states, (0 $_L$ and (1 $_L$), is

$$
\begin{aligned}
0_L \quad & 00000 \\
& 11000 \quad 01100 \quad 00110 \quad 00011 \quad 10001 \\
& 10100 \quad 01010 \quad 00101 \quad 10010 \quad 01001 \\
& 11110 \quad 01111 \quad 10111 \quad 11011 \quad 11101\ ,
\end{aligned}
$$

$$
\begin{aligned}
1_L \quad & 11111 \\
& 00111 \quad 10011 \quad 11001 \quad 11100 \quad 01110 \\
& 01011 \quad 10101 \quad 11010 \quad 01101 \quad 10110 \\
& 00001 \quad 10000 \quad 01000 \quad 00100 \quad 00010\ .
\end{aligned}
$$

Four measurement operators, $\mathbf{M}_0, \mathbf{M}_1, \mathbf{M}_3$ and $\mathbf{M}_4$, allow us to identify the qubit in error. They are tensor products of Pauli operators applied to the qubits of a codeword; the measurement operators are

$$\mathbf{M}_3 \quad X_0X_1Z_2Z_4, \quad \mathbf{M}_4 \quad X_1X_2Z_3Z_0, \quad \mathbf{M}_0 \quad X_2X_3Z_4Z_1, \quad \text{and} \quad \mathbf{M}_1 \quad X_3X_4Z_0Z_2.$$

The circuit in Figure 5.18 uses four groups, each consisting of four CNOT gates, to compute the eigenvalues of the measurement operators and then store the results in four ancilla qubits initially in the state, 0 . Table 5.10 lists the 16 possible errors, no-error, I_i, included. When a single error occurs, the nondemolition measurements uniquely identify the qubit in error and the type of error and drive unitary transformations, $\mathbf{U}$, used to restore the qubit affected by error to its original state.

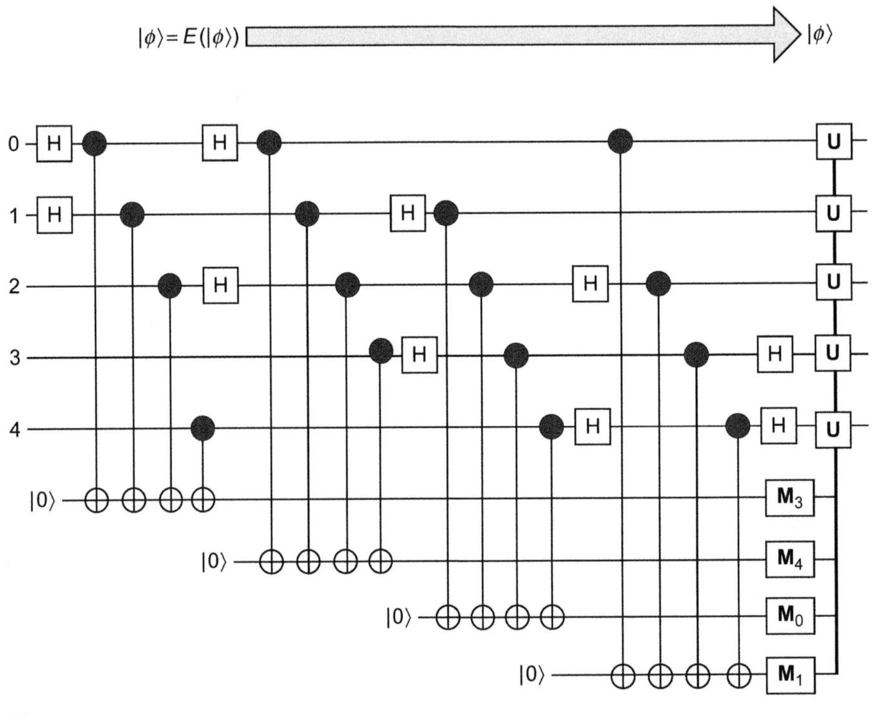

FIGURE 5.18

A circuit for the restoration of the qubit, $\varphi \quad \alpha_0\ 0\ _L \quad \alpha_1\ 1\ _L$, after a single error. The original qubit is encoded using ve physical qubits, labeled as 0, 1, 2, 3, and 4. The eigenvalues of four measurement operators, $\mathbf{M}_0, \mathbf{M}_1, \mathbf{M}_3$ and $\mathbf{M}_4$, are stored in four ancilla qubits initially in the state 0 . The $\mathbf{M}_3 \quad X_0X_1Z_2Z_4$, operator uses the rst group of four CNOT gates with the qubits $0, 1, 2$, and 4 as the source and the rst ancilla qubit as the target; the, $\mathbf{M}_4 \quad X_1X_2Z_3Z_0$, operator uses the second group of four CNOT gates with the qubits $0, 1, 2$, and 3 as the source and the second ancilla qubit as the target, and so on. When a single error occurs, the nondemolition measurements uniquely identify the qubit in error as well as the type of error and drive unitary transformations, $\mathbf{U}$, used to restore the physical qubits to their original state. Two Hadamard gates, H, on each line representing a physical qubit rotate the basis to measure the eigenvalues of X when necessary; the state of the qubit is not affected as $\mathbf{HH} \quad \mathbf{I}$.

Table 5.10 The Syndrome and the Error for the Restoration Circuit

M_3	M_4	M_0	M_1	E_i	M_3	M_4	M_0	M_1	E_i
0	0	0	0	$I_0I_1I_2I_3I_4$	1	0	0	0	$\mathbf{Z}_0I_1I_2I_3I_4$
0	0	0	1	$I_0I_1I_2I_3\mathbf{Z}_4$	1	0	0	1	$I_0I_1\mathbf{X}_2I_3I_4$
0	0	1	0	$I_0\mathbf{X}_1I_2I_3I_4$	1	0	1	0	$I_0I_1I_2I_3\mathbf{X}_4$
0	0	1	1	$I_0I_1I_2\mathbf{Z}_3I_4$	1	0	1	1	$I_0I_1I_2I_3\mathbf{Y}_4$
0	1	0	0	$I_0I_1I_2\mathbf{X}_3I_4$	1	1	0	0	$I_0\mathbf{Z}_1I_2I_3I_4$
0	1	0	1	$\mathbf{X}_0I_1I_2I_3I_4$	1	1	0	1	$\mathbf{Y}_0I_1I_2I_3I_4$
0	1	1	0	$I_0I_1\mathbf{Z}_2I_3I_4$	1	1	1	0	$I_0\mathbf{Y}_1I_2I_3I_4$
0	1	1	1	$I_0I_1I_2\mathbf{Y}_3I_4$	1	1	1	1	$I_0I_1\mathbf{Y}_2I_3I_4$

The analysis of this circuit in [128] is important because it establishes that errors in "all known quantum error-correcting codes can be corrected in a fault-tolerant way."

Error correction requires an increased number of physical qubits and quantum gates as well as an increased number of computational steps; thus, it increases the vulnerability of the system. For example, we have seen that to restore one logical qubit, we need nine physical qubits as well as 12 CNOT gates, 12 Hadamard gates, and several additional gates implementing Pauli transformations. This simple example shows that the overhead of error correction discussed in Section 5.3 cannot be ignored and raises the questions of whether error correction is really feasible, and whether the error correction introduces more errors and extends the number of steps and implicitly the computation time beyond the ability of physical qubits to maintain the state required by the computational steps of an algorithm.

A critical question addressed early on was if arbitrary long computations can be carried out with a quantum computing device whose state is subject to decoherence. The answer to this question was provided by the quantum accuracy threshold theorem [5, 239, 246, 342], which states that an arbitrarily long computation can be carried out reliably provided that all sources of "noise" affecting a quantum computer are weaker than the so-called "accuracy threshold." A lower bound on this threshold, $\epsilon > 1.9 \quad 10^{\ 4}$, was derived in [9].

5.15 QUANTUM CODES OVER $GF(p^k)$

The quantum codes we have constructed so far have been over $GF(2)$, the binary field with two elements. Codes over extension fields, $GF(p^k)$, have distinct advantages (see Section 4.15); this motivates us to explore the properties of linear codes over the finite field, $GF(p^k)$ [178].

The *trace*, $\mathrm{tr} : GF(p^k) \quad GF(p)$, is a linear mapping from $GF(p^k)$ to $GF(p)$:

$$\mathrm{tr}(\alpha) \quad \alpha_1^{p^0} \quad \alpha_2^{p^1} \qquad \alpha_k^{p^{k\ 1}}.$$

The elements, $\alpha_1, \alpha_2, \ldots, \alpha_k$, form a basis of, $GF(p^k)$, over $GF(p)$, if and only if

$$\det \begin{pmatrix} \alpha_1 & \alpha_2 & & \alpha_k \\ \alpha_1^p & \alpha_2^p & & \alpha_k^p \\ \vdots & \vdots & \ddots & \vdots \\ \alpha_1^{p^{k\ \ 1}} & \alpha_2^{p^{k\ \ 1}} & & \alpha_k^{p^{k\ \ 1}} \end{pmatrix} \quad 0.$$

The set, $\alpha_1, \alpha_2, \ldots, \alpha_k$, is a *trace orthogonal basis* if

$$\mathrm{tr}(\alpha_i \alpha_j) \quad 0,\ 1 \quad i,j \quad k, \quad i \quad j.$$

If, in addition, $\mathrm{tr}(\alpha_i^2) \quad 1$, then $\alpha_1, \alpha_2, \ldots, \alpha_k$ is a *self-dual basis*.

The finite field $GF(p^k)$ is a k-dimensional vector space over $GF(p)$; once we choose a basis, $\mathcal{B}$, there is a homomorphism, $\mathcal{B} : GF(p^k) \quad [GF(p)]^k$:

$$v \quad \mathcal{B}(v), \quad v \quad GF(q^k).$$

This homomorphism allows us to express any vector, $v \quad GF(q^k)$, in terms of the basis $\mathcal{B}$; this mapping is described by an $k \quad k$ matrix. Any linear transformation, $A \quad GL(n, GF(p^k))$, described by the $n \quad n$ matrix, $A \quad [a_{ij}], 1 \quad i,j \quad n$, can be expressed by substituting, a_{ij}, with $\mathcal{B}(a_{ij})$, and the change in the ground field, $GF(p)$, can be carried out before the mapping, A, or after the mapping, as shown by the commutative diagram, with $q \quad p^k$:

$$\begin{array}{ccc} & A & \\ GF(q^n) & - & GF(q^n) \\ \mathcal{B} & & \mathcal{B} \\ & \mathcal{B}(A) & \\ GF(p^{kn}) & - & GF(p^{kn}). \end{array}$$

If we consider two vectors, $v, w \quad GF(q^k)$, then the product, $v \quad w$, can be represented using a linear mapping, $\mathcal{M}_{\mathcal{B}}$, as

$$\mathcal{B}(v \quad w) \quad \mathcal{B}(w) \quad \mathcal{M}_{\mathcal{B}}(v).$$

Let $\mathcal{B} \quad (b_1, b_2, \ldots, b_k)$ be the dual of the basis, $\mathcal{B} \quad (b_1, b_2, \ldots, b_k)$; by definition

$$\mathrm{tr}(b_i b_j) \quad \delta_{ij},\ 1 \quad i,j \quad k.$$

Proposition. *Let C be the dual of the $[N,K]$ linear code, C, over the nite eld, $GF(2^k)$. Let $\mathcal{B}(C)$ be the binary expansion of the code, C, with respect to basis, $\mathcal{B}$, and let $\mathcal{B} \quad (C \quad)$ be the binary expansion of the dual of, C, with respect to the dual basis, $\mathcal{B}$; then*

$$\mathcal{B} \quad (C \quad) \quad [\mathcal{B}(C)] \quad .$$

Proof. The proof follows the one in [178]. Two arbitrary codewords, $c \quad (c_1 c_2 \ldots c_N) \quad \mathcal{C}$ and $d \quad (d_1 d_2 \ldots d_N) \quad \mathcal{C}$, are orthogonal:

$$c \quad d \quad 0 \qquad \sum_{i \quad 1}^{N} c_i d_i \qquad \sum_{i \quad 1}^{N} \left(\sum_{j \quad 1}^{k} c_{ij} b_j \right) \left(\sum_{l \quad 1}^{k} d_{il} b_l \right) \qquad 0,$$

with $c_{ij}, 1 \quad j \quad k$, the binary expression of $c_i \quad GF(p^k)$ and $d_{il}, 1 \quad l \quad k$, the binary expression of $d_i \quad GF(p^k)$; we wish to show that

$$\sum_{i \quad 1}^{N} \sum_{j \quad 1}^{k} c_{ij} d_{ij} \qquad 0.$$

If we take the trace of the transformations, $\mathcal{B}$ and $\mathcal{B}$, in the expression of, $c \quad d$, we observe that

$$\sum_{i \quad 1}^{N} \sum_{j \quad 1}^{k} \sum_{l \quad 1}^{k} c_{ij} d_{il} \operatorname{tr}(b_j b_l) \qquad \sum_{i \quad 1}^{N} \sum_{l \quad 1}^{k} c_{ij} d_{ij} \qquad 0$$

because $\operatorname{tr}(b_j b_l) \quad \delta_{jl}$. ■

This proposition allows us to conclude that the following diagram is commutative:

$$\begin{array}{ccc} \mathcal{C} & - & \mathcal{C} \\ \text{basis } \mathcal{B} & & \text{dual basis } \mathcal{B} \\ \mathcal{B}(\mathcal{C}) & - & \mathcal{B} \ (\mathcal{C} \ ). \end{array}$$

Quantum Circuits for Invertible Linear Transformations in *GF*(2^*n*)

We have seen that a linear mapping in $GF(2^n)$ can be reduced to mappings in $GF(2)$; thus, we first survey invertible linear transformations in $GF(2)$. The linear transformation, $P \quad GL(n, GF(2))$, that permutes the basis vectors in an n-dimensional vector space, $i \quad (\mathcal{H}_2) \quad ^n$, $i \quad P \, i$, can be implemented using at most, $3(n \quad 1)$, CNOT gates. Indeed, any permutation, P, that transforms an n-tuple over $GF(2)$ can be implemented as a product of at most, $(n \quad 1)$, transpositions, and a transposition, (i,j), can be implemented by a quantum circuit using three CNOT gates (see Figure 5.19); it follows that any permutation of the basis vectors in, $(\mathcal{H}_2) \quad ^n$, requires at most, $3(n \quad 1)$, CNOT gates.

Now we show that any linear transformation, $A \quad GL(n, GF(2))$, invertible in $GF(2^n)$, that carries out the transformation of basis vectors, $i \quad A \, i$, $i \quad (\mathcal{H}_2) \quad ^n$, can be implemented using at most, $(n \quad 1)(n \quad 3)$, CNOT gates. Indeed, any matrix, $A \quad GL(n, GF(2))$, can be decomposed as a product, $A \quad P \; L \; U$, with P, a permutation matrix and with L and U lower and, respectively, upper triangular matrices.

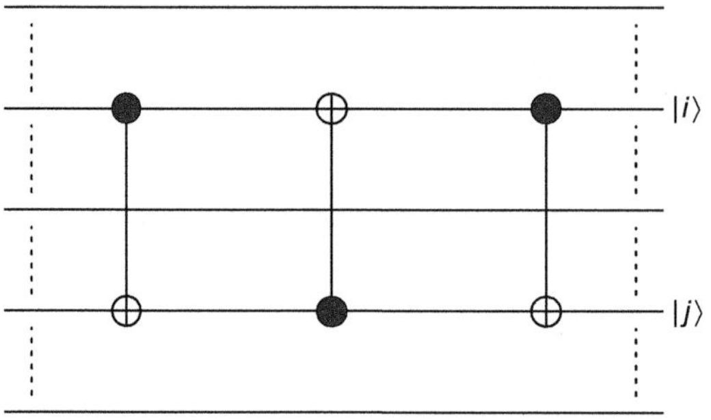

FIGURE 5.19

Any transposition, (i,j), can be implemented by a quantum circuit using three CNOT gates.

The lower triangular matrix, $L \quad [L_{p,q}], 1 \quad p,q \quad n$, can be factored as $L \quad L_1 L_2 \ldots L_i \ldots L_n$, with L_i obtained by replacing the i-th row of an identity matrix with the i-th row of L. Let the binary representation of the basis vector, k , be $k \quad (k_1, k_2, \ldots, k_i, \ldots, k_n)$. We multiply the vector, k and L_i,

$$kL_i \quad (k_1 \quad k_i L_{i,1}, \ldots, k_i \;{}_1 L_{i,i} \;{}_1, k_i, k_i \;{}_1, \ldots, k_n),$$

and notice that the j-th position of k is inverted if both, k_i and $L_{i,j}$, are equal to 1.

This operation requires at most, $(i \quad 1)$, CNOT gates with control qubit, i, and target qubit, j, whenever $L_{i,j} \quad 1$. Thus, the implementation of L requires, $n(n \quad 1)/2$, CNOT gates. A similar reasoning leads to the conclusion that implementation of, U, requires the same number of CNOT gates. The total number of CNOT gates required for the implementation of P, L, and U is

$$3(n \quad 1) \quad \frac{n(n \quad 1)}{2} \quad \frac{n(n \quad 1)}{2} \quad (n \quad 1)(n \quad 3).$$

Thus, the total number of CNOT gates required for the implementation of a linear transformation A is $\mathcal{O}(n^2)$.

To reduce a linear mapping in $GF(2^n)$ to a mapping in $GF(2)$, we first fix a basis, $\mathcal{B}$, in $GF(2^k)$. Then we extend the homomorphism, $\mathcal{B}$, and map n-dimensional vectors, v, with elements in $GF(2^k)$ to binary vectors, w, of length, kn, as shown earlier.

Cyclic discrete Fourier transform (*DFT*) over $GF(p^k)$. The Quantum Reed-Solomon codes discussed in the next section can be described by means of the cyclic, DFT, over $GF(p^k)$. If n divides $2^k \quad 1$ and β is an element of order n in $GF(p^k)$, then the cyclic, DFT, over $GF(p^k)$, is described by the matrix,

$$DFT \quad (\beta^{ij}), 0 \quad i,j \quad n \quad 1,$$

The condition that n divides $2^k \quad 1$ ensures that $GF(2^k)$ has a primitive element of order, n, and thus, $DFT \quad GL(n, GF(2^k))$. It is easy to see that the cyclic DFT, over $GF(p^k)$ can be implemented using $\mathcal{O}(n^2k^2)$ CNOT gates. This follows immediately from the fact that DFT is a linear transformation and, according to our observation, it requires, $\mathcal{O}(N^2)$, CNOT gates for vectors of length, $N \quad kn$. The inverse Fourier transform is denoted as $DFT \;{}^1$.

5.16 QUANTUM REED-SOLOMON CODES

Quantum Reed-Solomon (QRS) codes can be constructed as a family of CSS codes. In Section 4.15, we showed that a classical, $[N,K,d]$, Reed-Solomon code, over $GF(2^k)$, with length, $N \;\; 2^k \;\; 1$, distance, d, and dimension, $K \;\; N \;\; d \;\; 1$, is a cyclic code with the generator polynomial

$$g(x) \quad (x \quad \beta^j)(x \quad \beta^{j \;\; 1})\ldots(x \quad \beta^{j \;\; d \;\; 2}),$$

with β, a primitive element of $GF(2^k)$ of order N. A QRS code encodes, $k(2N \quad K)$, qubits in the states, φ_1 , φ_2 ,..., $\varphi_{k(2N \;\; K)}$, to kN qubits.

Proposition. *The $[N,K,d]$ Reed-Solomon code, C, is self-dual when $j \quad 0$ and $d > N/2 \quad 1$.*

Proof. If $g(x) \quad (x \quad \beta^0)(x \quad \beta^1)\ldots(x \quad \beta^{d \;\; 2})$ is the generator polynomial of C, then the generator polynomial of the dual code, C , is the reciprocal polynomial of $(x^N \quad 1)/g(x)$:

$$g \;\; (x) \quad (x \quad \beta^{\;\;(d \;\; 1)})(x \quad \beta^{\;\;d})\ldots(x \quad \beta^{\;\;(N \;\; 1)}).$$

$N \quad 2^k \quad 1$ and β is a primitive element of $GF(2^k)$; thus, $\beta^N \quad 1$, and we can write

$$g \;\; (x) \quad (x \quad \beta^1)(x \quad \beta^2)\ldots(x \quad \beta^{(N \;\; d \;\; 1)}).$$

The highest power of β in the expression of $g \;\; (x)$ satisfies the inequality, $N \quad d \quad 1 \quad d \quad 1$; indeed, $d > N/2 \quad 1 \quad\quad N \quad 2d \quad 2$ or $N \quad d \quad 1 \quad 2d \quad 2 \quad d \quad 1$. It follows that $g \;\; (x)$ is a divisor of $g(x)$; thus, $C \quad C$. ■

Reed-Solomon codes can be described by means of the cyclic, DFT, over $GF(2^k)$. The *spectrum*, *c*, of a vector, $c \quad \sum_i^n {}_1 \gamma_i c_i$, with the elements, $\gamma_i \quad GF(2^k)$, is

$$c \quad c \; DFT \qquad\qquad c \quad c \; DFT^{\;\;1}.$$

The following proposition will allow us to show that a classical Reed-Solomon code, C, can be described as the set of N-tuples, over $GF(2^k)$ with the first, d, components equal to zero:

$$C \quad w \; DFT^{\;\;1}, \text{ with } w \quad (w_1 \, w_2 \, \ldots \, w_d \, w_{d \;\; 1} \ldots w_N) \quad GF(2^k), \quad w_1 \quad w_2 \qquad\quad w_d \quad 0.$$

Proposition. *The spectrum c of a codeword, c, of a weakly self-dual, [N,K,d], Reed-Solomon code, over $GF(2^k)$, has its rst, $d \quad 1 \quad N \quad K$, components equal to zero, and the next, K, components with an arbitrary value; the spectrum, c , of a codeword of the dual code, $c \quad C$, has its rst component as well as the last, $d \quad 2$, components with an arbitrary value, while the other, K, components are equal to zero.*

Proof. The transformation carried out by the cyclic, DFT, over $GF(2^k)$, with a primitive element, β, is given by the $N \quad N$ matrix

$$DFT \quad (\beta^{ij})_{0 \ \ i,j \ \ N \ \ 1} \quad \begin{pmatrix} 1 & 1 & 1 & \dots & 1 & 1 \\ 1 & (\beta^1)^1 & (\beta^1)^2 & \dots & (\beta^1)^{N \ \ 2} & (\beta^1)^{N \ \ 1} \\ 1 & (\beta^2)^1 & (\beta^2)^2 & \dots & (\beta^2)^{N \ \ 2} & (\beta^2)^{N \ \ 1} \\ \vdots & \vdots & \vdots & \ddots & \vdots & \vdots \\ 1 & (\beta^{K \ \ 1})^1 & (\beta^{K \ \ 1})^2 & \dots & (\beta^{K \ \ 1})^{N \ \ 2} & (\beta^{K \ \ 1})^{N \ \ 1} \\ 1 & (\beta^K)^1 & (\beta^K)^2 & \dots & (\beta^K)^{N \ \ 2} & (\beta^K)^{N \ \ 1} \\ \vdots & \vdots & \vdots & \ddots & \vdots & \vdots \\ 1 & (\beta^{N \ \ 1})^1 & (\beta^{N \ \ 1})^2 & \dots & (\beta^{N \ \ 1})^{N \ \ 2} & (\beta^{N \ \ 1})^{N \ \ 1} \end{pmatrix}.$$

In Section 4.15, we introduced the generator and the parity-check matrix of a classical Reed-Solomon code. From that discussion, it follows that the transpose of the parity-check matrix of the weakly self-dual, $[N,K,d]$, Reed-Solomon code is the $N \quad (N \quad K)$ matrix

$$\mathrm{H}^T \quad \begin{pmatrix} 1 & 1 & \dots & 1 & 1 \\ (\beta^1) & (\beta^1)^2 & \dots & (\beta^1)^{N \ \ K \ \ 1} & (\beta^1)^{N \ \ K} \\ (\beta^2) & (\beta^2)^2 & \dots & (\beta^2)^{N \ \ K \ \ 1} & (\beta^2)^{N \ \ K} \\ \vdots & \vdots & \ddots & \vdots & \vdots \\ (\beta^{N \ \ 1}) & (\beta^{N \ \ 1})^2 & \dots & (\beta^{N \ \ 1})^{N \ \ K \ \ 1} & (\beta^{N \ \ 1})^{N \ \ K} \end{pmatrix}.$$

The cyclic, DFT, can then be expressed as

$$DFT \quad ([\mathrm{H}^T][A]),$$

with A, the $N \quad K$ matrix, given by

$$A \quad \begin{pmatrix} 1 & 1 & \dots & 1 & 1 \\ (\beta^1)^{N \ \ K} & (\beta^1)^{N \ \ K \ \ 1} & \dots & (\beta^1)^{N \ \ 2} & (\beta^1)^{N \ \ 1} \\ (\beta^2)^{N \ \ K} & (\beta^2)^{N \ \ K \ \ 1} & \dots & (\beta^2)^{N \ \ 2} & (\beta^2)^{N \ \ 1} \\ \vdots & \vdots & \ddots & \vdots & \vdots \\ (\beta^{N \ \ 1})^{N \ \ K} & (\beta^{N \ \ 1})^{N \ \ K \ \ 1} & \dots & (\beta^{N \ \ 1})^{N \ \ 2} & (\beta^{N \ \ 1})^{N \ \ 1} \end{pmatrix}.$$

If H is the parity-check matrix of a linear code, C and $c \quad C$, then

$$\mathrm{H}c^T \quad c\mathrm{H}^T \quad 0.$$

It follows that if C is a weakly self-dual $[N,K,d]$ Reed-Solomon code and $c \quad C$, then

$$c \quad c \ DFT \quad c \ [\mathrm{H}^T A].$$

The spectrum of any codeword has the first, $N \quad K \quad d \quad 1$, components equal to zero; the last, $N \quad d \quad 1$, components may have an arbitrary value (Figure 5.20a).

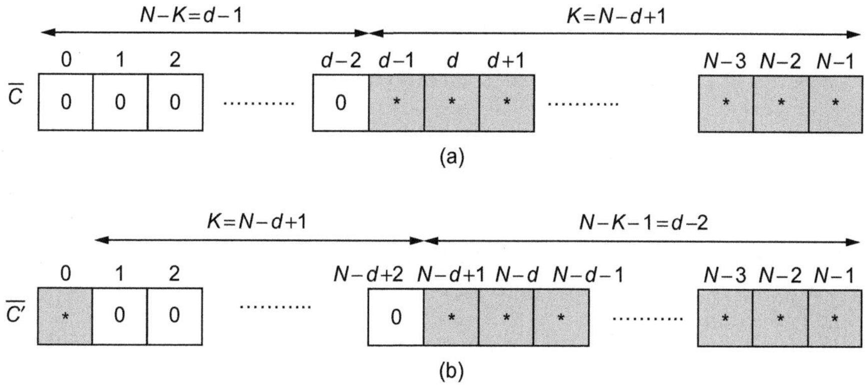

FIGURE 5.20

(a) The spectrum of codeword, $c \;\; \mathcal{C}$, has its rst, $d \;\; 1$, elements equal to zero and the last, $(N \;\; d \;\; 1)$, elements may take arbitrary values. (b) The spectrum of a codeword belonging to the dual code, $c \;\; \mathcal{C}$, allows its rst component to have an arbitrary value, the next $(N \;\; d \;\; 1)$ components are equal to zero, and the last $(d \;\; 2)$ components may have arbitrary values.

We can identify the zero components of a codeword of the dual code, $\mathcal{C}$, knowing that the generator of the original code, $\mathcal{C}$, is the parity-check matrix of the dual code. The transpose of, G, the generator matrix of the code, is the $N \;\; (N \;\; K)$ matrix,

$$\mathrm{G}^T \qquad \begin{pmatrix} 1 & 1 & 1 & \dots & 1 & 1 \\ 1 & (\beta^1)^1 & (\beta^2)^1 & \dots & (\beta^{K\;\;2})^1 & (\beta^{K\;\;1})^1 \\ 1 & (\beta^1)^2 & (\beta^2)^2 & \dots & (\beta^{K\;\;2})^2 & (\beta^{K\;\;1})^2 \\ \vdots & \vdots & \vdots & \ddots & \vdots & \vdots \\ 1 & (\beta^1)^{N\;\;1} & (\beta^2)^{N\;\;1} & \dots & (\beta^{K\;\;2})^{N\;\;1} & (\beta^{K\;\;1})^{N\;\;1} \end{pmatrix}.$$

The cyclic, DFT, can then be expressed as

$$DFT \qquad ([1][\mathrm{G}^T][B]),$$

with B, the $N \;\; (K \;\; 1)$ matrix

$$B \qquad \begin{pmatrix} 1 & 1 & \dots & 1 & 1 \\ (\beta^K)^1 & (\beta^{K\;\;1})^1 & \dots & (\beta^{N\;\;2})^1 & (\beta^{N\;\;1})^1 \\ (\beta^K)^2 & (\beta^{K\;\;1})^2 & \dots & (\beta^{N\;\;2})^2 & (\beta^{N\;\;1})^2 \\ \vdots & \vdots & \ddots & \vdots & \vdots \\ (\beta^K)^{N\;\;1} & (\beta^{K\;\;1})^{N\;\;1} & \dots & (\beta^{N\;\;2})^{N\;\;1} & (\beta^{N\;\;1})^{N\;\;1} \end{pmatrix}.$$

If G is the generator matrix of $\mathcal{C}$ and $c \;\;\; \mathcal{C}$, then

$$\mathrm{G} \;\; (c\,)^T \quad c \;\; \mathrm{G}^T \quad 0.$$

It follows that if $\mathcal{C}$ is a weakly self-dual $[N,K,d]$ Reed-Solomon code and $c \quad \mathcal{C}$, then

$$c \quad c \;\; DFT \quad c \;\; ([1][G^T][B]).$$

The *DFT* of any codeword of the dual code has a first component with an arbitrary value, the next $(N \quad d \quad 1)$ components are equal to zero, and the last $(d \quad 2)$ components may also have an arbitrary value (Figure 5.20b). ■

$\mathcal{C}$ is an $[N,K,d]$ self-dual binary code if and only if $\mathcal{C} \quad \mathcal{C}$. Let $w_j, 1 \quad j \quad 2^{N \quad 2K}$ be a representation of the cosets, $\mathcal{C} \quad /\mathcal{C}$; then the basis states of the quantum code, CSS, generated according to the rules discussed in Section 5.10, are

$$s \quad \frac{1}{\overline{\mathcal{C}}} \sum_{c \quad \mathcal{C}} c \quad w_j \; .$$

Let $\mathcal{C}$ be a classical $[N,K,d]$ Reed-Solomon code, such that $d > N/2 \quad 1$ and $j \quad 0$ and $\mathcal{B}$ be a self-dual basis of $GF(2^k)$, over $GF(2)$; then the *quantum Reed-Solomon code*, $\mathcal{Q}^{\mathcal{C}}$, is the CSS code of length, kN, derived from the weakly self-dual code, $\mathcal{B}(\mathcal{C})$.

Now we discuss the dimension of a CSS code constructed based on the weakly self-dual $[N,K,d]$ code, $\mathcal{C}$. If $\dim(\mathcal{C}) \quad K$, then $\dim(\mathcal{C} \quad) \quad N \quad K$ and

$$\dim(\mathcal{Q}^{\mathcal{C}}) \quad K \quad (N \quad K) \quad 2K \quad N.$$

The duality is a symmetric relationship[2]; $\mathcal{C}$ could be considered the original code and $\mathcal{C}$ its dual. The dimension must be a positive integer; therefore,

$$\dim(\mathcal{Q}^{\mathcal{C}}) \quad \begin{cases} 2K \quad N & \text{if } K > N/2, \\ N \quad 2K & \text{if } K < N/2. \end{cases}$$

When we assume that $K < N/2$, a CSS code, $\mathcal{Q}$, based on the weakly self-dual $[N,K,d]$ code, $\mathcal{C}$, is an $[N,N \quad 2K,d]$ code, with

$$d > N/2 \quad 1 \qquad d > (K \quad 1).$$

We conclude that in this case the CSS code, $\mathcal{Q}$, can detect at least, K, errors.

The transformation, $\mathcal{B}$, expands each $\alpha \quad GF(2^k)$ to k bits. It follows that, $\mathcal{Q}^{\mathcal{C}}$, is a $[kN, k(N \quad 2K), d] > K \quad 1$ Reed-Solomon code; thus, it encodes, $k(N \quad 2K)$, qubits as, kN, qubits (see Figure 5.21b), and it can detect at least, K, errors.

Let ψ_j be a codeword of the quantum Reed-Solomon code, $\mathcal{Q}^{\mathcal{C}}$. Then ψ_j can be expressed in $GF(2^k)$ as

$$\psi_j \quad \frac{1}{\mathcal{C}} \sum_{c \quad \mathcal{C}} (c \quad w_j) \; , \quad 1 \quad j \quad 2K \quad N.$$

---

[2]A relation, $\mathcal{R}$, is symmetric if $a\mathcal{R}b \qquad b\mathcal{R}a$.

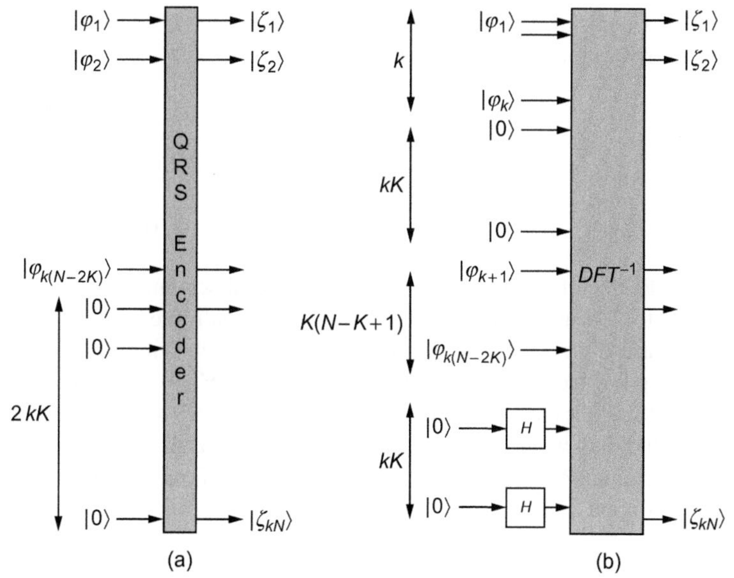

FIGURE 5.21

Encoders for quantum Reed-Solomon codes. (a) The basic scheme for a $[kN, k(2K \quad N), d]$ binary QRS code; it encodes, $k(N \quad 2K)$, qubits in the states, φ_1 , φ_2 ,..., $\varphi_{k(2N \quad K)}$, as kN qubits in the states, ζ_1 , ζ_2 ,..., ζ_{kN} . (b) A QRS encoder based on DFT.

We have $2K \quad N$ distinct values for w_j; thus, we can choose the last, K, components of w_j equal to zero; when we apply the transformation, $\mathcal{B}$, the codeword, ψ_j , can be expressed in binary as

$$\psi_j \quad \frac{1}{C} \sum_{c \quad C} \mathcal{B}(c \quad w_j) \,, \quad 1 \quad j \quad 2K \quad N.$$

The last, kK, components of $\mathcal{B}(w_j)$ are now zero; the spectrum of the codeword, $\psi_j \quad \mathcal{Q}^{\mathcal{C}}$, is

$$DFT(\psi_j) \quad \frac{1}{C} \sum_{c \quad C} DFT(\mathcal{B}(c)) \quad DFT(\mathcal{B}(w_j)) \quad \frac{1}{C} \sum_{c \quad C} c \quad w_j \,.$$

As we have shown earlier, the spectrum of any codeword of an $[N, K, d]$ Reed-Solomon code over $GF(2^k)$ has its first, $N \quad K$, symbols equal to zero; thus, in the binary representation, the code has its first, $k(N \quad K)$, components equal to zero and the next, kK, components, with an arbitrary value,

$$c \quad (0\,0 \ldots 0\, c_{k(N \quad K \quad 1)}\, c_{k(N \quad K \quad 2)} \ldots c_{kN}).$$

Now we apply a Walsh-Hadamard transform to the last kK qubits of the spectrum of ψ_i and leave the first $k(2K \quad N)$ components unchanged:

$$\zeta \quad I_2^{\;k(N \quad K)} \quad H_2^{\;kK} [DFT(\psi_j] \quad I_2^{\;k(N \quad K)} \quad H_2^{\;kK} c \quad w \,.$$

The first $k(N \quad K)$ components of ζ correspond to the DFT of w_j ; the desired result is produced by the circuit in Figure 5.21(b) implementing the transformation

$$\text{Enc} \quad DFT^{\;1} I_2^{\;k(N \quad K)} \quad H_2^{\;kK}$$

applied to the input state

$$\varphi \qquad \varphi_1 \qquad \varphi_k \quad 0 \quad \varphi_{k \quad 1} \qquad \varphi_{N \quad 2K} \quad 0 \qquad 0 \, .$$

Decoding quantum Reed-Solomon codes follows the general scheme for CSS codes discussed in Section 5.10 and illustrated in Figure 5.14; we compute first the syndrome for the bit-flip errors and then the syndrome for the phase-flip errors, as shown in Figure 5.22. Each symbol in $GF(2^k)$ is represented by k qubits, and encoding is done by applying the inverse cyclic DFT to a system of kN qubits, the last kK of them in the state 0 . After computing the bit-flip syndrome, we have to restore the original state; thus, we apply a DFT, and then we rotate the base by applying a Walsh-Hadamard transform, $H^{\;(kN)}$. The dimension of the original Reed-Solomon code, C, is K; therefore, the bit-flip and the phase-flip syndromes require, kK, qubits each.

Example. *The* [7,3,5] *Reed-Solomon code, C.* This code discussed in Section 4.15, with $N \quad 7, K \quad 3$, $d \quad 5$, and the generator polynomial,

$$g(x) \quad (x \quad \beta^0)(x \quad \beta^1)(x \quad \beta^2)(x \quad \beta^3),$$

has a dual, C , a [7,4,4] linear code, with the generator matrix, $g \;\; (x)$:

$$g \;\; (x) \quad \frac{x^6 \quad 1}{g(x)} \quad (x \quad \beta^{\;4})(x \quad \beta^{\;5})(x \quad \beta^{\;6}) \quad (x \quad \beta^3)(x \quad \beta^2)(x \quad \beta^1).$$

The corresponding quantum code, Q^C, is a $[21,3,d]$ code; indeed, $k \quad N \quad 7 \quad 3 \quad 21$, $k(N \quad 2K)$ $3(7 \quad 6) \quad 3$ and $d \quad N \quad K \quad 7 \quad 3 \quad 4$.

5.17 CONCATENATED QUANTUM CODES

Classical concatenated codes discussed in Section 4.18 have two simultaneous properties:

1. the probability of error decreases exponentially at all the rates below channel capacity, and
2. the decoding complexity increases only algebraically with the number of errors corrected by the code and implicitly with the length of a codeword.

Classical concatenated codes have a multilevel structure. The data are encoded first using an $[n,k,d]$ code; then each bit in a block is again encoded using this time an $[n_1,1,d_1]$ code, and the process continues for, say, l, levels. The result is an $[n_1 n_2 \ldots n_{l \quad 1}, k, d_1 d_2 \ldots d_{l \quad 1}]$, code. To decode the concatenated code, we first compute the error syndrome for the first level, the $[n_{l \quad 1}, 1, d_{l \quad 1}]$, code, for all the blocks of $n_{l \quad 1}$ bits in parallel; we then compute the error syndrome for the second level, the $[n_{l \quad 2}, 1, d_{l \quad 2}]$ code for all the blocks of $n_{l \quad 2}$ bits in parallel; and we continue for all the l levels. Thus, we compute

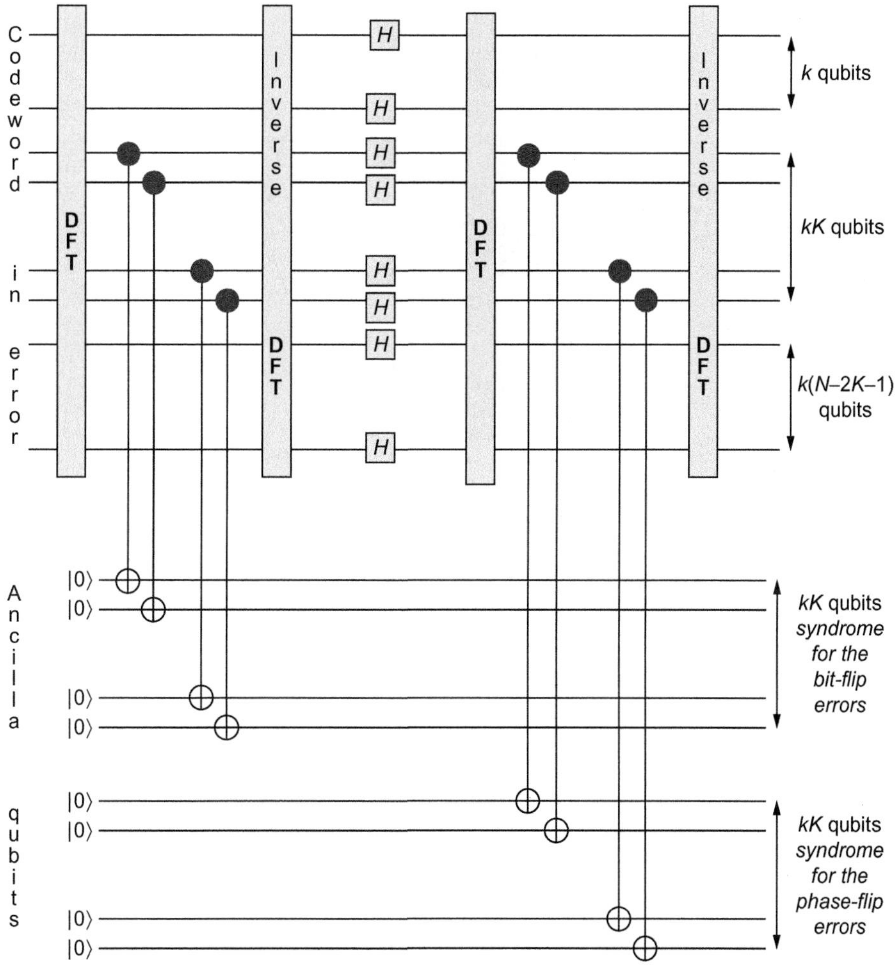

FIGURE 5.22

Quantum circuit to compute the syndrome for bit-flip and phase-flip errors for an $[N,K,d]$ quantum Reed-Solomon code.

the error syndrome for the entire code in a number of steps equal to the sum of the number of steps to compute the syndrome at each level.

Concatenated quantum codes were introduced by Knill and Laflamme in 1996 [244], and their performance and behavior was investigated by several groups including [348]. We consider two codes:

- An outer code. The M-qubit code, $\mathcal{C}^{out}$ E^{out}, D^{out}, with E^{out}, the encoding procedure and, D^{out}, the decoding procedure.
- An inner code. The N-qubit code, $\mathcal{C}^{in}$ E^{in}, D^{in}, with E^{in}, the encoding procedure and, D^{in}, the decoding procedure.

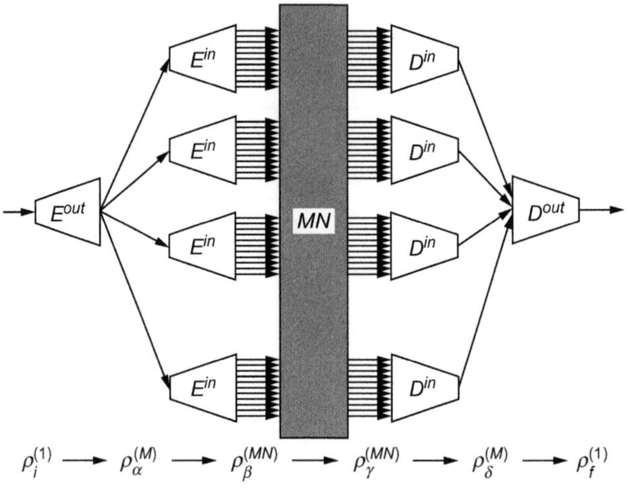

FIGURE 5.23

The encoding and decoding of a concatenated code, C C^{out}, C^{in} . The qubit carrying the information is rst encoded with the procedure, E^{out}, to, M, qubits, then each of the M qubits is encoded again to a block of, N, qubits, this time using the procedure, E^{in}. After passing through a noisy channel, the, MN, qubits are rst decoded block by block using the procedure, D^{in}, and then the decoding procedure, D^{out} is applied. The noise operator, $\mathcal{N}$, may represent a noisy communication channel, or noisy quantum circuits.

The qubit carrying the information is first encoded with the procedure, E^{out}, to, M, qubits, then each of the M qubits is encoded again to a block of N qubits, this time using the procedure, E^{in}. After passing through a noisy channel, the MN qubits are first decoded block by block using the procedure, D^{in}, and then the decoding procedure, D^{out}, is applied. The density matrix of the system evolves as follows:

$$\rho_i^1 \quad \rho_\alpha^{(M)} \quad \rho_\beta^{(MN)} \quad \rho_\gamma^{(MN)} \quad \rho_\delta^{(M)} \quad \rho_f^1,$$

with $\rho_i^{(1)}$, the density matrix of the original qubit; $\rho_\alpha^{(M)}$, the density matrix after the outer encoding procedure, E^{out}; $\rho_\beta^{(MN)}$, the density matrix after the inner encoding procedure, E^{in}; $\rho_\gamma^{(MN)}$, the density matrix before the inner decoding procedure, D^{in}; $\rho_\delta^{(M)}$, the density matrix before the outer decoding procedure, D^{out}; and ρ_f^1, the density matrix of the qubit after the error correction (Figure 5.23).

5.18 QUANTUM CONVOLUTIONAL AND QUANTUM TAIL-BITING CODES

Quantum convolutional codes (QCC) were introduced by Ollivier and Tillich [314]. More recently, Forney, Grassl, and Guha [155] showed how to extend the stabilizer formalism and to construct convolutional and tail-biting codes from classical self-dual codes with rates R $1/n$, over $GF(4)$. The authors show [155] that quantum convolutional codes compare favorably with quantum block codes; they have a better code rate and a lower decoding complexity, thus, a better performance.

This section builds on the concepts related to classical convolutional and tail-biting codes; therefore, a review of Section 4.16 is recommended. The D-transform is used in the literature of quantum convolutional and quantum tail-biting codes [155, 314]. We introduce the D-transform and use it to discuss convolutional encoders based on linear time-invariant systems. Then we discuss a finite-state representation of convolutional codes and review codes over $GF(4)$.

D-transform

We first review a few concepts related to generating functions and the Z-transform. A *generating function* is a formal power series whose coefficients encode information about a sequence, a_n, indexed by the natural numbers,

$$G(a_n;x) \quad a_n x^n.$$

For example,

$$G(n^2;x) \quad \sum_{n \quad 0} n^2 x^2 \quad \frac{x(x \quad 1)}{(1 \quad x)^3}.$$

Informally, a generating function is a "clothesline on which we hang up a sequence of numbers to display."

The *Z-transform* converts a discrete time-domain signal, a sequence of real numbers, a_n, to a complex frequency-domain representation, $A(z)$:

$$A(z) \quad \mathcal{Z}(a_n) \quad \sum_{n \quad 0} a_n z^{\quad n}.$$

The Z-transform allows us to compute the response of linear circuits. A linear circuit is characterized by a transfer function. When the circuit has one input, a_k, and one output, c_k, and the transfer function, also called an *impulse response*, is h_k, the output sequence is

$$c_k \quad \sum_{i>0} h_i a_{k \quad i}.$$

Let $A(z)$, $H(z)$, and $C(z)$ be the Z-transforms of a_k, h_k, and c_k, respectively; then

$$C(z) \quad H(z) \; A(z).$$

The convolution giving the response of the linear circuit in the time domain is transformed to a multiplication in the Z-domain.

Recall that a convolutional code is characterized by a triplet $[n,k,M]$, with n, the length of the output block c, k the length of the input block a, and M the memory of the code. A more general encoding rule for a convolutional code is

$$c_k \quad \sum_{i \;\; 0} G_i a_{k \;\; i},$$

where G_i are $k \quad n$ matrices; for $k \quad 1$, this expression coincides with the encoding rule given in Section 4.16, $c_k \quad \sum_{i \;\; 0} g_i a_{k \;\; i}$, with g_i, the components of the generating vector, g.

The analogy with linear circuits is clear. Now, instead of the indeterminate, $1/z$, used in the case of the Z-transform, we shall use the indeterminate, D. If $a(D), G(D)$, and $c(D)$ are the D-transforms, of the input, the generating matrix/vector, and the output, respectively, then the encoding rule can be written as a product of D-transforms:

$$c(D) \quad a(D) \;\; G(D).$$

The D-transform allows us to describe a convolutional code in a more compact format. We shall see that D can be viewed as a "delay" operator. For example, the expression, $c_k \quad a_k \quad a_{k \;\; 1} \quad a_{k \;\; 2}$, encoding the fact that, c_k, the output at time, k, is a function of the input at times k, $k \quad 1$, and $k \quad 2$, is represented by the polynomial, $(1 \quad D \quad D^2)$; when $c_k \quad a_k \quad a_{k \;\; 2}$, the polynomial is $(1 \quad D^2)$.

Consider a rate, $R \quad 1/2$, convolutional code, with generating vector, $g \quad (11\ 10\ 11)$. In this case, each input bit is encoded as two bits. The first bit is generated as a function of the first bit of each pair of the generating vector, (11 10 11), namely, 111; the second output bit is generated depending on the second bit of each pair, namely, 101. The D-transforms of the two sequences are

$$(111) \quad (1 \quad D \quad D^2) \quad \text{and} \quad (101) \quad (1 \quad D^2).$$

Thus, $G(D) \quad [(1 \quad D \quad D^2), (1 \quad D^2)]$.

Linear Time-Invariant (LTI) Systems Over Finite Fields

The encoders for convolutional codes are LTI systems over a finite field, $GF(q)$, and are characterized by an *impulse response* given by the D-transform, $G(D)$.

LTI systems are implemented using adders, multipliers, and delay/memory elements over $GF(q)$. They are linear systems because the output is a linear function of the input, the adders implement the addition in $GF(q)$, the multipliers implement the multiplication in $GF(q)$, and the delay/memory is also a linear operation. The systems are time-invariant because the relation between input and output is time-invariant; the output of a convolutional code, with generating vector, $g \quad (g_1, g_2, \ldots, g_s)$ at time, k, is $c_k \quad \sum_{i \;\; 0} g_i a_{k \;\; i} \quad k \quad \mathbb{Z}$.

Figure 5.24 shows an encoder for the rate, $R \quad 1/2$, block convolutional code in the previous example. The two output sequences are $(a_k \quad a_{k \;\; 1} \quad a_{k \;\; 2})$ and $(a_k \quad a_{k \;\; 2})$; thus, the encoder is characterized by the *impulse response*, $G(D) \quad [(1 \quad D \quad D^2), (1 \quad D^2)]$.

It is easy to see that this encoder generates a code capable of correcting two errors. Indeed, the distance of the code is $d \quad 5$. To show that, we observe that the minimum weight code is generated by the input sequence $\ldots, 0, 0, 1, 0, 0, \ldots$; in this case the sequence of the output bits is 111 for the upper bit stream and 101 for the lower bit stream. Thus, the minimum weight codeword is 5.

The encoder for a rate, $R \quad 1/n$, convolutional code is characterized by, n, polynomials in, D, one for each of the, n, output bits. The degree of each polynomial of a code, with a generating vector, $g \quad (g_0, g_1, \ldots, g_s)$, can be at most, s.

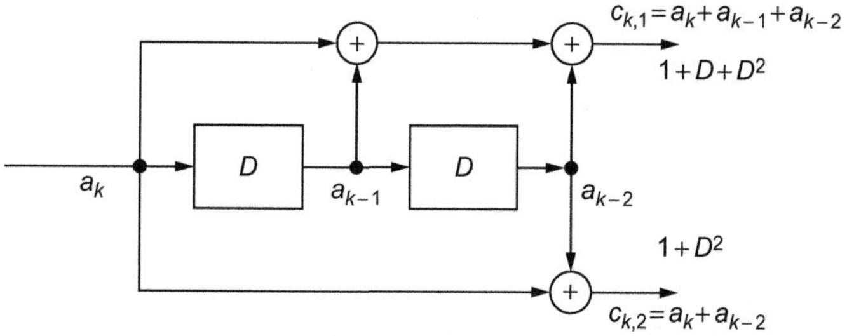

FIGURE 5.24

An encoder for a rate, $R \quad 1/2$, block convolutional code. The encoder generates two output bits for each input bit. The LTI system has two delay/memory elements and three adders. The generating vector is $g \quad (11.10.11)$. The upper output bit stream is generated according to the sequence, 111 (the rst bit of each pair of bits in g); thus, $c_{k,1} \quad (a_k \quad a_{k \ 1} \quad a_{k \ 2})$ and the corresponding polynomial is $(1 \quad D \quad D^2)$. The lower output bit stream is generated according to the sequence, 101 (the second bit of each pair of bits in g); thus, $c_{k,2} \quad (a_k \quad a_{k \ 2})$, and the corresponding polynomial is $(1 \quad D^2)$.

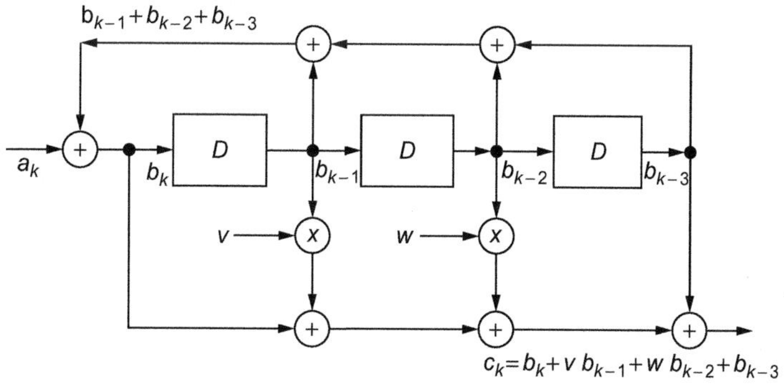

FIGURE 5.25

An encoder with a rational impulse response over $GF(2)$. The impulse response is in nite and the realization of the LTI system involves feedback. There are three memory elements (thus, the system has 3^2 states), six adders, and two multipliers with $v, w \quad GF(2)$.

Consider now an LTI system with, μ, delay/memory elements over, $GF(q)$, and characterize the state of the system by the contents of the, μ, memory elements. Then the system can be in one of, μ^q, states. It is easy to see that the impulse response, $G(D)$, must be a rational function. Indeed, the *autonomous* behavior corresponding to zero input must be periodic after time zero if the impulse response at time zero is $g_0(D) \quad 1$. Thus, $G(D)$ must be periodic—i.e., a rational function, $G(D) \quad G_1(D)/G_2(D)$.

Figure 5.25 shows an encoder with a rational impulse response over $GF(2)$. In this case, the output and the feedback are

$$G_1(D) \quad 1 \quad vD \quad wD^2 \quad D^3 \quad \text{and} \quad G_2(D) \quad 1 \quad D \quad D^2 \quad D^{\ 3}.$$

The impulse response is

$$G(D) \quad \frac{G_1(D)}{G_2(D)} \quad \frac{1 \quad vD \quad wD^2 \quad D^3}{1 \quad D \quad D^2 \quad D \ 3}.$$

Finite-State Representation of Convolutional Codes

The encoder of a convolutional code over $GF(q)$ is a finite-state machine. A state is labeled by the symbols stored by the memory elements; thus, a system with M memory elements, has M^q states. The transitions between states are labeled by the output streams. For example, the encoder for a rate, $R \quad 1/2$, block convolutional code over, $GF(2)$, in Figure 5.24 has, 2^2, states labeled, 00, 01, 10, 11, as shown in Figure 5.26. The trellis diagram in Figure 5.26(b) conveys the same information as the traditional state transition diagram in 5.26(a) and, in addition, it allows us to follow the path of an encoded sequence.

Linear Codes Over *GF*(4)

In this section we lay the groundwork to extend the stabilizer formalism and to construct convolutional and tail-biting codes from classical self-dual codes, with rates $R \quad 1/n$, over $GF(4)$.

Recall that $\mathcal{P}$, the set of Pauli operators, is not an Abelian group, but the set, Π, of equivalence classes of Pauli matrices, also called the *projective Pauli group*, is an Abelian group. The multiplication tables of $\mathcal{P}$ and Π are

	I	X	Y	Z
I	I	X	Y	Z
X	X	I	iZ	iY
Y	Y	iZ	I	iX
Z	Z	iY	iX	I

and

	[I]	[X]	[Y]	[Z]
[I]	[I]	[X]	[Y]	[Z]
[X]	[X]	[I]	[Z]	[Y]
[Y]	[Y]	[Z]	[I]	[X]
[Z]	[Z]	[Y]	[X]	[I]

.

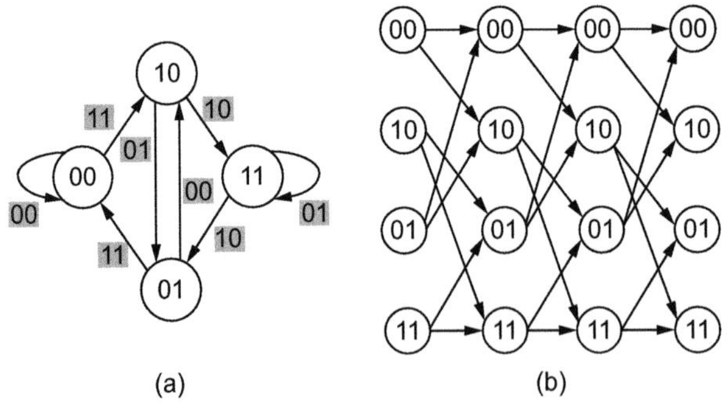

FIGURE 5.26

The states and transitions for the rate, $R \quad 1/2$, block convolutional code over, $GF(2)$, in Figure 5.24. A state is characterized by the contents of the two memory elements and a transition is labeled by the two output symbols. (a) State transition diagram. (b) The trellis diagram of the encoder; the labels on the transition are omitted.

We now consider the additive group, with four elements, $GF(4) \quad 0, 1, \omega, \omega$. The addition and multiplication tables of $GF(4)$ are, respectively,

	0	1	ω	ω
0	0	1	ω	ω
1	1	0	ω	ω
ω	ω	ω	0	1
ω	ω	ω	1	0

and

	0	1	ω	ω
0	0	0	0	0
1	0	1	ω	ω
ω	0	ω	ω	1
ω	0	ω	1	ω

.

Group isomorphisms allow us to discover and express properties of the elements of a group relative to another group. Recall that the elements of two isomorphic groups are in a one-to-one correspondence and respect the group operations. More precisely, two groups, $G \quad (S, \quad)$ and $G \quad (S, \quad)$, are isomorphic if there is a one-to-one mapping, $f : G \quad G$, such that

$$x_i, x_j \quad G f(x_i \quad x_j) \quad f(x_i) \quad f(x_j).$$

The projective Pauli group, Π, is isomorphic, with the additive group, $GF(4)$, $L : \Pi \quad GF(4)$; this implies that

$$L([A]) \quad L([B]) \quad L([A \quad B]), \qquad ([A],[B]) \quad [\Pi].$$

The correspondence between the respective elements of Π and $GF(4)$ is summarized in the following table:

Π	$GF(4)$
$[I]$	0
$[X]$	ω
$[Y]$	1
$[Z]$	ω

Let us now consider tensor products of, n, Pauli operators,

$$A \quad A_1 \quad A_2 \qquad A_i \qquad A_n \quad \text{and} \quad B \quad B_1 \quad B_2 \qquad B_i \qquad B_n,$$

with $A_i, B_i \quad \mathrm{I}, X, Y, Z$.

The mapping, L, of such tensor products induces a vector space, $\mathcal{V} \quad a, b, \ldots$, over, $GF(4)$, where

$$a \quad L(A), \quad b \quad L(B), \quad a \quad (a_1, a_2, \ldots, a_n), \quad b \quad (b_1, b_2, \ldots, b_n), \quad \text{and} \quad a_i \quad L(A_i), b_j \quad L(B_j).$$

It is easy to show that

$$L([A]) \quad L([B]) \quad L([A \quad B])$$

for any $[A], [B] \quad \mathcal{P}$. For example, consider $n \quad 1$, $A_1 \quad X$, and $B_1 \quad Y$; we have to show that

$$L([X]) \quad L([Y]) \quad \omega \quad 1 \quad \text{and} \quad L([X] \quad [Y]) \quad \omega.$$

From the correspondence table, it follows that

$$L([X]) \quad \omega,\ L([Y]) \quad 1, \qquad L([X]) \quad L([Y]) \quad \omega \quad 1 \quad \omega.$$

The last equality follows from the addition table in $GF(4)$; then

$$[X] \quad [Y] \quad [Z],\ L([Z]) \quad \omega \qquad L([X] \quad [Y]) \quad L([Z]) \quad \omega.$$

Now we introduce the Hermitian inner product and the trace inner product in, $\mathcal{V}$, and use them to study the commutativity/anticommutativity properties of elements of the Pauli group. The *Hermitian inner product*, a,b , and the *trace inner product*, tr a,b , of two elements, $a,b \quad GF(4)$, are defined as

,	0	1	ω	ω
0	0	0	0	0
1	0	1	ω	ω
ω	0	ω	1	ω
ω	0	ω	ω	1

and

tr ,	0	1	ω	ω
0	0	0	0	0
1	0	0	1	1
ω	0	1	0	1
ω	0	1	1	0

.

Note that $a,b, \quad a^{\dagger}b$, with $a^{\dagger}$, the adjoint of a. These definitions can be extended to $a,b \quad \mathcal{V}$.

The commutativity properties of matrices in the Pauli group, $\mathcal{P}$, can now be established based on the *Hermitian inner product* and the *trace inner product* of two elements, $a,b \quad \mathcal{V}$.

Thus, $A,B \quad \Pi$, and we have

$$A \quad B \quad (\quad 1)^{\mathrm{tr}\ L(A),L(B)} \ B \quad A.$$

We conclude that *two matrices,* $A,B \quad \Pi$, *commute when* tr $L(A),L(B) \quad 0$ *and anticommute when* tr $L(A),L(B) \quad 1$.

The Hermitian inner product in $\mathcal{V}$ is zero when the trace inner product is zero; indeed, if $(a,b) \quad \mathcal{V}$ and $\alpha \quad GF(4)$, then

$$\mathrm{tr}\ \alpha \quad a,b \quad \mathrm{tr}[\alpha \quad a,b\ ] \quad \alpha \quad \mathrm{tr}\ a,b\ .$$

It follows that the two matrices $A,B \quad \Pi$ commute if and only if the Hermitian inner product of their images in $\mathcal{V}$ is equal to zero.

To prove these properties, we have to consider all pairs of elements, $A,B \quad G$, and show that the property holds; for example, consider the case $n \quad 1$, $A_1 \quad X$, and $B_1 \quad Y$:

$$A_1 \quad B_1 \quad X \quad Y \quad iZ \quad \text{and} \quad B_1 \quad A_1 \quad Y \quad X \qquad iZ.$$

From the correspondence table, we have

$$L(X) \quad \omega,\ L(Y) \quad 1.$$

From the definition of the trace inner product, tr $\omega,1 \qquad 1$. We have thus verified that X and Y anticommute by showing that the trace inner product tr $L(X),L(Y) \qquad 1$. These results allow us to extend the stabilizer formalism and define quantum convolutional and quantum tail-biting codes.

Stabilizer Codes from Codes over *GF*(4)

Now it is a good time to review the basic concepts regarding stabilizer codes discussed in Section 5.12. Consider an $[n,k]$ code in $(\mathcal{H}_2)\ ^n$ stabilized by the set, $S\quad \mathbf{M}_1,\mathbf{M}_2,\ldots,\mathbf{M}_{n\ \ k}$; the $(n\quad k)$ generators $\mathbf{M}_i$, of the stabilizer, $\mathcal{S}$, are tensor products of, n, Pauli matrices. Each $\mathbf{M}_i$ has two eigenvalues, 1, and has two corresponding eigenspaces of dimension, $2^{n\ \ 1}$. There are, $(n\quad k)$, generators, thus, there are, $2^{n\ \ k}$, orthogonal eigenspaces, each of dimension, 2^k. The quantum code is the 2^k-dimensional eigenspace, such that all $(n\quad k)$ generators have eigenvalues equal to 1.

Table 5.11 summarizes the correspondence between stabilizer codes in the Hilbert space, $(\mathcal{H}_2)\ ^n$, and linear codes in, $\mathcal{V}$, induced by the mapping, L, and its inverse, $L\ ^1$, introduced in [155].

Error correction for a stabilizer code is done as follows. We measure $(n\quad k)$ generators, $\mathbf{M}_j$, simultaneously and map the eigenvalues of the error operator, $\mathbf{E}$, applied to the current state, ψ , to a binary $(n\quad k)$ tuple, $s\quad s_1,s_2,\ldots,s_{n\ \ k}$, called the syndrome. Bit s_i of the syndrome depends only on the eigenvalue of $\mathbf{E}$, (1) 0, (1) 1.

We can exploit the mapping discussed in this section to construct an $[n,k,d]$ stabilizer code from classical codes over, $GF(4)$, when $(n\quad k)$ is even. Given a self-dual, $[n,(n\quad k)/2,d]$, linear code, $\mathcal{C}$, and its $[n,(n\quad k)/2,d]$ dual, $\mathcal{C}$, the resulting $[n,k,d]$ stabilizer code will be able to correct, $(d\quad 1)/2$

Table 5.11 The Correspondence Between Stabilizer Codes in the Hilbert Space, $\mathcal{H}_2^n$, and Linear Codes in, $\mathcal{V}$, Induced by the Mapping, L

$(\mathcal{H}_2)^{\otimes n}$	$\xrightarrow{L}$	$\mathcal{V}$
$\mathbf{M}_i\quad \mathcal{P}, 1\quad i\quad (n\quad k)$ M_i tensor product of n Pauli operators generator of the stabilizer group $\mathcal{S}$		$m_i\quad L(M_i), 1\quad i\quad (n\quad k)$ m_i vector with elements from $GF(4)$ generator in $\mathcal{V}$
$\mathbf{M}_i\ \mathbf{M}_j\quad \mathbf{M}_j\ \mathbf{M}_i$ ($\mathbf{M}_i$ and $\mathbf{M}_j$ commute)		tr $m_i,m_j\quad 0$ (Hermitian inner product $m_i,m_j\quad 0$)
$S\quad \mathbf{M}_1,\mathbf{M}_2,\ldots,\mathbf{M}_{n\ \ k}$ $S\quad \Pi$ is an Abelian subgroup		$\Sigma\quad m_1,m_2,\ldots,m_{n\ \ k}$ $\Sigma\quad \mathcal{V}$ generates a classical code
$[n,k]$ stabilizer code subspace of $\mathcal{H}_2^n$ stabilized by $\mathcal{S}$		$L(S)$ set of linear combinations ωm_i and ωm_i $[n,(n\quad k)/2]$ block linear code over $GF(4)$ $m_i,m_j\quad 0\qquad L(S)$ is self-dual under the Hermitian inner product in $\mathcal{V}$
$N(S)$ normalizer group of stabilizer $\mathcal{S}$	$L\ ^1$	$L(S)$ classical code, the dual of $L(S)$ $[n,(n\quad k)/2]$ block linear code over $GF(4)$
error pattern $\mathbf{E}\quad \sigma_a\ ^n, a\quad I,x,y,z$ syndrome $s\quad s_1,s_2,\ldots,s_{n\ \ k}$		$e\quad L(E)$ syndrome $\gamma\quad \gamma_1,\gamma_2,\ldots,\gamma_{n\ \ k}, \gamma_j\quad L(s_j)$ $\gamma_j\quad$ tr e,m_j

errors. For example, the $[5,2,4]$ doubly-extended Reed-Solomon code generated by

$$\begin{array}{ccccc} 0 & \omega & \omega & \omega & \omega \\ \omega & 0 & \omega & \omega & \omega \end{array}$$

could be used to construct a $[5,1,3]$ single–error-correcting stabilizer code.

Convolutional Stabilizer Codes

A convolutional quantum code, $\mathcal{Q}$, with minimum distance, d, can be constructed from a classical convolutional, $[n,(n \quad k)/2,d]$, linear code, $\mathcal{C}$, and its dual, $\mathcal{C}$, an $[n,(n \quad k)/2,d]$. The rates of $\mathcal{C}, \mathcal{C}$, and $\mathcal{Q}$ are, respectively, (see Section 4.5)

$$R_{\mathcal{C}} \quad \frac{n \quad k}{2n}, \quad R_{\mathcal{C}} \quad \frac{n \quad k}{2n}, \quad \text{and} \quad R_{\mathcal{Q}} \quad \frac{k}{n}.$$

Consider for example $\mathcal{C}$, a rate $R \quad 1/3$, convolutional code over $GF(4)$, with the D-transform

$$m(D) \quad (1 \quad D, 1 \quad \omega D, 1 \quad \omega D).$$

A generator sequence is $m_j \quad (\ldots\ 000\ 111\ 1\omega\omega\ 000\ \ldots)$, and the generators are

$$\begin{array}{l} \ldots\ 000\ 111\ 1\omega\omega\ 000\ 000\ 000\ \ldots, \\ \ldots\ 000\ 000\ 111\ 1\omega\omega\ 000\ 000\ \ldots, \\ \ldots\ 000\ 000\ 000\ 111\ 1\omega\omega\ 000\ \ldots, \\ \ldots\ 000\ 000\ 000\ 000\ 111\ 1\omega\omega\ \ldots. \end{array}$$

$\mathcal{C}$ is self-dual; the generators are orthogonal, $m_j \quad m_j \quad 0$, according to the multiplication table in $GF(4)$. Then, $\mathcal{C}$ is a rate, $R \quad 2/3$, convolutional code generated by

$$m\ (D) \quad (\omega, \omega, 1).$$

It follows that $m_j \quad (\ldots\ 000\ 111\ \omega\omega 1\ 000\ \ldots)$. The generators of $\mathcal{C}$ are

$$\begin{array}{l} \ldots\ 000\ 111\ \omega\omega 1\ 000\ 000\ 000\ \ldots, \\ \ldots\ 000\ 000\ 111\ \omega\omega 1\ 000\ 000\ \ldots, \\ \ldots\ 000\ 000\ 000\ 111\ \omega\omega 1\ 000\ \ldots, \\ \ldots\ 000\ 000\ 000\ 000\ 111\ \omega\omega 1\ \ldots. \end{array}$$

A generator, M_j, of the stabilizer, $\mathcal{S}$, can be determined using the mapping between $GF(4)$ and Π:

$$M \quad F^{\ 1}[\ldots\ 000\ 111\ 1\omega\omega\ 000\ \ldots] \quad (\ldots\ \mathrm{III}\ XXX\ XZY\ \mathrm{III}\ \ldots].$$

All generators of $\mathcal{S}$ can be obtained by a cyclic shift of M to the right with an integral number of blocks

$$\begin{array}{l} \ldots \text{ III } XXX\ XZY \text{ III III III } \ldots, \\ \ldots \text{ III III } XXX\ XZY \text{ III III } \ldots, \\ \ldots \text{ III III III } XXX\ XZY \text{ III } \ldots, \\ \ldots \text{ III III III III } XXX\ XZY \ \ldots. \end{array}$$

Another quantum convolutional code is described in [314]. The generators of the stabilizer group are

$$\begin{array}{llllllllll} M_0 & X & Z & I & I & I & I & I & I & , \\ M_1 & Z & X & X & Z & I & I & I & I & , \\ M_2 & I & Z & X & X & Z & I & I & I & , \\ M_3 & I & I & Z & X & X & Z & I & I & , \\ M_4 & I & I & I & Z & X & X & Z & I & , \\ \vdots & & & & & & & & & \\ M_{4i\ j} & \mathrm{I}^{\ 5i} & M_j,\ 0 < i,\ 1 & j & 4, & & & & & \\ \vdots & & & & & & & & & \\ M & \ldots & I & I & I & I & Z & X. & & \end{array}$$

The code so defined has a particular structure: all generators, other than M_0 and M , can be grouped together to sets of four; each set acts on a set of seven consecutive qubits and each set has a fixed ovelap of two qubits with the set immediately before and after it.

Block Codes and Tail-Biting Codes

Two methods to make a block code out of a convolutional code, termination and tail-biting, are discussed in [155].

When using termination, we truncate the set of generators, Σ, of, $\mathcal{C}$, to create a block code, $\mathcal{B}$. For example, if we start with the R 2/3 code, $\mathcal{C}$, from the previous example, we can construct a [9,5,3] terminated block code, $\mathcal{B}$, using the generator matrix

$$G_t \quad \begin{pmatrix} \omega\omega 1 & 000 & 000 \\ 111 & 1\omega\omega & 000 \\ 000 & \omega\omega 1 & 000 \\ 000 & 111 & 1\omega\omega \\ 000 & 000 & \omega\omega 1 \end{pmatrix}.$$

The [9,4] orthogonal block code $\mathcal{B}$ is generated by the truncated generators of $\mathcal{C}$, which are not self-orthogonal; $\mathcal{B}$ is not self-orthogonal. The generator matrix is

$$G_t \quad \begin{pmatrix} 1\omega\omega & 000 & 000 \\ 111 & 1\omega\omega & 000 \\ 000 & 111 & 1\omega\omega \\ 000 & 000 & 111 \end{pmatrix}.$$

We can also construct a block code $\mathcal{B}$ from a concatenated code $\mathcal{C}$ by tail-biting by limiting the generators to those whose "starting time" lies in a given interval. We continue with the example from [154] and use the rate $R \quad 2/3$ code $\mathcal{C}$. The generator matrix in this case is

$$G_{tb} \quad \begin{pmatrix} \omega\omega 1 & 000 & 000 \\ 111 & 1\omega\omega & 000 \\ 000 & \omega\omega 1 & 000 \\ 000 & 111 & 1\omega\omega \\ 000 & 000 & \omega\omega 1 \\ 1\omega\omega & 000 & 111 \end{pmatrix}.$$

5.19 CORRECTION OF TIME-CORRELATED QUANTUM ERRORS

The quantum threshold theorem mentioned in Section 1.14 is based on the assumption that the quantum noise is Markovian and establishes the foundation for fault-tolerant computations in environments where spatial and temporal correlations decay exponentially. The "classical" theory of quantum error-correcting codes is based on the *independent qubit decoherence* model, which makes several assumptions:

- Errors are uncorrelated in space. An error affecting qubit, i, of an n-qubit register does not affect other qubits of the register. An error affecting one qubit is equally likely to be due to an interaction with the environment described by X, Y, or Z operators.
- Errors are uncorrelated in time. An error affecting qubit, i, at time, t, and corrected at time, $t \quad \Delta$, will have no further effect either on qubit i or on other qubits of the register.
- The error rate is constant. Moreover, whenever the probability of a qubit to be in error, ϵ, is very small, $\epsilon \quad 10^{\;4}$, then the probability of t or more errors is very small indeed, $\mathcal{O}(\epsilon^t)$, and we ignore such a case.

Different physical implementations of quantum devices reveal that the interactions of the qubits with the environment are more complex and force us to consider spatially as well as time-correlated errors. For example, if the qubits of an n-qubit register are confined to a three-dimensional structure, an error affecting one qubit will propagate to the qubits in a volume centered around the qubit in error.

Two major questions pertinent to error correction of correlated errors need to be addressed: First, can the quantum threshold theorem be extended for non-Markovian noise? Second, what are the practical means to deal with correlated errors? In this section we only address the second question and restrict our discussion on how to deal with time-correlated errors.

Correlated quantum errors could cause two or more errors to occur during an error correction cycle. There are two obvious approaches to deal with this problem:

1. Design a code capable of correcting, $(e_u \quad e_c)$, errors, where e_u is the number of uncorrelated errors and e_c is the expected number of time-correlated errors.
2. Use the classical information regarding past errors and quantum error-correcting codes capable of correcting a single error.

Spatially correlated errors and means to deal with the spatial noise are analyzed in recent papers [7, 100, 241]. Time-correlated errors occur after an error correction cycle and could affect the qubit restored to its original state after an error, or a different qubit. Figure 5.27 illustrates the evolution in time of two qubits, *i* and *j*, for two error-correction cycles and the effect of time-correlation. We expect that the correlations in the quantum system decay in time, and the latest error will influence the system most. A recent paper on the decoherence in a spin-boson model assumes that the qubits are perfect (thus, the only errors are due to dephasing) and considers a linear coupling to an ohmic bath [310];

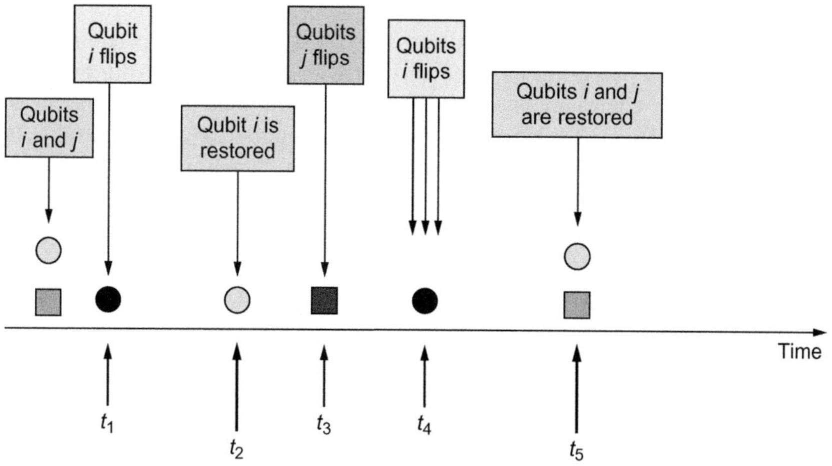

FIGURE 5.27

Time-correlated quantum errors. We follow the evolution in time of two qubits, *i* and *j*, for two error-correction cycles. The rst error-correction cycle starts at time, t_2, and the second at time, t_5. At time t_1 qubit, *i*, is affected by decoherence and flips; at time t_2 it is flipped back to its original state during the rst error-correction step; at time t_3 qubit, *j*, is affected by decoherence and is flipped; at time t_4, the correlation effect discussed in this section affects qubit *i* and flips it to an error state. The second error correction occurs at time, t_5, and qubits *i* and *j* are corrected.

then the Hamiltonian of the model can be written as

$$H \quad \frac{v_b}{2} \int dx[\partial_x \phi(x)]^2 \quad [\Pi(x)]^2 \quad \frac{\lambda}{2} \sum_n \partial_x \phi(n) \sigma_n^z,$$

with ϕ and $\Pi \quad \partial_x \theta$, canonic conjugate variables, and σ_n^z, the Pauli operators acting in the Hilbert space of the qubits. v_b is the velocity of the bosonic excitations, and the units are such that $\hbar \quad 1$. The exact time evolution between gates of a qubit can be expressed as the product of two vertex operators of the free bosonic theory,

$$U_n(t,0) \quad \exp[i\sqrt{\frac{\pi}{2}}\lambda(\theta(n,t) \quad \theta(n,0))\sigma_n^z],$$

with λ, the qubits' bosonic coupling strength; this model is then applied to the three-qubit Steane's code to calculate the probability of having errors in quantum error-correction cycles starting at times, t_1 and t_2. The conclusion of [310] is that the probability of errors consists of two terms; the first is the uncorrelated probability and the second is the contribution due to correlation between errors in different cycles (Δ is the error-correcting cycle):

$$P \quad \left(\frac{\epsilon}{2}\right)^2 \quad \frac{\lambda^4 \Delta^4}{8(t_1 \quad t_2)^4}.$$

This implies that the correlations in the quantum system decay in time, and the latest error will influence the system most.

An important question for the error model discussed in [310] is the independence of phase- and bit-flip errors—namely, can a phase-flip error be correlated to a posterior bit-flip error? The answer will depend on the particular environment one tries to model. For solid-state qubits, these errors are caused by different bath modes, which may not efficiently communicate with each other (such as for phonon-induced tunneling and bias fluctuations in quantum dot charge qubits, which cause bit- and phase-flip errors, respectively). In this case, errors of a different nature are expected to be very weakly correlated in time and space while correlated errors are mostly of the same kind.

An algorithm based on the stabilizer formalism for perfect as well as nonperfect quantum codes is discussed in [278]. The quantum code has a codeword consisting of n qubits and uses, k, ancilla qubits for syndrome measurement; the algorithm makes the standard assumptions for QEC:

- Quantum gates are perfect and operate much faster than the characteristic response of the environment.
- The states of the computer can be prepared with no errors.

It is also assumed that (1) there is no spatial-correlation of errors, and a qubit in error does not influence its neighbors; and (2) in each error-correcting cycle, in addition to a new error, $\mathbf{E}_a$, that occurs with a constant probability, ε_a, a time-correlated error, $\mathbf{E}_b$, may occur with the probability, $\varepsilon_b(t)$. As correlations decay in time, the qubit affected by error during the previous error correction cycle, has the highest probability to relapse.

QEC requires *nondemolition measurements of the error syndrome* in order to preserve the state of the physical qubits. In other words, a measurement of the probe (the ancilla qubits) should not influence the free motion of the signal system. The syndrome has to identify precisely the qubit(s) in error and the type of error(s). Thus, the qubits of the syndrome are either in the state with the density matrix, $0 \quad 0$, or in the state with the density matrix, $1 \quad 1$, which represent classical information. A quantum nondemolition measurement is used to construct the error syndrome, $\Sigma_{current}$. After examining the syndrome $\Sigma_{current}$, an error-correcting algorithm should be able to decide if:

1. no error has occurred, in which case no action should be taken;
2. one "new" error, $\mathbf{E}_a$, has occurred, in which case we apply the corresponding Pauli transformation to the qubit in error; or
3. two or more errors have occurred, in which case there are two distinct possibilities: (a) We have a "new" error as well as an "old" one, the time-correlated error. (b) There are two or more "new" errors. A quantum error-correcting code capable of correcting a single error will fail in both cases.

It is rather hard to distinguish the last two possibilities. For perfect codes, the syndrome S_{ab} corresponding to two errors, $\mathbf{E}_a$ and $\mathbf{E}_b$, is always identical to the syndrome, S_c, for some single error, $\mathbf{E}_c$. Thus, for perfect quantum codes the stabilizer formalism does not allow us to distinguish two errors from a single one; for a nonperfect code it is sometimes possible to distinguish the two syndromes and then using the knowledge regarding the time-correlated error it may be possible to correct both the "old" and the "new" error.

As we may recall from classical error-correcting codes, burst error-correcting codes are designed to deal with disturbances that affect a group of consecutive bits. Similar effects are noticeable in quantum computing and quantum communication when the correlation length of the source of decoherence is smaller than the separation in space of the qubits of a quantum register. In such cases, the independent qubit decoherence model is no longer valid; the source of decoherence affects an entire group of qubits (e.g., those within a volume of radius, r, in case of a three-dimensional structure). One method to deal with spatially correlated quantum errors is to consider burst-error-correcting codes.

The quantum error-correcting codes discussed in Sections 5.1 to 5.19 are subspaces of a Hilbert space and are collectively known as *subspace codes*; a code, $\mathcal{C}$, is a subspace of the Hilbert space, $\mathcal{H}$, defined with the aid of the projector, $\Pi_{\mathcal{C}}$; the correction of error, $\mathbf{E}$, produces the desired results only if, $\Pi_{\mathcal{C}} \mathbf{E} \Pi_{\mathcal{C}} \quad \alpha_{\mathcal{C}} \Pi_{\mathcal{C}}$, with $\alpha_{\mathcal{C}}$, a scalar. The detection of the error consists of the following steps:

1. Encode the information as a state, $\varphi \quad \mathcal{C}$.
2. Allow the error to occur.
3. Carry out a measurement of the resulting the state, $\mathbf{E} \ \varphi$:
   - If the outcome is $\Pi_{\mathcal{C}} \mathbf{E} \ \varphi$, then the state is in $\mathcal{C}$, and accept.
   - If the outcome is $(\mathbf{I} \quad \Pi_{\mathcal{C}}) \mathbf{E} \ \varphi$, then the state is in the orthogonal complement, and reject.

We now switch our focus to *subsystem codes*; when using them, we limit the quantum information restoration process of traditional subspace codes and encode the information in a noiseless subsystem; the subsystem codes are said to be based on a *passive* QEC strategy, while subspace codes are based on an *active* QEC strategy.

5.20 QUANTUM ERROR-CORRECTING CODES AS SUBSYSTEMS

The search for systematic ways to bypass decoherence was initiated as a study of the effects of pure dephasing; *decoherence-free subspaces* capture the behavior of systems decoupled from the environment and with completely unitary evolution; for example, two qubits possessing identical interactions with the environment that do not decohere are discussed in [317]. Noiseless quantum codes were introduced in [466], and noiseless encoding in quantum dot arrays was reported in [467].

Emmanuel Knill, Raymond Laflamme, and Lorenza Viola introduced *subsystem codes* in a paper on the theory of QEC for general noise [249]; a few years later David Kribs, Raymond Laflamme, and David Poulin developed operator quantum error-correcting (OQEC) codes [259], and Poulin proposed a stabilizer formalism for OQEC codes [337]. David Bacon introduced a class of subsystem codes referred to as Bacon-Shor codes [17, 18]. In this section and the next ones we follow the notations in the original papers to describe the evolution of ideas in the field of subsystem codes, and we drop the bold notations for operators.

The basic idea of subsystem codes is to use algebraic methods to classify quantum errors and, implicitly, to describe quantum error-correcting codes as tensor products of subspaces, or more accurately, as subsystems where the useful information is confined to one subsystem [249]. When the system consists of a number of qubits, each qubit is a subsystem; the approach of subsystem QEC is more general. It describes physical systems that cannot be canonically decomposed into qubits; for example, when the system consists of a number of photon modes, each photon mode is a subsystem.

Subspace Versus Subsystem Quantum Error Correction

A Hilbert space, $\mathcal{H}$, can be decomposed as the direct sum of two subspaces, $\mathcal{H} \quad \mathcal{H}^C \bigoplus \mathcal{H}^D$, or as the tensor product of two subspaces, $\mathcal{H} \quad \mathcal{H}^C \quad \mathcal{H}^D$; in case of a direct sum, $\dim(\mathcal{H}) \quad \dim(\mathcal{H}^C) \quad \dim(\mathcal{H}^D)$, while, $\dim(\mathcal{H}) \quad \dim(\mathcal{H}^C) \quad \dim(\mathcal{H}^D)$, for the tensor product. The direct sum and the tensor product obey a distributive law; thus, in the general case, we can *decompose, $\mathcal{H}$, as a direct sum of multiple tensor products of Hilbert spaces.* It follows immediately that a decomposition of the Hilbert space, $\mathcal{H}$, as a subspace, $\mathcal{H}^E$, and a perpendicular subspace, $(\mathcal{H}^E)$, is possible, with $(\mathcal{H}^E)$ expressed as a tensor product of two other subspaces, $\mathcal{H}^C$ and $\mathcal{H}^D$:

$$\mathcal{H} \quad (\mathcal{H}^E) \quad \bigoplus \mathcal{H}^E \quad \text{or} \quad \mathcal{H} \quad (\mathcal{H}^C \quad \mathcal{H}^D) \bigoplus \mathcal{H}^E.$$

The *subsystem encoding* for this decomposition requires the preparation of the quantum state,

$$\rho \quad (\rho^C \quad \rho^D) \quad 0^E,$$

and encoding the information into subsystem, $\mathcal{C}$, described in the Hilbert space, $\mathcal{H}^C$, for a fixed encoding into the subsystem, $\mathcal{D}$, described in the Hilbert space, $\mathcal{H}^D$. In this expression, ρ^C is the density matrix of the encoded quantum information, ρ^D is the density matrix of the arbitrary quantum information encoded into the subsystem $\mathcal{D}$, and 0^E is the all-zero matrix on the subspace $\mathcal{H}^E$.

$\mathcal{D}$ is referred to as a *gauge subsystem* and E as a *logical subsystem.* Regardless of the nontrivial transformation applied to gauge subsystem $\mathcal{D}$, the information encoded into subsystem $\mathcal{C}$ is not affected; *the QEC is restricted to the restoration of the information encoded into subsystem $\mathcal{C}$.*

The statement that subsystem QEC is the most general expression for QEC is justified: when $\mathcal{H}^D \quad 0$, we recognize the case of subspace codes; the subsystem codes correspond to the case, $\mathcal{H}^D \quad 0$, when we only need to correct errors modulo the subsystem structure. It follows that *the properties of subsystem codes are not fundamentally different than those of the subspace codes* we have analyzed extensively. As expected, the subsystem codes are more efficient than their subspace counterparts because (1) they require the computation of fewer syndrome bits to identify the error(s) and (2) the quantum circuits to implement the error correction are simpler.

Noiseless Subsystems

Consider a system, R, described in an N-dimensional Hilbert space, $\mathcal{H}_N$, with an environment, B, and let B_i be the linearly independent environment operators; then the interaction Hamiltonian of R and the environment B can be expressed as

$$J \quad \sum_i J_i \quad B_i.$$

When $\mathrm{tr}(J_i) \quad 0$ for all i, we are able to retain the internal interaction of the system R with the environment B and remove the internal evolution of B. Define the *noise strength parameter*, λ, to be the maximum eigenvalue of $\overline{JJ^\dagger}$.

Let $\mathcal{J}_1$ be the set of operators, $\mathcal{J}_1 \quad \mathrm{I}, J_1, J_2, \ldots$, with I the identity operator. A rigorous definition of interaction algebra, is given in Section 5.22; for now we only mention that the interaction algebra, $\mathcal{J}$ consists of linear combinations of products of operators in $\mathcal{J}_1$; for example, $\mathcal{J}_d \quad (\mathcal{J}_1)^d$ is the linear span of products of d or fewer operators from $\mathcal{J}_1$. When the system R consists of, n, qubits, the dimension of the Hilbert space is $N \quad 2^n$ and each operator, J_i, involves only a Pauli operator acting on a single qubit.

A *c-code* is a code intended only for transmission of classical information in some basis; an *e-error correcting code* is a c-code that permits correction of all errors described by operators in $\mathcal{J}_e$. To correct all errors in $\mathcal{J}_e$, a code, $\mathcal{C}$, detects the operators in $\mathcal{J}_e^\dagger \mathcal{J}_e$; indeed, the c-code $\mathcal{C}$ with an orthonormal basis $(c_1 \ , \ c_2 \ , \ldots)$ is able to detect error, E, if $c_i \ E \ c_j \quad \alpha_i \delta_{i,j}$. Thus, when we use the definition of distance it follows immediately that a c-code with the minimum distance, $d \quad 2e \quad 1$, is an e-error correcting code. The error amplitude of the information protected in an e-error correcting code is the amplitude of the part of the state orthogonal to the intended state; it is given as a function of time, t, and is at most, $(\lambda t)^{e \quad 1}/(e \quad 1)!$, with λ, the noise strength parameter [249]. The noise strength parameter is given as $\lambda \quad \mathcal{J}_e$, with $\mathcal{J}_e$, the maximum eigenvalue of $\sqrt{\mathcal{J}_e^\dagger \mathcal{J}_e}$. The minimum c-distance, d, implies the existence of an orthonormal basis, $(c_1 \ , \ c_2 \ , \ldots, \ c_k \ )$, such that every operator in the set, E_l, can be detected. It follows that $c_i \ E_l \ c_j \quad \alpha_{i,l}\delta_{i,j}$; in this expression, the operators in the set, $E_1 \quad \mathrm{I}, E_2, \ldots, E_D$, form a basis for $\mathcal{J}_{d \quad 1}$ and the dimension of this basis is D. Two results are proved in [249]:

1. There exist codes of dimension at least $\frac{N}{D}$ of R with minimum c-distance equal to d.
2. There exist codes of dimension at least $\frac{N}{D} \frac{1}{D \quad 1}$ of R with minimum distance equal to d.

Subsystem quantum error-correcting codes use a quantum operation before the errors occur instead of the classical approach of encoding first and then restoring the state affected by error. This quantum operation, described by a family, $\mathcal{A} \quad A_i$, of linear operators that transform the state, $\rho \quad \sum_i A_i \rho A_i^\dagger$, ensures the preservation of the information in a subsystem after the errors in $\mathcal{J}_e$ occur. The result of the state transformation of the system R due to the operator, $A_i \quad \mathcal{A}$, followed by the error, $E \quad \mathcal{J}_e$, is then expressed as the product of the two operators.

This treatment is possible because subsystems of R appear as factors of a tensor product of subspaces of the Hilbert space, $\mathcal{H}$. Let $\text{Mat}(\mathcal{H})$ be the set of all linear operators from $\mathcal{H}$ to itself and $Z(\mathcal{A})$ be the commutant of $\mathcal{A}$, the space of all operators commuting with those in $\mathcal{A}$. Then according to a theorem from representation theory, the system, R, can be described as a direct sum of tensor products of subsystems [249]

$$R \quad \bigoplus_i C_i \quad \mathcal{Z}_i, \quad \text{with} \quad \mathcal{A} \quad \bigoplus_i \text{Mat}(C_i) \quad I^{\mathcal{Z}_i} \quad \text{and} \quad Z(\mathcal{A}) \quad \bigoplus_i \mathbb{I}^{C_i} \quad \text{Mat}(\mathcal{Z}_i),$$

with $\mathcal{Z}_i(C_i)$, the state space of the subsystem, $Z_i(C_i)$. The states of the subsystem, Z_i in $\mathcal{Z}_i$, are immune to noise—the interaction operators act only on the cofactor, C_i; thus, Z_i is a *noiseless subsystem* and information in it is immune to noise.

5.21 BACON-SHOR CODE

To develop an intuition for the subsystem QEC formalism, we discuss first a self-correcting two-dimensional structure of n^2 qubits [17] and then a generalization of this idea known as Bacon-Shor code [18]. This code leads to simple and efficient methods for fault-tolerant error correction that does not require entangled ancilla states and can be implemented with nearest neighbor two-qubit measurements [9].

Recall from Section 5.12 that the stabilizer group, $\mathcal{S}$, is a subgroup of the Pauli group, $\mathcal{P}_n$, that does not contain $\ 1$. The code space, C, is the span of the vectors fixed by $\mathcal{S}$; if $\mathcal{S} \quad S_1, S_2, \ldots, S_s$, with $s < n$, then $S_i\ \varphi \quad \varphi\ , i \quad (1,s)$, and $\dim(C) \quad 2^{n \quad s}$.

If X_i and Z_i are the operators corresponding to the Pauli operators, X and Z, respectively, acting on qubit, i, we define $X_i, Z_i\ , 1 \quad i \quad n$, a set of operators in $\mathcal{P}_n$ isomorphic with X_i and Z_i, thus, satisfying the same commutation relations as X_i and Z_i. Several choices for $X_i, Z_i\ , 1 \quad i \quad n$ are possible as we can think about them as possible degrees of freedom.

Subsystem Error Correction for a Two-Dimensional Structure of n^2 Qubits

We consider a two-dimensional lattice of $n \quad n$ qubits and define a set of products of Pauli operators acting on the n^2 qubits. Then we partition the set into three subsets of operators; this process induces a partition of the Hilbert space, $(\mathcal{H}_2)\ ^{n^2}$, into subspaces. Lastly, we discuss the construction of a subsystem code and the error correction procedure for the code.

The Pauli operators, $P(a,b)$, acting on the n^2 qubits are defined with the aid of two binary configuration matrices, $a \quad [a_{i,j}]$ and $b \quad [b_{i,j}]$, with $(a_{i,j}, b_{i,j}) \quad 0,1\ ,1 \quad (i,j) \quad n$, as

$$P(a,b) \quad \prod_{i\ 1}^{n} \prod_{j\ 1}^{n} \begin{cases} X_{i,j} & \text{if } a_{i,j} \quad 1 \text{ and } b_{i,j} \quad 0, \\ Z_{i,j} & \text{if } a_{i,j} \quad 0 \text{ and } b_{i,j} \quad 1, \\ iY_{i,j} & \text{if } a_{i,j} \quad 1 \text{ and } b_{i,j} \quad 1. \end{cases}$$

We also define the error strings

$$e_j(a) \quad {}_{i\ 1}^{n} a_{i,j} \quad \text{and} \quad f_i(b) \quad {}_{j\ 1}^{n} b_{i,j},$$

with , the *XOR* operation. The set, $P(a,b)$, is partitioned in three subsets, $\mathcal{T}, \mathcal{S}$ $\mathcal{T}$, and $\mathcal{L}$, with the patterns of operators in each set depicted graphically in Figure 5.28:

- $\mathcal{T}$: operators in $P(a,b)$ with an even number of, $X_{i,j}$, operators in each column and an even number of, $Z_{i,j}$, operators in each row. If e_j f_i $0, 1$ (i,j) n, then $P(a,b)$ $\mathcal{T}$. The operators in $\mathcal{T}$ form a non-Abelian group under multiplication.
- $\mathcal{S}$: the stabilizers, operators in $P(a,b)$, with an even number of rows consisting entirely of, X, operators and an even number of columns consisting entirely of, Z, operators. If s_i^X and s_i^Z are the eigenvalues of the operators, S_i^X $\mathcal{S}$ and S_i^Z $\mathcal{S}$, respectively, call s^X and s^Z the corresponding binary vectors and recognize that the Hilbert space, $(\mathcal{H}_2)$ $^{n^2}$, can be represented as a direct sum of

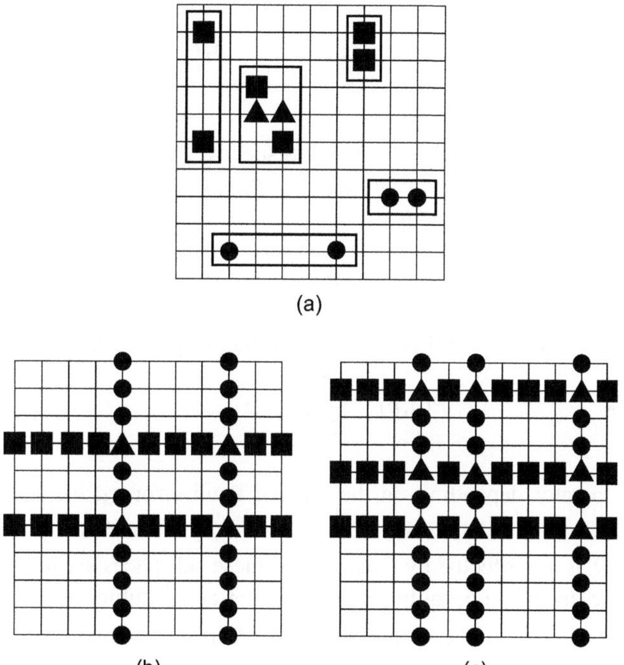

FIGURE 5.28

The operators in $\mathcal{T}, \mathcal{S}$, and $\mathcal{L}$. The qubits are located at the vertices of the $(11$ $11)$ two-dimensional lattice; a square on a vertex represents an, X, operator acting on the corresponding qubit, a circle denotes a, Z, operator, a triangle an, Y, operator, and the absence of any shape represents the identity operator, I. (a) $\mathcal{T}$: the subset of, $P(a,b)$, operators with an even number of, X, operators in each column and an even number of, Z, operators in each row; the rectangles mark such sets. (b) $\mathcal{S}$: the subset of, $P(a,b)$, operators with an even number of rows consisting entirely of, X, operators and an even number of columns consisting entirely of, Z, operators. (c) $\mathcal{L}$: the subset of, $P(a,b)$, operators with an odd number of rows consisting entirely of, X, operators and an odd number of columns consisting entirely of, Z, operators.

subspaces labelled by the binary vectors, s^X and s^Z:

$$(\mathcal{H}_2)\ \ ^{n^2}\quad \bigoplus_{s^X,s^Z} \mathcal{H}_{s^X,s^Z}.$$

The operators in $\mathcal{S}$ form a multiplicative Abelian subgroup of $\mathcal{T}$; every operator in $\mathcal{S}$ commutes with all other operators in $\mathcal{S}$ and also with all operators in $\mathcal{T}$.

- $\mathcal{L}$; operators in $P(a,b)$, with an odd number of rows consisting only of, $X_{i,j}$, operators and an odd number of columns consisting only of, $Z_{i,j}$, operators. The operators in $\mathcal{L}$ form a non-Abelian group under multiplication and commute with the operators in $\mathcal{S}$.

The operators in the sets, $\mathcal{T}$ and $\mathcal{L}$, must be represented on the subspaces, $\mathcal{H}_{s^X,s^Z}$, as subsystem actions; $\mathcal{T}$ and $\mathcal{L}$ induce a partitioning of the Hilbert spaces into direct sums of tensor products of subspaces, $\mathcal{H}_{s^X,s^Z}$, as follows:

$$(\mathcal{H}_2)\ \ ^{n^2}\quad \bigoplus_{s^X,s^Z} \mathcal{H}^{\mathcal{T}}_{s^X,s^Z}\quad \mathcal{H}^{\mathcal{L}}_{s^X,s^Z}.$$

The dimensions of the two spaces are, $2^{(n\ \ 1)^2}$ and 2, respectively, [17]; $\mathcal{H}^{\mathcal{T}}_{s^X,s^Z}$, is the gauge subsystem mentioned in Section 5.20. Operators, $T\quad \mathcal{T}$ and $L\quad \mathcal{L}$, act only on the first and, respectively, the second subspace of the two tensor products,

$$T\quad \bigoplus T_{s^X,s^Z}\quad \mathrm{I}_2 \quad \text{and} \quad L\quad \bigoplus \mathrm{I}_{2^{(n\ \ 1)^2}}\quad L_{s^X,s^Z}.$$

Under this decomposition, an operator, $P(a,b)$, is block diagonal and can be expressed as

$$P(a,b)\quad \bigoplus_{s^X,s^X} E_{s^X,s^Z}\quad \mathrm{I}_2,$$

with E_{s^X,s^Z}, an error operator depending on the type of Pauli error, (a,b), and the subspace labels, s^X and s^Z. The information encoded in the subspaces, $\mathcal{H}^{\mathcal{L}}_{s^X,s^Z}$, identified by $s^X\quad s^Z\quad 11\ldots11$, or, equivalently, $e_j\quad f_i\quad 0$, is not affected by these errors; these subspaces are noiseless.

The subsystem code encodes one qubit of information as n^2 qubits in the Hilbert space $\mathcal{H}^{\mathcal{L}}_{s^X,s^Z}$, with $s^X\quad s^Z\quad 11\ldots11$. To restore the system to $e_j(a)\quad f_i(a)$, we carry out simultaneously, $2(n\quad 1)$, measurements of the eigenvectors of S^X_i and S^Z_i. Measuring S^X_i and S^Z_i is equivalent to computing

$$\bigoplus_{j\ \ 1}^{n}(b_{i,j}\quad b_{i\ \ 1,j})\quad f_i(b)\quad f_{i\ \ 1}(b) \quad \text{and} \quad \bigoplus_{i\ \ 1}^{n}(a_{i,j}\quad a_{i,j\ \ 1})\quad e_j(b)\quad e_{j\ \ 1}(b).$$

Indeed, if $f_i(b)$ is treated as an n-bit codeword of a code with two logical codewords, $f_i(b)\quad 0$, $1\quad i\quad n$, and $f_i(b)\quad 1, 1\quad i\quad n$, then the measurement of the $(n\quad 1)$ eigenvalues of the stabilizer, S^X_i, is equivalent to computing the syndrome and then using this syndrome to correct the errors and restore, $f_i(b)\quad 1, 1\quad i\quad n$, to either 0 or 1 logical codewords. A similar reasoning can be carried out for $e_j(a)$.

Next, we discuss a stabilizer formalism for classical linear codes and review the basic ideas of Shor codes analyzed in Sections 5.5, 5.6, and 5.7; then we introduce the generalized Shor codes and the Bacon-Shor code.

A Stabilizer Formalism for a Binary Linear Code Used to Correct Bit-Flip or Phase-Flip Errors

Let C be an $[n,k,d]$ classical linear code, with a $k \quad n$ generator matrix, G $\quad [g_{i,j}]$, and an $(n \quad k) \quad n$ parity-check matrix, H $\quad [h_{i,j}]$. The k rows of G are linearly independent vectors in $GF(2^n)$, and $(n \quad k)$ rows of H are also linearly independent.

Call $G^c \quad [g^c_{i,j}]$ a $(n \quad k) \quad n$ matrix whose rows are linearly independent of each other and also of the rows of G; together, the rows of G and the rows of G^c form a basis in $GF(2^n)$. Similarly, $H^c \quad [h^c_{i,j}]$ is a $k \quad n$ matrix whose rows are linearly independent of each other and also of the rows of H; together the rows of H and the rows of H^c form a basis in $GF(2^n)$.

When we use, C, to correct bit-flip quantum errors, we define four sets of operators: stabilizers, S_i; logical X operators, X_i; logical Z operators, Z_i; and error operators, E_i.

The stabilizers are obtained from the rows of the parity-check matrix; the stabilizer generator S_i is a tensor product of n terms; each term is either the identity, I, or the Z operator; thus, the generators of the stabilizer group commute with each other. The Z operators in this product are selected by the elements $h_{i,j}$ of the parity-check matrix:

$$S_i \quad \bigotimes_{j \quad 1}^{n} Z^{h_{i,j}}.$$

To compute the syndrome, we measure the eigenvalues of the $(n \quad k)$ stabilizer generator S_i; indeed, the measurements are equivalent to computing the syndrome using the parity-check matrix H.

The k generators of *logical* X operators, X_i, for the k encoded qubits are selected by the rows of the generator matrix; each one is a tensor product of n terms each being an X operator selected by the elements of the generator matrix G:

$$X_i \quad \bigotimes_{j \quad 1}^{n} X^{g_{i,j}}.$$

It is easy to see that X_i commute, with S_j, $X_iS_j \quad S_jX_i$, and they are not in the stabilizer set, as they consist entirely of X operators. The k generators of the *logical* Z operator, Z_i, for the k encoded qubits are selected by the rows of the matrix H^c; each one is a tensor product of n terms, each being an X operator selected by the elements of the matrix $H^c \quad [h^c_{i,j}]$:

$$Z_i \quad \bigotimes_{j \quad 1}^{n} Z^{h^c_{i,j}}.$$

The *logical, Y,* operators are $Y_i \quad iX_iZ_i$. Lastly, the $(n \quad k)$ generators of error operators are selected by the rows of the matrix, $G^c \quad [g^c_{i,j}]$:

$$E_i \quad \bigotimes_{j \quad 1}^{n} X^{g^c_{i,j}}.$$

The products of, E_i, operators anticommute, with at least one S_i; thus, we can detect all error that can be expressed as a product of one or more E_i, with X_i logical operators.

To correct phase-flip errors, we interchange the Z operators with X operators in S_i and Z_i and the X operators with Z operators in X_i. The code for phase-flip error correction operates in the dual basis, $\quad , \quad$, with $\quad (0 \quad 1)/\ \overline{2}$ and $\quad (0 \quad 1)/\ \overline{2}$, instead of operating in the computational basis, $0 , 1$.

Generalized Shor Code

After reviewing concatenated codes. we recognize that the basic idea of the Shor code discussed in Section 5.7 is to concatenate an inner code, $\mathcal{C}_1 \quad [n_1,k_1,d_1]$, to correct bit-flip, or X-errors, with an outer code, $\mathcal{C}_2 \quad [n_2,k_2,d_2]$, to correct phase-flip, or Z-errors. The generator matrices of $\mathcal{C}_1$ and $\mathcal{C}_2$ are $G_1 \quad [g^1_{i,j}], 1 \quad i \quad k_1, 1 \quad j \quad n_1$ and $G_2 \quad [g^2_{i,j}], 1 \quad i \quad k_2, 1 \quad j \quad n_2$, respectively; the parity-check matrices of $\mathcal{C}_1$ and $\mathcal{C}_2$, are $H_1 \quad [h^1_{i,j}], 1 \quad i \quad n_1 \quad k_1, 1 \quad j \quad n_1$, and $H_2 \quad [h^2_{i,j}], 1 \quad i \quad n_2 \quad k_2$, $1 \quad j \quad n_2$, respectively.

The *generalized Shor code* denoted as $[n_1n_2, k_1k_2, \min(d_1,d_2)]$ is defined as follows [18]: arrange the n_1n_2 physical qubits in a two-dimensional array, with n_2 rows and n_1 columns; each row of n_1 qubits encodes, k_1, logical qubits; each column of n_2 qubits encodes, k_2, qubits; use $\mathcal{C}_1$ to correct bit-flip errors and $\mathcal{C}_2$ to correct phase-flip errors.

Phase errors on the qubits of a row of the array are phase errors on the qubits already encoded with $\mathcal{C}_1$; thus, we can use, $\mathcal{C}_2$, to correct phase-flip errors. Indeed, a tensor product of the identity, I, and the Z Pauli operators is either a stabilizer or a product of stabilizers; acting on a row of this array will result in a phase-flip error on some of the encoded qubits times an element of the stabilizer. We conclude that the $[n_1n_2, k_1k_2, \min(d_1,d_2)]$ code has distance, d_1, for bit-flip errors, distance, d_2, for phase-flip errors, and distance, $\min(d_1,d_2)$, for bit- and phase-flip errors; thus, the code can correct, $e \quad (\min(d_1,d_2) \quad 1)/2$, errors. Error correction for a generalized Shor code requires the measurement of $[(n_1 \quad k_1)n_2 \quad (n_2 \quad k_2)]$ eigenvalues of the stabilizer generators.

The Subsystem Code

The procedure follows closely the one discussed earlier in this section for the two-dimensional lattice of n^2 qubits. We use the rows of the parity-check matrix, H_1, of $\mathcal{C}_1$ to construct a stabilizer code on n_1 qubits and the rows of the parity matrix, H_2, of $\mathcal{C}_2$ to construct a stabilizer code on n_2 qubits. The first stabilizer group consists of $n_1 \quad k_1$ stabilizers, $\mathcal{S}_1 \quad S_1, S_2, \ldots, S_{n_1 \quad k_1}$, with $S_i \quad {}^{n_1}_{j \quad 1} Z^{h^1_{i,j}}$; the second stabilizer group consists of $n_2 \quad k_2$ stabilizers, $\mathcal{S}_2 \quad T_1, T_2, \ldots, T_{n_2 \quad k_2}$, with $T_i \quad {}^{n_2}_{j \quad 1} Z^{h^2_{i,j}}$. The stabilizers in $\mathcal{S}_1$ will be used on the columns and the ones in $\mathcal{S}_2$ on the rows to construct the non-Abelian group, $\mathcal{T}$. The group $\mathcal{T}$ consists of, $\mathcal{T}_1$, operators from, $\mathcal{S}_1$, acting on the columns of the array, and, $\mathcal{T}_2$, operators from, $\mathcal{S}_2$ acting on the rows of the array.

Next, we construct an Abelian subgroup, $\mathcal{S} \quad \mathcal{T}$, of operators from $\mathcal{S}_1$ and from $\mathcal{S}_2$, which commute with all the elements of $\mathcal{T}$. We take an operator from $\mathcal{S}_1$ and a codeword, $v \quad (v_1, v_2, \ldots, v_n) \quad \mathcal{C}_2$, and construct an operator acting on n qubits, with $S_1^{v_i}$, acting on each column, $j, 1 \quad j \quad n$. This operator is in $\mathcal{T}$ and commutes with every operator in $\mathcal{T}$; indeed, in a particular row the operators in $\mathcal{T}$ are constructed of Z operators and in a particular column they are made of operators in $\mathcal{S}_1$. Similarly, we take an operator from $\mathcal{S}_2$ and a codeword, $w \quad (w_1, w_2, \ldots, w_n) \quad \mathcal{C}_2$, and construct an operator on n qubits, with $S_2^{w_j}$ acting on each row $i, 1 \quad i \quad n$. This operator is in $\mathcal{T}$ and commutes with every operator in $\mathcal{T}$. We conclude that $\mathcal{S}$ is an Abelian invariant subgroup of $\mathcal{T}$, as all operators in $\mathcal{T}$ commute with all operators from $\mathcal{S}_1$ and $\mathcal{S}_2$.

$\mathcal{S}$ is the stabilizer group of the stabilizer code; the operators in $\mathcal{T}$ and $\mathcal{L}$ are in the normalizer set of this code and all their elements commute with the operators in $\mathcal{S}$. The subspaces of the Hilbert space, $(\mathcal{H}_2) \quad ^{n_1 n_2}$, are labeled with the eigenvalues of the $(n_1 \quad k_1)k_2$ stabilizer generators, which are tensor products of $\mathbb{I}$ and Z Pauli operators, and and with the eigenvalues of the $k_1(n_2 \quad k_2)$ stabilizer generators, which are tensor products of $\mathbb{I}$ and X Pauli operators.

Thus, we have (1) a stabilizer $\mathcal{S}$, with $(n_1 \quad k_1)k_2 \quad k_1(n_2 \quad k_2)$ generators; (2) $\mathcal{T}/\mathcal{S}$ generators, which act as a Pauli group on $(n_1 \quad k_1)(n_2 \quad k_2)$ qubits; and (3) encoded Pauli operators $\mathcal{L}$, which act on $k_1 k_2$ independent qubits that carry the information [18]. The total number of Pauli operators in these sets is $[(n_1 \quad k_1)k_2 \quad k_1(n_2 \quad k_2)] \quad [(n_1 \quad k_1)(n_2 \quad k_2)] \quad [k_1 k_2] \quad n_1 n_2$.

The Hilbert space can then be decomposed into tensor products of subspaces, $\mathcal{H}_s^{\mathcal{T}/\mathcal{S}}$, of dimension $2^{(n_1 \quad k_1)(n_2 \quad k_2)}$ and, $\mathcal{H}_s^{\mathcal{L}}$, of dimension $2^{k_1 k_2}$, with s an $[(n_1 \quad k_1)k_2 \quad k_1(n_2 \quad k_2)]$-tuple of eigenvalues equal to (1) of the stabilizer generators,

$$(\mathcal{H}_2) \quad ^{n_1 n_2} \qquad \bigoplus_{s \quad 0,1 \; (n_1 \quad k_1)k_2 \quad k_1(n_2 \quad k_2)} \qquad \mathcal{H}_s^{\mathcal{T}/\mathcal{S}} \quad \mathcal{H}_s^{\mathcal{L}}.$$

The quantum information is encoded into the subsystem described in $\mathcal{H}_s^{\mathcal{L}}$, a subspace where $s_i \quad 1$. The measurement of the generators of the stabilizer group $\mathcal{S}$ allows us to correct errors that can be corrected with $\mathcal{C}_1$ and $\mathcal{C}_2$ [18].

The concatenation for the generalized Shor codes led to an asymmetry between bit- and phase-flips and the Bacon-Shor code removes this asymmetry and produces a substantial saving in the number of measurements of the stabilizers for error correction; this number is reduced from $[(n_1 \quad k_1)n_2 \quad (n_2 \quad k_2)]$, for a generalized Shor code, to $[(n_1 \quad k_1)k_2 \quad k_1(n_2 \quad k_2)]$, for a subsystem code constructed from two classical codes [18]. Indeed, $k_1 \lll n_1$ and $k_2 \lll n_2$, and this reduction can be exploited successfully by fault-tolerant systems [9].

The operator quantum error formalism discussed next unifies the subspace codes based on the standard model for QEC, the decoherence-free subspaces model [317], and the noiseless subsystem model discussed in Section 5.20.

5.22 OPERATOR QUANTUM ERROR CORRECTION

In this section, we discuss a generalized formalism for quantum error correction, the operator quantum error correction (OQEC) [259, 260]. The roots of the operator quantum correction formalism can be traced back to Werner Heisenberg and Pascual Jordan and to concepts from modern functional analysis applied to quantum mechanics. OQEC is based on the concept of equivalence between states of composite systems expressed as tensor products; two states, $\rho^A \quad \rho^B$ and $\rho^A \quad (\rho \ )^B$, are said to be *equivalent* if they carry the same information even though ρ^B and $(\rho \ )^B$ are different.

Basic Concepts

A *Banach algebra*, A, is an associative algebra over, $\mathbb{R}$ or $\mathbb{C}$, the fields of real and complex numbers, respectively, when the multiplication obeys the rule

$$ab \qquad a \quad b \qquad a,b \quad \mathbb{R} \text{ or } a,b \quad \mathbb{C}.$$

A *C -algebra* is a Banach algebra over $\mathbb{C}$ together with an *involution*, a map $\ :A \qquad A$ with several properties [109]:

1. $(a \quad b) \qquad a \qquad b \ , \qquad a,b \quad A$
   $(ab) \qquad b \ a \ , \qquad a,b \quad A$
2. $(\lambda a) \qquad \lambda a \ , \qquad \lambda \quad \mathbb{C}$ and $a \quad A$
3. $(a \ ) \qquad a, \qquad a \quad A$
4. $\quad aa \qquad a \quad a \ , \qquad a \quad A$
   $\quad a \ a \qquad a \quad a \ , \qquad a \quad A$

Condition 4 is called the *C -identity* and can also be expressed as $aa \qquad a \ ^2$.

If A and B are two C -algebras, the mapping, $F : A \qquad B$, is a C *-homomorphism* when the following two conditions are satisfied:

1. $F(ab) \quad F(a)F(b), \qquad a,b \quad A$
2. $F(a \ ) \quad F(a) \ , \qquad a \quad A$

The C -identity implies that any homomorphism between C -algebras is *bounded*, its norm is smaller or equal to 1; a bijective C -homomorphism is a C *-isomorphism*.

Example. *A nite-dimensional C -algebra, also called a †-algebra.* Let $\mathcal{M}_m$ be the set of $m \quad m$ matrices, over $\mathbb{C}$, with the involution defined as the conjugate transpose. When the matrices in $\mathcal{M}_n$ are viewed as operators, over $\mathbb{C}^m$, with the operator norm, $\ . \ $, we have a finite-dimensional C -algebra.

The *nite direct sum* of two algebraic structures is the Cartesian product of their underlying sets. For example, given two Abelian groups, $(A, \ )$ and $(B, \ )$, then $A \bigoplus B$ is

$$(a_1,b_1) \quad (a_2,b_2) \qquad (a_1 \quad a_2),(b_1 \quad b_2),$$

with denoting the group operation. It is easy to show that a finite-dimensional C -algebra is isomorphic with a finite direct sum.

Given a Hilbert space, $\mathcal{H}$, the *algebra of bounded operators*, $\mathcal{B}(\mathcal{H})$, is a C -algebra, with $a \ $, the adjoint (dual) of the operator $a : \mathcal{H} \qquad \mathcal{H}$.

A quantum *channel*, also called a quantum operator, $\mathcal{E}$, is a completely positive and trace-preserving linear map, $\mathcal{E} : \mathcal{B}(\mathcal{H}) \quad \mathcal{B}(\mathcal{H})$, with an operator sum (Kraus) representation,

$$\mathcal{E}(\sigma) \quad \sum_a E_a \sigma E_a^\dagger, \qquad \sigma \quad \mathcal{B}(\mathcal{H}).$$

$\mathcal{E}$ is *unital* if

$$\mathcal{E}(\mathrm{I}) \quad \sum_a E_a E_a^\dagger \quad \mathrm{I}.$$

The algebra $\mathcal{A}$ associated with $\mathcal{E} \quad E_a$ is a $\dagger$-algebra called the *interaction algebra*; $\mathcal{A}$ is the algebra generated by E_a and $E_a^\dagger$. An orthonormal basis exists, and the matrix representation of the operators in $\mathcal{A}$ with respect to this basis is of the form

$$\mathcal{A} \quad \bigoplus_J (\mathcal{M}_{m_J} \quad \mathrm{I}_{n_J}).$$

This decomposition is unique up to a unitary equivalence.

The *noise commutant* associated with the channel, $\mathcal{E}$, is defined as

$$\mathcal{A} \quad \left\{\sigma \quad \mathcal{B}(\mathcal{H}) : E\sigma \quad \sigma E, \quad E \quad E_a E_a^\dagger \right\}.$$

The noise commutant is unitarily equivalent with the direct sum

$$\mathcal{A} \quad \bigoplus_J (\mathrm{I}_{m_J} \quad \mathcal{M}_{n_J}).$$

We see that elements of $\mathcal{A}$ are not affected by error of $\mathcal{A}$ when $\mathcal{E}$ is unital. It was proven that when $\mathcal{E}$ is unital the noise commutant coincides with the *xed point set*[3] of $\mathcal{E}$

$$\mathcal{A} \quad \mathrm{Fix}(\mathcal{E}) \quad \sigma \quad \mathcal{B}(\mathcal{H}) : \mathcal{E} \quad \sum_a E_a \sigma E_a^\dagger \quad \sigma .$$

The $\mathcal{A}$ may be used to produce noiseless subsystems for unital $\mathcal{E}$.

The interaction algebra, $\mathcal{A}$, due to its structure, induces a decomposition of the original Hilbert space

$$\mathcal{H} \quad \bigoplus_J \mathcal{H}_J^A \quad \mathcal{H}_J^B,$$

where the noisy subsystems, $\mathcal{H}_J^A$ and $\mathcal{H}_J^B$, have dimensions, m_J and n_J, respectively. We only discuss the case when the information is encoded in a single noiseless sector; then the decomposition is

$$\mathcal{H} \quad \left(\mathcal{H}^A \quad \mathcal{H}^B\right) \bigoplus \mathcal{K},$$

[3] x is a fix point of $f(x)$ if $f(x) \quad x$.

with $\dim(\mathcal{H}^A) \quad m$, $\dim(\mathcal{H}^B) \quad n$, and $\dim(\mathcal{K}) \quad \dim(\mathcal{H}) \quad mn$. σ^A and σ^B are the operators in $\mathcal{B}(\mathcal{H}^A)$ and $\mathcal{B}(\mathcal{H}^B)$, respectively. When $\sigma \quad \mathbb{I}^A$ is the identity element of $\mathcal{B}(\mathcal{H}^A)$, the restriction of the noise commutant, $\mathcal{A}$ to $\mathcal{H}^A \quad \mathcal{H}^B$, consists of the operators $\sigma \quad \mathbb{I}^A \quad \sigma^B$.

Generalized Noiseless Systems

We define a set of *matrix units* in, $\mathcal{A}$, for the decomposition, $\mathcal{H} \quad (\mathcal{H}^A \quad \mathcal{H}^B) \bigoplus \mathcal{K}$, as

$$P_{kl} \quad \alpha_k \quad \alpha_l \quad \mathbb{I}, \; 1 \quad k,l \quad m,$$

with α_k and β_k, basis vectors for $\mathcal{H}^A$ and $\mathcal{H}^B$, respectively, $\mathcal{H}^A$ span $\alpha_k \;{}^{m}_{k\;1}$ and $\mathcal{H}^B$ span $\beta_k \;{}^{n}_{k\;1}$; with this choice of the basis vectors, the matrix representation of the subalgebra of $\mathcal{A}$ is $\mathbb{I}^A \quad \mathcal{B}(\mathcal{H}^B)$. The following identities for a family of matrix elements are established in [259, 260]:

$$P_{kl} \quad P_{kk}P_{kl}P_{ll}, \quad 1 \quad k,l \quad m,$$

$$P^{\dagger}_{kl} \quad P_{lk}, \quad 1 \quad k,l \quad m,$$

$$P_{kl}P_{l\,k} \quad \begin{cases} P_{kk} & \text{if } l \quad l\,, \\ 0 & \text{if } l \quad l\,. \end{cases}$$

The projection, $P_{\mathcal{U}}$, and the superoperator, $\mathcal{P}_{\mathcal{U}}$, are defined as $P_{\mathcal{U}} \quad P_{11} \quad P_{22} \qquad P_{mm}$ and as the action $\mathcal{P}_{\mathcal{U}}(\,) \quad P_{\mathcal{U}}(\,)P_{\mathcal{U}}$, respectively.

For the fixed decomposition of $\mathcal{H}$ when the information is encoded in a single noiseless sector, we define the operators

$$\mathcal{U} \quad \sigma \quad \mathcal{B}(\mathcal{H}) : \sigma \quad \sigma^A \quad \sigma^B \text{ for some } \sigma^A, \sigma^B \;.$$

Given a quantum operation, $\mathcal{E}$, on $\mathcal{B}(\mathcal{H})$, a system is said to be *noiseless* for the decomposition, $\mathcal{H} \quad (\mathcal{H}^A \quad \mathcal{H}^B) \bigoplus \mathcal{K}$, if any one of the following three conditions are equivalent and define the properties of the noiseless subsystem B:

1. $\sigma^A, \; \sigma^B \quad \tau^A : \; \mathcal{E}(\sigma^A \quad \sigma^B) \quad \tau^A \quad \sigma_B$
2. $\sigma^A \quad \tau^A : \; \mathcal{E}(\mathbb{I}^A \quad \sigma^B) \quad \tau^A \quad \sigma_B$
3. $\sigma \quad \mathcal{U} \; : \; (\mathrm{tr}_A \quad \mathcal{P}_{\mathcal{U}} \quad \mathcal{E})(\sigma) \quad \mathrm{tr}_A(\sigma)$OQEC

Given a map, $\mathcal{E} \quad E_a$, and a semigroup, $\mathcal{U}$, of, $\mathcal{B}(\mathcal{H})$, the three equivalent necessary and sufficient conditions for existence of a noiseless subsystem are [259]:

1. The B-sector of $\mathcal{U}$ encodes a noiseless subsystem for $\mathcal{E}$ or a decoherence-free subspace when $m \quad 1$;
2. The subspace, $P_{\mathcal{U}}\mathcal{H} \quad \mathcal{H}^A \quad \mathcal{H}^B$, is invariant for the operation, E_a, and the restrictions, $E_a \; {}_{P_{\mathcal{U}}\mathcal{H}}$, belong to the algebra, $\mathcal{B}(\mathcal{H}^A) \quad \mathbb{I}^B$; and
3. Two conditions hold for any choice of the matrix units, $P_{kl} : 1 \quad k,l \quad m$, for $\mathcal{B}(\mathcal{H}^A) \quad \mathbb{I}^B$:
   a. $P_{kk}E_aP_{ll} \quad \lambda_{akl}P_{kl}, \quad a,k,l$ and for some set of scalars, (λ_{akl}), and
   b. $E_aP_{\mathcal{U}} \quad P_{\mathcal{U}}E_aP_{\mathcal{U}}, \quad a.$

The codes described by OQEC consist of operators semigroups and algebras, as we can see from the following definition: a *correctable B-sector of $\mathcal{U}$ for a channel, $\mathcal{E}$, and a recovery, $\mathcal{R}$,* is one when the $\mathcal{H}^B$ sector, $\mathcal{U}$, encodes a noiseless subsystem for the map, $\mathcal{R} \quad \mathcal{E}$, given the decomposition, $\mathcal{H} \quad (\mathcal{H}^A \quad \mathcal{H}^B) \bigoplus \mathcal{K}$; formally, we express this condition as

$$(\mathrm{tr}_A \quad \mathcal{P}_{\mathcal{U}} \quad \mathcal{R} \quad \mathcal{E})(\sigma) \quad \mathrm{tr}_A(\sigma), \ \sigma \quad \mathcal{U}.$$

The following proposition is proved in [260]:

Proposition. *Let $\mathcal{E} \quad E_a$ be a quantum operation on, $\mathcal{B}(\mathcal{H})$, and let the set of operators,*

$$\mathcal{U} \quad \sigma \quad \mathcal{B}(\mathcal{H}) : \sigma \quad \sigma^A \quad \sigma^B \text{ for some } \sigma^A, \sigma^B \ ,$$

be a semigroup. Then the B-sector of $\mathcal{U}$ is correctable for $\mathcal{E}$ if and only if a quantum recovery operation, $\mathcal{R} \quad \mathcal{B}(\mathcal{H})$, exists and

$$(\mathcal{R} \quad \mathcal{E})(\sigma) \quad \sigma, \ \sigma \quad \mathcal{U}.$$

We conclude this section with an example.

Example. *Two noiseless subsystems for a nonunital quantum channel* [260]. Let $\mathcal{E} \quad E_0, E_1$ be a channel on $\mathbb{C}^2 \quad \mathbb{C}^2$, with E_0 and E_1 given by

$$E_0 \quad \sqrt{1 \quad 2p}(\ 00 \ \ 00 \quad 11 \ \ 11\ ) \quad 01 \ \ 01 \quad 10 \ \ 10$$

and

$$E_1 \quad \overline{p}(\ 00 \ \ 00 \quad 10 \ \ 00 \quad 01 \ \ 11 \quad 11 \ \ 11\ ),$$

with $0 < p < 1$. Recall that a channel, $\mathcal{E} \quad E_a$, is *unital* if $\mathcal{E}(\mathbb{I}) \quad \sum_a E_a E_a^\dagger \quad \mathbb{I}$; the channel in this example is nonunital as $E_0 E_0^\dagger \quad E_1 E_1^\dagger \quad \mathbb{I}$.

The first subsystem decomposition is $\mathcal{H}^{A_1} \quad \mathbb{C}$ and $\mathcal{H}^{B_1} \quad$ span $\ 01 \ , \ 10 \ $; if subsystem B_1 is a noise-free subsystem, then the states immune to the effects of the channel $\mathcal{E}$ are the logical zero and the logical one, $0\ _L \quad 01$ and $1\ _L \quad 10$. For this decomposition, the condition 3a above for the existence of a noise-free subspace is satisfied for $Q \quad 01 \ \ 01 \quad 10 \ \ 10$. Indeed,

$$E_0 Q \quad Q \quad QE_0 \quad QE_0 Q \quad \text{and} \quad E_1 Q \quad 0 \quad QE_1 Q.$$

Alternatively, we can show that there exists, $\sigma \quad \mathbb{B}(\mathcal{H}^{B_1}) \quad \mathbb{B}(\mathcal{H}^{A_1} \quad \mathcal{H}^{B_1})$, such that $\mathcal{E}(\sigma) \quad \sigma$; let $a, b, c, d \quad \mathbb{C}$ and define

$$\sigma \quad a\ 01 \ \ 01 \quad b\ 01 \ \ 10 \quad c\ 10 \ \ 01 \quad d\ 10 \ \ 10\ .$$

We see immediately that the condition, $\mathcal{E}(\sigma) \quad \sigma$, is satisfied. Lastly, we observe that $E_1 \sigma \quad 0$ and $\sigma E_1 \quad 0$; thus, the noiseless subsystem, B_1, is not supported by a noise commutant for $\mathcal{E}$.

The second subsystem decomposition is supported by a noise commutant for $\mathcal{E}$. In this case: $\mathcal{H}^{A_2}$ span α_1 , α_2 and $\mathcal{H}^{B_2}$ span β_1 , β_2 . If B_2 is a noise-free subsystem, then the states immune to the effects of the channel $\mathcal{E}$ are the logical zero and the logical one, 0_L β_1 and 1_L β_2 . Indeed, B_2 is a noise-free subsystem as condition 3b above is satisfied; it is left as an exercise (see Section 5.28) to prove the following equalities:

$$P_{11}E_0P_{11} \quad \alpha P_{11}, \quad P_{11}E_1P_{11} \quad \beta P_{11}, \quad P_{11}E_0P_{22} \quad 0P_{12}, \quad P_{11}E_1P_{22} \quad 0P_{22}$$

and

$$P_{22}E_0P_{22} \quad P_{22}, \quad P_{22}E_1P_{22} \quad 0P_{22}, \quad P_{22}E_0P_{11} \quad 0P_{21}, \quad P_{22}E_1P_{22} \quad 0P_{21}.$$

Then we see that

$$\mathcal{E}(\mathbb{I}_2 \quad \sigma) \quad \begin{pmatrix} 1 \quad p & p \\ p & 1 \quad p \end{pmatrix} \quad \sigma, \quad \sigma \quad \mathbb{B}(\mathcal{H}^{B_2}).$$

We conclude the presentation of the subsystem codes with a discussion of stabilizers for OQEC in the next section, a topic analyzed in [337].

5.23 STABILIZERS FOR OPERATOR QUANTUM ERROR CORRECTION

OQEC is based on a fixed partition of the Hilbert space $\mathcal{H}$, $\mathcal{H}$ A B C , with C, the code space. When the information is encoded in subsystem, A, the density matrix of the resulting state is ρ^A ρ^B 0^C , with ρ^A $\mathcal{B}(A)$, the density matrix of the logical state carrying useful information and, ρ^B, the density matrix of an arbitrary state.

We consider a system consisting of n qubits in $\mathcal{H}_{2^n}$. We choose a set of $2n$ operators, X_j, Z_j j $1,n$ $\mathcal{P}_n$ [337]; the operators, X_j, Z_j j $1,n$, act on, n, virtual qubits in the *code space*, C, the same way the Pauli operators, X_i and Z_i, act on a single physical qubit.

The *stabilizer group* of the code, $\mathcal{S}$, is a Abelian subgroup of $\mathcal{P}_n$ that does not contain 1; the generators, S_j, in, $\mathcal{S}$ $S_1, S_2, \ldots, S_s$, s n, are independent commuting elements of $\mathcal{P}_n$. Thus, they can be diagonalized simultaneously; we can choose S_j Z_j. The code space, C, has dimension, $2^{n \quad s}$, and is the span of the vectors, φ , invariant to the transformation carried out by any stabilizer, S_j φ φ , 1 j s; the projector onto the code space, P, satisfies the relation, S_jP P, 1 j s.

Every pair of elements of the Pauli group $\mathcal{P}_n$ either commutes or anticommutes. $N(\mathcal{S})$ is the subgroup of $\mathcal{P}_n$ that commutes with every element of $\mathcal{S}$ and is called the *normalizer group*; the normalizer maps the code subspace to itself. In our case the normalizer is

$$N(\mathcal{S}) \quad i, Z_1, Z_2, \ldots, Z_n, X_1, X_2, \ldots, X_n \; .$$

We define an equivalence relation " " induced by a set of *gauge transformations*; two states with the density matrices, ρ and ρ , are said to be equivalent, ρ ρ , if and only if there exists a transformation, g, such that ρ $g\rho\, g^\dagger$; the gauge transformations, g, belong to the *gauge group*, $\mathcal{G}$.

The equivalence relation requires the gauge group, $\mathcal{G}$, to be a subgroup of the normalizer group, $\mathcal{G}$ $N(\mathcal{S})$, in order to keep the states, φ C, in the code subspace, C. The operators in $\mathcal{S}$, as well as the complex number, i, leave the states, φ C, invariant under conjugation and should be in $\mathcal{G}$. The gauge group can then be generated by the stabilizer generators, $S_j, 1 \quad j \quad s$, the complex number, i, and an arbitrary subset of X_{i_j} and Z_{i_j}, with $j \quad s$,

$$\mathcal{G} \quad S_1, S_2, \ldots S_s, X_{i_1}, \ldots, X_{i_\alpha}, Z_{i_1}, \ldots, Z_{i_\beta} \;.$$

Thus, $\mathcal{G}$ is a normal subgroup of $N(\mathcal{S})$. Then $\mathcal{L} \quad L_j \quad N(\mathcal{S})/\mathcal{G}$ is the quotient group[4] of $N(\mathcal{S})$ and $\mathcal{G} \quad \mathcal{L} \quad \mathcal{N}(\mathcal{S})$. In [468] it is shown that $[\mathcal{G}, \mathcal{L}] \quad 0$ and the operators, X_{i_j} and X_{i_j}, must appear in pairs; thus,

$$\mathcal{G} \quad S_1, S_2, \ldots, S_s, X_{s \; 1}, Z_{s \; 1}, \ldots, X_{s \; r}, Z_{s \; r} \;, \text{ with } S_j \quad Z_j, r \quad n \quad s, 1 \quad j \quad s$$

and

$$\mathcal{L} \quad X_{s \; r \; 1}, Z_{s \; r \; 1}, \ldots, X_n, Z_n \;.$$

The gauge group, $\mathcal{G}$, and the quotient group, $\mathcal{L}$, induce a subsystem structure and partition the code space, C, into subspaces, $C \quad A \quad B$, with $\dim(A) \quad 2^k$ and $\dim(B) \quad r$. The, n, virtual qubits are partitioned in: s, stabilizer qubits; r, gauge qubits of subsystem, B, that do not encode any useful information and are only used to absorb the transformation carried out by operators in $\mathcal{G}$; and k, logical qubits, with $n \quad r \quad s \quad k$. The actions of the operators in the two groups, $g \quad \mathcal{G}$ and $L \quad \mathcal{L}$, restricted to the code subspace are limited to one subsystem:

$$gP \quad \mathbb{I}^A_{2k} \quad g^B, g^B \quad \mathcal{B}(B) \quad \text{and} \quad LP \quad L^A \quad \mathbb{I}^B_{2rk}, L^A \quad \mathcal{B}(A).$$

The group, $\mathcal{L}_{\mathcal{B}}$, of Pauli operators acting on the, r, virtual qubits of subsystem B is generated by the operators, $Z_{s \; j}$ and $X_{s \; j}$, denoted as g^z_j and g^x_j, with $1 \quad j \quad r$. The, k, virtual qubits of subsystem, A, are acted upon by the operators in $\mathcal{L}$, a group generated by the operators, $Z_{s \; r \; j}$ and $X_{s \; r \; j}$, denoted as Z^j and X^j, with $1 \quad j \quad k$. The gauge group, $\mathcal{G}$, leaves the encode information invariant under conjugation, $\mathcal{G} \quad \mathcal{L}_{\mathcal{B}} \quad \mathcal{S} \quad (i)$. Thus, *the standard stabilizer formalism corresponds to an Abelian gauge group, while the OQEC corresponds to a non-Abelian* $\mathcal{G}$.

Our discussion of the error correction for OQEC codes starts with an analysis of a set of correctable errors and continues with the error detection and error recovery procedures. The projector, P, onto the code space, C, has the property that $PC \quad \mathcal{H} \quad A \quad B$; if $g^B_{ab} \quad \mathcal{B}(B)$ is an arbitrary operator, then the set of correctable errors, E_a , must satisfy the condition

$$PE^{\dagger}_a E_b P \quad \mathbb{I}^A \quad g^B_{ab} \,, \quad a, b.$$

When this condition is satisfied, a recovery map, $\mathcal{R} : \mathcal{B}(\mathcal{H}) \quad \mathcal{B}(\mathcal{H})$, exists; the recovery, $\mathcal{R}$, reverses the action of the transformation, $\mathcal{E} : \mathcal{B}(H) \quad \mathcal{B}(\mathcal{H})$, on the subsystem, A. Then $\mathcal{R} \quad \mathcal{E}(\rho^A \quad \rho^B) \quad \rho^A \quad \rho^{\,B}$, with $\mathcal{E}(\rho) \quad \sum_a E_a \rho E^{\dagger}_a$.

The *error detection* requires the measurement of the eigenvalues of the stabilizer generators, $S_j \quad \mathcal{S}, 1 \quad j \quad s$. The error syndromes can be $\quad 1$: an eigenvalue equal to $\quad 1$ indicates that the corresponding eigenvector is in the code subspace, C; detectable errors anticommute with at least one of the stabilizer generators, they correspond to an eigenvalue of $\quad 1$ and *do not have any effect on the useful information.*

---

[4] A quotient group is a group formed by identifying the elements of a larger group based on an equivalence relation.

The operator, $E_a E_b$ (the conjugation is irrelevant and it is omitted), is an element of either

- $\mathcal{P}_n \quad N(\mathcal{S})$ – the error can be corrected; there exists, $S \quad \mathcal{S}$, such that $E_a E_b, S \quad 0$ and $PE_aE_bP \quad PE_aE_bSP \quad PSE_aE_bP \quad PE_aE_bP \quad 0$, or
- $N(\mathcal{S}) \quad \mathcal{G}$ – the error cannot be corrected as $PE_aE_bP \quad L^A_{ab} \quad g^B_{ab}$ and $L^A_{ab} \quad \mathbb{I}^A$, or
- $\mathcal{G}$ – the error can be corrected; $PE_aE_bP \quad \mathbb{I}^A \quad g^B_{ab}$.

We conclude that an error is correctable if and only if $E_aE_b \quad N(\mathcal{S}) \quad \mathcal{G}, \quad (a,b)$.

The *error recovery* is based on the observation that equivalent errors have the same syndrome; then the measurement of the syndrome identifies the coset of $E_a \ /\mathcal{G}$ of the error and we recover the information by applying any element of the coset.

We conclude this section with the observation that a stabilizer OQEC code is characterized by $[n,k,r,d]$ with: n, the number of physical qubits; k, the number of logical qubits; r, the number of gauge qubits; and d, the distance of the code. Poulin [337] illustrates the possibility of simplifying existing $[n,k,0,d]$ codes by identifying gauge symmetries in their stabilizers.

After the discussion of subsystems codes in Sections 5.20 to 5.23, we discuss briefly an approach for correction of systematic errors proposed by Grover.

5.24 CORRECTION OF SYSTEMATIC ERRORS BASED ON FIXED-POINT QUANTUM SEARCH

In this section we overview a different approach to QEC based on Grover's fixed-point quantum search [190] discussed in Section 1.24. This strategy can be applied to the correction of systematic errors and of slow-varying random errors, when the parameters of the quantum circuit do not change in time. An error is *systematic* if the process is repetitive and perfectly reversible.

In Section 1.24 we concentrated on unitary transformations, $\mathbf{U}$, from a known initial state, γ , to a well-defined final state, τ , and considered the effect of an error operator, $\mathbf{E}$. Now, we do not have either a source or a target state, and the input can be any arbitrary state. Instead of the transformation $\mathbf{U}$ we apply $\mathbf{U} \quad \mathbf{UE}$. The reversibility implies that instead of $\mathbf{U}^\dagger$ we apply $\mathbf{U}^\dagger$. The transformation has a probability of error equal to ϵ; in other words, $\mathbf{U}$ applied to an arbitrary initial state γ reaches the corresponding desired state, τ , with the probability, $(1 \quad \epsilon)$. In this formulation it becomes clear that the fixed-point quantum search can be used to increase the probability of success from $(1 \quad \epsilon)$ to $(1 \quad \epsilon^3)$ when, instead of $\mathbf{U}$, we apply the composite operation, $\mathbf{UR}_\gamma\mathbf{U}^\dagger\mathbf{R}_\tau\mathbf{U}$. In this case, $\mathbf{R}_\tau$ and, respectively, $\mathbf{R}_\gamma$, rotate the state, with an angle, $\theta \quad \frac{\pi}{3}$.

The error operator, $\mathbf{E}$, associated with the linear transformation, $\mathbf{U}$, is restricted. The probability of error, the amount OF over-rotation of the initial state vector, is small, yet unknown; it is a function of the angle, θ, and the axis of rotation n, $\epsilon \quad \epsilon(\theta ,n)$. The error operator is

$$\mathbf{E} \quad e^{i\epsilon n\sigma},$$

with $\sigma \quad (X,Y,Z)$, the Pauli matrices. It is easy to observe that $\mathbf{E} \quad \mathbf{U}^\dagger\mathbf{U}$ because $\mathbf{UU}^\dagger \quad \mathbf{I}$.

Note that $\tau \quad \mathbf{U} \gamma$, thus, $\tau \quad \gamma \ \mathbf{U}^\dagger$. Then the over-rotation when we apply $\mathbf{U}$ to the arbitrary initial state is

$$\tau \ \mathbf{U} \ \gamma \qquad \gamma \ \mathbf{U}^\dagger\mathbf{U} \ \gamma \qquad \gamma \ \mathbf{E} \ \gamma \ .$$

Similarly,

$$\tau \ \mathbf{UR}_{\gamma}(\pi/3)\mathbf{U}^{\dagger}\mathbf{R}_{\tau}(\pi/3)\mathbf{U} \ \gamma \qquad \gamma \ \mathbf{U}^{\dagger}\mathbf{UER}_{\gamma}(\pi/3)E^{\dagger}\mathbf{UR}_{\tau}(\pi/3)\mathbf{UE} \ \gamma$$
$$\gamma \ \mathbf{ER}_{\gamma}(\pi/3)\mathbf{E}^{\dagger}\mathbf{R}_{\tau}(\pi/3)\mathbf{E} \ \gamma \ .$$

Reichardt and Grover [355] show that if the error is small then by concatenating the basic sequence

$$\mathbf{UR}_{\tau}(\pi/3)\mathbf{U}^{\dagger}\mathbf{R}_{\gamma}(\pi/3)\mathbf{U}$$

we obtain an even smaller error

$$\tau \ \mathbf{U} \ \gamma \qquad \gamma \ \mathbf{E} \ \gamma \qquad 1$$
$$\tau \ \mathbf{UR}_{\gamma}(\pi/3)\mathbf{U}^{\dagger}\mathbf{R}_{\tau}(\pi/3) \ \gamma \qquad \gamma \ \mathbf{ER}_{\gamma}(\pi/3)\mathbf{E}^{\dagger}\mathbf{R}_{\tau}(\pi/3)\mathbf{E} \ \gamma \qquad 1.$$

Figure 5.29 shows the effect of the transformation, $\mathbf{UR}_{\tau}(\pi/3)\mathbf{U}^{\dagger}\mathbf{R}_{\gamma}(\pi/3)\mathbf{U}$. We assume that the current state, γ , is along the, z , axis of the Bloch sphere and show the effect of the composite transformation on the error.

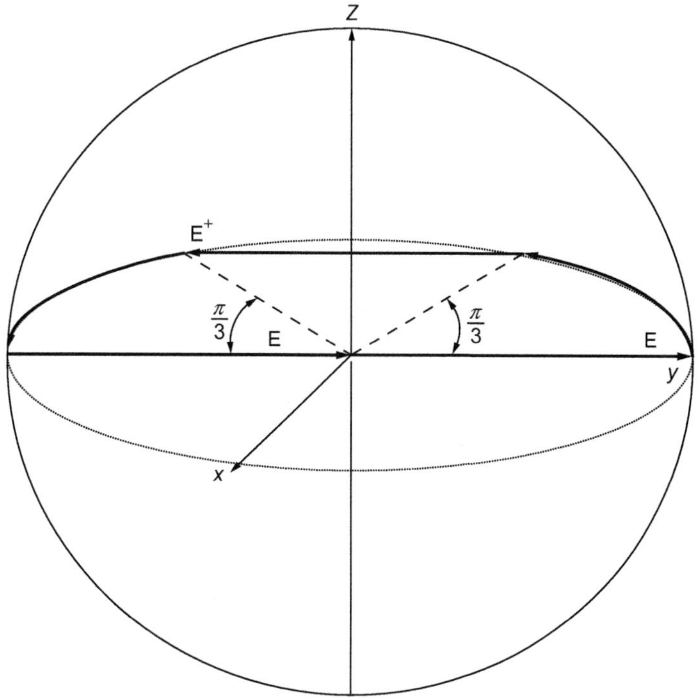

FIGURE 5.29

A Bloch sphere representation of error correction based on xed-point search. The current state is a vector along z. We show the effect of the transformation, $\mathbf{E}$, followed by a rotation of $\frac{\pi}{3}$ radians, $\mathbf{r}_{\tau}(\pi/3)$, followed by, $\mathbf{E}^{\dagger}$, followed by, another rotation of $\frac{\pi}{3}$ radians, $\mathbf{R}_{\gamma}(\pi/3)$, and nally followed by, $\mathbf{E}$.

Finally, we discuss an application of QEC to reliable quantum gates. Reliability of quantum systems is a vast subject, and the discussion in the next section only illustrates the role played by quantum error-correcting codes in the effort to build reliable quantum systems.

5.25 RELIABLE QUANTUM GATES AND QUANTUM ERROR CORRECTION

A reliable quantum computer can only be built with reliable quantum circuits. To design reliable quantum circuits, we have to consider "logical quantum gates" performing reliably the same function as the "physical quantum gates" we are accustomed to (e.g., CNOT and Toffoli gates); the logical gates should be built using quantum error-correcting codes.

Figure 5.30 shows a reliable CNOT gate based on the seven-qubit code of Steane. This circuit is able to correct one qubit in error after the application of the gate. This circuit is similar to the classical CNOT; it has two logical qubits and requires seven physical qubits to encode each logical qubit, as well as three ancilla qubits for syndrome calculation for each logical qubit.

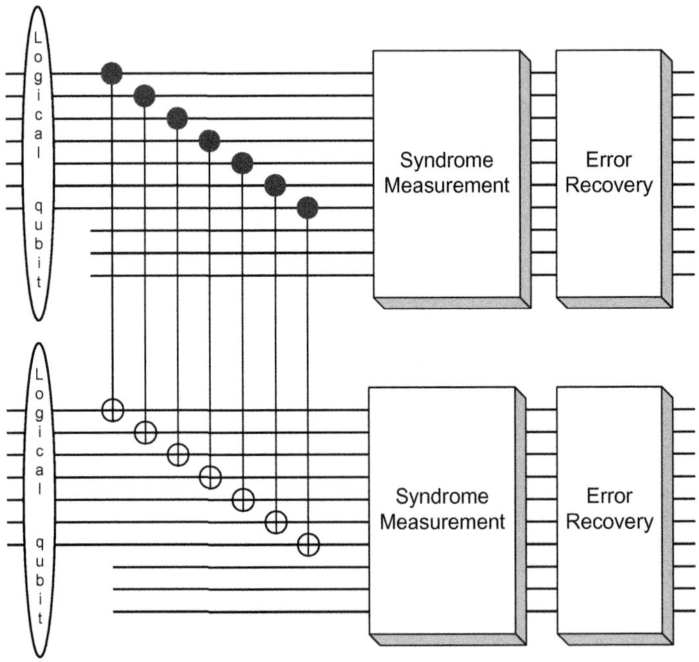

FIGURE 5.30

A reliable CNOT gate based on the seven-qubit quantum error-correcting code of Steane. Each one of the seven pairs of physical qubits traverses a CNOT gate; for each logical qubit, there are three ancilla qubits for syndrome calculation. The 14 physical qubits are subject to syndrome measurement and error recovery.

If an error affecting a register of quantum bits is described by a unitary transformation, $\mathbf{E}$, and if $\mathbf{G}_{cnot}$ is the transformation carried out by the CNOT gate, then

$$\mathbf{G}_{cnot}\mathbf{E} \quad \mathbf{E}^{\dagger}\mathbf{G}_{cnot}.$$

This equation shows that it makes no difference if an error occurs before, or after the application of a CNOT gate, and it is true for other types of quantum gates as well. This important property enables us to build reliable quantum gates consisting of a possibly large number of error-prone physical gates.

If p is the probability of an error affecting a single qubit, what is the probability, P_e, that the circuit in Figure 5.30 is affected by two or more errors? The answer to this question is critical for our ability to build a reliable CNOT gate based on a quantum error-correcting code (e.g., the seven-qubit Steane's code capable of correcting a single bit-flip or phase-flip error). We shall show that, $P_e \quad kp^2$, with k, a constant dependent on the quantum error-correcting code.

For Steane's code $k \quad k_s \quad 10^4$. If for a given ϵ we wish to achieve a probability of error, $P_e \lll \epsilon$, then p, the probability of a single qubit error, should satisfy the following condition:

$$p \quad \sqrt{\frac{P_e}{k}} \lll \sqrt{\frac{\epsilon}{k}}.$$

For example, when $k_s \quad 10^4$ (Steane's code) and when we require that $\epsilon \quad 10^{\ 10}$, then $p \lll 10^{\ 3}$ (e.g., $p \quad 10^{\ 5}$).

Let us now discuss the procedure to evaluate, k_s, when we assume that the probability of more than two errors in the circuit in Figure 5.30 is very low. This seems to be a reasonable assumption; indeed, if $p \quad 10^{\ 5}$, then $p^3 \quad 10^{\ 15}$. We evaluate several possible scenarios and compute the probability, $P_i \quad k_i p^2$, of an error caused by each scenario [287] and then compute

$$k_s \quad \sum_i k_i \quad \text{and} \quad P_e \quad k_s p^2.$$

There are several possibilities:

1. We have two independent errors, one affecting each logical qubit of the current gate. Each one of the seven physical qubits of a logical qubit is produced by a logical gate of a previous stage. For simplicity we assume that the previous gate is also a reliable CNOT based on Steane's code; thus there are 10 possible points of failure (seven qubits plus three ancilla qubits). Thus, $P_1 \quad (k_1 p)^2 \quad k_1 p^2$, with $k_1 \quad k_1^{\ 2}$ and $k_1 \quad 7 \quad 10 \quad 70$.
2. A single error enters the logical gate and an additional error occurs during syndrome calculations and error recovery. For each of the seven qubits, there are $2k_1$ possible pairs of failure points and $P_2 \quad (k_2 p) \quad p \quad k_2 p^2$, with $k_2 \quad 7 \quad (2k_1) \quad 7 \quad 140$.
3. Two failures occur during traversal of CNOT gates; then there are $10 \quad 10$ combinations for possible points of failure $P_3 \quad k_3 p^2$, with $k_3 \quad 10 \quad 10$.
4. Two failures occur during processing, one during the traversal of CNOT gates and one during the syndrome measurement; then $P_4 \quad k_4 p^2$, with $k_4 \quad 10 \quad 10$.
5. Two failures occur during syndrome measurement; $P_5 \quad k_5 p^2$, with $k_5 \quad k_1 \quad k_1 \quad 70 \quad 70$.

6. One failure occurs during syndrome measurement and another failure occurs during the recovery stage and then $P_6 \quad k_6 p^2$, with $k_6 \quad k_1 \quad 7 \quad 70 \quad 7$.
7. Two failures occur during the recovery stage and then $P_7 \quad k_7 p^2$, with $k_7 \quad 7 \quad 7$.

We conclude that $k_s \quad \sum_{i \;\; 1}^{7} k_i \quad 10^4$.

Quantum error correction is intimately related to the physical processes that cause decoherence. It is critical to have some understanding of the time scale for the decoherence phenomena for different qubit candidates. If τ_{gate} is the time required for a single gate operation and $\tau_{decoherence}$ is the decoherence time of a qubit, then n_{gates}, the number of gates that can be traversed by a register before it is affected by decoherence is given by

$$n_{gates} \qquad \frac{\tau_{decoherence}}{\tau_{gate}}.$$

Table 5.12 presents sample values of the time required for a single gate operation, τ_{gate}; the decoherence time of a qubit, $\tau_{decoherence}$; and n_{gates}, the number of gates that can be traversed before a register of qubits is affected by decoherence, for several qubit implementations [126]. We notice a fair range of values for the number of quantum gate operations that can be performed before decoherence affects the state, from about 10^3 for quantum dots to 10^{13} for a trapped indium ion.

The process of error correction introduces new errors at the same time it corrects others. The discussion in this section shows that the error correction may, or may not, reduce the error rate during a quantum computation. The error correction will help if and only if the gate error rate is low enough; otherwise it will produce more harm than benefits. Moreover, fault-tolerance may introduce additional overhead and limit the ability of a quantum computer to carry out long computations.

There is another aspect of QEC we have to pay attention to, *the complexity of decoding.* We can always use a code that corrects more errors, and presumably a quantum computer based on such codes will be able to withstand decoherence for longer periods of time and carry out longer computations. However, for most error-correcting codes, the complexity of decoding increases rapidly with the number of errors corrected by the code. The time to do error correction for such codes may in fact be so large that the performance of the quantum computer may decline when the number of errors corrected by the code increases above a certain threshold.

Thus, it is important to find quantum error-correcting codes where the time to measure the error syndrome and to carry out the error recovery and the complexity of the quantum circuits to carry

Table 5.12 The Time Required for a Single Gate Operation, τ_{gate}; the Decoherence Time of a Qubit, $\tau_{decoherence}$; and the Number of Gates that can be Traversed Before a Register of Qubits is Affected by Decoherence, n_{gates}

Qubit Implementation	$\tau_{decoherence}$ (sec)	τ_{gate} (sec)	n_{gates}
Nuclear spin	10^4	$10^{\;\,3}$	10^7
Trapped indium ion	$10^{\;\,1}$	$10^{\;\,14}$	10^{13}
Quantum dots/charge	$10^{\;\,9}$	$10^{\;\,12}$	10^3
Quantum dots/spin	$10^{\;\,6}$	$10^{\;\,9}$	10^3
Optical cavity	$10^{\;\,5}$	$10^{\;\,14}$	10^9

out these operations increase slowly with the error-correcting capability of the code. This problem was discussed in Section 4.18 when we showed that there exist polynomial time algorithms, $\mathcal{E} : 0,1^{K} \quad 0,1^{N}$, for encoding and $\mathcal{D} : 0,1^{N} \quad 0,1^{K}$ for decoding, such that Prob(decoding error) $< 1/N$.

5.26 HISTORY NOTES

In 1995, Peter Shor described a nine-qubit single error-correcting code [380]. Before this breakthrough, it was not clear that quantum error correction was even possible.

A seven-qubit quantum error-correcting code was found in 1995 by Steane [401] and Calderbank and Shor [80]; this was the first code of an entire class of quantum error-correcting codes, the so-called CSS (Calderbank, Shor, Steane) codes. In 1996, Bennett *et al* [40] and Laflamme *et al* [263] discovered an even more efficient five-qubit code.

The stabilizer formalism that provides a general theory of quantum error-correcting codes was developed soon thereafter by Gottesman and others [81, 173, 342].

Emmanuel Knill, Raymond Laflamme, and Lorenza Viola introduced subsystem codes in a paper on the theory of QEC for general noise [249]; David Kribs, Raymond Laflamme, and David Poulin developed operator quantum error correcting (OQEC) codes [259], and Poulin proposed a stabilizer formalism for OQEC codes [337]. David Bacon introduced a class of subsystem codes referred to as Bacon-Shor codes [17, 18].

The algorithmic approach to error correction is intimately tied to fault-tolerance. In 1996, Shor made a major contribution to fault-tolerant quantum computation on encoded states. Shor discusses the implementation of quantum gates to evolve the logical state of a quantum circuit [382].

Shor emphasizes the use of ancilla entangled states that are partially verified and the repetition of syndrome measurement. He also introduced a discrete universal set of quantum operations. The same year, DiVincenzo and Shor [128] generalized the fault-tolerant protocol for syndrome measurement for any stabilizer code. Daniel Gottesman also proposed fault-tolerant universal methods for stabilizer codes [173, 174].

5.27 SUMMARY AND FURTHER READINGS

The properties of quantum information make error correction very challenging. Quantum error correction requires a multi-qubit measurement that does not disturb the quantum information in the encoded state but retrieves information about the error. The measurement of the syndrome should reveal as much as possible about the error, but should not provide any information about the value that is stored in the logical qubit; otherwise, the measurement would destroy any quantum superposition of this logical qubit with other qubits in the quantum computer.

The measurement of the syndrome determines if a qubit has been corrupted and the type of error. The effect of the noise is either a bit-flip, or a phase-flip, or both and can be represented by transformations carried out by the Pauli operators X, Z, and Y. A puzzling question that comes to mind is how can we consider only a few distinct possibilities when the noise is arbitrary? The answer is that we perform a projective measurement of the syndrome; though the effect of the noise is arbitrary, it can be expressed as a superposition of basis operations given by the X, Y, and Z Pauli operators and the identity, when the qubit is not affected by error. The measurement of the syndrome "forces" the qubit to "choose" a specific "Pauli error" identified by the syndrome; to correct the error we apply to the corrupted qubit the same Pauli operator.

A quantum code is denoted by $\; n,k,d_1,d_2 \;$, with n, the number of qubits used to store or transmit, k, bits of information and allow correction of up to $(d_1 \quad 1)/2$ bit-flip errors and simultaneously up to $(d_2 \quad 1)/2$ phase-flip errors. The code, $\mathcal{C}$, is a subspace of dimension, 2^k, of a Hilbert space, H_{2^n}, and it is constructed by specifying a basis of it.

The following proposition [80] relates quantum and classical error-correcting codes: *If $\mathcal{C}_1$ and $\mathcal{C}_2$ are both $[n,k,d]$ codes, with* $0 \quad \mathcal{C}_2^{\;} \quad \mathcal{C}_1 \quad \mathcal{F}_2^n$, *then the quantum code,* $\mathcal{Q}_{\mathcal{C}_1,\mathcal{C}_1}$, *is a t-error-correcting code, with* $t \quad (d \quad 1)/2$.

The stabilizer formalism presented in [173] is very useful in QEC. An $[n,k]$ stabilizer code can be used to encode the state of a k-qubit system and correct a set Σ of error patterns. There are $(n \quad k)$ independent generators, $\mathbf{M}_i, 1 \quad i \quad (n \quad k)$, of the stabilizer group, S, which commute with each other.

The generators, $\mathbf{M}_i, 1 \quad i \quad (n \quad k)$, are n-tuples of Pauli operators (tensor products on n Pauli operators), and their eigenvectors span two orthogonal eigenspaces of dimension $2^{n \quad 1}$, corresponding to the two eigenvalues, $\quad 1$, respectively, of the generators. The stabilizer group S generated by $\mathbf{M}_i, 1 \quad i \quad (n \quad k)$, has a set of $2^{n \quad k}$ orthogonal eigenspaces, each of dimension 2^k. Each eigenspace corresponds to one of the $2^{n \quad k}$ combinations of values $\quad 1$, the eigenvalues of generators.

The *code subspace*, $\mathcal{C}$, is defined as the subspace of, $\mathcal{H}_{2^n}$, stabilized by S; the code subspace is a 2^k subspace for which the $(n \quad k)$ generators have eigenvalues equal to $\quad 1$.

The basic idea of subsystem codes is to use algebraic methods to classify quantum errors and, implicitly, to describe quantum error-correcting codes as tensor products of subspaces, or more accurately, as subsystems where the useful information is confined to one subsystem [249].

A Hilbert space, $\mathcal{H}$, can be decomposed as the direct sum of two subspaces, $\mathcal{H} \quad \mathcal{H}^C \bigoplus \mathcal{H}^D$, or as the tensor product of two subspaces, $\mathcal{H} \quad \mathcal{H}^C \quad \mathcal{H}^D$; in the case of a direct sum, $\dim(\mathcal{H}) \quad \dim(\mathcal{H}^C) \quad \dim(\mathcal{H}^D)$, while $\dim(\mathcal{H}) \quad \dim(\mathcal{H}^C) \quad \dim(\mathcal{H}^D)$ for the tensor product. A decomposition of the Hilbert space, $\mathcal{H}$, as a subspace, $\mathcal{H}^E$, and a perpendicular subspace, $(\mathcal{H}^E)$, is possible, with $(\mathcal{H}^E)$ expressed as a tensor product of two other subspaces, $\mathcal{H}^C$ and $\mathcal{H}^D$,

$$\mathcal{H} \quad (\mathcal{H}^E) \quad \bigoplus \mathcal{H}^E \quad \text{or} \quad \mathcal{H} \quad (\mathcal{H}^C \quad \mathcal{H}^D) \bigoplus \mathcal{H}^E.$$

The subsystem encoding for this decomposition requires the preparation of the quantum state

$$\rho \quad (\rho^C \quad \rho^D) \quad 0^E$$

and encoding the information into subsystem, $\mathcal{C}$, described in the Hilbert space $\mathcal{H}^C$ for a fixed encoding into the subsystem $\mathcal{D}$ described in the Hilbert space $\mathcal{H}^D$; in this expression, ρ^C, is the density matrix of the encoded quantum information, ρ^D, is the density matrix of the arbitrary quantum information encoded into the subsystem $\mathcal{D}$, and, 0^E, is the all-zero matrix on the subspace, $\mathcal{H}^E$. $\mathcal{D}$ is called a gauge subsystem and E a logical subsystem. Regardless of the nontrivial transformation applied to gauge subsystem $\mathcal{D}$, the information encoded into subsystem $\mathcal{C}$ is not affected; *the QEC is restricted to the restoration of the information encoded into subsystem $\mathcal{C}$.*

A different approach to QEC is taken by Lov Grover who in 2005 introduced fixed-point quantum search [189, 190] and then later discussed the application of fixed-point quantum search to error correction [355]. Grover's approach can be applied to correct small, systematic errors.

The most important readings on the subject of quantum error-correcting codes are the seminal papers of Shor [380]; Steane [402]; Calderbank and Shor [80]; Schumacher [368]; Knill, Laflamme, and Zurek [243]; and Knill and Laflamme [244]. A comprehensive presentation of the stabilizer formalism can be found in Gottesman's thesis [173] and in his later papers [176]. Forney, Grassl, and Guha have contributed to concatenated quantum codes [155]. The relevant references for OQEC are due to Knill, Laflamme, and Viola [249]; Kribs, Laflamme, and Poulin [259]; Poulin [337]; and Bacon [17, 18]. Fixed-point quantum search is discussed by Grover [190] and by Grover and Reichardt [355].

5.28 EXERCISES AND PROBLEMS

Problem 5.1. Let $\mathcal{C}$ be a linear $[n,k,d]$ code, capable of correcting, $\frac{d\ \ 1}{2}$, errors and $\mathcal{C}$ its dual. Assume that $\mathcal{C}$ $\mathcal{C}$. Let

$$s_v \quad 2^{\frac{n\ \ k}{2}} \sum_{w\ \ \mathcal{C}} v \quad w \ .$$

Prove that given, (v , u) $\mathcal{C}$, the two vectors are identical, s_v s_u , if and only if v u .

Problem 5.2. Show that the generators of a group provide a more compact description of the group than the enumeration of all the elements of the group.

Problem 5.3. Show that in the case of the nine-qubit error-correcting code any error operator, $\mathbf{E}^i_{x\ y\ z}$, $1 \quad i \quad 9$, anticommutes with at least one of the eight operators, $\mathbf{M}_j$, $1 \quad j \quad 8$; thus, any single qubit error can be corrected (see Section 5.12).

Problem 5.4. Show that all elements of the Pauli group, $\mathcal{P}$, are unitary, either Hermitian or anti-Hermitian, and either commute or anticommute.

Problem 5.5. Show that, S, is a normal subgroup of $N(S)$.

Problem 5.6. Show that, $NS \quad 4\ 2^{n\ \ k}$.

Hint: The Pauli group, $\mathcal{P}_n$, has $4^{n\ \ 1}$ elements including an overall phase of 1 and i. The factor of 4 accounts for the overall phase factor.

Problem 5.7. Show that it makes no difference if a qubit error occurs before or after the application of a CNOT gate. If $\mathbf{A}$ is a unitary transformation and $\mathbf{G}_{CNOT}$ is the transformation carried out by the CNOT gate, show that

$$\mathbf{G}_{CNOT}\mathbf{A} \quad \mathbf{A}^{\dagger}\mathbf{G}_{CNOT}.$$

Problem 5.8. Show that, S, is a stabilizer of a nontrivial Hilbert subspace, V_{2^n} $\mathcal{H}_{2^n}$, only if S $\mathbf{S}_1, \mathbf{S}_2, \ldots$ is an Abelian group:

$$\mathbf{S}_i\mathbf{S}_j \quad \mathbf{S}_j\mathbf{S}_i, \quad \mathbf{S}_i, \mathbf{S}_j \quad S, \ i \quad j.$$

Note: In Section 5.12, we showed that this condition is necessary.

Problem 5.9. Show that error operators for Steane's seven-qubit code anticommute with the generators listed in Table 5.4 from Section 5.12.

Problem 5.10. Establish the correspondences between the errors and the syndromes for Steane's seven-qubit code (see Table 5.5 from Section 5.12).

Problem 5.11. Show that error operators for the perfect quantum code discussed in Section 5.13 anticommute with the generators listed in Table 5.9.

Problem 5.12. Establish the correspondences between the errors and the syndromes for the perfect code discussed in Section 5.13 (see Table 5.8).

Problem 5.13. Construct a table showing the error operators that anticommute with each one of the four generators for the quantum restoration circuit in Section 5.14.

Problem 5.14. Establish the correspondences between the errors and the syndromes for the quantum restoration circuit in Section 5.14 (see Table 5.10).

Problem 5.15. Show that the following equalities given in the discussion of the example in Section 5.22 hold:

$$P_{11}E_0P_{11} \quad \alpha P_{11}, \quad P_{11}E_1P_{11} \quad \beta P_{11}, \quad P_{11}E_0P_{22} \quad 0P_{12}, \quad P_{11}E_1P_{22} \quad 0P_{12},$$

and

$$P_{22}E_0P_{22} \quad P_{22}, \quad P_{22}E_1P_{22} \quad 0P_{22}, \quad P_{22}E_0P_{11} \quad 0P_{21}, \quad P_{22}E_1P_{22} \quad 0P_{22}.$$

Hint: Show first that:

$$\alpha_1 \quad \beta_1 \quad \frac{00 \qquad 11}{\overline{2}}, \quad \alpha_1 \quad \beta_2 \quad \frac{00 \qquad 11}{\overline{2}}$$

and

$$\alpha_2 \quad \beta_1 \quad \frac{10 \qquad 01}{\overline{2}}, \quad \alpha_2 \quad \beta_2 \quad \frac{10 \qquad 01}{\overline{2}}.$$

Compute the matrix units for this decomposition using the expression,

$$P_{kl} \quad \alpha_k \ \alpha_l \quad \mathbb{I}^{B_2} \quad \alpha_k \ \alpha_l \quad (\ \beta_1 \ \beta_1 \quad \beta_2 \ \beta_2\ ), \quad 1 \quad k, l, \quad 2.$$

Then,

$$P_{11} \quad 00 \ 00 \quad 11 \ 11\ , \quad P_{12} \quad 00 \ 10 \quad 11 \ 01$$

and

$$P_{21} \quad 10 \ 00 \quad 01 \ 11\ , \quad P_{22} \quad 10 \ 10 \quad 01 \ 01\ .$$

CHAPTER

Physical Realization of Quantum Information Processing Systems

6

Reality is merely an illusion, albeit a very persistent one.
Albert Einstein

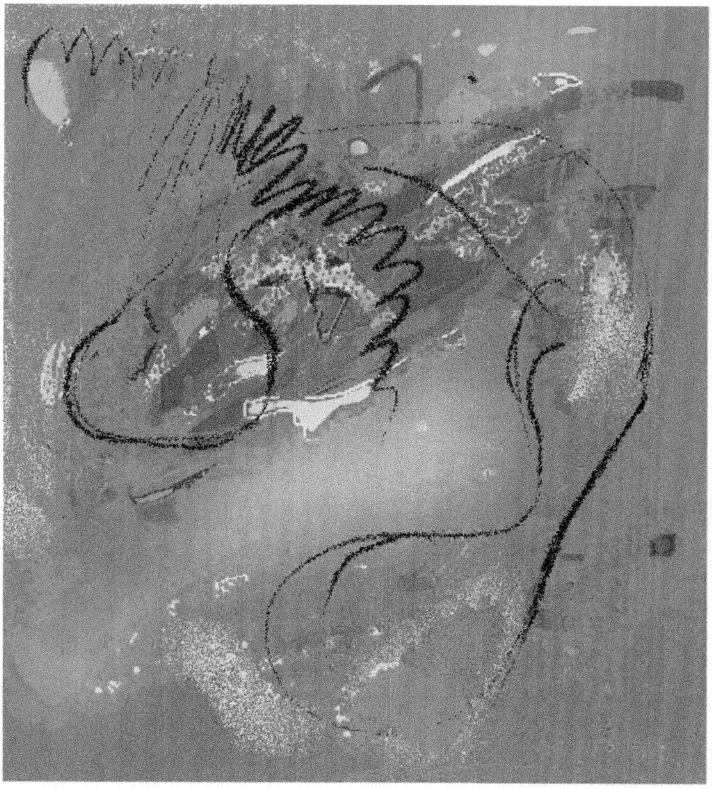

Throughout the previous chapters we limited ourselves to an abstract view of a quantum computer as a device suitable to conduct a special class of thoughts or Gedankenexperiments in computing and communication. It is now time to address the question of the physical implementation of quantum computing and quantum information theory concepts discussed earlier. We wish to see if indeed it is feasible to build quantum computers and, if so, what type of technological advances and scientific breakthroughs should occur before the powerful concepts presented earlier could have a practical impact.

The reason we deferred this topic to the very end of our journey is twofold. First, several physical implementations are possible, and at this time it is very difficult to assess in full the relative merits and potential of each of these approaches. Significant progress is reported in each of these areas, but a real breakthrough has not been achieved in any of them. Second, the physical phenomena involved in each of these areas are very complex and nontrivial to explain in detail.

The basic problems facing a physical implementation of a quantum computer are related to finding a system that is sufficiently insulated from unwanted interactions with the environment but, at the same time, can be controlled through some required interactions implementing logical quantum gates, and can be measured.

So far, there have been several serious proposals for quantum computing and quantum communication physical implementations, and they fall into several basic categories [473]: (1) nuclear magnetic resonance systems, (2) ion traps, (3) neutral atoms, (4) optical systems, (5) mesoscopic superconducting systems with tunnel junctions, and (6) solid-state systems (quantum dots, impurities). Among other unique qubit implementations, we mention incompressible higher order fractional quantum Hall states with non-Abelian braiding statistics that have been suggested as a good candidate for topological quantum computing.

Figure 6.1 summarizes the organization of this chapter, which starts with an analysis of the requirements for a physical implementation of a quantum information processing systems. From the multitude of proposals for the physical implementation of quantum information processing devices, we have selected only a subset. First, we discuss the cold ion traps, then the liquid nuclear magnetic resonance (NMR); we continue with quantum dots and with an in-depth discussion of the quantum Hall effect, and we conclude with a summary presentation of photonics.

6.1 REQUIREMENTS FOR PHYSICAL IMPLEMENTATIONS OF QUANTUM COMPUTERS

Quantum computing and communication devices must satisfy a set of generic requirements; these requirements guide the implementation efforts for quantum information processing systems. DiVincenzo formulated several criteria for realization of quantum computers in 2000; a decade later, in 2010, these criteria were reformulated and generalized for systems with low level of decoherence.

DiVincenzo s Criteria

In his seminal paper on the physical implementation of quantum computers [130], DiVincenzo formulated five criteria related to quantum computation plus two additional ones required for transfer of information:

First criterion: *The physical system has well-characterized qubits and is scalable.* The embodiment of a qubit is a quantum two-level system, such as the two spin states of a spin-½ particle, or the ground and some excited state of an atom, or the vertical and horizontal polarization of a single photon. The qubits can be in superposition states and/or entangled states, and these unique quantum mechanical characteristics are associated with the enormous computational power of quantum computing devices. These characteristics can be effective only if the qubits are well isolated from the environment; the process of decoherence due to the possible interaction with the environment must be very slow.

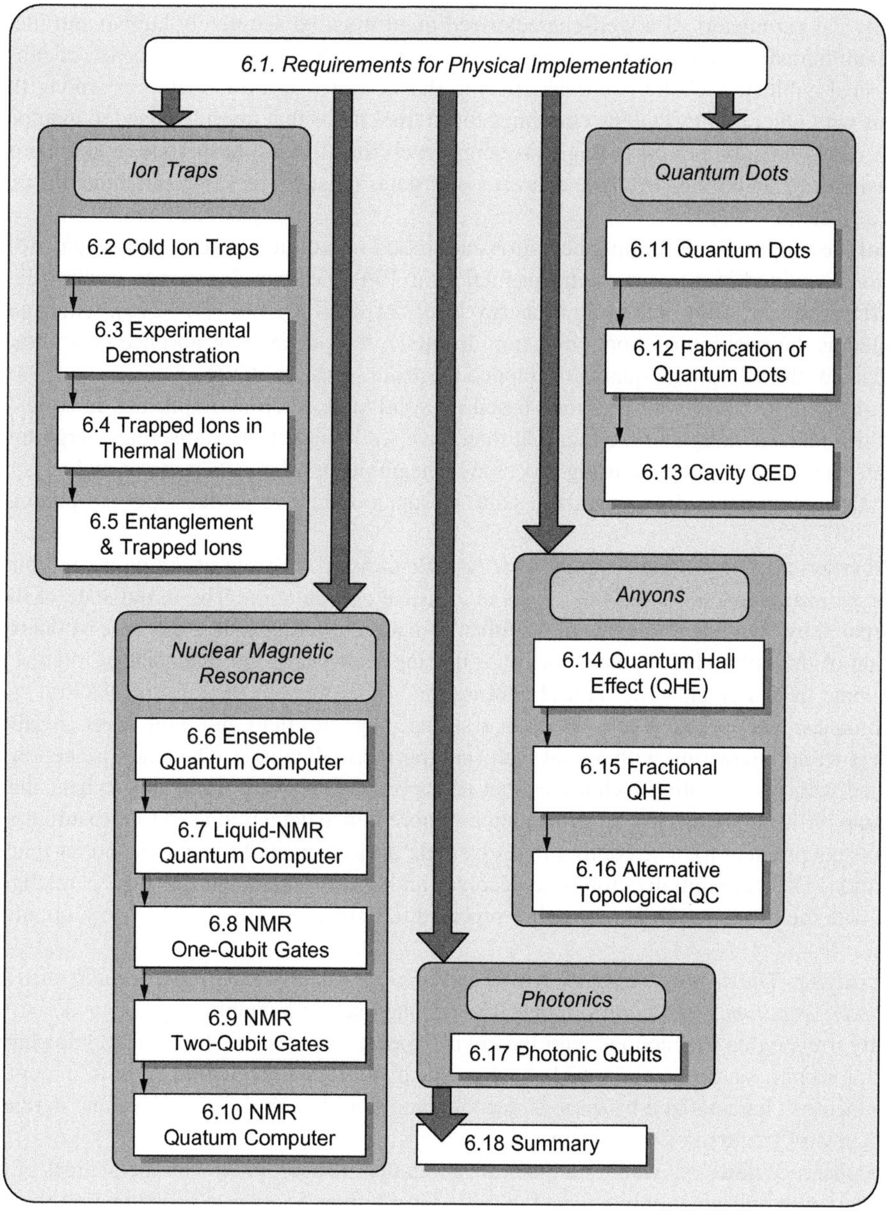

FIGURE 6.1

The organization of Chapter 6 at a glance.

The physical parameters of a well-characterized qubit must be accurately known and include the internal Hamiltonian (it determines the energy eigenstates of the qubit), the presence of other states of the physical qubit besides the two characteristic (basis) states and the couplings among them, the interactions with other qubits, and the couplings to external fields that might be used to manipulate the qubit. In general, the qubit has other (higher) energy levels than the two basis states and the probability of the transitions to such states from the characteristic states must be very low and under the control of the physical system.

Some of the proposed qubit implementations are based in atomic physics, such as pairs of energy levels of ions confined in a linear electromagnetic trap [96], Zeeman-degenerate ground states used in the NMR approach [104], and atomic energy levels of neutral atoms [75, 218]. Other qubits are implemented as the position of atoms in a trap or lattice, the presence or absence of a photon in an optical cavity, or the vibrational quanta of trapped electrons, ions, or atoms.

Some of the physical implementations based on solid state, such as impurities in solids and the quantum dots, take advantage of the fact that they have well-characterized discrete energy levels that can be controlled through manufacturing processes. The qubits in solid state systems include spin states or charge states of quantum dots, quantized states of superconductive devices, Cooper-pair charge, or the flux.

Second criterion: *The physical system must be able to have the state of the qubits initialized to a known low-entropy state, such as* 000 ... , as in classical computation. The initial state of the qubits can be prepared by "cooling," which can be either by natural cooling (when the state of interest is the ground state of the qubit Hamiltonian), or by projecting the system into the state of interest through the appropriate measurement. The natural cooling method is used in the case of electron spin resonance techniques: The electrons are placed in a strong magnetic field and their spins are allowed to align with it while interacting with a heat bath (natural thermalization). The state projection method is associated with laser cooling techniques that are applied to cooling ion states to near the ground state in a trap [96] and are usually accompanied by fluorescence detection used to measure the state of these ions (state projection). The cooling times by state projection could be much shorter than by natural relaxation. The state initialization time becomes an important factor if this time is relatively long compared with the gate operation time and/or repeated initialization is required by, say, quantum error correction.

Third criterion: *The decoherence times of the physical system implementing the qubit must be much longer than the quantum gate operation time.* The decoherence of a quantum system is due to thermodynamically irreversible interactions with the environment; it represents the principal mechanism for the transition from quantum to classical behavior. The evolution of a quantum system in contact with its environment is characterized by various decoherence times; each decoherence time is related to a different degree of freedom of the system.

The decoherence times relevant for a quantum computer are associated with the degrees of freedom that characterize the physical qubits (see Table 5.12 in Chapter 5); they also depend on the specifics of the qubits' couplings to these degrees of freedom. For example, the decoherence time of the spin of an impurity in a perfect semiconductor depends on its location, whether it is in the bulk of the solid or near the surface of the device and how close it is to the structures used to manipulate its quantum state [130]. The decoherence times must be long enough to allow the quantum system to evolve undisturbed until the computation is complete. No physical system can be completely free of decoherence, but decoherence times of the order of $(10^4 \quad 10^5)$ (*gate operation time*) are considered acceptable for fault-tolerant quantum computation,[1] in which case the error probabilities are lower than a critical threshold. The reality is that quantum systems with long decoherence times of the order mentioned above are relatively hard to find.

Fourth criterion: *A physical system as the embodiment of a quantum computer must have available a "universal" set of quantum logic gates.* A quantum computer executes a sequence of unitary transformations, $U_1, U_2, U_3, \ldots, U_n$, as specified by a quantum algorithm, with each transformation acting on one, two, or, at most, three qubits. All these unitary operations are implemented by a "universal" set of quantum gates; a convenient set of universal quantum gates contains one two-qubit gate plus a set of one-qubit gates.

The physical implementation of a quantum algorithm requires (1) identifying the Hamiltonians that generate the unitary transformations specified by the algorithm, given that

$$U_1 \quad e^{it\mathbf{H}_1/\hbar},\ U_2 \quad e^{it\mathbf{H}_2/\hbar},\ U_3 \quad e^{it\mathbf{H}_3/\hbar}, \ldots, U_n \quad e^{it\mathbf{H}_n/\hbar};$$

(2) designing the physical apparatus so that the interaction Hamiltonians, $\mathbf{H}_1, \mathbf{H}_2, \mathbf{H}_3, \ldots, \mathbf{H}_n$, are successively turned on at precise moments of time and for predetermined lengths of time. In reality, only some types of Hamiltonians can be turned on and off as requested by a quantum computation. Most of the physical implementations proposed so far consider only two-body interactions—i.e., two-qubit unitary transformations (gates) of the type CNOT. As we have seen in Section 1.17, three-qubit unitary transformations can be expressed in terms of sequences of one- and two-qubit interactions [16]. In case two-qubit interactions are not easily achievable, the CNOT gate can be constructed using the appropriate sequence of one-qubit interactions.

One-qubit gates act on either the phase or the excitation of a single qubit. A *phase gate* performs a unitary transformation of the type

$$U_z^{(1)}(\varphi) \quad e^{\ \ i\varphi\sigma_z}.$$

The state vector executes a rotation, with angle, φ, about the z-axis in a Bloch sphere representation; the characteristic states become

$$\begin{array}{lll} 0 & e^{i\varphi/2} & 0\,, \\ 1 & e^{\ \ i\varphi/2} & 1\,. \end{array}$$

An *excitation gate* performs the unitary transformation,

$$U_x^{(1)}(\theta) \quad e^{\ \ i\theta\sigma_x}.$$

The state vector executes a rotation, with angle, θ, about the x-axis in a Bloch sphere representation; the characteristic states become

$$\begin{array}{lll} 0 & \cos\theta\ 0 & i\sin\theta\ 1\,, \\ 1 & i\sin\theta\ 0 & \cos\theta\ 1\,. \end{array}$$

[1] *Gate operation time*, the time for the execution of an individual quantum gate represents the *clock time* of a quantum computer.

Two-qubit gates carry out controlled transformations of the second qubit state conditioned by the state of the first one. The controlled transformations can be of two types:

1. Controlled-NOT (CNOT):

$$U^{(2)}_{\text{CNOT}} \quad 0 \; 0 \quad I \quad 1 \; 1 \quad \sigma_x.$$

The gate changes the states of the two qubits in the well-known manner

$$\begin{array}{llll} 0,0 & 0,0\,, & 1,0 & 1,1\,, \\ 0,1 & 0,1\,, & 1,1 & 1,0\,. \end{array}$$

2. Controlled-Phase (C-Phase):

$$U^{(2)}_{\pi} \quad 0 \; 0 \quad I \quad 1 \; 1 \quad \sigma_x.$$

The states of the two qubits are transformed as follows:

$$\begin{array}{llll} 0,0 & 0,0\,, & 1,0 & 1,0\,, \\ 0,1 & 0,1\,, & 1,1 & 1,1\,. \end{array}$$

Here are two examples of universal sets of gates:

$$S_1 : \left\{ U^{(2)}_{\text{CNOT}}, U^{(1)}_x(\pi/4), U^{(1)}_z(\varphi), \varphi \quad [0, 2\pi] \right\},$$

$$S_2 : \left\{ U^{(2)}_{\pi}, U^{(1)}_z(\pi/4), U^{(1)}_x(\varphi), \varphi \quad [0, 2\pi] \right\}.$$

The time profile of a two-qubit interaction is usually described as a square pulse. In reality, the time profile is determined by requirements of each specific implementation of the interaction Hamiltonian. For example, when the interaction Hamiltonian could also couple the qubit state to other energy levels (states) of the quantum system (other than the levels involved in the embodiment of the qubit), then the desired transformation can be obtained by turning on and off the interaction slowly enough to make it compatible with an adiabatic approximation; the front and tail of the pulse are far from vertical. The adiabatic condition is satisfied if the duration of the pulse front is long enough.

The time duration of the interaction pulse is controlled by the maximum size attained by the matrix elements of, $\mathbf{H}(t)$, since the quantum gate action is determined only by the time integral, $\int \mathbf{H}(t)dt$. The maximum size of the matrix elements can be determined by fundamental constraints, such as the requirement that the linear approximation should be applicable to the system, and by practical constraints, such as the laser power that can be delivered to a particular ion. In the end, the clock time of the quantum computer is determined by the time interval needed to ensure that two consecutive pulses have negligible overlap.

The time profile of the interaction Hamiltonian, $\mathbf{H}(t)$, is physically controlled by "classical means," such as the intensity of a laser beam, the value of the gate voltage, or the current intensity in a wire, which are also quantum mechanical in nature at the atomic level. The fact that these control devices behave classically means that there should be no entanglement between their quantum states and the states of any physical implementation of a quantum computer.

For some physical implementations, such as the ion trap, there is no direct interaction between the atomic-level qubits. The two-qubit interaction is mediated by a special quantum subsystem that can interact with each of the qubits involved in a controlled gate: in the case of the ion trap, the collective vibrational state of the ion chain in the trap controls the interaction between qubits. Such auxiliary quantum systems introduce new sources of decoherence.

The time profile of the two-qubit interaction in an ion trap is influenced by the timing of the mediating subsystem and the decoherence effects occurring during gate operation. The fact that in some implementations there is a single subsystem that mediates each two-qubit interaction can affect the fully parallel gate operations required for successful error correction. We have to expect that in any physical implementation the quantum gates could not be implemented perfectly and the realization of the associated Hamiltonians would be accompanied by systematic errors (avoidable through careful calibration) and random errors (correctable with different techniques).

Fifth criterion: *The physical system must have a qubit-speci c measurement capability.* At the end of a computation the result is read out by measuring specific qubits. The measurement represents an interaction between qubits and the measurement apparatus, and this is an irreversible process. Realistic measurements of quantum systems are expected to have very low quantum efficiencies (100%); the probability of obtaining a certain value for an output qubit as a result of one measurement is very low. The efficiency of a measurement of a quantum computer output could be "increased" either by rerunning the computation for a number of times, or by copying the value of a single read-out qubit[2] to several ancilla qubits using CNOT gates and measuring each of them. In the case of a nondemolition measurement, a quantum efficiency of 1% could ensure reliable quantum computation if compounded with hundreds of "copy and remeasure" operations. For example, the bulk model of NMR implementation proposed [104] to perform successful quantum computations when using quantum efficiencies much lower than 1% and macroscopic quantities of copies of the same qubit (a large number of a certain molecule in solution). The final nondemolition measurement was done as an ensemble average over the whole macroscopic sample, while individual qubits were left almost undisturbed. If the measurement were performed on a time scale of 10 ^{4} of the decoherence time, its repeated execution during the quantum computation could simplify the process of error correction.

According to DiVincenzo [130], physical systems involved in both quantum communication and quantum computation must satisfy two additional requirements relevant for the task of transmitting intact qubits from one place to another:

First additional requirement: *The system must have the ability to interconvert stationary and ying qubits.* The term "flying qubits" refers to the physical qubits that are transmitted from place to place, while the "stationary qubits" represent the physical qubits for reliable local computation. The qubits encoded either in the polarization or in the spatial wave function of photons are the preferred flying qubits at this time. The light transmission through optical fibers is a well-developed technology and reliable enough for the transmission of qubits even at long distance. The unloading of the qubits from the quantum computer into a transmission system and loading them up through a reversed process presents technical difficulties.

Second additional requirement: *The ying qubits must be transmitted between speci ed locations without affecting their state.* The preservation of the photon quantum state during transmission through

---

[2]The state of a qubit cannot be copied, but the value of a qubit in a given basis can be copied.

optical fibers or through atmosphere has been the main concern of many experiments in quantum cryptography [215], but not of quantum computing yet.

The physical systems under consideration at this time as quantum computing and information processing devices fulfill most of these requirements, but not all of them. The systems based on trapped ions (atoms) and on cavity quantum electrodynamics, where the qubits are ions or atoms and their manipulation is performed with laser light, have the advantage that quantum phenomena can be observed very clearly; measurements performed with such systems (cavity-QED is at the core of atomic clocks) are the most precise that exits at this time.

Experimentalists have been able to implement certain quantum gates [269, 297], to create entanglement with trapped cold ions [192, 270, 363], and with cold atoms in optical lattices [281], to entangle up to six atoms [271], to demonstrate quantum teleportation with ions [315] and with atoms [20, 357], and to execute quantum error correction [89]. The disadvantage of these systems is the difficulty to scale them up for computations with a large number of atoms.

The solid state systems proposed as physical implementations have the advantage of being easily scalable and of using well-perfected manufacturing technologies. But their qubits (quantum dots or Cooper pairs) are difficult to protect against decoherence due to interactions with other atoms, with impurities, or phonons[3] present in the solid lattice environment. One- and two-qubit gates have been realized experimentally with quantum dots [276, 378].

In NMR systems the qubits are atoms within the same molecule and the NMR technique is used to manipulate them. The NMR technique is well established, and the state of the qubits can be controlled with precision. As a proof of the concept, Shor's algorithm was implemented on a system with seven qubits (five ^{19}F and two ^{13}C spin-½ nuclei) in a quantum computer molecule (the perfluorobutadienyl iron complex) and the number 15 was factorized [93]. However, the NMR systems have their disadvantages: The cooling of the molecules is necessary in order to preserve the benefits of quantum computation, and scaling up to a large number of atoms is limited by the difficulty of engineering molecules with larger and larger number of atoms (qubits).

Entanglement has also been realized with photonic systems using parametric down-conversion and conditional detection [318].

General Criteria for Systems with Low-Level Decoherence

The original criteria of DiVincenzo cannot be easily applied to some of the new emerging concepts; they can be rephrased [262] into three more general criteria based on the assumption that these can be achieved if the decoherence is kept at a low enough level. These new criteria are (1) scalability, (2) universal logic, and (3) correctability.

1. For any quantum system, *scalability* is achieved by adding new, well-characterized qubits. The qubit entanglement, allowed by quantum mechanics, makes the logic space available to a quantum system of N spin qubits very large; it is described by a group known as $SU(2^N)$. The space of

[3]Phonons are quanta of energy exchanged by collective vibrational modes in periodic solids. The atoms in the regular lattice of a solid material are connected through bonds that only allow them to vibrate about equilibrium positions like quantum harmonic oscillators. The atoms can not vibrate independently, and their vibrations take the form of collective modes that propagate through the material and exchange energy only in discrete amounts. Like the photons of electromagnetic energy, the phonons obey Bose-Einstein statistics.

N unentangled spins is described by a much smaller group, $SU(2)^{N}$. At the same time, the size and energy consumption of a quantum computer grows linearly with the number of qubits, N. The scalability of a technology depends on the scalability of the classical physical systems used to define the qubits; they may be microchips, lasers, microwave electronics, cryogenic refrigerators, or high vacuum systems.

2. *Universal logic.* This criterion requires the system to have a finite set of control operations. In the case of qubits, the set of universal logic gates may include nearly analogue single-qubit gates (such as spin-qubit arbitrary rotations) and any digital two-qubit entangling logic operation, such as the CNOT gate.

   The standard "circuit model" for quantum computing has been shown to have some promising alternatives such as the adiabatic and the cluster-state quantum computation.

   An adiabatic algorithm [145] follows the continuous-time evolution of a quantum system governed by a smoothly varying time-dependent Hamiltonian. In the adiabatic quantum computation, the answer to the calculation is defined as the ground state [294] of a whole network of interactions between qubits. By slowly turning on those interactions, the qubits are being adiabatically evolved from an initial Hamiltonian with a ground state easy to construct into the final Hamiltonian whose ground state is the solution of the computation. The state of the system will be very close to the desired final state if the evolution time is long enough; the required time depends on the energy difference between the two lowest states of the interpolating Hamiltonian. An adiabatic quantum computation does not require fast quantum logic operations and measurements. The universal logic criterion applies in this case with regard to the complexity of the available set of interactions, the length of time it takes to turn on those interactions, and how low is the temperature the system must be kept at.

   The other alternative, the cluster-state quantum computation [349], starts in one particular entangled quantum state, the so-called cluster state, of a large number of qubits; this represents the entire resource for the quantum computation. The information is written onto the cluster, processed by propagating it through the cluster. The computation is performed by changing the way in which the resulting wave function is measured; the qubits are measured in arbitrary bases and that provides the analogue component required by universal logic. The measurements are used to imprint a quantum circuit on the cluster state, but they are destroying its entanglements at the same time. The result of the computation is read out of the cluster using only one-particle measurements. The implementations for newer concepts, such as the adiabatic and the cluster state quantum computers could be simpler for some technologies; they are equivalent in computing power to gate-based quantum computers [294, 350].

3. *Correctability* requires that any quantum error correction protocol devised for any physical implementation should be able to maintain the desired state of the system by eliminating unwanted entropy introduced from the environment into the quantum computer while adding/dropping qubits required by encoding/decoding operations. That can be done through a combination of efficient state initialization and measurement. The initialization process cools the quantum system into a low-entropy state. The measurements help determine the state of the quantum system quickly and with the accuracy allowed by quantum mechanics. A quantum nondemolition measurement combines these two operations: It initializes the quantum system by projecting it to the measured (basis) state, and this state remains the same even after repeated measurements.

Actual quantum computers must have the capability of simultaneously controlling the quantum systems, measuring them, and switching between these operations, while preserving the quantum systems in high isolation from their environment.

Many physical systems proposed as candidates for a quantum information processing device, such as trapped ions, optical photons, quantum dots, and cavity quantum electrodynamics, show some promise in preserving quantum coherence long enough to allow the execution of the basic tasks of such a device.

6.2 COLD ION TRAPS

The cold ion trap was the first technology proposed for quantum information processing devices; it still is very promising and very popular. A set of N cold ions interacting with laser light and moving in a linear trap was proposed by Cirac and Zoller [96] and was demonstrated experimentally by Monroe *et al* [297] as a realistic physical system to implement a quantum computer. The ion traps use laser-induced coupling between the electron state and the vibrational motion of relatively heavy charged particles (ions) trapped in vacuum, at very low temperature.

The system consists of, N, spin-½ ions confined in the electrostatic field engineered inside the linear trap. The field is supposed to be sufficiently anisotropic and the ions sufficiently cold that they should lie "crystalized," aligned in a "string-of-pearls" formation along the longitudinal axis of the trap where the effective trapping potential is the weakest. This string of ions is the "register" of the quantum computer. The ions are laser (Doppler) cooled in all three dimensions to very low temperature of a few milliKelvin; at this temperature, they undergo very small oscillations around the equilibrium position. Their confined motion along the x-, y-, and z-directions can be described in terms of "normal" modes; each normal mode can be treated as an independent harmonic oscillator of frequency $\omega_x \quad \omega_y$ or $\omega_x \quad \omega_z$. There are a total of n such normal modes along the x-axis (the weak axis). The state of a string of N qubits in the trap can be expressed as

$$q_1, q_2, \ldots, q_j, \ldots, q_N \quad n \ ,$$

where the first term, $q_1, q_2, \ldots, q_j, \ldots, q_N$, represents the states of the logical qubits, ($q_j \quad 0$ or $q_j \quad 1$), and the second one, n , represents the normal-mode, vibrational state; the value of n, such as $n \quad 0, 1, 2, \ldots$, refers to the number of phonons present in the collective mode. For example, when $n \quad 0$ (no phonon in the normal mode), the ions are not vibrating and the state is

$$q_1, q_2, \ldots, q_j, \ldots, q_N \quad 0 \ .$$

In the lowest mode, $n \quad 1$ (one phonon present in the normal mode), the center-of-mass mode, the ions oscillate in lock step in the x-direction, the trap axis, with a frequency, ω_1, that coincides with, ω_x, of the harmonic trapping potential, ($\omega_1 \quad \omega_x$), and the state is

$$q_1, q_2, \ldots, q_j, \ldots, q_N \quad 1 \ .$$

Generally, the modes are numbered in order of increasing resonance frequency. The resonant frequencies of the ions' motion are of the order of a few megahertz, thus the wavelengths of the resonant radiation are larger than 100 meters; as the separation of the ions is of the order of $10\,\mu$m, the assumption that the external electric field is constant over the region occupied by the ion string is a good approximation. The normal mode degree of freedom is used as a "bus" to couple different ions together.

In the ion trap system, the ions themselves are the qubits. The two characteristic states of any qubit, j, ($j \quad 1, \ldots, N$), are identified with two internal states of the corresponding ion, such as the *ground state*, $g_j \quad 0_j$ and a particular long-lived metastable, nondegenerate, *excited state*, $e_j \quad 1_j$. The quantum state of each ion is a linear combination of the ground state, g , and the metastable excited state, e . A coherent linear combination of the two levels,

$$\alpha_0 \; g_j \qquad \alpha_1 e^{i\omega t} \; e_j,$$

can survive for a time comparable to the lifetime of the excited state. The relative phase oscillates because of the energy splitting, $\hbar\omega_0$, between the levels.

Qubit Manipulation in an Ion Trap

Individual qubits are independently manipulated by shining a different pulsed laser beam at each of the ions. If the laser beam directed to a particular ion is tuned to the frequency, ω_0, of the transition, $g \quad e_0$, then a periodic exchange of energy can occur between the electromagnetic light field and the two-level system. These periodic exchanges are called *Rabi oscillations*[4]; they are induced between the qubit states, 0 and 1 . During the Rabi cycle the atom/ion is usually in a superposition of ground and excited state. Any one-qubit unitary transformation can be applied, and any desired linear combination of 0 and 1 can be prepared by selecting an exact duration of the laser pulse and by appropriate choice of the laser phase.

The ions (qubits) can be read out by performing a measurement that projects their state onto the g , e_0 basis. That can be done in the following way: a laser is tuned to a transition from the ground state g to a short-lived excited state e , different from the metastable excited state e_0 . When the laser beam shines on the ions, each qubit with the value 0 absorbs and reemits repeatedly the laser light by oscillating between the g and the e levels. The ions emit fluorescent light by de-exciting from state e to the ground state; in this way qubits in state 0 become "visible." The qubits in state 1 remain "dark"; they are in the metastable excited state e_0 , which does not fluoresce. The other (excitation) levels of the ion are not affected by the laser light and do not disturb the computation process; some of them may be needed for implementing quantum gates. The laser acting on one ion will leave the internal states of all the other ions unaffected.

[4]*Rabi oscillations* occur when laser light interacts with a two-level atomic system, such as the ground state and the excited (nondegenerate) state of an atom/ion. They are associated with oscillations of the quantum-mechanical expectations values of the level populations in the ground and excited state and of the photon numbers. That can be interpreted as a periodic change between absorption and *stimulated emission* of photons. *Spontaneous emission* of photons, a competing process, is also possible and can prevent the observation of Rabi oscillations. Rabi oscillations can be described with two different models: the Rabi model, which assumes a quantum description of the atom and a classical description of the field, and the Jaynes-Cummings model, which assumes quantum descriptions of both atom and field.

Single-Qubit Rotations

Rotations of a single qubit, j, can be performed by tuning the laser frequency on resonance with the internal transition between ground and excited state of ion j, $g_j \quad e_{0\,j}$. The Hamiltonian for this interaction is

$$\mathbf{H}_j \quad \frac{\Omega}{2}\left(e_{0\,j\,j}\, g\, e^{\quad i\phi} \quad g_{\,j\,j}\, e_0\, e^{i\phi}\right),$$

where Ω is the Rabi frequency[5] and ϕ is the laser phase. If the laser shines on the ion for an interaction time $t \quad (k\pi/\Omega)$, where $k\pi$ is the length of the laser pulse, with k the laser wave vector, the evolution of the qubit is described by the unitary operator,

$$\mathbf{V}_j^k(\phi) \quad e^{\quad i\mathbf{H}t} \quad \exp\left[\quad ik\frac{\pi}{2}\left(e^{\quad i\phi}\, e_{0\,jj}\, g \quad h.c.\right)\right],$$

where $h.c.$ stands for Hermitian conjugate terms. As a result, the qubit performs a Rabi rotation and the two internal pure states evolve into superpositions:

$$g_j \quad \cos\left(\frac{k\pi}{2}\right) g_j \quad ie^{i\phi}\sin\left(\frac{k\pi}{2}\right) e_{0\,j},$$
$$e_{0\,j} \quad \cos\left(\frac{k\pi}{2}\right) e_{0\,j} \quad ie^{\quad i\phi}\sin\left(\frac{k\pi}{2}\right) g_j.$$

If the ion is in its ground state (qubit state 0), a laser π-pulse will make the electron wave function evolve through half of a Rabi oscillation period and leave the ion in its excited state (qubit state 1); the qubit rotates from the 0 to the 1 state. When the laser pulse duration is reduced by a factor of 2, a $\pi/2$-pulse, the ion is left in an equal superposition of the ground and excited states; the qubit rotates to the $\frac{1}{2}(0 \quad 1)$ state.

Entanglement of the Motional and Internal Degrees of Freedom of Trapped Ions

The coupling of the motion of the ions inside a trap is provided by the Coulomb repulsion between ions, which is much stronger than any other interaction when the typical separations between the ions is of a few optical wavelengths. For example, the average separation between calcium ions in a linear Paul trap is about $10\,\mu$m. The mutual Coulomb repulsion creates a spectrum of coupled normal modes of vibration for the trapped ions. The frequencies of different normal modes are well separated in the excitation spectrum.

The center-of-mass mode in which the ions oscillate in lock step in the harmonic potential well of the trap is the vibrational mode of lowest frequency, (ω_x), along the x-axis. Initially, the ions are laser cooled at a temperature much lower than that corresponding to ω_x to ensure that each vibrational mode occupies its quantum mechanical ground state. When an ion absorbs or emits a laser photon, the center of mass of the ion recoils.

[5]*Rabi frequency*, (Ω), is the rate at which the oscillations occur and is proportional to the amplitude (but not the intensity) of the light field and to the dipole moment of the level transition.

The laser pulses incident on the ion can be tuned to simultaneously cause internal spin transitions and vibrational (phonon) excitations; local (atomic) internal states are thus mapped into shared (collective) phonon states. For certain choices of the laser frequency, ω_L, the Hamiltonian describing the interaction between an ion and the electric field of the laser beam is resonant and the spin and the motion can be coupled efficiently (*sideband coupling*[6]) [459]. For example, for laser frequencies, $\omega_L \quad \omega_0 \quad \omega_x$, the interaction Hamiltonians take resonant forms and the spin transitions, , are accompanied by the corresponding motional mode transitions, $n \quad n \quad 1$. If the laser frequency takes the value, $\omega_L \quad \omega_0$, the Hamiltonian is such that the spin transitions do not change n (i.e., they are no longer accompanied by motional mode transitions).

In certain conditions, the so-called *Lamb-Dicke limit*, and for sufficiently low intensities, the laser beam will cause transitions that modify the state of only one of the collective modes. For example, if the laser is detuned from the resonance frequency, ω_0, by the quantity, $\Delta\omega_L^{(j)}$, which is equal to minus the center-of-mass frequency, ω_x, i.e. $\Delta\omega_L^{(j)} \quad \omega_x$ (*lower motional sideband*), the laser excites the ion internal transition *plus* the center-of-mass motion mode as the only collective vibrational mode.

If the laser addresses the j-th ion in a string of N ions with a frequency tuned to the lower motional sideband of the center-of-mass mode and the equilibrium position of the ion in the linear trap coincides with the node of the laser standing wave, an interaction takes place between the internal states of ion, j, and the center-of-mass mode; the corresponding Hamiltonian for this single ion [96] is

$$\mathbf{H}_{j,q} \quad \frac{\eta}{\overline{N}}\frac{\Omega}{2}\left(e_q\ j\ j\ g \quad a e \quad i\phi \quad g\ j\ j\ e_q \quad a^{\dagger} e^{i\phi}\right).$$

The terms, $a^{\dagger}$ and a, are the creation and annihilation operators of center-of-mass phonons, respectively; Ω is the Rabi frequency; ϕ is the laser phase; and η is the Lamb-Dicke parameter. The subscript, $q \quad 0, 1$, refers to the virtual electron level excited by the laser; the transition depends on the laser polarization. These are the intermediate (virtual, electronically excited) states of the Raman process.[7] The factor, $\overline{N}$, appears because for, N, ions the effective mass of the center-of-mass motion is NM (M is the mass of the ion), and the amplitude of the mode is proportional to $1/\ \overline{NM}$. It is assumed that $(\Omega/2\omega_x)^2\eta^2 \quad 1$ and in the Lamb-Dicke limit[8] $\eta \quad 1$.

If this laser shines on the ion for a time, $t \quad k\pi/(\Omega\eta/\ \overline{N})$ (a $k\pi$ laser pulse), the evolution of the system represents a mapping of the spin qubit to the motion qubit and is described by the

[6]Sideband coupling means that the frequency of the laser, ω_L, must be equal to the corresponding internal transition frequency, ω_0, minus/plus the frequency of the trap, ω_x; in this way it excites only a center-of-mass phonon. If $\omega_L \quad \omega_0 \quad \omega_x$ (*red-sideband coupling*), the spin transitions, , are accompanied by motional mode transitions, $n \quad n \quad 1$. If $\omega_L \quad \omega_0 \quad \omega_x$ (*blue-sideband coupling*), the spin transitions are accompanied by motional mode transitions $n \quad n \quad 1$. When $\omega_L \quad \omega_0$, the spin transitions do not change n.

[7]Raman process or Raman scattering refers to the *inelastic* scattering of photons (laser light) interacting with matter, ions in our case. A photon is absorbed and subsequently emitted via an intermediate electron state corresponding to a virtual energy level. An energy exchange occurs between the incident photon and the ion and, as a result, the energy of the emitted photon is equal to the difference between some vibrational/rotational electron energy levels of the ion.

[8]In the Lamb-Dicke limit, the vibrational amplitude of the ion is considered to be much smaller than the wavelength of the light. In our case, this condition ensures that a single ion is localized in a region much smaller than the wavelength of the laser light acting upon it.

unitary operator

$$\mathbf{U}_j^{k,q} \quad e^{\quad i\mathbf{H}_{j,q}t} \quad \exp\left[\quad ik\frac{\pi}{2}(\ e_q \ _j \ _j \ g \ ae^{\quad i\phi} \quad h.c.)\right].$$

The operator describing the evolution of the system is unitary if the laser pulse duration is proportional to $k\pi$, with k, the laser wave vector. This transformation does not change the system state, $g \ _j \ 0 \ _c$, but transforms coherently the states, $g \ _j \ 1 \ _c$ and $e_q \ _j \ 0 \ _c$, to

$$g \ _j \ 1 \ _c \quad \cos\left(\frac{k\pi}{2}\right) \ g \ _j \ 1 \ _c \quad ie^{i\phi}\sin\left(\frac{k\pi}{2}\right) \ e_q \ _j \ 0 \ _c,$$

$$e \ _j \ 0 \ _c \quad \cos\left(\frac{k\pi}{2}\right) \ e_q \ _j \ 0 \ _c \quad ie^{i\phi}\sin\left(\frac{k\pi}{2}\right) \ g \ _j \ 1 \ _c,$$

where $0 \ _c$ and $1 \ _c$ represent the center-of-mass states with zero phonon (vacuum), respectively, one phonon. According to the last transition, the laser pulse removes a bit of information stored as an internal excited state of the j-th ion and transfers it to the collective state of motion of all the ions. In this way, the state of motion of the i-th ion (where $i \quad j$) is going to be influenced by the change of internal state of the j-th ion. The entanglement between the spin of internal atomic states and motion is evident since the final state cannot be factored out into a product of spin and motional wave functions.

Two-Qubit Gate

This gate, by its very nature, requires the entanglement of two qubits and that is difficult to realize with laser-cooled trapped ions where the interaction between two two-level systems must be mediated by an auxiliary system (phonons or photons), which becomes a new source of decoherence.

The first model of a universal two-qubit gate proposed by Cirac and Zoller in 1995 [96] assumes two ions in a linear Paul trap[9] interacting with two lasers, and their state transformation is defined by,

$$\psi_1 \ _i \ \psi_2 \ _j \quad (\quad 1)^{\psi_1\psi_2} \ \psi_1 \ _i \ \psi_2 \ _j, \ (\ \psi_{1,2} \quad 0 \ , \ 1 \ ),$$

where ψ_1 and ψ_2 are two internal states (ground state and an excited state), respectively, for each of the ions, i and j, involved; they represent the qubit basis states. The implementation of this two-qubit model phase gate would require three steps:

1. A π laser pulse is applied to the lower motional (red) sideband of the transition, $g \ _i \quad e_0 \ _i$, of the i-th ion; the laser pulse has phase, $\phi \quad 0$, and polarization, $q \quad 0$. The evolution of ion, i, is described by the unitary operator, $\mathbf{U}_i^{1,0} \quad \mathbf{U}_i^{1,0}(0)$.

[9]The Paul trap now being used in many applications for storing ions is a three-dimensional version of the two-dimensional mass filter patented by Wolfgang Paul (Nobel Prize in physics 1989) and his student Steiwedel. An electric quadrupole made up of two pairs of electrodes, A-A and B-B (the symmetry axes of the pairs are perpendicular to each other), works as a two-dimensional mass filter if a D.C. voltage in addition of a radio-frequency oscillating, A.C. voltage is applied to the electrode pair A-A. The same voltages but of opposite signs are applied to the other electrode pair B-B. In a linear trap, the electrodes are parallel and create a saddle-shaped electric field; the ions are trapped by this electric field and kept levitated in the potential well created along the long axis of the electrodes. The electrode alignment and the machining requirements are critical constraints for a three-dimensional trap.

2. The laser directed on the j-th ion is next turned on for a time corresponding to a 2π pulse; this laser pulse has phase $\phi \quad 0$, polarization $q \quad 1$, and is applied to the lower motional (red) sideband of the transition from ground state to an intermediate excited state, $g_j \quad e_{1\,j}$, of the second ion, j. The lasers are applied to the lower motional sideband, $(\omega_L \quad \omega_0 \quad \omega_x)$, in order to excite a center-of-mass phonon only. The unitary operator corresponding to this action, $\mathbf{U}_j^{2,1}$, changes the sign of the state, $g_j\ 1_c$, which is rotated through the auxiliary state, $e_{1\,j}\ 0_c$ (the subscript, c, stands for center-of-mass mode); no other state is affected in the process.
3. The first step is repeated and a π laser pulse of phase, $\phi \quad 0$, and polarization, $q \quad 0$, is applied to the i-th ion; the unitary operator is, as above, $\mathbf{U}_i^{1,0}$.

The entire process can be described by the unitary operation

$$\mathbf{U}_{i,j} \quad \mathbf{U}_i^{1,0}\mathbf{U}_j^{2,1}\mathbf{U}_i^{1,0},$$

where the superscripts of each term reflect the laser pulse length (wave vector number k) and polarization. This unitary operation has the following effect on the possible states of the system:

$$\begin{array}{cccc}
 & \mathbf{U}_i^{1,0} & \mathbf{U}_{i,j}^{2,1} & \mathbf{U}_j^{1,0} \\
g_i\, g_j\, 0_c & g_i\, g_j\, 0_c & g_i\, g_j\, 0_c & g_i\, g_j\, 0_c, \\
g_i\, e_{0\,j}\, 0_c & g_i\, e_{0\,j}\, 0_c & g_i\, e_{0\,j}\, 0_c & g_i\, e_{0\,j}\, 0_c, \\
e_{0\,i}\, g_j\, 0_c & i\, g_i\, g_j\, 1_c & i\, g_i\, g_j\, 1_c & e_{0\,i}\, g_j\, 0_c, \\
e_{0\,i}\, e_{0\,j}\, 0_c & i\, g_i\, e_{0\,j}\, 1_c & i\, g_i\, e_{0\,j}\, 1_c & e_{0\,i}\, e_{0\,j}\, 0_c.
\end{array}$$

We notice that the sign of the state of the system of two ions coupled through the center-of-mass mode, c, changes (it acquires a phase) only when both ions, i and j, are initially excited. The state of the center-of-mass mode is restored to 0_c after the process.

A CNOT gate can be realized using this controlled-phase change coupled with an appropriate individual one-qubit rotation. This individual rotation acts on a single ion without modifying the center-of-mass motion involving both ions. The rotation can be performed using a laser with frequency on resonance with the internal transition between g_j and $e_{0\,j}$ of the second ion (laser detuning $\Delta\omega_L^{(j)} \quad 0$) and with polarization $q \quad 0$ while the equilibrium position of the ion coincides with the antinode of the laser standing wave. The Hamiltonian corresponding to the laser interaction with the second ion, j, is

$$\mathbf{H}_j \quad \frac{\Omega}{2}\left(e^{\quad i\phi}\ e_{0\,j\,j}\, g \quad e^{i\phi}\ g_{j\,j}\, e_0\right).$$

If we use a $k\pi$ laser pulse, the interaction time is, $t \quad k\pi/\Omega$, and the interaction process can be described by the unitary evolution operator

$$\mathbf{V}_j^k(\phi) \quad \exp\left[\quad ik\frac{\pi}{2}(e^{\quad i\phi}\ e_{0\,j\,j}\, g \quad h.c.)\right].$$

As a result of the interaction with the laser pulse, the states of the second ion, j, evolve into superpositions

$$g_j \qquad \cos\left(\frac{k\pi}{2}\right) g_j \quad ie^{i\phi}\sin\left(\frac{k\pi}{2}\right) e_{0\;j},$$

$$e_{0\;j} \qquad \cos\left(\frac{k\pi}{2}\right) e_{0\;j} \quad ie^{\quad i\phi}\sin\left(\frac{k\pi}{2}\right) g_j.$$

The complete CNOT gate for the states, $\psi_i \, \psi_j$, where $\psi \quad g, e_0$, is

$$G_{\text{CNOT}} \quad \mathbf{V}_j^{1/2}\left(\frac{\pi}{2}\right)\mathbf{U}_{i,j}\mathbf{V}_j^{1/2}\left(\quad\frac{\pi}{2}\right).$$

The gate has an intermediate step whereby the atomic amplitudes corresponding to the qubits are "hidden" in a third internal atomic level e_1 reached through a unitary transformation, $\mathbf{U}_j^{k\;1,q\;1}$. The rotations induced by the 2π laser pulse (unitary transformation $\mathbf{U}_j^{k\;2,q\;1}$) proceed through this state and selectively change the sign of the atomic amplitudes. In this scheme, the two-qubit gate was realized using the collective center-of-mass mode as an auxiliary quantum degree of freedom.

Three-Qubit Gate

The three-qubit gate can be implemented in the same way as the two-qubit gate, but through a five-step process involving ions i,j, and l:

1. A π laser pulse of phase, $\phi \quad 0$, and polarization, $q \quad 0$, excites the i-th ion (same as step 1 before).
2. A π laser pulse of phase, $\phi \quad 0$, and polarization, $q \quad 1$, is directed on the j-th ion (same as step 2 before, but with a π pulse).
3. A 2π laser pulse of phase, $\phi \quad 0$, and polarization, $q \quad 1$, is directed on the l-th ion (same as step 2 before, but with ion l).
4. A π laser pulse of phase, $\phi \quad 0$, and polarization, $q \quad 1$, is directed on the j-th ion (same as step 2 of this process).
5. A π laser pulse of phase, $\phi \quad 0$, and polarization, $q \quad 0$, excites the i-th ion (same as step 1 of this process).

In this case, the sign of the state is changed only if all three ions were initially excited. This procedure can be extended to the case of many ions.

The implementation of quantum gates following the procedure proposed by Cirac and Zoller [96] is based on two key ideas:

1. The non-local entanglement between individual qubits (ions) is achieved by transferring the internal atomic coherence (Rabi oscillation between ground and excited state) to and from the, c, center-of-mass motion involving all ions.
2. The gate has an intermediate step whereby the amplitudes of the atomic states corresponding to the qubits are "hidden" in a third internal atomic level e_1 ; the 2π laser pulse induces 2π rotations via this state and selectively changes the sign of the atomic state amplitudes. After the complete gate operation, no population would be left in the auxiliary atomic and center-of-mass levels.

Their procedure for implementing a two-qubit gate uses only product states as inputs and single-qubit measurements on the outputs [339]. In this way they avoid the preparation of Bell states to be used as inputs and performing Bell measurements on the outputs.

However, numerical calculations [339] have shown that only for Rabi frequencies much smaller than the trap frequency, $(\Omega \quad \omega_x)$, does the evolution of the system given by the Hamiltonians of Cirac and Zoller [96] correspond to their model two-qubit phase gate. For finite Rabi frequencies, the gate operation is not ideal; population is left in the auxiliary states, and we have an error in the gate operation. In fact, an error-detection procedure could be implemented by probing the population of these intermediate states with a laser inducing the fluorescence of such a state after each gate operation. The full characterization of the two-qubit gate implies the preparation of various initial states and the *quantum tomography* of the corresponding output states. The quantum tomography is the process that allows the complete characterization of an unknown quantum state that can be repeatedly prepared; it consists of finding an appropriate sequence of measurements that allows us to determine the complete density matrix, ρ, of the state.

Several research groups have shown that a controlled-NOT gate could be performed using two internal levels of a single ion and a phonon state [90, 298, 338, 399].

Advantages and Problems of Ion Traps

The ion traps have significant advantages: (1) many of the techniques required to prepare and manipulate quantum states have already been developed for precision spectroscopy work; (2) the decoherence rates associated with the decay of the excited ionic state and the damping of the ionic motion can be made small in the virtually perturbation-free environment of an ion trap (ultra-high vacuum, low electromagnetic noise level) and by using metastable atomic states; and (3) there is an experimentally demonstrated technique, quantum jumps, for reading out the result of a computation with high probability.

Major problems in ion traps, such as spontaneous transitions in the vibrational motion due to heating, internal radio-frequency (RF) transitions in the ions driven by thermal radiation, and experimental instabilities in the laser beam power, in trap-related RF voltages and mechanical vibrations, and fluctuating external magnetic fields were recognized early on [297, 406]. The accuracy of a quantum gate is decreased by the instabilities contributing to heating the ions and by the laser pulse not being exactly the right frequency and duration.

One basic feature of the Cirac and Zoller two-qubit gate implementation scheme—cooling to the ground state of motion of the trap—has proved to be very difficult to achieve. Initially, this basic step has been reliably realized in experiments for one ion [297] and two ions [426], which also were the first experimental demonstrations of a quantum logic gate.

Experimental Realization of Cold Ion Traps

Trapped ions have made significant experimental progress and demonstrated the essential characteristics as quantum registers of a quantum computer:

1. A trapped ion can be initialized to a known quantum state. Early experiments (1989) of Diedrich *et al* [122] showed that a single ion, $(^{198}\mathrm{Hg}\ \ )$, trapped in a Paul RF trap and laser cooled in the resolved sideband regime had its kinetic energy reduced to a value where the ion was spending

about 95% of its time in the ground state level of its confining potential well. A 99.9% probability of cooling to the ground state of vibration was measured for a single Ca ion in a Paul trap by Roos *et al* [360] and coherent quantum state manipulation was demonstrated for more than 30 periods of Rabi oscillations. A register of two ions was prepared in the ground state of their collective modes of motion almost nine years later by King *et al* [237]; two ^{9}Be ions trapped in an elliptical Paul RF trap were cooled and probed with laser beams. The two-qubit gate of Monroe *et al* [297] was implemented using a single trapped ion and a motional mode and two of the ion's internal states as the qubits.

2. Final states can be efficiently detected. The fluorescence of the ions was monitored to detect the internal state, or , with discrimination efficiency of almost 100% [360].
3. The decoherence times of qubits are relatively long. Coherence times longer than 1ms were observed for Rabi oscillations between the internal states, $S_{1/2}$ $D_{5/2}$, of a single trapped ion of ^{40}Ca prepared by sideband cooling in the lowest motional (vacuum) state by Schmidt-Kaler *et al* [364]. One logical qubit encoded into the decoherence-free subspace of a pair of trapped ^{9}Be ions is protected from collective dephasing, a major source of qubit decoherence for trapped ions; the storage time of the logical qubit increases by up to an order of magnitude [234] in comparison with a qubit encoded in a single ion.
4. Quantum logic gates of a universal set have been experimentally demonstrated. Single-qubit gates, rotations of a single physical qubit, are easy to implement [458] and much simpler than in other systems. Two-qubit gates have been realized based on slightly different concepts: entanglement of two qubits implemented as a single physical ion with conditional dynamics depending either on the wavelength of the applied laser radiation [298], or on the size of the atomic wave packet [112]; and entanglement of two qubits implemented as two ions with motion strongly coupled through their Coulomb repulsion and described in terms of the center-of-mass and *stretch*[10] normal modes [269, 365]. The qubit logical states are two hyperfine ground states of the ^{9}Be ion in [269]; in [365], the logical states of the qubits are the ground state and a metastable state of the ^{40}Ca ion and the phonon-number of the motional mode). Two-qubit gates with trapped ions, as implemented so far, have their limitations. Ideally, they should be independent of temperature and need not be cooled to their ground state for accurate operation, should allow ions to be packed close together for stronger interaction, and should be fast in order to minimize the effects of decoherence during gate operation and to speed up computation (gate time is limited by low laser intensity required for spectroscopic resolution of motional sidebands of the ions).
5. States of two and four qubits have been entangled [292, 363, 400, 426].

As first proposed, quantum computing with trapped ions involved a string of ions confined in a single trap, internal electron states of ions were used as qubit basis states, and quantum information was transferred between ions through their mutual Coulomb interaction. The ion trap system satisfied all the requirements mentioned above for a quantum register. The manipulation of a large number of ions in a single trap presents such serious technical difficulties that the scaling up of this scheme is limited to only tens of ions. A quantum computer based on the original Cirac-Zoller model that places all the ions in a single trap could not manipulate the 10^6 ions needed for a full-scale computation.

[10]In the *stretch* normal mode, two ions move in opposite directions with equal displacements at a frequency, ω_s $\overline{3}\omega_c$.

A trap containing 10^6 ions would be limited [120] to a center-of-mass phonon frequency ω_c 100 Hz, where ω_c $1/\ \overline{N}$. At such frequency, the 10^{11} operations required for the Shor algorithm without error correction would be executed in over 10^9 seconds, or about 30 years. In the trap array model of DeVoe [120] with some 64 000 traps, the N 64 ions per trap would have a phonon frequency ω_c 125 kHz; the computation could proceed 10^3 times faster than in the single-trap model. A logical solution to escape this limitation is to use arrays of micro-traps, eventually organized as a number of small ion-trap quantum registers, and quantum communication between them. Some proposals use photon coupling between traps [120, 322], others, spin-dependent Coulomb interactions [98].

Basic experimental steps have been made toward the development of scalable ion trap quantum computing devices with a larger number of ions that can be stored and moved to different regions to perform the gates required by quantum algorithms. A recent proposal of Kielpinski *et al* [235] of a "quantum charge-coupled device" architecture consists of a large number of interconnected ion traps and is based on techniques already demonstrated for manipulating small quantum registers. A few ions can be confined in each trap and manipulated using techniques already demonstrated experimentally or can be moved between traps by changing the voltages of the traps. The device has memory and interaction regions. Ions storing quantum information are confined to traps assigned to the memory regions. Logical operations are executed on the relevant ions that have been moved into the interaction region by applying appropriate transport voltages to the electrode segments in between. Once inside the interaction region, the ions are kept close together to allow the right amount of Coulomb coupling required by entangling gates. Gates are driven by laser beams focused on the interaction region. The trapping and transport potentials are realized through a combination of RF and quasi-static electric fields. The device can be constructed using standard microfabrication techniques. As a proof of concept, a pair of interconnected ion traps have been constructed at the National Institute of Standards and Technology. A qubit (^{40}Ca ion) was coherently transported between the traps through a Ramsey-type experiment without any heating of the ion motion or shortening of the ion lifetime in the trap. The ions used for quantum logic gates could be "sympathetically" cooled by another ion species used as a heat sink in the interaction region, as demonstrated experimentally [61]. Sources of decoherence during transport and positioning of ions could be reduced by several orders of magnitude by encoding each qubit into a decoherence-free subspace of two ions [234].

6.3 FIRST EXPERIMENTAL DEMONSTRATION OF A QUANTUM LOGIC GATE

The first experimental demonstration of a CNOT quantum gate operating on prepared quantum states of two qubits was reported in 1995 by Monroe, Meekhof, King, Itano, and Wineland [297] from the NIST in Boulder, Colorado. Their implementation followed the quantum logic gate scheme proposed by Cirac and Zoller [96]; the characteristic states of the two qubits are two internal and two external states, respectively, of a single, ^9Be , ion stored in a coaxial-resonator RF (Paul), ion trap. The *internal* states are two hyperfine ground states of the ionized beryllium atom, and the *external* states are two quantized harmonic oscillator states (usually denoted by n) associated with this ion's motion inside the trap.

The *target qubit* of the CNOT quantum logic gate, denoted $S^{(t)}$, is implemented as two $^2S_{1/2}$ hyperfine ground states of the ^9Be ion. The quantization axis for the hyperfine states is defined by a static magnetic field, **B** 0.2 mT, applied along the direction, $(\ 1/\ \overline{2})\ x \quad (1/2)(\ y \quad z\ )$, where x, y, z are the trap axes. The hyperfine ground states are the $2s^2S_{1/2}$ F $2, m_F$ 2 and $2s\ ^2S_{1/2}$ F $1, m_F$ 1,[11] usually represented as the spin-½ states, and , respectively. The separation in frequency between the two levels (the hyperfine splitting) is $\omega_0/2\pi$ 1.25 GHz.

Coherent transitions between these levels can be implemented with two laser beams that drive two-photon stimulated Raman transitions. In this case, the laser parameters are given by the differences between the parameters of the two laser beams at the mean position of the ion. Parameters such as the wave vector k, the frequency ω_L, and the phase ϕ become k k_{L1} k_{L2} (a vector assumed to be parallel to the x-axis), ω_L ω_{L1} ω_{L2}, and, respectively, ϕ ϕ_{L1} ϕ_{L2}.

The intermediate, electronically excited Raman states are $2p^2P_{1/2}$, corresponding to the state, e, and $2p^2P_{3/2}$, corresponding to the state, e. These two levels are separated by the fine structure splitting, ω_F, which is assumed to be much larger than the hyperfine splitting, ω_0, ω_0 ω_F; the hyperfine splitting of these two excited levels is considered negligible.

The *control qubit*, denoted $n^{(c)}$, is implemented as the first two quantized harmonic oscillator states 0 and 1 of the trapped ion. The separation in frequency of these two motional states is the vibrational frequency, $\omega_x/2\pi$ 11 MHz, of the trapped ion with a favored motion degree of freedom along the x-axis. The two-qubit register has four basis eigenstates, $n^{(c)}$ $S^{(t)}$ 0 , 0 , 1 , 1 . These states can be individually accessed in the following way: A pair of off-resonance laser beams generated from a single laser are directed to the ion and they are driving stimulated Raman transitions[12] between the states. The two laser beams, R_1, R_2, also called Raman beams, have right/left circular-polarization and linear-polarization, respectively. The beam, R_2, is shined along the direction, ($1/\overline{2}$) x (1/2)(y z), and the beam, R_1, is perpendicular to it [237]. In this way the wave vector difference between the Raman beams is parallel to the x-axis, and the transitions are sensitive to ion motion only along this direction. The Raman laser beams have about 1 mW power each at about 313 nm wavelength and are detuned about 50 MHz red off the $^2S_{1/2}$ - $^2P_{1/2}$ transition of the Be ion. The difference in frequency of the two Raman beams could be tuned in the range 1.20 to 1.30 GHz and the Raman pulse durations.

When the difference frequency, $\Delta\omega_L$, of the laser beams is set at:

1. $\Delta\omega_L$ ω_0 (the *carrier*), transitions $n^{(c)}$ - $n^{(c)}$ are coherently driven between the internal states, $S^{(t)}$, while preserving the external state, $n^{(c)}$; the carrier Rabi frequency for Be is $\Omega_0/2\pi$ 140 kHz.
2. $\Delta\omega_L$ ω_0 ω_x (the *red-sideband*), transitions, 1 - 0 , are coherently driven between the $n^{(c)}$ $S^{(t)}$ states; the red-sideband Rabi frequency is $\eta_x\Omega_0/2\pi$ 30 kHz, where the Lamb-Dicke parameter η_x 0.2.
3. $\Delta\omega_L$ ω_0 ω_x (the *blue-sideband*), transitions, 0 - 1 , are coherently driven between the $n^{(c)}$ $S^{(t)}$ states; the blue-sideband Rabi frequency is $\eta_x\Omega_0/2\pi$ 30 kHz.

---

[11] F represents the total angular momentum quantum number of the atom (nucleus electrons), and m_F is the corresponding magnetic quantum number. The Be nucleus has angular momentum 3/2; thus, the total angular momentum of the ion is F 3/2 1/2 2 and F 3/2 1/2 1 for spin states, and , respectively. The magnetic momentum takes values in the m_F $F, \ldots,$ F range.

[12] In a stimulated Raman transition a two-color (two-frequency) laser pulse transfers the atom/molecule from ground state to a vibrational excitation state if the difference in energy matches an allowed Raman transition.

In the last two cases, when $\Delta\omega_L$ is tuned to one of the sidebands, the stimulated Raman transitions entangle the internal spin, $S^{(t)}$, states with the external motional, $n^{(c)}$ states, and this feature is at the heart of the trapped ion CNOT quantum gate.

Before the actual operation of the CNOT gate, the ^{9}Be ion in the RF trap is Doppler precooled in all three dimensions with right circular-polarized laser beams D_1 and D_2 and then Raman cooled to the lowest ground state, 0 ; the two-qubit register is then prepared in an arbitrary state by applying to the ion the Raman laser pulses appropriately tuned and timed. For example, the eigenstate, 1 , is prepared by applying to the ion in the lowest ground state, 0 , a π pulse on the blue-sideband, $(\Delta\omega_L \quad \omega_0 \quad \omega_x)$, followed by a π pulse on the carrier, $(\Delta\omega_L \quad \omega_0)$, so that

$$0 \xrightarrow{\Delta\omega_L \quad \omega_0 \quad \omega_x} 1 \xrightarrow{\Delta\omega_L \quad \omega_0} 1 .$$

A third right circular-polarized laser beam prevents the optical pumping to the state, $F \quad 2, m_F \quad 1$.

The CNOT operation is implemented by sequentially shining three pulses of the Raman laser beams on the trapped ^{9}Be ion:

1. A $\pi/2$ pulse is applied on the carrier transition, $\Delta\omega_L \quad \omega_0$, and that causes the internal spin state, $S^{(t)}$, to go through a phase shift of $1/4$ of a complete Rabi cycle, while leaving the external state, $n^{(c)}$, unchanged.
2. A 2π pulse is applied on the blue-sideband transition between the internal state, , and an auxiliary atomic level, e , that is not part of any qubit. This auxiliary level is another ground state, the $^2S_{1/2}\ F \quad 2, m_F \quad 0$, Zeeman split from $^2S_{1/2}\ F \quad 2, m_F \quad 2$, the state, and is shifted by about 2.5 MHz when a magnetic field of 0.18 millitesla is applied. The auxiliary Rabi frequency is $\eta_x\Omega_e \quad 12$ kHz.

   This blue-sideband transition reverses the sign of any component of the quantum register in the 1 state by inducing a complete Rabi cycle through the auxiliary state: $1 \quad - \quad 0 \quad e \quad - \quad 1$.
3. A $\pi/2$ pulse is applied on the carrier transition, and the internal spin state $S^{(t)}$ undergoes another phase shift of $1/4$ of a complete Rabi cycle, but leaves the $n^{(c)}$ states unchanged.

As a result of this sequence of pulses, the components of the quantum register in the $n^{(c)} \quad 0$ state are not affected by the blue-sideband transition of step 2, and the effects of the two $\pi/2$ pulses of steps 1 and 3 cancel each other. However, the components of the quantum register in the 1 state change signs as a result of step 2, and the two Raman pulses of steps 1 and 3 add constructively and rotate the target spin by π radians (flip it).

Qubit State Detection

The change in the population of the register after such a sequence of operations can be detected through independent measurements of the control and target qubits. The probability that after a certain operation sequence the target qubit, $S^{(t)}$, is in the state , $\mathcal{P}\ S^{(t)}$, is estimated by measuring the ion fluorescence associated with the cycling transition from the state, $^2S_{1/2}\ F \quad 2, m_F \quad 2 \ - \ ^2P_{3/2}\ F \quad 3, m_F \quad 3$. The transition is induced by resonant excitation with the right circular-polarized D_2 laser beam of 19.4 MHz radiative line width at λ 313 nm. This radiation has a much lower probability of exciting the state state; thus, the detected fluorescence is considered to be proportional to $\mathcal{P}\ S^{(t)}$. The second measurement is supposed to evaluate the probability that the control qubit $n^{(c)}$ is in state 1 , $\mathcal{P}\ n^{(c)}$ 1 . The same operation sequence performed before the first measurement is repeated with the addition of a specific "preparation" Raman laser pulse before a new detection of the target qubit $S^{(t)}$. The "preparation" pulse has the role of mapping the $n^{(c)}$ onto $S^{(t)}$ in the following way:

1. If the first measurement finds the target qubit, $S^{(t)}$, to be in state, the initial operation sequence is repeated; a laser, π, pulse on the red-sideband is beamed at the ion before testing again the state of the target, $S^{(t)}$. If fluorescence is detected, then the control qubit, $n^{(c)}$, is in the state, 0 ; if fluorescence is not detected, the state $n^{(c)}$ is 1 .
2. If the first measurement finds the target qubit, $S^{(t)}$, to be in the state, the initial operation sequence is repeated and a laser, π, pulse on the blue-sideband is shone on the trapped ion before testing again the state of the control qubit, $n^{(c)}$. In this case, the presence of fluorescence indicates the $n^{(c)}$ state is 1 , while the absence of it indicates a state 0 . This procedure is applied before gate operation in order to verify the preparation of the states and to measure the register population and again after the operation of the CNOT gate to measure the population in the resulting register. When the control qubit $n^{(c)}$ is prepared in state 0 , the gate preserves the state of qubit $S^{(t)}$ with high probability; when the control qubit is prepared in state 1 , the CNOT gate flips the state of the target qubit $S^{(t)}$ with high probability.

The switching speed of this CNOT gate is limited at about 20 kHz by the 2π laser pulse in step 2; otherwise, the switching speed would be limited at about 50 MHz, the frequency separation between the control qubit vibrational energy levels. The decoherence rate was measured to be a few kHz and was due to several factors such as instabilities in the laser beam power and in the relative position of the ion and the laser beams; fluctuations of the external magnetic fields; and instabilities in the drive frequency and voltage amplitude of the RF ion trap.

Spontaneous Emission

A fundamental limitation of the coherence of atomic qubits is due to spontaneous emission; the ground state hyperfine levels that form the qubits have a very long radiative lifetime, but the decoherence rate due to spontaneous emission during the Raman transitions between these states is a real problem. The decoherence rate, Rate_{SE}, due to spontaneous emission from ${}^2P$ levels, is estimated as

$$\text{Rate}_{SE} \quad \sum_i \gamma_i p_i,$$

where p_i is the probability that an intermediate excited state is populated and γ_i is the decay rate of that excited state.

The probability of spontaneous emission during the time τ_π that a π laser pulse is acting on the carrier transition is given by

$$\text{Probability}_{SE} \quad \text{Rate}_{SE} \quad \tau_\pi .$$

The probability of spontaneous emission for the $^9\mathrm{Be}$ ion is evaluated [459] to be at least 0.001. For sideband transitions this probability will be increased by $1/\eta$,

$$\mathrm{Probability}_{SE}^{(\mathrm{sideband})} \quad \mathrm{Probability}_{SE}^{(\mathrm{carrier})} \quad \frac{1}{\eta},$$

with the Lamb-Dicke coefficient η 0.2 for the $^9\mathrm{Be}$ ion. Since the probabilities for errors during a gate operation must be on the order of 10 ^{4} or smaller to be able to perform error corrections during computations, the conclusion is that the $^9\mathrm{Be}$ ion would not be a good choice for the ultimate qubit. A good choice of ion should have a small ratio of spontaneous decay rate, γ, to fine structure splitting, ω_F; only then the effects of spontaneous emission could be greatly reduced. The probability of spontaneous emission during a laser carrier π pulse can be approximated as [459]

$$\mathrm{Probability}_{SE} \quad 2 \quad \overline{2\pi}\,\frac{\gamma}{\omega_F}.$$

The following approximations were made:

- The ions have half-integer nuclear spin, I, and a, π, pulse acts on the carrier of the transition

$$F \quad I \quad \frac{1}{2}, m_F \quad 0 \quad - \quad F \quad I \quad \frac{1}{2}, m_F \quad 0 \ ,$$

  for which the Rabi rate is independent of the spin I.
- The value of the decay rate, γ, corresponds to the relevant $^2P_{1/2}$ level.
- The hyperfine level splitting, ω_0, and the decay rate from the intermediate state, γ, are much smaller than the fine level splitting, ω_F, and the Raman lasers' detunings, Δ , respectively, $\omega_0, \gamma \quad \omega_F,\ \Delta$;
- The Raman beams have orthogonal linear polarization (experimentally a convenient choice).

The probabilities of spontaneous emission during a two-photon stimulated Raman transition for various ions of interest for quantum computing calculated in this approximation [459] are presented in Table 6.1. These values correspond to the carrier transition of the, $^2S_{1/2}$, levels

$$F \quad I \quad 1/2, m_F \quad 0 \quad - \quad F \quad I \quad 1/2, m_F \quad 0 \ .$$

From Table 6.1 we see that the spontaneous emission during Raman transitions is practically suppressed for heavier ions such as Cd   or $^{199}\mathrm{Hg}$. Nevertheless, even if heavy ions were to be used, fault-tolerance could be harder to reach for other transitions than the carrier transitions. Transitions such as sideband and two-qubit gates have a higher probability of spontaneous emission in the Lamb-Dicke limit.[13]

Among other factors, the probability of spontaneous emission is proportional to the time, τ_π, a π pulse takes to be executed (the probability is higher for longer τ_π). For sideband transitions and two-qubit gates, the spontaneous emission rate remains the same (as for the carrier transition), but τ_π

---

[13]Quantum logic gates with trapped ions generally require cooling to the Lamb-Dicke limit, η 1, where the spatial extent of the ion's motion is much smaller than the optical coupling wavelength, $\lambda/2\pi$ $1/k$.

Table 6.1 The probability of spontaneous emission during a two-photon stimulated Raman transition for ions of interest for quantum computing with trapped ions. (A π-pulse is applied on the carrier transition of the, ${}^2S_{1/2}$, levels, $F \quad I \quad 1/2, m_F \quad 0 \qquad F \quad I \quad 1/2, m_F \quad 0$.)

Ion	Nuclear Spin I	Carrier Transition (${}^2S_{1/2}$)	Probability of Spontaneous Emission
^{9}Be	3/2	$1, 0 \quad 2, 0$	$8.7 \quad 10^{\ 4}$
^{25}Mg	5/2	$2, 0 \quad 3, 0$	$1.4 \quad 10^{\ 4}$
^{43}Ca	7/2	$3, 0 \quad 4, 0$	$3.0 \quad 10^{\ 5}$
^{67}Zn	5/2	$2, 0 \quad 3, 0$	$2.6 \quad 10^{\ 5}$
^{87}Sr	9/2	$4, 0 \quad 5, 0$	$8.0 \quad 10^{\ 6}$
^{113}Cd	1/2	$0, 0 \quad 1, 0$	$5.3 \quad 10^{\ 6}$
^{199}Hg	1/2	$0, 0 \quad 1, 0$	$1.8 \quad 10^{\ 6}$

scales as the inverse of the Rabi frequency, which, in turn, scales as the carrier frequency, ω_0, times Lamb-Dicke parameter, η (i.e., $\Omega \quad \omega_0 \quad \eta$). Thus,

$$\tau_\pi \quad \frac{\pi}{2\Omega} \quad \frac{\pi}{2(\omega_0 \quad \eta)}.$$

Therefore, for fault-tolerant computation the values of η should not be too small compared to unity; in that case the Lamb-Dicke limit would no longer be a good approximation. Since the Rabi frequencies depend on the motional states, the precise control of the motion, for example, ground state cooling, is absolutely necessary. Error-resistant techniques proposed for fault-tolerant computation, such as adiabatic passage, spin echo, or composite pulses, like those used in NMR implementations, increase the probability of spontaneous emission and are not recommended.

Decoherence associated with unwanted motion in a scaled-up trapped ion quantum computer can be reduced by sympathetic cooling of multiple ion species (^{9}Be and cold ^{26}Mg) or multiple isotopes of the same ion species (^{112}Cd and cold ^{114}Cd) mediated by their Coulomb interactions; the coherence of the internal qubit is preserved during this type of cooling. As shown by Blinov *et al* [61], the Cd system (the pair ^{112}Cd as qubit and ^{114}Cd as refrigerator ion, or the pair ^{111}Cd as qubit and ^{116}Cd as refrigerator ion) is a convenient choice because the sympathetic cooling can be accomplished without extra lasers and without strong focusing.

6.4 TRAPPED IONS IN THERMAL MOTION

The original Cirac-Zoller model for an ion trap implementation of a quantum logic two-qubit gate [96] requires the cooling and maintenance of the trapped ions in their oscillatory center-of-mass quantum ground state. The fidelity of the quantum logic gates realized with trapped ions depends on the quantum state of the collective oscillatory degrees of freedom of these ions. The Coulomb force strongly couples the motion of the ions and, at the same time, makes them susceptible to interaction with external electric fields.

The ions in the center-of-mass mode get "heated" in the process of interacting with external electric fields and the mode becomes more and more excited; the wave function of the ions becomes more spatially spread, causing a random phase shift of the ions and, ultimately, decoherence. The solution to this problem is, on one hand, the understanding of and preempting the causes of the trapped ions heating, and, on the other hand, the use of alternative designs of quantum logic gates that operate without requiring the system to be in its quantum ground state of the phonon modes.

The simple method of avoiding the heating problem when quantum computing with trapped ions proposed by B. E. King *et al* [237] utilizes the "higher" modes ($m > 1$), such as $(x)_{stretch}$, $(xy)_{rocking}$, and $(xz)_{rocking}$, of the ions' collective oscillations inside the trap instead of the center-of-mass mode ($m \quad 1$) first proposed by Cirac and Zoller [96]. The heating rate of these higher motional modes is much smaller than that of the center-of-mass mode [237]. The *heating times*[14] measured experimentally for higher modes of a two-ion system were longer than 5 msec, while the heating times for center-of-mass modes were less than 0.1 msec. The higher modes seem to be well insulated from the influence of external heating fields and can be used as a reliable "quantum information bus."

The heating of the center-of-mass mode that is still present and causes a random phase shift of the ions by spatially spreading the wave function of the ions can be reduced and kept constant by cooling with a separate species of ion. The sequence of pulses required for implementing the CNOT gate in this case is the same as that recommended by Cirac and Zoller [96], but the laser frequencies employed (the sideband corresponding to the stretch mode in question) are different.

This scheme has its drawbacks: (1) the laser-ion coupling to the higher modes varies for different ions and, therefore, different pulse durations are required for different ions; (2) the energy levels of the higher modes are closer together in frequency and that makes them harder to resolve. It is possible that such laser control problems could make the quantum computer algorithm so complex that any speedup achieved through quantum parallelism would be nullified.

The proposal of Poyatos, Cirac, and Zoller [338] to realize a two-qubit logic gate between two ions in a linear trap at finite temperature is based on a concept different from that of the cold ion trap of Cirac and Zoller [96]. This "hot" ion trap implements the CNOT gate by splitting the atomic wave packets in the same way as in atomic interferometry and creates coherent states of the ions' collective oscillations. Under ideal conditions, the operation of the logic gate is independent of the state of motion of the ions without the restriction of the ground center-of-mass mode temperature.

The CNOT gate proposed by Poyatos *et al* is realized in three steps [340]:

1. *State-dependent displacement of the control ion* (ion **1**): A short laser pulse, appropriately tuned, with its wave vector, k, pointing along the main axis, x, of the trap is applied to ion **1** for a time, $t_L \quad \pi/\Omega \quad t_g$, which corresponds to the strong excitation regime, $\Omega \quad \omega_x$. Here, Ω is the Rabi frequency, and t_g is the period of the motion in the trapping potential when the frequencies for the reduced mass and center of mass are commensurable, $t_g \quad \omega_x^{\;1}$. The laser intensity is chosen in such a way that the internal state of the ion is flipped $0 \quad 1$, or $1 \quad 0$. Simultaneously, the laser pulse provides a momentum "kick" to the wave packet of the ion, which will move in

[14] The heating time is defined as the time it takes the occupation number of the mode to increase by one.

either one of two opposite directions depending on its internal state. Thus, when the ion **1** is in the state $|0\rangle$, it will be changed to the state $|1\rangle$ and, at the same time, it will move in one direction. When the ion **1** is in the state $|1\rangle$, it will be changed to the state $|0\rangle$ and it will move in the opposite direction. The choice of direction is associated with the absorption or emission of one phonon by the ion **1**. The wave packet of the control ion (ion **1**) will be split into two wave packets moving in opposite directions (right and left). Because of the strong ion–ion Coulomb interaction, the target ion (ion **2**) will also evolve into two spatially separated wave packets dependent on the internal state of the control ion (ion **1**). If the momentum kick provided by the initial laser pulse is sufficiently strong, then, after a short time, the wave packet associated with the state $|0\rangle$ will be spatially distinct from that associated with the state $|1\rangle$.

2. *Conditional flip* $|0\rangle \rightleftarrows |1\rangle$ *of ion* **2**: After a time, $t_0 = 2\pi/(3\omega_x)$, corresponding to the maximum distance between the wave packets, a laser beam is directed to the target ion (ion **2**) and is focused on the location of precisely that wave packet that arises if the control ion **1** were in the state $|0\rangle$; the laser interacts with that wave packet only. The laser beam does not induce any displacement of the ion; the setup is such that either one laser beam is used and it propagates perpendicular to the x-axis, the confinement axis of the trap, or two laser beams are used and they propagate in opposite directions along the x-axis, depending on the type of internal transition intended to be induced. But, for a given laser-ion interaction time, $t_1 \ll t_g$, an internal transition is induced on the target ion (ion **2**) dependent on the state of the control ion (ion **1**), such that $|0\rangle_2 \rightleftarrows |1\rangle_2$. After some time, the two ions go back to oscillate as free harmonic oscillators.
3. *State-dependent displacement of control ion* (ion **1**): After a time $(t_g - t_0)$, a short laser pulse is applied to the control ion (ion **1**) with a wave vector pointing in the direction of the x-axis as in the first step. The pulse has the same duration, $t_L = \pi/\Omega \ll t_g$, and the state after the interaction coincides with the fundamental two-qubit gate under ideal conditions. This pulse reverses the effect of the first laser pulse (the operators acting on the motional state cancel out), and the control ion is brought back to its original state.

This gate implementation is easy to extend to a system of three ions, but further scaling up is difficult. The fidelity of the gate is limited by factors, such as errors related to the finite size of the wave packets and the small distances between them; the focusing of the laser beam in the second step is extremely difficult. Random electric fields can induce a random relative phase shift for the two packets, as well as a heating of the ions, and that leads to an increase in the size of the ions' wave packets. Proper operation of such a "hot" ion gate could be maintained only with adequate cooling.

6.5 ENTANGLEMENT OF QUBITS IN ION TRAPS

Mølmer-Sørensen Entanglement Method

Mølmer and Sørensen [296] proposed the application of a simple interaction Hamiltonian to two or more ions (spin qubits) in an ion trap by simultaneously illuminating the ions with laser beams of two different colors (frequencies). The action of laser fields with two different frequencies, such that $\omega_L^1 + \omega_L^2 = 2\omega_0$, makes the two-photon process resonant, but neither of the frequencies is resonant with single excitations of the ions. Thus, it is possible to apply classical laser radiation, which cannot be absorbed by a single ion, but which induces two ions to undergo the, $|gg\rangle \to |ee\rangle$, transition simultaneously. If the laser frequencies are $\omega_L^1 \quad \omega_0 \quad \Delta$ and $\omega_L^2 \quad \omega_0 \quad \Delta$, with the detuning Δ close to, but not resonant with, the center-of-mass vibration frequency, there is a high probability that only the intermediate states with one excited ion and a vibrational quantum number raised or lowered by unity are involved in the interaction.

They showed that an interaction Hamiltonian composed of the center-of-mass vibrational energy of the ions string and the internal electronic energy of the ions can be used to create entangled states of multiple particles of the form

$$\psi \quad \frac{1}{2}\left(e^{i\phi_g}\; gg\ldots g \quad e^{i\phi_e}\; ee\ldots e\;\right),$$

where the product states, $gg\ldots g$ and $ee\ldots e$, with N terms each, describe N ions that are all in the same internal state, g or e.

Their method does not require the collective vibrational motion of the ions to be in its ground state; their scheme is based on two quantum mechanical "special effects": (1) the vibrational degrees of freedom used as intermediate states in the communication processes between ions have only a "virtual" role (they are not populated), and (2) the transition paths involving such unpopulated vibrational states interfere destructively, eliminating the dependence of decay rates and evolution frequencies on vibrational quantum number.

Sørensen and Mølmer proposed another scheme for a two-qubit (two-ion) gate where each ion in the ion trap is addressed with a single laser (like in the original scheme of Cirac and Zoller), but quantum logic operation of the gate is performed through off-resonant laser pulses as their entanglement method prescribes [399]. The two ions are illuminated with lasers detuned close to the upper and lower sidebands. The bichromatic light selects certain virtually excited intermediate states and, by choosing appropriate parameters, the desired internal state dynamics of the ions may ultimately be achieved, even if the vibrational collective motion used to couple the ions is not in its ground state.

The qubits are characterized by the states, gn and en, in each ion, where g and e are the internal ion states, ground and excited, respectively, and n is the collective motional state of the ion string, with n, the quantum number for the main vibrational mode of the trap; the vibrational mode can be other than the vibrational ground state and the trapped ions are called "hot." The intermediate vibrational states under consideration are $g(n \quad 1)$, $g(n \quad 1)$ and $e(n \quad 1)$, $e(n \quad 1)$. The laser addressing the first ion has a detuning close to the upper sideband, $\omega_L^{(1)} \quad \omega_0 \quad \delta$, which is close to resonance with the joint vibrational and internal excitation of the ion. The laser addressing the second ion has a detuning that is the negative of the detuning for the first ion, $\omega_L^{(2)} \quad \omega_0 \quad \delta$, and also close to resonance with the joint vibrational and internal excitation state of the ion. Here, ω_0 is the exact resonance frequency for the levels, gn and en, and is the same for the two ions; δ is the detuning interval. The two laser beams shining simultaneously on the two ions induce the coupling of the states, $g^{(1)}g^{(2)}n$ and $e^{(1)}e^{(2)}n$, using two equally probable pathways through the intermediate states, $g^{(1)}e^{(2)}(n \quad 1)$ and $e^{(1)}g^{(2)}(n \quad 1)$; i.e.,

$$g^{(1)}g^{(2)}n \qquad \left\{ e^{(1)}g^{(2)}(n \quad 1),\; g^{(1)}e^{(2)}(n \quad 1) \right\} \qquad e^{(1)}e^{(2)}n\,.$$

The detunings are chosen sufficiently close to the sideband in such a way that only one collective degree of vibrational excitation (the main vibrational mode of the trap) is taken into consideration, and the intermediate states, $e^{(1)}g^{(2)}(n \quad 1)$ and $g^{(1)}e^{(2)}(n \quad 1)$, are not populated in the process; they are "virtually" excited. Because of the opposite detunings, the factors containing n cancel out and the coherent evolution of the internal ionic state becomes insensitive to the vibrational quantum numbers; the coherent oscillation from gg to ee can be observed even when the ions are in any superposition or mixture of vibrational states and even when the vibrational motion exchanges energy with a thermal reservoir.

The heating of the vibrational motion is an important source of decoherence for the cold trapped ion systems. The Sørensen-Mølmer scheme can be made insensitive to the interaction with the environment. Their ion trap is assumed to be operating in the Lamb-Dicke limit where the vibrational quantum numbers n ensure that $\eta \ \overline{n \quad 1} \quad 1$, with η, the Lamb-Dicke parameter. This condition is fulfilled even for n values well higher than unity; the value $n \quad 1$ is associated with the vibrational ground state, the starting state for all the cold ion schemes.

Experimental Entanglement of Multiple Qubits in an Ion Trap

The entanglement of up to four spin qubits using the Sørensen-Mølmer technique [296, 399, 400] was realized experimentally by a group at NIST in Boulder, Colorado [363]. The experiment was performed with two and four, $^9\mathrm{Be}$, ions confined along the weak x-axis of a linear miniature radio-frequency trap [427]. Initially, both the center-of-mass and stretch modes were cooled to near their ground state. The characteristic states of the spin-½ qubit were two spectrally resolved ground state hyperfine sublevels,

$$^2S_{1/2}\ F \quad 2, m_F \quad 2 \text{ and } \quad ^2S_{1/2}\ F \quad 1, m_F \quad 1\ ,$$

with F, the total angular momentum quantum number, and m_F, the quantum number associated with the projection of the total angular momentum along the quantization axis defined by an externally applied magnetic field.

The ion trap used in the experiment was run at the center-of-mass frequency, ω_x, corresponding to the symmetric normal mode with the lowest energy. The next normal mode in frequency was the antisymmetric (stretch) mode[15]; the corresponding stretch mode frequencies were $\overline{3}\omega_x$ for two ions and $\overline{29/5}\omega_x$ for four ions. The Raman difference frequencies were tuned close to these asymmetric modes in the process of generating entanglement.

The ions were Doppler-cooled to the ground state temperature of both center-of-mass and stretch modes; at these temperatures the ions were strongly coupled by Coulomb interaction and formed a rigid crystal, a string lying along the trap main axis. The motional ground state of the ion string was prepared by repeating several times the following procedure: A pulse with $\theta \quad \pi$ driven at a difference frequency, $\Delta\omega_L \quad \omega_0 \quad \omega_i$ (the red sideband), was cooling down the ions while optical pumping was bringing them from state back to state [233]. The frequency ω_0 corresponds to the hyperfine splitting, and the frequencies ω_i characterize the normal modes of the quantized collective motion of the ions. The initial state was 0 for $N \quad 2$ ions and 0 for $N \quad 4$ ions.

The ions in the state were detected by illuminating all the ions with laser light with 313 nm wavelength and left circular polarization; this light induces the transition to the hyperfine sublevel, $^2\mathrm{P}_{3/2}\ F \quad 3, m_F \quad 3$. Dipole selection rules dictate that this level could decay only to the, , state, and transition to the other hyperfine sublevel and the qubit state was forbidden.

[15]In the antisymmetric *stretch mode*, alternating ions in a string oscillate out of phase in opposite directions.

The states, and , were coherently coupled through stimulated Raman transitions induced by shining two laser beams on the trapped ions. Controlled rotations of a single qubit were driven by laser beams tuned about midway between the states, $^2P_{3/2}$ and $^2P_{1/2}$.

The Rabi flopping, , was observed when the difference in frequency of the two Raman laser beams was close to the hyperfine splitting $\omega_0/2\pi$ 1.25 GHz; the state evolution during Rabi transitions was

$$\cos\frac{\theta}{2} \qquad e^{i\phi}\sin\frac{\theta}{2} \quad ,$$

$$\cos\frac{\theta}{2} \qquad e^{i\phi}\sin\frac{\theta}{2} \quad .$$

Here, the angle θ is proportional to the Raman pulse duration, and ϕ, the *ion phase*, is the phase difference between the Raman beams at the position of the ion. The phases ϕ_1 and ϕ_2 of two ions can be controlled independently by a combination of two techniques: changing the phase of the RF synthesizer that controls the Raman difference frequency and changing the strength of the ion trap field that controls the ion spacing. In this way, single-qubit rotations can be performed on one ion without affecting the other.

At the end of the interaction, the ions had evolved from the initial state, , into the maximally entangled states,

$$\varphi_e^2 \quad \frac{1}{\sqrt{2}}(\qquad i \qquad) \text{ for two ions}$$

and

$$\varphi_e^4 \quad \frac{1}{\sqrt{2}}(\qquad i \qquad) \text{ for four ions.}$$

The motional quantum state of the ions was controlled by driving stimulated Raman transitions that couple spin and motion. The difference wave vector resulting from the composition of the wave vectors of the two Raman beams was oriented along the string of ions (the trap main axis) and exerted a dipole force only along this direction.

The lasers were detuned about Δ 80 GHz blue-sideband of the, $2P_{1/2}$, excited state; their intensities give a Rabi oscillation frequency, $\Omega/2\pi$ 500 kHz. The stretch-mode frequency for the two- and four-ion experiments was 8.8 MHz and the corresponding Lamb-Dicke parameter was η $0.23/N^{1/2}$. The two driving frequencies required for the entanglement operation were generated by frequency-modulating one of the Raman beams. The Raman difference frequencies were chosen close to the motional mode which had equal amplitudes of motion for all ions; thus, the Raman Rabi frequencies were the same for all ions under uniform illumination.

The two-step transition, 0 to 0 , was driven by laser beams with frequencies, $(\omega_0 \quad \omega_i \quad \Delta)$ and $(\omega_0 \quad \omega_i \quad \Delta)$. When the detuning was sufficiently large in comparison with the transition linewidth, $\Delta > \eta\,\Omega$ (in this case, 80 GHz > 360 kHz), the intermediate states, 1 and 1 , were negligibly occupied (practically, no population of these states occurred through collective motion excitation). The interaction Hamiltonian in the rotating wave approximation[16] and the Lamb-Dicke limit is given by

$$\mathbf{H} \quad \frac{\hbar\overline{\Omega}}{2}(\qquad\qquad),$$

with $\overline{\Omega}$ $\eta^2\Omega^2/\Delta$, Ω, the Rabi frequency of the single-ion internal transition, $-$, and η, the Lamb-Dicke parameter. The Lamb-Dicke parameter is given by η $(\hbar k^2/2M\omega_i)^{1/2}$, with $\hbar k$, the momentum transfer and M the total particle mass involved in an excitation.

Entanglement is achieved by applying the Hamiltonian, H, for a time, t $\pi/2\overline{\Omega}$; it creates the spin state, ψ_2 (i)/ $\overline{2}$. In this process, the intermediate states, 1 and 1 , are negligibly populated if, as mentioned above, $\Delta > \eta\,\Omega$. On the other hand, from the expression of the Hamiltonian we see that the entanglement speed is maximized for small Δ. However, according to Sørensen and Mølmer [400], the technique still works for detuning values Δ $\eta\,\Omega$. For select values of the detuning Δ given by the condition,

$$\frac{\Delta}{\eta\Omega} \quad 2\ \overline{m}, \quad \text{for any integer}\, m,$$

the collective motion excitation disappears at the moment the entangled state is created. The NIST experiment was operated at a detuning

$$\Delta \quad 2\eta\Omega, \quad \text{for } m \quad 1.$$

The ion states after the entanglement operation were probed by illuminating them with a circularly polarized laser beam tuned to the cycling transition

$$2S_{1/2}(F \quad 2, m_F \quad 2) \quad 2P_{3/2}(F \quad 3, m_F \quad 3).$$

Thus, each ion in the state, $2S_{1/2}$, was identified by the bright fluorescent light emitted by the level, $2P_{3/2}$, excited through this transition; the ions in the state, , remained dark.

Each experiment was repeated 1000 times under the same conditions in order to determine with good accuracy the probabilities, P_j, that, j, ions are in the state, . The probabilities, $P_{\ldots}$ and $P_{\ldots}$, that all N ions (N 2,4) are in this very same state were large compared to the probabilities for the other cases; the probabilities for the intermediate cases were nonzero, and that indicated that the entangled states of N 2 and N 4 ions were not generated with perfect accuracy.

To prove entanglement, the *delity* of the experimental state defined as

$$F \quad \psi_e\, \rho\, \psi_e ,$$

where the density matrix, ρ, describes the experimental state and approximates the entangled state, ψ_e , must exceed $1/2$ as stated by Sackett *et al* [363]. The fidelity can be written as a sum of diagonal

[16]In the *rotating-wave approximation*, used in atom optics and magnetic resonance the terms in the interaction Hamiltonian that oscillate rapidly are neglected. This approximation is applied when the laser frequency is near an atomic resonance and its intensity is low.

and off-diagonal density matrix elements,

$$F \quad \frac{1}{2}(P(\ldots) \quad P(\ldots)) \quad \rho \quad \ldots \, , \, \ldots \quad .$$

Experimentally, the populations in the states, ... and ... , $P(\ldots)$ and $P(\ldots)$, respectively, are measured directly by detecting the number of ions in the state, ... ; the far-off-diagonal density matrix elements, $\rho_{\ldots , \ldots}$, are also measured experimentally after further manipulation of the ions. A Raman laser pulse with a certain duration and phase ϕ is applied and all ions are rotated together with an angle, $\theta \quad \pi/2$, and phases, $\phi_i \quad \phi$, relative to the entangling pulse. The parity Π of the ion string was measured as a function of the phase ϕ; for a string of N ions the parity was estimated to be given by

$$\Pi \quad 2 \, \rho_{\ldots , \ldots} \quad \cos N\phi$$

if only far-off-diagonal coherence elements are present (all the other off-diagonal coherence elements are absent). The value of the parity is $\Pi \quad 1$ if an even number of ions is in the state ... and $\Pi \quad 1$ if an odd number of ions is in the state ... .

In the case of two ions, the entanglement was estimated at $E \quad 0.5$. For two qubits Wootters [461] gives a formula to calculate "entanglement of formation" of an arbitrary state as a function of the density matrix of the system. This entanglement quantifies the resources necessary to create a given state. An entanglement of 0.5 indicates that two pairs of trapped ions entangled in the NIST experiment carry the same quantum information as a perfectly entangled pair. For the case of four ions, there is no similar formula and the entanglement was estimated at, $E_4 \quad 0.35$, from the term, $\rho_{\ldots , \ldots}$, measured experimentally.

The decoherence was stronger in the case of the four-ion experiment than in the two-ion one. The experiments showed that a four-qubit logic gate could be realized using trapped ions. The prevailing decoherence for the trapped ions is *collective dephasing*, which transforms each qubit state in the following way: - and - $e^{i\zeta}$, where ζ is an unknown phase. The states, and , of two qubits, if used as basis states, will protect the qubits against collective dephasing in a decoherence-free subspace; any superposition of these states is invariant under collective dephasing. A logical qubit can be defined using these two states in the decoherence-free subspace of two physical qubits.

A Single-Ion CNOT Gate Using Individual Addressing

The gate was realized by a group at Innsbruck University [365] using two ^{40}Ca ions in a linear Paul trap individually addressed with focused laser beams. They applied precise control of atomic phases and composite pulse sequences adapted from NMR techniques. The qubit basis states were the, $S_{1/2}$, ground state and the, $D_{5/2}$, metastable state (lifetime of about 1 second) as 0 and 1 , respectively. After an initial Doppler cooling, the electronic qubit states of the two ions were initialized by optical pumping and the collective motional breathing mode was prepared in the state, $n \quad 0$, using sympathetic sideband cooling; the ground state population was about 99%. The phonon number of the collective vibrational mode also forms a qubit, with n 1 and n 0 as the corresponding logical states, 0 and 1 . The sequence of operations implementing the gate follows the Cirac-Zoller proposal:

1. A π-pulse is applied on the blue sideband (all sideband pulses are on the blue sideband) of the first ion, which is the control ion, and maps the internal state of the ion to a corresponding state of the collective vibrational mode; thus, the quantum information carried by the control qubit is inscribed in the vibrational mode.
2. The second ion is addressed and a single-ion CNOT gate is performed between this ion and the collective vibrational mode. If the collective mode qubit is in state 1 (no vibration is present in this mode, n 0), the internal state of the second ion is flipped. The single-ion CNOT gate operation is the result of a sequence of pulses on the target ion: a carrier $\pi/2$-pulse (Ramsey pulse), a combination of four pulses implementing different rotation angles (determined by the lengths of Rabi pulses) about different rotation axes (determined by the phases of the exciting radiation), and ending with another carrier $\pi/2$-pulse. The composite pulse sequence replaces the 2π-rotation through an auxiliary excited level as proposed by Cirac and Zoller [96].
3. A π-pulse on the blue sideband applied to the first ion restores the control qubit and the collective mode to their initial states. The gate operation will flip the internal state of the second ion when the first ion was prepared in the, $D_{5/2}$, state and will leave it in its original state if the first ion was prepared in the, $S_{1/2}$, state. The state detection of each qubit in this experiment is performed with an efficiency of 98%. The fidelity of transferring the initially prepared eigenstates into the final target states is 80%, and the fidelity of creating an entangled state is F 0.71. Higher fidelity of more than 90% could be obtained for reduced laser frequency noise and improved control of the addressing beam; that requires more complex composite pulse techniques and quantum control techniques that correct experimental imperfections, such as residual thermal excitation and inaccuracies in pulse shaping.

Large-Scale Ion-Trap Quantum Computing Device

The original proposal and the subsequent studies of an ion-trap–based quantum computation device assumed a string of ions confined in a single ion trap; such a system proved that a quantum state could be efficiently prepared, manipulated, and read out and that gates could be executed with relatively high fidelity. In the end it was understood that manipulation of a large number of ions in a single trap was limited by technical difficulties involving, among other factors, precise control of electric and magnetic fields, very accurate tailoring of laser pulses length, and control of laser frequencies. The solutions proposed to avoid the limitations and to allow large-scale quantum computing systems involve small ion-trap quantum registers with quantum communication channels based on techniques such as photon coupling [120, 408], spin-dependent Coulomb interactions [98], or a combination of RF and quasistatic electric fields [235].

Cirac and Zoller [98] proposed a scalable trapped ion implementation of a quantum computer where the ions are placed in separated, but closely spaced harmonic oscillator wells and are individually addressed by laser beams. A laser pulse applies a momentum kick to the two ions dependent on the state of the first ion, as in the original proposal. But this time, the combination of the wave packets undergoing a displacement in the harmonic potential of each ion and their mutual Coulomb repulsion results in a state-dependent phase shift, and that can be used as a universal quantum gate. A scaled-up version would contain an array of separate, single-ion traps; a "head" ion that can be moved in the vicinity of any of the ions trapped in the array would be used as the control bit in a multiple CNOT gate.

The large-scale ion-trap array of Kielpinski, Monroe, and Wineland [235] has a "quantum charge-coupled device" (QCCD) type of architecture. It has a large number of interconnected traps; a few ions can be confined at a time in each trap or moved from storage (memory) to interaction registers just by changing the RF and the quasistatic electric (D.C.) voltages applied to the trap electrodes. The electrodes are segmented and arranged in three layers: two outer layers where RF voltages are applied and a middle layer for the quasistatic fields so that ions can be individually confined in a particular region or moved along the local trap axis. Ions encoding quantum information are stored in memory registers and are transported to an interaction region when a logical gate has to be performed. Gates are driven by laser beams focused through the interaction region where the ions are trapped close together to allow Coulomb coupling to play its role in entangling qubits.

The QCCD-type architecture proposed uses quantum manipulation techniques that have been experimentally demonstrated for small quantum registers [60]. Solid state ion traps have been built from alumina with gold evaporated electrodes [232] or with heavily doped silicon electrodes, using standard microfabrication techniques. Sympathetic cooling using a second species of ions as a heat sink in the interaction region will keep gate errors at a low level; decoherence during ion storage, gate execution, and ion transport can be reduced by several orders of magnitude by encoding each qubit into a decoherence-free subspace of two ions [234].

Trapped ions and atoms in small experimental systems have coherence times many orders of magnitude longer than times required for initialization, qubit control, and measurement. Preserving the high-fidelity control demonstrated in these small systems while scaling to larger architectures represents a serious challenge.

6.6 NUCLEAR MAGNETIC RESONANCE: ENSEMBLE QUANTUM COMPUTING

The magnetic induction by nuclear spins, the physical phenomenon on which *nuclear magnetic resonance* is based, was first observed in 1946 by the group of Edward Purcell [346] and by Felix Bloch [62]. Many atomic nuclei have a nonzero magnetic moment and behave as small bar magnets in response to applied external magnetic fields. The magnetic moment of a single nucleus is hard to detect by current technologies, but in a sufficiently large population of nuclei the contributions of the nuclear spins add up, and they can be observed as an ensemble using the nuclear magnetic resonance techniques. NMR has been used for more than 40 years to study chemical structure, dynamics, and reactions; it has become a relatively widely used medical imaging procedure under the name MRI (magnetic resonance imaging).

The very long coherence times associated with the nuclear spins motivated David Cory [103, 104] and, independently, Neil Gershenfeld and Isaac Chuang [170] to propose in the mid-1990s the NMR as a working concept for a quantum computer. The standard NMR spectroscopy and the NMR implementation of quantum computing use the same device and similar instrumentation, but are quite different in their goals. The liquid-state NMR has so far been used to demonstrate Shor's factoring algorithm by using a seven-qubit custom-synthesized molecule [93, 431] to factor the number 15.

One of the molecules used for quantum computing proof of concept is that of trichloroethylene, Cl_3HC_2, where the atom of the regular ^{12}C isotope (a spin-zero nucleus) is replaced with that of the ^{13}C isotope (a spin-½ nucleus). The hydrogen nucleus (a proton) of each trichloroethylene molecule has a strong magnetic moment; when the liquid sample that contains a very large number of Cl_3HC_2 molecules is placed in a powerful external magnetic field, the spin of each proton aligns itself with this magnetic field that determines the *z*-axis of the system.

The spins can be induced to change direction with RF pulses, and as soon as they tip off axis, the applied static magnetic field induces a rapid precession of the proton spins. The precession represents a rotation of the spin vector around the main axis, in this case, the *z*-axis determined by the external magnetic field; the precession frequency, also called Larmor frequency, ω_L, is related to the strength, $\mathbf{B}_{ext}$, of the external field,

$$\omega_L \quad \mu_p \mathbf{B}_{ext},$$

where μ_p 42.7 MHz/tesla is the magnetic moment of the proton. In a typical magnetic field of strength, $\mathbf{B}_{ext}$ 11.7 tesla, the precession frequency is ω_L 500 MHz.

The precession of the proton (hydrogen nucleus) spin induces oscillating currents tuned to the precession frequency in a coil placed around the sample. These oscillating currents induce a magnetic field that is detected, thus allowing the observation of the entire ensemble of protons in the sample. This is the magnetic induction by nuclear spin [62, 346] observed in 1946. At least 10^{15} molecules have to be present in the sample to be able to observe the magnetic induction signal. In general, the NMR sample contains about 10^{18} molecules.

NMR has important applications such as molecular structure determination, molecular dynamics studies in liquid, and solid state; the application of NMR to quantum computing is based on some of the same techniques as the molecular structure determination by NMR. But, instead of studying new molecules and trying to determine their structure, it uses synthesized molecules with well-defined nuclear spins that are manipulated as quantum information qubits; the molecule internal Hamiltonian and relaxation operator are precisely known and define the effective overall transformation produced by a complex series of operations applied to that sample.

In fact, the magnetic induction of spin had an early connection to information storage. In the mid-1950s, the magnetic induction was proposed to be used for classical information storage by Arthur Anderson and coworkers [11, 12] based on the *spin echo technique.* The spin echo technique corrects for the inhomogeneity of the applied static magnetic field. When the magnetic field is not constant across the sample, the precession frequency varies with the location of the spins. In this case, the induction magnetic fields produced by the spins are not aligned and they do not add up to produce a good magnetic induction signal; as a result, the induction signal vanishes.

With the spin echo technique, two pulses of the RF field are applied separated by a time interval *t*, and a strong nuclear induction signal is observed at a time, *t*, after the second pulse. The spin echo reverses the spins and, thus, reverses their precession, until their induction magnetic fields are all aligned again. The spin relaxation process, the main source of decoherence in this case, tends to realign the spins with the applied magnetic field (the transverse magnetization decays to zero in time), randomizes their phases, and leads to complete loss of the information encoded in the spins.

The spins, in general, are susceptible only to magnetic field action, but, since the spins of the nuclei are well shielded inside the atom from most sources of fluctuating magnetic fields, their relaxation times can be very long, of the order of thousands of seconds. In liquid state, the interplay between the weak nuclear interactions on one hand and the fast averaging effect due to the rapid tumbling motions of the molecules in the liquid on the other hand, makes for relaxation times of proton spins only of the order of tens of seconds, but long enough for quantum computation purposes.

6.7 LIQUID-STATE NMR QUANTUM COMPUTER

We discuss now the implementation of the qubits, the control Hamiltonian, and the qubit control for a liquid NMR computer.

The Qubits

Spin-½ nuclei are the preferred two-state quantum systems for implementing qubits in liquid-state nuclear magnetic resonance quantum computation. The application of the strong external magnetic field, $\mathbf{B}_0$, determines the z-axis direction, and the state space of the spin becomes a superposition of "up" and "down" states defined along this direction. The logical states, 0 and 1 , correspond to spin-up and spin-down states, respectively, and have different energies in the external static magnetic field; this is the so-called Zeeman splitting. The liquid NMR decoherence times are typically long because the nuclear spins interact only with magnetic fields and not with electric fields; moreover, the atomic/molecular electron cloud shields the nucleus spin from most sources of fluctuating magnetic fields.

Each molecule can be viewed as a single quantum computer, the state of which is determined by the orientation of its spins. Quantum logic gates are constituted of RF pulses; they manipulate spin orientations and spin couplings and perform unitary transformations on the state. The N-spin molecules in the NMR sample form an ensemble of N-qubit computers. This ensemble of independent quantum computers is supposed to be used in a global way without addressing them individually.

The sample containing about, 10^{18} N-spin, molecules is placed in a strong longitudinal static magnetic field, $\mathbf{B}_0$, and has a transverse RF magnetic field applied to it, the same way as in a conventional pulsed NMR device. The strong magnetic field, $\mathbf{B}_0$, induces the Zeeman splitting of the energy levels of the spin system in each molecule. For a two-level system, such as uncoupled protons, at thermal equilibrium the ratio of the populations in the higher and the lower energy levels is 0.999999 (i.e., it differs from unity by as little as 10 ^{6}). This is a small difference, but it produces a macroscopic precessing magnetization that can be detected in a pickup coil.

The time evolution of a spin-½ nucleus in the magnetic field, $\mathbf{B}_0$, along the z-axis is governed by the Hamiltonian operator,

$$\mathbf{H}_0 \quad \hbar\gamma\,\mathbf{B}_0\,\mathbf{I}_z \quad \hbar\omega_0\,\mathbf{I}_z \quad \frac{\hbar\omega_0}{2}\sigma_z \quad \begin{pmatrix} \hbar\omega_0/2 & 0 \\ 0 & \hbar\omega_0/2 \end{pmatrix},$$

where γ is the gyromagnetic ratio of the nucleus, $\mathbf{I}_z$ is the angular momentum of the nuclear spin in the z-direction, and $\omega_0/2\pi$ (where $\omega_0 \quad \gamma\,\mathbf{B}_0$) is the *precession frequency* or *Larmor frequency*. Sometimes ω_0 alone is called the Larmor frequency and the factor 2π is implicit. The three components of the nuclear magnetic moment, $\mathbf{I}_x$, $\mathbf{I}_y$, and $\mathbf{I}_z$, are related to the Pauli matrices,

$$\sigma_x \quad 2\mathbf{I}_x, \quad \sigma_y \quad 2\mathbf{I}_y, \quad \sigma_z \quad 2\mathbf{I}_z.$$

The energy of the spin up state, or 0 , given by

$$0\ \mathbf{H}_0\ 0 \quad \begin{pmatrix} 1 & 0 \end{pmatrix} \begin{pmatrix} \hbar\omega_0/2 & 0 \\ 0 & \hbar\omega_0/2 \end{pmatrix} \begin{pmatrix} 1 \\ 0 \end{pmatrix} \quad \begin{pmatrix} 1 & 0 \end{pmatrix} \begin{pmatrix} \hbar\omega_0/2 \\ 0 \end{pmatrix} \quad \frac{\hbar\omega_0}{2},$$

is lower than the energy of the spin down state, or 1 , given by

$$1\ \mathbf{H}_0\ 1 \quad \begin{pmatrix} 0 & 1 \end{pmatrix} \begin{pmatrix} \hbar\omega_0/2 & 0 \\ 0 & \hbar\omega_0/2 \end{pmatrix} \begin{pmatrix} 0 \\ 1 \end{pmatrix} \quad \begin{pmatrix} 0 & 1 \end{pmatrix} \begin{pmatrix} \hbar\omega_0/2 \\ 0 \end{pmatrix} \quad \frac{\hbar\omega_0}{2}.$$

The energy difference between the two states is

$$\Delta E \quad \hbar\omega_0/2 \quad (\quad \hbar\omega_0/2) \quad \hbar\omega_0.$$

The precession frequency is proportional to the energy difference, ΔE, between the "up" and "down" states

$$\omega_0 \quad \frac{\Delta E}{\hbar},$$

where $\hbar$ is the reduced Planck constant.

The time evolution of the state under the action of the external magnetic field [104] is described by

$$\psi_t \quad e^{\ \ i\omega_0 t/2}\alpha_0\ 0 \quad e^{i\omega_0 t/2}\alpha_1\ 1 \ ,$$

where the units for ω_0 are radian/second when the time, t, is given in seconds.

For liquid-state NMR the typical values for the static magnetic field $\mathbf{B}_0$ are in the range 5 15 tesla. The precession frequencies [433] depend on the nuclei's magnetic moments and can vary, for example, from 500 MHz for a proton (hydrogen nucleus) to 125 MHz for a ^{13}C nucleus in a trichloroethylene molecule in an external magnetic field of 11.74 tesla.

The differences in the precession frequencies help to distinguish between the types of nuclei placed in a given magnetic field; thus, they allow control and measurement of specific nuclear spins. Spins of the same nuclear species that are part of the same molecule can also have different precession frequencies. These differences between precession frequencies of the same nuclear species are called *chemical shifts* and they arise from the variable partial shielding of the applied magnetic field by the electron cloud surrounding a nucleus at a specific location inside the molecule. Different chemical environments inside a molecule determine different chemical shifts for nuclei of the same species. Large asymmetries in the molecular structure determine large chemical shifts. Typical chemical shifts range from a few tens to a few hundreds parts per million of the precession frequency. In the trichloroethylene molecule at 11.7 tesla magnetic field, the chemical shifts are between a few kHz and a few tens of kHz. They depend on the solvent, the molecule concentration in the solution, and the temperature.

The evolution in time of the initial state, ψ_0 , can be expressed in terms of the Pauli matrix, σ_z as

$$\psi_t \quad e^{i\frac{1}{2}\omega_0\sigma_z t}\ \psi_0 \ .$$

The operator, $\omega_0\sigma_z/2$, represents the internal Hamiltonian of the spin (i.e., the energy observable, here given in units for which the reduced Planck constant, $\hbar \quad h/(2\pi) \quad 1$). The Pauli matrix σ_z can be thought of as the observable for the nuclear spin along the z-axis, which is defined by the external static field. The observables for spin along the x- and y-axis are given by the corresponding Pauli matrices, σ_x and σ_y. The Pauli matrices are the spin observables in the laboratory frame and this representation is meaningful in the real space. In the Bloch sphere representation, the time evolution under the action of the Hamiltonian, $\mathbf{H}_0$, the operator associated with the static applied field, $\mathbf{B}_0$, is described as a processing motion (rotation) of the spin state vector about the z-axis, which is the direction of the magnetic field, $\mathbf{B}_0$. For a system of n uncoupled nuclei, the Hamiltonian $\mathbf{H}_0$ is

$$\mathbf{H}_0 \quad \sum_{i \; 1}^{n} \frac{1}{2}\hbar\omega_0^i\sigma_z^i.$$

The internal Hamiltonian of a molecule's nuclear spins is well approximated by

$$\mathbf{H} \quad \frac{1}{2}\sum_{i}\omega_i\sigma_z^i \quad \frac{\pi}{2}\sum_{i \; j} J_{ij}\sigma_z^i\sigma_z^j,$$

where the summation is over the nuclear spins, ω_i are resonance frequencies, and J_{ij} are scalar spin couplings. The dominant term of the internal Hamiltonian represents the action of a strong magnetic field applied to the sample. The Zeeman effect of this magnetic field determines an axis of quantization along which the spinors, σ_z, sum up; the Zeeman effect is very small and more than 10^{15} spins are necessary to produce an observable signal. A typical sample for liquid-state NMR contains about 10^{18} molecules; each molecule acts as an independent processor. An NMR spectrometer records the average state of the spins.

From the point of view of a quantum computer implementation, the single-particle terms in the Hamiltonian above are used to distinguish qubits, and the two-particle terms represent the building blocks of two-qubit CNOT gates. The resonance frequencies, ω_i, in the first term depend on the efficiency of screening of the nuclear spins from the applied magnetic field by the surrounding electrons; the scalar spin couplings, J_{ij}, in the second term are mediated by the electrons in molecular orbits that overlap both nuclear spins. Besides the interaction with the externally applied magnetic field, $\mathbf{B}_0$, the nuclear spins in molecules feel the effects of a direct coupling and an indirect coupling between them.

The *direct coupling* is due to a *magnetic dipole-dipole interaction* between two neighboring nuclear spins belonging either to the same molecule or to different molecules; the spins behave like bar magnets. For molecules in a liquid solution dipolar (direct), spin couplings average out due to the tumbling motions of the molecules and they have no effect on the Larmor precession. In solids the dipolar couplings can be averaged out by technical means, for example, by applying multiple-pulse sequences or by spinning the sample at a specific angle with respect to the magnetic field.

The *indirect coupling* between nuclear spins in a molecule, also called, J, *coupling* or *scalar coupling*, is mediated by the electrons shared in the chemical bonds between the atoms of the molecule; the shared electron wave functions of the two bonded atoms overlap. This is a so-called *Fermi contact interaction* and its *through-bond coupling strength*, J, depends on the nuclear species involved and the number of chemical bonds connecting the atoms. Typical values for the frequencies associated with the J coupling are a few hundred Hz for one-bond couplings and decrease to just a few Hz for three- or four-bond couplings. The effect of this scalar coupling on the nuclear spins is that of a static internal magnetic field along the (/)z-axis produced by neighboring nuclear spins, in addition to the externally applied static magnetic field $\mathbf{B}_0$.

The Hamiltonian operator associated with the J coupling between spins of nuclei i and j takes the form [433]

$$\mathbf{H}_J \quad \hbar\sum_{i,j}^{n} 2\pi J_{ij}\mathbf{I}_z^i\mathbf{I}_z^j \quad \hbar\sum_{i,j}^{n}\frac{\pi}{2}J_{ij}\left(\mathbf{I}_x^i\mathbf{I}_x^j \quad \mathbf{I}_y^i\mathbf{I}_y^j \quad \mathbf{I}_z^i\mathbf{I}_z^j\right), \quad i<j,$$

when the Larmor frequencies of different nuclear spin species satisfy the condition

$$\omega_0^i \quad \omega_0^j \qquad 2\pi\ J_{ij} \quad .$$

The Hamiltonian operator, $\mathbf{H}_J$, shifts the energy levels of J-coupled nuclei and modifies the Larmor frequency of spin, i, as a function of the state of the spin, j:

- If spin j is in the state 0 , the Larmor frequency of spin i is shifted by $J_{ij}/2$ and becomes $\omega_0^i \quad J_{ij}/2$.
- If spin j is in the state 1 , the Larmor frequency of spin i is shifted by $J_{ij}/2$ and becomes $\omega_0^i \quad J_{ij}/2$.

Here, ω_0 is the Larmor frequency of the uncoupled spin i. In the presence of an external magnetic field, the energy spectrum for an uncoupled spin i has a line centered at frequency, ω_0^i, while the energy spectrum of spin i in a coupled system with spin j shows two lines separated by an interval, J_{ij}, and centered around frequency, ω_0^i.

The Control Hamiltonian

In liquid-state NMR the state of a spin-½ nucleus in a static magnetic field, $\mathbf{B}_0$, can be manipulated by applying an RF electromagnetic field, $\mathbf{B}_1$, perpendicular to the z-axis determined by the field $\mathbf{B}_0$ and rotating in the x-y plane with a frequency, ω_{RF}, near or equal to the nuclear spin precession frequency ω_0.

The single-spin Hamiltonian operator, $\mathbf{H}_{\mathrm{RF}}$, associated with the RF field $\mathbf{B}_1$, is similar to, $\mathbf{H}_0$, the Hamiltonian associated with the static magnetic field $\mathbf{B}_0$,

$$\mathbf{H}_{\mathrm{RF}} \qquad \hbar\gamma B_1\left[\cos\left(\omega_{\mathrm{RF}}t \quad \phi\right)\mathbf{I}_x \quad \sin\left(\omega_{\mathrm{RF}}t \quad \phi\right)\mathbf{I}_y\right]$$

or

$$\mathbf{H}_{\mathrm{RF}} \qquad \frac{1}{2}\hbar\gamma B_1\left[\cos\left(\omega_{\mathrm{RF}}t \quad \phi\right)\sigma_x \quad \sin\left(\omega_{\mathrm{RF}}t \quad \phi\right)\sigma_y\right],$$

where γ, ϕ, and B_1 are the nuclear gyromagnetic ratio, the phase of the RF field, and its amplitude, respectively. The frequency of the RF field, ω_{RF}, is determined as $\omega_{\mathrm{RF}} \quad \gamma B_1$. The minus sign in front of the sine term shows that the spin evolution under the action of the RF field, $\mathbf{B}_1$, is in the same sense as its evolution under the field $\mathbf{B}_0$. In practice, $\mathbf{B}_1$ is a magnetic field applied perpendicular to the direction of the static magnetic field $\mathbf{B}_0$ and oscillating along a fixed axis in the laboratory. Such an oscillating field can be decomposed into two counter rotating fields: one field rotates with frequency ω_{RF} in the same direction as the spin precessing under field $\mathbf{B}_0$ and can be tuned at or near resonance with the spin, $\omega_{\mathrm{RF}} \quad \omega_0$; the other field component rotates in the opposite direction and is very far from resonance by as much as $2\omega_0$; thus, it has a negligible shift effect on the Larmor frequency.

The Larmor precession due to the nuclear spin interaction with a static external magnetic field and the couplings between spins of different nuclear species inside a molecule are characteristics of a particular system and can not be changed. However, the amplitude, B_1, the frequency, ω_0, and the phase, ϕ, of the RF field can be varied in time, and, thus, they are enabling our quantum control of an NMR system.

The effects of the static field and the resonant pulses are easier to understand if we use a "*rotating frame of reference*" instead of the laboratory frame. We imagine a coordinate system, x , y , z, attached to the NMR apparatus and the sample, and all of them rotating about the z-axis of the system, which is aligned with the static magnetic field, $\mathbf{B}_0$. The Hamiltonian for a spin placed in the static field, $\mathbf{B}_0$, and acted upon by the pulsed field, $\mathbf{B}_1$, can be expressed in the rotating frame as

$$\mathbf{H}_{\text{rot}} \quad \frac{1}{2}\hbar\left(\omega_0 \quad \omega_{\text{rotframe}}\right)\sigma_z \quad \frac{1}{2}\hbar\omega_{\text{RF}}\left[\cos\phi\,\sigma_x \quad \sin\phi\,\sigma_y\right].$$

When the frequency of rotation, ω_{rotframe}, is equal to the Larmor frequency, ω_0, the first term disappears in the expression above; thus, the effect of the static applied field in the rotating frame is cancelled, and the nuclear spin appears stationary (does not process) with respect to the field $\mathbf{B}_0$. In the rotating frame the spin precesses about the field $\mathbf{B}_1$, and this motion is called *nutation*.

In the case of a molecule containing several species of nuclear spins, we have to consider a rotating frame for each of them; at resonance the frequency of each rotating frame, $\omega^i_{\text{rotframe}}$, will be the characteristic Larmor frequency of the corresponding nucleus. In the aggregate rotating frame, the Hamiltonian of the system takes the simplified form

$$\mathbf{H}_{\text{system}} \quad \mathbf{H}_J \quad \hbar\sum_{i<j}^{n} 2\pi J_{ij}\mathbf{I}_z^i\mathbf{I}_z^j \quad \frac{1}{2}\hbar\pi\sum_{i<j}^{n} J_{ij}\sigma_z^i\sigma_z^j,$$

where the terms of the form, $\hbar\omega_0^i\sigma_z^i$, related to the static applied field are dropped and the remaining terms, $J_{ij}\sigma_z^i\sigma_z^j$, represent the couplings between the nuclear spins; they are invariant with respect to the magnetic fields applied.

In the rotating frame, the RF field, $\mathbf{B}_1$, applied along the x -axis and oscillating at a frequency equal to that of the rotating frame (i.e., the spin Larmor frequency), $\omega_{\text{RF}} \quad \omega_{\text{rotframe}} \quad \omega_0$, is at *resonance* and appears as constant. This field has enough energy to induce a rotation of the spin about its direction, the x -axis, and that represents a change of the spin state (qubit state). The angle of rotation, θ, is determined by the pulse length, τ (the length of time the field is on), and its magnitude, B_1:

$$\theta \quad 2\pi\gamma\tau B_1.$$

For example, assume a spin oriented parallel to the z-axis of its rotating frame after the application of the static field $\mathbf{B}_0$. A so-called, $\pi/2$, pulse of the $\mathbf{B}_1$ field rotates the spin by 90 about the x -axis and brings it parallel to the y -axis. Typically, the length (duration) of a $\pi/2$-pulse is 10 microseconds when applying 50 to 300 watts of power [432].

In the laboratory frame, the oscillating field $\mathbf{B}_1$ is decomposed into two components rotating in the $x\,y$ -plane, one in the same direction as the spin at frequency, ω_0, and the other in the opposite direction at frequency very far off resonance (by about $2\omega_0$); as a result, the spin will spiral around the z-axis down to the $x\,y$ -plane under the action of the resonant component. A so-called 180 pulse rotates the spin by 180 degrees about the x -axis and brings it along the z-axis.

The total NMR Hamiltonian for multiple nuclear spins (each with its rotating frame) and for multiple rotating RF fields (one RF field for each of the k nuclear spins) is made up of two terms:

$$\mathbf{H}_{\text{NMR}} \quad \mathbf{H}_{\text{system}} \quad \mathbf{H}_{\text{control}},$$

with

$$\mathbf{H}_{\text{system}} \quad \frac{1}{2}\hbar\pi \sum_{i,j} J_{ij}\sigma_z^i\sigma_z^j \ (i \quad j)$$

and

$$\mathbf{H}_{\text{control}} \quad \sum_{i,r} \quad \frac{1}{2}\hbar\omega_{\text{RF}}^k \left\{\cos\left[\left(\omega_{\text{rotframe}}^k \quad \omega_0^i\right)t \quad \phi^r\right]\sigma_x^i \quad \sin\left[\left(\omega_{\text{rotframe}}^k \quad \omega_0^i\right)t \quad \phi^k\right]\sigma_y^i\right\},$$

where the frequencies, ω_{RF}^k (with $\omega_{\text{RF}}^k \quad \gamma_i B_1^k$) and phases, ϕ^k, of the multiple RF fields, $\mathbf{B}_1^k$, can be controlled by the user.

Qubit Control

The qubit state changes only if the pulsed RF magnetic field is applied. The magnetic field of the resonant pulse appears like a constant field applied in the $x \quad y$ plane of the rotating frame, for example, along the (x)-axis of this frame. Now the nuclear spin rotates only about the x-axis, the axis of the pulse, and executes the following sequence of motions: it starts along the z-axis, tips toward the (y)-axis, continues to the (z)-axis, then to the $\;y$-axis and back to the z-axis. The spin/qubit state evolves in time as

$$\psi_t \quad e^{\;i\frac{1}{2}\omega_{\text{RF}}^{(x)}\sigma_x t}\,\psi_0\,,$$

where $\omega_{\text{RF}}^{(x)}$ is the nutation frequency (nutation about x-axis); it determines the strength (energy) of the pulse. If the frame rotates with a frequency $\omega_{\text{RF}}^x \quad \omega_0$, an observer in the rotating frame (Figure 6.2(a)) will see the spin precess about B_1, while an observer in the laboratory frame (Figure 6.2(b)) will see the spin rotate about the z-axis spiral down over the surface of the Bloch sphere.

If the pulse is turned on for a time interval, $t \quad \phi/\omega_{x\text{RF}}^{(x)}$, the qubit state rotates by an angle ϕ, which is called the *phase of the pulse* with respect to the x-axis. Similar rotations can be implemented around any axis of the rotating frame. Rotations around the z-axis are simpler to implement and can be executed exactly. General rotations on the Bloch sphere can be constructed as a series of rotations around the main axes of the rotating frame.

If the RF field is off resonance with respect to the spin precession (Larmor) frequency (i.e., $\omega_0 \quad \omega_{\text{RF}}^{(x)} \quad \Delta\omega$), the spin precesses in the rotating frame about an axis tilted away from the RF field axis, the x-axis in our example. The tilt angle is inverse proportional to $\Delta\omega$: If the RF pulse is far off resonance, the angle is very small. Therefore, if the Larmor frequencies are well separated, any one nuclear spin (qubit) can be selected and rotated without rotating the others.

All the axes are defined in the rotating frame or the *logical frame* with respect to the qubit implemented as nuclear spin. Each spin of interest in the molecule has its own rotating (logical) frame associated with it.

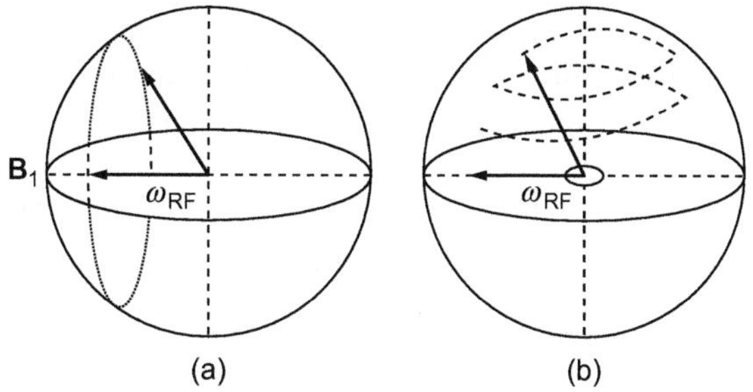

FIGURE 6.2

Spin nutation in a pulsed transverse RF field, $\mathbf{B}_1$. ω_{RF} is the nutation frequency. In the rotating frame, the RF field lies along a fixed axis. If the frame rotates with a frequency, ω_{RF} (also denoted ω_{RF}^x in the case where the RF field lies along the x-axis) and $\omega_{\mathrm{RF}}^x \quad \omega_0$, (a) an observer in the rotating frame will see the spin precess about $\mathbf{B}_1$, while (b) an observer in the laboratory frame will see the spin spiral down over the surface of the Bloch sphere.

The pulsed RF field applied to one nuclear spin is felt by other nuclear spins. The spins of other nuclear species have different precession frequencies and the difference at $B_0 \quad 11.7$ T is large, of the order of many MHz. A pulse that is in resonance with the target spin is not constant in the rotating frame of the non target nuclear species; it rotates rapidly and its effect is not significant. However, the spins of non target nuclei of the same species as the target will be affected by the resonant pulse, which appears as a constant field in their rotating frame; they will be rotated with the same angle as the target. The chemical shifts are too small (a few kHz) to make a difference.

How do we control a particular spin (qubit)? The practical solution to individually rotating nuclear spins of the same species is the following: first rotate all these spins by a given angle by applying a hard RF pulse for a time as short as possible; then rotate one spin that has a specific precession frequency by applying a soft pulse of sufficient duration. The power of this soft pulse is modulated in time to reduce the time needed for a rotation while minimizing cross-talk with other nuclear spins. For example, in the case of a two-spin system at equilibrium, a "hard" (nonselective) $\pi/2$-pulse along the y-axis yields $\mathbf{I}_x^1 \quad \mathbf{I}_x^2$, while a "soft" (spin-selective) $\pi/2$-pulse along the y-axis whose frequency spans the resonance frequency of only the first spin yields $\mathbf{I}_x^1 \quad \mathbf{I}_z^2$, leaving the second spin unaffected.

6.8 NMR IMPLEMENTATION OF SINGLE-QUBIT GATES

Single-qubit gates, spin state vector rotations in Hilbert space, can be directly implemented in the rotating frame by applying RF pulses to individual nuclear spins. When an RF pulse of amplitude, γB_1, is applied to a single nuclear spin, the rotating frame of which rotates at a frequency, $\omega_{\mathrm{RF}} \quad \omega_0$, the spin evolution is determined by the unitary transformation,

$$\mathbf{U} \quad \exp\left(-i\frac{1}{\hbar}\mathbf{H}_{\mathrm{control}}\, t_{\mathrm{pulse}}\right) \quad \exp\left(i\frac{1}{2}\gamma B_1 [\cos(\phi)\sigma_x - \sin(\phi)\sigma_y]\, t_{\mathrm{pulse}}\right),$$

where t_{pulse} is the RF pulse duration. In the Bloch sphere representation, this unitary transformation describes a rotation about an axis in the x-y plane. The direction of the axis of rotation is determined by the phase ϕ of the pulse, and the angle of rotation, θ, is proportional to the product between t_{pulse} and the nutation frequency $\omega_{\text{RF}} \quad \gamma B_1$.

A general rotation with an angle θ about an axis of rotation $\mathbf{n}$ (with $\mathbf{n}$ a three-dimensional vector) can be written as

$$\mathcal{R}_n(\theta) \quad \exp\left(\quad i\frac{\theta}{2}\mathrm{n} \quad \boldsymbol{\sigma}\right),$$

where $\sigma \quad \sigma_x x \quad \sigma_y y \quad \sigma_z z$ is a vector of Pauli matrices. For example, an RF pulse, with phase $\phi \quad \pi$ and $\omega_{\text{RF}}\, t_{\text{pulse}} \quad \pi/2$, will perform the unitary transformation

$$\begin{aligned}\mathbf{U} \quad & \exp\left\{i\frac{1}{2}\gamma B_1\, t_{\text{pulse}}\left[\cos(\phi)\,\sigma_x \quad \sin(\phi)\,\sigma_y\right]\right\}\\ & \exp\left\{i\frac{1}{2}\omega_{\text{RF}}\, t_{\text{pulse}}\left[\cos(\phi)\,\sigma_x \quad \sin(\phi)\,\sigma_y\right]\right\}\\ & \exp\left(\quad i\frac{1}{2}\frac{\pi}{2}\sigma_x\right),\end{aligned}$$

which is a 90 rotation about the x-axis, $\mathcal{R}_x(90)$; a pulse with the same phase, $\phi \quad \pi$, but twice as long, $\omega_{\text{RF}}\, t_{\text{pulse}} \quad \pi$, realizes a transformation

$$\mathbf{U} \quad \exp\left(i\frac{1}{2}\omega_{\text{RF}} t_{\text{pulse}}\left[\cos(\phi)\,\sigma_x \quad \sin(\phi)\,\sigma_y\right]\right) \quad \exp\left(\quad i\frac{1}{2}\pi\sigma_x\right),$$

i.e., a 180 rotation $\mathcal{R}_x(180)$. An RF pulse, with phase $\phi \quad 0$, determines a rotation, $\mathcal{R}_x(\quad 90)$, a negative rotation about the x-axis (i.e., a rotation with 90 about the (x)-axis).

RF pulses, with phases $\phi \qquad \pi/2$, implement positive and, respectively, negative rotations about the y-axis. For example, if $\phi \qquad \pi/2$ and $\omega_{\text{RF}} t_{\text{pulse}} \quad \pi/2$, then

$$\mathbf{U} \quad \exp\left(i\frac{1}{2}\omega_{\text{RF}} t_{\text{pulse}}\left[\cos(\phi)\,\sigma_x \quad \sin(\phi)\,\sigma_y\right]\right) \quad \exp\left(\quad i\frac{1}{2}\frac{\pi}{2}\sigma_y\right),$$

and that represents a transformation, $\mathcal{R}_y(\quad 90)$, a 90 rotation about the y-axis.

Thus, in practice, the RF field does not have to be applied along different axes in the lab frame. In the rotating frame, the relative phase of the RF field determines the direction of the nutation axis in the x-y plane. The absolute phase of the first RF pulse determines only a reference value against which the phases of all the subsequent RF pulses on the same spin, as well as the value of that spin, have to be estimated.

An arbitrary single-qubit unitary transformation, U, can be implemented as a sequence of rotations about only two axes, such as the x- and y-axis (see Section 4.3 [283]):

$$\mathbf{U} \quad e^{i\alpha}\mathcal{R}_x(\beta)\mathcal{R}_y(\gamma)\mathcal{R}_x(\delta),$$

where $\alpha, \beta, \gamma, \delta$ are real numbers. The rotations about the z-axis [170] can be decomposed in terms of rotations about the x- and y-axis:

$$\mathcal{R}_z(\theta) \quad \mathcal{R}_x(90)\mathcal{R}_y(\theta)\mathcal{R}_x(\;90)$$

and

$$\mathcal{R}_z(\theta) \quad \mathcal{R}_y(90)\mathcal{R}_x(\;\theta)\mathcal{R}_y(\;90).$$

Since rotations about the z-axis are easy to implement, they can be used to produce the respective rotations about either the x- or y-axis.

6.9 NMR IMPLEMENTATION OF TWO-QUBIT GATES

Two-qubit gates can be directly implemented through the action of the nuclear spin-spin coupling Hamiltonian, $\mathbf{H}_J$, estimated for molecules in a liquid; this Hamiltonian is invariant with respect to the laboratory or the rotating reference frame. The corresponding time evolution operator is

$$\mathbf{U}_J(t) \quad \exp\left(\; i\frac{1}{2}\pi J_{12}\sigma_z^{(1)}\sigma_z^{(2)}t\right).$$

In matrix form

$$U_J(t) \quad \begin{pmatrix} e^{\;i\pi J_{12}t/2} & 0 & 0 & 0 \\ 0 & e^{\;i\pi J_{12}t/2} & 0 & 0 \\ 0 & 0 & e^{\;i\pi J_{12}t/2} & 0 \\ 0 & 0 & 0 & e^{\;i\pi J_{12}t/2} \end{pmatrix}.$$

For a time interval, $t \quad 1/(2J_{12})$, the transformation under the action of the nuclear spin-spin coupling becomes

$$U_J\left(\frac{1}{2J_{12}}\right) \quad \begin{pmatrix} e^{\;i\pi/4} & 0 & 0 & 0 \\ 0 & e^{\;i\pi/4} & 0 & 0 \\ 0 & 0 & e^{\;i\pi/4} & 0 \\ 0 & 0 & 0 & e^{\;i\pi/4} \end{pmatrix}.$$

The transformation, $U_J(\frac{1}{2J_{12}})$, a two-qubit 90 rotation associated with, $\sigma_z^{(1)}\sigma_z^{(2)}$, is equivalent to a combination of two gates, each conditional on the logical state of qubit 1. The first gate applies a 90 rotation, $(\exp(\;i\sigma_z^{(2)}\pi/4))$, about the z-axis to qubit 2, conditional on qubit 1 being in the state, $\left|0\right._1$. The second gate applies a 90 rotation, $(\exp(i\sigma_z^{(2)}\pi/4))$, about the z-axis to qubit 2 if qubit 1 is in the state, $\left|1\right._1$. Now, after the two-qubit rotation, we apply a 90 rotation, $(\exp(i\sigma_z^{(2)}\pi/4))$, about the z-axis to qubit 2. As a result, the 90 rotation performed by the first gate if qubit 1, is in the state, $\left|0\right._1$, is canceled. If qubit 1 is in the state, $\left|1\right._1$, these rotations add to a 180 rotation, $(\exp(i\sigma_z^{(2)}\pi/4)\exp(i\sigma_z^{(2)}\pi/4) \quad \exp(i\sigma_z^{(2)}\pi/2) \quad i\sigma_z^{(2)})$, on qubit 2.

The Controlled-Phase (CPHASE) Gate

The transformation carried out by the *controlled-phase gate* (G_{CPHASE}) represents a 90 phase shift on each qubit and an overall phase

$$G_{\text{CPHASE}} \quad \overline{}iZ_1(\;90)\quad Z_2(\;90)U_J\left(\frac{1}{2J_{12}}\right) \quad \begin{pmatrix} 1 & 0 & 0 & 0 \\ 0 & 1 & 0 & 0 \\ 0 & 0 & 1 & 0 \\ 0 & 0 & 0 & \;1 \end{pmatrix}.$$

Here, the transformations, $Z_1(\;90)$ and $Z_2(\;90)$, represent phase shifts of spins 1 and 2 or rotations with 90 about their respective (z)-axes, or

$$Z_1(\;90)\quad Z_2(\;90)\quad \mathcal{R}_z\left(\;\frac{\pi}{2}\right)\quad \begin{pmatrix} e^{\;i(\;\frac{\pi}{4})} & 0 \\ 0 & e^{i(\;\frac{\pi}{4})} \end{pmatrix}\quad \begin{pmatrix} e^{i\frac{\pi}{4}} & 0 \\ 0 & e^{\;i\frac{\pi}{4}} \end{pmatrix}.$$

Now it is easy to calculate the transformation carried out by a G_{CPHASE} as

$$G_{\text{CPHASE}} \quad \overline{}i\begin{pmatrix} e^{i\frac{\pi}{4}} & 0 \\ 0 & e^{\;i\frac{\pi}{4}} \end{pmatrix}\begin{pmatrix} e^{i\frac{\pi}{4}} & 0 \\ 0 & e^{\;i\frac{\pi}{4}} \end{pmatrix}\begin{pmatrix} e^{\;i\pi/4} & 0 & 0 & 0 \\ 0 & e^{\;i\pi/4} & 0 & 0 \\ 0 & 0 & e^{\;i\pi/4} & 0 \\ 0 & 0 & 0 & e^{\;i\pi/4} \end{pmatrix}$$

$$\overline{}i\begin{pmatrix} e^{i\pi/2} & 0 & 0 & 0 \\ 0 & 1 & 0 & 0 \\ 0 & 0 & 1 & 0 \\ 0 & 0 & 0 & e^{\;i\pi/2} \end{pmatrix}\begin{pmatrix} e^{\;i\pi/4} & 0 & 0 & 0 \\ 0 & e^{\;i\pi/4} & 0 & 0 \\ 0 & 0 & e^{\;i\pi/4} & 0 \\ 0 & 0 & 0 & e^{\;i\pi/4} \end{pmatrix}$$

$$\overline{}i\begin{pmatrix} i & 0 & 0 & 0 \\ 0 & 1 & 0 & 0 \\ 0 & 0 & 1 & 0 \\ 0 & 0 & 0 & \;i \end{pmatrix}\frac{1}{\overline{}2}\begin{pmatrix} 1\;\;i & 0 & 0 & 0 \\ 0 & 1\;\;i & 0 & 0 \\ 0 & 0 & 1\;\;i & 0 \\ 0 & 0 & 0 & 1\;\;i \end{pmatrix}$$

$$\frac{\overline{}i}{\overline{}2}\begin{pmatrix} 1\;\;i & 0 & 0 & 0 \\ 0 & 1\;\;i & 0 & 0 \\ 0 & 0 & 1\;\;i & 0 \\ 0 & 0 & 0 & (1\;\;i) \end{pmatrix}\quad \frac{\overline{}i(1\;\;i)}{\overline{}2}\begin{pmatrix} 1 & 0 & 0 & 0 \\ 0 & 1 & 0 & 0 \\ 0 & 0 & 1 & 0 \\ 0 & 0 & 0 & \;1 \end{pmatrix}$$

$$\begin{pmatrix} 1 & 0 & 0 & 0 \\ 0 & 1 & 0 & 0 \\ 0 & 0 & 1 & 0 \\ 0 & 0 & 0 & \;1 \end{pmatrix}.$$

The CNOT Gate

A G_{CNOT} *gate* can be implemented as a controlled-phase transformation associated with a basis change of the target qubit and a phase shift on the control qubit [433],

$$G_{\text{CNOT}} \quad iZ_1(180)\, Y_2(\quad 90)\, G_{\text{CPHASE}}\, Y_2(90),$$

where

$$Z_1(180) \quad \mathcal{R}_z(180) \quad \begin{pmatrix} e^{\ i\frac{\pi}{2}} & 0 \\ 0 & e^{i\frac{\pi}{2}} \end{pmatrix} \quad i\begin{pmatrix} \ 1 & 0 \\ 0 & 1 \end{pmatrix},$$

$$Y_2(\quad 90) \quad \mathcal{R}_y(\quad 90) \quad \begin{pmatrix} \cos(\ \frac{\pi}{4}) & \sin(\ \frac{\pi}{4}) \\ \sin(\ \frac{\pi}{4}) & \cos(\ \frac{\pi}{4}) \end{pmatrix} \quad \frac{1}{\sqrt{2}}\begin{pmatrix} \ 1 & 1 \\ 1 & 1 \end{pmatrix},$$

$$Y_2(90) \quad \mathcal{R}_y(90) \quad \begin{pmatrix} \cos(\frac{\pi}{4}) & \sin(\frac{\pi}{4}) \\ \sin(\frac{\pi}{4}) & \cos(\frac{\pi}{4}) \end{pmatrix} \quad \frac{1}{\sqrt{2}}\begin{pmatrix} 1 & \ 1 \\ 1 & 1 \end{pmatrix}.$$

Then,

$$G_{\text{CNOT}} \quad i^2\begin{pmatrix} \ 1 & 0 \\ 0 & 1 \end{pmatrix} \frac{1}{\sqrt{2}}\begin{pmatrix} \ 1 & 1 \\ 1 & 1 \end{pmatrix}\begin{pmatrix} 1 & 0 & 0 & 0 \\ 0 & 1 & 0 & 0 \\ 0 & 0 & 1 & 0 \\ 0 & 0 & 0 & \ 1 \end{pmatrix}\begin{pmatrix} 1 & 0 \\ 0 & 1 \end{pmatrix} \frac{1}{\sqrt{2}}\begin{pmatrix} 1 & \ 1 \\ 1 & 1 \end{pmatrix}$$

$$\frac{1}{2}\begin{pmatrix} \ 1 & \ 1 & 0 & 0 \\ 1 & \ 1 & 0 & 0 \\ 0 & 0 & 1 & 1 \\ 0 & 0 & 1 & 1 \end{pmatrix}\begin{pmatrix} 1 & 0 & 0 & 0 \\ 0 & 1 & 0 & 0 \\ 0 & 0 & 1 & 0 \\ 0 & 0 & 0 & \ 1 \end{pmatrix}\begin{pmatrix} 1 & \ 1 & 0 & 0 \\ 1 & \ 1 & 0 & 0 \\ 0 & 0 & 1 & \ 1 \\ 0 & 0 & 1 & \ 1 \end{pmatrix}$$

$$\begin{pmatrix} 1 & 0 & 0 & 0 \\ 0 & 1 & 0 & 0 \\ 0 & 0 & 0 & 1 \\ 0 & 0 & 1 & 0 \end{pmatrix}.$$

A CNOT gate can be implemented as a CPHASE gate specified by a unitary matrix with the diagonal elements, $1, 1, 1, \ \ 1$, preceded and followed by $\mathcal{R}_y^2 \quad 90$. The single-spin rotations are executed by spin selective RF pulses while the two-spin rotations are implemented by allowing the system to evolve freely for a specific time interval.

Let us assume that in their initial states the two spins are oriented along their z-axes. First, a spin-selective RF pulse on spin 2 is applied along its y-axis. The frequency of the RF pulse is centered at $\omega_0^2/2\pi$ and has a spectral bandwidth that covers the frequency range, $\omega_0^2/2\pi \quad J_{12}/2$, but not the $\omega_0^1 \quad J_{12}/2$ frequency range which belongs to spin 1. The pulse rotates the spin 2 from z- to x-axis.

The duration of the RF pulse must be as short as possible, so that we can safely make the assumption that meanwhile the spin 1 has not changed its orientation under the action of the spin-spin coupling.

Next, the spin system is allowed to evolve freely for an interval of $1/(2J_{12})$ seconds during which the spins will execute 90 rotations about their respective z-axis. Spin 1 will remain parallel to its z-axis. As spin 2 is concerned, its precession frequency is shifted by $J_{12}/2$ depending on whether spin 1 is in the state, 1 or 0 ; therefore, it will arrive in $J_{12}/2$ seconds at either y- or $-y$-axis, depending on the state of spin 1. At this moment, a spin-selective pulse on spin 2 applied along its x-axis rotates spin 2 by 90 and brings it back along the z-axis if spin 1 is in the state, 0 , or brings it along the $-z$-axis if spin 1 is in the state, 1 .

The effects of the NMR pulse sequence that implements a CNOT gate can be visualized in the Bloch sphere representation for the logical initial states, 0 and 1 , only if the intermediate state are products of such states. If, for example, the initial state of the two spins is $1/\sqrt{2}(00 \quad 10)$, the spin 1 can be represented as pointing along the x-axis and spin 2 pointing along z-axis, but the spin-spin coupling, (J_{12}), leads to a superposition of states, a maximally entangled state, which can no longer be represented as a simple combination of arrows in the Bloch sphere.

If the input spin states are superpositions, then extra z-axis rotations have to be added to the sequence of pulses to give all elements in G_{CNOT} the same phase. If the spin-spin interaction Hamiltonian is not of the form, $\sigma_z^i\sigma_z^j$, but contains also transverse components (such as $\sigma_x^i\sigma_x^j$ and $\sigma_y^i\sigma_y^j$), the pulse sequences needed to implement the CPHASE and CNOT gates are more complicated.

When two spins are not directly coupled to each other, a CNOT gate can still be implemented if the two spins are connected by a network of couplings (Figure 6.3). If this is a nearest-neighbor coupling

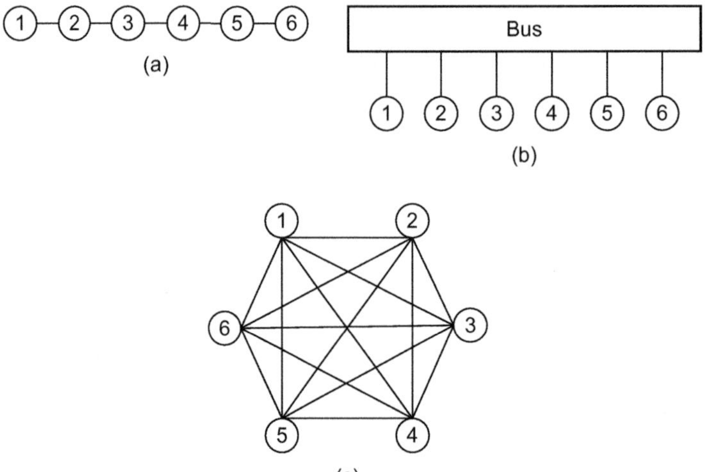

FIGURE 6.3

Possible coupling networks between six qubits. (a) A nearest-neighbor coupling network. Linear chains or two-dimensional such coupling networks are proposed for solid-state implementations. (b) A linear network coupling via a "bus" is used in ion trap implementations; since physical interactions decrease with distance, the "bus" degree of freedom ensures the good coupling only of a limited number of qubits. (c) A full coupling network; the size of such networks is also limited due to physical interactions becoming weaker with distance.

network, as the linear chain of qubits in Figure 6.3a, we can swap the states of two neighboring qubits at a time via a sequence of CNOT gates until the states of the two qubits we originally wanted to involve in a CNOT gate are in the immediate vicinity of each other. For example, we want to execute a CNOT gate between the qubits 2 (control) and 4 (target) that belong to the nearest-neighbor coupling chain of six qubits. Let us denote the gate as CNOT_{24}. The qubits 2 and 4 are not directly coupled to each other, but each of them is coupled to qubit 3.

The solution is to transfer the state of qubit 2 to the nearest neighbor of qubit 4 through permutations of qubit states. The states of qubits 2 and 3 can be swapped by performing a sequence of CNOT gates: CNOT_{23}, CNOT_{32}, CNOT_{23}. After that the required CNOT_{34} is applied, and the qubits 2 and 3 are swapped back (or just relabelled). The net effect is a CNOT_{24} gate. In a linear chain of n qubits with only nearest-neighbor couplings, we need to perform at most, $\mathcal{O}(n)$, permutations (SWAP operations) to perform a CNOT gate between any two qubits. In a network with qubits coupled via a "bus" degree of freedom (Figure 6.3b), a CNOT gate between any two qubits can be executed through a series of SWAP operations between the intermediary qubits.

In a full-coupling network where a qubit is coupled to several other qubits (Figure 6.3c), a CNOT between only two of them can be performed after removing the effect of the remaining couplings. In a quantum computer there are situations, such as maintaining a state in the memory, or selectively coupling certain qubits, when we have to control single spins in order to cancel the couplings' effects over a given time interval.

Spin-Refocusing Technique

The J-coupling between nuclear spins in a molecule cannot be physically turned off, but the unwanted spin couplings can be "removed" using the so-called *spin-refocusing* technique. The refocusing technique was developed by NMR spectroscopists before the advent of quantum computation. It works in the following way: A 180 RF pulse is applied to one of two coupled spins in the middle of the desired time interval.

We make the following assumptions: the two spins are in a weak-coupled regime with a positive coupling constant, J_{12}, and, if we consider the Bloch sphere representation, the spin 1 is in one of its logical states, 0 (spin up), or 1 (spin down) and spin 2 is in the xy-plane. The coupling between them can be interpreted as either an increase in precession frequency of spin 2 if spin 1 is up, or a decrease if spin 1 is down. Therefore, if spin 1 points along the z-axis, spin 2 in its own rotating frame precesses counterclockwise (from x-axis to y-axis) around spin 1.

If spin 1 points along the z-axis, spin 2 precesses clockwise (from x-axis to y-axis) around spin 1. If we wish to cancel the coupling effects over a period of 4 ms, we wait for 2 ms (2-ms delay) while spin 2 precesses counterclockwise/clockwise if spin 1 is up/down with the same angle. At the end of the 2 ms, we apply the refocusing RF pulse, which rotates spin 1 by 180 around the x-axis and produces an *inversion* of state for spin 1, respectively, a reversal in the direction of precession for spin 2. In the following 2 ms, spin 2 precesses clockwise/counterclockwise if spin 1 is down/up. The rotation angles in the first 2 ms and the next 2 ms are equal, but in opposite directions; as a result, spin 2 returns to its initial orientation. Another 180 RF pulse applied to spin 1 at the end of the 4-ms interval reverses the inversion of spin 1 and returns the two-spin system to its initial state.

The refocusing can be represented as

$$X_1(180)\,U_J(t)X_1(180)\,U_J(t) \quad U_J(\ \ t)\,U_J(t) \quad I.$$

We can replace $X_1(180)$ with $Y_1(180)$ with the same result:

$$Y_1(180)\,U_J(t)\,Y_1(180)\,U_J(t) \quad U_J(\ \ t)\,U_J(t) \quad I.$$

We can apply the pulses to either one of the two qubits to cancel the coupling between them, but if we apply pulses to both qubits the coupling is not removed.

Refocusing schemes have been specifically designed for the purpose of quantum computing; the pulse sequence can be simplified after a careful study of the quantum algorithm to be implemented. In actual implementations, control techniques are always complicated by experimental artifacts and undesired Hamiltonian terms that must be addressed. Two important factors, cross-talk and undesired couplings, are intrinsic to NMR. A pulse applied to a selected spin (qubit) will affect other spins (cross-talk) to some extent. The frequency selectivity of the RF pulses can be significantly improved if shaped and composite pulses are used. Unintended phase shifts, $\mathcal{R}_z(\theta)$, can be calculated in advance for each spin-pulse combination and can be corrected for with the corresponding, $\mathcal{R}_z(\ \ \theta)$, inserted in the pulse sequence.

The coupling terms, $J_{ij}\sigma_z^i\sigma_z^j$, act during time intervals of free evolution under the system's Hamiltonian; their effect can be removed using refocusing techniques, but only if the single spin (qubit) rotations are perfect and instantaneous. In practice, the RF pulses have a finite duration allowing the spin-spin coupling to distort the single-qubit rotations. Coupling effects can be neglected for short, high-power pulses used in heteronuclear spin systems where typically $J_{12} < 300$ Hz and $\omega_1 \quad 50$ kHz. But coupling effects can no longer be neglected for low-power pulses used in homonuclear spin systems for which J_{12} and ω_{RF} are of the same order of magnitude. Moreover, two spins could become entangled when acted upon by two simultaneous RF pulses. This problem can be corrected if the RF pulses are shaped and composite.

Due to its isolation inside the atom the nucleus has extremely long coherence times, up to thousands of seconds. Modern spectrometers enable complex operations and make an NMR quantum computer experimentally attractive. In an NMR quantum computer the qubit is embodied by the spin of an atomic nucleus and the unitary operations are the result of magnetic field pulses applied to individual varieties of nuclear spins in a strong magnetic field, while the couplings between spins are enabled by the chemical bonds between neighboring atoms. The initial state is prepared by polarizing the nuclear spins in a strong magnetic field and then by applying "effective pure state" preparation techniques. The result of a quantum computation will be read out by measuring the voltage signal induced by the precessing of the nuclear magnetic moment.

For liquid and solid NMR devices, the spins are identifiable based on the internal Hamiltonian determined by the chemistry of the sample. When the molecular system is carefully engineered, the spins (qubits) can be spatially localized.

In contrast to other physical implementations, NMR is an ensemble phenomenon and only an aggregate signal from many individual molecules can be practically observed. The initial states are not pure states, but thermal statistical mixtures. The techniques presently used to prepare an effective pure state reduce the signal exponentially in the number of qubits unless the initial polarization is high enough.

6.10 THE FIRST GENERATION NMR COMPUTER

The computational results from individual members of the ensemble are not accessible due to the averaging resulting from signals summation over the ensemble of a liquid NMR system. Nevertheless, Gershenfeld and Chuang [170] have shown in 1997 that NMR can be applied to perform quantum computation using ordinary liquids at room temperature and normal pressure and standard commercial instrumentation.

The first-generation devices based on liquid NMR and commercial technology have been used to manipulate systems of up to seven qubits [431]. The simplest instance of Shor's algorithm, the factorization of N 15 (prime factors 3 and 5), was implemented using a custom-synthesized "quantum computer" molecule, a perfluorobutadienyl iron complex. The qubits were five ^{19}F and two ^{13}C spin-½ nuclei. The experiments were performed in 2001 at IBM Almaden Research Center by a group of IBM scientists and Stanford University graduate students led by Isaac Chuang.

The defining step of the algorithm implementation was the computation of the function, $f(x)$ $a^x \bmod N$, for 2^n values of x in parallel. Like in the classic case, the function was computed using the identity, a^x $a^{2^{n\ 1}x_{n\ 1}}$ $a^{2x_1}a^{x_0}$, where x_k are the binary digits of x. Thus, the computation of $f(x)$ represents a serial multiplication by $a^{2^k} \bmod N$ for all k (where, 0 k n 1) for which x_k 1 . The powers of a^{2^k} can be computed on a classical machine by repeatedly squaring a; for N 15 the values of a may be 2, 4, 7, 8, 11, 13, or 14.

1. When a 2, 7, 8, or 13, $f(x)$ $a^4 \bmod N$ 1, with x 4 and k 2; then only two qubits, x_0 and x_1, control the multiplications leading to the computation of $f(x)$.
2. When a 4, 11, or 14, $f(x)$ $a^2 \bmod N$ 1, with x 2 and k 1; then only x_0 can be used for the computation of $f(x)$.

A register with n 3 qubits (though even n 2 qubits would have been enough) was used for detecting the periods of the modular exponentiation function and the quantum Fourier transform. Another register of m 4 qubits was used to store $f(x)$. As reported [431], the first three qubits were three fluorine, (^{19}F), nuclei and the second register of four qubits was made up of two fluorine, (^{19}F), and two carbon, (^{13}C), nuclei. This experiment was performed at room temperature (30 C) and the sample was in thermal equilibrium; the molecules and the initial state of each qubit was a statistical mixture of the spin states, 0 and 1 . This mixed state was converted into a seven-spin *effective pure state* through a variant [430] of the temporal averaging procedure [248].

An effective pure state (or pseudo-pure state as Cory *et al* [104] call it) is assumed to have a density matrix that can be shifted by adding a multiple of the unit matrix to it so as to obtain a scalar multiple of the density matrix of a pure state. In the original scheme [248], the temporal averaging for an n-qubit system involved the summation of 2^n 1 density matrices, each of them obtained from the density matrix, ρ_{eq}, of the mixed state at thermal equilibrium by cyclicly permuting all populations except the ground state population.

A new version takes a weighted sum of diagonal density matrices, ρ_i, obtained from ρ_{eq} by performing a few CNOT operations to rearrange the, 2^n 1, populations. The exact density matrix of the effective pure state, ρ_{effps}, may require up to 2^n 1 experiments, but a good approximation of ρ_{effps} may be obtained through much fewer experiments. The ρ_{effps} obtained with this new variant of temporal averaging proved to be a good approximation of the desired initial state of the seven qubits, $\psi_{initial}$ 0000001 .

The effective pure state of the seven qubits was created in the following way: the state of the five, ^{19}F, spins was made effective pure through the summation of nine experiments, where each experiment used a different sequence of two-qubit CNOT gates and single-qubit, spin-flip, NOT gates. These nine experiments were executed four times, each time with different additional CNOT and NOT gates, such that the result was an effective pure state of the two, ^{13}C, spins. The summation of the 36 experiments and a NOT operation on the seventh spin led to the effective pure state.

The algorithm was realized with a sequence of RF pulses selecting different spins; the time intervals between the pulses allowed for free evolution under the Hamiltonian of the system. Each pulse sequence was designed to produce a certain transformation of the spin state required by a computational step in the algorithm.

For a 7, the pulse sequence for the step calculating the function, $f(x)$ 7^x mod 15, was about 400 ms long, the inverse QFT was implemented in 120 ms, and the complete sequence of pulses for the Shor algorithm was about 720 ms long. After the execution of the complete sequence of pulses, the state of the first three qubits was estimated using NMR spectroscopy. The spectra showed that qubit 1 was in the state 0 , while qubits 2 and 3 were in a mixture of 0 and 1 . The three-qubit register was in a mixed state of 000 , 010 , 100 , and 110 which, in decimal notation, correspond to 0 , 2 , 4 , and 6 .

With the periodicity of these terms at 2, r $2^3/2$ 4 and the greatest common divisor (gcd), which can be efficiently computed on a classical computer, was $\gcd(7^{4/2}$ $1, 15)$ $3, 5$. A comparison between the measured NMR spectra and the spectra simulated based on a parameter-free decoherence model showed that the degree of unitary control in this first NMR quantum computation experiment was very high.

As the experiments have proven, NMR is a good tool for demonstration of experimental and theoretical techniques for precise control and modelling of quantum computers, but its scalability is limited by the size and stability of the custom-synthesized molecule containing an even larger number of spin-½ nuclei as qubits.

We continue our presentation of techniques for the implementation of quantum information processing systems with an analysis of quantum dots based on the solid state technology, which is well understood and very flexible.

6.11 QUANTUM DOTS

The physical implementations of quantum computation must achieve some very demanding conditions: (1) a well-defined two-level quantum system to play the role of the qubit, (2) precision control of the Hamiltonian operations on the qubit quantum system, and (3) a very high degree of quantum coherence. Atomic physics implementations, such as the ion trap and cavity quantum electrodynamics systems fulfil these conditions, but it is still unclear if they could be scaled up to the levels required by a full-scale quantum computer.

At this time, it is believed that solid-state systems could offer solutions for a scalable quantum computer technology. As noted by Burkard *et al* [78], in 2000 the number of proposals for solid-state implementations of quantum computers was almost equal to the other proposals put together. Solid state physics in itself is extremely versatile: On one hand, most phenomena possible in physics can be embodied in a solid-state system; on the other hand, solid state physics allied with computer technology has enabled the creation of a wide variety of artificial structures and devices.

Quantum dots, also known as nanocrystals or "artificial atoms," play a central role in all these solid-state proposals. Quantum dots are atom-like compound semiconductors structures where charge carriers are confined in all three spatial dimensions. A semiconductor quantum dot may contain from a few thousands to tens of thousands of atoms arranged in a nearly defect-free three-dimensional crystal lattice. The carriers in the quantum dot interact strongly with lattice vibrations and could be strongly influenced by defects, surfaces, or interfaces.

The size of a quantum dot is of the order of the Fermi wavelength in the host semiconductor material, typically ranging from 2 to 10 nanometers (10 50 atoms) in diameter. Quantum dots can be lithographically defined or self-assembled. The self-assembled quantum dots have smaller sizes and stronger confinement potentials than the lithographical ones. The confinement is usually defined by electrical gating of a two-dimensional electron gas.

The quantum dots have extraordinary tunability; their characteristics are extremely sensitive to the dot's size and composition and these can be controlled by specific engineering techniques. At such small sizes, semiconductors behave differently, which explains why quantum dots have unusual properties.

In bulk semiconductor material, the electrons have a range of energy levels so close together that they are described as *continuous*. Some of the energy levels are forbidden to electrons, and this energy region is called the *bandgap*. The overwhelming majority of the electrons of a bulk semiconductor occupy almost completely the *valence band* formed by the energy levels below the bandgap. Only a minuscule percentage of the semiconductor electrons may by found in the energy levels above the bandgap, in the so-called *conduction band.*

Electrons in the valence band could be stimulated to jump in the conduction band only if given enough energy through heat, voltage, or photon flux. The minimum energy required to raise an electron in the conduction band is equal to the energy bandgap. The pair formed by the exited electron (raised in the conduction band) and the hole left in the valence band is called an *exciton*. An electron in the conduction band falls back to its valence energy level after a very short time; the radiation emitted in this de-excitation process has a frequency corresponding to the transition energy. Most transitions are from the lowest levels of the conduction band to the highest levels of the valence band and the transition energy is equal to the bandgap. Since in a bulk semiconductor the bandgap is fixed, the transitions occur at fixed emission frequencies, the so-called *resonant frequencies*.

Quantum dots are also made of semiconductor material, but because their sizes are so small their electron energy levels can no longer be treated as continuous. The electron energy levels of a quantum dot must be treated as *discrete*, i.e., with a finite, small distance between them. This discrete level structure, through its *quantum con nement*[17] effects, determines properties in quantum dots totally different from bulk semiconductor material.

The addition/removal of a few atoms to/from the quantum dot changes the discrete electron energy levels and thus, alters the boundaries of the bandgap and changes its width. Because of the small size,

[17]Quantum confinement occurs when at least one dimension of a nanocrystal approaches the size of an exciton in bulk crystal, called the exciton Bohr radius. A quantum dot has all dimensions close to the Bohr exciton radius; in compound semiconductors the calculated Bohr exciton radius varies between 2.7 nm for CdS and 13 nm for GaAs.

the bandgap width can also be modified by changing the geometry of the quantum dot surface. The bandgap in a quantum dot will always be energetically larger than the bandgap in the corresponding bulk semiconductor. Since the transition energy is larger, the emission frequency is higher and, respectively, its wavelength is shorter or "blue shifted." By controlling the bandgap of a quantum dot containing a single electron, one can determine and manipulate the state of that electron.

The number of electrons in the conduction band of a quantum dot in a GaAs heterostructure can be precisely controlled one by one starting from zero [417]. This is the direct result of the so-called *Coulomb blockade*.[18] In the experiments of Tarucha *et al* [417], a gated vertical quantum dot (diameter of about 500 nm) was made from a double-barrier heterostructure with well-defined tunnel junctions and was cooled down to 50 millikelvin. The number of electrons was varied by changing the gate voltage, and the quantum dot displayed an atom-like shell structure associated with a two-dimensional harmonic potential. More than that, in the presence of a magnetic field applied parallel to the tunneling current, the electrons filling the shell structure showed pairing into spin-degenerate single-particle states like an atom.

Quantum dot systems are considered viable candidates for a quantum computer implementation. The qubit can be defined in several ways:

1. as a two-level system, the ground state and the excited state of a single-electron quantum dot controlled by optical resonant pulses [19],
2. as the spin of the excess electron in a single-electron quantum dot [276],
3. as two conduction band spin states of two electrons optically excited from two ground spin states in the presence of a magnetic field applied perpendicular to the confinement direction of the quantum dot[19][216], or
4. as the localization of the excess electron charge in one or the other of two coupled single-electron quantum dots [416].

A CNOT Gate Based on Dipole-Dipole Interaction Between Two Single-Electron Quantum Dots

Adriano Barenco, David Deutsch, and Arthur Ekert [19] were the first to propose an implementation of a quantum CNOT gate based on the electric dipole-dipole interaction between two single-electron quantum dots embedded in a semiconductor. In their simple model, the ground state and the excited state of the single electron in each quantum dot constitute the computation basis states, 0 and 1, of a qubit; the qubit quantum state can be controlled by optical resonant pulses.

The first quantum dot, the control qubit, and the second quantum dot, the target qubit, are assumed to have different resonant frequencies, ω_1 and ω_2, respectively, obtained be controlling their bandgaps through one of the methods mentioned above. Assume that we apply adiabatically an external static electric field, E_0, to the two qubits. In this way, electron transitions between levels are avoided and we observe only a quantum-confined Stark effect in both quantum dots: The charge distribution in

[18]The Coulomb blockade is a result of the discreteness of the electron charge and manifests itself as the increased resistance at small bias voltage of a low-capacitance tunnel junction. (See the glossary for more information.)

[19]The optical excitation also creates a single hole in the valence band spin state, which together with the two electrons in the conduction band are referred to as *trion* state.

the ground state (state 0) is shifted in the direction of the applied electric field while the charge distribution in the first excited state (state 1) is shifted in the direction opposite to the applied electric field.

This change in charge distribution gives rise to dipole moments corresponding to the state the single electron is in each of the two quantum dots. For example, if the qubit state is encoded in a single electron per quantum dot, we can assume that for selected coordinates, the dipole moments may be assigned the values d_i in the state 0 and d_i in the state 1 , where i 1,2 refers to the control and target dots, respectively.

The electron in each quantum dot creates an electric field that may shift the energy levels in the other dot without causing transitions. This is a good approximation because the dipole-dipole interaction term, $\mathbf{D}_{12}$, is much larger than the other terms in the total Hamiltonian of the two quantum dots system:

$$\mathbf{H} \quad \mathbf{H}_1 \quad \mathbf{H}_2 \quad \mathbf{D}_{12}.$$

The dipole-dipole interaction operator, $\mathbf{D}_{12}$, is diagonal in the four-dimensional state space spanned by the eigenstates of the free Hamiltonian of the system, $\mathbf{H}_1 \quad \mathbf{H}_2$, where $\mathbf{H}_1$ and $\mathbf{H}_2$ represent the Hamiltonians of the individual quantum dots.

The shift in energy levels due to the dipole-dipole interaction induced by the presence of a static electric field corresponds to a shift in the resonance frequencies of the transitions between ground level and excited level in the two quantum dots. The average shift in resonance frequencies, $\overline{\omega}$, is proportional to $d_1 d_2$, the product of the two dipole moments, and to $1/R^3$, the inverse of the cube power of the distance, R, between dots.

In the presence of a static electric field, the resonance frequency for transitions between the ground and excited state, respectively, between the states, 0 and 1 , of one quantum dot depends on the state of the neighboring dot. As seen in Figure 6.4, the resonant frequency for the first dot becomes, $\omega_1 \quad \overline{\omega}$ or $\omega_1 \quad \overline{\omega}$, as the second dot is in the state, 1 or 0 , respectively. At the same time, the resonant frequency of the second dot becomes, $\omega_2 \quad \overline{\omega}$ or $\omega_2 \quad \overline{\omega}$, if the first dot is in the state 1 or 0 . Thus, if we apply a, π, laser pulse at the frequency, $\omega_2 \quad \overline{\omega}$, the second dot will make a transition from the state, 0 to the state 1 , if and only if the first dot is in the state, 1 . At this resonant frequency, the first and the second quantum dots become the control qubit and, respectively, the target qubit of a CNOT gate.

A system of two single-electron quantum dots could be useful for quantum information processing if the decoherence time is greater than the time scale of the optical interaction. The decoherence time for resonant frequencies in the infrared regime is estimated at about $10^{\ 6}$s. Impurities and thermal vibration (phonons) can reduce it to about $10^{\ 9}$s, or even lower if their effects are not minimized by fabrication or by cooling the crystal. The time scale of the optical interaction is defined by the length of the π laser pulse at about $10^{\ 9}$s. The π pulse must by monochromatic and selective enough to initiate the desired transition. Thus, the length of the pulse must be greater than the inverse of the pulse carrier frequency and the inverse of the dipole-dipole interaction, $1/\omega$, which for this simple model is about $10^{\ 12}$s.

A Coupled Quantum Dot System

In the coupled quantum dot system proposed by Daniel Loss and David DiVincenzo [276], the qubit is realized as the spin of the excess electron in a single-electron quantum dot. The two-qubit gate operates by electrical gating of the tunneling barrier between neighboring quantum dots, not by spectroscopic

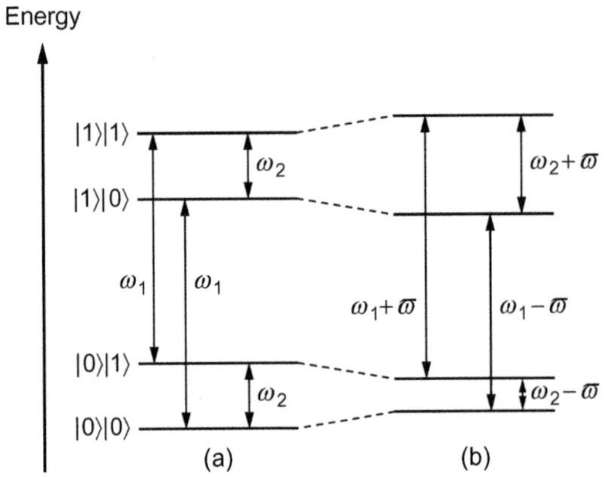

FIGURE 6.4

Energy levels of a two-quantum dot system. (a) In the absence of an external static electric field, the resonance frequency for the first dot is ω_1 and for the second dot is ω_2. (b) In the presence of a static electric field, the dipole-dipole interaction between the two quantum dots shifts their energy levels; the average shift in resonance frequencies is ϖ. The resonant frequency for the first dot becomes, $\omega_1 \quad \varpi$ or $\omega_1 - \varpi$, as the second dot is in the state, 1 or 0 , respectively. At the same time, the resonant frequency of the second dot becomes, $\omega_2 \quad \varpi$ or $\omega_2 - \varpi$, if the first dot is in the state, 1 or 0 .

manipulation as in the previous system. For single-electron quantum dots patterned as two-dimensional electron-gas structures, the gating of the tunneling barrier between neighbors can be controlled using a split-gate technique [272].

The control is achieved in the following way: If the barrier voltage is set "high," the tunneling is forbidden between the quantum dots and their states are preserved without evolution in time. If the barrier is pulsed to a "low" voltage, the spins of the electrons in neighboring quantum dots will be subject to a transient coupling (Heisenberg coupling) of the form,

$$\mathbf{H}_s(t) \quad J(t)S_1 \quad S_2,$$

where $J(t) \quad 4t_0^2(t)/u$ is the time-dependent exchange constant related to the turning on and off of the tunneling matrix element, $t_0(t)$; u is the charging energy of a single dot; and S_1 and S_2 are the spin-½ operators for dot 1 and 2, respectively. The expression above is based on several assumptions: (1) $\Delta E \quad kT$, where ΔE is the single-particle level spacing of the dots, k is Boltzmann's constant, and T is the temperature; thus the single-particle states of the dots higher than the ground state can be ignored. (2) The switching time, τ_s, for pulsing the gate potential "low" is $\tau_s > k/\Delta E$ to prevent transitions to higher orbital levels. (3) The charging energy, u, of a single dot is $u > t_0(t)$ for all t to ensure that the Heisenberg exchange approximation is correct. The quantum computation can be executed if (4) the decoherence time is much longer than the switching time, $\Gamma^{-1} \quad \tau_s$, where Γ is the decoherence rate. In quantum dots the decoherence times of the spin degrees of freedom, which are insensitive to environmental fluctuations of the electrical potential, are expected to be longer than the charge degrees of freedom. Nevertheless, the spin-½ degrees of freedom are affected by any magnetic coupling to the environment.

For a given duration of the spin-spin coupling, so that $J_0\tau_s \;\; \pi$, the effect of the pulsed Hamiltonian (pulsed voltage "low") is that of a unitary time evolution operator, $\mathbf{U}_s(t)$, when applied to the initial state of the two spins, $\psi(t) \;\; \mathbf{U}_s\psi(0)$. For a specific duration, $\int J(t)dt \;\; J_0\tau_s \;\; \pi(\mathrm{mod}\ 2\pi)$ of the pulsed Heisenberg spin-spin coupling, this operator is a spin "swap" operator: $\mathbf{U}_s(J_0\tau_s \;\; \pi) \;\; \mathbf{U}_{sw}$, such that $\mathbf{U}_{sw}\ ij \;\; ji$, where ij represent the basis states of the two spins in the, S^z, basis, with $i,j \;\; 0,1$. As such, this swap operator conserves the total angular momentum of the system and does not function as a useful two-qubit quantum gate. But, if the pulsed coupling is applied for half the given duration, the resulting operator is $\mathbf{U}_{sw}^{1/2}$. A sequence of $\mathbf{U}_{sw}^{1/2}$ and single-qubit operations act like a CNOT gate,

$$\mathbf{U}_{\mathrm{CNOT}} \;\; e^{i(\pi/2)S_1^z} e^{\;\; i(\pi/2)S_2^z} \mathbf{U}_{sw}^{1/2} e^{i\pi S_1^z} \mathbf{U}_{sw}^{1/2},$$

where terms of the form, $e^{\;\; i\pi} S_i^z, i \;\; 1,2$, are single-qubit operations and they can be realized by applying local magnetic fields, $\mathbf{B}_i$, pulsed exclusively onto spin, i. This expression for the CNOT gate is given in the basis where it has the form of a conditional phase-shift; a simple basis change for qubit 2 converts it to a standard CNOT.

An exchange constant of the order of $J_0 \;\; 80\,\mu$eV could be obtained with pulse durations of $\tau_s \;\; 25$ ps and decoherence times of $\Gamma^{\;\; 1} \;\; 1.4$ ns. Then, a $\pi/2$ single-qubit rotation could be realized over a time interval of the order of a pulse duration, τ_s, of 25 ps, by applying a magnetic field, $B \;\; 0.6$ T, a realistic value for solid state.

The two implementations discussed so far are based on electron spin coupling; now we discuss implantations based on electron charge.

Quantum Dot Qubit Based on Charge

The quantum dot implementation of qubits based only on the electrical effects [416] uses a field-effect-transistor structure for qubit state detection purposes. This implementation is closely related to conventional large-scale integrated, (Si), circuits and, in this sense, is easier to manufacture and operate, but the decoherence phenomena are considered extremely difficult to overcome.

The basic idea is to insert an excess electron into two coupled quantum dots and to consider the localization of this excess electron in one or the other of the quantum dots as the qubit states, 0 and 1 , respectively. A superposition state can be realized by varying the voltage on the gate electrodes attached to the quantum dots via oxide barriers.

The quantum dots are small in size (for example, diameter $d \;\; 10$ nm in silicon), and their electron energy levels are discrete. When the discrete energy level in one quantum dot matches that of the other quantum dot in a coupled quantum dot system, the corresponding electron can transfer from one dot to the other one inside the system. The electrons in two quantum dots of the coupled system are separated by an energy barrier that can be easily tunneled through when electron levels in the two dots match in energy. If this barrier is low and the coupling between the dots is strong, the two coupled quantum dots behave as a large single quantum dot where, possibly, only one electron resides. This large single quantum dot is itself characterized by a Coulomb blockade phenomenon like any regular single quantum dot. The electron tunneling in and out of a quantum dot is dominated by this classical effect related to the discrete nature of the electron charge. When the potential barrier delimiting the quantum dot is relatively high, the tunneling to and from the dot is weak, and the number of electrons on the dot is a well-defined integer, N. A current flowing as a result of a sequence of single electrons tunneling through the dot will change the electron number by one at a time.

Due to the Coulomb repulsion between the electrons in the dot, the addition of one electron is associated with an energy cost. The extra energy required is e^2/C, where e is the electron charge and C is the capacitance of the dot. No new current will flow until the extra energy is provided by increasing the source-drain voltage applied on the dot. This characteristic charging energy must be larger than the thermal energy of electrons for the Coulomb blockade to be observed. For a quantum dot of 1 femtofarad (10 ^{15}F) capacitance, this condition is satisfied at temperatures below 1 kelvin, which can be realized in a so-called dilution refrigerator.[20]

When an excess electron is injected into a system of two coupled quantum dots of different size, the system can be treated as a two-state system; the energy levels of the coupled system as a whole show the localized state of the wave function reflecting the different energy levels of the two independent isolated quantum dots. At a certain value of the applied gate bias voltage, the *resonant gate voltage*, the energy levels of the excess electron in each of the original quantum dots coincide and the electron transfers to the other dot through a process called *resonant tunneling*. In this way the coupling between the dots removes the degeneracy of the electron energy levels in the single-dot quantum states; the coupling creates new states associated with delocalization of the electron such that its even-parity and odd-parity wave functions spread over the two coupled dots.

The perfect localization of the electron in one or the other of the coupled dots represents the qubit basis states, (0 , 1). By adjusting the gate bias, the electron state can be controlled and changed from a well-localized one, (0 or 1), to an intermediate delocalized state where the electron transfers from one dot to the other within the coupled-dot system in a very short time (of the order of 1 picosecond). One dot is set close to the channel where the excess electron is inserted into the coupled-dot system, and the other dot is set near the gate electrode; the gate voltage controls the discrete energy levels of the coupled dots. The presence of an electron inside the dot on the channel side represents the qubit state, 1 , and the presence of the electron inside the dot on the gate electrode side represents the state, 0 .

The size of the two coupled dots is determined by the semiconductor material in which they are imbedded and by two requirements:

1. The electron, once inserted into the system, should be prevented from returning to the channel by applying a threshold gate voltage.
2. The discrete energy levels of the two dots should coincide at an applied gate voltage low enough to maintain the coherence of the coupled-dot state.

For coupled dots imbedded into the conventional silicon metal oxide semiconductor field effect transistor, these requirements translate into an asymmetry of the coupled dots with the larger dot close to the channel and the smaller dot near the gate electrode [416].

When two such coupled-dot qubits are positioned side by side, the electric field created by the charge distribution of the electron in a qubit changes the potential well profiles and the energy levels of the neighboring qubits. The electron state in a qubit is affected by the states of the electrons in the neighbor qubit, whether this is a state 0 , a state 1 , or an arbitrary superposition of 0 and 1 . The electric field generated by one coupled-dot qubit slightly changes the relative positions of the electron energy levels of the other coupled-dot qubit; it moves them off resonance and in this way

[20]The operation of the dilution refrigerator is based on the phase transition that a liquid mixture of two helium isotopes, ^{3}He and ^{4}He, undergoes at temperatures below 0.86 K, the triple-critical point. (For more information see the glossary.)

makes impossible the transfer of the localized excess electron between the dots of this coupled-dot qubit. Thus, by changing the charge distribution in a selective way in one coupled-dot qubit, we can control the total charge distribution of the electron, and respectively, the state of the electron in the other qubit.

A CNOT quantum gate can be easily implemented by positioning two coupled-dot qubits close to each other and connecting them to the same channel while keeping their gate electrodes separated. It is assumed that the electron transfer between two neighboring coupled-dot qubits can be neglected. Assume that the electron in the target qubit is localized in its larger dot at a lower gate bias while the control qubit is in the state, 1 (the electron charge is localized in the larger dot placed near the channel). As the gate voltage applied to the target qubit is raised, the energy level of the electron localization in the larger dot exceeds that of the smaller dot. When the gate voltage reaches the resonance value, the wave function of the lowest energy state and that of the excited state spread over the two dots (of the target qubit) with equal weight, and the electron transfers from the larger to the smaller dot, changing the state of the target qubit.

The energy levels of the single dots are split by the coupling of the dots and become separated by a small energy difference proportional with a frequency, ν, whose value is determined by the frequencies associated with the energy levels in the individual dots and by the coupling between the dots. This energy difference depends on the size of the dots and the semiconductor system (e.g., Si in SiO_2 matrix or GaAs in AlGaAs matrix) they are made of. When the control qubit is in the state, 1 , the resonance gate bias of the target qubit is shifted with this small energy difference toward a lower bias region, V_{res}^1; the target qubit moves between the dots, changing state. However, when the control qubit is in the state, 0 (the electron charge localized in the smaller dot of the control qubit), the energy level in the smaller qubit of the target qubit is raised and the resonance gate bias shifts toward a higher bias region, V_{res}^0, where

$$V_{res}^0 > V_{res}^1.$$

Thus, if we apply a gate bias, V_{res}^1, to the target qubit when the control qubit is in the state, 0 , the electron in the target qubit is prohibited from moving between the dots. The two coupled-dot qubits operate like a CNOT gate. It can be shown that the electron moves completely between the coupled dots of a pair with a period, $\pi/(2\nu)$. The charge transfer is realized when we apply a bias voltage at which the ground state and excited levels of the coupled dots come in the closest vicinity in the time period, $\pi/(2\nu)$.

6.12 FABRICATION OF QUANTUM DOTS

Quantum dot structures that contain carriers or excitons confined in all three dimensions to a nanometer-sized region of a semiconductor are possible to fabricate due to progresses in lithography, colloidal chemistry, and epitaxial growth. Quantum dots are made of different materials in different sizes: self-assembled quantum dots, semiconductor vertical or lateral dots, single molecules trapped between electrodes, metallic or superconducting nanoparticles, semiconducting nanowires, even carbon nanotubes between closely spaced electrodes.

The realization of an appropriate (two-level system) quantum dot circuit containing just a single conduction electron where electrical voltages are used to control one-electron quantum states is a real challenge. In reducing the number of electrons, the gates used to deplete the dots influence (decrease) the tunnel barriers; the current peaks associated with the Coulomb blockade become unmeasurably small, but not necessarily due to empty dots.

Few-electron quantum dots have been realized in self-assembled structures [328], in small semiconductor vertical pillars defined by etching [255], or as lateral quantum dots defined in a *two-dimensional electron gas* (2DEG) by surface metal gates on top of a semiconductor heterostructure. The first two types of quantum dots are hard to integrate into circuits with a controllable coupling between the elements. Multiple two-dimensional electron gas quantum dots can be integrated simply by increasing the number of gate electrodes, and the coupling between them can be controlled by setting the gate voltages.

Vertical Dots

The vertical dots can be lithographically defined; the technology is similar to the silicon-based very large system integration Tiwari *et al* used for manufacturing the single-electron memory device [420]. The quantum dot is inside the pillar and has the shape of a two-dimensional disc. The pillar is a three-terminal field-effect transistor where the current can be switched on and off by changing the gate voltage. In fact, there is a quasi-periodic set of voltages where the current switches. Since the switch can be triggered by a small fraction of a single electron charge, these devices are called *single-electron transistors.*

The vertical quantum dot single-electron transistors proposed by Kouwenhoven *et al* [255] are miniaturized resonant tunneling diodes. The pillars are etched from a semiconductor double-barrier heterostructure and a metal gate electrode deposited around the pillar in the region of the quantum dot. In lithographically patterned quantum dots, the confinement along the growth (vertical) direction is provided by the quantum potential well of the material, but the lateral confinement has to be provided by an electrostatically induced potential barrier. The gate potential together with the surface potential confine the electrons in the lateral x- and y-directions, while the double-barrier heterostructure defines the electron confinement in the z-direction. The confinement potential is repulsive, can be approximated by a harmonic potential, and has an, r^2-dependence, different from the, $r^{\ 1}$-dependence, of the attractive atomic potential.

The quantum dot, $12\,\mu$m thick, a few hundred nanometers in diameter, and containing up to 100 electrons, is created in the central well, which is made of undoped $In_{0.05}Ga_{0.95}As$. The tunnel barriers bordering the dot are made of undoped, $Al_{0.22}Ga_{0.78}As$, with thicknesses of 9.0 nm (upper one) and 7.5 nm (lower one). The conducting source and drain contacts are made of Si-doped n-GaAs with a variable concentration of Si dopant (increasing away from the tunnel barriers). The variation in the heterostructure material and in doping determines a particular conduction band profile for the quantum dot. For example, in this particular case, the introduction of 5% indium (In) atoms in the well lowers the bottom of its conduction band 32 meV bellow the Fermi level of the contacts.[21] The confinement potential in the z-direction (vertical) is such that the lowest quantum state and the second quantum state in the well are, respectively, 6 meV below and 63 meV above the Fermi level of the contacts and the

[21] The Fermi level is the last energy level occupied by electrons at the zero absolute temperature, T 0 K. According to the Pauli exclusion principle, at T 0 K the electrons of an atom fill all available levels starting with the lowest one. (For more information see the glossary.)

quantum dot is a good approximation of a two-state quantum system. Due to the presence of In, this quantum dot system contains electrons without applying voltages.

Self-Assembled Quantum Dots

Such structures form spontaneously during the epitaxial growth process and are characterized by smaller sizes. Their confinement is stronger and is provided in all three dimensions by a high bandgap in the surrounding material. The fabrication of self-assembled quantum dots is based on in-situ growth techniques such as molecular beam epitaxy or metalorganic chemical-vapor deposition. For example, a beam of indium (In) atoms (gallium or arsenic atoms can also be used) is sent onto an ultra-clean gallium arsenide (GaAs) surface heated at a high temperature [328]. The atoms diffuse across the clean surface and start arranging themselves to form a continuous epitaxial layer of $In_xGa_{1-x}As$.

The atomic arrangement on the surface develops so as to minimize the sum of surface, interfacial, and elastic energies of the epitaxial film. The elastic strain energy of the epitaxial film grows quadratically with the film thickness rapidly increasing the total energy of the film. If the epitaxial film has a lattice parameter even a few percent different from that of the GaAs substrate, nanometer-size islands can form on the surface of the film to minimize its total energy. Since the film relaxation is elastic, no defects are introduced during the island formation process. As the film strain due to lattice misfit between GaAs and $In_xGa_{1-x}As$ increases from 0% to 7%, the In concentration, x, increases from 0 to 1. Islands of InAs shaped as truncated pyramids nucleate randomly on the clean surface. They are converted into quantum dots by covering them with a GaAs epitaxial layer. The process can be repeated to create a three-dimensional array of quantum dots.

In general, the confining potential cannot be accurately estimated since the exact shape and composition of the islands are not well known and depend on the growth technique, the band structure of the strained material in the epitaxial film becomes more complex, and piezoelectric forces are present. Therefore, an *effective con ning potential* inside the quantum dots is often defined. This effective quantity is estimated from a combination of spectroscopic studies of the ground state and the excitation energies in quantum dots.

Capacitance spectroscopy and transmission spectroscopy in far-infrared can be performed on quantum dots fabricated in a special field-effect structure, based on the metal-insulator-semiconductor field-effect transistor (MISFET) design, in order to study the ground (charging) state and, respectively, the excitations of the n-electron state in the dot. The typical *capacitance* versus *voltage* characteristic of a quantum dot shows six maxima where each maximum corresponds to the insertion of one electron at a time, from one up to six, into the quantum dot.

The first two maxima correspond to the first two electrons occupying the lowest, spin-degenerate state, also called the, s, shell. After the insertion of the first electron, the second electron is harder to insert into the quantum dot due to its Coulomb interaction with the first electron (Coulomb blockade), which raises the energy of the first level. This raise in energy is compensated by a raise in the gate voltage. The addition of the third electron to the dot requires an energy equal to the quantum energy gap between the first and the second energy level of the dot plus the energy displacement of the first level.

The four electrons that occupy the second shell, the p shell, correspond to the second group of four maxima. These four maxima are equally spaced as a consequence of the shape of the quantum dot confining potential, which is symmetrical in the xy plane and has a stronger component in the z direction, the growth direction.

As long as only two electrons, (n 2), are inserted into the quantum dot, the far-infrared spectroscopy data agree well with a parabolic model of the confining potential, and its effective curvature can be precisely determined. The insertion of a third electron (a lithium atom-like quantum dot) changes the character of the electron energy spectrum, and the data can no longer be explained or predicted with the simple parabolic model. Direct measurements must be performed on specific quantum dot systems.

A fully tunable qubit quantum system with a combination of both a single-electron charge degree of freedom, which is controlled with electrical voltages, and the spin degree of freedom, which ensures a long coherence time, can be realized with a quantum dot integrated with a *quantum point contact*, QPC, as proposed by Elzerman *et al* [142]. The quantum dot is formed in the two-dimensional electron gas of a GaAs/AlGaAs heterostructure by applying negative voltages to the metal surface gates M, R, and T. The two-dimensional electron gas is depleted below the gates and a potential minimum, the quantum dot itself, is created in the middle.

The gate voltages can be tuned such that the dot contains either zero or one electron; the insertion of one electron is controlled by the voltage applied to the plunger gate P. The gates, T and R, create a tunnel barrier between the dot and the drain; this barrier can be made opaque to completely isolate the dot from the drain. The gates, T and M, set the barrier between the dot and the reservoir at a certain tunnel rate—for example, Γ (0.05 ms) ^{1}.

The QPC, defined by the gates, R and Q, is set to operate in the tunneling regime where the current flowing through the QPC is very sensitive to electrostatic changes in the vicinity. An electron tunneling into or off the dot changes the electrostatic potential around the dot and produces a change in the QPC current. The QPC can be used as a charge detector with a resolution much better than a single electron charge and a measurement timescale of about 8 μs, almost ten times shorter than $1/\Gamma$; it allows us to determine whether an electron is present in the dot or not.

The procedure for measuring the spin state when an electron of unknown spin orientation has been detected inside the quantum dot requires placing the quantum dot device inside a dilution refrigerator and applying a magnetic field of 10 T in the plane of the two-dimensional electron gas [255]; in this case, the Zeeman splitting of the electron levels in the dot was ΔE_Z 200 μeV, larger than the thermal energy, (25 μeV), but smaller than the orbital energy level spacing, (1.1 meV), and the charging energy, (2.5 meV). The procedure is controlled with voltage pulses on the plunger gate, P. When the dot is empty, both spin- and spin- electron energy levels, respectively, E and E , where $E > E$, are above the Fermi energy of the reservoir, E_F. A voltage pulse (about 10 mV) applied on P moves both levels below, E_F, and tunneling into the dot becomes energetically favorable for an electron of either spin or spin .

The tunneling can happen after a typical time interval of Γ ^{1}. For as long as the pulse amplitude is maintained at 10 mV, the electron stays trapped inside the quantum dot and the Coulomb blockade prevents a second electron from entering the dot. After a time interval, t_{wait}, the pulse amplitude is lowered to about 5 mV and the electron levels are repositioned, with respect to E_F, according to their spin.

If the electron spin is , its energy level is below E_F, and the electron stays trapped inside the quantum dot. If the electron spin is , its energy level is above E_F, and the electron tunnels out of the dot after a typical time of Γ ^{1}; the Coulomb blockade is canceled and an electron with spin can tunnel into the dot, usually after a time Γ ^{1}, where Γ Γ Γ. After a time, t_{read}, the pulse ends and the dot is emptied.

During t_{read}, the current at the QPC tracks the charge on the quantum dot: it goes up when an electron tunnels off the dot and goes down by the same amount when an electron tunnels into the dot. Therefore, the QPC current goes up and then down if a spin- electron is inside the dot at the beginning of t_{read}. If a spin- electron is inside the dot, the current at QPC should be flat. The spin measurement fidelity can be significantly improved by lowering the temperature of the electrons and by making the charge measurement faster. The single-spin energy relation time was estimated at about 1 ms for such quantum dots in an, 8 T, magnetic field.

Two such lateral quantum dots built next to each other and integrated with two QPCs as charge detectors can serve as a fully tunable two-qubit quantum system [141]. The QPCs provide information about the charge on the dot when an electron is pushed out of it by a more negative electrode voltage, even when no current flows through the dot. The QPCs are sensitive enough to detect inter-dot transitions. The charge read-out can affect the spin state indirectly via the spin-orbit interaction. This effect could be much reduced until spin-to-charge conversion is initiated if the charge detectors are switched only during the read-out stage of a cycle.

6.13 QUANTUM DOT ELECTRON SPINS AND CAVITY QED

Quantum computing implementation based on semiconductor quantum dot electron spins offers two significant advantages:

1. The quantum dot arrays could be scalable to 100 coupled qubits.
2. The spin decoherence times for conduction band electrons in III-IV and II-VI semiconductors are "very" long, as demonstrated by recent experiments. For example, in moderately n-doped GaAs spin conduction band electron spin decoherence times of 130ns were measured at temperature, T 5K, and magnetic field, B 0 T [236].

Microcavities are optical resonators (cavities) with dimensions in the micrometer or submicrometer range in all directions; while the cavities are very large compared to the optical wavelength, the microcavities are of the order of a wavelength. The coupling of the quantum dot electron spins through a cavity-QED mode could bring another advantage, that of long-distance, fast interactions between qubits, as first proposed by Pellizzari *et al* [322].

In the scheme of Sherwin *et al* [378], the qubits consist of the lowest electronic states of specially engineered quantum dots and are coupled by terahertz (THz) (10^{12} Hz) cavity photons. Each quantum dot contains exactly one electron. The lowest two energy levels form the states, 0 and 1 , of a qubit and the third energy level, denoted 2 , serves as an auxiliary state to perform conditional rotations of the state vector of the qubit. The spacing between and the absolute values of these electron energy levels of the semiconductor quantum dot are controlled with the voltages applied to the electrical gates of the dot via the Stark effect.[22] A large number of such quantum dots is enclosed in a three-dimensional microcavity with fundamental resonance of wavelength, λ_C, much longer than the

[22]In a semiconductor heterostructure like that used to manufacture the quantum dots, the effect of splitting and shifting of an atomic spectral line into several components in the presence of an electric field can be significantly enhanced by bound excitons.

diameter of a quantum dot. The electron in each quantum dot is acted upon with a continuous-wave laser of fixed wavelength λ_L, where $\lambda_L \quad \lambda_C$.

The THz photons in the cavity act as a data bus and can couple an arbitrary pair of quantum dots. The state of an electron in a quantum dot can be coherently manipulated by tuning the transition energies, E_{10} and E_{20}, between the dot states, 0 1 , and, respectively, 0 2 . The transition energies are tuned by varying the electric field across the quantum dot, i.e., by varying the electrode voltages. The transition, E_{10}, is tuned in and out of resonance, with $\hbar\omega_C$ and $\hbar\omega_L$, at electric fields, $\mathcal{E}_C$ and $\mathcal{E}_L$, respectively. The transition, E_{20}, is tuned in and out of resonance with the two-photon transition, with energy $\hbar\omega_L \quad \hbar\omega_C$, at the electric field, $\mathcal{E}_{L \quad C}$.

A CNOT Gate

The operation of a CNOT gate using quantum dot spins and cavity - QED is based on a series of voltage pulses applied across the electric gates of a pair of qubits. The pulses begin at a baseline value of the electric field $\mathcal{E} \quad 0$ and then rise to one of the target values $\mathcal{E}_C$, $\mathcal{E}_L$, or $\mathcal{E}_{L \quad C}$. The cavity always begins with no photons.

In the first step, a π pulse, with amplitude $\mathcal{E}_C$ and duration $\pi/(2g_{01})$, is applied to the control qubit, where the factor, g_{01}, represents the vacuum Rabi frequency for the transition, 0 1 .[23] If the control qubit is in the state, 0 , the pulse does not affect it, but if it is in the state, 1 , the pulse rotates it into the state, 0 , with phase i, and a single photon is transferred to the cavity.

In the second step, a 2π pulse of amplitude, $\mathcal{E}_{L \quad C}$, and duration, $\pi/\Omega(\mathcal{E}_{L \quad C})$, is applied to the target qubit; the two-photon effective Rabi frequency, $\Omega(\mathcal{E}_{L \quad C})$, refers to a Rabi oscillation between the states, 0 and 2 , involving both cavity and laser photons at electric field, $\mathcal{E}_{L \quad C}$. If the target qubit is in the state, 1 , it is not affected, but if the target qubit is in the state, 0 , it acquires a phase 1 if the cavity contains one photon.

In the third step, a π pulse of amplitude $\mathcal{E}_C$, identical to the one in the first step, is applied to the control qubit. If a photon is present in the cavity, the control qubit absorbs it, returns to the state, 1 , and acquires another phase, i. The end result after these three steps is a conditional two-qubit gate in which the state vector of the two-qubit system acquires a phase 1 if and only if both the control and the target qubit are initially in the state, 1 . A true CNOT gate is realized by applying to the target qubit, $\pi/2$, and, $3\pi/2$, pulses, with amplitude $\mathcal{E}_L$ and durations $\pi/(4\Omega_{L,01})$ and $3\pi/(4\Omega_{L,01})$,[24] respectively, before and after the sequence of π and 2π pulses previously discussed.

---

[23] The vacuum Rabi frequency for the transition, 0 1 , is

$$g_{01} \quad q z_{01} \mathcal{E}_{vac},$$

with q, the electron charge, z_{01}, the dipole matrix element of the, 0 1 , transition, and $\mathcal{E}_{vac}$, the amplitude of the vacuum electric field in the cavity. This amplitude is given by

$$\mathcal{E}_{vac} \quad \left(\frac{\hbar\omega_C}{2\varepsilon_0\varepsilon V}\right)^{1/2},$$

where ε and V are the dielectric constant and the volume of the cavity and ε_0 is the dielectric constant of the vacuum.

[24] The Rabi frequency, $\Omega_{L,01}$, refers to a Rabi oscillation between the states, 0 and 1 , involving one laser photon at electric field, $\mathcal{E}_L$.

The fidelity of such a CNOT gate can be ensured only if

1. all the state vector rotations occur while the electric field is at its required value, $\mathcal{E}$; therefore, the rise and fall times, δt, of the electric pulses must be short compared to the period of the Rabi oscillation at the respective field, $\mathcal{E}$;
2. the Hamiltonian changes are adiabatic in comparison with the rise and fall times of the electric pulses, $(\hbar/\mathcal{E}_{10} \quad \delta t)$, so that the probability of transition, $0 \qquad 1$, induced by the ramping of the electric field should be minimized;
3. the timing between the pulses can be adjusted to compensate for the quantum mechanical phases, $\exp(\quad i\mathcal{E}_{10}t/\hbar)$, accumulated by inactive qubits in their excited states;
4. the height and duration of the electric pulses are adjusted to account for the AC Stark shifts[25] in the energy levels of the quantum dots that are induced by the laser field.

If the dominant mechanism for decoherence is due to acoustic-phonon emission, calculations made by Sherwin *et al* [378] for a specific GaAs/al$_x$Ga$_{1 \quad x}$ quantum dot, a lossless dielectric cavity, a laser electric field of about 30 kV/m, and 2π and π electric voltage pulses of 25 ns and 3 ns, respectively, suggest that several thousand CNOT operations can be performed before the quantum computer decoheres. Here, the 2π pulse applied to the target qubit that is interacting with both a laser and a cavity photon is the longest operation; the π pulse applied to the control qubit is only 3 ns.

Self-Assembled InAs Quantum Dots

Imamoglu *et al* [216] have proposed self-assembled InAs quantum dots embedded in microdisk structures with a high cavity quality factor. The quantum dots have a strong confinement along the z-direction, which ensures that the lowest energy eigenstates consist of $m_z \qquad 1/2$ conduction band states and $m_z \qquad 3/2$ valence band states; the in-plane confinement is also strong enough to ensure that the electron will always be in the ground state orbital.

The quantum dots are doped such that for each of them the valence band is completely filled with electrons and the conduction band contains a single electron. A magnetic field is applied along the x-direction, B_x, and the quantum dot qubit is defined by the two spin states of the electron in the conduction band,

$$m_x \qquad 1/2 \qquad \text{and} \quad m_x \quad 1/2 \qquad ,$$

with energies $\hbar\omega$ and $\hbar\omega$, respectively. In this scheme each quantum dot is individually addressed by lasers and the two spin eigenstates are Raman coupled through strong laser fields and a single microcavity mode.

An arbitrary *single-qubit rotation* can be initiated by applying two laser fields, $\mathcal{E}_{L,x}(t)$ and $\mathcal{E}_{L,y}(t)$, polarized along the x and y directions, respectively. The laser frequencies, $\hbar\omega_{l,x}$, $\hbar\omega_{L,y}$, and the frequencies of their Rabi oscillations, $\Omega_{L,x}$ and $\Omega_{L,Y}$, satisfy exactly the Raman-resonance condition between the, and , states. The π/r (r any real number) laser field pulses have a short duration and the laser polarizations must have nonparallel components in order to create a nonzero Raman coupling.

[25]The shift of atomic levels is caused by a time-varying (A.C.) electric field of the order of 10^{10} V/cm, typically, that of a laser. In a two-level atom, the excited state shift is of equal magnitude but opposite sign of ground state shift.

Such arbitrary single-qubit rotations can be carried out in parallel on several quantum dots inside the microcavity.

A *two-qubit operation* can be implemented by a selective coupling between the control and the target qubit that is mediated by the microcavity field. A combination of a cavity mode with energy, $\hbar\omega_{\text{cav}}$, and x-direction polarization and a laser field with y-direction polarization could induce a Raman transition between the two conduction band states. As a first step in the implementation of a CNOT operation, two laser fields of frequencies, ω_L^i and ω_L^j, are shined onto two quantum dots to establish a near two-photon resonance condition for both the control, i, and target, j, qubits, respectively.

The *exact* two-photon resonance condition for the control quantum dot, i, is

$$\Delta\omega^i \quad \Delta\omega^i ,$$

with

$$\Delta\omega^i \quad \omega^i \quad \omega_v^i \quad \omega_{\text{cav}}^i$$

and

$$\Delta\omega^i \quad \omega^i \quad \omega_v^i \quad \omega_L^i,$$

where ω_v^i defines the energy, $\hbar\omega_v^i$, of the relevant valence band states for the control qubit.

The *exact* two-photon resonance condition for the target quantum dot, j, is given by a similar set of equations

$$\Delta\omega^j \quad \Delta\omega^j ,$$

with

$$\Delta\omega^j \quad \omega^j \quad \omega_v^j \quad \omega_{\text{cav}}^j \quad \text{and} \quad \Delta\omega^j \quad \omega^j \quad \omega_v^j \quad \omega_L^j.$$

The energy of the relevant valence band states for the target qubit, $\hbar\omega_v^j$, corresponds to ω_v^j.

A *near* two-photon resonance condition for both the control qubit (i) and the target qubit (j) can be realized if the two-photon detuning between the two qubits is

$$\Delta_{ij} \quad \Delta_i \quad \Delta_j \quad 0.$$

The detunings for the individual qubits are defined as

$$\Delta_i \quad \omega^i \quad \omega_{\text{cav}} \quad \omega_L^i \quad \text{and} \quad \Delta_j \quad \omega^j \quad \omega_{\text{cav}} \quad \omega_L^j,$$

with

$$\omega^i \quad \omega^i \quad \omega^i \quad \text{and} \quad \omega^j \quad \omega^j \quad \omega^j .$$

The detunings for each qubit must satisfy the following conditions:

$$\Delta_i \quad \omega^i \quad \text{and} \quad \Delta_j \quad \omega^j .$$

A fully controllable long-range transverse spin-spin interaction between electrons in two distant quantum dots can be established through Raman coupling via a common cavity mode if the laser field frequencies, ω_L^i and ω_L^j, are chosen such that the condition, Δ_{ij} 0, is fulfilled.

A two-qubit gate such as the *conditional phase- ip operation* between two spins, i and j, can be implemented by combining a transverse spin-spin coupling Hamiltonian with one-qubit rotations.

A critical issue for the quantum dots made with semiconductors in group III-V, such as GaAs and AlAs, is the presence of nuclear spins in the semiconductor substrate; they create an inhomogeneous magnetic field that reduces the decoherence time, and they produce decoherence through dynamic spin-diffusion induced by nuclear dipole-dipole interactions. The group IV semiconductors, such as Si and Ge, are free of nuclear spin effects. If the silicon semiconductor is used, the quantum dot can be replaced with a single impurity, such as a single phosphorous atom [228, 229], which binds a donor electron at room temperature. The quantum information can be encoded in the donor electron, or in the state of the single, ^{31}P, nuclear spin. The electrostatically defined quantum dots and silicon-based impurities have exchange interactions at extremely short range. Photonic connections between quantum dots assisted by cavity quantum electrodynamics with optical microcavities can help solve this problem. Progress in optical control and measurement of qubits in self-assembled quantum dots potentially enable very fast quantum computation. Though large wafers of spins trapped in quantum dots or impurities are routinely fabricated, the scaling of a system of coupled spins remains a difficult task.

6.14 QUANTUM HALL EFFECT

Topologically defined quantum gates represent a crucial development in quantum computation; they are considered to define not only very powerful fault-tolerant quantum error-correcting schemes among qubits, but also a very special method of quantum computation for which the appropriate physical implementation remains to be determined.

The anyons, a type of quantum excitation with fractional, non-Abelian statistics, are considered to play a part in topological condensed matter systems. Quantum logic implemented as the braiding of trajectories in a (2D   time)-space of such a quasiparticle is expected to offer a way to fault-tolerant quantum computation. Theoretical calculations and some the experimental indicate that some fractional quantum Hall effect states with filling factors such as, ν 5/2 and ν 12/5, could be states of non-Abelian anyons. These incompressible fractional quantum Hall effect states are the only experimentally established topological phases. Universal topological quantum computation requires both a topological phase and quasiparticles with non-Abelian braiding statistics and with topological properties in a special class.

We will review first some concepts related to the quantum Hall effect, then the fractional quantum Hall effect and some of its theoretical models. The quasiparticles introduced by these models to explain the fractional filling factors match the properties of the anyons, theoretical quasiparticles introduced by Frank Wilczek in 1982.

Quantum Hall Effect

The Hall effect was discovered by the American physicist Edwin Hall in 1879 while working on his doctoral thesis; the results were published in 1880 [195, 196]. This discovery was made almost 18 years *before* the electron was identified as an electric-charged particle.

The classical effect manifests itself as a potential difference, the *Hall voltage*, on the opposite sides of a conductor when an electric current is flowing through the conductor and a magnetic field is applied perpendicular to the current. A century later, in 1980, Klaus von Klitzing discovered the quantum Hall effect, QHE; he observed that the Hall resistance measured at a temperature of 1.5 kelvin and in a magnetic field of the order of 15 tesla, the Hall resistance variation with the strength of the applied magnetic field in semiconductor devices used as low-noise transistors, is no longer linear but "stepwise" [242].

In such conditions, the Hall effect is changing its characteristics. The devices contain electrons that, though trapped close to the interface between two distinct parts of the transistor (junction), are highly mobile along this surface. At very low temperature, the electrons move in two dimensions only, as if on a plane surface. The Hall resistance values where the steps occur do not depend on the properties of the material; they are given by a physical constant divided by an integer,

$$R_{\text{Hall}} \quad \frac{h}{e^2}\frac{1}{i} \quad 25\frac{1}{i}\,\text{k}\Omega, \quad \text{with} \quad i \quad 1,2,3,4,5,6,8,\text{and }10,$$

where h is the Planck constant and e is the electron charge. The Hall resistance is *quantized* and the effect becomes the *integer quantum Hall effect* because the factor, i, takes integer values. At the same time, it was observed that at magnetic field intensities where the Hall resistance takes quantized values the Ohmic resistance disappears, and the material behaves like a superconductor.

The behavior of the Hall resistance was explained by the fact that the degenerate electron gas in a very thin layer at the interface between the semiconductor and metal-oxide (the inversion layer in a metal-oxide-semiconductor field effect transistor) becomes fully quantized under the action of the electric and magnetic fields, when the device is operated at temperatures close to absolute zero and strong magnetic fields.

The electric field perpendicular to the surface orders the electron motion normal to the interface and the magnetic field produces the so-called *Landau quantization* of the motion parallel to the interface. In the absence of the magnetic field, the density of electron states in the interface plane is constant as a function of energy. In the presence of the magnetic field, the states group themselves in Landau levels separated by the cyclotron energy, $\hbar\omega_c$. The energy levels are identical to those of the quantum harmonic oscillator,

$$E_n \quad \hbar\omega_c\left(n \quad \frac{1}{2}\right), \quad \text{with} \quad n \quad 0,$$

where n is the main quantum number and ω_c is the cyclotron frequency, $\omega_c \quad eB/M$. Here, e and M are the charge and effective mass, respectively, of the particle rotating under the action of a magnetic field of strength, B. The presence of the Landau levels (the Landau quantization) can be felt when the mean thermal energy, k_BT (with k_B the Boltzmann constant and T the absolute temperature), becomes much smaller than the energy level separation, $\hbar\omega_c$.

The maximum number of electrons filling, n, Landau levels (assuming $T \quad 0$ K and the Fermi level lies just above the Landau level with quantum number n) is directly proportional to the strength of the applied magnetic field,

$$N_e^L \quad n\frac{\Phi}{\Phi_0},$$

where $\Phi \quad B \quad A$ is the magnetic flux through the two-dimensional electron gas plane of area, A, and Φ_0 is the "fundamental quantum of flux",[26] $\Phi_0 \quad h/e$, with h, the Planck constant and e, the electron charge. The ratio, $N_\Phi \quad \Phi/\Phi_0$, gives the number of flux quanta enclosed by the two-dimensional electron gas for a given value of the applied magnetic field.

The *lling factor*, ν, is defined as

$$\nu \quad \frac{N_e^L}{N_\Phi}$$

and is equal to the quantum number n of the maximum occupied Landau level when the position of the Fermi level is between n and $n \quad 1$.

The inverse of the filling factor,

$$\frac{1}{\nu} \quad \frac{N_\Phi}{N_e^L},$$

gives the average number of flux quanta enclosed by each electron at a given value of the applied magnetic field.

When the magnetic field increases, more electrons can fill a certain (higher energy) Landau level; thus, the Landau levels move relative to the Fermi level. When the Fermi level happens to be in the gap between two Landau levels, a region where there are no allowed electron states, the transport is dissipationless and the Ohmic resistance falls to zero, while the Hall resistance takes the value,

$$R_H \quad \frac{1}{\nu}\frac{h}{e^2}.$$

For filling factors, $\nu \quad$ *integer*, this value is the same as the classical Hall resistance. The difference is that while in the classical effect the Hall resistance varies monotonously with the strength of the applied magnetic field, in the QHE the Hall resistance exhibits relatively stable levels, or plateaus; it cannot change from its quantized value corresponding to each plateau for a range of magnetic field strengths for which the Fermi level stays in a gap between Landau levels. When the magnetic field reaches a value where the Fermi energy overlaps with a Landau level, electrons can scatter to new states, the Hall voltage changes, and the Hall resistance changes by a finite value.

The QHE plateaus correspond to gaps between electron states that extend across the sample and carry the current. At the same low temperatures, the plateaus in pure, high mobility semiconductor heterojunctions are much narrower than in dirtier, lower mobility samples where the gap between states is larger due to impurities.

6.15 FRACTIONAL QUANTUM HALL EFFECT

In 1982, Daniel Tsui, Horst Störmer, and Arthur Gossard [424] discovered experimental evidence of a new phenomenon called the fractional quantum Hall effect, FQHE. In experimental studies of

[26]The fundamental quantum of flux, or flux quantum, represents the smallest unit of magnetic flux that can be enclosed by an electron orbit.

the QHE performed at even lower temperatures and with even more powerful magnetic fields, they observed an unexpected new step in the Hall resistance of a GaAs – $Al_{(1\quad x)}Ga_xAs$ heterojunction; the two-dimensional electron gas resulting from donors located inside the AlGaAs was formed on the undoped, GaAs, side of the junction. The new quantized plateau was observed at Hall resistance

$$R_H \quad 3\frac{h}{e^2} \qquad \text{for a filling factor} \qquad \nu \quad \frac{1}{3}.$$

More such steps with heights obtained by dividing the constant, h/e^2, by different fractional numbers have been observed above and between the integer plateaus. Some of the fractional filling factors experimentally observed in the lowest Landau level are:

$$\frac{1}{3},\frac{1}{5},\frac{1}{7} \qquad \frac{2}{3},\frac{2}{5},\frac{2}{7} \qquad \frac{3}{5},\frac{3}{7},\frac{3}{13} \qquad \frac{4}{7},\frac{4}{9} \qquad \text{and} \qquad \frac{5}{9},\frac{5}{11}.$$

Quantum Fluid and Quasiparticles

Less than a year after the experimental discovery of the FQHE, Robert Laughlin [267] explained the new experimental facts as the result of the condensation of the two-dimensional electron gas in the heterojunction into a new type of collective ground state, a kind of quantum fluid incompressible due to long-range Coulomb interaction between electrons. Electrons are fermions and, as such, are not able to condensate as a fluid; to behave like the quantum fluid proposed by Laughlin, the electrons are assumed to combine first with *ux quanta* (the magnetic flux tubes they orbit around when placed in a magnetic field) and form *quasiparticles.*

The $\frac{1}{3}$ step observed by Tsui, Strömer, and Gossard was explained by assuming that each electron in the two-dimensional electron gas captured three flux quanta and thus formed a sort of composite particle that behaved like bosons and could be condensed at very low temperatures and very strong magnetic fields. The ground state of this electron gas forms a quantum fluid, and its elementary excitations are fractionally charged quasielectrons and quasiholes. This quantum fluid is incompressible for a given filling factor, $\nu \quad n/m$ (n and m are integers with m being odd), and has an energy gap for adding quasiparticles.

If only one electron is added to this ground state, it becomes excited and a number of quasiparticles are created. The quasiparticles are not true particles but manifestations of the interplay of the electrons in the quantum fluid.

Laughlin's quantum fluid explained the disappearance of the Ohmic resistance at the Hall resistance steps and all the fractional filling factors found experimentally. This quantum fluid is said to be *incompressible* because compressing or expanding it implies injecting particles with a cost in energy.

Assume the FQHE system is a full Landau level with collective excitation occurring at $\hbar\omega_c$. Its response to the compressive stress of an applied magnetic field is the following: first, the system generates Hall currents without compressing until the stress reaches a critical value and then it collapses by an area quantum, $m2\pi a_0$, and nucleates a quasiparticle that is surrounded by a vortex of Hall current rotating in a sense opposite to that induced by the stress.

The states in the n-th Landau level, $\; m,n \;$, are eigenstates of the total angular momentum, with eigenvalues $\mathcal{M} \quad 3m \quad 2n$. Laughlin explained the $\frac{1}{3}$-effect in terms of the states in the lowest Landau level with $n \quad 0$ [266]. The states of the lowest Landau level, $\; m,n \quad 0 \;$, are described qualitatively as states of cyclotron motion about the origin with angular momentum, m, and orbit of radius, $\;\overline{2m}$.

The quantization of the electron separation in the QHE has a physical origin: in a strong magnetic field, particles with the same charge sign do not repel one another, as if the Coulomb interaction were absent. But, they orbit each other with a speed proportional to the Coulomb force between them, at a distance, r, determined by a negative binding energy, e^2/r. The quantization of the interelectronic spacing follows from quantization of angular momentum.

Laughlin found that the first FQHE state observed experimentally at filling factor, $\nu \quad 1/3$, was correctly described by a many-body wave function he explicitly constructed; that wave function had a lower energy than the single-particle energy. The Laughlin wave function can be used to describe states at any filling factor with an odd number denominator of the type, $\nu \quad \frac{1}{(2m \quad 1)}$, such as $1/3, 1/5, 1/7$. As m increases, the energy difference between states is smaller and the plateaus become weaker. Laughlin also suggested that elementary excitations from the stable states are quasiparticles with fractional electric charge, $e \qquad e/m$. Thus, a quasihole is formed in the incompressible quantum fluid as a two-dimensional "bubble" of a size such that $1/m$ of an electron is removed. Laughlin's result was verified by others, such as Arovas *et al* [13]. The radius of a quasiparticle was estimated [267] as, $R_{\text{(quasiparticle)}} \qquad \overline{2}a_0$, assuming an ion disc of charge, $1/m$, where the magnetic length $a_0 \quad \overline{\hbar/m\omega_c} \quad \overline{\hbar c/eB}$. The energy required to create a quasiparticle, assuming accumulation of charge $1/m$ inside an ion disc, is approximately

$$\Delta E^{\text{disc}}_{\text{(quasiparticle)}} \qquad \frac{3}{2\ \overline{2\pi}}\,\frac{1}{m^2}\,\frac{e^2}{a_0}.$$

The energy to produce a quasiparticle does not depend on the position of the point where it is generated, as long as its distance from the boundary of the Hall sample is larger than its size. Elementary excitations of the quantum fluid are generated by the adiabatical transport of a flux quantum, $\Phi_0 \quad hc/e$, through the electron gas plane.

Nevertheless, states at filling factors, $\nu \quad p/q$, where $p > 1$ or where q is an even number, cannot be described with Laughlin's wave function.

Haldane [193] and Halperin [198] developed a *hierarchical model* to account for filling factors of the type $\nu \quad p/q$, with $p > 1$ and q odd. The model describes a hierarchy of incompressible quantum fluid states with fractional filling factors given by continued fractions of the type

$$\cfrac{1}{m \quad \cfrac{\alpha_1}{p_1 \quad \cfrac{\alpha_2}{\ddots \quad \cfrac{\alpha_n}{p_n}}}},$$

where

$$m \quad 1,3,5,\ldots, \qquad \alpha_i \quad 1, \qquad \text{and} \qquad p_i \quad 2,4,6,\ldots,$$

These filling factors are rational numbers with an odd denominator; thus, a given fluid state at filling factor, $\nu \quad [m, p_1, \ldots, p_n]$, cannot occur unless its parent state at, $\nu \quad [m, p_1, \ldots, p_n \quad 1]$, also occurs. Incompressible fluid states derive from parent incompressible states at simpler rational filling factors, ν. The most stable fluid states correspond to the simplest rational numbers with small values of m and p_i, where the fluid densities are highest.

In the hierarchical model, the Laughlin state at filling factor 1/3 is a ground state and a *parent state* for quasielectrons and quasiholes that are excited out of it and condensed into higher-order fractions, the so-called *daughter states*. For example, from the ν 1/3 parent state, one can obtain the daughter state with filling factor, ν 2/5, when quasielectrons are excited out of it, or the daughter state with filling factor, ν 2/7, when quasiholes are excited out. In turn, quasiparticles excited out of the daughter states condense into more daughter states and the process repeats itself with all the daughter states forming a hierarchy of states. The Haldane scheme is able to reproduce all FQHE steps corresponding to experimentally observed fractional filling factors with odd the denominators,

$$\nu \quad \frac{2}{7}, \frac{1}{3}, \frac{2}{5}, \frac{2}{3}, \frac{4}{5}, \frac{4}{3}, \text{ and } \frac{5}{3}.$$

They correspond to filling factors given (in Haldane notation) by [3,2], [3], [3, 2], [1,2, 2], [1,2], and [1,4], all derived from m 1 and m 3 hierarchies.

The hierarchical model has some problems: It does not account for the relative strength of quasielectron or quasihole states; after a few levels of the hierarchy, the number of quasiparticles will be larger than the number of electrons initially in the system; the system is not well defined between fractions; and the quasiparticles have to carry fractional charge. The FQHE efect represents only an indirect demonstration of the quantum fluid and its fractionally charged quasiparticles that carry the current.

The presence of quasiparticles has been observed directly in experiments like that of de-Picciotto *et al* [113] who measured the quantum shot noise, which probes the temporal behavior of the current and provides a direct way to measure the charge of the quasiparticles.

The charge transport in the FQHE regime can be explained based on the assumption that one-dimensional interacting electrons propagate along the edges of the plane where the two-dimensional electron gas is confined (the edges of the Hall sample) and behaves like a liquid [449]. The quantum shot noise generated by the weak backscattering of the current is proportional to the quasiparticle's charge and some other physical quantities given for a certain measurement. The shot noise measurements show that the current in the FQHE regime, at filling factor 1/3 is carried by quasiparticles with charge $e/3$.

Topological Order of FQHE States

The ground state degeneracy of the FQHE states depends on the topology of the space, and that suggests that the FQHE states possess what Wen [448] calls *topological orders*; i.e., *the number of degenerate ground states is a topological invariant for any given quantum Hall phase.* This topological invariance is related to the observation that when the bulk flow of the QHE fluid is restricted to the incompressible ground state, it becomes irrotational and incompressible.

The ground state degeneracy of the FQHE states does not change under arbitrary perturbations; therefore, it can be used to characterize different phases in phase space. Different FQHE states that are macroscopically distinct can give rise to the same Hall conductance; if that is the case, they can be distinguished, at least partially, by their different ground state degeneracies (as estimated on torus and Riemann surfaces). The ground state degeneracy of the FQHE state is directly determined by the fractional statistics of the quasiparticles and not by the filling factor, ν [448].

Alexei Kitaev [239] analyzes the apparent paradox: The FQHE ground state is characterized by topological quantum order in its degeneracy and, at the same time, is separated by a finite energy gap from the excited states, which causes all spatial correlation functions to decay exponentially. He shows that, though correlations can disappear at small scale, parts of the microscopic system keep track of the topology through long-range entanglement manifested in the excitation properties.

Fractional Statistics

The *adiabatic theorem in quantum mechanics* states that if we slowly vary the parameters in the Hamiltonian of a system, then a state occupying an isolated energy level, E, remains fully in that level; there are no "quantum jumps." If, after a sequence of adiabatic changes, the Hamiltonian returns to its starting value, the system returns to its original state, apart from a phase factor; the state vector (wave function) has a change in its phase.

Michael Berry showed [52] that in addition to the dynamical component given by $e^{iEt/\hbar}$, which keeps track of the evolution in time, t, of any stationary state, there is also a geometrical phase component, $e^{i\gamma}$, which remains finite in the adiabatic limit. This is the so-called *Berry phase* and is given by

$$\gamma \quad \oint i \; \psi(t) \; \frac{d\psi(t)}{dt} \; dt,$$

where the integral can be taken over time, or over any parameter used to describe the path of the state vector (wave function) in the state space. The integral depends only on the geometry of this path; it does not depend on the coordinate used to describe the path or on the way the evolution occurred in time (provided that the circuit is traveled slow enough for the adiabatic approximation to be applicable).

For a particle of spin, S, in a magnetic field, $\mathbf{B}$, the geometric phase factor depends only on the eigenvalue, s, of the spin component along, $\mathbf{B}$,

$$e^{i\gamma(C)} \quad e^{is\Omega(C)},$$

where $\Omega(C)$ is the solid angle subtended by, C, when $\mathbf{B}$ 0; the phase can be changed by varying the field $\mathbf{B}$ around a circuit C. For fermions, particles of spin-½ such as electrons, protons, or neutrons, a rotation, with angle Ω 2π (a whole turn of $\mathbf{B}$ around C) produces a phase factor 1. For bosons, particles of integer spin such as photons, the same rotation produces a phase factor 1. These quantum particles obey Fermi-Dirac and, respectively, Bose-Einstein statistics. Upon exchange of two particles, the two-particle state vector, $\psi(1,2)$, acquires a phase and becomes

$$\psi(2,1) \quad e^{i\pi\Theta} \; \psi(1,2) \; .$$

A new exchange brings the two particles to the initial state,

$$\psi(1,2) \quad e^{i\pi\Theta} \; \psi(2,1) \quad e^{i2\pi\Theta} \; \psi(1,2) \; .$$

Quantum state vectors are single valued, which requires

$$e^{i2\pi\Theta} \quad 1,$$

where Θ integer. In particular, for fermions Θ_{fermions} 1 and for bosons Θ_{bosons} 0. Such particles obey exchange statistics that are *integer statistics*.

In quantum mechanics the term *statistics* defines the effect of exchange of identical particles/quasi-particles. In three-dimensional space, under adiabatic conditions, the state vector of a quantum particle executing a full loop around another particle is not distinct from the state vector of the same particle executing a closed circuit without a particle inside; both acquire the same phase factor.

In two-dimensional space (a plane) the closed circuit with a particle inside is topologically distinct from a circuit without a particle inside. The evolution in time of particles exchanging places while confined to a plane represents *braiding* of their trajectories in (2D time). The interchange of the quasiparticles introduced by Wilczek in 1982 can give any phase; the factor, Θ, in the phase is no longer required to be an integer, it can be *any* real number and the particles obeying such *fractional statistics*, hence, their name, *anyons* [455].

Wilczek's quasiparticles are composites made of charged particles tied to magnetic flux tubes in two dimensions and obey fractional statistics: when the positions of two quasiparticles are interchanged, the state vector of the pair changes. It acquires a complex phase factor, $(\;\; 1)^{\;1/m_k}$ (similar to Berry's geometrical phase), with the sign depending on the sense of rotation as the quasiparticles pass each other and m being a rational number, m_k 1; this number is related to Θ above.

The Composite-Fermion Model

The model, introduced by J. K. Jain [217] in 1989, is a unified approach that explains the integer QHE and the fractional QHE as two different manifestations of the same type of correlations in the phase factors present at fractional filling factors, $\nu \;\; \frac{p}{2mp \;\; 1}$, with p and m integers, and present at integer filling factors, $\nu \;\; p$.

The composite-fermion consists of an electron, or hole, bound to an even number, $2p$, of magnetic flux quanta (vortices); a composite-fermion is interpreted as a particle. Electrons capture vortices to form composite fermions because they minimize their interaction energy in this way. The fractional QHE of electrons in external magnetic field, $\mathcal{B}$, is the manifestation of the integer QHE of composite-fermions in an "effective" magnetic field $\mathcal{B}$.

The composite-fermions have integer charge-like electrons, but because they move in an effective magnetic field, they appear to have a fractional topological charge. As composite-fermions move along closed paths, the vortices bound to them produce Berry phases that cancel part of the Aharonov-Bohm phases originating from the motion in the applied magnetic field. Composite-fermions sense an effective magnetic field, $\mathcal{B}$, much smaller than the applied field, $\mathcal{B}$; the effective field can even be zero. Now, the composite-fermions form Landau-like levels in the reduced $\mathcal{B}$ and occupy ν of them, which distinguishes the composite-fermions from electrons.

Composite-fermions are collective, topological quantum particles; even a single composite-fermion is a collective bound state of all the electrons. Their topological character is due to the fact that the quantum phase associated with a closed loop around a vortex is exactly, 2π, independent of the size or shape of the loop. Composite-fermions represent the non-perturbative reorganization that takes place when a two-dimensional electron gas is subjected to a strong magnetic field at very low temperature.

The composite-fermion model of the QHE correctly predicts all the observed fractions (with odd denominator), their relative intensities, and the order they appear in when the sample purity increases. This model predicts all these fractions at the first level, in contrast to the standard hierarchical model, which has to go down many levels to obtain some of the observed fractions; otherwise there is a close analogy between the states predicted by the two models and how they are obtained.

The statistics of composite particles is in general fractional [455] and is due to the special topology of the two-dimensional space where the electrons reside in the case of the fractional QHE. The relative motion of the particles confined to the two-dimensional plane is limited to going around each other, they cannot get out of the plane and move above or below each other.

During the evolution in time of the system, the paths (world lines) of the particles form *braids*[27] in the three dimensions defined by the two-dimensional plane time (see Section 1.18). Qubit braiding alone or together with some single-qubit gate and measurements could provide universal logic for quantum computation. The states of the FQHE were thought to have ordinary Abelian statistics associated with usual phase factors of 1 when braiding quasiparticles corresponding to FQHE states with odd denominator filling factors (similar to exchanging identical fermions or bosons among themselves).

Non-Abelian Quasiparticles

Considering the special topology of the quasiparticles, Moore and Read [299] suggested that some states could have non-Abelian statistics. Following the intuitive approach pioneered by Laughlin, they came up with several wave functions as verifiable solutions of the Hamiltonian for the quantum Hall system, and they showed that the quasiparticle excitations of the states associated with those wave functions were non-Abelian.

In (2 1) dimensions braiding sometimes changes the nature of the phase associated with the interchange of quasiparticles. The wave function of a set of excitations with specified positions and quantum numbers becomes a vector and each exchange of these quasiparticles gives rise to a unitary matrix transformation, instead of a simple alteration of its phase; that represents a *non-Abelian* action on this state vector.

In the case of quasiparticles with non-Abelian braiding statistics, the exchange of two particles does more than changing the phase of the wave function; it may rotate it into a different one in the space spanned by the state of the system. If we assume that a system of $a \quad 1,2,\ldots,m$ quasiparticles at positions, $x_1,x_2,\ldots,x_n$, is in a set of m, $m>1$ degenerate states, ψ_a, then by exchanging particles 1 and 2, the state vector becomes

$$\psi_a \quad A^{12}_{ab}\psi_b.$$

If particles 2 and 3 are exchanged, then the state vector becomes

$$\psi_a \quad A^{23}_{ab}\psi_b.$$

For some pairs of quasiparticles, the matrices, A^{12}_{ab} and A^{23}_{ab}, do not commute; such quasiparticles obey non-Abelian matrix statistics. If a system has a large set of non-Abelian states, the repeated application of the braiding transformation, A^{ij}_{ab}, would allow the approximation of any unitary transformation to arbitrary accuracy.

[27] The braid group, sometimes denoted by $\mathcal{B}_n$, replaces the permutation group in describing the particle statistics in two dimensions.

The simplest of the wave functions proposed by Moore and Read corresponds to a spin-polarized p-wave pairing Bardeen-Cooper-Schrieffer state for a fixed number of composite fermions. The non-Abelian nature of this Moore-Read state comes from the collective degeneracy of the quasiparticles in this state. If quasiparticles are moved around each other, the state of the entire collective ensemble changes in a way that depends only on the topology of the move and the result is a unitary transformation in Hilbert space. As we know, unitary transformations in Hilbert space are the operations executed by a quantum computation.

The physical system that may serve as a platform for topological quantum computation is the fractional QHE with Landau level filling factor, $\nu \quad 5/2$, associated with the quantized Hall plateau with transverse conductivity, $\sigma_{xy} \quad \frac{5}{2}\frac{e^2}{h}$, first observed in 1987 by Willett *et al* [456] at a temperature of about 100 mK. The state is observed on a regular basis in experiments on low-disorder samples of GaAs at very low temperature, ($T \quad 15$ mK), and magnetic field of about 5 tesla. The temperature dependence of the Hall resistance, R_{xx} at $\nu \quad 5/2$, measured in such experiments [135, 463] suggested an energy gap, $\Delta_{5/2} \quad 310$ mK. This state is assumed to belong to the first excited Landau level, with $N \quad 1$.

The state, with, $\nu \quad 5/2$, is easily destroyed and replaced by a strongly anisotropic phase when a strong magnetic field component, B , parallel to the two-dimensional electron gas plane is added to the applied perpendicular field. The application of the transverse magnetic field initiates a phase transition of the system from a gapped quantum Hall (incompressible) phase into an anisotropic compressible phase.

Shortly after its discovery, Moore and Reed [299] developed a theory predicting that the elementary excitations of the, $\nu \quad 5/2$, state are non-Abelian anyons. Then Nayak and Wilczek [303] found the corresponding braiding group representation.

Numerical calculations by Rezayi and Haldane [356] indicate that the 5/2 state belongs to the non-Abelian topological phase characterized by a so-called Pfaffian[28] quantum Hall wave function.

The existence of non-Abelian quasiparticles at filling factor, $\nu \quad 5/2$, depends on two important hypotheses:

1. The Coulomb repulsion in the second Landau level (where this state belongs) has a form-favoring pairing of the quasiparticles.
2. The electrons are fully polarized.

Numerical calculations (e.g., [326, 327]) offer a strong evidence that the first premise is satisfied, especially when the finite layer thickness is taken into account in the electron-electron Coulomb interaction; when the two-dimensional layer of electrons is assumed to have a thickness, $d \quad 4l$ (where $l \quad \overline{\hbar c/eB}$ is the magnetic length), the exact ground state of the Coulomb Hamiltonian is very well approximated by the Moore-Read Pfaffian wave function [326], which assumes pairing.

Experimental Evidence of Non-Abelian Statistics

Experiments set up to observe the quasiparticles with charge, $e/4$, support the first hypothesis, yet cannot distinguish and rule out some paired states that could have the same quasiparticle charge, but

[28]The Pfaffian is the square root of the determinant of an antisymmetric (skew-symmetric) matrix. The Pfaffian is non-vanishing only for $2n \quad 2n$ matrices.

are Abelian and not fully polarized. In a two-dimensional electron system under high magnetic field and at very low temperature where the FQH effect is present, electron-electron interactions lead to gaps in the bulk excitation spectra and the current can only be carried by states that propagate around the edges of the system.

A quantum point contact in a two-dimensional electron system represents a constriction where counter-propagating edge states are brought close enough together so that quasiparticles can tunnel between them. The weak quasiparticle tunneling depends [450, 452] strongly on the voltage difference between edges (or the current through the quantum contact point) and on temperature (the tunneling decreases slowly, as a power law, with increasing temperature). Its dependence on temperature provides a measurement of the effective quasiparticles charge, e , and the strength of the Coulomb interaction, g, both of them specific to the state. Quantum point contact measurements provide a discriminating probe of FQH effect wave functions.

The quantum point contact tunneling experiments on a high-mobility GaAs two-dimensional electron system [347] indicate that various candidate states at, ν $5/2$, when compared with the experimental tunneling data, predict a quasiparticle charge, $e/4$, but their Coulomb interaction strength parameter, g, can differ. The data are most consistent with charge, $e/4$, and factor, g $1/2$, values corresponding to two non-Abelian states, particle-hole conjugate of the Moore-Read state (anti-Pfaffian) and a $U(1)$ $SU_2(2)$ edge state [451]. But, at the same time, the Abelian state with charge, $e/4$, and, g $3/8$, is not excluded by these experiments.

Direct evidence of effective quasiparticle charge, $e/4$, in the state with fractional filling, ν $5/2$, in the second Landau level of the QHE was found in the experiments of Dolev *et al* [131]. In their measurements, the current tunnels across a constriction (quantum point contact) between two opposite edge states of a Hall bar GaAs/AlGaAs sample; the quasiparticle charge is estimated from the current shot noise. Their conclusion is that a quasiparticle charge of $e/4$ at filling ν $5/2$ indicates a paired composite fermion state, consistent with a spin-polarized Moore-Read Pffafian state [299]. This experiment is not a direct test of the non-Abelian nature of the quasiparticles, but if the Moore-Read state is the correct one for filling factor $5/2$, the observed quasiparticles with charge $e/4$ obey non-Abelian statistics.

Quasiparticle Charge

Experiments of quasiparticle interferometry are expected to allow direct observation of their fractional charge and, eventually, their non-Abelian statistics. The two point-contact interferometer as proposed in [87] consists of a Hall bar with two weak barriers and can be used to study quantum interference effects in strongly correlated (electron) systems. In the quantum Hall phase, the electrons are in the bulk of the Hall bar specimen while the low-energy excitations, the quasiparticles, lie on the edges. At the constrictions corresponding to the point-contacts, the quasiparticles carrying fractional charge and statistics can tunnel from one edge to another creating a tunneling current that modifies the quantum Hall conductance.

The presence of two tunneling sites results in interference between the tunneling events taking place at individual point-contacts (similar to double-slit experiments, see [283]); the tunneling current becomes sensitive to the phase of the tunneling events. The interference of the tunneling quasiparticles can be acted upon in three ways:

1. Changing the magnetic field at constant filling factor (it leads to Aharonov-Bohm oscillations).

2. Changing the number of quasiparticles enclosed by the interfering orbitals (it leads to fractional statistical oscillations).
3. Varying the source-drain voltage (it leads to so-called "Fabry-Pérot" oscillations).

Theoretically, it has been shown that these three effects can be studied separately and, in doing so, they provide means of measuring the fractional charge and fractional statistics of the quasiparticles.

Interference in integer QHE systems has been reported by several groups, the most recent with device designs analog to the optical Mach-Zehnder interferometer [82, 83, 219, 361]. Extensive measurements of fractional quantum Hall state interference performed recently [85, 86] have shown well-defined Hall conduction periodic structures over a range of integer and fractional QHE filling factors.

The proposed application of Fabry-Pérot interferometers to detection of non-Abelian braiding statistics will provide a verification of the microscopic ground state of certain observed fractional QHE states, such as the even-denominator, ν $5/2$. The interferometric measurement of charge has the advantage of a simple analysis, a direct comparison of the interference period at different quantum Hall state filling factors to that at the state, $5/2$.

Interferometer devices constructed to accommodate the less robust fractional quantum Hall states [457] will permit interference of fractional state excitations with small coherence lengths; that is considered a crucial step toward experimentally proving the non-Abelian nature of the ν $5/2$ state specifically associated with charge $e/4$. A charge $e/2$ is a potentially possible value for a ν $5/2$ state with Abelian properties.

A two-point contact interferometer has the geometry of a Hall bar with two constrictions, such that the quasiparticle tunneling occurs between two edge currents on the opposite sides of the bar. The tunneling current is sensitive to the total topological charge of anyons trapped inside the interferometer loop. Such an interferometer could form a universal gate set together with the short-range interaction between two anyons transported sufficiently close to each other where they could exchange a virtual quasiparticle, and with topological quantum computation operations.

Electron Spin Polarization

As for the second hypothesis concerning fully polarized electrons, calculations [123, 300] indicate that a fully spin-polarized state is lower in energy than an unpolarized one. That is seen as excluding the hypothesis that the ν $5/2$ state with charge $e/4$ could be a completely unpolarized state. However, a partially polarized state cannot be excluded [123], and that could be either Abelian or non-Abelian in nature.

So far, there is little experimental evidence to support the fully polarized state hypothesis. Experiments set up to directly measure the spin polarization at, ν $5/2$, through the resistively detected NMR technique have so far been unsuccessful. This technique relies on the hyperfine interaction in the following way: A finite nuclear spin polarization in the nuclear species present in the FQHE sample creates an effective magnetic field, B_N, that modifies the Zeeman energy, E_{Zeeman} $g\mu_B(\mathcal{B}$ $B_N)$, and with it the spectrum of hyperfine levels. For example, for electrons in GaAs (g 0.44), the effective magnetic field induced is B_N 5.3 T if the three spin-3/2 nuclear species, ^{69}Ga, ^{71}Ga, and ^{75} as present in the sample are all fully polarized.

Similar experiments at ν $1/2$ have shown that the system remains partially polarized up to magnetic fields of about 8 tesla [423]; the $5/2$ state is typically observed at half the magnetic field. However, the results of all these experiments at low magnetic fields could also be explained assuming fully or partially spin-polarized, even unpolarized states; that leaves us with a confusing picture of the nature of the 5/2 state.

For the quantum Hall state with filling factor ν 12/5 recently observed experimentally [463], there is no direct evidence yet that it has non-Abelian statistics, though calculations indicate that it could be a state of non-Abelian anyons (a Read-Rezayi state). At the same time, it is also possible that it belongs to the conventional Abelian hierarchy of odd-denominator states. If the ν 12/5 state proves to be a Read-Rezayi ($\mathcal{Z}_3$ parafermion) state [351], then it has braiding statistics that allow universal topological quantum computation. Calculations show that the state is close to a phase transition between the Abelian hierarchy state and the non-Abelian Read-Rezayi (parafermion) state and the system could be pushed over the phase boundary from an Abelian to a non-Abelian phase by engineering minor changes in the experimental Hall sample. However, besides the well-quantized plateau observed in the original experiments [463], very little is known about the ν 12/5 state and its spin polarization and quasiparticle charge.

Application of FQHE to Quantum Computing

The quasiparticles in ν 5/2 state cannot support universal computation; the transformations generated by braiding operations with such quasiparticles are not sufficient to implement all possible unitary transformations [159]. The braiding of two such quasiparticles has the effect of a 90 rotation in the Hilbert space of a system of multi-quasiparticles. Arbitrary unitary transformations cannot be constructed through composition of 90 rotations. However, braiding together with a single-qubit $\pi/8$ phase gate and a two-qubit measurement form a universal gate set. The two extra gates can be implemented through some nontopological operations [74] and are not topologically protected.

A *single-qubit phase gate* can be implemented by bringing two quasiparticles (out of the four making a qubit) together at a given distance and waiting for some time before pulling them apart. The choice of distance and waiting time determines the phase gate. The other gate under consideration for the universal set is a nondemolition measurement of the total topological charge of any four quasiparticles out of the eight forming two qubits [159]. This is done with an interference measurement, and depending on the total topological charge of the four quasiparticles, the two possible trajectories interfere with a phase 1. In this way we measure the total parity of the two qubits. A non-Abelian topological, ν 12/5, state would have the advantage of supporting universal topological quantum computation through quasiparticle braiding alone.

6.16 ALTERNATIVE PHYSICAL REALIZATIONS OF TOPOLOGICAL QUANTUM COMPUTERS

In the previous section, we saw that quantum information can be processed by braiding the anyons in a two-dimensional structure with an exponentially large topologically protected Hilbert space. Topological quantum computing introduced by Kitaev in 1997 and further developed by Freedman, Kitaev, Larsen, and Wang [157, 158] has the potential to revolutionize fault-tolerance and justifies the interest in alternative physical realizations of such systems.

Universal topological computation requires both a topological phase and quasiparticles with non-Abelian braiding statistics and with topological properties in a special class. A *topological phase* of a physical system is defined [304] as the system's ground state in the presence of multiple quasiparticles or in a nontrivial topology with a stable degeneracy resistant to weak (but finite) local perturbations. The stability of the ground state degeneracy to local perturbations requires the existence of an excitation gap.

The possible non-Abelian fractional quantum Hall states, such as ν $5/2$ and $12/5$, are the first candidates. Incompressible fractional quantum Hall states are, so far, the only experimentally established topological phases. There are other physical systems, besides the fractional quantum Hall states, which are assumed to contain topological phases, such as transition metal oxides and ultra-cold atoms in optical traps.

The transition metal oxides are already known for displaying strong collective phenomena, (high-T_c, superconductivity, huge magnetoresistance, and thermoelectricity). At least for one material in this family, the Sr_2RuO_4 (strontium ruthenate), there is strong evidence that it is an odd-parity, spin-triplet (p-wave) paring superconductor at low temperature, T_c 1.5 K. Half-quantum vortices in a thin film of this metal oxide superconductor are expected to exhibit non-Abelian statistics.

The ultra-cold atoms in optical traps represent another type of physical systems where topological phases could be present. The Hamiltonian can be controlled by tuning the lasers that define an optical lattice or by tuning through a Feshbach resonance. Topological phases can be generated in ultra-cold gases using two techniques: (1) by fast rotating dilute boson gases to make quantum Hall systems of bosons and (2) by using a gas of ultra-cold fermions, with a p-wave Feshbach resonance to form a polarized chiral p-wave superfluid. Topological phases are hard to detect directly in the transition metal oxide or in ultra-cold atomic system unless the phase breaks parity and time-reversal symmetries spontaneously or in an external magnetic field.

CNOT gate based on parity measurement. In a system of ν $5/2$ anyons, two quasiparticles form a two-level system equivalent to a spin-½ and can encode a qubit. That is the most efficient way of encoding a logical qubit, but in this case, due to charge superselection rules, a qubit cannot be prepared in a superposition of basis states (e.g., 0 1). The solution adopted by [74] is to represent a logical qubit by a group of four quasiparticles and to identify the logical basis states, 0_{logical} and 1_{logical}, with the physical states, 0,0 and 1,1 . In the four-quasiparticles representation, a qubit can be prepared in a superposition of basis states.

However, any two-qubit logical state that can be prepared by topological operations has the form of a product of two logical one-qubit states; therefore, in the four-quasiparticle representation, no entangled states of two logical qubits can be prepared topologically. This apparent drawback could be circumvented: a state approximating the logical entangled target state is prepared using noisy nontopological operations and then it is purified following a purification protocol involving only topological quantum computing operations [74]. The entanglement of two logical qubits could be implemented through topological braiding operations and a nondestructive measurement of the total topological charge of four quasiparticles without destroying their pairwise correlations. In the case of ν $5/2$ anyons, the total charge measurement is equivalent to a parity measurement.

The Pauli operators can be implemented by braiding operations; for example, if the qubit is represented by four quasiparticles, 1,2,3,4 in succession, the action of the Pauli operator, σ_z, corresponds to braiding quasiparticle 1 around quasiparticle 2, or to braiding quasiparticle 3 around quasiparticle 4. The braiding of quasiparticle 2 around quasiparticle 3 implements the action of the

Pauli operator, σ_x. The action of Pauli operator, σ_y corresponds to braiding quasiparticle 1 around quasiparticle 3. Any four-quasiparticle braiding gate commutes with the parity operator.

A two-qubit CNOT gate for such multi-quasiparticle qubits was proposed to be implemented based on parity measurements [469]. In the case of ν 5/2 anyons, what is measured is in fact the charge of four quasiparticles, and that is equivalent to measuring their spin parity. The total topological charge of the quasiparticles can be measured using an interferometric technique. This implementation of a CNOT gate uses an ancilla qubit besides the target and control qubits.

The CNOT operation is executed through spin parity nondemolition measurements on groups of four neighboring quasiparticles each: two quasiparticles of the control qubit and the neighboring two quasiparticles of the ancilla qubit and then the other two quasiparticles of the ancilla qubit and the neighboring two quasiparticles of the target qubit. For example, assume anyons 1, 2, 3, 4 form the control qubit, anyons 5, 6, 7, 8 form the ancilla qubit, and anyons 9, 10, 11, 12 form the target qubit. The parity measurement of anyons 3, 4, 5, 6 gives the parity of the control and ancilla qubits while the parity measurement of anyons 7, 8, 9, 10 gives the parity of the ancilla and target qubits. The Hadamard rotations required by the CNOT operation are implemented through topological braiding of anyons of the ancilla and, respectively, target qubit. The last measurement is of anyons 5 and 6 of the ancilla.

This implementation uses only local braiding operations between the nearest-neighbor anyons composing a qubit, while that of [74] requires long-range anyon braiding operations between qubits. Both implementations would require some unprotected operations, but that could be avoided if ν 12/5 non-Abelian anyons were used.

The topological quantum computation tries to use the emergent properties of many-particle systems to encode and manipulate quantum information in a way that is fault-tolerant. The physics of topological phases evolves alongside topological quantum computation and it is influenced by it. So far, the incompressible fractional quantum Hall states, such as ν 5/2 and 12/5, are the only experimentally established topological phases. Universal topological quantum computation has an additional requirement, which is that quasiparticles with non-Abelian braiding statistics and with topological properties belonging to a special class are needed.

6.17 PHOTONIC QUBITS

Over the past several years, photons and an all-optical architecture have emerged as a leading approach to a physical implementation of quantum information processing. An optical qubit is often realized as one photon in two optical modes, such as horizontal and vertical polarization. But qubits can be encoded in any of the other degrees of freedom of single photons, such as time bin or path. Quantum information may also be encoded in the continuous phase and amplitude variables of many-photon laser beams. Single photons are relatively free of decoherence.

Single-Photon Source

Single photons can be prepared by conditional measurements on a biphoton state generated in the process of *parametric down-conversion*, a process investigated experimentally in 1986 [180, 213]. The parametric down-conversion is the elementary quantum process of the spontaneous decay of a photon of frequency ω_0 into two new photons of frequencies, ω_1 and ω_2, such that $\omega_0 \quad \omega_1 \quad \omega_2$.

The emission of the two photons is simultaneous and the process obeys the conservation relations for momentum and energy. The two photons are in a highly entangled quantum state known as biphoton, have the same polarization, and travel in well-defined directions.

In the 1986 experiment, a coherent beam of light from an argon-ion laser (351.1 nm line) was traversing a (8 cm long) nonlinear crystal of potassium dihydrogen phosphate. Some incident photons split into two lower frequency photons, the so-called *signal* and *idler* photons. The signal and the idler photons were directed by two mirrors to pass through a beam splitter, and the superposed beams interfered and were detected by two photodetectors.

The rate at which the photons were detected in coincidence was measured when the beam splitter was displaced from its symmetry position by various small distances $c\delta\tau$. It was observed that the signal and idler photons (1) had no definite phase; (2) were mutually incoherent: they exhibited no second-order interference when brought together at either one of the two photodetectors; and (3) exhibited fourth-order interference effects as demonstrated by the coincidence counting rate between the two photodetectors, the so-called *cosine modulation.*

The individual frequencies, ω_1 and ω_2, had large uncertainties (though ω_0 was very well defined). The large uncertainties were determined by the pass bands of the interference filters, which were of the order of $5 \quad 10^{12}$ corresponding to a coherence time for each photon of 100 fsec.[29] The two-photon probability amplitudes at the two photodetectors were expected to interfere only if they overlap to this accuracy. Detection of a photon in one of the emission channels causes the collapse of the non-local photon pair and the projection of the quantum state of the other channel into a single-photon well-defined state. Most single-photon sources use entangled photons produced through parametric down-conversion.

Linear Optics Quantum Computation

Single-qubit logic gates acting as polarization rotations are easy to implement using birefringent wave plates; polarizing beam splitters can achieve conversion between polarization and path encoding. One of the earliest proposals for quantum information processing uses photons to implement a quantum logic (Fredkin) gate [293]. In fact, optical systems represent the only realistic implementation of long-distance quantum communication; they are applied to quantum cryptography (quantum key distribution) [215].

The realization of entangling, two-qubit logic gates required for universal quantum computation, is a major difficulty. An optical CNOT gate can be implemented only if a π phase shift is applied to the target qubit when the control qubit has value 1; this conditional phase shift is applied after a Hadamard gate creates a superposition state for the target qubit and a second Hadamard gate flips its value. Normally, the phase shift would require optical nonlinearities stronger than what is available in conventional nonlinear materials. Otherwise, electromagnetically induced transparency and atom-photon interactions enhanced by an optical cavity (cavity quantum electrodynamics) were considered the only available option.

In 2001, Knill, Laflamme, and Milburn (KLM) [250] showed that quantum information processing could efficiently (with polynomial resources) be performed with passive linear optics. What linear optics quantum computation requires are single-photon sources, beam splitters, phase shifters, photo-detectors, and feedback from photo-detector outputs. Photonic (bosonic) qubits are defined by states

[29]A femtosecond is $10 \quad ^{15}$ seconds.

of optical modes, each mode being characterized by the number of photons. For example, the basic states of a photonic qubit encoded in two modes, m_1 and m_2, are given by $\mathbf{0}$ $\quad 0_{\ m_1}\ 1_{\ m_2}$ and $\mathbf{1}$ $\quad 1_{\ m_1}\ 0_{\ m_2}$. As any other implementation, an optical quantum computer must have the means to prepare state, apply quantum gates, and extract the result. The initial state is a vacuum state, a state in which there are no photons in any of the used optical modes. Photons are added to the initial state using a single photon[30] source. Quantum states containing a definite number of photons (Fock states) are required in quantum cryptography for increased capacity and security in communication channels and for efficient linear quantum computation.

With the injection of a single photon in a well-defined state, the state of any given mode can be set to the one-photon state, $\mathbf{1}$, in this way. The state preparation should be non-deterministic; i.e., its probability of success is nonzero and it is known whether it succeeded or not.

Simple optical elements like beam splitters and phase shifters generate evolutions that preserve the total photon number, and their effects on the creation operator of each mode can be described by unitary matrices. The unitary matrix associated with phase shifter, P_θ is $u_{(P_\theta)}\quad e^{i\theta}$, and the unitary matrix associated with beam splitter, $B_{\theta,\phi}$, is

$$u_{(B_{\theta,\phi})}\quad \begin{pmatrix} \cos(\theta) & e^{i\phi}\sin(\theta) \\ e^{\ \ i\phi}\sin(\theta) & \cos(\theta) \end{pmatrix}.$$

A sequence of beam splitters and phase shifters is able to implement the operation associated with any such unitary matrix up to a global phase. The state space of a photonic qubit is preserved in the interaction with phase shifters and beam splitters; their effect can be expressed in the qubit basis using Pauli matrices, σ_x, σ_y, and σ_z. For example, a phase shifter, $P_\theta^{(1)}$, corresponds to the application of the, $e^{\ \ i\sigma_z(\theta/2)}$, operation up to a global phase and a beam splitter, $B_\theta^{(12)}$, applies the operation, $e^{\ \ i\sigma_y\theta}$. Thus, one-qubit rotations can be implemented with linear optics [250].

A two-qubit gate such as the conditional sign flip in one mode was proposed to be implemented by using two ancilla modes with one prepared photon and post-selection based on measuring the ancillas. The output is accepted only if the ancilla photon counters detect one photon in only one of the ancilla modes and none in the other. For any other detection pattern, the CNOT gate is not applied. Knill, Laflamme, and Milburn showed that measurement-induced nonlinearity was sufficient to implement a CNOT gate and to obtain efficient quantum computation.

In the case of a nondeterministic CNOT gate, the control and target qubits, encoded in photon polarization, for example, and two ancilla photons simultaneously enter an optical network of beam splitters (a beam splitter implements a Hadamard gate) where the paths of the four photons are combined. The CNOT operation is considered to have been applied to the state of the emerging control and target qubits conditional on a single photon being detected at both ancilla detectors. As such, the probability of a successful CNOT gate was estimated to be $P < 1/16$; the probability of a successful computation decreases exponentially with the number of applied CNOT gates.

The success probability of a two-qubit gate can be increased by reducing it to a state preparation problem by using quantum teleportation. The control qubit (mode 1) in an unknown state and one

[30] The single photon character is usually stated by showing that the amount of energy flowing during a certain characteristic time, such as coherence time, or time-of-flight between source and detector, is small compared to $h\nu$ [180]. The detection process appears discrete.

of the ancilla (mode 2) photons are subjected to a Bell measurement. Before the measurement, the control qubit is in the unknown state, $c \quad \alpha_0\, 0^{\,(1)}$, and the ancilla (mode 2) photon is prepared in a maximally entangled state, $\varphi^{\,(23)} \quad 01^{\,(23)} \quad 10^{\,(23)}$, with the second (mode 3) ancilla. The measurement is performed in the Bell basis, $01^{\,(12)} \quad 10^{\,(12)}, 00^{\,(12)} \quad 11^{\,(12)}$. In the process, the unknown input state of the control and target qubits can be teleported to the first and second ancilla, respectively.

The nondeterministic CNOT can be independently repeated on the ancillas until the gate succeeds and then we proceed with the teleportation [175]. A successful gate could be detected by measuring the ancilla photons in some appropriate state. The control and target qubits are preserved until the gate works and CNOT operation can be implemented deterministically. Quantum teleportation had been realized experimentally with single photons in 1997 [68].

The techniques applied include near-deterministic nondestructive parity measurements to assess photon loss and sign, a method for creating entanglement by local measurements of uncorrelated photons shared with beam splitters, and nearly unconditional quantum teleportation and Bell state measurements with linear optics. The KLM scheme for efficient linear optics quantum computation with photonic (bosonic) qubits shows that these techniques can be implemented with several types of operations: (1) preparation of X, Y, or Z eigenstate (eigenvalue 1 or 1); (2) measurements as X, Y, and Z; (3) rotations X_{180}, Y_{180}, and Z_{180}; rotation X_{ϕ}; (4) rotation Z_{90}; (5) rotation $(Z^{(1)}Z^{(2)})_{90}$ on two physical qubits that encode one logical qubit. The 180 rotations of logical qubits can be applied by using the corresponding 180 rotations directly on the physical qubits.

The computation result is read out by measuring a mode with a photo-detector associated with a photon counter; the measurement determines if one or more photons are present in that mode but destroys it/them in the process. The sources of errors for linear optics quantum computation are photon loss at the single-photon source or during processing, and detector inefficiency, which can be considered as loss of photons and phase errors.

In principle, the probability of photon loss can be estimated from the characteristics of the optical devices used by the implementation, but the loss itself introduces a so-called "erasure" failure, whereby we no longer know what happened to the state of the qubit. The probability for erasure as well as phase errors can be reduced if several photons are used to encode logical qubits and error-correcting codes are applied. A linear optics implementation of a quantum computer as proposed by [250] requires good mode matching, good synchronization of pulses or delay lines that can be rapidly controlled, tunable phase shifters and beam splitters, single-photon sources, and fast, high efficiency single-photon detectors.

The linear optics quantum computation scheme proposed by Knill, Laflamme, and Milburn was not economical in its use of optical components or ancilla photons. Several groups have been working on improving the theoretical efficiency of the gates and reducing their complexity [209, 252, 330]; improved CNOT gates have been demonstrated experimentally in the following years by Pittman *et al* [331], O'Brien *et al* [311], and Gasparoni *et al* [168], among others. Encoding against measurement errors was also demonstrated experimentally.

A proof-of-principle of a high-fidelity realization of a two-qubit Z-measurement quantum error correction scheme was provided for a two-qubit code of the type,

$$\varphi_L \quad \alpha\, 0_L \quad \beta\, 1_L \quad \alpha(\,00 \quad 11\,) \quad \beta(\,01 \quad 10\,),$$

where the encoded state is prepared from an arbitrary input state [312]. The input into the target mode of a nondeterministic optical CNOT gate was a single qubit prepared in an arbitrary state, $\varphi \quad \alpha 0 \quad \beta 1$, using a half-wave plate and a quarter-wave plate. An ancilla qubit prepared in real equal superposition, 0 1 , with a half-wave plate is input into the control mode of the gate. The resulting two-qubit encoded state, φ_L, generated from the possible inputs, $\varphi \quad 0 , 1 , 0 \quad 1 , 0 \quad i 1$, was measured using two-qubit quantum state tomography and found with high fidelity. The two-qubit measurements were performed using a pair of analyzers each consisting of a quarter-wave plate and a half wave plate followed by a polarizing beam splitter and a single-photon counter; each analyzer could perform any one-qubit projective measurement.

A combination of two-qubit coding and fast feed-forward control techniques was used by Pittman *et al* [332] to demonstrate QEC for measurement errors in linear optics quantum computation. In this experiment the qubit state, φ , and the ancilla state, 0 , were represented by the polarization states of two single photons from a parametric down-conversion pair. The logical state, φ_L, was encoded probabilistically using linear optics and post-selection. The feed-forward-controlled bit-flip was performed with an electro-optic polarization rotator triggered by the output of single-photon detectors. An intentional, Z, measurement was performed on one of the photons, and the success of the error-correction procedure was verified by comparing the corrected polarization state of the remaining photon with the initial state, φ .

The encoding was done with a single polarizing beam splitter (PBS), which transmits horizontally, (H), polarized single photons and reflects the vertically, (V), polarized ones. A qubit photon in the arbitrary state, $\varphi \quad 0 \quad 1$, was sent into one of the input ports of the PBS, and the ancilla in the state, 0 , was sent into the other input. Coincidence basis measurements on the two outputs of PBS made sure that one photon will exit each output port. Single photons from pairs produced by a down-conversion source were used as qubit and ancilla. The encoding operation for an arbitrary qubit state φ can be understood as a two-photon quantum interference effect that uses a beam splitter and post-selection to generate polarization entanglement, in the coincidence basis, from an initial product state of two single photons.

The fidelity of the logical qubit state, φ_L, was determined by the quality of the two-photon interference effects. The Z measurement was performed with a second PBS controlled in such a way that a photon with polarization corresponding to state 0 would be transmitted to a designated single-photon detector, while photons with polarization corresponding to state 1 would be reflected to another designated detector. The detector recording state 1 photons triggered the application of a high-voltage pulse on the polarization rotator whose special axis orientation would cause a bit-flip in the computational basis. The polarization state exiting was measured by a different detector. The coincidences between events in this detector and the events in Z-measurement detectors represented the coincidence-basis operation of this specific purpose encoding device, which was proven to be able to accomplish two-qubit encoding.

Some New Contributions to Linear Optics Implementations

In 2004, Michael Nielsen [309] showed that efficient linear optical quantum computation is possible without the teleportation and Z-measurement error correction steps used by the KLM scheme. Nielsen combines KLM nondeterministic gates with fewer optical components with the cluster-state model proposed by Raussendorf and Briegel [349].

The cluster state is a massively entangled state of a network of many qubits; the computation is executed as a sequence of single-qubit measurements with the measurement outcome determining the basis for the measurement on the next qubit; the answer state is left in the output qubits in the network. The preparation of the cluster state can be probabilistic and nondeterministic linear optics CNOT gates can be used in the process. The advantage is that much of the large overhead required for error encoding used to make nondeterministic CNOT gates is no longer necessary.

A first experimental demonstration of an all-optical one-way implementation of Deutsch algorithm was performed in 2006 on a four-qubit cluster state [415]. The entangling operations required by the computation were realized by using the cluster resource and the nonlinearity induced by detection. The qubit states, $0\ _i$, and $1\ _i$, were embodied by the horizontal and the vertical polarization state, respectively, of one photon populating a spatial mode, $i \quad 1, \ldots, 4$. The input photons were produced in pairs in entangled states of the form, $(1/\ \overline{2})(\ 00 \quad 11\ )_{ab}$ and $(1/\ \overline{2})(\ 00 \quad 11\ )_{ab}$, following two passages of an ultraviolet pump-laser beam through a nonlinear beta-barium-borate crystal. The cluster state, $\varphi\ _c \quad (1/2)(\ 0000 \quad 0011 \quad 1100 \quad 1111\ )_{1234}$, was prepared through post-selection, a four-photon coincidence event at the single-photon detectors facing each spatial mode.

The algorithm was executed by using quarter-wave plates, half-wave plates, PBSs, and photo-counter pairs for polarization measurements in arbitrary bases of the photons in mode, i. In this implementation of Deutsch's algorithm, the application of a CNOT gate is replaced with a series of rotations, Hadamard gates, and a controlled-phase gate. On a linear cluster the two-qubit operations CNOT or CPHASE (controlled-PHASE) can be implemented by changing the order of measurements. Given a perfect cluster state and feed-forward control based on measurements made on ancilla qubits could ensure a deterministic photonic quantum computation.

With active feed-forward control, whereby a later measurement depends on an earlier measurement and its fed-forward result, Prevedel *et al* [345] have demonstrated "error-free" single-qubit and two-qubit gate operations in an implementation of Grover's search algorithm. The resources required by a linear optics implementation can be reduced by several orders of magnitude if CNOT gates are replaced with other techniques.

Recent developments show that a hybrid approach, such as storing quantum information in the single-photon sources, which are themselves inherently quantum mechanical, may have advantages; the photonic qubits are easy to transport and are robust against decoherence, but the execution of gate operations required to prepare photon entanglement encounters difficulties. Even the more improved methods of preparing photon entanglement based on measurement-induced nonlinearities have to contend with probabilistic interactions and most often require photon detection based on number discrimination. For large entangled multi-photon states, the result is long preparation times and a large number of resources for photon storage, which limits the eventual use of on-chip devices

A photonic module proposed by Devitt *et al* [119] would rapidly prepare a large class of entangled photon states for quantum computation and communication applications in a completely deterministic way without the need for number-discriminated photon detection. The internal construction of such a module is independent of the state being prepared, no single photon detection is necessary, and the coherence time required for the atomic qubit is limited by the time it takes to measure specific stabilizers (of the Gottesman formalism) describing the state. The module is a generic resource for preparation of any geometric graph state, including states for optical cluster state computation; Bell states for quantum cryptographic protocols, quantum dense coding, purification protocols, and quantum repeaters; and N-photon Greenberger-Horne-Zeilinger states for loss protection scheme for optical quantum computation.

The operation of the module is based on atom/cavity interaction, where the atomic system used in the cavity can be rubidium or nitrogen-vacancy (NV) diamond; they have coherence times long enough to perform two-photon parity measurements, or to prepare highly entangled states very quickly (NV diamond). Single photons of known polarity from a single-photon source pass as a two-photon train through the module where they interact with an atom/cavity system and must induce a nondemolition bit-flip on the two-level atomic qubit. The atom/cavity qubit is initially prepared in the state 0 . After the two photons have passed through the module, the state of the atom/cavity qubit is measured and the two output photons are projected into an entangled Bell state of either even parity, if the atom qubit is in the state 0 , or odd parity if the atom qubit is in the state 1 .

The measurement of the atom/cavity system never collapses the two photons to an unentangled state. The number of entangled photons that are prepared by the module depends only on the number of photons sent through the module; no change in the internal structure of the module is necessary. The stabilizer formalism used for describing large entangled states is closely linked to the concept of parity measurement. Any N-photon stabilized state can be prepared by performing a parity check on the, N, stabilizers describing the state.

Assuming that the individual photons within a train can be selectively routed (using the time tag of each pulse) and local operations are applied to them, the parity measurement performed by the module prepares a stabilizer state. For example, a cluster state intended for quantum computation has a well-known stabilizer structure [349] with a maximum parity-weight of five. The optical module has to maintain a decoherence time of the atom/cavity system long enough to permit the passage of five photons between initialization and measurement. In the end, multiple modules combined with an appropriate single-photon source, optical wave plates and classical routing can be used to build a static on-chip system for fast preparation of specific entangled states such as large cluster states for quantum computation or multiple Bell pairs in sequence for quantum communications and cryptography.

The most important challenge to optical quantum computing, by far more serious than photon sources, detectors, and nonlinearities, and similar to decoherence in matter-based qubits, is the photon loss. However, the advances in photonic quantum computing support photonic qubits and benefit photonic quantum communication between matter qubits such as trapped ions, quantum dots, and solid-state impurities/dopants.

6.18 SUMMARY AND FURTHER READINGS

Any physical implementation of a quantum information processing system must be based on a system insulated from unwanted interactions with the environment, must be controlled through some required interactions implementing logical quantum gates, and must be measured. The physical implementation must satisfy a set of generic requirements formulated by DiVincenzo in 2000 and recently reformulated and generalized for systems with a low level of decoherence.

The first technology for quantum information processing devices, proposed by Cirac and Zoller [96] and demonstrated experimentally by Monroe *et al* [297] was the cold ion trap. The ion traps use laser-induced coupling between electron spin states and the vibrational motion (phonon) of relatively heavy charged particles, ions, trapped in high vacuum at very low temperature.

The main advantages of ion traps are that some of the techniques have already been developed, the decoherence rates associated with the decay of excited ionic state and the damping of the ionic motion can be made small, and quantum jumps can be used for reading out the result of a computation with high probability. The major problems are spontaneous transitions in the vibrational motion due to heating; internal transitions in the ions driven by thermal radiation; experimental instabilities in the laser beam power, RF voltages; mechanical vibrations; and fluctuating external magnetic fields.

The first experimental demonstration of a CNOT quantum gate was reported by a group at the National Institute of Standards and Technology (NIST) in Boulder, Colorado [297]. The characteristic states of the two qubits are two internal and two external states, respectively, of a single, ^{9}Be , ion stored in a coaxial-resonator RF ion trap.

The very long coherence times associated with the nuclear spins stimulated David Cory [103, 104] and Neil Gershenfeld and Isaac Chuang [170] to propose the use of nuclear magnetic resonance (NMR) as an alternative implementation of a quantum computer. Single-qubit gates, spin state vector rotations in Hilbert space, can be directly implemented in the rotating frame by applying RF pulses to individual nuclear spins. The liquid-state NMR has been used to demonstrate Shor's factoring algorithm by using a seven-qubit custom-synthesized molecule [93, 431] to factor the number 15.

Solid state is extremely versatile and could offer solutions for a scalable quantum computer technology as most phenomena possible in physics can be embodied in a solid-state system. Solid-state physics has enabled the creation of a wide variety of artificial structures and devices; thus, it is no wonder that in 2000 the number of proposals for solid-state implementations of quantum computers was almost equal to all the other proposals put together.

Quantum dots are atom-like compound semiconductor structures where charge carriers are confined in all three spatial dimensions. A semiconductor quantum dot may contain from a few thousand to tens of thousands of atoms arranged in a nearly defect-free three-dimensional crystal lattice. The carriers in the quantum dot interact strongly with lattice vibrations and could be strongly influenced by defects, surfaces, or interfaces. The size of a quantum dot is of the order of the Fermi wavelength. Quantum dots can be lithographically defined or self-assembled.

Quantum computing implementation based on semiconductor quantum dot electron spins has two significant advantages: The quantum dot arrays could be scalable to 100 coupled qubits, and the spin decoherence times for conduction band electrons in III-IV and II-VI semiconductors are "very" long.

In 1982, Tsui, Störmer, and Gossard [424] discovered experimental evidence of a new phenomenon called the fractional quantum Hall effect, FQHE. Laughlin [267] explained the new experimental facts as the result of the condensation of the two-dimensional electron gas in the heterojunction into a new type of collective ground state, a kind of quantum fluid incompressible due to long-range Coulomb interaction between electrons.

The ν $1/3$ step observed experimentally was explained by assuming that each electron in the two-dimensional electron gas captured three flux quanta and thus formed a sort of composite particles that behaved like bosons and could be condensed at very low temperatures and very strong magnetic fields. The ground state of this electron gas forms a quantum fluid, and its elementary excitations are fractionally charged quasiparticles and quasiholes. When only one electron is added to the ground state of the two-dimensional electron gas, it becomes excited and a number of quasiparticles are created; the quasiparticles are not true particles but manifestations of the interplay of the electrons in the quantum fluid.

Topological quantum computation introduced by Kitaev in 1997 [204] and further developed by Friedman, Kitaev, Larsen, and Wang [157, 158] has the potential to revolutionize fault-tolerance and justifies the interest in alternative physical realizations of such systems. Universal topological computation requires both a topological phase and quasiparticles with non-Abelian braiding statistics and with topological properties in a special class. The system's ground state should have a stable degeneracy resistant to weak local perturbations in the presence of multiple quasiparticles or in a nontrivial topology. So far, the incompressible fractional quantum Hall states, such as ν $5/2$ and $12/5$, are the only experimentally established topological phases. The topological quantum computation tries to use the emergent properties of many-particle systems to encode and manipulate quantum information in a way that is fault-tolerant. There are other physical systems, besides the fractional quantum Hall states, that are assumed to contain topological phases, such as transition metal oxides and ultra-cold atoms in optical traps.

After almost two decades of quantum information science, we are prepared to make an objective examination of different complete architectures of quantum computers, the resources required to implement algorithms of useful sizes with negligible error. In practice, we may discover that imperfections in the coherent control of the qubits may hinder the good performance of the quantum computer more than the decoherence itself. A large-scale, fault-tolerant quantum computer belongs to a new form of technological reality.

New research papers are published at an overwhelming rate in major science journals, such as *Physical Review A, Physical Review B, and Physical Review Letters, Nature, Applied Physics Letters, New Journal of Physics, Journal of the Optical Society of America B, and Proceedings of the Royal Society London*. The *Virtual Journal of Quantum Information* distributed online by the American Institute of Physics (AIP) and the American Physical Society (APS) contains citations of all recent publication in the field. The site *http://arxiv.org*, owned and operated by Cornell University, gives free access to e-prints of research papers in several fields of science; of special interest, papers related to the physical realization of quantum information processing devices, are published under identifiers such as quantum physics, condensed matter.

Appendix: Observable Algebras and Channels

In this appendix, we present a formal framework for some of the concepts we are familiar with from quantum mechanics and quantum information theory, while avoiding the introduction of more intricate concepts from operator theory. Our presentation closely follows the notations and conventions of Keyl [230].

A physical system, be it a classical or a quantum system, is characterized by its *state*, ρ, and the *effect*, A, of a measurement of a physical property, or an observable; an effect is associated with an operator, and A denotes both the effect and the operator. We assume that the experiment to observe the system produces statistical results; the effect A may be observed or not in the state ρ. The formalism is based upon two sets—$\mathcal{S}$, the set of states, and $\mathcal{E}$, the set of effects/operators—and a map:

$$(\rho, A) \in \mathcal{S} \times \mathcal{E} \mapsto \rho(A).$$

The functional,[1] $\rho(A) \in [0,1]$, is the probability of observing effect A (or applying the operator A) to a system in the state ρ.

Let $\mathcal{H}$ be a Hilbert space and $\mathbb{B}_{\mathcal{H}}$ be the algebra of bounded operators[2] in $\mathcal{H}$ and $\mathbb{I}$ the identity operator in this algebra. If ψ is an element of $\mathbb{B}_{\mathcal{H}}$ and A is an operator, then the expression, $\langle \psi, A\psi \rangle$, is the scalar product of the elements before and after the transformation, A.
An *observable algebra*, $\mathcal{A}$, is a linear subspace of $\mathbb{B}_{\mathcal{H}}$, with the following properties:

- $\mathcal{A}$ contains the identity, $\mathbb{I} \in \mathcal{A}$.
- $\mathcal{A}$ is closed under multiplication of operators:

$$(A, B) \in \mathcal{A} \Rightarrow AB \in \mathcal{A}.$$

- A^*, the adjoint of operator A is also in $\mathcal{A}$:

$$A \in \mathcal{A} \Rightarrow A^* \in \mathcal{A}.$$

The *norm* of operator, $A \in \mathcal{A}$, inherited from $\mathbb{B}_{\mathcal{H}}$, is defined as

$$||A|| = \sup_{||\psi||-1} = ||A\psi||, \quad \forall \psi \in \mathcal{H}.$$

The *operator ordering*, inherited from $\mathbb{B}_{\mathcal{H}}$, is defined as

$$A \geq B \Leftrightarrow \langle \psi, A\psi \rangle \geq \langle \psi, B\psi \rangle, \quad \forall \psi \in \mathcal{H}.$$

---

[1] A functional is a real-valued function on a vector space of functions.

[2] A bounded operator is a linear transformation, L, between vector spaces, X and Y, with the property that the ratio of the norms of $L(x)$ and $x \in X$ is bounded by the same number k:

$$||L(x)||_Y \leq k||x||_X, \quad \text{with} \quad k > 0, \quad \forall x \in X, \quad x \neq 0.$$

The *dual* space of $\mathcal{A}$, $\mathcal{A}^*$ is the set of all linear forms on $\mathcal{A}$. The set of the states of the system described by the algebra $\mathcal{A}$, $\mathcal{S}(\mathcal{A})$, is:

$$\mathcal{S}(\mathcal{A}) = \{\rho \in \mathcal{A}^* | \rho \geq 0, \rho(\mathbb{I}) = 1\}.$$

In this expression the positivity of the functional, ρ, means: $\rho \geq 0 \Leftrightarrow \rho(A) \geq 0, \forall A \in \mathcal{A}$. The set of the effects of the system described by the algebra $\mathcal{A}$, $\mathcal{E}(\mathcal{A})$, is:

$$\mathcal{E}(\mathcal{A}) = \{A \in \mathcal{A} | A \geq 0, A \leq \mathbb{I}\}.$$

Both spaces are *convex*; if $0 \leq \lambda \leq 1$, then

$$\forall \rho, \sigma \in \mathcal{S}(\mathcal{A}) \Rightarrow \lambda\rho + (1-\lambda)\sigma \in \mathcal{S}(\mathcal{A}) \quad \text{and} \quad \forall A, B \in \mathcal{E}(\mathcal{A}) \Rightarrow \lambda A + (1-\lambda)B \in \mathcal{S}(\mathcal{A}).$$

The extreme points[3] of the two sets have special properties:

- The extreme points of $\mathcal{S}(\mathcal{A})$ are called *pure states*.
- The extreme points of $\mathcal{E}(\mathcal{A})$ are called *propositions*; they are effects that occur with probability one.

When $\mathcal{A} = \mathbb{B}_{\mathcal{H}}$, the observable algebra, $\mathcal{A}$, coincides with, $\mathbb{B}_{\mathcal{H}}$, and then the correspondence between the concepts from the algebra of observables and the concepts we are familiar with from quantum mechanics is straightforward:

- $\mathcal{A} \mapsto \mathcal{H}_d$, Hilbert space of dimension d.
- Vectors $\psi \in \mathcal{A} \mapsto$ ket vectors $|\psi\rangle \in \mathcal{H}_d$.
- Vectors $\psi^* \in \mathcal{A}^* \mapsto$ bra vectors $\langle\psi| \in \mathcal{H}_d$.
- Scalar product $\langle A, B\rangle \mapsto \text{tr}(A^*B)$.
- Effect $A \in \mathcal{E}(\mathcal{A}) \mapsto$ positive operator, A, bounded by $\mathbb{I}$.
- Proposition $P \in \mathcal{E}(\mathcal{A}) \mapsto$ projector P.
- States $\mapsto$ density matrices, positive, and normalized self-adjoint operators.
- State space $\mapsto$ the set of density matrices.
- When $d = 2$ the state space $\mathcal{S}(\mathcal{H}_2) \mapsto$ the Bloch ball.
- Extreme points of $\mathcal{S}(\mathcal{H}_2) \mapsto$ pure states, points on the Bloch sphere.

The Hilbert space of dimension, d, is a bounded algebra of $d \times d$ complex matrices, $\mathbb{B}_{\mathcal{H}_d}$. The positivity of the operator, ρ, is a consequence of the positivity of the functional, ρ:

$$\rho(|\psi\rangle\langle\psi|) \geq 0 \Leftrightarrow \langle\psi, \rho(\psi)\rangle \geq 0.$$

The normalization condition, $\rho(\mathbb{I}) = 1$, translates into $\text{tr}(\rho) = 1$:

$$\rho(\mathbb{I}) = 1 \Leftrightarrow \text{tr}(\rho) = 1.$$

[3]The extreme point of a convex set, S, is a point in the set, S, that does not lie in any open line segment joining two points of S.

Consider now the dual space, $\mathbb{B}^*_{\mathcal{H}_d}$, and call $\mathbb{I}^\perp \in \mathbb{B}^*_{\mathcal{H}_d}$ the dual of the unit operator $\mathbb{I}$; let $\sigma_1, \sigma_2, \ldots, \sigma_{d^2-1}$ be an orthonormal basis in $\mathbb{I}^\perp$. Then each self-adjoint operator ρ, $\text{tr}(\rho) = 1$, can be expressed as

$$\rho = \frac{\mathbb{I}}{d} + \frac{1}{2}\sum_{i=1}^{d^2-1} x_i \sigma_i =: \frac{\mathbb{I}}{d} + \frac{1}{2}\bar{x}\bar{\sigma}, \quad \text{with} \quad \bar{x} \in \mathbb{R}^{d^2-1}.$$

This expression has a geometric interpretation. When $d = 2$ we choose the Pauli matrices as a basis, $\bar{\sigma} = (\sigma_I, \sigma_x, \sigma_y, \sigma_z)$, and the state space, $\mathcal{S}(\mathcal{H}_2)$, coincides with the Bloch ball, $\bar{x} \in \mathbb{R}^3, |\bar{x}| \leq 1$, because the positivity of ρ implies $|\bar{x}| \leq 1$. In fact,

$$\rho \geq 0 \Leftrightarrow |\bar{x}| \leq 1.$$

The pure states, the extremes of $\mathcal{S}(\mathcal{H}_2)$, are the states on the Bloch sphere $\bar{x} = 1$.

We now consider a finite set of values, $x \in X$, and an observable, E; the effect, $E_x \in \mathcal{E}(\mathcal{A})$, means that, x, is the outcome of observing E in algebra, $\mathcal{A}$, and it is true if we observe the value x and is false otherwise. If the system is in the state, ρ, then $p_x = \rho(E_x)$ is the probability to obtain, x, as the outcome of a measurement.

Given the observable algebra $\mathcal{A}$, a *positive operator valued measure* (POVM) on the finite set, X, is a family of effects,

$$E = E_x : x \in X, \quad 0 \leq E_x \leq \mathbb{I},$$

if and only if $\sum_{x \in X} E_x = \mathbb{I}$ holds.

We are now concerned with composite systems and their representation in an algebra of observables. Consider two Hilbert spaces, $\mathcal{H}$ and $\mathcal{K}$, and vectors, $|\psi_1\rangle, |\varphi_1\rangle \in \mathcal{H}$ and $|\psi_2\rangle, |\varphi_2\rangle \in \mathcal{K}$. The bilinear form, $|\psi_1\rangle \otimes |\psi_2\rangle$, called the tensor product is defined as $|\psi_1\rangle \otimes |\psi_2\rangle(\varphi_1, \varphi_2) = \langle \psi_1 | \varphi_1 \rangle \rangle \langle \psi_2 \rangle | \varphi_2 \rangle$. To describe composite systems, we introduce the tensor product of two bounded operators, $A_1 \in \mathbb{B}_{\mathcal{H}}$ and $A_2 \in \mathbb{B}_{\mathcal{K}}$, applied to the tensor product, $|\psi_1\rangle \otimes |\psi_2\rangle$:

$$A_1 \otimes A_2(|\psi_1\rangle \otimes |\psi_2\rangle) = (A_1|\psi_1\rangle) \otimes (A_2|\psi_2\rangle).$$

The space, $\mathbb{B}_{\mathcal{H}\otimes\mathcal{K}}$, is the span of all tensor products, $A_1 \otimes A_2$. The tensor product can be extended to more than two systems.

The *partial trace* of $\rho \in \mathbb{B}_{\mathcal{H}\otimes\mathcal{K}}$ is

$$\text{tr}_{\mathcal{H}}(\text{tr}_{\mathcal{K}}(\rho)A)) = \text{tr}_{\mathcal{H}\otimes\mathcal{K}}(\rho A \otimes \mathbb{I}), \quad \forall A \in \mathbb{B}_{\mathcal{H}}.$$

Proposition. *For each element, $|\Psi\rangle$, of the tensor product, $\mathcal{H} \otimes \mathcal{K}$, there exist two orthonormal sets of vectors, $|\psi_j\rangle \in \mathcal{H}, 1 \leq j \leq n$ and $|\varphi_k\rangle \in \mathcal{K}, 1 \leq k \leq n$, such that*

$$|\Psi\rangle = \sum_j \sqrt{\lambda_j} |\psi_j\rangle \otimes |\varphi_j\rangle,$$

with $\sqrt{\lambda_j}$ called the Schmidt coefficients. We do not require either set of vectors to be a basis; in other words, $n \leq \dim(\mathcal{H})$ and $n \leq \dim(\mathcal{K})$.

The following proposition is a consequence of the Schmidt decomposition.

Proposition. *Each state, $\rho \in \mathbb{B}^*_{\mathcal{H}}$, can be extended to a pure state, $|\Psi\rangle$, on a larger system with the Hilbert space, $\mathcal{H} \otimes \mathcal{K}$, such that $\rho = |\Psi\rangle\langle\Psi|$.*

In Section 1.15, we observed that quantum information processing requires several types of procedures, including free-time quantum evolution, controlled-time evolution, preparation, and measurement. The process of encoding classical information into a quantum system involves a preparation step, and the process of retrieving classical information from a quantum system requires a measurement. The mathematical description of each of these steps requires the concept of *channel* or map.

A *channel* converts input systems characterized by the input observable algebra, $\mathcal{A}$, into output systems characterized by an output observable algebra, $\mathcal{B}$; sometimes $\mathcal{A}$ and $\mathcal{B}$ coincide. A channel, T, is thus a mapping, $T : \mathcal{A} \mapsto \mathcal{B}$, with several properties:

- T maps effects to effects; thus, it must be positive, $T(A) \geq 0, \forall A \geq 0$, and bounded, $T(\mathbb{I}) \leq \mathbb{I}$.
- Given two channels, $T_1 : \mathcal{A}_1 \mapsto \mathcal{B}_1$ and $T_2 : \mathcal{A}_2 \mapsto \mathcal{B}_2$, the channel, $T = T_1 \otimes T_2$, maps composite systems of the type, $\mathcal{A}_1 \otimes \mathcal{A}_2$, into, $\mathcal{B}_1 \otimes \mathcal{B}_2$. The channel, T, must be positive as well.

The *dual channel/map*, T^*, is defined for all states, $\rho \in \mathcal{B}^*$, and all operators, $A \in \mathcal{A}$:

$$T^* : \mathcal{B}^* \mapsto \mathcal{A}^* \Rightarrow T^*\rho(A) = \rho(TA).$$

There are two ways to describe a channel:

- *The Heisenberg picture:* We invoke first the channel, $T : \mathcal{A} \mapsto \mathcal{B}$, and then measure the effect, A, on the output systems. Thus, T measures the set of effects on the input systems:

$$T : \mathcal{E}(\mathcal{B}) \mapsto \mathcal{E}(\mathcal{A}).$$

- *The Schrödinger picture:* We consider the states and interpret the channel as a transformation of $\mathcal{A}$ systems in the state, $\rho \in \mathcal{S}(\mathcal{A})$, into $\mathcal{B}$ systems in the state, $T^*(\rho)$:

$$T^* : \mathcal{S}(\mathcal{A}) \mapsto \mathcal{S}(\mathcal{B}).$$

The dual map, T^*, is called the Schrödinger representation of the channel, T.

Let T be a liner map (channel) from an observable algebra, $\mathcal{A}$, to another, $\mathcal{B}$.

- $T : \mathcal{A} \mapsto \mathcal{B}$ is positive if $T(A) \geq 0, \quad \forall A \in \mathcal{A}, \quad A \geq 0$.
- $T : \mathcal{A} \mapsto \mathcal{B}$ is completely positive if:

$$T \otimes \mathbb{I}_{\mathbb{B}} : \mathcal{A} \otimes \mathbb{B}(\mathbb{C}^n) \mapsto \mathbb{B}_{\mathcal{H}} \otimes \mathbb{B}(\mathbb{C}^n), \quad \forall n \in \mathbb{N},$$

  with $\mathbb{I}_{\mathbb{B}}$, the identity element in $\mathbb{B}(\mathbb{C}^n)$. The dual map, T^*, is completely positive if T is.
- $T : \mathcal{A} \mapsto \mathcal{B}$ is unital if $T(\mathbb{I}) = \mathbb{I}$.

Theorem. *(Stinespring dilation theorem) Every completely positive map*

$$T : \mathcal{A} \mapsto \mathcal{B} : \quad A \in \mathcal{A}, \mathcal{A} = \mathbb{B}_{H_1}, \dim(\mathcal{H}_1) = d_1, \quad B \in \mathcal{B}, \mathcal{B} = \mathbb{B}_{H_2}, \dim(\mathcal{H}_2) = d_2$$

has a unique decomposition of the form

$$T(A) = V^* (A \otimes \mathbb{I}_{\mathcal{K}}) V,$$

where $\mathcal{K}$ is an additional Hilbert space and V is an operator:

$$V : \mathcal{H}_2 \mapsto \mathcal{H}_1 \otimes \mathcal{K}.$$

The Hilbert space, $\mathcal{K}$, and the, V, are chosen, such that $\forall|\varphi\rangle \in \mathcal{H}_2$, the span of $(A \otimes \mathbb{I}_{\mathcal{K}}) V|\varphi\rangle$, is dense in $\mathcal{H}_1 \otimes \mathcal{K}$. The minimal $\mathcal{K}$ satisfies the condition:

$$\dim(\mathcal{K}) \leq d_1^2 d_2.$$

The Stinespring dilation theorem is a critical result for quantum information theory and quantum error correction, as we saw when we discussed the role of measurements carried out on ancilla qubits in Chapters 3 and 5.

We conclude this presentation with a theorem regarding the purification of mixed states; the result is based upon the fact that a $d \times d$ matrix represents a bilinear form as well as an operator on a vector space V. The proof of the theorem is presented by Keyl [230].

Theorem. *Given a bipartite system in $\mathcal{H} \otimes \mathcal{H}_1$, for every state, $\rho \in \mathcal{H} \otimes \mathcal{H}_1$, there exists*

- *a Hilbert space $\mathcal{K}$;*
- *a pure state, $\sigma \in \mathcal{H} \otimes \mathcal{K}$; and*
- *a channel, $T : \mathbb{B}_{H_1} \mapsto \mathbb{B}_{\mathcal{K}}$, such that*

$$\rho = \left(\mathbb{I}_{\mathcal{B}^*} \otimes T^*\right) \sigma,$$

with $\mathbb{I}_{\mathcal{B}^*}$, the identity element in $\mathcal{B}^*_{\mathcal{H}}$.

The pure state, σ, is a purification of the state, ρ, and can be chosen so that $\mathrm{tr}_{\mathcal{K}}(\sigma)$ does not have zero eigenvalues, and in this case, T and σ, are unique (up to a unitary equivalence).

Glossary

adiabatic Term frequently used in thermodynamics meaning without transfer of heat (without change in temperature). For example, if we open the valve of a gas canister, the gas rushes out and expands without having the time to equalize its temperature with the environment. The rushing gas feels cool.

Aharonov-Bohm effect Given a magnetic field, **B**, consisting of a single field line with flux, Φ_B, at positions, R, not on the flux line, the field is $\mathbf{B} = 0$, but there must be a vector potential, $\mathcal{B}(R)$, satisfying the expression

$$\oint_C \mathcal{B}(R) \cdot dR = \Phi$$

for closed contours, C, threaded by the flux line. Aharonov and Bohm had shown [4] that in quantum mechanics such vector potentials, though corresponding to a zero-value field, have physical significance. Berry proved [52] that the effect of such vector potentials on a system of particles with charge, q, clustered inside a box at a zero-field position, R, and moved around the magnetic field line can be interpreted as a geometrical phase factor *Berry phase* acquired by the system's single-valued state vector after completion of the closed circuit, C. The geometrical phase factor for any state, $|n(R)\rangle$, of energy, E_n, independent of position R and unaffected by the vector potential is given as

$$\gamma_n(C) = \frac{q}{\hbar} \oint_C \mathcal{B}(R) \cdot dR = \frac{q}{\hbar} \Phi.$$

Notice that the geometrical phase factor is independent of the state $|n\rangle$ and even of the contour, C, if the system makes a complete rotation about the flux line. The geometrical phase factor can be observed by interference between the particles inside the rotated box and particles in a box that was not moved around the closed contour.

amplitude, probability amplitude Given a set of basis or logical states, $|i\rangle$, a quantum system may be in a superposition of these states, $|\psi\rangle = \sum_i \alpha_i |i\rangle$. The complex numbers, α_i, are called amplitudes or probability amplitudes.

ancillas Auxiliary qubits used to assist in a computation.

anti-commutator of two operators Given two operators, **A** and **B**, the *anticommutator* of the two operators is $\{\mathbf{A}, \mathbf{B}\} = \mathbf{AB} + \mathbf{BA}$. We say **A** anticommutes with **B** if $\{\mathbf{A}, \mathbf{B}\} = 0$.

anyons Indistinguishable particles in two dimensions that are neither bosons nor fermions. Consider a gas of electrons squeezed between two slabs of semiconductor materials such that the movement of electrons is restricted to two dimensions only. At very low temperature and in a strong magnetic field, the two-dimensional electron gas has a strongly entangled ground (lowest energy) state separated from all other states by an energy gap. This lowest-energy state carries an electric charge that is not an integer multiple of the electron charge and does not have the quantum numbers associated with electrons [344]. The properties of the anyons manifest themselves as the fractional quantum hall (FQH) effect. Anyons are suggested as candidates to explain high temperature superconductors.

atomic radiation cascade (ARC) A system that produces two polarization entangled photons that move in opposite directions. An atom, e.g., Ca^{40}, laser pumped to an excited, S, the upper level of the cascade state, emits two photons that appear to have the same polarization.

basis states of a quantum system Linearly independent, complete subset of orthonormal state vectors in the vector space of the quantum system. Two orthonormal states, such as $|0\rangle$ and $|1\rangle$, could be the basis states of a qubit.

Bayes rule Probability rule. Assume that it is known that event, $\mathcal{A}$, occurred, but it is not known which one of the set of mutually exclusive and collectively exhaustive events, $\mathcal{B}_1, \mathcal{B}_2, \ldots, \mathcal{B}_n$, has subsequently occurred. Then the conditional probability that one of these events, $\mathcal{B}_j$, occurs, given that $\mathcal{A}$ occurs is

$$P(\mathcal{B}_j|\mathcal{A}) = \frac{P(\mathcal{A}|\mathcal{B}_j)P(\mathcal{B}_j)}{\sum_i P(\mathcal{A}|\mathcal{B}_i)P(\mathcal{B}_i)},$$

with $P(\mathcal{B}_j|\mathcal{A})$, the *a posteriori probability*.

BB84 A quantum key distribution protocol proposed by Charles Bennett and Gilles Brassard in 1984.

BCH (Bose-Chaudhuri-Hocquenghem) bound Let $g(x)$ be the generator polynomial of the cyclic $[n,k]$ code, $\mathcal{C}$, over $GF(q)$. Let $\beta \in GF(q)$ be a primitive element of order n. If $g(x)$ has among its zeros, $\beta^{\alpha}, \beta^{\alpha+1}, \ldots, \beta^{\alpha+d-2}$, then the minimum distance of the cyclic code is $d_{min}(\mathcal{C}) \geq d$.

BCH codes Cyclic codes independently discovered by A. Hocquenghem [208] and by R. C. Bose and R. K. Ray-Chaudhuri [67]. Given the prime number q, and positive integers m and d, the $BCH_{q;m;d}$ is obtained as follows: let $n = q^m$ and $GF(q^m)$ be an extension of $GF(q)$ and let C' be the (extended) $[n; n-(d-1); d]_{q^m}$ Reed-Solomon code obtained by evaluating polynomials of degree at most, $(n-d)$, over $GF(q^m)$ at all the points of $GF(q^m)$. Then the code $BCH_{q;m;d}$ is the $GF(q)$-subfield subcode of C'. A BCH code has dimension at least $n - m(d-1)$ [412].

Bell, John Stewart (1928–1990) British physicist whose work led to the possibility of exploring seemingly philosophical questions in quantum mechanics, such as the nature of reality, directly through experiments. Bell started from locality and argued for the existence of deterministic hidden variables. He was able to show that the behavior predicted by quantum theory could not be duplicated by a hidden variable theory if the hidden variables acted locally. Bell's inequality is a constraint on the sum of averages of measured observables, based on assumptions of *local realism* (i.e., of locality and of existence of hidden variables).

Bell states Four distinct quantum states of a two-particle system with a very strong coupling of the individual states of the particles. These states form an orthonormal basis:

$$|\beta_{00}\rangle = \frac{|00\rangle + |11\rangle}{\sqrt{2}}, \quad |\beta_{01}\rangle = \frac{|01\rangle + |10\rangle}{\sqrt{2}},$$

$$|\beta_{10}\rangle = \frac{|00\rangle - |11\rangle}{\sqrt{2}}, \quad |\beta_{11}\rangle = \frac{|01\rangle - |10\rangle}{\sqrt{2}}.$$

Bennett, Charles Distinguished American scientist working at IBM Research who has made outstanding contributions to quantum information theory. He built on the work of Rolf Landauer to show that general-purpose computation can be performed by a logically and thermodynamically reversible apparatus; he proposed a reinterpretation of Maxwell's demon, attributing its inability to break the second law to the thermodynamic cost of destroying, rather than acquiring, information; together, with several collaborators, he discovered quantum teleportation.

Berry phase See the *Aharonov-Bohm effect*.

black box A computational device whose inner structure is unknown and that implements a set of operations. An experiment with a black box, called a *query*, consists of supplying an input and observing the output. The smallest number of queries necessary to determine the operation is called the *query complexity*.

Bloch sphere A geometrical representation of the pure state space of a two-level quantum mechanical system. We represent impure states as points inside a Bloch sphere; thus, sometimes we talk about a Bloch ball.

block code A code where a group of information symbols is encoded into a fixed-length code word by adding a set of parity-check or redundancy symbols.

boson Quantum particles whose spin quantum number can be $s = 1$, $s = -1$, or $s = 0$ (e.g., photons and mesons). The other type of quantum particles are fermions, particles whose spin quantum number can be $s = 1/2$ or $s = -1/2$ (e.g., electrons, protons, and neutrons). The spin is the quantum number characterizing the intrinsic angular momentum of the quantum particle.

bounded distance decoding The strategy when the $[n,k,d]$ code, C, corrects patterns of at most, e, and no other errors. When

$$e = \left\lfloor \frac{d-1}{2} \right\rfloor$$

the bounded distance decoding is identical to minimum distance decoding. See also *minimum distance decoder*.

Bounded-error probabilistic polynomial time (BPP) Term used in computational complexity theory to describe a class of probabilistic algorithms. BPP consists of polynomial time algorithms that produce the correct result with a probability at least $2/3$ (or other values strictly between $1/2$ and 1). We may get the wrong answer when we run such an algorithm once, but by repeatedly running the algorithm and then taking a majority vote, the probability of obtaining the correct answer can be made arbitrarily close to 1, while maintaining the polynomial running time. See also *polynomial time algorithm*.

Bounded-error quantum polynomial (BQP) Term used in computational complexity theory to describe a class of probabilistic algorithms. BQP functions are all functions with the domain being the set of binary strings computable by uniform quantum circuits whose number of gates is polynomial in the number of input qubits, and which give the correct answer at least $2/3$ of the time.

bounded operator Linear transformation, L, between vector spaces, A and B, with the property that the ratio of the norms of $L(a)$ and $a \in A$ is bounded by the same number, k, is $||L(a)||_B \leq k||a||_A$, with $k > 0, \forall a \in A, a \neq 0$.

braid group B_n, the braid group on n strands, is an infinite group with applications in knot theory and topological quantum computing. B_n has an intuitive geometric representation. It shows how n strands can be laid out to connect two groups of n objects.

capacity of a classical memoryless communication channel A discrete memoryless channel is characterized by the triplet, $(X, Y, \{p(y|x), x \in \{X, y \in Y\})$, with X the input channel alphabet, Y the output channel alphabet, and $\{p(y|x), x \in X, y \in Y\})$ the probability of observing output symbol y when the input symbol is x. The capacity of the discrete memoryless channel, $(X, Y, \{p(y|x), x \in X, y \in Y\})$ is $C = \max_{p(x)} I(X;Y)$, with $p(x)$, the input distribution and $I(X, Y)$ the mutual information between input and output.

capacity of a quantum memoryless communication channel A quantum communication channel has four distinct capacities [47]: (1) C, the ordinary capacity for transmitting classical information; (2) Q, the ordinary capacity for transmitting quantum states, typically $Q < C$; (3) Q_2, the classically-assisted quantum capacity for transmitting quantum states with the assistance of a two-way classical channel, $Q \leq Q_2 \leq C$; and (4) C_E, the entanglement-assisted classical capacity, $Q \leq C \leq C_E$.

cardinality of a finite set Let G be a finite set. The cardinality of G, denoted as $|G|$, represents the number of elements in the set.

cavity quantum electrodynamics (CQED) In CQED systems individual photons circulating in a high-finesse resonator can interact strongly via their mutual coupling to a single atom positioned inside the cavity. The atom is strongly coupled to the cavity mode in a manner that allows for efficient transfer of electromagnetic fields from input to output (photon) channels. The atom-cavity system may be viewed as a quantum-optical device.

Cauchy sequence A sequence, $\{a_n\}$, is Cauchy if for any $\epsilon > 0$ there exists, $N \in \mathbb{R}$, such that $||a_k - a_r|| < \epsilon$ for $k, r > N$.

cavity Cavities are optical resonators with dimensions much larger than the optical wavelength. They are arrangements of optical components in an "open" setup where the light is allowed to circulate either by bouncing back and forth between two end mirrors (linear cavities) or by doing round trips in two opposite directions (ring cavities). A laser cavity contains a gain medium that can compensate for the cavity losses in each round trip of the light.

cavity modes Cavity modes are self-consistent electric field distributions inside a cavity that are self-reproducing after each cavity round trip of the light. A cavity is *stable* if it has such modes (the modes do not always exist). The stability depends on the cavity parameters, such as curvature of reflecting surfaces and distance between optical components. Laser frequencies correspond precisely to certain mode frequencies. A laser can operate on a single mode of its cavity and then is called a single-frequency laser. Single mode operation is required, for example, to drive resonant cavities, for high resolution spectroscopy, for coherent beam combining of laser outputs. While the modes of an empty cavity are mutually orthogonal, the modes of a laser cavity are no longer so due to transversely varying gain and loss. The optical frequencies for which the optical phase is self-consistently reproduced after each round trip of the cavity (i.e., the optical phase shift is an integer multiple of 2π) are called *mode frequencies* or *resonance frequencies*. The *Q factor* is the ratio of resonance frequency and bandwidth. Bandwidth is defined as the width of the resonance frequency line.

Church-Turing principle Every function that can be regarded as computable can be computed by a mechanical device equivalent to a Turing machine (Turing's thesis) or by the use of Church's λ-calculus. David Hilbert formulated in the 1930s the Entscheidungs problem, which asked if there was a mechanical procedure for separating mathematical truths from mathematical falsehoods. While studying this problem, Alonzo Church and Stephen Kleene introduced the λ-definable functions. Church proved that "every effectively calculable function (effectively decidable predicate) is general recursive." Turing proved that every function that could be regarded as computable is computable by a Turing machine.

closed quantum system An idealization of a system of quantum particles. The system is assumed to be isolated and the interaction of the particles with the environment is non-existent. In reality, we can only construct quantum systems with a very weak interaction with the environment.

CNOT, controlled-NOT gate A two-qubit gate. It has as input a control qubit and a target qubit. The control input is transferred directly to the control output of the gate. The target output is equal to the target input if the control input is $|0\rangle$, and it is flipped if the control input is not $|0\rangle$.

coherent states Minimum uncertainty states. For such states, $\Delta(X)\Delta(P_X) = \hbar/2$, rather than $\Delta(X)\Delta(P_X) \geq \hbar/2$. Let $|e_i\rangle$ be a set of basis sets. A quantum system in the state, $|\varphi\rangle = \sum_i \alpha_i |e_i\rangle \in \mathcal{H}_n$, is in a *coherent superposition* of basis states, $|e_i\rangle$, if its density matrix, ρ, is not diagonal, $\rho \neq 1/nI_n$; if, in addition, the system is in a pure state, then it is said to be *completely coherent*. If ρ is diagonal, $\rho = 1/nI_n$, the system is in an incoherent superposition of states; in particular, a qubit is in an incoherent superposition of states if

$$\rho = \frac{1}{2}\begin{pmatrix} 1 & 0 \\ 0 & 1 \end{pmatrix}.$$

In case of an incoherent superposition, the basis states are prepared independently and there is no definite phase relationship between the bases states.

commutator of two operators Given two operators, $\mathbf{A}$ and $\mathbf{B}$, the *commutator* of the two operators is $[\mathbf{A}, \mathbf{B}] = \mathbf{A}\,\mathbf{B} - \mathbf{B}\,\mathbf{A}$. We say $\mathbf{A}$ commutes with $\mathbf{B}$ if $[\mathbf{A}, \mathbf{B}] = 0$.

complementarity A principle enounced in 1928 by Niels Bohr. This basic tenant of quantum mechanics says that quantum systems exhibit a wave-particle duality. The behavior of such phenomena as light and electrons is sometimes wavelike and sometimes particle-like, and it is impossible to observe simultaneously both aspects.

complete positivity The linear transformation, $\mathcal{E}(\rho)$, enjoys *complete positivity* if there exist operators $\mathbf{A}_i$, such that

$$\mathcal{E}(\rho) = \sum_i \mathbf{A}_i \rho \mathbf{A}_i^\dagger, \quad \text{with} \quad \sum_i \mathbf{A}_i^\dagger \mathbf{A}_i \leq \mathbf{I}.$$

When $\sum_i \mathbf{A}_i^\dagger \mathbf{A}_i = \mathbf{I}$, then $\mathrm{tr}\,(\mathcal{E}(\rho)) = 1$.

complete vector space A vector space is complete if any Cauchy sequence is convergent. See also *Cauchy sequence*.

concatenated codes Codes introduced by David Forney based upon serial concatenation of two or more codes. For example, the *outer code* (applied first and removed last) could be a Reed-Solomon code followed by a convolutional *inner code* (applied last, removed first). Concatenated codes encode first a block of, k, information symbols into, n, symbols using the outer code and then encode each one of the n symbols using the inner code. Turbo codes are a refinement of concatenated codes; they use the encoding structure of concatenated codes and an iterative algorithm for decoding. See also *convolutional codes*.

conservative logic gate A logic gate that conserves the number of 1s at its input. For example, the Fredkin gate is a conservative gate.

convex set A convex set, S, in a vector space, V, over the field, $\mathbb{R}$, is one when the line segment joining any two points in S lies entirely in S.

convex subspace of a vector space A subset that contains all the points of a straight line connecting any two points in the subset. The set of density operators in a Hilbert space, $\mathcal{H}_n$, is a convex linear subset of $n \times n$ Hermitian matrices.

convolutional codes Concatenated codes do not encode blocks of input symbols but a continuous input steam. They are primarily used to maximize the transmission rate in case of satellite communication and remote sensing.

Copenhagen interpretation of quantum mechanics The Copenhagen school is the name given to the group of theoreticians who shared the views of Niels Bohr regarding quantum mechanics interpretations.

correspondence rule in Copenhagen interpretation Heuristic principle requiring that in such cases when the influence of Planck's constant could be neglected, the numerical values predicted by quantum mechanics should be the same as if they were predicted by classical radiation theory.

coset of a group For a subgroup, H of a group, G, $H \subset G$, and an element $x \in G$; define xH to be the set, $\{xh : h \in H\}$, and Hx to be the set $\{hx : h \in H\}$. A subset of G of the form, xH, for some $x \in G$ is said to be a *left coset of H*, and a subset of the form Hx is said to be a *right coset of H*. For any subgroup, H, we can define an equivalence relation, $(\equiv)$ by $x \equiv y$, if $x = yh$ for some $h \in H$. The equivalence classes of this equivalence relation are exactly the left cosets of H, and an element $x \in G$ is in the *equivalence class xH*.

Coulomb blockade A result of the discreteness of the electron charge, it manifests itself as the increased resistance at small bias voltage of a low-capacitance tunnel junction. The transfer by tunneling of one electron through the junction barrier increases the tunnel junction capacitor with one elementary charge; that causes a voltage buildup, $V = e/C$, and an increase in electrostatic energy of the system by an amount $E = e^2/(2C)$, where e is the electron charge of 1.6×10^{-19} coulomb and C is the capacitance of the junction. For a small capacitance, the voltage buildup can be large enough to prevent tunneling by another electron. The electric current is suppressed and the resistance of the junction is no longer constant for as long as the bias voltage is raised but does not reach the buildup value, V. The Coulomb blockade can be observed experimentally only at low temperature where the charging energy, E, is larger than the thermal energy, kT, of the charge carriers; i.e., $e^2/2C \gg kT$, where k is Boltzmann's constant and T is the absolute temperature. For typical capacitances of $C \leq 10^{-15}$ farad this regime can be reached at temperatures of the order of millikelvin and the discreteness of the electron energy spectrum has to be taken into account.

CSS, Calderbank-Shore-Steane, codes Quantum codes formed from two classical codes.

cyclic code Code invariant under the cyclic permutation of bits or qubits.

cyclic discrete Fourier transform, *DFT* If n divides $2^k - 1$ and β is an element of order n in $GF(2^k)$, then the cyclic DFT over $GF(p^k)$ is described by the matrix

$$DFT = (\beta^{ij})_{0 \leq i,j \leq n-1}, \quad \text{with} \quad \beta^n = 1.$$

The inverse cyclic discrete Fourier transform is denoted as DFT^{-1}.

cyclic redundancy check (CRC) Error-detecting code; the parity-check symbols are computed over the characters of the message and are then appended to the packet by the networking hardware.

data processing inequality Informally, the data processing inequality states that one cannot get more information by processing a set of data than the information provided by the data to begin with. If the random variables, $X \mapsto Y \mapsto Z$, form a Markov chain, then

$$I(X;Z) \geq I(X;Y).$$

Random variables, X, Y, and Z, form a Markov chain if the conditional probability distribution of Z depends only upon Y and it is independent of X:

$$p_{XYZ}(x,y,z) = p_X(x)p_{Y|X}(y|x)p_{Z|Y}(z|y) \Longrightarrow X \mapsto Y \mapsto Z.$$

de Broglie, Louis Victor Pierre Raymond, duke (1892–1987) Distinguished French physicist who speculated that nature did not single out light as being the only matter that exhibits a wave-particle duality. He proposed that ordinary "particles" such as electrons, protons, or bowling balls could also exhibit wave characteristics in certain circumstances. Quantitatively, he associated a wavelength, λ, to a particle of mass, m, moving at speed, v, the two being related by *de Broglie's equation*:

$$p = \frac{h}{\lambda},$$

with p, the momentum of the particle and h, Planck's constant.

decoherence The most significant cause of errors in quantum systems. The destruction of the superposition of pure quantum states due to the interaction of the quantum system with the environment. On a large scale, the world obeys the laws of classical physics due to decoherence.

degenerate code A quantum code for which linearly independent correctable errors acting on the coding space produce linearly dependent states. Degenerate codes have the potential to be more efficient than nondegenerate codes because they are better than many known bounds on efficiency suggest.

density matrix of a quantum system The density matrix or operator is a simplifying notation to represent pure or mixed states. For a pure state $|\varphi\rangle$, the density operator is $\rho = |\varphi\rangle\langle\varphi|$; for mixed states, it is a probabilistic expression, $\rho = \sum_i \mu_i |\varphi_i\rangle\langle\varphi_i|$, with $\sum_i \mu_i = 1$, and with more than one, $\mu_i \neq 0$.

depolarizing channel model The quantum error process is memoryless, and the probability of error is p; thus, $\text{Prob}(\sigma_I) = 1 - p$ and $\text{Prob}(\sigma_x) = \text{Prob}(\sigma_y) = \text{Prob}(\sigma_z) = p/3$.

Descartes, René (1556–1650) Very influential French mathematician and philosopher of the seventeenth century. He founded analytical geometry and invented the Cartesian coordinate system. Descartes is also a major promoter of rationalism, a philosophical method later embraced by Baruch Spinoza and Gottfried Leibnitz. Famous also for the statement "Cogito ergo sum," Latin for "I think, therefore I exist."

Deutsch, David Distinguished physicist born in Haifa, Israel, and educated at Cambridge and Oxford Universities in the United Kingdom. Member of the Quantum Computation and Cryptography group at Clarendon Laboratory at Oxford University and author of the theory of parallel universes.

Deutsch's principle Extends the Church-Turing principle. Every finitely realizable physical system can be perfectly simulated by a universal computing machine operating by finite means.

diameter of the electron According to high-energy electron-electron scattering, the diameter of the electron is $< 10^{-18}$ m.

dilution refrigerator The operation of the dilution refrigerator is based on the phase transition a liquid mixture of two helium isotopes, ^{3}He and ^{4}He, undergoes at temperatures below 0.86 K, the triple critical point, the point in the phase diagram of liquid He where the three boundary lines between "normal liquid," "superfluid liquid," and "unstable composition" meet. When cooled below this critical point, the He mixture undergoes a spontaneous phase separation and forms a ^{3}He-rich phase and a ^{3}He-poor phase. The transport of ^{3}He atoms from the ^{3}He-rich phase into the ^{3}He-poor phase requires energy. If atoms are made to continuously cross this boundary, they cool the mixture. Temperatures as low as 2 millikelvin (mK) can be reached with the best dilution systems.

Dirac, Paul Adrien Maurice (1902–1984) Distinguished British mathematician. His work has been concerned with the mathematical and theoretical aspects of quantum mechanics. He used a noncommutative algebra for calculating atomic properties leading to the relativistic theory of the electron (1928) and the theory of holes (1930). This latter theory required the existence of a positive particle having the same mass and charge as the known (negative) electron. This particle, the positron, was discovered experimentally at a later date (1932) by C. D. Anderson. The importance of Dirac's work lies essentially in his famous wave equation, which introduced special relativity into Schrödinger's equation. In 1932, he became Lucasian Professor of Mathematics at Cambridge. In 1933, Dirac shared the Nobel Prize for Physics with Schrödinger.

Dirac's ket **and** bra **notation** Notation used in quantum mechanics for state vectors.

discrete memoryless classical communication channel Channel over a finite alphabet and with a finite number of states. The memoryless property of the communication channel implies that the output of the channel is a Markov process; it is affected only by the current input and not by the history of the channel states.

discrete memoryless quantum communication channel The quantum systems are described as vectors in finite-dimensional Hilbert spaces. The memoryless property of the communication channel implies that the output of the channel is affected only by the current input and not by the history of the channel states.

distinguishable states of a quantum system Two states of a quantum system are distinguishable if they are orthogonal. If two states are distinguishable, then a measurement exists that guarantees to determine which one of the two states the system is in.

dual operator/matrix $A^\dagger$, the dual, or the adjoint of a matrix A, is obtained by transposing the matrix and then taking the complex conjugate of each element. The order of the two operations can be reversed. See also *adjoint matrix.*

efficient computation A computation is efficient if it requires, at most, polynomial resources as a function of the size of the input. For example, if a computation returns the value of the function, $f(x)$, with x as a bit string of length, n, then the computation of $f(x)$ is efficient if the number of steps it requires is bounded by n^k for some k.

eigenvalue A scalar, λ_i, associated with an *eigenvector*, $|\psi_i\rangle$, of a linear operator (observable), $\mathbf{O}$, as

$$\mathbf{O}|\psi_i\rangle = \lambda_i|\psi_i\rangle.$$

In quantum mechanics, the eigenvalues of an operator represent those values of the corresponding observable that have nonzero probability of occurring. The set of all the eigenvalues is called the operator (matrix) *spectrum*.

eigenvector A state vector, $|\psi_i\rangle$, is an eigenvector of a linear operator (observable), $\mathbf{O}$, if when operated on by the operator the result is a scalar multiple of itself as

$$\mathbf{O}|\psi_i\rangle = \lambda_i|\psi_i\rangle.$$

The scalar, λ_i, is called the *eigenvalue* associated with the eigenvector. If $|\psi_i\rangle$ is an eigenvector with the eigenvalue, λ_i, then any nonzero multiple of $|\psi_i\rangle$ is also an eigenvector with eigenvalue λ_i. If the set of state vectors, $|\psi_0\rangle, |\psi_1\rangle, \ldots, |\psi_{n-1}\rangle$, are eigenvectors to *different* eigenvalues, $\lambda_0, \lambda_1, \ldots, \lambda_{n-1}$, then the state vectors $|\psi_0\rangle, |\psi_1\rangle, \ldots, |\psi_{n-1}\rangle$ are necessarily linear independent.

empiricism Theory of knowledge whose main tenet is that scientific knowledge is related to experience and that experiments are essential for the development of scientific concepts; all hypotheses and theories must be tested against observations of the natural world. The sophists in the fifth century BC, Aristotle (384–322 BC), the stoics and the epicureans a generation later, followed by Thomas Aquinas and Roger Bacon in the thirteenth century, and John Locke, George Berkeley, David Hume, in the seventeenth and eighteenth centuries, attribute sensory perception and a posteriori observation a major role in the formation of human knowledge. They believed that human knowledge of the natural world is grounded in sense experience. See also *rationalism*.

encoding of a logical qubit into multiple physical qubits We encode the state of a qubit, $|\psi\rangle = \alpha_0|0\rangle + \alpha_1|1\rangle$ as $|\varphi\rangle = \alpha_0|0_L\rangle + \alpha_1|1_L\rangle$, with 0_L and 1_L resilient to errors. For example, we encode $|0\rangle$ as $|000\rangle$ and $|1\rangle$ as $|111\rangle$. In this case, a random error can cause departures from the subspace spanned by $|000\rangle$ and $|111\rangle$. We can correct small errors because the component that was $|000\rangle$ is likely to remain in a subspace spanned by the four vectors $|000\rangle, |001\rangle, |010\rangle$, and $|100\rangle$, while the component that was $|111\rangle$ is likely to remain in a subspace spanned by the four vectors $|111\rangle, |110\rangle, |101\rangle$, and $|011\rangle$.

entanglement Nonlocal, nonclassical correlation between two quantum systems that gives quantum computers and quantum communication systems their power. The translation of the German term *Verschränkung* used by Schrödinger, who was the first to recognize this quantum effect. It means that a two-particle quantum system is in a state that cannot be written as a tensor product of the states of the individual particles. The two quantum particles share a joint state, and it is not possible to describe one of the particles in isolation.

entanglement fidelity A measure of the overlap between the initial purification and the joint state of the quantum system, Q, and the reference system, R, after the transmission of Q through a quantum communication channel,

$$F_e(\rho, \mathcal{E}) = \langle\varphi^{(Q,R)}|\,(\mathcal{I}_R \otimes \mathcal{E})\left(|\varphi^{(Q,R)}\rangle\langle\varphi^{(Q,R)}|\right)|\varphi^{(Q,R)}\rangle,$$

with $|\varphi^{(Q,R)}\rangle$, the joint state of the quantum system Q and the reference system R before the transmission and $\mathcal{I}_R$ the identity transformation of the reference system R.

entropy Measure of the uncertainty, or the degree of disorder of a system.

entropy exchange If $\mathcal{E}$ is the quantum transformation of the input state ρ, carried out during transport through a quantum communication channel, then the entropy exchange is defined as the von Neumann entropy of the environment after the operation [275, 370]:

$$S_e(\rho, \mathcal{E}) = S\left(\rho^{(E)}_{final}\right).$$

Initially, the environment, E, is in a pure state as a result of a purification operation; $\rho^{(E)}_{final}$ is the density matrix of E after the operation.

Entscheidungs problem German term for the "decision problem" posed by Hilbert: "Could there exist, at least in principle, a definite method or process, by which it could be decided whether any mathematical assertion was provable?"

EPR paradox Experiment proposed by Einstein, Podolsky, and Rosen in 1935 to show that quantum mechanics is not a complete theory.

ergodic process Stochastic process for which time averages and set averages are equal to one another.

ergodicity An attribute of stochastic systems; generally, a system that tends in probability to a limiting form that is independent of the initial conditions.

Euler, Leonhard (1707–1783) Eminent mathematician born in Basel, Switzerland, who introduced new concepts in analysis and revised almost all of the branches of pure mathematics known at that time. In 1748, he wrote *Introductio in Analysin Infinitorum*, an introduction to pure analytical mathematics, and in 1755, the *Institutiones Calculi Differentialis*. Euler's formulae:

$$\sin\theta = \frac{1}{2i}\left(e^{i\theta} - e^{-i\theta}\right), \quad \cos\theta = \frac{1}{2}\left(e^{i\theta} + e^{-i\theta}\right).$$

extreme point of a convex set The extreme point of a convex set, S, is a point in the set S that does not lie in any open line segment joining two points of S.

fanout A measure of the ability of a logic gate output to drive a number of inputs of other logic gates of the same type to form more complex circuits.

fast Fourier transform (FFT) An algorithm proposed by J. W. Cooley and J. W. Tukey in 1965, which reduces the number of operations to compute the Fourier transform from $2n^2$ to $2n\log_2 n$. A similar idea is used for the quantum Fourier transform. The algorithm decomposes the transformation for $n = 2^m$, recursively, into two transforms of length, $n/2$, using the identity

$$\begin{aligned}\sum_{j=0}^{n-1} a_j\, e^{-i2\pi jk/n} &= \sum_{j=0}^{n/2-1} a_{2j}\, e^{-i2\pi(2j)k/n} + \sum_{j=0}^{n/2-1} a_{2j+1}\, e^{-i2\pi(2j+1)k/n} \\ &= \sum_{j=0}^{n/2-1} a_j^{even} e^{-i2\pi jk/(n/2)} + e^{-i2\pi k/n} \sum_{j=0}^{n/2-1} a_j^{odd} e^{-i2\pi jk/(n/2)}.\end{aligned}$$

Fermi, Enrico (1901–1954) Distinguished Italian-born physicist who discovered in 1926 the statistical laws, nowadays known as the "Fermi statistics," governing the particles subject to Pauli's exclusion principle (now referred to as *fermions*, in contrast with *bosons*, which obey the Bose-Einstein statistics). Fermi directed a classical series of experiments that ultimately led to the atomic pile and the first controlled nuclear chain reaction. He played an important part in solving the problems connected with the development of the first atomic bomb.

Fermi level, Fermi energy The Fermi level is the last energy level occupied by electrons at the zero absolute temperature, $T = 0\,\text{K}$. According to the Pauli exclusion principle, at $T = 0\,\text{K}$ the electrons of an atom fill all available levels starting with the lowest one. The top of this sea of electrons, also called the "Fermi sea," defines the *Fermi level*, or the *Fermi energy*. The Fermi energy, E_F, is the maximum energy occupied by an electron at $T = 0\,\text{K}$. The Fermi energy is very large in comparison with the energy that an electron could gain by ordinary physical interactions with the environment. For example, the Fermi energy for copper is $E_F = 7\,\text{eV}$, while the thermal energy acquired by an electron at $T = 300\,\text{K}$, the room temperature is $kT = 0.026\,\text{eV}$, where k is Boltzmann's constant.

fermions Quantum particles with a spin multiple of 1/2 (e.g., electrons, protons, and neutrons). The spin quantum number of fermions can be $s = 1/2$, $s = -1/2$. The spin of a complex system of fermions is a multiple of $\pm 1/2$. See also *bosons*.

Feynman, Richard Phyllips (1918–1988) Distinguished American physicist who received his doctorate from Princeton in 1942 under J. A. Wheeler (he was also advised by E. Wigner) and worked on the atomic bomb project at Princeton University (1941–1942) and then at Los Alamos (1943–1945). Feynman's main contribution was to quantum mechanics; he introduced diagrams (now called Feynman diagrams) that are graphic analogues of the mathematical expressions needed to describe the behavior of systems of interacting particles. He was awarded the Nobel Prize in 1965, jointly with J. Schwinger and S-I. Tomonoga, for fundamental work in quantum electrodynamics and physics of elementary particles; his later work led to the current theory of quarks, fundamental in pushing forward an understanding of particle physics. He made significant contributions to the field of quantum computing as well. According to his obituary published in the Boston Globe, "He was widely known for his insatiable curiosity, gentle wit, brilliant mind and playful temperament."

fidelity of a pure quantum state Let $|\varphi\rangle$ be a pure state of a quantum system. The fidelity of a possibly mixed state with density matrix, ρ and $|\varphi\rangle$, is defined as

$$F = \langle\varphi|\rho|\varphi\rangle.$$

fidelity of two operators The fidelity of positive operators, **A** and **B**, is defined as $F(A,B) = \text{tr}\sqrt{A^{1/2}BA^{1/2}}$.

fidelity of two probability density functions The fidelity of two probability density functions, $p_X(x)$ and $p_Y(x)$, is defined as $F(p_X(x), p_Y(x)) = \sum_x \sqrt{p_X(x)p_Y(x)}$.

forward classical communication costs FCCC measures the capacity of the classical channel needed to simulate a quantum channel.

Fredkin gate A three-qubit gate with two target inputs, a, b, and one control input, c, and three outputs, a', b', and $c' = c$. When the control input is not set, $c = 0$, and the target inputs appear unchanged at the output, $a' = a$ and $b' = b$. When the control is set, $c = 1$, and the target inputs are swapped,

$a' = b$ and $b' = a$. The control input is always transferred to the output unchanged. There are both classical and quantum versions of the Fredkin gate.

functional A real-valued function on a vector space, V.

fundamental theorem of algebra Any polynomial equation of degree n has n solutions.

fundamental theorem of arithmetic Any positive integer, $n > 1$, can be expressed as a product of prime numbers with unique non-negative exponents as

$$n = 2^{n_2} \cdot 3^{n_3} \cdot 5^{n_5} \cdot \ldots = \prod_{p \text{ prime}} p^{n_p}.$$

Galois, Évariste (1811–1832) French mathematician who developed the Galois theory and articulated first the concept of finite field; he also introduced the term *group* for the well-known algebraic structure.

Galois field A finite field with q elements is also called a *Galois field*, and it is denoted as $GF(q)$. In particular, $\mathbb{Z}_p$ is denoted as $GF(p)$. It is easy to prove that a finite field has, p^n, elements, with p, a prime number.

gate (quantum gate) Building block of a quantum circuit. A physical system capable of transforming one or more qubits.

Gauss-Markov process A stochastic process, $X(t)$, with three properties:

1. If $h(t)$ is a nonzero scalar function of t, then $Z(t) = h(t)X(t)$ is also a Gauss-Markov process.
2. If $f(t)$ is a nondecreasing scalar function of t, then $Z(t) = X(f(t))$ is also a Gauss-Markov process.
3. There exists a nonzero scalar function, $h(t)$, and a nondecreasing scalar function, $f(t)$, such that $X(t) = h(t)W(f(t))$, where $W(t)$ is the standard Wiener process. Every Gauss-Markov process can be synthesized from the standard Wiener process.

See also the *standard Wiener process*.

Gedankenexperiment "Gedanken" is the German word for "thought." A thought experiment enables us to prove or disprove a conjecture when the experiment enabling us to study the physical phenomena is not feasible. We construct the result of the thought experiment according to a set of assumptions and a model of the system, all based upon universally accepted laws of physics.

generating function A formal power series whose coefficients encode information about a sequence, a_n, indexed by natural numbers, $G(a_n;x) = a_n x^n$. For example,

$$G(n^2;x) = \sum_{n=0}^{\infty} n^2 x^2 = \frac{x(x+1)}{(1-x)^3}.$$

Informally, a generating function is a "clothesline on which we hang up a sequence of numbers to display."

general linear group Multiplicative group of nonsingular $n \times n$ matrices. Denoted as $GL(n)$ or as $GL(n,\mathbb{R})$, $GL(n,\mathbb{C})$, $GL(n,F)$ if matrices have real, complex, or elements from a field, F, $a_{ij} \in F$, respectively.

generators of a group A set of μ elements, $\langle g_1, g_2, \ldots, g_\mu \rangle$, is called the generator of the group G if the element, g_i, can be written as a product of (possibly repeated) elements from the list,

$\langle g_1, g_2, \ldots, g_\mu \rangle$, $\forall g_i \in G$. We write $G = \langle g_1, g_2, \ldots, g_\mu \rangle$. The number, μ, of generators of group, G, satisfies the inequality, $\mu \leq \log(|G|)$.

generators of a stabilizer group Let $\mathbf{M}_1, \mathbf{M}_2, \ldots, \mathbf{M}_q$ be the generators of the stabilizer group, S, of a quantum code, $\mathcal{Q}$. The eigenvectors of the generators, $\{\mathbf{M}_1, \mathbf{M}_2, \ldots, \mathbf{M}_q\}$, have special properties: those corresponding to eigenvalues of $+1$ are the codewords of $\mathcal{Q}$ and those corresponding to eigenvalues of -1 are codewords affected by errors. If $\mathbf{M}_j|\psi_i\rangle = (+1)|\psi_i\rangle$, then $|\psi_i\rangle$ is a codeword, $|\psi_i\rangle \in \mathcal{Q}$. This justifies the name given to the set, S, any operator in S leaves the state of a word unchanged. On the other hand, if $\mathbf{M}_j|\varphi_k\rangle = (-1)|\varphi_k\rangle$, then $|\varphi_k\rangle = \mathbf{E}_\alpha|\psi_k\rangle$; the state, $|\varphi_k\rangle$, is the result of an error affecting the codeword, $|\psi_k\rangle \in \mathcal{Q}$.

Gilbert-Varshamov bound If C is an $[n,k,d]$ linear code over $GF(q)$ and if $M(n,d) = |C|$ is the largest possible number of codewords for a given n and d values, then the code satisfies the Gilbert-Varshamov bound:

$$M(n,d) \geq \frac{q^n}{\sum_{k=0}^{d-1} \binom{n}{k} (q-1)^k}.$$

Gleason theorem Given a Hilbert space, $\mathcal{H}_n$, of dimension $n \geq 3$, the only possible measure of the probability of the state associated with a particular linear subspace $\mathcal{H}_m, m < n$, of the Hilbert space $\mathcal{H}_n$ is the trace of the product of the projection operator, $\mathbf{E}$, and the density matrix, ρ, of the system, $\mathrm{tr}[\rho\mathbf{E}]$.

Hadamard, Jacques (1865–1963) Distinguished French mathematician who proved the prime number theorem, developed Hadamard matrices, and worked on the calculus of variations.

Hadamard gate The H gate describes a unitary quantum "fair coin flip" performed upon a single qubit. For example, it transforms an input qubit in the state, $|0\rangle$, into a superposition state, $(|0\rangle + |1\rangle)/\sqrt{2}$, or $(|0\rangle - |1\rangle)/\sqrt{2}$. The transfer matrix of a Hadamard gate is

$$H = \frac{1}{\sqrt{2}} \begin{pmatrix} 1 & 1 \\ 1 & -1 \end{pmatrix}.$$

Hall, Edwin (1855–1938) American physicist who discovered the Hall effect in 1879.

Hall effect The classical effect, discovered by Edwin Hall in 1879, manifests itself as a potential difference, the *Hall voltage*, on the opposite sides of a conductor when an electric current is flowing through the conductor and a magnetic field is applied perpendicular to the current. Electrons moving along a direction perpendicular to an applied magnetic field are acted upon by the Lorentz force oriented normal to both the magnetic field direction and the electron motion direction. In the absence of the magnetic field, the electrons flow in a straight line; when the magnetic field is applied, the electrons are deflected; their new path is curved so that the moving charges accumulate on one face of the conductor. Equal charges of opposite sign accumulate on the other face. The separation of charges results in the buildup of an electric potential drop (Hall voltage) across the sample.

The *Hall voltage* for a metal that is characterized by only one type of charge carrier, the electrons, is given by

$$V_{\text{Hall}} = -\frac{IB}{dne},$$

where I is the current flowing along the length of the sample, B is the magnetic flux density, d is the sample thickness, n is the density of the charge carriers (electrons), and e is the electron charge (1.602×10^{-19} *Coulomb*). Other Hall effect characteristics are the *Hall coefficient* and the *Hall resistance*, defined, respectively, as

$$R_{\text{Hall}} = \frac{V_{\text{Hall}} d}{IB} = -\frac{1}{ne} \quad \text{and} \quad \rho_{\text{Hall}} = \frac{V_{\text{Hall}}}{I}.$$

The Hall effect has the ability to differentiate between positive and negative charges that move in opposite directions and at one time it was the first proof that the electric current in metals is carried by electrons and not by protons. The classical Hall effect can be observed at room temperature and at moderate magnetic fields of less than one tesla (T). The Hall resistance, ρ_{Hall}, varies linearly with the strength of the magnetic field, B.

The Hall effect is also present in semiconductors where it differentiates between the two types of carriers, electrons and holes. The Hall voltage has opposite signs for these two carriers, such as negative for n-type semiconductors and positive for p-type semiconductors. The Hall coefficient for moderate magnetic fields takes into account the contributions of both carriers, electrons and holes, which have different mobilities and are present in different concentrations; it is given by

$$R_{\text{Hall}} = \frac{-n\mu_e^2 + p\mu_h^2}{e(n\mu_e + p\mu_h)^2},$$

with n being the electron concentration, μ_e the electron mobility, p the hole concentration, μ_p the hole mobility, mobility, and e the electron charge. The *Hall mobility* can determined from the equation

$$\mu = \frac{|V_{\text{Hall}}|}{R_{\text{sheet}} IB},$$

where all the quantities involved can be measured experimentally. The Hall coefficient for large applied magnetic fields takes a simpler form, similar to that for a single type of carrier

$$R_{\text{Hall}} = \frac{1}{(n-p)e}.$$

Hamiltonian The name *Hamiltonian* honors the Irish mathematician and astronomer William R. Hamilton who, personally, had nothing to do with quantum mechanics, but in the 1830s happened to restate Lagrange's equations of motion with emphasis on momenta instead of forces. Hamilton's equations are first-order differential equations with respect to time and involve the Hamiltonian function, $\mathbf{H}$, which is the total energy expressed as a function of the generalized coordinates for position, q_i, and momenta, p_i,

$$\frac{d}{dq_i} = \frac{\partial \mathbf{H}}{\partial p_i}, \quad \frac{d}{dp_i} = -\frac{\partial \mathbf{H}}{\partial q_i}.$$

Hamming, Richard Wesley (1915–1998) American mathematician best known for his work on error-detecting and error-correcting codes. His fundamental paper on this topic appeared in 1950.

Hamming bound Given an $[n,k,d]$ linear code, with n the length of the codeword, k the number of information symbols, $r = n - k$ the number of parity-check symbols, and d the distance of the code, the Hamming bound gives the minimum number of parity-check bits necessary to construct a code with a certain error-correction capability (e.g., a code capable of correcting all single-bit errors):

$$2^r \geq \sum_{i=0}^{\lfloor (d-1)/2 \rfloor} \binom{n}{i}.$$

Hamming distance The number of positions where two code words differ.

Hamming sphere All n-tuples of a block code that are not codewords and are at distance, $d \leq e$, from a codeword.

Heisenberg, Werner (1901–1976) Eminent German physicist; the founder of quantum mechanics. Heisenberg and his fellow student Wolfgang Pauli started their study of theoretical physics under Arnold Sommerfeld in 1920. In 1924–1925, he worked with Niels Bohr at the University of Copenhagen. In 1925, Heisenberg formulated matrix mechanics, the first coherent mathematical version of quantum mechanics. Matrix mechanics was further developed in 1926 in a paper coauthored with M. Born and P. Jordan. He is perhaps best known for the *uncertainty principle*, discovered in 1927. In 1928, Heisenberg published *The Physical Principles of Quantum Theory*. In 1932, he was awarded the Nobel Prize in Physics for the creation of quantum mechanics.

Heisenberg uncertainty principle Intrinsic property of the quantum systems: the precise knowledge of some basic physical properties such as position and momentum is simply forbidden. The *uncertainty principle* states that the uncertainty in determining the position, ΔX, and the uncertainty in determining the momentum, ΔP_X, at position, X, are constrained by the inequality

$$\Delta X \, \Delta P_X \geq \frac{\hbar}{2},$$

where $\hbar = h/2\pi$ is a modified Planck constant.

Hertz, Heinrich (1857–1894) German physicist, the first to demonstrate experimentally the production and detection of Maxwell electromagnetic waves. The photoelectric effect was first discovered accidentally in 1887 by Hertz, while carrying on investigations on the electromagnetic waves.

hidden variable theory A theory based on the assumption that there is a variable or variables that determine the real properties of quantum particles. These variables have definite values from the moment the particle is created, and they determine the result of the measurement performed on that property of the quantum particle.

Hilbert, David (1862–1943) Eminent German mathematician; in 1899 he published *Grundlagen der Geometrie*, putting geometry in a formal axiomatic setting. He delivered the speech "The Problems of Mathematics" at the Second International Congress of Mathematicians in Paris, challenging mathematicians to solve fundamental questions such as the continuum hypothesis, the well ordering of the reals, Goldbach conjecture, the transcendence of powers of algebraic numbers, the Riemann hypothesis, and the extension of the Dirichlet principle. Hilbert's work in integral equations led to the research in functional analysis and established the basis for his work on infinite-dimensional space, later called Hilbert space.

Hilbert space, n-dimensional An n-dimensional complex vector space denoted as $\mathcal{H}_n$, with an inner product and, thus, with a norm. An n-dimensional Hilbert space is isomorphic with $\mathbb{C}^n$.

Holevo bound A source generates letters, X, from an alphabet with the probabilities, p_i; Alice encodes the random variable, X, and transmits it through a quantum channel; Bob performs a measurement of the quantum system; and the result of this measurement is a random variable, Y. The mutual information, $I(X;Y)$, measures the amount of information about X that Bob can infer from the measurement. Regardless of the measurement performed by Bob, the mutual information is limited by the Holevo bound:

$$I(X;Y) \leq S\left(\sum_i p_i \rho_i\right) - \sum_i p_i S(\rho_i).$$

Holevo-Schumacher-Westmoreland (HSW) noisy quantum channel encoding theorem The HSW theorem was proven independently by Holevo [211] and by Schumacher and Westmorland [373].

impure states of a single qubit Impure, or mixed states, are represented by points inside the Bloch sphere. This implies that the trace of the square of their density matrix is less than one; $\text{tr}(\rho^2) < 1$. See also *pure state of a single qubit.*

index of a subgroup Let G be a group with cardinality $|G|$, and $H \subset G$ be a subgroup of G with cardinality $|H|$. Then the index of the subgroup H is

$$I_{G/H} = \frac{|G|}{|H|}.$$

indistinguishability in quantum mechanics Principle of quantum mechanics: all quantum particles of the same type are alike. For example, we cannot distinguish an electron from another. Therefore the operation of swapping the position of two electrons in a systems with many electrons leaves the system's state unchanged, or, in other words, the operation is symmetric and it is represented by a unitary transformation acting upon the wave function. In three dimensions an exchange of two bosons is represented by an identity operator; the wave function is invariant and we say that the particles obey the Bose statistics. The exchange of two fermions in three dimensions changes the sign of the wave function (it is represented by multiplication with -1); the particles are said to obey Fermi statistics.

interference The interaction of optic, acoustic, or electromagnetic waves that are correlated, or coherent with each other; interference occurs if the waves come from the same source, or if the waves have about the same frequency. In quantum mechanics, consider a transition between two states in the Hilbert space, $\mathcal{H}_n$, with an orthonormal basis, $\{|e_i\rangle\}$:

$$|\varphi\rangle = \sum_i \alpha_i |e_i\rangle \quad \mapsto \quad |\psi\rangle = \sum_i \beta_i |e_i\rangle.$$

The probability of this transition is the square of the modulus of the inner product of the two states

$$\text{Prob}\,(|\varphi\rangle \mapsto |\psi\rangle) = |\langle\varphi|\psi\rangle|^2 = |\sum_i \alpha_i^* \beta_i|^2 = \sum_{i,j} \alpha_i^* \alpha_j \beta_j^* \beta_i.$$

The last sum can be separated into two terms; thus, the expression becomes

$$\text{Prob}\,(|\varphi\rangle \mapsto |\psi\rangle) = \sum_i |\alpha_i|^2|\beta_i|^2 + \sum_{i,j,i\neq j} \alpha_i^*\alpha_j\beta_j^*\beta_i.$$

The first term corresponds to a classical interference; the second term reflects the quantum interference between $i \neq j$ paths of quantum evolution and the quantum probability rule.

irreversible/noninvertible gate A gate characterized by the fact that, knowing the output, we cannot determine the input for all possible combinations of input values. The irreversibility of classical gates, other than NOT, means that there is an irretrievable loss of information, and this has very serious consequences regarding the energy consumption of classical gates.

isomorphism The word derives from the Greek *iso*, meaning "equal," and *morphosis*, meaning "to form" or "to shape." Formally, an isomorphism is bijective morphism. Informally, an isomorphism is a map that preserves sets and relations among elements. A space isomorphism is a vector space in which addition and scalar multiplication are preserved. Two groups, G_1 and G_2, with binary operators "+" and "×," are isomorphic if there exists a map, $f : G_1 \mapsto G_2$, which satisfies the relation, $f(x+y) = f(x) \times f(y)$. An isomorphism preserves the identities and inverses of a group. See also *automorphism and homomorphism.*

Kant, Immanuel (1724–1804) German philosopher, the last major exponent of the *Enlightenment*, and author of the *Critique of Pure Reason*. His work bridges the gap between rationalism and empiricism. He defined the Enlightenment as an age shaped by the motto "Sapere aude," in Latin, meaning "Dare to know." He expressed the belief that we are never able to transcend the bounds of our own mind; we cannot access the "Ding an sich," in German, the "thing-in-itself."

kernel of a linear transformation Ker($\mathcal{L}$), the kernel of a linear transformation, $\mathcal{L} : \mathcal{H}_a \mapsto \mathcal{H}_b$, is the set of all vectors, $|\zeta\rangle \in \mathcal{H}_a$, such that $\mathcal{L}(|\zeta\rangle) = 0$.

key distribution Mechanism for distribution of cryptographic keys—in particular, of public keys.

Lagrange theorem If H is a subgroup of a finite group, G, then $|H|$ divides $|G|$.

Landauer principle The Landauer principle traces the energy consumption in a computation to the act of erasing information. When a computer erases one bit of information, the amount of energy dissipated into the environment is at least, $k_B T \ln(2)$, with k_B (Boltzmann's constant) and T (the temperature of the environment). An equivalent formulation: the entropy of the environment increases by at least $k_B \ln(2)$ when one bit of information is erased.

Laurent series A complex function, $f(z)$, can be represented as a power series with terms of negative degree as

$$f(z) = \sum_{n=-\infty}^{n=+\infty} a_n(z-c)^n.$$

The power series is defined with respect to point, c, and a path of integration, γ. The path of integration is counterclockwise and must lie in an annulus inside of which, $f(z)$, is holomorphic.

The constant, a_n, is

$$a_n = \frac{1}{2\pi i} \oint_{\gamma} \frac{f(z)dz}{(z-c)^{n+1}}.$$

The Laurent series expansion is used when the Taylor series expansion cannot be applied.

light A visible form of electromagnetic radiation. An electromagnetic field consists of an electric and a magnetic field perpendicular to each other and oscillating in a plane perpendicular to the direction of propagation of the electromagnetic wave.

linear code A subspace of a vector space; a selected subset of n-tuples satisfying some distance properties.

manifold, smooth manifold A smooth manifold is a geometrical object that is locally the same as the Euclidian space, $\mathbb{R}^n$ [268]. Although manifolds resemble Euclidean spaces near each point, the global structure of a manifold may be more complicated. A manifold is typically endowed with a differentiable structure that allows calculus and a Riemannian metric to measure distances and angles. For example, a line and a circle are one-dimensional manifolds, a plane and sphere (the surface of a ball) are two-dimensional manifolds; every point of an n-dimensional manifold has a neighborhood homeomorphic to the n-dimensional space $\mathbb{R}^n$.

matrix exponentiation If β is a real number and if matrix, A, is such that $A^2 = I$, then

$$e^{i\beta A} = \cos(\beta)I + i\sin(\beta)A.$$

matrix in a Hilbert space An $n \times m$ matrix, A, is regarded as a linear operator from an n-dimensional Hilbert space, $\mathcal{H}_n$, to an m-dimensional Hilbert space, $\mathcal{H}_m$, namely,

$$A : \mathcal{H}_n \longrightarrow \mathcal{H}_m.$$

maximal quantum test Consider a quantum system and assume that the maximum number of different outcomes obtainable in a test of the system is N; any test that has exactly, N, different outcomes is a maximal quantum test. For example, when atoms with spin, s, are used in a Stern-Gerlach experiment, $2s+1$ different outcomes are observed, regardless of the orientation of the magnetic field.

maximum likelihood decoding Decoding strategy when a received n-tuple is decoded into the code word to minimize the probability of errors.

Maxwell, James Clerk (1831–1879) Eminent Scottish mathematician and theoretical physicist. His most significant achievement was aggregating a set of equations in electricity, magnetism, and inductance, the Maxwell's equations. Maxwell demonstrated that electric and magnetic fields travel through space, in the form of waves, and at the constant speed of light. He also developed the Maxwell distribution, a statistical means to describe aspects of the kinetic theory of gases.

Maxwell demon In 1871, Maxwell proposed the following "thought" experiment. Imagine the molecules of a gas in a cylinder divided in two by a wall that has a slit; the slit is covered with a door controlled by a little demon. The demon examines every molecule of gas and determines its velocity; those of high velocity on the left side are allowed to migrate to the right side and those with low velocity on the right side are allowed to migrate to the left side. As a result of these measurements, the demon separates hot from cold in blatant violation of the second law of thermodynamics.

measurement of a quantum system The process that makes a connection between the quantum and the classical worlds; generally considered as an irreversible operation that destroys quantum information about an observable (property) of a quantum system and replaces it with classical information. In quantum mechanics, the measurement of an observable of a quantum system (such as momentum, energy, or spin) is associated with a Hermitian operator, **A**, on the Hilbert space of state vectors of the system. If v is an eigenvector of **A**, with the eigenvalue λ, then measuring the system in a state described by state vector v will always give the result λ. If the state vector is not an eigenvector of **A**, the measurement process forces the system to jump (collapse) randomly to a state corresponding to a state vector, v_i, an eigenvector of **A**. The result of the measurement is the corresponding eigenvalue of **A**, λ_i. See also *projective measurements*.

memoryless channel A *discrete memoryless classical channel* is the triplet, $(X, Y, \{p(y|x), x \in \{X, y \in Y\})$, with:

- X the input channel alphabet,
- Y the output channel alphabet, and
- $\{p(y|x), x \in X, y \in Y\})$ the probability of observing output symbol, y, when the input symbol is x. The channel is memoryless if the probability distribution of the output depends only upon the input at that time and not upon the past history.

A *discrete memoryless quantum channel* means that the state of the quantum system is a vector in a finite-dimensional Hilbert space; the memoryless property has the same meaning as for the classical information; the mean of the output of the quantum communication channel is only determined by the current input.

minimum distance decoder Minimum distance (also called nearest neighbor) decoder for a code, $\mathcal{C}$, with distance, d, is a decoder that, given a received word, c_r, selects the code word, $c \in \mathcal{C}$, that satisfies, $d(c_r, c) < d/2$, if such a codeword exists; otherwise, it declares failure.

mixed state A quantum system whose state, $|\psi\rangle$, is not known precisely is said to be in a mixed state. A mixed state is a superposition of different pure states; the system is in a state, $|\varphi_i\rangle$, with the probability, p_i. The density operator of a quantum system in a mixed state is $\rho = \sum_i p_i |\varphi_i\rangle\langle\varphi_i|$. The trace of the density operator is $\text{tr}(\rho^2) < 1$. A mixed state is also called an *impure state*. See also *pure state of a single qubit*.

monic polynomial A polynomial, $f(x) = x^n + a_{n-1}x^{n-1} + \cdots + a_1 x + 1$, in which the coefficient of the highest order term is 1.

mutual information The mutual information measures the reduction in uncertainty of a random variable, X, due to another random variable, Y:

$$I(X;Y) = \sum_x \sum_y p_{XY}(x,y) \log \frac{p_{XY}(x,y)}{p_X(x) p_Y(y)}.$$

Newmark theorem The theorem [6] states that one can extend the Hilbert space, $\mathcal{H}_n$, in which the POVM operators, A_i, are defined to a larger Hilbert space, $\mathcal{H}_N$, with $N > n$, such that in $\mathcal{H}_N$ there exist a set of orthogonal projectors, P_i, with two properties:

- $\sum_i P_i = I$.
- A_i is the result of projecting $P_i \in \mathcal{H}_N$ into $\mathcal{H}_n$.

nondegenerate code A quantum code for which linearly independent correctable errors acting on the coding space produce linearly independent states. See also, *degenerate codes.*

nondegenerate stabilizer code A nondegenerate stabilizer code $[n,k,d]$ is one when none of the elements of the stabilizer, S (except the identity), has weight less than d; in case of a degenerate stabilizer code, at least one element of the stabilizer S other than the identity has a weight less than d.

nondemolition measurement If the density matrix of a qubit is

$$\rho = p|0\rangle\langle 0| + (1-p)|1\rangle\langle 1| + \alpha|0\rangle\langle 1| + \alpha^*|1\rangle\langle 0|,$$

with $\alpha \in \mathbb{C}$, then an "ideal measurement" should provide an outcome of 0, with the probability p and an outcome of 1, with the probability $(1-p)$, regardless of the value of α, regardless of the state of the neighboring qubits, and without changing the state of the quantum computer. A "nondemolition" measurement also leaves the qubit in the state, $|0\rangle\langle 0|$, after reporting the outcome 0 and leaves it in the state, $|1\rangle\langle 1|$, after reporting the outcome 1. Quantum error correction requires nondemolition measurements of the error syndrome to preserve the state of the physical qubits.

norm of a vector in a Hilbert space A non-negative function that measures the "length" of a vector. If $|\psi\rangle \in \mathcal{H}_n$, then the norm, $\| \, |\psi\rangle \|$, can be computed from the inner product of the state vector with itself, $\| \, |\psi\rangle \|^2 = |\langle \psi | \psi \rangle|$.

normal operator in a Hilbert space An operator, $\mathbf{U} \in \mathcal{H}_n$, with the property that $[\mathbf{U},\mathbf{U}^\dagger] = \mathbf{U}\mathbf{U}^\dagger - \mathbf{U}^\dagger\mathbf{U} = 0$.

normal unitary basis of an n-dimensional Hilbert space, $\mathcal{H}_n$ A set of n vectors, $|\psi_0\rangle, |\psi_2\rangle, \ldots, |\psi_i\rangle, \ldots, |\psi_{n-1}\rangle$, where each vector has the norm (or length) equal to 1, $\||\psi_0\rangle\| = \||\psi_1\rangle\| = \cdots = \||\psi_{n-1}\rangle\| = 1$, and any two of them are orthogonal, $\langle \psi_i | \psi_j \rangle = 0$, for $(i \neq j)$.

normalization condition in quantum mechanics Requirement that the square of the modulus of the projections of a state vector on the orthonormal basis sum is equal to 1. This translates into the condition that the sum of probabilities of all possible outcomes of a measurement of a quantum system must be equal to 1.

normalizer of a stabilizer S. The normalizer $N(S)$ of a quantum code, $\mathcal{Q}$, with stabilizer, $S \in \mathcal{G}_n$, is the set of elements in $\mathcal{G}_n$ that fix the stabilizer, S, under conjugation. $\mathcal{G}$ is the n-qubit Pauli group. S is a normal subgroup of $N(S)$, $S \subset N(S)$.

observable A physical property of a quantum system that can be measured by an external observer. Mathematically, each observable X has an associated Hermitian operator, $\mathcal{M}_X$, with a complete set of eigenvectors.

operator A function used to transform the state of a system.

operator-sum representation An operator-sum representation, also called Kraus representation, for transformation, $\mathcal{E}(\rho)$, for a system initially in the state, ρ, is $\mathcal{E}(\rho) = \sum_k A_k \rho A_k^\dagger$, with A_k, a set of tracepreserving operators; in other words, $\sum_k A_k^\dagger A_k = I$.

oracle The name "oracle" comes form the Latin verb "orare," meaning "to speak." An oracle is a source of wise pronouncements or prophetic opinions with some connection to an infallible spiritual authority. In the ancient world, the Greeks believed that they could read the future with the help of oracles. Pythia, the priestess of the temple of Apollo at Delphi, is probably the most famous oracle, but lesser ones are known. According to Herodotus, when Croesus, the king of Lydia, demanded to know if he should go to war against Cyrus the Great (of Persia), Pythia responded in a truly Delphian style, "If you do, you will destroy a great empire." Believing the response favorable, Croesus attacked, but it was his own empire that was ultimately destroyed by the Persians in 547 BC.

oracle models An oracle implements a computational model and solves a problem for which there is no efficient solution. Oracle models are used to compare the power of two computational models. For example, in 1994, D. Simon showed that a quantum computer with a specific oracle, $\mathcal{O}$, efficiently could solve problems that classical computers having access to the same oracle $\mathcal{O}$ could not solve efficiently. See also *efficient computation*.

order of an element in a finite field $GF(q)$ Let $GF(q)$ be a finite field, with q elements, and $a \in F$. The order of a, ord(a), is the smallest integer, s, such that $a^s = 1 \mod q$.

orthogonal state vectors Two vectors, $|\psi_a\rangle$ and $|\psi_b\rangle$ in $\mathcal{H}_n$, are *orthogonal*, and we write $|\psi_a\rangle \perp |\psi_b\rangle$ if their inner product is zero, $\langle\psi_a|\psi_b\rangle = 0 \Longrightarrow |\psi_a\rangle \perp |\psi_b\rangle$.

orthonormal basis in a Hilbert space, $\mathcal{H}_n$ A basis consisting of a set of n orthonormal vectors, such as $\{|0\rangle, |1\rangle, \ldots, |i\rangle, \ldots, |n-1\rangle\}$. Any two vectors from the set are orthogonal, and all have a norm equal to 1. See also *normal unitary basis of an n-dimensional Hilbert space, $\mathcal{H}_n$*.

outer product in a Hilbert space, $\mathcal{H}_n$ *The outer product* of a ket *vector and a* bra *vector,* $|\psi_a\rangle\langle\psi_b|$, *is a linear operator. For example, in $\mathcal{H}_4$, we have*

$$|\psi_a\rangle\langle\psi_b| = \begin{pmatrix} \alpha_0 \\ \alpha_1 \\ \alpha_2 \\ \alpha_3 \end{pmatrix} \begin{pmatrix} \beta_0^* & \beta_1^* & \beta_2^* & \beta_3^* \end{pmatrix} = \begin{pmatrix} \alpha_0\beta_0^* & \alpha_0\beta_1^* & \alpha_0\beta_2^* & \alpha_0\beta_3^* \\ \alpha_1\beta_0^* & \alpha_1\beta_1^* & \alpha_1\beta_2^* & \alpha_1\beta_3^* \\ \alpha_2\beta_0^* & \alpha_2\beta_1^* & \alpha_2\beta_2^* & \alpha_2\beta_3^* \\ \alpha_3\beta_0^* & \alpha_3\beta_1^* & \alpha_3\beta_2^* & \alpha_3\beta_3^* \end{pmatrix}.$$

parametric down-conversion A nonlinear crystal splits incoming photons into pairs of photons of lower energy whose combined energy and momentum are equal to the energy and momentum of the original photon; the energy and momentum must be conserved. The photon pair is entangled in the frequency domain due to phase matching.

partial trace Let $\mathcal{C} = \mathcal{AB}$ be a composite system consisting of two subsystems, $\mathcal{A}$ and $\mathcal{B}$, described by the density operator, $\rho_{\mathcal{C}}$. The partial trace of $\rho_{\mathcal{C}}$, over system $\mathcal{B}$, is:

$$\mathrm{tr}_{\mathcal{B}}(\rho_{\mathcal{C}}) = \mathrm{tr}_{\mathcal{B}}(|a_1\rangle\langle a_2| \otimes |b_1\rangle\langle b_2|) = |a_1\rangle\langle a_2| \; \mathrm{tr}_{\mathcal{B}}(\, |b_1\rangle\langle b_2| \,) = |a_1\rangle\langle a_2|\langle b_1|b_2\rangle,$$

with $|a_1\rangle, |a_2\rangle$, any two vectors in the state space of $\mathcal{A}$ and $|b_1\rangle, |b_2\rangle$, any two vectors in the state space of $\mathcal{B}$.

Pauli, Wolfgang Ernst (1900–1958) Distinguished Austrian-born physicist who proposed in 1924 a quantum spin number for electrons. Best known for his (Pauli) exclusion principle, proposed in 1925, which states that no two electrons in an atom can have the same four quantum numbers. He predicted mathematically, in 1931, that conservation laws in beta decay required the existence of a new particle, electrically neutral and with very low mass and named it "neutron." In 1933, he published his prediction and he made the claim that the particle had zero mass. The particle, which we now know as the neutron, has a nonzero mass and was discovered by Chadwick in 1932.

Pauli's particle was named the "neutrino" by Fermi in 1934, and at that time, he correctly stated that it was not a constituent of the nucleus of an atom. The neutrino was found experimentally in 1956 by C. Cowan and F. Reines.

Pauli exclusion principle Two electrons on the same orbital around the nucleus of an atom cannot be in identical states, including the spin; they must have their spins oriented in opposite directions, because they already share three quantum numbers. If, as a result of an experiment, one of the electrons is made to change the orientation of its spin, then a simultaneous measurement of the other finds it in a state with the *opposite* spin.

Pauli group The one-qubit Pauli group, $\mathcal{G}_1$, consists of the Pauli matrices, σ_I, σ_x, σ_y, and σ_z, together, with the multiplicative factors, ± 1 and $\pm i$:

$$\mathcal{G}_1 \equiv \{\pm\sigma_I, \pm i\sigma_I, \pm\sigma_x, \pm i\sigma_x, \pm\sigma_y, \pm i\sigma_y, \pm\sigma_z, \pm i\sigma_z\}.$$

The cardinality of $\mathcal{G}$ is $|\mathcal{G}_1| = 16 = 2^4$. The n-qubit Pauli group, $\mathcal{G}_n$, consists of the 4^n tensor products of σ_I, σ_x, σ_y, and σ_z and an overall phase of ± 1 or $\pm i$. The group has 4^{n+1} elements.

Pauli matrices Matrices, $\sigma_I, \sigma_x, \sigma_y, \sigma_z$, describing transformations (rotations) of a single qubit:

$$\sigma_I = \begin{pmatrix} 1 & 0 \\ 0 & 1 \end{pmatrix}, \quad \sigma_x = \begin{pmatrix} 0 & 1 \\ 1 & 0 \end{pmatrix}, \quad \sigma_y = i\begin{pmatrix} 0 & -1 \\ 1 & 0 \end{pmatrix}, \quad \text{and} \quad \sigma_z = \begin{pmatrix} 1 & 0 \\ 0 & -1 \end{pmatrix}.$$

Pauli matrices are Hermitian $\left(\sigma_I^\dagger = \sigma_I, \sigma_x^\dagger = \sigma_x, \sigma_y^\dagger = \sigma_y, \sigma_z^\dagger = \sigma_z\right)$ and unitary $\left(\sigma_x\sigma_x^\dagger = \sigma_x^2 = \sigma_I, \sigma_y\sigma_y^\dagger = \sigma_y^2 = \sigma_I, \sigma_z\sigma_z^\dagger = \sigma_z^2 = \sigma_I\right)$. Their multiplication table is

$\times$	σ_I	σ_x	σ_y	σ_z
σ_I	σ_I	σ_x	σ_y	σ_z
σ_x	σ_x	σ_I	$i\sigma_z$	$-i\sigma_y$
σ_y	σ_y	$-i\sigma_z$	σ_I	$i\sigma_x$
σ_z	σ_z	$i\sigma_y$	$-i\sigma_x$	σ_I.

perfect code capable of correcting *e* errors Linear code where Hamming spheres of radius e about each code word are disjoint and exhaust the entire space of n-tuples.

perfect quantum code A quantum code achieving the quantum Hamming bound; see *quantum Hamming bound.*

photon From the Greek word "photos," meaning light; a light particle.

Plotkin bound If C is an $[n,k,d]$ linear code, if $2d > n$ and $M = |C|$, then the code satisfies the Plotkin bound:

$$M \leq 2\left\lfloor \frac{d}{2d-n} \right\rfloor.$$

polarization filter A partially transparent material that transmits light of a particular polarization.

polarization measurements Polarization is measured by passing a photon through a polarizer. If the polarizer axis is oriented parallel to the polarization of the photon, the photon passes through unimpeded. If it is oriented perpendicular to the polarization of the photon, the photon is absorbed. At an intermediate angle, the photon will have a certain probability of being transmitted.

polarization of light As an *electromagnetic radiation*, light consists of an electric and a magnetic field perpendicular to each other and, at the same time, perpendicular to the direction the energy is transported by the electromagnetic (light) wave. The electric field oscillates in a plane perpendicular to the direction of light, and the way the electric field vector travels in this plane defines the polarization of the light. When the electric field oscillates along a straight line, we say that the light is *linearly polarized.* When the end of the electric field vector moves along an ellipse, the light is *elliptically polarized.* When the end of the electric field vector moves around a circle, the light is *circularly polarized.* If the light comes toward us and the end of the electric field vector moves around in a counterclockwise direction, we say that the light has *right-hand polarization*; if the end of the electric field vector moves in a clockwise direction, we say that the light has *left-hand polarization.*

polynomial time algorithm Term in computational complexity theory to describe algorithms that require a number of steps, $T(n)$, bounded by a polynomial, when n is the size of the input. $\mathcal{P}$ is the class of polynomial time algorithms. $\mathcal{NP}$ is the class of polynomials that require a number of steps not bounded by a polynomial.

positive operator A positive operator, **A**, is any Hermitian operator with non-negative eigenvalues; $\sqrt{A}$ denotes the unique positive root of **A**.

positive operator-valued measure (POVM) A set of generalized measurement positive operators concerned only with the statistics of the measurement; POVM operators are not necessarily orthogonal or commutative and allow the possibility of measurement outcomes associated with non-orthogonal states. The number of POVM operators may differ from the dimension of the Hilbert space, while the number of projective operators is precisely equal to the dimension of the Hilbert space.

primitive element of a finite field Given a finite field, $F(q)$, with q elements, there is an element, $\alpha \in F$, such that the set on nonzero elements of F is

$$F - \{0\} = \{\alpha^1, \alpha^2, \ldots, \alpha^{q-2}\} \quad \text{and} \quad \alpha^{q-1} = 1.$$

projection operator, projector The outer product of a unit vector with itself as

$$\mathbf{P}_a = |\Psi_a\rangle\langle\Psi_a|.$$

It has the defining property: $\mathbf{P}_a^2 = \mathbf{P}_a$. A complete set of orthogonal projectors in $\mathcal{H}_n$ is a set, $\{\mathbf{P}_0, \mathbf{P}_1, \ldots, \mathbf{P}_i, \ldots, \mathbf{P}_{m-1}\}$, such that $\sum_{i=0}^{m-1} \mathbf{P}_i = 1$.

projective measurement In a von Neumann-type projective measurement, the measurement operators are Hermitian and idempotent. The number of such operators is equal to the dimension of the Hilbert space. Since orthogonal measurement operators commute, they correspond to simultaneous observables. A projective measurement is characterized by a set of projectors, $\mathcal{M}_i$, such that $\sum_i \mathcal{M}_i = I$ and $\mathcal{M}_i \mathcal{M}_j = \delta_{ij} \mathcal{M}_i$. The outcome of the measurement is the one with the index, i, associated with $\mathcal{M}_i$. The probability of outcome, i, for a system in the state, $|\psi\rangle$, is $p_i = |\mathcal{M}_i|\psi_i\rangle|^2$. Given the outcome i, the quantum state "collapses" to the state, $M_i|\psi_i\rangle/\sqrt{p_i}$.

pure state of a single qubit Pure states are characterized by maximum knowledge; they can be expressed as superpositions of basis vectors of an orthonormal basis. Pure states of a single qubit are represented by points on the Bloch sphere. This implies that the trace of the square of their density matrix is one, $\text{tr}(\rho^2) = 1$. See also *mixed state* or *impure states of a single qubit*.

purification Purification is the process of finding a reference system, $\mathcal{B}$, given a quantum system, $\mathcal{A}$, in a mixed state, $\rho_{\mathcal{A}}$. If $|\psi_{\mathcal{C}}\rangle$ is the pure state of the composite system, $\mathcal{C} = \mathcal{AB}$, we wish that $\rho_{\mathcal{A}} = \text{tr}_{\mathcal{B}}(|\psi_{\mathcal{C}}\rangle\langle\psi_{\mathcal{C}}|)$. Also, a purification is any pure state in the extended Hilbert space, $\mathcal{H}_{\mathcal{C}} = \mathcal{H}_{\mathcal{A}} \otimes \mathcal{H}_{\mathcal{B}}$, having $\rho_{\mathcal{A}}$ as the reduced state for the subsystem.

quadratic form Let $A(a;b)$ be a symmetric bilinear function. Then $A(a;a)$ is called a quadratic form.

quantitative version of the Church-Turing principle "Any physical computing device can be simulated by a Turing machine in a number of steps polynomial in the resources used by the computing device." This thesis allows us to relate the behavior of the abstract model of computation provided by the Turing machine concept with the physical computing devices used to carry out a computation.

quantum Latin word meaning "some defined quantity." In physics, it is used with the same meaning as *discrete* in mathematics.

quantum accuracy threshold theorem An arbitrarily long computation can be carried out reliably provided that all sources of "noise" affecting a system are weaker than the accuracy threshold; a lower bound on this threshold, $\epsilon > 1.9 \times 10^{-4}$, was derived in [9].

quantum circuit complexity The smallest number of quantum gates necessary to implement an operation on a fixed number of qubits.

quantum communication channel Physical media allowing two parties to exchange quantum information. For example, an optic fiber allowing photons with a certain polarization to circulate from a source to a destination.

quantum data processing inequality Consider a quantum system in a state characterized by the density matrix, ρ, transmitted through a quantum channel (Figure 3.20); the resulting state, $\mathcal{E}_1(\rho)$, becomes the input to a second channel, and the resulting state is $\mathcal{E}_2 \circ \mathcal{E}_1(\rho)$. The two transformations $\mathcal{E}_1$ and $\mathcal{E}_2$, are trace-preserving. Then the coherent information between the original input and the output of the second channel, $I(\rho, \mathcal{E}_2 \circ \mathcal{E}_1(\rho))$, cannot be larger than the coherent information between the original input and the output of the first channel, $I(\rho, \mathcal{E}_1(\rho))$, and, in turn, $I(\rho, \mathcal{E}_1(\rho))$ cannot be larger than $S(\rho)$, the von Neumann entropy of the source; $S(\rho) \geq I(\rho, \mathcal{E}_1) \geq I(\rho, \mathcal{E}_2 \circ \mathcal{E}_1)$.

quantum erasure A phenomenon caused by entanglement; the state of a system is characterized jointly by the density matrix and the information regarding the components of a composite system.

quantum errors The most general type of error of a qubit can be described as a combination of two basic types of quantum errors: amplitude, or *bit-flip*, and phase, or *phase-flip*. Classical errors correspond to quantum bit-flip errors, changes of the form $|0\rangle \leftrightarrow |1\rangle$; the Pauli, σ_x, transformation applied to a qubit leads to a bit-flip. Phase-flips are changes of the form, $|0\rangle + |1\rangle \leftrightarrow |0\rangle - |1\rangle$; the Pauli, σ_z, transformation applied to a qubit leads to a phase-flip. There is no correspondence of quantum phase-flip in classical error correction.

quantum Fano inequality Relates $S_e(\rho, \mathcal{E})$, the entropy exchange, and $F_s(\rho, \mathcal{E})$, the entanglement fidelity, $S_e(\rho, \mathcal{E}) \leq h(p)F_s(\rho, \mathcal{E}) + [1 - F_s(\rho, \mathcal{E})]\log(d^2 - 1)$. Here, d is the dimension of the Hilbert space of the quantum system Q, and $h(p)$ is the binary entropy associated to the probability p, as $h(p) = -p \log p - (1-p) \log(1-p)$.

quantum Hamming bound Gives the smallest number n of physical qubits required to encode a logical qubit $2(3n+1) \leq 2^n$. Thus, $n = 5$ is the smallest number of qubits required to encode the two superposition states $|0_L\rangle$ and $|1_L\rangle$, and then be able to recover them regardless of the qubit in error and the type of error.

quantum parallelism Term capturing the fact that a quantum computer can manipulate an exponential set of inputs simultaneously.

quantum particle Particle obeying the laws of quantum mechanics.

qubit Quantum bit; mathematical abstraction for a quantum system capable of storing one bit of information.

qudit Multilevel quantum states used for quantum information processing; for example, for $d = 3$, we have a quantum system with three basis states.

Rabi oscillation The oscillation of the level population of a two-state quantum system in the presence of an oscillatory driving field. The optical Rabi oscillation is the result of a coherent nonlinear light-matter interaction; the oscillation results from the quantum interference between the probability amplitudes of two atomic dipole eigenstates involved in the interaction. When an atom or some other two-level system is illuminated by a coherent beam of photons, it will cyclically absorb photons and re-emit them by stimulated emission; such a cycle is a *Rabi cycle*, and the inverse of its duration is the *Rabi frequency* of the photon beam.

Raman transition The result of inelastic scattering of monochromatic light, usually laser light in the visible, near infrared or near ultraviolet range; in a Raman process, the parent atom or molecule is excited from the ground state to a virtual energy state and relaxes into a vibrational excited state by emitting a photon. The emitted photon is shifted up or down in energy in relation to the incident one; this shift in energy (Raman shift) is equal to the vibrational level involved in the transition and gives information about the phonon modes in the system. In a stimulated Raman transition, a two-color (frequency) laser pulse transfers the atom/molecule from ground to a vibration excitation state if the difference in energy matches an allowed Raman transition.

rank of a linear transformation The dimension of the range of the transformation, where the range (image) is the set of vectors the transformation maps to.

rank of a matrix The maximum number of linearly independent rows or columns of matrix, A.

rationalism, continental rationalism Theory of knowledge based upon the idea that the criterion of truth is not sensory but intellectual and deductive. The origins of this philosophy whose main credo is self-sufficiency of reason can be traced back to Pythagoras and Plato and was later embraced by Descartes, Spinoza and Leibnitz, who advocated the introduction of mathematical methods in philosophy. Spinoza and Leibnitz believed that, in principle, all knowledge, including scientific knowledge, could be gained through the use of reason alone. Rationalism is often contrasted to empiricism; see also *empiricism*.

ray in a Hilbert space A mathematical abstraction that exhibits only direction. It can be represented as a straight line through the origin of the coordinate system.

reciprocal polynomial Given a polynomial of degree m, $h(x) = \sum_{k=0}^{m} c_k x^k, c_m \neq 0$, its reciprocal is the polynomial

$$\bar{h}(x) = \sum_{k=0}^{m} c_{m-k} x^k.$$

Reed-Solomon code A Reed-Solomon code over $GF(q)$ of length, $n = 2^k - 1$, dimension k, and distance, d, is a cyclic code with the generator polynomial,

$$g(x) = \left(x - \beta^j\right)\left(x - \beta^{j+1}\right)\ldots\left(x - \beta^{j+d-2}\right),$$

with β being a primitive element of order $n = 2^k - 1$ of the finite field $GF(2^k)$.

reduced density operator Let $\mathcal{C} = \mathcal{AB}$ be a composite system described by the density operator, $\rho_{\mathcal{C}}$. The reduced density operator of subsystem, $\mathcal{A}$, is $\rho^{\mathcal{A}} = \text{tr}_{\mathcal{B}}\left(\rho_{\mathcal{C}}\right)$.

relative entropy The relative entropy between two distributions, $p(x)$ and $q(x)$, quantifies how close the two distributions are from each other

$$H(p(x) \mid\mid q(x)) = \sum p(x) \log \frac{p(x)}{q(x)}.$$

If the real probability distribution of events is $p(x)$, but we erroneously assume that it is $q(x)$, then the surprise when an event occurs is $\log(1/q(x))$ and the average surprise is $\sum p(x) \log \frac{1}{q(x)}$.

reversible physical process A physical process is said to be reversible if it can evolve forward as well as backward in time; a reversible system can always be forced back to its original state from a new state reached during an evolutionary process.

Schmidt decomposition Let $|\psi_{\mathcal{C}}\rangle$ be the state of a composite system, $\mathcal{C} = \mathcal{AB}$, with $\mathcal{A} \in \mathcal{H}_A$ and $\mathcal{B} \in \mathcal{H}_B$. There are two orthonormal bases, called Schmidt bases, $\{|e_j^A\rangle\} \in \mathcal{H}_A$ and $\{|e_j^B\rangle\} \in \mathcal{H}_B$, such that $|\psi_{\mathcal{C}}\rangle = \sum_j \lambda_j |e_j^A\rangle \otimes |e_j^B\rangle$; the Schmidt coefficients, λ_j, are real numbers and $\sum_j \lambda_j^2 = 1$.

Schmidt number Consider a bipartite pure state, $|\psi_{\mathcal{C}}\rangle$, of a composite system, $\mathcal{C} = \mathcal{AB}$. The number of nonzero eigenvalues, λ_j, of the density matrix, $\rho_{\mathcal{C}}$, thus, the number of terms in the Schmidt decomposition of $|\psi_{\mathcal{C}}\rangle$, is called the Schmidt number of the state, $|\psi_{\mathcal{C}}\rangle$. The Schmidt number provides a clear distinction between the bipartite pure states, $|\psi_{\mathcal{AB}}\rangle$, that are separable and those that are entangled: *a bipartite pure state, $|\psi_{\mathcal{AB}}\rangle$, is entangled if its Schmidt number is greater than 1; otherwise, it is separable.*

Schrödinger, Erwin (1887–1961) Eminent Austrian-born physicist. He is one of the founders of quantum physics and has made significant contributions to statistical mechanics and the general theory of relativity. Schrödinger transferred the idea of a wave associated to a particle, predicted by de Broglie, to Bohr's atomic model. In 1927, he showed the mathematic equivalence of his equation, $\mathbf{H}\Psi = \mathbf{E}\Psi$, and Heisenberg's matrix mechanics. That same year he became Max Planck's successor for the theoretical physics chair at the University of Berlin. He was awarded the Nobel Prize for Physics in 1933 (together with Paul Dirac) for his contributions to the development of quantum mechanics.

Schumacher compression We can compress the quantum information produced by the quantum source, $\mathcal{A}$, with density matrix, $\rho^{\mathcal{A}}$, up to $\log\left[2^{nS(\rho^{\mathcal{A}})+\mathcal{O}(n)}\right] \approx nS\left(\rho^{\mathcal{A}}\right)$ qubits.

Schwartz inequality Inequality satisfied by any state vectors, $|\psi_a\rangle, |\psi_b\rangle \in \mathcal{H}_n$, where

$$\langle\psi_a|\psi_a\rangle\langle\psi_b|\psi_b\rangle \geq |\langle\psi_a|\psi_b\rangle|^2.$$

second law of thermodynamics The entropy of a system is a nondecreasing function of time.

self-adjoint operator or matrix A linear operator or a matrix, A, with the property that $A = A^\dagger$, where $A^\dagger$ is the dual, or the adjoint of A. See also *adjoint matrix*.

self-dual code A linear code, $\mathcal{C}$, is called weakly self-dual if $\mathcal{C}^\perp \subset \mathcal{C}$. When $\mathcal{C} = \mathcal{C}^\perp$, the code is called strictly self-dual.

self-dual basis See *trace-orthogonal basis*.

Shannon, Claude Elwood (1916–2001) Eminent American mathematician and electrical engineer; founder of the modern information theory. In 1948, Shannon published his landmark work *A Mathematical Theory of Communication*, and in 1949, he published the *Communication Theory of Secrecy Systems*, generally credited with transforming cryptography from an art to a science.

Shannon entropy Given a random variable, X, with a probability density function, $p_X(x)$, the entropy is a positive real number, $H(X) = -\sum_x p_X(x) \times \log_2 p_X(x)$.

Shor, Peter Distinguished American mathematician, well known for his quantum factoring algorithm and for his pioneering work on quantum error-correcting codes.

singlet electron state The antisymmetric state of a pair electrons with *antiparallel* spins, $1/\sqrt{2}(|\uparrow\downarrow\rangle - |\downarrow\uparrow\rangle)$. The electrons have different spin quantum numbers, $+1/2$ and $-1/2$, and the total spin of the state is zero. See also *triplet electron state*.

single value decomposition theorem An $n \times n$ matrix, Γ, over a field, F, of real or complex numbers, can be decomposed into a product of three $n \times n$ matrices over F: $\Gamma = A\Delta B^\dagger$, with $A = [\alpha_{ij}]$ and $B = [\beta_{ij}]$ unitary matrices, and $\Delta = [\delta_{ij}]$ a diagonal matrix.

Singleton bound Every $[n,k,d]$ linear code, $\mathcal{C}$, over the field, $GF(q)$, satisfies the Singleton bound, $k + d \leq n + 1$.

Solovay-Kitaev theorem $\mathcal{G} \subset SU(d)$ is a universal family of gates if (1) $\mathcal{G}$ is closed under the inverse operation, $\forall g \in \mathcal{G} \leftrightarrow g^{-1} \in \mathcal{G}$, and (2) $\mathcal{G}$ generates a dense subset of $SU(d)$; then

$$\forall U \in SU(d), \epsilon > 0, \exists g_1, g_2, \ldots, g_q \in \mathcal{G} : ||U - U_{g_1,g_2,\ldots g_q}|| \leq \epsilon \quad \text{and} \quad q = \mathcal{O}(\log^2 1/\epsilon),$$

where $||U||$ is the norm of the linear operator U and $U_{g_1,g_2,\ldots,g_q}$ is an implementation of U using only gates from the set $\mathcal{G}$. For example, if a set $\mathcal{G}$ of single-qubit quantum gates generates a dense subset of $SU(2)$, then it is possible to construct good approximations of any gate using short sequences of the set $\mathcal{G}$ of one-qubit gates.

special unitary group of degree n, $SU(n)$ The multiplicative group of $n \times n$ unitary matrices with determinant equal to 1. $SU(n)$ is a subgroup of the *unitary group*, $U(n)$, of all $n \times n$ unitary matrices. In turn, $U(n)$ is a subgroup of the *general linear group*, $GL(n, \mathbb{C})$.

spectral decomposition of a normal operator in a Hilbert space In $\mathcal{H}_n$, every normal operator, $\mathbf{A}$, has n eigenvectors, $|a_i\rangle$, and, correspondingly, n eigenvalues, $\lambda_i, 1 \leq i \leq n$. If $\mathbf{P}_i$ are the projectors corresponding to these eigenvectors, $\mathbf{P}_i = |a_i\rangle\langle a_i|$, then the operator, $\mathbf{A}$, has the spectral decomposition, $\mathbf{A} = \sum_\mathbf{i} \lambda_\mathbf{i} \mathbf{P}_\mathbf{i}$.

spectrum of a vector in an extension finite field The spectrum, $\bar{c}$, of a vector, $c = \sum_{i=1}^{n} \alpha_i c_i$, with elements, $\alpha_i \in GF(2^k)$, is $\bar{c} = DFT(c)$, with DFT, the cyclic discrete Fourier transform.

spin The observable associated with the intrinsic rotation of a quantum particle is the intrinsic angular momentum, also called the *spin angular momentum*. The "spin" is the quantum number characterizing the intrinsic angular momentum of the quantum particles. There are two types of particles: *bosons*, particles whose spin quantum number can be $s = 1$, $s = -1$, or $s = 0$ (e.g., photons and mesons), and *fermions*, particles whose spin quantum number can be $s = 1/2$ or $s = -1/2$ (e.g., electrons, protons, and neutrons).

stabilizer A complete description of a particular class of quantum error-correcting codes by the set of tensor products of Pauli matrices that fix every state in the coding space.

stabilizer code A class of quantum error-correcting codes described by an Abelian subgroup of the group, $\mathcal{G}$, of Pauli matrices.

standard Wiener process A standard Wiener process (often called Brownian motion) on the interval, $[0,T]$, is a continuous random variable, $W(t)$ on $t \in [0,T]$, such that $W(0) = 0$, and $W(t) - W(s) \sim \sqrt{t-s}N(0,1), 0 \leq s < t \leq T$, also $W(t) - W(s)$ and $W(v) - W(u)$ are independent for $0 \leq s < t < u < v \leq T$. $N(0,1)$ is a normal distribution with zero mean and unit variance. The process is also referred to as Gaussian.

Stark effect Splitting and shifting of an atomic spectral line into several components in the presence of an electric field, discovered by Johannes Stark in 1913; analogous to the Zeeman effect, the splitting of an atomic spectral line into several components in the presence of a magnetic field. The effect is due to the interaction of the electric dipole moment with the external electric field. In a semiconductor heterostructure where a small bandgap material is sandwiched between layers of a material with larger bandgap, like that used to manufacture the quantum dots, the effect can be significantly enhanced by bound excitons. The electron and the hole forming the excitons are pushed in opposite directions by the electric field, but are confined in the smaller bandgap material and cannot be pulled apart.

stationary, ergodic source of classical information Source that emits symbols with a probability that does not change over time; an *ergodic* source emits information symbols with a probability equal to the frequency of their occurrence in a long sequence. Stationary, ergodic sources have a finite but arbitrary and potentially long correlation time.

Stern-Gerlach experiment Experiment revealing the spin of quantum systems.

Stinespring dilation theorem If A is a unital, C^*, algebra and $B(\mathcal{H})$ is a bounded operator on the Hilbert space, $\mathcal{H}$, then for every completely positive map, $T : A \mapsto B(\mathcal{H})$, there exists another Hilbert space, $\mathcal{K}$, and a unital *-homomorphism, $U : A \mapsto B(\mathcal{K})$, such that $T(a) = VU(a)V^*, \forall a \in A$ if $V : \mathcal{K} \mapsto \mathcal{H}$ is a bounded operator. Then $||T(1)|| = |V||^2$ [411]. The interpretation of Stinespring's dilation theorem is that any completely positive and trace-preserving map in a Hilbert space can be constructed from three operations: (1) tensoring with a second Hilbert system in a specified state, (2) a unitary transformation on the larger space obtained as a result of step (1), and (3) reduction to a subsystem.

Stirling approximation $n! \approx \sqrt{2\pi n}\left(\frac{n}{e}\right)^n$.

strong superposition principle Any complete set of mutually orthogonal vectors has a physical realization as a maximal quantum test [324].

superposition probability rule In quantum mechanics, if an event may occur in two or more indistinguishable ways, then the probability amplitude of the event is the sum of the probability amplitudes of each case considered separately.

superposition state of a quantum system If the states, $|0\rangle, |1\rangle, \ldots, |n-1\rangle$, of a quantum system are distinguishable and if the complex numbers, α_i, satisfy the condition, $\sum_i |\alpha_i|^2 = 1$, then the state, $\sum_i \alpha_i |i\rangle$, is a valid quantum state called a superposition state. See also *basis states of a quantum system*.

super selection rule (SSR) An SSR is a restriction on the allowed local operations on a system, not on its allowed states, and it is associated with a group of physical transformations [24]. Such restrictions could be imposed by the properties of the underlying theory, or arise due to physical restrictions. The operations it applies to include unitary transformations, $\mathbf{O}\rho = U\rho U^\dagger$, and measurements, $\mathbf{O}_r\rho = \mathcal{M}_r\rho\mathcal{M}_r^\dagger$, with $\sum \mathcal{M}_r\mathcal{M}_r^\dagger = 1$. SSRs prohibit entangled states involving different particle numbers; thus, a two-qubit system consisting of two quantum dots and an electron in a superposition state as being on one or the other quantum dot would not satisfy the first requirement [130].

support of a normal operator $\text{supp}[\rho]$, the *support of a normal operator* $\rho \in \mathcal{L}(\mathcal{H})$, is defined as the subspace of $\mathcal{H}$ spanned by the eigenvectors of ρ having nonzero eigenvalues, $\text{supp}[\rho] = \text{spann}\{|\xi_i\rangle : 1 \le i \le n,\ \lambda_i \ne 0\}$, where $\rho = \sum_{i=1}^n \lambda_i |\xi_i\rangle\langle\xi_i|$ is a spectral decomposition of ρ.

symmetric bilinear function (form) Let $A(a;b)$ be a bilinear function. We say that $A(a;b)$ is a symmetric bilinear function if $A(a;b) = A(b;a)$.

symmetry A symmetry is a transformation of the state of a quantum system that leaves all the observables of the system unchanged. To be a symmetry, the mapping should preserve the absolute value of inner products:

$$|\varphi^{(i)}\rangle \mapsto |\psi^{(i)}\rangle \text{ and } |\varphi^{(j)}\rangle \mapsto |\psi^{(j)}\rangle \Longrightarrow |\langle\varphi^{(i)}|\varphi^{(j)}\rangle| = |\langle\psi^{(i)}|\psi^{(j)}\rangle| \quad \forall(\varphi^{(i)}, \psi^{(i)}) \in \mathcal{H}_n.$$

Szilárd, Leó (1898–1964) Distinguished Hungarian-born physicist who proposed the idea of nuclear chain reactions before nuclear fission was discovered. In 1939 Szilárd and Fermi conducted a simple experiment at Columbia University and discovered significant neutron multiplication in uranium, proving that the chain reaction was possible and opening the way to nuclear weapons. Szilárd was directly responsible for the creation of the Manhattan Project. In 1929, Leó Szilárd stipulated that information is physical while trying to explain Maxwell's demon paradox [414].

Szilárd's engine Imaginary engine powered by Maxwell's demon.

teleportation In a science fiction context, making an object or person disintegrate in one place and have it reembodied as the same object or person somewhere else. In the context of quantum information theory, "a way to scan out part of the information from an object A, which one wishes to teleport, while causing the remaining, unscanned, part of the information to pass, via the Einstein-Podolsky-Rosen effect, into another object C, which has never been in contact with A. Later, by applying to C a treatment depending on the scanned-out information, it is possible to maneuver C into exactly the same state as A was in before it was scanned" (see http://www.research.ibm.com/quantuminfo/teleportation). In this process, the original state is destroyed.

thermodynamic entropy The thermodynamic entropy of a gas, S, quantifies the notion that a gas is a statistical ensemble and it measures the randomness (or the degree of disorder) of the ensemble. The entropy is larger when the vectors describing the individual movements of the molecules of gas are in a higher state of disorder than when all of them are well organized and moving in the same direction with the same speed. Ludwig Boltzmann postulated that $S = k_B \ln \Omega$, with k_B, the Boltzmann's constant, $k_B = 1.3807 \times 10^{-23}$ joules per degree kelvin, and Ω, the number of microstates.

thermodynamics Thermodynamics is the study of energy, its ability to carry out work, and the conversion between various forms of energy, such as the internal energies of a system, heat, and work. Thermodynamic laws are derived from statistical mechanics. C. P. Snow, the English physicist and novelist, summarized the three laws of thermodynamics as follows:

1. You cannot win; matter and energy are conserved, thus, you cannot get something for nothing.
2. You cannot break even; you cannot return to the same energy state because there is always an increase in disorder. The entropy always increases.
3. You cannot get out of the game; absolute zero is unattainable.

Toffoli gate A three-qubit gate with two control inputs, a and b, and one target input, c. The outputs are $a' = a$, $b' = b$ and c'. When $c = 1$, then $c' = 1 \oplus (a \text{ AND } b) = \text{NOT}(a \text{ AND } b)$; otherwise, $c' = c$. The Toffoli gate is a universal gate and it is reversible. There are both classical and quantum versions of the Toffoli gate.

trace distance between two operators The trace distance between **A** and **B** is

$$D(A,B) = \frac{1}{2}\text{tr}|A - B| = \frac{1}{2}\text{tr}\sqrt{(A-B)^\dagger (A-B)},$$

with A and B, the matrices corresponding to operators, **A** and **B**, respectively.

trace distance between two probability density functions The trace distance between $p_X(x)$ and $p_Y(x)$ is denoted by $D(p_X(x), p_X(x))$; also called Kolmogorov, or L1, distance, it is defined as $D(p_X(x), p_Y(x)) = \frac{1}{2}\sum_x |p_X(x) - p_Y(x)|$.

trace of a linear operator The trace of an operator is the trace of the matrix associated with the operator.

trace of a matrix The sum of the diagonal elements of matrix, A, $\text{tr}(A) = \sum_{i=1}^{n} a_{ii}$; for two matrices, A and B, $\text{tr}(A + B) = \text{tr}(A) + \text{tr}(B)$.

trace of an extension Galois field $GF(p^k)$ over $GF(p)$ A linear mapping from $GF(p^k)$ to $GF(p)$, $\text{tr}: GF(p^k) \mapsto GF(p)$, defined as $\text{tr}(\alpha) = \alpha + \alpha^k + \cdots + \alpha^{p^{k-1}}$.

trace-orthogonal basis The elements, $\alpha_1, \alpha_2, \ldots, \alpha_k$, form a basis of $GF(p^k)$, over $GF(p)$, if and only if

$$\det \begin{pmatrix} \alpha_1 & \alpha_2 & \cdots & \alpha_k \\ \alpha_1^p & \alpha_2^p & \cdots & \alpha_k^p \\ \vdots & \vdots & \ddots & \vdots \\ \alpha_1^{p^{k-1}} & \alpha_2^{p^{k-1}} & \cdots & \alpha_k^{p^{k-1}} \end{pmatrix} \neq 0.$$

The set, $\{\alpha_1, \alpha_2, \ldots, \alpha_k\}$, is a *trace orthogonal basis* if $\text{tr}(\alpha_i \alpha_j) = 0, 1 \le i, j \le k, i \ne j$. If, in addition, $\text{tr}(\alpha_i^2) = 1$, then $\{\alpha_1, \alpha_2, \ldots, \alpha_k\}$ is a *self-dual basis*.

trace-preserving operation Informally, a trace-preserving quantum operation takes place when a quantum system interacts with the environment and when no measurement is performed either on the quantum system or the environment. When classical information about the quantum system is made available through a measurement, the operation is non-trace-preserving. If the operation is denoted by, $\mathcal{E}$, and the quantum system is in a state characterized by the density matrix, ρ, then a trace-preserving operation requires that $\text{tr}(\mathcal{E}(\rho)) = 1$. A non-trace-preserving quantum operation A_k is characterized by the inequality, $\sum_k A_k^\dagger A_k \leq I$.

triplet electron state The state of a pair of electrons with *parallel* spins, $|\uparrow\uparrow\rangle$ or $|\downarrow\downarrow\rangle$, or in a symmetric superposition of antiparallel spins, $(1/\sqrt{2})(|\uparrow\downarrow\rangle + |\downarrow\uparrow\rangle$. The total spin of a triplet state is $+1$.

turbo codes Turbo codes are a refinement of concatenated codes; they use the encoding structure of concatenated codes and an iterative algorithm for decoding.

Turing, Alan Mathison (1912–1954) Eminent British mathematician regarded as the founder of modern computer science. In 1936, he published the paper "*On Computable Numbers, with an Application to the Entscheidungsproblem.*" Turing defined a computable number as a real number whose decimal expansion could be produced by a Turing machine starting with a blank tape. In March 1946, he submitted a report proposing the automatic computing engine (ACE), an original design for a modern computer. In 1948, he moved to Manchester, and in 1950, he published a paper, "*Computing Machinery and Intelligence in Mind*" where he proposed the Turing test—the test used today to answer the question of whether a computer can be intelligent.

uniform family of (quantum) circuits A set of circuits, $\{C_n\}$, with one circuit for each input of length n. The *uniformity condition* requires that the description of a circuit, C_n, should be computed in time, *poly(n)*, on some Turing machine.

unitary matrix A matrix, $A = [a_{ij}]$, with complex elements; $a_{ij} \in \mathbb{C}$ is said to be unitary if $A^\dagger A = I$. Here, $A^\dagger$ is the *adjoint* of A, a matrix obtained from A by first constructing A^T, the *transpose* of A, and then taking the complex conjugate of each element (or by first taking the complex conjugate of each element and then transposing the matrix). The determinant of a unitary matrix is 1.

unitary operator A linear operator, $\mathbf{A}$, on a Hilbert space that preserves the inner product, thus, the distance. See also *unitary matrix.*

universal set of quantum gates A set of gates with the property that there exists a network of them capable of implementing every single unitary operation.

von Neumann, John (1903–1957) Eminent Hungarian-born mathematician who has made contributions to quantum physics, functional analysis, set theory, topology, economics, computer science, numerical analysis, hydrodynamics (of explosions), statistics, and many other mathematical fields. von Neumann was a pioneer of the application of operator theory to quantum mechanics [441] and the cocreator of game theory and the concepts of cellular automata and the universal constructor. John von Neumann was instrumental in the development of the theory of thermonuclear reactions and the development of the hydrogen bomb, along with Edward Teller and Stanislaw Ulam. John von Neumann was one of the pioneers of computing; the "von Neumann architecture" implements a universal Turing machine and the common "referential model" of specifying sequential architectures. The model uses a processing unit and a single separate storage structure to hold both instructions and data, thus, the term "stored-program computer."

von Neumann entropy The *von Neumann entropy*, S, of a quantum system, $\mathcal{A}$, is a function of the density matrix, $S(\rho^{\mathcal{A}}) = -\mathrm{tr}(\rho^{\mathcal{A}} \log \rho^{\mathcal{A}})$.

wave function Function describing the state of a stationary system and the evolution in time of a nonstationary system. See also *Schrödinger equation*.

weakly self-dual code See *self-dual code*.

Wigner, Eugene (1902–1995) Distinguished Hungarian-born physicist and mathematician. Wigner laid the foundation for the theory of symmetries in quantum mechanics and in 1927 introduced what is now known as the Wigner D-matrix. He and Hermann Weyl introduced group theory into quantum mechanics. He received the Nobel Prize in Physics in 1963. In 1960, Wigner discussed the power of mathematics in his best-known essay outside physics, the "*Unreasonable Effectiveness of Mathematics in the Natural Sciences*."

References

[1] S. Aaronson, D. Gottesman, Improved simulation of stabilizer codes, Phys. Rev. A, 70 (2004) 052328–052342.

[2] E. S. Abers, Quantum Mechanics, Prentice Hall, Upper Saddle River, NJ, 2003.

[3] D. Aharonov, W. van Dam, J. Kempe, Z. Landau, S. Loyd, O. Regev, Adiabatic quantum computation is equivalent to standard quantum computation. Preprint, http://arxiv.org/abs/quant-ph/0405098, v2, 2005.

[4] Y. Aharonov, D. Bohm, Significance of electromagnetic potentials in the quantum theory, Phys. Rev. 115 (3) (1959) 485–490.

[5] D. Aharonov, M. Ben-Or, Fault-tolerant quantum computation with constant error rate. Preprint, http://arxiv.org/abs/quant-ph/9906129, v1, 1999.

[6] N. I. Akhiezer, I. M. Glazman, Theory of Linear Operators in Hilbert Spaces, vol. 2, Ungar, New York, 1963.

[7] R. Alicki, M. Horodecki, P. Horodecki, R. Horodecki, Dynamical description of quantum computing: generic nonlocality of quantum noise, Phys. Rev. A, 65 (2002) 062101.

[8] P. Aliferis, D. Gottesman, J. Preskill, Quantum accuracy threshold for concatenated distance-3 codes, Quantum Inf. Comput. 6 (2006) 97–165.

[9] P. Aliferis, A. W. Cross, Subsystem fault tolerance with the Bacon-Shor code, Phys. Rev. Lett., 98 (2007) 220502.

[10] D. Z. Albert, Quantum Mechanics and Experience, Harvard University Press, Cambridge, MA, 1992.

[11] A. G. Anderson, R. Garwin, E. L. Hahn, J. W. Horton, G. L. Tucker, Spin echo serial storage memory, J. Appl. Phys. 26 (1955) 1324.

[12] A. G. Anderson, E. L. Hahn, Spin echo storage technique, U. S. Patent # 2,174,714, 1955.

[13] D. Arovas, J. R. Schrieffer, F. Wilczek, Fractional statistics and the quantum Hall effect, Phys. Rev. Lett. 53 (7) (1984) 722–724.

[14] V. I. Arnold, Mathematical Methods of Classical Mechanics, Springer Verlag, Heidelberg, 1997.

[15] R. B. Ash, Information Theory, Dover Publishing House, New York, 1965.

[16] A. Barenco, C. H. Bennett, R. Cleve, D. P. DiVincenzo, N. Margolus, P. Shor, et al, Elementary gates for quantum computation, Phys. Rev. A, 52 (1995) 3457–3570.

[17] D. Bacon, Operator quantum error correcting subsystems for self-correcting quantum memories, Phys. Rev. A, 73 (2006) 012340.

[18] D. Bacon, A. Casaccino, Quantum error subsystem codes from two classical linear codes. Preprint, http://arxiv.org/abs/quant-ph/0610088, v2, 2006.

[19] A. Barenco, D. Deutsch, A. Ekert, Conditional quantum dynamics and logic gates, Phys. Rev. Lett., 74 (20) (1995) 4083–4086.

[20] M. D. Barrett, J. Chiaverini, T. Schaetz, J. Britton, W. M. Itano, J. D. Jost, et al, Deterministic quantum teleportation of atomic qubits, Nature. 429 (2004) 737–739.

[21] H. Barnum, C. M. Caves, C. A. Fuchs, R. Jozsa, B. Schumacher, Noncommuting states cannot be broadcast, Phys. Rev. Lett. 76 (15) (1996) 2818–2821.

[22] H. Barnum, M. A. Nielsen, B. Schumacher, Information transmission through a noisy quantum channel, Phys. Rev. A, 57 (6) (1998) 4153–4175.

[23] J. D. Barrow, P. C. W. Davis, C. L. Harper (Eds.), Science and Ultimate Reality. Quantum Theory, Cosmology and Complexity, Cambridge University Press, Cambridge, UK, 2004.

[24] S. D. Bartlett, H. M. Wiseman, Entanglement constrained by superselection rules, Phys. Rev. Lett. 91 (9) (2003) 097903.

[25] J. S. Bell, On the Einstein-Podolsky-Rosen paradox, Physics, 1 (1964) 195–200.

[26] J. S. Bell, Speakable and Unspeakable in Quantum Mechanics: Collected Papers on Quantum Philosophy, Cambridge University Press, Cambridge, UK, 1987.
[27] P. Benioff, The computer as a physical system: a microscopic quantum mechanical Hamiltonian model of computers as represented by Turing machines, J. Stat. Phys. 22 (1980) 563–591.
[28] P. Benioff, Quantum mechanical models of Turing machines that dissipate no energy, Phys. Rev. Lett. 48 (1982) 1581–1584.
[29] P. Benioff, Quantum mechanical models of Turing machines, J. Stat. Phys. 29 (1982) 515–546.
[30] C. H. Bennett, Logical reversibility of computation, IBM J. Res. Dev. 17 (1973) 525–535.
[31] C. H. Bennett, The thermodynamics of computation—A review, Int. J. Theor. Phys. 21 (1982) 905–928.
[32] C. H. Bennett, G. Brassard, Quantum cryptography: public key distribution and coin tossing, in: Proceedings of the IEEE Conference on Computers, Systems, and Signal Processing, IEEE Press, Los Alamitos, CA, 1984, pp. 175–179.
[33] C. H. Bennett, S. J. Wiesner, Communication via one- and two-particle operators on Einstein-Podolsky-Rosen states, Phys. Rev. Let. 69 (1992) 2881–2884.
[34] C. H. Bennett, G. Brassard, C. Crépeau, R. Jozsa, A. Peres, W. K. Wootters, Teleporting an unknown state via dual classical and Einstein-Podolsky-Rosen channels, Phys. Rev. Lett., 70 (13) (1993) 1895–1899.
[35] C. H. Bennett, G. Brassard, R. Jozsa, D. Mayers, A. Peres, B. Schumacher, et al, Reduction of quantum information entropy by reversible extraction of classical information, J. Mod. Opt. 41 (12) (1994) 2307–2314.
[36] C. H. Bennett, H. J. Bernstein, S. Popescu, B. Schumacher, Concentrating partial entanglement by local operations. Preprint, http://arxiv.org/abs/quant-ph/9511030, 1994.
[37] C. H. Bennett, Quantum information and computation, Phys. Today, 48 (10) (1995) 24–30.
[38] C. H. Bennett, Quantum information processing, in: Computer Science: Reflections on the Field, Reflections from the Field. National Research Council. Committee on the Fundamentals of Computer Science: Challenges and Opportunities, National Academies Press, Wasington, DC, 2004, pp. 51–56.
[39] C. H. Bennett, G. Brassard, S. Popescu, B. Schumacher, J. A. Smolin, W. K. Wootters, Purification of noisy entanglement and faithful teleportation via noisy channels, Phys. Rev. Lett. 76 (1996) 722–725.
[40] C. H. Bennett, D. DiVincenzo, J. A. Smolin, W. K. Wootters, Mixed states entanglement and quantum error correction, Phys. Rev. A, 54 (1996) 3824–3851.
[41] C. H. Bennett, C. A. Fuchs, J. A. Smolin, Entanglement-enhanced classical communication on a noisy quantum channel. Preprint, http://arxiv.org/abs/quant-ph/9611006, 1996.
[42] C. H. Bennett, E. Bernstein, G. Brassard, U. Vazirani, Strengths and weaknesses of quantum computation, SIAM J. Comput. 26 (1997) 1510–1523.
[43] C. H. Bennett, D. DiVincenzo, J. A. Smolin, Capacities of quantum erasure channels, Phys. Rev. Lett. 78 (16) (1997) 3217–3220.
[44] C. H. Bennett, P. W. Shor, Quantum information theory, IEEE Trans. Inf. Theory 44 (6) (1998) 2724–2742.
[45] C. H. Bennett, P. W. Shor, J. A. Smolin, A. V. Thapliyal, Entanglement-assisted classical capacity of noisy quantum channels. Preprint, http://arxiv.org/abs/quant-ph/9904023, August, 1999.
[46] C. H. Bennett, D. P. DiVicenzo, Quantum information and computation, Science. 404 (2000) 247–255.
[47] C. H. Bennett, P. W. Shor, J. A. Smolin, A. V. Thapliyal, Entanglement-assisted capacity of a quantum channel and the reverse Shannon theorem, IEEE Trans. Inf. Theory 48 (10) (2002) 2637–2655.
[48] C. H. Bennett, T. Mor, J. A. Smolin, The parity bit in quantum cryptography. Preprint, http://arxiv.org/abs/quant-ph/9604040, 2002.
[49] E. Berlekamp, Algebraic Coding Theory, McGraw-Hill, New York, 1968.
[50] E. Bernstein, U. Vazirani, Quantum complexity theory, SIAM J. Comput. 26 (1997) 1411–1473.
[51] C. Berrou, A. Glavieux, P. Thitimajshima, Near Shannon limit error-correcting codes and decoding: turbocodes, in: Proceedings of the IEEE International Conference on Communication, IEEE Press, Los Alamitoa, CA, 1993, pp. 1064–1070.

[52] M. V. Berry, Quantal phase factors accompanying adiabatic changes, Proc. R. Soc. Lond. A392 (1984) 45–57.
[53] A. Berthiaume, G. Brassard, The quantum challenge to structural complexity theory, in: Proceedings of the 7-th Annual Conference on Structure in Complexity Theory, IEEE Press, Los Alamitos, CA, 1992, pp. 132–137.
[54] A. Berthiaume, G. Brassard, Oracle quantum computing, in: Proceedings of the Workshop on Physics of Computation, IEEE Press, Los Alamitos, CA, 1992, pp. 195–199.
[55] I. Bialynicki, J. Mycielski, J. H. Eberly, A reply to the comment by David Farrelly, Ernestine Lee, and T. Uzer, Commun. Math. Phys. 44 (1975) 129.
[56] G. Birkhoff, J. von Neumann, The logic of quantum mechanics, Ann. Math. 37 (4) (1936) 823–843.
[57] G. Birkhoff, S. Mac Lane, A Survey of Modern Algebra, Macmillan Publishing, New York, 1965.
[58] D. Biron, O. Biham, E. Biham, M. Grassl, D. A. Lidar, Generalized Grover search algorithm for arbitrary initial amplitude distribution. Preprint, http://arxiv.org/abs/quant-ph/9801066, 1998.
[59] E. Biham, O. Biham, D. Biron, M. Grassl, D. A. Lidar, D. Shapira, Analyis of generalized Grover search algorithms using recursion equations, Phys. Rev. A, 63 (2001) 012310.
[60] R. B. Blakestad, C. Ospelkaus, A. P. VanDevender, J. M. Amini, J. Britton, D. Leibfried, et al, High-fidelity transport of trapped-ion qubits through an X-junction trap array, Phys. Rev. Lett. 102 (2009) 153002.
[61] B. B. Blinov, L. Deslauriers, P. Lee, M. J. Masden, R. Miller, C. Monroe, Sympathetic cooling of Cd^+ isotopes, Phys. Rev. A 65 (4) (2002) 040304.
[62] F. Bloch, Nuclear induction, Phys. Rev. 70 (1946) 460.
[63] D. Bohm, Quantum Theory, Prentice Hall, Upper Saddle, NJ, 1951.
[64] G. Boole, An Investigation of the Laws of Thought, Dover Publications, New York, 1958.
[65] M. Born, V. A. Fock, Beweis des Adiabatensatzes, Z. Phys. A: Hadrons Nucl. 51 (3–4) (1928) 165–180.
[66] M. Born, The statistical interpretations of quantum mechanics, Nobel Lectures, Physics 1942–1962, 256–267. December 11, 1954. Also: http://nobelprize.org/nobel_prizes/physics/laureates/1954/born-lecture.pdf.
[67] R. C. Bose, R. K. Ray-Chaudhuri, On a class of error-correcting binary codes, Inf. Control. 3 (1960) 68–79.
[68] D. Bouwmeester, J.-W. Pan, K. Mattle, M. Eibl, H. Weinfurter, A. Zeilinger, Experimental quantum teleportation, Nature. 390 (1997) 575–579.
[69] D. Bouwmeester, A. Ekert, A. Zeilinger (Eds.), The Physics of Quantum Information, Springer Verlag, Heidelberg, 2003.
[70] H. E. Brandt, Positive operator valued measure in quantum information processing, Am. J. Phys. 5 (1998) 434–479.
[71] H. E. Brandt, Positive operator valued measure in quantum measurement with a postive operator-valued measure, J. Opt. B Quantum Semiclassical Opt. 5 (2003) S266–S270.
[72] G. Brassard, Searching a quantum phone book, Science. 275 (1997) 627–628.
[73] G. Brassard, P. Hoyer, M. Mosca, A. Tapp, Quantum amplitude amplification and estimation. Preprint, http://arxiv.org/abs/quant-ph/0005055, v1, 2000.
[74] S. Bravyi, Universal quantum computation with the $\nu = 5/2$ fractional quantum hall state, Phys. Rev. A, 73 (4) (2006) 042313.
[75] G. K. Brennen, C. M. Caves, P. S. Jessen, I. H. Deutsch, Quantum logic gates in optical lattices, Phys. Rev. Lett. 82 (5) (1999) 1060–1063.
[76] L. de Broglie, The wave nature of the electron, Nobel Lectures, Physics 1922–1941, pp. 244–256. December 12, 1929. Also: http://nobelprize.org/nobel_prizes/physics/laureates/1929/broglie-lecture.pdf.
[77] J. Brown, The Quest for the Quantum Computer, Simon and Schuster, New York, 1999.
[78] G. Burkard, H. A. Engel, D. Loss, Spintronics and quantum dots for quantum computing and quantum communication, Fortschr. Phys. 48(9–11) (2000) 965–986.
[79] A. W. Burks, H. H. Goldstine, J. von Neumann, Preliminary discussion of the logical design of an electronic computer instrument. Report to the U.S. Army Ordnance Department, 1946. Also in: W. Asprey, A. W. Burks (Eds.), Papers of John von Neumann, MIT Press, Cambridge, MA, 1987, pp. 97–146.

[80] A. R. Calderbank, P. W. Shor, Good quantum error-correcting codes exist, Phys. Rev. A 54 (42) (1996) 1098–1105.
[81] A. R. Calderbank, E. M. Rains, P. W. Shor, N. J. A. Sloan, Quantum error correction and orthogonal geometry, Phys. Rev. Lett. 78 (1997) 405–408.
[82] F. E. Camino, W. Zhou, V. J. Goldman, Aharonov-Bohm electron interferometer in the integer quantum Hall regime, Phys. Rev. B, 72 (2005) 155313.
[83] F. E. Camino, W. Zhou, V. J. Goldman, Aharonov-Bohm superperiod in a Laughlin quasiparticle interferometer, Phys. Rev. Lett. 95 (2005) 246802.
[84] F. E. Camino, W. Zhou, V. J. Goldman, Experimental realization of a primary-filling $e/3$ quasiparticle interferometer, Preprint, http://arxiv.org/abs/condmat/0611443, 2006.
[85] F. E. Camino, W. Zhou, V. J. Goldman, Quantum transport in electron Fabry-Pérot interferometers, Phys. Rev. B, 76 (2007) 155305.
[86] F. E. Camino, W. Zhou, V. J. Goldman, Experimental realization of Laughlin quasiparticle interferometers, Physica E. 40 (5) (2008) 949–953. Proceedings of the Conference on Electronic Properties of 2D Systems (EP2DS-17), Genoa, Italy, 2007.
[87] C. de C. Chamon, D. E. Freed, S. A. Kivelson, S. L Sondhi, X. G. Wen, Two point-contact interferometer for quantum Hall systems, Phys. Rev. B, 55 (4) (1997) 2331–2343.
[88] N. J. Cerf, C. Adami, Accesible information in quantum mesaurements. Preprint, http://arxiv.org/abs/quant-ph/9611032, 1996.
[89] J. Chiaverini, D. Leibfried, T. Schaetz, M. D. Barrett, R. B. Blakestad, J. Britton, et al, Realization of quantum error correction, Nature. 432 (2004) 602–605.
[90] A. M. Childs, I. L. Chuang, Universal quantum computation with two-level trapped ions, Phys. Rev. A, 63 (1) (2001) 012306.
[91] I. L. Chuang, Y. Yamamoto, Quantum bit regeneration, Phys. Rev. Lett. 76 (1996) 4281–4284.
[92] I. L. Chuang, M. A. Nielsen, Prescription for experimental determination of the dynamics of a quantum black box, J. Mod. Opt. 44 (11/12) (1997) 2455–2467.
[93] I. L. Chuang, L. M. K. Vandersypen, X. Zhou, D. W. Leung, S. Lloyd, Experimental realization of a quantum algorithm, Nature. 393 (6681) (1998) 143–146.
[94] I. L. Chuang, IBM's test-tube quantum computer makes history, http://www.spaceref.com/news/viewpr.html?pid=6949, 2001.
[95] A. Church, A note on the Entscheidungsproblem, J. Symbolic Logic. 1 (1936) 40–41.
[96] J. I. Cirac, P. Zoller, Quantum computation with cold trapped ions, Phys. Rev. Lett. 74 (20) (1995) 4091–4094.
[97] J. I. Cirac, L. Duan, P. Zoller, Quantum optical implementation of quantum information processing. Preprint, http://arxiv.org/abs/quant-ph/0405030, v1, 2004. See also Experimental Quantum Computation and Information, P. De Martini, C. Monroe (Eds.), IOS Press, Amsterdam, 2002.
[98] J. I. Cirac, P. Zoller, A scalable quantum computer with ions in an array of microtraps, Nature. 404 (2000) 579–581.
[99] J. F. Clauser, M. A. Horne, A. Shimony, R. A. Holt, Proposed experiment to test local hidden-variable theory, Phys. Rev. Lett. 23 (1969) 880.
[100] J. P. Clemens, S. Siddiqui, J. Gea-Banacloche, Quantum error correction against correlated noise, Phys. Rev. A, 69 (2004) 062313.
[101] R. Cleve, A. Ekert, L. Henderson, C. Macchiavello, M. Mosca, On quantum algorithms. Preprint, http://arxiv.org/abs/quant-ph/9903061 v1, 1999.
[102] G. Collins, Computing with quantum knots, Sci. Am. 296 (2006) 56–63.
[103] D. G. Cory, A. F. Fahmy, T. F. Havel, Nuclear magnetic resonance spectroscopy: An experimentally accessible paradigm for quantum computing, in: T. Toffoli, M. Biafore, J. Leao (Eds.), Proceedings of the PhysComp96, New England Complex Systems Institute, Cambridge, MA, 1996, pp. 87–91.

[104] D. G. Cory, A. F. Fahmy, T. F. Havel, Ensemble quantum computing by NMR spectroscopy, Proc. Natl. Acad. Sci. 94 (5) (1997) 1634–1639.

[105] D. G. Cory, M. D. Price, T. F. Havel, Nuclear magnetic resonance spectroscopy: an experimentally accessible paradigm for quantum computing, Physica D. 120 (1998) 82–101.

[106] T. M. Cover, J. A. Thomas, Elements of Information Theory, Wiley, New York, 1991.

[107] A. W. Cross, D. P. DiVincenzo, B. M. Terhal, A comparative study for quantum fault tolerance. Preprint, http://arxiv.org/abs/0711.1556, 2007.

[108] W. van Dam, A universal quantum cellular automaton, in: T. Toffoli, M. Biafore, J. Leao (Eds.), Proceedings of the Worshop on Physics and Computation (PhysComp96), New England Complex Systems Institute, Cambridge, MA, 1996, pp. 323–331.

[109] K. Davidson, C^*—Algebras by Example, Fields Institute Monographs, American Math. Society, Providence, RI, 1996.

[110] E. B. Davies, Information and quantum measurement, IEEE Trans. Inf. Theory IT. 24 (5) (1978) 596–599.

[111] C. M. Dawson, M. A. Nielsen, The Solovay-Kitaev algorithm. Preprint, http://arxiv.org/abs/quant-ph/0505030, 2005.

[112] B. DeMarco, A. Ben-Kish, D. Leibfried, V. Meyer, M. Rowe, B. M. Jelencović, et al, Experimental demonstration of a controlled-NOT wave-packet gate, Phys. Rev. Lett. 89 (26) (2002) 267901.

[113] R. de-Picciotto, M. Reznikov, M. Heiblum, V. Umansky, G. Bunin, D. Mahalu, Direct observation of a fractional charge, Nature. 389 (1997) 162–164.

[114] D. Deutsch, Quantum theory, the Church-Turing principle and the universal quantum computer, Proc. R. Soc. Lond. A, 400 (1985) 97–117.

[115] D. Deutsch, Quantum computational networks, Proc. R. Soc. Lond. A, 425 (1989) 73–90.

[116] D. Deutsch, R. Jozsa, Rapid solution of problems by quantum Computations, Proc. R. Soc. Lond. A, 439 (1992) 553–558.

[117] D. Deutsch, A. Barenco, A. Ekert, Universality in quantum computation, Proc. R. Soc. Lond. A, 449 (1995) 669–677.

[118] D. Deutsch, A. Ekert, R. Jozsa, C. Macchiavello, S. Popescu, A. Sanpera, Quantum privacy amplification and the security of quantum cryptography over noisy channels, Phys. Rev. Lett. 80 (1996) 2818–2821. Erratum: quantum privacy amplification and the security of quantum cryptography over noisy channels, Phys. Rev. Lett. 80 (1998) 2022.

[119] S. J. Devitt, A. D. Greentree, R. Ionicioiu, J. L. Q'Brien, W. J. Munro, L. C. L. Hollenberg, The photonic module: An on-demand resource fro photonic entanglement, Phys. Rev. A, 76 (2007) 052312.

[120] R. G. DeVoe, Elliptical ion traps and trap arrays for quantum computation, Phys. Rev. A, 58 (2) (1998) 910–914.

[121] D. Dhar, L. L. Grover, S. M. Roy, Preserving quantum states using inverting pulses: a super-Zeno effect, Phys. Rev. Lett. 96 (10) (2006) 100405.

[122] F. Diedrich, J. C. Bergquist, W. M. Itano, D. J. Wineland, Laser cooling to the zero-point energy of motion, Phys. Rev. Lett. 62 (4) (1989) 403–406.

[123] I. Dimov, B. I. Halperin, C. Nayak, Spin order in paired quantum Hall states, Phys. Rev. Lett. 100 (12) (2008) 126804.

[124] P. A. M. Dirac, Theory of electrons and positrons. Nobel Lecture, December 12, 1933. Also: http://nobelprize.org/nobel_prizes/physics/laureates/1933/dirac-lecture.pdf.

[125] P. A. M. Dirac, The Principles of Quantum Mechanics, fourth ed., Sec. 2, 4–7, Clarendon Press, Oxford, 1967.

[126] D. P. DiVincenzo, Quantum computation, Science. 270 (1995) 255–261.

[127] D. P. DiVincenzo, Two-bit gates are universal for quantum computation, Phys. Rev. A, 51 (1995) 1015–1022.

[128] D. P. DiVincenzo, P. W. Shor, Fault-tolerant error correction with efficient quantum codes, Phys. Rev. Lett. 77 (1996) 3260–3263.

[129] D. P. DiVincenzo, Quantum gates and circuits, Philos. Trans. R. Soc. Lond. A 454 (1998) 261–276. Also: Proceedings Mathematical, Physical and Engineering Sciences, Vol. 454, No. 1969, Quantum Coherence and Decoherence (January 1998), pp. 261–276. Also: Preprint, http://arxiv.org/abs/quanth-ph/9705009, May, 1997.

[130] D. P. DiVincenzo, The physical implementation of quantum computation, Fortschr. Phys. 48 (9–11) (2000) 771–783.

[131] M. Dolev, M. Heilblum, V. Umansky, A. Stern, D. Mahalu, Towards identification of a non-Abelian state: observation of a quarter of electron charge at $\nu = 5/2$ quantum state, Preprint, http://arxiv.org/abs/0802.0930, 2008.

[132] A. Einstein, Elektrodynamik bewegter Körper, Ann. Phys. 891 (1905) 921.

[133] A. Einstein, B. Podolsky, N. Rosen, Can quantum-mechanical description of physical reality be considered complete? Phys. Rev. 47 (1935) 777.

[134] A. Einstein, Autobiographycal notes, in: P. A. Schilpp (Ed.), Albert Einstein: Philosopher-Scientist, Open Court, La Salle, IL, 1970, pp. 1–96.

[135] J. P. Eisenstein, K. B. Cooper, L. N. Pfeiffer, K. W. West, Insulating and fractional quantum Hall states in the first excited Landau level, Phys. Rev. Lett. 88 (7) (2002) 076801.

[136] A. K. Ekert, B. Huttner, G. M. Palma, A. Peres, Eavesdropping on quantum-cryptographical systems, Phys. Rev. A, 50 (1996) 1047–1057.

[137] A. K. Ekert, C. Machiavello, Quantum error-correction for communication, Phys. Rev. Lett. 77 (1996) 2585–2588.

[138] A. K. Ekert, R. Jozsa, Quantum algorithms: entanglement enhanced information processing, Proc. R. Soc. Lond. A 356 (1743) (1998) 1769–1782.

[139] A. K. Ekert, P. Hayden, H. Inamori, Basic concepts in quantum computing. Preprint, http://arxiv.org/abs/quant-ph/0011013 v1, 2000.

[140] P. Elias, Coding for noisy channels, IRE Conv. Rec. 4 (1955) 37–46.

[141] J. M. Elzerman, R. Hanson, J. S. Greidanus, L. H. Willems van Beveren, S. De Francaschi, L. M. K. Vandersypen, et al, Tunable few-electron double quantum dots with integrated charge read-out, Physica E. 25 (2004) 135–141.

[142] J. M. Elzerman, R. Hanson, L. H. Willems van Beveren, B. Witkamp, L. M. K. Vandersypen, L. P. Kouwenhoven, Single-shot read-out of an individual electron spin in a quantum dot, Nature. 430 (2004) 431–435.

[143] R. M. Fano, Transmission of Information: A Statistical Theory of Communication, Wiley, New York, 1961.

[144] R. M. Fano, A Heuristic discussion of probabilistic decoding, IEEE Trans. Inf. Theory IT. 9 (1963) 64–74.

[145] E. Farhi, J. Goldstone, S. Gutmann, M. Sipser, Quantum computation by adiabatic evolution, Preprint, http://arxiv.org/abs/quant-ph/0001106, v1, 2000.

[146] A. Feinstein, Foundation of Information Theory, McGraw-Hill, New York, 1958.

[147] W. Feller, An Introduction to Probability Theory and Its Applications, Wiley, New York, 1957.

[148] R. P. Feynman, R. B. Leighton, M. Sands, The Feynman Lectures on Physics, Volumes 1, 2, and 3, Addison-Wesley, Reading, MA, 1977.

[149] R. P. Feynman, Simulating physics with computers, Int. J. Theor. Phys. 21 (1982) 467–488.

[150] R. P. Feynman, Quantum mechanical computers, Found Phys. 16 (1986) 507–531.

[151] R. P. Feynman, QED: The Strange Theory of Light and Matter, Princeton University Press, Princeton, NJ, 1985.

[152] R. P. Feynman, Lectures on Computation, Addison-Wesley, Reading, MA, 1996.

[153] D. L. Forney, Concatenated Codes, MIT Press, Cambridge, MA, 1966.

[154] D. L. Forney, Performance and complexity, 1995 Shannon Lecture, Proceedings of the IEEE International Symposium on Information Theory, IEEE Press, Los Alamitos, CA, 1995.

[155] G. D. Forney, M. Grassl, S. Guha, Convolutional and tail-bitting quantum error-correcting codes. Preprint, http://arxiv.org/abs/quant-ph/0511016, v1, 2005.

[156] E. Fredkin, Digital machines: an informational process based on reversible universal cellular automata, Physica D. 45 (1990) 254–270.

[157] M. H. Freedman, M. Larsen, Z. Wang, A modular functor which is universal for quantum computation. Preprint, http://arxiv.org/abs/quant-ph/0001108, v2, 2000.

[158] M. H. Freedman, A. Kitaev, Z. Wang, Simulation of topological field theories by quantum computers, Commun. Math. Phys. 227 (2000) 587–603.

[159] M. Freedman, C. Nayak, K. Walker, Towards universal topological quantum computation in the $\nu = \frac{5}{2}$ fractional quantum Hall state, Phys. Rev. B, 73 (24) (2006) 245307.

[160] H. Fritzsch, An Equation That Changed the World, University of Chicago Press, Chicago, IL, 2004.

[161] C. A. Fuchs, Distinguishability and accessible information in quantum theory. Preprint, http://arxiv.org/abs/quant-ph/9601020, v1, 1996.

[162] C. A. Fuchs, Nonorthogonal quantum states maximize classical channel capacity. Preprint, http://arxiv.org/abs/quant-ph/9703043, v1, 1997.

[163] W. Fulton, J. Harris, Representation Theory: A First Course, third ed., Springer-Verlag, Heidelberg, 1997.

[164] B. Furrow, A panoply of quantum algorithms. Preprint, http://arxiv.org/abs/quant-ph/0606127, 2006.

[165] P. Galison, Einstein's Clocks, Poincare's Maps, W. W. Norton & Co., London, 2003.

[166] R. G. Gallagher, A simple derivation of the coding theorem and some applications, IEEE Trans. Inf. Theory IT. 11 (1965) 3–18.

[167] R. G. Gallagher, Information Theory and Reliable Communication, Wiley, New York, 1968.

[168] S. Gasparoni, J.-W. Pan, P. Walter, T. Rudolph, A. Zeilinger, Realization of a photonic controlled-NOT gate sufficient for quantum computation, Phys. Rev. Lett. 93 (2) (2004) 020504.

[169] I. M. Gelfand, Lectures on Linear Algebra, Dover Publications, New York, NY, 1989.

[170] N. A. Gershenfeld, I. L. Chuang, Bulk spin-resonance quantum computation, Science. 275 (5298) (1997) 350–356.

[171] A. M. Gleason, Measures on a close subspace of a Hilbert space, J. Math. Mech. 6 (1957) 885–893.

[172] D. Gorenstein, N. Zierler, A class of error-correcting codes in p^m symbols, J. Soc. Ind. Appl. Math. (SIAM) 9 (1961) 207–214.

[173] D. Gottesman, Stabilizer codes and quantum error correction, Ph.D. Thesis, California Institute of Technology. Preprint, http://arxiv.org/abs/quant-ph/9705052, v1 May 1997.

[174] D. Gottesman, Theory of fault-tolerant computation, Phys. Rev. A 57 (1998) 127.

[175] D. Gottesman, I. L. Chuang, Demonstrating the viability of universal quantum computation using teleportaion and single-qubit operations, Nature. (London) 402 (1999) 390–393.

[176] D. Gottesman, An introduction to quantum error correction, in: S. J. Lomonaco (Ed.), Quantum Computation: A Grand Challenge for the 21st Century and the Millenium, Am. Mat. Soc., Providence, RI, 2002, pp. 221–235. Also Preprint, http://arxiv.org/abs/quant-ph/0004072, v1, 2000.

[177] R. M. Gray, Entropy and Information Theory, Springer Verlag, Heidelberg, 1990.

[178] M. Grassl, W. Gieselman, T. Beths, Quantum Reed-Solomon codes, in: Applied Algebra, Algebraic Logarithms and Error-Correcting Codes, Lecture Notes in Computer Science, Springer Verlag, Heidelberg, vol. 1719, 1999, pp. 231–244.

[179] I. S. Gradshteyn, I. M. Ryzhik, Table of Integrals, Series, and Products, Academic Press, Orlando, FL, 1980.

[180] P. Grangier, G. Roger, A. Aspect, Experimental evidence for a photon anticorrelation effect on a beam splitter: a new light on single-photon interferences, Europhys. Lett. 1 (4) (1986) 173–179.

[181] R. B. Griffiths, C.-S. Niu, Semiclassical Fourier transform for quantum computation, Phys. Rev. 76 (1996) 3228–3231.

[182] L. K. Grover, A fast quantum algorithm for database search, in: Proceedings of the ACM Symposium on Theory of Computing, ACM Press, New York, 1996, pp. 212–219. Also updated version: Preprint, http://arxiv.org/abs/quant-ph/9605043, 1996.
[183] L. K. Grover, Quantum mechanics helps in searching for a needle in a haystack, Phys. Rev. Lett. 78 (1997) 325–328.
[184] L. K. Grover, A framework for fast quantum mechanical algorithms, in: Proceedings of the Symposium on Theory of Computing, ACM Press, New York, 1998, pp. 53–62.
[185] L. K. Grover, Quantum computers can search rapidly using almost any transformation, in: Proceedings of the Symposium on Theory of Computing, ACM Press, New York, 1998, 53–62.
[186] L. K. Grover, Searching with quantum computers, Phys. Rev. Lett. 80 (19) (1998) 4329–4332.
[187] L. K. Grover, From Schrödinger's equation to the quantum search algorithm, Am. J. Phys. 69 (7) (2001) 769–777.
[188] L. K. Grover, A. M. Sengupta, Classical analog of quantum search, Phys. Rev. A, 65 (3) (2002) 032319.
[189] L. K. Grover, Fixed-point quantum search, Phys. Rev. Lett. 95 (15) (2005) 150501.
[190] L. K. Grover, A different kind of quantum search. Preprint, http://arxiv.org/abs/quant-ph/0503205, v1, 2005.
[191] S. Guiasu, Information Theory and Applications, McGraw-Hill, New York, 1976.
[192] H. Häffner, W. Hänsel, C. F. Roos, J. Benhelm, D. Chek-al-kar, M. Chwalla, et al, Scalable multiparticle entanglement of trapped ions, Nature 438 (2005) 643–646.
[193] F. D. M. Haldane, Fractional quantization of the Hall effect: a hierarchy of incompressible quantum fluid states, Phys. Rev. Lett. 51 (7) (1983) 605–608.
[194] P. C. Haljan, P. J. Lee, K-A. Brickman, M. Acton, L. Deslauriers, C. Monroe, Entanglement of trapped-ion clock states, Phys. Rev. A, 72 (2005) 062316.
[195] E. H. Hall, On the new action of the magnet on a electric currents, Am. J. Sci. 19 (3) (1880) 200–205; Philos. Mag. 9 (5) (1880) 225–230.
[196] E. H. Hall, On the new action of magnetism on a permanent electric current, Am. J. Sci. 20 (3) (1880) 161–186; Philos. Mag. 10 (5) (1880) 301–328.
[197] B. I. Halperin, Quantized Hall conductance, current-carrying edge states, and the existence of extended states in a two-dimensional disordered potential, Phys. Rev. B, 25 (4) (1982) 2185–2190.
[198] B. I. Halperin, Statistics of quasiparticles and the hierarchy of fractional quantized Hall states, Phys. Rev. Lett. 52 (18) (1984) 1583–1586.
[199] R. W. Hamming, Error detecting and error correcting codes, Bell Syst. Tech. J. 29 (1950) 147–160.
[200] L. Hardy, Why is nature described by quantum theory, in: J. D. Barrow, P. C. W. Davis, C. L. Harper (Eds.), Science and Ultimate Reality. Quantum Theory, Cosmology and Complexity, Cambridge University Press, Cambridge, UK, 2004, pp. 45–71. Also Quantum theory from five reasonable axioms. Preprint, http://arxiv.org/abs/quant-ph/0101012, v4, 2001.
[201] Y. Hardy, W.-H. Steeb, Classical and Quantum Computing, Birkhäuser, Boston, MA, 2001.
[202] A. W. Harrow, B. Recht, I. L. Chuang, Efficient discrete approximations of quantum gates, J. Math. Phys. 43 (2002) 4445–4450.
[203] M. B. Hastings, Superadditivity of communication capacity using entangled inputs, Nat. Phys. 5 (2009) 255–257.
[204] P. Hausladen, R. Jozsa, B. Schumacher, M. Westmoreland, W. K. Wootters, Classical information capacity of a quantum channel, Phys. Rev. A, 54 (1) (1996) 1869–1876.
[205] M. Hayashi, H. Nagaoka, General formulas for capacity of classical-quantum channels, IEEE Trans. Inf. Theory. 49 (7) (2003) 1753–1768.
[206] W. Heisenberg, The development of quantum mechanics, Nobel Lectures, Physics 1922–1942, pp. 290–301, December 11, 1933. Also: http://nobelprize.org/nobel_prizes/physics/laureates/1932/heisenberg-lecture.pdf.

[207] K.-E. Hellwig, K. Kraus, Pure operations and measurements, Commun. Math. Phys. 11 (1969) 214–220. See also Operations and Measurements II. Commun. Math. Phys. 16 (1970) 142–147.
[208] A. Hocquenghem, Codes correcteurs d'erreurs, Chiffres (Paris) 2 (1959) 147–156.
[209] H. F. Hofmann, S. Takeuchi, Quantum phase gate for photonic qubits using only beam splitters and postselection, Phys. Rev. A 66 (2) (2001) 024308.
[210] A. S. Holevo, Statistical problems in quantum physics, in: G. Marayama, J. V. Prohorov (Eds.), proceedings of the Second Japan-URSS Symposium on Probability Theory, Lecture Notes in Mathematics, vol. 330, Springer Verlag, 104–119, 1973.
[211] A. S. Holevo, The capacity of quantum channel with general signal states, IEEE Trans. Inf. Theory. 44 (1998) 269–273.
[212] J. P. Home, M. J. McDonnell, D. M. Lucas, G. Imreh, B. C. Keitch, D. J. Szwer, et al, Deterministic entanglement and tomography of ion-spin qubits, New J. Phys. 8 (2006) 188.
[213] C. K. Hong, L. Mandel, Experimental realization of a localized one-photon state, Phys. Rev. Lett. 56(1) (1986) 58–60.
[214] P. Hoyer, On arbitrary phases in quantum amplitude purification, Phys. Rev. Lett. A, 62 (2000) 052304.
[215] R. J. Hughes, G. I. Morgan, C. G. Peterson, Quantum key distribution over a 48 km long optical fibre network, J. Mod. Opt. 47 (2000) 533–547.
[216] A. Imamoglu, D. D. Awschalom, G. Burkard, D. P. DiVincenzo, D. Loss, M. Sherwin, et al, Quantum information processing quantum dot spins and cavity QED, Phys. Rev. Lett. 83 (1999) 42–4.
[217] J. K. Jain, Composite-fermion approach for the fractional quantum Hall effect, Phys. Rev. Lett. 63 (2) (1989) 199–202.
[218] D. Jaksch, H.-J. Briegel, J. I. Cirac, P. Zoller, Entanglement of atoms via cold controlled collisions, Phys. Rev. Lett. 82 (9) (1999) 1975–1978.
[219] Y. Ji, Y. Chung, D. Sprinzak, M. Heilblum, D. Mahalu, H. Shtrikman, An electronic Mach-Zehnder interferometer, Nature. 422 (2003) 415–418.
[220] R. Jozsa, Fidelity for mixed quantum states, J. Mod. Opt. 41 (12) (1994) 2315–2323.
[221] R. Jozsa, B. Schumacher, A new proof of the quantum noiseless coding theorem, J. Mod. Opt. 41 (12) (1994) 2343–2349.
[222] R. Jozsa, Entanglement and quantum computation, in: S. A. Huggett, L. J. Mason, K. P. Tod, S. T. Tsou, N. M. J. Woodhouse (Eds.), Geometric Universe: Science, Geometry and the Work of Roger Penrose, Oxford University Press, New York, 1998, pp. 369–379.
[223] R. Jozsa, Quantum algorithms and the Fourier transform. Preprint, http://arxiv.org/abs/quant-ph/9707033, v1, 1997.
[224] R. Jozsa, Searching in Grover's algorithm. Preprint, http://arxiv.org/abs/quant-ph/9901021, v1, 1999.
[225] R. Jozsa, Quantum factoring, discrete logarithms, and the hidden subgroup problem. Preprint, http://arxiv.org/quant-ph/0012084 v1, 2000.
[226] R. Jozsa, Illustrating the concept of quantum information. Preprint, http://arxiv.org/abs/quant-ph/0305114, v1, 2003.
[227] J. Justensen, T. Hoholdt, A Course in Error Correcting Codes, European Mathematical Society, Zuerich, Switzerland, 2004.
[228] B. E. Kane, Silicon-based nuclear spin quantum computer, Nature. 393 (1998) 133–137.
[229] B. E. Kane, Silicon-based quantum computation. Preprint, http://arxiv.org/abs/quant-ph/0003031, v1, 2000.
[230] M. Keyl, Fundamentals of quantum information. Preprint, http://arxiv.org/abs/quant-ph/0202122 v1, 2002.
[231] A. Ya. Khinchin, Mathematical Foundations of Information Theory, Dover Publications, Mineola, NY, 1957.
[232] D. Kielpinski, Entanglement and Decoherence in a Trapped-Ion Quantum Register, Ph.D. Thesis, University of Colorado, Boulder, 2001.

[233] D. Kielpinski, A. Ben-Kish, J. Britton, V. Meyer, M. A. Rowe, C. A. Sackett, et al, Recent results in trapped-ion quantum computing. Preprint, http://arxiv.org/abs/quant-ph/0102086, v1, 2001.
[234] D. Kielpinski, V. Meyer, M. A. Rowe, C. A. Sackett, W. M. Itano, C. Monroe, et al, A decoherence-free quantum memory using trapped ions, Science. 291 (5506) (2001) 1013–1015.
[235] D. Kielpinski, C. Monroe, D. J. Wineland, Architecture for a large-scale ion-trap quantum computer, Nature. 417 (2002) 709–711.
[236] J. M. Kikkawa, D. D. Awschalom, Resonant spin amplification in n-type GaAs, Phys. Rev. Lett. 80 (19) (1998) 4313–4316.
[237] B. E. King, C. S. Wood, C. J. Myatt, Q. A. Turchette, D. Leibfried, W. M. Itano, et al, Cooling the collective motion of trapped ions to initialize a quantum register, Phys. Rev. Lett. 81 (1998) 1525–1528.
[238] A. Yu. Kitaev, Quantum measurements and the Abelian stabilizer problem. Preprint, http://arxiv.org/abs/quant-ph/9511026, v1, 1995.
[239] A. Yu. Kitaev, Quantum computations and error correction, Russ. Math. Surv. 52 (1997) 1191–1249.
[240] A. Yu. Kitaev, Fault-tolerant quantum computations by anions, Ann. Phys. 303 (2003) 2–30.
[241] R. Klesse, S. Frank, Quantum error correction in spatially correlated quantum noise, Phys. Rev. Lett. 95 (2005) 230503.
[242] K. von Klitzing, G. Dorda, M. Pepper, New method for high-accuracy determination of the fine-structure constant based on quantized Hall resistance, Phys. Rev. Lett. 45 (6) (1980) 494–497.
[243] E. Knill, R. Laflamme, W. H. Zurek, Threshold accuracy for quantum computation. Preprint, http://arxiv.org/abs/quant-ph/9610011, 1996.
[244] E. Knill, R. Laflamme, Concatenated quantum codes. Preprint, http://arxiv.org/abs/quant-ph/9608012, 1996.
[245] E. Knill, R. Laflamme, Theory of quantum error-correcting codes, Phys. Rev. A 55 (2) (1997) 900–911.
[246] E. Knill, R. Laflamme, W. H. Zurek, Resilient quantum computation: error models and thresholds, Proc. R. Soc. Lond. A 454 (1998) 365–384.
[247] E. Knill, R. Laflamme, W. H. Zurek, Resilient quantum computation, Science. 279 (1998) 342.
[248] E. Knill, I. Chuang, R. Laflamme, Effective pure states for bulk quantum computation, Phys. Rev. A, 57 (5) (1998) 3348–3363.
[249] E. Knill, R. Laflamme, L. Viola, Theory of quantum error correction for general noise, Phys. Rev. Lett. 84 (11) (2000) 2525–2528.
[250] E. Knill, R. Laflamme, G. J. Milburn, A scheme for efficient quantum computation with linear optics, Nature. 409 (2001) 46–52.
[251] E. Knill, R. Laflamme, H. Barnum, D. Dalvit, J. Dziarmaga, J. Gubernatis, et al, Quantum information processing, Kluwer Encyclopedia of Mathematics, Supplement III, 2002.
[252] M. Koashi, T. Yamamoto, N. Imoto, Probabilistic manipulation of entangled photons, Phys. Rev. A, 63 (3) (2001) 030301.
[253] A. N. Kolmogorov, Logical basis for information theory and probability theory, IEEE Trans. Inf. Theory IT. 14 (1968) 662–664.
[254] A. N. Kolmogorov, S. V. Fomin, Elements of the Theory of Functions and Functional Analysis, Dover Publications, Mineola, NY, 1999.
[255] L. P. Kouwenhoven, D. G. Austing, S. Tarucha, Few-electron quantum dots, Rep. Prog. Phys. 64 (6) (2001) 701–736.
[256] K. Kraus, States, Effects, and Operators: Fundamental Notions of Quantum Theory, Springer Verlag, Berlin, 1983.
[257] D. Kretschmann, D. Schlingemann, R. F. Werner, The information–disturbance tradeoff and the continuity of Stinespring's representation. Preprint, http://arxiv.org/abs/quant-ph/0605009, 2006.
[258] E. Kreyszig, Advanced Engineering Mathematics, Wiley, New York, 1998.

[259] D. W. Kribs, R. Laflamme, D. Poulin, A unified and generalized approach to quantum error correction, Phys. Rev. Lett. 94 (2005) 180–501.
[260] D. W. Kribs, R. Laflamme, D. Poulin, M. Lesosky, Operator quantum error correction. Preprint, http://arxiv.org/abs/quant-ph/0504189, v3, 2006.
[261] B. O. Küppers, Information and the Origin of Life, MIT Press, Cambridge, MA, 1990.
[262] T. D. Ladd, F. Jelezko, R. Laflamme, Y. Nakamura, C. Monroe, J. L. O'Brien, Quantum computers, Nature. 464 (2010) 4552.
[263] R. Laflamme, C. Miquel, J.-P. Paz, W. H. Zurek, Perfect quantum-error correcting code, Phys. Rev. Lett. 77 (1996) 198–201.
[264] R. Landauer, Irreversibility and heat generation in the computing process, IBM J. Res. Dev. 5 (1961) 182–192.
[265] R. B. Laughlin, Quantized Hall conductivity in two dimensions, Phys. Rev. B, 23 (10) (1981) 5632–5633.
[266] R. L. Laughlin, Quantized motion of three two-dimensional electrons in a strong magnetic field, Phys. Rev. B, 27 (6) (1983) 3383–3389.
[267] R. L. Laughlin, Anomalous quantum Hall effect: an incompressible quantum fluid with fractionally charged excitations, Phys. Rev. Lett. 50 (18) (1983) 1395–1398.
[268] J. M. Lee, Introduction to Smooth Manifolds, Springer Verlag, Heidelberg, 2002.
[269] D. Leibfried, B. DelMarco, V. Meyer, D. Lucas, M. Barrett, J. Britton, et al, Experimental demonstration of a robust, high-fidelity, geometric two ion-qubit phase gate, Nature. 422 (2003) 412–415.
[270] D. Leibfried, M. D. Barrett, T. Schaetz, J. Britton, J. Chiaverini, W. M. Itano, et al, Toward Heisenberg-limited spectroscopy with multiparticle entangled states, Science. 304 (2004) 1476–1478.
[271] D. Leibfried, E. Knill, S. Seidlin, J. Britton, R. B. Blakestad, J. Chiaverini, et al, Creation of a six-atom 'Schrödinger Cat' state, Nature. 438 (2005) 639–642.
[272] C. Livermore, C. H. Crouch, R. M. Westervelt, K. L. Campman, A. C. Grossard, The Coulomb blockade in coupled quantum dots, Science. 274 (5291) (1996) 1332–1335.
[273] S. Lloyd, A potentially realizable quantum computer, Science. 261 (1993) 1569–1571.
[274] S. Lloyd, Almost any quantum logic gate is universal, Phys. Rev. Lett. 75 (1995) 346–349.
[275] S. Lloyd, Capacity of a noisy communication channel, Phys. Rev. A, 55 (1997) 1613–1622.
[276] D. Loss, D. P. DiVincenzo, Quantum computation with quantum dots, Phys. Rev. A, 57 (1) (1998) 120–126.
[277] F. Lu, D. C. Marinescu, An $R \,||\, C_{max}$ quantum scheduling algorithm, Quantum Inf. Process. 6 (3) (2007) 159–178.
[278] F. Lu, D. C. Marinescu, Quantum error correction of time-correlated errors, Quantum Inf. Process. 6 (4) (2007) 273–293.
[279] D. J. C. MacKay, Information Theory, Inference, and Learning Algorithms, Cambridge University Press, Cambridge, UK, 2003.
[280] F. J. MacWilliams, N. J. A. Sloane, The Theory of Error Correcting Codes, North Holland, Amsterdam, 1983.
[281] O. Mandel, M. Greiner, A. Widera, T. Rom, T. W. Hänsch, I. Bloch, Controlled collisions for multi-particle entanglement of optically trapped atoms, Nature. 425 (6961) (2003) 937–940.
[282] Y. Manin, Classical computing, quantum computing, and Shor's algorithm, Talk at the Bourbaki seminar, June 1999. Preprint, http://arxiv.org/abs/quant-ph/9903008, v1, 1999.
[283] D. C. Marinescu, G. M. Marinescu, Approaching Quantum Computing, Prentice Hall, Upper Saddle River, NJ, 2004.
[284] D. C. Marinescu, G. M. Marinescu, Quantum information: a glimpse at the strange and intriguing future of information, Comput. J. 50 (5) (2007) 505–521.
[285] S. McCartney, ENIAC, The Triumphs and Tragedies of the World's First Computer, Walker and Company Publishing House, New York, 1999.

[286] B. Mc.Millan, The basic theorems of information theory, Ann. Math. Stat. 24 (1953) 196–212.
[287] Z. Meglicki, Quantum Computing with Magic, MIT Press, Cambridge, MA, 2008.
[288] N. D. Mermin, The topologival theory of defects in ordered media, Rev. Mod. Phys. 51 (3) (1979) 591–648.
[289] N. D. Mermin, What's wrong with these elements of reality?, Phys. Today. 43 (6) (1990) 9–11.
[290] N. D. Mermin, Quantum Computer Science: An Introduction, Cambridge University Press, Cambridge, England, 2007.
[291] E. Mertzbacher, Quantum Mechanics, third ed., Wiley, New York, 1998.
[292] V. Meyer, M. A. Rowe, D. Kielpinski, C. A. Sackett, W. M. Itano, C. Monroe, et al, Experimental demonstration of entanglement-enhanced rotation angle estimation using trapped ions, Phys. Rev. Lett. 86 (26) (2001) 5870–5873.
[293] G. J. Milburn, Quantum optical Fredkin gate, Phys. Rev. Lett. 62 (1988) 2124–2127.
[294] A. Mizel, D. A. Lidar, M. Mitchell, Simple proof of equivalence between adiabatic quantum computation and the circuit model, Phys. Rev. Lett. 99 (7) (2007) 070502.
[295] C. Mochon, Anyons from non-solvable finite groups Are sufficient for universal quantum computation. Preprint, http://arxiv.org/abs/quant-ph/0206128, 2003.
[296] K. Mølmer, A. Sørensen, Multiparticle entanglement of hot trapped ions, Phys. Rev. Lett. 82 (9) (1999) 1835–1838.
[297] C. Monroe, D. M. Meekhof, B. E. King, W. M. Itano, D. J. Wineland, Demonstration of a fundamental quantum logic gate, Phys. Rev. Lett. 74 (25) (1995) 4714–4718.
[298] C. Monroe, D. Leibfried, B. E. King, D. M. Meekhof, W. M. Itano, D. J. Wineland, Simplified quantum logic with trapped ions, Phys. Rev. A 55 (1997) 2489–2491.
[299] G. Moore, N. Read, Nonabelions in the fractional quantum Hall effect, Nucl. Phys. B, 360 (2–3) (1991) 362–369.
[300] R. H. Morf, Transition from quantum Hall to compressible states in the second Landau level, Phys. Rev. Lett. 80 (7) (1998) 1505–1508.
[301] M. Mosca, A. Ekert, The hidden subgroup problem and eigenvalue estimation on a quantum computer. Preprint, http://arxiv.org/abs/quant-ph/9903071, v1, 1999.
[302] D. E. Muller, Application of Boolean algebra to switching circuit design and to error detection, IEEE Trans. Comput. 3 (1954) 612.
[303] C. Nayak, F. Wilczek, 2n-Quasihole states realize 2^{n-1}-dimensional spinor braiding statistics in paired quantum Hall states, Nucl. Phys. B, 479 (1996) 529–553.
[304] C. Nayak, S. H. Simon, A. Stern, M. Freedman, S. Das Sarma, Non-abelian anyons and topological quantum computation, Rev. Mod. Phys. 80 (2008) 1083–1159.
[305] M. A. Nielsen, I. L. Chuang, Programmable quantum gate arrays, Phys. Rev. Lett. 79 (1997) 321.
[306] M. A. Nielsen, Quantum Information Theory, Ph.D. Thesis, University of New Mexico. Preprint, http://arxiv.org/abs/quant-ph/0011036, v6, 2000.
[307] M. A. Nielsen, I. L. Chuang, Quantum Computing and Quantum Information, Cambridge University Press, Cambridge, UK, 2000.
[308] M. A. Nielsen, Quantum computation by measurement and quantum memory, Phys. Lett. A, 308 (2–3) (2003) 96–100.
[309] M. A. Nielsen, Optical quantum computation using cluster states. Preprint, http://arxiv.org/abs/quant-ph/0402005, v1, 2004.
[310] E. Novais, H. U. Baranger, Decoherence by correlated noise and quantum error correction, Phys. Rev. Lett. 97 (2006) 040501.
[311] J. L. O'Brien, G. J. Pryde, A. Gilchrist, D. F. V. James, N. K. Langford, T. C. Ralph, et al, Quantum process tomography of a controlled-NOT gate, Phys. Rev. Lett. 93 (8) (2004) 080502.
[312] J. L. O'Brien, G. J. Pryde, A. G. White, T. C. Ralph, High-fidelity Z-measurement error encodeing of optical qubits, Phys. Rev. A, 71 (6) (2005) 060303.

[313] B. W. Ogburn, J. Preskill, Topological quantum computation, in: Quantum Computing and Quantum Communication, Lecture Notes in Computer Science, vol. 1509, Springer-Verlag, Heidelberg, 1999, pp. 341–359.

[314] H. Ollivier, J-P. Tillich, Description of a quantum convolutional code. Preprint, http://arxiv.org/abs/quant-ph/0304189, v2, 2003.

[315] S. Olmschenk, D. N. Matsukevich, P. Maunz, D. Hayes, L.-M. Duan, C. Monroe, Quantum teleportation between distant matter qubits, Science. 323 (5913) (2009) 486–489.

[316] R. Omnès, The Interpretation of Quantum Mechanics, Princeton University Press, Princeton, NJ, 1994.

[317] G. M. Palma, K.-A. Suominen, A. Ekert, Quantum computers and dissipation, Proc. R. Soc. A, 452 (1996) 567.

[318] J. W. Pan, D. Bouwmeester, M. Daniell, H. Weinfurter, A. Zeilinger, Experimental test of quantum nonlocality in three-photon Greenberger-Horne-Zeilinger entanglement, Nature. 403 (6769) (2000) 515–519.

[319] C. M. Papadimitriou, Computational Complexity, Addison-Wesley, Reading, MA, 1994.

[320] W. Pauli, Exclusion principle and quantum mechanics. Nobel Lectures, Physics 1942–1962, pp. 27–43, December 13, 1946. Also: http://nobelprize.org/nobel_prizes/physics/laureates/1945/pauli-lecture.pdf.

[321] W. Peterson, E. J. Weldon, Error Correcting Codes, second ed., MIT Press, Cambridge, MA, 1972.

[322] T. Pellizzari, S. A. Gardiner, J. I. Cirac, P. Zoller, Decoherence, continuous oservation, and quantum computing: a cavity QED model, Phys. Rev. Lett. 75 (21) (1995) 3788–3791.

[323] R. Penrose, The Road to Reality: A Complete Guide to the Laws of the Universe, Vintage Books, London, 2007.

[324] A. Peres, W. K. Wootters, Optimal detection of quantum information, Phys. Rev. Lett. 66 (9) (1991) 1119–1122.

[325] A. Peres, Quantum Theory: Concepts and Methods, Kluwer Academic Press, Boston, MA, 1995.

[326] M. R. Peterson, T. Jolicoeur, S. Das Sarma, Finite layer thickness stabilizes the Pfaffian state for the 5/2 fractional quantum Hall effect: wavefunction overlap and topological degeneracy, Phys. Rev. Lett. 101 (1) (2008) 016807.

[327] C. A. Petri, Nets, time, and space, Theor. Comput. Sci. 153 (1996) 3–48.

[328] P. M. Petroff, A. Lorke, A. Imamoglu, Epitaxially self-assembled quantum dots, Phys. Today. 45 (5) (2001) 46–52.

[329] D. Petz, Entropy, von Neumann and the von Neumann entropy. Preprint, http://arxiv.org/abs/math-ph/0102013, v1, 2001.

[330] T. B. Pittman, B. C. Jacobs, J. D. Franson, Probabilistic quantum logic operations using polarizing beam splitters, Phys. Rev. A, 64 (6) (2001) 062311.

[331] T. B. Pittman, M. J. Fitch, B. C. Jacobs, J. D. Franson, Experimental controlled-NOT logic gate for single photons in the coincidence basis, Phys. Rev. A, 68 (3) (2003) 032316.

[332] T. B. Pittman, B. C. Jacobs, J. D. Franson, Demonstration of quantum error correction using linear optics, Phys. Rev. A, 71 (5) (2005) 052332.

[333] M. K. E. L. Planck, The genesis and present state of development of the quantum theory. Nobel Lectures, Physics 1901–1922. June 2, 1920. Also: http://nobelprize.org/nobel_prizes/physics/laureates/1918/planck-lecture.pdf.

[334] M. B. Plenio, V. Vitelli, The physics of forgetting: Landauer's erasure principle and information theory. Preprint, http://arxiv.org/abs/quant-ph/0103108, v1, 2001.

[335] S. Popescu, Bell's inequalities and density matrices. Revealing 'Hidden' Nonlocality. Preprint, http://arxiv.org/abs/quant-ph/9502005, v1, 1995.

[336] A. O. Pittinger, An Introduction to Quantum Algorithms, Birkhäuser, Boston, MA, 1999.

[337] D. Poulin, Stabilizer formalism for operator quantum error correction, Phys. Rev. Lett. 95 (2005) 230504.

[338] J. F. Poyatos, J. I. Cirac, P. Zoller, Quantum gates with 'hot' trapped ions, Phys. Rev. Lett. 81 (6) (1998) 1322–1325.
[339] J. F. Poyatos, J. I. Cirac, P. Zoller, Complete characterization of a quantum process: the two-bit quantum gate, Phys. Rev. Lett. 78 (2) (1997) 390–393.
[340] J. F. Poyatos, J. I. Cirac, P. Zoller, Schemes of quantum computations with trapped ions, Fortschr. Phys. 48 (9–11) (2000) 785–799.
[341] J. Preskill, Fault tolerant quantum computation. Preprint, http://arxiv.org/abs/quant-ph/9712048, v1, 1997.
[342] J. Preskill, Lecture Notes for Physics 229: Quantum Information and Computing, California Institute of Technology, Pasadena, CA, 1998.
[343] J. Preskill, Quantum clock synchronization and quantum error correction. Preprint, http://arxiv.org/abs/quant-ph/0010098, v1, 2000.
[344] J. Preskill, Topological quantum computation, Chapter 9 of Class Notes on Quantum Computing. http://www.theory.caltech.edu/~preskill/ph219/topological.pdf, 2004.
[345] R. Prevedel, P. Walter, F. Tiefenbach, P. Böhi, R. Kaltenbaek, T. Jennewein, et al, High-speed linear optics quantum computing using active feed-forward, Nature. 445 (2007) 65–69.
[346] E. M. Purcell, H. C. Torrey, R. V. Pound, Resonance absorption by nuclear magnetic moments in a solid, Phys. Rev. 69 (1946) 37.
[347] I. P. Radu, J. B. Miller, C. M. Marcus, M. A. Kastner, L. N. Pfeiffer, K. W. West, Quasiparticle tunneling in the fractional quantum Hall state at $\nu = 5/2$. Preprint, http://arxiv.org/abs/0803.3530, v1, 2008.
[348] B. Rahn, A. C. Doherty, H. Mabuchi, Exact performance of concatenated quantum codes. Preprint, http://arxiv.org/abs/quant-ph/0206061, v1, 2002.
[349] R. Raussendorf, H. J. Briegel, One-way quantum computer, Phys. Rev. Lett. 86 (22) (2001) 5188–5191.
[350] R. Raussendorf, D. E. Browne, H. J. Briegel, Measurement-based quantum computation on cluster states, Phys. Rev. A, 68 (2) (2003) 022312.
[351] N. Read, E. Rezayi, Quasiholes and fermionic zero modes of paired fractional quantum Hall states: the mechanism for non-abelian statistics, Phys. Rev. B, 54 (23) (1999) 16864.
[352] I. S. Reed, A class of multiple-error-correcting codes and the decoding scheme, IEEE Trans. Inf. Theory. 4 (1954) 3849.
[353] I. S. Reed, G. Solomon, Polynomial codes over certain finite fields, SIAM J. Appl. Math. 8 (1960) 300–304.
[354] M. Reed, B. Simon, Methods of Modern Mathematical Physics, vol. 1: Functional Analysis, Academic Press, New York, 1972.
[355] B. W. Reichardt, L. K. Grover, Quantum error correction of systematic errors using a quantum search framework, Phys. Rev. A, 72 (4) (2005) 042326.
[356] E. H. Rezayi, F. D. M. Haldane, Incompressible paired Hall state, stripe order, and the composite fermion liquid phase in half-filled Landau levels, Phys. Rev. Lett. 84 (20) (2000) 4685–4688.
[357] M. Riebe, H. Häffner, C. F. Roos, W. Hänsel, J. Benhelm, G. P. T. Lancaster, et al, Deterministic quantum teleportation with atoms, Nature. 429 (2004) 734–737.
[358] M. Riebe, K. Kim, P. Schindler, T. Monz, P. O. Schmidt, T. K. Kärber, et al, Process tomography of ion trap quantum gates, Phys. Rev. Lett. 97 (2006) 220407.
[359] E. Rieffel, W. Polak, An introduction to quantum computing for non-physicists, ACM Comput. Surv. 32 (3) (2000) 300–335.
[360] C. Roos, T. Zeigler, H. Rohde, H. C. Nägerl, J. Eschner, D. Leibfried, et al, Quantum state engineering on an optical transition and decoherence in a Paul trap, Phys. Rev. Lett. 83 (23) (1999) 4713–4716.
[361] P. Roulleau, F. Portier, P. Roche, A. Cavanna, G. Faini, U. Gennser, et al, Direct measurement of the coherence length of edge states in the integer quantum Hall regime, Phys. Rev. Lett. 100 (2008) 126802.
[362] M. A. Rowe, D. Kielpinski, V. Meyer, C. A. Sackett, W. M. Itano, C. Monroe, et al, Experimental violation of a Bell's inequality with efficient detection, Nature. 409 (6822) (2001) 791–794.

[363] C. A. Sackett, D. Kielpinski, B. E. King, C. Langer, V. Meyer, C. J. Myatt, et al, Experimental entanglement of four particles, Nature. 404 (2000) 256–259.

[364] F. Schmidt-Kaler, C. Roos, H. C. Nägerl, H. Rohde, S. Gulde, A. Mundt, et al, Ground state cooling, quantum state engineering and study of decoherence of ions in Paul traps, J. Mod. Opt. 47 (14/15) (2000) 2573–2582.

[365] F. Schmidt-Kaler, H. Häffner, M. Riebe, S. Guide, G. P. T. Lancaster, T. Deuschle, et al, Realization of the Cirac-Zoller controlled-NOT quantum gate, Nature. 422 (2003) 408–411.

[366] E. Schrödinger, The fundamental idea of wave mechanics, Nobel Lectures, Physics 1922–1941, pp. 305–314. December 12, 1933. Also: http://nobelprize.org/nobel_prizes/physics/laureates/1933/schrodinger-lecture.pdf.

[367] E. Schrödinger, The present situation in quantum mechanics, Die Naturwissenschaften, 23 (1935) 807–812; 823–828; 944–849. Also Proc. Cambridge Philos. Soc. 31 (1936) 555–563, 1935 and 32:446–452.

[368] B. W. Schumacher, Quantum coding, Phys. Rev. A, 51 (4) (1995) 2738–2747.

[369] B. W. Schumacher, M. D. Westmoreland, W. K. Wootters, Limitations on the amount of accessible information in a quantum channel, Phys. Rev. Lett. 76 (1996) 3452–3455.

[370] B. W. Schumacher, Entropy exchange and coherent quantum information, in: Proceedings of the Fourth Workshop on Physics and Computation, New England Complex Systems Institute, Cambridge, MA, 1996, pp. 292–296

[371] B. W. Schumacher, M. A. Nielsen, Quantum data processing and error correction, Phys. Rev. A, 54 (4) (1996) 2629–2636.

[372] B. W. Schumacher, Sending quantum entanglement through noisy quantum channels, Phys. Rev. A, 54 (4) (1996) 2614–2628.

[373] B. W. Schumacher, M. D. Westmoreland, Sending quantum information via noisy quantum channels, Phys. Rev. A, 56 (1) (1997) 131–138.

[374] C. E. Shannon, A mathematical theory of communication, Bell Syst. Tech. J. 27 (1948) 379–423 & 623–656.

[375] C. E. Shannon, Communication in the presence of noise, Proc. IRE, 37 (1949) 10–21.

[376] C. E. Shannon, Certain results in coding theory for noisy channels, Inf. Control. 1 (1) (1957) 6–25.

[377] C. E. Shannon, W. Weaver, A Mathematical Theory of Communication, University of Illinois Press, Urbana, IL, 1963.

[378] M. S. Sherwin, A. Imamoglu, T. Montroy, Quantum computation with quantum dots and terahertz cavity quantum electrdynamics, Phys. Rev. A, 60 (5) (1999) 3508–3514.

[379] P. W. Shor, Algorithms for quantum computation: discrete log and factoring, Proceedings of the 35 Annual Symposium on Foundations of Computer Science, IEEE Press, Piscataway, NJ, 1994, pp. 124–134.

[380] P. W. Shor, Scheme for reducing decoherence in quantum computer memory, Phys. Rev. A, 52 (4) (1995) 2493–2496.

[381] P. W. Shor, Polynomial-time algorithms for prime factorization and discrete logarithms on a quantum computer. SIAM J. Comput. 26 (1997) 1484–1509. Also Preprint, http://arxiv.org/abs/quant-ph/9508027, v2, 1996.

[382] P. W. Shor, Fault-tolerant quantum computation, 37th Annual Symposium on Foundations of Computer Science, IEEE Press, Piscataway, NJ, 1996, pp. 56–65.

[383] P. W. Shor, J. A. Smolin, Quantum error-correcting codes need not completely reveal the error syndrome. Preprint, http://arxiv.org/abs/quant-ph/9604006, 1996.

[384] P. W. Shor, J. Preskill, Simple proof of security of the BB84 quantum key distribution protocol. Preprint, http://arxiv.org/abs/quant-ph/0003004, 2000.

[385] P. W. Shor, Introduction to quantum algorithms. Preprint, http://arxiv.org/abs/quant-ph/0005003, 2001.

[386] P. W. Shor, Why haven't more quantum algorithms been found? J. ACM. 50 (1) (2003) 87–90.

[387] P. W. Shor, Capacities of quantum channels and how to find them, Math. Program. 97 (2003) 311–335.
[388] P. W. Shor, Quantum channels, http://www-math.mit.edu/~shor/NewDirections/channels.pdf, 2005.
[389] P. W. Shor, Equivalence of additive questions in quantum information theory, Commun. Math. Phys. 246 (2004) 453–472.
[390] P. W. Shor, The bits don't add up, Nat. Phys. 5 (2009) 247–248.
[391] D. R. Simon, On the power of quantum computation, SIAM J. Comput. 26 (1997) 1474–1483.
[392] L. Sklar, Philosophy of Physics, Westview Press, Boulder, CO, 1992.
[393] B. Sklar, A primer on turbo code concepts, IEEE Comm. Mag. 35 (4) (1997) 94–102.
[394] B. Sklar, Digital Communication: Fundamentals and Applications, second ed., Prentice Hall, Upper Saddle River, NJ, 2001.
[395] T. Sleator, H. Weinfurter, Realizable universal quantum logic gates, Phys. Rev. Lett. 74 (1995) 4087–4090.
[396] D. Slepian, J. K. Wolf, Noiseless coding of correlated information sources, IEEE Trans. Inf. Theory IT-19 (1973) 471–480.
[397] G. Smith, J. Smolin, A. Winter, The quantum capacity with symmetric side-channels, IEEE Trans. Inf. Theory IT. 54 (9) (2008) 4208–4217.
[398] G. Smith, J. Yard, Quantum communication with zero-capacity channels, Science. 321 (2008) 1812–1815.
[399] A. Sørensen, K. Mølmer, Quantum computation with ions in thermal motion, Phys. Rev. Lett. 82 (9) (1999) 1971–1974.
[400] A. Sørensen, K. Mølmer, Entanglement and quantum computation with ions in thermal motion, Phys. Rev. A 62 (2000) 022311.
[401] A. M. Steane, Multiple particle interference and quantum error correction, Proc. R. Soc. Lond. A, 452 (1996) 2551.
[402] A. M. Steane, The ion trap quantum information processor. Preprint, http://arxiv.org/abs/quant-ph/9608011, v2, 1996.
[403] A. M. Steane, Quantum computing, Rep. Prog. Phys. 61 (1998) 117.
[404] A. M. Steane, Error correcting codes in quantum theory, Phys. Rev. Lett. 77 (1996) 793–797.
[405] A. M. Steane, Active stabilization, quantum computation, and quantum state synthesis, Phys. Rev. Lett. 78 (1997) 2252.
[406] A. Steane, The ion trap quantum information processor, Appl. Phys. B, 64 (1997) 623–642.
[407] A. M. Steane, Efficient fault-tolerant quantum computing, Nature (London), 399 (1999) 124.
[408] A. M. Steane, D. M. Lucas, Quantum computing with trapped ions, atoms and light, Fortschr. Phys. 48 (2000) 839–858.
[409] A. M. Steane, Overhead and noise threshold of fault-tolerant quantum error correction, Phys. Rev. A 68 (2003) 042322.
[410] O. Stern, The method of molecular rays, Nobel Lectures, Physics 1942–1961, pp. 8–16. Also: http://nobelprize.org/nobel_prizes/physics/laureates/1943/stern-lecture.pdf.
[411] W. F. Stinespring, Positive functions on C^*-algebras, Proc. Am. Math. Soc. 6 (1955), 211–216.
[412] M. Sudan, Essential Coding Theory. Lecture Notes. At http://people.csail.mit.edu/madhu/FT04/, 2004.
[413] N. S. Szabo, R. I. Tanaka, Residue Arithmetic and Its Applications to Computer Technology, McGraw-Hill, New York, 1967.
[414] L. Szilárd, Über die Entropieverminderung in einem thermodynamichen System bei Eingriffen intelligenter Wesen, Z. Phys. 53 (1929) 840–856.
[415] M. S. Tame, R. Prevedel, M. Paternostro, P. Böhi, M. S. Kim, A. Zeilinger, Experimental realization of Deutsch's algorithm in a one-way Wuantum computer. Phys. Rev. Lett. 98 (14) (2007) 140501.
[416] T. Tanamoto, Quantum gates by coupled quantum dots and measurement procedure in field-effect-transistor structure, Fortschr. Phys. 48 (9–11) (2000) 1005–1021.
[417] S. Tarucha, D. G. Austing, T. Honda, R. J. van der Hage, L. P. Kouwenhoven, Shell filling and spin effects in a few electron quantum dots, Phys. Rev. Lett. 77 (17) (1996) 3613–3616.

[418] B. M. Terhal, Is entanglement monogamous? IBM J. Res. Dev., 48 (1) (2004) 71–78.
[419] B. M. Terhal, A. C. Doherty, D. Schwab, Symmetric extensions of quantum states and local hidden variable theories, Phys. Rev. Lett. 90 (15) (2003) 157903.
[420] S. Tiwari, F. Rana, H. Hanafi, A. Hartstein, E. F. Crabbé, K. Chan, A silicon nanocrystals based memory, Appl. Phys. Lett. 68 (10) (1996) 1377–1379.
[421] T. Toffoli, Reversible computing, in: J. W. de Bakker, J. van Leeuwen (Eds.), Automata, Languages, and Programming, Spinger Verlag, New York, 1980, pp. 632–644.
[422] A. C. de la Tore, A. Daleo, I. Garcia-Mata, The photon-box Bohr-Einstein debate demythologized. Preprint, http://arxiv.org/abs/quant-ph/9910040 v1, 1999.
[423] L. A. Tracy, J. P. Eisenstein, L. N. Pfeiffer, K. W. West, Spin transition in the half-filled Landau level, Phys. Rev. Lett. 98 (8) (2007) 086801.
[424] D. C. Tsui, H. L. Störmer, A. C. Gossard, Two-dimensional magnetotransport in the extreme quantum limit, Phys. Rev. Lett. 48 (22) (1982) 1559–1562.
[425] T. Tulsi, L. K. Grover, A. D. Patel, A new algorithm for directed quantum search. Preprint, http://arxiv.org/abs/quant-ph/0505007, 2005.
[426] Q. A. Turchette, C. S. Wood, B. E. King, C. J. Myatt, D. Leibfried, W. M. Itano, et al, Deterministic entanglement of two ions, Phys. Rev. Lett., 81 (1998) 3631–3634.
[427] Q. A. Turchette, D. Kielpinski, B. E. King, D. Leibfried, D. M. Meekhof, C. J. Myatt, et al, Heating of trapped ions from the quantum ground state, Phys. Rev. A, 61 (6) (2000) 063418.
[428] A. M. Turing, On computable numbers with an application to the Entscheidungsproblem, Proc. Lond. Math. Soc. 2, 42 (1936) 230.
[429] A. Uhlmann, The transition probability in the state space of C^*-algebra, Rep. Math. Phys. 9 (1976) 273–279.
[430] L. M. K. Vandersypen, M. Steffen, G. Breyta, C. S. Yannoni, R. Cleve, I. L. Chuang, Experimental realization of an order-finding algorithm with an NMR quantum computer, Phys. Rev. Lett. 85 (25) (2000) 5452–5455.
[431] L. M. K. Vandersypen, M. Steffen, G. Breyta, C. S. Yannoni, M. H. Sherwood, I. S. Chuang, Experimental realization of Shor's quantum factoring algorithm using nuclear magnetic resonance, Nature. 414 (2001) 883–887.
[432] L. Vandersypen, I. Chuang, NMR computing - lessons for the future, Quantum Inf. Comput. 1 (2001) 134–142.
[433] L. M. K. Vandersypen, I. S. Chuang, NMR techniques for quantum control and computation, Rev. Mod. Phys. 76 (2004) 1037–1069.
[434] S. J. Van Enk, J. I. Cirac, P. Zoller, H. J. Kimble, H. Mabuchi, Quantum state transfer in a quantum network: a quantum-optical implementation, J. Mod. Opt. 44 (10) (1997) 1727–1736.
[435] S. A. Vanstone, P. C. van Oorschot, An Introduction to Error Correcting Codes with Applications, Kluwer, Boston, MA, 1989.
[436] V. Vedral, The role of entropy in quantum information theory. Preprint, http://arxiv.org/abs/quant-ph/0102094, v1, 2001.
[437] S. Verdù, T. S. Han, A general formula for classical channel capacity, IEEE Trans. Inf. Theory. 40 (1994) 1147–1157.
[438] G. Vidal, Efficient classical simulation of slightly entangled quantum computations, Phys. Rev. Lett. 91 (2003) 14792–14976.
[439] A. J. Viterbi, Error bounds for convolutional codes and an asymptotically optimum decoding algorithm, IEEE Trans. Inf. Theory IT. 13 (1967) 260–269.
[440] J. von Neumann, Thermodynamik quantummechanischer Gesamtheiten (The Thermodynamics of Quantum Mechanical Ensembles), Nachrichten von der Gesellschaft der Wissenschaften zu Göttingen, Matematisch-Physikalische Klasse 1 (1927) 273–291. Also, http://www-gdz.sub.uni-goettingen.de/cgi-bin/digbib.cgi?PPN252457811.

[441] J. von Neumann, Mathematical Foundations of Quantum Mechanics, Trans. R. T. Bayer. Princeton University Press, Princeton, NJ, 1955.

[442] J. von Neumann, Probabilistic logic and synthesis of reliable organisms from unreliable components, in: C. E. Shannon, J. McCarthy (Eds.), Automata Studies, Princeton University Press, Princeton, NJ, 1956, pp. 43–98.

[443] J. von Neumann, Theory of Self-Reproduced Automata, Fourth University of Illinois lecture, edited and completed by A. W. Burks, University of Illinois Press, Urbana, IL, 1966.

[444] J. Watrous, Lecture Notes: Theory of Quantum Information, University of Waterloo. At: http://www.cs.uwaterloo.ca/~watrous/quant-info/, 2007.

[445] A. Wehrl, General properties of entropy, Rev. Mod. Phys. 50 (2) (1978) 221–260.

[446] C. F. von Weizsäcker, Die Einheit der Natur (The Unity of Nature), Farrar, Straus, and Giroux, New York, 1980.

[447] C. F. von Weizsäcker, E. von Weizsäcker, Wideraufname der begrifflichen Frage: was ist Information?, (Revisting the fundamental question: what is information?), Nova Acta Leopold. 37 (1) (1972) 535.

[448] X. G. Wen, Q. Niu, Ground-state degeneracy of the fractional quantum Hall states in the presence of a random potential and on high-Genus Riemann surfaces, Phys. Rev. B, 41 (13) (1990) 9377–9396.

[449] X. G. Wen, Chiral luttinger liquid and the edge excitations in the fractional quantum Hall effect, Phys. Rev. B, 41 (14) (1990) 12838–12844.

[450] X. G. Wen, Edge transport properties of the FQH states and impurity scattering of a one-dimensional charge-density wave, Phys. Rev. B, 44 (1991) 5708–5719.

[451] X. G. Wen, Non-abelian statistics in the fractional quantum Hall states, Phys. Rev. Lett. 66 (6) (1991) 802–805.

[452] X. G. Wen, Topological orders and edge excitations in FQH states, Adv. Phys. 44 (1995) 405–473.

[453] R. F. Werner, Quantum states with Einstein-Podolsky-Rosen correlations admitting a hidden-variable model, Phys. Rev. A, 40 (1989) 4277–4281.

[454] E. P. Wigner, Gruppentheorie und ihre Anwendungen auf die Quantenmechanik der Atomspektren, Vieweg Verlag, Braunschweig, 1931. Translated into English: J. J. Griffin, Group Theory and Its Application to the Quantum Mechanics of Atomic Spectra, Academic Press, New York, 1959.

[455] F. Wilczek, Quantum mechanics of fractional-spin particles, Phys. Rev. Lett. 49 (14) (1982) 957–959.

[456] R. Willett, J. P. Eisenstein, H. L. Störmer, D. C. Tsui, A. C. Gossard, J. H. English, Observation of an even-denominator quantum number in the fractional quantum hall effect, Phys. Rev. Lett. 59 (15) (1987) 1776–1779.

[457] R. I. Willett, M. J. Manfra, I. N. Pfeiffer, K. W. West, Interferometric measurement of filling factor 5/2 quasiparticle charge. Preprint, http://arxiv.org/abs/0807.0221, 2008.

[458] D. J. Wineland, C. Monroe, W. M. Itano, D. Leibfried, B. E. King, D. M. Meekhof, Experimental issues in coherent quantum state manipulation of trapped atomic ions, J. Res. NIST 103 (1998) 259–328.

[459] D. J. Wineland, M. Barrett, J. Britton, J. Chiaverini, B. DeMarco, W. M. Itano, et al, Quantum information processing with trapped ions. Preprint, http://arxiv.org/abs/quant-ph/0212079, v2, 2003.

[460] W. K. Wootters, W. H. Zurek, A single quantum cannot be cloned, Nature. 299 (1982) 802–803.

[461] W. K. Wootters, Entanglement of formation of an arbitrary state of two qubits, Phys. Rev. Lett. 80 (10) (1998) 2245–2248.

[462] J. M. Wozencraft, B. Reiffen, Sequential Decoding, MIT Press, Cambridge, MA, 1961.

[463] J. S. Xia, W. Pan, C. L. Vincete, E. D. Adams, N. S. Sullivan, H. L. Stormer, et al, Electron correlation in the second Landau level: a competition between many nearly degenerate quantum phases, Phys. Rev. Lett. 93 (17) (2004) 176809.

[464] A. Yao, Quantum circuit complexity, Proceedings of the 34th Ann. Symp. Found. Comput. Sci. (1993) 352–361.

[465] C. Zalka, Grover's quantum searching algorithm is optimal, Phys. Rev. A 60 (1999) 2746–2751.
[466] P. Zanardi, M. Rasatti, Noiseless quantum codes, Phys. Rev. Lett. 79 (1998) 3306–3308.
[467] P. Zanardi, F. Rossi, Quantum information in semiconductors: noiseless encoding in a quantum-dot array, Phys. Rev. Lett. 81 (1998) 4752–4755.
[468] P. Zanardi, D. A. Lidar, S. Lloyd, Quantum tensor product structures Are observable induced, Phys. Rev. Lett. 92 (2004) 060402.
[469] O. Zilberberg, B. Braunecker, D. Loss, Controlled-NOT for multiparticle qubits and topological quantum computation based on parity measurements. Preprint, http://arxiv.org/abs/0708.1062, 2007.
[470] W. H. Zurek, Decoherence and the transition from quantum to classical, Phys. Today. 44 (10) (1991) 36–44.
[471] W. H. Zurek, Decoherence, Einselection, and the quantum origins of the classical. Preprint, http://arxiv.org/abs/quant-ph/0105127, 2003.
[472] W. H. Zurek, Probabilities from entanglement, Born's rule from envariance. Preprint, http://arxiv.org/abs/quant-ph/0405161, 2004.
[473] Quantum information science and technology roadmap, ARDA - Quantum Information Science and Technology Project, Version 2.0, 2004. Preprint, http://qist.lanl.gov/qcomp_map.shtml.
[474] Semiconductor Industry Association Roadmap 2000–2001. http://public.itrs.net, 2010.
[475] Stanford Encyclopedia of Philosophy. http://plato.stanford.edu/entries/qm-copenhagen, 2008.

Index

Page numbers followed by "*f*" indicate figures, "*t*" indicate tables and "*n*" indicate footnotes.

C

D

E

I

J

K

N

Printed and bound by CPI Group (UK) Ltd, Croydon, CR0 4YY
08/05/2025
01864869-0002